D1555709

COINCIDENCE TABLES FOR ATOMIC SPECTROSCOPY

Principal Lines for Emission Spectral Analysis
with Coincidences from 2,000 to 10,000 Å

Scientific Editor Prof. Dr. Ing. František Čůta,
Corresponding Member
of the Czechoslovak Academy of Sciences

Reviewer Prof. Dr. Ing. Josef Knop,
Corresponding Member
of the Czechoslovak Academy of Sciences

Coincidence Tables for Atomic Spectroscopy

JOSEF KUBA
LUDVÍK KUČERA
FRANTIŠEK PLZÁK
MILOSLAV DVOŘÁK
JAN MRÁZ

ELSEVIER PUBLISHING COMPANY
AMSTERDAM—LONDON—NEW YORK

1965

Sole distributors for the United States and Canada:
Elsevier Publishing Company, Inc.
52 Vanderbilt Avenue, New York, N. Y. 10017

Sole distributors for the United Kingdom:
Elsevier Publishing Company Ltd.
Rippleside Commercial Estate
Barking, Essex

In coedition with

Publishing House
of the Czechoslovak Academy of Sciences,
Prague

Elsevier Publishing Company,
335 Jan van Galenstraat, Amsterdam, The Netherlands

Library of Congress Catalog Card Number: 64-66151

Printed in Czechoslovakia

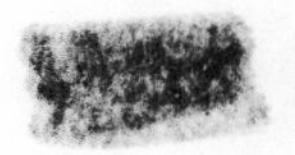

PREFACE

To solve difficult analytical tasks, tables of atomic spectroscopy have been prepared which contain the principal lines for emission spectral analysis together with coincidences within the wavelength range 2,000 to 9,250 Å.

When drafting the structure and especially when arranging and calculating the tables, the authors had in mind the practical tasks of qualitative atomic spectroscopy. They know from experience the difficulties which arise in the evaluation of more complex spectra taken with spectrographs of medium dispersion. Comparison patterns are inadequate for a more exacting analysis.

The hitherto generally known and used tables arranged by Kayser, the tables of the Soviet authors based on the MIT wavelength tables drawn up by G. R. Harrison, as the authors (Zajdel, Prokofjev, Rajskij) themselves admit, are merely an enumeration, a summary of spectral lines according to wavelength without any critical comparison and without any evaluation from the point of view of the angle of coincidence. It is quite natural that the basic Soviet tables, as well as the tables composed by Kayser and Harrison, will continue to be indispensable in every spectroscopic laboratory. In principle, the coincidence tables are complementary tables which are necessary for evaluation of the most complex spectrograms when the normally used tables prove inadequate. The enclosed coincidence tables do not contain absolutely all the coincidences, it is true, but they do include the great majority of them. That the tables contain the great majority of existing lines is proved by the fact that the lines given for the first coincidence region epresent more than 99% of the whole energy radiated by the normally applied specti oscopic excitation. The tables do not contain only the simple and principal coincidences found in the generally used tables, but practically all the coincidences appearing in the given region. When establishing the united coincidence region, the authors set out from the dispersion curves of the most widely used spectrographs (Q 24, KSA 1, FDU) which represent the main group of apparatus used in Czechoslovak and European spectroscopic laboratories. The first coincidence region was determined and based on the well known triplet of iron, Fe 3,100, which is generally considered to be the practical criterion of the capacity of resolution of an instrument: it should be noted, for instance, that for the Q 24 apparatus more suitable lines than this triplet are not known. A more detailed description of the selection of the regions will be found in the following chapters.

A suitable supplement to the spectral lines are the photographs of spectra which will prove especially useful for rapid orientation in the spectrograms: moreover, they show in an instructive way the principal coincidence regions of the respective spectrogram.

In conclusion, it may be said that the enclosed tables, although not a universal help, will be of great assistance in such undoubtedly responsible work as the evaluation of complex spectrograms. The tables will help those concerned with atomic spectroscopy to effect a change from current spectrographic practice to independent creative spectroscopic work. It is hardly necessary to mention the fact that slight errors could not be completely avoided in these newly conceived tables and therefore the authors will be grateful to all spectrographic workers who draw their attention to inaccuracies, thus enabling their elimination in future editions.

The authors would like to express their sincere thanks to Prof. Čůta, corresponding member of the Czechoslovak Academy of Sciences, Prof. Knop, corresponding member of the Czechoslovak Academy of Sciences, and Dr. Polej, for their kind advice. Our thanks are also due to M. Frauenterka for drawing the spectra, to Miss L. Kmínková for arranging and writing the tables, and especially to the compositors and foremen of the printing works for the great understanding they showed in fulfilling such a complex task. The editors of the Publishing House of the Czechoslovak Academy of Sciences assisted us right from the beginning to the end of our task and without their kindness these tables would never have been accomplished.

Authors

CONTENTS

INTRODUCTION

In the course of only two decades spectral analysis has become a well proved analytical method applied in the testing and research laboratories of our national economy. Spectroscopic apparatus can be found in iron works and the laboratories of the engineering industry as well as in those of agricultural and food production, in geological institutes and in medical research institutes. It would be necessary to state the name of 180 works and laboratories which are members of the "Association of Research Workers in Spectral Analysis" attached to "The Committee for Spectroscopy of the Czechoslovak Academy of Sciences" in order to give at least an approximate idea of the wide use of this modern method.

The wide use of spectral analysis in the Czechoslovak Socialist Republic is closely connected with the tasks imposed on our industry and our economy as a whole and may be explained by the great advantages of the spectroscopic method. With a comparison spectroscope (steeloscope) it is possible to determine, within a few minutes, not only the kind of material, but also its approximate chemical composition. The analysis may be carried out on finished products without their incurring any damage. Spectroscopy is, therefore, one of the group of non-destructive testing methods.

The advantage of spectroscopy consists, in the first place, in the high degree of sensitivity and high speed of the analysis; moreover, only a small quantity of the material to be tested is required.

In recent years radioactive isotopes have enabled enhancement of the sensitivity of analytical methods up to the limit of the given possibilities. In most cases, however, exceptional methods are concerned which are rarely suitable for industrial practice. Spectroscopy is therefore, still today one of the most sensitive analytical methods. In one thousandth of a gram of a sample impurities even in thousandths of one per cent may be identified. Using the flame spectrum, 10^{-10} g of sodium may be identified and some authors quote an even higher sensitivity, especially in cases where scintillation photomultipliers were used. On the whole, spectroscopy can be considered a very speedy and sensitive method of qualitative analysis. A single spectrum is usually sufficient for determination of all the components of a sample without their difficult separation which is unavoidable when chemical and physical-chemical methods are used. In spectroscopy differentiation of the elements is achieved by a grating or by the prism of a spectrograph in which the physical laws of reflection and of the refraction of light rays of different wavelengths are applied. In this case difficulties naturally arise which do not generally occur during a classic chemical analysis. These difficulties arise due to the actual nature of qualitative spectroscopy and, in particular, to the laws which give rise to the spectral lines.

For every chemical element there is a corresponding characteristic spectrum line which, under normal conditions, always lies in the same place in the spectrum. Atoms of elements in the gaseous state emit spectral lines, either neutral atoms (arc spectrum), or ionized atoms (spark spectrum). That is why these spectra are called atomic spectra and the whole branch atomic spectroscopy in contradistinction to molecular spectroscopy. The spectral lines of the different chemical elements differ to a very considerable extent. The method of excitation also influences the number and character of the lines. The spark-spectrum is much richer in lines than the arc spectrum or the flame spectrum. This fact will be dealt with in greater detail in the second part of the book entitled Spectral Analysis. We are publishing here only the most concise summary obtained from Bohr's model of an atom which does not, it is true, provide an answer to a number of questions (for instance, why are some lines thicker and others thinner?), but which does facilitate comprehension of the way in which spectral lines arise. In accordance with Bohr's model let us suppose that the atom is not excited. The electrons thus move in their stable levels. If we supply the atom with energy from an external source (for instance, by heating), the electron will be pushed to one of the more distant levels from the nucleus and the atom will become excited. This state of excitation is not stable and after a short while (10^{-8} sec) the electron falls from the higher level to a lower one which lies nearer the nucleus. The surplus energy, set free by this transition, is emitted in the form of a quantity of light of a certain wavelength corresponding to one spectral line. Generally there is a great number of electrons and several "permitted" levels and that is why the spectrum has a large number of lines. The most important are those which correspond to the transitions to the lowest level. These lines are of the greatest importance for qualitative analysis and are therefore given the greatest attention in these tables. In consequence of their diminishing concentration, these lines will disappear last of all and that is why they are called persistent lines. When investigating a spectrogram for the first time it is just these basic lines that we look for. Being most sensitive from the analytical point of view, inversion frequently takes place and in the literature they are therefore called absorption lines or resonance lines. In the following tables they are indicated by the letter *R*.

If a sufficient supply of energy is applied to the atom the electron is completely removed from the atom shell and the neutral atom is ionized, i.e. it becomes an ion. By a further supply of energy it is possible to excite the electrons of an ionized atom in which case a somewhat different spectrum from the neutral spectrum is emitted.

Its lines are also given in the tables, but they are indicated by the symbols which express the stage of ionization. There exist lines of an atom which has lost even more electrons. As an ionized atom has one or more electrons less, its spark spectrum resembles the arc spectrum of the element which lies one or two places to the left in the Periodic Chart of Elements. Ionization takes place as a rule in spark discharges. That is why by spark spectra the spectra of ionized atoms are usually understood, whereas the spectra of neutral atoms are called arc spectra. Quantum and wave mechanics have changed a great deal of Bohr's original conception of the atom model; however, one of the most important facts, upon which the whole of spectral analysis is established, has remained unchanged, viz. the fact that for every element there is a certain characteristic spectrum the lines of which always lie, under normal conditions, in the same place, which means that they have a certain wavelength which today is defined with great accuracy. These lines are shown in the spectral tables, so that by determining the wavelengths of the lines in the spectrum and looking them up in the tables a qualitative analysis is basically effected. But in practice difficulties arise which were mentioned at the beginning of the introduction. Thousands and tens of thousands of lines of atoms and ions have wavelengths which lie so close to one another that the lines in the spectrum practically amalgamate — coincide — so that it may happen that the lines are interchanged when routine analyses are carried out. To preclude such interchanges the following tables have been composed, as will be seen from the following chapters.

1. PURPOSE AND SCOPE OF THE TABLES

The tables are intended for the identification of elements when evaluating emission spectrograms. They contain, in the first place, a set of analytical lines of all the elements which may be found under the usual conditions prevailing in spectral laboratories. A list of the elements treated will be found in Table 1. This table does not contain data in regard to the following elements: Tc (43), Pm (61), Po (84), At (85), Fr (87), Bk (97), Cf (98), Es (99), Fm (100), Md (101), and No (102). Table II and Table III contain a survey of the principal Coincidence Tables.

Tables IV and V show the analytical lines of rare earth elements. The emission spectra of some actinides are contained in Table VI.

The analytical lines of gases will be found in Tables VII and VIII. The spark spectrum of the air is given in Table IX.

The principal part of the tables consists of the Coincidence Tables XVI containing 683 analytical lines of the 73 elements.

For each element such a set of lines has been chosen as ensures complete certainty even when the most complex spectrograms in the region from 2,000 to 10,000 Å and at a linear dispersion of 3—30 Å/mm are investigated.

To each line there are attached the coincidence lines which may be produced in accordance with the principle that all lines in an optical region have been given within the limits of a possible photographic differentiation of two thin lines which are contained in Harrison's atlas of lines for the given narrow range of the spectrum. Those coincidence lines whose minimum intensity is 1/1 or 1/2 of the intensity of the line being investigated should be considered as lying outside this region (see Chapter 2.2. and Chapter 2.2.1.) For the sake of convenience the lines in the individual intervals of coincidence are differentiated by the type and size of the letters. The first coincidence region will prove satisfactory for work with spectrographs of a dispersion of 0.5—5 Å/mm. For work with spectrographs of medium dispersion (2 to 30 Å/mm) the coincidence regions 1 and 2 should be applied; when working with spectographs of a dispersion of 30 Å/mm (for instance, with the spectrograph Q 12) the lines of all three ranges should be used.

In Table XVII (the rare earth elements) only coincidences in the first interval are given because the elements emit complex spectra and therefore spectrographs yielding a greater dispersion would in principle be used. That is why only two or possibly three of the most sensitive analytical lines have been selected.

In Table X a set of lines of iron is given in the range from 2,000 to 10,000 Å, selected in such a way that at least one line is approximately in the span of every 25 Å.

This table is intended for precision wavelength measurement of lines subject to investigation with the aid of a dispersion curve or with the use of different interpolation methods. In Chapter 3, concise instructions are given how to proceed when a wavelength is to be determined with the use of linear interpolation of the values of two given lines.

Table XI shows the wavelength standards for the lines of Cd, Kr, and Fe, accurate at least to the third decimal place.

Tables XII and XIII indicate the band heads of inorganic molecules which appear in emission spectrograms. A survey of the excitation potentials for different stages of ionization is given in Table XIV. In Fig. 1 the relative values of excitation potentials for neutral atoms are arranged in the form of a periodic table.

2. TABLES OF THE ANALYTICAL LINES OF ELEMENTS AND OF THEIR COINCIDENCES

2.1. Analytical Lines

The coincidence tables contain the analytical lines of the 73 elements in the range of wavelengths from 2,000 to 10,000 Å.

As analytical lines have been chosen, on the one hand, the resonance lines emitted by the transitions of the electrons to their normal positions (indicated in the tables by different types of letters) and on the other hand, some intensive lines suitable for analytical purposes, the reason for this being as follows: The resonance lines of elements marked by high ionization tension (exceeding 10 eV) are in the region lying below 2,100 Å and are therefore not perceivable by normal spectroscopic procedures in an air medium. This applies, for instance, to sulphur whose analytical lines are represented by the lines of a twice and three times ionized atom due to their being the most intensive ones under the discharge conditions generally used for its excitation. In some cases (V, Ti, Mn) the selectivity of the lines was the decisive criterion in regard to the other analytical lines.

Table 1

List of Elements in the Coincidence Tables

Symbol	Element	Number of Analytical Lines	Symbol	Element	Number of Analytical Lines
Ag	Silver	15	N	Nitrogen	8
Al	Aluminium	8	Na	Sodium	6
Ar	Argon	9	Nb	Niobium	10
As	Arsenic	14	Ne	Neon	10
Au	Gold	7	Ni	Nickel	16
B	Boron	3	O	Oxygen	12
Ba	Barium	9	Os	Osmium	7
Be	Beryllium	8	P	Phosphorus	6
Bi	Bismuth	7	Pb	Lead	7
Br	Bromine	9	Pd	Palladium	6
C	Carbon	8	Pt	Platinum	7
Ca	Calcium	10	Ra	Radium	7
Cd	Cadmium	12	Rb	Rubidium	8
Ce	Cerium	7	Re	Rhenium	7
Cl	Chlorine	10	Rh	Rhodium	9
Co	Cobalt	14	Rn	Radon	9
Cr	Chromium	14	Ru	Ruthenium	7
Cs	Cesium	4	S	Sulphur	8
Cu	Copper	11	Sb	Antimony	6
F	Fluorine	10	Sc	Scandium	9
Fe	Iron	17	Se	Selenium	8
Ga	Gallium	5	Si	Silicon	7
Ge	Germanium	9	Sn	Tin	8
H	Hydrogen	9	Sr	Strontium	7
He	Helium	5	Ta	Tantalum	6
Hf	Hafnium	12	Te	Tellurium	5
Hg	Mercury	11	Th	Thorium	8
In	Indium	8	Ti	Titanium	12
Ir	Iridium	12	Tl	Thallium	7
I	Iodine	13	U	Uranium	7
K	Potassium	12	V	Vanadium	10
Kr	Krypton	13	W	Tungsten	8
La	Lanthanum	10	Xe	Xenon	6
Li	Lithium	10	Y	Yttrium	7
Mg	Magnesium	9	Zn	Zinc	11
Mn	Manganese	12	Zr	Zirconium	7
Mo	Molybdenum	10			

In the selection we applied our experience of many years and the experience represented by the tabular works of other authors (W. Gerlach, E. Riedel: Die Chemische Emissionsspektralanalyse, III. Teil — Tabellen zur qualitativen Analyse. Leipzig 1939. — A. N. Zajdel, V. K. Prokofjev, S. M. Rajskij: Tablicy spektralnych linij. Moskva 1960. — G. R. Harrison: Wavelength Tables. New York 1960. — J. Kuba, M. Dvořák: Spektrální analysa kvalitativní. Praha 1957).

A list of elements with the number of analytical lines contained in the coincidence tables is given in Table 1.

2.1.1. Location of Analytical Lines in the Spectrum

On pages 61 to 91 the position of the main lines in the range of the spectrum from 2,000 to 10,000 Å is indicated graphically. The spectrum is drawn in linear dispersion, i.e. practically as it is produced by the gratings of a spectrograph.

In the horizontal direction the spectrum is divided into three bands. The lower band shows the lines corresponding to neutral atoms, while the upper band shows the lines produced by ionized atoms. The central band is common to all lines, therefore containing atom lines, ion lines and also lines of which the stage of excitement is unknown.

The central band therefore illustrates the density of the most intensive element lines and serves as a visual aid facilitating general determination of the spectrum region suitable for the given analytical case as well as for fixing the minimum dispersion of a spectral instrument.

Indicated at the sides of the lines are the chemical symbols of the elements and their respective wavelength in angstroms given with an accuracy of two decimal places.

For the sake of convenience, the upper spectrum band is divided into sections of 100 Å each.

The lines are classified according to thickness into five grades representing the intensity regions on Harrison's intensity scale:

1. Grade lines of an intensity below 50
2. Grade lines of an intensity from 50 to 100
3. Grade lines of an intensity from 100 to 500
4. Grade lines of an intensity from 500 to 2,000
5. Grade lines of an intensity from 2,000 to 9,000

This means that the lines of the smallest density, starting with those which may still be traced, and others up to the value 50 on Harrison's intensity scale are indicated by the thinnest lines. The region of relative intensities from 50 to 2,000 is divided into a further three grades. The most intensive lines from 2,000 to 9,000 are indicated with the fifth grade line, which is the thickest.

Comparing the frequency of atom lines with that of ion lines and taking into consideration the intensity rating, we are able to decide not only on the spectrum region and spectral dispersion of the apparatus, but also the suitability of the excitation conditions.

Classification of analytical lines of elements in the region from 2,000 to 10,000 Å

1. Grade	indicates lines of an intensity below	50
2. Grade	indicates lines of an intensity	50 to 100
3. Grade	indicates lines of an intensity	100 to 500
4. Grade	indicates lines of an intensity	500 to 2,000
5. Grade	indicates lines of an intensity	2,000 to 9,000

2.2. Coincidence of Analytical Lines

To each analytical line a list of possible coincidence lines is attached. The left-hand part contains coincidences in the direction towards the shortwave portion, the right-hand part those in the direction towards the longwave portion. To each line the rating of its relative intensity in arc and in spark are attached. Below is an example taken from the table on page 488 and concerning a spark line of potassium, wavelength 2,240.89 Å (K 2,240.89), the spark intensity of which is 40 (*J*—40). It is obvious from the table that in that portion of the spectral region which is directed towards the wavelengths shorter than the length of the line (K 2,240.89) under observation, it may be a case of coincidence with the lines of the following elements: W, Nb, Ir, Fe, V. As can be seen from the table, concerned here is first of all the nearest line (W 2,240.85) which lies, in the direction towards the shorter wavelengths, only 0.04 Å from the potassium line which is being studied. However, the tungsten line is of very weak density (its spark intensity is 5 and arc intensity 4) so that this coincidence is of importance only in the case of a high content of tungsten. At a somewhat greater distance is the line for niobium (Nb 2,240.65) of a spark intensity of 15 and an arc intensity of 10. A hundredth of an Å farther in the direction towards the shorter wavelengths lies the iridium line (Ir 2,240.64) of the same spark intensity as the above mentioned niobium line (*J*—15). The following two coincidence lines have a higher intensity: the iron line (Fe 2,240.63) has a spark intensity equal to 3 and the vanadium line (V 2,240.62) is of an intensity as high as 100. All the intensities are of the conventional type in accordance with Harrison's tables and are therefore of very approximate comparison value, which is, however, sufficient for orientation in a certain spectral line region.

In the second (right) part of the table of our example there are similarly given the possible coincidences of the spectral lines of longer wavelengths exceeding the wavelength of the potassium line (K 2,240.89) which is being studied. In the first place, the krypton line (Kr 2,240.89) should be noted. Its intensity lies almost on the border of distinguishability (its spark intensity is 2). The wavelengths of both lines are almost coincidental from the practical point of view and cannot be differentiated even numerically by the third decimal place. In the spectrum, however, the krypton line follows the potassium line in the direction towards the lines of longer wavelength, the same sequence being observed in the table.

A longer wavelength is indicated by the other line (Ir 2,240.96) belonging to iridium of a spark intensity equal to 3. Further there follow the lines of cadmium, niobium and tungsten with gradually increasing wavelengths.

Example:

K 2,240.89				J—40			
Element	λ	I		Element	λ	I	
		J	O			J	O
W	2,240.85	5	4	Kr	2,240.89	(2)	—
Nb	2,240.65	15	10	Ir	2,240.96	3*h*	5
Ir	2,240.64	15	—	Cd	2,240.99	(3)	—
Fe	2,240.63	3*h*	20	Nb	2,241.02	6	—
V	2,240.62	100	—	W	2,241.08	15	10

The first page for each element contains data of its boiling (t_1) and melting (t_0) points as well as the values of its ionization potentials.

2.2.1. *Selection of Coincidence Regions*

When selecting the coincidence regions, we proceeded on the basis of the following considerations:

1. It may be assumed that two lines will be maximally distinguishable on a photographic plate only if the centres of the lines lie at a distance of 0.5 mm from one another, as is e.g. the case of the lines of the characteristic triplet of Fe at 3,100 Å* (Fig. 2, page 4, and Fig. 3, page 4).

* Fe 3,099.897 Å This is, in effect, a quadruplet, but a spectrograph of medium dispersion
Fe 3,099.971 Å does not distinguish the line 3,099.897 and the line 3,099.971 Å.
Fe 3,100.304 Å
Fe 3,100.666 Å

This is not connected in any way with the capacity of resolution of the spectral instrument. If in the case of the quoted triplet

$$D = \frac{\lambda}{\Delta\lambda} = \frac{3{,}100}{0.36\ \text{Å}} = \sim 8{,}600\,,$$

then for equally near lines in the short wave portion, e.g. Fe 2,148.394 Å and Fe 2,148.50 Å,

$$D = \frac{2{,}148}{0.106} = \sim 20{,}000$$

the value being twice as large.

2. From the dispersion curves for the quartz prism spectrograph Q 24 for the ultraviolet region, for the spectrograph FDU for the visible region, and for the spectrograph KSA 1 for the range exceeding 6,000 Å, we deduced the coincidence intervals $\pm\Delta\lambda_A$ for each analytical line λ_A in such a way that the value of the line, increased or decreased by the increment $\Delta\lambda_A$, delimited the interval $d = 0.05$ mm on the plate. In view of the gentle course of the dispersion curve the deduction was always carried out for the ranges of 100 Å, and in the long wavelength region for the ranges of 400—500 Å, see Table 2.

3. For the analytical line λ_A the lines within the region $\pm\lambda_A$ are considered to be coincident, and in the Tables there are entered all the wavelengths of the lines which are contained in that region ($\lambda_A \pm \Delta\lambda_A$) in Harrison's tables. In this connection it must be stressed that Harrison's tables contain about one half of all the spectral lines known in the studied region. Apart from the lines of atoms in various stages of ionization, the lines of relative intensity 1 were not included. In spite of this, however, all the lines given there represent 99% of the radiation emitted by atoms in the region of the wavelengths from 2,000—10,000 Å. For the purposes of qualitative evaluation, when a certain element is identified by its most intensive lines, the possibility of coincidence with a line that is not given arises only if the sought element occurs in hardly traceable quantities and the disturbing element forms the principal part of the investigated sample. But even in such a case it is improbable that such a situation would arise and effect all of the 5—10 lines which are given in the tables of analytical lines for evaluation and control.

4. For more disturbing lines (i.e. for a higher concentration of the elements in the discharge plasma) whose image on the plate under the given conditions is of a width not exceeding 0.205 mm and could therefore coincide even in the case of greater wavelength differences, the values $\Delta'\lambda_A$ for $d = 0.5$ mm and $\Delta''\lambda_A$ for $d = 1$ mm were determined (see the third and fourth columns of the table).

5. For the coincidence region from $(\lambda_A \pm \Delta\lambda_A)$ to $(\lambda_A \pm \Delta'\lambda_A)$ only the lines of relative intensity exceeding half the intensity of the investigated analytical line are given as possible disturbing lines (see enclosure).

6. For the coincidence from $(\lambda_A \pm \Delta'\lambda_A)$ to $(\lambda_A \pm \Delta''\lambda_A)$ only the lines of relative intensity equal to or exceeding the intensity of the studied analytical line are given as possible disturbing lines.

7. In all the coincidence ranges there are listed, besides the disturbing lines, all the spectral lines that belong to the investigated element.

2.3. Intensity Ratings of Lines

Attached to each wavelength are the arc intensity value of the line, and the spark intensity value. The indication of intensity has also been adopted from Harrison's MIT Wavelength Tables. The intensities are determined by visual estimation of the blackening of the lines on the photographic plate using the scale 1—9,000. The value 9,000 represents the maximum intensity. The line on the borderline of identification is of an intensity equal to 1. These very weak lines are left out in Harrison's original tables with the exception of a few lines of Fe. The entire intensity range in the 1—9,000 system is graduated into 25 different values. The advantage of this scale does not consist in that the intensity ratings are given with greater accuracy than would be the case with a linear scale, e.g. from 1 to 25, but in the fact that the scale 1—9,000 corresponds better to the relative intensities. The estimations were established by means of comparisons of the lines of all the elements, but in each case only for an individual spectral sector. The individual spectral sectors were brought into harmony by multiplying by an estimated factor, which may be a considerable source of errors. This means that it is hardly probable that a line of an intensity equal to 5,000 in the region about 8,000 Å will have the same intensity as another line of an intensity equal to 5,000 in the region about 3,000 Å. Another source of the disharmony of intensities is the varying sensitivity of photographic emulsion with respect to the different spectral regions.

When evaluating the intensity ratings, the applied source of excitation must also be taken into account. Harrison used the same arc, burning in air, connected to a mains 220 V with stabilising impedance and a resistance. The long Pfund arc was applied for the region below 4,500 Å, and the short arc (about 8 mm long) for the region $>$ 4,500 Å.

The estimation in the column "Spark" relates to a condensed spark of 20,000 V. The gap between the sparking points is about 5 mm.

Table 2

A. Coincidence intervals $\Delta\lambda_A$ for the ultraviolet region of the spectrum for dispersion of the Q 24 spectrograph

Wave-length Å	$\Delta\lambda_A$ for d=0.05 mm Å	$\Delta'\lambda_A$ for d=0.5mm Å	$\Delta''\lambda_A$ for d=1.0 mm Å	Wave-length Å	$\Delta\lambda_A$ for d=0.5 mm Å	$\Delta'\lambda_A$ for d=0.05 mm Å	$\Delta''\lambda_A$ for d=1.0 mm Å
2,000	±0.20	±2.00	± 4.00	3,100	±0.75	± 7.50	±15.00
2,100	±0.23	±2.30	± 4.60	3,200	±0.85	± 8.50	±17.00
2,200	±0.26	±2.60	± 5.20	3,300	±0.92	± 9.25	±18.50
2,300	±0.30	±3.00	± 6.00	3,400	±1.00	±10.00	±20.00
2,400	±0.34	±3.45	± 6.90	3,500	±1.07	±10.75	±21.50
2,500	±0.39	±3.90	± 7.80	3,600	±1.15	±11.50	±23.00
2,600	±0.44	±4.40	± 8.80	3,700	±1.25	±12.50	±25.00
2,700	±0.49	±4.90	± 9.80	3,800	±1.37	±13.75	±27.50
2,800	±0.55	±5.50	±11.00	3,900	±1.50	±15.00	±30.00
2,900	±0.62	±6.25	±12.50	4,000	±1.62	±16.25	±32.50
3,000	±0.67	±6.75	±13.50	4,100	±1.70	±17.00	±35.00

B. Coincidence intervals $\Delta\lambda_A$ for the visible region of the spectrum for dispersion of the FDU spectrograph

Wave-length Å	$\Delta\lambda_A$ for d=0.05 mm Å	$\Delta'\lambda_A$ for d=0.5 mm Å	$\Delta''\lambda_A$ for d=1.0 mm Å	Wave-length Å	$\Delta\lambda_A$ for d=0.05 mm Å	$\Delta'\lambda_A$ for d=0.5 mm Å	$\Delta''\lambda_A$ for d=1.0 mm Å
4,200	±1.05	±10.50	±21.00	5,200	±1.90	±19.00	±38.00
4,300	±1.12	±11.25	±22.50	5,300	±2.00	±20.00	±40.00
4,400	±1.20	±12.02	±24.04	5,400	±2.10	±21.00	±42.00
4,500	±1.29	±12.87	±25.75	5,500	±2.20	±22.00	±44.00
4,600	±1.37	±13.75	±27.50	5,600	±2.30	±23.00	±46.00
4,700	±1.46	±14.62	±29.25	5,700	±2.42	±24.25	±48.50
4,800	±1.55	±15.50	±31.00	5,800	±2.55	±25.50	±51.00
4,900	±1.64	±16.37	±32.75	5,900	±2.70	±27.00	±54.00
5,000	±1.72	±17.25	±34.50	6,000	±2.85	±28.50	±57.00
5,100	±1.81	±18.12	±36.25				

C. Coincidence intervals $\Delta\lambda_A$ for the infrared region of the spectrum for dispersion of the KSA 1 spectrograph

Wave-length Å	$\Delta\lambda_A$ for d=0.05 mm Å	$\Delta'\lambda_A$ for d=0.5 mm Å	$\Delta'\lambda_A$ for d=0 Å	Wave-length Å	$\Delta\lambda_A$ for d=0.05 mm Å	$\Delta'\lambda_A$ for d=0.5 mm Å	$\Delta''\lambda_A$ for d=1.0mm Å
6,200	±1.25	±12.50	±25.000	7,800	±1.81	±18.15	±36.30
6,600	±1.45	±14.50	±29.00	8,500	±1.92	±19.20	±38.40
7,000	±1.53	±15.35	±30.70	9,000	±2.08	±20.80	±41.60
7,400	±1.66	±16.60	±33.3				

2.3.1. Excitation and Ionization Tensions

We imaginė that the excitation of atoms to the state when they emit light arises due to the shifting, through the introduction of outside energy, of one electron of the atom from its normal level (E_1) (in the interpretation of Bohr's theory) to a higher energy level (E_2). This state of excitation is not stable and after a very short time (10^{-8} sec) the electron returns to its normal level. It can pass through stages whose energy is less than E_2, unless these transitions are not allowed by the selection rules. The surplus of energy ΔE is radiated in the form of a quantum of light. In the case of a direct return to the normal level the wavelength of the emitted line of the shortest wavelength is expressed by the relation

$$\Delta E = E_2 - E_1 = \boldsymbol{h}\nu\,, \tag{1}$$

where $\boldsymbol{h}$ is Planck's constant ($\boldsymbol{h} = 6.624 \,.\, 10^{-27}$ erg. sec.), and ν the frequency of the emitted line.

As $\nu = \boldsymbol{c}/\lambda$, we can determine the wavelength from the equation (1) in the following way

$$\lambda = \frac{\boldsymbol{hc}}{\Delta E}\,.$$

The energy ΔE which must be supplied to the atom to produce radiation of a certain wavelength will be expressed in electronvolts.*

In microscopical terminology it is called excitation tension. For example, to produce the Mg radiation of wavelength 3,838.26 Å the atom must be supplied with energy equal to 5.94 eV. Similarly, ionization tension is denoted and rated in electronvolts as that quantity of energy which must be supplied to an atom in order to remove one electron from it and thus bring the desired ion into existence. If the rating corresponds to the removal of the first electron from the neutral atom, or to the removal of the second or the third electron, the tension is called primary, secondary, and teıtiary ionization tension respectively. Since the neutral state of an atom is indicated as a rule by the Roman numeral I following the symbol of the element (e.g. Al I), the numeral II (III, IV) and so on is attached to indicate the fact that the atom is in the first (second, third and so on) stage of ionization.

The total energy which is required to achieve multiple ionization is calculated by adding the different energy values required to push the electron to the various stages of ionization. For instance, the first stage of ionization tension for aluminium is 5.985 eV, the second stage of ionization tension being 18.824 eV. It is therefore necessary to apply energy of (5.985 + 18.824) = 24.809 electronvolts if an atom

* Numerically 1 eV equals $1 \,.\, 60 \,.\, 10^{-12}$ erg and represents the energy supplied to an electron in an electric field when crossing a difference of potentials of 1 V.

of aluminium is to be converted into the ion Al^{2+}. In the same way the excitation energy ratings of the analytical lines given in the coincidence tables are based on the normal states of atoms and ions. This means that it is necessary to add to the excitation energy ratings of the lines in the spectrum given for the ion lines the value of the atom ionization energy if the value of the excitation energy of the lines of the ion spectrum is to be calculated, see Table XIV.

A knowledge of line excitation tensions, or of atom ionization tensions, is of considerable importance in practical spectroscopy. Fig. 1 shows the primary ionization tensions of the elements given in the Periodic Chart of Elements. It is obvious at first glance that the periodicity in the atomic structure is reflected also in the periodicity of the ionization tensions. In conformity with the experience that they may be excited in the easiest way, alkali metals need the lowest ionization tensions, whereas rare gases, on the contrary, need the highest ionization tensions.

Most frequently, however, we make use of the knowledge of the spectral line tensions when seeking suitable doublets of lines for quantitative analysis. Those lines should be selected that have approximately the same excitation tensions because such doublets show high stability of the ratio of intensities even under uncontrollable variations of the excitation conditions.

It is also possible to determine whether the line under study is liable to self-inversion when the low level energy is known. In this case we set out from the fact that those lines are subjected to the absorption phenomenon which are in the basic state when their energy $E_1 = 0$, or in the unstable state when $E_1 = 1$ eV. On the other hand lines having energy E_1 of the value of several eV's in their low levels show no inclination to self-inversion and are, therefore, suitable for quantitative determination.

The low level energy E_1 is determined from the relation

$$E_2 - E_1 = h\nu \, .$$

Knowing the excitation tension of the upper level expressed in electronvolts (E_2), the following rearranged relation

$$E_1 = E_2 - \frac{12.395}{\lambda}$$

should be used where λ is rated in Å.

2.4. Characteristics of the Lines

The spectral line wavelengths given in the tables are given in Å with an accuracy to two decimal places in all cases in which wavelength ratings of such accuracy could be found in the literature. In the case of the wavelength coincidence of two or more lines, differentiation was achieved by the addition of the third decimal place. In spite of this, however, it was impossible to eliminate cases in which the wavelengths of some lines could not be distinguished numerically. In those cases the sequence of lines in the tables is the same as their sequence in the spectrum.

The stage of ionization is indicated for every wavelength if it is contained in Harrison's tables. Atomic lines are indicated by I, while ion lines, in accordance with their stage of ionization, are followed by the numerals II, III and so forth.

In conformity with international practice, the letter *R* indicates the lines inclined to self-inversion (self-absorption or the so-called inverted lines). The letter *h* indicates diffuse lines and in square brackets there are given the lines whose intensity was rated on the ground of the spectrum produced by a discharge in low pressure chambers. All these markings are used with the intensity ratings. Below is a list of the letters used to characterise coincidence lines in the tables.

O arc
J spark
bh band bead
h diffuse, dull line
I interferometric measurement
JS international primary standard
l asymmetric line (shifted to the longer wavelength portion)
r weak resonance line showing inversion
R strong resonance line showing inversion
s asymmetric line (shifted towards the shorter wavelength region)
S international secondary standard
W wide or composite line
I line emitted by neutral atom
II line emitted by an atom in the first stage of ionization
m average value of several rating (intensity rating for lines emitted in discharge tubes, e.g. in Geissler tubes or in discharge tubes with a hollow cathode). Only spark lines are selected as in coincidence lines
* the case of the ratings of relative intensity that, in view of their low intensity, lines of the relevant conditions of excitation are not given in the coincidence tables
*P** links up with the preceding table
*N** links up with the following table
t_0 melting point
t_1 boiling point

2.5. Evalution with the Help of Coincidence Tables

The coincidence tables of analytical lines prove extremely valuable when qualitative spectrograms are evaluated. The method of application of the tables can be shown schematically as follows:

First of all, the spectrogram should be evaluated with the use of comparison patterns with photographed spectra of iron, provided with indices of the location of the individual analytical lines as normally supplied with spectrographs. The provisional results should be classified into two groups.

1. Those elements which were unequivocally determined, i.e. the presence of the element is evidenced by at least three analytical lines. Elements of this group are indicated in the following explanation by the letter N.

2. Those elements whose identification is hampered by real or assumed coincidences. One element of this group is designated by the letter X.

Now the analytical lines pertaining to the element X must be found in the tables and it must be checked whether the first, second or third region does not contain elements pertaining to the group N, i.e. lines of elements which were found and identified with all certainty in the investigated spectrum.

The checking procedure should be carried out on such a line which does not present any possibility of coincidence of the element X with the element of the group N in the coincidence regions.

If a spectrograph of medium dispersion is used and the line is well developed (not extending to the sides), it will be sufficient to consider the first region.

Should it happen that with each analytical line there appears in the coincidence region under consideration an element of the group N as an interfering element, the identification must, in its further stages, be effected according to the relation of the intensity of the "interfering" line of some elements N to the intensity of the line X. For every element an adequate number of analytical lines are given and it is therefore possible to find such characteristically distinct intensities for every possible combination that even in the quoted (rare) case, when all the analytical lines of the assumed element X contain in their coincidence regions all the elements of the group N, unequivocal identification can be achieved when the intensities of the elements of both groups are compared.

3. MEASUREMENT OF WAVELENGTHS OF LINES

Direct determination of the line wavelength in a spectrogram can be realized in several ways. The simplest method consists in the direct interpolation of the wavelength of the two lines which are known and between which the investigated lines lie. The most suitable lines for the interpolation are the lines of iron, arc or spark lines, which occur in large numbers along the whole spectrum region. This fact makes it possible to divide the dispersion curve into sufficiently small sectors which may be considered, at the first approach, to be a straight line. The lines of iron selected for this purpose are given in Table X which cover, at intervals of 25 Å, the whole of the emission spectrum from 2,000 to 9,000 Å. If the wavelength of an unknown line λ_x must be determined, the line must first be approximately located, using the scale, and then in Table X must be found the nearest two lines of iron (λ_1 and λ_2) between which the investigated line lies (Fig. 6 and Fig. 7). The given lines λ_1 and λ_2 must be found in the spectrum of iron correlated to the spectrum which is being analysed, and their positions (d_2 and d_1) measured by means of a measuring microscope or a line microphotometer. If the position of the unknown line is d, then the wavelength of that line is given by the following equation:

$$\lambda_x = \lambda_1 + \frac{\lambda_2 - \lambda_1}{d_2 - d_1}(d - d_1)\,. \qquad (1)$$

4. LIST OF TABLES

IX. Spark Spectrum of Air (Page 47).

X. Selected Lines of Iron in the Region from 2,000 to 10,000 Å (Page 50).

XI. Wavelength Standards (Page 52).

XII. Molecule Bands of Inorganic Compounds which Occur in Emission Spectra in so far as they are Useful for the Analysis. The bands are listed in alphabetical order according to the molecule symbols (Page 55).

XIII. Molecule Bands of Inorganic Compounds Occurring in Emission Spectra in so far as they are Useful for the Analysis. The bands are listed according to wavelength (Page 56).

XIV. Ionization Potentials of the Neutral Atoms of the Elements and of their Ions (Page 57).

XV. Classification of Electrons by their Quantum Numbers into Levels of the Electron Shell (Page 59).

XVI. Coincidence of Analytical Lines for the Region from 2,000 Å to 10,000 Å (Page 93).

XVII. Coincidence of the Analytical Lines of the Rare Earth Elements (Page 1083).

5. CONCLUSION

Arrangement of the tables of the spectral lines represented an extraordinarily difficult task for the authors. It is quite understandable that it was impossible to check by experiments tens of thousands of data and therefore already published data had to be used. On this occasion it should be stated that in the whole of world literature on this subject not one single fundamental tabular work containing all spectral lines together with their intensity rating under strictly defined excitation conditions can be found. Also lacking is a uniting angle of view when the line intensities are being evaluated, this being felt particularly where different spectral regions are concerned. However, in this connection it should be stressed that the intensity ratings of spectral lines are an integral part of the tables serving for qualitative analysis. Apart from accurate measurement of the wavelength in angstroms, the intensity rating is absolutely essential for evaluation of a spectral line.

In this respect the best spectroscopic tables to date are G. R. Harrison's MIT Wavelength Tables, the latest edition of which appeared in 1960. They contain more than 110,000 spectral lines, 75,000 of which were checked by the author and his co-workers, corrected and listed in those coincidence tables. G. R. Harrison's Wavelength Tables contain lines representing, when intensities are being investigated, 99% of the entire light flux passing through the spectrograph. We certainly cannot maintain that the tables contain all the lines. Indeed, the number of lines is constantly rising due to continuously increasing scientific knowledge and possibilities. An important part is played in this respect by the scientific or technical importance of an element in connection with new techniques and technology. As examples we can mention uranium, zirconium, boron, titanium and other elements.

In the preface to his tables Harrison mentions similar facts, although the preface was written as long ago as in the year 1939. It is obviously that the number of lines as well as more precise knowledge of their properties are proportional to the time spent in studying them.

Harrison's tables show, however, certain deficiences, the most important of them being the omission of intensity ratings. The electrical conditions under which they were obtained are not, on the one hand, unequivocally specified and, on the other hand (judging by the comparison of intensities of lines measured under normal conditions prevailing in spectroscopic practice), they do not conform to the conditions under which a routine spectral analysis is usually carried out. And thus it sometimes happens that the intensity ratings of a series of lines which are well known to the experienced analyst are in contradistinction to the intensity ratings in the scale given by Harrison. Nevertheless, Harrison's work is still the best and, in particular, the most complete summary of spectral lines. Proof of this lies also in the fact that the spectral line tables published in the USSR (later translated and published in the G. D. R.) by the foremost Soviet scientists (Prokofjev, V. K., Rajskij S. M., Zajdel A. N.) were based upon Harrison's data.

In conclusion, a further explanation should be given of the reasons that induced the authors to make certain corrections. When starting this work it was agreed that the tables should be used with spectrographs of all dispersions. This made a third coincidence range essential. The said range is necessary and useful for spectrographs of small dispersion (such as Q 12). By the introduction of a third coincidence range the total number of lines was increased only slightly (by approximately 5%) as for this range there were selected only lines of an intensity equal to or exceeding the intensity of the investigated line.

The wavelengths of the lines are given with an accuracy up to two decimal places with the following exceptions:

a) in cases in which the two measured wavelengths differ only in the third decimal place, also this third decimal is given;

b) in all cases in which the wavelengths of the selected lines of iron (in Table X) were measured with an accuracy of three decimal places, those values were left unchanged.

The reason for this was that it is just those lines that serve as wavelength standards for calculations of the wavelengths of unknown lines either by the method of interpolation or by plotting of a part of the dispersion curve. The last decimal places in the measurements of wavelengths were corrected in conformity with the latest sources available, but it should not be assumed that these data are final and correct.

6. BIBLIOGRAPHY

Eder J. M., Valenta E.: Atlas typischer Spektren. Wien 1924.

Gatterer A., Junkers J.: Atlas der Restlinien von 30 chemischen Elementen. Castel Gandolfo 1937.

Gatterer A., Junkers J.: Funkenspektrum des Eisens von 4,650—2,242 Å. Castel Gandolfo 1935.

Gatterer A., Junkers J.: Arc spectrum of Iron from 8,388—2,242 Å. Castel Gandolfo 1935.

Gerlach W., Riedel E.: Chemische Emissionsspektralanalyse, III. Teil — Tabellen zur quantitativen Analyse. Leipzig 1939.

Harrison G. R.: Massachusetts Institute of Technology. "Wavelength Tables". John Wiley & Sons. New York 1960.

Kayser H., Ritschel K.: Tabelle der Hauptlinien der Linienspektren aller Elemente. II. ed. Leipzig 1939.

Lowe F.: Atlas der Analysenlinien der wichtigsten Elemente. 2. ed. Dresden—Leipzig 1936.

Q 24 Eisenspektrum von 4,555 Å bis 2,227 Å. VEB, Karl Zeiss, Jena 1957.

Kalinin S. K., Javnel A. A., Najmark L. E.: Atlas dugovogo i iskrovogo spektrov železa. Moskva 1953.

Knop J.: Chem. listy *50* (1956).

Hartmann J.: Über eine einfache Interpolationsformel für prismatisches Spektrum. Publ. Potsch. Obs. 12. Anhang (1898) Astrophys. *8* (1898).

Harmy M.: Comptes Rendus *160* (1915).

Salet P.: Bull. de l'Observ. de Lyon 1921, Comptes Rendus *160* (1915), *175* (1922).

Kayser H.: Handbuch der Spektroskopie. Leipzig 1900.

Zajdel' A. N., Prokofjev V. K., Rajskij S. M.: Tablicy spektralnych linij. Moskva—Leningrad 1952.

Lange N. A.: Handbook of chemistry. New York, Toronto, London 1961.

ILLUSTRATIONS

Period	Ia	IIa	IIIb	IVb	Vb	VIb	VIIb	VIIIb			Ib	IIb	IIIa	IVa	Va	VIa	VIIa	VIIIa
1																	1 H 13.6	2 He 24.6
2	3 Li 5.4	4 Be 9.3											5 B 8.3	6 C 11.3	7 N 14.54	8 O 13.6	9 F 17.4	10 Ne 21.6
3	11 Na 5.1	12 Mg 7.6											13 Al 6.0	14 Si 8.1	15 P 11.0	16 S 10.3	17 Cl 13.0	18 Ar 15.7
4	19 K 4.3	20 Ca 6.1	21 Sc 6.6	22 Ti 6.8	23 V 6.7	24 Cr 6.8	25 Mn 7.4	26 Fe 7.9	27 Co 7.9	28 Ni 7.6	29 Cu 7.7	30 Zn 9.4	31 Ga 6.0	32 Ge 8.1	33 As 9.8	34 Se 9.7	35 Br 11.8	36 Kr 14.0
5	37 Rb 4.2	38 Sr 5.7	39 V 6.6	40 Zr 6.9	41 Nb 6.8	42 Mo 7.4	43 Tc	44 Ru 7.55	45 Rh 7.7	46 Pd 8.3	47 Ag 7.6	48 Cd 9.0	49 In 5.8	50 Sn 7.3	51 Sb 8.6	52 Te 9.0	53 J 10.4	54 Xe 12.1
6	55 Cs 3.9	56 Ba 5.2	57 La 5.6	72 Hf 5.5	73 Ta 6.0	74 W 8.0	75 Re 7.9	76 Os 8.7	77 Ir 9.2	78 Pt 9.0	79 Au 9.2	80 Hg 10.4	81 Tl 6.1	82 Pb 7.4	83 Bi 8	84 Po	85 At	86 Rn 10.7
7	87 Fr	88 Ra 5.3	89 Ac	90 Th	91 Pa	92 U 4	93 Np	94 Pu	95 Am	96 Cm	97 Bk	98 Cf	99 E	100 Fm	101 Mv	102 No		
6	**Lanthanides**				58 Ce 6.6	59 Pr 5.8	60 Nd 6.3	61 Pm	62 Sm 5.6	63 Eu 5.7	64 Gd 6.2	65 Tb 6.7	66 Dy 6.8	67 Ho	68 Er	69 Tm	70 Yb 6.25	71 Lu 5.0

Fig. 1. The relative values of excitation potentials.

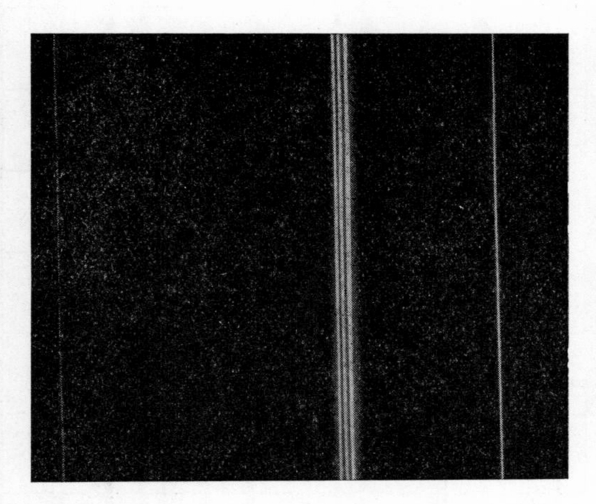

Fig. 2. Triplet Fe at 3,000 Å. Spectrograph Q 24. Magnification 25 ×.

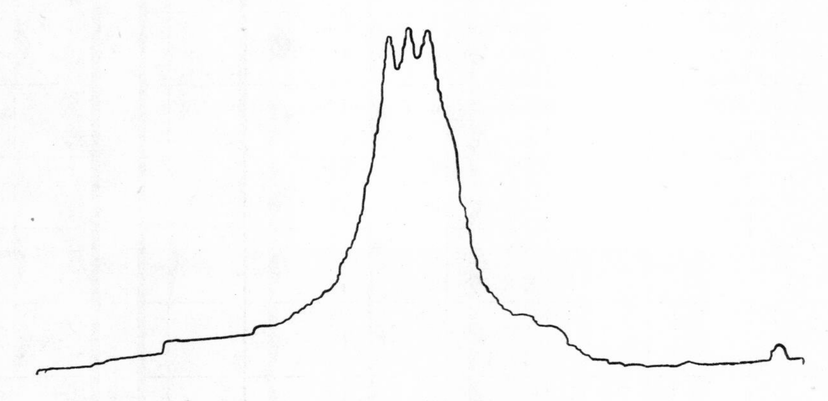

Fig. 3. Photometric record of the triplet Fe at 3,100 Å.

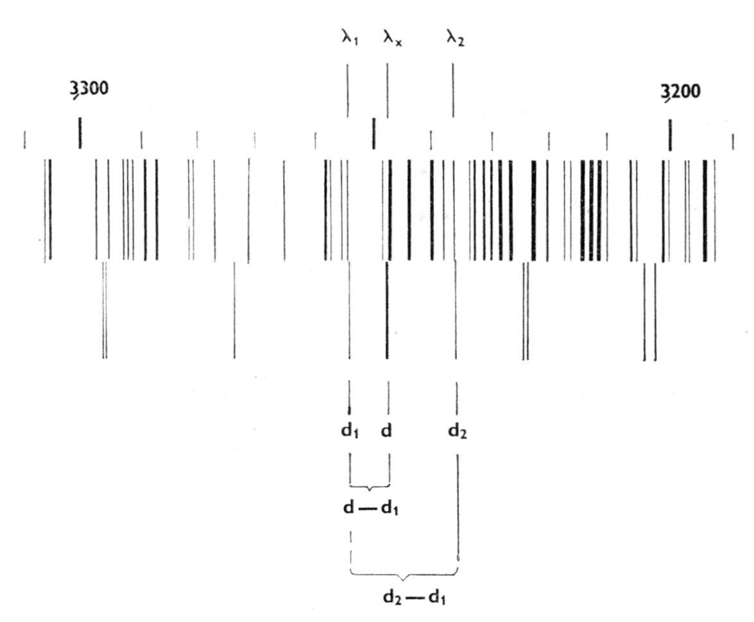

Fig. 4. An example of wavelength determination of an unknown line using interpolation of two lines the wavelengths of which are known, Fe 3,254.36 Å and Fe 3,236.22 Å. The distance ($d_2 - d_1$) equals (10.32 mm), the distance ($d - d_1$) is 3.87 mm.
These values when put in the equation (1) will thus give λ_x 3,247.56 Å. In the Wavelength Tables it will be found that this value is corresponding to the line of copper λ 3,247.54 Å.

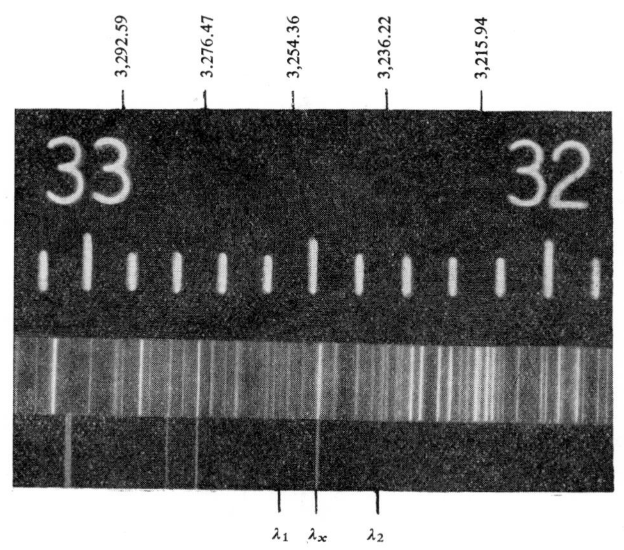

Fig. 5. Section of the spectrum in the region 3,200 – 3,300 Å. The upper part is the spectrum of iron, the lower part contains the lines of copper.

TABLES I—XV

Table I

List of Elements Given in the Tables

Symbol	Table	Symbol	Table
Ar	II, III, VII, VIII, IX, XVI	N	II, III, VII, VIII, IX, XVI
Ac	VI	Na	II, III, XVI
Ag	II, III, XVI	Nb	II, III, XVI
Al	II, III, XVI	Nd	II, IV, V, XVII
Am	VI	Ne	II, III, VII, VIII, IX, XVI
As	II, III, XVI	Ni	II, III, XVI
Au	II, III, XVI	Np	VI
B	II, III, XVI	O	II, III, VII, VIII, IX, XVI
Ba	II, III, XVI	Os	II, III, XVI
Be	II, III, XVI	P	II, III, XVI
Bi	II, III, XVI	Pa	VI
Br	II, III, XVI	Pb	II, III, XVI
C	II, III, XVI	Pd	II, III, XVI
Ca	II, III, XVI	Pr	II, IV, V, XVII
Cd	II, III, XVI	Pt	II, III, XVI
Ce	II, III, XVI	Pu	VI
Cl	II, III, VII, VIII, XVI	Ra	II, III, XVI
Co	II, III, XVI	Rb	II, III, XVI
Cr	II, III, XVI	Re	II, III, XVI
Cs	II, III, XVI	Rh	II, III, XVI
Cu	II, III, XVI	Rn	II, III, VI, VIII, XVI
Dy	II, IV, V, XVII	Ru	II, III, XVI
Er	II, IV, V, XVII	S	II, III, XVI
Eu	II, IV, V, XVII	Sb	II, III, XVI
F	II, III, VII, VIII, XVI	Sc	II, III, XVI
Fe	II, III, X, XVI	Se	II, III, XVI
Ga	II, III, XVI	Si	II, III, XVI
Gd	II, IV, V, XVII	Sm	II, IV, V, XVII
Ge	II, III, XVI	Sn	II, III, XVI
H	II, III, VII, VIII, XVI	Sr	II, III, XVI
He	II, III, VII, VIII, IX, XVI	Ta	II, III, XVI
Hf	II, III, IV, V, XVI	Tb	II, IV, V, XVII
Hg	II, III, XVI	Te	II, III, XVI
Ho	II, IV, V, XVII	Th	II, III, XVI
In	II, III, XVI	Ti	II, III, XVI
Ir	II, III, XVI	Tl	II, III, XVI
J	II, III, XVI	Tm	II, IV, V, XVII
K	II, III, XVI	U	II, III, XVI
Kr	II, III, VII, VIII, IX, XVI	V	II, III, XVI
La	II, III, IV, V, XVI	W	II, III, XVI
Li	II, III, XVI	Xe	II, III, VII, VIII, IX, XVI
Lu	II, IV, V, XVII	Y	II, III, XVI
Mg	II, III, XVI	Yb	II, IV, V, XVII
Mn	II, III, XVI	Zn	II, III, XVI
Mo	II, III, XVI	Zr	II, III, XVI

Table II

Table of Analytical Lines of the Elements Given in the Coincidence Tables

Symbol	Atomic No.	λ	*J*	*O*	eV
Ag	47	**II 2,246.41**	**300**	**25**	**10.63**
		I 2,375.06	300	300	8.96
		II 2,413.18	300	50	10.18
		II 2,437.79	**500**	**60**	**9.93**
		II 2,614.59	300	—	15.51
		II 2,711.21	300	1	15.71
		I 2,824.37	200	150	8.13
		II 2,938.55	200	200	14.98
		I 3,280.68	**1000*R***	**2000*R***	**3.76**
		I 3,382.89	**700*R***	**1000*R***	**3.76**
		3,968.22	60	100	—
		I 4,055.26	500*R*	800*R*	6.72
		I 4,668.48	70	200	6.44
		I 5,209.07	**1000*R***	**1500*R***	**6.04**
		I 5,465.49	**500*R***	**1000*R***	**6.04**
Al	13	I 2,373.36	100*R*	200*R*	5.22
		II 2,669.17	**100**	**3**	**4.64**
		2,816.18	**100**	**15**	**11.82**
		I 3,082.16	**800**	**800**	**4.02**
		I 3,092.71	**1000**	**1000**	**4.0**
		I 3,944.03	**1000**	**2000**	**3.14**
		I 3,961.53	**2000**	**3000**	**3.14**
		II 6,243.36	**80**	—	**15.06**
Ar	18	II 3,545.58	(300)	—	23.25
		I 3,606.52	(1000)	—	15.06
		I 3,948.98	(2000)	—	14.68
		I 4,266.29	(1200)	—	14.53
		I 4,702.32	(1200)	—	14.46
		I 6,965.43	**(400)**	—	**13.32**
		I 7,067.22	**(400)**	—	**13.29**
		I 7,503.87	**(700)**	—	**13.47**
		I 8,115.31	**(5000)**	—	**13.07**
As	33	**I 2,288.12**	**5**	**250**	**6.7**
		I 2,349.84	**18**	**250**	**6.6**
		I 2,456.53	**8**	**100*R***	**6.5**
		I 2,780.20	**75**	**75*R***	**6.7**
		I 2,860.45	**50**	**50*R***	**6.6**
		I 3,032.84	70	125*R*	6.4
		II 3,116.63	150	—	—
		II 3,749.77	100	—	—
		II 4,352.25	200	—	13.11
		4,474.60	200	—	—
		II 4,539.97	200	—	—
		II 4,630.14	200	—	—
		II 4,407.82	200	—	—
		II 5,331.54	200	—	12.43
Au	79	**I *2,427.95**	**100**	**400*R***	**5.10**
		I 2,675.95	**100**	**250*R***	**4.63**
		I 3,122.78	5	500	5.10
		II 3,804.00	150	25	—

Symbol	Atomic No.	λ	J	O	eV
		I 4,792.60	60	200	7.69
		I 5,837.40	10	400	—
		I 6,278.18	20	700	—
B	5	**I 2,496.77**	**300**	**300**	**4.96**
		I *2,497.73	**400**	**500**	**4.96**
		II 3,451.41	**100**	—	**12.69**
Ba	56	**II 2,335.27**	**100*R***	**60*R***	**6.83**
		I 3,071.59	**50*R***	**100*R***	**4.04**
		I 3,501.12	**20**	**1000**	**3.54**
		II *4,554.04	**200**	**1000*R***	**2.72**
		II 4,934.09	**400**	**400**	**2.51**
		I 5,424.62	**30*R***	**100*R***	**3.81**
		I 5,519.12	**60**	**200**	**3.82**
		I *5,535.55	**200*R***	**1000*R***	**2.24**
		I 5,777.67	**100*R***	**500*R***	**3.82**
Be	4	**I *2,348.61**	**50**	**2000*R***	**5.28**
		II *3,130.42	**200**	**200**	**3.95**
		II 3,131.07	**150**	**200**	**3.95**
		I 3,321.01	—	**50**	**6.45**
		I 3,321.09	**15**	**100**	**6.45**
		I 3,321.34	**30**	**1000**	**6.45**
		4,672.2	100	—	—
		II 4,673.46	100	—	14.81
Bi	83	**I 2,897.97**	**500*R***	**500*R***	**5.7**
		I 2,938.30	**300**	**300**	**6.1**
		I 2,989.03	**100**	**250**	**5.5**
		I *3,067.72	**2000**	**3000*R***	**4.04**
		I 4,722.55	**100**	**1000**	**4.04**
		II 5,209.29	600	—	5.72
		5,552.35	100	500	—
Br	35	2,389.69	(70)	—	—
		2,541.45	(40)	—	—
		2,660.49	(25)	—	—
		2,872.59	(25)	—	—
		4,542.93	(250)	—	—
		II *4,704.86	**(250)**	—	**14.4**
		II 4,785.50	**(400)**	—	**14.4**
		II 4,816.71	**(300)**	—	**14.4**
		5,506.78	(300)	—	—
C	6	III 2,296.89	(200)	—	—
		I *2,478.57	**(400)**	**400**	**7.69**
		II 2,512.03	(400)	—	18.66
		II 2,836.71	**200**	—	**16.34**
		II 4,267.02 }	**350**	—	**20.94**
		II 4,267.27 }	**500**	—	**20.94**
		I 5,793.51	—	30	10.08
		II 6,578.03	500	—	16.33
Ca	20	I 2,398.56	20	100*R*	—
		II 2,573.09	150	3	—
		II 3,158.87	**300**	**100**	**7.05**

Symbol	Atomic No.	λ	J	O	eV
		II 3,179.33	**400**	**100**	**7.05**
		I 3,644.41	15	200	5.30
		II *3,933.67	**600*R***	**600*R***	**3.15**
		II 3,968.47	**500*R***	**500*R***	**3.12**
		I *4,226.73	**50**	**500*R***	**2.93**
		I 4,425.44	**20**	**100**	**4.68**
		I 4,434.96	**25**	**150**	**4.63**
Cd	48	**II 2,265.02**	**300**	**25**	**5.47**
		I *2,288.02	**300*R***	**1500*R***	**5.41**
		II 2,312.84	**200**	**1**	**11.14**
		II 2,748.58	**200**	**5**	**10.28**
		I 3,261.06	**300**	**300**	**3.80**
		I 3,403.65	**500**	**800**	**7.37**
		I 3,466.20	**500**	**1000**	**7.37**
		I 3,610.51	**500**	**1000**	**7.37**
		I 4,678.16	200	200	6.39
		I 4,799.92	300	300	6.39
		I 5,085.82	500	1000	6.39
		I 6,438.47	**1000**	**2000**	**7.34**
Ce	58	2,225.10	100	—	—
		II 3,539.09	10	100	4.4
		II 3,560.80	2	300	4.8
		II 3,577.46	12	300	4.5
		II 4,012.39	**20**	**60**	**4.2**
		II *4,186.60	**25**	**80**	**3.34**
		5,522.99	—	100	—
Cl	17	II 3,329.12	(150)	—	20.05
		II 3,353.39	(125)	—	18.03
		II 3,827.62	(150)	—	21.48
		II 3,860.83	(150)	—	19.17
		II 4,132.48	(200)	—	19.00
		II 4,740.40	(150)	—	20.00
		II 4,768.68	(150)	—	19.68
		II 4,794.54	**(250)**	—	**15.95**
		II 4,810.06	**(200)**	—	**15.94**
Co	27	**II 4,819.46**	**(200)**	—	**15.94**
		II *2,286.16	**300**	**40**	**5.84**
		II 2,378.62	**50**	**25**	**5.62**
		II 2,519.82	**200**	**40**	**6.24**
		I 3,405.12	**150**	**2000*R***	**4.07**
		I 3,412.34	100	1000*R*	4.16
		I *3,453.50	**200**	**3000*R***	**4.02**
		I 3,465.80	**25**	**2000*R***	**3.57**
		I 3,502.28	20	2000*R*	3.97
		I 3,529.81	**20**	**1000*R***	**4.03**
		I 3,995.31	20	1000*R*	4,0
		I 4,118.77	—	1000*R*	4.06
		I 4,121.32	25	1000*R*	3.93
		I 4,530.96	8	1000	5.66
		I 4,581.60	10	1000	5.66
		I 4,780.01	500	500	5.87
		I 4,813.48	6	1000	5.79

Symbol	Atomic No.	λ	J	O	eV
Cr	24	**II *2,835.63**	**400R**	**100**	**12.6**
		II 2,843.25	**400R**	**125**	**12.6**
		II 2,849.84	**150R**	**80**	**12.6**
		II 2,855.68	**200**	**60**	**12.5**
		II 2,860.93	**100**	**60**	**12.5**
		I 3,578.69	400R	500R	3.46
		I 3,593.49	400	500	3.44
		I 3,605.33	400R	500R	3.43
		I *4,254.35	**1000R**	**5000R**	**2.91**
		I 4,274.80	**800R**	**4000R**	**2.90**
		I 4,289.72	**800R**	**3000R**	**2.89**
		I 5,204.52	**100**	**400R**	**3.32**
		I 5,206.04	**200**	**500R**	**3.32**
		I 5,208.44	**100**	**500R**	**3.32**
Cs	55	I 3,876.39	—	300	3.17
		I 3,888.65	10	150	3.16
		I 4,555.36	**100**	**2000R**	**2.72**
		I 4,593.18	**50R**	**1000R**	**2.70**
Cu	29	**II 2,192.26**	**500**	**25**	**5.76**
		II 2,246.99	**500**	**30**	**5.51**
		I 2,824.37	300	1000	5.78
		I *3,247.54	**2000R**	**5000R**	**3.82**
		I 3,273.96	**1500R**	**3000R**	**3.78**
		I 4,539.70	80	100	7.88
		I 4,586.95	80	250	7.80
		I 5,105.54	—	**500**	**3.82**
		I 5,153.24	—	**600**	**6.19**
		I 5,218.20	—	**700**	**6.19**
		I 5,782.13	—	1000	3.79
F	9	II 3,501.42	(200)	—	28.66
		II 3,503.10	(400)	—	28.66
		II 3,505.61	(600)	—	28.66
		II 3,847.09	(800)	—	25.12
		II 3,849.99	(600)	—	25.11
		II 3,851.67	(200)	—	25.11
		II 4,246.16	(300)	—	31.56
		I 6,239.64	(300)	—	14.68
		I 6,856.02	**(1000)**	—	**14.50**
		I 6,902.46	**(500)**	—	**14.52**
Fe	26	**II *2,382.04**	**100R**	**40R**	**5.20**
		II 2,395.62	**100**	**50**	**5.22**
		I 2,483.27	50	500R	4.99
		II 2,599.40	1000	1000	4.77
		II 2,755.74	100	300	5.48
		I 3,020.64	600R	1000R	4.11
		I 3,581.19	600R	1000R	4.32
		I *3,719.93	**700**	**1000R**	**3.32**
		I 3,734.87	600	1000R	4.18
		I 3,737.13	**600**	**1000R**	**3.35**
		I 3,745.56	**500**	**500**	**3.39**
		I 3,748.26	**200**	**500**	**3.41**
		I 4,271.76	700	1000	4.39
		I 4,307.91	800R	1000R	4.44

Symbol	Atomic No.	λ	J	O	eV
		I 4,325.76	700	1000	4.47
		I 4,383.55	800	1000	4.31
		I 4,404.75	700	1000	4.37
Ga	31	**I 2,874.24**	**15*R***	**10**	**4.29**
		I 2,943.64	**20*R***	**10**	**4.29**
		I 4,032.98	**500*R***	**1000*R***	**3.07**
		I *4,172.06	**1000*R***	**2000*R***	**3.07**
		I 6,396.61	20	—	4.98
Ge	32	***2,651.18**	**20**	**40**	**4.8**
		I 2,651.58	**20**	**30**	**4.7**
		2,709.63	**20**	**30**	**4.6**
		2,754.59	20	30	—
		I 3,039.06	**1000**	**1000**	**4.96**
		I 3,124.82	80	200	4.85
		I 3,269.49	**300**	**300**	**4.67**
		I 4,226.58	**50**	**200**	**4.96**
		II 4,814.80	200	—	12.41
		II 5,131.7	100	—	—
		II 5,893.46	100	—	9.84
H	1	3,797.91	(20)	—	13.45
		3,835.40	(40)	—	13.42
		3,889.06	(60)	—	13.38
		3,970.07	(80)	—	13.31
		4,101.74	(100)	—	13.21
		4,340.47	(200)	—	13.01
		4,861.33	**(500)**	—	**12.74**
		6,562.73	**(1000)**	—	—
		6,562.85	**(2000)**	—	—
He	2	I 3,187.74	(200)	—	23.70
		II 3,203.14	(100)	—	52.23
		I *3,888.65	**(1000)**	—	**23.00**
		II 4,685.75	**(300)**	—	**51.00**
		I 5,875.62	**(1000)**	—	**23.06**
Hf	72	**II 2,513.03**	**70**	**25**	—
		II 2,516.68	**100**	**35**	—
		II *2,641.41	**125**	**40**	—
		II 2,820.22	**100**	**40**	—
		I 2,916.48	**15**	**50**	—
		I 2,940.77	**12**	**60**	—
		I 3,072.88	**18**	**80**	**4.04**
		II 3,134.72	**125**	**80**	**4.33**
		II 5,040.82	150	100	3.94
		II 5,298.06	100	80	4.23
		II 5,311.60	150	100	4.12
		II 6,644.60	200	100	3.64
Hg	80	**I 2,536.52**	**1000*R***	**2000*R***	**4.88**
		I 3,125.66	150	200	8.85
		I 3,131.55	300	400	8.82
		I 3,131.83	100	200	8.85
		I 3,650.15	**500**	**200**	**8.86**
		I 3,663.28	**400**	**500**	**8.85**

Symbol	Atomic No.	λ	J	O	eV
		I 4,046.56	**300**	**200**	**7.73**
		I 4,358.35	**500**	**3000**	**7.73**
		I 5,460.74	**(200)**	—	**7.73**
		I 5,769.59	200	600	8.85
		I 5,790.65	(1000)	—	8.85
In	49	I 2,560.23	50*R*	150*R*	4.81
		I 2,710.27	200*R*	800*R*	4.81
		I 2,753.88	300	300*R*	5.01
		I 3,039.36	**500*R***	**1000*R***	**4.1**
		I 3,256.09	**600*R***	**1500*R***	**4.1**
		I 3,258.56	**300*R***	**500*R***	**4.1**
		I 4,101.77	**1000*R***	**2000*R***	**3.02**
		I *4,511.32	**4000*R***	**5000*R***	**3.02**
Ir	77	I 2,543.97	100	200	>4.9
		I 2,639.71	15	100	>4.7
		I 2,664.79	50	200	>4,6
		I 2,694.23	50	150	4,6
		I 2,849.73	**20**	**40**	**4.3**
		I 3,133.32	2	40	>4,0
		I *3,220.78	**30**	**100**	**4.20**
		I 3,513.65	**100**	**100**	**3.52**
		I 4,268.10	15	200	3.79
		4,311.50	10	300	—
		4,399.47	100	400	—
		4,426.27	10	400	—
J	53	**2,062.38**	**(900)**	—	—
		II 2,566.26	(300)	—	—
		II 2,582.81	(400)	—	—
		II 2,878.64	(400)	—	15.19
		3,055.37	(350)	—	—
		3,078.77	(350)	—	—
		3,931.01	(400)	—	—
		3,940.24	(500)	—	—
		II 4,452.88	(700)	—	14.76
		I 4,862.31	(700)	—	9.47
		I 5,119.29	(500)	—	9.34
		II 5,161.19	**(300)**	—	**12.11**
		5,464.61	**(900)**	—	—
K	19	2,240.89	40	—	—
		I 3,102.03	—	50*R*	3.99
		I 3,217.02	20	100*R*	3.86
		I 3,217.50	25	50*R*	3.86
		I 3,446.39	100*R*	150*R*	3.60
		I 3,447.39	75*R*	100*R*	3.60
		I 4,044.14	**400**	**800**	**3.06**
		I 4,047.20	**200**	**400**	**3.06**
		II 4,388.13	40	—	23.45
		II 4,608.43	40	—	—
		I 5,339.67	—	40	3.94
		I 5,782.60	—	60	3.76
		***7,664.91**	**400**	**9000*R***	**1.62**
		7,698.98	**200**	**5000*R***	**1.61**

Symbol	Atomic No.	λ	J	O	eV
Kr	36	II 2,464.77	(100)	—	22.8
		II 2,712.40	(80)	—	23.2
		II 2,833.00	(100)	—	22.79
		II 3,607.88	(100)	—	22.97
		II 3,631.87	(200)	—	21.89
		II 3,653.97	(250)	—	21.92
		II 3,718.02	(300)	—	23.83
		II 3,920.14	(200)	—	21.93
		I 4,319.58	(1000)	—	12.78
		II 4,355.48	(3000)	—	21.92
		II 4,658.87	(2000)	—	19.30
		I 5,570.29	**(2000)**	—	**12.14**
		I 5,870.92	**(3000)**	—	**12.14**
La	57	III 3,171,67	1000	2	5.59
		II 3,337.49	300	800	4.12
		II 3,344.56	200	300	3.94
		II 3,916.05	400	400	3.40
		II *3,949.11	**800**	**1000**	**3.54**
		II 4,077.34	**400**	**600**	**3.27**
		II 4,123.23	**500**	**500**	**3.32**
		I 5,455.15	**1**	**200**	**2.40**
		I 5,930.65	—	**250**	**2.09**
		I *6,249.93	—	**300**	**2.49**
Li	3	I 2,562.54	15	150	4.83
		I 2,741.31	—	200	4.52
		I 3,232.61	**500**	**1000*R***	**3.83**
		I 3,915.0	—	200	5.02
		I 4,132.29	—	400	4.85
		I 4,273.28	100	200*R*	4.75
		I 4,602.86	—	**800**	**4.54**
		I 4,971.99	—	500	4.34
		I 6,103.64	**300**	**2000*R***	**3.87**
		I *6,707.84	**200**	**3000*R***	**1.90**
Mg	12	**II *2,795.53**	**300**	**150**	**4.43**
		II 2,802.70	**300**	**150**	**4.42**
		I *2,852.13	**100*R***	**300*R***	**4.34**
		I 3,829.35	**150**	**100**	**5.94**
		I 3,832.31	**200**	**250**	**5.94**
		I 3,838.26	**200**	**300**	**5.94**
		I 5,167.34	**50**	**100**	**5.11**
		I 5,172.70	**100**	**200**	**5.11**
		I 5,183.62	**300**	**500**	**5.11**
Mn	25	**II *2,576.10**	**2000*R***	**300*R***	**4.81**
		II 2,593.73	**1000*R***	**200*R***	**4.77**
		II 2,605.69	**500*R***	**100*R***	**4.75**
		2,794.82	5	1000*R*	—
		2,798.27	80	800*R*	—
		2,801.06	60	600*R*	—
		I* 4,030.76	**20**	**500*R***	**3.08**
		I 4,033.07	**20**	**400*R***	**3.08**
		I 4,034.49	**20**	**250*R***	**3.08**
		I 4,783.42	60	400	4.89
		I 4,823.52	80	400	4.89
		I 5,341.07	100	200	4.44

Symbol	Atomic No.	λ	J	O	eV
Mo	42	**II *2,816.15**	**300**	**200**	**11.9**
		II 2,848.23	**200**	**125**	**11.8**
		II 2,871.51	**100**	**100**	**11.7**
		II 2,890.99	**50**	**30**	**11.7**
		II 2,909.12	**40**	**25**	**11.6**
		I 3,170.35	25*R*	1000*R*	3.91
		I *3,798.25	**1000*R***	**1000*R***	**3.26**
		I 3,864.11	**500*R***	**1000*R***	**3.20**
		I 3,902.96	**500*R***	**1000*R***	**3.17**
		I 5,506.49	100	200*R*	3.58
N	7	II 3,006.86	(50)	—	24.53
		II 3,995.00	(300)	—	21.60
		I 4,099.94	**150)**	—	**13.69**
		I 4,109.98	**(1000)**	—	**13.70**
		I 4,935.03	(250)	—	13.19
		II 5,666.64	**(300)**	—	**20.62**
		II 5,676.02	**(100)**	—	**20.62**
		II 5,679.56	**(500)**	—	**20.66**
Na	11	**I 3,302.32**	**300*R***	**600*R***	**3.75**
		I 3,302.99	**150*R***	**300*R***	**3.75**
		I 5,682.66	—	**80**	**4.29**
		I 5,688.22	—	**300**	**4.29**
		I *5,889.95	**1000*R***	**9000*R***	**2.11**
		I 5,895.92	**500*R***	**5000*R***	**2.10**
Nb	41	**II *3,094.18**	**1000**	**100**	**4.52**
		II 3,130.79	**100**	**100**	**4.40**
		II 3,194.98	**300**	**30**	**4.21**
		II 3,225.48	**800*R***	**150**	**4.14**
		I 3,580.27	300	100	3.59
		I *4,058.94	**400**	**1000**	**3.18**
		I 4,079.73	**200**	**500**	**3.12**
		I 4,100.92	**200**	**300**	**3.07**
		I 4,123.81	**125**	**200**	**3.02**
		5,344.17	200	400	—
Ne	10	I 3,369.81	(500)	—	20.29
		I 3,369.91	(700)	—	20.29
		I 3,520.47	(1000)	—	20.36
		I 4,537.75	(1000)	—	21.11
		I 4,884.92	(1000)	—	20.65
		I 4,884.92	(1000)	—	21.60
		I 4,957.03	(1000)	—	21.10
		I 5,400.56	**(2000)**	—	**18.95**
		I 5,852.49	**(2000)**	—	**18.96**
		I 6,402.25	**(2000)**	—	**18.56**
Ni	28	**II 2,253.86**	**300**	**100**	**6.81**
		II 2,264.46	**400**	**150**	**6.72**
		II 2,270.21	**400**	**100**	**6.61**
		II *2,287.08	**500**	**100**	**7.28**
		I 3,002.49	100	1000*R*	4.16
		I 3,050.82	—	1000*R*	4.09
		I *3,414.77	**50**	**1000*R***	**3.65**
		I 3,461.65	50	800*R*	3.60

Symbol	Atomic No.	λ	J	O	eV
		I 3,492.96	**100**	**1000*R***	**3.65**
		I 3,515.05	**50**	**1000*R***	**3.63**
		I 3,524.54	**100**	**1000*R***	**3.54**
		I 3,619.39	150	2000*R*	3.85
		I 4,401.55	30	1000	6.00
		I 4,714.42	8	1000	6.00
		I 4,786.54	2	300	6.00
		I 5,035.37	5	300	6.09
O	8	I 3,823.47	(125)	—	15.77
		II 3,911.95	(150)	—	28.83
		I 3,947.33	(300)	—	12.28
		II 3,973.27	(125)	—	26.56
		II 4,072.16	(300)	—	28.69
		II 4,075.87	(800)	—	28.80
		I 4,368.30	(1000)	—	12.36
		II 4,414.89	(300)	—	26.24
		I 5,330.66	(500)	—	13.06
		I *7,771.93	**(1000)**	—	**10.73**
		I 7,774.14	**(300)**	—	**10.73**
		I 7,775.43	**(100)**	—	**10.73**
Os	76	**I *2,909.06**	**400**	**500*R***	**4.2**
		I 3,058.66	**500**	**500*R***	**5.05**
		I 3,262.29	**50**	**500*R***	**4.32**
		I 3.267.95	**30**	**400*R***	**3.80**
		I 3,301.56	50	500*R*	3.76
		I 4,260.85	200	200	2.91
		I 4,420.47	**100**	**400*R***	**2.80**
P	15	**I 2,534.01**	**(20)**	**50**	**7.2**
		I 2,535.65	**(30)**	**100**	**7.2**
		I 2,553.28	**(20)**	**80**	**7.1**
		I 2,554.93	**(20)**	**60**	**7.1**
		III 4,222.15	(150)	300	17.54
		II 5,296.09	(300)	—	13.14
Pb	82	**I 2,169.99**	**1000*R***	**1000*R***	**5.67**
		II *2,203.50	**5000*R***	**50**	**7.25**
		I 2,614.18	**80**	**200*R***	**5.68**
		I 2,833.07	**80*R***	**500*R***	**4.4**
		I 3,639.58	**50*h***	**300**	**4.38**
		I 3,683.47	**50**	**300**	**4.34**
		I *4,057.82	**300*R***	**2000*R***	**4.38**
Pd	46	**I *3,404.58**	**1000*R***	**2000*R***	**4.46**
		I 3,421.24	**1000*R***	**2000*R***	**4.58**
		I 3,516.94	**500*R***	**1000*R***	**4.49**
		I 3,609.55	**700*R***	**1000*R***	**4.40**
		I 3,634.70	**1000*R***	**2000*R***	**4.23**
		I 4,212.95	300	500	4.40
Pt	78	**I 2,659.45**	**500*R***	**2000*R***	**4.6**
		I 2,830.29	**600*R***	**1000*R***	**4.4**
		I 2,929.79	**200**	**800*R***	**4.2**
		I 2,997.97	**200*R***	**1000*R***	**4.23**
		I *3,064.71	**300*R***	**2000*R***	**4.05**

Symbol	Atomic No.	λ	J	O	eV
		I 4,442.55	25	800	—
		I 5,301.02	10	150	6.90
Ra	88	II 2,708.96	(200)	—	—
		II 2,813.76	(400)	—	—
		II 3,649.55	(1000)	—	—
		II *3,814.42	**(2000)**	—	**3.25**
		II 4,682.28	**(800)**	—	**2.65**
		I *4,825.91	**(800)**	—	**2.57**
		I 5,660.81	(1000)	—	—
Rb	37	I 3,350.89	150	—	3.71
		II 3,461.57	200	—	4.14
		3,492.77	300	—	—
		I 3,587.08	40	200	3.47
		I 4,201.85	**500**	**2000*R***	**2.95**
		I 4,215.57	**300**	**1000*R***	**2.94**
		I *7,800.23	—	**9000*R***	**1.59**
		I 7,947.60	—	**5000*R***	**1.56**
Re	75	2,275.25	300*R*	300*R*	—
		I 3,424.60	—	300	—
		I 3,451.81	—	100	3.59
		I *3,460.47	—	**1000**	**3.58**
		I 3,464.72	—	100	3.57
		I 4,889.17	—	**2000**	**2.53**
		I 5,275.53	—	500	2.35
Rh	45	I 3,283.57	—	150*R*	4.10
		I 3,323.09	**200*R***	**1000*R***	**3.92**
		I 3,396.85	**500*R***	**1000*R***	**3.64**
		I *3,434.89	**200*R***	**1000*R***	**3.60**
		I 3,657.99	**200**	**500**	**3.57**
		I 3,692.36	**150**	**500**	**3.35**
		I 4,128.87	150	300	3.97
		I 4,374.80	500	1000	—
		I 4,528.73	60	500*R*	—
Rn	86	3,006.8	(300)	—	—
		3,054.3	(250)	—	—
		3,621.0	(250)	—	—
		3,634.8	(250)	—	—
		I 4,349.60	(5000)	—	—
		4,625.48	(500)	—	—
		4,680.83	(500)	—	—
		I 7,055.42	**(400)**	—	**8.4**
		7,450.00	**(600)**	—	**8.5**
Ru	44	**2,678.76**	**300**	**100**	**>11.0**
		2,712.41	**300**	**80**	**>11.0**
		I 3,436.74	**150**	**300*R***	**3.75**
		I *3,498.94	**200**	**500*R***	**3.54**
		I 3,596.18	**100**	**30**	**3.70**
		I 4,554.51	200	1000*R*	3.53
		I 4,584.45	80	150*R*	3.70

Symbol	Atomic No.	λ	J	O	eV
S	16	IV 3,097.46	—	50	26.49
		III 3,497.34	100	—	—
		II 4,153.10	(600)	—	18.88
		I 4,694.13	**(50)**	—	**9.16**
		I 4,695.45	**(30)**	—	**9.16**
		I 9,212.91	**(200)**	—	**7.87**
		I 9,228.11	**(200)**	—	**7.86**
		I 9,237.49	**(200)**	—	**7.86**
Sb	51	**I *2,068.38**	**3**	**300*R***	**5.98**
		I 2,175.89	**40**	**300**	**5.69**
		I 2,528.53	**200**	**300*R***	**6.12**
		I 2,877.91	**150**	**250**	**5.36**
		I 3,232.50	**250**	**150**	**6.12**
		I 3,267.50	**150**	**150**	**5.82**
		I 5,568.09	200	6	7.92
		II 6,004.6	200	—	11.46
Sc	21	**II 3,613.84**	**70**	**40**	**3.45**
		II 3,630.74	**70**	**50**	**3.42**
		II 3,642.78	**50**	**60**	**3.40**
		I 3,907.48	**25**	**125**	**3.17**
		I *3,911.81	**30**	**150**	**3.19**
		I 4,020.40	**20**	**50**	**3.08**
		I 4,023.69	**25**	**100**	**3.10**
		II 4,670.40	300	100	4.01
		I 5,700.23	—	400*R*	3.60
Se	34	**I 2,039.85**	**(1000)**	—	**6.3**
		I 2,062.79	**(800)**	—	**6.3**
		2,591.41	(125)	—	—
		I 4,730.78	**(1000)**	—	**>2.6**
		I 4,739.03	**(800)**	—	**>2.6**
		I 4,742.25	**(500)**	—	**>2.6**
		5,142.14	(500)	—	—
		5,175.98	(600)	—	—
Si	14	**I 2,506.90**	**200**	**300**	**4.95**
		I 2,516.12	**500**	**500**	**4.95**
		I 2,528.52	**500**	**400**	**4.93**
		I *2,881.58	**400**	**500**	**5.08**
		I 2,987.65	100	100	4.93
		I 5,754.26	—	40	7.11
		I 7,423.54	—	500	7.29
Sn	50	**I *2,839.99**	**300*R***	**300*R***	**4.78**
		I 2,863.33	**300*R***	**300*R***	**4.32**
		I 3,009.15	**200**	**300**	**4.33**
		I 3,034.12	**150**	**200**	**4.30**
		I 3,175.02	**400**	**500**	**4.33**
		I 3,262.33	**300**	**400**	**4.87**
		I 4,524.74	**50**	**500**	**4.87**
		I 5,631.69	200	50	4.33
Sr	38	I 3,351.25	15	300	5.54
		II 3,380.71	**200**	**150**	**6.61**
		II 3,464.46	**200**	**200**	**6.62**

Symbol	Atomic No.	λ	J	O	eV
		II *4,077.71	**500**	**400R**	**3.04**
		II 4,215.52	**400**	**300R**	**2.94**
		I *4,607.33	**50R**	**1000R**	**2.69**
		I 4,832.08	**8**	**200**	**4.34**
Ta	70	I 2,653.27	15	200	>4.7
		2,675.90	200	150	>4.6
		I 2,714.67	8	200	5.05
		***3,311.16**	**70**	**300**	**>3.7**
		I 4,551.95	8	400	3.21
		I 4,780.94	200	50	3.91
Te	52	**I *2,142.75**	—	**600**	—
		I 2,383.25	**(300)**	**500**	**5.8**
		I 2,385.76	**(300)**	**600**	**5.8**
		4,866.22	800	—	—
		5,755.87	250	—	—
Th	90	**3,290.59**	**40**	—	**>7.3**
		3,538.75	**50**	—	**>3.5**
		3,601.04	10	8	>3.4
		3,741.19	80	80	>3.3
		II 4,019.14	**8**	**8**	**>3.08**
		4,381.86	20	30	>2.8
		II 4,391.11	40	50	>3.37
		II 4,919.81	20	50	3.28
Ti	22	3,234.52	500R	100	3.88
		II *3,349.04	**800R**	**125**	**4.31**
		II 3,361.21	**600R**	**100**	**3.71**
		II 3,372.80	**400R**	**80**	**3.68**
		II 3,383.76	**300R**	**70**	**3.66**
		I 3,635.46	**100**	**200**	**3.4**
		I 3,642.68	**125**	**300**	**3.42**
		I 3,653.50	**200**	**500**	**3.44**
		I *4,981.73	**125**	**300**	**3.33**
		I 4,991.07	**100**	**200**	**3.32**
		I 4,999.51	**80**	**200**	**3.30**
		I 5,007.21	**40**	**200**	**3.29**
Tl	81	**I 2,767.87**	**300R**	**400R**	**4.44**
		I 2,918.32	**200R**	**400R**	**5.18**
		I 3,229.75	**800**	**2000**	**4.77**
		I 3,519.24	**1000R**	**2000R**	**4.49**
		I 3,775.72	**1000R**	**3000R**	**3.28**
		I *5,350.46	**2000R**	**5000R**	**3.28**
		I 6,549.77	50	300	5.14
U	92	**3,552.17**	**12**	**8**	**>3.5**
		II 3,670.07	18	15	3.49
		3,672.58	**15**	**8**	**>3.4**
		II 3,859.58	30	20	3.24
		II 4,090.14	40	25	3.24
		II 4,241.67	**50**	**40**	**3.49**
		II 5,492.97	50	60	**2.26**

Symbol	Atomic No.	λ	J	O	eV
V	23	II *3,093.11	400R	100R	4.40
		II 3,102.30	300R	70	4.36
		II 3,110.71	300R	70	4.33
		II 3,118.38	200R	70	4.31
		II 3,125.28	200R	80	4.29
		I 3,183.41	100R	200R	3.91
		I 3,183.98	400R	500R	3.90
		I 3,185.40	400R	500R	3.96
		I *4,379.24	200R	200R	3.13
		I 4,389.97	60R	80R	3.10
		I 4,408.51	20R	30	3.08
W	74	II 2,397.09	30	18	5.56
		II 2,589.17	25	15d	5.55
		I 2,944.40	20	30	4.57
		I 2,946.98	18	20	4.57
		II 3,613.79	30	10	5.24
		I 4,008.75	45	45	3.45
		I 4,294.61	50	50	3.25
		I *4,302.11	60	60	3.24
Xe	54	II 2,979.32	(200)		17.35
		III 3,624.05	(600)		—
		II 3,922.53	(500)		—
		I 4,500.98	(500)		11.07
		I 4,624.28	(1000)		11.0
		I 4,671.23	(2000)		10.97
Y	39	II 3,242.28	100	60	4.01
		II 3,600.73	300	100	3.62
		II 3,633.12	100	50	3.41
		II *3,710.29	150	80	3.52
		II 3,774.33	100	12	3.41
		I 4,643.70	100	50	2.67
		I *4,674.85	100	80	2.72
Zn	30	I *2,138.56	500	800R	5.80
		II 2,502.00	400	20	10.95
		II 2,557.96	300	10	10.95
		I 3,282.33	300	500R	7.78
		I 3,302.59	300	800	7.78
		I 3,302.94	300R	700R	7.78
		I 3,345.02	300	800	7.78
		I 4,680.14	200	300	6.66
		I 4,722.16	300	400	6.66
		I 4,810.53	300	400	6.66
		I 6,362.35	500	1000	7.74
Zr	40	II *3,391.97	400	300	3.82
		II 3,438.23	200	250	3.69
		I 3,547.68	12	200	3.56
		I *3,601.19	15	400	3.59
		I 4,687.80	—	125	3.4
		I 4,739.48	—	100	3.26
		I 4,772.31	—	100	3.22

Table III

Analytical Lines of the Elements Cointained in the Coincidence Tables including the Lines of the Rare Earth Elements, Arranged according to Wavelength

λ	Symbol	J	O	eV
I 2,039.85	Se	(1000)	—	6.3
2,062.38	J	(900)	—	—
I 2,062.79	Se	(800)	—	6.3
I 2,068.38*	Sb	3	300R	5.98
I 2,138.56*	Zn	500	800R	5.80
I 2,142.75*	Te	—	600	—
I 2,169.99	Pb	1000R	1000R	5.67
I 2,175.89	Sb	40	300	5.69
II 2,192.26	Cu	500	25	5.76
II 2,203.51*	Pb	5000R	50	7.25
2,225.10	Ce	100	—	—
2,240.89	K	40	—	—
II 2,246.41	Ag	300	25	10.36
II 2,247.00	Cu	500	30	5.51
II 2,253.86	Ni	300	100	6.81
II 2,264.46	Ni	400	150	6.72
II 2,265.02	Cd	300	25	5.47
II 2,270.21	Ni	400	100	6.61
2,275.25	Re	300R	300R	—
II 2,286.16*	Co	300	40	5.84
II 2,287.08*	Ni	500	100	7.28
I 2,288.02*	Cd	300R	1500R	5.41
I 2,288.12	As	5	250R	6.7
II 2,312.84	Cd	200	1	11.14
II 2,335.27	Ba	100R	60R	6.83
I 2,348.61*	Be	50	2000R	5.28
I 2,349.84	As	18	250R	6.6
I 2,373.36	Al	100R	200R	5.22
I 2,375.06	Ag	300	300	8.96
II 2,382.04*	Fe	100R	40R	5.20
I 2,383.25	Te	300	500	5.8
I 2,385.76	Te	(300)	600	5.8
2,389.69	Br	(70)	—	—
II 2,395.62	Fe	100	50	5.22
II 2,397.09	W	30	18	5.56
I 2,398.56	Ca	120	100R	—
II 2,413.18	Ag	300	50	10,18
I 2,427.95*	Au	100	400R	5.10
II 2,437.79*	Ag	500	60	9.93
I 2,456.53	As	8	100R	6.5
II 2,464.77	Kr	(100)	—	22.8
I 2,478.57*	C	(400)	400	7.69
I 2,483.27	Fe	50	500R	4.99
I 2,496.78	B	300	300	4.96
I 2,497.73*	B	400	500	4.96
II 2,502.00	Zn	400	20	10.95
I 2,506.90	Si	200	300	4.95
II 2,512.03	C	400	—	18.66
II 2,513.03	Hf	70	25	—
I 2,516.12	Si	500	500	4.95
II 2,516.88	Hf	100	35	—
I 2,528.52	Si	500	400	4.93
I 2,520.54	Sb	200	300R	6.12

λ	Symbol	J	O	eV
I 2,534.01	**P**	**(20)**	**50**	**7.2**
I 2,535.65	**P**	**(30)**	**100**	**7.2**
I 2,536.52	**Hg**	**1000*R***	**2000*R***	**4.88**
2,541.45	Br	(40)	—	—
I 2,543.97	Ir	100	200	4.9
I 2,553.28	**P**	**(20)**	**80**	**7.1**
I 2,554.93	**P**	**(20)**	**60**	**7.1**
II 2,557.96	**Zn**	**300**	**10**	**10.95**
I 2,560.23	In	50*R*	150*R*	4.81
I 2,562.54	Li	15	150	4.83
I 2,566.26	J	(300)	—	—
II 2,573.09	Ca	150	3	—
II 2,576.10*	**Mn**	**2000*R***	**300*R***	**4.81**
II 2,582.81	J	(400)	—	—
II 2,589.17	**W**	**25**	**15**	**5.5**
2,591.41	Se	(150)	—	—
II 2,593.73	**Mn**	**1000*R***	**200*R***	**4.77**
II 2,599.40	Fe	1000	1000	4.77
II 2,605.69	**Mn**	**500*R***	**100*R***	**4.75**
I 2,614.18	**Pb**	**80**	**200*R***	**5.68**
II 2,614.59	Ag	300	—	15.51
2,615.42*	**Lu**	**250**	**100**	—
I 2,639.71	Ir	15	100	4.7
II 2,641.41*	**Hf**	**125**	**40**	—
2,651.18	**Ge**	**20**	**40**	**4.8**
I 2,651.58*	**Ge**	**20**	**30**	**4.7**
2,653.27	Ta	15	200	4.7
I 2,659.45	**Pt**	**500*R***	**2000*R***	**4.6**
2,660.49	Br	(25)	—	—
I 2,664.79	Ir	50	200	4.6
II 2,669.17*	**Al**	**100**	**3**	**10.6**
2,675.90	Ta	200	150	4.6
I 2,675.95	**Au**	**100**	**250*R***	**4.63**
2,678.76	**Ru**	**300**	**100**	**11.0**
I 2,694.23	Ir	50	150	4.6
II 2,708.96	Ra	(200)	—	—
2,709.63	Ge	20	30	4.6
I 2,710.27	In	200*R*	800*R*	4.81
II 2,711.21	Ag	300	1	15.71
II 2,712.40	Kr	(80)	—	23.2
2,712.41	**Ru**	**300**	**80**	**11.0**
I 2,714.67	Ta	8	200	5.05
I 2,741.31	Li	—	200	4.52
II 2,748.58	**Cd**	**200**	**5**	**10.28**
I 2,753.88	In	300	300*R*	5.01
2,754.59	Ge	20	30	—
II 2,755.74	Fe	100	300	5.48
I 2,767.87	**Tl**	**300*R***	**400*R***	**4.44**
I 2,780.20	**As**	**75**	**75*R***	**6.7**
2,794.82	Mn	5	1000*R*	—
II 2,795.53*	**Mg**	**300**	**150**	**4.43**
2,798.27	Mn	80	800*R*	—
2,801.06	Mn	60	600*R*	—
II 2,802.70	**Mg**	**300**	**150**	**4.42**
II 2,813.76	Ra	(400)	—	—
II 2,816.15*	**Mo**	**300**	**200**	**11.9**
II 2,816.18	**Al**	**100**	**10**	**17.8**

λ	Symbol	*J*	*O*	eV
II 2,820.22	**Hf**	**100**	**40**	—
I 2,824.37	Ag	200	150	8.15
I 2,830.30	**Pt**	**600*R***	**1000*R***	**4.4**
II 2,833.00	Kr	(100)	—	22.79
I 2,833.07	**Pb**	**80**	**500*R***	**4.4**
II 2,835.63*	**Cr**	**400*R***	**100**	**12.6**
II 2,836.71	**C**	**200**	—	**16.34**
I 2,839.99*	**Sn**	**300*R***	**300*R***	**4.78**
II 2,843.25	**Cr**	**400*R***	**125**	**12.6**
II 2,848.23	**Mo**	**200**	**125**	**11.8**
I 2,849.73	**Ir**	**20**	**40**	**4.3**
II 2,849.84	**Cr**	**150*R***	**80**	**12.6**
I 2,852.13*	**Mg**	**100*R***	**300*R***	**4.34**
II 2,855.68	**Cr**	**200**	**60**	**12.5**
I 2,860.45	**As**	**50**	**50*R***	**6.6**
II 2,860.93	**Cr**	**100**	**60**	**12.5**
I 2,863.33	**Sn**	**300*R***	**300*R***	**4.32**
II 2,871.51	**Mo**	**100**	**100**	**11.7**
2,872.59	Br	(25)	—	—
I 2,874.24	**Ga**	**15*R***	**10**	**4.29**
I 2,877.92	**Sb**	**150**	**250**	**5.36**
II 2,878.64	J	(400)	—	15.19
I 2,881.58*	**Si**	**400**	**500**	**5.08**
II 2,890.99	**Mo**	**50**	**30**	**11.7**
2,891.38	Yb	100	50	—
2,894.84	**Lu**	**200**	**60**	—
I 2,897.98	**Bi**	**500*R***	**500*R***	**5.6**
I 2,909.06*	**Os**	**400**	**500*R***	**4.2**
II 2,909.12	**Mo**	**40**	**25**	**11.6**
2,911.39	**Lu**	**300**	**100**	—
I 2,916.48	**Hf**	**15**	**50**	—
I 2,918.32	**Tl**	**200*R***	**400*R***	**5.18**
I 2,929.79	**Pt**	**200**	**800*R***	**4.2**
2,936.77	**Ho**	**1000*R***	—	—
I 2,938.30	**Bi**	**300**	**300**	**6.1**
II 2,938.55	Ag	200	200	14.98
I 2,940.77	**Hf**	**12**	**60**	—
I 2,943.64	**Ga**	**20*R***	**10**	**4.29**
I 2,944.40	W	20	30	4.57
I 2,946.98	W	18	20	4.57
II 2,979.32	Xe	(200)	—	17.35
I 2,987.65	Si	100	100	4.93
I 2,989.03	**Bi**	**100**	**250**	**5.5**
I 2,997.97	**Pt**	**200*R***	**1000*R***	**4.23**
I 3,002.49	Ni	100	1000*R*	4.16
3,006.8	Ra	(300)	—	—
II 3,006.86	N	(50)	—	24.53
I 3,009.15	**Sn**	**200**	**300**	**4.33**
I 3,020.64	Fe	600*R*	1000*R*	4.11
II 3,032.84	As	70	125*R*	6.4
I 3,034.12	**Sn**	**150**	**200**	**4.30**
I 3,039.06	Ce	1000	1000	4.96
I 3,039.36	**In**	**500*R***	**1000*R***	**4.1**
I 3,050.82	Ni	—	1000*R*	4.09
3,054.3	Rn	(250)	—	—
3,055.37	J	(350)	—	—
I 3,058.66	**Os**	**500**	**500*R***	**4.05**

λ	Symbol	J	O	eV
I 3,064.71*	Pt	300R	2000R	4.05
I 3,067.72	Bi	2000	3000R	4.04
I 3,071.59	Ba	50R	100R	4.04
I 3,072.88	Hf	18	80	4.04
II 3,077.60	Lu	200	100	5.57
3,078.77	J	(350)	—	—
I 3,082.15	Al	800	800	4.0
I 3,092.71	Al	1000	1000	4.0
II 3,093.11*	V	400R	100R	4.40
II 3,094.18*	Nb	1000	100	4.52
IV 3,097.46	S	—	50	26.49
I 3,102.03	K	—	50R	3.99
II 3,102.30	V	300R	70	4.36
II 3,110.71	V	300R	70	4.33
II 3,116.63	As	150	—	—
II 3,118.38	V	200R	70	4.31
I 3,122.78	Au	5	500	5.10
I 3,124.82	Ge	80	200	4.85
II 3,125.28	V	200R	80	4.29
I 3,125.66	Hg	150	200	8.85
II 3,130.42*	Be	200	200	3.95
II 3,130.79	Nb	100	100	4.40
II 3,131.07	Be	150	200	3.95
3,131.26	Tm	500	400	—
I 3,131.55	Hg	300	400	8.82
I 3,131.83	Hg	100	200	8.85
I 3,133.32	Ir	2	40	4.0
II 3,134.72	Hf	125	80	4.33
II 3,158.87	Ca	300	100	7.05
I 3,170.35	Mo	25R	1000R	3.91
III 3,171.67	La	1000	2	5.59
I 3,175.02	Sn	400R	500	4.33
II 3,179.33	Ca	400	100	7.05
I 3,183.41	V	100R	200R	3.91
I 3,183.98	V	400R	500R	3.90
I 3,185.40	V	400R	500	3.96
I 3,187.4	He	(200)	—	23.70
II 3,194.98	Nb	300	30	4.21
II 3,203.14	He	(100)	—	52.23
I 3,217.02	K	20	100R	3.86
I 3,217.50	K	25	50R	3.86
I 3,220.78*	Ir	30	100	4.20
II 3,225.48	Nb	800R	150	4.14
I 3,229.75	Tl	800	2000	4.77
I 3,232.50	Sb	250	150	6.12
I 3,232.61	Li	500	1000R	3.83
II 3,234.52	Ti	500R	100	3.88
II 3,242.28	Y	100	60	4.01
I 3,247.54*	Cu	2000R	5000R	3.82
I 3,256.09	In	600R	1500R	4.1
I 3,258.56	In	300R	500R	4.1
I 3,261.06	Cd	300	300	3.80
I 3,262.29	Os	50	500R	4.32
I 3,262.33	Sn	300	400	4.87
I 3,267.50	Sb	150	150	5.82
I 3,267.95	Os	30	400R	3.80
I 3,269.49	Ge	300	300	4.67

λ	Symbol	J	O	eV
I 3,273.96	**Cu**	**1500R**	**3000R**	**3.78**
I 3,280.68*	**Ag**	**1000R**	**2000R**	**3.78**
I 3,282.33	**Zn**	**300**	**500R**	**7.78**
I 3,283.57	Rh	—	150R	4.10
II 3,289.37	**Yb**	**1000R**	**500R**	**3.77**
3,290.59	**Th**	**40**	—	**7.3**
I 3,301.56	Os	50	500R	3.76
I 3,302.32	**Na**	**300R**	**600R**	**3.75**
I 3,302.59	**Zn**	**300**	**800**	**7.78**
I 3,302.94	**Zn**	**300R**	**700R**	**7.78**
I 3,302.99	**Na**	**150R**	**300R**	**3.75**
3,311.16*	**Ta**	70	**300**	**3.7**
I 3,321.01	**Be**	—	**50**	**6.45**
I 3,321.09	**Be**	**15**	**100**	**6.45**
I 3,321.34	**Be**	**30**	**1000**	**6.45**
I 3,323.09	**Rh**	**200R**	**1000R**	**3.92**
II 3,329.12	Cl	(150)	—	20.05
II 3,337.49	La	300	800	4.12
II 3,344.56	La	200	300	3.94
I 3,345.02	**Zn**	**300**	**800**	**7.78**
II 3,349.04*	**Ti**	**800R**	**125**	**4.31**
II 3,350.48	Gd	180	150	3.85
I 3,350.89	Rb	—	150	3.71
I 3,351.25	Sr	15	300	5.54
II 3,353.39	Cl	(125)	—	18.03
II 3,361.21	**Ti**	**600R**	**100**	**3.71**
I 3,369.81	Ne	(500)	—	20.29
I 3,369.91	Ne	(700)	—	20.29
3,372.75	Er	20	35	—
II 3,372.80	**Ti**	**400**	**80**	**3.68**
II 3,380.71	**Sr**	**200**	**150**	**6.61**
I 3,382.89	**Ag**	**700R**	**1000R**	**3.66**
II 3,383.76	**Ti**	**300R**	**70**	**3.66**
II 3,391.98*	**Zr**	**400**	**300**	**3.82**
I 3,396.85	**Rh**	**500R**	**1000R**	**3.64**
II 3,397.07	**Lu**	**20R**	**50**	**5.11**
3,398.98	Ho	60	40	—
I 3,403.65	**Cd**	**500**	**800**	**7.37**
I 3,404.58*	**Pd**	**1000R**	**2000R**	**4.46**
I 3,405.12	**Co**	**150**	**2000R**	**4.1**
I 3,412.34	Co	100	1000	4.2
I 3,414.77*	**Ni**	**50**	**1000**	**3.65**
I 3,421.24	**Pd**	**1000R**	**2000R**	**4.58**
II 3,422.47	Gd	100	80	3.86
I 3,424.60	Re	—	300	—
I 3,434.89*	**Rh**	**200R**	**1000R**	**3.60**
I 3,436.74	**Ru**	**150**	**300R**	**3.75**
II 3,438.23	**Zr**	**200**	**250**	**3.69**
I 3,446.39	K	100R	150R	3.60
I 3,447.39	K	75R	100R	3.60
II 3,451.41	**B**	**100**	—	**12.69**
I 3,451.81	Re	—	100	3.59
I 3,453.50*	**Co**	**200**	**3000R**	**4.0**
I 3,460.47*	**Re**	—	**1000**	**3.58**
II 3,461.57	Rb	200	—	4.14
I 3,461.65	Ni	50	800R	3.60
3,462.20	**Tm**	**200**	**250**	—

λ	Symbol	J	O	eV
II 3,464.46	**Sr**	**200**	**200**	**6.62**
I 3,464.72	Re	—	100	3.57
I 3,465.80	**Co**	**25**	**2000*R***	**3.6**
I 3,466.20	**Cd**	**500**	**1000**	**7.37**
3,492.77	Rb	300	—	—
I 3,492.96	Ni	100	1000*R*	3.65
III 3,497.34	S	(100)	—	—
I 3,498.94*	**Ru**	**200**	**500*R***	**3.54**
3,499.10	Er	15	18	—
I 3,501.12	Ba	20	1000	3.54
II 3,501.42	F	(200)	—	28.66
I 3,502.28	Co	20	2000	4.0
II 3,503.10	F	(400)	—	28.66
II 3,505.61	F	(600)	—	28.66
3,509.17	**Tb**	**200**	**200**	—
I 3,513.65	**Ir**	**100**	**100**	**3.52**
I 3,515.05	**Ni**	**50**	**1000*R***	**3.63**
I 3,516.94	**Pd**	**500*R***	**1000*R***	**4.49**
I 3,519.24	**Tl**	**1000*R***	**2000*R***	**4.49**
I 3,520.47	Ne	(1000)	—	20.36
I 3,524.54	**Ni**	**100**	**1000*R***	**3.54**
I 3,529.81	**Co**	**30**	**1000*R***	**4.0**
3,531.71	Dy	100	100	—
3,538.75	**Th**	**50**	—	**3.5**
II 3,539.09	Ce	10	100	4.4
II 3,545.58	Ar	(300)	—	23.25
I 3,547.68	**Zr**	**12**	**200**	**3.56**
3,552.17	**U**	**12**	**8**	**3.5**
II 3,554.43	**Lu**	**150**	**50**	**5.63**
II 3,560.80	Ce	2	300	4.8
3,561.74	**Tb**	**200**	**200**	—
II 3,568.26	Sm	50	40	3.96
II 3,577.46	Ce	12	300	4.5
I 3,578.69	Cr	400*R*	500*R*	3.46
I 3,580.27	Nb	300	100	3.59
I 3,581.19	Fe	600*R*	1000*R*	4.32
II 3,584.96	Gd	100	100	3.60
I 3,587.08	Rb	40	200	3.47
3,590.35	Nd	300	400	—
II 3,592.60	Sm	50	40	3.83
I 3,593.49	Cr	400	500	3.44
I 3,596.18	**Ru**	**100**	**30**	**3.70**
II 3,600.73	**Y**	**300**	**100**	**3.62**
3,601.04	**Th**	**10**	**8**	**3.4**
I 3,601.19*	**Zr**	**15**	**400**	**3.59**
I 3,605.33	Cr	400*R*	500*R*	3.43
I 3,606.52	Ar	(1000)	—	15.06
II 3,607.88	Kr	(100)	—	22.97
II 3,609.48	Sm	100	60	3.71
I 3,609.55	**Pd**	**700*R***	**1000*R***	**4.40**
I 3,610.51	**Cd**	**500**	**1000**	**7.37**
II 3,613.79	**W**	**30**	**10**	**5.24**
II 3,613.84*	**Sc**	**70**	**40**	**3.45**
I 3,619.39	Ni	150	2000*R*	3.85
3,621.0	Rn	(250)	—	—
III 3,624.05	Xe	(600)	—	—
II 3,630.74	**Sc**	**70**	**50**	**3.42**

λ	Symbol	J	O	eV
II 3,631.87	Kr	(200)	—	21.89
II 3,633.12	**Y**	**100**	**50**	**3.41**
II 3,634.27	Sm	25	100	3.59
I 3,634.70	**Pd**	**1000**	**2000*R***	**4.23**
3,634.8	Rn	(250)	—	—
I 3,635.46	**Ti**	**100**	**200**	**3.40**
I 3,639.58	**Pb**	**50**	**300**	**4.38**
I 3,642.68	**Ti**	**125**	**300**	**3.42**
II 3,642.79	**Sc**	**50**	**60**	**3.40**
I 3,644.41	Ca	15	200	5.30
3,645.42	Dy	100	300	—
II 3,646.20	**Gd**	**150**	**200**	**3.63**
II 3,649.55	Ra	(1000)	—	—
I 3,650.15	**Hg**	**500**	**200**	**8.86**
I 3,653.50	**Ti**	**200**	**500**	**3.44**
II 3,653.97	Kr	(250)	—	21.92
I 3,657.99	**Rh**	**200**	**500**	**3.57**
I 3,663.28	**Hg**	**400**	**500**	**8.85**
II 3,670.07	U	18	15	3.49
3,672.58	**U**	**15**	**8**	**3.4**
I 3,683.47	**Pb**	**50**	**300**	**4.34**
I 3,692.36	**Rh**	**150**	**500**	**3.35**
3,692.65	**Er**	**12**	**20**	—
II 3,694.20*	**Yb**	**1000*R***	**500*R***	**3.35**
II 3,710.29*	**Y**	**150**	**80**	**3.52**
II 3,718.02	Kr	(300)	—	23.83
I 3,719.93*	**Fe**	**700**	**1000*R***	**3.32**
II 3,724.99	Eu	50	250	3.22
I 3,734.87	Fe	600	1000*R*	4.18
I 3,737.13	**Fe**	**600**	**1000*R***	**3.35**
3,741.19	Th	80	80	3.33
I 3,745.56	**Fe**	**500**	**500**	**3.39**
3,748.17	**Ho**	**40**	**60**	—
I 3,748.26	**Fe**	**200**	**500**	**3.41**
II 3,749.77	As	100	—	—
II 3,764.92	**Tm**	**120**	**200**	**3.29**
II 3,761.33	**Tm**	**150**	**250**	**3.29**
II 3,768.40	**Gd**	**20**	**20**	**3.36**
II 3,774.33	**Y**	**100**	**12**	**3.41**
I 3,775.72	**Tl**	**1000**	**3000**	**3.28**
3,797.91	H	(20)	—	13.45
I 3,798.25*	**Mo**	**1000*R***	**1000*R***	**3.26**
II 3,804.00	Au	150	25	—
II 3,814.42*	**Ra**	**(2000)**	—	**3.25**
II 3,819.66	Eu	500	500	3.24
I 3,823.47	O	(125)	—	15.77
II 3,827.62	Cl	(150)	—	21.48
I 3,829.35	**Mg**	**150**	**100**	**5.94**
I 3,832.31	**Mg**	**200**	**250**	**5.94**
3,835.40	H	(40)	—	13.42
I 3,838.26	**Mg**	**200**	**300**	**5.94**
II 3,847.09	F	(800)	—	25.12
3,848.75	**Tb**	**200**	**100**	—
II 3,849.99	F	(600)	—	25.11
II 3,851.67	F	(200)	—	25.11
II 3,859.58	U	30	20	3.24
II 3,860.83	Cl	(150)	—	19.17

λ	Symbol	J	O	eV
I 3,864.11	**Mo**	**500*R***	**1000*R***	**3.20**
II 3,872.12	Dy	150	300	—
3,874.18	**Tb**	**200**	**200**	—
I 3,876.39	Cs	—	300	3.17
I 3,888.65*	**He**	**(1000)**	—	**23.00**
I 3,888.65	Cs	10	150	3.16
3,889.06	H	(60)	—	13.38
3,891.02	**Ho**	**40**	**200**	—
II 3,898.54	Dy	—	100	—
3,899.19	Tb	100	200	—
I 3,902.96	**Mo**	**500*R***	**1000*R***	**3.17**
II 3,905.89	Nd	30	40	—
II 3,906.32	**Er**	**12**	**25**	—
II 3,907.11	Eu	500*R*	1000*R*	3.37
I 3,907.48	**Sc**	**25**	**125**	**3.17**
II 3,908.03	Pr	50	100	3.72
II 3,908.43	Pr	60	100	3.17
I 3,911.81*	**Sc**	**30**	**150**	**3.19**
II 3,911.95	O	(150)	—	28.83
I 3,915.0	Li	—	200	5.02
II 3,916.05	La	400	400	3.40
II 3,920.14	Kr	(200)	—	21.93
III 3,922.53	Xe	(500)	—	—
3,931.01	J	(400)	—	—
II 3,933.67*	**Ca**	**600*R***	**600*R***	**3.15**
3,940.24	J	(500)	—	—
II 3,941.51	Nd	30	60	3.20
I 3,944.03	**Al**	**1000**	**2000**	**3.1**
II 3,944.69	Dy	150	300	—
I 3,947.33	O	(300)	—	12.28
I 3,948.98	Ar	(2000)	—	14.68
II 3,949.11*	**La**	**800**	**1000**	**3.54**
I 3,961.53*	**Al**	**2000**	**3000**	**3.1**
II 3,964.83	Pr	80	125	3.18
II 3,965.26	Pr	50	100	3.33
3,968.22	Ag	60	100	—
3,968.40	Dy	—	300	—
II 3,968.47	**Ca**	**500*R***	**500*R***	**9.2**
3,970.07	H	(80)	—	13.31
II 3,971.99	Eu	—	1000*R*	3.32
II 3,973.27	O	(125)	—	26.56
3,976.82	Tb	200	150	—
I 3,987.99*	**Yb**	**500*R***	**1000*R***	**3.10**
II 3,994.83	Pr	25	300	3.15
II 3,995.00	N	(300)	—	21.60
I 3,995.31	Co	20	1000*R*	4.03
II 4,000.45	**Dy**	**300**	**400**	—
I 4,007.97	Er	70	35	—
I 4,008.75	**W**	**45**	**45**	**3.45**
II 4,012.25	Nd	40	80	3.71
I 4,012.39	**Ce**	**20**	**60**	**4.2**
II 4.019.14	**Th**	**8**	**8**	**3.08**
I 4,020.40	**Sc**	**20**	**50**	**3.08**
I 4,023.69	**Sc**	**25**	**100**	**3.10**
I 4,030.76*	**Mn**	**20**	**500*R***	**3.08**
I 4,032.98	**Ga**	**500*R***	**1000*R***	**3.07**
I 4,033.07	**Mn**	**20**	**400*R***	**3.08**

λ	Symbol	J	O	eV
I 4,034.49	**Mn**	**20**	**250*R***	**3.08**
I 4,044.14	**K**	**400**	**800**	**3.06**
4,045.43	Ho	80	200	—
4,045.98	**Dy**	**12**	**150**	—
I 4,046.56	**Hg**	**300**	**200**	**7.73**
I 4,047.20	**K**	**200**	**400**	**3.06**
4,053.92	Ho	200	400	—
I 4,055.26	Ag	500*R*	800*R*	6.72
I 4,057.82*	**Pb**	**300*R***	**2000*R***	**4.38**
I 4,058.94*	**Nb**	**400**	**1000**	**3.18**
II 4,072.16	O	(300)	—	28.69
II 4,075.87	O	(800)	—	28.80
II 4,077.34	**La**	**400**	**600**	**3.27**
II 4,077.71*	**Sr**	**500**	**400*R***	**3.04**
4,077.97	**Dy**	**100**	**150*R***	—
I 4,079.73	**Nb**	**200**	**500**	**3.12**
4,087.64	Er	1	20	—
II 4,090.14	U	40	25	3,24
4,094.18	Tm	30	300	—
II 4,098.91	Gd	100	100	3.62
I 4,099.94*	**N**	**(150)**	—	**13.69**
II 4,100.75	Pr	50	200	3.57
I 4,100.92	**Nb**	**200**	**300**	**3.07**
4,101.74	H	(100)	—	13.21
I 4,101.77	**In**	**1000*R***	**2000*R***	**3.02**
4,103.84	Ho	400	400	—
4,105.84	Tm	30	300	—
I 4,109.98	**N**	**(1000)**	—	**13.70**
I 4,118.77	Co	—	1000*R*	4.1
I 4,121.32	Co	25	1000*R*	3.9
II 4,123.23	**La**	**500**	**500**	**3.32**
I 4,123.81	**Nb**	**125**	**200**	**3.02**
I 4,128.87	Rh	150	300	3.97
II 4,129.74	**Eu**	**50**	**150*R***	**3.00**
I 4,132.29	Li	—	400	4.85
II 4,132.48	Cl	(200)	—	19.00
II 4,141.26	Pr	50	150	3.54
II 4,143.14	Pr	50	200	3.36
4,151.11	Er	3	20	—
II 4,153.10	S	(600)	—	18.88
4,163.03	Ho	100	100	—
4,167.97	**Dy**	**12**	**50**	—
I 4,172.06*	**Ga**	**1000*R***	**2000*R***	**3.07**
4,179.42*	**Pr**	**40**	**200**	—
II 4,184.25	Lu	200	100	5.11
II 4,184.26	Gd	150	150	3.45
II 4,186.60*	**Ce**	**25**	**80**	**3.34**
I 4,186.81	Dy	12	100	—
4,187.62	Tm	30	300	—
II 4,189.52	**Pr**	**50**	**100**	**3.33**
I 4,201.85	**Rb**	**500**	**2000*R***	**2.95**
II 4,205.05*	**Eu**	**50**	**200*R***	**2.94**
I 4,211.72	**Dy**	**15**	**200**	—
I 4,212.95	Pd	300	500	4.40
II 4,215.52	**Sr**	**400**	**300*R***	**2.94**
I 4,215.56	**Rb**	**300**	**1000*R***	**2.94**
III 4,222.15	P	(150)	300	17.54

λ	Symbol	J	O	eV
II 4,222.98	Pr	40	125	2.99
II 4,225.33	**Pr**	**40**	**50**	**2.93**
I 4,225.85	Gd	50	150	3.14
I 4,226.58	**Ge**	**50**	**200**	**4.96**
I 4,226.73*	**Ca**	**50**	**500*R***	**2.9**
II 4,241.67	U	**50**	**40**	**3.49**
II 4,242.15	Tm	100	500	2.95
II 4,246.16	F	(300)	—	31,56
II 4,251.74	Gd	10	300	3.29
I 4,254.35*	**Cr**	**1000*R***	**5000*R***	**2.91**
II 4,256.40	Sm	150	150	3.29
I 4,260.85	Os	200	200	2.91
II 4,262.10	Gd	10	150	3.63
II 4,262.68	Sm	150	200	3.58
I 4,266.29	Ar	(1200)	—	14.53
II 4,267.02	**C**	**(350)**	—	**20.94**
II 4,267.72	**C**	**(500)**	—	**20.94**
I 4,268.10	Ir	15	200	3.79
I 4,271.76	Fe	700	1000	4.39
I 4,273.28	Li	100	200*R*	4.75
I 4,274.80	**Cr**	**800*R***	**4000*R***	**2.90**
4,278.51	Tb	100	200	—
II 4,279.67	Sm	100	100	3.17
II 4,280.78	Sm	200	200	3.38
I 4,289.72	**Cr**	**800*R***	**3000*R***	**2.89**
I 4,294.61	**W**	**50**	**50**	**3.25**
I 4,302.11*	**W**	**60**	**60**	**3.24**
4,303.57*	**Nd**	**40**	**100**	—
I 4,307.91	Fe	800*R*	1000*R*	4.44
4,311.50	Ir	10	300	—
4,318.85	Tb	30	150	—
I 4,319.58	Kr	(1000)	—	12.78
I 4,325.76	Fe	700	1000	4.57
4,325.83	Tb	—	100	—
4,326.48	Tb	4	150	—
4,340.47	H	(200)	—	13.01
I 4,346.46	Gd	60	150	2.97
I 4,349.60	Rn	(5000)	—	—
II 4,352.25	As	200	—	13.11
II 4,355.48	Kr	(3000)	—	21.92
I 4,358.35	**Hg**	**500**	**3000**	**7.73**
I 4,359.93	Tm	30	300	2.84
I 4,368.30	O	(1000)	—	12.36
I 4,374.80	Rh	500	1000	—
I 4,379.24*	V	**200*R***	**200*R***	**3.13**
4,381.86	Th	30	30	2.8
I 4,383.55	Fe	800	1000	4.31
II 4,388.13	K	40	—	23.45
I 4,389.97	V	60*R*	80*R*	3.10
II 4,390.87	**Sm**	**150**	**150**	**3.00**
II 4,391.11	Th	40	50	3.37
4,399.47	Ir	100	400	—
I 4,401.55	Ni	30	1000	6.00
I 4,404.75	Fe	700	1000	4.37
I 4,408.51	V	20*R*	30	3.08
II 4,408.84	Pr	100	125	3.36
I 4,414.63	Cd	200	—	—

λ	Symbol	J	O	eV
II 4,414.89	O	(300)	—	26.24
4,419.61	Er	4	25	—
I 4,420.47	**Os**	**100**	**400*R***	**2.80**
II 4,424.34*	**Sm**	**300**	**300**	**3.28**
I 4,425.44	**Ca**	**20**	**100**	**4.7**
4,426.27	Ir	10	400	—
II 4,434.32	**Sm**	**200**	**200**	**3.17**
I 4,434.96	**Ca**	**25**	**150**	**4.7**
II 4,435.60	Eu	100	400*R*	3.00
I 4,442.55	Pt	25	800	—
II 4,451.57	Nd	50	100	3.16
II 4,452.88	J	(700)	—	14.76
II 4,467.34	Sm	200	200	3.43
4,474.60	As	(200)	—	—
I 4,500.98	**Xe**	**(500)**	—	**11.07**
I 4,511.32*	**In**	**4000*R***	**5000*R***	**3.02**
4,518.57	**Lu**	**40**	**300**	—
I 4,524.74	**Sn**	**50**	**500**	**4.87**
I 4,528.73	Rh	60	500*R*	—
I 4,530.96	Co	8	1000	5.66
I 4,537.75	Ne	(1000)	—	21.11
I 4,539.70	Cu	80	100	7.88
II 4,539.97	As	200	—	—
4,542.93	Br	(250)	—	—
I 4,551.95	Ta	8	400	3.21
II 4,554.04*	**Ba**	**200**	**1000**	**2.72**
I 4,554.51	Ru	200	1000*R*	3.53
I 4,555.36	**Cs**	**100**	**2000*R***	**2.72**
I 4,581.60	Co	10	1000	5.66
I 4,584.45	Ru	80	150*R*	3.70
I 4,586.95	Cu	80	250	7.80
I 4,593.18	**Cs**	**50*R***	**1000*R***	**2.70**
I 4,594.02*	**Eu**	**200**	**500**	**2.70**
I 4,602.86	**Li**	—	**800**	**4.54**
I 4,607.33*	**Sr**	**50*R***	**1000*R***	**2.69**
II 4,608.43	K	40	—	—
I 4,624.28	**Xe**	**(1000)**	—	**11.00**
4,625.48	Rn	(500)	—	—
I 4,627.23	Eu	15	50	2.68
II 4,630.14	As	200	—	—
I 4,643.70*	**Y**	**100**	**50**	**2.67**
II 4,658.87	Kr	(2000)	—	19.30
I 4,668.48	Ag	70	200	6.44
II 4,670.40	Sc	300	100	4.01
I 4,671.23	**Xe**	**(2000)**	—	**10.97**
4,672.2	Be	100	—	—
II 4,673.46	Be	100	—	14.81
I 4,674.85	**Y**	**100**	**80**	**2.72**
4,675.62	Er	4	15	—
I 4,678.16	Cd	200	200	6.39
4,680.14	**Zn**	**200**	**300**	**6.66**
4,680.83	Rn	(500)	—	—
II 4,682.28	**Ra**	**(800)**	—	**2.65**
II 4,687.80	**Zr**		**125**	**9.16**
I 4,695.45	**S**	**(30)**	—	**9.16**
I 4,702.32	Ar	(1200)	—	14.46
II 4,704.86*	**Br**	**(250)**	—	**14.4**

λ	Symbol	J	O	eV
II 4,707.82	As	200	—	—
I 4,714.42	Ni	8	1000	6.00
I 4,722.16	**Zn**	**300**	**400**	**6.66**
I 4,722.55	**Bi**	**100**	**1000**	**4.04**
I 4,730.78	**Se**	**(1000)**	—	**2.6**
I 4,739.03	**Se**	**(800)**	—	**2.6**
I 4,739.48	**Zr**	—	**100**	**3.26**
II 4,740.40	Cl	(150)	—	20.00
I 4,742.25	**Se**	**(500)**	—	**2.6**
4,752.52	Tb	80	100	—
II 4,768.68	Cl	150	—	19.68
I 4,772.31	**Zr**	—	**100**	**3.22**
I 4,780.01	Co	500	500	5.87
I 4,780.94	Ta	200	50	3.91
I 4,783.42	Mn	60	400	4.89
II 4,785.50	**Br**	**(400)**	—	**14.4**
I 4,786.54	Ni	2	300	6.00
I 4,792.60	Au	60	200	7.69
II 4,794.54	**Cl**	**(250)**	—	**28.9**
I 4,799.92	Cd	300	300	6.39
II 4,810.06	**Cl**	**(200)**	—	**28.9**
I 4,813.48	Co	6	1000	5.79
II 4,816.71	**Br**	**(300)**	—	**14.4**
II 4,819.46	**Cl**	**(200)**	—	**28.9**
I 4,823.52	Mn	80	400	4.89
I 4,825.91*	**Ra**	**(800)**	—	**2.57**
I 4,832.08	**Sr**	**8**	**200**	**4.34**
4,861.33	**H**	**(500)**	—	**12.74**
I 4,862.31	J	(700)	—	9.47
4,866.22	Te	(800)	—	—
I 4,884.92	Ne	(1000)	—	20.65
I 4,884.92	Ne	(1000)	—	21.10
I 4,889.17	**Re**	—	**2000**	**2.53**
II 4,919.81	Th	20	50	3.28
II 4,934.09	**Ba**	**400**	**400**	**2.51**
I 4,935.03	N	(250)	—	13.19
4,935.50	Yb	10	200	—
I 4,957.03	Ne	(1000)	—	21.10
4,957.36	Dy	3	20	—
I 4,971.99	Li	—	500	4.34
I 4,981.73*	**Ti**	**125**	**300**	**3.33**
I 4,991.07	**Ti**	**100**	**200**	**3.32**
I 4,999.51	**Ti**	**80**	**200**	**3.30**
I 5,007.21	**Ti**	**40**	**200**	**3.29**
I 5,035.37	Ni	5	300	6.09
II 5,040.82	Hf	150	100	3.94
I 5,085.82	Cd	500	1000	6.39
I 5,105.54	**Cu**	—	**500**	**3.82**
I 5,119.29	J	(500)	—	9.34
5,142.14	Se	(500)	—	—
I 5,153.24	**Cu**	—	**600**	**6.19**
II 5,161.19	**J**	**300**	—	**12.11**
I 5,167.34	**Mg**	**50**	**100**	**5.11**
I 5,172.70	**Mg**	**100**	**200**	**5.11**
5,175.98	Se	(600)	—	—
I 5,183.62	**Mg**	**300**	**500**	**5.11**
I 5,204.52	**Cr**	**100**	**400*R***	**3.32**

λ	Symbol	*J*	*O*	eV
I **5,206.04**	**Cr**	**200**	**500*R***	**3.32**
I **5,208.44**	**Cr**	**100**	**500*R***	**3.32**
I **5,209.07**	**Ag**	**1000*R***	**1500*R***	**6.04**
II 5,209.29	Bi	600	—	5.72
I **5,218.20**	**Cu**	—	**700**	**6.19**
II 5,259.74	Pr	3	125	2.99
I 5,275.53	Re	—	500	2.35
II 5,296.09	P	(300)	—	13.14
II 5,298.06	Hf	100	80	4.23
5,301.02	Pt	10	150	6.90
II 5,311.60	Hf	150	100	4.12
I 5,330.66	O	(500)	—	13.06
II 5,331.54	As	200	—	12.43
5,335.16	Yb	400	150	—
I 5,339.67	K	—	40	3.94
I 5,341.07	Mn	100	200	4.44
5,344.17	Nb	200	400	—
I **5,350.46***	**Tl**	**2000**	**5000**	**3.28**
I **5,400.56**	**Ne**	**(2000)**	—	**18.95**
I **5,424.62**	**Ba**	**30*R***	**100*R***	**3.81**
5,449.29	Yb	100	20	—
I **5,455.15**	**La**	**1**	**200**	**2.40**
I **5,460.74**	**Hg**	**2000**	—	**7.73**
5,464.61	**J**	**(900)**	—	—
I **5,465.49**	**Ag**	**500*R***	**1000*R***	**6.04**
II 5,492.97	U	50	60	2.26
I 5,506.49	Mo	100	200*R*	**3.58**
5,506.78	Br	(300)	—	—
I **5,519.12***	**Ba**	**60*R***	**200*R***	**3.82**
5,522.99	Ce	—	100	—
I **5,535.55**	**Ba**	**200*R***	**1000*R***	**2.24**
5,552.35	Bi	100	500	—
I 5,568.09	Sb	200	6	7.92
I **5,570.29**	**Kr**	**2000**	—	**12.14**
5,631.69	Sn	200	50	4.33
I 5,660.81	Ra	(1000)	—	—
II **5,666.64**	**N**	**300**	—	**20.62**
II **5,676.02**	**N**	**(100)**	—	**20.62**
II **5,679.56**	**N**	**(500)**	—	**20.66**
I **5,682.66**	**Na**	—	**80**	**4.29**
I **5,688.22**	**Na**	—	**300**	**4.29**
I 5,700.23	Sc	—	400*R*	3.60
I 5,754.26	Si	—	40	7.11
5,755.87	Te	250	—	—
I 5,769.59	Hg	200	600	8.85
I **5,777.67**	**Ba**	**100*R***	**500*R***	**3.82**
I 5,782.13	Cu	—	1000	3.79
I 5,782.60	K	—	60	3.76
I 5,790.65	Hg	(1000)	—	8.85
I 5,793.51	C	—	30	10.08
I 5,837.40	Au	10	400	—
I **5,852.49**	**Ne**	**(2000)**	—	**18.96**
I **5,870.92**	**Kr**	**(3000)**	—	**12.14**
I **5,875.62**	**He**	**(1000)**	—	**23.06**
I **5,889.95***	**Na**	**1000**	**9000**	**2.11**
II 5,893.46	Ge	100	—	9.84
I **5,895.92**	**Na**	**500*R***	**5000*R***	**2.10**

λ	Symbol	J	O	eV
I 5,930.63	La	—	150	2.09
II 6,004.6	Sb	200	—	11.46
I 6,103.64	Li	300	2000*R*	3.87
II 6,221.87	Lu	1000	500	3.53
I 6,239.64	F	(300)	—	14.68
II 6,243.36	Al	80	—	15.06
I 6,249.93*	La	—	300	2.49
I 6,278.18	Au	20	700	—
I 6,362.35	Zn	500	1000	7.74
I 6,396.61	Ga	20	—	4.98
I 6,402.25	Ne	(2000)	—	18.56
I 6,438.47	Cd	1000	2000	7.34
I 6,549.77	Tl	50	300	5.14
6,562.73	H	(1000)	—	—
6,562.85	H	(2000)	—	12.09
II 6,578.03	C	500	—	16.33
II 6,644.60	Hf	200	100	3.64
I 6,707.84*	Li	200	3000*R*	1.90
I 6,856.02	F	(1000)	—	14.50
I 6,902.46	F	(500)	—	14.52
I 6,965.43	Ar	(400)	—	13.32
I 7,055.42	Rn	(400)	—	8.4
I 7,067.22	Ar	(400)	—	13.29
I 7,423.54	Si	—	500	7.29
I 7,450.00	Rn	(600)	—	8.5
I 7,503.87	Ar	(700)	—	13.47
I 7,664.91*	K	400	9000*R*	1.62
I 7,698.98	K	200	5000*R*	1.61
I 7,771.93*	O	(1000)	—	10.73
I 7,774.14	O	(300)	—	10.73
I 7,775.43	O	(100)	—	10.73
I 7,947.60	Rb	—	5000*R*	1.56
I 8,115.31	Ar	(5000)	—	13.07
I 8,521.11*	Cs	—	5000*R*	1.46
I 8,943.50	Cs	—	2000*R*	1.39
I 9,212.91	S	(200)	—	7.87
I 9,228.11	S	(200)	—	7.86
I 9,237.49	S	(200)	—	7.86

Table IV

Analytical Lines of the Rare Earth Elements in Alphabetical Order of the Element Symbols

Symbol	Atomic Mass	Atomic No.	Wavelengths of characteristic lines by Å
Dy	162.50	66	3,531.71
			3,645.42
			II 3,872.12
			II 3,898.54
			II 3,994.69
			3,968.40
			II 4,000.45
			4,045.98
			4,077.97
			4,167.97
			I 4,186.81
			I 4,211.72
			4,957.36
Er	167.26	68	3,499.10
			3,372.75
			3,692.65
			II 3,906.32
			I 4,007.97
			4,087.64
			4,151.11
			4,419.61
			4,675.62
Eu	151.96	63	3,724.99
			II 3,819.66
			II 3,907.11
			II 3,971.99
			II 4,129.74
			II 4,205.05
			4,435.60
			I 4,594.02
			I 4,627.23
Gd	157.25	64	3,350.48
			3,422.47
			3,584.96
			3,646.20
			II 3,768,41
			II 4,098.61
			II 4,184.26
			I 4,225.85
			II 4,251.74
			II 4,262.10
			I 4,346.46
Hf	178.49	72	2,516.88
			2,641.41
			2,820.22
			2,898.26
			2,916.48
			3,072.88
			3,134.72
			II 3,719.28
			II 3,918.09
			II 4,093.16
			II 4,232,44
			I 4,800.50
			5,181.86
			II 5,311.60
			5,719.18
Ho	164.930	67	2,936.77
			3,453.13
			3,398.98
			3,748.17
			3,891.02
			4,045.43
			4,053.92
			4,103.84
			4,163.03
La	138.91	57	3,337.49
			II 3,794.77
			II 3,949.11
			II 4,077.34
			II 4,086.71
			4,123.23
			II 4,333.73
			II 4,429.90
			II 4,522.37
			II 4,921.78
			II 5,183.42
			6,249.93
			I 6,709.50
			II 7,066.21
Lu	174.97	71	2,615.42
			2,894.84
			2,911.39
			3,077.60
			3,397.07
			3,554.43
			II 4,184.25
			I 4,518.57
			II 6,221.87
Nd	144.24	60	3,951.15
			3,905.89
			3,941.51
			II 4,012.25
			II 4,061.08
			II 4,109.073
			II 4,109.455

Symbol	Atomic Mass	Atomic No.	Wavelengths of charakteristiclines by Å
			II 4,156.08
			II 4,303.57
			4,451.57
Pr	140.907	59	3,908.03
			3,908.43
			3,964.82
			3,965.26
			3,994.83
			II 44,100.75
			4,141.26
			II 4,143.14
			II 4,179.42
			II 4,189.52
			II 4,222.98
			II 4,225.33
			II 4,408.84
			II 5,259.74
Sm	150.35	62	3,568.26
			3,592.60
			3,609.49
			3,634.27
			II 4,256.40
			II 4,262.68
			II 4,279.67
			II 4,280.78
			4,390.86
			II 4,424.34
			4,434.32
			II 4,467.34
Tb	158.924	65	3,509.17
			3,561.74
			3,848.75
			3,874.18
			3,899.19
			3,976.82
			4,278.51
			4,318.94
			4,325.83
			4,326.48
			4,752.52
Tm	168.934	69	3,131.26
			3,462.20
			3,761.33
			3,761.92
			I 4,094.18
			I 4,105.84
			I 4,187.62
			II 4,242.15
			I 4,359.93
Yb	173.04	70	2,891.38
			3,298.37
			3,694.20
			I 3,975.30
			3,987.99
			4,149.07
			4,935.50
			5,335.16
			I 5,449.29

Table V

Table of the Analytical Lines of the Rare Earth Elements according to Wavelength

Symbol	λ in Å	Symbol	λ in Å	Symbol	λ in Å
Hf	2,516.88	Er	II 3,906.32	Eu	**II 4,205.05**
Lu	2,615.42	Eu	**II 3,907.11**	Dy	**I 4,211.72**
Hf	2,641.41	Pr	3,908.03	Pr	**II 4,222.98**
Hf	2,820.22		3,908.43	Pr	II 4,225.33
Yb	2,891.38	Hf	II 3,918.09	Gd	I 4,225.85
Lu	2,894.84	Nd	3,941.51	Hf	II 4,232.44
Hf	2,898.26	Dy	**II 3,944.69**	Tm	II 4,242.15
Lu	2,911.39	La	**II 3,949.11**	Gd	II 4,251.74
Hf	2,916.48	Nd	3,951.15	Sm	II 4,256.40
Ho	2,936.77	Pr	3,964.83	Gd	II 4,262.10
Hf	3,072.88	Pr	3,965.26	Sm	II 4,262.68
Lu	3,077.60	Dy	3,968.40	Tb	**4,278.51**
Tm	3,131.26	Eu	II 3,971.99	Sm	II 4,279.67
Hf	3,134.72	Yb	**I 4,975.30**	Sm	II 4,280.78
Yb	3,289.37	Tb	3,976.82	Nd	II 4,303.57
La	3,337.49	Yb	3,987.99	Tb	4,318.94
Gd	3,350.48	Pr	3,994.83	Tb	4,325.83
Er	3,372.75	Dy	II 4,000.45		4,326.48
Lu	3,397.07	Er	**I 4,007.97**	La	**II 4,333.73**
Ho	3,398.98	Nd	II 4,012.25	Gd	I 4,346.46
Gd	3,422.47	Ho	4,045.43	Tm	I 4,359.93
Ho	3,453.13	Dy	4,045.98	Sm	4,390.86
Tm	3,462.20	Ho	4,053.92	Pr	II 4,408.84
Er	3,499.10	Nd	II 4,061.08	Er	4,419.61
Tb	3,509.17	La	II 4,077.34	Sm	**II 4,424.34**
Dy	3,531.71	Dy	4,077.97	La	II 4,429.90
Lu	3,554.43	La	II 4,086.71	Sm	4,434.32
Tb	3,561.74	Er	4,087.64	Eu	4,435.60
Sm	3,568.26	Hf	**II 4,093.16**	Nd	4,451.57
Gd	3,584.96	Tm	I 4,094.18	Sm	II 4,467.34
Sm	3,592.60	Gd	**II 4,098.61**	Lu	I 4,518.57
Sm	3,609.48	Pr	II 4,100.75	La	II 4,522.37
Sm	3,634.27	Ho	**I 4,103.84**	Eu	I 4,594.02
Dy	3,645.42	Tm	I 4,105.84	Eu	I 4,627.23
Gd	3,646.20	Nd	II 4,109.07	Er	4,675.62
Er	3,962.65		II 4,109.45	Tb	4,752.52
Yb	3,694.20	La	4,123.23	Hf	I 4,800.50
Hf	II 3,719.28	Eu	**II 4,129.74**	La	II 4,921.78
Eu	3,724.93	Pr	II 4,141.26	Yb	4,935.50
Ho	3,748.17	Pr	II 4,143.14	Dy	4,957.36
Tm	3,761.92	Yb	4,149.07	Hf	5,181.86
	3,761.33	Er	**4,151.11**	La	II 5,183.42
Gd	II 3,768.41	Nd	II 4,156.08	Pr	II 5,259.74
La	II 3,794.77	Ho	4,163.03	Hf	II 5,311.60
Eu	II 3,819.66	Dy	4,167.97	Yb	5,335.16
Tb	3,848.75	Pr	**II 4,179.42**	Yb	I 5,449.29
Dy	II 3,872.12	Lu	**II 4,184.25**	Hf	5,719.18
Tb	3,874.18	Gd	**II 4,184.26**	Lu	II 6,221.87
Ho	3,891.02	Dy	I 4,186.81	La	6,249.93
Dy	II 3,898.54	Tm	**I 4,187.62**	La	I 6,709.50
Tb	**3,899.19**	Pr	II 4,189.52	La	II 7,066.21
Nd	3,905.89				

Table VI

Emission Spectra of Some of the Actinides

Ac 89

λ	I		λ	I		λ	I	
	J	O		J	O		J	O
4,088.37	100	—	4,359.09	30	—	4,413.17	100	—
4,168.40	100	—	4,386.37	100	—	4,812.25	60	—
4,179.93	60	—						

Pa 91

λ	J	O	λ	J	O	λ	J	O
2,640.3	100	—	2,970.7	20	—	3,092.6	20	—
2,644.9	20	—	2,972.4	20	—	3,093.2	50	—
2,651.6	20	—	2,974.0	50	—	3,103.8	20	—
2,670.5	50	—	2,977.2	20	—	3,105.9	50	—
2,672.1	50	—	2,980.5	100	—	3,111.7	50	—
2,697.5	50	—	2,981.9	20	—	3,117.7	20	—
2,732.2	100	—	2,982.9	20	—	3,131.2	20	—
2,743.2	100	—	2,983.4	20	—	3,136.9	20	—
2,743.9	200	—	2,984.7	50	—	3,157.6	20	—
2,755.9	20	—	2,987.0	20	—	3,170.8	20	—
2,793.3	50	—	2,987.9	100	—	3,171.5	20	—
2,796.2	100	—	2,991.6	20	—	3,178.2	20	—
2,808.0	20	—	2,992.7	50	—	3,179.6	20	—
2,811.5	50	—	2,994.9	50	—	3,204.2	20	—
2,815.4	20	—	2,997.7	50	—	3,205.4	20	—
2,826.6	50	—	3,004.6	100	—	3,214.0	50	—
2,832.6	50	—	3,005.7	50	—	3,221.9	20	—
2,835.5	50	—	3,009.5	50	—	3,240.5	20	—
2,843.3	50	—	3,015.2	50	—	3,332.8	20	—
2,845.5	50	—	3,023.5	50	—	3,545.0	50	—
2,855.4	20	—	3,025.3	100	—	3,632.4	50	—
2,870.0	50	—	3,033.6	50	—	3,643.8	20	—
2,871.4	50	—	3,034.3	50	—	3,663.2	20	—
2,873.3	50	—	3,036.7	20	—	3,674.9	20	—
2,891.0	20	—	3,039.0	20	—	3,679.8	20	—
2,893.0	100	—	3,042.0	100	—	3,692.8	20	—
2,906.3	20	—	3,042.7	50	—	3,716.8	20	—
2,906.9	50	—	3,045.6	50	—	3,729.3	20	—
2,907.5	50	—	3,051.8	20	—	3,735.7	20	—
2,909.0	50	—	3,053.5	100	—	3,737.8	20	—
2,909.6	100	—	3,054,6	100	—	3,754.4	20	—
2,922.8	50	—	3,057.9	20	—	3,765.7	20	—
2,928.3	50	—	3,059.1	20	—	3,780.9	20	—
2,932.9	20	—	3,060.3	20	—	3,795.6	20	—
2,934.3	50	—	3,060.9	20	—	3,802.8	20	—
2,940.2	100	—	3,066.4	50	—	3,807.4	20	—
2,943.0	50	—	3,067.7	20	—	3,814.0	20	—
2,943.5	100	—	3,070.3	20	—	3,818.2	20	—
2,954.6	50	—	3,075.6	20	—	3,823.6	20	—
2,959.7	100	—	3,076.7	20	—	3,828.4	20	—
2,968.0	50	—	3,082.3	20	—	3,849.9	50	—
2,969.0	50	—	3,083.2	50	—	3,856.9	50	—

λ	I		λ	I		λ	I	
	J	O		J	O		J	O
3,868.5	20	—	3,983.3	20	—	4,129.3	20	—
3,875.9	50	—	3,988.3	20	—	4,139.6	20	—
3,892.6	20	—	4,012.9	50	—	4,160.1	50	—
3,905.0	20	—	4,018.2	50	—	4,161.0	20	—
3,908.0	50	—	4,029.9	20	—	4,161.5	20	—
3,909.1	20	—	4,046.9	20	—	4,164.2	50	—
3,909.6	20	—	4,047.6	20	—	4,176.1	20	—
3,911.3	20	—	4,051.8	20	—	4,192.1	50	—
3,921.5	20	—	4,056.1	50	—	4,196.9	50	—
3,928.7	50	—	4,070.4	50	—	4,197.8	50	—
3,944.0	20	—	4,074.3	20	—	4,205.3	20	—
3,945.9	20	—	4,087.8	20	—	4,210.4	20	—
3,952.6	20	—	4,089.4	20	—	4,217.2	100	—
3,956.6	20	—	4,090.4	20	—	4,230.6	50	—
3,957.8	100	—	4,099.2	20	—	4,248.1	100	—
3,961.5	50	—	4,102.1	50	—	4,286.7	50	—
3,967.3	20	—	4,107.9	20	—	4,291.3	50	—
3,970.0	50	—	4,108.4	20	—	4,299.4	20	—
3,975.4	20	—	4,128.7	20	—	4,371.7	20	—
3,982.2	20	—						

Np 93

λ	J	O	λ	J	O	λ	J	O
2,655.0	20	—	2,971.0	20	—	3,722.3	20	—
2,669.6	20	—	2,974.3	50	—	3,794.9	20	—
2,678.2	20	—	2,974.9	50	—	3,829.2	50	—
2,733.7	20	—	3,026.4	50	—	3,832.3	20	—
2,734.5	20	—	3,027.3	20	—	3,865.2	20	—
2,785.2	20	—	3,029.1	20	—	3,888.2	20	—
2,789.8	20	—	3,052.0	50	—	3,949.2	20	—
2,833.6	20	—	3,057.5	20	—	3,987.0	20	—
2,841.0	20	—	3,070.3	20	—	3,988.6	20	—
2,841.5	20	—	3,078.0	50	—	3,989.8	20	—
2,864.1	20	—	3,084.7	20	—	4,028.9	20	—
2,864.4	20	—	3,090.4	20	—	4,108.4	20	—
2,865.5	20	—	3,092.0	20	—	4,164.5	50	—
2,866.0	20	—	3,092.5	20	—	4,256.7	20	—
2,867.6	20	—	3,171.6	20	—	4,258.1	20	—
2,869.8	20	—	3,402.5	20	—	4,279.6	20	—
2,873.3	20	—	3,590.6	20	—	4,290.9	50	—
2,956.6	50	—	3,665.6	20	—	4,307.8	20	—
2,957.9	50	—	3,708.3	20	—	4,336.6	20	—
2,968.2	20	—						

Pu 94

λ	J	O	λ	J	O	λ	J	O
2,677.0	20	—	2,787.2	20	—	2,826.2	20	—
2,684.8	20	—	2,806.0	20	—	2,833.2	50	—
2,693.3	20	—	2,808.4	50	—	2,835.5	100	—
2,781.3	50	—	2,809.0	20	—	2,898.0	50	—
2,784.4	50	—	2,815.6	50	—	2,899.8	20	—

λ	I		λ	I		λ	I	
	J	O		J	O		J	O
2,904.3	20	—	3,069.3	20	—	3,904.1	50	—
2,905.0	50	—	3,091.2	20	—	3,907.2	100	—
2,910.3	50	—	3,091.9	20	—	3,910.3	20	—
2,912.7	20	—	3,092.7	50	—	3,912.5	50	—
2,913.7	20	—	3,093.3	20	—	3,913.4	50	—
2,914.2	20	—	3,104.1	20	—	3,928.0	20	—
2,914.9	20	—	3,105.0	20	—	3,928.5	20	—
2,915,6	50	—	3,105.9	20	—	3,931.2	20	—
2,918.0	20	—	3,123.8	20	—	3,949.2	20	—
2,925.2	100	—	3,136.5	20	—	3,953.0	20	—
2,925,8	50	—	3,159.2	50	—	3,958.9	20	—
2,926.3	20	—	3,179.3	20	—	3,961.5	50	—
2,928.2	50	—	3,221.4	20	—	3,962.8	20	—
2,930.9	20	—	3,231.8	20	—	3,963.6	20	—
2,932.4	20	—	3,232.6	20	—	3,967.3	20	—
2,933.2	20	—	3,312.6	20	—	3,972.2	50	—
2,933.8	20	—	3,401.0	20	—	3,975.4	50	—
2,937.6	50	—	3,585.9	20	—	3,976.0	20	—
2,938.9	100	—	3,632.1	20	—	3,980.4	20	—
2,941.3	50	—	3,709.1	20	—	3,985.5	100	—
2,945.2	50	—	3,718.1	20	—	3,988.5	20	—
2,945.9	50	—	3,720.2	20	—	3,989.7	100	—
2,950.0	20	—	3,720.6	20	—	3,991.5	20	—
2,951.6	50	—	3,721.5	20	—	3,992.2	20	—
2,954.3	100	—	3,726.0	50	—	4,015.9	20	—
2,964.5	20	—	3,726.8	50	—	4,021.5	50	—
2,966.7	20	—	3,732.1	20	—	4,064.6	20	—
2,967.3	20	—	3,753.6	20	—	4,066.8	20	—
2,968.9	50	—	3,755.9	20	—	4,078.0	20	—
2,970.2	50	—	3,758.3	20	—	4,101.0	20	—
2,972.3	100	—	3,773.6	20	—	4,102.0	20	—
2,977.1	20	—	3,775.7	20	—	4,116.4	20	—
2,977.8	50	—	3,792.2	20	—	4,141.2	20	—
2,978.4	20	—	3,803.6	20	—	4,196.2	50	—
2,980.0	50	—	3,810.2	20	—	4,229.7	20	—
2,981.2	20	—	3,812.3	20	—	4,254.4	20	—
2,986.9	20	—	3,814.9	20	—	4,254.8	20	—
2,988.1	100	—	3,823.9	20	—	4,256.0	20	—
2,991.5	20	—	3,827.5	20	—	4,257.9	20	—
2,993.9	100	—	3,835.5	20	—	4,273.3	100	—
2,996.4	100	—	3,836.9	20	—	4,280.3	20	—
3,000.4	100	—	3,852.7	50	—	4,289.1	20	—
3,008.9	50	—	3,874.2	20	—	4,330.6	20	—
3,009.6	50	—	3,878.6	20	—	4,337.2	20	—
3,028.8	20	—	3,887.4	20	—	4,352.7	50	—
3,042.5	20	—	3,892.7	20	—	4,358.1	20	—
3,043.0	20		3,895.9	20	—			

Am 95

λ	J	O	λ	J	O	λ	J	O
2,661.6	20	—	2,691,4	100	—	2,725.4	20	—
2,664.8	20	—	2,706.4	50	—	2,727.6	20	—
2,668.8	20	—	2,716.5	50	—	2,728.7	200	—
2,686.9	20	—	2,720.5	20	—	2,731.9	20	—

λ	I	
	J	O
2,732.1	20	—
2,735.4	100	—
2,739.6	100	—
2,742.3	50	—
2,743.2	100	—
2,746.6	100	—
2,747.0	100	—
2,749.4	500	—
2,749.8	20	—
2,753.4	20	—
2,755.3	200	—
2,755.9	200	—
2,756.7	500	—
2,776.0	50	—
2,776.5	50	—
2,796.8	50	—
2,812.9	200	—
2,815.3	100	—
2,816.0	50	—
2,816.4	50	—
2,832.3*P*	2000	—
2,852,2	100	—
2,861.6	50	—
2,873.3	100	—
2,881.6	50	—
2,888.5	200	—
2,893.3	100	—
2,899.6	100	—
2,911.2	500	—
2,920.6	500	—
2,928.0	100	—
2,937.0	100	—
2,957.0	200	—
2,966.8	500	—
2,969.4	500	—
2,972.6	200	—
2,982.2	100	—
2,993.5	50	—
2,994.4	20	—
3,004.3	200	—
3,020,7	100	—
3,021.1	50	—
3,028.0	200	—
3,028.8	20	—
3,037.4	20	—
3,037.8	100	—
3,047.6	20	—
3,057.4	20	—
3,059.1	50	—
3,071.3	100	—
3,082.8	100	—
3,162.1	500	—
3,162.9	100	—
3,203.3	100	—
3,314.0	100	—
3,332.1	100	—
3,362.5	50	—
3,392.0	50	—
3,428.0	20	—
3,431.6	20	—
3,439.8	200	—
3,446.0	50	—
3,448.2	50	—
3,452.0	100	—
3,509.5	50	—
3,510.2	100	—
3,547.9	100	—
3,559.6	50	—
3,561.8	50	—
3,562.6	200	—
3,569.2	100	—
3,570.0	100	—
3,573.1	100	—
3,573.6	100	—
3,581.2	200	—
3,584.7	100	—
3,593.7	100	—
3,615.8	20	—
3,616.5	20	—
3,617.4	20	—
3,631.4	50	—
3,639.5	20	—
3,673.1	200	—
3,683.4	50	—
3,696.4	200	—
3,706.2	100	—
3,707.9	100	—
3,710.3	100	—
3,712.9	50	—
3,720.0	50	—
3,720.5	50	—
3,721.2	20	—
3,725.4	50	—
3,733.0	20	—
3,734.7	20	—
3,737.1	20	—
3,737.6	50	—
3,740.4	50	—
3,748.3	20	—
3,749.4	50	—
3,751.0	50	—
3,752.6	100	—
3,753.2	100	—
3,757.7	200	—
3,761.6	500	—
3,767,1	20	—
3,769.0	20	—
3,774.3	20	—
3,777.4	500	—
3,788.7	50	—
3,789.0	20	—
3,801.7	100	—
3,848.9	100	—
3,871.5	500	—
3,892.9	20	—
3,896.0	20	—
3,901.5	50	—
3,904.2	100	—
3,914.3	100	—
3,916.0	500	—
3,921.5	500	—
3,926.2	1000	—
3,936.2	100	—
3,942.7	20	—
3,943.8	50	—
3,944.0	100	—
3,944.5	50	—
3,945.2	20	—
3,946.0	50	—
3,948.0	20	—
3,950.3	20	—
3,952,5	500	—
3,963.3	50	—
3,973.1	100	—
3,975.2	50	—
3,982.6	20	—
3,984.6	50	—
3,991.9	50	—
4,000.6	50	—
4,003.0	20	—
4,006.2	20	—
4,009.2	100	—
4,013.7	100	—
4,017.9	20	—
4,019.7	100	—
4,020.2	50	—
4,021.3	20	—
4,027.4	50	—
4,027.8	50	—
4,032.9	20	—
4,033.5	20	—
4,036.4	200	—
4,057.8	50	—
4,069.8	50	—
4,079.7	20	—
4,080.5	20	—
4,089.3	1000	—
4,099.0	20	—
4,102.3	20	—
4,107.0	20	—
4,108.8	20	—
4,109.5	20	—
4,109.9	20	—
4,124.5	20	—
4,127.4	20	—
4,128.3	20	—
4,128.9	20	—
4,130.7	20	—
4,136.7	20	—
4,137.7	20	—
4,144.0	20	—

λ	I		λ	I		λ	I	
	J	O		J	O		J	O
4,148.0	20	—	4,226.8	100	—	4,302.5	20	—
4,149.9	20	—	4,233.6	20	—	4,302.8	20	—
4,154.7	50	—	4,234.0	20	—	4,309.7	100	—
4,182.6	20	—	4,248.4	500	—	4,309.9	50	—
4,184.0	20	—	4,249.4	20	—	4,324.6	500	—
4,184.3	20	—	4,249.9	50	—	4,332.8	100	—
4,186.6	50	—	4,252.0	20	—	4,335.6	50	—
4,188.2	500	—	4,261.8	50	—	4,337.2	20	—
4,188.8	20	—	4,262.4	20	—	4,341.8	20	—
4,189.6	20	—	4,265.6	200	—	4,348.3	50	—
4,197.6	20	—	4,272.2	20	—	4,349.3	50	—
4,198.7	20	—	4,278.2	20	—	4,366.7	100	—
4,200.0	50	—	4,289.3	100	—	4,372.4	50	—
4,219.4	20	—	4,289.9	20	—	4,374.9	50	—
4,221.0	20	—	4,300.9	20	—			

Table VII

Analytical Lines of Gases in Alphabetical Order of the Element Symbols

Symbol	Atomic Weight	Atomic No.	λ in Å		
Ar	39.944	18	8,115.31	4,806.07	3,850.57
			7,503.87	4,348.11	3,729.29
			7,067.22	4,277.55	3,588.44
			6,965.43	4,103.91	2,516.81
Cl	35.457	17	4,819.46	3,861.34	
			4,810.06	3,602.10	
			4,794.54	3,340.36	
			4,132.48	3,191.40	
F	19.00	9	6,902.46	3,505.61	
			6,856.02	3,503.10	
			*5,291.0	3,501.42	
			4,446.71	3,472.96	
			4,246.16		
H	1.0080	1	6,562.85	3,835.40	
			6,562.72	3,691.55	
			4,861.33	3,666.07	
			4,340.47	3,656.71	
			4,101.74		
He	4.003	2	5,875.62	3,888.65	
			5,015.68	3,203.14	
			4,685.75	3,187.74	
				2,733.32	
Kr	83.80	36	5,870.92	3,778.09	
			5,570.29	3,741.69	
			4,671.61	3,653.97	
			4,273.97		
N	14.008	7	5,679.56	3,995.00	
			5,676.02	3,955.85	
			5,666.64	3,437.16	
			4,109.98	3,006.86	
Ne	20.183	10	6,402.25	3,713.08	
			5,852.49	3,694.20	
			5,400.56	3,520.47	
				3,369.91	
O	16	8	7,775.43	5,330.66	3,947.33
			7,774.14	4,414.89	3,823.47
			7,771.93	4,368.30	2,881.70
				4,075.87	2,445.55
Rn	222.00	86	7,450.00	4,644.18	
			7,055.42	4,349.60	
			5,084.48	4,203.23	
Xe	131.30	54	4,671.23	3,950.92	
			4,624.28	2,605.54	
			4,500.98	2,475.89	

Table VIII

Table of the Analytical Lines of the Gases according to Wavelength

Symbol	λ in Å	Symbol	λ in Å	Symbol	λ in Å
O	2,445.55	H	3,835.40	Cl	4,810.06
Xe	2,475.89	Ar	3,850.57	Cl	4,819.46
Ar	2,515.60	Cl	3,861.34	Xe	4,844.33
Ar	2,516.81	He	3,888.65	H	4,861.33
Xe	2,605.54	O	3,947.33	He	5,015.68
He	2,733.32	Xe	3,950.92	Rn	5,084.48
O	2,881.70	N	3,955.85	F	5,291.0
N	3,006.86	N	3,995.00	O	5,330.66
He	3,187.74	O	4,075.87	Ne	5,400.56
Cl	3,191.40	H	4,101.74	Kr	5,570.29
He	3,203.14	Ar	4,103.91	N	5,666.64
Cl	3,340.36	N	4,109.98	N	5,676.02
Ne	3,369.91	Cl	4,132.48	N	5,679.56
N	3,437.16	Rn	4,203.23	Ne	5,852.49
F	3,472.96	F	4,246.16	Kr	5,870.92
F	3,501.42	Kr	4,273.97	He	5,875.62
F	3,503.10	Ar	4,277.55	Ne	6,402.25
F	3,505.61	H	4,340.46	H	6,562.73
Ne	3,520.47	Ar	4,348.11	H	6,562.85
Ar	3,588.44	Rn	4,349.60	F	6,856.02
Cl	3,602.10	O	4,368.30	F	6,902.46
Kr	3,653.97	O	4,414.89	Ar	6,965.43
H	3,656.71	F	4,446.71	Rn	7,055.42
H	3,666.07	Xe	4,500.98	Ar	7,067.22
H	3,691.55	Xe	4,624.28	Rn	7,450.00
Ne	3,694.20	Rn	4,644.18	Ar	7,503.87
Ne	3,713.08	Xe	4,671.23	O	7,771.93
Ar	3,729.20	Kr	4,671.61	O	7,774.14
Kr	3,741.69	He	4,685.75	O	7,775.43
Kr	3,778.09	Cl	4,794.54	Ar	8,115.31
O	3,823.47	Ar	4,806.07		

Table IX

Spark Spectrum of the Air

λ in Å	I	Symbol	λ in Å	I	Symbol	λ in Å	I	Symbol
2,287.9	1	N	3,609.8	1	—	4,105.00	3	O
2,318.5	1	O	3,639.6	3	—	4,110.84	2	O
2,382.1	2	—	3,702.9	1	—	4,112.09	1	O
2,395.62	1	—	3,707.3	1	O	4,114.0	0	O
2,399.0	1	—	3,709.2	1	O	4,119.3	4	O
2,404.9	2	—	3,712.7	2	O	4,120.5	2	O
2,406.9	1	—	3,727.34	4	O	4,121.5	2	O
2,433.6	1	O	3,729.3	1	N	4,124.1	2	O
2,445.5	1	O	3,749.51	5	O	4,129.5	1	O
2,507.2	2	—	3,754.5	1	O	4,132.88	2	O
2,514.5	1	—	3,759.8	1	O	4,133.70	2	N
2,599.5	2	—	3,770.9	1	N	4,142.2	2	O
2,739.8	1	—	3,804.0	1	O	4,143.7	1	O
2,746.7	1	—	3,830.7	1	N	4,145.90	3	N
2,749.0	1	—	3,839.1	2	N	4,153.5	3	O
2,755.9	2	—	3,842.8	1	N	4,169.36	1	O
2,795.5	1	—	3,845.1	0	N	4,176.2	2	N
2,858.3	1	—	3,848.04	1	O	4,185.5	4	O
2,927.5	1	—	3,850.6	1	N	4,189.8	6	O
3,007.0	1	O	3,851.2	1	O	4,199.3	0	N
3,047.0	1	—	3,856.7	1	N	4,206.7	2	N
3,059.15	2	—	3,864.6	1	O	4,211.1	1	N
3,130.1	1	—	3,882.3	2	O	4,233.3	1	N
3,135.3	1	O	3,893.3	1	N	4,228.0	2	N
3,139.3	2	O	3,907.6	1	O	4,236.8	3	N
3,158.7	1	—	3,909.1	1	N	4,241.75	2	N
2,265.2	1	O	3,912.1	3	O	4,253.7	2	N
3,288.9	1	—	3,919.10	6	N	4,266.4	2	N
3,301.9	1	—	3,933.6	9	?	4,275.9	1	N
3,312.5	1	O	3,940.2	1	N	4,303.7	1	O
3,318.8	1	—	3,945.1	1	O	4,317.11	3	O
3,320.7	2	O	3,947.45	1	O	4,319.62	3	O
3,325.0	1	O	3,954.4	1	O	4,325.7	1	O
3,329.5	2	N	3,955.9	4	N	4,327.5	1	O
3,331.8	2	N	3,968.4	1	A(?)	4,328.5	1	O
3,334.8	1	—	3,973.30	4	O	4,331.04	1	N
3,354.08	1	O	3,982.76	2	O	4,331.9	1	O
3,365.8	1	N	3,995.1	10	N	4,336.8	2	O
3,367.3	1	N	4,014.0	1	O	4,345.54	3	O
3,370.9	1	N	4,025.7	1	N	4,347.44	2	O
3,374.0	2	N	4,034.9	2	N	4,348.0	2	N
3,377.2	2	O	4,041.3	3	N	4,349.40	4	O
3,390.3	2	O	4,057.8	1	N	4,351.3	2	O
3,408.3	2	O	4,063.2	1	N	4,361.6	0	N
3,437.32	3	N	4,069.90	8	O	4,366.87	3	N
3,450.9	1	—	3,069.90	8	O	4,369.2	1	O
3,471.2	2	—	4,072.25	8	O	4,371.4	1	N
3,491.9	2	—	4,075.93	8	O	4,379.6	1	N
3,514.8	1	—	4,078.9	2	O	4,392.4	0	(?)
3,560.6	1	—	4,085.20	2	O	4,396.0	1	O
3,570.3	1	—	4,089.1	1	O	4,401.2	1	N
3,577.2	1	—	4,093.00	2	O	4,414.9	6	O
3,589.0	1	—	4,097.2	3	N	4,417.0	5	O
3,594.6	1	—	4,103.3	2	N	4,425.9	1	N

λ in Å	I	Symbol	λ in Å	I	Symbol	λ in Å	I	Symbol
4,430.1	1	N	4,879.7	1	N	5,530.2	3	N
4,432.4	2	N	4,890.9	0	O	5,535.2	5	N
4,434.0	0	N	4,895.3	1	N	5,543.4	3	N
4,443.3	1	O	4,906.8	1	O	5,552.0	2	N
4,447.04	6	N	4,924.6	2	O	5,566.0	0	N
4,452.4	2	O	4,934.8	1	N	5,592.3	0	O
4,460.1	1	N	4,941.0	1	N	5,645.6	1	N
4,465.4	2	O	4,942.5	1	N	5,666.6	5	N
4,467.8	2	O	4,943.0	1	O	5,675.9	3	N
4,469.4	1	O	4,955.0	1	O	5,679.5	10	N
4,477.7	1	N	4,964.7	0	N	5,686.2	3	N
4,507.62	2	N	4,987.4	1	N	5,710.7	2	N
4,514.8	2	N	4,991.3	1	N	5,730.6	2	N
4,529.9	2	N	4,994.4	3	N	5,747.5	1	N
4,544.8	1	N	5,001.4	6	N	5,767.4	2	N
4,552.5	2	N	5,005.2	6	N	5,927.8	4	N
4,590.93	3	O	5,007.4	3	N	5,931.8	7	N
4,596.12	3	O	5,010.6	2	N	5,940.5	1	N
4,601.48	4	N	5,013.9	0	...	5,941.6	10	N
4,607.14	4	N	5,016.4	2	N	5,952.4	4	N
4,609.4	1	N	5,022.9	1	N	6,158.1	0	O
4,613.84	3	N	5,025.7	2	N	6,171.0	2	O
4,621.30	4	N	5,032.0	0	...	6,284.3	1	N
4,630.53	10	N	5,045.1	2	N	6,341.5	0	N
4,634.0	1	N	5,061.8	0	N	6,358.1	0	N
4,638.8	2	O	5,073.5	0	N	6,370.7	0	...
4,640.5	1	N	5,136.0	0	...	6,379.3	2	N
4,641.8	3	O	5,143.6	0	O	6,456.0	0	N
4,643.1	4	N	5,150.0	0	...	6,482.0	5	N
4,649.1	4	O	5,160.1	0	O	6,563.2	3	H
4,650.8	2	O	5,172.0	1	N	6,610.4	6	N
4,654.5	1	N	5,173.4	1	N	6,640.7	0	...
4,661.6	5	O	5,175.9	2	N	6,654.8	2	...
4,674.9	1	N	5,179.4	1	N	6,721.3	1	...
4,676.1	3	O	5,183.2	0	O	6,811.9	0	...
4,697.6	0	N(?)	5,185.1	0	N	6,864.0	0	...
4,699.2	3	O	5,190.6	1	N	6,887.6	1	...
4,703.1	0	O	5,206.5	1	O	6,950.0	ON	...
4,705.1	1	N	5,250.6	1	N(?)	6,965.9	1	A
4,705.4	3	O	5,263.0	0	...	7,067.6	0	A
4,709.9	2	O	5,281.7	0	N	7,157.4	9	O(?)
4,718.4	2	N	5,320.5	1	N	7,384.5	1	A
4,735.7	1	N	5,325.1	0	O	7,424.0	8	N
4,751.2	1	N	5,328.6	0	N	7,432.9	0	...
4,764.6	1	N	5,338.7	1	N	7,442.7	10	N
4,774.2	1	N	5,341.2	1	N	7,458.7	0	...
4,779.8	2	N	5,351.2	0	N	7,468.7	10	N
4,781.2	0	N(?)	5,356.4	0	N	7,479.0	0	O
4,788.2	4	N	5,411.5	1	N	7,505.9	0	A
4,793.7	2	N	5,432.1	0	N(?)	7,515.2	0	A
4,803.3	5	N	5,452.1	1	N	7,635.7	1	A
4,805.9	1	N	5,454.1	1	N	7,772.1	10	O
4,810.3	2	N	5,462.8	1	N	7,774.3	7	O
4,847.7	1	N(?)	5,478.1	0	N	7,775.6	6	O
4,856.8	1	O	5,480.1	1	N	7,947.8	4	O
4,860.3	1	N	5,495.7	2	N	7,951.1	3	O
4,871.6	0	O	5,526.2	2	N	7,952.3	2	O

λ in Å	I	Symbol	λ in Å	I	Symbol	λ in Å	I	Symbol
8,185.3	4	N	8,230.2	0	O	8,683.7	1	N
8,188.4	4	N	8,242.8	4	N	8,686.4	0	N
8,200.7	1	N	8,446.8	5	O	8,692.0	0	...
8,211.1	2	N	8,594.0	0	...	8,703.8	0	N
8,216.7	7	N	8,630.0	0	...	8,712.0	0	N
8,223.5	4	N	8,680.6	2	N	8,719.2	0	N

Table X

Selected Lines of Iron in the Region of from 2,000 to 10,000 Å

The spectral lines in the range of up to 6,000 Å are situated at an approximate distance of 25 Å, in the range of from 6,000 to 10,000 Å at a distance of approximately 60 Å from one another.

λ	I		λ	I		λ	I	
	J	*O*		*J*	*O*		*J*	*O*
2,083.51	8	—	3,153.21	80	100	4,325.76	700	1000
2,108.202	4	5	3,175.45	200	200	4,375.93	(200)	500
2,125.01	—	30	3,191.66	150	200	4,404.75	700	1000
2,125.23	3	—	3,215.94	150	300	4,427.31	(200)	500
2,146.04	10	1	3,236.22	200	300	4,447.722	100	200
2,175.447	25	8	3,254.36	150	200	4,466.55	300	500
2,201.117	—	10	3,276.47	50	(100)	4,482.75	2	20
2,201.408	2	—	3,292.59	150	300	4,494.568	150	400
2,228.80	20	—	3,314.74	200	200	4,528.619	200	600
2,229.07	—	6	3,335.77	50	50	4,547.85	100	200
2,262.68	12	6	3,355.23	100	100	4,581.52	2	60
2,272.07	3	15	3,370.79	200	300	4,602.94	100	300
2,301.68	5	15	3,394.59	80	150	4,625.055	12	100
2,313.10	3	25	3,424.29	150	200	4,647.437	40	25
2,332.80	40	15	3,453.02	15	30	4,678.852	100	150
2,351.19	7	2	3,485.34	50	100	4,707.28	12	100
2,374.52	5	6	3,506.50	30	50	4,733.596	1	15
2,402.60	20	10	3,536.56	200	300	4,745.806	1	8
2,423.09	—	15	3,554.93	300	400	4,772.817	4	10
2,423.21	40	—	3,575.37	80	80	4,786.810	—	150
2,457.59	30	70	3,612.07	50	80	4,791.248	200	200
2,475.1	10	5	3,623.19	80	100	4,821.047	200	200
2,502.39	60	30	3,655.47	25	25	4,859.748	40	150
2,523.66	10	15	3,679.91	300	500	4,878.218	4	80
2,555.44	35	10	3,697.43	60	100	4,903.317	2	500
2,570.84	100	70	3,724.38	150	200	4,924.776	—	100
2,603.56	3	20	3,749.49	700	1000	4,957.609	150	300
2,604.05	4	—	3,765.54	150	200	4,973.108	—	100
2,620.41	40	70	3,795.00	400	500	4,994.133	—	200
2,651.71	60	60	3,836.33	60	100	5,001.871	40	300
2,669.50	25	50	3,869.56	80	100	5,012.071	—	300
2,703.99	400	30	3,886.28	400	600	5,041.759	—	300
2,723.58	200	300	3,906.48	200	300	5,083.342	—	200
2,754.43	20	70	3,922.91	400	600	5,102.200	—	80
2,773.24	40	90	3,949.96	100	150	5,127.363	—	100
2,804.52	200	300	3,977.74	150	300	5,150.843	—	150
2,823.28	300	200	4,005.25	200	250	5,171.599	60	300
2,843.98	300	300	4,021.87	100	200	5,202.339	10	300
2,875.30	200	125	4,045.81	300	400	5,227.192	60	400
2,901.91	40	125	4,076.64	50	80	5,250.650	—	150
2,923.85	70	100	4,107.49	100	120	5,270.360	80	400
2,941.34	300	600	4,118.55	100	200	5,307.365	—	125
2,987.29	200	300	4,134.68	100	150	5,328.534	35	150
3,009.09	60	80	4,156.80	80	100	5,341.026	15	200
3,024.03	200	300	4,175.64	80	100	5,371.493	—	700
3,055.26	150	200	4,202.03	300	400	5,397.131	50	400
3,075.72	400	400	4,233.61	150	250	5,405.778	70	400
3,098.19	60	70	4,260.48	300	400	5,429.699	40	500
3,116.59	150	—	4,282.41	300	600	5,455.613	30	300
3,116.63	—	150	4,294.13	400	700	5,474.917	—	100

λ	I		λ	I		λ	I	
	J	O		J	O		J	O
5,497.519	5	150	6,705.12	12	12	8,232.35	—	10
5,525.553	2	40	6,726.78	—	15	8,248.15	—	10
5,554.887	—	100	6,750.16	18	50	8,264.27	—	20
5,572.849	25	300	6,810.25	18	15	8,327.06	2	40
5,586.763	50	400	6,828.61	25	18	8,339.43	—	18
5,615.652	300	400	6,855.18	80	60	8,365.64	—	15
5,624.549	125	150	6,885.77	8	10	8,387.78	1	35
5,658.826	80	100	6,916.70	5	35	8,468.41	—	20
5,679.022	4	5	6,945.21	20	60	8,514.07	—	10
5,701.556	25	50	6,978.85	12	60	8,611.81	—	10
5,731.770	3	10	6,999.90	—	25	8,661.91	—	100
5,753.136	20	40	7,016.44	25	100	8,674.75	—	50
5,775.091	2	12	7,038.25	20	7	8,688.63	—	150
5,806.730	5	10	7,068.41	30	40	8,710.29	—	20
5,825.691	—	6	7,090.40	25	50	8,757.19	—	50
5,855.130	—	5	7,130.94	80	100	8,764.00	—	100
5,873.219	2	8	7,164.47	100	200	8,790.62	—	10
5,883.848	10	15	7,187.34	300	500	8,793.38	—	25
5,905.683	8	12	7,207.41	300	300	8,804.62	—	10
5,930.186	10	30	7,239.88	20	25	8,824.23	—	200
5,952.362	—	3	7,261.54	10	18	8,838.43	—	30
5,975.358	10	10	7,288.76	20	30	8,866.96	—	150
6,003.034	15	30	7,293.07	50	100	8,919.95	—	10
6,027.057	12	6	7,311.10	25	60	8,945.20	—	20
6,065.487	30	50	7,351.56	7	18	8,975.41	—	15
6,089.564	—	2	7,386.40	25	40	8,999.56	—	100
6,103.333	40	—	7,411.18	—	100	9,012.10	—	30
6,109.318	—	4	7,445.78	—	150	9,024.47	—	15
6,137.696	—	100	7,476.30	—	12	9,088.33	—	40
6,147.849	6	5	7,491.68	—	20	9,089.41	—	30
6,173.339	—	18	7,495.09	—	200	9,118.89	—	20
6,191.562	20	100	7,507.28	—	40	9,258.31	—	20
6,230.728	50	60	7,531.17	—	80	9,259.06	—	15
6,252.561	25	60	7,586.04	—	10	9,350.44	—	10
6,265.140	5	12	7,653.76	—	80	9,401.14	—	10
6,301.517	50	50	7,710.39	—	10	9,414.14	—	20
6,318.022	25	40	7,711.73	15	25	9,433.98	—	10
6,335.335	20	50	7,748.28	—	25	9,513.24	—	10
6,358.690	6	8	7,780.59	—	25	9,569.96	—	40
6,393.605	80	100	7,832.22	1	30	9,626.56	—	30
6,430.851	80	100	7,937.17	1	40	9,653.14	—	20
6,462.729	7	20	7,945.88	2	30	9,738.62	—	200
6,494.985	150	400	7,998.97	1	35	9,753.13	—	10
6,518.374	7	10	8,028.34	—	100	9,763.45	—	15
6,546.245	50	150	8,046.07	1	25	9,763.91	—	15
6,575.022	15	12	8,085.20	1	20	9,800.33	—	20
6,592.919	80	150	8,198.95	—	20	9,861.79	—	30
6,627.566	8	4	8,207.77	—	100	9,889.08	—	0
6,677.93	100	250						

Table XI

Wavelength Standards

At the General Conference on Weights and Measures held in Paris from the October 11 – 20, 1960, the meter was newly defined as equalling 1,650,763.73 times the wavelength of the orange line emitted by the Krypton Kr 86 isotope which is produced by the radiation corresponding to the leap of the electron from the level marked spectrographically 2 p_{10} to the level marked 5 d_5. (Hollow cathode discharge, in sealed cryostat, at a temperature of about – 210°C (63°K) – triple point of nitrogen).

Primary Standard (up to the year 1960)

The wavelength of the Cadmium red line in the air, at an atmospheric pressure of 760 mm Hg and at a temperature of 15°C. Measurements carried out by Benoit, Fabry, and Perot in the year 1907.

6,438.4696 Å

Secondary Standards

As secondary standards are given the lines of iron measured interferrometrically from 2,100.795Å to 3,497.8418 Å by Meggers and Humphrey (Jour. Research of B. of S. 18, 543, 1937). The region of from 3,513.821 to 6,750.157 by C. Fabry (International Critical Tables, 1929 – selected from the Handbook of Chemistry and Physics, 37 ed. 1955 – 1956). The infrared wave length range of from 7,164.469 Å to 10,216.351 Å contains lines determined by Meggers (Jour. of B. of S. 14, 33, 1935). All measurements relate to atmospheric pressure and to 15°C.

Table XI

Å	Å	Å	Å	Å
2,100.795	2,279.922	2,474.8131	2,874.1722	3,284.5892
2,102.349	2,283.653	2,487.0643	2,877.3003	3,286.7538
2,108.955	2,284.087	2,496.5324	2,894.5050	3,298.1328
2,110.233	2,287.2477	2,507.8987	2,895.0352	3,305.971
2,112.966	2,287.632	2,519.6279	2,899.4156	3,306.356
2,115.168	2,291.122	2,530.6938	2,912.1581	3,314.7421
2,130.962	2,292.5227	2,542.1007	2,920.6906	3,323.7374
2,132.015	2,293.8454	2,551.0936	2,929.0081	3,328.8669
2,135.957	2,294.4059	2,562.5348	2,941.3430	3,337.6655
2,138.589	2,296.9247	2,575.7442	2,953.9400	3,340.5659
2,139.695	2,297.785	2,576.1033	2,957.3654	3,347.9262
2,141.715	2,299.2180	2,584.5349	2,959.9924	3,355.2285
2,145.188	2,300.1397	2,585.8753	2,965.2551	3,370.7845
2,147.787	2,301.6818	2,598.3689	2,981.4448	3,380.1111
2,150.182	2,303.4225	2,611.8725	2,987.2919	3,383.9808
2,151.099	2,303.576	2,613.8240	2,990.3923	3,396.9772
2,153.004	2,308.9971	2,617.6160	2,999.5123	3,399.3343
2,154.458	2,313.1022	2,621.6690	3,003.0311	3,401.5196
2,157.792	2,320.3561	2,625.6663	3,009.5698	3,407.4608
2,161.577	2,327.3940	2,628.2923	3,015.9129	3,413.1335
2,163.366	2,331.3067	2,635.8082	3,024.0330	3,427.1207
2,163.860	2,332.7972	2,643.9972	3,030.1491	3,443.8774
2,164.547	2,338.0052	2,647.5576	3,037.3891	3,445.1506
2,165.860	2,344.2802	2,651.7059	3,040.4281	3,465.8622
2,172.581	2,354.8888	2,662.0563	3,047.6059	3,476.7035
2,173.211	2,359.1039	2,673.2127	3,055.2631	3,485.3415
2,176.837	2,359.997	2,679.0608	3,057.4452	3,490.5746
2,180.866	2,360.294	2,689.2117	3,059.0874	3,497.8418
2,183.979	2,362.019	2,699.1060	3,067.2433	3,513.821
2,186.890	2,364.8269	2,706.5812	3,075.7204	3,556.882
2,187.191	2,366.592	2,711.6548	3,083.7419	3,606.682
2,191.202	2,368.595	2,714.413	3,091.5777	3,640.392
2,196.039	2,370.497	2,718.4352	3,116.6329	3,676.314
2,200.7227	2,371.4285	2,723.5770	3,125.653	3,677.630
2,201.117	2,374.517	2,727.540	3,134.1113	3,724.381
2,207.068	2,375.193	2,735.473	3,143.9896	3,753.615
2,210.686	2,379.2756	2,739.5467	3,157.0388	3,805.346
2,211.234	2,380.7591	2,746.4833	3,160.6582	3,843.261
2,228.1704	2,384.386	2,746.9823	3,175.4465	3,850.820
2,231.211	2,388.6270	2,749.325	3,178.0137	3,865.527
2,240.627	2,389.9713	2,755.7366	3,184.8948	3,906.482
2,245.651	2,399.2396	2,763.1078	3,191.6583	3,907.937
2,248.858	2,404.430	2,767.5208	3,196.9288	3,935.816
2,249.177	2,406.6593	2,778.2205	3,200.4741	3,977.744
2,253.1251	2,410.5172	2,781.8347	3,205.3992	4,021.870
2,255.861	2,411.0663	2,797.7751	3,215.9398	4,076.638
2,259.511	2,413.3087	2,804.5200	3,217.3796	4,118.549
2,260.079	2,431.025	2,806.9840	3,222.0682	4,134.680
2,264.3894	2,438.1811	2,813.2861	3,225.7883	4,147.673
2,265.053	2,442.5674	2,823.2753	3,236.2226	4,191.436
2,270.8601	2,443.8707	2,832.4350	3,239.4362	4,233.606
2,271.781	2,447.7086	2,838.1193	3,244.1887	4,282.406
2,272.0670	2,453.4746	2,845.5945	3,254.3628	4,315.087
2,274.0085	2,457.5956	2,851.7970	3,257.5937	4,375.933
2,276.0247	2,465.1479	2,863.864	3,271.0014	4,427.313
2,277.098	2,468.8782	2,869.3075	3,280.2613	4,466.556

Å	Å	Å	Å	Å
4,494.568	5,586.763	7,418.674	8,232.347	8,975.408
4,531.152	5,615.652	7,445.776	8,239.130	8,999.561
4,547.851	5,658.825	7,495.088	8,248.151	9,012.098
4,592.655	5,709.396	7,511.045	8,293.527	9,079.599
4,602.945	5,763.013	7,531.171	8,327.063	9,088.326
4,647.437	5,857.759 Ni	7,586.925	8,331.941	9,089.413
4,691.414	5,892.882 Ni	7,583.796	8,339.431	9,118.888
4,707.282	6,027.058	7,568,044	8,360.822	9,147.800
4,736.782	6,065.489	7,620.538	8,365.642	9,210.030
4,789.654	6,137.697	7,661.223	8,387.781	9,258.30
4,878.219	6,191.563	7,664.302	8,439.603	9,350.46
4,903.318	6,230.729	7,710.390	8,468.413	9,359.420
4,919.001	6,265.141	7,748.281	8,514.075	9,362.370
5,001.872	6,318.023	7,780.586	8,526.685	9,372.900
5,012.072	6,335.338	7,832.224	8,582.267	9,430.08
5,049.825	6,393.606	7,912.866	8,611.807	9,513.24
5,083.343	6,430.852	7,937.166	8,621.612	9,569.960
5,110.414	6,494.985	7,945.878	8,661.908	9,626.562
5,167.491	6,546.245	7,994.473	8,674.751	9,653.143
5,192.353	6,592.920	7,998.972	8,688.633	9,753.129
5,232.948	6,677.994	8,025.200	8,757,192	9,763.450
5,266.564	6,750.157	8,028.341	8,764.000	9,763.913
5,371.493	7,164.469	8,046.073	8,793.376	9,800.335
5,405.779	7,187.341	8,080.668	8,804.624	9,861.793
5,434.527	7,207.406	8,096.874	8,824.227	9,889.082
5,455.613	7,389.425	8,198.951	8,838.433	10,065.080
5,497.520	7,401.689	8,207.767	8,866.961	10,145.601
5,506.783	7,411.178	8,220.406	8,945.204	10,216.351
5,569.626				

Table XII

Molecule Bands of Inorganic Compounds which Occur in Emission Spectra is so far as they are Useful for the Analysis. The bands are listed in alphabetical order by the molecule symbols

a	*b*	*c*	*d*	*e*
Molecule	λ in Å	Band direction	Sensitivity in carbon arc	Note
AlO	4,821.1	K	1	At high concentration of Al only
BaF	4,950.8	K	—	At higher concentration of F
BeO	4,708.7	K	2	At high concentration of Be only
C_2	4,737.1 5,165.2	L L		
CN	3,883.35 4,215.97	L L		Intensive, always found in the spectrum of carbon electrodes
CaCl	5,934.0	K	3	Narrow band, of low intensity ~1% when Cl in minerals is being identified
CaF	5,291.0	K	3	~0.1% may be used in the determination of F in ores (1)
	6,064.4	L		Overlapping with the band of CaO; not used for analytical purposes
CaO	5,473.0	K	3	Very characteristic; from 5,473 to 5,560 Å
LaO	4,371.9 4,418.1 5,599.9 7,379.8	K K K K	4 4 4 4	0.1 – 0.5%
	7,403.5 7,871.2 7,910.5	K K K	4 3 4	
ScO	6,036.2	K	5	0.1%
SrF	5,772.0	K	—	At higher concentration of F
YO	5,972.2 6,132.1	K K	5 5	0.1 – 0.03%

Explanations: L – slow decrease in the intensity of the band in the direction towards the shortwave portion

K – slow decrease of the intensity of the band in the direction towards the longwave portion

Estimation of the relative intensities in the column *d* is established on the decimal scale.

Table XIII

Molecule Bands of Inorganic Compounds Occurring in Emission Spectra in the Case they may be Used in Analysis

The bands are listed by order of wavelength

a	*b*	*c*	*d*
λ in Å	Molecule	Band direction	Sensitivity in arc
3,883.35	CN	L	—
4,215.97	CN	L	—
4,371.9	LaO	K	4
4,418.1	LaO	K	4
4,708.7	BeO	K	2
4,737.1	C_2	L	—
4,821.1	AlO	K	1
4,950.8	BaF	K	—
5,165.2	C_2	L	—
5,291.0	CaF	K	5
5,473.0	CaO	K	3
5,599.9	LaO	K	4
5,772.0	SrF	K	—
5,934.0	CaCl	K	3
5,972.2	YO	K	5
6,036.2	SeO	K	5
6,064.4	CaF	L	—
6,132.1	YO	K	5
7,379.8	LaO	K	4
7,403.5	LaO	K	4
7,877.2	LaO	K	3
7,910.5	LaO	K	4

Table XIV

Ionization Potentials of the Neutral Atoms of the Elements and of their Ionts

Symbol	I	II	III	IV	V
1. H	13.596	—	—	—	—
2. He	24.581	54.405	—	—	—
3. Li	5.390	75.622	122.427	—	—
4. Be	9.321	18.207	153.85	217.671	—
5. B	8.296	25.119	37.921	259.31	340.156
6. C	11.265	24.377	47.866	64.478	392.0
7. N	14.545	29.606	47.609	77.4	97.87
8. O	13.615	35.082	55.118	77.28	113.7
9. F	17.422	34.979	62.647	87.142	114.22
10. Ne	21.559	40.958	63.427	96.897	126.43
11. Na	5.138	47.292	71.650	—	—
12. Mg	7.645	15.032	80.119	109.533	—
13. Al	5.985	18.824	28.442	119.961	154.28
14. Si	8.149	16.339	33.489	45.131	166.5
15. P	10.977	19.653	30.157	51.356	65.01
16. S	10.357	23.405	35.048	47.294	62.2
17. Cl	12.959	23.799	39.905	54.452	67.8
18. Ar	15.736	27.619	40.68	61	78
19. K	4.340	31.811	45.7	—	—
20. Ca	6.112	11.868	51.209	67.2	—
21. Sc	6.56	12.9	24.753	73.913	91.8
22. Ti	6.835	13.6	27.5	43.237	99.84
23. V	6.738	14.2	26.5	48.5	64
24. Cr	6.761	16.7	—	—	73.0
25. Mn	7.429	15.636	—	—	76.0
26. Fe	7.86	16.240	30.6	—	—
27. Co	7.876	17.4	—	—	—
28. Ni	7.633	18.2	—	—	—
29. Cu	7.723	20.283	—	—	—
30. Zn	9.392	17.960	39.7	—	—
31. Ga	5.997	20.509	30.7	64.1	—
32. Ge	8.126	15.93	34.216	45.7	93.43
33. As	9.81	20.2	27.297	50.123	62.61
34. Se	9.750	21.691	34.078	42.900	73.11
35. Br	11.844	19.2	35.888	—	—
36. Kr	13.996	26.5	36.94	68	—
37. Rb	4.176	27.499	47	80	—
38. Sr	5.693	11.026	—	—	—
39. Y	~6.6	12.4	20.5	—	~77
40. Zr	6.951	14.03	24.10	33.972	—
41. Nb	6.77	—	24.332	—	~50
42. Mo	7.383	—	—	—	61.12
44. Ru	~7.5	—	—	—	—
45. Rh	~7.7	—	—	—	—
46. Pd	8.334	19.9	—	—	—
47. Ag	7.574	21.960	36.10	—	—
48. Cd	8.991	16.904	38.217	—	—
49. In	5.785	18.867	28.030	58.037	—
50. Sn	7.332	14.629	30.654	40.740	81.13
51. Sb	8.64	~18.6	24.825	44.147	55.69
52. Te	9.007	21.543	30.611	37.817	60.27
53. J	10.44	19.010	—	—	—
54. Xe	12.127	21.204	32.115	~46	~76
55. Cs	3.893	32.453	~35	~51	~58

Symbol	I	II	III	IV	V
56. Ba	5.2097	10.001	—	—	—
57. La	5.614	11.43	19.17	—	—
58. Ce	~6.57	—	19.70	36.715	—
59. Pr	~5.76	—	—	—	—
60. Nd	~6.31	—	—	—	—
62. Sm	5.6	~11.4	—	—	—
63. Eu	5.67	11.24	—	—	—
64. Gd	6.16	—	—	—	—
65. Tb	~6.74	—	—	—	—
66. Dy	~6.82	—	—	—	—
70. Yb	6.25	12.11	—	—	—
71. Lu	~5.0	—	—	—	—
72. Hf	~5.5	—	—	—	—
73. Ta	~6.0	—	—	—	—
74. W	7.98	—	—	—	—
75. Re	7.87	—	—	—	—
76. Os	~8.7	—	—	—	—
77. Ir	9.2	—	—	—	—
78. Pt	8.96	~19.3	—	—	—
79. Au	9.223	20.1	—	—	—
80. Hg	10.434	18.752	34.5	~72	~82
81. Tl	6.106	20.423	29.8	50.8	—
82. Pb	7.415	15.04	32.1	38.97	69.7
83. Bi	~8	16.7	25.56	45.3	56.0
86. Rn	10.746	—	—	—	—
88. Ra	5.278	10.145	—	—	—
90. Th	—	—	29.5	—	—
92. U	~4	—	—	—	—

Table XV

Classification of Electrons according to their Quantum Numbers into Levels of the Electron Shell

	K	*L*		*M*			*N*				*O*				*P*			*Q*
n =	1	2		3			4				5				6			7
l =	0	0	1	0	1	2	0	1	2	3	0	1	2	3	0	1	2	0
	s	*s*	*p*	*s*	*p*	*d*	*s*	*p*	*d*	*f*	*s*	*p*	*d*	*f*	*s*	*p*	*d*	*s*
	2	2	6	2	6	10	2	6	10	14	2	6	10	14	2	6	10	2
1. H	1																	
2. He	2																	
3. Li	2	1																
4. Be	2	2																
5. B	2	2	1															
6. C	2	2	2															
7. N	2	2	3															
8. O	2	2	4															
9. F	2	2	5															
10. Ne	2	2	6															
11. Na	2	2	6	1														
12. Mg	2	2	6	2														
13. Al	2	2	6	2	1													
14. Si	2	2	6	2	2													
15. P	2	2	6	2	2													
16. S	2	2	6	2	4													
17. Cl	2	2	6	2	5													
18. Ar	2	2	6	2	6													
19. K	2	2	6	2	6		1											
20. Ca	2	2	6	2	6		2											
21. Sc	2	2	6	2	6	1	2											
22. Ti	2	2	6	2	6	2	2											
23. V	2	2	6	2	6	3	2											
24. Cr	2	2	6	2	6	5	1											
25. Mn	2	2	6	2	6	5	2											
26. Fe	2	2	6	2	6	6	2											
27. Co	2	2	6	2	6	7	2											
28. Ni	2	2	6	2	6	8	2											
29. Cu	2	2	6	2	6	10	1											
30. Zn	2	2	6	2	6	10	2											
31. Ga	2	2	6	2	6	10	2	1										
32. Ge	2	2	6	2	6	10	2	2										
33. As	2	2	6	2	6	10	2	3										
34. Se	2	2	6	2	6	10	2	4										
35. Br	2	2	6	2	6	10	2	5										
36. Kr	2	2	6	2	6	10	2	6										
37. Rb	2	2	6	2	6	10	2	6			1							
38. Sr	2	2	6	2	6	10	2	6			2							
39. Y	2	2	6	2	6	10	2	6	1		2							
40. Zr	2	2	6	2	6	10	2	6	2		2							
41. Nb	2	2	6	2	6	10	2	6	4		1							
42. Mo	2	2	6	2	6	10	2	6	5		1							
43. Tc	2	2	6	2	6	10	2	6	6		1							
44. Ru	2	2	6	2	6	10	2	6	7		1							
45. Rh	2	2	6	2	6	10	2	6	8		1							
46. Pd	2	2	6	2	6	10	2	6	10									

	K	L		M			N				O				P			Q
n =	1	2		3			4				5				6			7
l =	0	0	1	0	1	2	0	1	2	3	0	1	2	3	0	1	2	0
	s	s	p	s	p	d	s	p	d	f	s	p	d	f	s	p	d	s
	2	2	6	2	6	10	2	6	10	14	2	6	10	14	2	6	10	2
47. Ag	2	2	6	2	6	10	2	6	10		1							
48. Cd	2	2	6	2	6	10	2	6	10		2							
49. In	2	2	6	2	6	10	2	6	10		2	1						
50. Sn	2	2	6	2	6	10	2	6	10		2	2						
51. Sb	2	2	6	2	6	10	2	6	10		2	3						
52. Te	2	2	6	2	6	10	2	6	10		2	4						
53. J	2	2	6	2	6	10	2	6	10		2	5						
54. Xe	2	2	6	2	6	10	2	6	10		2	6						
55. Cs	2	2	6	2	6	10	2	6	10		2	6			1			
56. Ba	2	2	6	2	6	10	2	6	10		2	6			2			
57. La	2	2	6	2	6	10	2	6	10		2	6	1		2			
58. Ce	2	2	6	2	6	10	2	6	10	1	2	6	1		2			
59. Pr	2	2	6	2	6	10	2	6	10	2	2	6	1		2			
60. Nd	2	2	6	2	6	10	2	6	10	3	2	6	1		2			
61. Pm	2	2	6	2	6	10	2	6	10	4	2	6	1		2			
62. Sm	2	2	6	2	6	10	2	6	10	5	2	6	1		2			
63. Eu	2	2	6	2	6	10	2	6	10	6	2	6	1		2			
64. Gd	2	2	6	2	6	10	2	6	10	7	2	6	1		2			
65. Tb	2	2	6	2	6	10	2	6	10	8	2	6	1		2			
66. Dy	2	2	6	2	6	10	2	6	10	9	2	6	1		2			
67. Ho	2	2	6	2	6	10	2	6	10	10	2	6	1		2			
68. Er	2	2	6	2	6	10	2	6	10	11	2	6	1		2			
69. Tu	2	2	6	2	6	10	2	6	10	12	2	6	1		2			
70. Yb	2	2	6	2	6	10	2	6	10	13	2	6	1		2			
71. Lu	2	2	6	2	6	10	2	6	10	14	2	6	1		2			
72. Hf	2	2	6	2	6	10	2	6	10	14	2	6	2		2			
73. Ta	2	2	6	2	6	10	2	6	10	14	2	6	3		2			
74. W	2	2	6	2	6	10	2	6	10	14	2	6	4		2			
75. Re	2	2	6	2	6	10	2	6	10	14	2	6	5		2			
76. Os	2	2	6	2	6	10	2	6	10	14	2	6	6		2			
77. Ir	2	2	6	2	6	10	2	6	10	14	2	6	7		2			
78. Pt	2	2	6	2	6	10	2	6	10	14	2	6	8		2			
79. Au	2	2	6	2	6	10	2	6	10	14	2	6	10		1			
80. Hg	2	2	6	2	6	10	2	6	10	14	2	6	10		2			
81. Tl	2	2	6	2	6	10	2	6	10	14	2	6	10		2	1		
82. Pb	2	2	6	2	6	10	2	6	10	14	2	6	10		2	2		
83. Bi	2	2	6	2	6	10	2	6	10	14	2	6	10		2	3		
84. Po	2	2	6	2	6	10	2	6	10	14	2	6	10		2	4		
85. At	2	2	6	2	6	10	2	6	10	14	2	6	10		2	5		
86. Rn	2	2	6	2	6	10	2	6	10	14	2	6	10		2	6		
87. Fr	2	2	6	2	6	10	2	6	10	14	2	6	10		2	6		1
88. Ra	2	2	6	2	6	10	2	6	10	14	2	6	10		2	6		2
89. Ac	2	2	6	2	6	10	2	6	10	14	2	6	10		2	6	1	2
90. Th	2	2	6	2	6	10	2	6	10	14	2	6	10		2	6	2	2
91. Pa	2	2	6	2	6	10	2	6	10	14	2	6	10		2	6	3	2
92. U	2	2	6	2	6	10	2	6	10	14	2	6	10		2	6	5	1

DISLOCATION OF MAIN LINES IN THE REGION OF THE SPECTRUM 2,000—10,000 Å

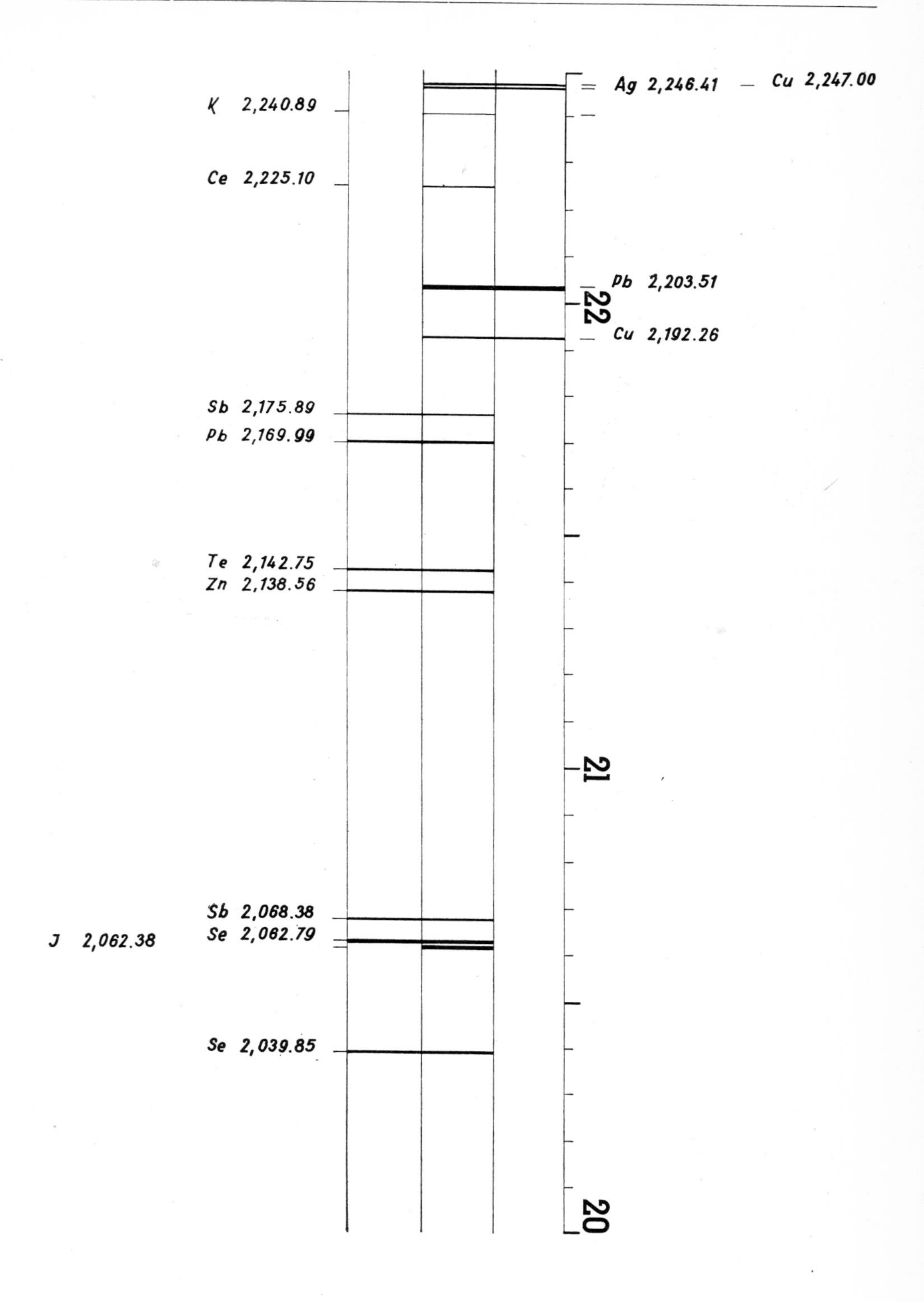

Ag 2,246.41
Cu 2,247.00
K 2,240.89
Ce 2,225.10
Pb 2,203.51
22
Cu 2,192.26
Sb 2,175.89
Pb 2,169.99
Te 2,142.75
Zn 2,138.56
21
Sb 2,068.38
J 2,062.38
Se 2,062.79
Se 2,039.85
20

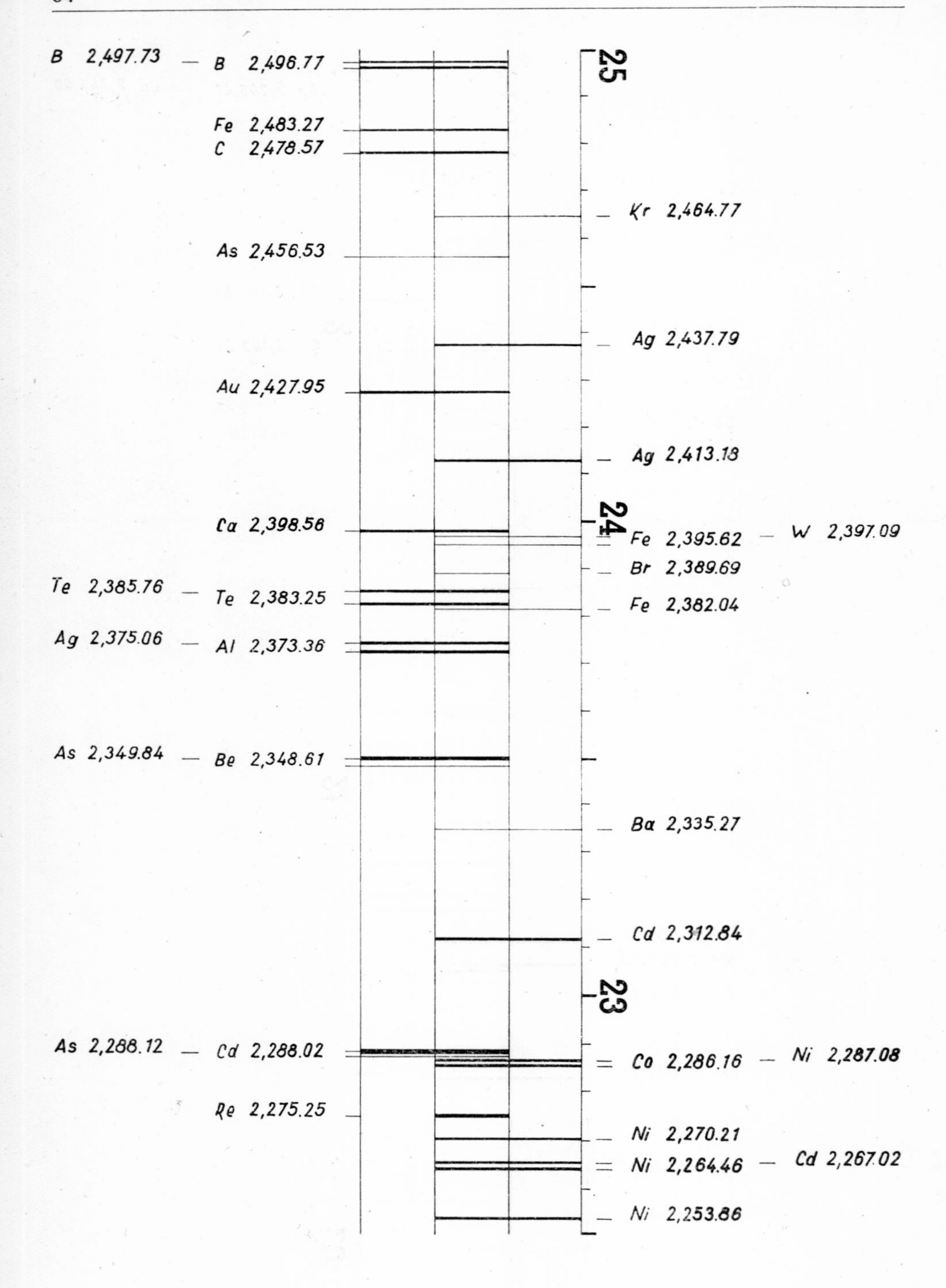
B 2,497.73 — B 2,496.77
Fe 2,483.27
C 2,478.57
Kr 2,464.77
As 2,456.53
Ag 2,437.79
Au 2,427.95
Ag 2,413.18
Ca 2,398.56
Fe 2,395.62 — W 2,397.09
Br 2,389.69
Te 2,385.76 — Te 2,383.25
Fe 2,382.04
Ag 2,375.06 — Al 2,373.36
As 2,349.84 — Be 2,348.61
Ba 2,335.27
Cd 2,312.84
As 2,288.12 — Cd 2,288.02
Co 2,286.16 — Ni 2,287.08
Re 2,275.25
Ni 2,270.21
Ni 2,264.46 — Cd 2,267.02
Ni 2,253.86
25
24
23

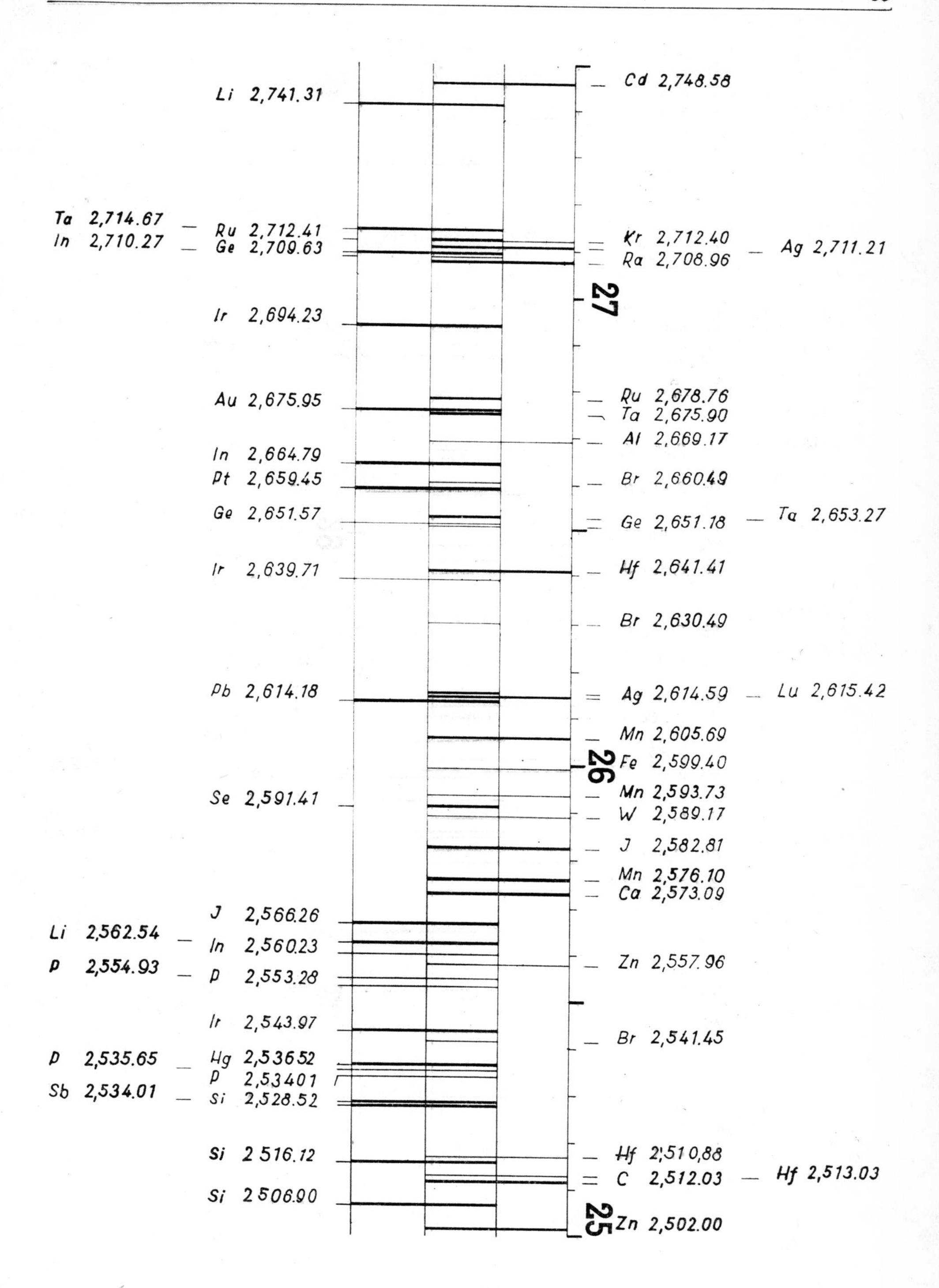

Li 2,741.31
Cd 2,748.58
Ta 2,714.67
In 2,710.27
Ru 2,712.41
Ge 2,709.63
Kr 2,712.40
Ra 2,708.96
Ag 2,711.21
27
Ir 2,694.23
Au 2,675.95
Ru 2,678.76
Ta 2,675.90
Al 2,669.17
In 2,664.79
Pt 2,659.45
Br 2,660.49
Ge 2,651.57
Ge 2,651.18
Ta 2,653.27
Ir 2,639.71
Hf 2,641.41
Br 2,630.49
Pb 2,614.18
Ag 2,614.59
Lu 2,615.42
Mn 2,605.69
26
Fe 2,599.40
Se 2,591.41
Mn 2,593.73
W 2,589.17
J 2,582.81
Mn 2,576.10
Ca 2,573.09
J 2,566.26
Li 2,562.54
In 2,560.23
P 2,554.93
P 2,553.28
Zn 2,557.96
Ir 2,543.97
Br 2,541.45
P 2,535.65
Hg 2,536.52
P 2,534.01
Sb 2,534.01
Si 2,528.52
Si 2 516.12
Hf 2,510.88
C 2,512.03
Hf 2,513.03
Si 2 506.90
25
Zn 2,502.00

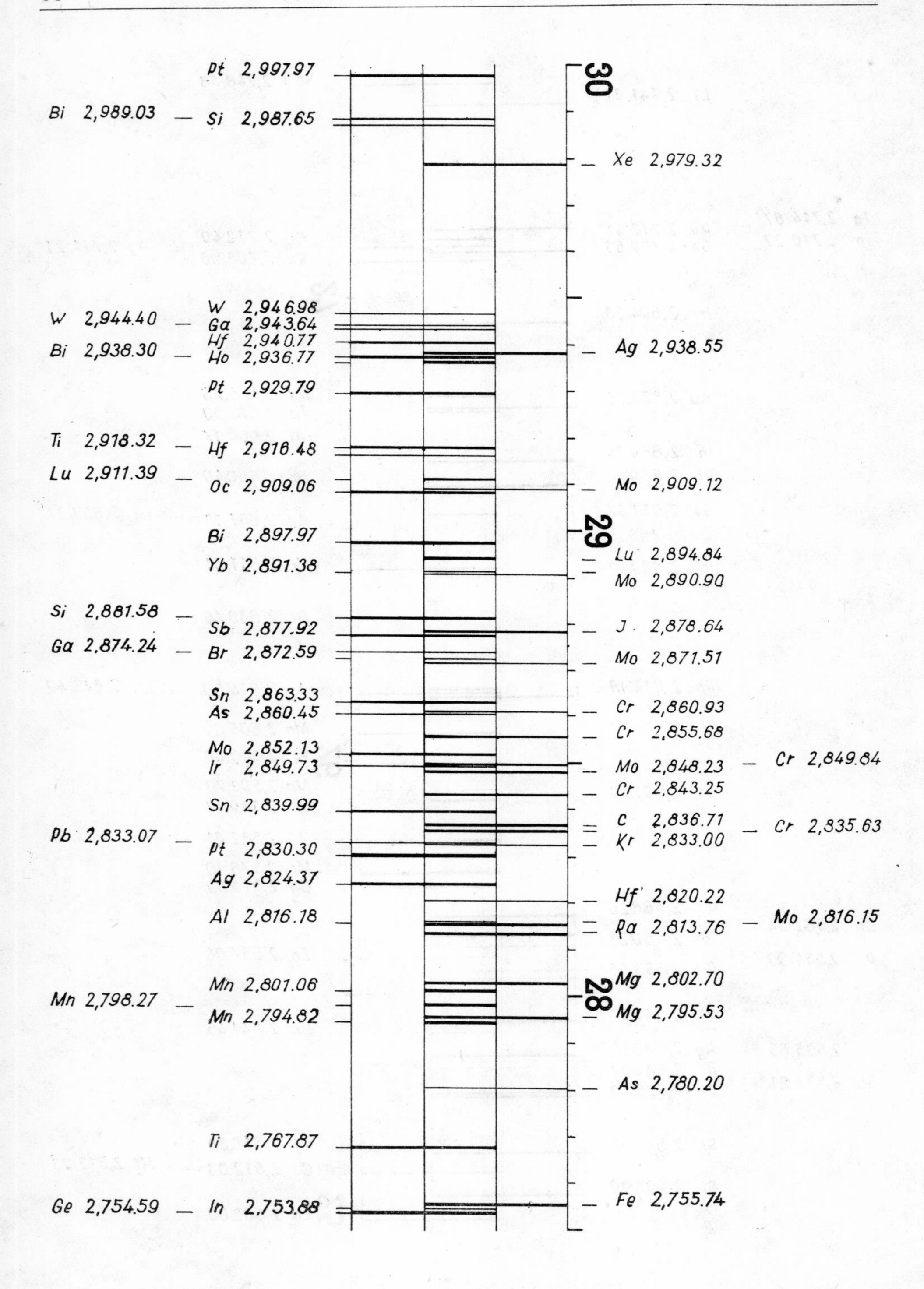

Pt 2,997.97
Bi 2,989.03
Si 2,987.65
30
Xe 2,979.32
W 2,944.40
W 2,946.98
Ga 2,943.64
Hf 2,940.77
Bi 2,938.30
Ho 2,936.77
Ag 2,938.55
Pt 2,929.79
Ti 2,918.32
Hf 2,916.48
Lu 2,911.39
Oc 2,909.06
Mo 2,909.12
Bi 2,897.97
29
Lu 2,894.84
Yb 2,891.38
Mo 2,890.90
Si 2,881.58
Sb 2,877.92
J 2,878.64
Ga 2,874.24
Br 2,872.59
Mo 2,871.51
Sn 2,863.33
As 2,860.45
Cr 2,860.93
Cr 2,855.68
Mo 2,852.13
Ir 2,849.73
Mo 2,848.23
Cr 2,849.84
Cr 2,843.25
Sn 2,839.99
C 2,836.71
Cr 2,835.63
Pb 2,833.07
Pt 2,830.30
Kr 2,833.00
Ag 2,824.37
Hf 2,820.22
Al 2,816.18
Ra 2,813.76
Mo 2,816.15
Mn 2,801.06
Mg 2,802.70
28
Mn 2,798.27
Mn 2,794.82
Mg 2,795.53
As 2,780.20
Ti 2,767.87
Ge 2,754.59
In 2,753.88
Fe 2,755.74

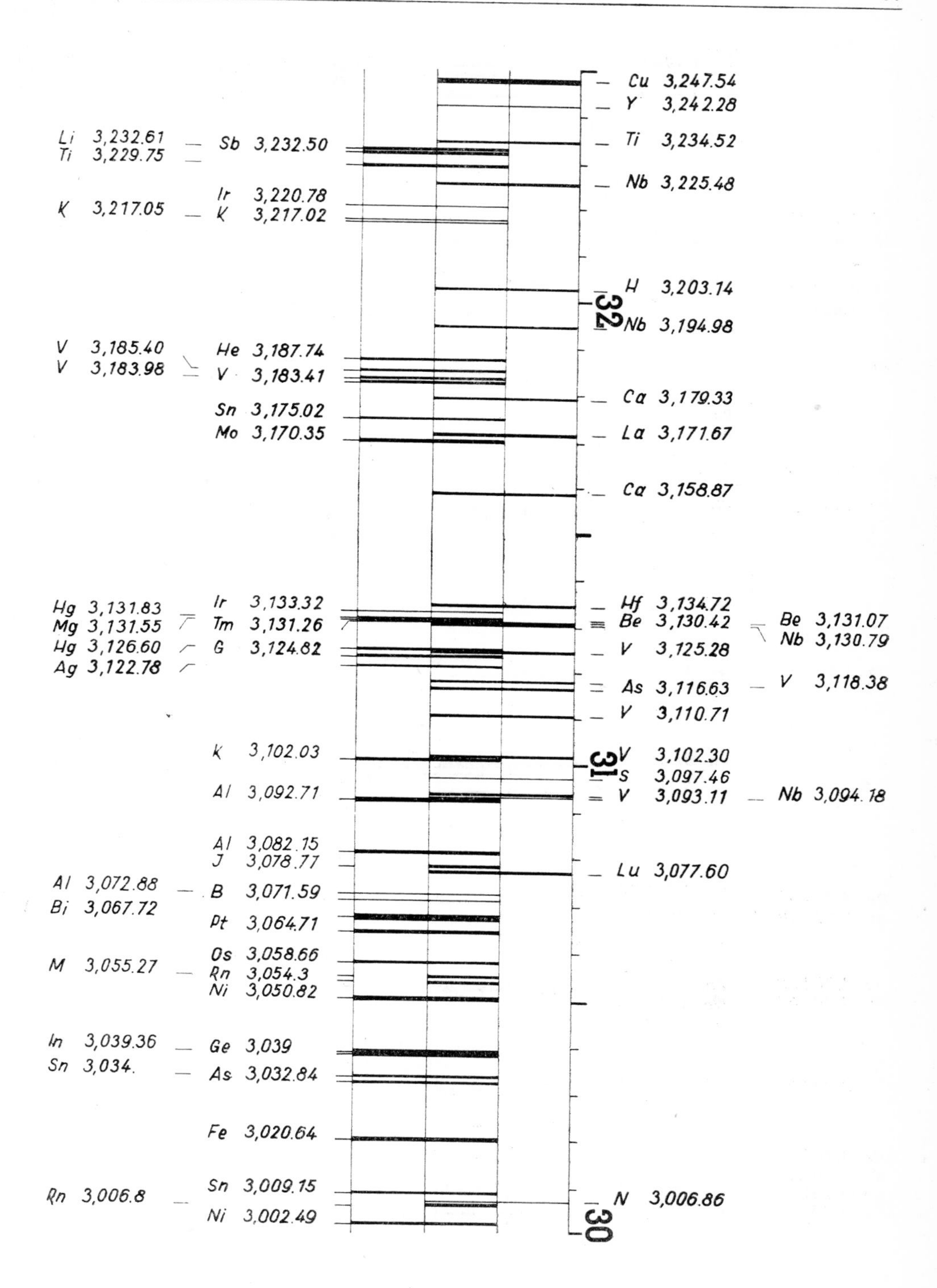
Cu 3,247.54
Y 3,242.28
Li 3,232.61
Ti 3,229.75
Sb 3,232.50
Ti 3,234.52
Nb 3,225.48
Ir 3,220.78
K 3,217.05
K 3,217.02
H 3,203.14
32
Nb 3,194.98
V 3,185.40
V 3,183.98
He 3,187.74
V 3,183.41
Ca 3,179.33
Sn 3,175.02
Mo 3,170.35
La 3,171.67
Ca 3,158.87
Hg 3,131.83
Mg 3,131.55
Hg 3,126.60
Ag 3,122.78
Ir 3,133.32
Tm 3,131.26
G 3,124.82
Hf 3,134.72
Be 3,130.42
Be 3,131.07
Nb 3,130.79
V 3,125.28
As 3,116.63
V 3,118.38
V 3,110.71
K 3,102.03
V 3,102.30
31
S 3,097.46
Al 3,092.71
V 3,093.11
Nb 3,094.18
Al 3,082.15
J 3,078.77
Lu 3,077.60
Al 3,072.88
B 3,071.59
Bi 3,067.72
Pt 3,064.71
M 3,055.27
Os 3,058.66
Rn 3,054.3
Ni 3,050.82
In 3,039.36
Ge 3,039
Sn 3,034.
As 3,032.84
Fe 3,020.64
Rn 3,006.8
Sn 3,009.15
N 3,006.86
Ni 3,002.49
30

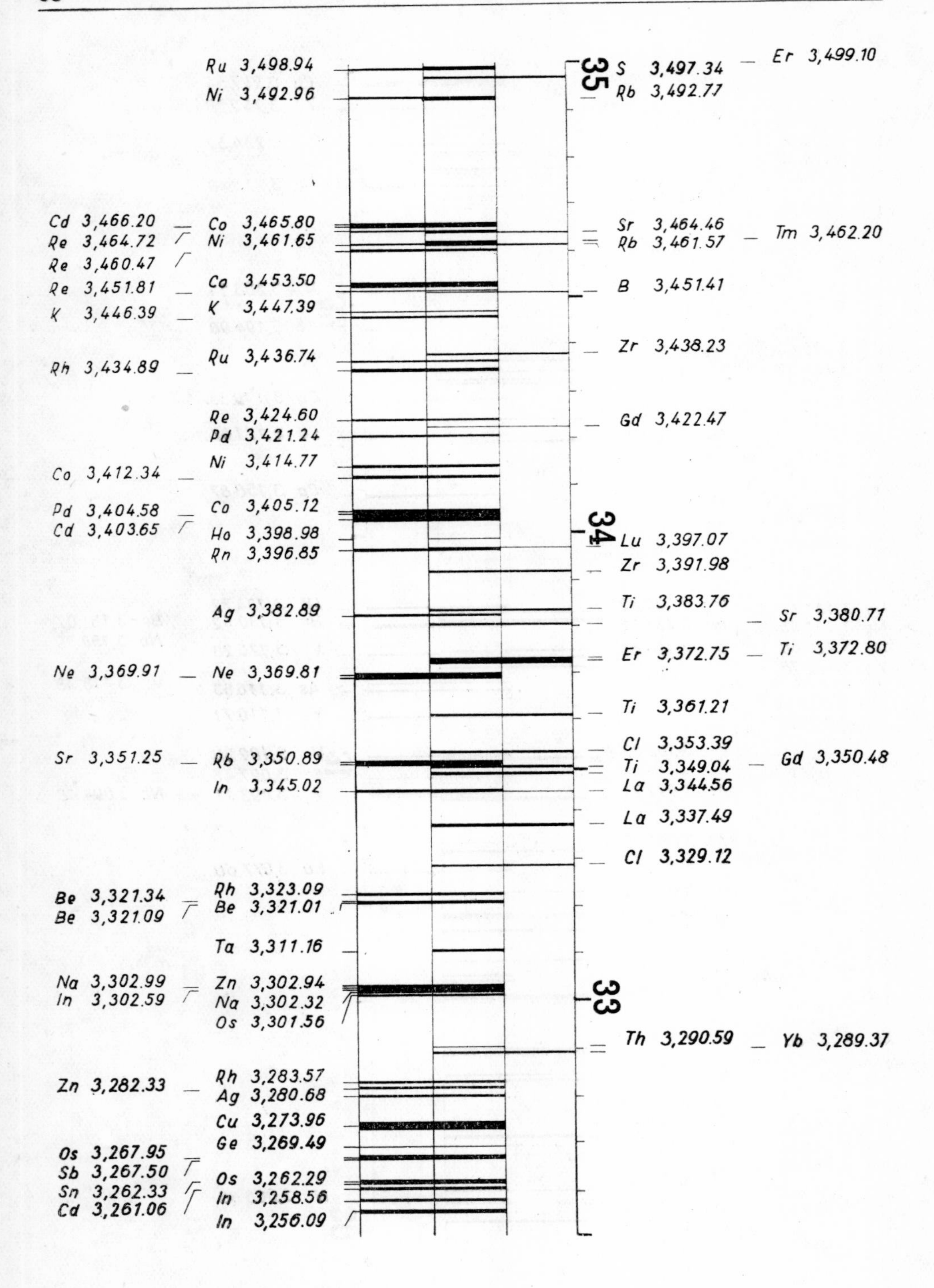
Ru 3,498.94
Ni 3,492.96
35
S 3,497.34
Rb 3,492.77
Er 3,499.10
Cd 3,466.20
Re 3,464.72
Re 3,460.47
Re 3,451.81
K 3,446.39
Rh 3,434.89
Co 3,465.80
Ni 3,461.65
Co 3,453.50
K 3,447.39
Ru 3,436.74
Sr 3,464.46
Rb 3,461.57
Tm 3,462.20
B 3,451.41
Zr 3,438.23
Re 3,424.60
Pd 3,421.24
Ni 3,414.77
Gd 3,422.47
Co 3,412.34
Pd 3,404.58
Cd 3,403.65
Co 3,405.12
Ho 3,398.98
Rn 3,396.85
34
Lu 3,397.07
Zr 3,391.98
Ti 3,383.76
Ag 3,382.89
Sr 3,380.71
Er 3,372.75
Ti 3,372.80
Ne 3,369.91
Ne 3,369.81
Ti 3,361.21
Cl 3,353.39
Sr 3,351.25
Rb 3,350.89
In 3,345.02
Ti 3,349.04
Gd 3,350.48
La 3,344.56
La 3,337.49
Cl 3,329.12
Be 3,321.34
Be 3,321.09
Rh 3,323.09
Be 3,321.01
Ta 3,311.16
Na 3,302.99
In 3,302.59
Zn 3,302.94
Na 3,302.32
Os 3,301.56
33
Th 3,290.59
Yb 3,289.37
Zn 3,282.33
Rh 3,283.57
Ag 3,280.68
Cu 3,273.96
Ge 3,269.49
Os 3,267.95
Sb 3,267.50
Sn 3,262.33
Cd 3,261.06
Os 3,262.29
In 3,258.56
In 3,256.09

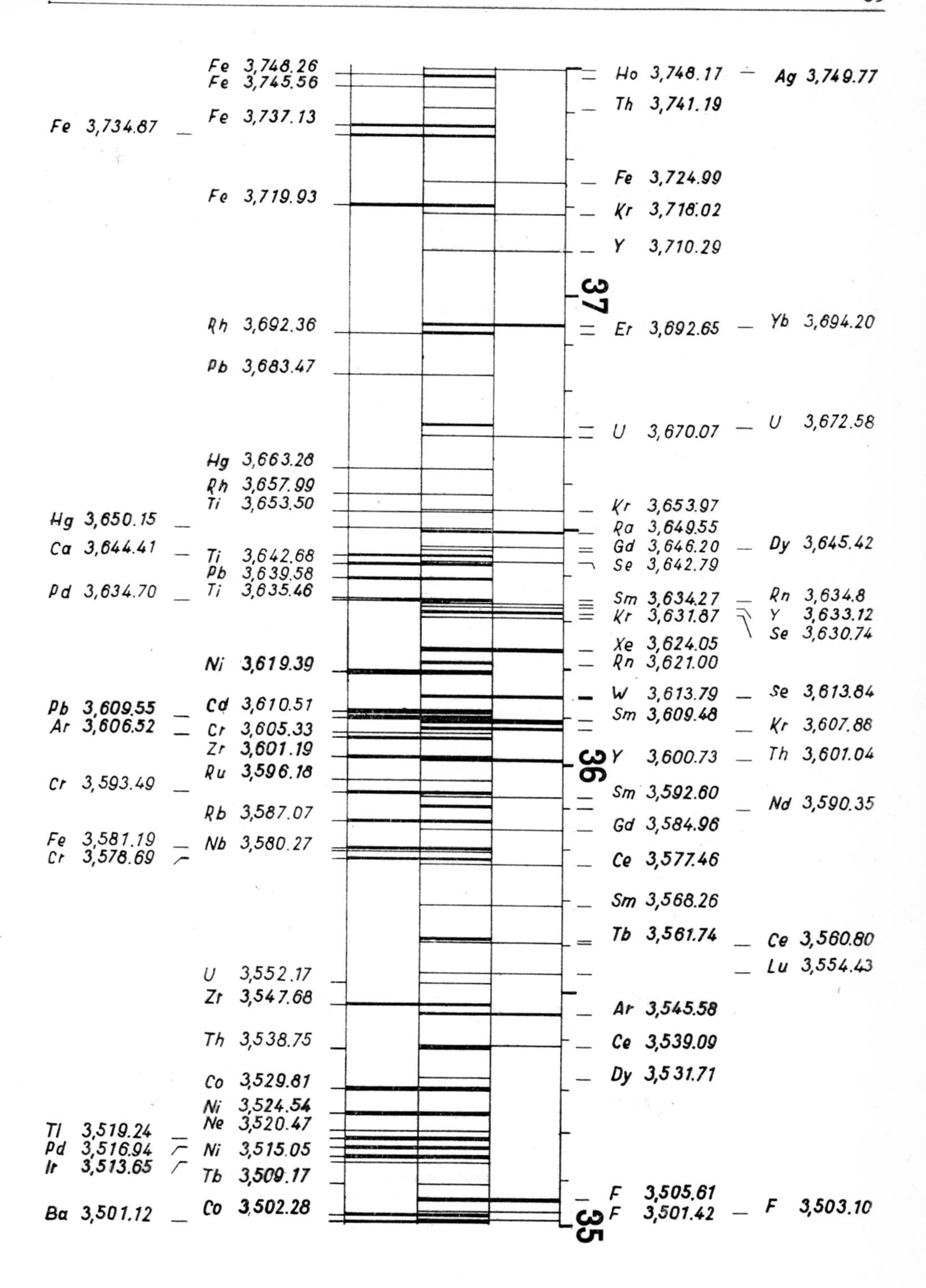
Fe 3,748.26
Fe 3,745.56
Fe 3,737.13
Fe 3,734.87
Fe 3,719.93
Rh 3,692.36
Pb 3,683.47
Hg 3,663.28
Rh 3,657.99
Ti 3,653.50
Hg 3,650.15
Ca 3,644.41
Ti 3,642.68
Pb 3,639.58
Pd 3,634.70
Ti 3,635.46
Ni 3,619.39
Pb 3,609.55
Ar 3,606.52
Cd 3,610.51
Cr 3,605.33
Zr 3,601.19
Ru 3,596.18
Cr 3,593.49
Rb 3,587.07
Fe 3,581.19
Cr 3,578.69
Nb 3,580.27
U 3,552.17
Zr 3,547.68
Th 3,538.75
Co 3,529.81
Ni 3,524.54
Ne 3,520.47
Tl 3,519.24
Pd 3,516.94
Ir 3,513.65
Ni 3,515.05
Tb 3,509.17
Ba 3,501.12
Co 3,502.28
Ho 3,748.17
Ag 3,749.77
Th 3,741.19
Fe 3,724.99
Kr 3,718.02
Y 3,710.29
37
Er 3,692.65
Yb 3,694.20
U 3,670.07
U 3,672.58
Kr 3,653.97
Ra 3,649.55
Gd 3,646.20
Dy 3,645.42
Se 3,642.79
Sm 3,634.27
Rn 3,634.8
Kr 3,631.87
Y 3,633.12
Se 3,630.74
Xe 3,624.05
Rn 3,621.00
W 3,613.79
Se 3,613.84
Sm 3,609.48
Kr 3,607.88
36
Y 3,600.73
Th 3,601.04
Sm 3,592.60
Nd 3,590.35
Gd 3,584.96
Ce 3,577.46
Sm 3,568.26
Tb 3,561.74
Ce 3,560.80
Lu 3,554.43
Ar 3,545.58
Ce 3,539.09
Dy 3,531.71
F 3,505.61
F 3,501.42
F 3,503.10
35

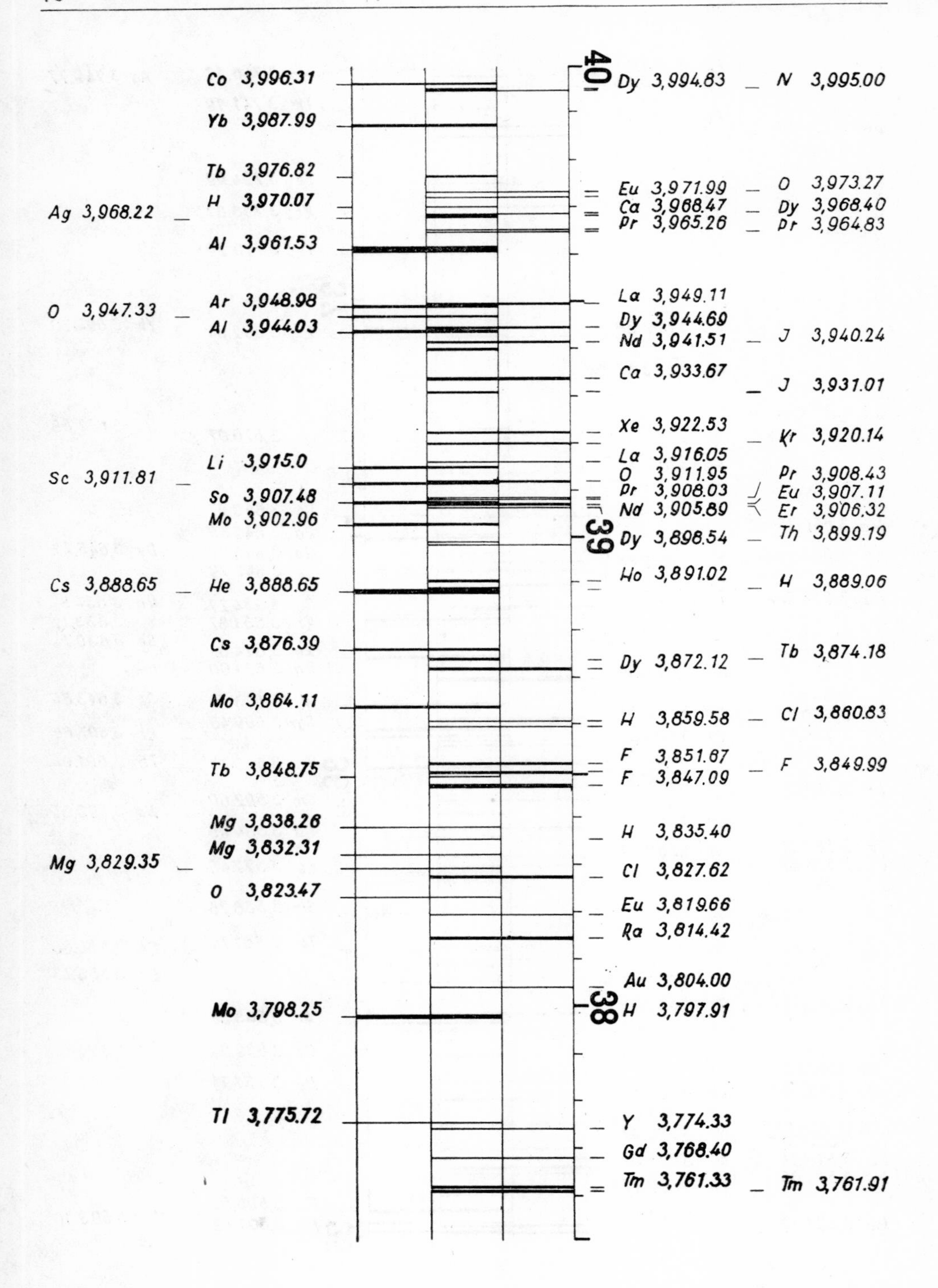

Co 3,996.31
Yb 3,987.99
Tb 3,976.82
H 3,970.07
Ag 3,968.22
Al 3,961.53
O 3,947.33
Ar 3,948.98
Al 3,944.03
Li 3,915.0
Sc 3,911.81
So 3,907.48
Mo 3,902.96
Cs 3,888.65
He 3,888.65
Cs 3,876.39
Mo 3,864.11
Tb 3,848.75
Mg 3,838.26
Mg 3,832.31
Mg 3,829.35
O 3,823.47
Mo 3,798.25
Tl 3,775.72
40
39
38
Dy 3,994.83
N 3,995.00
Eu 3,971.99
O 3,973.27
Ca 3,968.47
Dy 3,968.40
Pr 3,965.26
Pr 3,964.83
La 3,949.11
Dy 3,944.69
Nd 3,941.51
J 3,940.24
Ca 3,933.67
J 3,931.01
Xe 3,922.53
Kr 3,920.14
La 3,916.05
O 3,911.95
Pr 3,908.43
Pr 3,908.03
Eu 3,907.11
Nd 3,905.89
Er 3,906.32
Dy 3,898.54
Th 3,899.19
Ho 3,891.02
H 3,889.06
Dy 3,872.12
Tb 3,874.18
H 3,859.58
Cl 3,860.83
F 3,851.67
F 3,849.99
F 3,847.09
H 3,835.40
Cl 3,827.62
Eu 3,819.66
Ra 3,814.42
Au 3,804.00
H 3,797.91
Y 3,774.33
Gd 3,768.40
Tm 3,761.33
Tm 3,761.91

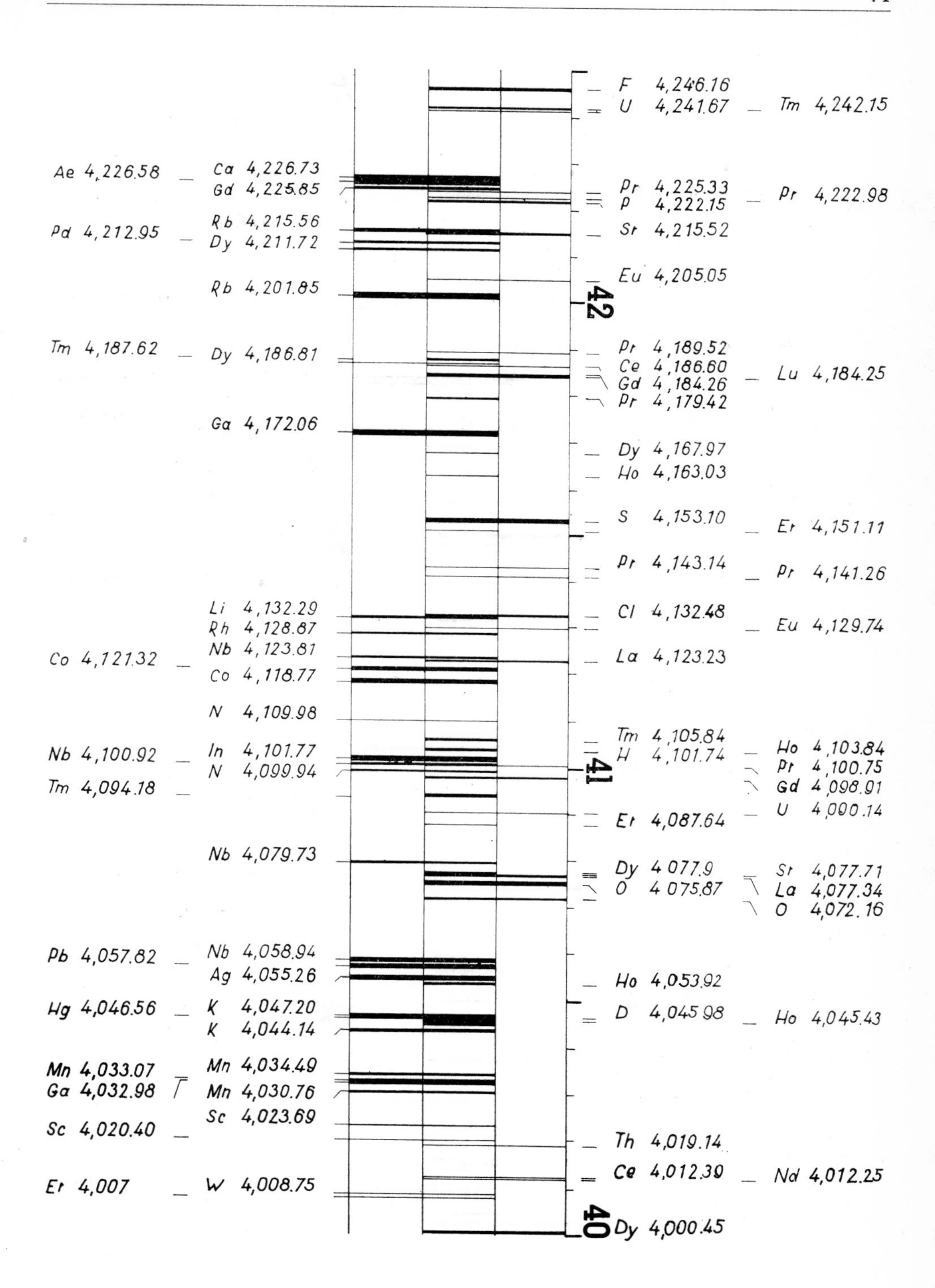
F 4,246.16
U 4,241.67
Tm 4,242.15
Ae 4,226.58
Ca 4,226.73
Gd 4,225.85
Pr 4,225.33
P 4,222.15
Pr 4,222.98
Pd 4,212.95
Rb 4,215.56
Dy 4,211.72
Sr 4,215.52
Eu 4,205.05
Rb 4,201.85
42
Tm 4,187.62
Dy 4,186.81
Pr 4,189.52
Ce 4,186.60
Gd 4,184.26
Pr 4,179.42
Lu 4,184.25
Ga 4,172.06
Dy 4,167.97
Ho 4,163.03
S 4,153.10
Er 4,151.11
Pr 4,143.14
Pr 4,141.26
Li 4,132.29
Rh 4,128.87
Cl 4,132.48
Eu 4,129.74
Co 4,121.32
Nb 4,123.81
Co 4,118.77
La 4,123.23
N 4,109.98
Tm 4,105.84
H 4,101.74
Ho 4,103.84
Pr 4,100.75
Gd 4,098.91
U 4,090.14
Nb 4,100.92
In 4,101.77
N 4,099.94
Tm 4,094.18
41
Er 4,087.64
Nb 4,079.73
Dy 4 077.9
O 4 075.87
Sr 4,077.71
La 4,077.34
O 4,072.16
Pb 4,057.82
Nb 4,058.94
Ag 4,055.26
Ho 4,053.92
Hg 4,046.56
K 4,047.20
K 4,044.14
D 4,045.98
Ho 4,045.43
Mn 4,033.07
Ga 4,032.98
Mn 4,034.49
Mn 4,030.76
Sc 4,023.69
Sc 4,020.40
Th 4,019.14
Ce 4,012.39
Nd 4,012.25
Er 4,007
W 4,008.75
40
Dy 4,000.45

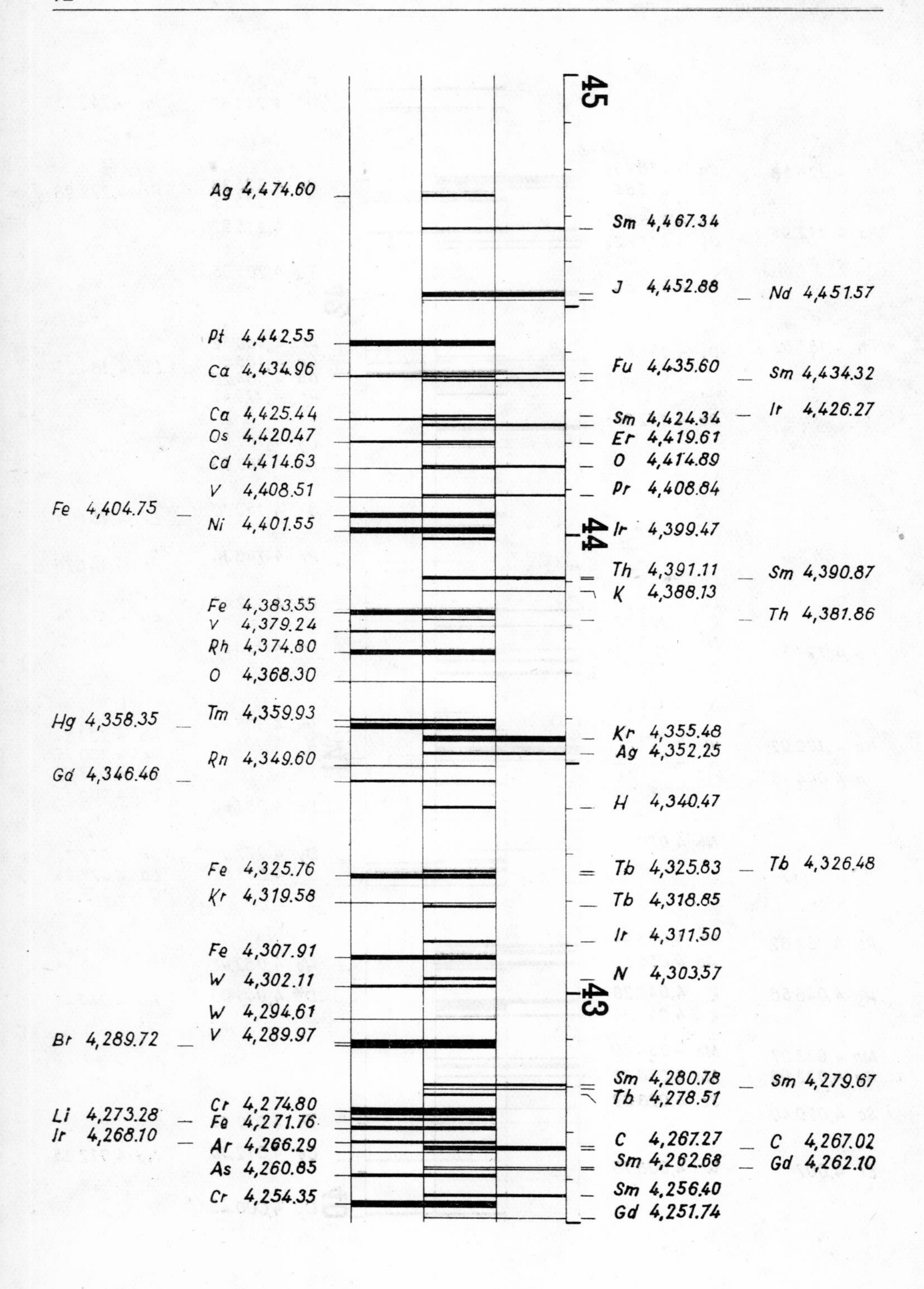
45
44
43
Ag 4,474.60
Sm 4,467.34
J 4,452.88
Nd 4,451.57
Pt 4,442.55
Ca 4,434.96
Fu 4,435.60
Sm 4,434.32
Ca 4,425.44
Sm 4,424.34
Ir 4,426.27
Os 4,420.47
Er 4,419.61
Cd 4,414.63
O 4,414.89
V 4,408.51
Pr 4,408.84
Fe 4,404.75
Ni 4,401.55
Ir 4,399.47
Th 4,391.11
Sm 4,390.87
K 4,388.13
Fe 4,383.55
V 4,379.24
Th 4,381.86
Rh 4,374.80
O 4,368.30
Tm 4,359.93
Hg 4,358.35
Kr 4,355.48
Ag 4,352.25
Rn 4,349.60
Gd 4,346.46
H 4,340.47
Fe 4,325.76
Tb 4,325.83
Tb 4,326.48
Kr 4,319.58
Tb 4,318.85
Ir 4,311.50
Fe 4,307.91
N 4,303.57
W 4,302.11
W 4,294.61
Br 4,289.72
V 4,289.97
Sm 4,280.78
Sm 4,279.67
Tb 4,278.51
Cr 4,274.80
Li 4,273.28
Fe 4,271.76
Ir 4,268.10
C 4,267.27
C 4,267.02
Ar 4,266.29
Sm 4,262.68
Gd 4,262.10
As 4,260.85
Sm 4,256.40
Cr 4,254.35
Gd 4,251.74

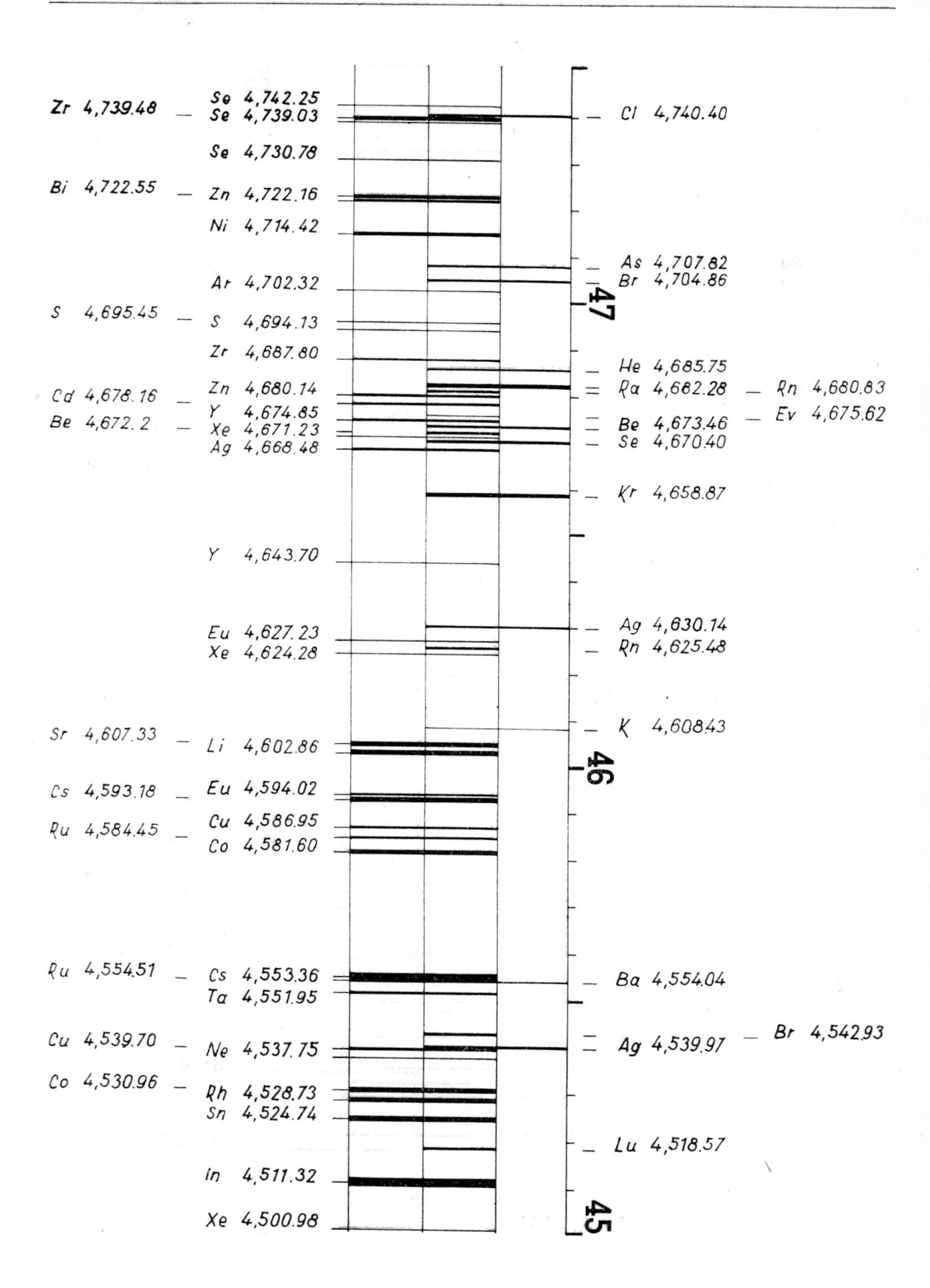
Zr 4,739.48
Bi 4,722.55
S 4,695.45
Cd 4,678.16
Be 4,672.2
Sr 4,607.33
Cs 4,593.18
Ru 4,584.45
Ru 4,554.51
Cu 4,539.70
Co 4,530.96
Se 4,742.25
Se 4,739.03
Se 4,730.78
Zn 4,722.16
Ni 4,714.42
Ar 4,702.32
S 4,694.13
Zr 4,687.80
Zn 4,680.14
Y 4,674.85
Xe 4,671.23
Ag 4,668.48
Y 4,643.70
Eu 4,627.23
Xe 4,624.28
Li 4,602.86
Eu 4,594.02
Cu 4,586.95
Co 4,581.60
Cs 4,553.36
Ta 4,551.95
Ne 4,537.75
Rh 4,528.73
Sn 4,524.74
In 4,511.32
Xe 4,500.98
47
46
45
Cl 4,740.40
As 4,707.82
Br 4,704.86
He 4,685.75
Ra 4,682.28
Be 4,673.46
Se 4,670.40
Kr 4,658.87
Ag 4,630.14
Rn 4,625.48
K 4,608.43
Ba 4,554.04
Ag 4,539.97
Lu 4,518.57
Rn 4,680.83
Ev 4,675.62
Br 4,542.93

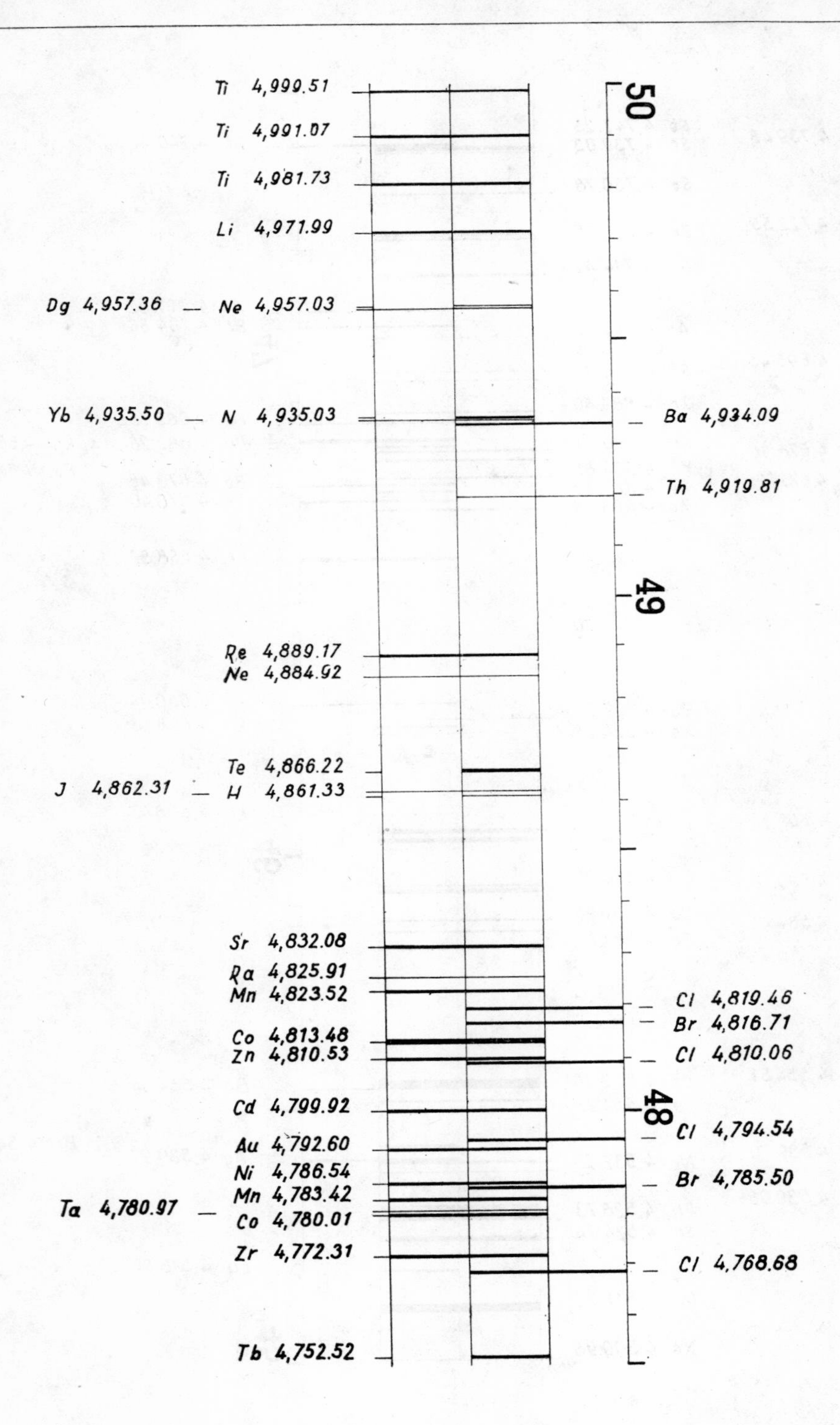
Ti 4,999.51
Ti 4,991.07
Ti 4,981.73
Li 4,971.99
Dg 4,957.36
Ne 4,957.03
Yb 4,935.50
N 4,935.03
Ba 4,934.09
Th 4,919.81
Re 4,889.17
Ne 4,884.92
Te 4,866.22
J 4,862.31
H 4,861.33
Sr 4,832.08
Ra 4,825.91
Mn 4,823.52
Cl 4,819.46
Br 4,816.71
Co 4,813.48
Zn 4,810.53
Cl 4,810.06
Cd 4,799.92
Cl 4,794.54
Au 4,792.60
Ni 4,786.54
Br 4,785.50
Mn 4,783.42
Ta 4,780.97
Co 4,780.01
Zr 4,772.31
Cl 4,768.68
Tb 4,752.52
50
49
48

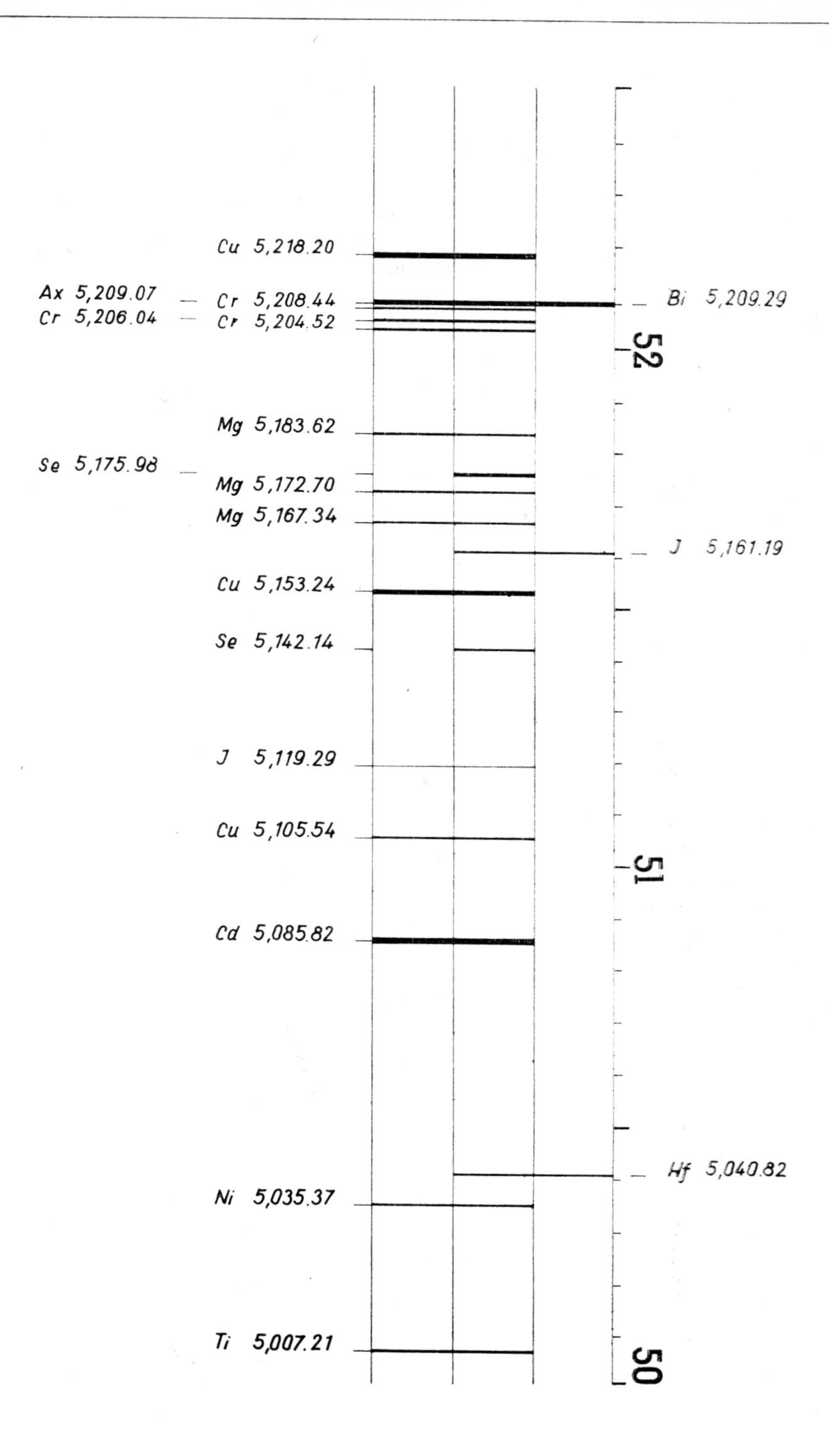
Cu 5,218.20
Ax 5,209.07
Cr 5,208.44
Bi 5,209.29
Cr 5,206.04
Cr 5,204.52
52
Mg 5,183.62
Se 5,175.98
Mg 5,172.70
Mg 5,167.34
J 5,161.19
Cu 5,153.24
Se 5,142.14
J 5,119.29
Cu 5,105.54
51
Cd 5,085.82
Hf 5,040.82
Ni 5,035.37
Ti 5,007.21
50

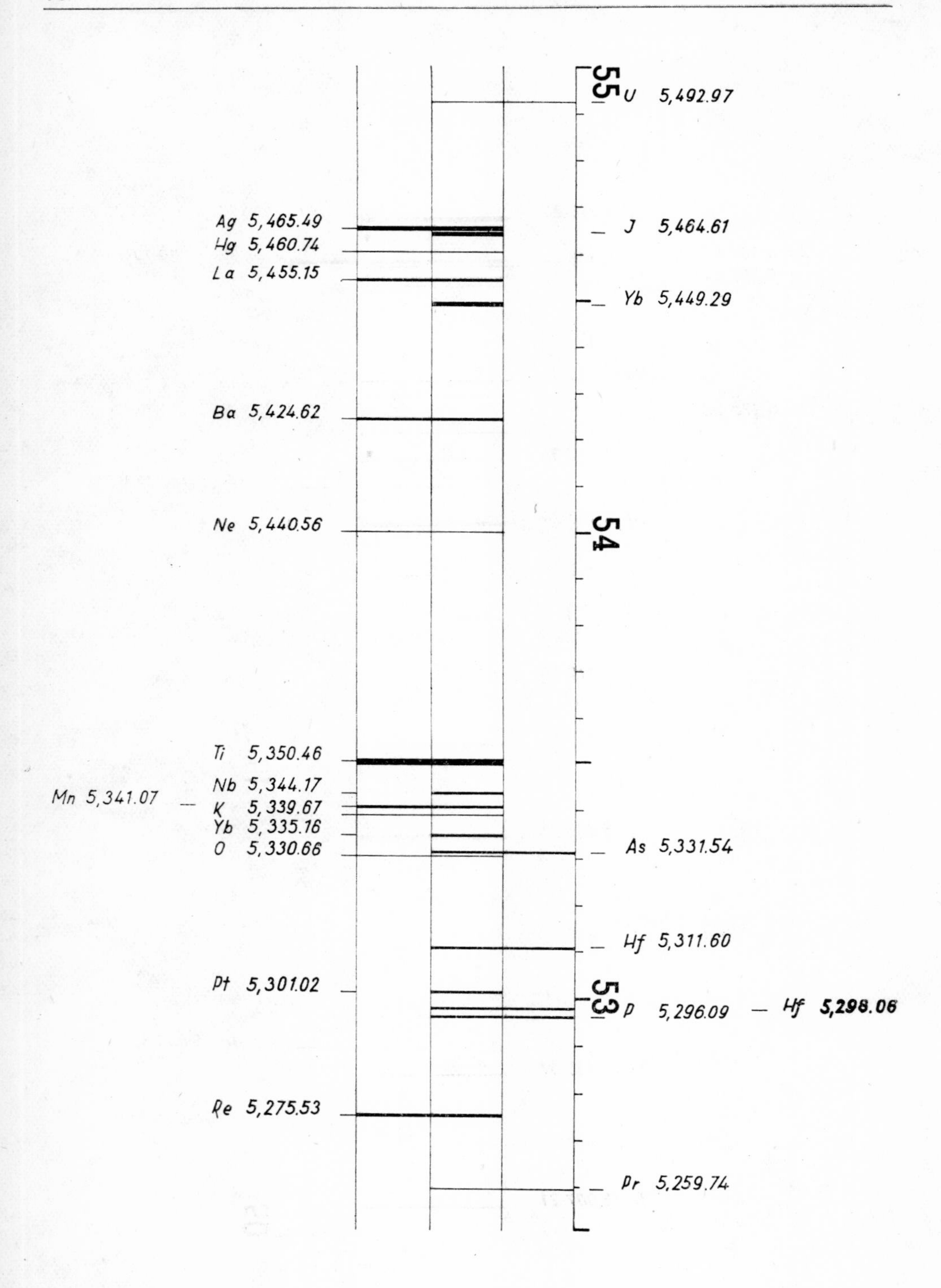

55
U 5,492.97
Ag 5,465.49
J 5,464.61
Hg 5,460.74
La 5,455.15
Yb 5,449.29
Ba 5,424.62
Ne 5,440.56
54
Ti 5,350.46
Nb 5,344.17
Mn 5,341.07
K 5,339.67
Yb 5,335.16
O 5,330.66
As 5,331.54
Hf 5,311.60
Pt 5,301.02
53
P 5,296.09
Hf 5,298.06
Re 5,275.53
Pr 5,259.74

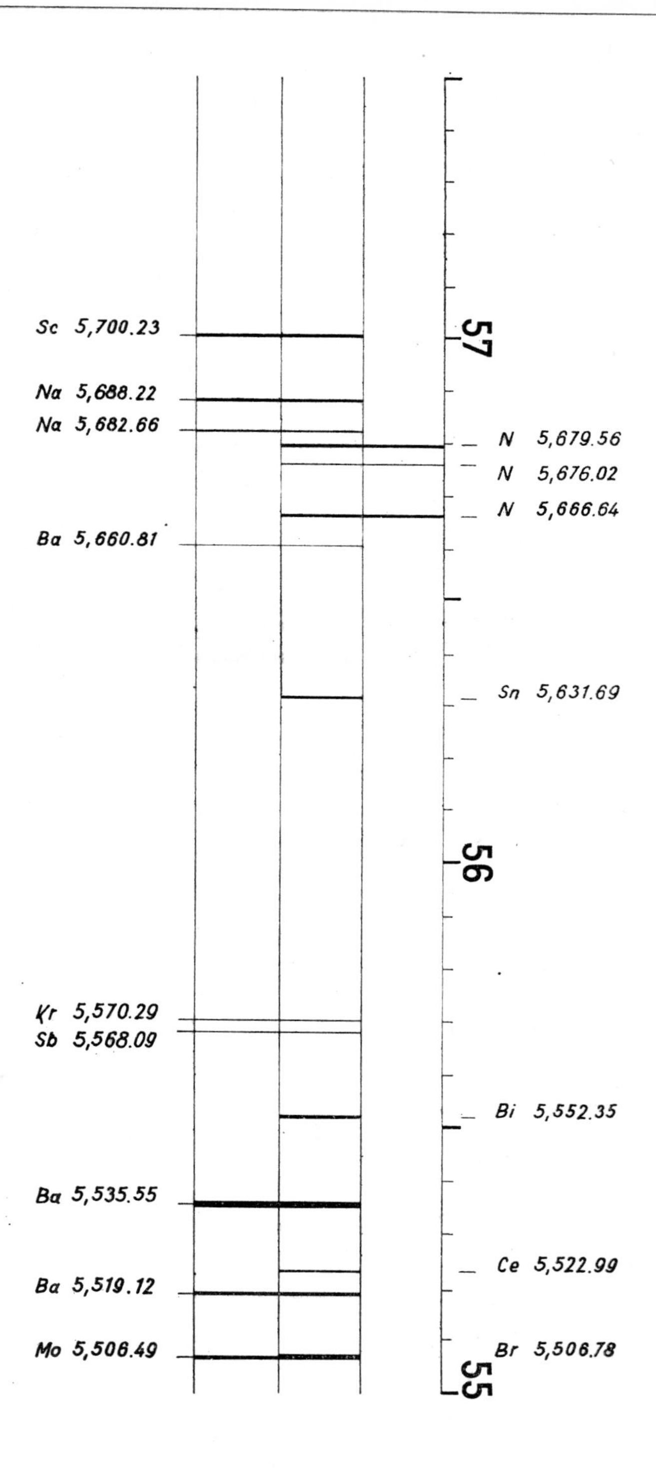
Sc 5,700.23
Na 5,688.22
Na 5,682.66
Ba 5,660.81
Kr 5,570.29
Sb 5,568.09
Ba 5,535.55
Ba 5,519.12
Mo 5,506.49
N 5,679.56
N 5,676.02
N 5,666.64
Sn 5,631.69
Bi 5,552.35
Ce 5,522.99
Br 5,506.78
57
56
55

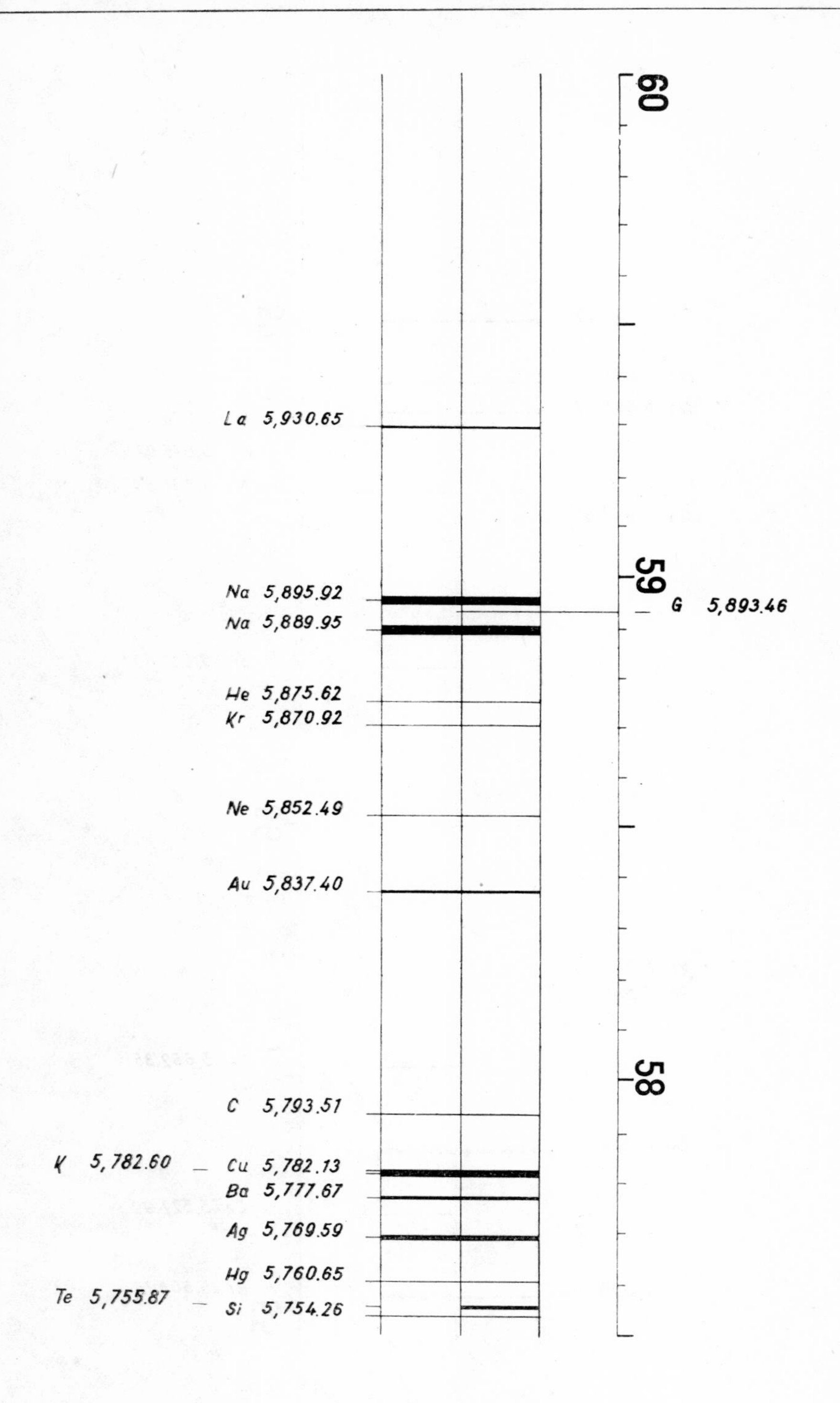
60
La 5,930.65
59
G 5,893.46
Na 5,895.92
Na 5,889.95
He 5,875.62
Kr 5,870.92
Ne 5,852.49
Au 5,837.40
58
C 5,793.51
K 5,782.60
Cu 5,782.13
Ba 5,777.67
Ag 5,769.59
Hg 5,760.65
Te 5,755.87
Si 5,754.26

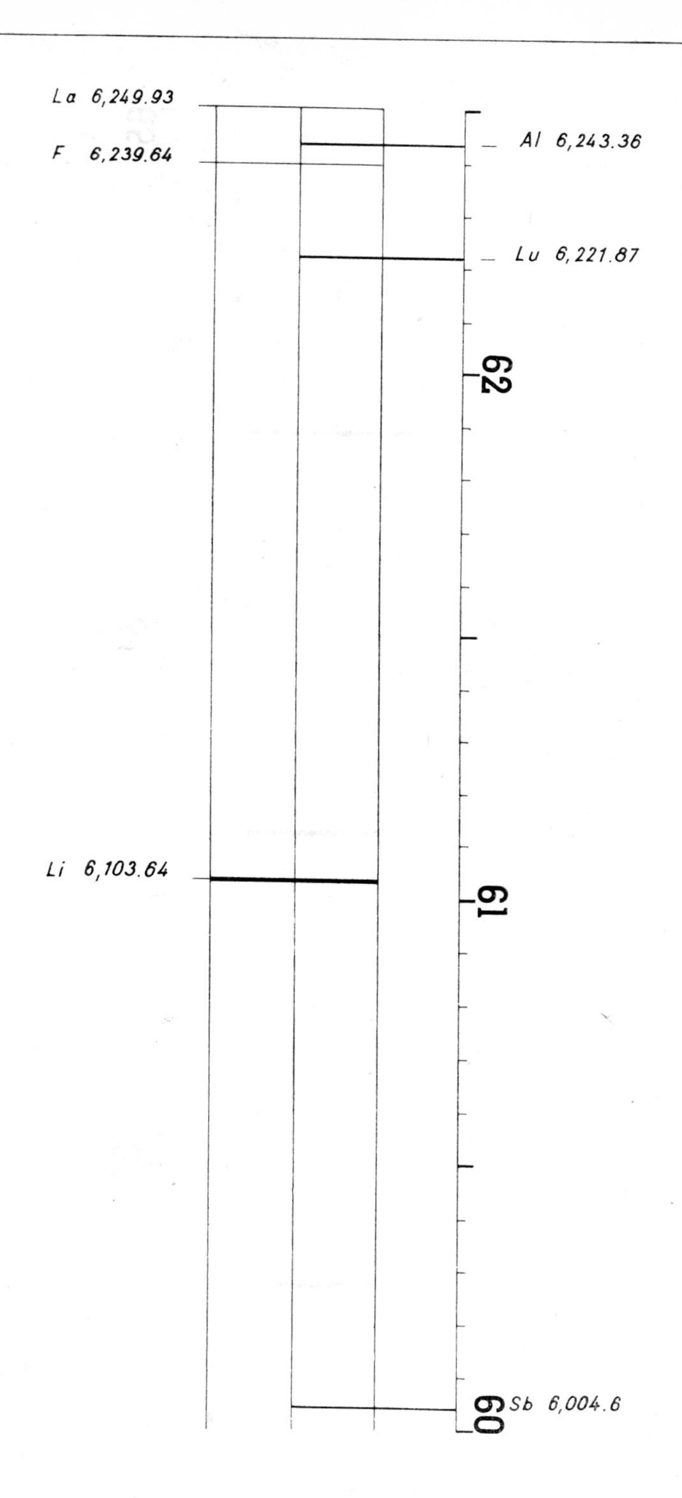
La 6,249.93
F 6,239.64
Al 6,243.36
Lu 6,221.87
62
Li 6,103.64
61
Sb 6,004.6
60

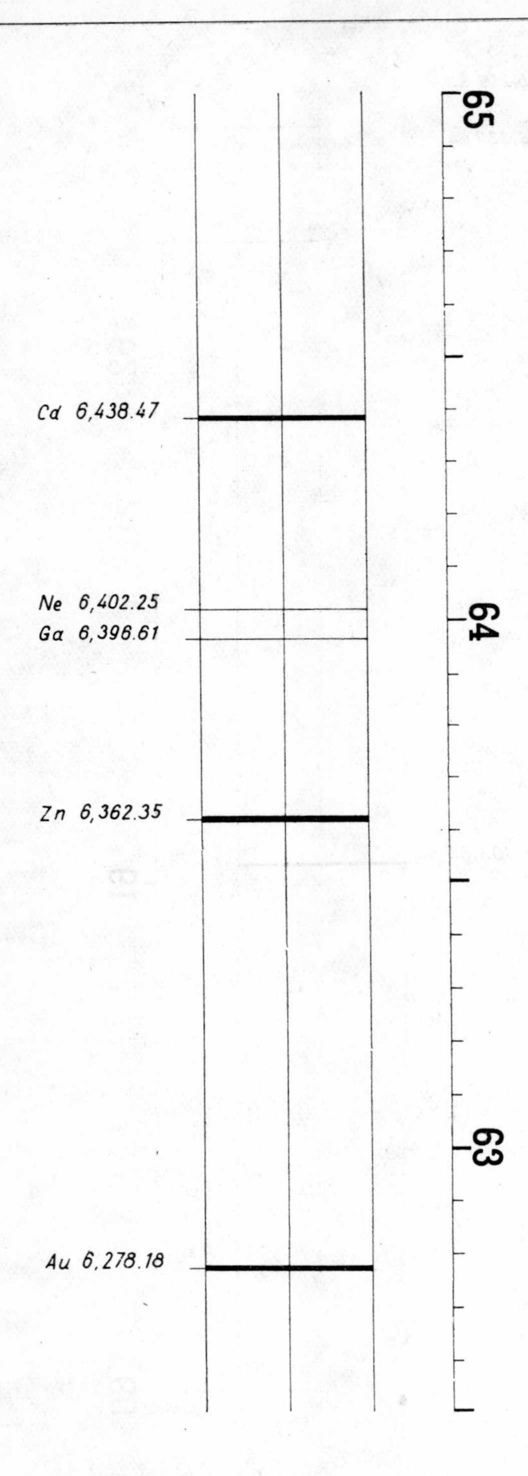

65
Cd 6,438.47
Ne 6,402.25
Ga 6,396.61
64
Zn 6,362.35
63
Au 6,278.18

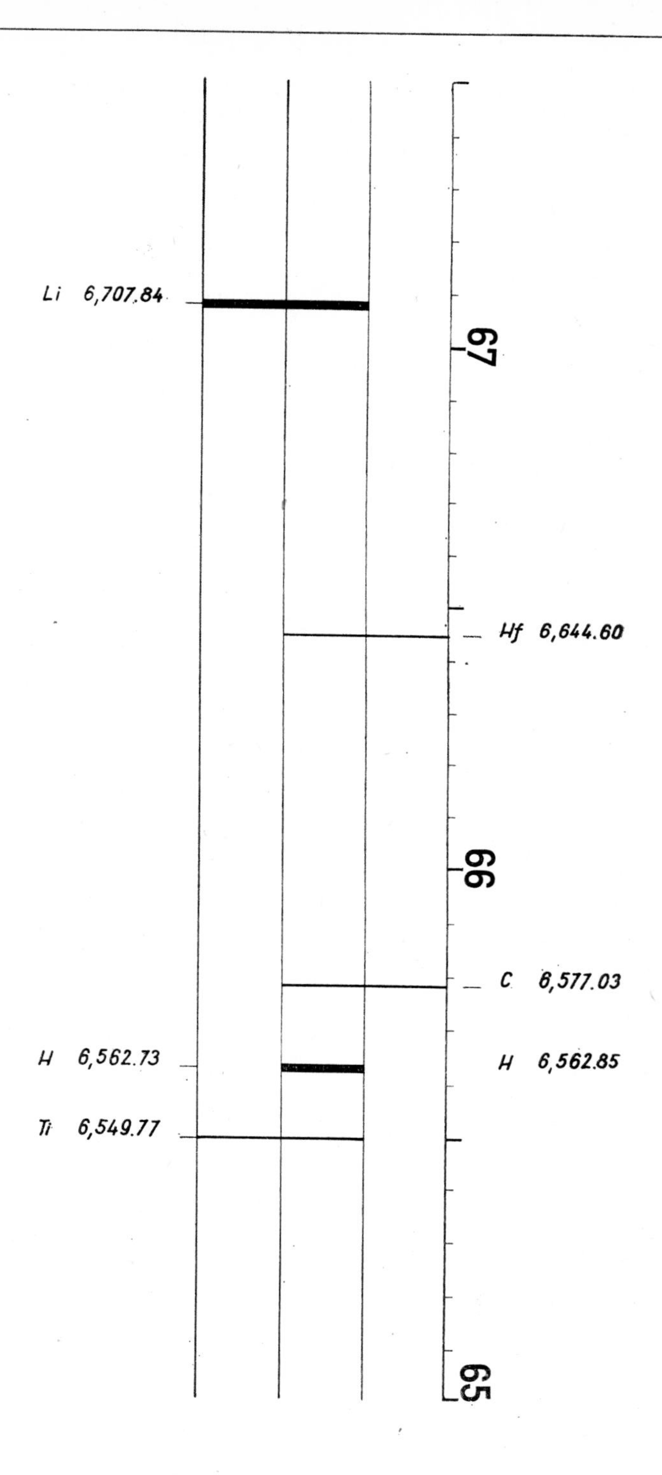
Li 6,707.84
67
Hf 6,644.60
66
C 6,577.03
H 6,562.73
H 6,562.85
Ti 6,549.77
65

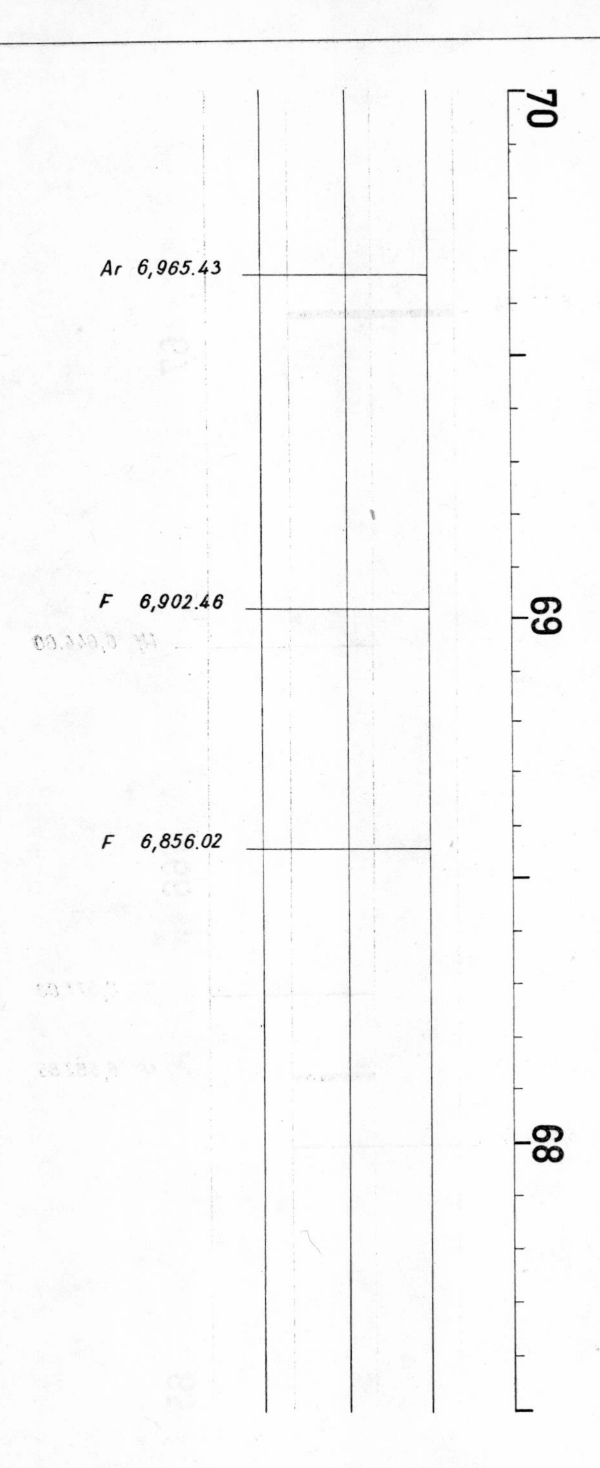
Ar 6,965.43
F 6,902.46
F 6,856.02
70
69
68

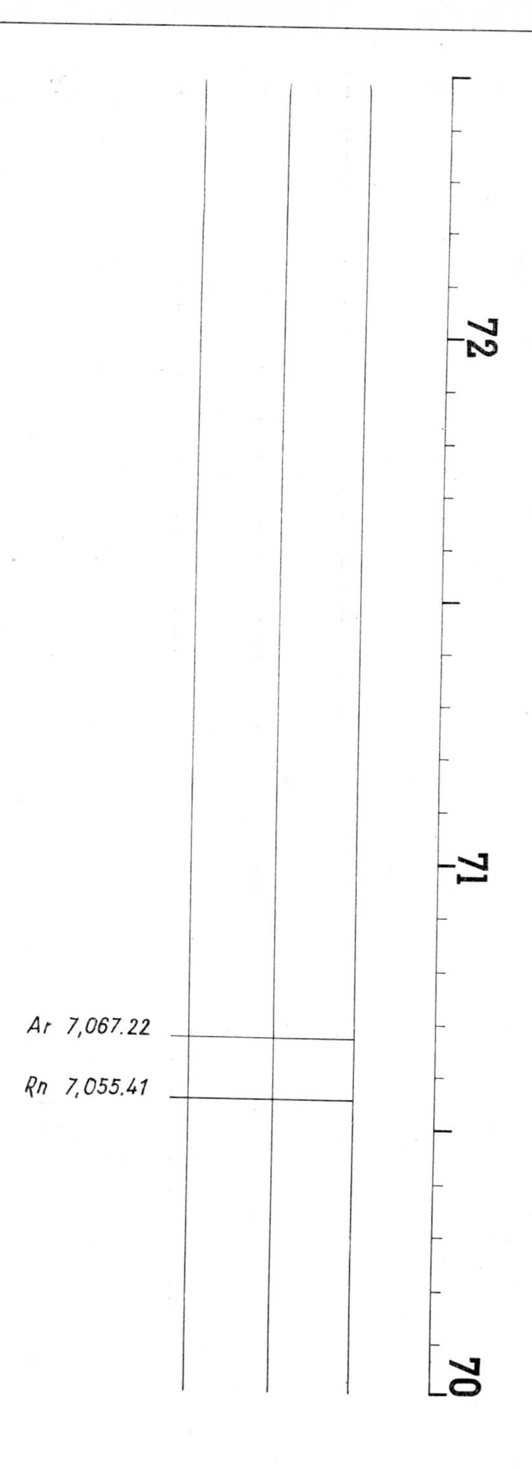
72
71
70
Ar 7,067.22
Rn 7,055.41

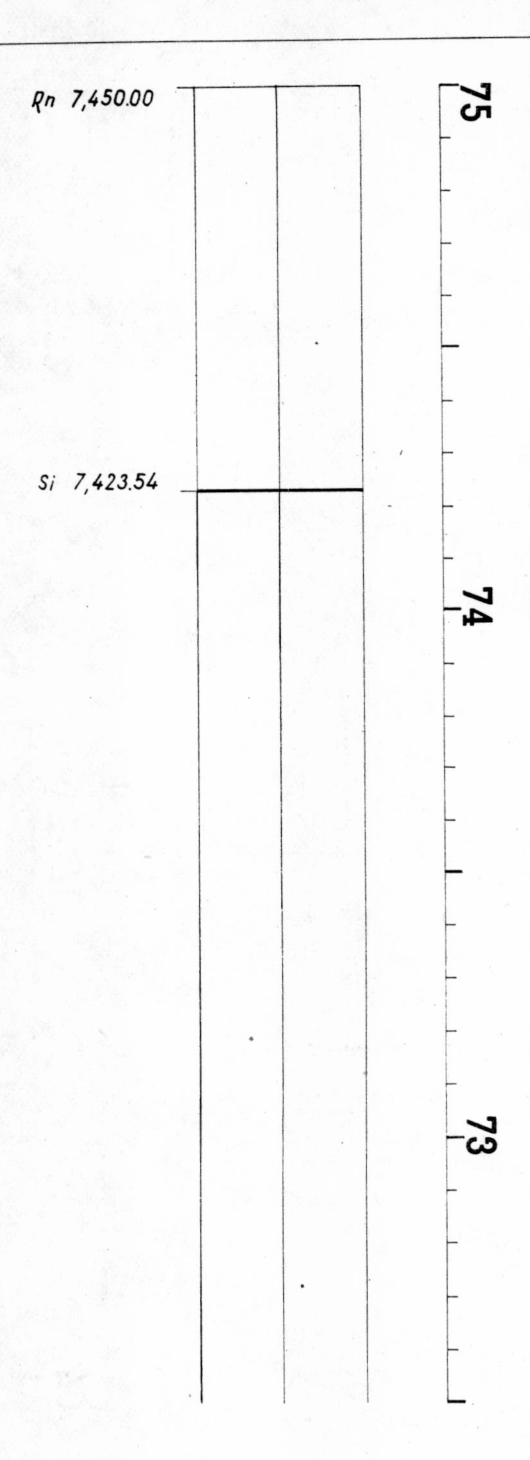

Rn 7,450.00
Si 7,423.54
75
74
73

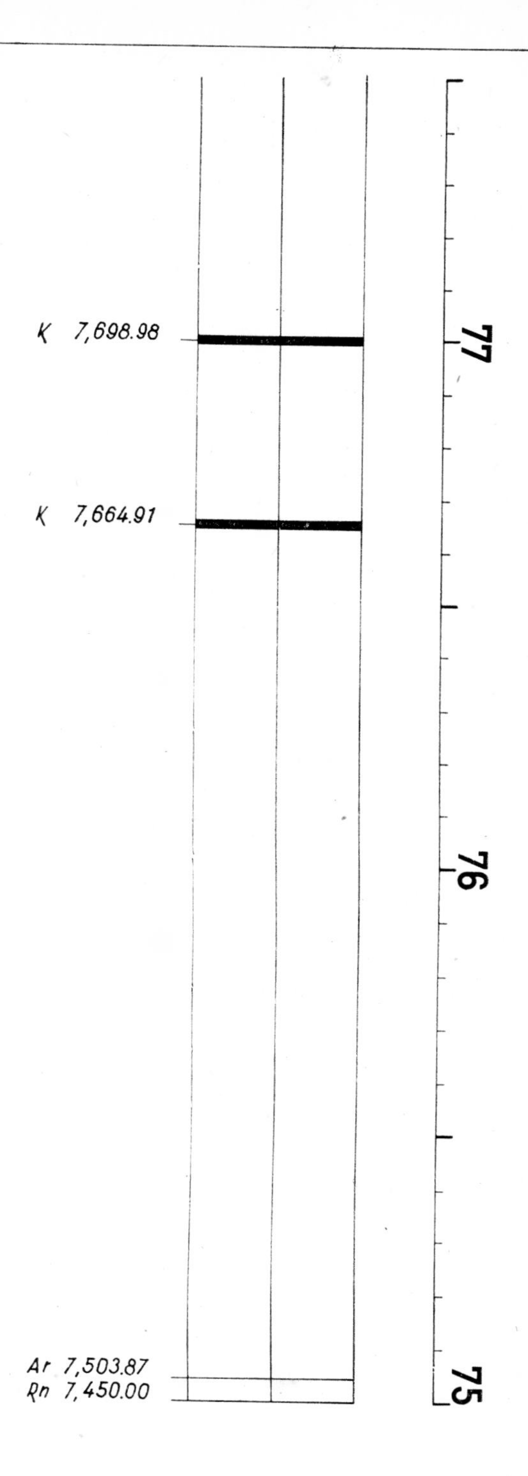
K 7,698.98
K 7,664.91
Ar 7,503.87
Rn 7,450.00
77
76
75

Rb 7,947.60
O 7,774.14
O 7,775.43
O 7,771.93
80
79
78

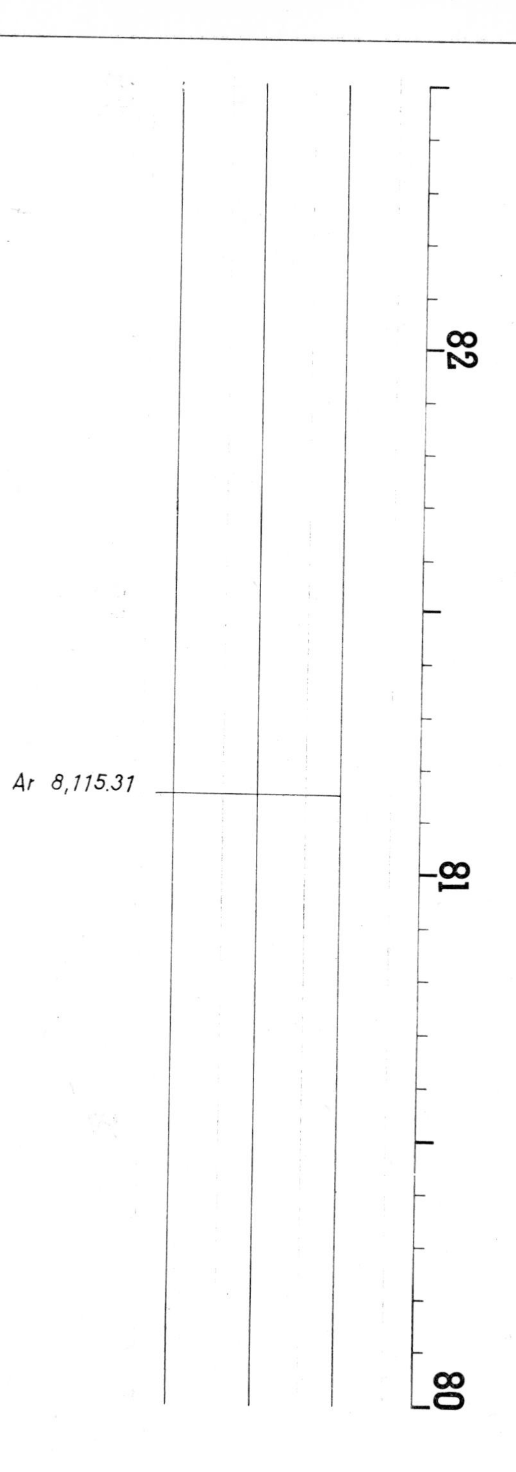
82
Ar 8,115.31
81
80

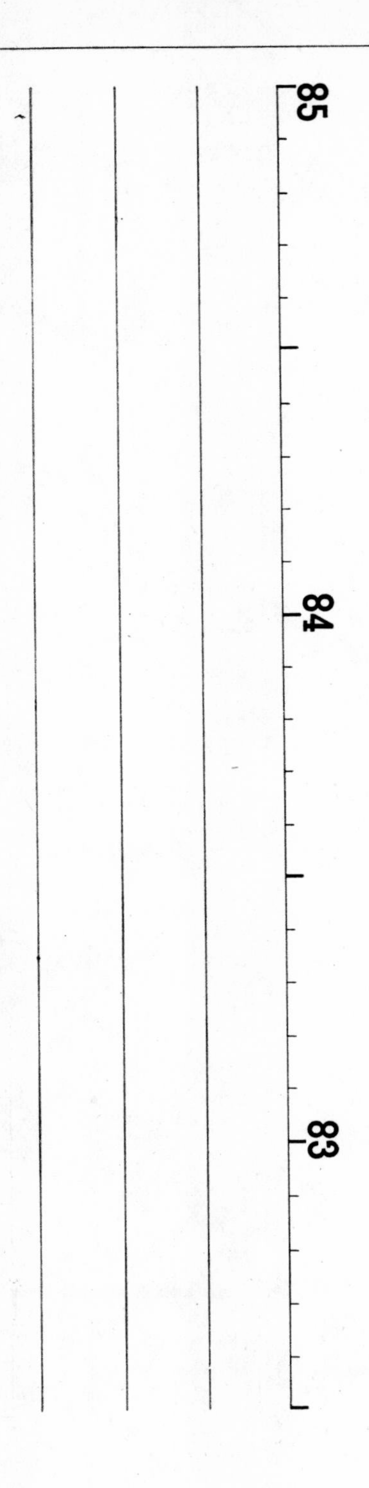
85
84
83

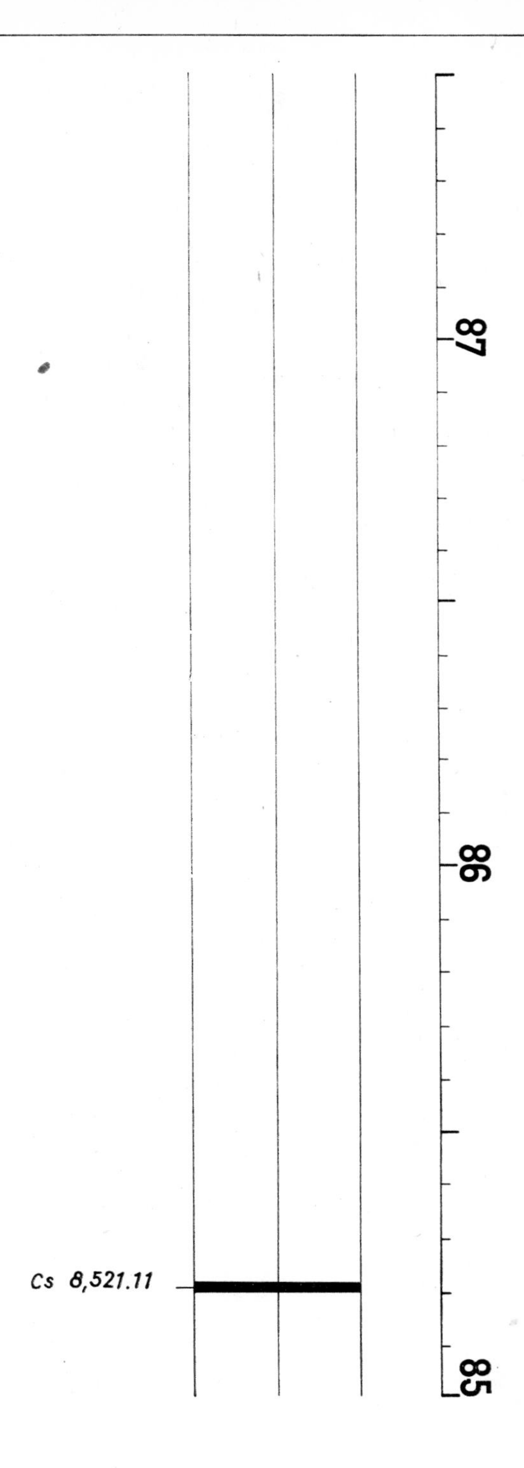
87
86
85
Cs 8,521.11

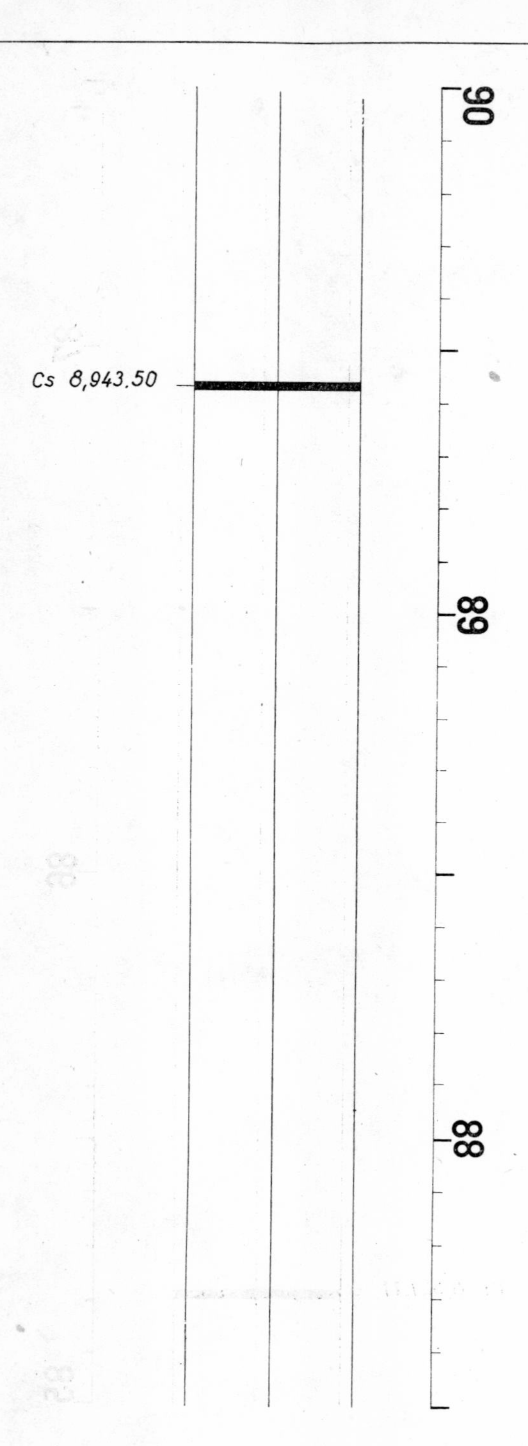
Cs 8,943.50
90
89
88

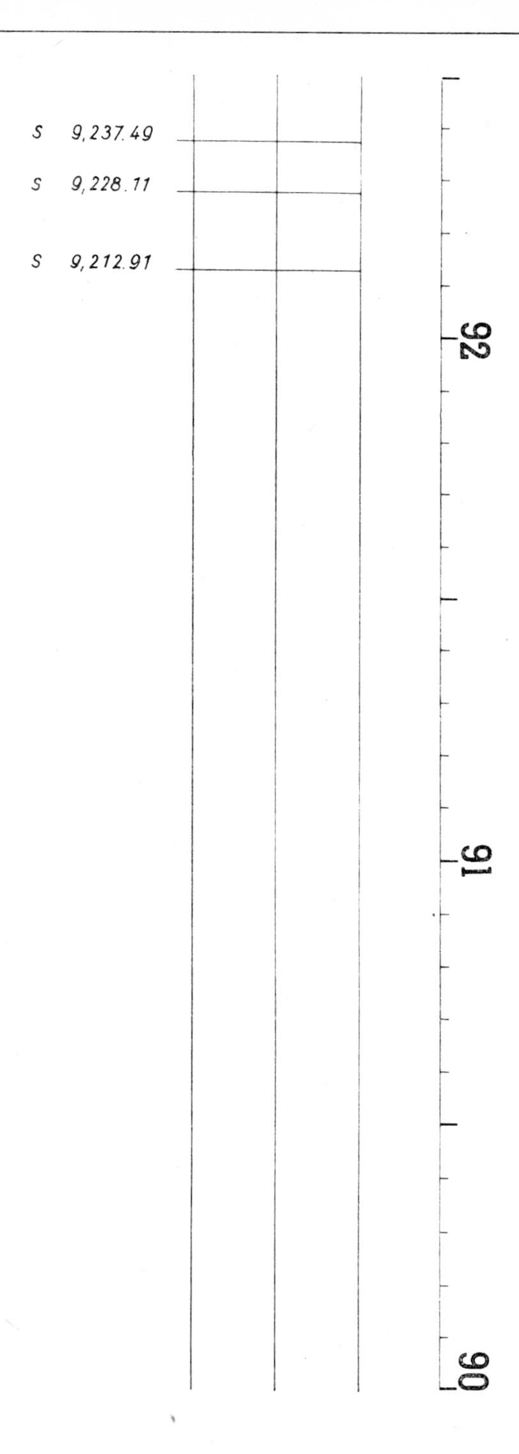
S 9,237.49
S 9,228.11
S 9,212.91
92
91
90

TABLE XVI

COINCIDENCE OF ANALYTICAL LINES FOR THE REGION OF FROM 2,000 TO 10,000 Å

Ag 47 107.870

t_0 960.5°C

t_1 2,212.0°C

I.	II.	III.	IV.	V.
7.574	21.960	36.10	—	—

λ	I		eV
	J	*O*	
II 2,246.41	300	25	10.36
I 2,375.06	300	300	8.96
II 2,413.18	300	50	10.18
II 2,437.79	500	60	9.93
II 2,614.59	300	—	15.51
II 2,711.21	300	1	15.71
I 2,824.37	200	150	8.13
II 2,938.55	200	200	14.98
I 3,280.683	1000*R*	2000*R*	3.78
I 3,382.891	700*R*	1000*R*	3.66
3,968.22	60	100	—
I 4,055.26	500*R*	800*R*	6.72
I 4,668.48	70	200	6.44
I 5,209.07	1000*R*	1500*R*	6.04
I 5,465.49	500*R*	1000*R*	6.04

Ag 2,246.41 *J* 300 *O* 25

	λ	*I*			λ	*I*	
		J	*O*			*J*	*O*
Rh	**2,246.38**	**50**	**—**	**Bi**	**2,246.418**	**2**	**—**
W	**2,246.36**	**6**	**4**	**Nb**	**2,246.42**	**—**	**3*h***
V	**2,246.33**	**2**	**—**	**Au**	**2,246.43**	**15**	**—**
K	**2,246.32**	**(5)**	**—**	**Re**	**2,246.47**	**5**	**25**
U	**2,246.23**	**4**	**—**	**Nd**	**2,246.47**	**12**	**—**
Nb	**2,246.17**	**—**	**10**	**Nb**	**2,246.49**	**4**	**—**
Co	**2,246.15**	**3*w***	**—**	**Pt**	**2,246.50**	**10**	**2**
				W	**2,246.63**	**12**	**6**
				Au	**2,246.68**	**18**	**—**
Sn	2,246.053	100*R*	100*R*				
Ir	2,245.76	150	10				
Fe	2,245.65	—	12	Pb	2,246.89	100*R*	30*R*
Ba	2,245.61	12	12	Ir	2,246.90	100	10
Pt	2,245.52	30	25	Cu	2,246.99	500	30
				*bh*C	2,247.5	—	12
Re	2,245.14	2*h*	15	Os	2,247.68	6	40
Co	2,245.13	35	15				
Pt	2,244.97	10	25	Zr	2,248.05	—	40
Ni	2,244.53	—	40	W	2,248.26	15	12
Ni	2,244.48	3	15	Ag	2,248.74	150*wh*	15
				Rh	2,248.74	3	20
Cu	2,244.26	9	25	W	2,248.75	25	20
Re	2,244.12	4	15				
				Fe	2,248.86	—	35
Yt	*2.243.06*	*35w*	*25*				
Ir	*2,242.68*	*300*	*50*	*Pt*	*2,249.30*	*12*	*25*
Cu	*2,242.61*	*50h*	*25*	*Pt*	*2,249.90*	*8*	*25*
Re	*2,242.41*	*10*	*25*	*Zr*	*2,250.75*	*—*	*35*
Os	*2,242.10*	*5*	*30*	*Sn*	*2,251.15*	*50*	*25*
Rh	*2,242.05*	*25*	*25*				
V	*2,241.84*	*—*	*25*				
Os	*2,241.62*	*2w*	*25*				
V	*2,241.53*	*200*	*—*				
Ru	*2,241.07*	*—*	*60*				

Ag 2,375.06 *J* 300 *O* 300

	λ	*J*	*O*		λ	*J*	*O*
W	**2,375.03**	**12**	**—**	**Os**	**2,375.06**	**25**	**10**
Pd	**2,374.97**	**(3)**	**—**	**Re**	**2,375.07**	**7**	**25**
Te	**2,374.94**	**(30)**	**—**	**Th**	**2,375.07**	**4**	**8**
Fe	**2,374.873**	**—**	**2**	**Ir**	**2,375.09**	**20**	**10**
Mo	**2,374.87**	**—**	**8**	**Co**	**2,375.18**	**20**	**9**
Rh	**2,374.84**	**10**	**—**	**Fe**	**2,375.19**	**15**	**—**
Ru	**2,374.78**	**4**	**—**	**Ru**	**2,375.27**	**5**	**80**
W	**2,374.76**	**—**	**10**	**Tm**	**2,375.31**	**15**	**—**
Co	**2,374.75**	**—**	**3**	**W**	**2,375.39**	**2*h***	**—**
Ta	**2,374.72**	**12*h***	**2**				

	λ	I J	I O
Al	2,373.36	100R	200R
Al	2,373.13	30	100R
V	2,373.06	200	—
Ir	2,372.77	40	100
V	2,372.18	100	—
V	*2,371.07*	*500*	—

	λ	I J	I O
Ru	2,375.63	80	50
Nb	2,376.40	60	8
Fe	2,376.44	50h	3
Os	2,377.03	30	50
Os	2,377.61	15	50
Rh	2,377.81	50	—
Tl	*2,379.69*	*200R*	*100R*
Rb	*2,380.44*	*(125)*	—
Rb	*2,381.30*	*(100)*	—
Ru	*2,381.99*	*150*	*50*

Ag 2,413.18 *J* 300 *O* 50

	λ	I J	I O
Ir	**2,413.10**	**5h**	—
Ni	**2,413.05**	**50**	—
Pt	**2,413.04**	**10**	**60**
V	**2,413.03**	**20h**	**20**
Mo	**2,413.01**	**30h**	**15**
Co	**2,412.89**	**1**	**6**
In	**2,412.83**	**(2)**	—
Ir	**2,412.81**	**5h**	—
V	2,412.69	20h	25
Nb	2,412.46	100	5
Rh	2,411.94	50	2
Pb	2,411.73	15	75
Co	2,411.62	50	250R
Ag	2,411.35	150h	25
Fe	2,411.07	70h	35
Ru	2,410.89	—	30
Fe	2,410.52	70h	50
Co	2,410.51	—	40w
Hf	2,410.14	50	25
Mo	2,410.09	60h	—
Pd	*2,408.74*	*100*	—
Cr	*2,408.62*	*2r*	*150r*
Ru	*2,407.92*	*50*	*60*
Rh	*2,407.88*	*5*	*60*
Co	*2,407.25*	*2*	*100*
Pd	*2,406.74*	*150*	*1*
Cu	*2,406.67*	*50*	*150*
Fe	*2,406.66*	*1wh*	*50*
Ta	*2,406.55*	—	*60*

	λ	I J	I O
Co	**2,413.19**	—	**15**
Fe	**2,413.309**	**100h**	**60**
Ir	**2,413.31**	—	**25**
Hf	**2,413.33**	**8**	**5**
Ru	**2,413.38**	**4**	—
Pd	**2,413.39**	**40**	—
Rn	**2,413.4**	**(3)**	—
Th	**2,413.41**	—	**5**
Be	**2,413.45**	**(25)**	—
Th	**2,413.49**	**20**	**5**
Zr	**2,413.50**	—	**8**
Se	**2,413.52**	**(125)**	—
Co	**2,413.58**	**2**	**15**
Nb	2,413.94	300	—
Co	2,414.46	15	40R
Nb	2,414.48	100	—
Os	2,414.52	3d	25
Pd	2,414.73	150	—
Ru	2,414.82	12	25
Fe	2,415.06	50	1
Co	2,415.30	18	40R
V	2,415.333	25h	25
Rh	2,415.84	200	100
Ni	2,416.14	250h	40
V	2,416.75	40h	40r
Ta	*2,416.89*	*150*	*100*
Nb	*2,416.99*	*200*	*8*
Fe	*2,417.87*	*100*	*10*
Pt	*2,418.06*	*50*	*300*
Nb	*2,418.69*	*500*	*5*
Ag	*2,420.07*	*100hw*	—

Ag 2,437.79 *J* 500 *O* 60

	λ	I J	I O		λ	I J	I O
Pd	**2,437.76**	3*h*	—	**Pd**	**2,437.80**	15*h*	—
Cl	**2,437.74**	(3)	—	**Ni**	**2,437.89**	200	40*w*
Mo	**2,437.736**	80*w*	10	**Mn**	**2,437.91**	25*wh*	—
Nb	**2,437.72**	5*wh*	—	**Pd**	**2,437.93**	15*wh*	—
U	**2,437.67**	4*h*	4	**W**	**2,437.96**	—	6
Fe	**2,437.666**	3*h*	1	**La**	**2,438.00**	20	—
J	**2,437.61**	(20)	—	**V**	**2,438.04**	15*h*	—
Th	**2,437.56**	4	5				
W	**2,437.48**	9	—				
Mn	**2,437.42**	40*wh*	—	Fe	2,438.18	4	30
				Ta	2,438.64	4	40
Nb	**2,437.41**	100	4	Si	2,438.78	20	30
				Rh	2,438.79	100	2
				Fe	2,439.30	100	15
*bh*B	2,437.1	—	250				
Pt	2,436.69	20	300	Pt	2,440.06	100*wh*	800*w*
Ni	2,436.67	20	30	Ti	2,440.98	—	35
Co	2,436.66	25	50*R*				
W	2,435.96	10	30				
				Cu	*2,441.64*	*100*	*200*
V	2,435.52	35	35*r*	*Fe*	*2,442.57*	*10*	*70*
Si	2,435.16	80	150	*Rh*	*2,443.71*	*100*	*4*
*bh*C	2,435.00	—	30	*Pb*	*2,443.84*	*15*	*100w*
Ru	2,434.88	—	50	*Ta*	*2,443.94*	—	*60*
Ni	2,434.42	—	40*w*				
				Rh	*2,444.06*	*100*	*2*
				Ag	*2,444.21*	*80w*	—
Nb	*2,433.79*	*100*	*3*	*Rh*	*2,444.27*	*3*	*100w*
Ni	*2,433.57*	*80*	—	*Ta*	*2,444.67*	*200*	*50*
O	*2,433.53*	*(250)*	—				
Ru	*2,432.93*	—	*60*				
Ta	*2,432.70*	*400*	*300r*				
Re	*2,432.17*	—	*100w*				
Ir	*2,431.94*	*50*	*50*				

Ag 2,614.59 *J* 300 —

	λ	J	O		λ	J	O
Hf	**2,614.59**	3*h*	—	**Cs**	**2,614.62**	(8)	—
Ru	**2,614.586**	4	50	**Cr**	**2,614.63**	6	—
Er	**2,614.55**	2	3	**Cl**	**2,614.64**	(3)	—
Fe	**2,614.49**	5	40	**Sb**	**2,614.666**	3	4
W	**2,614.44**	5*s*	6	**Mg**	**2,614.67**	2*h*	5
Cu	**2,614.41**	5*w*	—	**J**	**2,614.70**	(30)	—
V	**2,614.40**	35	1	**Nb**	**2,614.76**	30	—
W	**2,614.38**	6	—	**Ru**	**2,614.86**	30	—
Co	**2,614.36**	60*w*	6	**Fe**	**2,614.87**	10	—
Nb	**2,614.31**	20	2	**Cd**	**2,614.96**	(2)	—
In	**2,614.31**	2	—	**Ir**	**2,614.98**	5	25
Zn	**2,614.30**	(10)	—	**J**	**2,615.01**	(12)	—

	λ	I J	I O
Hf	2,614.29	6	6
Ne	2,614.26	(5)	—
Ir	2,614.20	2	10
Fe	2,614.18	3	—
Pb	2,614.178	80	200r
Sn	2,614.178	5	5
Fe	2,613.82	400	400
Hf	2,613.60	80	20
Pd	2,613.43	100	—
Lu	2,613.40	100	30
Ru	2,612.51	80	—
Fe	2,611.87	500	500
Na	2,611.81	(80)	—
Ni	2,611.65	125	—
Ru	2,611.51	80	3
Fe	2,611.07	80	20
La	2,610.33	150	10
Mn	2,610.20	100h	15
Ni	2,610.09	900h	—
Pd	2,609.86	200	—
Rh	2,609.17	200	3
Zn	2,608.64	100	300
Ta	2,607.84	150	20h
Fe	2,607.09	400	300
Ni	2,606.39	600h	—
Ag	2,606.16	200wh	10
Mn	2,605.69	500R	100R
Ru	2,615.09	100	60
Ni	2,615.19	900h	—
Lu	2,615.42	250	100
V	2,616.25	70	4
Fe	2,617.62	400	300
Mn	2,618.14	100h	50
Cu	2,618.37	100	500w
Fe	2,619.08	150	5
Lu	2,619.26	100	30
Nb	2,620.45	200	3
Yb	2,621.12	100	2
Fe	2,621.67	400	100

Ag 2,711.21 J 300 O 1*

	λ	I J	I O
Kr	2,711.11	(2)	—
U	2,711.10	4	8
Mo	2,710.93	25	1
Pt	2,710.922	15	1
Cr	2,710.92	70	1
W	2,710.78	15	6d
Mo	2,710.74	1	20
Ta	2,710.72	10h	1
Mn	2,710.33	40h	12
In	2,710.27	2000Rh	800R
Ru	2,710.23	100	50
V	2,710.16	60	6
N	2,709.82	(50)	—
Cr	2,709.31	60	2
P	2,711.28	(5)	—
W	2,711.31	4	—
Nb	2,711.37	5	1
Th	2,711.456	4	6
Fe	2,711.463	3	12
Mo	2,711.49	25	1
Zr	2,711.51	20	40
U	2,711.55	6w	—
Ir	2,711.57	6h	—
Mn	2,711.58	125h	2
Cs	2,711.6	(2)	—
Te	2,711.61	(50)	—
Xe	2,711.65	(2)	—
Fe	2,711.654	50	100
J	2,711.66	(20)	—

	λ	I	
		J	O
Ta	2,709.27	150	40
Tl	2,709.23	200R	400R
Fe	2,709.06	100	3
Ra	2,708.96	(200)	—
Ni	2,708.79	500	—
Fe	2,708.57	50	80
Mn	2,708.45	50h	15
V	2,707.86	150	70
Nb	2,707.83	80	3
Mn	2,707.53	50h	10
Co	2,707.50	100wh	—
Ru	2,707.29	60	—
Fe	2,707.13	70	—
Co	2,706.74	100wh	—
Hf	2,706.73	50	10
V	2,706.70	200R	60
Fe	2,706.58	150	150
Sn	2,706.51	150R	200R
V	2,706.17	400R	100
Pt	2,705.89	200wh	1000wh
Co	2,705.85	100w	15w
Rh	2,705.63	300wh	100
Ru	2,704.57	100	—
Fe	2,703.99	400	30
Cu	2,703.18	200	10
Pt	2,702.40	300	1000
Nb	2,702.20	100	10
V	2,702.19	300R	80
Eu	2,701.89	200	300W
Lu	2,701.71	150	40
Mo	2,701.42	100	20
Cu	2,700.96	400	20
V	2,700.94	500R	125

	λ	I	
		J	O
V	2,711.74	150R	50
Fe	2,711.84	110	4
Ag	2,712.06	200h	3
J	2,712.23	(100)	—
Cr	2,712.31	70	30
Fe	2,712.39	100	2
Kr	2,712.40	(80h)	—
Ru	2,712.41	300	80
Hf	2,712.42	50	25
V	2,713.05	80	40
Ru	2,713.07	80	—
Cu	2,713.50	300w	50
Ru	2,713.58	100	—
In	2,713.93	125wh	200R
V	2,714.20	100	60
Pd	2,714.32	150	—
Fe	2,714.41	400	200
Co	2,714.42	200W	12
Pd	2,714.90	200	—
Rh	2,715.31	500wh	50
Nb	2,715.34	100	2
V	2,715.69	300R	50
Nb	2,715.88	100	2
Co	2,715.99	75	75w
Ru	2,716.12	100	—
Fe	2,716.22	150	20
Ti	2,716.25	70	5
Ru	2,716.58	80	—
Nb	2,716.62	200	10
Eu	2,716.97	300	300
Mo	2,717.35	100	20
Ru	2,717.40	60	80
Cu	2,718.77	300w	40
Fe	2,719.02	300r	500r
Pt	2,719.04	100W	1000W
Tm	2,721.19	100	60

Ag 2,824.37 *J* 200 *O* 150

	λ	J	O
Cu	2,824.369	300	1000
Co	2,824.36	—	2
Er	2,824.32	—	3
W	2,824.30	8	1
U	2,824.28	30	25
Re	2,824.25	—	20
Ho	2,824.19	3	20
Pt	2,824.41	2	2
V	2,824.44	15	2
Ir	2,824.45	15	20
Cr	2,824.537	10	—
Eu	2,824.54	—	2w
Ce	2,824.629	—	2
U	2,824.63	6	10

	λ	I	
		J	O
Mo	2,824.17	3	—
Os	2,824.166	4	20
J	2,824.15	(4)	—
Cs	2,824.12	(8)	—
Ta	2,824.05	—	4
Ce	2,824.03	—	15
In	2,823.99	3	—
Ag	2,823.95	3h	—
Nb	2,823.88	10	1
Fe	2,823.28	300	200
Pb	2,823.19	40	150R
Ru	2,823.18	80	20
Au	2,822.72	80	—
Hf	2,822.68	90	30
Ru	2,822.55	150	30
V	2,822.44	70h	4
Cr	2,822.37	100	20
Sc	2,822.13	20	50
Ru	2,822.03	5	50
Cr	2,822.01	80	10
Ru	2,821.424	100	5
Ti	2,821.42	70wh	—
Ni	2,821.29	125	125
Eu	2,820.77	200W	200W
Hf	2,820.22	100	40
Co	2,820.01	—	50
Ti	2,820.00	70wh	—
Re	2,819.955	150W	—
Au	2,819.95	150	—
Ta	2,819.37	5	100
Ru	2,818.95	3	50
Yb	2,818.75	80	—
Cr	2,818.36	80	8
Pt	2,818.25	4	70
Ti	2,817.87	200	10
Fe	2,817.51	60	100
Ta	2,817.50	10	80
Ta	2,817.10	100	80d
Mo	2,816.15	300h	200
Ag	2,815,54	80wh	3
Ta	2,815.12	4	100
Ta	2,815.01	15	150
Hg	2,814.93	(200)	—
Zr	2,814.90	1	70
Ta	2,814.80	5	125
Eu	2,813.95	300wh	300w

	λ	I	
		J	O
Fe	2,824.67	2d	2d
Th	2,824.68	25d	—
Ru	2,824.77	—	10
Ho	2,824.79	10h	—
Ta	2,824.81	5h	60W
Zr	2,824.82	—	4
J	2,824.84	(8)	—
U	2,824.86	8	12
Ce	2,824.88	—	10
Yb	2,824.96	10	2
Ru	2,825.06	60	—
Co	2,825.15	—	75w
Ni	2,825.236	125	—
Co	2,825.24	200	5
Ru	2,825.46	80	—
Fe	2,825.56	150	150
Fe	2,825.69	60	70
V	2,825.87	70h	7
Tl	2,826.16	100R	200R
Ta	2,826.18	5	60
Ru	2,826.22	80	—
Rn	2,826.5	(70)	—
Mo	2,826.55	5	40
Rh	2,826.67	50d	100
Ru	2,826.68	100	—
Cr	2,826.75	3	70
Co	2,826.80	—	50W
Tm	2,827.02	50	20
Nb	2,827.08	50	8
Ta	2,827.18	10	200
Ti	2,827.21	80wh	—
Rh	2,827.31	—	50
Ta	2,827.55	100d	3d
Fe	2,827.89	50	70
Tm	2,827.92	100	50
Ti	2,828.15	200h	2
Ta	2,828.58	100	75
Fe	2,828.63	80	—
Eu	2,828.69	150	200
Mn	2,828.76	—	50wh
Fe	2,828.81	60	100
Ti	2,838.9	150wh	—
Ru	2,829.16	8	50
Pt	2,830.29	600r	1000R
Cr	2,830.47	80h	15
Fe	2,831.56	500	1

	λ	I J	I O
Ra	2,813.76	(400)	—
Ru	2,813.71	125	50
Fe	2,813.29	400	400
Ti	2,832.16	100	25
Fe	2,832.44	200	300
Cr	2,832.46	125	2
Kr	2,833.00	(100)	—
Pb	2,833.07	80R	500R
Ta	2,833.64	40w	300w
Re	2,834.06	—	100r
Rh	2,834.12	30	70
Cd	2,834.19	(100)	—
Cr	2,834.26	125	—
Pt	2,834.71	5	80
Nb	2,835.12	100	5d
Fe	2,835.46	100	100
Cr	2,835.63	400r	100

Ag 2,938.55 *J* 200 *O* 200

	λ	I J	I O
Ti	**2,938.539**	—	**2**
Mg	**2,938.538**	—	**25**
Cs	**2,938.5**	**(20)**	—
W	**2,938.499**	**6**	**8**
Ir	**2,938.47**	**12**	**18*h***
Ta	**2,938.43**	**3**	**50**
Ce	**2,938.32**	—	**2**
Mo	**2,938.30**	**30**	**1**
Bi	**2,938.298**	**300*w***	**300*w***
V	**2,938.25**	**60**	**2**
Ce	**2,938.22**	—	**5**
Yb	**2,938.18**	**3**	**1**
Th	**2,938.11**	**4**	**5**
Nb	**2,938.07**	**5**	**3**
Ce	**2,938.051**	—	**5**
Fe	**2,938.05**	**1**	**2**
Ir	**2,937.96**	—	**5**
Mn	**2,937.92**	—	**25*wh***
Fe	2,937.81	150	300
Hf	2,937.79	100	50
Na	2,937.73	(40)	6
Fe	2,936.90	500*r*	700*r*
Ho	2,936.77	1000*R*	—
Ti	2,936.17	100*wh*	—
Tm	2,936.00	300	80
Ru	2,935.52	80	10
*bh*B	2,934.9	—	100
Ag	2,934.23	200*h*	10
Ta	**2,938.56**	—	**10**
Mo	**2,938.59**	**10**	**1**
V	**2,938.67**	**2**	**12**
Ce	**2,938.68**	—	**3**
Ti	**2,938.70**	**100*wh***	—
In	**2,938.71**	**10**	—
Fe	**2,938.73**	—	**2**
Ir	**2,938,76**	—	**10**
Mo	**2,938.77**	**5**	**2**
Pt	**2,938.81**	**2**	**15**
Sm	**2,938.82**	—	**6**
Cr	**2,938.85**	—	**6*wh***
W	**2,938.852**	**9**	**1**
Mo	**2,938.89**	**3**	—
Hg	**2,939.03**	**(10)**	—
U	**2,939.04**	**2**	**6**
W	**2,939.043**	**4**	**9**
Fe	**2,939.08**	**20**	**80**
Ru	**2,939.13**	—	**12**
W	**2,939.18**	**2**	**8**
Ta	2,939.28	40*h*	200
Mn	2,939.30	—	50
Ta	2,940.06	40*w*	100
Ta	2,940.21	50	150
Ru	2,940.36	3	50
Fe	2,940.59	80	200
Hf	2,940.77	12	60
In	2,941.05	(80)	—
Fe	2,941.34	300	600

	λ	I			λ	I	
		J	O			J	O
Ta	2,933.55	150	400	V	2,941.37	300*r*	40
Ru	2,933.24	150	20	V	2,941.49	150*r*	12
Mn	2,933.06	15	80	Nb	2,941.54	300	50
Ta	2,932.69	80*w*	400	Ti	2,941.99	150	100
In	2,932.62	300	500	Ta	2,942.14	40	150
V	2,932.32	80	12	Te	2,942.16	(100*h*)	—
Rh	*2,931.94*	*20*	*80*	Ru	2,942.25	100	30
Ti	*2,931.26*	*150wh*	—	V	2,942.35	20*h*	80*r*
V	*2,930.81*	*150r*	*30*	Ar	2,942.90	(100)	—
Co	*2,930.43*	*150wh*	—	Ti	2,943.13	60*wh*	—
Pt	*2,929.79*	*200w*	*800R*	Re	2,943.14	—	60
Co	*2,929.51*	—	*75*	Co	2,943.15	100*wh*	—
Rh	*2,929.11*	*10*	*100*	Fe	2,944.40	600	70
Fe	*2,929.01*	*100*	*150*	V	2,944.57	300*r*	50
Ho	*2,928.79*	*100*	—	*Fe*	*2,945.05*	*30*	*100*
Mg	*2,928.75*	*100*	*25*	*He*	*2,945.10*	*(100)*	—
Nb	*2,927.81*	800*R*	*200*	*Ti*	*2,945.47*	*100wh*	—
Ru	*2,927.54*	*200*	*50*	*Ru*	*2,945.67*	*300*	*60*
Re	*2,927.40*	—	*125w*	*Nb*	*2,945.88*	*100*	*2*
Fe	*2,926.59*	*400*	*150*	*Yt*	*2,945.95*	*100*	*2*
Ta	*2,926.46*	*10*	*100*	*Ta*	*2,946.91*	*10*	*150*
				Fe	*2,947.66*	*100*	*10*
				Fe	*2,947.88*	*200*	*600r*
				Ti	*2,948.25*	*30*	*100*
				Fe	*2,948.43*	*70*	*80*
				Pb	*2,948.72*	*125*	—
				Mn	*2,949.20*	*30*	*100*
				Te	*2,949.52*	*(100)*	—
				Fe	*2,950.24*	*300*	*700*
				V	*2,950.35*	*100r*	*25*
				Nb	*2,950.88*	*200*	*150*

Ag 3,280.683 *J* 1000*R* *O* 2000*R*

Eu	**3,280.682**	—	**1000*R***	**Cu**	**3,280.685**	**2**	**10**
Co	**3,280.681**	—	**2**	**In**	**3,280.69**	**3**	—
Pd	**3,280.68**	2*h*	—	**Th**	**3,280.74**	2*d*	2*d*
Mo	**3,280.67**	—	**3**	**Zr**	**3,280.75**	**2**	**3**
Ce	**3,280.668**	—	**6**	**Mn**	**3,280.76**	**30**	**60**
U	**3,280.61**	**2**	**1**	**Sm**	**3,280.84**	**6**	**20**
Cl	**3,280.60**	**(8)**	—	**Ta**	**3,280.87**	**2**	**3**
Rh	**3,280.55**	**10**	**30*R***	**Mo**	**3,280.88**	**25**	—
Lu	**3,280.50**	—	**10**	**Yt**	**3,280.91**	**12**	**8**
Ce	**3,280.485**	**1**	**15**	**Os**	**3,280.92**	—	**5**
Xe	**3,280.48**	**(4)**	—	**Th**	**3,281.03**	**5**	**2**

	λ	I	
		J	O
U	3,280.40	1	5
Ti	3,280.39	—	3
Th	3,280.37	12	10
Mo	3,280.32	5	5
Tb	3,280.28	15	30
Fe	3,280.26	150	150
Er	3,280.22	3	12
Sm	3,280.218	—	4
Nd	3,280.202	4h	2h
P	3,280.20	(30h)	—
Dy	3,280.10	5	70
U	3,280.00	8	8
Ti	3,279.99	40	10
Hf	3,279.98	25	25
Yb	3,279.98	—	5
Nb	3,279.97	5	—
Gd	3,279.92	—	2
Pb	3,279.85	10	—
V	3,279.845	8	20
Ce	3,279.84	5	30
Nb	3,279.83	—	4
Cu	3,279.82	30	25
Ti	3,278.92	150	40
Fe	3,278.73	60	100
Ti	3,278.29	100	25
Fe	3,277.35	200	40
Fe	3,276.47	50	100
V	3,276.12	200R	50
Yb	3,275.81	100	12
Os	3,275.20	15	200
Ta	3,274.95	35W	200
Cu	3,273.96	1500R	3000R
Nb	3,273.89	100W	20r
Ti	3,272,08	100	25
Fe	*3,271.00*	*300*	*300*
Ge	*3,269.49*	*300*	*300*
Os	*3,269.21*	*20*	*200*
Os	*3,267.94*	*30*	*400R*
Sb	*3,267.50*	*150Wh*	*150*
Pd	*3,267.35*	*200h*	—
La	*3,265.67*	*200*	*300*
Fe	*3,265.62*	*300*	*300*
Fe	*3,265.05*	*150*	*200*
Hg	*3,264.06*	*(200)*	—
Rh	*3,263.14*	*40*	*200*
Sn	*3,262.33*	*300h*	*400h*
Os	*3,262.29*	*50*	*500R*

	λ	I	
		J	O
Mo	3,281.07	3	25
Ce	3,281.09	—	18
U	3,281.115	1	2
V	3,281.115	50	3
Ho	3,281.17	4h	—
Xe	3,281.26	(8h)	—
Th	3,281.28	5	4
Fe	3,281.30	100	15
Mo	3,281.34	1	5
Tb	3,281.40	15	50
Th	3,281.41	10	8
Ca	3,281.48	4	1h
Nd	3,281.487	6	4
Rb	3,281.49	(20)	—
Ba	3,281.50	—	25
U	3,281.55	1	5
Ce	3,281.587	—	3
Co	3,281.588	—	7
Cd	3,281.61	5	5
Ti	3,282.33	150	30
Zn	3,282.333	300	500R
Ni	3,282.70	—	100
Nb	3,283.46	100	2
Sn	3,283.51	100h	—
Rh	3,283.57	—	150
Fe	3,284.59	125	200
Na	3,285.748	(100)	40
Fe	3,286.755	400	500
Ni	3,286.95	1	100
Pd	3,287.25	25	300w
Ti	3,287.65	200	40
Rh	3,289.14	50	150
Yb	*3,289.37*	*1000R*	*500R*
Yb	*3,289.85*	—	*1000*
Os	*3,290.26*	*20*	*200*
Cr	*3,291.76*	*200*	*10*
Fe	*3,292.02*	*125*	*150*
Mo	*3,292.31*	*300*	*10*
Fe	*3,292.59*	*150*	*300*
Ru	*3,294.11*	*200*	*60*
Cr	*3,295.43*	*200*	*10*
Fe	*3,298.13*	*150*	*200*

Ag 3,382.891   *J* 700*R*   *O* 1000*R*

	λ	*I* *J*	*I* *O*
Ce	3,382.89	—	8
Yt	3,382.83	3	3
Nd	3,382.81	10	200
Tb	3,382.80	8	15
Ce	3,382.70	—	8
Cr	3,382.683	200	35
O	3,382.68	(12)	—
U	3,382.67	4	4
Pr	3,382.66	2	10
W	3,382.61	12	10
Pd	3,382.57	2*h*	—
V	3,382.53	125	—
Ce	3,382.51	—	8
Mo	3,382.48	15	15
Eu	3,382.412	—	2*h*
Fe	3,382.409	10	50
Sm	3,382.407	40	100
Nb	3,382.407	40*h*	5
Ti	3,382.312	7	30
Ce	3,382.308	—	8
U	3,382.304	1	3
Mo	3,382.29	6	10
Bi	3,382.28	—	2
W	3,382.10	7	8
Nd	3,382.09	—	4
Cr	3,382.08	1	30
Er	3,382.07	1	15
Au	3,382.00	8	6
U	3,381.95	6*d*	4*d*
Co	3,381.50	—	100*W*
Nb	3,380.94	200	—
La	3,380.91	100*h*	200
Ni	3,380.89	12	200
Sr	3,380.71	200	150
Pd	3,380.67	2*h*	150*W*
Ni	3,380.57	100	600*R*
Ti	3,380.28	150*r*	25
Fe	3,380.11	25	200
Cr	3,379,82	100	15
Cr	3,379.37	100	6
Nb	3,379.30	100	1
Fe	3,378.68	80	150
Cr	3,378.34	150	25
Co	3,377.06	—	100
Lu	3,376.50	10	100
La	3,376.33	50	100
Yb	3,375.48	100	30
Er	3,382.892	1	18
Zr	3,382.90	—	3
Yt	3,383.05	3	3
Th	3,383.120	10	5
W	3,383.12	18	—
Sb	3,383.14	50	40
Ce	3,383.28	—	8
Pr	3,383.38	3	15
Ce	3,383.39	1	12
U	3,383.40	3	6
Nd	3,383.45	—	2
Mo	3,383.55	1	5
Ce	3,383.69	2	20
Fe	3,383.70	70	100
Pr	3,383.73	3	12
Ti	3,383.76	300*R*	70
Tb	3,383.78	—	8
Er	3,383.78	1	12
Nb	3,383.80	5	15
Pt	3,383.82	8	1
J	3,383.86	(3)	—
Co	3,383.92	—	60
Ce	3,383.925	—	8
Fe	3,383.98	100	200
Co	3,385.22	15	250*R*
Hg	3,385.25	(200)	—
Nb	3,386.24	100	5
Os	3,387.836	15	100
Ti	3,387.837	125	60
Zr	3,387.87	100	100
Co	3,388.17	12	250*R*
O	3,390.26	(100)	—
Ni	3,391.05	40	400
Cr	3,391.43	150	4
Zr	3,391.97	400	300
Fe	*3,392.31*	*80*	*125*
Ru	*3,392.54*	*40*	*100*
Fe	*3,392.66*	*200*	*300*
Ar	*3,392.81*	*(100)*	—
Ni	*3,392.99*	—	*600R*
Ar	*3,393.75*	*(250)*	—
Cr	*3,393.84*	*125*	*15*
Cr	*3,394.30*	*150*	*15*
Ti	*3,394.58*	*200*	*70*
Fe	*3,394.59*	*80*	*150*

	λ	I J	I O		λ	I J	I O
Ni	3,374.22	6	400	*Co*	*3,395.38*	*50*	*400R*
Ar	3,373.48	(300)	—	*Rh*	*3,396.85*	*500*	*1000W*
				Re	*3,399.30*	—	*200w*
Pd	*3,373.00*	*500wh*	*800r*	*Fe*	*3,399.34*	*200*	*200*
Ti	*3,372.80*	*400R*	*80*	*Rh*	*3,399.70*	*60*	*500*
Nb	*3,372.56*	*200*	*10h*				
Rh	*3,372.25*	*200*	*300*	*Fe*	*3,401.52*	*90*	*150*
Sc	*3,372.15*	*150*	*7*	*Os*	*3,401,86*	*20*	*200*
				Fe	*3,402.26*	*150*	*150*
Ni	*3,371.99*	*10*	*400*	*Os*	*3,402.51*	*15*	*200*
Fe	*3,370.79*	*200*	*300*	*Cd*	*3,403.65*	*500h*	*800*
Os	*3,370.59*	*30*	*300R*				
Ne	*3,369.91*	*(700)*	—				
Ne	*3,369.81*	*(500)*	—				
Ni	*3,369.57*	*100*	*500R*				
Fe	*3,369.55*	*200*	*300*				
Rh	*3,368.37*	*50*	*300*				
Co	*3,367.11*	*30*	*300R*				
Ni	*3,366.17*	*12*	*400W*				
Ni	*3,365.77*	*12*	*400W*				

Ag 3,968.22 *J* 60 *O* 100

	λ	J	O		λ	J	O
W	**3,968,17**	**8**	**8**	**Zr**	**3,968.26**	**4**	**100**
Rh	**3,968.164**	—	**2**	**Gd**	**3,968.35**	—	**20**
Ta	**3,968.16**	—	**4***h*	**Ar**	**3,968.36**	**(200)**	—
Pr	**3,968.158**	**25**	**10**	**Fe**	**3,968.370**	—	**2**
Tb	**3,968.15**	—	**2**	**U**	**3,968.374**	**2**	**1**
V	**3,968.09**	**40**	**25**	**Dy**	**3,968.39**	—	**300**
Hg	**3,968.03**	**(50)**	—	**Yt**	**3,968.43**	**30**	**10**
Yb	**3,968.03**	**3**	—	**Ru**	**3,968.461**	**200**	**12**
Hf	**3,968.01**	**1**	**5**	**Lu**	**3,968.464**	—	**50**
U	**3,968.007**	**6***h*	**4**	**Ca**	**3,968.468**	**500***R*	**500***R*
Fe	**3,967,97**	**15**	**60**	**Ce**	**3,968.469**	**35***w*	**35**
Ce	**3,967.91**	—	**3**	**Nb**	**3,968.471**	**10**	**3**
Sm	**3,967.78**	**2**	**6**	**Ir**	**3,968.475**	—	**25**
Nd	**3,967.70**	**6**	**20**	**W**	**3,968.59**	**6**	**6**
Yt	**3,967.69**	**10***h*	**3**	**Br**	**3,968.66**	**(8)**	—
Sm	**3,967.67**	**6**	**8**	**Zr**	**3,968.72**	—	**3**
Pr	**3,967.66**	**3**	**10**	**Tb**	**3,968.73**	—	**2**
Tb	**3,967.65**	—	**6**	**Mo**	**3,968.75**	**50**	**8**
Ce	**3,967.644**	—	**3**	**Eu**	**3,968.87**	—	**3***wh*
U	**3,967.639**	**2**	—	**Nd**	**3,968.88**	**4**	**20**
Xe	**3,967.541**	**(200)**	—	**Th**	**3,969.003**	**5**	**10**
P	**3,967.54**	**(15)**	—	**Gd**	**3,969.005**	—	**40**
Ce	**3,967.53**	**1**	**5**	**Mo**	**3,969.01**	**3**	**3**
Dy	**3,967.52**	—	**8**	**U**	**3,969.02**	**8**	**5**
Ir	**3,967.51**	—	**3**	**Cr**	**3,969.06**	**50**	**80**

	λ	I	
		J	O
U	3,967.48	—	10
Re	3,967.424	—	25
Fe	3,967.423	100	125
Th	3,967.41	1	2
Nb	3,967.37	50*h*	—
Nd	3,967.31	2	10
Cs	3,967.30	(4)	—
Th	3,967.214	8	8
Tb	3,967.211	—	20
Ce	3,967.18	2	6
Pr	3,967.131	25*d*	40*d*
Eu	3,967.13	—	25*W*
Nd	3,967.07	10	30
Ce	3,967.05	6	35
U	3,967.01	2	2
Th	2,966.97	5	5
Gd	3,966.85	—	8
W	3,966.75	3	5
Th	3,966.72	1	3
K	3,966.69	(30)	—
Zr	3,966.66	3	—
Fe	3,966.63	40	80
Eu	3,966.61	—	5*W*
Pr	3,966.573	70*d*	100*d*
U	3,966.567	30	20
Pt	3,966.36	40	80
Nb	3,966.25	30	10
Fe	3,966.07	70	100
Sm	3,966.05	50	60
Pr	3,965.26	50	100
Os	3,964.96	12	60
Ru	3,964.90	40	50
Eu	3,964.89	—	40*wd*
Kr	3,964.89	(30*hl*)	—
Pr	3,964.82	80*d*	125*d*
He	3,964.73	(50)	—
Fe	3,964.52	25	80
Nb	3,964.28	50	1
Ti	3,964.27	40	80
Pr	3,964.26	50	60
Cr	3,963.69	300	300
Gd	3,963.65	60	50
Os	3,963.63	50	500
Fe	3,963.11	50	125
Sm	3,963.00	40	50
Ti	3,962.85	35	80
Re	3,962.48	—	100
Pr	3,962.44	50	60
Co	3,969.12	6	100*w*
In	3,969.13	15	—
Nb	3,969.135	20*h*	—
Ce	3,969.16	—	2
Ir	3,969.17	10	30
Te	3,969.18	(10)	—
W	3,969.20	10	12
Eu	3,969.23	—	20*W*
Dy	3,969.233	—	6
Gd	3,969.261	—	200
Fe	3,969.261	400	600
Sr	3,969.261	—	30
Th	3,969.34	2*w*	3*w*
U	3,969.42	—	5
Er	3,969.43	1	6
Pr	3,969.51	3	8
Th	3,969.53	5	5
Fe	3,969.63	5	5
Nd	3,969.666	4	20
Os	3,969.67	100	100
Cr	3,969.748	90	200
Pr	3,969.75	3	12
Ru	3,969.79	4	8
Th	3,969.83	5	5
Eu	3,969.90	—	8*W*
Tb	3,969.92	—	3
H	3,970.07	(80)	—
Ta	3,970.10	40	100
Fe	3,970.39	30	50
Ni	3,970.50	—	40*w*
Pr	3,971.16	60	100
Cr	3,971.25	50	80
Pb	3,971.3	(30)	—
Fe	3,971.33	125	200
Sm	3,971.39	30	50
Rn	3,971.67	(80)	—
Pr	3,971.69	40	60
Eu	3,971.90	—	100*Rwh*
Eu	3,971.99	—	1000*Rwh*
Pr	3,972.16	80	125
Ni	3,972.17	6	100
Co	3,972.53	6	100
K	3,972.55	(30)	—
Cr	3,972.69	12	60
Co	3,973.15	6	150*w*
Ta	3,973.18	400*W*	1

	λ	I	
		J	O
Zr	3,961.59	8	500
Al	3,961.53	2000	3000
Mo	3,961.50	500	5
Eu	3,961.14	—	50*W*
Os	3,961.02	20	125
Co	3,960.99	10	60
Pr	3,960.60	25*h*	50
Os	3,960.51	15	50
Hg	3,960.24	(30)	—
Sm	3,959.53	40	50
Rh	3,958.86	100	200
Pd	3,958.64	200	500*w*
Tb	3,958.36	15	100*w*
Zr	3,958.22	150	500
Ti	3,958.21	100	150
Tm	3,958.10	40	200
Tb	3,957.97	15	60*d*
Co	3,957.93	—	100*R*
Dy	3,957.80	—	70
Gd	3,957.68	200	300*W*
P	3,957.62	—	(100)
Nd	3,957.46	40*s*	60
Ca	3,957.05	3	80
Fe	3,957.03	15	50
Fe	3,956.68	150	150
Fe	3,956.46	100	100
Ti	3,956.34	50	100
N	3,955.85	(35)	—
Eu	3,955.75	50*W*	—
K	3,955.21	(30)	—
Kr	3,954.78	(90*wh*)	—
O	3,954.60	(40)	—
O	3,954.38	(100)	—
Nd	3,953.525	60	60
Pr	3,953.516	100	150
Gd	3,953.37	50	100
Cr	3,953.16	12	60
Fe	3,953.15	40	80
Co	*3,952.92*	*75*	*100*
Mn	*3,952.84*	*75*	*60*
Ru	*3,952.68*	*30*	*20*
Fe	*3,952.61*	*50*	*80*
Ce	*3,952.54*	*30*	*60*
Cr	*3,952.40*	*18*	*60*
Nb	*3,952.37*	*50*	*3*
Gd	*3,952.01*	*60*	*100*
V	*3,951.97*	*50*	*35*
Mn	*3,951.96*	*50*	*40*
O	3,973,27	(125)	—
Ni	3,973.56	10	800
V	3,973.64	40	25
Ca	3,973.71	15	200
Gd	3,974.07	80	100
Xe	3,974.42	(40)	—
Co	3,974.73	10	100
Nd	3,975.20	30	40
Zr	3,975.29	1	50
Os	3,975.44	12	50
Mn	3,975.89	50	40
Pr	3,976.29	10	50
Ir	3,976.31	70	10
Nb	3,976.51	80*h*	—
Fe	3,976.61	35	8
Cr	3,976.66	300	300
Tb	3,976.82	200	150
Mn	3,977.08	100	50
Os	3,977.23	40	300
Fe	3,977.74	150	300
Rb	3,978.21	(40)	—
Ru	3,978.44	70	60
Dy	3,978.57	15	200
Co	3,978.65	—	100
Cr	3,978.68	40	80
Ce	3,978.89	50	50
V	3,979.14	8	50
Sm	3,979.19	50	50
Ta	3,979.28	3*h*	50*h*
Gd	3,979.34	20	100
Hf	3,979.40	40	6
Ru	3,979.42	60	60
Nd	3,979.48	30	40
Co	3,979.52	12	150*w*
Cr	3,979.80	20	80
S	3,979.86	(35)	—
V	3,980.52	35	40
Ta	3,981.01	40	2
Cr	3,981.23	50	100
Rn	3,981.68	(150)	—
Ti	3,981.76	70	100
Fe	3,981.77	100	150
Tb	3,981.88	200	80
Dy	3,981.94	100	150
Pr	3,982.06	100	125
Ti	3,982.48	30	80
Mn	3,982.58	30	20
Yt	3,982.59	100	60

	λ	I J	I O
P	3,951.50	(70)	—
Fe	3,951.17	125	150
Xe	3,950.92	(125)	—
Yt	3,950.36	100	60
Fe	3,949.96	100	150
Pr	3,949.44	100	150
La	3,949.11	800	1000
Ar	3,948.98	(2000)	—
Fe	3,948.78	100	150
Hg	3,948.29	(100)	—
Xe	3,948.16	(60)	—
Fe	3,948.11	50	125
Pr	3,947.63	60d	125d
Ar	3,947.50	(1000)	—
O	3,947.33	(300)	—
Tb	3,946.87	30	150
Ru	3,945.57	100	50
Gd	3,945.54	150	200W
Co	3,945.33	15	200
Hg	3,945.09	(100)	—
Dy	3,944.69	150	300
Al	3,944.03	1000	2000
Mn	3,942.85	75	75
Hg	3,942.59	(100)	—
Fe	3,942.44	70	100
Hg	3,942.24	(100)	—
Co	3,941.73	—	200wh
Cr	3,941.49	60	200r
Mo	3,941.48	150	5
Co	3,940.89	—	100
Fe	3,940.88	80	150
Rb	3,940.57	(200)	—
J	3,940.24	(500)	—
Tb	3,939.60	200	200
Os	3,938.59	20	125
Nb	3,938.55	100h	—
La	3,936.22	50	100
Nb	3,936.02	200	5
Co	3,935.97	15	400R
Pr	3,935.82	50	125
Fe	3,935.81	8	100

	λ	I J	I O
Mn	3,982.91	30	20
Gd	3,983.03	40	40
Sm	3,983.14	60	100
Dy	3,983.66	8	150
Cr	3,983.91	60	200
Fe	3,983.96	125	200
Hg	3,983.98	(400)	—
Dy	3,984.23	—	80
Cr	3,984.34	60	80
Ru	3,984.86	70	60
Mn	3,985.24	100	75
Fe	3,985.39	40	125
Li	3,985.79	—	100
Fe	3,986.17	8	125
Mn	3,986.83	75	40
Mn	3,987.10	60	30
Gd	3,987.22	100	100
Er	3,987.95	20	100
Yb	3,987.99	500R	1000R
La	3,988.52	800	1000
Zn	3,989.23	(100)	—
Pr	3,989.72	125	200
Ti	3,989.76	100	150
V	3,990.57	40	125
Cr	3,991.12	60	200
Zr	3,991.13	60	100
Cr	3,991.67	50	100
Ir	3,992.12	60	150
Mn	3,992.49	75	40
Cr	3,992.84	70	150
Ba	3,993.40	50r	100R
Kr	3,994.83	(100)	—
Pr	3,994.834	25	300
N	3,994.99	(300)	—
Co	3,995.31	20	1000R
Tm	3,995.58	—	100
La	3,995.75	300	600
Ta	3,996.17	30h	100
Gd	3,996.32	100	100
Tm	3,996.52	40	200
Dy	3,996.70	80	200
Pr	3,997.05	40	100
P	3,997.16	(70)	—
Fe	3,997.40	150	300
Co	3,997.91	20	200
Kr	3,997.95	(100wh)	—
Fe	3,998.05	100	150

	λ	I J	I O		λ	I J	I O
				Ti	*3,998.64*	*100*	*150*
				V	*3,998.73*	*25*	*100*
				S	*3,998.79*	*(60)*	*—*
				Dy	*4,000.45*	*300*	*400*

Ag 4,055.26 *J* 500*R* *O* 800*R*

	λ	J	O		λ	J	O
Eu	4,055.25	—	2	Sm	4,055.28	3	5
W	4,055.23	6	7	Rn	4,055.3	(5)	—
Mn	4,055.21	5	10	Zr	4,055.31	—	3
Tb	4,055.20	—	3	Tb	4,055.32	—	5
Dy	4,055.159	8	5	Ir	4,055.35	—	6
Ce	4,055.158	1	8	Er	4,055.47	8	12
Fe	4,055.04	10	40	Os	4,055.496	3	30
Ir	4,055.034	—	4	Ho	4,055.50	2	—
Er	4,055.03	—	3	Mn	4,055.543	80	80
Zr	4,055.030	5	100	Mo	4,055.545	10	2
Ti	4,055.02	30	80	W	4,055.65	5	6
Dy	4,055.01	—	2*h*	Nd	4,055.67	5	10
Ce	4,054.99	6	12	Zr	4,055.71	3	25
Tb	4,054.96	—	2*h*	Ir	4,055.72	—	7
Fe	4,054.88	5	25	U	4,055.73	5	—
Th	4,054.86	3	5	Pr	4,055.76	3	6*wh*
Pr	4,054.845	40	50	Ce	4,055.84	—	12
U	4,054.84	4	1	Er	4,055.85	—	6
Fe	4,054.83	5	25	U	4,055.98	—	8
Pt	4,054.77	2	5	Er	4,055.99	—	5
Gd	4,054.734	20	80	Mo	4,056.01	30	25
Ru	4,054.732	—	5	Gd	4,056.04	—	3
Nd	4,054.72	4	12	Cr	4,056.05	8	30
Ce	4,054.66	—	5	U	4,056.07	5	—
Ca	4,054.64	2	2	Tb	4,056.11	—	2
Co	4,054.62	—	2	Pr	4,056.13	2	5
Tb	4,054.60	—	2	Yb	4,056.18	10	2
Sc	4,054.55	9	10	Ti	4,056.21	2	3
Ar	4,054.52	(80)	—	Ce	4,056.25	—	3
Sm	4,054.51	2	—	V	4,056.26	3	1
Ho	4,054.49	2	3	U	4,056.29	—	5
Lu	4,054.45	3	25	Mo	4,056.318	15	15
Zr	4,054.43	—	20	J	4,056.321	(15)	—
Ta	4,054.39	—	4	Ce	4,056.338	—	4
U	4,054.31	15	12	Rh	4,056.34	2	2
Fe	4,054.183	—	2	W	4,056.46	5	2
Cr	4,054.183	3	1	Ir	4,056.47	2	12
Cl	4,054.18	(9)	—	Zr	4,056.51	—	6
Tb	4,054.12	1	9	Pr	4,056.54	60	100
Ce	4,054.10	—	2	Kr	4,056.57	(3)	—

	λ	I	
		J	O
Ru	**4,054.05**	**100**	**40**
Br	**4,054.03**	**(4)**	—
Tb	**4,054.00**	—	**5**
Cs	**4,053.96**	**(15)**	—
W	**4,053.94**	**10**	**9**
Co	**4,053.922**	**2**	**6**
Ho	**4,053.92**	**200**	**400**
Er	**4,053.88**	**1**	**20**
Dy	**4,053.86**	—	**3**
Ti	**4,053.837**	**8**	**25**
Nd	**4,053.837**	**3**	**12**
V	**4,053.66**	—	**2**
Cu	**4,053.65**	*2h*	—
Gd	**4,053.648**	**40**	**100**
V	**4,053.59**	**70**	—
U	**4,053.57**	**5**	**5**
Gd	4,053.30	80	100
Ru	4,051.40	200	125
Gd	4,049.90	60	100
Te	4,048.89	(70)	—
Cr	4,048.78	50	80
Gd	4,047.85	50	150
K	4,047.20	200	400
Hg	4,046.56	300	200
Dy	4,045.98	12	150
Ar	4,045.97	(150)	—
Fe	4,045.81	300	400
Ho	4,045.43	80	200
Co	4,045.39	—	400
P	4,044.49	(150*w*)	—
Ar	4,044.42	(1200)	—
K	4,044.14	400	800
La	4,042.91	300	400
Os	4,041.92	6	100
Mn	4,041.36	50	100
Ho	*4,040.84*	*30*	*150*
Cr	*4,039.10*	*40*	*100*
Gd	*4,037.91*	*30*	*100*
Xe	*4,037.59*	*(100)*	—
Co	*4,035.55*	*3*	*150*
Mn	*4,034.49*	*20*	*250*
Mn	*4,033.07*	*20*	*400r*
Tb	*4,033.04*	*5*	*125*
Ga	*4,032.98*	*500R*	*1000R*
S	*4,032.81*	*(125)*	—
La	*4,031.69*	*300*	*400*

	λ	I	
		J	O
Sc	**4,056.58**	**2**	**5**
In	**4,056.59**	**(5)**	—
Sr	**4,056.67**	**4**	**4**
Cu	**4,056.7**	—	**8*wh***
U	**4,056.74**	**2**	**1**
In	**4,056.75**	**(50)**	—
In	**4,056.78**	**(30)**	—
Cr	**4,056.79**	**3**	**15**
Al	**4,056.8**	**(2)**	—
Nd	**4,056.84**	**8**	**8**
Ce	**4,056.90**	**2**	**15**
In	**4,056.936**	**(500)**	—
Nb	**4,056.94**	**5**	**3**
Co	**4,056.98**	**2**	**20**
Kr	4,057.01	(300*hs*)	—
In	4,057.07	(100)	—
Co	4,057.20	—	100
Xe	4,057.46	(100*wh*)	—
Pb	4,057.82	300*R*	2000*R*
Co	4,058.19	—	100
Gd	4,058.23	60	100
Co	4,058.60	—	100
Nb	4,058.94	400*w*	1000*w*
P	4,059.27	(100)	—
Fe	4,062.44	100	120
Cu	4,062.70	20	500*w*
Pr	4,062.82	50	150
Mn	4,063.53	60	100
Fe	4,063.60	300	400
Zr	4,064.15	6	100
V	4,065.08	100	2
Kr	4,065.11	(300)	—
Os	4,066.69	100	100
Fe	4,066.98	80	100
La	4,067.39	150	80
Ta	4,067.91	40	100
Fe	4,067.98	100	150
Co	4,068.54	100	150
O	4,069.90	(125)	—
Se	4,070.16	(500)	—
J	4,070.75	(150)	—
Fe	*4,071.74*	*200*	*300*
Ar	*4,072.01*	*(150)*	—
O	*4,072.16*	*(300)*	—
In	*4,072.40*	*200wh*	—
Te	*4,073.57*	*(300)*	—

	λ	I J	I O		λ	I J	I O
Mn	*4,030.75*	*20*	*500r*	*O*	*4,075.87*	*(800)*	—
Fe	*4,030.49*	*60*	*120*	*La*	*4,077.34*	*400*	*600*
Se	*4,030.07*	*(150)*	—	*Sr*	*4,077.71*	*500W*	*400r*
S	*4,028.79*	*(200)*	—	*Hg*	*4,077.81*	*150*	*150*
				Dy	*4,077.97*	*100*	*150r*
Co	*4,027.04*	*4*	*200*				
F	*4,025.49*	*(300)*	—	*Ti*	*4,078.47*	*50*	*125*
F	*4,025.01*	*(150)*	—	*Nb*	*4,079.73*	*200w*	*500w*
F	*4,024.73*	*(500)*	—	*P*	*4,080.04*	*(150)*	—
Co	*4,023.40*	—	*200*	*Ru*	*4,080.60*	*300*	*125*
				Zr	*4,081.21*	*7*	*150*
Cu	*4,022.66*	*25*	*400*				
Fe	*4,021.87*	*100*	*200*	*Fe*	*4,084.50*	*80*	*120*
Co	*4,020.90*	—	*500w*	*Co*	*4,086.31*	*15*	*400*
				La	*4,086.71*	*500*	*500*
				Pd	*4,087.34*	*100*	*500*
				Kr	*4,088.33*	*(500)*	—

Ag 4,668.48 *J* 70 *O* 200

	λ	J	O		λ	J	O
W	**4,668.46**	**3**	**20**	**Xe**	**4,668.49**	**(50)**	—
Pr	**4,668.45**	—	**6**	**S**	**4,668.58**	**(50)**	—
Ti	**4,668.36**	**1**	**8**	**Na**	**4,668.60**	**100**	**200**
Pr	**4,668.230**	**1**	**10**	**Tm**	**4,668.69**	—	**10**
Gd	**4,668.23**	—	**12**	**Mo**	**4,668.80**	**5**	**5**
Dy	**4,668.20**	**2**	**2**	**W**	**4,668.911**	—	**5**
J	**4,668.15**	**(10)**	—	**La**	**4,668.914**	**300r**	**200r**
Th	**4,668.149**	—	**3**	**U**	**4,668.94**	**3**	**3**
Eu	**4,668.14**	—	**15w**	**Ir**	**4,668.99**	—	**20**
Fe	**4,668.14**	**10**	**125**	**Ne**	**4,669.02**	**(50)**	—
Co	**4,667.87**	—	**10**	**Nd**	**4,669.13**	—	**5**
Se	**4,667.80**	**(70)**	—	**Ru**	**4,669.138**	—	**15**
Dy	**4,667.79**	**2**	**2**	**S**	**4,669.14**	**(35)**	—
Ni	**4,667.77**	—	**100**	**Ta**	**4,669.143**	**15**	**300**
Er	**4,667.59**	**2**	**3**	**Fe**	**4,669.18**	**2**	**15**
Ti	**4,667.588**	**8**	**150**	**Hf**	**4,669.24**	**6**	**6**
Os	**4,667.54**	—	**5**	**V**	**4,669.308**	**8**	**10**
Yt	**4,667.462**	—	**7**	**U**	**4,669.309**	**15**	**8**
Fe	**4,667.459**	**20**	**150**	**Cr**	**4,669.34**	**20**	**50**
Mo	**4,667.42**	**20**	—	**Se**	**4,669.38**	**(10)**	—
Eu	**4,667.41**	—	**3**	**Sm**	**4,669.390**	**35**	**40**
Ce	**4,667.364**	—	**2**	**Tb**	**4,669.39**	—	**3**
Ne	**4,667.356**	**(100)**	—	**Dy**	**4,669.40**	**2**	**3**
In	**4,667.32**	**10**	—	**Ce**	**4,669.50**	—	**4**
Cu	**4,667.30**	**5**	—	**Er**	**4,669.52**	—	**3**
N	**4,667.28**	**(5)**	—	**Mo**	**4,669.637**	**3**	**2**
Tb	**4,667.28**	—	**3**	**Ce**	**4,669.638**	—	**3**
Tm	**4,667.28**	—	**15**	**Sm**	**4,668.65**	**40**	**50**
Nb	**4,667.223**	**10**	**15**	**N**	**4,669.77**	**(10)**	—
U	**4,667.217**	**5**	**3**	**Nb**	**4,669.87**	**2**	**3**

	λ	I	
		J	O
Zr	**4,667.14**	—	**5**
Ni	4,666.99	—	50
U	4,666.86	40	25
Ne	4,666.65	(50)	—
J	4,666.52	(250)	—
Te	4,665.33	(70)	—
Se	4,664.98	(150)	—
Na	4,664.86	—	80
Cr	4,664.80	20	70
Pr	4,664.65	15	100
Te	4,664.34	(800)	—
Se	4,664.20	(150)	—
Hf	4,664.12	100	50
Os	4,663.82	5	100
La	4,663.76	200	100
Co	4,663.41	—	700*W*
Ne	4,663.09	(40)	—
Mo	4,662.76	40	40
La	4,662.51	200	150
Eu	4,661.88	—	100
O	4,661.65	(125)	—
Ta	4,661.12	5*h*	300
Ne	4,661.10	(150)	—
Hg	4,660.28	(200)	—
W	4,659.87	70	200
K	4,659.32	(40)	—
Kr	4,658.87	(2000)	—
P	4,658.11	(100)	—
Lu	4,658.02	15	100
Ar	4,657.94	(150)	—
Sb	4,657.91	30	—
Co	4,657.39	35	100
In	4,656.81	(35)	—
S	4,656.74	(80)	—
Ti	4,656.47	70	150
Ne	4,656.39	(300)	—
In	4,655.79	(100)	—
In	4,655.66	(50)	—
In	4,655.52	(45)	—
La	4,655.50	300	150
In	4,655.41	(45)	—
O	4,655.36	(50)	—
Te	4,654.38	(800)	—
Ru	4,654.31	—	125
Ne	4,653.70	(50)	—
Cr	*4,652.16*	*150*	*200R*

	λ	I	
		J	O
Ru	**4,669.98**	—	**40**
Te	**4,670.11**	**(30)**	—
Sc	4,670.40	300*wh*	100
V	4,670.49	40*r*	60*R*
Ne	4,670.88	(70)	—
Xe	4,671.23	(2000)	—
U	4,671.41	30	20
Mn	4,671.69	5	100
La	4,671.83	150	100
Mo	4,671.90	30	30
Pr	4,672.08	25*w*	100
Nb	4,672.09	100	150
Be	4,672.2	100	—
Xe	4,672.20	(50*hl*)	—
As	4,672.70	50	—
O	4,672.75	(30)	—
Be	4,673.46	(100)	—
O	4,673.71	(30)	—
Sm	4,674.59	40	80
Cu	4,674.76	30*W*	200
Yt	4,674.85	100	80
Rh	4,675.03	50	100
Nb	4,675.37	30*w*	50*w*
J	4,675.53	(50)	—
P	4,675.78	(70)	—
O	4,676.25	(125)	—
Xe	4,676.46	(100*wh*)	—
Sm	4,676.91	50	100
J	4,676.94	(80)	—
Cd	4,678.16	200*W*	200*W*
Ne	4,678.22	(300)	—
Ne	4,678.60	(50)	—
Br	4,678.69	(200)	—
Fe	4,678.85	100	150
P	4,678.94	(100)	—
Ne	4,679.13	(150)	—
Zn	4,680.14	200*h*	300*w*
Ne	4,680.36	(100)	—
Kr	4,680.41	(500)	—
W	4,680.52	40	150
Rn	4,680.83	(500)	—
In	4,681.11	(200)	—
Ne	4,681.20	(50)	—
Ru	4,681.79	—	100
Ta	4,681.87	50	200
Ti	4,681.92	100	200
In	4,682.00	250*W*	—

	λ	I J	I O		λ	I J	I O
Xe	*4,651.94*	*(100)*	—	Ra	4,682.28	(800)	—
Pr	*4,651.52*	*40w*	*125*	Yt	4,682.32	100	60
Cr	*4,651.28*	*100*	*100*	Co	4,682.38	—	500
Cu	*4,651.13*	*40*	*250*				
O	*4,650.85*	*(70)*	—	*Gd*	*4,683.34*	50	*100*
Ne	*4,649.90*	*(70)*	—	*Ru*	*4,684.02*	—	*100*
O	*4,649.15*	*(300)*	—	*Ta*	*4,684.87*	*2*	*100*
Ni	*4,648.66*	*3*	*400w*	*Pr*	*4,684.94*	*10w*	*125*
Se	*4,648.44*	*(800)*	—	*In*	*4,685.22*	*(100)*	—
Ru	*4,647.61*	—	*125*	*He*	*4,685.75*	*(300)*	—
Fe	*4,647.44*	*40*	*125*	*Ni*	*4,686.22*	*1*	*200*
Sb	*4,647.32*	*(80)*	—	*Te*	*4,686.95*	*(300)*	—
Cr	*4,646.17*	*150*	*100*	*Sm*	*4,687.18*	—	*100*
Ne	*4,645.42*	*(300)*	—	*Ne*	*4,687.67*	*(100)*	—
Ti	*4,645.19*	*10*	*100*	*Zr*	*4,687.80*	—	*125*
Ru	*4,645.09*	*30*	*100*	*Eu*	*4,688.23*	*2*	*100*
Rn	*4,644.18*	*(300)*	—	*Xe*	*4,690.97*	*(100)*	—
Yt	*4,643.69*	*100*	*50*	*Kr*	*4,691.28*	*(100)*	—
N	*4,643.11*	*(100)*	—	*Ti*	*4,691.34*	*25*	*125*
Sm	*4,642.23*	*40*	*100*	*Ba*	*4,691.62*	*40*	*100*
Ar	*4,642.15*	*(80)*	—	*Ta*	*4,691.90*	*5h*	*400*
O	*4,641.83*	*(150)*	—	*La*	*4,692.50*	*300*	*200*
Te	*4,641.19*	*(70)*	—	*Co*	*4,693.21*	*25*	*500*
Zr	*4,640.6*	—	*150*	*Ta*	*4,693.35*	*3*	*150*
Ne	*4,640.44*	*(70)*	—	*S*	*4,694.13*	*(50)*	—
				Kr	*4,694.44*	*(200hl)*	—
				Xe	*4,697.02*	*(300)*	—
				Gd	*4,697.49*	*4*	*100*

Ag 5,209.97 *J* 1000*R* *O* 1500*R*

	λ	J	O		λ	J	O
Pd	**5,208.91**	—	**10**	**W**	**5,209.10**	—	**10**
Ce	**5,208.90**	—	**2**	**Bi**	**5,209.29**	**600h**	—
Ne	**5,208.86**	**(70)**	—	**Zr**	**5,209.30**	—	**5**
Sb	**5,208.80**	**(8)**	—	**Tb**	**5,209.34**	—	**15**
Bi	**5,208.8**	**70**	**5*wh***	**Cs**	**5,209.44**	**(15)**	—
Fe	**5,208.60**	**8**	**200**	**Ru**	**5,209.50**	—	**7**
Pt	**5,208.59**	—	**2**	**Cs**	**5,209.62**	**(15)**	—
Cr	**5,208.44**	**100**	**500*R***	**Nd**	**5,209.90**	—	**2**
Kr	**5,208.32**	**(500)**	—	**Sm**	**5,209.93**	—	**20**
Ar	**5,208.04**	**(10)**	—	**Co**	**5,210.06**	—	**100**
W	**5,207.974**	—	**5**	**Er**	**5,210.28**	—	**12**
Tb	**5,207.97**	—	**15**	**Ti**	**5,210.39**	**35**	**200**
Pr	**5,207.90**	**1**	**15**	**Mo**	**5,210.44**	**4**	**15**
Eu	**5,207.89**	—	**20*h***	**Ar**	**5,210.49**	**(200)**	—
Ti	**5,207.87**	—	**25**	**Sc**	**5,210.52**	—	**20**
***bh*F**	**5,207.8**	—	**2**	**Ne**	**5,210.57**	**(50)**	—
U	**5,207.79**	—	**4**	**Sb**	**5,210.69**	**4*h***	—

	λ	I	
		J	O
V	**5,207.68**	**8**	**8**
Sm	**5,207.64**	—	**3**
Ce	**5,207.34**	—	**8**
Ar	**5,207.17**	**(10)**	—
Sm	**5,207.15**	—	**25*d***
Cu	**5,207.13**	**20**	—
Cr	5,206.04	200	500*R*
Ra	5,205.93	(250)	—
Fe	5,204.58	—	125
Cr	5,204.52	100	400*R*
La	5,204.15	300	50
Ne	5,203.89	(150)	—
Fe	5,202.34	10	300
Eu	5,200.92	—	300
Sm	5,200.59	1	200
Yt	5,200.41	150	60
Eu	5,199.85	—	500
Fe	5,195.48	—	100*h*
Ru	5,195.02	—	100
Fe	5,194.95	15	200
Ne	5,193.22	(150)	—
Rh	5,193.14	3	200
Ne	5,193.13	(150)	—
Nb	5,193.07	20	100
V	5,192.99	75*h*	100
Ti	5,192.97	25	150
Fe	5,192.36	50	400
Co	5,192.35	—	100*w*
Fe	5,191.47	35	400
P	5,191.40	(100)	—
Xe	5,191.37	(200)	—
Ti	*5,188.70*	*100*	*80*
Ne	*5,188.61*	*(150)*	—
La	*5,188.23*	*500*	*50*
Ar	*5,187.75*	*(800)*	—
In	*5,184.44*	*(300)*	—
Mg	*5,183.62*	*300*	*500wh*
La	*5,183.42*	*400*	*300*
Zn	*5,181.99*	*2*	*200*
Nb	*5,180.31*	*15*	*150*
La	*5,177.31*	*30*	*150*
Co	*5,176.08*	—	*500r*
Se	*5,175.98*	*(600)*	—
Rh	*5,175.97*	*1*	*200*
In	*5,175.56*	*(150)*	—
In	*5,175.42*	*(300)*	—

	λ	I	
		J	O
Sm	**5,210.75**	—	**8**
Hg	**5,210.79**	**(60)**	—
Co	**5,210.84**	—	**50**
Gd	**5,210.99**	—	**12**
Co	5,211.82	—	100
La	5,211.87	5	300*r*
Co	5,212.71	—	300*w*
Eu	5,213.37	—	150
Ar	5,214.77	(200)	—
Eu	5,215.09	—	1000
Fe	5,215.18	5	200
Fe	5,216.28	10	300
Eu	5,217.02	—	125
Fe	5,217.40	3	150
Cu	5,218.20	—	700
Ar	5,221.27	(500)	—
Eu	5,223.48	—	700
Fe	5,226.87	15	200
Cs	5,227.00	(200)	—
Fe	5,227.19	60	400
Se	5,227.51	(600)	—
Fe	*5,229.87*	*15h*	*200*
Co	*5,230.22*	—	*300R*
Fe	*5,232.94*	*150*	*800*
La	*5,234.27*	*8W*	*200r*
Co	*5,247.93*	—	*500w*

	λ	I J	I O
In	5,175.29	(400)	—
Ti	5,173.75	20	125
Mg	5,172.70	100wh	200wh
Fe	5,171.60	60h	300
Ru	5,171.03	—	150

Ag 5,465.49 J 500R O 1000R

	λ	J	O
bhZr	5,465.4	—	20
Ce	5,465.34	—	20
Te	5,465.17	(15)	—
J	5,464.61	(900)	—
W	5,464.46	—	4
Tb	5,464.39	—	10
La	5,464.38	30	25
Fe	5,464.28	—	6
U	5,464.24	—	5
Ce	5,464.20	—	5
Sb	5,464.08	(100)	—
Zr	5,464.06	—	2
Cr	5,463.97	—	15s
Sm	5,463.79	—	2
Sn	5,463.6	2	2h
Hf	5,463.38	10	10
Fe	5,463.28	—	100
Gd	5,463.277	—	8
Hg	5,460.74	(2000)	—
Xe	5,460.39	(200)	—
Ar	5,457.37	(200)	—
S	5,453.88	(750)	—
Eu	5,452.96	—	1000s
Ar	5,451.65	(500)	—
Eu	5,451.53	—	1000s
Co	5,444.57	—	400w
Ar	5,442.22	(500)	—
Ar	5,439.97	(500)	—
Xe	5,438.96	(400)	—
S	5,432.83	(600)	—
Ar	5,421.35	(500)	—
Lu	5,465.50	—	3
Tm	5,465.53	—	20
Mo	5,465.57	5	20
U	5,465.69	8	12
Cs	5,465.9	—	5wh
bhZr	5,466.0	—	8
Br	5,466.23	(150)	—
Fe	5,466.41	—	25
Er	5,466.46	—	12
Yt	5,466.47	20	150
Sm	5,466.73	—	80
La	5,466.93	—	3
Fe	5,466.95	5	—
Eu	5,467.05	—	10W
Ar	5,467.13	(6)	—
Sm	5,467.21	—	8
Nd	5,467.50	—	2
W	5,467.55	—	8
Kr	5,468.17	(200hs)	—
Ag	5,471.55	100	500h
Eu	5,472.33	—	1000s
Xe	5,472.61	(500)	—
Ar	5,473.44	(500)	—
S	5,473.63	(750)	—
Lu	5,476.69	1000	500
Co	5,483.34	—	500w
Eu	5,488.65	—	500
Ar	5,495.87	(1000)	—
J	5,496.92	(900)	—
Ar	5,506.11	(500)	—

Al 13 26.9815

t_0 660°C

t_1 2,056°C

I.	II.	III.	IV.	V.
5.985	18.824	28.442	119.961	154.28

λ	*I*		eV
	J	*O*	
I 2,373.36	100*R*	**200R**	5.22
II 2,669.17	100	3	10.6
2,816.18	100	10	17.8
I 3,082.15	800	800	4.0
I 3,092.71	1000	1000	4.0
I 3,944.03	1000	2000	3.1
I 3,961.53	2000	3000	3.1
II 6,243.36	80	—	21.0

Al 2,373.36 *J* 100*R* *O* 200*R*

	λ	*I*	
		J	*O*
Pt	**2,373.32**	**10**	—
Re	**2,373.29**	**3**	**7**
Pr	**2,373.25**	**8**	—
Rb	**2,373.22**	**(70)**	—
Ir	**2,373.172**	**10*wh***	**5**
Nd	**2,373.17**	**7**	—
Au	**2,373.14**	**5**	—
Al	**2,373.13**	**30**	**100*R***
Os	**2,373.12**	**10**	—
Ba	**2,373.11**	**2*w***	**12**
Co	**2,373.09**	**6**	**4**
Nb	**2,373.07**	—	**5**
Yb	**2,373.06**	**5**	**1**
V	2,373.06	200	—
In	2,373.04	(10*h*)	—
Ir	2,372.77	40	100
Nb	2,372.72	50	2
Ce	2,372.37	50	—
V	2,372.17	100	—
Pd	2,372.15	50	5
V	2,371.07	500	—
Te	2,369.92	(50)	—
Ir	*2,368.04*	*125*	*25*
Al	*2,367.06*	*50R*	*150R*

	λ	*I*	
		J	*O*
Mn	**2,373.37**	**50**	**4**
Co	**3,373.38**	**2**	**20**
Cs	**2,373.4**	**(2)**	—
W	**2,373.43**	**4**	**4**
Re	**2,373.52**	**15**	**7**
Os	**2,373.62**	**5**	**4**
Sb	**2,373.623**	**25**	**75**
Fe	**2,373.624**	—	**2**
Kr	**2,373.68**	**(4)**	—
Ag	**2,373.70**	**2*h***	—
Fe	**2,373.73**	**15**	**6**
Cr	**2,373.73**	—	**60**
Ag	2,375.06	300*wh*	300*wh*
Ru	2,375.63	80	50
Nb	2,376.40	60	8
Fe	2,376.43	50	3
Tl	*2,379.69*	*200R*	*100R*

Al 2,669.17 *J* 100 *O* 3*

	λ	*J*	*O*
Ne	**2,669.13**	**(3)**	—
Hf	**2,669.00**	**8**	**15**
Ir	**2,668.99**	**5**	**50**
W	**2,668.95**	**7**	**3**
V	**2,668.89**	**4**	**2**
Cs	**2,668.76**	**(8)**	—
Yb	**2,668.74**	**20**	**2**
Cr	**2,668.71**	**12**	**10**
In	**2,668.68**	**(50)**	—
In	**2,668.623**	**(30)**	—
Ta	2,668.619	100	1
Eu	2,668.33	400	300*W*
Ru	2,667.79	80	—
Ru	2,667.39	150	10
Mn	2,667.00	50*h*	5

	λ	*J*	*O*
U	**2,669.174**	**4**	**6**
Rh	**2,669.20**	**15**	**1**
Tm	**2,669.26**	**8**	**2**
Ti	**2,669.263**	**1**	**15**
Tb	**2,669.29**	**3**	**10**
W	**2,669.30**	**30*d***	**15**
Mn	**2,669.31**	**20**	—
Ne	**2,669.36**	**(5)**	—
Cr	**2,669.37**	**1**	**6**
Ru	**2,669.42**	**100**	—
Ir	**2,669.46**	**5**	**10**
Zr	**2,669.494**	**2**	—
Fe	**2,669.496**	**25**	**50**
Os	**2,669.53**	**3**	**8**
Hf	**2,669.56**	**4**	**2**
J	**2,669.59**	**(20)**	—

	λ	I J	I O
Yb	2,666.97	150	5
Mn	2,667.77	50*h*	8
Fe	2,666.64	80	5
Nb	2,666.59	50	5
Yb	2,666.08	150	5
Nb	2,665.251	300	3
Ni	2,665.249	125	—
Yb	2,665.03	60	10
Ir	2,664.79	50	200*h*
Fe	2,664.66	300	20
Ta	*2,664.23*	*100h*	*10h*
V	*2,663.25*	*100*	*12*
Ru	*2,661.61*	*150*	*80*
Ru	*2,661.17*	*100*	*20*
Pd	*2,661.14*	*100*	—
Mo	*2,660.58*	*125*	*25*
Ag	*2,660.46*	*150*	*30*
Pt	*2,659.45*	*500R*	*2000R*
Ti	**2,669.60**	**12**	**60**
Sb	**2,669.64**	**3**	—
Co	2,669.91	100*Wh*	—
V	2,670.23	70	2
Ni	2,670.33	80	—
Mn	2,671.80	50	—
Na	2,671.829	(60)	12
Mo	2,671.834	100	1
Nb	2,671.93	200	20
V	2,672.00	300*R*	50
Cl	2,672.19	(50)	—
Mn	2,672.59	125*h*	15
Yb	2,672.65	80	20
Mo	2,672.84	100	15
Ru	2,673.01	50	8
V	2,673.23	60	10
Mo	2,673.27	100	1
Mn	2,673.37	50	—
Eu	2,673.41	50	125
Nb	*2,673.57*	*500*	*10*
Rh	*2,674.44*	*200W*	*1*
Ne	*2,675.24*	*(150)*	—
Ne	*2,675.64*	*(150)*	—
Ta	*2,675.90*	*200*	*150*
Nb	*2,675.94*	*100*	*10*
Au	*2,675.95*	*100*	*250R*
Ru	*2,676.19*	*100*	*8*
Rh	*2,676.25*	*100*	*1*
Cl	*2,676.95*	*(150)*	—
Pt	*2,677.15*	*200W*	*800W*
Cr	*2,677.16*	*300r*	*35*
V	*2,677.80*	*300R*	*70*
Eu	*2,678.28*	*100*	*150*
V	*2,678.57*	*150R*	*30*
Zr	*2,678.63*	*100*	*80*
Ru	*2,678.76*	*300*	*100*

Al 2,816.79 *J* 100 *O* 10*

	λ	J	O
Mo	**2,816.15**	**300*h***	**200**
Th	**2,864.08**	**5**	**10**
Ba	**2,816.07**	**30**	—
Hf	**2,816.069**	**6**	**10**
U	**2,815.98**	**6**	**8**
Eu	**2,816.18**	**50**	**50*W***
Mn	**2,816.32**	**2**	**1**
Ca	**2,816.33**	**3**	—
Yb	**2,816.33**	**2**	—
Ho	**2,816.38**	**10**	—

	λ	I J	I O		λ	I J	I O
Os	**2,815.78**	**4**	**40**	**Dy**	**2,816.39**	**2**	**5**
U	**2,815.76**	**8**	**6**	**U**	**2,816.42**	**6*h***	**3**
				Kr	**2,816.46**	**(60)**	**—**
				O	**2,816.52**	**(25)**	**—**
Ag	2,815.54	80 *Wh*	3	**Tm**	**2,816.56**	**10**	**2**
Mn	2,815.02	75	25				
Hg	2,814.93	(200)	—	**Fe**	**2,816.66**	**4**	**5**
Ta	2,814.31	50	50*r*	**W**	**2,816.67**	**1**	**5**
Eu	2,813.95	300 *Wh*	300 *Wh*	**Nb**	**2,816.68**	**50**	**2**
Ra	2,813.76	(400)	—				
Ru	2,813.71	125	50	Ta	2,817.10	100	80*d*
Fe	2,813.61	60	5	V	2,817.50	50	18
Sn	2,813.58	50	50	Fe	2,817.51	60	100
Ru	2,813.30	75	—	Ti	2,817.87	200	10
				Cr	2,818.36	80	8
Fe	2,813.29	400	400				
Cr	2,812.00	80	8	Yb	2,818.75	80	—
Ta	2,811.72	150	1	Au	2,819.95	150	—
Co	2,810.86	75*h*	5	Ti	2,820.00	70 *Wh*	—
Nb	2,810.81	100	3	Hf	2,820.22	100	40
				Eu	2,820.77	200 *W*	200 *W*
Ru	*2,810.55*	*200*	*50*	Lu	2,821.32	50*h*	2
Ti	*2,810.30*	*150*	*6*	Ni	2,821.29	125	125
Fe	*2,809.81*	*100h*	*1*	Ti	2,821.42	70 *Wh*	—
Bi	*2,809.62*	*100*	*200 W*	Ru	2,821.424	100	5
La	*2,808.39*	*150*	*10d*				
Fe	*2,806.98*	*200*	*200*	*Cr*	*2,822.37*	*100*	*20*
Ru	*2,806.74*	*100*	*50*	*Ru*	*2,822.55*	*150*	*30*
Ni	*2,805.67*	*200h*	—	*Fe*	*2,823.28*	*300*	*200*
				Cu	*2,824.369*	*300*	*1000*
				Ag	*2,824.37*	*200 W*	*150 Wh*
				Ni	*2,825.236*	*125*	—
				Co	*2,825.24*	*200*	*5*
				Fe	*2,825.56*	*150*	*150*
				Tl	*2,826.16*	*100R*	*200R*
				Ru	*2,826.68*	*100*	—

Al 3,082.15 *J* 800 *O* 800

	λ	J	O		λ	J	O
V	**3,082.11**	**2*h***	**80*r***	**Yt**	**3,082.17**	**10*d***	**4**
Er	**3,082.08**	**3**	**12**	**Th**	**3,082.18**	**10**	**10**
Mn	**3,082.05**	**—**	**50**	**Mo**	**3,082.22**	**40**	**4**
Th	**3,082.03**	**8**	**12**	**Sm**	**3,082.26**	**2*h***	**1**
U	**3,082.02**	**8**	**8**	**Rb**	**3,082.27**	**—**	**10**
Y	**3,082.01**	**2*h***	**15**	**Rn**	**3,082.320**	**(10)**	**—**
Gd	**3,082.00**	**60**	**100**	**Ce**	**3,082.304**	**2**	**20**
Ce	**3,081.984**	**—**	**5**	**Tb**	**3,082.36**	**8**	**3**
S	**3,081.98**	**(15)**	**—**	**Re**	**3,082.43**	**—**	**100 *W***
Mo	**3,081.95**	**—**	**25**	**Ta**	**3,082.45**	**1*h***	**15*h***

	λ	I J	I O		λ	I J	I O
W	3,081.86	7	10	Dy	3,082.51	5	15
Ta	3,081.85	5	50	V	3,082.52	30	3
Fe	3,081.84	2	2	Sc	3,082.56	2*h*	2
Nb	3,081.77	15	1	Tm	3,082.57	—	10
U	3,081.67	4	6	U	3,082.59	4	6
J	3,081.664	(100)	—	Ti	3,082.615	—	6
Th	3,081.662	5	10	Co	3,082.618	50	150*R*
Fe	3,081.656	2*h*	2*h*	Xe	3,082.62	(12)	—
Mo	3,081.65	20	—	Cd	3,082.68	—	30
Yt	3,081.60	2	—	Mn	3,082.70	12	12
Cd	3,081.58	2	—	Re	3,082.77	—	2
Ti	3,081.57	40*Wh*	—	Co	3,082.84	—	35
Fe	3,081.55	2	2	Nb	3,082.86	5	3
Sc	3,081.548	2*h*	1	J	3,082.87	(25)	—
Ca	3,081.547	—	2				
Lu	3,081.47	8	80	Fe	3,083.74	500	500
O	3,081.46	(5*h*)	—	Ti	3,088.02	500*R*	70
La	3,081.42	4	7	*N**			
Ni	3,080.75	60	200				
J	3,078.77	(350)	—				
Ti	3,078.64	500*R*	60				
Fe	3,077.17	300	1				
Fe	3,075.72	400	400				
Ti	3,075.22	300*R*	40				
Bi	*3,067.72*	*2000Wh*	*3000hR*				

Al 3,092.713 *J* 1000 *O* 1000

	λ	I J	I O		λ	I J	I O
Yt	3,092.712	—	8	U	3,092.720	3	5
Mo	3,092.70	2	20	V	3,092.720	50*r*	100*r*
Yb	3,092.56	30	—	Ce	3,092.724	—	4
Sc	3,092.52	4	3	Na	3,092.73	(200)	50
Ta	3,092.44	15	50	Fe	3,092.78	30	50
Se	3,092.42	—	3	Ce	3,092.82	1	4
Xe	3,092.41	(10)	—	Al	3,092.84	18	50*R*
Ir	3,092.401	—	5	Nb	3,092.89	5	1*h*
Fe	3,092.399	5	4	Ne	3,092.91	(4)	—
Cd	3,092.393	15	10	Nd	3,092.915	6	8
Te	3,092.32	(15)	—	Mo	3,092.92	10*d*	—
W	3,092.28	3	7	Mg	3,092.991	20	125
Hf	3,092.245	20	20	Ta	3,092.994	1	18
Zr	3,092.24	—	3	Cl	3,093.00	(4)	—
Cl	3,092.22	(50)	—	Rn	3,093.00	(18)	—
Ce	3,092.20	—	3	U	3,093.01	20	20

	λ	I			λ	I	
		J	O			J	O
Mo	3,092.07	100	30	Th	3,093.05	6	12
Gd	3,092.064	2	3	Dy	3,093.108	5	15
Ta	3,092.057	1	15	V	3,093.108	400*R*	100*R*
Ce	3,092.03	—	2	Tm	3,093.12	60	30
				Ce	3,093.24	—	6
P*				Si	3,093.28	6	—
				Au	3,093.30	5	—
				Zr	3,093.32	—	3
				Ce	3,093.34	—	12
				Fe	3,093.36	40	70
				U	3,093.37	3*d*	4*d*
				Ar	3,093.41	(50)	—
				Yb	3,093.44	2	—
				Ru	3,093.45	—	4
				Ca	3,093.46	3	2
				Nb	3,094.18	1000	100
				Ni	*3,101.55*	*150*	*1000R*

Al 3,944.03 *J* 1000 *O* 2000

	λ	J	O		λ	J	O
Rh	3,944.019	4	5	Ir	3,944.06	—	2
Re	3,944.016	—	15	Se	3,944.07	(20)	—
Er	3,944.014	—	2	Ce	3,944.09	—	8
Eu	3,944.01	—	8*h*	Ni	3,944.126	—	5*Wh*
Ce	3,943.89	15	40	U	3,944.13	15	8
U	3,943.82	5	35	Pr	3,944.14	5	9
Pb	3,943.80	5	—	La	3,944.15	2	—
Pr	3,943.75	8	10	Ru	3,944.19	4	10
Th	3,943.69	10	10	Tb	3,944.20	—	6
Nb	3,943.67	50	20	Th	3,944.25	1	3
V	3,943.664	18	50	Ar	3,944.27	(50)	—
Tb	3,943.66	—	8	Ga	3,944.29	2	—
Gd	3,943.63	20	20	F	3,944.33	(20)	—
Sm	3,943.62	3	8	Re	3,944.35	—	15
Cr	3,943.61	4	18	Ir	3,944.37	4	20
Eu	3,943.60	—	3	Nd	3,944.42	—	20
Fe	3,943.59	—	2*h*	Er	3,944.42	—	12
Xe	3,943.57	(10)	—	Sb	3,944.58	8	—
Mo	3,943.51	4	5	Eu	3,944.59	—	5*W*
U	3,943.498	10	6	Pr	3,944.617	3	8
Ce	3,943.497	1	6	U	3,944.62	2	8
As	3,943.47	3	—	Yt	3,944.687	—	3
Pr	3,943.37	3	5	Dy	3,944.69	150	300
Fe	3,943.348	8	40	Re	3,944.73	—	30

	λ	I	
		J	O
Gd	3,943.246	40	40
Sm	3,943.23	25	40
Ir	3,943.20	—	6
Er	3,943.19	—	10
Ce	3,943.14	3	12
Mo	3,943.09	10	10
Eu	3,943.08	15	50
Hf	3,943.06	—	3
Mo	3,943.04	6	10
Ce	3,943.00	—	3
U	3,942.95	2	4
Ag	3,942.95	10*h*	5*h*
Tb	3,942.94	—	6
Eu	3,942.938	—	15*W*
Kr	3,942.93	20*Wh*	—
Pr	3,942.92	6	15
Mn	3,942.85	75	75
Ca	3,942.84	3	2
U	3,942.83	8	20
Ce	3,942.75	20	50
Rh	3,942.72	25	60
Mo	3,942.71	20	—
Co	3,942.692	—	8
Eu	3,942.69	—	2*W*
Sm	3,942.66	2	10
Th	3,942.65	8	10
Gd	3,942.64	30	60
Nd	3,942.63	25	25
Hg	3,942.59	(100)	—
Re	3,942.56	—	10
U	3,942.55	10	8
Dy	3,942.54	5	30
Er	3,942.53	—	3
J	3,940.24	500	—
Ca	3,933.67	600*R*	600*R*
J	3,931.01	(400)	—
Eu	3,930.50	400*R*	1000*R*
Fe	3,930.30	400	600
Sm	3,944.74	5	5
Fe	3,944.75	1	4
Ru	3,944.78	—	4
W	3,944.80	6	7
Ce	3,944.83	—	5
Fe	3,944.896	8	15
Pr	3,944.899	12	30
Ce	3,944.92	—	6
Co	3,944.95	2	5*h*
O	3,945.04	(20)	—
Hg	3,945.09	(100)	—
Fe	3,945.13	10	30
Eu	3,945.16	—	5
V	3,945.17	—	2
W	3,945.21	6	5
Mo	3,945.25	6	10
Co	3,945.33	15	200
Hf	3,945.36	6*h*	8
Ce	3,945.364	—	2
Tb	3,945.39	—	8
Pr	3,945.42	4	12
Kr	3,945.48	(5*h*)	—
Cr	3,945.49	15	50
Ce	3,945.507	1	3
Th	3,945.515	15	20
J	3,945.52	(2)	—
O	3,947.33	(300)	—
Ar	3,947.50	(1000)	—
Ar	3,948.98	(2000)	—
La	3,949.11	800	1000
*N**			

Al 3,961.53 *J* 2000 *O* 3000

	λ	J	O
Ru	3,961.52	4	6
U	3,961.51	1	8
Mo	3,961.50	500	5
Ce	3,961.39	—	2
Ag	3,961.33	5	15
Yb	3,961.54	—	30
S	3,961.55	(10)	—
Zr	3,961.59	8	500
Nb	3,961.62	10*h*	—
Cl	3,961.62	(12)	—

	λ	I J	I O		λ	I J	I O
Pr	3,961.28	5	10	U	3,961.66	3	1
Er	3,961.21	—	6	Ce	3,961.661	2	6
W	3,961.18	9	—	W	3,961.76	5	5
Fe	3,961.15	7	25	Sm	3,961.81	10	25
Eu	3,961.14	—	50*W*	Tb	3,961.97	—	3
U	3,961.06	—	6	U	3,961.98	2	6
Re	3,961.03	—	30	Mo	3,961.988	—	3
Os	3,961.02	20	125	Eu	3,961.99	—	10*W*
Co	3,960.99	10	60	La	3,962.03	3	2
Nb	3,960.98	5	8	In	3,962.04	(25)	—
Th	3,960.95	1	2	Ce	3,962.09	4	15
Ce	3,960.91	8	40	Gd	3,962.10	25	25
Fe	3,960.89	3*h*	3	Ni	3,962.12	—	10*h*
W	3,960.87	10	—	Sm	3,962.14	3	5
U	3,960.80	8	5	Nb	3,962.15	—	3
Cr	3,960.76	8	40	In	3,962.159	(18)	—
Eu	3,960.73	—	3	U	3,962.16	—	5
Tb	3,960.69	—	5	Nb	3,962.163	10*h*	—
Pr	3,960.60	25*h*	50	Zr	3,962.17	—	2
Re	3,960.58	—	15	Pr	3,962.186	3	9
Ir	3,960.55	—	4	Cr	3,962.187	1	8
Sb	3,960.55	18*h*	—	Nd	3,962.22	20	30
U	3,960.52	10	6	Sm	3,962.24	5	10
Sm	3,960.51	3	5	U	3,962.271	2	4
Os	3,960.506	15	50	In	3,962.275	(25)	4
Br	3,960.47	(6)	—	Ca	3,962.28	2	3*W*
Ca	3,960.46	2*h*	3	W	3,962.33	9	9
U	3,960.39	8	6	Kr	3,962.34	(10*h*)	—
Ce	3,960.38	2	10	mh	3,962.346	8	8
Th	3,960.34	15	10	Fe	3,962.35	1	7
Tb	3,960.30	—	2	In	3,962.42	(5)	—
Fe	3,960.28	6	30	Ir	3,962.443	—	6
Hg	3,960.24	(30)	—	Pr	3,962.445	50	60
Zr	3,960.20	1	2	U	3,962.47	4	1
U	3,960.17	5	6	Re	3,962.48	—	100
Mo	3,960.14	5	—	In	3,962.59	(25)	—
Sm	3,960.12	3	5	Dy	3,962.60	—	8
Gd	3,960.11	20	20	In	3,962.609	(25)	—
Tb	3,960.09	3	3	Tb	3,962.609	—	5
Mo	3,959.95	2	3	Sr	3,962.61	—	10*h*
Er	3,959.908	1	3	Fe	3,962.72	—	2*h*
Ta	3,959.906	—	5	Sn	3,962.75	5	—
				Ir	3,962.78	—	20
				U	3,962.79	8	15
				Ti	3,962.85	35	80
P*				Ce	3,962.90	—	3
				Tb	3,962.94	—	4
				Sm	3,963.0	40	50
				U	3,963.037	2	—

	λ	I			λ	I	
		J	O			J	O
				La	**3,963.04**	**5**	—
				Fe	**3,963.109**	**50**	**125**
				Nd	**3,963.114**	**25**	**30**
				S	**3,963.13**	**(10)**	—
				Dy	**3,963.16**	—	**3**
				Eu	3,971.99	—	1000*RWh*

Al 6,243.36 *J* 80

	λ	J	O		λ	J	O
J	**6,243.28**	**(15)**	—	**Ar**	**6,243.39**	**(6)**	—
Te	**6,243.24**	**(15)**	—	**Kr**	**6,243.53**	**(2*W*)**	—
Ar	**6,243.13**	**(15)**	—	**V**	**6,243.55**	**2**	**10**
V	**6,243.10**	**4**	**35**	**Ba**	**6,243.85**	**2**	**6**
V	**6,242.811**	**10**	**20**	**J**	**6,244.11**	**(15)**	—
J	**6,242.81**	**(30)**	—	**Pr**	**6,244.34**	**1**	**10**
Cl	**6,242.66**	**(2)**	—	**J**	**6,244.55**	**(30)**	—
N	**6,242.52**	**(70)**	—	**Se**	**6,244.61**	**(30)**	—
Lu	**6,242.34**	**200**	**40**				
Hg	**6,242.24**	**(10)**	—				
				Te	6,245.61	(150)	—
Se	**6,242.21**	**(80)**	—	Ne	6,246.73	(100)	—
				Yb	6,246.97	60	40
				Ra	6,247.16	(50)	—
F	6,239.64	(300)	—	Hf	6,248.95	100	80
J	6,236.40	(50)	—				
Lu	6,235.36	100	25				
J	6,232.85	(50)	—	*Ti*	*6,258.10*	*100*	*200*
Al	6,231.76	(35)	—	*Ti*	*6,258.70*	*250*	*300*
				Ne	*6,258.80*	*(100)*	—
				Ti	*6,261.10*	*100*	*300*
Te	*6,230.80*	*(300)*	—	*La*	*6,262.30*	*150*	*125*
Al	*6,226.19*	*(25)*	—				
Lu	*6,221.87*	*1000*	*500*	*Se*	*6,266.23*	*(200)*	—
				Ne	*6,266.49*	*(1000)*	—

Ar $^{18}_{39.948}$

t_0 −189.2°C t_1 −185.7°C

I.	II.	III.	IV.	V.
15.736	27.619	40.68	61	78

λ	I	eV
II 3,545.58	(300)	23.25
I 3,606.52	(1000)	15.06
I 3,948.98	(2000)	14.68
I 4,266.29	(1200)	14.53
I 4,702.32	(1200)	14.46
I 6,965.43	(400)	13.32
I 7,067.22	(400)	13.29
I 7,503.87	(700)	13.47
I 8,115.31	(5000)	13.07

Ar 3,545.58 (300)

	λ	I J	I O
Ca	3,545.58	4	3
J	3,545.48	(10)	—
Tb	3,545.40	3	8
Nb	3,545.396	5*w*	3*w*
Rb	3,545.37	(2)	—
Th	3,545.35	3	3
Th	3,545.29	3	5
W	3,545.23	10	12
V	3,545.20	300*R*	40
Hg	3,545.06	2	—
Co	3,545.04	30	2
Ti	3,544.99	2	10
Sm	3,544.983	2	5
Gd	3,544.981	3	3
Yt	3,544.980	5	7
W	3,544.97	6	7
Cu	3,544.963	6	35
Pr	3,544.96	10	30
Th	3,544.95	2	5
W	3,544.80	5	7
Eu	3,544.77	1	10*w*
Ba	3,544.71	5	20
Nb	3,544.65	15	20
Fe	3,544.63	6	50
Mo	3,544.62	2	40*h*
Os	3,544.58	10	12
Kr	3,544.54	(30*wh*)	—
Fe	3,541.09	200	200
Nb	3,540.96	500	15
Fe	3,536.56	200	300
Ar	3,535.33	(15)	—
Nb	3,535.30	500	300
Tl	*3,529.43*	*800*	*1000*

	λ	I J	I O
Ce	3,545.60	1	10
N	3,545.62	(5)	—
Fe	3,545.64	70	90
U	3,545.67	8	3
Dy	3,545.74	2	6
As	3,545.77	5	—
Ce	3,545.78	1	8
Gd	3,545.79	125	125
Fe	3,545.83	1	2
Rn	3,545.84	(5)	—
Ar	3,545.84	(125)	—
Ce	3,545.91	1	5
Ho	3,545.97	20	20
Mo	3,546.00	25	—
Yt	3,546.01	2*h*	2
Pr	3,546.02	1	8
Nb	3,546.03	8	5
Tb	3,546.05	3	15
U	3,546.13	6	12
Os	3,546.13	5	8
Nb	3,546.16	3	2
Ce	3,546.19	2	20
Th	3,546.26	3	5
Pr	3,546.28	2	6
U	3,546.38	15	1
Cu	3,546.45	1*h*	2
Kr	3,546.46	(3)	—
Nb	3,546.49	5	5
W	3,546.49	7	3
Tb	3,546.52	8	50
Ti	3,546.54	1	4
U	3,546.55	3	5
Hg	3,549.42	(200)	—
Ar	3,553.58	(15)	—
Ar	3,554.31	(300)	—
Lu	3,554.43	150	50
Fe	3,554.93	300	400
Ar	3,555.97	(100)	—
Fe	*3,558.52*	*300*	*400*
Ar	*3,559.51*	*(15)*	—
Ar	*3,561.04*	*(15)*	—
Ar	*3,563.26*	*(100)*	—
Ar	*3,564.27*	*(100)*	—
Fe	*3,565.38*	*300*	*400*

Ar 3,606.52 (1000)

	λ	I J	I O
Nb	3,606.49	5	3*h*
Yb	3,606.47	60	15
La	3.606.43	3*h*	2
Nd	3,606.40	2	12
Nb	3,606.35	3*h*	—
W	3,606.34	8	10
Sm	3,606.33	1	2
U	3,606.32	12	8
Mo	3,606.29	10	1
Nd	3,606.28	4	12
Nb	3,606.27	5*wh*	2
Eu	3,606.21	1*h*	2*h*
Th	3,606.20	5*h*	—
Ru	3,606.15	1	2
Dy	3,606.13	100	200
Tb	3,606.12	15*d*	30
U	3,606.116	6	1
Ho	3,606.09	4	—
Er	3,606.08	4	15*d*
W	3,606.07	10	12
Ti	3,606.06	1	12
Pr	3,605.96	3	10
Fe	3,605.91	1	3
Ar	3,605.89	(15)	—
Rh	3,605.86	30	25
U	3,605.82	8	8
Hg	3,605.80	(200)	—
U	3,605.75	4	1
Th	3,605.66	6	8
Ru	3,605.64	9	2
Gd	3,605.62	15	20
Rn	3,605.6	(10)	—
V	3,605.59	20*h*	30
Cs	3,605.53	(4)	—
U	3,605.48	4	1
Yt	3,605.46	6*h*	2*h*
Fe	3,605.46	150	300
Cr	3,605.33	400*R*	500*R*
Ar	3,599.67	(20)	—
Cr	*3,593.49*	*400R*	*500R*
Ar	*3,588.44*	*(300)*	—
Br	3,606.66	(15)	—
Mo	3,606.67	20	8
W	3,606.68	2	—
Fe	3,606.682	150	200
V	3,606.69	70	80
Th	3,606.72	2	1
Ti	3,606.79	4	12
F	3,606.80	(10)	—
Nb	3,606.81	1	3
Mo	3,606.91	30	—
Hg	3,606.92	(30)	—
U	3,606.95	1	3
Nb	3,607.01	10	1
Gd	3,607.066	20	15
W	3,607.066	12	10
Ti	3,607.13	5	25
Pr	3,607.22	2	10
F	3,607.32	(6)	—
Nb	3,607.33	5	2
Tm	3,607.36	10	8
Zr	3,607.38	9	8
Th	3,607.39	4	3
Ta	3,607.406	35	70
Xe	3,607.41	(5)	—
Mo	3,607.412	4	4
J	3,607.51	(18)	—
U	3,607.52	2	—
Mn	3,607.54	40	75
Au	3,607.54	20	—
Hg	3,607.60	(18)	—
Ce	3,607.62	8	15
Fe	3,608.86	400	500
Pd	3,609.55	700*R*	1000*R*
Cd	3,610.51	500	1000
Cd	3,612.87	500	800
Ar	*3,622.15*	*(15)*	—

Ar 3,948.98 (2000)

	λ	I			λ	I	
		J	O			J	O
Th	**3,948.97**	**30**	**30**	**La**	**3,949.11**	**800**	**1000**
Ca	**3,948.90**	**15**	**40**	**Eu**	**3,949.12**	**1**	**25***w*
Cr	**3,948.85**	**2**	**25**	**Fe**	**3,949.15**	**1**	**4**
Se	**3,948.80**	**(25)**	—	**Gd**	**3,949.206**	**20**	**10**
Fe	**3,948.779**	**100**	**150**	**W**	**3,949.213**	**3**	—
Nd	**3,948.778**	**8**	—	**Tm**	**3,949.27**	**5**	**50**
As	**3,948.74**	**50**	—	**U**	**3,949.31**	**6**	**2**
Xe	**3,948.72**	**(10)**	—	**Nb**	**3,949.33**	**2***h*	**3**
Ti	**3,948.67**	**40**	**80***h*	**Ce**	**3,949.38**	**3**	**20**
Mo	**3,948.65**	**10**	**6**	**Ru**	**3,949.42**	**5**	**10**
Pr	**3,948.62**	**2**	**9**	**Pr**	**3,949.438**	**100**	**150**
Pt	**3,948.39**	**5**	**60**	**Ag**	**3,949.44**	**8**	**3**
Tb	**3,948.35**	**15**	**20**	**Nb**	**3,949.45**	**50**	**1**
Nd	**3,948.325**	**10**	**20**	**Tb**	**3,949.509**	**3**	**10**
Ir	**3,948.317**	**3**	**10**	**Ba**	**3,949.51**	**(20)**	—
Hg	**3,948.29**	**(100)**	—	**U**	**3,949.52**	**8**	**2**
W	**3,948.20**	**5**	—	**Ca**	**3,949.57**	**2**	—
Xe	**3,948.16**	**(60)**	—	**Cr**	**3,949.58**	**1**	**5**
C	**3,948.15**	**2**	—	**Os**	**3,949.78**	**10**	**50**
Sm	**3,948.108**	**50**	**50**	**Sm**	**3,949.84**	**2**	**2**
Fe	**3,948.107**	**50**	**125**	**J**	**3,949.90**	**(20)**	—
Th	**3,948.04**	**1**	**3**	**Ca**	**3,949.92**	**2***r*	—
W	**3,947.98**	**9**	**10**	**Nb**	**3,949.93**	**3**	**4**
Ce	**3,947.97**	**3**	**20**	**Fe**	**3,949.957**	**100**	**150**
U	**3,947.96**	**8**	**2**	**Cl**	**3,949.96**	**(10)**	—
Te	**3,947.93**	**(10)**	—	**Ru**	**3,950.00**	**8**	**10**
Sm	**3,947.83**	**8**	**15**	**W**	**3,950.12**	**12**	**9**
Ti	**3,947.77**	**35**	**70**	**U**	**3,950.13**	**6**	**3**
Kr	**3,947.66**	**(5***h***)**	—	**Ru**	**3,950.21**	**15**	**12**
Th	**3,947.65**	**1**	**3**	**V**	**3,950.23**	**10**	**20**
Pr	**3,947.63**	**60***d*	**124***d*	**Mo**	**3,950.26**	**4**	**4**
O	**3,947.61**	**(18)**	—	**Er**	**3,950.35**	**10***h*	**30**
Fe	**3,947.532**	**20**	**70**	**Yt**	**3,950.36**	**100**	**60**
Nb	**3,947.528**	**10**	**10**	**Th**	**3,950.39**	**30**	**30**
O	**3,947.51**	**(50)**	—	**Ru**	**3,950.41**	**10**	**10**
U	**3,947.509**	**10**	**10**	**Nd**	**3,950.417**	**10**	**20**
Ar	**3,947.50**	**(1000)**	—	**Ce**	**3,950.424**	**3**	**10**
Ar	3,946.10	(25)	—	Ar	3,952.74	(15)	—
Ar	3,944.27	(50)	—	Al	3,961.53	2000	3000
Al	3,944.03	1000	2000				
Ar	*3,932.55*	*(25)*	—	*Ar*	*3,968.36*	*(200)*	—
Ar	*3,931.24*	*(15)*	—	*Ar*	*3,974.76*	*(15)*	—
Ar	*3,928.62*	*(125)*	—				

Ar 4,266.29 (1200)

	λ	I J	I O
Mo	4,266.18	20	25
Ho	4,266.04	2	5
Nb	4,266.02	20	15
Mn	4,265.92	50	100
Dy	4,265.83	2	6
U	4,265.63	2	3
Ar	4,265.52	(2)	—
Ir	4,265.30	2	60
V	4,265.17	15	30
As	4,265.13	30	—
Ra	4,265.12	(6)	—
Ar	4,259.36	(1200)	—
Cr	*4,254.35*	*1000R*	*5000R*
Ar	*4,251.18*	*(800)*	—
Ar	*4,249.37*	*(20)*	—

	λ	I J	I O
Sm	4,266.31	1	9
Tb	4,266.35	1	30
Lu	4,266.40	2h	—
Dy	4,266.52	2h	2
Ar	4,266.53	(200)	—
W	4,266.54	8	15
Mo	4,266.61	15h	2
Rb	4,266.62	10	—
U	4,266.67	3	3
Cr	4,266.82	1	30
Fe	4,266.97	30	70
Yb	4,266.99	8h	4
Gd	4,267.016	40	40
C	4,267.02	350	—
As	4,267.24	3	—
C	4,267.27	500	—
Ne	4,267.29	(2)	—
Fe	4,271.76	700	1000
Ar	4,272.17	(1200)	—
Kr	4,273.97	(1000)	—
Cr	4,274.80	800r	4000R
Ar	*4,277.55*	(80)	—
Ar	*4,282.90*	*(40)*	—

Ar 4,702.32 (1200)

	λ	I J	I O
Er	4,702.17	2	3
Nb	4,702.053	4	3h
U	4,702.047	18	8
Ra	4,701.97	(4)	—
Hg	4,701.8	(5)	—
Cs	4,701.79	(25)	—
Rn	4,701.70	(50)	—
Ho	4,701.69	2	2
In	4,701.48	35	—
Br	4,701.40	(5)	—
Ta	4,701.32	2h	150
Al	4,701.30	6	—
O	4,701.16	(20)	—
Ho	4,701.16	1h	2
Mn	4,701.159	5	100
Yt	4,701.02	2	4

	λ	I J	I O
Th	4,702.32	2	4
Gd	4,702.323	100	50
J	4,702.46	(8)	—
W	4,702.47	5	15
U	4,702.52	20	10
Ne	4,702.53	(150)	—
La	4,702.64	3	9
Mg	4,702.02	3	8r
Zr	4,703.03	3	2
Gd	4,703.136	20	50
O	4,703.14	(30)	—
La	4,703.28	300r	200r
Ar	4,703.36	(10)	—
Dy	4,703.47	2	5
Hf	4,703.61	12	4

	λ	I J	I O		λ	I J	I O
U	**4,700.98**	**12**	**5**	Ne	4,704.39	(1500)	—
Nb	**4,700.91**	**5**	**3*h***	Ne	4,708.85	(1200)	—
				Ar	4,709.50	(30)	—
				Ne	4,710.06	(1000)	—
				Ne	4,712.06	(1000)	—
				Ne	4,715.43	(1500)	—
				S	4,716.23	(600)	—
				Ar	*4,719.94*	*(20)*	—
				Ar	*4,726.91*	*(200)*	—
				Se	*4,730.78*	*(1000)*	—

Ar 6,965.43 (400)

	λ	J	O		λ	J	O
K	**6,964.69**	**(5)**	—	**Hf**	**6,965.80**	**3**	**1**
K	**6,964.18**	**(3)**	—	**F**	**6,966.35**	**(70)**	—
Tl	**6,964.10**	**5**	—	**Tl**	**6,966.43**	**40**	—
Yb	**6,963.11**	**5**	**2**	**Nb**	**6,966.89**	**3**	**15**
				Cl	**6,966.95**	**(2)**	—
Ar	6,960.23	(20)	—	Ra	6,980.22	(1000)	—
J	6,958.78	(1000)	—				
Ar	6,951.46	(20)	—				
Xe	*6,942.11*	*(400wh)*	—	*Xe*	*6,990.88*	*(700)*	—
Ar	*6,937.67*	*(100)*	—				

Ar 7,067.22 (400)

	λ	J	O		λ	J	O
Nb	**7,066.41**	**2**	**8**	**J**	**7,067.24**	**(15)**	—
La	**7,066.21**	**150**	**400**	**Fe**	**7,067.44**	**10*h***	—
He	**7,065.70**	**(10)**	—	**Fe**	**7,068.41**	**30**	**40**
				Ar	**7,068.73**	**(30)**	—
Se	7,062.06	(1000)	—				
Ne	7,059,11	(200)	—				
Rn	7,055.42	(400)	—				

Ar 7,503.87 (700)

	λ	J	O		λ	J	O
Xe	**7,503.00**	**(3*h*)**	—	**Ar**	**7,505.13**	**(100)**	—
Ne	7,488.87	(500)	—	Ar	7,514.65	(200)	—
Ar	*7,484.24*	*(15)*	—				

Ar 8,115.31 (5000)

	λ	I				λ	I	
		J	O				J	O
Kr	**8,112.90**	**(5000)**	—		**Xe**	**8,115.94**	**(30*h*)**	—
Kr	**8,104.36**	**(5000)**	—		*Ar*	*8,119.18*	*(50)*	—
Kr	**8,104.02**	**(500)**	—					
Ar	**8,103.69**	**(2000)**	—					
Ar	8,094.06	(20)	—					
Ar	8,079.68	(20)	—					

As 33 74 9216

t_0 814°C t_1 610°C

I.	II.	III.	IV.	V.
9.81	20.2	27.297	50.123	62.61

λ	*I*		eV
	J	*O*	
I 2,288.12	5	250*R*	6.7
I 2,349.84	18	260*R*	6.6
I 2,456.53	8	100*R*	6.5
I 2,780.20	75	75*R*	6.7
I 2,860.45	50	50*R*	6.6
I 3,032.84	70	125*R*	6.4
II 3,116.63	150	—	—
II 3,749.77	100	—	—
II 4,352.25	200	—	13.11
4,474.60	200	—	—
II 4,539.97	200	—	—
II 4,630.14	200	—	—
II 4,707.82	200	—	—
II 5,331.54	200	—	12.43

As 2,288.12 *J* 5* *O* 250*R*

	λ	*I* J	*I* O		λ	*I* J	*I* O
V	**2,288.09**	**15**	**3*h***	**Sb**	**2,288.14**	—	**5**
Cd	**2,288.02**	**300*R***	**1500*R***	**Pt**	**2,288.19**	**30**	**15**
Ir	**2,287.88**	**5**	**20**	**Ni**	**2,288.39**	**2**	**12**
Ta	**2,287.84**	—	**5**	**Mn**	**2,288.42**	**3**	**6**
Co	**2,287.81**	**4**	**12*d***				
				Pt	2,292.39	100	400
Ni	2,287.08	500	100				
Zr	2,285.23	—	100				

As 2,349.84 *J* 18* *O* 250*R*

	λ	*I* J	*I* O		λ	*I* J	*I* O
W	**2,349.82**	**10**	**6**	**Cu**	**2,349.85**	—	**5**
Os	**2,349.81**	**6**	**40**	**U**	**2,349.85**	**8**	**2*h***
V	**2,349.806**	**150**	**3*w***	**Sb**	**2,349.853**	**14**	**8**
Mo	**2,349.78**	—	**10*w***	**Cd**	**2,349.86**	—	**2**
Yt	**2,349.70**	—	**8**	**Ti**	**2,349.94**	**12**	**3**
Rh	**2,349.68**	**125**	**3**	**Nb**	**2,350.03**	—	**2**
Ir	**2,349.66**	—	**5**	**Os**	**2,350.04**	—	**12**
Dy	**2,349.63**	—	**3**	**Ir**	**2,350.05**	**5*wh***	**15**
Re	**2,349.61**	**9**	**12**	**Zr**	**2,350.20**	**2*h***	**2**
Zr	**2,349.59**	**1**	**10**				
				Hf	2,351.21	150	100
Be	2,348.61	50	2000*R*	Sn	2,354.84	150*R*	150*R*

As 2,456.53 *J* 8* *O* 100*R*

	λ	*I* J	*I* O		λ	*I* J	*I* O
Os	**2,456.46**	**3**	**12*s***	**W**	**2,456.531**	**5**	**15**
Tl	**2,456.45**	—	**5*R***	**Zr**	**2,456.54**	—	**3**
Ru	**2,456.44**	**50**	**60**	**Ru**	**2,456.57**	**50**	**60**
Th	**2,456.29**	**4**	**6**	**Nb**	**2,456.67**	**5*w***	**1*w***
Co	**2,456.24**	**7**	**20*w***	**Ir**	**2,456.77**	—	**2**
W	**2,456.23**	—	**8**	**Th**	**2,456.86**	**2**	**4**
Rh	**2,456.18**	**150**	**5**	**Ru**	**2,456.95**	—	**10**
Ru	2,455.53	100*r*	80	Fe	2,457.59	30	70
Ru	2,454.92	5	60	Ru	2,458.62	3	60
Ta	2,454.48	—	60	Rh	2,458.90	300	50
Ta	2,454.21	—	50	Ta	2,460.55	—	50
Ni	*2,450.99*	*30*	*100h*	*Re*	*2,461.18*	—	*125*
Pt	*2,450.97*	*10*	*400*	*Fe*	*2,462.64*	*50*	*200R*

As 2,780.20 *J* 75 *O* 75*R*

	λ	I *J*	I *O*
Ga	2,780.15	(40)	—
K	2,780.1	(2)	—
V	2,780.097	10	—
Ne	2,780.06	(5)	—
Fe	2,780.045	20	—
Yb	2,780.04	2	—
U	2,780.040	8	8
Mo	2,780.036	100h	60
Ce	2,780.005	—	15
Mn	2,779.998	—	25
Fe	2,779.91	40	—
Cs	2,779.9	(8)	—
Co	2,779.84	3	—
Mg	2,779.83	50	40
Ce	2,779.825	—	2
Sn	2,779.82	100	80
La	2,779.78	10	1
W	2,779.724	7	8
Nb	2,779.72	4	5
Ta	2,779.704	4	30r
Pd	2,779.701	5	—
Fe	2,779.70	—	4
Ir	2,779.65	—	3
Rh	2,779.54	6	100
Fe	2,779.30	300	25
B	2,779.26	100	—
Ta	2,779.10	5h	150w
Ru	2,778.99	50	50
Fe	2,778.84	40	70
Co	2,778.82	8	75
V	2,778.58	60h	—
Mn	2,778.56	—	60
Ru	2,778.38	150	—
Fe	2,778.22	80	100
Rh	2,778.15	100	2
Cr	2,778.06	60	12
Rh	2,778.05	3	100
V	2,777.73	100R	40h
Cr	2,777.67	1	40
Ru	2,777.50	50	5
Ru	2,777.39	50	—
Fe	2,776.40	30	100
Yb	2,776.27	40	6
Mn	2,776.23	—	80
Fe	2,776.17	40	—
Ru	2,775.91	—	50

	λ	I *J*	I *O*
Ta	2,780.21	—	4
La	2,780.23	10h	20
Nb	2,780.24	200r	30
Tl	2,780.25	(20)	—
Cd	2,780.28	(25)	—
J	2,780.28	(20)	—
W	2,780.283	20	10
Cr	2,780.30	100	—
Ir	2,780.41	—	3
Bi	2,780.52	100	200W
Eu	2,780.53	—	20
Fe	2,780.54	2	10
Ti	2,780.56	60wh	—
Fe	2,780.70	15	30
Cr	2,780.703	15	600R
Ta	2,781.37	4	50
Gd	2,781.40	40	30
Re	2,781.448	—	40
V	2,781.454	125h	4
Rh	2,781.80	150	1
Fe	2,781.83	60	90
Eu	2,781.89	100w	100w
Ru	2,782.21	1	50
Cr	2,782.35	35	—
Os	2,782.55	15	40
Mn	2,782.73	—	50
Rh	2,783.03	10	150
Re	2,783.57	—	150w
Fe	2,783,70	400	30
Cr	2,783.84	35	—
V	2,784,27	50h	4
Ru	2,784.53	100	60
Ta	2,784.97	100	50
Mo	2,784.99	200	100
Sn	2,785.03	60	60
Tm	2,785.08	30	60
B	2,785.14	35	—
Fe	2,785.21	40h	—
Ba	2,785.26	—	50
Ru	2,785.65	200	60
Cr	2,785.70	80	5
Ta	*2,787.69*	*40*	*400R*
Ru	*2,787.83*	*150*	*60*
Fe	*2,788.10*	*150*	*150*

	λ	I	
		J	O
Ta	2,775.88	30	200
Rh	2,775.77	125	5
V	2,775.76	70*h*	12
Ru	2,775.63	150	50
Co	2,775.58	—	50
Mo	2,775.40	100*h*	80
In	2,775.36	30	80
Ta	2,775.35	15	80
Ni	2,775.33	250*wh*	—
Ru	2,775.185	—	50
Co	2,775.181	100*h*	30*h*
Ta	2,775.11	80	100*w*
Cd	2,775.05	20	50
Tm	2,774.99	40	20
Co	2,774.96	—	50
Ta	2,774.88	3	100
Pt	2,774,78	100*wh*	10
Fe	2,774.73	10	80
V	2,774.72	50*h*	20
Ho	2,774.70	300	—
Fe	2,774.69	50	—
Cr	*2,774.44*	*100*	—
V	*2,774.28*	*100R*	*25*
Fe	*2,773.24*	*40*	*90*
Lu	*2,772.58*	*150h*	*5*
Ru	*2,772.45*	*150*	—
Fe	*2,772.11*	*300*	*300*
V	*2,772.01*	*80h*	*2*
Ta	*2,771.83*	*100*	*3h*
Pt	*2,771.67*	*15*	*500*
Rh	*2,771.51*	*8*	*100*
Ru	*2,771.07*	*100*	—
Zn	*2,770.98*	*150*	*300*
Zn	*2,770.86*	*25*	*300*
Sb	*2,769.94*	*75*	*100*
Cr	*2,769.91*	*40*	*400r*
Mo	*2,769.76*	*100*	*10*
Cu	*2,769.67*	*400*	*5h*
Fe	*2,769.30*	*10*	*90*
Ta	*2,788.30*	*3*	*150*
Ta	*2,790.71*	*10*	*150*
Mg	*2,790.79*	*80*	*40*
Rn	*2,791.16*	*14*	*100*

As 2,860.45 *J* 50 *O* 50*R*

Yb	**2,860.40**	**8**	**1**
Ru	**2,860.37**	—	**3**
Hf	**2,860.312**	**30**	**15**
Lu	**2,860.31**	**3**	—
Ti	**2,860.28**	—	**7**
U	**2,860.47**	**30**	**35**
In	**2,860.52**	**5**	—
Tm	**2,860.55**	—	**10**
Ce	**2,860.555**	—	**3**
Hf	**2,860.56**	**2**	**20**

	λ	I J	I O
Er	**2,860.26**	**—**	**4**
Fe	**2,860.21**	**—**	**2**
Dy	**2,860.17**	**—**	**2**
W	**2,860.16**	**9**	**5**
Tm	**2,860.13**	**20**	**15**
Os	**2,860.06**	**10**	**25**
Ru	**2,860.02**	**12**	**60**
V	**2,859,97**	**10**	**50**
Nb	**2,859,96**	**3**	**5**
O	**2,859.89**	**(5*d*)**	**—**
W	**2,859.86**	**2**	**4**
Er	**2,859.81**	**10**	**25**
Yb	**2,859.80**	**30**	**18**
Eu	2,859.67	—	40
Co	2,859.66	—	40
Na	2,859.48	(40)	2
Nb	2,859.04	50	1
V	2,858.98	20	40
Cr	2,858.91	80*wh*	50
U	2,858.90	25	35
Fe	2,858.897	80	100
Cu	2,858.73	2*h*	30
Mn	2,858.66	—	50
Cr	2,858.65	30	18
Ta	2,858.43	300	100
Fe	2,858.34	200	3
Te	2,858.29	(100)	2
Cu	2,858.22	2*h*	30
Cr	2,857.97	40	2
V	2,857.94	7*h*	50
Ti	2,857.81	70*wh*	—
Ru	2,857.78	60	4
Pd	2,857.73	100	—
Cr	2,857.40	80	20
Ta	2,857.28	5	60
Ru	2,857.22	25	—
Fe	2,857.17	30	—
Cr	2,856.77	60	20
Ru	2,856.55	50	—
Ti	2,856.24	100*wh*	—
Rh	2,856.16	30*h*	60
Fe	2,856.14	40	1
Mo	2,855.99	25	2
La	2,855.90	50*hl*	3
Cr	2,855.68	200*Wh*	60
Fe	2,855.67	200	2
Os	2,855.34	8	25
V	2,855.29	35	1
Pb	**2,860.64**	**2**	**—**
Ce	**2,860.643**	**—**	**2**
Ir	**2,860.66**	**2**	**12**
Rh	**2,860.67**	**—**	**30**
Pt	**2,860.678**	**150*h***	**30**
W	**2,860.679**	**9**	**6**
Dy	**2,860.70**	**—**	**2**
Cl	**2,860.71**	**(5)**	**—**
Ar	**2,860.73**	**(5)**	**—**
Rh	**2,860.76**	**10*h***	**30**
V	**2,860.80**	**12**	**15**
Ti	**2,860.84**	**25*wh***	**—**
Ce	**2,860.85**	**(8)**	**—**
Zr	**2,860.851**	**—**	**15**
Ce	**2,860.86**	**—**	**2**
J	**2,860.92**	**(12)**	**—**
Cr	**2,860.93**	**100**	**60**
Os	**2,860.96**	**25**	**100**
Na	**2,861.011**	**(2)**	**1**
Hf	**2,861.012**	**90**	**40**
Te	**2,861.03**	**(25)**	
W	**2,861.05**	**5**	**1**
Nb	2,861.093	100	10
Ta	2,861.12	—	30
Fe	2,861.19	30	1
Yb	2,861.21	30	3
Yb	2,861.31	25	3
Ru	2,861.41	35	60
Hf	2,861.70	125	50
Tm	2,861.74	30	20
Ta	2,861.98	5	30
Ti	2,861.99	100*wh*	—
Ta	2,862.02	10	40
Ti	2,862.32	40	20
Fe	2,862.50	40	100
Eu	2,862.57	70	100*w*
Cr	2,862.571	300*R*	80
Co	2,862.61	—	50
Ru	2,862.88	60	6
Rh	2,862.93	60	150
Ru	2,863.32	80	30
Sn	2,863.33	300*R*	300*R*
Fe	2,863.43	80	100
Ni	2,863.70	250	—
Bi	2,863.75	18	80*w*
Tm	2,863.76	40	15
Mo	2,863.81	100*h*	30

	λ	I J	I O		λ	I J	I O
V	2,855.22	1	50	Fe	2,863.86	100	125
Cr	2,855.07	100	4	Ni	2,864.15	300wh	—
Ce	2,854.88	—	25s	Pb	2,864.26	60	—
Ru	2,854.72	60	2	Mo	2,864.31	2	40
Ce	2,854.67	—	30s	V	2,864.36	25r	40
Pd	2,854.58	500h	4	Rh	2,864.40	10	70
Xe	2,854.53	(30)	—	Ta	2,864.50	30	125
V	2,854.34	100R	20	Pb	2,864.51	60	—
Ru	2,854.07	35	60	V	2,864.52	35	6
				Mo	2,864.66	3	40
Mo	2,853.23	100h	25	Xe	2,864.73	(100)	—
Cr	2,853.22	100R	5	Fe	2,864.97	50	—
Na	2,853.03	15	80R	Pt	2,865.05	80h	20
V	2,852.87	7h	60	Cr	2,865.11	200R	60
Na	2,852.83	20	100R	Nb	2,865.61	30	5
Ta	2,852.35	100	5	U	2,865.679	50	30
Fe	2,852.13	80	150	In	2,865.684	(50)	—
Mg	2,852.129	100R	300R	Ru	2,866.08	30	—
Ce	2,852.12	1	50d	Ru	2,866.25	30	—
Hf	2,852.01	50	20	Hf	2,866.37	12	50
Fe	2,851.80	150	200	V	2,866.42	—	35
Cr	2,851.36	80	20	V	2,866.59	6	25
Hf	2,851.21	50	25	Fe	2,866.63	80	125
Yb	2,851.12	50	10	Ru	2,866.64	25	60
Sb	2,851.11	45	50				
Ti	2,851.10	80	20	Cr	2,866.74	125R	80
Ta	2,850.98	150	400	Fe	2,867.31	30	60
Os	2,850.76	25	75	Ta	2,867.41	150	54
Sn	2,850.62	100wh	80	Fe	2,867.56	30	60
bhB	2,850.6	—	50	Cr	2,867.65	100R	80
Ta	2,850.49	100	200	Cd	2,868.26	80	100
Co	2,850.04	—	75	Fe	2,868.45	40	80
Cr	2,849.84	150r	80	Al	2,868.52	80	—
Ta	2,849.82	50W	15Ws	Nb	2,868.525	300	15
Tl	2,849.80	(300)	—	Ta	2,868.65	40	150
Fe	2,849.61	—	50	Rn	2,868.7	(70)	—
Nb	2,849.56	100w	2w	Te	2,868.86	(100)	—
Mo	2,849.38	5	50	Fe	2,868.87	60	5
Ru	2,849.29	100	3	V	2,869.13	150r	25
Hf	2,849.21	100	30	Tm	2,869.22	300	100
V	2,849.05	50	7	Fe	2,869.31	70	300
Fe	2,848.72	30	60	Co	2,870.03	50wh	—
Ru	2,848.58	3	50	Ti	2,870.04	100wh	—
Ta	2,848.53	50	300	Cr	2,870.44	300W	25
Zr	2,848.52	—	100	V	2,870.55	20r	50r
Co	2,848.37	60h	—	Ru	2,870.55	50	—
Mo	2,848.23	200h	125	Co	2,871.24	100	—
Ta	2,848.05	15	150	Rh	2,871.35	10h	100
Fe	2,848.046	70	—	Ta	2,871.42	50	200
Hg	2,847.83	100	15	Cr	2,871.45	80	—

	λ	I J	I O		λ	I J	I O
				Mo	*2,871.51*	*100h*	*100*
				Cr	*2,871.63*	*2*	*50*
				Ru	*2,871.64*	*5*	*50*
				Re	*2,871.81*	*—*	*50*
				Fe	*2,872.34*	*50*	*150*
				Os	*2,872.40*	*8*	*50*
				Mo	*2,872.88*	*50*	*2*
				J	*2,872.89*	*(60)*	*—*

As 3,032.84 *J* 70 *O* 125*R*

	λ	I J	I O		λ	I J	I O
Tb	**3,032.83**	**8**	**8**	**Gd**	**3,032.85**	**100**	**100**
Os	**3,032.81**	**15**	**50**	**Sm**	**3,032.86**	—	**15**
Re	**3,032.79**	—	**20**	**Cr**	**3,032.93**	**100**	**10**
Sn	**3,032.78**	**20**	**50**	**U**	**3,032.98**	**2**	**4**
Kr	**3,032.77**	(5*wh*)	—	**Ce**	**3,033.05**	—	**10**
Nb	**3,032.768**	**300**	**3**	**Fe**	**3,033.10**	**20**	**40**
U	**3,032.76**	**2**	**2**	**Xe**	**3,033.11**	**(3)**	—
Ce	**3,032.73**	—	**10**	**Ce**	**3,033.12**	—	**8**
Ru	**3,032.67**	—	**3**	**Au**	**3,033.18**	**30*h***	**25**
Mo	**3,032.54**	—	**5**	**Dy**	**3,033.18**	**1**	**8**
Rn	**3,032.5**	**(40)**	—	**U**	**3,033.19**	**12**	**12**
O	**3,032.47**	**(34)**	—	**Rn**	**3,033.2**	**(10)**	—
U	**3,032.44**	**4*h***	**1**	**Mo**	**3,033.23**	**10**	**8**
Ce	**3,032.41**	**(4)**	—	**J**	**3,033.27**	**(5)**	—
Ir	**3,032.41**	**1**	**50**	**Pr**	**3,033.31**	**4*h***	—
Ce	**3,032.33**	—	**8**	**Mo**	**3,033.33**	**30**	**1**
B	**3,032.28**	**10**	—	**Te**	**3,033.35**	**(5)**	—
Pd	**3,032.20**	**100*h***	**2**	**Ta**	**3,033.392**	—	**2**
V	**3,032.19**	**8**	—	**Sm**	**3,033.393**	**24**	**1**
				Nb	**3,033.393**	**2**	**2**
Ni	3,031.87	—	200	**Ra**	**3,033.44**	**(150)**	—
Fe	3,031.64	200	200	**U**	**3,033.444**	**12**	**1**
Cr	3,031.35	30	40	**Fe**	**3,033.445**	**2**	**4**
Pt	3,031.22	40*h*	10	**V**	**3,033.448**	**40**	**20**
Fe	3,031.21	150	150	**Ru**	**3,033.45**	**10**	**70**
Hf	3,031.16	90	70	**Zr**	**3,033.46**	**2**	—
Yb	3,031.11	30	100*h*	**Bi**	**3,033.5**	**(15)**	—
Os	3,030.69	40	500				
Re	3,030.45	—	100				
Ne	3,030.31	(50)	—	V	3,033.82	90*r*	20
Cr	3,030.24	150	200*r*	Gd	3,034.059	60	100
Fe	3,030.15	300	300	Ru	3,034.060	5	60
Sb	3,029.81	200*wh*	100	Sn	3,034.12	150*wh*	200*wh*
Ti	3,029.73	150	12	Cr	3,034.19	60	200*r*
V	3,029.58	40	—	Co	3,034.43	2	80
				Fe	3,034.54	40	70
Yb	3,029.55	40*h*	1	Nb	3,034.95	100*w*	1
Zr	3,029.51	5	60	Bi	3,035.18	—	60*h*

	λ	I	
		J	O
Ir	3,029.36	3	60
Fe	3,029.23	60	80
Cr	3,029.16	50	70
Na	3,029.07	(60)	—
Nb	3,028.44	200	50
Rh	3,028.43	—	80
Cr	3,028.12	125	2
V	3,028.04	50	2
Pd	3,027.91	200h	150
Ru	3,027.79	50	12
Gd	3,027.61	60	100
Ta	3,027.51	35w	125
Cr	3,026.65	125	8
Fe	3,026.46	200	200
Co	3,026.37	40	100
Fe	*3,025.84*	*300r*	*400r*
Fe	*3,025.64*	*100*	*100*
Nb	*3,024.74*	*200*	*10w*
Bi	*3,024.63*	*50*	*250wh*
Cr	*3,024.35*	*125*	*300r*
Fe	*3,024.03*	*200*	*300*
Ti	*3,023.86*	*100wh*	—
Mo	*3,023.30*	*100*	*5*
Te	*3,023.29*	*(100)*	—
Ti	*3,022.82*	*150Wh*	—
Nb	*3,022.74*	*100*	*5w*
Cr	*3,021.56*	*200r*	*300r*
Fe	*3,021.07*	*300r*	*700R*
Cr	*3,020.67*	*100*	*200r*
Fe	*3,020.64*	*600r*	*1000R*
Lu	*3,020.54*	*100*	—
Fe	*3,020.49*	*300r*	*300r*

	λ	I	
		J	O
Ru	3,035.47	4	60
Fe	3,035.74	60	100
Zn	3,035.78	100	200
V	3,036.08	35h	—
Cu	3,036.10	50	200
Pt	3,036.45	10	200
Ru	3,036.47	150	50
Yt	3,036.59	40	10
Rn	3,036.8	(60)	—
Cr	3,037.044	100	200r
Na	3,037.07	(40)	8
Fe	3,037.39	400r	700R
Ta	3,037.50	100	8h
Rn	3,037.7	(40)	—
Ni	3,037.93	100	800R
Cl	3,037.98	(35)	—
Ru	3,038.18	5	80
Se	3,038.66	(60)	—
Ti	3,038.71	40	2
Ge	3,039.06	1000	1000
In	3,039.36	500R	1000R
Co	3,039.57	—	70
Nb	*3,039.81*	*300*	*5*
Fe	*3,040.43*	*400*	*400*
Sb	*3,040.67*	*(400wh)*	—
Cr	*3,040.85*	*200*	*500R*
Os	*3,040.90*	*100*	*200*
Fe	*3,041.64*	*80*	*80*
Fe	*3,041.738*	*80*	*100*
Cr	*3,041.74*	*125*	*2*
Fe	*3,042.02*	*100*	*125*
Ta	*3,042.06*	*100*	*5*
V	*3,042.26*	*80*	*6*
Pt	*3,042.64*	*250R*	*200R*
Fe	*3,042.66*	*200*	*300*
Cr	*3,042.79*	*100*	*1*
Pb	*3,043.90*	*100*	—
Co	*3,044.00*	—	*400R*
Ni	*3,045.01*	*10*	*200*
Fe	*3,045.08*	*100*	*150*
Ta	*3,045.96*	*50w*	*150*

As 3,116.63 J 150

	λ	I J	I O
Ar	3,116.63	(3)	—
Se	3,116.63	(20)	—
J	3,116.60	(7)	—
Fe	3,116.59	150	—
Nb	3,116.57	4wh	1
Yb	3,116.49	3	1
Th	3,116.48	10	10
Os	3,116.47	15	50
U	3,116.43	4	6
Nb	3,116.36	3	8
Ce	3,116.35	12	50
Rn	3,116.3	(5)	—
Te	3,116.30	(5)	—
Th	3,116.29	10	5d
Fe	3,116.25	4	5
Hg	3,116.24	(100)	—
W	3,116.21	3	7
Nd	3,116.14	4	10
Mo	3,116.09	60	1
Yb	3,116.06	3	1
V	3,116.05	10h	—
Ce	3,115.98	—	5
U	3,115.93	15	8
Ta	3,115.86	18w	50
U	3,115.79	4	6
Th	3,115.74	10	10
Zr	3,115.73	2	2
Bi	3,115.42	500	—
Fe	3,114.29	80	—
Pd	3,114.04	500w	400w
V	3,113.57	100	7
Ti	3,112.05	70	7
Ta	3,110.815	70w	—
V	3,110.71	300R	70
Ti	3,110.67	100	10
V	3,109.37	70	1
Hf	3,109.12	100	50
Ti	3,106.23	150	25
Ti	3,103.80	200	20
V	3,102.30	300R	70
Ni	3,101.88	150	400R
Zr	3,116.68	4h	—
Cr	3,116.74	35	—
Xe	3,116.78	(2)	—
V	3,116.781	25	—
W	3,116.869	3	9
Dy	3,116.869	2	4
Er	3,116.947	2	20
Hf	3,116.950	20	20
Sm	3,117.186	1	4
Fe	3,117.191	2	2
Cr	3,117.258	30	1
Tb	3,117.26	3	8
U	3,117.28	4	6
Ir	3,117.33	2	2
W	3,117.38	5	6
Ti	3,117.67	200	15
Pb	3,117.7	(100)	—
V	3,118.38	200R	70
Cr	3,118.65	200	35
Fe	3,119.49	80	100
Ti	3,119.80	150	4
Cr	3,220.37	150	40
Fe	3,120.43	80	100
V	3,120.73	80	12
V	3,121.14	200r	60
Xe	3,121.87	(150)	—
Mo	3,122.00	150	5
Cr	3,122.60	80	10
V	3,122.89	30r	12
V	3,125.28	200R	80
Fe	3,125.65	300	400
Hg	3,125.66	150	200
Ne	3,126.19	(150)	—
Cr	3,128.70	150	30
V	3,130.27	200r	50
Be	3,130.42	200	200
Be	3,131.07	150	200
Tm	3,131.26	500	400
Hg	3,131.55	300	400
Hg	3,131.55	300	400

As 3,749.77 *J* 100

	λ	*I*	
		J	*O*
Yb	3,749.69	10	2
W	3,749.66	8	7
Ce	3,749.49	24	2
Fe	3,749.487	700	1000*r*
Rn	3,749.48	(8)	—
O	3,749.47	(125)	—
U	3,749.16	8	2
Pb	3,749.15	2	—
Os	3,749.07	2	4
Ni	3,749.04	5	50
Cr	3,749.00	125*R*	125*R*
Th	3,748.97	5	8
Fe	3,748.969	20	35
Pr	3,748.82	3	9
U	3,748.68	25	15
Sm	3,748.63	5	10
Cr	3,748.61	30	40
Nb	3,748.55	10	10
Fe	3,748.26	200	500
Rh	3,748.22	100	200
Mo	3,748.13	50	1
Ir	3,747.20	60	100
Nb	3,746.91	80	20
Fe	3,745.90	100	150
V	3,745.80	600	35
Fe	3,745.56	500	500
Kr	3,744.80	(150*hs*)	—
Mo	3,744.37	80	20
P	3,744.21	(70)	—
Ar	3,743.76	(100)	—
Fe	3,743.36	150	200
Ru	3,742.78	50	50
Nb	3,742.39	50	30
Ru	3,742.28	100	70
Kr	3,741.69	(200)	—
Ti	3,741.64	200	30
Th	3,741.19	80	80
Nb	3,740.72	50	3
Gd	3,740.05	50	50
Pb	3,739.95	60*h*	150
Nb	3,739.79	200	100
Gd	3,739.76	50	100
Fe	3,738.31	100	100
Fe	*3,737.13*	*600*	*1000r*
Fe	*3,734.87*	*600*	*1000r*
Sm	3,749.79	34	—
Pr	3,749.80	2	5
Nd	3,749.85	8	6
U	3,749.86	8	—
Yt	3,749.89	4	3
Co	3,749.93	5	60
Cl	3,750.00	(30)	—
Ce	3,750.08	3	12
Pr	3,750.12	15	30
Eu	3,750.13	2*wh*	8*wh*
Th	3,750.15	5	20
H	3,750.152	(10)	—
Nb	3,750.22	5	3
Nd	3,750.31	10	8
Dy	3,750.33	4	8
Ca	3,750.35	3	20
Pr	3,750.49	10	20
Ar	3,750.50	(5)	—
Cr	3,750.56	12*h*	—
Os	3,750.59	5	15
Nb	3,750.63	4	4
Zr	3,750.64	3	4
Sm	3,750.67	8	10
Th	3,750.69	5	10
Ce	3,750.73	2	2
Nd	3,750.74	6	10
Mn	3,750.76	30	60
Ir	3,750.77	2	15
W	3,750.78	12	—
Os	3,750.80	5	15
Cl	3,750.81	(8)	—
V	3,750.87	20	10
Pr	3,751.001	30	40
Ce	3,751.002	5	15
V	3,751.23	100	4
Co	3,751.63	60	100
Os	3,752.52	100	400*R*
Th	3,752.65	50	40
Ti	3,752.86	80	200
Ru	3,753.53	60	30
Fe	3,753.61	100	150
Rn	3,753.65	(50)	—
Ne	3,754.22	(50)	—
Kr	3,754.24	(80)	—
Tb	3,755.24	100	50
Mo	3,755.54	50	1

	λ	I J	I O
V	*3,732.76*	*500R*	*70R*
Fe	*3,732.40*	*150*	*200*
Mn	*3,731.93*	*100*	*75*
Gd	*3,730.86*	*100*	*100w*
Ti	*3,729.81*	*150*	*500*
Ar	*3,729.29*	*(200)*	—
Mn	*3,728.89*	*100*	*75*
Sm	*3,728.47*	*100*	*100*
V	*3,728.34*	*150*	*20*
Ru	*3,728.03*	*150*	*100*
Fe	*3,727.62*	*150*	*200*
V	*3,727.34*	*200*	*40*
Ne	*3,727.08*	*(125)*	—
Ru	*3,726.93*	*150*	*100*
Nb	*3,726.24*	*100*	*30*
Ru	3,755.93	60	30
Fe	3,756.94	60	80
Dy	3,757.37	50	200
Cr	3,757.66	50	50
Ti	3,757.69	100	30
Fe	3,758.23	700	700
La	3,759.08	150	400
Ti	3,759.29	400*R*	100
Nb	3,759.55	50	40
Ru	3,760.03	50	20
Fe	3,760.05	100	150
Fe	3,760.53	70	100
Ti	3,761.32	300*r*	100
Tm	3,761.33	150	250
Pr	3,761.87	100	150
Tm	3,761.92	120	200
Fe	*3,763.79*	*400*	*500*
Tb	*3,765.14*	*100*	*70*
Fe	*3,765.54*	*150*	*200*
Fe	*3,767.19*	*400*	*500*
Ar	*3,770.37*	*(400)*	—
Yt	*3,774.33*	*100*	*12*

As 4,352.25 *J* 200

	λ	I J	I O
Eu	4,352.24	2	4
Ar	4,352.23	(30)	—
P	4,352.22	(30)	—
Sm	4,352.10	100	125
Pr	4,352.05	2	4
Mg	4,351.91	2	15
Pr	4,351.85	60	80
Cr	4,351.77	300	300
Fe	4,351.76	30	30
Nb	4,351.57	20	10
Fe	4,351.55	5	30
Mo	4,351.549	15	15
Os	4,351.53	2	9
Pb	4,351.5	(3)	—
Ce	4,351.39	3*wh*	2
Kr	4,351.36	(100)	—
Nd	4,351.29	10	30
O	4,351.27	125	—
Br	4,351.22	(20)	—
Nd	4,351.185	5	15
Sb	4,352.25	6*h*	—
V	4,352.44	6	5
Nd	4,352.49	2	10
Rb	4,352.5	(8)	—
Pr	4,352.51	1	5
Ir	4,352.56	2	50
Hf	4,352.57	6	10
Pb	4,352.7	(10)	—
Ce	4,352.71	5	40
Fe	4,352.74	150	300
V	4,352.87	6	10
Mo	4,352.88	3	5
As	4,353.02	100	—
Nd	4,353.03	1	5
W	4,353.11	4	8
B	4,353.17	2	—
Pr	4,353.18	1	10
Nb	4,353.27	5	34
W	4,353.30	2	7
Mo	4,353.31	25	25

	λ	I	
		J	O
Cr	**4,351.05**	**150**	**100**
Mo	**4,351.02**	**5*h***	**5*h***
Kr	**4,351.02**	**(40*wh*)**	—
Ti	**4,350.834**	**30**	**6**
Th	**4,350.834**	**2**	**6*h***
V	**4,350.82**	**4**	**8**
Sm	**4,350.81**	**1**	**3**
Mo	**4,350.73**	**15**	**40**
Hf	**4,350.51**	**40**	**20**
Sm	4,350.46	150	150
Pr	4,350.40	25	70
Ba	4,350.37	20	40
Mo	4,350.34	40	50
Rn	4,349.60	(5000)	—
O	4,349.43	(300)	—
Ar	4,348.12	(500*h*)	—
Gd	4,347.32	100	100
Sm	4,345.85	100	100
O	4,345.57	(125)	—
Ar	4,345.17	(1000)	—
Cr	4,344.51	300	400*r*
Cl	4,343.62	(100)	—
Gd	4,342.19	200	200
Gd	4,341.29	125	200
Ra	4,340.64	(1000)	—
H	4,340.46	(600)	—
Cr	*4,339.72*	*150*	*150*
Se	*4,339.59*	*(200)*	—
Cr	*4,339.45*	*300*	*300r*
Ti	*4,337.92*	*125*	*70*
Cr	*4,337.57*	*300*	*500*
Fe	*4,337.06*	*150*	*400*
As	*4,336.85*	*100*	—
Ar	*4,335.34*	*(800)*	—
Sm	*4,334.15*	*200*	*200*
Pr	*4,333.91*	*100*	*150*
La	*4,333.73*	*500*	*800*
Ar	*4,333.56*	*(1000)*	—
Ar	*4,331.25*	*(2000)*	—
Xe	*4,330.52*	*(500wh)*	—
Se	*4,330.28*	*(200)*	—
Sm	*4,329.02*	*300*	*300*
Se	*4,328.70*	*(200)*	—
Fe	*4,325.76*	*700*	*1000*
Gd	*4,325.69*	*250*	*500R*
Kr	*4,319.58*	*(1000)*	—
V	**4,353.33**	**5**	**6**
Hf	**4,353.34**	**6*h***	**8**
Ce	**4,353.37**	**2**	**6**
Th	**4,353.40**	**5**	**8**
Eu	**4,353.52**	**1**	**2**
Nd	**4,353.61**	**1**	**5**
Pr	**4,353.80**	**2**	**10**
Co	**4,353.82**	**2**	**4**
Kr	**4,353.90**	**(2)**	—
Cr	**4,353.98**	**3**	**20**
Ti	**3,454.06**	**5**	**25**
Ru	**4,354.13**	**20**	**25**
Kr	**4,354.23**	**(2)**	—
U	**4,354.36**	**6**	**4**
La	4,354.40	100	80
Kr	4,355.48	(3000)	—
N	4,358.27	(250)	—
Hg	4,358.35	500	3000*w*
Cr	4,359.63	150	200
Sm	4,362.03	150	150
Kr	4,362.64	(500)	—
Mo	4,363.64	200	5
Te	4,364.02	(400)	—
Br	*4,365.60*	*(200)*	—
O	*4,366.91*	*(100)*	—
O	*4,368.30*	*(1000)*	—
Xe	*4,369.20*	*(100wh)*	—
Kr	*4,369.69*	*(200)*	—
Fe	*4,369.774*	*100*	*200*
Gd	*4,369.775*	*150*	*250*
Cr	*4,371.28*	*150*	*200*
Ru	*4,372.21*	*100*	*125*
Rh	*4,374.80*	*500*	*1000w*
Sm	*4,374.97*	*200*	*200*
Fe	*4,375.93*	*200*	*500*
Kr	*4,376.12*	*(800)*	—
Mo	*4,377.76*	*200*	*5*
V	*4,379.24*	*200R*	*200R*
Se	*4,382.87*	*(800)*	—
Fe	*4,383.55*	*800*	*1000*
Cr	*4,384.98*	*200*	*150*
Kr	*4,386.54*	*(300h)*	—
Xe	*4,393.20*	*(200wh)*	—

	λ	I: J	I: O
Sm	4,318.93	300	300
Kr	3,318.55	(400)	—
Kr	4,317.81	(500wh)	—
Fe	4,315.09	300	500

As 4,474.60 J 200

	λ	I: J	I: O
Mo	4,474.564	125	125
U	4,474.564	2h	2h
Gd	4,474.14	150	150
Th	4,474.08	6	10
V	4,474.05	20	30
W	4,474.04	5	12
La	4,474.03	20	3
Xe	4,473.85	(24)	—
Pr	4,473.83	6	30
Cr	4,473.78	1	25
Pd	4,473.59	6	60
Mo	4,473.58	2	3
Ta	4,473.51	10	3
Er	4,473.50	1	10
U	4,473.49	3	2
Pt	4,473.46	2	2
J	4,473.44	(80)	—
Sm	4,473.01	150	150
Br	4,472.62	(125)	—
Sm	4,472.42	100	100
He	4,471.48	(100)	—
Fe	4,469.38	100	200
Pr	4,468.71	100	125
Ti	4,468.47	150	80
Se	4,467.60	(300)	—
Sm	4,467.34	200	200
Fe	4,466.554	300	500
Gd	4,466.553	150	200
S	4,464.42	(100)	—
Kr	4,463.69	(800)	—
S	4,463.58	(200)	—
Xe	4,462.19	(500wh)	—
Rn	4,459.25	(250)	—
Fe	4,459.12	200	400
Sm	4,458.51	200	150
Kr	4,453.92	(600)	—
J	4,452.88	(700)	—
Sm	4,452.71	200	200
Se	4,449.15	(300)	—
Nb	4,474.62	5h	—
Ce	4,474.69	1	12
V	4,474.71	20	25
Nd	4,474.72	1	5
Ar	4,474.77	(20)	—
U	4,474.836	3	3
Pr	4,474.84	1	9
Ti	4,474.85	30	80
Kr	4,475.00	(800hs)	—
Se	4,475.02	(12)	—
Ne	4,475.13	(5)	—
Sm	4,475.16	10	10
Th	4,475.23	2	5h
Mo	4,475.25	3	1
P	4,475.26	—	(150)
Ne	4,475.66	(100)	—
Fe	4,476.02	300	500
Br	4,477.75	(200W	—
Sm	4,478.66	100	100
Te	4,478.73	(800)	—
Xe	4,480.86	(200wh)	—
Gd	4,481.07	100	70
Ne	4,483.19	(150)	—
S	4,483.42	(100)	—
Mg	4,487.48	(300)	—
Ne	4,488.09	(300)	—
Kr	4,489.88	(400h)	—
As	4,494.59	200	—
Cr	4,496.86	200	200

As 4,539.97 J 200

	λ	I J	I O
Os	4,539.92	2	100
Cr	4,539.79	25	40
Hf	4,539.76	6	3
Ce	4,539.75	10	20
W	4,539.70	1	4
Cu	4,539.695	80W	100W
Eu	4,539.69	3	3w
Mo	4,539.64	6	6
Yb	4,539.29	2	12
Pr	4,539.288	3w	30d
U	4,539.28	3h	—
Nd	4,539.26	2	10
Mo	4,539.21	3	4
Ne	4,539.17	(50)	—
Ti	4,539.10	4	15
Cs	4,538.94	(30)	—
Mo	4,538.88	4	5
La	4,538.87	8h	—
Br	4,538.77	(15)	—
Dy	4,538.76	2	5
Ne	4,538.31	(300)	—
Gd	4,537.82	100	150
Ne	4,537.75	(1000)	—
Ne	4,537.68	(300)	—
Ne	4,536.31	(150)	—
Pr	4,535.92	100	125
Cr	4,535.72	100	125
Ti	4,533.97	150	30
Hf	4,533.11	(300)	—
Xe	4,532.49	(100)	—
P	4,530.78	(150)	—
Cr	4,530.74	125	150
Fe	4,528.62	200	600
Xe	*4,524.68*	*(400)*	—
Kr	*4,523.14*	*(400h)*	—
Eu	*4,522.60*	*200*	*200R*
Tm	*4,522.57*	*300*	*200*
La	*4,522.37*	*400*	*200*
Ar	*4,522.32*	*(800)*	—

	λ	I J	I O
V	4,540.01	12	15
Gd	4,540.02	200	80
Nb	4,540.08	1	5
Pr	4,540.15	1	5
P	4,540.20	(70)	—
Cu	4,540.21	2	—
W	4,540.31	—	7
Cu	4,540.23	3	—
Ne	4,540.38	(50)	—
Th	4,540.42	8	8
Ti	4,540.48	1	7
Cr	4,540.50	40	40d
J	4,540.63	(8)	—
Eu	4,540.66	4	4W
La	4,540.71	25	4
Cr	4,540.72	40	40d
Mo	4,540.75	25d	25d
Rb	4,540.77	10	—
Ti	4,540.87	1	6
Xe	4,540.89	(200h)	—
Hf	4,540.93	2	50
Mo	4,541.02	4	4
Cu	4,541.03	5	—
Cr	4,541.07	8	30
P	4,541.12	(70)	—
Pd	4,541.14	3	15
Br	4,542.93	(250)	—
As	4,543.76	200	—
Ar	4,545.08	(200)	—
Xe	4,545.23	(200wh)	—
Fe	4,547.85	100	200
As	4,549.23	125	—
Fe	4,549.47	100	100
Ti	4,549.63	200	100
S	4,552.38	(200)	—
Ba	*4,554.04*	*200*	*1000R*
Ru	*4,554.51*	*200*	*1000R*
Kr	*4,556.61*	*(200h)*	—
Te	*4,557.84*	*(300)*	—
La	*4,558.46*	*200*	*100*
Cr	*4,558.66*	*600wk*	*20*
Ti	*4,563.77*	*200*	*100*
Se	*4,563.95*	*(200)*	—

As 4,630.14 J 200

	λ	I	
		J	O
Hg	4,630.14	(30)	—
Fe	4,630.13	2	10
Nb	4,630.11	20	30
Mo	4,630.02	12	15
Se	4,629.82	(4)	—
U	4,629.72	5	34
Al	4,629.7	(4)	—
Ca	4,629.69	2	—
Br	4,629.42	(15)	—
Co	4,629.38	5	600 W
Ti	4,629.34	7	70
Fe	4,629.33	8	7
W	4,629.26	1	5
Ta	4,629.19	3h	5w
U	4,629.14	2	1
Dy	4,629.13	2	4h
Ho	4,629.10	4	6
Ar	4,628.44	(1000)	—
Ne	4,628.31	(150)	—
As	4,627.80	200	—
In	4,627.38	(150)	+
Cr	4,626.19	125	100
Rn	4,625.48	(500)	—
Xe	4,624.28	(1000)	—
Se	4,623.77	(150)	—
Br	4,622.75	(200)	—
In	4,620.24	(200)	—
La	4,619.88	200	150
Kr	4,619.15	(1000)	—
Se	4,618.77	(100)	—
Ti	4,617.27	100	200
In	4,617.16	(200)	—
Cr	*4,616.14*	*200*	*300r*
Tm	*4,615.93*	*300*	*200*
Kr	*4,615.28*	*(500)*	—
Xe	*4,611.89*	*(700)*	—
Ar	*4,609.60*	*(300)*	—
Rn	*4,609.38*	*(250)*	—
As	*4,607.46*	*200*	—
Rn	*4,606.40*	*(200)*	—
Se	*4,604.34*	*(300)*	—
Xe	*4,603.03*	*(300h)*	—
As	*4,602.73*	*200*	—

	λ	I	
		J	O
C	4,630.52	2d	—
N	4,630.55	(300)	—
Se	4,630.56	(12)	—
Te	4,630.57	(50)	—
Er	4,630.88	3	15
Mo	4,630.92	3	3
Si	4,631.22	4	—
Ca	4,631.39	44	—
Al	4,631.5	(2)	—
Dy	4,631.51	2	4
Kr	4,633.88	(800)	—
Se	4,636.65	(150)	—
In	4,638.10	(200)	—
In	4,638.24	(125)	—
O	4,641.83	(150)	—
N	4,643.11	(100)	—
Yt	4,643.69	100	50
Rn	*4,644.18*	*(300)*	—
Ne	*4,645.42*	*(300)*	—
Se	*4,648.44*	*(800)*	—
O	*4,649.15*	*(300)*	—
Te	*4,654.38*	*(800)*	—
La	*4,655.50*	*300*	*150*
Ne	*4,656.39*	*(300)*	—

As 4,707.82 J 200

	λ	I J	I O
Cr	4,707.75	1	8
Pr	4,707.54	10w	80
V	4,707.44	5	6
Fe	4,707.28	12	100
Mo	4,707.25	125	125
Ce	4,707.00	1	2
Se	4,706.97	5	10
U	4,706.76	3	2
V	4,706.57	15	20
Nd	4,706.54	5	50
Te	4,706.53	(70)	—
U	4,706.40	4	3
O	4,705.32	(300)	—
Br	4,704.86	(250)	—
Mg	4,704.63	(200)	—
Ne	4,704.39	(1500)	—
La	4,703.28	300r	200r
Ne	4,702.53	(150)	—
Gd	4,702.32	100	50
Ar	4,702.316	(1200)	—
La	4,699.63	200r	200r
O	4,699.21	(100)	—
Xe	4,698.01	(150h)	—
Xe	4,697.02	(300)	—
Kr	4,694.44	(200h)	—
La	*4,692.50*	*300*	*200*
Te	*4,686.95*	*(300)*	—
He	*4,685.75*	*(300)*	—
Ra	*4,682.28*	*(800)*	—
In	*4,682.00*	*250W*	—
In	*4,681.08*	*(200)*	—
Rn	*4,680.83*	*(500)*	—
Kr	*4,680.41*	*(500)*	—
Zn	*4,680.14*	*200h*	*300w*
Br	*4,678.69*	*(200)*	—

	λ	I J	I O
J	4,707.83	(8)	—
Gd	4,707.89	30	30
Pr	4,707.94	10w	50
Cr	4,708.04	150	200
Pr	4,708.15	10w	50
U	4,708.17	2	—
La	4,708.19	2	25
Xe	4,708.21	(5)	—
Mo	4,708.22	30	30
Nb	4,708.29	30	50
Ar	4,708.46	(2)	—
Ti	4,708.663	20	2
Ta	4,708.663	5d	2
Ca	4,708.81	5	—
Hf	4,708.84	4	2
Ne	4,708.85	(1200)	—
Yt	4,708.86	1	3
Xe	4,708.92	(8h)	—
Ba	4,708.94	(80)	—
Fe	4,708.96	50	50
Ar	4,709.08	(10)	—
Fe	4,709.10	1	20
Dy	4,709.23	2	4
Ne	4,710.06	(1000)	—
Sb	4,711.26	(100)	—
Ne	4,712.06	(1000)	—
La	4,712.93	150	100r
Ne	4,713.13	(100)	—
Ne	4,715.34	(1500)	—
S	4,716.23	(600)	—
La	4,716.44	200	100
Cr	4,718.43	150	200
La	4,719.95	300	200r
Co	4,721.41	100h	8
Rn	4,721.76	(150)	—
Zn	4,722.16	300h	400w
Gd	*4,723.73*	*200*	*100*
Yb	*4,726.07*	*200*	*45*
Ar	*4,726.91*	*(200)*	—
La	*4,728.42*	*300*	*400r*
Se	*4,730.78*	*(1000)*	—
Gd	*4,732.61*	*300*	*300*
Xe	*4,734.15*	*(600)*	—
Ar	*4,735.94*	*(400)*	—

As 5,331.54 *J* 200

	λ	I	
		J	O
Pr	5,331.48	1	6
Co	5,331.47	80	500w
Nb	5,331.19	2h	—
Kr	5,331.08	(2)	—
Ne	5,330.78	(600)	—
Br	5,330.57	(15)	—
O	5,330.66	(500)	—
J	5,330.16	(2)	—
Sr	5,329.82	2	40
Rh	5,329.74	2	30
Cr	5,329.72	2	5
O	5,329.59	(150)	—
O	5,328.98	(100)	—
Fe	5,328.05	100	400
Ra	5,320.29	(250)	—
Fe	5,316.61	150	—
P	5,316.07	(150w)	—
Xe	5,313.87	(500)	—
Hf	5,311.60	150	100
Ar	5,309.52	(200)	—
Kr	5,308.66	(200)	—
Se	5,305.35	(500)	—
La	5,301.98	200	300r
P	5,296.09	(300w)	—
Xe	5,292.22	(800)	—

	λ	I	
		J	O
U	5,331.85	2	2
Br	5,332.04	(100)	—
Sn	5,332.36	(20)	—
V	3,233.66	5	—
Sn	5,333.21	1	24
Ne	5,333.32	(50)	—
Kr	5,333.41	(500h)	—
La	5,333.42	5	3
Yb	5,335.16	400	150
J	5,338.19	(300)	—
Xe	5,339.38	(500)	—
La	5,340.67	100	80
Mn	5,341.06	100	200
Ne	5,341.09	(1000)	—
Ne	5,343.28	(600)	—
Nb	5,344.17	200	400
P	5,344.72	(150w)	—
J	5,345.15	(300)	—
Yb	5,345.67	100	20
Yb	5,347.20	200	40
Ar	5,347.41	(200)	—
Ne	5,349.21	(150)	—
Tl	5,350.46	2000R	5000R
Yb	5,352.96	250	100
Sb	5,354.24	(200)	—
Cs	5,358.53	(500)	—

Au 79 196.967

t_0 1,063°C t_1 2,966°C

I.	II.	III.	IV.	V.
9.223	20.1	—	—	—

λ	*I*		eV
	J	*O*	
I 2,427.95	100	400*R*	5.10
I 2,675.95	100	250*R*	4.60
I 3,122.78	5	500	5.10
II 3,804.00	150	25	—
I 4,792.60	60	200	7.69
I 5,837.40	10	400	—
I 6,278.18	20	700	—

Au 2,427.95 *J* 100 *O* 400*R*

	λ	I			λ	I	
		J	*O*			*J*	*O*
Os	2,427.90	40	8	Ir	2,427.96	—	5
W	2,427.81	7	1	Mn	2,427.98	50*wh*	—
Cl	2,427.79	(20)	—	Th	2,427.99	25	3
Mn	2,427.75	50*wh*	—	Cl	2,428.02	(10)	—
V	2,427.745	6	6	Pt	2,428.03	10	100
Ru	2,427.74	12	2	Sr	2,428.09	—	10*h*
Al	2,427.70	(15)	—	W	2,428.17	12	5
Cs	2,427.65	(20)	—	Mo	2,428.18	15	—
Ta	2,427.64	1	150	Ag	2,428.196	40*wh*	1
U	2,427.62	2	12	Fe	2,428.20	—	9
Ir	2,427.61	2	25	Pt	2,428.203	20	100
				Ti	2,428.23	—	15
				V	2,428.28	20	30*r*
Mn	2,427.41	50*wh*	—	Fe	2,428.286	10	1
Rh	2,427.34	50	2	Co	2,428.29	20	10
Pd	2,426.87	50	—				
Te	2,426.39	(100)	—				
Co	2,424.93	—	250*R*	Sn	2,429.49	250*R*	200*R*
				Ag	2,429.63	150*wh*	—
Pt	2,424.87	100	50	V	2,430.04	70	—
				Fe	2,430.07	70	15
Pd	*2,424.48*	*100*	—				
Cu	*2,424.44*	*200*	*4*	*Ta*	*2,432.70*	*400*	*300r*
Sn	*2,421.69*	*200R*	*150R*	*O*	*2,433.53*	*(250)*	—
				Nb	*2,433.79*	*100*	*3*

Au 2,675.95 *J* 100 *O* 250*R*

	λ	I			λ	I	
		J	*O*			*J*	*O*
Nb	2,675.94	100	10	V	2,675.97	—	6
Ta	2,675.90	200	150	Co	2,675.98	5	10*W*
U	2,675.88	10	15	Co	2,676.01	10*h*	—
W	2,675.87	6	12	Tl	2,676.03	(30)	—
V	2,675.761	2	12	V	2,676.04	15*h*	—
Tl	2,675.76	(30)	—	Eu	2,676.05	5	—
Ce	2,675.733	—	2	Ti	2,676.08	—	6
W	2,675.73	12	—	Rh	2,676.110	5	10
Cr	2,675.68	15	1	Fe	2,676.11	7*w*	15*w*
Th	2,675.67	4	8	Ce	2,676.12	—	2
La	2,675.65	5	2	Nb	2,676.125	10*h*	2
Ne	2,675.64	(150)	—	Yb	2,676.13	2	—
Ta	2,675.54	—	2	Ru	2,676.19	100	8
Ru	2,675.52	50	—	Rh	2,676.25	100	1
Mn	2,675.51	25	—	P	2,676.28	(10)	—

	λ	I	
		J	O
Ne	2,675.24	(150)	—
J	2,674.80	(60)	—
Pt	2,674.57	10	200
Rh	2,674.44	200*w*	1
Ru	2,674.19	50	—
Nb	2,673.57	500	10
Eu	2,673.41	50	125
Mo	2,673.27	100	1
V	2,673.23	60	10
Ru	2,673.01	50	8
Mo	2,672,84	100	15
Yb	2,672.65	80	20
Mn	2,672.59	125*h*	15
Cl	2,672.19	(50)	—
V	2,672.00	300*R*	50
Nb	2,671.93	200	20
Mo	2,671.83	100	1
Na	2,671.829	(60)	12
Mn	2,671.80	50	—
Co	*2,669.91*	*100wh*	—
Ru	*2,669.42*	*100*	—
Al	*2,669.17*	*100*	*3*
Ta	*2,668.62*	*100*	*1*
Eu	*2,668.33*	*400*	*300w*
Ru	*2,667.39*	*150*	*10*
Yb	*2,666.97*	*150*	*5*
W	**2,676.31**	**6**	—
Mn	**2,676.33**	—	**15**
V	**2,676.35**	**18*h***	—
Ru	**2,676.353**	**3**	**50**
Ce	**3,676.36**	—	**2**
U	**2,676.41**	**4**	**4**
W	**2,676.412**	**1**	**9**
Cl	2,676.95	(150)	—
Pt	2,677.15	200*w*	800*w*
Cr	2,677.16	300*r*	35
Nb	2,677.66	50	2
V	2,677.80	300*R*	70
Eu	2,678.28	100	150
V	2,678.57	150*R*	30
Zr	2,678.63	100	80
Ru	2,678.76	300	100
Cr	2,678.79	80	10
Fe	2,679.06	200	200
Ni	2,679.24	500*wh*	—
V	2,679.32	300*R*	70
Tm	2,679.57	50	30
Pd	2,679.58	100	—
Nb	2,680.06	80	4
Ta	2,680.06	50	30
Ru	2,680.54	50	5
Mo	*2,681.36*	*100*	*10*
Ag	*2,681.38*	*100wh*	—
Rh	*2,681.60*	*100*	—
V	*2,682.87*	*200R*	*50*
V	*2,683.09*	*150R*	*35*
Mo	*2,683.23*	*150*	*20*
Hf	*2,683.35*	*100*	*15*
Rh	*2,683.56*	*200*	*1*
Mo	*2,684.14*	*150*	*40*
Zn	*2,684.16*	*6*	*300*
Rh	*2,684.21*	*150*	*2*
Ni	*2,684.41*	*600wh*	—
Fe	*2,684.75*	*400*	*3*
Pd	*2,684.76*	*100*	—
Eu	*2,685.65*	*100*	*150*

Au 3,122.78 *J* 5* *O* 500

	λ	*I* *J*	*I* *O*		λ	*I* *J*	*I* *O*
Mo	**3,122.76**	**1**	**5**	**Tb**	**3,122.83**	**3**	**8**
Er	**3,122.65**	**8**	**12**	**Ce**	**3,122.85**	—	**3**
Nb	**3,122.646**	**3**	**8**	**V**	**3,122.89**	**300*r***	**12**
Ce	**3,122.62**	—	**10**	**Dy**	**3,122.946**	**1*h***	**3**
U	**3,122.61**	**4**	**5**	**Sc**	**3,122.954**	**8**	**2**
Cr	**3,122.60**	**80**	**10**	**Th**	**3,122.96**	**20**	**20**
Hf	**3,122.547**	**1**	**8**	**Pr**	**3,123.00**	**2**	**10**
Fe	**3,122.547**	**4**	**4**	**Sm**	**3,123.03**	**1**	**2**
Sc	**3,122.542**	—	**2*h***	**Tb**	**3,123.05**	**3**	**15**
Tm	**3,122.54**	—	**10**	**Nd**	**3,123.07**	**6**	**8**
Eu	**3,122.49**	—	**5*d***	**Ti**	**3,123.074**	**15**	**35**
Cu	**3,122.43**	**1**	**7**	**Er**	**3,123.09**	**1**	**12**
Ir	**3,122.38**	**1**	**25**	**Re**	**3,123.17**	—	**25**
Sm	**3,122.36**	**1**	**4**	**U**	**3,123.18**	**1**	**2**
Fe	**3,122.30**	**20**	**70**	**Ir**	**3,123.22**	—	**2**
Ce	**3,122.21**	—	**2**	**Sr**	**3,123.30**	**2**	**2**
Ti	**3,122.06**	**50**	**2**	**Tm**	**3,123.30**	**20**	**10**
Dy	**3,122.03**	**5*h***	**5**	**Fe**	**3,123.346**	**4**	**10**
				Ce	**3,123.349**	—	**8**
Ni	3,114.12	50	300	**Mo**	**3,123.36**	**25**	**5**
Pd	3,114.04	500*w*	400*w*	**Yb**	**3,123.51**	**2**	**1**
				Fe	3,125.65	300	400
				Tm	*3,131.26*	*500*	*400*
				Hg	*3,131.55*	*300*	*400*
				Mo	*3,132.59*	*300R*	*1000R*
				Ni	*3,134.11*	*150*	*1000R*

Au 3,804.00 *J* 150 *O* 25*

	λ	*I* *J*	*I* *O*		λ	*I* *J*	*I* *O*
Th	**3,803.99**	**5**	**8**	**Fe**	**3,804.01**	**10**	**40**
Sm	**3,803.94**	**2**	**15**	**W**	**3,804.08**	**5**	**7**
V	**3,803.90**	**3**	**10**	**Nd**	**3,804.102**	**10*d***	**15*d***
Nb	**3,803.88**	**20**	**30**	**Th**	**3,804.103**	**2**	—
Mo	**3,803.80**	**5**	**3**	**Ce**	**3,804.156**	**2**	**4**
V	**3,803.78**	**3**	**10**	**Hf**	**3,804.16**	**4**	**5**
W	**3,803.68**	**15*w***	—	**Nb**	**3,804.20**	**3*h***	**5**
J	**3,803.49**	**(5*d*)**	—	**Eu**	**3,804.26**	**10*W***	**10*W***
V	**3,803.474**	**40**	**50**	**Mo**	**3,804.52**	**20**	**20**
Nd	**3,803.474**	**20**	**20**	**Hf**	**3,804.53**	**2**	**3**
Mo	**3,803.41**	**20*w***	—	**W**	**3,804.55**	**8*h***	—
U	**3,803.35**	**8**	**8**	**V**	**3,804.59**	**2**	**5**
W	**3,803.32**	**6*h***	—	**Kr**	**3,804.67**	**(30*h*)**	—

	λ	I	
		J	O
Ru	3,803.20	8	10
Ar	3,803.19	(15)	—
O	3,803.13	(20)	—
Er	3,803.12	1d	8d
Pr	3,803.11	20	50d
Eu	3,803.10	10w	9w
Ce	3,803.097	5	35
U	3,803.093	8	1
Sm	3,803.09	3	10
Th	3,803.08	15	15
Ho	3,803.07	4h	—
W	3,802.93	6	7
Nb	3,802.92	50	50
V	3,802.88	8	20
U	3,802.86	5	3
Ce	3,802.80	4	3
Th	3,802.79	5	10
Pr	3,802.76	15	30d
Eu	3,802.75	5w	5w
Dy	3,802.68	2	4
Ir	3,802.678	1h	10
Nb	3,802.64	2	3
P	3,802.07	(100)	—
Sn	3,801.00	150h	200h
Ir	3,800.12	100	150
Fe	3,799.55	300	400
Ru	3,799.35	100	70r
Rh	3,799.31	100	25
Pd	3,799.19	150	200w
Ru	3,798.90	100	70
Fe	3,798.51	300	400
Mo	3,798.25	1000R	1000R
Nb	3,798.12	80	50
Fe	3,797.52	200	300
Gd	3,796.39	150	150w
Tm	3,795.76	150	250
Fe	3,795.00	400	500
La	3,794.77	200	400
Os	3,793.91	300	125
Bi	3,792.8	500h	—
B	3,792.5	(500)	—
Nb	3,791.21	80	80
La	3,790.82	300	400
Ru	3,790.51	150	70
Mn	3,790.21	125	100
Fe	*3,787.88*	*300*	*500*

	λ	I	
		J	O
Ce	3,804.697	3	3
Th	3,804.698	8	10
Nb	3,804.74	50	20
Mn	3,804.757	5	5
La	3,804.76	2h	—
Nd	3,804.77	10	10
Er	3,804.78	1	10
Cr	3,804.80	30	100
Pr	3,804.85	50w	25w
U	3,804.87	4	12r
V	3,804.92	9	25
W	2,804.95	6	7
U	3,805.06	4	1
Cs	3,805.10	(25)	—
Gd	3,805.11	10	10
Ti	3,805.12	1	8
Sm	3,805.19	1	6
Cl	3,805.24	(75)	—
Cu	3,805.30	2h	20
Fe	3,805.34	300	400
Nd	3,805.36	30	50
Dy	3,806.28	80	25
Hg	3,806.38	(200)	—
Fe	3,806.70	150	200
Fe	3,807.54	100	150
Mn	3,809.59	150	150
Bi	3,811.14	(150)	—
Gd	3,812.02	200h	200R
Fe	3,812.96	300	400
Ra	3,814.42	(2000)	—
V	3,815.39	150h	1
Bi	3,815.8	(300)	—
Fe	3,815.84	700	700
Nb	*3,818.86*	*300*	*8*
Eu	*3,819.66*	*500wd*	*500wd*
Fe	*3,820.43*	*600*	*800*
Fe	*3,825.88*	*400*	*500*
P	*3,827.44*	*(150)*	—
Cl	*3,827.62*	*(150)*	—
Fe	*3,827.82*	*200*	*200*
Mg	*3,829.35*	*150*	*100w*
N	*3,830.39*	*(150)*	—

	λ	I J	I O		λ	I J	I O
Kr	*3,783.13*	*(500h)*	—				
Os	*3,782.19*	*200*	*400R*				
Nb	*3,781.38*	*200*	*5*				
Ar	*3,781.36*	*(300)*	—				
Kr	*3,778.09*	*(500h)*	—				

Au 4,792.60 *J* 60 *O* 200

	λ	J	O		λ	J	O
Eu	**4,792.57**	**2**	**20**	**Xe**	**4,792.62**	**(150)**	—
Cr	**4,792.51**	**40**	**200**	**Nd**	**4,792.623**	—	**2**
Ti	**4,792.49**	**12**	**70**	**Se**	**4,792.73**	**(20)**	—
La	**4,792.465**	—	**3**	**Mo**	**4,792.74**	**30**	**40**
U	**4,792.463**	—	**2**	**W**	**4,792.82**	—	**3**
Tb	**4,792.38**	—	**2**	**Sc**	**4,792.84**	**2**	**3**
Sn	**4,792.22**	**(2)**	—	**Co**	**4,792.86**	**5**	**600*W***
Ar	**4,792.12**	**(20)**	—	**Dy**	**4,792.89**	—	**4*h***
P	**4,792.06**	**(70)**	—	**Pr**	**4,792.92**	—	**10*h***
Cl	**4,792.04**	**(12)**	—	**V**	**4,792.96**	**3**	**4**
S	**4,792.02**	**(40)**	—	**Rn**	**4,793.0**	**(10)**	—
Pr	**4,791.95**	—	**8**	**Mn**	**4,793.004**	—	**8**
Dy	**4,791.89**	—	**2**	**Nb**	**4,793.06**	—	**3**
Eu	**4,791.84**	—	**10*W***	**Ir**	**4,793.18**	—	**2**
U	**4,791.83**	—	**3**	**Tb**	**4,793.24**	—	**2**
Mo	**4,791.826**	**5**	**5**	**Zr**	**4,793.28**	—	**3**
Br	**4,791.81**	**(2)**	—	**Mo**	**4,793.41**	**30**	**30**
Gd	**4,791.60**	—	**150**	**Gd**	**4,793.45**	—	**5*h***
Pr	**4,791.597**	—	**3**	**Re**	**4,793.57**	—	**2**
Sm	**4,791.58**	—	**150**	**N**	**4,793.66**	**(5)**	—
Au	**4,791.54**	**2**	—	**Cu**	**4,793.80**	**2*wh***	**5*wh***
Se	**4,791.50**	**6**	**12**	**Tb**	**4,793.82**	—	**2**
Rh	**4,791.48**	—	**2*h***	**Mo**	**4,793.821**	**15**	**15**
Re	**4,791.42**	—	**200*w***	**Os**	**4,793.99**	**6**	**300**
La	**4,791.40**	—	**6**				
Dy	**4,791.30**	**2**	**8**	Cl	4,794.54	(250)	—
Pr	**4,791.27**	—	**6**	Co	4,795.85	—	100
Fe	**4,791.25**	**200*R***	**200**	Te	4,796.10	(70)	—
Eu	**4,791.16**	—	**3**	Cr	4,796.17	1*h*	125
Ar	**4,791.15**	**(2)**	—	Kr	4,796.33	(60*h*)	—
Kr	**4,791.15**	**(3)**	—	Co	4,796.37	—	100
U	**4,791.10**	—	**3*s***	Mo	4,796.52	40	40
K	**4,791.08**	—	**2**	Hg	4,797.01	(300)	—
				Ar	4,798.74	(30)	—
				Gd	4,798.87	60	25
Ne	4,790.73	(30)	—				
Cr	4,790.34	1	100	Cd	4,799.92	300*w*	300*w*
Ne	4,790.22	(50)	—	Gd	4,800.11	80	30
Fe	4,789.65	—	100	Cr	4,801.03	70	200
Ne	4,789.60	(100)	—	Gd	4,801.08	200	200

	λ	I J	I O
Cr	4,789.38	100	300
Ne	4,788.93	(300)	—
Pd	4,788.17	4*h*	200*h*
Xe	4,787.77	(50)	—
Fe	4,786.81	—	150
Yb	4,786.60	200	50
Ni	4,786.54	2	300*W*
Sm	4,785.87	—	100
Br	4,785.50	(400)	—
Cl	4,785.44	(50)	—
Lu	4,785.42	200	100
Mo	4,785.12	30	30
Te	4,784.85	(70)	—
Gd	4,784.64	50	100
Sb	4,784.03	(70)	—
Mn	4,783.42	60	400
Pr	4,783.35	10*w*	125
Mo	4,782.94	40	40
Gd	4,781.93	50	200
Cl	4,781.82	(50)	—
Co	4,781.43	2*h*	400
Cl	4,781.32	(75)	—
Ta	4,780.93	200	50
Ne	4,780.88	(30)	—
Ne	4,780.34	(50)	—
Br	4,780.31	(125)	—
Co	4,780.01	500	500*w*
Ti	4,779.95	100*h*	10
Sc	4,779.35	40	80
Xe	4,779.18	(50)	—
Cl	4,778.93	(45)	—
P	4,778.80	(30)	—
Co	4,778.25	—	100
Sm	4,777.84	—	100
Br	*4,776.42*	*(200)*	—
Rb	*4,776.41*	*(100)*	—
Co	*4,776.32*	—	*300*
O	*4,773.76*	*(70)*	—
Co	*4,771.11*	—	*500w*
Xe	*4,769.05*	*(100)*	—
Cl	*4,768.68*	*(150)*	—
Ar	*4,768.67*	*(150)*	—
Rn	*4,768.59*	*(100)*	—
Co	*4,768.08*	*10*	*300*
Br	*4,767.10*	*(200)*	—
Te	*4,766.03*	*(150)*	—
Kr	*4,765.74*	*(1000)*	—
Cr	*4,764.89*	*(150)*	—
Sb	4,802.01	(40)	—
O	4,802.20	(30)	—
Gd	4,802.58	40*h*	100
O	4,803.00	(50)	—
N	4,803.27	(30)	—
Gd	4,803.55	40	100
La	4,804.04	150	150
Ti	4,805.10	125	15
Mo	4,805.58	30	30
Gd	4,805.82	80	200
Ar	4,806.07	(500)	—
Ni	4,807.00	1	150*W*
Xe	4,807.02	(500)	—
Gd	4,807.46	40	100
V	4,807.53	30*h*	40*h*
La	*4,809.01*	*150*	*150*
Cl	*4,810.06*	*(200)*	—
Ne	*4,810.063*	*(150)*	—
Zn	*4,810.53*	*300h*	*400w*
Ne	*4,810.63*	*(100)*	—
Nd	*4,811.34*	*60*	*60*
Kr	*4,811.76*	*(300)*	—
Ac	*4,812.25*	*60*	—
Co	*4,813.48*	*6*	*1000W*
Ge	*4,814.80*	*200*	—
S	*4,815.51*	*(800)*	—
Sm	*4,815.81*	*80*	*125*
Br	*4,815.71*	*(300)*	—
Rn	*4,817.15*	*(100)*	—
Ne	*4,817.64*	*(300)*	—
Xe	*4,818.02*	*(100)*	—
Ne	*4,818.79*	*(150)*	—
Mo	*4,819.25*	*60*	*80*
Cl	*4,819.46*	*(200)*	—
Ne	*4,819.94*	*(70)*	—
Yb	*4,820.24*	*60*	*15*
Fe	*4,821.05*	*200h*	*200h*
Gd	*4,821.71*	*80*	*150*
Ne	*4,821.92*	*(300)*	—
Ne	*4,823.17*	*(100)*	—
Xe	*4,823.41*	*(150h)*	—
Mn	*4,823.52*	*80*	*400*

	λ	I J	I O
Cr	*4,764.29*	*35*	*200*
Se	*4,763.65*	*(800)*	—
J	*4,763.38*	*(80)*	—
Kr	*4,762.43*	*(300)*	—
Hg	*4,762.22*	*(100)*	—

Au 5,837.40 *J* 10* *O* 400

	λ	I J	I O
Ta	**5,837.27**	—	**4**
*bh***Yt**	**5,837.2**	—	**6**
Yb	**5,837.16**	**150**	**50**
Nd	**5,837.14**	—	**2**
Tm	**5,836.72**	—	**5**
Ta	**5,836.62**	—	**2***h*
Sm	**5,836.59**	—	**2**
Fe	**5,836.56**	—	**2**
*bh***Sc**	**5,836.5**	—	**6**
Sm	**5,836.36**	—	**100***d*
*bh***F**	**5,836.3**	—	**30**
U	**5,836.05**	—	**30**
*bh***F**	**5,835.9**	—	**20**
Ce	**5,835.84**	—	**25**
Er	**5,835.83**	—	**8**
Mo	**5,835.59**	—	**20**
Re	**5,835.48**	—	**2**
Fe	**5,835.281**	—	**3**
Nd	**5,835.28**	—	**2**
*bh***F**	**5,835.2**	—	**30**
Pr	**5,835.13**	—	**15**
Nb	**5,834.90**	**10**	**15**
Re	*5,834.32*	—	*200*
Eu	*5,831.05*	—	*2000 W*
Eu	*5,818.74*	—	*1000*
*bh***F**	**5,837.5**	—	**30**
Fe	**5,837.70**	—	**3**
U	**5,837.71**	**1**	**30**
*bh***V**	**5,837.8**	—	**7**
Eu	**5,838.03**	—	**10**
Tb	**5,838.05**	—	**15**
*bh***Yt**	**5,838.1**	—	**10**
Nb	**5,838.15**	**2**	**15**
Ce	**5,838.16**	—	**15**
Sm	**5,838.43**	—	**5**
Nb	**5,838.64**	**100**	**200**
Cr	**5,838.68**	—	**4**
Tm	**5,838.77**	**40**	**20**
Hf	**5,838.87**	—	**2**
*bh***F**	**5,838.9**	—	**30**
Th	**5,838.95**	**1**	**12**
W	**5,838.99**	—	**25**
U	**5,839.05**	—	**2**
Nd	**5,839.13**	—	**2**
Ho	**5,839.14**	—	**12**
*bh***F**	**5,839.3**	—	**10**
Ce	**5,839.38**	—	**10**
Ho	**5,839.47**	—	**30**
Sm	**5,839.54**	—	**3**
*bh***Sc**	**5,839.6**	—	**10**
La	**5,839.79**	—	**3**
*bh***Zr**	**5,839.8**	—	**2**
Sm	**5,839.88**	—	**5***d*
Ba	5,853.68	100*h*	300
*bh*C	5,858.2	—	400
Mo	5,858.27	200	200
Ho	5,860.28	—	200

Au 6,278.18 *J* 20* *O* 700

	λ	I J	I O
Rh	**6,278.08**	—	**3**
Ta	**6,277.56**	—	**2**
Ti	**6,277.52**	—	**2**
Rh	**6,277.46**	—	**15**
Nd	**6,277.29**	—	**5**
Th	**6,277.25**	—	**10**
Ce	**6,277.11**	—	**3**
Sm	**6,276.77**	—	**10*d***
Rh	**6,276.66**	—	**8**
Co	**6,276.63**	—	**40**
*bh***F**	**6,276.5**	—	**20**
Sm	**6,276.48**	—	**2**
Ce	**6,276.46**	—	**10**
Sc	**6,276.31**	**8**	**10**
Sn	**6,275.78**	—	**2**
Nb	**6,275.42**	**2*h***	**6**
Co	**6,275.13**	—	**25**
Sm	**6,275.08**	—	**4**
*bh***Yt**	**6,275.0**	—	**15**
Er	**6,274.96**	—	**8**
Os	**6,274.94**	—	**3**
Ho	**6,274.90**	—	**2**
Yb	6,274.79	—	100
Eu	6,262.28	—	600
Ni	6,256.36	10	600*w*
Sm	**6,278.23**	—	**9*d***
Nd	**6,278.25**	—	**2**
Ta	**6,278.34**	—	**15**
Mo	**6,278.43**	—	**2**
Pr	**6,278.67**	—	**9**
Re	**6,278.76**	—	**20**
Sm	**6,279.06**	—	**2**
Th	**6,279.17**	**2**	**20**
Sm	**6,279.49**	—	**10**
U	**6,279.63**	—	**4**
Sc	**6,279.76**	**25**	**10**
Zr	**6,279.77**	—	**6**
Hf	**6,279.84**	**20**	**15**
Sc	**6,280.17**	**3**	**5*h***
U	**6,280.19**	—	**10**
Rh	**6,280.67**	—	**3**
*bh***F**	**6,280.9**	—	**20**
V	**6,280.91**	—	**2**
Pr	**6,281.31**	**1**	**25**
Ta	**6,281.33**	—	**50**
Er	**6,281.40**	—	**4**
Co	6,282.63	—	300*W*
Rb	6,298.33	150	1000
Eu	6,299.76	—	500*W*
Eu	*6,303.39*	—	*700*

Ba 56 137.34

t_0 850°C

t_1 1,140°C

I.	II.	III.	IV.	V.
5,2097	10,001	—	—	—

λ	I		eV
	J	*O*	
II 2,335.27	100*R*	60*R*	12.0
I 3,071.59	50*R*	100*R*	4.0
I 3,501.12	20	1000	3.54
II 4,554.04	200	1000*R*	7.9
II 4,934.09	400	400	7.7
I 5,424.62	30*R*	100*R*	3.8
I 5,519.11	60*R*	200*R*	3.8
I 5,535.55	200*R*	1000*R*	2.2
I 5,777.66	100*R*	500*R*	3.8

Ba 2,335.27 *J* 100*R* *O* 60*R*

	λ	*I* *J*	*I* *O*
Hf	2,335.25	2	—
W	2,335.20	8	6
Pt	2,335.19	25	8
Co	2,335.11	—	15
Re	2,335.03	—	10
Tm	2,335.00	8	—
Ir	2,334.96	25	—
Sn	2,334.81	100*R*	100*R*
Rh	2,334.77	500	25
In	2,334.57	(50*h*)	—
V	2,334.21	250	—
Ir	2,334.84	10	40
V	2,333.60	50	—
Rb	2,333.38	(80)	—
Rh	2,333.31	125	8
Hf	2,332.97	50	40
Pb	2,332.43	30	60
V	*2,331.77*	*300*	—
Zr	*2,331.57*	*100w*	—
Ag	*2,331.37*	*150wh*	*18*
V	*2,330.46*	*300*	*8*
Rh	2,335.28	15	—
Ir	2,335.285	—	5
Nb	2,335.31	10*w*	3*h*
V	2,335.34	3	2
Mo	2,335.38	5	—
Yb	2,335.419	15*h*	3
Rb	2,335.42	(5)	—
Xe	2,335.42	(2)	—
Pd	2,335.43	(3)	—
J	2,335.45	(60)	—
Th	2,335.49	3	—
V	2,335.492	50	3
J	2,335.52	(15*d*)	—
W	2,335.56	3	—
Pd	2,335.57	(5)	—
Re	2,335.75	9	40
V	2,336.10	50	2
Ni	2,336.59	150	—
Os	2,336.80	80	50
Rh	2,336.84	125	1
Rb	2,337.05	(125)	—
V	2,337.13	100	—
V	2,337.32	80	—
Ar	2,337.79	(60)	—
Ru	*2,340.69*	*4*	*60*

Ba 3,071.59 *J* 50*R* *O* 100*R*

	λ	*I* *J*	*I* *O*
Yb	3,071.59	3	—
Ca	3,071.587	1	5
Cr	3,071.57	12	—
Nb	3,071.56	50	10
Ru	3,071.50	—	20
U	3,071.483	1	4
Nd	3,071.482	—	2
Mo	3,071.44	3	25
Nd	3,071.43	2	4*d*
Xe	3,071.39	(3*h*)	—
Nb	3,071.36	—	2
Cl	3,071.35	(40)	—
U	3,071.33	1	5
Cr	3,071.30	1	15
Sm	3,071.27	4	8
vzduch	3,071.60	5	—
Ce	3,071.62	1	18*s*
Cd	3,071.65	2	—
W	3,071.72	15	8
U	3,071.721	4	5
Re	3,071.75	—	10*h*
U	3,071.82	2	3
P	3,071.83	(30)	—
Dy	3,071.92	4	25
Pt	3,071.94	15	60
Co	3,071.96	2	80
Fe	3,072.05	3	6
Zn	3,072.06	125	200
Ta	3,072.07	—	2*wh*
Ti	3,072.11	125	25

	λ	I J	I O
Fe	**3,071.267**	**2**	**1**
Ti	**3,071.24**	**70**	**12**
Sn	**3,071.2**	**2*wh***	**—**
Nb	**3,071.18**	**15**	**3**
Re	**3,071.16**	**—**	**50**
Fe	**3,071.15**	**2**	**2**
Ce	**3,071.11**	**1**	**20*s***
Ne	**3,071.08**	**(4)**	**—**
Rh	**3,071.03**	**—**	**20**
U	**3,070.99**	**8**	**8**
Ir	**3,070.92**	**—**	**3**
Nb	**3,070.90**	**15**	**3**
Mo	**3,070.895**	**3**	**40**
V	**3,070.892**	**2*h***	**10**
Sm	**3,070.87**	**2**	**2**
Co	**3,070.86**	**—**	**5**
Mo	3,070.62	25	—
Mn	3,070.27	25	100
V	3,070.12	50	—
Re	3,069.95	—	125
Os	3,069.94	15	125
Ta	3,069.24	70	150
Te	3,068.94	(50)	—
Rn	3,068.9	(100)	—
Gd	3,068.65	50	50
Ru	3,068.26	8	60
Fe	3,068.18	150	150
Bi	3,067.72	2000*wh*	3000*hR*
Mo	3,067.64	50	10
Re	3,067.39	—	60
Rh	3,067.30	1	80
Fe	3,067.24	300	300
Cr	3,067.16	40	25
Ge	3,067.01	40	60
S	3,066.89	(35)	—
Fe	3,066.48	40	60
Ru	3,066.40	25	10
V	3,066.37	125*r*	400*r*
Ti	3,066.35	40	15
Al	3,066.16	25	25
Os	3,066.12	10	50
Pd	3,066.10	2	150
Nb	3,066.097	30	2
Mn	3,066.02	—	75
V	3,065.60	35	2
Fe	3,065.31	60	—

	λ	I J	I O
Th	**3,072.12**	**8**	**10**
Er	**3,072.13**	**1**	**6**
Ga	**3,072.14**	**3**	**—**
Nb	**3,072.18**	**5**	**1**
J	**3,072.25**	**(7)**	**—**
Nb	**3,072.30**	**1**	**2**
Yt	**3,072.334**	**8**	**8**
Fe	**3,072.335**	**7**	**8**
Ru	**3,072.335**	**40**	**5**
U	**3,072.336**	**6**	**6**
Co	3,072.34	100	200*R*
Gd	3,072.57	25	20
V	3,072.71	40*r*	70*r*
Hf	3,072.88	18	80
Ti	3,072.97	200*r*	35
Tm	3,073.08	150	60
Mn	3,073.13	20	75
Ru	3,073.34	5	50
Tm	3,073.50	60	25
Ru	3,073.51	80	10
Te	3,073.53	(100)	—
Cr	3,073.68	25	35
Cu	3,073.80	20	70
V	3,073.82	20*r*	60
Tm	3,073.85	50	15
Fe	3,073.98	25	40
Os	3,074.08	20	125
Os	3,074.15	25	40
Na	3,074.33	(60)	8
Mo	3,074.37	15	60
Fe	3,074.44	25	40
V	3,074.658	30	—
Al	3,074.665	(50)	—
Os	3,074.96	20	125
Ti	3.075.22	300*R*	50
Ru	3,075.31	40	10
As	3,075.32	35	60
Fe	3,075.72	400	400
Zn	3,075.90	50	150
S	3,076.20	(25)	—
Ta	3,076.38	25*ws*	35
Te	3,076.57	(25)	—
Ru	3,076.78	3	50
Nb	3,076.87	50	10*w*
Gd	3,076.971	25	25
Ne	3,076.971	(150)	—
Os	3,077.056	12	100
Ru	3,077.063	40	30

	λ	I J	I O
Pd	3,065.306	100	10
Nb	3,065.26	200	10
Sc	3,065.11	25	12d
Cr	3,065.07	50	20
Yb	3,065.05	30	4
Ru	3,064.84	60	70
Pt	3,064.71	300R	2000R
Hf	3,064.68	30	10
Ni	3,064.62	50	200r
Rn	3,064.6	(40)	—
Nb	3,064.53	200	5w
Co	3,064.37	—	100
Mo	3,064.28	10	80
Ne	*3,063.69*	*(150)*	*—*
Cu	*3,063.41*	*50*	*300*
V	*3,063.25*	*80r*	*30*
Fe	*3,062.23*	*400*	*2*
Os	*3,062.19*	*30*	*100*
Co	*3,061.82*	*125*	*200R*
Ru	*3,060.49*	*50*	*8*
V	*3,060.46*	*100r*	*150r*
Os	*3,060.30*	*30*	*100*
Ta	*3,060.29*	*35W*	*125*
Ru	*3,060.23*	*50*	*20*
Co	*3,060.05*	*1*	*150*
F	*3,059.96*	*(60)*	*—*
Cr	*3,059.52*	*60*	*9*
Pd	*3,059.43*	*150w*	*—*
Fe	*3,059.09*	*400*	*600r*
Os	*3,058.66*	*500*	*500R*
Ti	*3,058.09*	*70*	*12*
Lu	*3,057.90*	*150h*	*3*
Ni	*3,057.64*	*125*	*400R*
Fe	*3,057.45*	*400*	*400*
Ne	*3,057.39*	*(250)*	*—*
Ta	*3,057.12*	*125*	*25w*
Ru	*3,056.86*	*150*	*12*
Ti	*3,056.74*	*70*	*12*
Lu	*3,056.72*	*100*	*50*
Ta	*3,056.61*	*70*	*1wh*
Fe	3,077.17	300	1
Ta	3,077.24	50w	150w
Cr	3,077.25	40	—
Os	3,077.44	8	80
W	3,077.52	40wh	2
Lu	3,077.60	200	100
Fe	3,077.64	25	60
Mo	3,077.66	125	20
Rn	3,077.7	(30)	—
V	3,077.718	30	10
Os	3,077.720	30	100
Cr	3,077.83	125r	25
Fe	3,078.02	80	100
Os	3,078.11	15	125
Ta	3,078.23	5	50
Na	3,078.31	(60)	12
Os	3,078.38	15	125
Fe	3,078.43	50	80
Ti	3,078.64	500R	60
J	3,078.77	(350)	—
Th	3,078.83	25h	10
Tb	3,078.86	80	30
Ne	3,078.87	(75)	—
Ne	3,079.17	(75)	—
Mn	*3,079.63*	*40*	*125*
Nb	*3,080.35*	*100*	*8*
Hf	*3,080.66*	*100*	*30*
Ni	*3,080.75*	*60*	*200*
Cd	*3,080.83*	*100*	*150*
V	*3,081.25*	*50*	*5*
Ru	*3,081.38*	*50*	*4*
J	*3,081.66*	*(100)*	*—*
Gd	*3,082.00*	*60*	*100*
Al	*3,082.15*	*800*	*800*
Re	*3,082.43*	*—*	*100w*
Co	*3,082.62*	*50*	*150R*
V	*3,083.21*	*50*	*2*
Fe	*3,083.74*	*500*	*500*
Rh	*3,083.96*	*2*	*150*
Mo	*3,085.61*	*25*	*125*

Ba 3,501.12 *J* 20* *O* 1000

	λ	*I* *J*	*I* *O*
Ce	**3,501.03**	—	5
U	**3,501.00**	—	8
Fe	**3,500.86**	—	2
Mn	**3,500.852**	—	5
Ni	**3,500.852**	80	500*wh*
Th	**3,500.851**	3	5
Tb	**3,500.84**	15	70
Ce	**3,500.83**	—	4
V	**3,500.82**	25	35
Ce	**3,500.68**	—	18
Yt	**3,500.64**	—	5
Fe	**3,500.57**	20	50
Th	**3,500.55**	5	6
Dy	**3,500.50**	4	2
Sm	**3,500.498**	15	30
V	**3,500.35**	2	2
Ti	**3,500.34**	35	15
U	**3,500.333**	10	3
Pr	**3,500.328**	—	6
Cu	**3,500.324**	2*h*	20
Pt	**3,500.318**	—	2*h*
W	**3,500.28**	10	10
Tb	**3,500.27**	—	50
Th	**3,500.269**	2	2
W	**3,500.23**	—	5
Zr	**3,500.15**	3	4
Nb	**3,500.11**	1	5
U	**3,500.08**	2	15
Ru	3,498.94	200	500*R*
Rh	3,498.73	60	500
Co	3,495.69	25	1000*R*
Ni	3,492.96	100*h*	1000*R*

	λ	*I* *J*	*I* *O*
Er	**3,501.14**	—	9
Os	**3,501.16**	15	100
Tb	**3,501.19**	—	8
Sm	**3,501.23**	1	2
Cu	**3,501.338**	1*h*	5
Ce	**3,501.339**	—	4
U	**3,501.34**	3*d*	5*d*
Nb	**3,501.341**	30	3
Zr	**3,501.347**	1	15
Ru	**3,501.354**	3	30
Dy	**3,501.44**	3	10
Ce	**3,501.45**	3	18*w*
Th	**3,501.46**	10	10
V	**3,501.48**	20	25
Zr	**3,501.49**	3	12
Cu	**3,501.529**	1*h*	2*h*
Nd	**3,501.535**	4	10*d*
Ce	**3,501.59**	—	12
U	**3,501.65**	2	4
Ag	**3,501.68**	20*h*	8*h*
Os	**3,501.69**	5	30
Co	**3,501.72**	100	5
Ce	**3,501.81**	—	8
Dy	**3,501.87**	1	4
Ag	**3,501.94**	2	5
Yt	**3,501.96**	2	3
Ta	**3,501.97**	1	3
Nd	**3,502.01**	4	8
Dy	**3,502.09**	4	6
Tb	**3,502.10**	—	8
Co	3,502.28	20	2000*R*
Rh	3,502.52	150	1000
Rh	3,507.32	125	500
Ni	3,510.34	50*h*	900*R*
Ni	*3,515.05*	*50h*	*1000R*
Pd	*3,516.94*	*500R*	*1000R*
Tl	*3,519.24*	*1000R*	*2000R*

Ba 4,554.04 *J* 200 *O* 1000*R*

	λ	*J*	*O*
Ce	**4,554.03**	—	35*s*
Mo	**4,554.03**	4	1
Zr	**4,553.97**	12	4
Cr	**4,553.95**	3	20

	λ	*J*	*O*
Dy	**4,554.23**	—	2
Ar	**4,554.32**	(15)	—
Ce	**4,554.33**	—	2
Ne	**4,554.41**	(10)	—

	λ	I	
		J	O
U	4,553.86	1	4
Th	4,553.85	2	3
Nb	4,553.84	8	5
Mo	4,553.80	20	20
Hf	4,553.78	—	10
Ce	4,553.75	—	2
*bh*Pb	4,553.7	—	6
Ta	4,553.69	2	200
W	4,553.66	1	6
Yb	4,553.56	60	20
Mo	4,553.503	25	—
Pr	4,553.498	—	4
Ce	4,553.42	—	2
Ti	4,553.41	—	6
Co	4,553.33	—	25
Mo	4,553.32	4	12
Mn	4,553.31	—	12*h*
Eu	4,553.3	—	3*w*
Ca	4,553.27	4*h*	—
Pr	4,553.26	—	5*h*
Hf	4,553.24	—	5*h*
Mo	4,553.22	6	12
Ni	4,553.17	—	15*r*
Ne	4,553.16	(50)	—
Er	4,553.13	—	5
*bh*C	4,553.1	—	—
Ce	4,553.06	—	8
V	4,553.05	15	20
Th	4,553.04	3*h*	3
Zr	4,553.01	—	10
Pd	4,552.892	—	2
Hg	4,552.89	(30)	—
Pr	4,552.83	—	5
Mo	4,552.80	5	5
Kr	4,552.77	(3)	—
Sm	4,552.66	40	80
S	4,552.38	(200)	—
Ta	4,551.95	8	400
Co	4,549.66	—	600
Ti	4,549.63	200	100
Fe	4,549.47	100	100
As	4,549.23	125	—
Cr	4,545.96	125	200
Xe	4,545.23	(200*wh*)	—
Ar	4,545.08	(200)	—
Co	4,543.81	—	500*W*
As	4,543.76	200	—

	λ	I	
		J	O
Sm	4,554.44	—	60
Yt	4,554.46	6	3
Pr	4,554.50	—	2
Ru	4,554.51	200	1000*R*
Ce	4,554.557	—	6
Ne	4,554.561	(5)	—
Ca	4,554.590	2	—
Pt	4,554.593	5	10
Tm	4,554.65	—	5
W	4,554.68	—	4
Ir	4,554.78	—	4
Pr	4,554.79	1	4
P	4,554.80	(100)	—
Ne	4,554.82	(40)	—
Cr	4,554.83	2	25
Nd	4,554.97	1	5
Gd	4,554.99	2	6
Cr	4,555.03	40*h*	—
Th	4,555.07	—	3
Ti	4,555.08	2	12
Cr	4,555.092	50	15
U	4,555.095	40	20
Zr	4,555.13	—	15
Nd	4,555.14	—	15
Dy	4,555.24	—	4
Tm	4,555.26	50	25
Te	4,555.27	(30)	—
Cr	4,555.30	—	15
Yt	4,555.30	2	2
W	4,555.33	1	7
Cs	4,555.35	100	2000*R*
Eu	4,555.38	4	4
Ne	4,555.39	(30)	—
Xe	4,555.94	(100*wh*)	—
Kr	4,556.61	(200*h*)	—
Te	4,557.84	(300)	—
P	4,558.03	(100)	—
La	4,558.46	200	100
Cr	4,558.66	600*wh*	20
La	4,559.29	150	100
Ti	4,563.77	200	100
Se	4,563.95	(200)	—
V	4,564.59	150	—
P	4,565.21	(100)	—
Co	4,565.59	12	800*W*
Ti	*4,571.98*	*300*	*150*

	λ	I	
		J	O
Br	4,542.93	(250)	—
Xe	4,540.89	(200*h*)	—
Gd	4,540.02	200	80
As	4,539.97	200	—
Ne	4,538.31	(300)	—
Ne	4,537.75	(1000)	—
Ne	4,537.68	(300)	—
Ra	4,533.11	(300)	—
Co	4,530.96	8	1000*w*
Fe	4,528.62	200	600
La	*4,574.87*	*300*	*300*
Ne	*4,575.06*	*(300)*	—
Kr	*4,577.20*	*(800)*	—
Rn	*4,577.72*	*(250)*	—
Co	*4,581.60*	*10*	*1000W*

Ba 4,934.086 *J* 400 *O* 400

	λ	I	
		J	O
Er	**4,934.074**	—	**18**
Pr	**4,934.071**	—	**3**
Co	**4,934.06**	—	**25**
Fe	**4,934.02**	—	**40**
Ni	**4,934.00**	—	**3**
Th	**4,933.85**	**1**	**8**
Dy	**4,933.84**	—	**2**
W	**4,933.82**	—	**12**
Re	**4,933.74**	—	**15*w***
Mo	**4,933.73**	**4**	**12**
U	**4,933.66**	**8**	**8**
Zr	**4,933.64**	—	**4**
Fe	**4,933.63**	**70**	**2**
Ta	**4,933.53**	**1**	**5*s***
Ir	**4,933.50**	—	**2**
Mo	**4,933.46**	**2**	**4**
Tb	**4,933.38**	—	**2**
Fe	**4,933.35**	**30**	**50**
Mo	**4,933.33**	**3**	**15**
Sm	**4,933.30**	—	**25**
F	**4,933.25**	**(30)**	—
Ar	**4,933.24**	**(30)**	—
Nb	**4,933.22**	**6*w***	**5*h***
Mo	**4,933.10**	**15**	**30**
U	**4,933.06**	—	**6**
Co	**4,932.88**	—	**5**
Eu	**4,932.82**	—	**8**
Pr	**4,932.81**	—	**2**
W	**4,932.79**	**1*h***	**12**
Co	4,928.28	—	200*W*
Fe	4,925.29	50*r*	1000*R*
Xe	4,923.15	(500)	—
Th	**4,934.088**	**2**	**4**
Mn	**4,934.15**	**5**	**25**
Sc	**4,934.24**	**8**	—
Hf	**4,934.45**	**50**	**40**
Kr	**4,934.48**	**(4*h*)**	—
Ru	**4,934.61**	—	**4**
La	**4,934.82**	**100**	**150**
N	**4,935.03**	**(250)**	—
Pr	**4,935.14**	—	**2**
Co	**4,935.22**	—	**2**
*bh***Mg**	**4,935.3**	—	**4**
Pr	**4,935.38**	—	**2**
Sm	**4,935.46**	—	**20**
Er	**4,935.498**	—	**35**
Yb	**4,935.502**	**10**	**200**
P	**4,935.55**	**15**	—
La	**4,935.62**	—	**8**
Ru	**4,935.63**	—	**10**
*bh***C**	**4,935.7**	—	—
Sc	**4,935.73**	**1**	**5**
Cr	4,936.33	5	200
Ni	4,937.34	—	400*w*
Eu	4,938.31	—	250*W*
Fe	4,938.82	1	300
Kr	4,945.59	(300)	—
La	4,949.77	—	200
Ne	*4,957.03*	*(1000)*	—

	λ	I J	I O		λ	I J	I O
Cr	4,922.27	40	200				
La	4,921.78	400	500				
Xe	4,921.48	(500)	—				
La	4,920.97	400	500				
Fe	4,920.50	125	500				
Fe	4,919.00	50	300				
Ni	4,918.36	1	200*W*				
Xe	*4,916.51*	*(500)*	—				
Ni	*4,904.41*	*1*	*400W*				
Fe	*4,903.32*	*2*	*500*				

Ba 5,424.62 *J* 30*R** *O* 100*R*

	λ	J	O		λ	J	O
Yt	**5,424.36**	—	**5**	**W**	**5,424.64**	—	**3**
Dy	**5,424.25**	—	**3**	**Gd**	**5,424.65**	—	**7**
Th	**5,424.15**	—	**3**	**Ni**	**5,424.654**	—	**30**
V	**5,424.082**	**25**	**25**	**Rh**	**5,424.72**	**1**	**25**
Fe	**5,424.076**	**20**	**400**	**W**	**5,424.93**	—	**2**
Nd	**5,424.068**	—	**3**	**Tb**	**5,425.00**	—	**15**
Rh	**5,424.068**	**2**	**100**	**Rh**	**5,425.45**	**1**	**25**
Sm	**5,423.96**	**1**	**2**	**Ba**	**5,425.55**	—	**10**
W	**5,423.93**	—	**10**	**Sc**	**5,425.57**	—	**3**
La	**5,423.82**	**4**	**3*h***	**Co**	**5,425.62**	—	**4**
Re	**5,423.80**	—	**6**	**Sm**	**5,425.64**	—	**9**
Gd	**5,423.61**	—	**8**	**Th**	**5,425.68**	**3**	**15*d***
Re	**5,423.60**	—	**5**	**Mo**	**5,425.99**	**2**	**3**
Nd	**5,423.55**	—	**3**	**Ti**	**5,426.26**	**30**	**15**
Zr	**5,423.51**	—	**2**	**Zr**	**5,426.36**	—	**3**
W	**5,423.45**	—	**5**	**Ce**	**5,426.37**	—	**6**
Ce	**5,423.42**	—	**12**	**Sm**	**5,426.39**	—	**3**
Dy	**5,423.32**	—	**5**	**Tb**	**5,426.43**	—	**15**
Rh	**5,423.29**	—	**15**	**Dy**	**5,426.71**	—	**5**
Re	**5,423.27**	—	**2**				
Tb	**5,423.24**	—	**15**	Eu	5,426.93	—	200
Sm	**5,422.97**	—	**4*wh***	Fe	5,429.70	40	500
W	**5,422.89**	—	**9**	Rb	5,431.53	—	100
				Ta	5,431.66	—	60*w*
				V	5,434.17	50	50
Er	5,422.80	—	30				
Lu	5,421.90	5	50	Fe	5,434.53	35	300
Eu	5,421.08	—	125	Ta	5,435.27	—	80
Mn	5,420.36	—	60	Ni	5,435.87	—	50
Ta	5,419.13	—	80*R*	Os	5,443.31	—	50
				Eu	5,443.56	—	125
Os	5,416.69	—	50				
Os	5,416.34	—	80	Co	5,444.57	—	400*w*
Sm	5,416.05	—	100	Fe	5,445.04	—	150
V	5,415.26	75	75				

	λ	I J	I O		λ	I J	I O
Fe	5,415.21	20	500	*Fe*	*5,446.92*	*35*	*300*
				Eu	*5,451.53*	—	*1000s*
Er	5,414.64	—	50	*Eu*	*5,452.96*	—	*1000s*
Eu	5,411.84	—	80	*Sm*	*5,453.02*	—	*100*
Fe	5,410.91	10	200	*Co*	*5,454.55*	—	*300w*
Cr	5,409.79	30	300*R*				
Ti	5,409.61	1	50	*Ru*	*5,454.82*	—	*100*
				La	*5,455.15*	*1*	*200*
Ce	5,409.22	—	50	*Fe*	*5,455.61*	*30*	*300*
Co	5,407.51	—	100	*Fe*	*5,463.28*	—	*100*
Mn	5,407.42	—	60	*Ag*	*5,465.49*	*500R*	*1000R*
Fe	5,405.78	70	400				
Sm	5,405.24	—	80	*Yt*	*5,466.47*	*20*	*150*
Ta	5,404.95	—	80				
Rh	5,404.73	1	50				
Fe	5,404.15	35	300				
Eu	5,402.79	—	1000				
Lu	*5,402.57*	*10*	*150*				
Co	*5,401.98*	—	*100w*				
V	*5,401.93*	*100*	*100*				
Ru	*5,401.04*	—	*125*				
Fe	*5,400.50*	—	*125*				
Fe	*5,397.13*	*50*	*400*				
Fe	*5,393.182*	*10*	*150*				
V	*5,393.177*	—	*100*				
Eu	*5,392.91*	—	*150*				
Rh	*5,390.44*	*3*	*125*				
Ta	*5,389.30*	—	*100W*				
Fe	*5,383.37*	*40*	*400h*				

Ba 5,519.11 *J* 60*R** *O* 200*R*

	λ	J	O		λ	J	O
Th	**5,518.99**	—	**4**	**W**	**5,519.16**	—	**7**
Ta	**5,518.91**	—	**100W**	**Nd**	**5,519.35**	—	**5**
Sm	**5,518.87**	—	**4**	**Re**	**5,519.52**	—	**3*h***
*bh***Mg**	**5,518.8**	—	**3**	*bh***Zr**	**5,519.6**	—	**5**
W	**5,518.79**	—	**4**	**Sm**	**5,519.62**	—	**7**
Tb	**5,518.77**	—	**10**	**Eu**	**5,519.64**	—	**3**
Er	**5,518.75**	—	**12**	**Th**	**5,519.90**	—	**10**
Nd	**5,518.68**	—	**10**	**Nd**	**5,520.03**	—	**2**
Ce	**5,518.49**	—	**12**	**Mo**	**5,520.04**	**12**	**20**
Nd	**5,518.39**	—	**5**	**Re**	**5,520.06**	—	**15*h***
Ti	**5,518.21**	—	**6**	**Pr**	**5,520.30**	**4*w***	—
U	**5,518.16**	—	**2**	**Sc**	**5,520.50**	—	**80**
Er	**5,518.12**	—	**12**	**Nd**	**5,520.60**	—	**15**
Zr	**5,518.05**	—	**2**	**Mo**	**5,520.64**	**12**	**20**
Ta	**5,517.73**	—	**2*s***	**W**	**5,521.01**	—	**10**

	λ	I J	I O
Mo	**5,517.43**	**1**	**8**
Ce	**5,517.395**	**—**	**10**
Nb	**5,517.388**	**1**	**3**
La	**5,517.34**	**—**	**30**
Zr	**5,517.11**	**—**	**4**
Sm	5,516.14	—	200
Ti	5,512.53	12	125
Ru	5,510.71	—	100
Eu	5,510.55	—	300*s*
La	5,510.34	—	200
Fe	5,506.78	10	150
Mo	5,506.49	100	200*r*
La	5,503.81	—	100
Fe	5,501.47	—	150
La	5,501.34	50	200
Fe	5,497.52	5	150
Eu	*5,488.65*	—	*500*
Co	*5,483.34*	—	*500w*
Ni	*5,476.91*	*8*	*400w*
Lu	*5,476.69*	*1000*	*500*
U	**5,521.04**	**—**	**6**
Re	**5,521.10**	**—**	**20**
Ta	**5,521.15**	**—**	**4**
Mo	**5,521.17**	**2**	**3**
Ce	5,522.99	—	100
Co	5,523.29	—	300*w*
Os	5,523.53	—	100
Sc	5,526.81	300*wh*	100
Yt	5,527.64	15	100
Co	5,530.77	—	500
Re	5,532.66	—	100
Mo	5,533.05	100	200
Ba	5,535.55	200*R*	1000*R*
Yb	5,539.06	5	200
Eu	*5,547.45*	—	*1000*
Bi	*5,552.35*	*100*	*500wh*
Yb	*5,556.48*	*50*	*1500*

Ba 5,535.55 *J* 200*R* *O* 1000*R*

	λ	J	O
Sm	**5,535.50**	**—**	**3**
Nd	**5,535.48**	**—**	**3**
Fe	**5,535.41**	**—**	**50**
N	**5,535.39**	**(70)**	**—**
V	**5,535.382**	**2**	**2**
Tb	**5,535.38**	**—**	**15**
Nd	**5,535.27**	**—**	**2**
Ce	**5,535.24**	**—**	**15**
Pr	**5,535.18**	**2**	**10**
Gd	**5,535.16**	**—**	**8**
Rh	**5,535.04**	**1**	**80**
Cu	**5,534.98**	**3**	**—**
Fe	**5,534.86**	**10**	**—**
Sr	**5,534.81**	**15**	**20**
Fe	**5,534.66**	**—**	**20**
Tb	**5,534.58**	**—**	**10**
Gd	**5,534.55**	**—**	**8**
Mo	**5,534.54**	**4**	**5**
Ar	**5,534.45**	**(60)**	**—**
Gd	**5,534.29**	**—**	**8**
*bh***La**	**5,534.2**	**—**	**8**
Rh	**5,533.84**	**—**	**3**
Nd	**5,533.840**	**—**	**10**
La	**5,535.67**	**100**	**50**
U	**5,535.80**	**2**	**6**
In	**5,535.94**	**(70)**	**—**
C	**5,536.0**	**5*h***	**—**
K	**5,536.01**	**(10)**	**—**
Eu	**5,536.13**	**—**	**30**
Sm	**5,536.20**	**—**	**2**
Er	**5,536.26**	**—**	**8**
Tb	**5,536.27**	**—**	**25**
Br	**5,536.30**	**(50)**	**—**
*bh***La**	**5,536.4**	**—**	**10**
Br	**5,536.40**	**(20)**	**—**
Nd	**5,536.44**	**—**	**2**
In	**5,536.55**	**(70)**	**—**
Zr	**5,536.68**	**—**	**2**
Te	**5,536.73**	**(25)**	**—**
Ta	**5,536.74**	**—**	**2*h***
Eu	**5,536.83**	**—**	**3**
In	**5,537.03**	**(50)**	**—**
Sm	**5,537.07**	**—**	**50**
Th	**5,537.13**	**1**	**10**
Tm	**5,537.27**	**—**	**3**
Nd	**5,537.28**	**—**	**3**

	λ	I	
		J	O
V	5,533.838	18	18
Ne	5,533.68	(75)	—
Er	5,533.64	—	8
Nd	5,533.40	—	15
Mo	5,533.05	100	200
Xe	5,531.07	(300)	—
Co	5,530.77	—	500
Sc	5,526.81	300wh	100
Ar	5,524.93	(300)	—
In	5,523.91	(100)	—
Rb	5,522.79	100	—
Se	5,522.42	(750)	—
In	5,519.36	(500)	—
Br	*5,506.78*	*(300)*	—
Ar	*5,506.11*	*(500)*	—
Ra	*5,501.98*	*(250)*	—
As	*5,497.98*	*200*	—
J	*5,496.92*	*(900)*	—
Ar	*5,495.87*	*(1000)*	—
W	5,537.30	—	8
Ti	5,537.32	—	10
Ar	5,537.39	(2)	—
Zr	5,537.42	—	5
Ti	5,537.52	—	15
Ce	5,537.53	—	10
W	5,537.74	—	12
Eu	5,547.45	—	1000
Bi	5,552.35	100	500wh
Kr	5,552.99	(100whs)	—
Ra	5,553.57	(250)	—
O	5,554.94	(100h)	—
Ra	5,555.85	(500)	—
In	5,556.04	(100)	—
Yb	5,556.48	50	1500
As	5,558.31	200	—
Ar	*5,558.70*	*(500)*	—
Ar	*5,559.62*	*(200)*	—
N	*5,560.37*	*(200)*	—
Kr	*5,562.22*	*(500)*	—
Ne	*5,562.77*	*(500)*	—
N	*5,564.37*	*(200)*	—
Se	*5,566.93*	*(500)*	—
Sb	*5,568.09*	*200wh*	*6*
Kr	*5,570.29*	*(2000)*	—
Eu	*5,570.36*	—	*1000*
Ar	*5,572.55*	*(500)*	—
In	*5,576.75*	*(300)*	—
Eu	*5,577.13*	—	*1000*

Ba 5,777.66 *J* 100*R* *O* 500*R*

	λ	J	O
bhF	5,777.6	—	100
Pr	5,777.28	—	4
Te	5,777.25	(15)	—
Zn	5,777.11	15	10
U	5,776.90	—	2
Re	5,776.84	—	300w
Ta	5,776.77	—	80
V	5,776.68	25	50
Xe	5,776.39	(150)	—
Nd	5,776.12	—	15
Nb	5,776.07	3	30
Gd	5,776.03	—	10
Pr	5,775.92	—	4h
Kr	5,777.72	(2wh)	—
U	5,777.81	1	2h
Mo	5,778.19	—	12
Ra	5,778.28	(35)	—
Ir	5,778.28	—	2
Re	5,778.33	—	2
Sm	5,778.331	—	50
Ce	5,778.41	—	4
bhZr	5,778.5	—	60
Tb	5,778.94	—	10
Sm	5,779.25	—	50
Pr	5,779.29	—	50
Mo	5,779.36	—	20

	λ	I	
		J	O
Ce	**5,775.80**	—	**3**
Kr	**5,775.56**	**(2)**	—
Zn	**5,775.50**	—	**4**
*bh***Mg**	**5,775.5**	—	**5**
Lu	**5,775.40**	**5**	**50**
*bh***Sc**	**5,775.3**	—	**10**
J	5,775.11	(30)	—
Ti	5,774.05	50	70*W*
Ar	5,772.12	(100)	—
Kr	5,771.41	(100)	—
Ne	5,770.31	(50)	—
Hg	5,769.59	200	600
La	5,769.07	60	30
Ti	5,766.35	50	70*W*
Te	5,765.25	(70)	—
Eu	5,765.20	—	2000
Ne	5,764.42	(700)	—
J	5,764.33	(100)	—
Ti	5,762.27	50	70*h*
Ne	5,760.58	(70)	—
Xe	5,758.65	(150)	—
Te	5,755.87	(250)	—
Kr	5,752.98	(60)	—
Mo	*5,751.40*	*100*	*125*
Xe	*5,751.03*	*(200)*	—
Ne	*5,748.30*	*(500)*	—
Ar	*5,739.52*	*(500)*	—
V	*5,737.06*	*100*	*100*
V	*5,731.25*	*100*	*250*
V	*5,727.03*	*150*	*150*
Xe	*5,726.91*	*(200)*	—
*bh***F**	**5,779.4**	—	**200**
La	**5,779.91**	**4**	—
Tb	**5,779.91**	—	**15**
Ho	**5,780.01**	—	**12**
Ta	**5,780.02**	—	**60**
Mo	**5,780.11**	—	**12**
Mn	**5,780.19**	—	**10**
Ce	**5,780.21**	—	**2**
Cu	5,782.13	—	1000
Hg	5,789.66	(500)	—
Hg	5,790.65	(1000)	—
Mo	5,791.88	60	100
La	5,797.59	150	80
Te	5,803.07	(50)	—
Ne	*5,804.45*	*(500)*	—
La	*5,805.78*	*120*	*60*
Ne	*5,811.42*	*(300)*	—
Ra	*5,813.63*	*(500)*	—
Eu	*5,818.74*	—	*1000*
Yb	*5,819.43*	*100*	*7*
Ne	*5,820.15*	*(500)*	—
Xe	*5,823.89*	*(300)*	—
Xe	*5,824.80*	*(150)*	—

Be $^{4}_{9.0122}$

t_0 1,350°C

t_1 1,530°C

I.	II.	III.	IV.	V.
9.321	18.207	153.85	217.671	—

λ	I		eV
	J	O	
I 2,348.61	50	2000*R*	5.28
II 3,130.42	200	200	3.95
II 3,131.07	150	200	3.95
I 3,321.01	—	50	6.45
I 3,321.09	15	100	6.45
I 3,321.34	30	1000	6.45
4,672.20	100	—	—
II 4,673.46	100	—	14.81

Be 2,348.610 *J* 50 *O* 2000*R*

	λ	*I* *J*	*I* *O*
Lu	2,348.60	2	—
Ta	2,348.59	12*s*	—
Zr	2,348.586	1	15
Mo	2,348.58	9	—
J	2,348.56	(12)	—
W	2,348.56	—	2*h*
Pt	2,348.55	20	3
Co	2,348.46	2*w*	—
Nd	2,348.46	20	—
Ru	2,348.33	—	50
V	2,348.30	25	—
Pd	2,347.76	(25)	—
Ba	2,347.58	40	30
Hf	2,347.44	125	80
Co	2,347.39	25	10
Pd	2,347.31	25*w*	3
V	2,347.15	150	—
V	2,346.88	25	—
Rh	2,346.77	30	2
Pd	2,346.46	25	2
Rh	2,346.44	100	2
V	2,346.34	125	—
J	2,346.22	(30)	—
Cu	2,346.13	25	2
Rb	2,345.35	(100)	—
Cr	2,345.33	60	8
Fe	*2,343.49*	*50*	*10*
Hf	*2,343.32*	*80*	*60*
V	*2,343.11*	*250*	—

	λ	*I* *J*	*I* *O*
Os	2,348.61	—	3
Nb	2,348.66	5*wh*	—
Ni	2,348.74	1	10
Te	2,348.74	(5)	—
Nb	2,348.75	—	3*w*
Re	2,348.80	—	9
Cu	2,348.82	20	15*d*
Mn	2,348.83	15	3
Mo	2,348.84	9	3
La	2,348.86	2	2*h*
Tm	2,348.89	8	—
U	2,348.91	2	2
Rh	2,349.68	125	3
Rb	2,349.80	(80)	—
V	2,349.81	150	3*w*
Mo	2,349.89	25	—
Te	2,350.12	(25)	—
Ce	2,350.15	50	—
Os	2,350.23	50	30
Rh	2,350.35	50	—
Ir	2,350.61	25	—
Pd	2,350.70	25*h*	6
Hf	2,351.21	150	100
Te	2,351.22	(25)	—
V	2,351.26	50	2
Pd	2,351.34	60	10
V	2,351.54	50	—
V	*2,352.18*	*200*	*5*
Ir	*2,353.12*	*50w*	*4*

Be 3,130.416 *J* 200 *O* 200
3,131.072 *J* 150 *O* 200

Xe	3,130.40	(2*wh*)	—
P	3,130.38	(30)	—
Ho	3,130.38	4*h*	—
Ti	3,130.376	1	2
U	3,130.37	—	3*h*
Ce	3,130.23	1	30
Ta	3,130.29	1	15
Ir	3,130.285	—	4
Fe	3,130.278	4	5
V	3,130.27	200*r*	50

W	3,130.46	8	10
Si	3,130.48	5	—
Ce	3,130.52	—	5
U	3,130.564	15	1
Cr	3,130.565	12	1
Fe	3,130.567	4	4
Ba	3,130.570	3	2
Ta	3,130.578	35	100*W*
Ir	3,130.578	—	3
Cs	3,130.7	(4)	—

	λ	I	
		J	O
La	3,130.25	9	3
Sm	3,130.23	—	5
Ce	3,130.20	—	15
Ti	3,130.17	—	2
Dy	3,130.16	2h	4
W	3,130.15	4	5
Se	3,130.12	(8)	—
Zr	3,130.063	—	3
Mo	3,130.060	20	—
Ag	3,130.01	15h	25h
Os	3,130.00	8	30
Gd	3,129.96	1	2
Sm	3,129.950	—	5
Ta	3,129.946	8	50
Yt	3,129.93	50	8
Th	3,129.91	8	10
Ru	3,129.84	4	60
Zr	3,129.76	10	10
U	3,129.73	15	8
Fe	3,129.33	60	100
Ni	3,129.31	—	125
Re	3,128.95	—	100W
Pr	3,128.70	150	30
Ru	3,127.91	100	10
Ta	3,127.76	100	18w
Co	3,127.25	—	100
V	3,126.21	100R	60
Ne	3,126.19	(150)	—
Fe	3,126.17	70	150
Hg	3,125.66	150	200
Fe	3,125.65	300	400
V	3,125.28	200R	80
Cr	3,124.98	125	20
Os	3,124.94	10wh	150h
Ge	3,124.82	80	200
Rh	3,123.70	2	150
V	3,122.89	300r	12
Au	3,122.78	5	500h
Mo	*3,122.00*	*150*	*5*
Xe	*3,121.87*	*(150)*	*—*
V	*3,121.14*	*200r*	*60*
Cr	*3,120.37*	*150*	*40*
Ti	*3,119.80*	*150*	*4*
Cr	*3.118.65*	*200*	*35*
V	*3,118.38*	*200R*	*70*
Re	*3,118.19*	*—*	*200*
Ti	*3,117.67*	*200*	*15*

	λ	I	
		J	O
S	3,130.71	(15)	—
U	3,130.73	6	10
Eu	3,130.74	100	100W
Nb	3,130.786	100	100
Rh	3,130.790	2	60
Ti	3,130.800	100	24
Ar	3,130.80	(20)	—
Gd	3,130.81	1	3
Ce	3,130.87	2	30
Ce	3,130.92	—	5
Ho	3,130.99	6	6
Th	3,131.070	10	12
Er	3,131.07	—	3
Be	3,131.072	150	200
Os	3,131.11	30	125
Ta	3,131.22	25h	5h
Tm	3,131.28	500	400
Hg	3,131.55	300	400
Hg	3,131.83	100	200
Cr	3,132.06	125	25
Mo	3,132.59	300R	1000R
Ta	3,132.64	25	250w
Cd	3,133.17	300	200
V	3,133.33	200r	50
Tm	3,133.89	200	200
Ni	3,134.108	150	1000R
Fe	3,134.111	125	200
Hf	3,134.72	125	80
O	3,134.79	(100)	—
Ru	3,134.80	100	10
V	3,134.93	150r	30
Fe	3,135.36	100	1
Dy	3,135.37	50	100
Ru	3,135.80	80	10
Ta	3,135.89	100	35
V	3,136.51	200	20
Co	3,137.33	—	150r
Rh	3,137.71	—	100
Pb	3,137.83	100	—
Pt	*3,139.39*	*80*	*300*
V	*3,139.74*	*150*	*15*
Pd	*3,142.81*	*100*	*300*
Fe	*3,143.99*	*150*	*200*
Ni	*3,145.72*	*3*	*200*

	λ	I J	I O
As	*3,116.63*	*150*	—
Fe	*3,116.59*	*150*	—
Bi	*3,115.42*	*500*	—
Ni	*3,114.12*	*50*	*300*
Pd	*3,114.04*	*500w*	*400w*
Eu	*3,111.43*	—	*200*
V	*3,110.71*	*300R*	*70*
Ni	*3,105.47*	*35*	*200*

3,321.01	—	*O* 50
Be 3,321.09	*J* 15*	*O* 100
3,321.34	*J* 50*	*O* 1000 r

	λ	I J	I O
Pd	**3,220.99**	—	**15**
Ce	**3,320.94**	—	**10**
U	**3,320.92**	**2**	**1**
Mo	**3,320.90**	**80**	**3**
Nb	**3,320.81**	**100**	**3**
Ce	**3,320.79**	—	**8**
Fe	**3,320.779**	**12**	**30**
Ni	**3,320.779**	—	**10**
Sc	**3,320.71**	—	**5**
Mn	**3,320.69**	**30*h***	**60**
Re	**3,320.651**	—	**2**
Fe	**3,320.650**	**10**	**20**
Sm	**3,320.59**	**2**	**4**
Ce	**3,320.55**	—	**5**
Gd	**3,320.44**	**3**	**3**
Os	**3,320.425**	**8**	**10**
Ce	**3,320.423**	—	**12**
Sc	**3,320.422**	**12**	**5**
W	**3,320.37**	**8**	**9**
Ir	**3,320.36**	—	**2**
U	**3,320.34**	**1**	**2**
Yb	**3,320.31**	**20**	**3**
Th	**3,320.30**	**12**	**10**
Mo	**3,320.28**	**2**	**8**
Ni	3,320.26	15	400*w*
Ho	3,320.24	8	6
Sm	3,320.16	8	40
Au	3,320.15	5	20
V	3,320.141	3	12
Eu	3,320.14	2*h*	3
Dy	3,319.89	9	150
Co	3,319.82	2	35
Be	**3,321.09**	**15**	**100**
W	**3,321.13**	**5**	**6**
Tb	**3,321.15**	**30**	**30**
Sm	**3,321.18**	**15**	**50**
Cr	**3,321.19**	**1**	**20**
Mo	**3,321.20**	**15**	**1**
Ni	**3,321.24**	—	**4**
Ru	**3,321.25**	—	**12**
Ce	**3,321.27**	—	**3**
Be	**3,321.34**	**30**	**1000*r***
Nd	**3,321.39**	—	**8**
Th	**3,321.45**	**12**	**10**
Re	**3,321.46**	—	**10**
V	**3,321.54**	**150**	**3**
W	**3,321.56**	**7**	**8**
Ti	**3,321.59**	**15**	**15**
V	**3,321.68**	—	**12**
Ti	**3,321.70**	**125**	**15**
U	**3,321.71**	**1**	**5**
Eu	**3,321.86**	**5**	**30**
Co	**3,321.91**	—	**25**
Os	**3,322.05**	**10**	**10**
U	**3,322.12**	**12**	**18**
Mo	**3,322.17**	**30**	**4**
Ce	**3,322.175**	—	**2**
Co	3,322.199	—	100*W*
Re	3,322.205	—	25*W*
Ru	3,322.226	—	8
Sr	3,322.231	8	100
W	3,322.25	12*d*	10
Eu	3,322.26	2	20
Ni	3,322.31	10	400

	λ	I J	I O
Cu	3,319.68	20	60
Co	3,319.48	—	80
Yb	3,319.41	—	25
Fe	3,319.25	50	70
Co	3,319.16	—	60
Zr	3,319.02	6	25
Ta	3,318.84	35	125
Ru	3,318.82	8	50
Re	3,318.67	—	30
Ta	3,318.53	3	70
Co	3,318.40	—	35
Cr	3,318.08	—	80 *Wh*
Ti	3,318.02	125	60
Hf	3,317.99	18	25
Ru	3,317.96	—	30
Ta	3,317.93	25 *W*	200
Ru	3,317.89	12	50
Mn	3,317.30	30*h*	100
Cu	3,317.22	20	60
Fe	3,317.121	80	100
Dy	3,317.119	4	35
Tm	3,316.87	20	60
Os	3,316.69	15	30
Sm	3,316.58	10	25
Ru	3,316.39	—	80
Dy	3,316.32	4	50
Os	3,315.69	15	40
Ni	3,315.66	20	400*R*
Ru	3,315.44	5	30
Os	3,315.42	15	50
Ru	3,315.23	25	60
Ru	3,315.047	12	50
Pt	3,315.046	10	200
Mn	3,314.90	—	35
Fe	3,314.74	200	200
Ce	3,314.72	3	25
Mn	3,314.424	—	30*h*
Ti	3,314.423	20	40
Co	3,314.08	—	100
Re	3,313.95	—	40
Cr	3,313.72	1	30
Eu	3,313.32	40	35
Mn	3,313.22	—	50
Hf	3,312.86	10	30
Dy	3,312.73	5	50
Nb	3,312.60	50	40
Er	3,312.42	15	25
Ni	3,312.32	2	70
Fe	3,322.477	100	150
Re	3,322.478	—	150
Ir	3,322.60	18	30
Ba	3,322.87	—	30*r*
Nd	3,322.936	2	25
Ti	3,322.937	300*R*	80
Rh	3,323.09	200	1000
Er	3,323.20	4	25
Cr	3,323.25	—	25
Fe	3,323.74	150	150
Pt	3,323.79	10	150
Mo	3,323.95	25	40
Os	3,324.33	15*d*	50*d*
Tb	3,324.40	50	70
Fe	3,324.54	80	100
Re	3,324.93	—	25
Ru	3,324.99	12	60
Co	3,325.24	3	80
Ce	3,325.33	1	25*s*
Fe	3,325.46	80	100
Tb	3,325.52	—	30
Mo	3,325.67	25	50
Ta	3,325.74	3	50
Co	3,326.56	—	60
Cr	3,326.59	18	40
Zr	3,326.80	100	100
Co	3,326.99	—	100
Mo	3,327.30	20	40
Os	3,327.42	15	80
Ru	3,327.71	6	50
Yt	3,327.87	60	60
Sm	3,327.90	20	50
Co	3,328.21	—	40
Fe	3,328.87	100	150
Cr	3,329.05	6	30
Tb	3,329.08	8	30
Ti	3,329.46	200*r*	80
Co	3,329.47	—	80
Fe	3,329.53	6	35
Cu	3,329.64	10	60
Er	3,329.66	3	25
V	3,329.85	40	100
Mg	3,329.93	8	80
Sr	3,329.99	10	100
Sn	3,330.59	100*h*	100*h*
Cr	3,330.60	—	80
Mn	3,330.67	—	75
Rh	3,331.09	10	50

	λ	I J	I O		λ	I J	I O
Re	3,312.29	—	25	*Gd*	*3,331.39*	*80*	*100*
Ce	3,312.21	5	30	*Fe*	*3,331.61*	*70*	*125*
Co	3,312.15	2	60	*Ru*	*3,332.05*	*10*	*60*
Ir	3,312.13	15	25	*Mg*	*3,332.15*	*25*	*100*
				Ta	*3,332.41*	*3*	*50*
Lu	3,312.11	10	100				
Mn	3,311.90	—	75	*Ru*	*3,332.64*	*5*	*60*
				Co	*3,333.39*	—	*100*
				Cr	*3,333.60*	*3h*	*125*
Ta	*3,311.16*	*70w*	*300w*	*Co*	*3,334.14*	—	*250R*
Os	*3,310.91*	*30*	*200*	*Fe*	*3,334.22*	*100h*	*150h*
Tm	*3,310.59*	*20*	*60*				
Fe	*3,310.49*	*40*	*50*	*Eu*	*3,334.33*	*5*	*50*
Fe	*3,310.34*	*80*	*100*	*Cr*	*3,334.69*	—	*150wh*
				Ti	*3,335.19*	*150*	*60*
Ni	*3,310.20*	—	*50*	*Cu*	*3,335.21*	*15*	*60*
Tm	*3,309.80*	*60*	*80*	*Re*	*3,335.37*	—	*100*
Ta	*3,309.78*	*5*	*70*				
Ti	*3,309.50*	*25*	*60*	*Fe*	*3,335.77*	*100*	*125*
Ti	*3,308.39*	*10*	*50*	*Os*	*3,336.15*	*50*	*200R*
				Ru	*3,336.64*	*4*	*50*
Au	*3,308.31*	*15*	*50*	*Mg*	*3,336.68*	*60*	*125*
Ru	*3,307.98*	—	*50*	*Co*	*3,337.17*	*2*	*60*
Cu	*3,307.95*	*30*	*60*				
Sr	*3,307.53*	*10s*	*200*	*La*	*3,337.49*	*300wh*	*800*
Fe	*3,307.23*	*60*	*80*	*Fe*	*3,337.67*	*100*	*125*
				Ta	*3,337.80*	*18s*	*100*
Co	*3,307.15*	—	*80*	*Ru*	*3,337.82*	*8*	*60*
Sm	*3,306.37*	*40*	*100*	*Cu*	*3,337.84*	*50*	*70*
Fe	*3,306.35*	*150*	*200*				
Zr	*3,306.28*	*80*	*80*	*Re*	*3,338.17*	—	*150*
Os	*3,306.23*	*12*	*80*	*Rh*	*3,338.54*	*50*	*200*
				Fe	*3,338.64*	*25*	*70*
Ru	*3,306.17*	*12*	*60*	*Tb*	*3,339.00*	*8*	*50*
Fe	*3,305.97*	*300*	*400*	*Fe*	*3,339.19*	*50*	*80*
bhB	*3,305.4*	—	*50*				
Ru	*3,304.82*	*3*	*50*	*Ru*	*3,339.55*	*60*	*100*
Ta	*3,304.37*	*15*	*70*	*Co*	*3,339.78*	—	*150w*
Ru	*3,303.99*	*8*	*60*				
Co	*3,303.88*	—	*60r*				
Fe	*3,303.57*	*10*	*70*				
La	*3,303.11*	*150*	*400*				
Na	*3,302.99*	*150R*	*300R*				
Zn	*3,302.94*	*300R*	*700R*				
Ta	*3,302.76*	*14*	*50*				
Zn	*3,302.59*	*300*	*800*				
Bi	*3,302.55*	—	*150*				

Be 4,672.2 *J* 100

4,673.46 *J* 100

Br	**4,672.11**	**(4)**	—	**Xe**	**4,672.20**	**(50*h*)**	—
Nb	**4,672.091**	**100**	**150**	**Br**	**4,672.56**	**(12)**	—
Kr	**4,672.09**	**(2*wh*)**	—	**Nd**	**4,672.69**	**2**	**5**

	λ	I J	I O
Pr	4,672.08	25*w*	100
Tm	4,671.98	20	15
Mo	4,671.90	30	30
La	4,671.83	150	100
Li	4,671.8	(4)	—
Cu	4,671.693	10	—
Mn	4,671.688	5	100
W	4,671.65	1	12
Kr	4,671.61	(10)	—
Er	4,671.58	1	3
U	4,671.41	30	20
Xe	4,671.23	(2000)	—
Eu	4,671.18	1	30
Dy	4,671.10	4	4
Er	4,671.09	1	2
Hf	4,670.91	2	5
Ne	4,670.88	(70)	—
Gd	4,670.85	2*h*	3
Yt	4,670.83	2	3
Se	4,670.40	300*wh*	100
Ne	4,669.02	(50)	—
La	4,668.91	300*r*	200*r*
Na	4,668.60	100	200
S	4,668.58	(50)	—
Xe	4,668.49	(50)	—
Ag	4,668.48	70	200
Se	4,667.80	(70)	—
Ne	4,667.36	(100)	—
Ne	4,666.65	(50)	—
J	4,666.52	(250)	—
Te	4,665.33	(70)	—
Se	4,664.98	(150)	—
Te	4,664.34	(800)	—
Se	4,664.20	(150)	—
Hf	4,664.12	100	50
La	4,663.76	200	100
La	4,662.51	200	150
O	4,661.65	(125)	—
Ne	4,661.10	(150)	—
Hg	4,660.28	(200)	—
W	4,659.87	70	200
Kr	4,658.87	(2000)	—
P	*4,658.11*	*(100)*	—
Ar	*4,657.94*	*(150)*	—
Ne	*4,656.39*	*(300)*	—
In	*4,655.79*	*(100)*	—

	λ	I J	I O
Hg	4,672.7	(5)	—
As	4,672.70	50	—
O	4,672.75	(30)	—
Mo	4,673.03	3	3
Fe	4,673.17	2	20
Br	4,673.38	(4)	—
Be	4,673.46	(100)	—
Cu	4,673.55	6	—
Ng	4,673.59	5	2
Dy	4,673.61	8	10
Ba	4,673.62	5	40
Th	4,673.66	2	6
O	4,673.71	(30)	—
U	4,673.74	3*h*	—
In	4,673.77	(5)	—
Kr	4,673.80	(3)	—
W	4,673.99	1	2
U	4,674.23	8	8
Xe	4,674.56	(25)	—
Sm	4,674.592	40	80
Nd	4,674.595	10	50
Ho	4,674.62	3	4
Cu	4,674.76	30*W*	200
Yt	4,674.848	100	80
Er	4,674.849	15	50
Cs	4,674.89	(10)	—
N	4,674.98	(5)	—
Rh	4,675.03	50	100
J	4,675.53	(50)	—
P	4,675.78	(70)	—
O	4,676.25	(125)	—
Xe	4,676.46	(100*wh*)	—
Sm	4,676.91	50	100
J	4,676.94	(80)	—
Gd	4,678.16	200*W*	100*W*
Ne	4,678.22	(300)	—
Ne	4,678.60	(50)	—
Br	4,678.69	(200)	—
Fe	4,678.85	100	150
P	4,678.94	(100)	—
Ne	4,679.13	(150)	—
Zn	4,680.14	200*h*	300*h*
Ne	4,680.36	(100)	—
Kr	4,680.41	(500)	—
Rn	4,680.83	(500)	—
In	4,681.11	(200)	—
Ne	4,681.20	(50)	—

	λ	I	
		J	O
La	4,655.50	300	150
Te	4,654.38	(800)	—
Cr	4,652.16	150	200R
Xe	4,651.94	(100)	—
Cr	4,651.28	100	100
O	4,649.15	(300)	—
Sc	4,648.44	(800)	—
Cr	4,646.17	150	100
Ne	4,645.42	(300)	—
In	4,644.54	(125)	—
Rn	4,644.18	(300)	—
Yt	4,643.69	100	50
N	4,643.11	(100)	—
Ta	4,681.87	50	200
Ti	4,681.92	100	200
In	4,682.00	250W	—
Ra	4,682.28	(800)	—
Yt	4,682.32	100	60
Gd	4,683.34	50	100
In	4,685.22	(100)	—
He	4,685.75	(300)	—
Te	4,686.95	(300)	—
Ne	4,687.67	(100)	—
Xe	4,690.97	(100)	—
Kr	4,691.28	(100)	—
La	4,692.50	300	200
Kr	4,694.51	(200h)	—
Xe	4,697.02	(300)	—
Xe	4,698.01	(150h)	—
O	4,699.21	(100)	—
La	4,699.63	200r	200r
Ar	4,702.316	(1200)	—
Gd	4,702.323	100	50
Ne	4,702.53	(150)	—

Bi 83 208.980

t_0 271°C t_1 1,470°C

I.	II.	III.	IV.	V.
~8	16.7	25.56	45.3	56.0

λ	I		eV
	J	O	
I 2,897.97	500R	500R	5.6
I 2,938.30	300	300	6.1
I 2,989.03	100	250	5.5
I 3,067.72	2000	3000R	4.0
I 4,722.55	100	1000	4.0
II 5,209.29	600	—	5.72
5,552.35	100	500	—

Bi 2,897.97 *J* 500*R* *O* 500*R*

	λ	*I*			λ	*I*	
		J	*O*			*J*	*O*
V	**2,897.90**	**25**	**1**	**Mn**	**2,897.99**	—	**15**
Pt	**2,897.87**	**15**	**400**	**U**	**2,898.01**	**6**	**6**
Fe	**2,897.85**	—	**2**	**Fe**	**2,898.06**	—	**2**
J	**2,897.83**	**(20)**	—	**W**	**2,898.09**	**3**	—
Nb	**2,897.81**	**150**	**15**	**Be**	**2,898.19**	—	**15**
Mn	**2,897.80**	—	**15**	**Tb**	**2,898.20**	**10**	—
La	**2,897.76**	**5*hI***	**2**	**Ru**	**2,898.22**	**60**	—
Ru	**2,897.71**	**60**	**6**	**W**	**2,898.25**	**3**	**8**
Cr	**2,897.70**	**25**	**3**	**Zr**	**2,898.256**	—	**4**
Yt	**2,897.68**	**12**	**3**	**Hf**	**2,898.259**	**12**	**50**
Ir	**2,897.65**	—	**3**	**Th**	**2,898.267**	**3**	**4**
Fe	**2,897.64**	**1**	**4**	**Be**	**2,898.27**	—	**20**
U	**2,897.636**	**6*h***	**3*h***	**Yb**	**2,898.33**	**10**	—
Rh	**2,897.633**	**10**	**2**	**Ce**	**2,898.34**	—	**6**
Mo	**2,897.628**	**25**	**20**	**Ir**	**2,898.350**	—	**10**
Re	**2,897.59**	—	**10*d***	**Fe**	**2,898.355**	**30**	**100**
Er	**2,897.52**	**2**	**12**	**U**	**2,898.37**	**4**	**4**
N	**2,897.49**	**(15*h*)**	—	**Mo**	**2,898.39**	—	**2**
Tb	**2,897.46**	**20*h***	**3**	**Ta**	**2,898.42**	**5**	**30**
U	**2,897.458**	**4*h***	**6**	**Cl**	**2,898.45**	**(8)**	—
Mn	**2,897.43**	—	**8**	**Fe**	**2,898.47**	—	**2**
Mo	**2,897.42**	**20**	—	**Mo**	**2,898.48**	**15**	—
Ho	**2,897.36**	**20*h***	—	**Cr**	**2,898.536**	**40**	**12**
Nb	**2,897.35**	**2**	**3**	**Ru**	**2,898.538**	**4**	**20**
				U	**2,898.56**	**4**	**8**
Te	2,895.49	(300*h*)	—				
Pt	2,893.86	25	500	Nb	2,899.24	500	20
V	2,893.32	300*r*	50	Ta	2,902.05	200	1000*w*
Ta	*2,891.84*	*100*	*500w*	*Ta*	*2,904.07*	*40*	*300w*
Tb	*2,891.41*	*500*	*3*	*Os*	*2,909.06*	*400*	*500R*

Bi 2,938.30 *J* 300 *O* 300

	λ	*J*	*O*		λ	*J*	*O*
V	**2,938.25**	**60**	**2**	**Mo**	**2,938.30**	**30**	**1**
Ce	**2,938.22**	—	**5**	**Ce**	**2,938.32**	—	**2**
Yb	**2,938.18**	**3**	**1**	**Ta**	**2,938.43**	**3**	**50**
Th	**2,938.11**	**4**	**5**	**Ir**	**2,938.47**	**12**	**18*h***
Nb	**2,938.07**	**5**	**3**	**W**	**2,938.50**	**6**	**8**
Ce	**2,938.051**	—	**5**	**Cs**	**2,938.5**	**(20)**	—
Fe	**2,938.05**	**1**	**2**	**Mg**	**2,938.538**	—	**25**
Ir	**2,937.96**	—	**5**	**Ti**	**2,938.539**	—	**2**
Mn	**2,937.92**	—	**25*Wh***	**Ag**	**2,938.55**	**200*wh***	**200**
Te	**2,937.90**	**(15)**	—	**Ta**	**2,938.56**	—	**10**
Re	**2,937.814**	—	**8**	**Mo**	**2,938.59**	**10**	**1**
Fe	**2,937.811**	**150**	**300**	**V**	**2,938.67**	**2**	**12**

	λ	I J	I O
Hf	2,937.80	100	50
Zr	2,937.730	2	1
Na	2,937.725	(40)	6
Nb	2,937.707	20wh	—
Ce	2,937.707	—	6
V	2,937.69	10	20w
Fe	2,936.90	500r	700r
Ho	2,936.77	1000R	—
Tm	2,935.00	300	80
Ag	2,934.23	200h	10
Ta	2,933.55	150	400
Ru	2,933.24	150	20
Ta	2,932.69	80w	400
In	2,932.62	300	500
Pt	*2,929.79*	*200w*	*800R*
Nb	*2,927.81*	*800r*	*200*
Fe	*2,926.59*	*400*	*150*

	λ	I J	I O
Ce	2,938.68	—	3
Ti	2,938.70	100wh	—
In	2,938.71	10	—
Fe	2,938.73	—	2
Ir	2,938.76	—	10
Mo	2,938.77	5	2
Pt	2,938.81	2	15
Sm	2,938.82	—	6
Cr	2,938.85	—	6wh
W	2,938.852	9	1
Mo	2,938.89	3	—
Ta	2,939.28	40h	200
Ta	2,940.21	50	150
Fe	2,940.59	80	200
Fe	2,941.34	300	600
V	2,941.37	300r	40
V	2,941.49	150r	12
Nb	2,941.54	300	50
Ti	2,941.99	150	100
Ta	2,942.14	40	150
Fe	2,944.40	600	70
V	2,944.57	300r	50
Ru	*2,945.67*	*300*	*60*
Fe	*2,947.88*	*200*	*600r*
Fe	*2,950.24*	*300*	*700*

Bi 2,989.03 *J* 100 *O* 250

	λ	J	O
Ce	2,989.02	—	2h
U	2,989.00	—	2
Ca	2,988.98	—	5
Ir	2,988.979	—	35
Te	2,988.97	(10)	—
Sc	2,988.952	2d	20s
Ru	2,988.948	100	250
Fe	2,988.94	1	2
Sm	2,988.92	4h	4
W	2,988.885	12	6
Ce	2,988.878	—	3
Sm	2,988.84	5	6
Nb	2,988.79	2	2
Zr	2,988.77	6wh	—
Dy	2,988.715	—	2
U	2,988.71	4	5d
Kr	2,988.69	(3wh)	—

	λ	J	O
Ta	2,989.05	5h	40w
U	2,989.11	1	2d
Os	2,989.13	40	12
Cr	2,989.19	90	10
Lu	2,989.27	4	50
Er	2,989.29	1	18
V	2,989.300	20	1
Ca	2,989.30	6	—
Ce	2,989.31	—	4
Ru	2,989.33	1	12
U	2,989.39	2	5r
Ta	2,989.497	15	200
W	2,989.502	4	—
Os	2,989.53	4	5
Er	2,989.55	1	5d
Ce	2,989.57	—	6
Co	2,989.588	30	75R

	λ	I	
		J	O
Mo	**2,988.68**	**2**	**25**
Cr	**2,988.65**	**150**	**200r**
Ca	**2,988.60**	**7**	**—**
Tb	**2,988.59**	**5**	**10**
Ta	**2,988.58**	**20**	**40**
Li	**2,988.5**	**15**	**—**
Re	**2,988.48**	**—**	**40**
Fe	**2,988.47**	**30**	**60**
F	**2,988.45**	**(10r)**	**—**
U	**2,988.42**	**1**	**2**
J	**2,988.39**	**(15)**	**—**
Rh	**2,988.36**	**2**	**2**
V	2,988.02	80	10
Si	2,987.65	100	100
Fe	2,987.29	200	300
Co	2,987.16	50	75R
Ta	2,986.81	100	20d
Tm	2,986.52	150	50
Cr	2,986.47	125	125r
Fe	2,986.46	60	100
Rh	2,986.20	60	150
Mo	2,986.16	50	4
Fe	2,985.55	300	80
Cr	2,985.32	60	10
V	2,985.17	60	1
Nb	2,985.05	50	2h
Fe	2,984.83	400	200r
Na	2,984.43	(80)	20
Yb	2,983.98	70	10
Fe	2,983.57	400r	1000R
V	2,983.55	60	10
V	2,982.75	50	4
Ne	2,982.66	(250)	—
Tm	*2,981.49*	*100*	*60*
Fe	*2,981.45*	*200*	*300*
Sb	*2,980.96*	*(125hd)*	*—*
Pd	*2,980.65*	*200R*	*—*
Cd	*2,980.63*	*500*	*1000R*
Fe	*2,979.35*	*100*	*20*
Xe	*2,979.32*	*(200)*	*—*
Ti	*2,979.20*	*100wh*	*—*
Ru	*2,978.641*	*150*	*50*
Th	*2,978.64*	*100h*	*1*
Ta	*2,978.18*	*150*	*—*
Nb	*2,977.68*	*300*	*1*
Ru	*2,976.586*	*200*	*60*

	λ	I	
		J	O
In	**2,989.59**	**2**	**—**
V	**2,989.595**	**40**	**8**
U	**2,989.60**	**—**	**3**
Ru	**2,989.65**	**2**	**30**
Ti	2,990.16	80Wh	—
Nb	2,990.26	200	5
Au	2,990.28	50	—
Fe	2,990.39	100	150
Ru	2,991.62	100	50
Fe	2,991.64	80	100
Cr	2,991.89	60	125r
Nb	2,991.95	100	1
Ne	2,992.44	(150)	—
Bi	2,993.34	100wh	200wh
J	2,993.86	(70)	—
Nb	2,993.96	50	—
Cr	2,994.07	50	150
Fe	2,994.43	600r	1000R
Ni	2,994.46	10	125R
Nb	2,994.73	300	100
Yb	2,994.80	80	10
Au	2,994.99	100	—
Cr	2,995.10	75	200r
Ti	2,995.75	70wh	10h
Cr	*2,996.58*	*125*	*300r*
Cu	*2,997.36*	*30*	*300*
Pt	*2,997.97*	*200r*	*1000R*
Ru	*2,998.89*	*100*	*50*
Fe	*2,999.51*	*300*	*500*
Pd	*2,999.55*	*100h*	*—*
Cr	*3,000.89*	*125*	*150*
Fe	*3,000.95*	*300r*	*800R*
V	*3,001.20*	*200r*	*20*
Ni	*3,002.49*	*100*	*1000R*

	λ	I J	I O		λ	I J	I O
Ta	*2,976.26*	*150*	*2wh*				
Te	*2,975.91*	*(100)*	—				
Hf	*2,975.88*	*100*	*80*				

Bi 3,067.72 *J* 2000 *O* 3000*R*

	λ	I J	I O		λ	I J	I O
Sm	**3,067.672**	**15**	**15**	**In**	**3,067.73**	**3**	—
Eu	**3,067.672**	—	**8**	**Th**	**3,067.734**	**20**	**12**
Mo	**3,067.64**	**50**	**10**	**Sn**	**3,067.75**	**15**	**10**
W	**3,067.57**	**12**	**3**	**U**	**3,067.76**	**6**	**8**
Nb	**3,067.53**	**15**	—	**Eu**	**3,067.78**	—	**7**
Ce	**3,067.44**	—	**4**	**Cs**	**3,067.8**	**(4)**	—
Hf	**3,067.414**	**10**	**30**	**Lu**	**3,067.83**	**3*h***	—
W	**3,067.41**	**5**	**2**	**W**	**3,067.87**	**12**	**3**
Re	**3,067.39**	—	**60**	**Ce**	**3,067.89**	—	**6**
Yb	**3,067.37**	**3**	—	**Ir**	**3,067.939**	—	**2*h***
Rh	**3,067.305**	**1**	**80**	**Fe**	**3,067.944**	**10**	**25**
Xe	**3,067.30**	**(20)**	—	**Mo**	**3,068.00**	**1**	**30**
U	**3,067.25**	**5**	—	**Zr**	**3,068.02**	**2**	**2**
Fe	**3,067.24**	**300**	**300**	**J**	**3,068.05**	**(5)**	—
Ne	**3,067.21**	**(5)**	—	**Nb**	**3,068.06**	**10**	—
Cs	**3,067.2**	**(4)**	—	**Fe**	**3,068.18**	**150**	**150**
Cr	**3,067.16**	**40**	**25**	**Ru**	**3,068.26**	**8**	**60**
Ge	**3,067.13**	—	**2**	**Yb**	**3,068.28**	**3**	—
Zr	**3,067.124**	**4**	**5**	**Pb**	**3,068.5**	**(10)**	—
Fe	**3,067.120**	**6**	**6**				
V	**3,067.116**	—	**15**				
Ca	**3,067.009**	**2**	**6**				
Ge	**3,067.007**	**40**	**60**				
Fe	**3,067.00**	**2**	**2**				
Dy	**3,066.99**	**4**	**20**				
W	**3,066.98**	**12**	**4**				
Pt	3,064.71	300*R*	2000*R*				

Bi 4,722.55 *J* 100 *O* 1000

	λ	I J	I O		λ	I J	I O
Sr	**4,722.28**	—	**30**	**Ti**	**4,722.62**	**8**	**80**
Bi	**4,722.19**	**5**	**10**	**Sm**	**4,722.63**	—	**3*h***
Kr	**4,722.16**	**(3)**	—	**Pr**	**4,722.67**	—	**5**
Zn	**4,722.159**	**300*h***	**400*w***	**Er**	**4,722.70**	—	**12**
Ne	**4,722.150**	**(5)**	—	**Ne**	**4,722.71**	**(15)**	—
La	**4,722.14**	**2*h***	—	**U**	**4,722.73**	**50**	**40**
Th	**4,722.11**	—	**2**	**Hg**	**4,722.8**	**(5)**	—

	λ	I	
		J	O
Er	**4,722.02**	—	**4**
Pr	**4,721.91**	—	**3**
Ir	**4,721.88**	—	**2**
Rn	**4,721.76**	**(150)**	—
Hf	**4,721.71**	**4**	**10**
Pr	**4,721.67**	—	**3**
Ar	**4,721.62**	**(10)**	—
Ne	**4,721.54**	**(70)**	—
V	**4,721.51**	**12**	**15**
Pr	**4,721.471**	—	**2**
Gd	**4,721.466**	**20*h***	**20**
Cl	**4,721.43**	**(25)**	—
Co	**4,721.41**	**100*h***	**8**
Sm	**4,721.40**	—	**20**
U	**4,721.32**	**2*h***	—
Os	**4,721.284**	—	**12**
Cl	**4,721.28**	**(6)**	—
V	**4,721.25**	**2**	**3**
Dy	**4,721.23**	**5**	**12**
La	4,719.95	300	200*r*
Br	4,719.77	(80)	—
Cr	4,718.43	150	200
Mo	4,717.92	50	50
Ne	4,717.61	(70)	—
La	4,716.44	200	100
S	4,716.23	(600)	—
Ne	4,715.34	(1500)	—
Xe	4,715.18	(80)	—
Ni	4,714.42	8	1000
Ne	4,714.34	(70)	—
Fe	4,714.07	50	50
Ne	4,713.13	(100)	—
La	4,712.93	150	100*r*
Ne	4,712.06	(1000)	—
Sb	4,711.26	(100)	—
Te	4,711.16	(70)	—
Ne	4,710.06	(1000)	—
O	4,710.00	(60)	—
Ru	4,709.48	80	150
Fe	4,708.96	50	50
Ba	4,708.94	(80)	—
Ne	4,708.85	(1200)	—
Cr	4,708.04	150	200
As	4,707.82	200	—
Mo	4,707.25	125	125
Bi	**4,722.83**	**5**	**10**
V	**4,722.86**	**15**	**20**
Ta	**4,722.88**	—	**200**
Sm	**4,722.95**	—	**3*h***
Mo	**4,723.06**	**20**	**10**
W	**4,723.09**	—	**4**
Cr	**4,723.10**	**8**	**125**
Nb	**4,723.15**	**3*h***	**1**
Dy	**4,723.167**	**2**	**2**
Ti	**4,723.168**	**7**	**40**
Ru	**4,723.22**	—	**6**
Er	**4,723.24**	—	**3**
Mo	**4,723.31**	**4**	**4**
Mo	**4,723.449**	**4**	**4**
Th	**4,723.452**	**4**	**5**
Sm	**4,723.69**	—	**20**
La	**4,723.72**	—	**6**
Gd	**4,723.73**	**200**	**100**
Nb	**4,723.795**	**5**	**5**
Th	**4,723.795**	**10**	**12**
Ne	**4,723.81**	**(20)**	—
Pr	**4,723.899**	—	**4**
Zr	**4,723.900**	—	**3**
Eu	**4,723.91**	—	**20**
Dy	**4,723.92**	—	**3**
Pd	**4,724.00**	**2*h***	**15**
P	4,724.25	(75)	—
Ne	4,725.14	(70)	—
Yb	4,726.07	200	45
Gd	4,726.74	100	40
Ar	4,726.91	(200)	—
P	4,727.46	(100)	—
Co	4,727.94	—	300
La	4,728.42	300	400*r*
Gd	4,728.47	100	150
Sc	4,729.23	50*h*	100
Te	4,729.83	(50)	—
Cr	4,730.71	50	100
Se	4,730.78	(1000)	—
As	4,730.92	125	—
Xe	4,731.19	(50*h*)	—
Te	4,731.27	(70)	—
Mo	4,731.44	100	100
U	4,731.60	50	40
Gd	4,732.61	300	300
Sc	4,734.09	60*h*	100
Xe	4,734.15	(600)	—

	λ	I J	I O
O	4,705.32	(300)	—
Br	4,704.86	(250)	—
Hg	4,704.63	(200)	—
Ne	4,704.39	(1500)	—
La	4,703.28	300r	200r
Ne	4,702.53	(150)	—
Gd	4,702.323	100	50
Ar	4,702.320	(1200)	—
La	4,699.63	200r	200r
O	4,699.21	(100)	—
Xe	4,698.01	(150h)	—
Xe	4,697.02	(300)	—
Kr	4,694.44	(200h)	—
Gd	4,735.76	150	150
Ar	4,735.93	(400)	—
Fe	4,736.78	50	125
Kr	4,739.00	(3000)	—
Se	4,739.03	(800)	—
Ta	4,740.16	100	100R
La	4,740.28	300	150
Cl	4,740.40	(150)	—
Eu	4,740.52	2	500
Se	4,740.97	(600)	—
Sc	4,742.25	(500)	—
Br	4,742.70	(200)	—
La	4,743.08	300	300r
Gd	4,743.65	300	300
La	4,748.73	200	100
Ne	4,749.57	(300)	—
Co	4,749.68	100h	500

Bi 5,209.29 *J* 600

	λ	J	O
Ag	5,209.07	1000R	1500R
Ne	5,209.86	(70)	—
Sb	5,208.80	(8)	—
Bi	5,208.8	70	5wh
Fe	5,208.60	8	200
Cr	5,208.44	100	500R
Kr	5,208.32	(500)	—
Ar	5,208.04	(10)	—
Pr	5,207.90	1	15
V	5,207.68	8	8
La	5,204.15	300	50
Ar	5,187.75	(800)	—
Se	5,175.98	(600)	—
Cs	5,209.44	(15)	—
Cs	5,209.62	(15)	—
Ti	5,210.39	35	200
Mo	5,210.44	4	15
Ar	5,210.49	(200)	—
Ne	5,210.57	(50)	—
Sb	5,210.69	4h	—
Hg	5,210.79	(60)	—
Ar	5,221.27	(500)	—
Se	5,227.51	(600)	—

Bi 5,552.35 *J* 100 *O* 500

	λ	J	O
Ce	5,552.30	—	6
Sc	5,552.25	4	—
Gd	5,552.21	—	2
Mo	5,552.19	3	12
Hf	5,552.12	5	40
Mn	5,551.98	—	10
N	5,551.95	(30)	—
Xe	5,552.38	(80)	—
Mn	5,552.45	—	4
U	5,552.62	3	4
P	5,552.63	(15)	—
Ar	5,552.76	(10)	—
Nd	5,552.86	—	8
Os	5,552.88	—	12

	λ	I J	I O
*bh*Zr	**5,551.7**	—	**60**
J	**5,551.65**	**(15)**	—
Er	**5,551.50**	—	**8**
Xe	**5,551.50**	**(2*whs*)**	—
*bh*B	**5,551.5**	—	**8**
Sm	**5,551.45**	—	**3**
U	**5,551.44**	**5**	**8**
Ce	**5,551.41**	—	**8**
Th	**5,551.37**	—	**10**
Nb	**5,551.35**	**10**	**30**
U	**5,551.26**	**1**	**2**
W	**5,551.02**	—	**5**
Ti	**5,550.96**	—	**10**
Hf	**5,550.60**	**5**	**30**
Sc	**5,550.40**	—	**2*h***
Sm	**5,550.399**	—	**125**
Zr	**5,550.37**	—	**3*h***
W	**5,550.33**	—	**8**
Gd	**5,550.22**	—	**8**
Nd	**5,550.09**	**1**	**10**
Ce	**5,550.04**	—	**6*w***
Eu	5,547.45	—	1000
Yt	5,544.61	80	10
P	5,544.49	(50)	—
P	5,541.18	(50)	—
Ne	5,538.64	(50)	—
In	5,537.03	(50)	—
In	5,536.55	(70)	—
Br	5,536.30	(50)	—
In	5,535.94	(70)	—
La	5,535.67	100	50
Ba	5,535.55	200*R*	1000*R*
N	5,535.39	(70)	—
Ar	5,534.45	(60)	—
Ne	5,533.68	(75)	—
Mo	5,533.05	100	200
Xe	5,531.07	(300)	—
Co	5,530.77	—	500
N	5,530.27	(50)	—
Sc	*5,526.81*	*300wh*	*100*
Ar	*5,524.93*	*(300)*	—
In	*5.523.91*	*(100)*	—
Rb	*5,522.79*	*100*	—
Se	*5,522.42*	*(750)*	—
In	*5,519.36*	*(500)*	—
In	*5,512.92*	*(100)*	—

	λ	I J	I O
Kr	**5,552.99**	**(100*whs*)**	—
Sm	**5,553.01**	—	**3**
*bh*Zr	**5,553.1**	—	**60**
Xe	**5,553.10**	**(3*h*)**	—
Nb	**5,553.12**	**1**	**3**
Ho	**5,553.14**	—	**30**
Er	**5,553.16**	—	**12**
Ti	**5,553.328**	—	**15**
Nd	**5,553.33**	—	**2**
Pr	**5,553.399**	—	**5*d***
Ar	**5,553.40**	**(2)**	—
Tb	**5,553.40**	—	**10**
Ra	**5,553.57**	**(250)**	—
Fe	**5,553.58**	—	**6**
Sc	**5,553.59**	—	**5**
Ni	**5,553.69**	—	**8**
Ar	**5,554.07**	**(5)**	—
W	**5,554.11**	—	**6**
Cr	**5,554.28**	—	**3*s***
Tb	**5,554.42**	—	**25**
O	5,554.94	(100*h*)	—
In	5,555.43	(70)	—
Ra	5,555.85	(500)	—
In	5,556.04	(100)	—
Yb	5,556.48	50	1500
As	5,558.31	200	—
Ar	5,558.70	(500)	—
Ar	5,559.62	(200)	—
N	5,560.37	(200)	—
Kr	5,562.23	(500)	—
Ne	5,562.44	(150)	—
Ne	5,562.77	(500)	—
Cs	5,563.02	(125)	—
Ne	5,563.05	(75)	—
N	5,564.37	(200)	—
S	5,564.93	(150)	—
Xe	5,566.61	(100)	—
Se	5,566.93	(500)	—
Sb	5,568.09	200*wh*	6
Kr	5,568.65	(100)	—
Fe	5,569.62	15	300
Kr	5,570.29	(2000)	—
Eu	5,570,36	—	1000
Mo	5,570.45	100	200
Xe	5,572.19	(50)	—
Ar	5,572.55	(500)	—
Fe	5,572.85	25	300

	λ	I J	I O
In	5,512.82	(150)	—
Br	5,506.78	(300)	—
Mo	5,506.49	100	200r

	λ	I J	I O
Te	5,576.40	(100)	—
In	5,576.75	(300)	—
In	5,576.91	(150)	—
In	5,577.04	(100)	—
Eu	5,577.13	—	1000
Rn	5,582.4	(200)	—
Yb	5,588.47	100	30
Ar	5,588.69	(500)	—
Br	5,589.93	(250)	—
Co	5,590.73	—	500
Se	5,591.16	(500)	—
Al	5,593.23	(200)	—
Ar	5,597.46	(500)	—

Br $^{35}_{79.909}$

t_0 −7.2°C t_1 58.8°C

I.	II.	III.	IV.	V.
11.844	19.2	35.888	—	—

λ	I	eV
2,389.69	(70)	—
2,541.45	(40)	—
2,660.49	(25)	—
2,872.59	(25)	—
4,542.93	(250)	—
II 4,704.86	(250)	14.4
II 4,785.50	(400)	14.4
II 4,816.71	(300)	14.4
5,506.78	(300)	—

Br 2,389.69 (70)

	λ	I J	I O		λ	I J	I O
Ta	**2,389.60**	**12**	**3**	**V**	**2,389.70**	**100**	**5**
Mn	**2,389.59**	7*w*		**Cr**	**2,389.76**	**25**	—
Tm	**2,389.56**	6*h*	—	**Er**	**2,389.79**	**3**	—
Co	**2,389.54**	**20**	**12**	**Te**	**2,389.79**	**(25)**	—
Pt	**2,389.533**	**18**	**25**	**W**	**2,389.80**	**8**	—
Zr	**2,389.53**	**6**	**2**	**La**	**2,389.84**	3*h*	—
Ge	**2,389.47**	**1**	**2**	**Rh**	**2,389.84**	**10**	—
Zr	**2,389.405**	**4**	**2**	**Pd**	**2,389.86**	**2**	—
Fe	**2,389.402**	**2**	**2**	**Nd**	**2,389.92**	**20**	—
Se	**2,389.38**	**(10)**	—	**Zn**	**2,390.04**	**(5)**	—
Co	2,388.91	35	10	Ag	2,390.54	80*h*	—
Pd	2,388.29	40	2	Rh	2,390.62	50*w*	10
Ar	2,387.96	(40)	—	Lu	2,392.19	100	30
Ta	2,387.06	50	20	Rh	2,392.42	40	—
Mn	2,387.02	35	2*h*				
Mo	2,386.96	35	—	*Hf*	*2,393.36*	*80*	*50*
				V	*2,393.57*	*500*	—
				Pb	*2,393.79*	*1000*	*2500*
V	*2,385.82*	*100*	—	*Hf*	*2,393.83*	*100*	*80*
Te	*2,385.76*	*(300)*	*(600)*	*Fe*	*2,395.62*	*100wh*	*50*
Ir	*2,384.81*	*80*	*4*				
Te	*2,383.25*	*300*	*500*	*Rh*	*2,396.55*	*200*	*2*
V	*2,383.00*	*80*	*8*				

Br 2,541.45 (40)

	λ	I J	I O		λ	I J	I O
Nb	**2,541.42**	**80**	**3**	**Ir**	**2,541.47**	**2**	**12**
U	**2,541.37**	**2**	**6**	**Ca**	**2,541.49**	**6**	—
Cr	**2,541.353**	**3**	**60**	**W**	**2,541.51**	**3**	—
Pt	**2,541.352**	**3**	**15**	**U**	**2,541.52**	**2**	**3**
Rh	**2,541.15**	**15**	**2**	**La**	**2,541.59**	**4**	—
Mn	**2,541.11**	**80**	—	**In**	**2,541.62**	**2**	—
Fe	**2,541.10**	**15**	**1**	**Cd**	**2,541.64**	**(1)**	**2**
Nb	**2,541.09**	3*h*	—	**W**	**2,541.69**	**4**	**10**
W	**2,541.06**	**7**	**1**	**V**	**2,541.77**	**2**	**7**
				Fe	**2,541.83**	**5**	**2**
Fe	2,540.67	30	6				
Co	2,540.65	40	6	Co	2,541.94	300*h*	40
Nb	2,540.61	150	2	Zr	2,542.10	50	100
Ru	2,540.30	100	10	Rh	2,542.16	50	1
Al	2,540.12	(30)	—	V	2,542.44	20*wh*	1
				Mo	2,542.67	25	20
Tb	2,540.12	50	3				
Fe	2,539.98	20*h*	3	Fe	2,542.73	40	1
W	2,539.90	20	2	Mn	2,542.92	100	1
Ru	2,539.72	100	12	V	2,542.93	35	2
Mo	2,539.44	20	1	Ru	2,543.25	150	50

	λ	I	
		J	O
Mn	2,539.40	20Wh	—
Pd	2,539.36	50wh	—
Pt	2,539.20	20	400
Ni	2,539.10	250w	—
Fe	2,539.00	20	10
Fe	2,538.81	30	15
Yb	2,538.67	20	10
Mo	2,538.45	125	30
Cr	2,538.29	25	—
Fe	2,538.20	35	2
Mn	2,538.05	25	—
Pd	2,537.97	100	—
Te	2,537.80	(300)	—
Rh	2,537.73	50	4
V	2,537.62	25	—
Pd	*2,537.17*	*100*	—
Rh	*2,537.04*	*100*	*15*
Hg	*2,536.52*	*1000R*	*2000R*
Co	*2,535.96*	*40*	*10r*
Ti	*2,535.87*	*60*	*20*
Mn	*2,535.64*	*80*	—
Ru	*2,535.59*	*100*	—
Ru	*2,534.94*	*40*	*3*
Ar	*2,2534.74*	*(40)*	—
Ti	*2,534.62*	*80*	*25*
Pd	*2,534.60*	*100*	—
Rh	*2,534.57*	*100w*	*2w*
V	*2,534.52*	*80*	*10*
Nb	*2,534.44*	*40*	*1*
Fe	*2,534.42*	*50*	*7*
Ru	*2,534.00*	*80*	*4*
Co	*2,533.81*	*60*	*5r*
Fe	2,543.38	50	5
Mn	2,543.45	100	4
Sb	2,543.84	25	—
Fe	2,543.920	20	40
Cs	2,543.92	(20)	—
Rh	2,543.94	100r	15
Ir	2,543.97	100	200h
Co	2,544.25	100	50r
Ar	2,544.72	(40)	—
Cu	2,544.802	700R	—
Nb	2,544.803	300	5
Pd	2,544.83	200	—
Co	2,545.04	30	3
Th	2,545.10	20	—
Mn	2,545.16	25	—
Sc	2,545.20	20	15
Fe	2,545.22	25	4
Rh	2,545.35	150	8
Al	*2,545.60*	*(50)*	—
Nb	*2,545.64*	*150*	*1*
Ni	*2,545.90*	*200h*	*20*
Sn	*2,546.55*	*100*	*100*
Co	*2,546.74*	*50*	*1*
Ni	*2,547.19*	*100*	—
Ru	*2,547.67*	*80*	*6*
Se	*2,547.98*	*(60)*	—
Co	*2,548.34*	*75*	*20*
Nb	*2,548.63*	*80*	*2*
V	*2,548.69*	*80*	*10*
Mn	*2,548.74*	*150*	*8*
Ru	*2,549.18*	*150*	*5*
Se	*2,549.19*	*(50)*	—

Br 2,660.49 (25)

	λ	J	O
Ag	**2,660.46**	**150**	**30**
Fe	**2,660.402**	**15**	**40**
Hg	**2,660.4**	**(2)**	**5h**
Cd	**2,660.40**	**(5)**	**50h**
Al	**2,660.39**	**60**	**150R**
Mn	**2,660.34**	**5**	—
Pd	**2,660.19**	**(5)**	—
U	**2,660.14**	**10**	**15**
Tm	**2,660.09**	**10**	**30**
Nb	**2,660.03**	**30**	**2**
Yb	**2,660.01**	**3**	**1**
Hf	**2,660.52**	**2**	—
W	**2,660.522**	**8**	**10**
Mo	**2,660.58**	**125**	**25**
Mn	**2,660.62**	**10**	—
Ti	**2,660.64**	**1**	**12**
J	**2,660.75**	**(20)**	—
Mg	**2,660.755**	**6**	**40**
Mg	**2,660.82**	**6**	**40**
Os	**2,660.92**	**1**	*5*
Kr	**2,660.97**	**(8h)**	—

	λ	I J	I O
Ga	2,659.87	12	5
W	2,659.70	12	—
Ru	2,659.615	12	80
V	2,659.606	40	9
Pt	2,659.45	500*R*	2000*R*
J	2,659.27	(20)	—
Rh	2,659.11	100	—
Mn	2,659.08	25	—
Nb	2,659.05	30	34
V	2,658.97	40	10
Tb	2,658.91	500	—
Ta	2,658.86	50	25
Cl	2,658.74	(100)	—
Pd	2,658.72	300	20
Cr	2,658.59	35	18
V	2,658.51	15	—
Rh	2,658.36	25	2
Fe	2,658.25	80	—
Ru	2,658.24	35	—
Ta	2,658.14	15*h*	3*h*
W	2,658.04	20	10
Nb	2,658.03	200	—
Fe	2,657.92	20	—
Hf	2,657.84	25	20
Lu	2,657.80	150	50
Pd	2,657.56	200	—
Ne	2,657.52	(15)	—
Hf	2,657.50	20	10
Rh	2,657.32	70	—
V	2,657.29	35	1
Ru	2,657.19	50	—
Mo	2,656.98	25	—
Ag	2,656.92	20*h*	—
Br	2,656.83	(25)	—
Fe	2,656.80	25	50
Ag	2,656.66	15*h*	1*h*
Kr	2,656.38	(15*h*)	—
Ru	2,656.25	150	20
Mn	2,656.17	12	—
Fe	2,656.15	40	70
Nb	2,656.08	200	8
Mn	2,655.914	20	5
Ni	2,655.907	500*Wh*	—
J	2,655.84	(20)	—
Mo	2,655.81	20	—
V	2,655.68	100*h*	10
Ni	*2,655.47*	*400wh*	—

	λ	I J	I O
Na	2,661.00	(80)	—
Te	2,661.13	(30)	—
Pd	2,661.14	100	—
Ru	2,661.17	100	20
J	2,661.22	(12)	—
Sh	2,661.25	80	100
V	2,661.42	80	100
Ru	2,661.61	150	80
W	2,661.85	12	—
Hf	2,661.87	40	25
Ir	2,661.98	15	150*h*
Se	2,662.05	(25)	—
Fe	2,662.06	40	70
Te	2,662.11	(25)	—
Ru	2,662.16	30	—
Mn	2,662.54	50	—
Fe	2,662.56	15*h*	2*h*
In	2,662.58	(30)	—
In	2,662.68	(30)	—
S	2,662.82	(15)	—
Ru	2,662.86	40	—
Pb	2,663.17	40	300*wh*
V	2,663.25	100	12
Co	2,663.53	60*w*	15*w*
Mn	2,664.03	25	—
U	2,664.15	20	18
Ta	2,664.23	100*h*	10*h*
W	2,664.32	20	10
Rh	2,664.49	30	1
Fe	2,664.66	300	20
Ir	2,664.79	50	200*h*
Yb	2,665.03	60	10
Ga	2,665.05	(40)	—
Mn	2,665.18	15	—
Ni	2,665.249	125	—
Nb	2,665.251	300	3
J	2,665.31	(30)	—
Ta	*2,665.60*	*40h*	*80d*
Hf	*2,665.97*	*35*	*20*
Yb	*2,666.08*	*150*	*5*
Nb	*2,666.59*	*50*	*5*
Fe	*2,666.63*	*80*	*5*
Mn	*2,666.77*	*50h*	*8*
Yb	*2,666.97*	*150*	*5*
Mn	*2,667.01*	*50h*	*5*
Cl	*2,667.36*	*(40)*	—
Ru	*2,667.39*	*150*	*10*

	λ	I	
		J	O
Hg	*2,655.12*	*40*	*80*
Rh	*2,654.77*	*25*	—
In	*2,654.76*	*(30)*	—
Ru	*2,653.95*	*80*	—
Yb	*2,653.75*	*200*	*50*
Co	*2,653.70*	*40*	*20*
Hg	*2,653.68*	*40*	*80*
Cr	*2,653.59*	*35*	*12*
Mo	*2,653.35*	*150*	*25*
V	*2,652.78*	*40h*	—
Rh	*2,652.66*	*25*	*100*
Sb	*2,652.61*	*75*	*50*
Fe	*2,652.57*	*40*	*1*
Al	*2,652.49*	*60*	*150R*
Mn	*2,652.48*	*100*	*3*
Yb	*2,652.23*	*60*	*2*
Hg	*2,652.04*	*60*	*100*
Yb	*2,651.71*	*60*	*2*
Fe	*2,651.706*	*60*	*60*
Ta	*2,651.22*	*80*	—
Hf	*2,651.16*	*40*	*15*
Nb	*2,651.12*	*200*	*3*
Ne	*2,651.01*	*(50)*	—
Mn	*2,650.99*	*150*	—
Pt	*2,650.86*	*100*	*700*
Nb	*2,667.76*	*30*	*3*
Ru	*2,667.79*	*80*	—
Eu	*2,668.33*	*400*	*300W*
Ta	*2,668.619*	*100*	*1*
In	*2,668.623*	*(30)*	—
In	*2,668.68*	*(50)*	—
Al	*2,669.17*	*100*	*3*
W	*2,669.30*	*30d*	*15*
Ru	*2,669.42*	*100*	—
Fe	*2,669.50*	*25*	*50*
Co	*2,669.91*	*100wh*	—
Fe	*2,669.93*	*25*	—
V	*2,670.23*	*70*	*2*

Br 2,872.59 (25)

	λ	J	O
Zr	**2,872.53**	**3**	**2**
W	**2,872.50**	**4**	**5**
Os	**2,872.40**	**8**	**50**
Fe	**2,872.38**	**20**	**2**
Cs	**2,872.35**	**(8)**	—
Fe	**2,872.34**	**50**	**150**
Te	**2,872.18**	**(5)**	—
Ir	**2,872.111**	**3*h***	**2**
U	**2,872.110**	**4**	**3**
Se	**2,872.08**	**(3)**	—
U	**2,871.99**	**4*h***	**6*h***
Mo	2,871.51	100*h*	100
Ru	2,871.47	30	—
Cr	2,871,45	80	—
Ta	2,871.42	50	200
F	2,871.40	(25)	—
Na	2,871.27	(40)	6
Xe	2,871.24	(25*hs*)	—
Ne	**2,872.66**	**(35)**	—
Nb	**2,872.80**	**10**	**1**
Er	**2,872.84**	**1**	**6**
Mo	**2,872.88**	**30**	**2**
J	**2,872.89**	**(60)**	—
Mn	**2,872.91**	**3**	**1**
Ne	**2,873.00**	**(10)**	—
Pb	**2,873.0**	**(20)**	—
U	**2,873.005**	**6**	**10**
Tm	**2,873.01**	**20**	**15**
Si	**2,873.10**	**(2)**	—
F	**2,873.13**	**(5)**	—
La	**2,873.18**	**2**	**1**
V	**2,873.182**	**50**	**4**
Zn	**2,873.19**	**(3)**	—
Hg	2,873.24	(20)	—
Ru	2,873.31	60	4
Pb	2,873.32	60	100*R*
Ta	2,873.36	40*h*	200*W*

	λ	I			λ	I	
		J	O			J	O
Co	2,871.24	100	—	Fe	2,873.40	300	—
Fe	2,871.13	20	—				
Fe	2,871.06	40	—	Cr	2,873.48	125	30
				Ta	2,873.56	50	150
U	2,870.97	20	18	Ag	2,873.65	100*wh*	3
W	2,870.90	12	9	Ru	2,873.73	12	—
Fe	2,870.60	15	—	Cr	2,873.82	40	20
Ru	2,870.548	50	—				
V	2,870.547	20*r*	50*r*	Ta	2,874.17	15	150
				Fe	2,874.17	200	300
Cr	2,870.44	300*W*	25	V	2,874.21	20	7
Th	2,870.41	20	18	Ga	2,874.24	15*r*	10
V	2,870.11	12	1	F	2,874.80	(15)	—
Ti	2,870.04	100*wh*	—				
Co	2,870.03	50*wh*	—	Mo	2,874.85	60	2
				Fe	2,874.88	20	60
V	2,869.96	20	7	Os	2,874.95	15	50
Hf	2,869.82	20	25	Ru	2,874.98	50	80
Zr	2,869.81	30	30	U	2,875.20	12	18
Fe	2,869.31	70	300				
Tm	2,869.22	300	100	J	2,875.25	(12)	—
				Fe	2,875.30	50	125
Mo	2,869.217	15	—	Fe	2,875.35	70	—
V	2,869.13	150*r*	25	Nb	2,875.39	300	50*r*
Fe	2,868.87	60	5	Ir	2,875.60	15	25
Te	2,868.86	(100)	—				
J	2,868.77	(30)	—	V	2,875.69	40	10
				Ti	2,875.79	40*wh*	—
Ti	2,868.74	25	15	Pt	2,875.85	80*h*	20
W	2,868.73	25	6	Ir	2,875.98	15	25
Rh	2,868.7	(70)	—	Cr	2,875.99	80*wh*	30
Ta	2,868.65	40	150				
Nb	2,868.525	300	15	Cr	2,876.24	80*wh*	25
				Hf	2,876.33	100	30
Al	2,868.52	(80)	—	Ne	2,876.43	(18)	—
Fe	2,868.45	40	80	Cr	2,876.66	20	—
Mo	2,868.32	20	2	Fe	2,876.80	100	—
Cd	2,868.28	80	100				
Mo	2,868.11	20	3	W	2,876.93	12	8
				V	2,876.94	25	4
V	2,868.20	30*r*	40	Nb	2,876.95	500*W*	40*W*
Tm	2,868.01	40	15	U	2,877.05	12	6
Mo	2,867.82	40*h*	—	Fe	2,877.30	125	200
Cr	2,867.65	100*R*	80				
Fe	2,867.56	30	60	Ti	2,877.44	100	30
				Pt	2,877.52	200*h*	40
Ta	2,867.41	150	5*h*	Ta	2,877.686	80*h*	15
Fe	2,867.31	30	60	V	2,877.688	100*R*	15
Cr	2,867.10	35	20	Cu	2,877.689	20	5
Ru	2,867.09	18	—				
Yb	2,867.04	40	4	Pd	2,877.87	15*wh*	—
				Sb	2,877.91	150	250*W*
Pt	2,866.89	15	6	Cr	2,877.98	100	30
Cr	2,866.74	125*R*	80	Ru	2,878.04	15	—
Mo	2,866.69	30	30	Ta	2,878.20	15*h*	4
Ru	2,866.64	25	60				
Fe	2,866.63	80	125	Tm	2,878.21	20	3
				Tm	2,878.36	20	10

	λ	I	
		J	O
In	2,866.46	(18)	—
Hf	2,866.37	12	50
Ru	2,866.25	30	—
Ru	2,866.08	30	—
In	2,865.684	(50)	—
U	2,865.679	50	30
Nb	2,865.61	30	5
Cr	2,865.11	200R	60
Pt	2,865.05	80h	20
Fe	2,864.97	50	—
Xe	2,864.73	(100)	—
V	2,864.52	35	6
Pb	2,864.51	60	—
Ta	2,864.50	30	125
V	2,864.36	25r	40
Pb	2,864.26	60	—
Ni	2,864.15	300wh	—
Fe	2,863.86	100	125
Mo	2,863.81	100h	30
Tm	2,863.76	40	15
Ni	2,863.70	250	—
Fe	2,863.43	80	100
Sn	2,863.33	300R	300R
Ru	2,863.32	80	30
Rh	2,862.93	60	150
Ru	2,862.88	60	6
Cr	2,862.571	300R	80
Eu	2,862.57	70	100W
Fe	2,862.50	50	100
Ti	2,862.32	40	20
V	2,862.30	25	1
Ti	2,861.99	100wh	—
Tm	2,861.74	30	20
Hf	2,861.70	125	50
Ru	2,861.41	35	50
Yb	2,861.31	25	3
Yb	2,861.21	30	3
Fe	2,861.19	30	1
Nb	2,861.09	100	10
Te	2,861.03	(25)	—
Hf	2,861.01	90	40
Os	2,860.96	25	100
Cr	2,860.93	100	60
Ti	2,860.84	25hw	—
Pt	2,860.68	150h	30
U	2,860.47	30	35
As	2,860.45	50	50r
Hf	2,860.31	30	15
Os	2,878.40	12	40
Cr	2,878.449	80	20
Sm	2,878.449	25	4
J	2,878.64	(400)	—
Mo	2,879.05	100h	15
Ru	2,879.06	60	—
V	2,879.16	35	50
Fe	2,879.24	25	—
Mo	2,879.69	25	1
N	2,879.73	(25h)	—
Ta	2,880.02	50	150
V	2,880.03	150r	25
Ru	2,880.08	25	—
Co	2,880.29	50wh	—
Se	2,880.31	(25)	—
La	2,880.64	40	4
Nb	2,880.71	50	4w
Fe	2,880.76	50	15
Cd	2,880.77	125	200R
Fe	2,880.83	25	1
Cr	2,880.87	25	20
Na	2,881.14	(60)	8
Cd	2,881.23	(30)	50R
Al	2,881.46	(30)	—
Si	2,881.58	400	500
Cr	2,881.93	30	1
Ru	2,882.12	200	30
Zn	2,882.15	(25)	—
Ta	2,882.33	80	3
Mo	2,882.38	25	1
V	2,882.50	200r	35
Nb	2,883.18	800R	100
Fe	2,883.70	300	—
Rn	2,883.8	(25)	—
P	2,883.90	(25)	—
Ti	2,884.11	125	35
Al	2,884.20	(30)	—
Cu	2,884.38	30	—
As	2,884.51	25	—
Fe	2,884.779	25h	—
V	2,884.785	200r	40
Tb	2,885.14	70	—

Br 4,542.93 (250)

	λ	I J	I O
W	4,542.887	3	15
Mo	4,542.886	4	3
Ir	4,542.87	2wh	3
Nb	4,542.80	5	5
Gd	4,542.73	3	3
Cr	4,542.62	8	30
Nd	4,542.60	5	50
U	4,542.59	2	—
Pr	4,542.54	8	20
Mn	2,542.44	5	80
Fe	4,542.42	1	3
U	4,542.11	3	1
Nd	4,542.05	5	50
Yt	4,542.04	2	3
Gd	4,542.03	50	60
La	4,541.79	5	30
Hf	4,541.72	2	12
U	4,541.705	12	8
Dy	4,541.698	8	6
Er	4,541.67	1	5
Xe	4,540.89	(200h)	—
Gd	4,540.02	200	80
As	4,539.97	200	—
Ne	4,538.31	(300)	—
Ne	4,537.75	(1000)	—
Ne	4,537.68	(300)	—
Ne	4,536.31	(150)	—
Ti	4,533.97	150	30
Ra	4,533.11	(300)	—
P	4,530.78	(150)	—
Cr	4,530.74	125	150
Xe	*4,524.68*	*(400)*	—
Kr	*4,523.14*	*(400h)*	—
Tm	*4,522.57*	*300*	*200*
La	*4,522.37*	*400*	*200*
Ar	*4,322.32*	*(800)*	—

	λ	I J	I O
Hf	4,543.01	2	5
U	4,543.03	2h	—
Th	4,543.21	2	3
W	4,543.28	1	7
Mo	4,543.40	15	2
W	4,543.51	10	25
Pr	4,443.53	3	25w
U	4,543.63	80	50
Cs	4,543.71	(10)	—
Cr	4,543.74	2	20
As	4,543.76	200	—
J	4,543.86	(10)	—
Yb	4,543.93	10	—
Sm	4,543.94	50	100
Pr	4,543.96	3	10
Ti	4,544.01	20	5
Hf	4,544.02	2	20
Ne	4,544.11	(5)	—
Gd	4,544.25	6h	6h
Rh	4,544.27	4	25
J	4,544.29	(8)	—
Yt	4,544.30	3	5
Ar	4,545.08	(200)	—
Xe	4,545.23	(200wh)	—
As	4,549.23	125	—
Ti	4,549.63	200	100
S	4,552.38	(200)	—
Ba	4,554.04	200	1000R
Ru	4,554.51	(200)	1000R
Te	*4,557.84*	*(300)*	—
Cr	*4,558.66*	*600wh*	*20*

Br 4,704.86 (250)

	λ	I J	I O
Xe	4,704.67	(8wh)	—
Hg	4,704.63	(200)	—
Cu	4,704.60	50	200
Sb	4,704.48	10	—
U	4,704.476	3	3

	λ	I J	I O
Fe	4,704.96	1	10
V	4,705.09	12	15
O	4,705.32	(300)	—
Bi	4,705.35	50	—
Kr	4,705.44	(2h)	—

	λ	I J	I O
Ne	4,704.39	(1500)	—
Rh	4,704.08	4	10
Th	4,703.99	1	3
Nb	4,703.93	5h	3h
Hf	4,703.61	12	4
Dy	4,703.47	2	5
La	4,703.28	300r	200r
Ne	4,702.53	(150)	—
Ar	4,702.32	(1200)	—
La	4,699.63	200r	200r
Xe	4,698.01	(150h)	—
Xe	4,697.02	(300)	—
Kr	4,694.44	(200h)	—
La	4,692.50	300	200
Te	*4,686.95*	*(300)*	—
He	*4,685.75*	*(300)*	—
Ra	*4,682.28*	*(800)*	—
In	*4,682.00*	*250W*	—
Rn	*4,680.83*	*(500)*	—
Kr	*4,680.41*	*(500)*	—
Ne	*4,678.22*	*(300)*	—

	λ	I J	I O
Th	4,705.76	6	10
Os	4,705.94	1	15
Mo	4,706.06	25	25
Ta	4,706.09	2	200
Cr	4,706.10	1	30
Nb	4,706.14	50	50
V	4,706.16	12	15
W	4,706.17	1	15
Mo	4,706.20	4	4
Th	4,706.22	4	5
Kr	4,706.31	(3)	—
Mo	4,707.25	125	125
As	4,707.82	200	—
Cr	4,708.04	150	200
Ne	4,708.85	(1200)	—
Ne	4,710.06	(1000)	—
Ne	4,712.06	(1000)	—
La	4,712.93	150	100r
Ne	4,715.34	(1500)	—
S	4,716.23	(600)	—
La	4,716.44	200	100
Cr	4,718.43	150	200
La	*4,719.95*	*300*	*200r*
Zn	*4,722.16*	*300h*	*400w*
La	*4,728.42*	*300*	*400r*
Se	*4,730.78*	*(1000)*	—
Gd	*4,732.61*	*300*	*300*

Br 4,785.50 (400)

	λ	I J	I O
Cl	4,785.44	(50)	—
Lu	4,785.42	200	100
Dy	4,785.31	2	3
Br	4,785.19	(20)	—
Mo	4,785.12	30	30
Dy	4,784.92	2	4
Te	4,784.85	(70)	—
J	4,784.76	(5)	—
Sb	4,784.76	24	—
Gd	4,784.64	50	100
V	4,784.47	10	12
Mo	4,784.41	5	5
B	4,784.29	4	—
Nb	4,784.28	1	3
Sb	4,784.03	(70)	—

	λ	I J	I O
Nb	4,785.70	2	3
Ar	4,786.19	(2)	—
Cs	4,786.36	(15)	—
Mo	4,786.46	25	25
V	4,786.51	25h	30
Ni	4,786.54	2	300w
Yt	4,786.58	25	15
Yb	4,786.598	200	50
Er	4,786.603	2	2
Xe	4,786.65	(8h)	—
Gd	4,786.81	2	40
Yt	4,786.88	15	15
Dy	4,786.93	4	8

	λ	I			λ	I	
		J	O			J	O
Ta	4,780.93	200	50	Ne	4,788.93	(300)	—
Co	4,780.01	500	500*w*	Fe	4,791.25	200*R*	200
Br	4,776.42	(200)	—	Cl	4,794.54	(250)	—
				Hg	4,797.01	(300)	—
				Cd	4,799.92	300*w*	300*w*
Kr	*4,765.74*	*(1000)*	—	*Ar*	*4,806.07*	*(500)*	—
Se	*4,763.65*	*(800)*	—	*Xe*	*4,807.03*	*(500)*	—
				S	*4,815.51*	*(800)*	—

Br 4,816.71 (300)

	λ	J	O		λ	J	O
Yb	**4,816.40**	**1**	**20**	**Gd**	**4,816.84**	**20**	**50**
Nb	**4,816.38**	**50**	**50**	**Mo**	**4,816.96**	**3**	**3**
W	**4,816.11**	**1**	**10**	**Pd**	**4,817.01**	**2*h***	**4**
Co	**4,815.89**	**1**	**4**	**Rn**	**4,817.15**	**(100)**	—
Sm	**4,815.81**	**80**	**125**	**Hf**	**4,817.21**	**40**	**15**
S	**4,815.51**	**(800)**	—	**Xe**	**4,817.22**	**(20*wh*)**	—
Ta	**4,815.12**	**15**	**5**	**C**	**4,817.33**	**(5)**	—
				Pd	**4,817.51**	**8**	**40**
				Ne	**4,817.64**	**(300)**	—
Ge	4,814.80	200	—	**W**	**4,817.69**	**1**	**5**
Kr	4,811.76	(300)	—				
Zn	4,810.53	300*h*	400*w*	**Mo**	**4,817.70**	**25**	**25**
Ne	4,810.063	(150)	—	**Xe**	**4,818.02**	**(100)**	—
Cl	4,810.06	(200)	—				
La	4,809.01	150	150	Ne	4,818.80	(150)	—
Xe	4,807.02	(500)	—	Cl	4,819.46	(200)	—
Ar	4,806.07	(500)	—	Fe	4,821.05	200*h*	200*h*
La	4,804.04	150	150	Ne	4,821.92	(300)	—
				Xe	4,823.41	(150*h*)	—
Gd	*4,799.92*	*300w*	*300w*	La	4,824.07	150	150
Hg	*4,797.01*	*(300)*	—	Kr	4,825.18	(300)	—
Ne	*4,788.93*	*(300)*	—	Ra	4,825.91	(800)	—
				Ne	4,827.34	(1000)	—
				Ne	4,827.59	(300)	—
				Xe	4,829.71	(400)	—
				Te	4,831.29	(800)	—
				Kr	4,832.07	(800)	—
				Ne	*4,837.31*	*(500)*	—
				Se	*4,840.63*	*(800)*	—
				Xe	*4,843.29*	*(300)*	—
				Xe	*4,844.33*	*(1000)*	—
				Se	*4,844.96*	*(800)*	—
				Kr	*4,846.60*	*(700)*	—

Br 5,506.78 (300)

	λ	I	
		J	O
In	5,506.71	(30)	—
Mo	5,506.49	100	200r
Ar	5,506.11	(500)	—
V	5,505.86	10	10
Yb	5,505.501	2	40
Ra	5,505.50	(25)	—
Ar	5,505.18	(10)	—
V	5,504.87	15	15
J	5,504.72	(60)	—
Nb	5,504.58	34	30w
Ra	5,501.98	(250)	—
P	5,499.71	(150)	—
As	5,497.98	200	—
J	5,496.92	(200)	—
Ar	5,495.87	(1000)	—
Br	5,495.06	(150)	—
Lu	5,476.69	1000	500
S	5,473.63	(750)	—
Ar	5,473.44	(500)	—
Xe	5,472.61	(500)	—
Ag	5,465.49	500R	1000R
J	5,464.61	(900)	—

	λ	I	
		I	O
Fe	5,506.782	(10)	150
In	5,506.82	(15)	—
S	5,507.01	(25)	—
In	5,507.11	(30)	—
P	5,507.13	(70)	—
Cs	5,507.17	(15)	—
In	5,507.33	(70)	—
Ne	5,507.34	(25)	—
Xe	5,507.46	(10)	—
Ar	5,507.63	(10)	—
V	5,075.75	60	60
Br	5,508.38	(5)	—
V	5,508.62	5	5
Pr	5,508.78	2	7
In	5,512.82	(150)	—
In	5,519.36	(500)	—
Se	5,522.42	(750)	—
Ar	5,524.93	(300)	—
Se	5,526.81	300wh	100
Xe	5,531.07	(300)	—

C 6 12.01115

t_0 3,527°C

t_1 4,200°C

I.	II.	III.	IV.	V.
11.265	24.377	47.866	64.478	392.0

λ	*I*		eV
	J	*O*	
III 2,296.89	200	—	
I 2,478.57	(400)	400	7.7
II 2,512.03	400	—	18.66
II 2,836.71	200	—	27.6
II 4,267.02	350	—	32.2
II 4,267.27	500	—	32.2
I 5,793.51	—	30	10.08
II 6,578.03	500	—	16.33

C 2,296.89 *J* 200

	λ	*I*			λ	*I*	
		J	*O*			*J*	*O*
Ta	**2,296.86**	**25**	—	**Fe**	**2,296.92**	**2**	**15**
Pb	**2,296.85**	**3**	—	**W**	**2,296.97**	**6**	**3**
K	**2,296.79**	**(5)**	—	**U**	**2,297.02**	**4**	—
Co	**2,296.71**	**2**	**18**	**In**	**2,297.09**	**10**	—
Ru	**2,296.65**	**5**	—	**Ni**	**2,297.138**	**18**	**15**
				Ir	**2,297.138**	**20**	**2**
V	2,294.99	100	5	**Cr**	**2,297.19**	**50**	**10**
V	*2,292.85*	*250*	—	Lu	2,297.41	100	15
				Pr	2,297.77	100	—
				La	2,297.78	150	2*h*
				V	2,297.85	100	—
				Rh	2,298.26	150	—
				Tl	2,298.95	150	30

C 2,478.573 *J* (400) *O* 400

	λ	*J*	*O*		λ	*J*	*O*
Tm	**2,478.57**	**15**	**15**	**Cr**	**2,478.59**	**3*w***	—
Fe	**2,478.569**	**40**	**4**	**Pb**	**2,478.594**	**2**	—
Nb	**2,478.566**	**4*wh***	—	**V**	**2,478.62**	**20**	**2**
Pd	**2,478.565**	**25**	—	**Ti**	**2,478.65**	**50**	—
Hf	**2,478.563**	**300*wh***	**100**	**Hg**	**2,478.66**	**15*h***	**15*h***
Fe	**2,478.45**	**5*h***	—	**Pd**	**2,478.80**	**15*h***	—
Ru	**2,478.40**	**5*wh***	**8**	**Xe**	**2,478.82**	**(2)**	—
W	**2,478.315**	**10**	—	**Kr**	**2,478.85**	**(3)**	—
Sb	**2,478.311**	**100**	**75**	**W**	**2,478.87**	**8**	—
Nb	**2,478.29**	**8**	**4**	**Ru**	**2,478.93**	**60**	**80**
P	**2,478.22**	**(10)**	—				
Ta	**2,478.219**	—	**60**	Fe	2,479.78	30	200*R*
Mo	**2,478.219**	**4**	—				
Co	**2,478.21**	**20**	**4**				
				Fe	*2,483.27*	*50*	*500rh*
Nb	2,477.38	200	5				
Pd	2,476.42	50	300*r*				
Ni	*2,473.15*	*500*	*80*				
Fe	*2,472.91*	—	*1000*				

C 2,512.03 *J* 400

	λ	*J*	*O*		λ	*J*	*O*
Hg	**2,512.03**	**5**	—	**Ta**	**2,512.036**	**800**	**5**
Ru	**2,511.99**	**40**	**20**	**Mo**	**2,512.05**	**10**	—
Nb	**2,511.97**	**50*h***	—	**Yb**	**2,512.06**	**50**	**10**

	λ	I J	I O
Cr	2,511.96	1	25
V	2,511.95	2h	35
W	2,511.83	8	—
Nb	2,511.760	10	—
Fe	2,511.759	100	25
Kr	2,511.74	(3)	—
C	2,511.71	60	—
Ta	2,511.69	100d	2d
Hg	2,511.64	(1)	2h
Ni	2,510.87	250h	50h
Rh	2,510.65	200wh	5
C	2,509.11	200	—
Cu	*2,506.27*	*500r*	—

	λ	I J	I O
Cs	2,512.1	(2)	—
Ir	2,512.103	25h	—
Fe	2,512.105	30	—
W	2,512.19	4	—
In	2,512.25	(25)	—
Fe	2,512.36	1	12
In	2,512.37	(30)	—
Cr	2,512.40	6	—
Si	2,514.33	200	300
Pd	2,514.48	200	—
Si	*2,516.12*	*500*	*500*

C 2,836.71 J 200

	λ	I J	I O
Mo	2,836.70	25h	1
Hg	2,836.67	5	—
Ti	2,836.64	100wh	—
Ta	2,836.62	2	80r
Ti	2,836.61	2	5
Ru	2,836.57	1h	30
V	2,836.52	80	20
Fe	2,836.51	12	3
Cr	2,836.48	20	3
Ra	2,836.46	(25)	—
Th	2,836.44	2	5
Ir	2,836.40	10	25
O	2,836.35	(5)	—
Fe	2,836.32	5	8
Rn	2,836.3	(25)	—
Mo	2,836.29	10	2
Br	2,836.27	(2)	—
W	2,836.25	10s	10
Nb	2,836.24	5	3
K	2,836.2	(10)	—
Fe	2,836.19	10	3
Cr	2,835.63	400r	100
Fe	2,835.46	100	100
Nb	2,835.12	100	5d
Cr	2,834.26	125	—
Cd	2,834.19	(100)	—

	λ	I J	I O
Fe	2,836.72	20	—
Ga	2,836.87	2	—
Cd	2,836.91	80	200
U	2,836.918	10h	5
In	2,836.919	80	80
J	2,836.922	(40)	—
Fe	2,837.03	2	1
Pd	2,837.12	(40)	—
U	2,837.19	8	10
Se	2,837.21	(35)	—
Cu	2,837.55	250	—
Fe	2,838.12	150	150
Ta	2,838.24	150	2
Os	2,838.63	100	100R
Ti	2,839.80	100wh	—
Pd	2,839.89	100	—
Sn	2,839.99	300R	300R
Cr	2,840.02	125	25
Pd	2,841.03	100	—
Ru	2,841.12	125	—
Nb	2,841.15	100	10
Ru	2,841.68	200	50
Ti	2,841.94	125	40
Rn	2,842.1	(150)	—
Cr	*2,843.25*	*400r*	*125*
Fe	*2,843.98*	*300*	*300*

	λ	I J	I O
Kr	2,833.00	(100)	—
Cr	2,832.46	125	2
Fe	2,832.44	200	300
Ti	2,832.16	100	25
Fe	2,831.56	500	1
Pt	*2,830.29*	*600r*	*1000R*
Ti	*2,828.15*	*200h*	*2*
Co	*2,825.24*	*200*	*5*

	λ	I J	I O
Ta	*2,844.46*	*200*	*200*
Hg	*2,847.67*	*(300)*	—

C 4,267.02 *J* 350

4,267.27 *J* 500

	λ	J	O
Gd	**4,267.016**	**40**	**40**
Yb	**4,266.99**	**8*h***	**4**
Fe	**4,266.97**	**30**	**70**
Cr	**4,266.82**	**1**	**30**
U	**4,266.67**	**3**	**3**
Rb	**4,266.62**	**10**	—
Mo	**4,266.61**	**15*h***	**2**
W	**4,266.54**	**8**	**15**
Ar	**4,266.53**	**(200)**	—
Dy	**4,266.52**	**2*h***	**2**
Lu	**4,266.40**	**2*h***	—
Tb	**4,266.35**	**1**	**30**
Sm	**4,266.31**	**1**	**9**
Ar	**4,266.29**	**(1200)**	—
Mo	**4,266.18**	**20**	**25**
Ho	**4,266.04**	**2**	**5**
Nb	**4,266.02**	**20**	**15**
Mn	**4,265.92**	**50**	**100**
Dy	**4,265.83**	**2**	**6**
U	**4,265.63**	**2**	**3**
Ar	**4,265.52**	**(2)**	—
Ir	**4,265.30**	**2**	**60**
V	**4,265.17**	**15**	**30**
As	**4,265.13**	**30**	—
Ra	**4,265.12**	**(6)**	—
Mo	**4,265.117**	**15**	**15**
Sm	**4,265.07**	**25**	**60**
La	4,263.58	150	150
Sm	4,262.68	150	200
Te	4,261.08	(300)	—
Os	4,260.85	200	200
Fe	4,260.48	300	400

	λ	J	O
As	**4,267.24**	**3**	—
C	***4,267.27***	***500***	—
Ne	**4,267.29**	**(2)**	—
Ta	**4,267.298**	**20**	**1**
U	**4,267.303**	**10**	**12**
Nd	**4,267.49**	**200*wh***	—
Nb	**4,267.63**	**10*h***	—
Ne	**2,267.72**	**(5)**	—
S	**4,267.80**	**(60)**	—
Fe	**4,267.83**	**60**	**125**
U	**4,267.93**	**4**	**15**
Ba	**4,267.95**	**(80)**	—
Ne	**4,268.01**	**(70)**	—
Zr	**4,268.02**	**1**	**40**
W	**4,268.05**	**5**	**10**
Mo	**4,268.07**	**20**	**20**
Hf	**4,268.10**	**5**	—
Ir	**4,268.101**	**15**	**200**
Ta	**4,268.25**	**15*h***	**50**
Co	**4,268.44**	**2**	**2**
Kr	**4,268.57**	**(60*wh*)**	—
V	**4,268.64**	**20**	**40**
Nb	**4,268.67**	**10**	**5**
Gd	**4,268.74**	**40**	**40**
Fe	**4,268.76**	**10**	**30**
Cr	**4,268.79**	**3**	**30**
Kr	**4,268.81**	**(100*wh*)**	—
U	**4,268.85**	**10**	**12**
Cs	**4,268.89**	**(10)**	—
Ir	**4,268.94**	**2**	**10**
C	**4,268.99**	**(10)**	—
U	**4,269.08**	**1**	**3**
Pr	**4,269.10**	**4**	**12**
Ce	**4,269.25**	**2**	**5**

	λ	I: J	I: O
Ar	4,259.36	(1200)	—
Sm	4,256.40	150	150
Cr	*4,254.35*	*1000R*	*5000R*
Sm	*4,251.79*	*200*	*200*
Ar	*4,251.18*	*(800)*	—
Fe	*4,250.79*	*250*	*400*
Kr	*4,250.58*	*(150)*	—
Fe	*4,250.13*	*150*	*250*
Sc	*4,246.83*	*500*	*80*
Pt	**4,269.26**	**1*h***	**2**
La	4,269.49	150	150
Fe	4,271.16	300	400
Fe	4,271.76	700	1000
Ar	4,272.17	(1200)	—
Kr	4,273.97	(1000)	—
Cr	4,274.80	800*r*	4000*R*
La	4,275.64	500	40
Se	*4,280.36*	*(150)*	—
Sm	*4,280.78*	*200*	*200*
Fe	*4,282.41*	*300*	*600*
Sm	*4,285.48*	*200*	*200*
La	*4,286.97*	*300*	*400*
Cr	*4,289.72*	*800r*	*3000r*
Kr	*4,292.92*	*(600)*	—
Fe	*4,292.13*	*400*	*700*
Fe	*4,298.04*	*400*	*100*
Fe	*4,299.24*	*400*	*500*
Ar	*4,300.10*	*(1200)*	—

C 5,793.51 *O* 30

	λ	J	O
Sm	**5,793.28**	**2**	**15**
Si	**5,793.13**	—	**18**
*bh***F**	**5,793.1**	—	**100**
W	**5,793.07**	—	**20**
Pr	**5,792.95**	—	**3**
Rh	**5,792.77**	—	**40**
Eu	**5,792.72**	—	**30**
Sm	**5,792.53**	—	**3**
Th	**5,792.43**	—	**5**
Nd	**5,792.41**	—	**2**
*bh*F	**5,792.4**	—	**20**
Mo	**5,791.85**	**60**	**100**
Cr	**5,791.781**	—	**5**
Sm	**5,791.776**	—	**3**
U	**5,791.77**	—	**4**
Ce	**5,791.68**	—	**6**
Tb	**5,791.64**	—	**10**
Re	**5,791.60**	—	**25**
Gd	**5,791.39**	—	**10**
Pr	**5,791.38**	—	**8**
W	**5,791.36**	—	**6**
La	**5,791.345**	—	**200**
Nd	**5,793.98**	—	**2*h***
Nb	**5,794.24**	**10**	**15**
Eu	**5,794.58**	—	**4*wh***
Ce	**5,794.78**	—	**4**
Nd	**5,795.17**	—	**10**
Sm	**5,795.29**	—	**3**
*bh*F	**5,795.4**	—	**10**
*bh***Cr**	**5,795.5**	—	**10**
Tb	**5,795.64**	—	**25**
Mo	**5,795.77**	—	**20**
Rh	**5,795.79**	—	**8**
*bh*F	**5,795.9**	—	**100**
Ce	**5,796.06**	—	**6**
Ni	**5,796.08**	—	**3**
W	5,796.51	—	20
La	5,797.59	150	80
Zr	5,797.74	—	50
Si	5,797.91	—	25
U	5,798.55	1	35
*bh*F	5,798.9	—	80
V	5,799.90	2	40

	λ	I J	I O		λ	I J	I O
Zr	**5,791.338**	—	**2**	*bh*Yt	5,800.0	—	15
Ce	**5,791.32**	—	**5**	Eu	5,800.27	—	200
Er	**5,791.12**	—	**20**	Ba	5,800.28	20	100
Fe	**5,791.04**	**2***h*	**6***h*	Mo	5,800.46	—	25
Cr	**5,791.00**	—	**40***wh*	Sm	5,800.50	—	80
				Lu	5,800.59	2	30
				Os	5,800.60	—	50
*bh*F	5,790.3	—	100	C	5,801.17	—	15
*bh*F	5,789.3	—	30				
La	5,789.25	—	125	Sm	5,801.67	—	25*d*
V	5,788.56	—	25*W*	*bh*F	5,801.8	—	80
Sm	5,788.39	—	30	K	5,801.96	20	50*h*
				U	5,802.13	—	20
Nd	5,788.22	—	30	Mo	5,802.67	—	30
Ce	5,788.13	—	25				
Cr	5,787.99	—	50*wh*	Sm	5,802.82	—	80
*bh*F	5,787.6	—	100	Tb	5,803.15	—	40
Nb	5,787.54	15	80	Yb	5,803.44	—	15
				Nd	5,804.02	—	100
Sm	5,787.15	—	15	Nb	5,804.03	5	15
Cr	5,787.04	—	15				
Sm	5,786.99	—	200	Ti	5,804.26	50	100*h*
*bh*F	5,786.4	—	30	Ce	5,804.41	—	25
V	5,786.16	—	75	*bh*F	5,804.7	—	80
				W	5,804.87	12	25*w*
Ti	5,785.98	60	100*W*	Ni	5,805.23	—	50
Cr	5,785.82	—	15				
Tb	5,785.18	—	40	Eu	5,805.68	—	20
Cr	5,785.00	—	20	Ba	5,805.69	—	70
Nd	5,784.96	—	20	La	5,805.78	120	60
				Mo	5,806.19	—	15
*bh*F	5,784.8	—	100	Mo	5,806.69	—	20
V	5,784.38	30	50				
Cr	5,783.93	—	30*h*	Rh	5,806.91	1	100
Eu	5,783.71	—	150*s*	Re	5,806.98	—	20
V	5,783.50	—	30	V	5,807.14	40	75
				*bh*F	5,807.7	—	80
*bh*F	5,783.4	—	30	La	5,808.33	60	25
Mo	5,783.33	—	20*h*				
Cr	5,783.11	—	30*h*	*bh*Zr	5,809.2	—	30
V	5,782.61	—	30	Tb	5,809.49	—	15
K	5,782.60	—	60	Hf	5,809.50	30	20
				*bh*Sc	5,809.8	—	20
Cu	5,782.13	—	1000	*bh*F	5,810.60	—	50
*bh*F	5,782.1	—	150				
Sm	5,781.89	—	100	Ce	5,810.72	—	15
Cr	5,781.81	—	20	Ta	5,811.10	—	100
Cr	5,781.19	—	18	Nd	5,811.57	1	15
				*bh*Se	5,811.6	—	15
Os	5,780.81	—	50	K	5,812.52	—	30
Ti	5,780.78	20	20				
Ta	5,780.71	—	80	Sm	5,812.67	—	25
U	5,780.61	—	40	Ce	5,812.93	—	40
*bh*F	5,780.5	—	50	*bh*F	5,813.7	—	50
				Nd	5,813.89	—	30
Si	5,780.45	—	15	Sm	5,814.87	—	60
Ta	5,780.02	—	60				

	λ	I	
		J	O
Tb	5,779.91	—	15
*bh*F	5,779.4	—	200
Mo	5,779.36	—	20
Pr	5,779.29	—	50
Sm	5,779.25	—	50
*bh*Zr	5,778.5	—	60
Sm	5,778.33	—	50
Ba	5,777.66	100*R*	500*R*
*bh*F	5,777.6	—	100
Re	5,776.84	—	300*w*
Ta	5,776.77	—	80
V	5,776.68	25	50
Nd	5,776.12	—	15
Nb	5,776.07	3	30
Lu	5,775.40	5	50
*bh*F	5,774.8	—	100
Mo	5,774.55	—	20
Ti	5,774.05	50	70*W*
Sm	5,773.77	—	100
Ce	5,773.12	—	30
Ce	5,772.88	—	15
Sc	5,772.7	—	15
V	5,772.42	25	50
Si	5,772.26	—	30
*bh*F	5,771.9	—	100
Yb	5,771.67	50	30
Mo	5,771.05	—	15
V	5,770.55	2	18
Nd	5,770.50	—	20
La	5,770.01	—	25
Er	5,769.93	—	20
Hg	5,769.59	200	600
La	5,769.35	—	70
La	5,769.07	60	30
Ce	5,768.89	—	15
Ta	*5,767.91*	—	*100w*
Ta	*5,766.56*	—	*80w*
Ti	*5,766.35*	*50*	*70W*
Eu	*5,765.20*	—	*2000*
Tm	*5,764.29*	—	*50*
Pt	*5,763.57*	—	*30*
Fe	*5,763.01*	*35*	*80*
Er	*5,762.79*	—	*30*
Ti	*5,762.27*	*50*	*70h*
La	*5,761.84*	—	*60*
Ni	*5,760.85*	—	*50*
Nb	*5,760.34*	*30*	*30*

	λ	I	
		J	O
Ru	5,814.98	—	25
*bh*Ti	5,815.0	—	20
Pr	5,815.18	1	15
Th	5,815.43	2	18*d*
Mo	5,815.52	—	20
Re	5,815.87	—	50*w*
Fe	5,816.38	10*h*	15*h*
Tm	5,816.51	—	15
Ta	5,816.510	—	40*w*
*bh*F	5,816.7	—	50
V	5,817.06	—	50
V	5,817.53	—	100*h*
*bh*Yt	5,818.6	—	15
Eu	*5,818.74*	—	*1000*
bhF	*5,819.9*	—	*30*
Eu	*5,820.00*	—	*30*
Sm	*5,820.67*	—	*40*
Eu	*5,820.91*	—	*30*
La	*5,822.00*	—	*40*
bhF	*5,823.0*	—	*30*
Ti	*5,823.71*	*50*	*35*
Pr	*5,823.72*	*1*	*60w*
Ba	*5,826.29*	—	*150wh*
Er	*5,826.79*	—	*50*
Sm	*5,829.72*	—	*35*
V	*5,830.72*	*80*	*100*
bhF	*5,830.8*	—	*30*
Sm	*5,831.01*	—	*80*
Eu	*5,831.05*	—	*2000W*
bhF	*5,831.1*	—	*50*
bhF	*5,831.5*	—	*50*
Rh	*5,831.58*	*1*	*80*
Sm	*5,831.77*	—	*40r*
bhF	*5,832.00*	—	*50*
K	*5,832.09*	—	*50*
bhF	*5,832.6*	—	*50*
bhF	*5,833.4*	—	*30*
Yb	*5,834.01*	*1*	*60h*
bhF	*5,834.2*	—	*30*
Re	*5,834.33*	—	*200*
bhF	*5,835.2*	—	*30*
U	*5,836.05*	—	*30*
bhF	*5,836.3*	—	*30*
Sm	*5,836.36*	—	*100d*
Yb	*5,837.16*	*150*	*50*
Au	*5,837.40*	*10*	*400h*
bhF	*5,837.5*	—	*30*
U	*5,837.71*	*1*	*30*

	λ	I J	I O
Sm	5,759.50	—	60
Er	5,757.62	—	30
Ta	5,755.81	—	40
Ni	5,754.67	—	150w
Si	5,754.26	—	40
Fe	5,753.14	20	40
Re	5,752.95	—	200w
Mo	5,751.40	100	125
Ho	5,751.12	—	30
V	5,750.65	—	50W
Ni	5,748.34	—	40
bhZr	5,748.1	—	100
Tb	5,747.58	—	60
Ta	5,746.71	—	60
La	5,744.41	—	80
V	5,743.45	20	60
Nd	5,742.92	—	40
Bi	5,742.55	10	30
Nb	5,838.64	100	200
bhF	5,838.9	—	30
Ho	5,839.47	—	30
Pt	5,840.12	—	80
bhF	5,840.4	—	30
bhF	5,842.0	—	30
Hf	5,842.23	80	50
Sm	5,842.26	—	40
bhF	5,843.8	—	30

C 6,578.03 J 500

	λ	J	O
Bi	**6,577.2**	**20**	**2h**
Yt	**6,576.88**	**6**	**10**
Kr	**6,576.42**	**(20)**	—
Fe	**6,575.02**	**15h**	**12h**
Nb	**6,574.73**	**2**	**12**
Te	**6,574.50**	**(30)**	—
J	6,566.48	(400)	—
H	6,562.85	(2000)	—
H	6,562.72	(1000)	—
Ne	6,506.53	(1000)	—
In	**6,578.16**	**12**	—
Te	**6,578.63**	**(15)**	—
Br	**6,579.20**	**(20)**	—
Sn	**6,579.26**	**15wh**	—
F	**6,580.38**	**(30)**	—
J	**6,580.58**	**(15)**	—
Hf	**6,581.15**	**2**	**2**
Ar	**6,581.60**	**(2)**	—
J	**6,581.68**	**(15)**	—
Ra	6,593.34	(500)	—
Xe	6,595.01	(400)	—
Ba	6,595.32	300	1000
Ne	6,598.95	(1000)	—
N	6,644.96	(500)	—

Ca $^{20}_{40.08}$

t_0 851°C t_1 1,487°C

I.	II.	III.	IV.	V.
6.112	11.868	51.209	67.2	—

λ	*I*		eV
	J	*O*	
I 2,398.56	120	100*R*	—
II 2,573.09	150	3	—
II 3,158.87	300	100	13.1
II 3,179.33	400	100	13.1
II 3,644.41	15	200	5.30
II 3,933.67	600*R*	600*R*	9.2
II 3,968.47	500*R*	500*R*	9.2
I 4,226.73	50	500*R*	2.9
I 4,425.44	20	100	4.7
I 4,434.96	25	150	4.7

Ca 2,398.56 *J* 20* *O* 100*R*

	λ	*I* *J*	*I* *O*		λ	*I* *J*	*I* *O*
U	2,398.55	2	2	V	2,398.68	—	10
Co	2,398.55	—	4	Re	2,398.73	—	12
*bh*B	2,398.5	—	200	Ir	2,398.75	150	10
Nb	2,398.48	30	10	V	2,398.87	—	2
Co	2,398.37	18	2	Eu	2,398.91	5*h*	4
W	2,398.27	—	3	Cr	2,399.06	2	50
V	2,398.268	—	15				
Ta	2,398.24	—	6				
Fe	2,398.21	—	3	Bi	2,400.88	100	200*R*
				Pt	2,401.87	30	300
				Pb	2,401.95	40	50
Co	2,396.774	—	90				
Os	2,396.77	12	100				
Ru	2,396.71	80	60	*Ru*	*2,402.72*	*150r*	*100*
Ta	2,396.30	—	80	*Pt*	*2,403.09*	*50*	*400*
Fe	2,395.62	100*Wh*	50	*Cu*	*2,403.33*	*300*	*100*
				Re	*2,505.05*	—	*100*
Sb	2,395.20	15	50				
Pb	*2,393.79*	*1000*	*2500*				

Ca 2,573.09 *J* 150 *O* 3*

	λ	*I* *J*	*I* *O*		λ	*I* *J*	*I* *O*
Cs	2,537.05	20	—	Nb	2,573.13	20	1
Nb	2,573.02	10	2	Br	2,573.14	(5)	—
Fe	2,572.97	15	1	Yb	2,573.14	10	3
Hf	2,572.96	3	5	Te	2,573.14	(15)	—
U	2,572.94	4*h*	3	U	2,573.20	2	2
In	2,572.8	2	—	Fe	2,573.21	40	—
Mn	2,572.76	50	200	W	2,573.250	4*d*	—
Ir	2,572.70	5	25	As	2,573.25	5	—
Ru	2,572.66	8	—	Co	2,573.40	12	40
Ti	2,572.651	40	10	La	2,573.48	2*h*	—
U	2,572.647	15*h*	—				
				Hf	2,573.90	100	25
				Fe	2,574.37	150	50
Cu	2,571.74	150	—	V	2,574.52	80	9
Hf	2,571.67	80	30	Nb	2,574.84	100	2
Sn	2,571.59	125	100	Al	2,575.10	80*R*	200*R*
Zr	2,571.39	400*R*	300*R*				
Nb	2,571.33	100	4	O	2,575.30	(100)	—
				Pd	2,575.49	100	—
Lu	2,571.23	100	30	Mn	2,576.10	2000*R*	300*R*
Ru	2,571.08	100	6	Pd	2,576.40	100	—
Mn	2,570.94	80	—	Pd	2,576.55	(100)	—
Fe	2,570.84	100	70				
Pd	2,569.55	150	20	Pd	2,577.10	150	3
				Ta	2,577.37	150*d*	80*d*
Rh	2,569.07	125	3				

	λ	I J	I O		λ	I J	I O
Zr	2,568.87	200	100	*Yb*	*2,579.58*	*200*	*5*
Rh	2,568.83	100	2	*Ag*	*2,580.74*	*150wh*	*1*
				Rh	*2,581.69*	*150*	—
Yb	*2,567.63*	*150*	*5*				
Fe	*2,566.91*	*150*	*60*				
J	*2,566.26*	*(300)*	—				
Ni	*2,566.08*	*600h*	—				

Ca 3,158.869 *J* 300 *O* 100

Cr	**3,158.83**	**1**	**6**	**Th**	**3,158.871**	**1*h***	**4**
				Ce	**3,158.88**	—	**5**
Ce	**3,158.812**	—	**15**				
W	**3,158.806**	**6**	**7**	**Ru**	**3,158.889**	**12**	**60**
Co	**3,158.77**	—	**150r**	**U**	**3,158.89**	**4*d***	**3*d***
Mn	**3,158.74**	—	**8**	**Mo**	**3,158.94**	**10**	**2**
Tb	**3,158.66**	—	**8**	**Fe**	**3,159.02**	**3**	**7**
				Ta	**3,159.05**	**3**	**25*w***
Pr	**3,158.65**	**2**	**12**				
Cu	**3,158.62**	**2**	—	**Ce**	**3,159.07**	—	**8**
Th	**3,158.61**	**4**	**10**	**Pt**	**3,159.08**	**15**	**3**
C	**3,158.6**	**(6)**	—	**Cr**	**3,159.10**	**25**	**4**
Sc	**3,158.542**	—	**2**	**Zr**	**3,159.12**	**2**	**2**
				Ir	**3,159.15**	**2*h***	**50*r***
B	**3,158.54**	**2**	**5**				
U	**3,158.53**	**4**	**1**	**W**	**3,159.18**	**10**	**10**
Ba	**3,158.514**	—	**2**	**Nd**	**3,159.22**	**4**	**6**
Fe	**3,158.511**	—	**1**	**Mo**	**3,159.23**	**25**	—
Ce	**3,158.43**	—	**20**	**Rh**	**3,159.25**	**10**	**5**
				Dy	**3,159.309**	—	**4**
Ag	**3,158.40**	**1*h***	**3**				
Fe	**3,158.400**	**3**	**6**	**Te**	**3,159.31**	**(10*h*)**	—
In	**3,158.396**	**(100)**	—	**Re**	**3,159.311**	—	**15**
Re	**3,158.31**	—	**200**	**Nb**	**3,159.32**	**3**	**2**
Yb	**3,158.30**	**18**	**2**	**U**	**3,159.33**	**2*h***	**2**
				Mo	**3,159.34**	**5**	**20**
Ce	**3,158.26**	—	**2**				
Te	**3,158.23**	**(10*h*)**	—	**Os**	**3,159.356**	**5**	**20**
U	**3,158.20**	**1**	**2**	**V**	**3,159.364**	**40**	**1**
Gd	**3,158.18**	**2**	—	**Ce**	**3,159.38**	—	**3**
Mo	**3,158.165**	**30*r***	**300*R***	**Tb**	**3,159.39**	—	**15**
				W	**3,159.42**	**3**	**4**
Fe	**3,158.163**	**3**	—				
Rb	**3,158.105**	—	**8**	**Ni**	**3,159.521**	—	**10**
Nb	**3,158.102**	**5**	**1**	**Ir**	**3,159.522**	**1**	**7**
Cr	**3,158.02**	**35**	**1**	**U**	**3,159.53**	**2**	**4**
				Cr	**3,159.59**	**20*h***	**60*wh***
				Co	**3,159.66**	**2*h***	**100**
Fe	3,157.89	100	100				
Tm	3,157.34	150	200	**Ho**	**3,159.67**	—	**6**
Os	3,157.24	2	100	**Ce**	**3,159.71**	—	**4**
Fe	3,157.04	100	150				
Ru	3,156.82	—	50				
				Ru	3,159.92	25	70

	λ	I J	I O
Os	3,156.77	3	100
Cu	3,156.63	15	50
Pt	3,156.56	50	150
Dy	3,156.52	20	50
Fe	3,156.27	100	125
Os	3,156.25	15*wh*	500*R*
Rh	3,155.775	2	150
In	3,155.771	(200)	—
Nb	3,154.81	200*w*	3*w*
Co	3,154.79	4*h*	100
Co	3,154.68	—	100
Fe	3,154.21	400	—
Ru	4,153.82	12	60
Re	3,153.79	—	80
Os	3,153.61	20	125
Fe	3,153.21	80	100
Co	3,152.71	—	100
Os	3,152.67	18	150
Rh	3,152.60	3	80
Os	3,152.07	15	80
Re	3,151.63	—	150*w*
Rh	3,151.36	2	80
Fe	3,151.35	150	300
V	3,151.32	150*w*	8*w*
Tm	3,151.03	200	200
Ta	3,150.85	35*w*	50*w*
Ca	3,150.74	2	50
Ru	3,150.69	60	60
Fe	*3,148.41*	*40*	*100*
Mn	*3,148.18*	*40*	*150*
Co	*3,147.06*	—	*150R*
Cu	*3,146.82*	*20*	*100*
Ni	*3,145.72*	*3*	*200*
Fe	*3,144.49*	*100*	*150*
Fe	*3,143.99*	*150*	*200*
Pd	*3,142.81*	*100*	*300*
La	*3,142.76*	*50*	*150*
Fe	*3,142.45*	*100*	*125*
Pt	*3,141.66*	*5h*	*150*

	λ	I J	I O
Fe	3,160.20	50	70
Fe	3,160.66	125	150
Mn	3,161.04	50	150
Fe	3,161.37	60	80
Os	3,161.44	12	80
Co	3,161.65	—	60
Os	3,161.73	10	100
Ti	3,161.77	150	35
Fe	3,161.95	150	200
Fe	3,162.33	50	70
Ti	3,162.57	200*r*	50
Ta	3,162.72	7	70
Dy	3,162.82	60	80
Ta	3,163.13	35	70*r*
Os	3,164.61	12	60
Fe	3,165.01	60	100
Fe	3,165.86	80	100
Fe	3,166.44	80	100
Os	3,166.51	20	200
Fe	*3,167.92*	*30*	*100*
Co	*3,168.06*	—	*100*
Os	*3,168.28*	*15*	*100*
Re	*3,168.38*	—	*150w*
Ti	*3,168.521*	*300r*	*70*
Ru	*3,168.525*	*25r*	*100*
Co	*3,169.77*	—	*100*
Dy	*3,169.98*	*50*	*100*
Ta	*3,170.29*	*35*	*250w*
Mo	*3,170.35*	*25r*	*1000R*
Fe	*3,171.35*	*80*	*100*
La	*3,171.67*	*1000wh*	*2h*
Fe	*3,172.07*	*100*	*100*
Tm	*3,172.82*	*200*	*200*
Os	*3,173,20*	*15*	*100*
Sn	*3,175.02*	*400hr*	*500h*
Fe	*3,175.45*	*200*	*200*

Ca 3,179.33 *J* 400 *O* 100

	λ	J	O
Mo	**3,179.322**	**5**	**6**
Hg	**3,179.32**	**2**	—
Ti	**3,179.29**	**2*h***	*10*
Cr	**3,179.28**	**10*h***	**100**
Ru	**3,179.264**	**50*r***	**50**
Th	**3,179.338**	**2**	**5**
Ce	**3,179.341**	—	**5**
B	**3,179.35**	**100**	**5**
U	**3,179.38**	**6*h***	**8*r***
Pd	**3,179.410**	**1*h***	**15**

	λ	I	
		J	O
Os	**3,179.259**	**4**	**20**
Ag	**3,179.24**	**15*h***	**2**
Nb	**3,179.23**	**10**	**2**
Ir	**3,179.20**	**2*h***	**12**
Pr	**3,179.15**	**1**	**10**
Ag	**3,179.14**	**2**	**2**
W	**3,179.06**	**8**	**10**
Na	**3,179.055**	**(40)**	**6**
Th	**3,179.047**	**10**	**10**
U	**3,179.04**	**6**	**10**
Dy	**3,179.03**	—	**3**
Pt	**3,179.00**	**8**	**1**
Fe	**3,178.97**	**15**	**30**
Nd	**3,178.92**	—	**6**
U	**3,178.91**	**5**	**1**
Mo	**3,178.87**	**4**	—
Th	**3,178.785**	**5**	**5**
Cr	**3,178.778**	**10**	—
Pd	**3,178.77**	**10**	—
Ce	**3,178.75**	—	**15**
Ru	**3,178.73**	—	**30**
Ir	**3,178.69**	**10**	**10**
Nb	**3,178.630**	**5*wh***	**2**
Ti	**3,178.630**	**25*wh***	—
Re	**3,178.612**	—	**30**
Os	**3,178.61**	**(10)**	—
Fe	**3,178.55**	**6**	**10**
Zn	**3,178.54**	**(2)**	—
Mn	**3,178.495**	**50**	**150**
Re	**3,178.492**	—	**15**
Ce	**3,178.46**	—	**12**
Os	3,178.24	10	80
Os	3,178.06	20	150
Fe	3,178.01	150	300
Re	3,177.71	—	80
Fe	3,177.53	300	5
Co	3,177.27	—	100
Ru	3,177.05	200	60
Ru	3,176.29	3	50

	λ	I	
		J	O
V	**3,179.415**	**25**	**1**
Yt	**3,179.418**	**30**	**20**
W	**3,179.425**	**12**	**6**
Cr	**3,179.45**	**10**	—
Ir	**3,179.48**	—	**2**
Fe	**3,179.51**	**6**	—
Ta	**3,179.54**	**2*h***	**15*h***
Er	**3,179.61**	**4**	**4**
Rh	**3,179.73**	—	**50**
Zr	**3,179.75**	—	**2**
Mo	**3,179.77**	**5**	**6**
La	**3,179.78**	**2**	**8**
Cu	**3,179.79**	**3**	—
U	**3,179.83**	**10**	**12**
Tb	**3,179.84**	—	**8**
W	**3,179.97**	**20**	**6*d***
Cd	**3,180.01**	**2**	—
Fe	**3,180.16**	**15**	**10**
Ir	**3,180.17**	—	**6**
Fe	3,180.23	300	300
Nb	3,180.29	200	5
Fe	3,180.75	100	100
Ta	3,180.95	35	100
Fe	3,181.52	70	80
Ni	3,181.74	1*h*	50
Os	3,181.88	12	100
Fe	3,182.06	80	80
Co	3,182.12	2*h*	80
Os	3,182.567	15	100
Ta	3,182.571	18	70*r*
Re	3,182.87	—	100
Fe	3,182.97	70	125
V	3,183.41	1000	2000
Sm	3,183.92	40	60
V	3,183.98	400*R*	500*R*
Ni	3,184.37	3	150
Ta	3,184.55	18	70
Fe	3,184.62	40	60
Re	3,184.75	—	150
Fe	3,184.90	150	200
Os	3,185.33	12	150
V	3,185.40	400*R*	500*R*
Re	3,185.56	—	200
Rh	3,185.59	20	100
Ru	3,186.04	25	80
Co	3,186.35	—	70
Ti	3,186.45	80	150
Fe	3,186.74	300	20

	λ	I J	I O		λ	I J	I O
				Os	3,186.98	15	100
				Os	3,187.33	12	80
				He	3,187.74	(200)	—
				Cr	3,188.01	60*h*	150*h*
				Co	*3,188.37*	*2h*	*100*
				Fe	*3,188.57*	*100*	*150*
				Fe	*3,188.82*	*100*	*150*
				Rh	*3,189.05*	*20*	*100*
				Os	*3,189.46*	*15*	*125*
				Nb	*3,191.10*	*300w*	*100w*
				Rh	*3,191.19*	*50*	*300*
				Fe	*3,191.66*	*150*	*200*
				Ti	*3,191.99*	*20*	*100*
				Fe	*3,192.80*	*8*	*150*
				Fe	*3,193.23*	*70*	*100*
				V	*3,193.92*	*20*	*100*
				Mo	*3,193.97*	*50r*	*1000r*
				Os	*3,194.23*	*15*	*125*
				Fe	*3,194.42*	*70*	*100*
				Os	*3,195.38*	*12*	*100*
				Ni	*3,195.57*	—	*125*
				Fe	*3,196.13*	—	*100*

Ca 3,644.41 *J* 15* *O* 200

	λ	J	O		λ	J	O
Hf	**3,644.355**	**50**	**25**	**Ti**	**3,644.46**	—	**3**
Th	**3,644.347**	**4**	**5**	**Eu**	**3,644.50**	**1***wh*	**6***w*
Ce	**3,644.30**	**1**	**8**	**Pr**	**3,644.54**	**3**	**8**
Eu	**3,644.27**	—	**5***W*	**Ce**	**3,644.55**	**1**	**5**
U	**3,644.24**	**2**	**18**	**Fe**	**3,644.59**	**1**	**2**
Fe	**3,644.18**	—	**4**	**Nd**	**3,644.66**	**4**	**10**
Tb	**3,644.13**	**3**	**15**	**Sc**	**3,644.686**	—	**2**
Eu	**3,644.08**	**1**	**2**	**Cr**	**3,644.693**	**18**	**5**
Sm	**3,643.99**	—	**8**	**Ti**	**3,644.70**	**5**	**35**
Ni	**3,643.94**	—	**2**	**V**	**3,644.71**	**50**	**80**
Dy	**3,643.89**	**6**	**15**	**Th**	**3,644.72**	**4**	**5**
V	**3,643.86**	**30**	**40**	**Er**	**3,644.73**	**2**	**3**
Fe	**3,643.81**	**1**	**7**	**Ca**	**3,644.76**	—	**30**
Tb	**3,643.76**	**3**	**30**	**Fe**	**3,644.80**	**6**	**20**
Nb	**3,643.72**	**15**	**15**	**U**	**3,644.85**	**6**	**4**
Sm	**3,643.715**	**3**	**5**	**Pr**	**3,644.90**	**2**	**5**
Fe	**3,643.713**	**4**	**6**	**Tb**	**3,644.93**	—	**15**
Tm	**3,643.65**	**40**	**60**	**Nb**	**3,644.935**	**10**	**5**
Cu	**3,643.631**	—	**5**	**Ce**	**3,644.97**	—	**2**
Nd	**3,643.630**	**1**	**5**	**Eu**	**3,644.99**	—	**5***w*

	λ	I	
		J	O
Fe	3,643.626	8	20
Nb	3,643.52	2	2
Th	3,643.51	1	5
Dy	3,643.50	—	3
Ir	3,643.48	—	2
Mo	3,643.47	20	1h
Ce	3,643.45	—	5
Nb	3,643.34	10	5
Pr	3,643.32	8	25
W	3,643.31	5	7
Tb	4,643.26	3	15
Re	3,642.99	—	100
Ti	3,642.67	125	300
Ta	3,642.06	18	125
V	3,641.10	30wh	100h
Fe	3,640.39	200	300
Os	3,640.33	40	200
Pb	3,639.58	50h	300
Rh	3,639.51	70	125
Co	3,639.44	20	200
Pt	3,638.79	10	250
Fe	3,638.30	80	100
Zr	3,636.45	30	200
Ti	3,635.46	100	200
Mo	3,635.14	10	100h
Pd	3,634.69	1000R	2000R
Sm	3,634.27	25	100
Fe	*3,631.46*	*300*	*500*
Pt	*3,628.11*	*20*	*300w*
Co	*3,627.81*	—	*200*

	λ	I	
		J	O
U	3,645.03	15	8
Fe	3,645.080	8	20
Ca	3,645.080	2h	3
Pr	3,645.11	2	5
Nd	3,645.167	2	6
Eu	3,645.17	5wh	10w
Co	3,645.19	3	60
Fe	3,645.22	—	9
Ce	3,645.228	1	5
Cu	3,645.232	5	20
Sm	3,645.29	7	9
Ir	3,645.30	2h	25
Se	3,645.31	50	50
Nb	3,645.36	5	5
Tb	3,645.38	15	50
Sm	3,645.39	6	8
Er	3,645.399	12	25s
Yt	3,645.403	4	4
Ho	3,645.41	8	8
La	3,645.414	60	100
Dy	4,645.42	100	300
Eu	3,645.43	—	5
Ce	3,645.45	2	10w
U	3,645.46	2	1
Fe	3,645.49	7	15
Pr	3,645.54	8	20
Gd	3,646.20	150	200w
Co	3,647.66	8	100
Lu	3,647.77	5	100
Fe	3,647.84	400	500
Os	3,648.81	10	100
Co	3,649.35	4	200
Sm	3,649.506	30	100
Fe	3,649.508	100	200
Hg	3,650.15	500	200
La	3,650.17	60	100
Fe	3,651.47	200	300
Co	3,652.54	—	200r
Tl	3,652.95	50	150
Ti	3,653.50	200	500
Cr	3,653.91	25	100
Os	3,654.49	15	100
Ti	3,654.59	40	100
Gd	3,654.64	200	200W
Gd	*3,656.16*	*200*	*200W*
Rh	*3,657.99*	*200W*	*500W*

	λ	I J	I O		λ	I J	I O
				Gd	*3,662.27*	*200*	*200w*
				Hg	*3,663.28*	*400*	*500*
				Ni	*3,664.09*	*30*	*300*
				Gd	*3,664.62*	*200*	*200w*

Ca 3,933.67 *J* 600*R* *O* 600*R*

	λ	J	O		λ	J	O
Hf	**3,933.664**	**15**	**20**	**Eu**	**3,933.677**	—	**10**
Ir	**3,933.664**	—	**20**	**Ru**	**3,933.680**	**200**	**5**
U	**3,933.662**	**10**	**2**	**Ce**	**3,933.73**	**60**	**60**
Co	**3,933.65**	—	**80**	**Ir**	**3,933.901**	**8**	**20**
Ag	**3,933.62**	**80**	**80**	**Tb**	**3,933.905**	—	**4**
Fe	**3,933.60**	**200**	**200**	**Co**	**3,933.91**	—	**60**
Sm	**3,933.59**	**200*h***	**200**	**U**	**3,933.98**	—	**6**
Tb	**3,933.47**	—	**6**	**V**	**3,934.01**	**30**	**100**
Nb	**3,933.39**	**3**	**3**	**Ce**	**3,934.08**	—	**6**
Sc	**3,933.38**	**60**	**60**	**Nd**	**3,934.09**	—	**20**
P	**3,933.37**	**(50)**	—	**Nd**	**3,934.107**	—	**5**
Pr	**3,933.30**	**2**	**9**	**Br**	**3,934.11**	**(2)**	—
S	**3,933.29**	**(80)**	—	**Zr**	**3,934.121**	**12**	**20**
Zr	**3,933.18**	—	**9**	**In**	**3,934.123**	**(10)**	—
U	**3,933.03**	**10**	**5**	**Nb**	**3,934.14**	**15**	**5**
Nb	**3,933.01**	**3**	**3**	**Dy**	**3,934.17**	—	**12**
Dy	**3,932.98**	—	**10**	**Rh**	**3,934.23**	**2**	**100**
Ce	**3,932.978**	**1**	**6*s***	**Fe**	**3,934.23**	—	**2**
Pr	**3,932.978**	**8**	**25**	**Ti**	**3,934.24**	**2**	**30**
Ir	**3,932.971**	—	**6**	**Pr**	**3,934.257**	**2**	**12**
Gd	**3,932.97**	**10**	**10**	**Mo**	**3,934.260**	**5**	**5**
Sm	**3,932.966**	**3**	**6**	**Yb**	**3,934.30**	**4**	**2**
Fe	**3,932.921**	**4**	**8**	**Eu**	**3,934.39**	—	**5**
Cu	**3,932.917**	—	**10**	**Tb**	**3,934.40**	—	**10**
Th	**3,932.915**	**3**	**10**	**Nb**	**3,934.41**	**5**	**5**
Ce	**3,932.79**	—	**3**	**In**	**3,934.43**	**(5)**	—
Fe	**3,932.63**	**40**	**80**	**K**	**3,934.46**	**(20)**	—
Tb	**3,932.55**	—	**3**	**U**	**3,934.47**	**1**	**2**
Ar	**3,932.55**	**(25)**	—	**Ce**	**3,934.51**	—	**2**
La	**3,932.52**	**2*h***	—	**W**	**3,934.60**	**5**	**5**
Hf	**3,932.40**	**10**	**3**	**Co**	**3,934.71**	**2**	**5**
Ce	**3,932.39**	—	**3**	**Ce**	**3,934.75**	—	**10**
Tb	**3,932.37**	—	**12**	**Zr**	**3,934.79**	**15**	**20**
S	**3,932.30**	**(10)**	—	**Gd**	**3,934.80**	**50**	**100**
Ru	**3,932.29**	—	**3**	**Er**	**3,934.812**	—	**5**
Eu	**3,932.28**	—	**3*w***	**Nb**	**3,934.813**	**5*h***	**1*h***
Fe	**3,932.27**	—	**3**	**Sm**	**3,934.818**	**8**	**4**
Er	**3,932.25**	**5**	**20**	**Nd**	**3,934.823**	**30**	**60**
Th	**3,932.230**	**10**	**15**	**Ir**	**3,934.84**	**50**	**200**
Dy	**3,932.228**	—	**30**	**Pr**	**3,934.88**	**2**	**8**

	λ	I J	I O		λ	I J	I O
J	3,931.01	(400)	—	**U**	**3,934.89**	—	**10**
Eu	3,930.50	400*R*	1000*R*	**Mo**	**3,934.96**	**6**	**5**
Fe	3,930.30	400	600	**Rh**	**3,934.98**	**2**	**3**
La	3,929.22	300	400	**F**	**3,935.00**	**(6)**	—
Fe	3,927.92	300	500	**Mo**	**3,935.02**	**5**	**5**
Fe	3,922.91	400	600	**W**	**3,935.04**	**10**	**12**
La	3,921.53	200	400	**Pr**	**3,935.13**	**3*d***	**10*d***
Fe	3,920.26	300	500	**V**	**3,935.14**	**25**	**40**
Cr	3,919.16	125	300*r*	**Br**	**3,935.15**	**(15)**	—
Eu	*3,907.11*	*500R*	*1000RW*	Co	3,935.97	15	400*R*
				J	3,940.24	(500)	—
				Al	3,944.03	1000	2000
				Dy	3,944.69	150	300
				O	3,947.33	(300)	—
				Ar	3,947.50	(1000)	—
				Ar	*3,948.98*	*(2000)*	—
				La	*3,949.11*	*800*	*1000*
				Al	*3,961.53*	*2000*	*3000*

Ca 3,968.468 *J* 500*R* *O* 500*R*

	λ	I J	I O		λ	I J	I O
Lu	**3,968.464**	—	**50**	**Ce**	**3,968.469**	**35*w***	**35**
Ru	**3,968.461**	**200**	**12**	**Nb**	**3,968.471**	**10**	**3**
Yt	**3,968.43**	**30**	**10**	**Ir**	**3,968.475**	—	**25**
Dy	**3,968.39**	—	**300**	**W**	**3,968.59**	**6**	**6**
U	**3,968.374**	**2**	**1**	**Br**	**3,968.66**	**(8)**	—
Fe	**3,968.370**	—	**2**	**Zr**	**3,968.72**	—	**3**
Ar	**3,968.36**	**(200)**	—	**Tb**	**3,968.73**	—	**2**
Gd	**3,968.35**	—	**20**	**Mo**	**3,968.75**	**50**	**8**
Zr	**3,968.26**	**4**	**100**	**Eu**	**3,968.87**	—	**3*wh***
Ag	**3,968.22**	**60**	**100**	**Nd**	**3,968.88**	**4**	**20**
W	**3,968.17**	**8**	**8**	**Th**	**3,969.003**	**5**	**10**
Rh	**3,968.164**	—	**2**	**Gd**	**3,969.005**	—	**40**
Ta	**3,968.16**	—	**4*h***	**Mo**	**3,969.01**	**3**	**3**
Pr	**3,968.158**	**10**	**25**	**U**	**3,969.02**	**8**	**5**
Tb	**3,968.15**	—	**2**	**Cr**	**3,969.06**	**50**	**80**
V	**3,968.09**	**40**	**25**	**Co**	**3,969.12**	**6**	**100*w***
Hg	**3,968.03**	**(50)**	—	**In**	**3,969.13**	**15**	—
Yb	**3,968.03**	**3**	—	**Nb**	**3,969.135**	**20*h***	—
Hf	**3,968.01**	—	**5**	**Ce**	**3,969.16**	—	**2**
U	**3,968.007**	**4**	**6*h***	**Ir**	**3,969.17**	**10**	**30**
Fe	**3,967.97**	**15**	**60**	**Te**	**3,969.18**	**(10)**	—
Ce	**3,967.91**	—	**3**	**W**	**3,969.20**	**10**	**12**
Sm	**3,967.78**	**2**	**6**	**Eu**	**3,969.23**	—	**20*W***
Nd	**3,967.70**	**6**	**20**	**Dy**	**3,969.233**	—	**6**

	λ	I J	I O
Yt	**3,967.69**	**10***h*	**3**
Sm	**2,967.67**	**6**	**8**
Pr	**3,967.66**	**3**	**10**
Tb	**3,967.65**	—	**6**
Ce	**3,967.644**	—	**3**
U	**3,967.639**	**2**	—
Xe	**3,967.541**	**(200)**	—
P	**3,967.54**	**(15)**	—
Ce	**3,967.53**	**1**	**5**
Dy	**3,967.52**	—	**8**
Ir	**3,967.51**	—	**3**
U	**3,967.48**	—	**10**
Re	**3,967.424**	—	**25**
Fe	**3,967.423**	**100**	**125**
Th	**3,967.41**	**1**	**2**
Nb	**3,967.37**	**50***h*	—
Nd	**3,967.31**	**2**	**10**
Cs	**3,967.30**	**(4)**	—
Th	**3,967.214**	**8**	**8**
Tb	**3,967.211**	—	**20**
Ce	**3,967.18**	**2**	**6**
Pr	**3,967.131**	**25***d*	**40***d*
Eu	**3,967.13**	—	**25***W*
Nd	**3,967.07**	**10**	**30**
Ce	**3,967.05**	**6**	**35**
U	**4,967.01**	**2**	**2**
Th	**3,966.97**	**5**	**5**
Gd	**3,966.85**	—	**8**
Cr	3,963.69	300	300
Os	3,963.63	50	500
Zr	3,961.59	8	500

	λ	I J	I O
Gd	**3,969.261**	—	**200**
Fe	**3,969.261**	**400**	**600**
Sr	**3,969.261**	—	**30**
Th	**3,969.34**	**2***w*	**3***w*
U	**3,969.42**	—	**5**
Er	**3,969.43**	—	**6**
Pr	**3,969.51**	**3**	**8**
Th	**3,969.53**	**5**	**5**
Fe	**3,969.63**	**5**	**5**
Nd	**3,969.666**	**4**	**20**
Os	**3,969.671**	**100**	**100**
Cr	**3,969.748**	**90**	**200**
Pr	**3,969.755**	**3**	**12**
Ru	**3,969.79**	**4**	**8**
Th	**3,969.83**	**5**	**5**
Eu	**3,969.90**	—	**8***W*
Tb	**3,969.92**	—	**3**
N	**3,969.95**	**(2)**	—
Ce	**3,970.041**	**3**	**12**
Sr	**3,970.043**	—	**20**
Hf	**3,970.05**	**3**	**10**
Pt	**3,970.06**	**15**	**4**
Pr	**3,970.070**	**4**	**15**
H	**3,970.074**	**(80)**	—
Eu	3,971.99	—	1000*Rwh*
Ta	3,973.18	400*W*	1
Ni	3,973.56	10	800
Cr	3,976.66	300	300
Os	3,977.23	40	300
Fe	3,977.74	150	300
Hg	3,983.98	(400)	—
Yb	*3,987.99*	*500R*	*1000R*
Lu	*3,988.52*	*800*	*1000*
Co	*3,995.31*	*20*	*1000R*
La	*3,995.75*	*300*	*600*

*P**

Ca 4,226.728 *J* 50 *O* 500*R*

	λ	J	O
Yt	**4,226.726**	**15**	**5**
Mo	**4,226.726**	**2**	**15**
Ru	**4,226.66**	—	**15**
Ar	**4,226.65**	**(2)**	—
Ir	**4,226.63**	—	**5**

	λ	J	O
Ce	**4,226.734**	**30**	**50**
Ir	**4,226.735**	—	**30**
Cr	**4,226.76**	**30**	**125**
Al	**4,226.81**	**(35)**	—
W	**4,226.91**	**3**	**15**

	λ	I	
		J	O
V	4,226.62	3	8
U	4,226.60	2	1
Ge	4,226.57	50	200
Mo	4,226.55	5	10
Os	4,226.53	6	12
Tb	4,226.44	—	50
Cl	4,226.43	(10)	—
Fe	4,226.430	25	80
Se	4,226.37	(20)	—
W	4,226.34	3	10
Ce	4,226.33	—	2
Mo	4,226.29	20	20
Nb	4,226.25	3	4
Nb	4,226.21	5	3*w*
Ir	4,226.205	—	2
Ce	4,226.196	—	3
Sm	4,226.17	2	5
Tb	4,226.15	—	3
Br	4,226.15	(4)	—
B	4,226.15	2	4
Rb	4,226.1	(8)	—
U	4,226.065	8	6
Rn	4,226.06	(50)	—
Fe	4,225.96	30	80
As	4,225.87	10	—
Gd	4,225.85	50	150
U	4,225.75	3	1
Ce	4,225.746	8	6
Te	4,225.70	(50)	—
K	4,225.60	(40)	—
Se	4,125.58	—	4
Nd	4,225.56	2	3
J	4,225.54	(15)	—
Ir	4,225.50	2	15
Fe	4,225.465	20	80
Zr	4,225.463	—	6
U	4,225.37	8	8
Pr	4,225.33	40	50
Sm	4,225.32	30	40
Gd	4,225.264	—	3
Zr	4,225.263	—	4
Mo	4,225.25	3	5
V	4,225.22	10	3
Dy	4,225.153	8	40
Gd	4,225.148	10	20
Ho	4,225.13	2	3
F	4,225.12	(20*h*)	—
Co	4,225.11	—	5
Hf	4,225.10	5	—
Nd	4,226.99	5	6
Ar	4,227.02	(10)	—
Mo	4,227.08	25	3
Cs	4,227.10	(50)	—
Gd	4,227.14	20	50
In	4,227.16	(50*h*)	—
Hg	4,227.29	(100)	—
Rn	4,227.3	(15)	—
U	4,227.33	8	6
Al	4,227.406	(8)	—
Ce	4,227.412	—	10
Fe	4,227.43	250	300
Re	4,227.46	—	200 *W*
Al	4,227.50	(30)	—
Eu	4,227.58	—	8
Ti	4,227.65	2	18
Th	4,227.66	1	3
Nd	4,227.72	8	20
V	4,227.74	5	10
Ce	4,227.75	5	40
Zr	4,227.76	8	150
N	4,227.83	(10*h*)	—
Al	4,227.86	(2)	—
Hg	4,227.87	(70)	—
Al	4,227.92	(6)	—
Cu	4,227.936	7	—
Yb	4,227.94	10	—
In	4,227.98	(10)	—
Al	4,227.982	(20)	—
Nd	4,228.02	5	12
Eu	4,228.04	—	3*w*
Hf	4,228.08	2	8
Ar	4,228.18	(40)	—
Nd	4,228.20	6	15
Ca	4,228.23	3	—
C	4,228.28	(5)	—
Ce	4,228.30	—	15
Cs	4,228.35	(35)	—
As	4,228.42	10	—
U	4,228.43	1	10
Te	4,228.46	(50)	—
Pr	4,228.50	3*w*	15*w*
Tm	4,228.55	1	2
Nd	4,228.57	2	10
Ta	4,228.61	10	25
Nb	4,229.15	100	50
Sm	4,229.70	30	40
Se	4,230.05	(40)	—

	λ	I	
		J	O
Ru	**4,225.09**	—	**25**
Gd	**4,225.03**	—	**15**
W	**4,224.99**	**12**	—
J	**4,224.98**	**(8)**	—
Mo	**4,224.93**	**5**	**5**
Cl	**4,224.92**	**(15)**	—
Eu	**4,224.88**	—	**4**
Fe	4,224.18	80	200
Br	4,223.99	(80)	—
Tl	4,223.05	(25)	—
N	4,223.04	(25)	—
Xe	4,223.00	(200*h*)	—
Pr	4,222.98	40	125
K	4,222.97	(40)	—
O	4,222.78	(50)	—
Fe	4,222.22	200	200
P	4,222.15	(150*w*)	300
Cr	4,221.57	35	80
J	4,220.96	(80)	—
Sm	4,220.65	100	100
Te	4,220.42	(100)	—
Fe	4,220.35	40	80
In	4,219.83	(50)	—
Ne	4,219.76	(100)	—
In	4,219.50	(30)	—
Fe	4,219.36	200	250
Sb	4,219.07	30	—
Yb	4,218.57	50	3
Nb	4,217.94	50	50
Cr	4,217.63	70	150
Fe	4,217.555	100	200
La	4,217.554	100	200
S	4,217.23	(30)	—
Gd	4,217.19	100	100
Ne	4,217.15	(30)	—
O	4,217.09	(30)	—
Hg	4,216.72	(50*h*)	—
Cr	4,216.36	25	60
Fe	4,216.19	100	200
Ba	4,216.04	(25)	—
Xe	4,215.60	(100)	—
Rb	4,215.56	300	1000*R*
Sr	4,215.52	400*W*	300*r*
Gd	4,215.024	150	200
Se	4,215.02	(150)	—
N	4,214.73	(25)	—
Nb	4,214.67	100	—

	λ	I	
		J	O
Cr	4,230.64	(30)	—
La	4,230.95	50	150
S	4,230.98	(35)	—
Ne	4,231.60	(50)	—
Te	4,231.74	(30)	—
Cs	4,232.19	(25)	—
Mo	4,232.59	100	125
Fe	4,233.17	100	100
O	4,233.32	(100)	—
Fe	4,233.61	150	250
Ne	4,233.86	(30)	—
Cl	4,234.09	(50)	—
Sm	4,234.57	40	60
Mn	4,235.29	100	80
J	4,235.47	(25)	—
Cl	4,235.49	(25)	—
Eu	4,235.60	—	400*r*
Fe	4,235.942	200	300
Yt	4,235.943	30	60
Kr	4,236.64	(100*h*)	—
Sm	4,236,74	50	60
Br	4,236.88	(25)	—
N	4,236.98	(30*h*)	—
Ar	4,237.23	(40)	—
Sm	4,237.66	50	60
Xe	4,238.25	(200*h*)	—
La	4,238.38	300	500
Gd	4,238.78	200	200
Fe	4,238.82	100	200
Mn	4,239.72	50	100
Cr	4,240.70	30	200
Mo	4,240.83	25	30
Cl	4,241.38	(60)	—
U	4,241.67	50	40
Au	4,241.77	30	40
N	4,241.80	(100*h*)	—
Tm	4,242.15	100	500
Cr	4,242.38	50	4
Rb	4,242.6	(150)	—
Ru	4,243.06	40	100
Mo	4,243.14	25	—
As	4,243.26	100	—
Gd	4,243.84	100	60
U	4,244.37	25	25
Rb	4,244.44	25	—
P	4,244.55	(30)	—
Sm	4,244.70	80	100

	λ	I	
		J	O
Ru	4,214.44	40	100
Xe	4,213.72	(200h)	—
Fe	4,213.65	60	100
Cs	4,213.13	(30)	—
In	4,213.10	(50)	—
Pd	4,212.95	300W	500W
Se	4,212.58	(200)	—
Hg	4,212.53	(50)	—
Hg	4,212.22	(30)	—
Ru	4,212.06	80	125
Gd	4,212.02	50	150
Pr	4,211.86	25d	50d
Os	4,211.85	50	150
Se	4,211.83	(200)	—
Cr	4,211.35	30	100
Rh	4,211.14	200	15
Ag	4,210.94	30h	200h
Kr	4,210.67	(25wh)	—
Fe	4,210.35	200	300
Mo	4,209.65	80	4
Xe	4,209.47	(100h)	—
Cr	4,209.37	40	100
Zr	4,208.98	25	30
Fe	4,208.61	50	100
Xe	4,208.48	(200h)	—
Ta	4,208.44	30h	2
Cr	4,208.36	25	100
Cl	4,208.03	(30)	—
As	4,208.00	30	—
F	*4,207.16*	*(50)*	—
Pr	*4,206.74*	*50*	*50*
In	*4,205.08*	*(50)*	—
Eu	*4,205.05*	*50*	*200R*
Fe	*4,203.99*	*120*	*200*
Xe	*4,203.69*	*(50)*	—
Rn	*4,203.23*	*(200)*	—
Fe	*4,202.03*	*300*	*400*
Rb	*4,201.85*	*500*	*2000R*
Ar	*4,200.67*	*(1200)*	—
Ru	*4,199.90*	*300*	*150*
Fe	*4,199.10*	*200*	*300*
Ru	*4,198.87*	*100*	*60*
Ar	*4,198.32*	*(1200)*	—
Fe	*4,198.31*	*150*	*250*
Ne	*4,198.10*	*(70)*	—
Gd	*4,197.70*	*50*	*40*
Ru	*4,197.58*	*100*	*100*
Ru	*4,196.87*	*50*	*60*
Ra	4,244.72	(40)	—
Mo	4,244.80	80	4
Fe	4,245.26	40	80
Xe	4,245.38	(200h)	—
Mo	4,246.02	30	30
Fe	4,246.09	30	80
F	4,246.16	(300h)	—
Mo	4,246.62	25	—
Se	*4,246.83*	*500*	*80*
P	*4,246.88*	*(150w)*	*70*
Fe	*4,247.43*	*100*	*200*
Se	*4,248.00*	*(100)*	—
P	*4,249.57*	*(100)*	—
La	*4,249.99*	*50*	*100*
Fe	*4,250.13*	*150*	*250*
Kr	*4,250.58*	*(150)*	—
Ne	*4,250.68*	*(50)*	—
Mo	*4,250.69*	*125*	*5*
Fe	*4,250.79*	*250*	*400*
Te	*4,251.15*	*(70)*	—
Ar	*4,251.18*	*(800)*	—
Xe	*4,251.57*	*(50wh)*	—
Sm	*4,251.79*	*200*	*200*
Mo	*4,251.87*	*60*	*60*
Kr	*4,252.67*	*(50hs)*	—
Nb	*4,252.97*	*50*	*30*
Cl	*4,253.51*	*(75)*	—
Gd	*4,253.62*	*50*	*50*
O	*4,253.74*	*(50h)*	—
O	*4,253.98*	*(100h)*	—
Cr	*4,254.35*	*1000R*	*5000R*
Kr	*4,254.85*	*(100h)*	—
Nb	*4,255.44*	*50*	*30*
Te	*4,256.10*	*(50)*	—
Sm	*4,256.40*	*150*	*150*
Ar	*4,259.36*	*(1200)*	—
Kr	*4,259.44*	*(800hs)*	—
Bi	*4,259.62*	*60wh*	—
Fe	*4,260.48*	*300*	*400*
Os	*4,260.85*	*200*	*200*
Te	*4,261.08*	*(300)*	—
Cr	*4,261.35*	*50*	*125*
Hg	*4,261.88*	*(70)*	—
Sm	*4,262.68*	*150*	*200*
Cr	*4,263.14*	*80*	*125*
La	*4,263.58*	*150*	*150*

	λ	I J	I O		λ	I J	I O
La	*4,196.55*	*150*	*200*				
Rh	*4,196.50*	*50*	*100*				
Fe	*4,196.21*	*50*	*100*				
Se	*4,195.51*	*(100)*	—				
Fe	*4,195.34*	*100*	*150*				
Sb	*4,195.17*	*50*	—				
Se	*4,194.55*	*(50)*	—				
Ra	*4,194.09*	*(80)*	—				
Xe	*4,193.53*	*(150)*	—				
Xe	*4,193.15*	*(200h)*	—				
La	*4,192.36*	*50*	*50*				
Fe	*4,191.44*	*100*	*200*				
Ar	*4,191.03*	*(1200)*	—				
Ar	*4,190.71*	*(600)*	—				
O	*4,189.79*	*(500)*	—				
S	*4,189.71*	*(250)*	—				
Pr	*4,189.52*	*50*	*100*				

Ca 4,425.44 *J* 20* *O* 100

U	**4,425.41**	—	**8**	**Pr**	**4,425.50**	—	**3**
Ce	**4,425.33**	—	**3***s*	**Ce**	**4,425.61**	—	**6**
Tb	**4,425.25**	—	**2***h*	**V**	**4,425.71**	**7**	**9**
Pr	**4,425.22**	—	**2**	**Ir**	**4,425.76**	—	**10**
U	**4,425.20**	**5**	**2**	*bh***Ca**	**4,425.8**	—	**3**
Cr	**4,425.13**	**1**	**15**	**Re**	**4,425.77**	—	**3**
Ce	**4,425.12**	—	**2**	**Dy**	**4,425.82**	—	**3**
Gd	**4,425.01**	**4**	**8**	**Ti**	**4,425.83**	**1**	**10**
Sm	**4,424.99**	**2**	**3**	**Pr**	**4,425.86**	—	**2**
Ta	**4,424.96**	**3***h*	**10**	**U**	**4,425.87**	**2**	**4**
W	**4,424.91**	**2**	**8**	**W**	**4,425.91**	**6**	**15**
Ru	**4,424.78**	—	**25**	**Ce**	**4,425.92**	—	**4**
Ir	**4,424.75**	—	**3**	**Tm**	**4,425.96**	—	**3**
Pr	**4,424.59**	**35**	**90**	**Sm**	**4,425.98**	**7**	**9**
Er	**4,424.57**	—	**10**	**Th**	**4,425.99**	**4**	**6**
V	**4,424.56**	**15**	**20**	**V**	**4,426.00**	**15***h*	**25***h*
Ce	**4,424.54**	—	**3**	**Ru**	**4,426.01**	—	**10**
Tb	**4,424.46**	—	**3**	**Ti**	**4,426.05**	**25**	**80**
Ti	**4,424.39**	**2**	**15**	**Ce**	**4,426.08**	—	**4**
Nd	**4,424.343**	**50**	**50**	**Nd**	**4,426.10**	—	**5**
Sm	**4,424.342**	**300**	**300**	**U**	**4,426.14**	**1**	**3**
Ce	**4,424.31**	—	**6**	**Gd**	**4,426.15**	**2**	**50**
Cr	**4,424.28**	**35**	**25**	**Hf**	**4,426.18**	**8**	**3***h*
				Ir	**4,426.27**	**10**	**400***w*
				Tb	**4,426.30**	—	**5**
Ti	4,422.82	25	80				
Yt	4,422.59	60	60	**Ce**	**4,426.301**	—	**4**
Fe	4,422.57	125	300	**Pr**	**4,426.32**	—	**3**

	λ	I J	I O
Gd	4,422.41	40	100
Ti	4,421.76	15	60
Ru	4,421.46	—	60
Gd	4,421.24	8	100
Pr	4,421.23	35*w*	100
Sm	4,421.14	150	150
Sm	4,420.53	200	200
Os	4,420.47	100	400*R*
Mn	4,419.78	20	100
Pr	4,419.67	50	100
Pr	4,419.06	30	80
Gd	4,419.04	200	200
*bh*La	4,418.2	—	50
Sm	4,417.58	80	80
Ti	4,417.28	20	80
Eu	4,417.25	—	60*w*
Ti	4,416.54	10	70
Se	4,415.56	25	100
Fe	4,415.12	400	600
Mn	4,414.88	60	150
Gd	4,414.73	50	100
Gd	4,414.16	60	100
Pr	4,413.76	40	90
Mn	*4,411.88*	*20*	*100*
Gd	*4,411.16*	*50*	*100*
Ru	*4,410.03*	*80*	*150*
Sm	*4,409.34*	*100*	*100*
Pr	*4,408.84*	*100*	*125*
Fe	*4,408.42*	*60*	*125*
Gd	*4,408.26*	*150*	*100*
Fe	*4,407.72*	*50*	*100*
Pr	*4,405.85*	*100*	*100*
Fe	*4,404.75*	*700*	*1000*
Ir	*4,403.78*	*10*	*300*
Pr	*4,403.60*	*40*	*100*
Gd	*4,403.14*	*100*	*100*
Ta	*4,402.50*	*20h*	*100*
Gd	*4,401.85*	*100*	*200*
Ni	*4,401.55*	*30*	*1000W*

	λ	I J	I O
Tm	**4,426.34**	—	**4**
Eu	**4,426.42**	**2**	**4***W*
Pr	**4,426.51**	—	**3**
Ti	4,427.10	60	125
Fe	4,427.31	200	500
Ru	4,428.46	—	125
Pr	4,428.24	125	200
La	4,429.90	300	200
Fe	4,430.62	8	200
Gd	4,430.63	40	150
Sc	4,431.37	3	50
Gd	4,431.77	1	60
Ba	4,431.90	30	60
Pr	4,432.34	10*w*	80
Fe	4,433.22	20	150

*N**

Ca 4,434.960 *J* 25* *O* 150

	λ	J	O
Ir	**4,434,958**	—	**3**
Mo	**4,434.953**	**80**	**80**
Th	**4,434.953**	—	**4**
Tb	**4,435.00**	—	**4**
Dy	**4,435.02**	—	**2**
Nd	**4,435.09**	**4**	**15**

	λ	I			λ	I	
		J	*O*			*J*	*O*
Ce	4,434.952	—	8	Fe	4,435.15	3	70
U	4,434.94	—	8	Pr	4,435.25	—	5
Ir	4,434.93	—	3	W	4,435.44	—	6
Pr	4,434.85	3*w*	25	Ce	4,435.47	—	2
Eu	4,434.80	2	20*w*	Eu	4,435.53	—	2000
V	4,434.60	12	20	U	4,435.533	6	8
Hf	4,434.52	—	6*h*	Tb	4,435.55	—	12
Tb	4,434.48	—	20	Ir	4,435.59	—	8
Ir	4,434.47	—	5	Eu	4,435.60	100	400*R*
Ce	4,434.37	—	3	Ce	4,435.62	—	5
Sm	4,434.32	200	200	Ca	4,435.69	15	100
Tb	4,434.06	—	2	Pr	4,435.73	—	15*w*
W	4,434.05	—	3	W	4,435.74	4	10
Ti	4,434.00	50	100	Dy	4,435.78	—	4
Mg	4,433.99	—	8	Zr	4,435.84	—	3
Cr	4,433.97	1	10	La	4,435.85	3	5
Ir	4,433.91	—	10	Mn	4,436.02	—	12
U	4,433.89	12	15	Th	4,436.05	4	8
Sm	4,433.88	200	200	Gd	4,436.10	—	4
Fe	4,433.79	2	30	Tb	4,436.11	—	15
				V	4,436.14	25*h*	25*h*

*P** *N**

Ca 4,454.781 *J* 5* *O* 200

	λ	*J*	*O*		λ	*J*	*O*
Th	4,454.778	—	4	Zr	4,454.80	1	7
U	4,454.776	—	2	Tb	4,454.89	—	4
Ce	4,454.76	—	5	Ce	4,454.98	1	12
Gd	4,454.70	—	3	Mn	4,455.01	10	25
Pr	4,454.69	10	30	Fe	4,455.035	1	20
Re	4,454.67	—	100	Pr	4,455.035	—	10*d*
Sm	4,454.63	100	100	Th	4,455.04	4	4
Ce	4,454.53	—	3	La	4,455.21	—	3
Th	4,454.52	4	8	Sm	4,455.23	—	3
Fe	4,454.383	80	200	Fe	4,455.26	3	2
Pr	4,454.382	15	60	Mo	4,455.30	12	10
W	4,454.06	—	3	Mn	4,455.32	15	25
Ce	4,454.05	—	2	Ti	4,455.33	80	150
Tm	4,454.04	1	20	Zr	4,455.43	—	6
Tb	4,453.96	—	2	Cr	4,455.458	—	8
Gd	4,453.93	10	10	Ce	4,455.461	—	3
Re	4,453.875	—	20	Pr	4,455.462	—	8
Mo	4,453.87	5	4	W	4,455.465	5	15
La	4,453.85	3	3	Dy	4,455.49	—	10
Ce	4,453.77	—	3	Ho	4,455.59	—	5
Ti	4,453.71	40	80	Nd	4,455.62	3	10

	λ	I J	I O
W	4,453.62	—	5
Dy	4,453.61	—	2h
Ti	4,453.32	70	150
Sm	4,452.71	200	200
Mn	4,451.59	100	125
Nd	4,451.57	50	100
Ti	4,450.90	60	150
Pr	4,449.87	80	125
Ru	4,449.34	100	125
Ti	4,449.15	80	150
Fe	4,447.72	100	200
Os	4,447.35	3	200
Nd	4,446.39	50	100
Co	4,445.71	2	125
Sm	4,444.26	100	100
P	4,443.87	(50)	—
Fe	*4,443.20*	*100*	*200*
O	*4,443.04*	*(50)*	—
Pt	*4,442.55*	*25*	*800*
Fe	*4,442.34*	*200*	*400*

	λ	I J	I O
Ce	4,455.66	1	18
Pr	4,455.69	—	8
La	4,455.79	25	40
Mn	4,455.82	15	25
U	4,455.886	8	6
Ca	4,455.887	75	100
Ce	4,455.90	—	3
Tb	4,456.02	—	2
Eu	4,456.07	—	4w
Ti	4,457.43	100	150
Pr	4,458.34	10w	90
Sm	4,458.51	200	150
Ni	4,459.04	20	400
Fe	4,459.12	200	400
Ru	4,460.03	80	150
Fe	4,461.65	125	300
Ni	4,462.46	20	150
Ti	4,465.81	40	100
Pr	4,465.98	30	90
Gd	*4,466.553*	*150*	*200*
Fe	*4,466.554*	*300*	*500*
Co	*4,466.89*	*5*	*300*
Sm	*4,467.34*	*200*	*200*
Fe	*4,469.38*	*100*	*200*
Co	*4,469.55*	*5*	*300*
Fe	*4,476.02*	*300*	*500*
Ir	*4,478.48*	*10*	*200*
Cu	*4,480.36*	*20*	*200*
Ta	*4,480.93*	*10h*	*200w*
Tm	*4,481.27*	*50*	*400*

P*

Ca 6,122.218 *J* 100 *O* 100

	λ	I J	I O
Br	6,122.12	(50)	—
*bh*C	6,122.1	—	—
Zr	6,121.91	—	60
Sm	6,121.81	—	30d
Dy	6,121.64	—	2
Se	6,121.54	(40)	—
Th	6,121.44	—	6h
Ti	6,121.01	—	35
Zr	6,120.83	—	12
Th	6,120.55	3	15
Yb	6,120.38	20h	—
La	6,120,34	1	2

	λ	I J	I O
Co	6,122.218	—	40
Nd	6,122.224	—	20
Pr	6,122.24	—	5
Au	6,122.28	—	2
Hg	6,122.3	(5)	—
Mn	6,122.44	(80)	—
Co	6,122.65	—	125
Sm	6,122.74	—	10
Mn	6,122.80	(15)	—
Nd	6,122.96	—	3
Yb	6,122.99	—	4
Ir	6,123.01	—	3

	λ	I	
		J	O
Sm	**6,120.26**	—	*10d*
K	**6,120.22**	**(60)**	—
***bh*Zr**	**6,120.1**	—	**3**
Tb	**6,119.98**	—	**4**
K	**6,119.95**	**(10)**	—
Ce	**6,119.79**	—	**4**
Ni	**6,119.78**	—	**2**
Ar	**6,119.67**	**(2)**	—
Dy	**6,119.63**	—	**2**
Kr	**6,119.56**	**(10*w*)**	—
Cu	**6,119.55**	—	**25**
V	**6,119.52**	**20**	**30**
W	**6,119.35**	—	**3**
Eu	6,118.78	—	400*W*
Co	6,116.98	—	80
Cd	6,116.19	—	50
Ni	6,116.18	—	150
Xe	6,115.08	(50)	—
Ar	6,114.92	(100)	—
La	6,111.72	—	50
Cd	6,111.52	—	100
Ba	6,110.78	60	200*Wh*
As	6,110.66	150	—
As	6,110.30	150	—
In	6,108.99	(60)	—
In	6,108.65	(50)	—
La	6,108.49	—	70
Eu	6,108.13	—	150
Ni	6,108.12	—	200
Ar	6,105.64	(60)	—
Li	6,103.64	300	2000*R*
Ca	6,102.721	50	80
Rh	6,102.721	—	100
S	6,102.26	(50)	—
Se	6,101.96	(2000)	—
Ta	6,101.58	—	150
Xe	6,101.43	(200)	—
Se	6,101.81	(50)	—
Eu	6,099.38	—	600
Cd	6,099.18	—	300
Ar	6,099.81	(60)	—
Ti	6,098.67	—	60
Xe	6,097.59	(600)	—
Ne	6,096.16	(300)	—
Se	6,096.12	(50)	—
In	6,095.96	(80)	—
In	6,095.85	(50)	—
Cl	6,094.65	(100)	—
Yb	**6,123.03**	**4**	—
Hg	**6,123.27**	**(15)**	—
Ar	**6,123.38**	**(6)**	—
Se	**6,123.49**	**(60)**	—
Mo	**6,123.53**	—	**12**
Sm	**6,123.60**	—	**50*d***
Ce	**6,123.67**	—	**15**
Mo	**6,123.69**	—	**12*s***
La	**6,123.76**	—	**5**
Xe	**6,123.91**	**(5)**	—
Eu	**6,124.68**	—	**150**
Zr	**6,124.84**	—	**40**
Si	**6,124.85**	—	**2*h***
Sm	**6,124.88**	—	**40**
Sm	**6,124.95**	—	**2**
Si	**6,125.03**	—	**4*h***
J	6,125.53	(100)	—
Mn	6,125.85	(50)	—
Ti	6,126.21	60	150
Zr	6,127.44	—	500
J	6,127.46	(125)	—
Cu	6,127.73	—	80
Ne	6,128.45	(100)	—
In	6,129.70	(60)	—
Sb	6,129.98	150*h*	10*h*
*bh*Yt	6,132.1	—	200
In	6,132.74	(50)	—
J	6,132.94	(50)	—
La	6,134.39	—	70*R*
Zr	6,134.55	—	300
Bi	6,134.82	30	50
Se	6,135.04	(70)	—
Fe	6,136.62	—	100
Te	6,136.81	(70)	—
In	6,137.19	(50)	—
Fe	6,137.70	—	100
Se	6,138.46	(60)	—
S	6,138.98	(50)	—
Ba	6,141.72	2000*wh*	2000*wh*
Ne	6,142.51	(100)	—
La	6,142.98	—	50
Ne	6,143.06	(1000)	—
Zr	6,143.20	—	300
In	6,143.23	(80)	—
Ta	6,144.56	—	50
Ar	6,145.43	(100)	—
Re	6,145.80	—	50*w*

	λ	I	
		J	O
Xe	6,093.56	(150)	—
Co	6,093.13	—	200
Ti	6,091.17	25	125
J	6,086.77	(150)	—
Ni	6,086.29	—	100
Ti	6,085.23	60	100
Eu	6,083.87	—	500
J	6,082.46	(1000)	—
Co	6,082.43	—	300W
Pb	6,081.5	(200)	—
V	6,081.44	10	100
Sb	6,079.55	100h	20
bhSc	6,079.3	—	100
Eu	6,077.37	—	100W
V	6,077.36	2	300h
Pb	6,075.8	(200)	—
Eu	6,075.59	—	300
Ne	6,074.34	(1000)	—
bhSc	6,072.7	—	100
Rb	6,070.75	50	600
bhF	6,064.4	—	200
Ba	6,063.12	60	200wh
bhF	6,062.3	—	150

	λ	I	
		J	O
Ti	6,146.22	—	400
Xe	6,146.45	(50)	—
Re	6,146.82	—	50
bhYt	6,148.4	—	100
Br	6,148.62	(200)	—
Hg	6,149.50	(200)	—
In	6,149.67	50	15
Ne	6,150.30	(100)	—
Te	6,151.14	(50)	—
Na	6,154.23	100	500
Ta	6,154.50	—	200
O	6,155.99	(150)	—
O	6,156.78	(300)	—
O	6,158.20	(1000)	—
Rb	6,159.62	—	400
Lu	6,159.94	200	50
Na	6,160.76	100	500
Ni	6,163.42	—	100
Ne	6,163.59	(1000)	—
La	6,165.70	—	100
Te	6,166.84	(100)	—
Ar	6,170.18	(100)	—
As	6,170.47	150	—
Eu	6,173.05	—	600
Ar	6,173.11	(100)	—
Ni	6,175.42	—	300
Ni	6,176.81	—	400w
Xe	6,178.30	(150)	—
Eu	6,178.75	—	120
Xe	6,179.66	(125)	—
Tl	6,179.98	(100)	—
Ne	6,182.15	(150)	—

Cd 48 112.40

t_0 320.9°C

t_1 765°C

I.	II.	III.	IV.	V.
8.991	16.904	38.217	—	—

λ	I		eV
	J	O	
II 2,265.02	300	25	5.47
I 2,288.02	300R	1500R	5.41
II 2,312.84	200	1	11.14
II 2,748.58	200	5	10.28
I 3,261.06	300	300	3.80
I 3,403.65	500	800	7.37
I 3,466.20	500	1000	7.37
I 3,610.51	500	1000	7.37
I 4,414.63	200	—	—
I 4,678.16	200	200	6.39
I 4,799.92	300	300	6.39
I 5,085.82	500	1000	6.39
I 6,438.4696	1000	2000	7.34

Cd 2,265.017 *J* 300 *O* 25*

	λ	*I*			λ	*I*	
		J	*O*			*J*	*O*
Mo	2,264.95	4	—	Te	2,265.02	(10)	—
Cr	2,264.92	10	—	K	2,265.04	(30)	—
Ce	2,264.89	20	—	In	2,265.06	6	—
Mo	2,264.74	18	—	Ir	2,265.16	50	5
				Mo	2,265.17	2	—
Ni	2,264.46	400	150	W	2,265.35	10	5
Rh	2,263.43	200	5				
				Zn	2,266.00	(250)	—
				Ni	*2,270.21*	*400*	*100*

Cd 2,288.02 *J* 300*R* *O* 1500*R*

	λ	*J*	*O*		λ	*J*	*O*
V	2,287.93	15	—	Ir	2,288.09	5*w*	—
J	2,287.89	(12)	—	V	2,288.095	15	3*h*
Ir	2,287.88	5	20	As	2,288.12	5	250*R*
Ce	2,287.85	10	—	Sb	2,288.14	—	5
Ta	2,287.84	—	5	Pr	2,288.15	7	—
Te	2,287.84	(5)	—	Cl	2,288.16	(4)	—
Co	2,287.81	4	12*d*	Pt	2,288.19	30	15
Ar	2,287.8	(60)	—	Au	2,288.25	2*h*	—
U	2,287.80	2	—	Rh	2,288.25	40	—
Kr	2,287.79	(30)	—	U	2,288.30	2	—
Ni	2,287.08	500	100				
Co	2,286.16	300	40	Rh	2,290.03	500	25

Cd 2,312.84 *J* 200 *O* 1*

	λ	*J*	*O*		λ	*J*	*O*
Re	2,312.82	5	15	W	2,312.91	5	—
Tm	2,312.71	10	—	Ni	2,312.92	18	—
Mn	2,312.69	40	—	Re	2,312.98	3	15
Rh	2,312.65	100	2	Lu	2,313.03	5	—
Ta	2,312.60	30	5	Pt	2,313.04	20	3
U	2,312.56	12	2	O	2,313.05	(7*h*)	—
Yb	2,312.56	2	1	Fe	2,313.10	3	25
Co	2,312.55	12	3				
				V	2,314.19	100	—
Rb	2,312.45	(100)	—				
V	2,311.35	150	—				
				V	*2,318.07*	*250*	—
V	*2,309.84*	*125*	—				
Ag	*2,309.64*	*200h*	*150h*				

Cd 2,748.58 *J* 200 *O* 5*

	λ	*I* *J*	*I* *O*
W	**2,748.576**	**3**	**10**
U	**2,748.45**	**15**	**18**
W	**2,748.43**	**9**	—
Mo	**2,748.41**	**10**	—
La	**2,748.32**	**8**	**1**
W	**2,748.31**	**12**	**10**
Cr	**2,748.29**	**5**	**300**
Au	**2,748.26**	**80**	**40**
Cs	**2,748.18**	**(20)**	—
Ru	2,747.97	100	50
Rh	2,747.63	100	1
Fe	2,746.98	300*wh*	200
Ti	2,746.71	150*wh*	—
Fe	2,746.48	300*wh*	150
Cu	2,745.45	150	8
Ru	2,745.25	150	12
Ru	2,743.94	100	50
Ni	*2,742.99*	*500*	—
Rh	*2,739.92*	*300*	*10*
Fe	*2,739.55*	*300h*	*200*
Yb	**2,748.66**	**30**	**5**
In	**2,748.72**	**25***wh*	—
Ta	**2,748.78**	**50**	**400**
Nb	**2,748.849**	**10**	**10**
W	**2,748.849**	**10**	**12**
Au	**2,748.85**	**5***r*	—
Al	**2,748.86**	**(30*h*)**	—
Os	**2,748.863**	**5**	**12**
Cr	**2,748.98**	**200**	**35**
W	**2,749.000**	**4**	**9**
S	**2,749.00**	**(8)**	—
Fe	2,750.14	100	300*h*
Yb	2,750.48	150	20
Cr	2,750.73	150	30
Ti	2,751.70	200*wh*	—
Cr	2,751.87	125	20
Ta	2,752.49	300	300
Ru	2,752.77	150	50
Fe	2,753.29	150*h*	25
V	2,753.40	200*R*	50
In	*2,753.88*	*300wh*	*300R*
Ag	*2,756.51*	*200*	5

Cd 3,261.06 *J* 300 *O* 300

	λ	*J*	*O*
vzduch	**3,261.0**	**8**	—
Pb	**3,261.0**	**(50)**	—
Ir	**3,260.998**	—	**2*h***
Ce	**3,260.97**	**3**	**25**
Ca	**3,260.93**	**3**	**1*h***
Th	**3,260.922**	**10**	**8**
Zr	**3,260.916**	—	**2*h***
Eu	**3,260.88**	—	**5**
Tb	**3,260.83**	**3**	**15**
Co	**3,260.82**	**4**	**70**
B	**3,260.74**	**10**	**4**
J	**3,260.69**	**(10)**	—
Dy	**3,260.69**	**2**	**10**
Nd	**3,260.65**	**4**	**10**
Se	**3,260.64**	**(8)**	—
Os	**3,260.57**	**5**	**15**
Nb	**3,260.56**	**300**	**15**
U	**3,260.53**	**1**	**2**
Pt	**3,261.07**	—	**3**
Zr	**3,261.078**	—	**2*h***
V	**3,261.081**	**10**	**15**
Th	**3,261.11**	**4**	**5**
Ce	**3,261.12**	—	**2**
Ru	**3,261.13**	**3**	**30**
W	**3,261.16**	**8**	**10**
Pb	**3,261.21**	**2**	—
Dy	**3,261.22**	**2**	**5**
Ce	**3,261.24**	—	**8**
Nd	**3,261.33**	**4*h***	—
Fe	**3,261.333**	**7**	**25**
Yb	**3,261.509**	**18**	**5**
Ta	**3,261.511**	**35*h***	—
Ce	**3,261.52**	—	**2**
Th	**3,261.54**	**10**	**8**
Re	**3,261.55**	—	**50**
Kr	**3,261.58**	**(8*h*)**	—

	λ	I J	I O
Mo	**3,260.48**	**5**	**10**
Ru	**3,260.35**	**50**	**100**
Th	**3,260.33**	—	**2**
Ce	**3,260.32**	—	**3**
Os	**3,260.30**	**10**	**60**
Co	**3,260.29**	—	**5**
Sm	**3,260.265**	**2**	**9**
Fe	**3,260.261**	**15**	**20**
Ti	**3,260.259**	**30**	**12**
Mn	**3,260.23**	**50**	**75**
Ne	**3,260.22**	**(15)**	—
Ta	**3,260.18**	**18***w*	**125**
Ru	**3,260.17**	—	**12**
Nb	**3,260.14**	**5**	**5**
Fe	3,259.99	100	150
Fe	3,259.05	200	1
Pd	3,258.78	200*h*	300
Fe	3,258.77	150	—
In	3,258.56	300*R*	500*R*
In	3,256.09	600*R*	1500*R*
Fe	3,254.36	150	200
Lu	3,254.31	150	50
Co	3,254.21	—	300*R*
Nb	3,254.07	300	20
Ti	3,242.91	200*r*	60
Cd	3,242.52	300	300
Ti	3,251.91	150	50
Pd	*3,251.64*	*500*	*200*
Fe	*3,251.23*	*150*	*300*
La	*3,249.35*	*80*	*300*
Cu	*3,247.54*	*2000R*	*5000R*
La	*3,245.12*	*300*	*400*
Fe	*3,244.19*	*200*	*300*
Ni	*3,243.06*	*15*	*400R*
Pd	*3,242.70*	*100R*	*2000wh*
Eu	**3,261.59**	—	**2***h*
S	**3,261.60**	**(8)**	—
Tl	**3,261.605**	**300***r*	**70**
Ce	**3,261.63**	**2***h*	**2**
Tm	**3,261.66**	**100**	**30**
Pt	**3,261.692**	**2**	**8**
Nb	**3,261.695**	**50**	**1**
Yb	**3,261.70**	**5**	—
U	**3,261.72**	**10**	**15**
Pd	**3,261.73**	**2***h*	—
Tb	**3,261.74**	**3**	**15**
Mo	**3,261.84**	—	**2**
Nb	**3,261.88**	**3**	**10**
Ba	**3,261.96**	—	**40**
Os	3,262.29	50	500*R*
Sn	3,262.33	300*h*	400*h*
Rh	3,263.14	40	200
Nb	3,263.37	500	3
Hg	3,264.06	(200)	—
Fe	3,265.05	150	200
Fe	3,265.62	300	300
La	3,265.67	200	300
Pd	3,267.35	200*h*	—
Sb	3,267.50	150*Wh*	150
Gs	3,267.94	30	400*R*
Os	3,269.21	20	200
Ge	3,269.49	300	300
Fe	*3,271.00*	*300*	*300*
Cu	*3,273.96*	*1500R*	*3000R*

Cd 3,403.65 *J* 500 *O* 800

	λ	J	O
Ce	**3,403.60**	—	**15**
Cr	**3,403.59**	**3**	**35**
Pr	**3,403.57**	**2**	**8**
U	**3,403.55**	**10**	**10**
Nb	**3,403.49**	**5**	—
Nd	**3,403.46**	**4**	**10**
Tb	**3,403.66**	—	**8**
Er	**3,403.678**	**1**	**9**
Zr	**3,403.684**	**15**	**15**
In	**3,403.70**	**18**	—
Ce	**3,403.73**	—	**3**
Nb	**3,403.75**	**4**	**5**

	λ	I J	I O
Dy	3,403.45	4	4
Ni	3,403.43	—	40
Ti	3,403.37	2	12
V	3,403.36	15	30
Mo	3,403.35	3	20
Cr	3,403.322	200	30
Fe	3,403.32	3	7
Th	3,403.275	3d	3d
Dy	3,403.27	5	6
U	3,403.22	2	4
Ce	3,403.18	—	10
Eu	3,403.16	2	5
Nd	3,403.15	15	20
Sm	3,403.08	8	15
Eu	3,403.03	—	5
Nb	3,403.01	80w	20w
Ta	3,402.99	25w	—
Zr	3,402.873	10	10
Er	3,402.87	1	8
Mo	3,402.81	100	5d
Bi	3,402.80	—	3
Ir	3,402.791	2	10
Kr	3,402.79	(2)	—
Tb	3,402.78	3	8
U	3,402.775	1	4
Hg	3,402.77	(18)	—
W	3,402.767	4	5
Th	3,402.70	15	10
Ir	3,402.68	—	3
Rh	3,399.70	60	500
Rh	3,396.85	500	1000w
Co	3,395.37	50	400R
Ar	3,393.75	(250)	—

	λ	I J	I O
Ru	3,403.77	—	8
W	3,403.79	3	4
Ce	3,403.85	—	8
U	3,403.89	2	2
Sb	3,403.91	10h	2
Yb	3,404.10	30	9
Ce	3,404.130	—	18
Er	3,404.13	—	8
In	3,404.131	(18)	—
W	3,404.14	4	—
Ta	3,404.16	1	5
W	3,404.22	7	8
Tb	3,404.24	3	15
K	3,404.24	(30)	—
In	3,404.297	(18)	—
Fe	3,404.304	25	25
P	3,404.33	(50)	—
Mo	3,404.34	25	20
Fe	3,404.36	50	100
V	3,404.42	50h	—
Ce	3,404.43	—	12
In	3,404.45	(10)	—
La	3,404.52	2	9
Pd	3,404.58	1000R	2000R
Th	3,404.65	4	4
Co	3,450.12	150	2000R
Fe	3,407.46	400	400
Co	3,409.18	125	1000R
Co	3,412.34	100	1000R
Co	3,412.63	40	1000R
Fe	3,413.13	300	400
Ni	3,413.48	15	500
Ni	*3,414.76*	*50wh*	*1000R*
Ne	*3,417.90*	*(500)*	*—*
Pd	*3,421.24*	*1000R*	*2000R*

Cd 3,466.20 *J* 500 *O* 1000

	λ	I J	I O
U	3,466.14	—	4
Dy	3,466.12	1	2
Ce	3,466.08	—	5
Ce	3,466.03	—	3
Re	3,465.983	—	20
Tb	3,465.98	—	30

	λ	I J	I O
Pr	3,466.237	1h	2
In	3,466.24	18	—
U	3,466.30	2	8
Mn	3,466.336	(18)	—
Ar	3,466.34	(10)	—
Eu	3,466.42	2	30

	λ	I J	I O
Th	3,465.93	5	5
Hf	3,465.92	2	5
Nb	3,465.865	40	30
Fe	3,465.863	400	500
Mo	3,465.86	5	5
Ar	3,465.80	(3)	—
Co	3,465.800	25	2000R
Th	3,465.77	10	10
Pr	3,465.76	4	9
Mg	3,465.75	2	—
Sn	3,465.73	(3)	—
Mo	3,465.66	6	6
Zr	3,465.63	—	3
Cr	3,465.58	8	30
Ti	3,465.56	60	6
Ta	3,465.505	15h	—
Ce	3,465.503	—	5
Sm	3,465.46	—	5
Nd	3,465.44	—	6wh
Os	3,465.43	12	60
Ce	3,465.42	—	8
Kr	3,465.41	(6wh)	—
W	3,465.408	4	4
Ca	3,465.40	2h	10
Ce	3,465.302	—	8
Dy	3,465.30	—	2
Ru	3,465.29	2h	5
Cr	3,465.250	30	35
V	3,465.249	25	2
Ir	3,465.22	1	12
Cs	3,465.20	(4)	—
Mn	3,465.19	—	2
Er	3,465.12	—	8d
Co	3,462.80	80	1000R
Rh	3,462.04	150	1000
Ni	3,461.65	50h	800R
Ar	3,461.08	(300)	—
Pd	3,460.77	600h	300r
Re	3,460.47	—	1000W
Mn	3,460.33	500	60
Ni	3,458.47	50h	800R
Co	*3,455.23*	*10*	*2000R*
Co	*3,453.50*	*200*	*3000R*
Ni	*3,446.26*	*50h*	*1000R*

	λ	I J	I O
Fe	3,466.500	70	30
U	3,466.503	—	3
Th	3,466.54	1	5
Tb	3,466.57	—	8
Ne	3,466.58	(150)	—
V	3,466.59	30	—
Ce	3,466.646	—	2
Th	3,466.647	2	5
Sm	3,466.74	—	3
Pr	3,466.75	2	10
Sm	3,466.78	—	4
Ce	3,466.80	—	10
Mo	3,466.826	10	6
W	3,466.828	—	2hd
Ta	3,466.85	18h	2
Eu	3,466.88	1	20
Fe	3,466.89	4	10
Th	3,466.90	3	3
Tb	3,466.92	—	8
Ce	3,466.95	—	10
Gd	3,466.96	10	15
Mo	3,466.97	8	5
Cr	3,467.02	20	50
Pr	3,467.04	2	8
Ru	3,467.05	3	50
Ho	3,467.07	6	8
Dy	3,467.07	1	3
Sb	3,467.24	2h	1
Ti	3,467.260	6	25
Tb	3,467.26	3	3
Gd	3,467.28	100	100
Cd	3,467.656	400	800
Rh	3,470.66	125	500
Ni	3,472.54	40	800R
Ne	3,472.57	(500)	—
Sb	3,473.91	300wh	3
Co	3,474.02	100	3000R
Mn	3,474.13	400	12
Rh	3,474.78	125	700
Fe	3,475.45	300	400

Cd 3,610.510 *J* 500 *O* 1000

	λ	*I*			λ	*I*	
		J	*O*			*J*	*O*
In	**3,610.508**	**18**	—	**Eu**	**3,610.57**	**2***h*	—
Se	**3,610.50**	**(35)**	—	**Ta**	**3,610.60**	**15**	—
Re	**3,610.493**	—	**40**	**Mo**	**3,610.62**	**5**	**3**
U	**3,610.487**	**8**	**4**	**Pr**	**3,610.68**	**2**	**3**
Ni	**3,610.46**	—	**1000r**	**U**	**4,610.69**	**12***d*	**3***d*
Ce	**3,610.45**	—	**2**	**Fe**	**3,610.70**	**3**	**10**
Th	**3,610.40**	**4**	**8**	**Nb**	**3,610.76**	**5**	**3**
Xe	**3,610.32**	**(15)**	—	**Gd**	**3,310.766**	**6**	**25**
Mn	**3,610.30**	**40**	**60**	**Dy**	**3,610.77**	**2**	**4**
Ce	**3,610.26**	—	**5**	**Th**	**3,610.79**	**4**	**8**
La	**3,610.24**	**5**	**7**	**Cu**	**3,610.81**	**6**	**25**
Fe	**3,610.162**	**90**	**100**	**Pt**	**3,610.909**	**1**	**4**
Ti	**3,610.156**	**70**	**100**	**Ce**	**3,610.914**	—	**10**
Cr	**3,610.05**	**8**	**20**	**U**	**3,611.001**	**2**	**2***h*
Th	**3,610.04**	**1**	**3**	**Ba**	**3,611.002**	**3**	**10**
Cl	**3,610.02**	**(4)**	—	**Eu**	**3,611.02**	**1**	**10***w*
Nb	**3,610.00**	**5**	**3**	*Yt*	**3,611.047**	**60**	**40**
Eu	**3,609.94**	—	**3***w*	**Gd**	**3,611.049**	**5***h*	**5**
Ta	**3,609.93**	**18**	**1***h*	**Sm**	**3,611.06**	**2**	**4**
Ce	**3,609.89**	—	**2**	**La**	**3,611.1**	—	**6***h*
Tb	**3,609.88**	—	**8**	**La**	**3,611.103**	**3**	—
Nd	**3,609.79**	**10**	**15**	**Ta**	**3,611.13**	**1**	**25**
Ir	**3,609.77**	**25**	**30**	**Pr**	**3,611.159**	**2**	**5**
Co	**3,609.76**	**3**	**5**	**Dy**	**3,611.16**	**1**	**3**
Cl	**3,609.74**	**(2)**	—	**U**	**3,611.24**	**10**	**5**
Ce	**3,609.69**	**10**	**40**	**Nb**	**3,611.28**	**5***h*	**3**
U	**3,609.68**	**12**	**15**	**Yb**	**3,611.31**	**50**	**12**
Zr	**3,609.64**	—	**3**	**Tb**	**3,611.33**	**8**	**50**
Ti	**3,609.59**	**2**	**12**	**Eu**	**3,611.332**	—	**10***h*
Tb	**3,609.55**	—	**15**	**Ce**	**3,611.34**	—	**10**
Pd	**3,609.548**	**700***R*	**1000***R*	**U**	**3,611.39**	**1**	**12**
Tm	**3,609.54**	**25**	**15**	*bh***La**	**3,611.4**	—	**4***d*
Nd	**3,609.495**	**30**	**25**	**Tb**	**3,611.41**	**8**	**8**
Mo	**3,609.493**	**10**	**5**	**Cs**	**3,611.52**	—	**200**
Sm	**3,609.484**	**100**	**60**	**Eu**	**3,611.580**	**5**	**40**
Cr	**3,609.479**	**12**	**20**	**V**	**3,611.58**	**15***wh*	—
Th	**3,609.45**	**10**	**12**	**Ce**	**3,611.65**	**2***w*	**20***w*
Er	**3,609.44**	**1***h*	**6***d*				
Nb	**3,609.360**	**5**	**1**				
Ta	**3,609.357**	**1**	**8**	Cd	3,612.87	500	800
				Fe	3,618.77	400	400
				Ni	3,619.39	150*h*	2000*R*
Fe	3,608.86	400	500	Nb	3,619.73	300	3
Ar	3,606.52	(1000)	—	Rn	3,621.0	(250)	—
Cr	3,605.33	400*R*	500*R*				
Yt	3,600.73	300	100				
Ni	*3.597.70*	*50h*	*1000r*				

	λ	I J	I O
Ne	3,593.53	(500)	—
V	3,589.76	600R	80

Cd 4,414.63 J 200

	λ	I J	I O
P	4,414.60	(70)	—
Zr	4,414.544	3	4
V	4,414.541	5	10
Nd	4,414.43	3	12
Pr	4,414.40	4	20
Mo	4,414.35	4	5
P	4,414.28	(100)	—
Pt	4,414.25	1	2
Gd	4,414.16	60	100
Cr	4,413.87	15	25
W	4,413.86	1*h*	5
Ce	4,413.80	1	20
Nd	4,413.78	1	5
Pr	4,413.76	40	90
Ca	4,413.74	3	—
Ba	4,413.683	3	10
V	4,413.678	10	15
Mo	4,413.67	6	6
As	4,413.64	50	—
Ne	4,413.56	(15)	—
Gd	4,413.47	2	5
Ac	4,413.17	100	—
Ti	4,411.08	100	7
Sm	4,409.34	100	100
Ne	4,409.30	(150)	—
J	4,408.96	(250)	—
Pr	4,408.84	100	125
Gd	4,408.26	150	100
Xe	4,406.68	(100*wh*)	—
Gd	4,406.67	200	70
Pr	4,405.85	100	100
Fe	4,404.75	700	1000
Gd	4,403.14	100	100
Kr	4,399.97	(200)	—
Hg	4,398.62	(300)	—
Xe	4,395.77	(200*wh*)	—
Xe	4,395.20	(200*wh*)	—
Eu	4,414.65	2	6
Gd	4,414.73	50	100
V	4,414.74	34	6
Xe	4,414.84	(150)	—
Mn	4,414.879	60	150
Nb	4,414.881	3*w*	4*w*
O	4,414.888	(300)	—
Zr	4,414.893	2	3
V	4,415.06	4	7
W	4,415.07	6	15
Fe	4,415.12	400	600
Ne	4,415.14	(5)	—
U	4,415.24	12	12
Yt	4,415.39	2	3
Sb	4,415.40	4	—
Gd	4,415.43	1	34
Se	4,415.56	25	100
Mo	4,415.67	3	2
In	4,415.69	15*h*	—
Cd	4,415.70	20	1
W	4,415.71	3	10
Ta	4,415.74	10	40
O	4,416.97	(150)	—
Gd	4,419.04	200	200
Os	4,420.47	100	400R
Sm	4,420.53	200	200
Sm	4,421.14	150	150
Ne	4,422.52	(300)	—
Fe	4,422.57	125	300
Kr	4,422.70	(100*hs*)	—
Sm	4,424.34	300	300
Ne	4,424.80	(300)	—
Kr	4,425.19	(100)	—
Ne	4,425.40	(150)	—
Ar	4,426.01	(300)	—
Fe	4,427.31	200	500
As	4,427.38	200	—
La	4,429.90	300	200
Kr	4,431.67	(500)	—
As	4,431.73	200	—

	λ	I	
		J	O
Sm	*4,433.88*	*200*	*200*
Sm	*4,434.32*	*200*	*200*
Rn	*4,435.05*	*(200)*	—
Ra	*4,436.27*	*(200)*	—
Kr	*4,436.81*	*(600)*	—

Cd 4,678.16 *J* 200 *O* 200

	λ	I	
		J	O
Sm	**4,678.12**	—	**3**
Tl	**4,678.1**	—	**8**
Ta	**4,678.02**	**2**	**40**
J	**4,677.94**	**(10)**	—
N	**4,677.94**	**(10*w*)**	—
U	**4,677.90**	**4**	—
Ag	**4,677.87**	**1*h***	**2*h***
Tm	**4,677.85**	**2**	**50**
Pb	**4,677.8**	**(7)**	—
Pr	**4,677.78**	—	**2**
Cl	**4,677.76**	**(4)**	—
W	**4,677.69**	**3**	**25**
Gd	**4,677.62**	—	**5**
Ti	**4,677.48**	—	**3**
Pd	**4,677.46**	**2**	**8**
Rh	**4,677.41**	**2**	**6**
Tb	**4,677.31**	—	**2**
Co	**4,677.25**	—	**4**
Tm	**4,677.19**	—	**5**
Pr	**4,676.99**	—	**3**
Gd	**4,676.98**	—	**8**
Ti	**4,676.97**	—	**10**
J	**4,676.94**	**(80)**	—
Sm	**4,676.91**	**50**	**100**
*bh***Yt**	**4,676.9**	—	**3**
Tb	**4,676.89**	—	**25**
Te	**4,676.88**	**(15)**	—
Xe	**4,676.75**	**(5*h*)**	—
Pr	**4,676.72**	—	**5**
Xe	4,676.46	(100*wh*)	—
O	4,676.25	(125)	—
Rh	4,675.03	50	100
Yt	4,674.85	100	80
Cu	4,674.76	30*W*	200
Be	4,673.46	(100)	—
Be	4,672.2	100	—
Nb	4,672.09	100	150
Pr	4,672.08	25*w*	100
La	4,671.83	150	100
Pr	**4,678.168**	**2**	**12**
In	**4,678.17**	**30**	—
Mo	**4,678.20**	**5**	**3**
Ne	**4,678.218**	**(300)**	—
Th	**4,678.220**	—	**2**
Gd	**4,678.25**	**10*h***	**10*h***
Xe	**4,678.31**	**(2*h*)**	—
Sr	**4,678.33**	—	**20*h***
Yt	**4,678.35**	—	**2**
Cb	**4,678.43**	**5**	**5**
Ce	**4,678.51**	—	**2**
Cb	**4,678.52**	**5**	**5**
*bh***C**	**4,678.6**	—	—
Ne	**4,678.60**	**(50)**	—
Ce	**4,678.62**	—	**2**
Br	**4,678.69**	**(200)**	—
Tb	**4,678.81**	—	**2**
Fe	**4,678.85**	**100**	**150**
Nd	**4,678.91**	**1**	**2*h***
P	**4,678.94**	**(100)**	—
J	**4,678.98**	**(8)**	—
Pr	**4,679.039**	—	**5**
W	**4,679.041**	**2**	**15**
Er	**4,679.07**	**2**	**6**
Pr	**4,679.11**	**3**	**8**
Dy	**4,679.12**	**2**	**3**
Ne	**4,679.13**	**(150)**	—
Gd	**4,679.18**	**25**	**25**
Xe	**4,679.45**	**(3*h*)**	—
Re	**4,679.47**	—	**20**
Eu	**4,679.48**	—	**4*w***
Pr	**4,679.50**	—	**2**
Tb	**4,679.62**	—	**2**
Zn	4,680.14	200*h*	300*w*
Ne	4,680.36	(100)	—
Kr	4,680.41	(500)	—
W	4,680.52	40	150
Rn	4,680.83	(500)	—

	λ	I J	I O
Mn	4,671.69	5	100
Xe	4,671.23	(2000)	—
Se	4,670.40	300*wh*	100
Ta	4,669.14	15	300
La	4,668.91	300*r*	200*r*
Na	4,668.60	100	200
Ag	4,668.48	70	200
Fe	4,668.14	10	125
Ni	4,667.77	—	100
Ti	4,667.59	8	150
Fe	4,667.46	20	150
Ne	4,667.36	(100)	—
J	4,666.52	(250)	—
Se	4,664.98	(150)	—
Pr	4,664.65	15	100
Te	4,664.34	(800)	—
Se	4,664.20	(150)	—
Hf	4,664.12	100	50
Os	4,663.82	5	100
La	4,663.76	200	100
Co	*4,663.41*	—	*700W*
La	*4,662.51*	*200*	*150*
Ta	*4,661.12*	*5h*	*300*
Hg	*4,660.28*	*(200)*	—
W	*4,659.87*	*70*	*200*
Kr	*4,658.87*	*(2000)*	—
Ne	*4,656.39*	*(300)*	—
La	*4,655.50*	*300*	*150*
Te	*4,654.38*	*(800)*	—
Cr	*4,652.16*	*150*	*200R*
Cu	*4,651.13*	*40*	*250*
O	*4,649.15*	*(300)*	—
Ni	*4,648.66*	*3*	*400w*

	λ	I J	I O
In	4,681.11	(200)	—
Ru	4,681.79	—	100
Ta	4,681.87	50	200
Ti	4,681.92	100	200
In	4,682.00	250*W*	—
Ra	4,682.28	(800)	—
Yt	4,682.32	100	60
Co	4,682.38	—	500
Gd	4,683.34	50	100
Ru	4,684.02	—	100
Ta	4,684.87	2	100
Pr	4,684.94	10*w*	125
In	4,685.22	(100)	—
He	4,685.75	(300)	—
Ni	4,686.22	1	200
Te	4,686.95	(300)	—
Sm	4,687.18	—	100
Ne	4,687.67	(100)	—
Zr	4,687.80	—	125
Eu	4,688.23	2	100
Xe	4,690.97	(100)	—
Kr	4,691.28	(100)	—
Ti	4,691.34	25	125
Ba	4,691.62	40	100
Ta	4,691.90	5*h*	400
La	4,692.50	300	200
Co	*4,693.21*	*25*	*500*
Kr	*4,694.44*	*(200h)*	—
Xe	*4,697.02*	*(300)*	—
Eu	*4,698.14*	*2*	*300*
Co	*4,698.38*	*8*	*300*
La	*4,699.63*	*200r*	*200r*
Ar	*4,702.32*	*(1200)*	—
La	*4,703.28*	*300r*	*200r*
Ni	*4,703.81*	—	*200*
Ne	*4,704.39*	*(1500)*	—
Sm	*4,704.41*	—	*200*
Cu	*4,704.60*	*50*	*200*
Hg	*4,704.63*	*(200)*	—
Br	*4,704.86*	*(250)*	—
O	*4,705.32*	*(300)*	—
Ta	*4,706.09*	*2*	*200*

Cd 4,799.918 *J* 300 *O* 300

	λ	*I*	
		J	*O*
Gd	**4,799.87**	**60**	**25**
Ti	**4,799.80**	**15**	**80**
V	**4,799.77**	**12**	**15**
As	**4,799.68**	**10**	—
Br	**4,799.61**	**(8)**	—
Re	**4,799.50**	—	**3**
Xe	**4,799.45**	**(10*h*)**	—
Nd	**4,799.423**	—	**15**
Pr	**4,799.418**	—	**4**
Eu	**4,799.38**	—	**5**
Yt	**4,799.30**	**3**	**5**
Re	**4,799.11**	—	**10**
Eu	**4,798.92**	—	**5**
Sm	**4,798.87**	—	**50*r***
Pr	**4,798.75**	—	**25**
Ar	**4,798.74**	**(30)**	—
Rh	**4,798.67**	—	**3**
Tb	**4,798.66**	—	**2**
Ti	**4,798.53**	**15**	**2**
Pb	**4,798.52**	**20**	—
Ru	**4,798.443**	—	**25**
Dy	**4,798.436**	—	**2**
Pb	**4,798.4**	**5**	—
Cl	**4,798.40**	**(15)**	—
Hg	4,797.01	(300)	—
Cl	4,794.64	(250)	—
Os	4,793.99	6	300
Co	4,792.86	5	600*W*
Xe	4,792.62	(50)	—
Au	4,792.60	60	200*W*
Cr	4,792.51	40	200
Gd	4,791.60	—	150
Sm	4,791.58	—	150
Re	4,791.42	—	200*w*
Fe	4,791.25	200*R*	200
Cr	4,789.38	100	300
Ne	4,788.93	(300)	—
Pd	4,788.17	4*h*	200*h*
Fe	4,786.81	—	150
Yb	4,786.60	200	50
Ni	4,786.54	2	300*W*
Br	4,785.50	(400)	—
Lu	4,785.42	200	100
Mn	*4,783.42*	*60*	*400*
Co	*4,781.43*	*2h*	*400*

	λ	*I*	
		J	*O*
W	**4,799.919**	**10**	**50**
La	**4,800.00**	—	**15**
In	**4,800.01**	**9**	**5**
J	**4,800.07**	**(8)**	—
Ne	**4,800.111**	**(15)**	—
Gd	**4,800.111**	**80**	**30**
K	**4,800.16**	—	**3**
Th	**4,800.18**	**2**	**3**
La	**4,800.26**	—	**15**
Tb	**4,800.44**	—	**3**
Hf	**4,800.50**	**6**	**50**
Fe	**4,800.66**	—	**15**
Dy	**4,800.68**	—	**4**
Eu	**4,800.82**	—	**2*w***
Mo	**4,801.01**	**4**	**3**
Cr	**4,801.03**	**70**	**200**
Gd	**4,801.08**	**200**	**200**
Pr	**4,801.15**	**3**	**40**
Ru	**4,801.18**	—	**10**
Ce	**4,801.36**	—	**2**
Tm	**4,801.41**	**5**	**10**
La	4,804.04	150	150
Gd	4,805.82	80	200
Ar	4,806.07	(500)	—
Ni	4,807.00	1	150*w*
Xe	4,807.02	(500)	—
Ca	4,809.01	150	150
Cl	4,810.06	(200)	—
Ne	4,810.063	(150)	—
Zn	4,810.53	300*h*	400*h*
Kr	4,811.76	(300)	—
Ta	4,812.75	5	150
Co	4,813.48	6	1000*W*
Ge	4,814.80	200	—
S	*4,815.51*	*(800)*	—
Br	*4,816.71*	*(300)*	—
Ne	*4,817.64*	*(300)*	—
Ne	*4,821.92*	*(300)*	—
Mn	*4,823.52*	*80*	*400*
Kr	*4,825.18*	*(300)*	—
Ra	*4,825.91*	*(800)*	—
Ne	*4,827.34*	*(1000)*	—
Ne	*4,827.59*	*(300)*	—
Ni	*4,829.03*	*2h*	*300w*

	λ	I			λ	I	
		J	O			J	O
Co	*4,780.01*	*500*	*500w*	*Xe*	*4,829.71*	*(400)*	—
Co	*4,776.32*	—	*300*				
Co	*4,771.11*	—	*500w*				

Cd 5,085.82 *J* 500 *O* 1000

	λ	J	O		λ	J	O
Yb	**5,085.74**	**4*h***	—	**U**	**5,085.86**	**6**	**10**
Eu	**5,085.60**	—	**8**	**W**	**5,085.90**	—	**6*d***
Sc	**5,085.55**	**70**	**80**	**In**	**5,085.98**	**5**	—
Rh	**5,085.52**	—	**10**	**La**	**5,806.24**	—	**3**
Ni	**4,085.48**	—	**10**	**Kr**	**5,086.52**	**(250*h*)**	—
Ti	**5,085.34**	—	**20**	**Ce**	**5,086.57**	—	**6**
Zr	**5,085.26**	—	**10**	**La**	**5,086.71**	**2**	**4**
Th	**5,085.12**	**1**	**2**	**Fe**	**5,086.76**	**100**	**2**
Al	**5,085.02**	**(25)**	—	**Nb**	**5,086.83**	**1*h***	**5*h***
Nb	**5,084.84**	**1**	**5**	**Ir**	**5,086.85**	—	**2*h***
Ce	**5,084.790**	—	**5**	**W**	**5,086.92**	—	**3**
Sm	**5,084.788**	**1**	**2**	**Sc**	**5,086.95**	**25**	**60**
W	**5,084.67**	—	**3*h***	**Tl**	**5,086.99**	**4**	—
Rn	**5,084.48**	**(300)**	—	**Ti**	**5,087.068**	**1**	**70**
Mo	**5,084.23**	**3**	**10**	**Sm**	**5,087.073**	—	**30**
K	**5,084.21**	—	**20**	**Ar**	**5,087.09**	**(60)**	—
Ce	**5,084.17**	—	**12**	**Pr**	**5,087.11**	—	**25**
Ni	**5,084.08**	**2**	**300*w***	**Sc**	**5,087.15**	**20**	**40**
				Eu	**5,087.18**	—	**4**
				Ta	**5,087.37**	—	**60**
Xe	5,080.62	(500)	—	**Yt**	**5,087.42**	**100**	**50**
Se	5,068.65	(250)	—	**Tb**	**5,087.48**	—	**15**
Ar	*5,060.08*	*(500)*	—	Se	5,096.57	(350)	—
				Ra	5,097.56	(250)	—
				J	*5,119.28*	*(500)*	—

Cd 6,438.4696 *J* 1000 *O* 2000

	λ	J	O		λ	J	O
Br	**6,438.08**	**(2)**	—	**Ra**	**6,438.9**	**(30)**	—
W	**6,438.03**	**1**	**3**	**In**	**6,438.96**	**5**	—
Eu	**6,437.69**	—	**700**	**Zr**	**6,439.03**	—	**8**
Ar	**6,437.63**	**(4)**	—	**Ca**	**6,439.07**	**50**	**150**
Sm	**6,437.630**	—	**10**	**Co**	**6,439.171**	—	**80**
In	**6,437.54**	**(12)**	—	**Nd**	**6,439.171**	—	**20**
Ta	**6,427.36**	—	**2**	**Sm**	**6,439.32**	—	**3**
Yt	**6,437.16**	**5**	**4**	**Sm**	**6,329.72**	—	**10*d***

	λ	I	
		J	O
W	**6,439.720**	**1**	**6**
Co	**6,439.83**	—	**2*h***
Rn	**6,439.83**	**(2)**	—
Nd	**6,439.86**	—	**3**
Ra	6,446.20	(1000)	—
Co	6,450.24	—	1000

Ce 58 140.12

t_0 640°C t_1 1,400°C

I.	II.	III.	IV.	V.
6.57	—	19.70	36.715	—

λ	I		eV
	J	O	
2,225.10	100	—	—
II 3,539.09	10	100	4.4
II 3,560.80	2	300	4.8
II 3,577.46	12	300	4.5
II 4,012.39	20	60	4.2
II 4,186.60	25	80	3.34
5,522.99	—	100	—

Ce 2,225.10 *J* 100

	λ	*I*			λ	*I*	
		J	*O*			*J*	*O*
Nb	**2,225.96**	**5***wh*	—	**Sb**	**2,225.15**	**(10)**	—
Rh	**2,224.97**	**30**	—	**Kr**	**2,225.18**	**(2)**	—
Sb	**2,224.94**	**25**	**30**	**Cd**	**2,225.20**	**(10)**	—
Co	**2,224.87**	**5**	—	**W**	**2,225.23**	**4**	**2**
Ni	**2,224.867**	**25**	**15**	**Os**	**2,225.27**	**20**	**8**
				J	**2,225.27**	**(30)**	—
Ar	2,222.93	(60)	—	**Pd**	**2,225.29**	**5**	**15**
V	2,222.70	100	—	**Co**	**2,225.350**	**2**	**12**
				Hf	**2,225.35**	**7**	**6**
Ce	*2,222.04*	*100*	—				
Ir	*2,221.07*	*100*	*2*	Ar	2,227.70	(60)	—
				V	*2,228.30*	*100*	—
				J	*2,229.97*	*(400)*	—

Ce 3,539.09 *J* 10* *O* 100

	λ	*J*	*O*		λ	*J*	*O*
Zr	**3,539.01**	**3**	**4**	**Nb**	**3,539.12**	**15**	**1**
U	**3,538.98**	**3**	**1**	**Nd**	**3,539.18**	**4**	**10**
Mo	**3,538.97**	—	**4**	**Fe**	**3,539.20**	**1***h*	**1***h*
Ce	**3,538.96**	—	**3**	**U**	**3,539.21**	**4**	**4**
Mo	**3,538.92**	**4**	**3**	**Eu**	**3,539.24**	—	**2**
Tb	**3,538.90**	**3**	**15**	**Sm**	**3,539.25**	**1**	**8**
Sm	**3,538.864**	**1**	**10**	**Ru**	**3,539.26**	**5**	**30**
Mg	**3,538.86**	—	**8**	**Er**	**3,539.29**	—	**3**
Nd	**3,538.858**	**6**	**20**	**Th**	**3,539.32**	**5**	**5**
Pr	**3,538.85**	**2**	**9**	**Re**	**3,539.329**	—	**25**
Th	**3,538.84**	—	**4***d*	**W**	**3,539.330**	**3**	**6**
Fe	**3,538.795**	—	**1***h*	**Ru**	**3,539.369**	**15**	**60**
Ce	**3,538.792**	**2**	**3**	**Nd**	**3,539.372**	**4**	**4**
Ce	**3,538.76**	**2**	**5**	**Dy**	**3,539.38**	**2***h*	**18**
W	**3,538.63**	**9**	**8**	**W**	**3,539.458**	**7**	**3**
Fe	**3,538.55**	—	**1***h*	**Mo**	**3,539.465**	**3**	**3**
Yt	**5,538.53**	**3**	**10**	**Th**	**3,539.59**	**8**	**8**
Dy	**3,538.523**	**40**	**150**	**Er**	**3,539.60**	—	**9**
Er	**3,538.519**	**9**	**18**	**Dy**	**3,539.62**	**4**	**6**
Dy	**3,538.50**	**2**	**5**	**Pr**	**3,539.63**	**2**	**5**
Tb	**3,538.50**	**3**	**3**	**Nb**	**3,539.648**	**15**	**15**
Ce	**3,538.46**	**1**	**2**	**U**	**3,539.654**	**2**	**12**
Th	**3,538.44**	—	**3***d*	**Eu**	**3,539.76**	—	**3**
U	**3,538.42**	**8**	**2**	**Tb**	**3,539.81**	**3**	**15**
Ce	**3,538.41**	**1**	**2**	**Ce**	**3,539.84**	—	**5**
Pr	**3,538.31**	**2**	**3**	**Os**	**3,539.86**	**3**	**1**
Ce	**3,538.30**	—	**2**	**Dy**	**3,539.88**	—	**4**
Fe	**3,538.29**	**1***h*	**2***h*	**Sm**	**3,539.89**	—	**10**

	λ	I	
		J	O
Ag	**3,538.27**	**3**	**10**
Rh	**3,538.26**	**4**	**50**
V	**3,538.24**	**100**	**10**
Th	**3,538.223**	**3**	**8**
U	**3,538.221**	**6**	**8**
Ir	**3,538.15**	**1*h***	**18**
Rh	**3,538.14**	**10**	**100**
Eu	**3,538.09**	**10*h***	**20*w***
Ru	3,537.95	25	70
Fe	3,537.90	25	50
Re	3,537.47	—	80*w*
Tm	3,536.57	20	60
Ru	3,536.567	—	50
Fe	3,536.56	200	300
Dy	3,536.02	10	125
Ru	3,535.83	12	60
Tm	3,535.52	25	80
Nb	3,535.30	500	300
Dy	3,534.96	—	125
Cu	3,533.75	15	50
Co	3,533.36	—	200*w*
Fe	3,533.20	50	50
Na	3,533.010	(200)	50
Fe	3,533.010	75	50
Ru	3,532.81	12	60
Os	3,532.80	20	100
Mn	3,532.12	30	50*h*
Mn	3,532.00	8	50*h*
Dy	3,531.71	100	100
Ru	3,531.39	9	60
Fe	3,530.389	25	50
Cu	3,530.386	20	50
Os	3,530.06	20	100
Fe	3,529.82	80	125
Co	3,529.81	30	1000*R*
Tl	3,529.43	800	1000
Co	3,529.03	—	200*R*
Ru	3,528.68	12	60
Os	3,528.60	50	400*R*
Rh	*3,528.02*	*150*	*1000w*
Ni	*3,527.98*	*15*	*200*
Fe	*3,527.80*	*80*	*100*
Co	*3,526.85*	*25*	*300R*
Ni	*3,524.54*	*100wh*	*1000R*
Os	*3,523.64*	*30*	*150*
Zr	**3,539.91**	**7**	**8**
Pr	**3,539.92**	**5**	**25**
W	**3,539.93**	**4**	**6**
Re	**3,539.95**	—	**15**
Ce	**3,540.117**	—	**3**
Fe	**3,540.118**	**60**	**100**
Tb	3,540.24	50	50
Fe	3,541.09	200	200
Nb	3,541.25	5	50
Ru	3,541.63	10	60
Rh	3,541.91	10	50
Fe	3,542.08	100	150
Dy	3,542.33	20	90
Os	3,542.71	10	150
V	3,543.50	50	50
Fe	3,543.67	30	60
Tb	3,543.86	15	50
Rh	3,543.95	40	150
Fe	3,544.63	6	50
Fe	3,545.64	70	90
Gd	3,545.79	125	125
Tb	3,546.52	8	50
Dy	3,546.84	10	50
Zr	3,547.68	12	200
Sr	3,548.08	—	50
Ni	3,548.18	25	400
Gd	3,549.36	125	125
Rh	3,549.54	50	150
Dy	*3,550.23*	*100*	*200*
Co	*3,550.59*	—	*200*
Pd	*3,553.08*	*15wh*	*100R*
Fe	*3,553.74*	*100*	*100*
Fe	*3,554.93*	*300*	*400*
Nd	*3,555.72*	*15*	*100*
Fe	*3,556.88*	*150*	*300*
Fe	*3,558.518*	*300*	*400*
Gd	*3,558.519*	*50*	*100r*
*N**			

	λ	I J	I O
Ni	*3,523.44*	*—*	*100*
Co	*3,523.43*	*25*	*100r*
Co	*3,521.57*	*25*	*200r*
Fe	*3,521.26*	*200*	*300*
Co	*3,520.08*	*—*	*100W*
Ni	*3,519.77*	*30*	*500h*
Zr	*3,519.60*	*10*	*100*
Tl	*3,519.24*	*1000R*	*2000R*
Os	*3,518.72*	*30*	*200*
Co	*3,518.35*	*100*	*200W*

Ce 3,560.80 *J* 2* *O* 300

	λ	I J	I O
Nd	**3,560.729**	**20**	**30**
Yb	**3,560.727**	**100**	**8**
Fe	**3,560.70**	**15**	**50**
V	**3,560.60**	**50**	**10**
Eu	**3,560.57**	**1*h***	**7*w***
Os	**3,560.47**	**5*h***	**1**
Nb	**3,560.34**	**2**	**3**
Yb	**3,560.33**	**50**	**20**
U	**3,560.308**	**2*h***	**4**
Co	**3,560.308**	**3**	**18*h***
Sm	**3,560.27**	**3**	**25**
Ho	**3,560.15**	**8**	**6**
Dy	**3,560.149**	**20**	**40**
W	**3,560.07**	**4**	**5**
Th	**3,559.96**	**8**	**6**
Ni	**3,559.93**	**—**	**5**
Er	**3,559.90**	**7**	**15**
Mo	**3,559.88**	**5**	**5**
Os	**3,559.79**	**50**	**150**
Cr	**3,559.78**	**10**	**15**
Tb	**3,559.76**	**8**	**15**
W	**3,559.71**	**8**	**10**
*P**			
Os	**3,560.85**	**100**	**150*R***
Th	**3,560.86**	**4**	**10**
Fe	**3,560.887**	**—**	**2**
Co	**3,560.893**	**25**	**200**
Cr	**3,560.90**	**1*h***	**2*h***
Tm	**3,560.92**	**—**	**8**
Zr	**3,561.11**	**—**	**2**
Nb	**3,561.14**	**10**	**5**
Ir	**3,561.17**	**2**	**5**
Hg	**3,561.20**	**6*h***	**10**
W	**3,561.25**	**4**	**8**
Pr	**3,561.27**	**4**	**25**
Er	**3,561.273**	**—**	**12s**
Eu	**3,561.28**	**1*h***	**3**
Mo	**3,561.36**	**20**	**1**
Ir	**3,561.37**	**—**	**5**
U	**3,561.41**	**—**	**12**
Ce	**3,561.54**	**—**	**6**
Ti	**3,561.57**	**20**	**10**
Sm	**3,561.58**	**4**	**15**
Nd	**3,561.59**	**2**	**8**
Hf	**3,561.66**	**35**	**20**
Pr	**3,561.68**	**3**	**6**
Nb	**3,561.69**	**10**	**4**
Tb	**3,561.74**	**200**	**200**
Ni	**3,561.75**	**12**	**70**
U	**3,561.80**	**30**	**12**
Th	**3,561.809**	**1**	**3**
Fe	**3,561.812**	**—**	**4**
Er	**3,561.891**	**—**	**8**
Ru	**3,561.894**	**5**	**2**
Zr	**3,561.90**	**2**	**3**
Ti	**3,561.91**	**12**	**3**

	λ	I J	I O
Dy	3,563.15	100	200
Co	3,564.95	—	150*d*
Fe	3,565.38	300	400
Ni	3,566.37	100*wh*	2000*R*
Co	3,569.38	100	400*R*
Fe	3,570.10	300	300
Rh	3,570.18	150	400*r*
Dy	3,570.25	—	300
Ni	3,571.87	40*h*	1000*R*

*N**

Ce 3,577.458 *J* 12* *O* 300

	λ	I J	I O
Re	3,577.350	—	2
U	3,577.349	1	6
Co	3,577.26	—	3
Sr	3,577.243	—	2
Ni	3,577.240	—	4
Nb	3,577.23	3	4
Th	3,577.22	2	1
Ir	3,577.09	—	3
Tb	3,577.08	3	8
U	3,577.076	3	15
Yb	3,577.05	20	7
Dy	3,576.87	50	200
Zr	3,576.854	25	15
Mo	3,576.851	3	1
U	3,576.84	1	2
Tb	3,576.83	3	15
Pr	3,576.83	—	2
Ni	3,576.762	40*h*	2
Fe	3,576.760	40	80
Gd	3,576.759	3*h*	5
Ir	3,576.74	2	10
Th	3,576.56	3	3
W	3,576.38	5	7
Se	3,576.34	45	18
Pr	3,576.32	3	10
Dy	3,576.25	—	300
Co	3,575.36	25	200*r*
Co	3,574.96	25	200
Dy	3,574.16	100	200
Pb	3,572.73	20	200

*P**

	λ	I J	I O
Pr	3,577.465	8	30
Os	3,577.487	—	3
U	3,577.487	—	3
Tm	3,577.51	8	5
Zr	3,577.55	1	12
Ba	3,577.61	—	4
Ir	3,577.62	—	4
Co	3,577.69	—	2*h*
Nb	3,577.72	15	10
Fe	3,577.75	—	2
Sm	3,577.77	3	20
U	3,577.78	1	4
Ce	3,577.80	2	2
Pr	3,577.85	—	6
V	3,577.87	40	50
Mn	3,577.88	25	50
U	3,577.92	—	10
Dy	3,577.99	50	150
Tb	3,578.00	—	8
Co	3,578.07	—	18
Pr	3,578.106	5	10
Eu	3,578.110	1	6
Zr	3,578.226	6	5
Nb	3,578.234	5	1
Ti	3,578.27	1	8
U	3,578.327	1	8
Er	3,578.33	—	3*w*
Gd	3,578.34	5	5
Fe	3,578.38	5	40
Tb	3,578.40	—	8
Pr	3,578.42	3	9
Dy	3,578.47	2	4
Eu	3,578.49	1	2
Pr	3,578.57	1	2
Gd	3,578.58	—	8

	λ	I J	I O		λ	I J	I O
				Nb	**3,578.584**	**1**	**15**
				Cr	3,578.69	400*r*	500*R*
				Fe	3,581.19	600*r*	1000*R*
				Rh	3,583.10	125	200
				Dy	3,585.07	100	300
				Fe	3,585.32	100	150
				Dy	3,585.78	100	150
				Fe	3,586.99	150	200
				Rb	3,587.08	40	200
				Co	3,587.19	50*h*	200*r*
				Ni	3,587.93	12	200
				Nd	*3,590.35*	*300 W*	*400 W*
				Cr	*3,593.49*	*400 R*	*500 R*
				Ni	*3,597.70*	*50h*	*1000r*
				Os	*3,598.11*	*30*	*300*

Ce 4,012.388 *J* 20 *O* 60

	λ	J	O		λ	J	O
Mo	**4,012.27**	**3**	**3**	**Ti**	**4,012.391**	**50**	**35**
Re	**4,012.26**	—	**25**	**Tb**	**4,012.45**	—	**5**
Er	**4,012.253**	—	**12**	**Cr**	**4,012.47**	**60**	**70**
Zr	**4,012.252**	—	**20**	**Th**	**4,012.50**	**15**	**15**
Nd	**4,012.250**	**40**	**80**	**Mo**	**4,012.51**	**3**	**3**
Nb	**4,012.17**	**100**	—	**Dy**	**4,012.52**	—	**4**
U	**4,012.161**	**4**	**6**	**Er**	**4,012.58**	**2**	**4**
Co	**4,012.160**	—	**2**	**Nd**	**4,012.70**	**10**	**15**
Ce	**4,012.14**	—	**4**	**U**	**4,012.71**	**2**	**2*h***
Ta	**4,012.11**	**2*h***	**5*h***	**Mo**	**4,012.80**	**2**	**3**
K	**4,012.10**	**(20)**	—	**Ti**	**4,012.81**	**1**	**12**
W	**4,012.096**	**5**	—	**Eu**	**4,012.816**	**25**	**20**
Mo	**4,011.97**	**25**	**25**	**Dy**	**4,012.82**	**2**	**5**
Hf	**4,011.94**	—	**3**	**Tb**	**4,012.87**	**1**	**6**
Mn	**4,011.90**	**10**	**12**	**Pr**	**4,012.91**	**1**	**8**
Rb	**4,011.9**	**(15)**	—	**Se**	**4,012.96**	**(150)**	—
Se	**4,011.88**	**(200)**	—	**In**	**4,012.96**	**(10)**	—
W	**4,011.81**	**3**	**4**	**U**	**4,013.03**	—	**8**
U	**4,011.78**	**5**	**10**	**W**	**4,013.18**	**5**	**6**
Ce	**4,011.77**	—	**2**	**Ta**	**4,013.19**	**1**	**5**
Th	**4,011.75**	**15**	**15**	**Mo**	**4,013.21**	**40**	**1**
Ru	**4,011.729**	—	**7**	**Nd**	**4,013.22**	**5**	**10**
Fe	**4,011.728**	—	**2**	**Pr**	**4,013.23**	**4**	**10**
Sm	**4,011.72**	**4**	**8**	**Th**	**4,013.264**	**5**	**5**
Eu	**4,011.683**	—	**25**	**Ce**	**4,013.265**	—	**2**
Te	**4,011.68**	**(30)**	—	**Nb**	**4,013.27**	**5**	**3**
Ce	**4,011.56**	**3**	**15**	**Tb**	**4,013.28**	—	**20*d***

	λ	I	
		J	O
Ti	4,011.534	—	4
Pr	4,011.532	1	5
Mn	4,011.531	15	15
Re	4,011.51	—	35
U	4,011.45	10	8
Fe	4,011.41	1	5
V	4,011.31	2	9
Ce	4,011.30	—	4
Dy	4,011.24	8	12
Ar	4,011.23	(5)	—
Nd	4,011.07	10	15
As	4,010.66	10	—
Cs	4,010.54	(10)	—
Mo	4,010.30	10	—
C	4,009.90	10	—
Fe	4,009.72	100	120
Co	4,009.71	10	5
Ti	4,009.66	25	60
Mo	4,009.37	25	20
He	4,009.27	(10)	—
Gd	4,009.21	2	50
U	4,009.17	15	8
Ti	4,008.93	35	80
Br	4,008.76	(20)	—
Nd	4,008.754	10	12
W	4,008.753	45	45
Pr	4,008.71	50	150
Mo	4,008.67	20	—
Kr	4,008.48	(10*wh*)	—
Gd	4,008.331	10	15
Sm	4,008.330	10*h*	8
Nb	4,008.28	10	5
Ru	4,008.27	20	20
V	4,008.17	10	2
Kr	4,008.08	(25)	—
Ti	4,008.06	7	50
Er	4,007.97	7	35
Se	4,007.90	(150)	—
In	4,007.61	(10)	—
In	4,007.54	(15)	—
Ru	4,007.53	10	20
Sm	4,007.49	25	50
Nd	4,007.43	20	20
Br	4,007.33	(10)	—
Fe	4,007.27	50	80
Th	4,007.03	20	20
Ta	4,006.83	20	30

	λ	I	
		J	O
U	4,013.29	2	2
Nd	4,013.36	3	3
Gd	4,013.430	—	5
Pr	4,013.43	8	15
U	4,013.431	2*h*	2
In	4,013.49	(30)	—
Ru	4,013.50	12	15
Ta	4,013.54	—	5
Eu	4,013.55	—	2*W*
Ti	4,013.58	7*h*	70*h*
Fe	4,013.647	1	8
Ce	4,013.647	—	3
Ru	4,013.74	5	10
J	4,013.794	(15)	—
Fe	4,013.795	40	80
Mg	4,013.80	—	2
P	4,013.80	(30)	—
Gd	4,013.817	3	25
Fe	4,013,824	—	200
Dy	4,013.83	2	12
Ar	4,013.87	(200)	—
Ca	4,013.88	2*h*	2
Gd	4,013.92	—	10
In	4,013.93	(80)	—
Nd	4,013.935	3	5
Ce	4,013.941	—	3
Co	4,013.945	—	300
Ne	4,013.99	(2)	—
Ir	4,014.00	—	4
Br	4,014.32	(25)	—
Fe	4,014.53	100	200
Cr	4,014.57	8	40
Se	4,014.77	(70)	—
Ce	4,104.90	12	60
Cs	4,014.99	(10)	—
Cl	4,015.06	(10)	—
W	4,015.22	30	25
Ti	4,015.38	10*h*	70*h*
Pr	4,015.389	30	50
La	4,015.393	2*h*	100
Au	4,016.05	15	10
Mo	4,016.06	10	—
Nb	4,016.08	30	5
In	4,016.24	(50)	—
Ti	4,016.28	5	30
W	4,016.53	12	10
Pr	4,016.75	20	25

	λ	I	
		J	O
Cs	4,006.77	(10)	—
Fe	4,006.63	15	20
Ru	4,006.60	15	25
Cs	4,006.54	(30)	—
Te	4,006.50	(100)	—
Th	4,006.39	10	10
As	4,006.34	50	—
Fe	4,006.31	35	60
Hg	4,006.27	(30)	—
Mo	4,006.07	20	20
Ti	4,005.97	3	35
V	4,005.71	30	10
Ru	4,005.64	30	25
Kr	4,005.57	(30*h*)	—
Tb	4,005.55	125	100*d*
Th	4,005.549	30*w*	20*w*
W	4,005.40	10	8
Fe	4,005.25	200	250
Os	4,005.15	20	35
Mo	4,005.12	20	—
Re	4,004.93	—	30
In	4,004.83	(10)	—
Pr	4,004.714	25	20
In	4,004.709	(15)	—
In	4,004.53	(30)	—
Nd	4,004.26	10	15
U	4,004.06	20	15
Os	4,004.02	6	50
Nd	4,004.01	15	20
Cr	4,003.92	12	30
Ti	4,003.81	70	50
Ce	4,003.771	18	40
Fe	4,003.767	80	30
Eu	4,003.71	25	18
Os	4,003.48	6	50
Sm	4,003.45	20	30
U	4,003.40	10	6
Cr	4,003.33	20	—
Th	4,003.32	15	15
Th	4,003.11	10	10
Mg	4,003.10	(20)	—
Se	4,003.08	(60)	—
Cu	4,003.04	1*h*	40
Mo	4,002.97	20	—
V	4,002.94	80	6
Tb	4,002.58	5	50
Eu	4,002.56	2	30
Ti	4,002.49	5	40
Xe	4,002.35	(40*wh*)	—

	λ	I	
		J	O
V	4,016.82	(15*wh*)	—
Fe	4,017.15	50	80
J	4,017.21	(25)	—
V	4,017.29	15*h*	—
K	4,017.51	(15)	—
Nb	4,017.56	10	3
Eu	4,017.58	25	25*W*
U	4,017.72	25	25
Rn	4,017.75	(150)	—
Ti	4,017.77	8	70*h*
Mn	4,018.10	60	80
Cr	4,018.20	8	35
Os	4,018.26	4	60
Fe	4,018.27	7	50
Se	4,018.52	(70)	—
Nd	4,018.83	10	15
U	4,018.97	15	25
Tb	4,019.12	5	40
W	4,019.23	15	18
Co	4,019.30	—	80
P	4,019.45	(50)	—
Se	4,019.50	(10)	—
Se	4,019.72	(20)	—
Gd	4,019.73	10	15
Sm	4,019.98	15	30
Ir	4,020.03	100	80
Cl	4,020.06	(15)	—
Nb	4,020.24	10*h*	2
Sc	4,020.40	20	50
Mo	4,020.45	10	10
Nd	4,020.87	15	15
Co	4,020.90	—	500*w*
Pr	4,020.991	30	40
Ru	4,020.995	12	15
Mo	4,021.01	25	15
Se	4,021.26	(20)	—
Nd	4,021.33	12	12
Tm	4,021.39	10	4
In	4,021.66	(50)	—
Nd	4,021.79	10	12
Ti	4,021.83	20	100
Fe	4,021.87	100	200
In	4,021.99	(10)	—
Th	4,022.09	15	20
W	4,022.12	10	12
Ru	4,022.16	100	40
Cr	4,022.26	40	80
Nb	4,022.39	10	2

	λ	I	
		J	O
U	4,002.34	18	10
Tb	4,002.18	2	40*w*
Se	4,002.07	(60)	—
Mn	4,001.91	10	15
W	4,001.89	15	—
Cs	4,001.68	(20)	—
Fe	4,001.67	50	80
Cr	4,001.44	80	200
Gd	4,001.24	3	80
K	4,001.20	(40)	—
Nb	4,001.13	15	10
W	4,000.69	10	12
Nb	4,000.60	50	2
Dy	4,000.454	300	400
Fe	4,000.452	10	35
Er	4,000.45	6	35
Pr	4,000.19	25	—
N	3,999.98	(15)	
Cr	3,999.68	10	40
Tb	3,999.41	15	12
Ti	3,999.36	5	30
Ta	3,999.28	20*d*	30
Ce	3,999.24	20	80
V	3,999.191	40*h*	—
Pr	3,999.188	40*d*	50*d*
Nb	3,999.18	10	5
Zr	3,998.97	30	30
Os	3,998.93	12	80
S	3,998.79	(60)	—
V	3,998.73	25	100
Nd	3,998.69	15	40
Ti	3,998.64	100	150
Mo	3,998.63	25	—
Ho	3,998.28	6	40
U	3,998.24	18	5
Nd	3,998.15	12	20
Fe	3,998.05	100	150
Kr	3,997.95	(100*wh*)	—
Nd	3,997.93	10	20
Co	3,997.91	20	200
Th	3,997.86	10	10
Gd	3,997.77	30	20
Fe	3,997.40	150	300
Mn	3,997.21	25	12
P	3,997.16	(70)	—
Br	3,997.13	(12)	—
V	3,997.12	40	25
Pr	3,997.05	40	100
Fe	3,996.97	20	40

	λ	I	
		J	O
Cu	4,022.66	25	400
Nd	4,023.00	15	15
Gd	4,023.15	10	20
Sm	4,023.223	25	30
Sc	4,023.223	—	60
Se	4,023.23	(20)	—
Gd	4,023.350	10	20
Re	4,023.353	—	40*w*
V	4,023.39	30	10
Co	4,023.40	—	200
Mo	4,023.56	25	—
Cs	4,023.58	(10)	—
La	4,023.59	15	50
Sc	4,023.69	25	100
Cr	4,023.74	15	40
In	4,023.76	(15)	—
Ru	4,023.83	60	25
Zr	4,023.98	2	30
Br	4,024.04	(20)	—
Tb	4,024.07	1	40*w*
Mo	4,024.09	25	30
Ti	4,024.57	35	80
F	4,024.73	(500)	—
Fe	4,024.74	30	120
Nd	4,024.78	10	20
In	4,024.83	100*wh*	—
K	4,024.92	(15)	—
F	4,025.010	(150)	—
Cr	4,025.012	25	100
J	4,025.08	(30)	—
Ti	4,025.14	25	15
Xe	4,025.19	(15)	—
F	4,025.49	(300)	—
Pr	4,025.55	25	40
W	4,025.60	10	—
Th	4,025.61	20	20
Cs	4,025.67	(10)	—
La	4,025.88	50	50
Hg	4,025.95	(20)	—
Mo	4,025.99	30*h*	30*h*
U	4,026.02	25	25
N	4,026.09	(10*h*)	—
Th	4,026.16	10	10
Cr	4,026.17	35	100
Hr	4,026.19	(70)	—
Nb	4,026.32	10	—
Mn	4,026.43	40	50
Al	4,026.5	(30)	—
Ti	4,026.54	10	70

	λ	I	
		J	O
Os	3,996.80	10	50
Dy	3,996.70	80	200
Sc	3,996.61	10	40
Pt	3,996.57	—	50
Tm	3,996.52	40	200
Al	3,996.38	(10)	—
Gd	3,996.32	100	100
Ta	3,996.17	30*h*	100
Al	3,996.16	18	—
Rh	3,996.15	10	25
Th	3,996.07	10	15
Fe	*3,995.99*	*20*	*60*
Ru	*3,995.98*	*30*	*30*
Al	*3,995.86*	*(30)*	—
La	*3,995.75*	*300*	*600*
Tm	*3,995.58*	—	*100*
Co	*3,995.31*	*20*	*1000R*
K	*3,995.10*	*30*	—
N	*3,994.99*	*(300)*	—
U	*3,994.980*	*(20wh)*	*8*
J	*3,994.979*	*(35)*	—
Os	*3,994.93*	*5*	*30*
Pr	*3,994.834*	*25*	*300*
Kr	*3,994.83*	*(100)*	—
Nd	*3,994.68*	*40*	*80*
Co	*3,994.54*	—	*60*
Cr	*3,993.97*	*20*	*60*
S	*3,993.53*	*(50)*	—
Ba	*3,993.40*	*50r*	*100R*
Sm	*3,993.30*	*25*	*25*
Cr	*3,992.84*	*70*	*150*
V	*3,992.80*	*20*	*60*
Nd	*3,992.57*	*20*	*30*
Mn	*3,992.49*	*75*	*40*
Br	*3,992.39*	*(20)*	—
Ir	*3,992.12*	*60*	*150*
Ar	*3,992.06*	*(25)*	—
Nd	*3,991.74*	*40*	*60*
Co	*3,991.69*	*6*	*60*
Nb	*3,991.68*	*20*	*15*
Cr	*3,991.67*	*50*	*100*
Mn	*3,991.60*	*25*	*20*
Zr	*3,991.13*	*60*	*100*
Cr	*3,991.12*	*60*	*200*
Kr	*3,991.08*	*(20)*	—
S	*3,990.94*	*(40)*	—
Yb	*3,990.89*	*20*	*60*
Rb	4,026.9	(25)	—
Ta	4,026.94	30	40
Co	4,027.04	4	200
Cr	4,027.10	30	80
Zr	4,027.20	4	100
Ti	4,027.48	3	30
In	4,027.79	(50*h*)	—
Nb	4,027.98	10	5
Cr	4,028.02	—	35
Ti	4,028.34	80	20
Ce	4,028.41	8	35
Au	4,028.48	10	—
S	*4,028.79*	*(200)*	—
Fe	*4,029.636*	*25*	*80*
Re	*4,029.639*	—	*80*
Mo	*4,029.94*	*30*	*1*
Se	*4,030.07*	*(150)*	—
Fe	*4,030.49*	*60*	*120*
Ti	*4,030.51*	*18*	*80*
Cr	*4,030.68*	*30*	*40*
Mn	*4,030.75*	*20*	*500v*
La	*4,031.69*	*300*	*400*
Pr	*4,031.75*	*30*	*50*
Fe	*4,031.96*	*50*	*80*
Ru	*4,032.20*	*20*	*20*
Nb	*4,032.52*	*50*	*30*
Fe	*4,032.63*	*15*	*80*
S	*4,032.81*	*(125)*	—
Br	*4,032.85*	*(20)*	—
Ar	*4,032.97*	*(20)*	—
Ga	*4,032.98*	*500R*	*1000R*
Tb	*4,033.04*	*5*	*125*
Ta	*4,033.069*	*10*	*100*
Mn	*4,033.073*	*30*	*400r*
Sb	*4,033.54*	*60*	*70*
Ir	*4,033.76*	*25*	*100*
Ar	*4,4033.83*	*(30)*	—
Pr	*4,033.86*	*35*	*50*
Mn	*4,034.49*	*20*	*250r*
Ar	*4,035.47*	*(30)*	—
Co	*4,035.55*	*3*	*150*
V	*4,035.63*	*80*	*40*
Mo	*4,035.66*	*25*	*3*
Mn	*4,035.73*	*60*	*50*
J	*4,036.08*	*(50)*	—
V	*4,036.78*	*40*	*8*
Xe	*4.037.29*	*(50)*	—

	λ	I J	I O
Mo	3,990.84	30	—
V	3,990.57	40	125
U	3,990.42	20	18
Fe	3,990.38	25	70
Xe	3,990.33	(30wh)	—
Co	3,990.30	10	80
Cl	3,990.19	(20)	—
Nd	3,990.10	20	40
Sm	3,990.00	25	40
Cr	3,989.99	40	80
Ti	3,989.76	100	150
Pr	3,989.72	125	200
Mo	3,989.51	25h	—
Gd	3,989.25	25	10
Zn	3,989.23	(100)	—
V	3,988.83	35	70
La	3,988.52	800	1000
Th	3,988.01	30	50
Yb	3,987.99	500R	100R
Er	3,987.95	20	100r
Gd	3,987.84	25	50
Ru	3,987.79	50	3
Kr	3,987.78	(25)	—
Nd	3,987.43	20	20
Gd	3,987.22	100	100
Co	3,987.11	—	80
Mn	3,987.10	60	30
Mn	3,986.83	75	40
Nd	3,986.23	20	30
Mo	3,986.20	20	5
Fe	3,986.17	8	125
P	3,985.86	(30)	—
U	3,985.795	30	25
Li	3,985.79	—	100
V	3,985.790	40	1
Fe	3,985.39	40	125
Mn	3,985.24	100	75
Xe	3,985.20	(30)	—
Ag	3,985.19	20	2
Ru	3,984.86	70	60
Cr	3,984.34	60	80
Dy	3,984.23	—	80
Mn	3,984.18	20	20
Hg	3,983.98	(400)	—
Fe	3,983.96	125	200
J	3,983.95	(25)	—
Cr	3,983.91	60	200
Dy	3,983.66	8	150
W	3,983.29	25	12
Cr	4,037.294	12	80
Gd	4,037.34	30	100
Xe	4,037.59	(100)	—
Nb	4,037.66	20	—
Kr	4,037.83	(30)	—
Os	4,037.84	4	80
Gd	4,037.91	30	100
Se	4,038.31	(40)	—
Ar	4,038.82	(40)	—
Cr	4,039.10	40	100
Ru	4,039.21	50	25
Pr	4,039.36	20	50
Nb	4,039.53	50	30
Cs	4,039.84	(50)	—
Hg	4,040.40	(20)	—
Ce	4,040.76	5	70
Nd	4,040.80	40	40
Ho	4,040.84	30	150
Au	4,040.94	40	50
N	4,041.32	(20h)	—
Mn	4,041.36	50	100
Os	4,041.92	6	100
K	4,042.59	(30)	—
Ne	4,042.64	(50)	—
Ar	4,042.91	(80)	—
La	4,042.911	300	400
Cs	4,043.42	(20)	—
Cu	4,043.50	25	—
J	4,043.88	(20)	—
K	4,044.14	400	800
U	4,044.416	25	18
Ar	4,044.418	(1200)	—
P	4,044.49	(150w)	—
Fe	4,044.61	35	70
Kr	4,044.67	(80)	—
Pr	4,044.82	35	50

	λ	I J	I O		λ	I J	I O
Sm	*3,983.14*	*60*	*100*				
Gd	*3,983.03*	*40*	*40*				
Mn	*3,982.91*	*30*	*20*				
O	*3,982.72*	*(20)*	*—*				
Yt	*3,982.59*	*100*	*60*				
Mn	*3,982.58*	*30*	*20*				
Ti	*3,982.48*	*30*	*80*				
As	*3,982.45*	*25*	*—*				
Mn	*3,982.17*	*25*	*12*				
Pr	*3,982.06*	*100*	*125*				
Dy	*3,981.94*	*100*	*150*				
Tb	*3,981.88*	*200*	*80*				
Fe	*3,981.77*	*100*	*150*				
Ti	*3,981.76*	*70*	*100*				
Rn	*3,981.68*	*(150)*	*—*				
Ag	*3,981.64*	*20*	*30*				
Cr	*3,981.23*	*50*	*100*				
Th	*3,981.11*	*20*	*20*				
Ta	*3,981.01*	*40*	*2*				
V	*3,980.52*	*35*	*40*				
Br	*3,980.39*	*(25)*	*—*				

Ce 4,186.599 *J* 25 *O* 80

U	**4,186.48**	**5**	**6**	**Tb**	**4,186.60**	**—**	**2**
Eu	**4,186.42**	**—**	**6**	**Zr**	**4,186.69**	**3**	**3**
Pr	**4,186.39**	**3**	**12**	**Er**	**4,186.71**	**2**	**8**
Cr	**4,186.36**	**10**	**50**	**Zr**	**4,186.78**	**—**	**3**
Sb	**4,186.34**	**4**	**—**	**U**	**4,186.79**	**5**	**6**
Nd	**4,186.311**	**3**	**25**	**Dy**	**4,186.81**	**12**	**100*w***
Tm	**4,186.31**	**8**	**5**	**Ho**	**4,186.84**	**3**	**3**
Mo	**4,186.281**	**12**	**15**	**Ce**	**4,186.86**	**—**	**2**
Ir	**4,186.280**	**—**	**2**	**Yb**	**4,186.90**	**10*h***	**—**
Tb	**4,186.24**	**1*h***	**10**	**U**	**4,186.98**	**4**	**10**
K	**4,186.23**	**(60)**	**—**	**Fe**	**4,187.04**	**200**	**250**
Ti	**4,186.12**	**40**	**100**	**Tb**	**4,187.16**	**—**	**15**
Nb	**4,186.10**	**8**	**5**	**Co**	**4,187.25**	**3**	**50**
U	**4,186.04**	**1**	**6**	**La**	**4,187.316**	**40**	**50**
Nd	**4,186.03**	**4**	**8**	**Ce**	**4,187.323**	**35**	**15**
W	**4,186.02**	**2**	**12**	**Zr**	**4,187.47**	**—**	**4**
S	**4,185.95**	**(15)**	**—**	**Zr**	**4,187.56**	**3**	**9**
Tb	**4,185.89**	**—**	**8**	**Fe**	**4,187.59**	**1**	**3**
Mo	**4,185.825**	**40**	**40**	**Mo**	**4,187.61**	**3**	**3**
Pr	**4,185.82**	**5**	**15**	**Tm**	**4,187.62**	**30**	**300**
U	**4,185.78**	**6**	**6**				
Nd	**4,185.77**	**8*s***	**15**	Fe	4,187.80	150	200
Er	**4,185.724**	**—**	**15*w***	Rn	4,187.81	(35)	—

	λ	I	
		J	O
Re	**4,185.722**	—	*2h*
Nb	**4,185.67**	*2h*	**2**
Ir	**4,185.66**	—	**25**
Cl	**4,185.61**	**(20)**	—
Nb	**4,185.54**	**30**	**1**
O	4,185.45	(150)	—
Kr	4,185.12	(50)	—
Fe	4,184.89	80	100
Kr	4,184.47	(20)	—
Nb	4,184.44	50	20
Ti	4,184.33	20	8
Gd	4,184.264	150	150
Se	4,184.26	(25)	—
Lu	4,184.25	200	100
Te	4,183.99	(30)	—
Sm	4,183.76	15	10
V	4,183.43	20	3
Zr	4,183.32	1	40
Ir	4,183.21	4	40
Re	4,182.98	—	150*r*
Ir	4,182.47	6	50
Ru	4,182.46	30	20
Fe	4,182.39	30	80
Ar	4,181.88	(1000)	—
Fe	4,181.76	150	200
Ta	4,181.15	25	40
Sm	4,181.10	25	8
Mo	4,181.05	15	25
Hg	4,180.95	(100)	—
In	4,180.94	15	—
Se	4,180.94	(800)	—
Ti	4,180.87	20*h*	100
Yb	4,180.83	100	10
Xe	4,180.10	(500*h*)	—
Ac	4,179.93	60	—
Nb	4,179.75	20	10
Br	4,179.64	(40)	—
Kr	4,179.58	(20*wh*)	—
Pr	4,179.42	40	200
Ar	4,179.31	(20)	—
Cr	4,179.26	40	100
Te	4,179.24	(70)	—
Ge	4,179.04	25*wh*	—
Tb	4,178.97	2*h*	50*d*
Ar	4,178.39	(20)	—
P	4,178.36	(300*w*)	—
Mo	4,178.27	20	25
P	4,188.07	(30*h*)	—
Sm	4,188.12	25	10
Mo	4,188.32	80	100
P	4,189.08	(30)	—
Pr	4,189.52	50	100
S	4,189.71	(250)	—
O	4,189.79	(500)	—
Mn	4,189.99	40	80
Mo	4,190.00	15	20
Cr	4,190.13	15	40
Gd	4,190.15	—	100*w*
Yb	4,190.30	30	7
Ar	4,190.71	(600)	—
Gd	4,190.78	40	100
Nb	4,190.88	30	20
Ar	4,191.028	(1200)	—
Mo	4,191.03	30	—
Gd	4,191.08	—	100
Cr	4,191.27	15	70
Fe	4,191.44	100	200
Cl	4,191.59	(15)	—
Pr	4,191.61	25	40
Gd	4,191.62	15	40
Dy	4,191.63	2	40
Cr	4,191.75	6	50
Sm	4,191.92	15	8
Nb	4,192.07	20	20
Cr	4,192.10	15	40
Mo	4,192.29	40	—
La	4,192.36	50	50
Pt	4,192.43	2	100
O	4,192.53	(15)	—
Xe	4,193.01	(20)	—
Rb	4,193.10	40	—
Xe	4,193.15	(200*h*)	—
Se	3,193.32	(20)	—
Br	4,193.46	(25)	—
S	4,193.51	(15)	—
Xe	4,193.53	(150)	—
Cr	4,193.66	25	100
Nb	4,193.80	15	—
Ra	4,194.09	(80)	—
Ho	4,194.34	15	30
La	4,194.35	15	15
Se	4,194.55	(50)	—
Mo	4,194.56	30	50
Re	4,194.67	—	40
Dy	4,194.83	12	50

	λ	I J	I O
Hg	4,178.02	(50h)	—
Ta	4,177.919	15	20
Mo	4,177.917	15h	—
Cr	4,177.90	1h	40h
Cu	4,177.75	1	60
Fe	4,177.60	25	100
Yt	4,177.55	50	50
Nd	4,177.32	25	15
Mo	4,177.26	20	20
Ta	4,176.99	40	15
Mn	4,176.60	40	100
Fe	4,176.57	50	100
Ar	4,176.33	(20)	—
Br	4,175.79	(50)	—
Fe	4,175.64	80	100
Os	4,175.63	4	100
W	4,175.59	25	7
Ne	4,175.49	(40)	—
Se	4,175.32	(800)	—
Ne	4,175.22	(60)	—
Ta	4,175.21	40h	100
Fe	4,174.92	25	100
Cr	4,174.79	40	100
Ne	4,174.37	(70)	—
S	4,174.30	(150)	—
Yt	4,174.13	8	100
S	4,174.04	(50)	—
J	4,173.79	(30)	—
Gd	4,173.56	30	30
Ti	4,173.55	40	12
Os	4,173.23	6	100
Ir	4,172.56	12	150
Pr	4,172.27	40	75
Fe	4,172.13	50	80
Ga	4,172.06	100R	2000R
Ti	4,171.903	70	15
Mo	4,171.90	25	—
Pr	4,171.82	40	75
U	4,171.59	30	30
Fe	4,170.91	40	80
J	4,170.47	(25)	—
Gd	4,170.11	50	50
Ru	4,170.05	25	20
Pd	4,169.842	50	200
Cr	4,169.838	25	80
Te	4,169.77	(100)	—
Sm	4,169.48	25	15
O	4,169.23	(50)	—
Cr	4,194.95	25	70
Nb	4,195.09	20	20
Cl	4,195.11	(18)	—
Os	4,195.14	1	100s
Sb	4,195.17	50	—
Fe	4,195.34	100	150
Se	4,195.51	(100)	—
Fe	4,196.21	50	100
Se	4,196.24	(20)	—
Ne	4,196.41	(15)	—
Rh	4,196.50	50	100
La	4,196.55	150	200
Ru	4,196.87	50	60
Cr	4,197.23	25	70
Ru	4,197.58	100	100
As	4,197.61	30	—
Gd	4,197.70	50	40
Se	4,198.05	(40)	—
Ne	4,198.10	(70)	—
Fe	4,198.31	150	250
Ar	4,198.32	(1200)	—
Cr	4,198.52	30	100
Ru	4,198.87	100	60
Mo	4,198.91	25	1
Fe	4,199.10	200	300
Ru	4,199.90	300	150
Tm	4,199.92	20	100
Cr	4,200.10	8	80
Ar	4,200.67	(1200)	—
Fe	4,200.93	20	80
Kr	4,201.42	(30wh)	—
Rb	4,201.85	500	2000R
Ar	4,201.99	(20)	—
Fe	2,402,03	300	400
Os	4,202.06	4	100
P	4,202.24	(30h)	—
Br	4,202.50	(25)	—
Rn	4,203.23	(200)	—
Cr	4,203.59	20	100
Xe	4,203.69	(50)	—
Tm	4,203.73	25	250
Fe	4,203.99	120	200
La	4,204.04	25	200
Cr	4,204.47	30	80
Eu	4,205.05	50	200R
In	4,205.08	50)	—
In	4,205.15	(30)	—
Ta	4,205.88	30	100

	λ	I J	I O		λ	I J	I O
S	*4,168.41*	*(50)*	—	*Ru*	*4,206.02*	*40*	*100*
Ac	*4,168.40*	*100*	—	*Hg*	*4,206.10*	*(30)*	—
Nb	*4,168.13*	*80*	*100*	*Fe*	*4,206.70*	*25*	*125*
Ru	*4,167.51*	*150*	*100*	*Pr*	*4,206.74*	*50*	*50*
Ru	*4,166.88*	*25*	*20*	*Cr*	*4,206.90*	*25*	*80*
				Fe	*4,207.13*	*40*	*80*
				F	*4,207.16*	*(50)*	—
				F	*4,207.44*	*(30)*	—

Ce 5,522.99 *O* 100

	λ	J	O		λ	J	O
Pr	**5,522.80**	**1**	**15*w***	**Er**	**5,523.16**	—	**8**
Fe	**5,522.461**	—	**8**	**Co**	**5,523.19**	—	**300*w***
Ce	**5,552.459**	—	**15**	**Sm**	**5,523.31**	—	**2**
Tb	**5,522.30**	—	**15**	**Re**	**5,523.39**	—	**10*w***
Nd	**5,522.16**	—	**10**	**Tb**	**5,523.44**	—	**10**
Nb	**5,521.88**	**1**	**5*h***	**Os**	**5,523.53**	—	**100**
Sr	**5,521.83**	**10**	**50**	**Nb**	**5,523.57**	**10**	**30**
Th	**5,521.77**	—	**4**	**Zr**	**5,523.87**	—	**2**
Gd	**5,521.71**	—	**8**	**Nd**	**5,523.94**	—	**5**
Yt	**5,521.70**	**40**	**4**	**Ta**	**5,523.98**	—	**2*s***
Pt	**5,521.68**	—	**4**	**Tb**	**5,524.12**	—	**40**
Sm	**5,521.64**	—	**3**	**Pr**	**5,524.14**	—	**10*W***
Yt	**5,521.63**	—	**5**	**Th**	**5,524.21**	—	**8**
Mo	**5,521.17**	**2**	**3**	**Hf**	**5,524.35**	**50**	**40**
Ta	**5,521.15**	—	**4**	**Yb**	**5,524.55**	—	**10**
Re	**5,521.10**	—	**20**	**Gd**	**5,524.57**	—	**10**
U	**5,521.04**	—	**6**	**Th**	**5,524.61**	—	**8*w***
W	**5,521.01**	—	**10**	**Co**	**5,524.98**	—	**25*w***
Sc	5,520,50	—	80	Eu	5,526.62	—	60
Ba	5,519.11	60	200*wh*	Sc	5,526.81	300*wh*	100
Ta	5,518.91	—	100*W*	Yt	5,527.54	15	100
Mn	5,516.77	—	50	Mg	5,528.46	30	60
Tb	5,516.24	—	50	Co	5,530.77	—	500
Sm	5,516.14	—	200	Re	5,532.66	—	100
W	5,514.70	8	50*w*	Mo	5,533.05	100	200
Fe	5,514.63	10	50	Rh	5,535.04	1	80
Ti	5,514.542	15	80	Fe	5,535.41	—	50
Tb	5,514.54	—	50	Ba	5,535.55	200*R*	1000*R*
Ti	5,514.35	10	70	La	5,535.67	100	50
Sc	5,514.21	—	60	Sm	5,537.07	—	50
Ti	5,512.53	12	125	Fe	5,538.57	—	50
Sm	5,512.10	—	80	Yb	5,539.06	5	200
Ce	5,512.08	—	50*s*	La	5,541.26	—	50*W*
Ru	5,510.71	—	100	Eu	5,542.54	—	80
Eu	5,510.55	—	300*s*	Pd	5,542.80	2	100

	λ	I J	I O
La	5,510.34	—	200
Pr	5,509.15	2	50
V	5,507.75	60	60
Fe	5,506.78	10	150
Mo	5,506.49	100	200r
La	5,506.00	—	50
Sr	5,504.17	25	60
Ti	5,503.90	3	60
La	5,503.81	—	100
Fe	5,501.47	—	150
La	5,501.34	50	200
Fe	*5,597.52*	*5*	*150*
Eu	*5,595.17*	—	*125*
Co	*5,489.65*	—	*150w*
Eu	*5,488.65*	—	*500*
Co	*5,483.96*	—	*150w*
Co	*5,483.34*	—	*500w*
Sr	*5,480.84*	*30*	*100h*
Rh	5,544.58	1	50
Eu	*5,547.45*	—	*1000*
Sm	*5.550.40*	—	*125*
Bi	*5,552.35*	*100*	*500wh*
Fe	*5,554.89*	—	*100*
Yb	*5,556.48*	*50*	*1500*
Re	*5,563.21*	—	*150w*
Fe	*5,563.60*	*5*	*100*
Ho	*5,566.52*	—	*100*

Cl $^{17}_{35.453}$

t_0 −101.6°C

t_1 −34.6°C

I.	II.	III.	IV.	V.
12.959	23.799	39.905	54.452	67.8

λ	I	eV
II 3,329.12	(150)	20.05
II (3,353.39)	(125)	18.03
II 3,827.62	(150)	21.48
II 3,860.83	(150)	19.17
II 4,132.48	(200)	19.00
II 4,740.40	(150)	20.00
II 4,768.68	(150)	19.68
II 4,794.54	(250)	28.9
II 4,810.06	(200)	28.9
II 4,819.46	(200)	28.9

Cl 3,329.12 (150)

	λ	I J	I O
Tb	3,329.08	8	30
La	3,329.07	4	4
Fe	3,329.054	3	8
Cr	3,329.053	6	30
Fe	3,328.87	100	150
N	3,328.79	(15)	—
Eu	3,328.69	4	3
Mo	3,328.56	20	1
V	3,328.402	1	7
Cr	3,328.35	40	20
Ta	3,328.30	1*h*	10
Nd	3,328.270	6	15
U	3,328.270	2	2
Th	3,328.25	4	4
Hf	3,328.21	15	20
Zr	3,326.80	100	100
Ti	3,326.76	125	15
Ar	3,325.50	(100)	—
Fe	3,325.46	80	100
Fe	3,324.54	80	100
Fe	3,323.74	150	150
Rh	3,323.09	200	1000
Fe	3,323.07	100	—
Ti	3,322.94	300*R*	80
Fe	3,322.48	100	150
Ti	3,321.70	125	15
V	3,321.54	150	3
Mo	3,320.90	80	3
Nb	3,320.81	100	3
Ar	*3,319.24*	*(300)*	—
Cl	*3,316.86*	*(50)*	—
Cl	*3,315.44*	*(100)*	—
Fe	*3,314.74*	*200*	*200*
Cl	*3,312.78*	*(15)*	—
Cr	*3,310.65*	*200*	*2*

	λ	I J	I O
Os	3,329.13	5	20
Nb	3,329.16	8*h*	—
Ne	3,329.20	(12)	—
Mo	3,329.21	100	2
U	3,329.30	2	6
Gd	3,329.35	2	2
Nb	3,329.36	15	10
Cs	3,329.43	(10)	—
Ti	3,329.46	200*r*	80
vzduch	3,329.50	10	—
Fe	3,329.532	6	35
Eu	3,329.533	5	9
Ta	3,329.535	1	18
Nb	3,329.19	3	3
Sm	3,329.623	2	8
Cu	3,329.64	10	60
Er	3,329.59	3	25
Ho	3,329.66	44	—
Ag	3,329.83	2*h*	—
V	3,329.85	40	100
Rb	3,329.91	(5)	—
U	3,329.92	2	10*h*
Mg	3,329.93	8	80
Sr	3,329.99	10	100
Ru	3,330.00	(40)	—
Sn	3,330.59	100*h*	100*h*
Ta	3,331.01	200*w*	18
Gd	3,331.39	80	100
Ti	3,332.11	125	40
Mo	3,332.52	80	—
Fe	3,334.22	100*h*	150*h*
Ne	3,334.87	(250)	—
Ti	3,335.19	150	60
Fe	3,335.77	100	125
Cr	3,336.33	80	18
La	3,337.49	300*wh*	800
Fe	3,337.67	100	125
V	3,337.85	150	2
Cr	*3,339.80*	*150*	*25*
Ti	*3,341.87*	*300R*	*100*
La	*3,344.56*	*200wh*	*300*
Zn	*3,345.02*	*300*	*800*

Cl 3,353.39 (125)

	λ	I J	I O
Nb	3,353.36	3h	5w
Ce	3,353.33	30	10
Ru	3,353.31	15	30
Fe	3,353.27	5	10
Cr	3,353.13	50	15
Cr	3,353.03	30	20
Ce	3,353.99	2	10
W	3,352.95	12	10
Ti	3,352.94	2	25
Fe	3,352.93	3	5
Tb	3,352.89	8	30
Nb	3,352.87	2h	10
Ir	3,352.84	2h	—
Nb	3,352.83	5h	—
Co	3,352.80	30	—
W	3,352.77	12	4
Dy	3,352.70	5	20
U	3,352.68	2	4
Eu	3,352.592	2	2h
Nb	3,352.590	10	15
Sn	3,352.43	100h	—
W	3,352.39	10	3
Sn	3,351.97	(60)	—
Fe	3,351.74	60	80
Fe	3,351.52	60	70
Gd	3,350.48	180	150
Ti	3,349.41	400R	100
Nb	3,349.35	100	5
Nb	3,349.06	100	80
Ti	3,349.03	800R	125
Fe	3,347.93	100	150
Cr	3,347.84	125	35
Rb	3,347.00	(60)	—
Nb	3,346.75	80	5
Cr	3,346.74	80r	150R
Ti	3,346.73	60	60
V	3,345.90	125	—
Zn	3,345.57	100	500
Zn	3,345.02	300	800
La	3,344.56	200wh	300
Pt	3,343.90	80	100
Ti	3,343.77	70	60

	λ	I J	I O
Pr	3,353.49	1	6
Nb	3,353.50	20	10
Nd	3,353.52	6	10
W	3,353.55	7	8
Ho	3,353.57	8	6
Nd	3,353.592	3	5
Dy	3,353.595	5	35
U	3,353.60	14	2
Ne	3,353.63	(4)	—
Ru	3,353.65	4	50
Se	3,353.67	(20)	—
Nb	3,353.69	2	5h
Sc	3,353.73	60	50
Yb	3,353.74	8	5
W	3,353.74	8	9
V	3,353.77	100	2
Cs	3,353.88	(4)	—
Os	3,353.91	15	40
Rb	3,353.98	(30)	—
vzduch	3,352.05	12	—
Fe	3,354.06	40	40
Ca	3,354.14	2	1
Sm	3,354.182	5	10
Th	3,354.185	8	5
Sm	3,354.28	2	5
F	3,354.34	(3)	—
Zr	3,354.39	10	10
Fe	3,355.23	100	100
V	3,356.35	60	125
Cr	3,357.41	125	6
Nb	3,358.42	100	100
Cr	3,358.50	200	40
Gd	3,358.63	100	100
Cr	3,360.29	200	50
Ti	3,361.21	600R	100
V	3,361.51	200	—
Cr	3,361.77	100	10
Nb	3,362.17	100	—
Gd	3,362.24	180	150
Tm	3,362.61	200	250
Xe	*3,366.72*	*(150h)*	—
Cr	*3,368.05*	*125*	*35*
Fe	*3,369.55*	*200*	*300*
Ne	*3,369.81*	*(500)*	—
Ne	*3,369.91*	*(700)*	—

	λ	I J	I O		λ	I J	I O
				Fe	*3,370.79*	*200*	*300*
				Se	*3,372.15*	*150*	*7*
				Rh	*3,372.25*	*200*	*300*
				Nb	*3,372.56*	*200*	*10h*
				Ti	*3,372.80*	*400R*	*80*
				Pd	*3,373.00*	*500wh*	*800r*
				Ar	*3,373.48*	*(300)*	—

Cl 3,827.62 (150)

	λ	J	O		λ	J	O
Ce	**3,827.60**	**3**	**3**	**Ti**	**3,827.67**	**1**	**5**
Fe	**3,827.57**	**2**	**4**	**Fe**	**3,827.82**	**200**	**200**
Ru	**3,827.55**	**1***wh*	**4**	**Ce**	**3,827.85**	**1**	**8**
U	**3,827.48**	**1**	**4**	**Pb**	**3,827.89**	**2**	—
P	**3,827.44**	**(150)**	—	**Nd**	**3,827.99**	**8**	**6**
Cd	**3,827.41**	**5**	—	**Ti**	**3,828.02**	**1**	**6**
Ce	**3,827.37**	**10**	**8**	**Sm**	**3,828.05**	**8**	—
Gd	**3,827.365**	**10**	**10**	**U**	**3,828.06**	**8**	—
Sm	**3,827.356**	**8**	**10**	**Th**	**3,828.14**	**8**	**8**
W	**3,827.35**	**3**	**4**	**Nd**	**3,828.17**	**40**	**50**
Eu	**3,827.29**	**5***W*	**4***W*	**Sc**	**3,828.179**	**30**	**30**
Ce	**3,827.21**	**1**	**6**	**Er**	**3,828.18**	**3***w*	**15***w*
Rb	**3,827.20**	**(40)**	—	**Dy**	**3,828.19**	**3**	**10**
Pb	**3,827.20**	**(20)**	—	**Ti**	**3,828.193**	**15**	**35**
Mo	**3,827.16**	**25**	**25**	**Nb**	**3,828.24**	**15**	**5**
Os	**3,827.14**	**20**	**50**	**Th**	**3,828.388**	**5**	**5**
Pd	**3,827.13**	**25***wh*	**4**	**Ce**	**3,828.389**	**2**	**3**
Nb	**3,827.01**	**5**	**5**	**Sm**	**3,828.41**	**2**	**8**
V	**3,826.969**	**25**	**1**	**Rh**	**3,828.48**	**60**	**100**
Ti	**3,826.969**	**2**	**12**	**Br**	**3,828.49**	**(6)**	—
Ir	**3,826.96**	**1***h*	**8**	**Fe**	**3,828.50**	**1**	**2**
Th	**3,826.95**	**8**	**10**	**Br**	**3,828.55**	**(12)**	—
W	**3,826.93**	**4**	**5**	**V**	**3,828.56**	**5**	**20**
Nd	**3,826.91**	**25***w*	**40***w*	**Ru**	**3,828.71**	**8**	**30**
Cu	**3,826.908**	**2***h*	—	**As**	**3,828.74**	**10**	—
Xe	**3,826.86**	**(15)**	—	**U**	**3,828.81**	**5**	**2**
Ta	**3,826.849**	**3**	**25**	**V**	**3,828.836**	**3**	**15**
Fe	**3,826.846**	**2**	**6**	**Nd**	**3,828.842**	**20**	**20**
Ar	**3,826.83**	**(15)**	—	**Mo**	**3,828.87**	**30**	**40**
Er	**3,826.82**	**2***d*	**12***d*	**Eu**	**3,828.93**	**8***wh*	**6***wh*
V	**3,826.77**	**3**	**12**	**Ta**	**3,828.95**	**10**	**25**
Pr	**3,826.708**	**4**	—				
Rb	**3,826.708**	**(15)**	—				
Ce	**3,826.708**	**6**	**4**	Mg	3,829.35	150	100*w*
Mo	**3,826.69**	**25**	**30**	N	3,830.39	(150)	—
				Nb	3,831.84	300	5
Eu	**3,826.68**	**10***W*	**15***W*	Pd	3,832.29	150	150
Os	**3,826.63**	**12**	**30**	Mg	3,832.31	200	250

	λ	I J	I O
Hg	3,826.61	(30)	—
Sm	3,826.56	3	2
Cr	3,826.425	20	40
Nd	3,826.416	20	10
Pr	3,826.29	50d	80d
Xe	3,826.27	(2h)	—
Fe	3,825.88	400	500
Fe	3,824.44	100	150
Mn	3,823.51	75	75h
O	3,823.47	(125)	—
Rh	3,822.26	100	100
Fe	3,821.18	100	100
Fe	3,820.43	600	800
Cl	3,820.25	(100)	—
Eu	3,819.66	500wd	500wd
Nb	3,818.86	300	8
Pr	3,818.28	100	125
Fe	3,815.84	700	700
Bi	3,815.80	(300)	—
V	3,815.39	150h	1
Ra	3,814.42	(2000)	—
Fe	*3,812.96*	*300*	*400*
Gd	*3,812.02*	*200h*	*200R*
Bi	*3,811.14*	*(150)*	—
Mn	*3,809.59*	*150*	*150*
Fe	*3,806.70*	*150*	*200*
Hg	*3,806.38*	*(200)*	—
Fe	*3,805.34*	*300*	*400*
Au	*3,804.00*	*150*	*25*
Sn	*3,801.00*	*150h*	*200h*

	λ	I J	I O
Yt	3,832.89	80	30
Ta	3,833.74	200	40
Mn	3,833.86	75	75
Fe	3,834.22	400	400
Mn	3,834.36	75	75r
Ar	3,834.68	(800)	—
Nd	3,836.54	100	80
Mo	3,837.29	100w	—
Mg	3,838.26	200	300
Fe	3,839.26	75	100
Mn	3,839.78	125	100
Fe	3,840.44	300	400
Fe	3,841.05	400	500
Cr	3,841.28	80	150
F	*3,847.09*	*(800)*	—
Tm	*3,848.02*	*250*	*400*
Tb	*3,848.75*	*200*	*200*
La	*3,849.01*	*150*	*200*
Fe	*3,849.97*	*400*	*500*
F	*3,849.99*	*(600)*	—
Ar	*3,850.57*	*(400)*	—
Pr	*3,851.62*	*150w*	*200w*
F	*3,851.67*	*(200)*	—

Cl 3,860.83 (150)

	λ	I J	I O
Rb	3,860.80	(5)	—
In	3,860.73	(5)	—
Ru	3,860.723	4	20
Eu	3,860.72	6	5d
S	3,860.64	(15)	—
U	3,860.630	15h	1
Ce	3,860.626	15	8
Sm	3,860.62	10	25
Cu	3,860.46	7	30
Ce	3,860.40	3	6
La	3,860.31	2	—
Sm	3,860.28	5	4

	λ	I J	I O
Nb	3,860.86	10	5
Hf	3,860.910	6	6
Fe	3,860.915	2	1
Nd	3,860.94	4	4
Cl	3,860.99	(100)	—
Ce	3,860.993	2	2
Sm	3,861.059	4	15d
W	3,861.062	10	10
Ta	3,861.077	1	3h
Hg	3,861.08	(10h)	—
Ti	3,861.084	3	10h
Co	3,861.16	15	300R

	λ	I	
		J	O
Ce	3,860.18	4	2
S	3,860.15	(8)	—
Sm	3,860.14	1	5
W	3,859.98	30	15
Ce	3,859.94	5	3
Fe	3,859.91	600	1000*r*
Th	3,859.83	10	5
Ta	3,859.80	2	10
Ru	3,859.71	15	6
Ca	3,859.675	5*h*	6
Nd	3,859.670	25	25
U	3,859.58	30	20
Nd	3,859.422	12	10
Pr	3,859.416	3	3
Au	3,859.37	3	—
V	3,859.34	8	20
Al	3,859.33	(10)	—
Fe	3,859.22	100	100
Fe	3,856.37	300	500
V	3,855.84	200	200
Pb	3,854.05	100	—
Fe	3,852.57	100	150
F	3,851.67	(200)	—
Pr	3,851.62	150*w*	200*w*
Cl	3,851.42	(75)	—
Cl	3,851.02	(100)	—
Fe	3,850.82	75	200
Ar	3,850.57	(400)	—
F	3,849.99	(600)	—
Fe	3,849.97	400	500
La	3,849.01	150	200
Th	3,848.75	200	100
Tm	3,848.02	250	400
F	3,847.09	(800)	—
Fe	3,846.80	100	125
Fe	3,846.41	75	50
Bi	3,846.03	100	—
Bi	3,845.80	(100)	—

	λ	I	
		J	O
Eu	3,861.18	30*w*	30*w*
Sm	3,861.183	7	15
Sn	3,861.20	2	—
Rb	3,861.20	(2)	—
W	3,861.24	8	8
Mo	3,861.29	4	—
Pr	3,861.31	5	15
Ir	3,861.32	2	3
Cl	3,861.34	(50)	—
Fe	3,861.342	50	80
K	3,861.41	(10)	—
Cs	3,861.49	(4)	—
Ce	3,861.58	4*h*	3*h*
Sm	3,861.59	8	15
Fe	3,861.60	2	3
V	3,861.61	3*h*	10*h*
Ho	3,861.68	20	40
*bh*C	3,861.7	—	—
U	3,861.73	6	8
Ti	3,861.74	3	10
Cu	3,861.75	2	50
Sm	3,861.79	6	10
Cl	3,861.88	(20)	—
Ce	3,861.93	1	3
Ir	3,861.94	10	15
Zr	3,861.95	2	4
Eu	3,861.96	1*h*	2
U	3,862.05	1	8
Sm	3,862.051	9	4
V	3,862.22	20	80
Sm	3,862.24	9	3
V	3,862.28	2	3
Bi	3,863.90	(100)	—
Mo	3,864.11	500*R*	1000*R*
Bi	3,864.20	150*h*	—
La	3,864.49	150	100
Nb	3,865.02	200*h*	—
Pr	3,865.46	125*r*	200*r*
Os	3,865.47	200	125
Fe	3,865.53	400	600
Fe	3,867.22	100	150
S	3,867.56	(150)	—
Fe	3,869.56	80	100
Dy	3,872.12	150	300
Fe	3,872.50	300	300
Co	3,873.11	80	500*R*
Fe	3,873.76	80	125

	λ	I J	I O
Co	3,873.95	80	400*R*
Tb	3,874.18	200	200
Kr	3,875.44	150*wh*	—
C	*3,876.19*	*125*	—
Fe	*3,878.02*	*300*	*400*
Fe	*3,878.57*	*300*	*300 R*
Nb	*3,879.35*	*300*	*5*
P	*3,885.17*	*(150)*	—
Fe	*3,886.28*	*400*	*600*
La	*3,886.37*	*200*	*400*
He	*3,888.65*	*(1000)*	—

*P**

Cl 4,132.48 (200)

	λ	I J	I O
Cr	**4,132.44**	**2**	—
Ba	**4,132.43**	**4**	**10**
Ce	**4,132.31**	**2*s***	**8**
Gd	**4,132.28**	**20**	**25**
Eu	**4,132.24**	**14**	**4**
Mo	**4,132.23**	**20**	**20**
Pr	**4,132.23**	**8**	**15**
W	**4,132.21**	**8**	**6**
Fe	**4,132.06**	**200**	**300**
V	**4,132.02**	**10**	**12**
Cs	**4,132.00**	**(10)**	—
Mo	**4,131.92**	**10**	**20**
Ce	**4,131.85**	**2**	**5**
Pr	**4,131.82**	**4**	**12**
U	**4,131.78**	**6**	**6*h***
La	**4,131.74**	**3**	**3**
Ar	**4,131.73**	**(80)**	—
Nb	**4,131.53**	**10**	**2**
Gd	**4,131.48**	**50**	**50**
Se	4,129.15	(200)	—
Rh	4,128.87	150	300
Se	4,126.57	(150)	—
Nb	4,123.81	125	200
La	4,123.23	500	500
Nb	4,119.28	200	2
O	4,119.22	(300)	—
Ru	*4,112.74*	*200*	*125*
Ca	**4,132.497**	**2**	—
La	**4,132.501**	**104**	—
Nd	**4,132.55**	**5**	**4**
Ce	**4,132.64**	**24**	**8**
Tm	**4,132.69**	**15**	**12**
Mo	**4,132.75**	**10**	**12**
Th	**4,132.751**	**12**	**10**
Se	**4,132.76**	**(200)**	—
C	**4,132.82**	**(100)**	—
Dy	**4,132.85**	**2**	**9**
Ir	**4,132.90**	**15**	—
Sc	**4,133.01**	**3**	**8**
Eu	**4,133.11**	**1*h***	**4**
Sm	**4,133.17**	**4**	**4**
U	**4,133.20**	**8**	**8**
J	**4,133.25**	**(15)**	—
La	**4,133.33**	**25**	**15**
Nd	**4,133.36**	**10**	**15**
Dy	**4,133.37**	**4**	**10**
Nb	**4,133.42**	**52*h***	**2**
Fe	4,134.68	100	150
Rh	4,135.27	150	300
Se	4,136.28	(100)	—
Kr	4,139.11	(100*wh*)	—
Hg	4,140.38	(200)	—
La	4,141.74	200	200
S	4,142.29	(150)	—
Fe	4,143.87	260	400
Ru	4,144.16	200	150
S	4,145.10	(250)	—
Kr	4,145.12	(250)	—

	λ	I			λ	I	
		J	O			J	O
				N	4,151.46	(1000)	—
				La	4,151.95	300	200

Cl 4,740.40 (150)

	λ	J	O		λ	J	O
Mo	4,740.36	5	5	Th	4,740.517	15	20
U	4,740.285	10	3	Eu	4,740.524	2	500
La	4,740.277	300	150	Nb	4,740.61	3	3
Ta	4,740.16	100	100R	Cl	4,740.68	(10)	—
Ra	4,740.07	(4)	—	Dy	4,740.93	2	3
Eu	4,739.96	2	2	Se	4,740.97	(600)	—
La	4,739.79	8	4	Sc	4,741.02	60h	100
Cs	4,739.66	(20)	—	U	4,741.28	2	1
P	4,739.49	(30)	—	Yt	4,741.40	3	2
Cl	4,739.42	(10)	—	W	4,741.52	2	12
Rh	4,739.22	3	15	Fe	4,741.53	1	12
Mn	4,739.11	15	150	Dy	4,741.54	2	3
Se	4,739.03	(800)	—	O	4,741.71	(20)	—
Kr	4,739.00	(3000)	—	Cd	4,741.78	3	—
Cr	4,737.35	80	200	Se	4,742.25	(500)	—
Ar	4,735.93	(400)	—	Br	4,742.70	(200)	—
Gd	4,735.76	150	150	La	4,743.08	300	300r
Fe	4,734.15	(600)	—	Gd	4,743.65	300	300
Gd	4,732.61	300	300	Ar	4,746.82	(80)	—
Mo	4,731.44	100	100	La	4,748.73	200	100
As	4,730.92	125	—	Ne	4,749.57	(300)	—
Se	4,730.78	(1000)	—	Co	4,749.68	1004	500
Gd	4,728.47	100	150	Kr	4,752.02	(100h)	—
La	4,728.42	300	400r	Br	4,752.27	(100)	—
P	4,727.46	(100)	—	Th	4,752.52	80	100
Ar	4,726.91	(200)	—	Ne	4,752.73	(1000)	—
Gd	4,726.74	100	40	Ar	4,752.94	(150)	—
Yb	4,726.07	200	45	Ne	4,754.44	(100)	—
Cr	4,723.73	200	100	Ne	4,758.73	150	—
Zn	4,722.16	300h	400x	Kr	4,762.43	(300)	—
Rn	4,721.76	(150)	—	Se	4,763.65	(800)	—
La	4,719.95	300	200r	Ar	4,764.89	(150)	—
Cr	4,718.43	150	200	Kr	4,765.74	(1000)	—
La	4,716.44	200	100	Te	4,766.03	(150)	—
S	4,716.23	(600)	—	Br	4,767.10	(200)	—
Ne	4,715.34	(1500)	—	Ar	4,768.67	(150)	—
La	4,712.93	150	100				
Ne	4,712.06	(1000)	—				

Cl 4,768.68 (150)

	λ	I J	I O		λ	I J	I O
Ar	**4,768.67**	**(150)**	—	**Ta**	**4,768.98**	**5**	**150**
Rh	**4,768.59**	**(100)**	—	**Xe**	**4,769.05**	**(100)**	—
Cs	**4,768.41**	**(10)**	—	**U**	**4,769.26**	**15**	**6**
J	**4,768.16**	**(20)**	—	**Dy**	**4,769.63**	**2**	**4**
Co	**4,768.08**	**10**	**300**	**Te**	**4,769.73**	**(30)**	—
Cr	**4,767.86**	**8**	**100**	**Ti**	**4,769.77**	**2**	**12**
W	**4,767.78**	**1**	**12**	**C**	**4,770.00**	**(10)**	—
Gd	**4,767.25**	**25**	**100**				
Br	4,767.10	(200)	—	Rh	4,776.41	(100)	—
Te	4,766.03	(150)	—	Br	4,776.42	(200)	—
Kr	4,765.74	(1000)	—	Ti	4,779.95	100*h*	10
Ar	4,764.89	(150)	—	Co	4,780.01	500	500*x*
Se	4,763.65	(800)	—	Br	4,780.31	(125)	—
J	4,763.38	(80)	—	Ta	4,780.93	200	50
Kr	4,762.43	(300)	—	Cl	4,781.32	(75)	—
Hg	4,762.22	(100)	—				
Mo	4,760.19	125	125	*Lu*	*4,785.42*	*200*	*100*
Ne	4,758.73	(150)	—	*Br*	*4,785.50*	*(400)*	—
Cr	4,756.11	100	300	*Yb*	*4,786.60*	*200*	*50*
Ne	4,754.44	(100)	—	*Ne*	*4,788.93*	*(300)*	—
				Fe	*4,791.25*	*200R*	*200*
				Xe	*4,792.62*	*(150)*	—

*P**

Cl 4,794.54 (250)

	λ	I J	I O		λ	I J	I O
Co	**4,794.48**	**2**	—	**Mo**	**4,794.60**	**10**	**12**
Os	**4,793.99**	**6**	**300**	**V**	**4,795.10**	**4**	**5**
Mo	**4,793.82**	**15**	**15**	**Br**	**4,795.23**	**(5)**	—
Cu	**4,793.80**	**2*wh***	**5*wh***	**Mo**	**4,795.37**	**3**	**4**
N	**4,793.66**	**(5)**	—	**Xe**	**4,795.40**	**(2*h*)**	—
Mo	**4,793.41**	**30**	**30**	**Er**	**4,795.50**	**1**	**9**
Rn	**4,793.00**	**(10)**	—	**Ne**	**4,795.62**	**(15)**	—
				Ir	**4,795.67**	**2*wh***	**2*wh***
				Hf	**4,795.96**	**2**	**3**
Xe	4,792.62	(150)	—	**Te**	**4,796.10**	**(70)**	—
Fe	4,791.25	200*R*	200				
Ne	4,788.93	(300)	—				
Yb	4,786.60	200	50	Hg	4,797.01	(300)	—
Br	4,785.50	(400)	—	Cd	4,799.92	300*w*	300*w*
Lu	4,785.42	200	100	Gd	4,801.08	200	200
Ta	4,780.93	200	50	La	4,804.04	150	150
Br	4,780.31	(125)	—	Ti	4,805.10	125	15
Co	4,780.01	500	500*w*	Ar	4,806.07	(500)	—
				Xe	4,807.02	(500)	—
				La	4,809.01	150	150

*P** *N**

Cl 4,810.06 (200)

	λ	I J	I O
W	4,809.65	1	2
Se	4,809.61	(4)	—
Ne	4,809.50	(10)	—
Nb	4,809.37	8	5
Eu	4,809.29	2	10
Hf	4,809.18	10	8
Cl	4,809.05	(9)	—
La	4,809.01	150	150
Eu	4,808.62	2	1
Ti	4,808.53	2	12
Yb	4,808.51	4*h*	—
Xe	4,807.02	(500)	—
Ar	4,806.07	(500)	—
Ti	4,805.10	125	15
La	4,804.04	150	150
Gd	4,801.08	200	200
Cd	4,799.92	300*x*	300*x*
Hg	4,797.01	(300)	—
Ne	4,810.063	(150)	—
N	4,810.29	(5)	—
Rb	4,810.49	25	15
Nd	4,810.509	2*wh*	2
Kr	4,810.51	(3)	—
Zn	4,810.53	300*h*	400*w*
Nb	4,810.60	10	100
Ne	4,810.63	(100)	—
Mo	4,811.06	50	50
Ti	4,811.08	1	12
Hf	4,811.14	2	6
V	4,811.14	4	—
Nb	4,811.30	2	15
Nd	4,811.34	60	60
Ce	4,811.57	(12)	—
Au	4,811.62	15	50
Kr	4,811.76	(300)	—
Ge	4,814.80	200	—
S	4,815.51	(800)	—
Br	4,816.71	(300)	—
Rn	4,817.15	(100)	—
Ne	4,817.64	(300)	—
Xe	4,818.02	(100)	—

*P** *N**

Cl 4,819.46 (200)

	λ	I J	I O
P	4,819.34	(15)	—
Mo	4,819.25	60	80
Dy	4,819.04	2	7
V	4,819.05	2	3
Ti	4,819.03	2	40
Hf	4,818.87	4	25
Ne	4,818.79	(150)	—
Th	4,818.64	5	5
Cl	4,818.55	(4)	—
Br	4,818.41	(5)	—
Yb	4,818.38	20	3
Xe	4,818.02	(100)	—
U	4,819.54	12	12
Si	4,819.57	8	—
S	4,819.60	(25)	—
Yt	4,819.64	4	4
Cl	4,819.79	(25)	—
Se	4,819.80	(25)	—
Ne	4,819.94	(70)	—
Yb	4,820.24	60	15
Er	4,820.34	2	25
Ho	4,820.36	1	2
Ti	4,820.41	30	125
Eu	4,820.49	2	3
Ga	4,820.59	2	—
Th	4,820.90	2	3
Fe	4,821.05	200*h*	200*h*
Ne	4,821.92	(300)	—

*P**

	λ	I	
		J	O
Ne	4,823.17	(100)	—
Xe	4,823.41	(150*h*)	—
La	4,824.07	150	150
Kr	4,825.18	(300)	—
Ra	4,825.91	(800)	—
Ne	4,827.59	(300)	—
K	4,829.21	(100)	—
Xe	4,829.71	(400)	—
Mo	4,830.51	100	125
Te	4,831.29	(800)	—
Kr	4,832.07	(800)	—
Ne	*4,837.31*	*(500)*	—
Se	*4,840.63*	*(800)*	—
Xe	*4,843.29*	*(300)*	—
Xe	*4,844.33*	*(1000)*	—
Se	*4,844.96*	*(800)*	—
Kr	*4,846.60*	*(700)*	—

Co $^{27}_{58.9332}$

t_0 1,480°C t_1 3,000°C

I.	II.	III.	IV.	V.
7.876	17.4	—	—	—

λ	I		eV
	J	*O*	
II 2,286.16	300	40	5.84
II 2,378.62	50	25	5.62
II 2,519.82	200	40	6.24
I 3,405.12	150	2000*R*	4.1
I 3,412.34	100	1000*R*	4.2
I 3,453.50	200	3000*R*	4.0
I 3,465.80	25	2000R	3.6
I 3,502.28	20	2000*R*	4.0
I 3,529.81	30	1000*R*	4.0
I 3,995.31	20	1000*R*	4.0
I 4,118.77	—	1000*R*	4.1
I 4,121.32	25	1000*R*	3.9
I 4,530.96	8	1000	5.66
I 4,581.60	10	1000	5.66
I 4,780.01	500	500	5.87
I 4,813.48	6	1000	5.79

Co 2,286.16 *J* 300 *O* 40

	λ	*J*	*O*		λ	*J*	*O*
Fe	**2,286.15**	**2**	—	**Ti**	**2,286.18**	**6**	**2**
V	**2,286.11**	**2*W***	—	**Pt**	**2,286.19**	**3*W***	—
Tl	**2,285.95**	**2**	—	**W**	**2,286.29**	**5**	**2**
Xe	**2,285.94**	**(4)**	—	**Nb**	**2,286.35**	**4*W***	—
Nd	**2,286.86**	**3**	—	**Mo**	**2,286.42**	**9**	**2*h***
				Ag	**2,286.46**	**12**	—
Rh	2,284.08	200	—				
Co	2,283.52	15	10	Ni	2,287.08	500	100
				Cd	2,288.02	300*R*	1.500*R*
Co	*2,281.88*	*25*	*3*				
				Rh	*2,290.03*	*500*	*25*

Co 2,378.62 *J* 50 *O* 25

	λ	*J*	*O*		λ	*J*	*O*
W	**2,378.60**	**12**	**5**	**Mo**	**2,378.65**	**3**	**6**
U	**2,378.60**	**8**	—	**Pd**	**2,378.72**	**9*h***	—
Ca	**2,378.59**	**2**	—	**Os**	**2,378.74**	**5**	**25**
Os	**2,378.54**	**5*w***	**25**	**Hf**	**2,378.81**	**2*h***	—
Fe	**2,378.53**	**3**	—	**Er**	**2,378.85**	**6**	—
Ta	**2,378.52**	—	**2**	**Co**	**2,378.91**	—	**5**
Re	**2,378.42**	—	**4**	**Rh**	**2,378.93**	—	**8**
Al	**2,378.41**	**20**	**40**	**Nb**	**2,378.94**	—	**2*h***
Hg	**2,378.33**	**20**	**20**	**Fe**	**2,378.980**	**2**	**2**
Ta	**2,378.31**	**18**	**5**	**Pr**	**2,378.98**	**10*s***	—
Ir	**2,378.30**	—	**2**				
				Fe	2,379.27	15	12
				Ir	2,379.38	8	25
U	2,378.16	35	—	Os	2,379.39	9	40
In	2,378.147	—	15	La	2,379.42	60	—
Os	2,378.14	5	30	Os	2,379.64	5	15
J	2,377.99	(30)	—				
Ir	2,377.98	15	30	Tl	2,379.69	200*R*	100*R*
				Ni	2,379.72	2	15
Pd	2,377.92	30	—	Re	2,379.79	8	15
Rh	2,377.81	50	—	Os	2,379.84	8	25
Os	2,377.61	15	50	Ce	2,380.16	50	—
Re	2,377.38	6	20				
Pt	2,377.275	25	15	Re	2,380.24	5	15
				Hf	2,380.30	60	30
Ir	2,377.275	5*w*	25	Tl	2,380.34	60	20
Co	2,377.21	2	12	Mo	2,380.41	—	12
Mo	2,377.12	30	—	Rb	2,380.44	(125)	—
Os	2,377.03	30	50				
Fe	2,376.43	50*h*	3	Co	2,380.48	10	20*d*
				Zr	2,380.56	—	15
Nb	2,376.40	60	8	Fe	2,380.76	15	12
Cu	2,376.38	30	3	Os	2,380.82	20	30
Os	2,376.29	5	25	Re	2,380.90	5	12

	λ	I J	I O
Cu	2,376.27	25	—
Au	2,376.24	3	25
Ni	2,376.03	5	15
Re	2,375.83	7	18
Ru	2,375.63	80	50
Ir	2,375.58	—	12
Ni	2,375.42	30	10
Ru	2,375.27	5	80
Re	*2,375.07*	*7*	*25*
Ag	*2,375.06*	*300wh*	*300wh*
Os	*2,374.51*	—	*25*
Os	*2,374.33*	*3*	*25*
Cr	*2,373.73*	—	*60*
Sb	*2,373.62*	*25*	*75*
Mn	*2,273.37*	*50*	*4*
Al	*2,373.36*	*100R*	*200R*
Rb	*2,373.22*	*(70)*	—
Al	*2,373.13*	*30*	*100R*
V	*2,373.06*	*200*	—
Os	*2,372.91*	*3*	*25*
Cr	*2,372.89*	*3*	*40*
Ir	*2,372.77*	*40*	*100*
Nb	*2,372.72*	*50*	*2*
Ce	*2,372.37*	*50*	—
Mo	*2,372.27*	*6*	*25*
V	*2,372.17*	*100*	—
Pd	*2,372.15*	*50*	*5*
V	2,380.92	50	6
Hf	2,381.00	40	20
Pd	2,381.02	9	25
Ta	2,381.13	40	15
Re	2,381.14	7	40
As	2,381.18	4	75
Rb	2,381.30	(100)	—
Cr	2,381.48	25	3
Ta	2,381.52	40	8
Ir	2,381.62	2h	30
Lu	2,381.69	30h	—
Ir	2,381.82	50	8
Ru	2,381.99	150	50
bhC	2,382.0	—	30
Fe	2,382.04	100R	40r
Os	*2,382.46*	*5*	*30*
V	*2,382.47*	*100w*	—
Rh	*2,382.89*	*5*	*50*
V	*2,383.00*	*80*	*8*
Te	*2,383.25*	*300*	*500*
Rh	*2,383.40*	*10*	*50*
Pd	*2,383.40*	*50wh*	—
Re	*2,383.48*	*5*	*25*
Sb	*2,383.63*	*20*	*75*
Pt	*2,383.64*	*20*	*30*
Mn	*2,384.05*	*3*	*40*
Zr	*2,384.17*	*1h*	*25*
Os	*2,384.62*	*5*	*30*
Rh	*2,384.65*	—	*25*
Ir	*2,384.81*	*80*	*4*
Pd	*2,385.01*	*50wh*	—

Co 2,519.82 *J* 200 *O* 40

	λ	J	O
Ti	**2,519.81**	**7**	**3**
Tm	**2,519.80**	**50**	**1**
Os	**2,519.79**	**5**	**12**
Ta	**2,519.78**	—	**50**
Nb	**2,519.69**	**15w**	**2w**
Fe	**2,519.63**	**20**	**30**
V	**2,519.62**	**50**	**125r**
Cr	**2,519.51**	**6**	**150r**
W	**2,519.44**	**12**	**2**
Ru	2,519.208	80	20
Si	2,519.207	300	300
Nb	**2,519.84**	**3**	**1**
W	**2,519.88**	**2**	**15**
Ru	**2,519.95**	**1**	**12**
Zr	**2,519.98**	—	**6**
Re	**2,520.01**	—	**50**
Th	**2,520.13**	**2**	**5**
Sb	**2,520.18**	**5**	**20**
W	**2,520.19**	**3**	—
Rh	2,520.53	1000wh	10
Ti	2,520.54	4	40
Fe	2,520.88	—	80

	λ	I J	I O		λ	I J	I O
Os	2,518.44	5	20	Co	2,521.36	150	75
Fe	2,518.10	50	200R	Nb	2,521.40	100	3
Mo	2,517.83	—	20				
				Re	2,521.59	—	100R
Fe	2,517.66	12	20	Ru	2,521.61	1	60
Ru	2,517.62	—	50	Fe	2,522.85	50	300R
Rh	2,517.52	150	—				
Mo	2,517.46	—	25				
Ti	2,517.43	30	20	*Sn*	*2,523.91*	*60*	*60*
				Si	*2,524.12*	*400*	*400*
Tl	2,517.41	—	30R	*Ni*	*2,524.22*	—	*50R*
Ru	2,517.32	80	60	*Fe*	*2,524.29*	*50*	*100R*
V	2,517.14	25	35r	*Bi*	*2,524.49*	*25*	*100*
Cr	2,516.92	2	35				
Hf	2,516.88	100	35	*Mo*	*2,524.81*	—	*40h*
				Co	*2,524.96*	*700*	*50W*
Si	2,516.123	500	500	*Ni*	*2,525.39*	*300wh*	—
V	2,516.118	100	25	*Ta*	*2,526.02*	*80*	*100*
Re	2,516.116	—	125	*V*	*2,526.21*	*150R*	*150*
				Ta	*2,526.35*	—	*100*
Zn	*2,515.81*	*20*	*150W*	*Ta*	*2,526.45*	—	*150*
Rh	*2,515.74*	*10*	*60*	*Cu*	*2,526.59*	*200*	—
Bi	*2,515.69*	*25*	*100*	*Ta*	*2,526.66*	—	*50*
Pt	*2,515.58*	*20*	*500*	*Ru*	*2,526.83*	*20*	*50*
Ru	*2,515.28*	*2*	*60*				
				Fe	*2,527.43*	*50*	*200R*
Os	*2,515.04*	*5*	*40s*	*Mn*	*2,527.44*	*12*	*150*
Pt	*2,515.03*	*20*	*150*				
Pd	*2,514.48*	*200*	—				
Si	*2,514.33*	*200*	*300*				
Pt	*2,514.07*	*10h*	*150*				
Ru	*2,513.32*	*80*	*50*				
Os	*2,513.25*	*8*	*50*				
Os	*2,512.87*	*3*	*40*				
Ru	*2,512.81*	*2*	*80*				
Ta	*2,512.65*	—	*100*				
Ta	*2,512.04*	*800*	*5*				
C	*2,512.03*	*400*	—				

Co 3,405.12 *J* 150 *O* 2000*R*

Ti	**3,405.09**	**2**	**20**	**V**	**3,405.160**	**15**	**30**
Ag	**3,405.03**	**2*h***	**3**	**Kr**	**3,405.16**	**80*wh***	—
Dy	**3,404.99**	**4**	**4**	**Mo**	**3,405.20**	**5**	**8**
Ti	**3,404.97**	**2*d***	—	**Cr**	**3,405.22**	**1**	**12**
V	**3,404.96**	**1**	**8**	**Ru**	**3,405.277**	—	**3**
U	**3,404.93**	**2**	**3**	**W**	**3,405.277**	**6**	**7**
Au	**3,404.92**	**3*h***	—	**Bi**	**3,405.33**	**10**	**40**
Ce	**3,404.91**	**2**	**18**	**Nb**	**3,405.41**	**50**	**80**
Mo	**3,404.86**	**4**	**6**	**Ce**	**3,405.44**	—	**6**
Zr	**3,404.83**	**35**	**40**	**Th**	**3,405.56**	**4**	**3**

	λ	I	
		J	O
W	3,404.80	6	8
Ne	3,404.77	(12)	—
Sm	3,404.767	2	2
Nd	3,404.76	6	4
Fe	3,404.75	—	2
Re	3,404.72	—	100
Tb	3,404.71	8	3
Th	3,404.65	4	4
Pd	3,404.58	1000R	2000R
La	3,404.52	2	9
In	3,404.45	(10)	—
Ce	3,404.43	—	12
V	3,404.42	50h	—
Fe	3,404.36	50	100
Mo	3,404.34	25	20
P	3,404.33	(50)	—
Fe	3,404.304	25	25
In	3,404.297	(18)	—
K	3,404.24	(30)	—
Tb	3,404.24	3	10
W	3,404.22	7	8
Ta	3,404.16	1	5
W	3,404.14	4	—
In	3,404.131	(18)	—
Er	3,404.13	—	8
Ce	3,404.130	—	18
Cd	3,403.65	500h	800
Cr	3,403.32	200	30
Nb	3,403.01	80x	20x
Mo	3,402.81	100	5d
Ti	3,402.42	90	15
Cr	3,402.40	80	25
Fe	3,402.26	150	150
Fe	3,401.52	90	150
Hf	3,399.79	100	60
Fe	3,399.34	200	200
Rh	3,396.85	500	1000x
Nb	3,396.37	150	2
Cr	3,395.61	100	2
Ti	3,394.57	200	70
Cr	3,394.29	150	15
Ar	3,393.75	(250)	—
Fe	3,392.66	200	300
Zr	3,391.97	400	300
Cr	3,391.43	150	4
Hg	3,385.25	(200)	—

	λ	I	
		J	O
Fe	3,405.58	—	1
Ce	3,405.63	—	4
Bi	3,405.660	—	60
Dy	3,405.663	5	10
Eu	3,405.67	—	6w
Mo	3,405.68	3d	1d
Nd	3,405.681	—	4
U	3,405.75	—	12
Ce	3,405.81	—	10
Co	3,405.82	—	30R
Fe	3,405.83	1	3
Ru	3,405.88	2	50
Re	3,405.89	—	150
Mo	3,405.94	25	25
Ce	3,405.977	3	25
F	3,405.980	(10)	—
Tb	3,406.01	8	15
Pd	3,406.04	2	3
V	3,406.06	3	—
W	3,406.094	8	—
Ce	3,406.095	—	5
Eu	3,406.12	4	15W
N*			

Co 3,412.339 *J* 100 *O* 1000*R*

	λ	*I*	
		J	*O*
Hf	**3,412.338**	**—**	**10**
Ce	**3,412.334**	**—**	**10**
Mo	**3,412.29**	**20**	**1**
Rh	**3,412.27**	**60**	**300**
Eu	**3,412.26**	**4**	**2*h***
U	**3,412.09**	**10**	**6**
Ru	**3,412.08**	**—**	**30**
W	**3,412.05**	**5*d***	**5*d***
F	**3,412.04**	**(3)**	**—**
Mo	**3,412.02**	**5**	**5**
Gd	**3,412.01**	**2**	**2**
Sr	**3,411.94**	**—**	**4**
Er	**3,411.89**	**1*h***	**6**
Ce	**3,411.83**	**—**	**12*s***
Bi	**3,411.80**	**(15*h*)**	**—**
Zr	**3,411.791**	**—**	**3**
Th	**3,411.789**	**6**	**6**
La	**3,411.76**	**25*h***	**5**
Tb	**3,411.73**	**—**	**8**
Ta	**3,411.72**	**2**	**5**
Ti	**3,411.68**	**2**	**10**
F	**3,411.66**	**(6)**	**—**
Ru	**3,411.64**	**20**	**80**
Tm	**3,411.58**	**5**	**8**
Ir	**3,411.577**	**1**	**7**
Ce	**3,411.57**	**—**	**12**
Ho	**3,411.56**	**6**	**6**
U	**3,411.53**	**8**	**10**
Dy	**3,411.52**	**2**	**3**
Ce	**3,411.43**	**—**	**10**
Th	**3,411.362**	**3**	**3**
Fe	**3,411.356**	**30**	**80**
Zr	3,410.25	50	50
Hf	3,410.17	60	25
Nb	3,409.19	100	10
Co	3,409.18	125	1000*R*
Cr	3,408.76	100	35
Nb	3,408.68	50	5
Pt	3,408.13	60	250*W*
Gd	3,407.60	100	100*r*
Fe	3,407.46	400	400
Ti	3,407.20	50	12
P	3,406.93	(50)	—
Fe	3,406.80	60	100

	λ	*I*	
		J	*O*
U	**3,412.36**	**12*d***	**12*d***
Th	**3,412.41**	**3**	**2**
Yb	**3,412.45**	**5**	**15**
Yt	**3,412.469**	**4**	**4**
Dy	**3,412.47**	**2**	**2**
Nb	**3,412.48**	**20*h***	**2*h***
Sm	**3,412.57**	**1**	**4**
Ir	**3,412.59**	**2*h***	**9**
Tm	**3,412.60**	**2**	**10**
Co	**3,412.63**	**40**	**1000*R***
Fe	**3,412.64**	**—**	**8**
U	**3,412.69**	**—**	**6**
Eu	**3,412.73**	**8**	**20**
Os	**3,412.740**	**—**	**15**
W	**3,412.743**	**12**	**2**
Ru	**3,412.80**	**4**	**50**
*bh***Sr**	**3,412.8**	**—**	**4**
Sn	**3,412.80**	**5*Wh***	**—**
Ho	**3,412.86**	**4**	**—**
Ta	**3,412.89**	**3**	**20**
Nb	**3,412.93**	**150**	**5**
Mo	**3,412.95**	**15**	**—**
W	**3,412.96**	**9**	**10**
Th	**3,413.02**	**2**	**5**
Sn	**3,413.03**	**(15)**	**—**
Ne	**3,413.13**	**(7)**	**—**
Fe	**3,413.135**	**300**	**400**
Ni	3,413.48	15	500
P	3,413.51	(70)	—
Ta	3,414.14	100*W*	18*W*
Ni	3,414.76	50*wh*	1000*R*
Nb	3,415.97	50	50
Ti	3,416.96	50	7
Ru	3,417.35	70	1
Fe	3,417.84	100	150
Ne	3,417.90	(500)	—
Ne	3,418.01	(50)	—
Fe	3,418.51	100	150
P	3,419.24	(100)	—
Nb	3,420.63	50	5
Nb	3,421.16	50*w*	10*w*
Cr	3,421.21	200	50
Pd	3,421.24	1000*R*	2000*R*
Gd	3,422.47	100	80
Cr	3,422.74	125	35
Fe	3,424.29	150	200

*P**

	λ	I J	I O
P	*3,424.87*	*(100)*	—
Tm	*3,425.08*	*300*	*200*
Nb	*3,425.42*	*300*	*30r*
Nb	*3,426.57*	*200*	*5*
Ru	*3,428.31*	*100*	*100*
Tm	*3,429.97*	*100*	*100*
Bi	*3,430.83*	*(200)*	—
Bi	*3,431.23*	*(150)*	—
Co	*3,431.57*	*40*	*500R*

Co 3,453.505 *J* 200 *O* 3000*R*

	λ	J	O
Re	**3,453.502**	—	**40**
Eu	**3,453.47**	**2**	**3**
Kr	**3,453.46**	**(3*h*)**	—
Tb	**3,453.46**	**3**	**15**
Ta	**3,453.35**	**10*h***	—
Cr	**3,453.33**	**35**	**35**
Re	**3,453.28**	—	**20**
Ce	**3,453.24**	—	**8**
Sm	**3,453.23**	**4*W***	**4**
La	**3,453.17**	**40**	**50**
Ho	**3,453.13**	**20**	**30**
Dy	**4,453.12**	**2**	**5*Wh***
Er	**3,453.10**	**1**	**12**
Ne	**3,453.10**	**(7)**	—
V	**3,453.08**	**60**	—
Os	**3,453.05**	**10**	**20*s***
Er	**3,453.04**	**1**	**12**
Fe	**3,453.02**	**15**	**30**
Pb	**3,453.00**	**(2)**	—
Ta	**3,452.97**	**2**	**5**
U	**3,452.96**	—	**2**
Pr	**3,452.93**	**1**	**6**
Ru	**3,452.90**	**6**	**60**
Ni	**3,452.89**	**50**	**600*R***
Ce	**3,452.81**	—	**5**
Mo	**3,452.79**	**20**	—
Sm	**3,452.77**	**1**	**4**
U	**3,452.73**	**3*W***	—
Th	**3,452.68**	**10**	**10**
Nb	**3,452.65**	**5**	**15**
Ce	**3,452.623**	—	**5**
W	**3,452.622**	**4**	**7**
Mo	**3,452.60**	**8**	**10**
Ce	**3,452.54**	—	**5**
W	**3,452.51**	**12**	**4**
Ti	**3,452.47**	**100**	**12**
Ti	**3,453.53**	—	**5*h***
Sm	**3,453.55**	**4**	**15**
U	**3,453.57**	**8**	**5**
Ce	**3,453.64**	—	**2**
Ti	**3,453.65**	—	**3**
Tm	**3,453.66**	**80**	**150**
Cr	**3,453.74**	**25**	**30**
Ce	**3,453.76**	—	**3**
U	**3,453.780**	**3**	**4**
Pr	**3,453.784**	—	**6**
Tl	**3,453.83**	**(10)**	—
Pt	**3,453.86**	**8**	**1**
Eu	**3,453.88**	—	**2*W***
W	**3,453.88**	**10**	—
Th	**3,453.92**	**4**	**4**
Nb	**3,453.97**	**100**	—
Ce	**3,454.02**	—	**3**
Se	**3,454.05**	**(8)**	—
Tb	**3,454.06**	**30**	**80**
Yb	**3,454.07**	**250**	**40**
Ar	**3,454.10**	**(10)**	—
Pt	**3,454.13**	**2**	**2**
Gd	**3,454.149**	**15**	**15**
Eu	**3,454.15**	**1**	**3**
Ni	**3,454.161**	**2**	—
Ti	**3,454.165**	—	**15**
Yt	**3,454.18**	**2**	**5**
Ne	**3,454.19**	**(75)**	—
Th	**3,454.21**	**8**	**8**
Mo	**3,454.22**	**5**	**5**
U	**3,454.23**	**2**	**2**
Ce	**3,454.26**	—	**2**
Er	**3,454.32**	**8**	**20**
Dy	**3,454.33**	**10**	**100**
Nd	**3,454.39**	**2**	**6**
Pr	**3,454.47**	**1*h***	**6**

	λ	I J	I O		λ	I J	I O
Nb	3,452.35	200	5	**Ce**	**3,454.475**	**40**	**10**
Hg	3,451.69	(200)	—	**Bi**	**3,454.50**	**(5)**	**—**
B	3,451.41	100	—	**Dy**	**3,454.52**	**2**	**18**
Pd	3,451.35	400*h*	—	**W**	**3,454.54**	**9**	**—**
Gd	3,450.38	100	100				
				Zr	**3,454.58**	**4**	**7**
Co	3,449.44	125	500*R*	**U**	**3,454.62**	**—**	**10**
Co	3,449.17	125	500*R*				
Ne	3,447.70	(150)	—				
Ta	3,446.91	150*W*	2*W*	Cr	3,454.99	100	—
K	3,446.72	100*R*	150*R*	Co	3,455.23	10	2000*R*
				Bi	3,455.27	100*h*	—
Ni	3,446.26	50*h*	1000*R*	Ti	3,456.39	125	25
Fe	3,445.15	150	300	V	3,457.15	150	2
Ti	3,444.31	150	60				
Fe	3,443.88	200	400	Cr	3,457.63	125	4
Co	3,443.64	100	500*R*	Yb	3,458.28	100	12
				Mn	3,460.33	500	60
				Pd	3,460.77	600*h*	300*r*
Fe	*3,440.99*	*200*	*300*	Ar	3,461.08	(300)	—
Fe	*3,440.61*	*300*	*500*				
Rb	*3,439.34*	*(200)*	*—*	Ti	3,461.50	125	80
Zr	*3,438.23*	*200*	*250*	Rb	3,461.57	(200)	—
Ta	*3,437.37*	*300Wh*	*7*	Rh	3,462.04	150	1000
				Tm	3,462.20	200	250
Rh	*3,434.89*	*200r*	*1000r*	Co	3,462.80	80	1000*R*
Pd	*3,433.45*	*500h*	*1000h*				
				Gd	3,463.98	125	100
				Sr	*3,464.46*	*200*	*200*
				Co	*3,465.80*	*25*	*2000R*
				Fe	*3,465.86*	*400*	*500*
				Cd	*3,466.20*	*500*	*1000*
				Cd	*3,467.66*	*400*	*800*
				Ne	*3,472.57*	*(500)*	*—*
				Sb	*3,473.91*	*300Wh*	*3*
				Co	*3,474.02*	*100*	*3000R*
				Mn	*3,474.13*	*400*	*12*

Co 3,465.80 *J* 25* *O* 2000*R*

	λ	J	O		λ	J	O
Th	**3,465.77**	**10**	**10**	**Mo**	**3,465.86**	**5**	**5**
Pr	**3,465.76**	**4**	**9**	**Fe**	**3,465.863**	**400**	**500**
Mo	**3,465.66**	**6**	**6**	**Nb**	**3,465.865**	**40**	**30**
Zr	**3,465.63**	**—**	**3**	**Hf**	**3,465.92**	**2**	**5**
Cr	**3,465.58**	**8**	**30**	**Th**	**3,465.93**	**5**	**5**
Ti	**3,465.56**	**60**	**6**	**Tb**	**3,465.98**	**—**	**30**
Ce	**3,465.50**	**—**	**5**	**Re**	**3,465.983**	**—**	**20**
Sm	**3,465.46**	**—**	**5**	**Ce**	**3,466.03**	**—**	**3**
Nd	**3,465.44**	**—**	**6*Wh***	**Ce**	**3,466.08**	**—**	**5**
Os	**3,465.43**	**12**	**60**	**Dy**	**3,466.12**	**1**	**2**

	λ	I J	I O
Ce	3,465.42	—	8
W	3,465.41	4	4
Cu	3,465.40	2h	10
Ce	3,465.302	—	8
Dy	3,465.30	—	2
			35
Ru	3,465.29	2h	5
Cr	3,465.250	30	35
V	3,465.249	25	2
Ir	3,465.22	1	12
Mn	3,465.19	—	2
Er	3,465.12	—	8d
Th	3,465.02	5	5
Ce	3,464.99	—	12
U	3,464.95	4	4
Nd	3,464.93	2h	10
Fe	3,464.92	—	2
Os	3,464.88	5	15
U	3,464.87	—	2
Ce	3,464.86	—	12
Cr	3,464.84	3	30
Ce	3,464.72	—	2
Co	3,462.80	80	1000R
Rh	3,462.04	150	1000
Ni	3,461.65	50h	800R
Re	3,460.47	—	1000W
Ni	3,458.47	50h	800R
Co	3,455.23	10	2000R
Co	*3,453.50*	*200*	*3000R*
U	3,466.14	—	4
Cd	3,466.20	500	1000
Pr	3,466.24	1h	2
U	3,466.30	2	8
Eu	3,466.42	2	30
Fe	3,466.500	70	30
U	3,466.503	—	3
Th	3,466.54	1	5
Tb	3,466.57	—	8
Ce	3,466.646	—	2
Th	3,466.647	2	5
Sm	3,466.74	—	3
Pr	3,466.75	2	10
Sm	3,466.78	—	4
Ce	3,466.80	—	10
Mo	3,466.826	10	6
W	3,466.828	—	2hd
Ta	3,466.85	18h	2
Eu	3,466.88	1	20
Cd	3,467.66	400	800
Ni	3,472.54	40	800R
Co	3,474.02	100	3000R

Co 3,502.279 *J* 20* *O* 2000R

	λ	J	O
U	3,502.24	—	5
Tb	3,502.10	—	8
Dy	3,502.09	4	6
Nd	3,502.01	4	8
Ta	3,501.97	1	3
Yt	3,501.96	2	3
Ag	3,501.94	2	5
Dy	3,501.87	1	4
Ce	3,501.81	—	8
Co	3,501.72	100	5
Os	3,501.69	5	30
Ag	3,501.68	20h	8h
U	3,501.65	2	4
Mn	3,502.280	—	2h
Zr	3,502.306	—	5
Cr	3,502.307	6	35
Ru	3,502.42	4	20
Eu	3,502.46	2	10W
Tb	3,502.49	—	15
Ta	3,502.495	2h	7
Rh	3,502.52	150	1000
Ni	3,502.59	—	100
Ir	3,502.616	—	2
Co	3,502.624	5	60
Fe	3,502.63	—	2
Ce	3,502.65	—	8

	λ	I J	I O		λ	I J	I O
Ce	3,501.59	—	12	Nd	3,502.70	4	8*d*
Nd	3,501.54	4	10*d*	Rc	3,502.73	—	20
Co	3,501.53	1*h*	2*h*	Th	3,502.783	8	8
Zr	3,501.49	3	12	Er	3,502.783	1	10
V	3,501.48	20	25	Eu	3,502.81	5	15*d*
Th	3,501.46	10	10	Ta	3,502.87	3	7
Ce	3,501.45	3	18*W*	Ce	3,502.89	—	8*s*
Dy	3,501.44	3	10	Ir	3,502.94	4	8
Ru	3,501.354	3	30	Ce	3,502.98	—	5
Zr	3,501.347	1	15	W	3,503.038	7	8
Nb	3,501.341	30	3	U	3,503.041	1	2
U	3,501.34	3*d*	5*d*	Re	3,503.059	—	80
Ce	3,501.339	—	4	Pr	3,503.060	4	15
Cu	3,501.338	1*h*	5	Ce	3,503.08		10
Sm	3,501.23	1	2	V	3,503.176	4	10
				Dy	3,503.179	3*h*	8
				Nb	3,503.20	5	10
Ba	3,501.12	20	1000				
Co	3,495.69	25	1000*R*	Eu	3,503.23	—	5
Ni	3,492.96	100*h*	1000*R*	Sm	3,503.29	—	20
				Zr	3,503.318	—	2
				U	3,503.324	—	2
				Ni	3,510.34	50*h*	900*R*

*N**

Co 3,529.81 *J* 30* *O* 1000*R*

	λ	J	O		λ	J	O
U	3,529.77	5	2	Re	3,529.816	—	10
Tb	3,529.76	3	15	Fe	3,529.820	80	125
Pr	3,529.74	2	5	Ca	3,529.97	3	2
V	3,529.735	2	20	Zr	3,529.98	6*h*	2
Ce	3,529.732	—	3	Sm	3,530.00	—	15
U	3,529.62	6	2	Ce	3,530.02	3	18
Ru	3,529.61	—	3	Os	3,530.06	20	100
W	3,529.56	20	10	Nb	3,530.09	5	5
Ce	3,529.54	2	5	Ce	3,530.14	2	4
Fe	3,529.53	—	2	Zr	3,530.22	1	10
Dy	3,529.52	—	4	Er	3,530.36	1	6*W*
Ba	3,529.49	—	4	Eu	3,530.36	1	12*W*
Tl	3,529.43	800	1000	Tb	3,530.38	—	15
Nb	3,529.39	5	3	Cu	3,530.386	20	50
Er	3,529.34	—	7*W*	Fe	3,530.389	25	50
Ru	3,529.28	3	30	Ce	3,530.40	—	2
Ce	3,529.27	2	8	Th	3,530.50	2*d*	4*d*
Re	3,529.21	—	25	U	3,530.540	—	6
Nd	3,529.19	2	8	Dy	3,530.54	4	6
U	3,529.08	3	3	Co	3,530.55	—	2

	λ	I J	I O		λ	I J	I O
Nd	**3,529.05**	—	**10*d***	**Ti**	**3,530.58**	**1**	**15**
Ce	**3,529.041**	**2**	**12**	**Ni**	**3,530.59**	—	**30**
Dy	**3,529.036**	—	**5**	**Sm**	**3,530.60**	**4**	**8**
Co	**3,529.03**	—	**200*R***	**Ce**	**3,530.63**	—	**10**
Th	**3,528.95**	**6**	**6**	**Tb**	**3,530.64**	—	**8**
Dy	**3,528.93**	**10**	**2**	**La**	**3,530.66**	**6**	**15**
Nb	**3,528.90**	**40**	**2**	**Ir**	**3,530.74**	**2**	**10**
Ni	**3,528.89**	—	**15**	**W**	**3,530.76**	**7**	**8**
Th	**3,528.82**	**3**	**4**	**V**	**3,530.77**	**100**	**40**
Pd	**3,528.72**	—	**10**	**Mo**	**3,530.81**	**3**	**1**
				Nb	**3,530.82**	**15**	**3**
Os	3,528.60	50	400*R*	**Pr**	**3,530.84**	**2**	**9**
Rh	3,528.02	150	1000*W*	**Zr**	**3,530.85**	**5**	**5**
Ni	3,524.54	100*Wh*	1000*R*	**V**	**3,530.869**	**20**	**1**
Ni	3,519.77	30	500*h*	**Hf**	**3,530.874**	**1**	**5**
Tl	3,519.24	1000*R*	2000*R*				
				Dy	**3,530.88**	—	**4**
				Ce	**3,530.95**	**1**	**10**
Pd	*3,516.94*	*500R*	*1000R*				
Ni	*3,515.05*	*50h*	*1000R*				

*P**

Co 3,995.31 *J* 20* *O* 1000*R*

	λ	I J	I O		λ	I J	I O
Ir	**3,995.291**	—	**6**	**Ce**	**3,995.42**	—	**4**
Ta	**3,995.290**	—	**3**	**Tb**	**3,995.43**	—	**2**
Er	**3,995.27**	—	**6**	**Mo**	**3,995.48**	**2**	**3**
Nd	**3,995.24**	—	**20**	**U**	**3,995.53**	—	**4**
Fe	**3,995.200**	—	**10**	**Tm**	**3,995.58**	—	**100**
Ir	**3,995.196**	—	**3**	**Sm**	**3,995.59**	**3**	**8**
Tb	**3,995.14**	—	**3**	**Rh**	**3,995.61**	**10**	**15**
Eu	**3,995.07**	—	**4*W***	**Ba**	**3,995.66**	**5**	**18**
U	**3,994.98**	**20*Wh***	**8**	**Re**	**3,995.68**	—	**10*h***
Os	**3,994.93**	**5**	**30**	**La**	**3,995.750**	**300**	**600**
V	**3,994.89**	—	**35**	**Ce**	**3,995.752**	—	**6**
Er	**3,994.86**	—	**5**	**Er**	**3,995.754**	—	**4**
Pr	**3,994.83**	**25**	**300**	**Eu**	**3,995.755**	—	**12**
Ce	**3,994.81**	—	**3**	**U**	**3,995.77**	**3**	**4**
W	**3,994.76**	**2**	**4**	**Tb**	**3,995.79**	—	**8**
U	**3,994.72**	**2**	**2**	**Pr**	**3,995.85**	**2**	**4**
Ti	**3,994.70**	—	**25**	**U**	**3,995.97**	**8**	**10**
Nd	**3,994.684**	**40**	**80**	**Ru**	**3,995.98**	**30**	**30**
Eu	**3,994.68**	—	**5**	**Fe**	**3,995.989**	**20**	**60**
Mo	**3,994.63**	**3**	**3**	**Eu**	**3,995.994**	**2**	**3**
Ce	**3,994.573**	—	**6**	**Th**	**3,996.07**	**10**	**15**
U	**3,994.572**	—	**4**	**Rh**	**3,996.15**	**10**	**25**
Ru	**3,994.56**	—	**5**	**Ta**	**3,996.17**	**30*h***	**100**
Th	**3,994.55**	**10**	**30**	**Mo**	**3,996.25**	**2**	**3**
Co	**3,994.54**	—	**60**	**Cr**	**3,996.27**	—	**2**

	λ	I			λ	I	
		J	O			J	O
Dy	**3,994.53**	—	**5**	**Gd**	**3,996.32**	**100**	**100**
Nd	**3,994.46**	—	**2**	**Ce**	**3,996.36**	—	**2*h***
Nb	**3,994.43**	—	**4**	**Ir**	**3,996.45**	**2**	**8**
Pr	**3,994.34**	—	**4**	**Ce**	**3,996.49**	—	**10**
U	**3,994.29**	**6**	**6**	*bh***Ca**	**3,996.5**	—	**4**
Gd	**3,994.18**	**6**	**8**	**Ru**	**3,996.51**	**4**	**10**
Re	**3,994.14**	—	**5*h***	**Tm**	**3,996.52**	**40**	**200**
Fe	**3,994.12**	**10**	**25**	**Pt**	**3,996.57**	—	**50**
Nd	**3,994.10**	**2**	**20**	**Tb**	**3,996.59**	—	**3**
Tb	**3,994.04**	—	**4**	**Sc**	**3,996.61**	**10**	**40**
Pr	**3,994.01**	**3**	**20**	**Ti**	**3,996.65**	—	**12**
Cr	**3,993.97**	**20**	**60**	**Pr**	**3,996.686**	**4**	**20**
Ni	**3,993.95**	—	**30*h***	**Er**	**3,996.695**	—	**25**
Mo	**3,993.934**	**5**	**5**	**Tb**	**3,996.696**	**3**	**9**
Eu	**3,993.928**	**2**	**8**	**Dy**	**3,996.699**	**80**	**200**
W	**3,993.90**	**6**	**7**	**Ho**	**3,996.70**	**4**	**6**
Ce	**3,993.82**	**6**	**50**	**Ce**	**3,996.77**	—	**4**
U	**3,993.81**	**4**	**12**	**Hf**	**3,996.80**	**5**	**1**
Yb	**3,993.77**	—	**3**	**Os**	**3,996.805**	**10**	**50**
Th	**3,993.72**	**5**	**8**	**Sm**	**3,996.91**	**2**	**2**
Ho	**3,993.71**	**4**	**10**	**Re**	**3,996.92**	—	**3*h***
La	3,988.52	800	1000	Dy	4,000.45	300	400
Yb	3,987.99	500*R*	1000*R*				
Eu	*3,971.99*	—	*1000Rwh*				

Co 4,118.77 *O* 1000*R*

	λ	J	O		λ	J	O
Pt	**4,118.69**	**10**	**400**	**Fe**	**4,118.90**	**1**	**2**
Ir	**4,118.66**	—	**8**	**Hf**	**4,118.91**	—	**3**
V	**4,118.64**	**3**	**9**	**Mo**	**4,118.96**	**8**	**10**
Hf	**4,118.60**	**1**	**3**	**Ce**	**4,119.01**	**3**	**25**
Fe	**4,118.549**	**100**	**200**	**V**	**4,119.10**	**2**	**2**
Sm	**4,118.546**	**60**	**50**	**Gd**	**4,119.20**	—	**3**
Ru	**4,118.50**	—	**40**	**Nb**	**4,119.28**	**200**	**2**
Pr	**4,118.48**	**50*d***	**250*d***	**Yb**	**4,119.29**	—	**4**
Tb	**4,118.43**	—	**5**	**Dy**	**4,119.32**	**4**	**15**
U	**4,118.39**	**8**	**3**	**Er**	**4,119.329**	**2*Wh***	**18*W***
Tb	**4,118.20**	—	**5**	**Eu**	**4,119.33**	**3**	**10**
Ce	**4,118.19**	—	**3**	**Mo**	**4,119.35**	**2*h***	**3**
W	**4,118.184**	**10**	**9**	**Pr**	**4,119.37**	**2**	**5**
V	**4,118.182**	**5**	**10**	**Gd**	**4,119.38**	—	**8**
Ce	**4,118.144**	**8**	**25**	**Fe**	**4,119.39**	**1**	**2**
U	**4,118.144**	**2**	**2**	**Yb**	**4,119.43**	**20**	**10**
Ta	**4,118.07**	**4*h***	**5**	**V**	**4,119.46**	**6**	**12**

	λ	I J	I O
W	**4,118.05**	**9**	**8**
Dy	**4,118.01**	—	**4**
Ce	**4,117.99**	**1**	**3**
Fe	**4,117.86**	**1**	**6**
Ce	**4,117.83**	**1*h***	**4**
Co	4,110.53	—	600
In	4,101.77	1000*R*	2000*R*

	λ	I J	I O
Tb	**4,119.49**	—	**6**
Sm	**4,119.57**	**5**	**4**
Ir	**4,119.60**	—	**15**
Mo	**4,119.63**	**50**	**5**
Rh	**4,119.68**	**25**	**100**
U	**4,119.69**	**4**	**8**
Nb	**4,119.72**	**20**	**1*h***
Pr	**4,119.80**	**1**	**5**

*N**

Co 4,121.32 *J* 25* *O* 1000*R*

	λ	I J	I O
Ce	**4,121.28**	—	**2**
Cr	**4,121.26**	**8**	**35**
U	**4,121.23**	**3**	**3**
Gd	**4,121.07**	**1**	**3**
Ce	**4,121.069**	**2**	**2**
Tb	**4,121.01**	—	**5**
Ru	**4,120.99**	**30**	**25**
Pr	**4,120.96**	**1**	**3**
W	**4,120.86**	**8**	**9**
Er	**4,120.84**	—	**8**
Ce	**4,120.83**	**6**	**25**
Ce	**4,120.80**	—	**4**
V	**4,120.77**	—	**2**
Mo	**4,120.70**	—	**8**
U	**4,120.69**	**1*h***	**3**
Dy	**4,120.66**	—	**5**
Nd	**4,120.65**	**4**	**6**
Cr	**4,120.61**	**10**	**40**
V	**4,120.54**	**4**	**15**
Tb	**4,120.52**	**2**	**12**
Ce	**4,120.51**	**2**	**2**
Tb	**4,120.43**	—	**2**

	λ	I J	I O
Sm	**4,121.35**	**15**	**15**
Zr	**4,121.46**	**2**	**30**
Cr	**4,121.47**	**1**	**10**
Bi	**4,121.53**	**50**	**125*Wh***
Sm	**4,121.54**	**4**	**10*h***
Tb	**4,121.57**	—	**2**
Ce	**4,121.60**	**1**	**8**
Re	**4,121.63**	—	**50**
Ti	**4,121.64**	**3**	**15**
Zr	**4,121.67**	—	**5**
Rh	**4,121.68**	**50**	**150**
Cu	**4,121.74**	—	**20**
Fe	**4,121.806**	**40**	**100**
Eu	**4,121.81**	—	**2*W***
Cr	**4,121.82**	**10**	**40**
Bi	**4,121.85**	**2**	**5**
Ce	**4,121.91**	—	**3**
Nd	**4,121.94**	**8**	**4**
Sm	**2,122.017**	**4**	**2**
W	**4,122.017**	**8**	**7**
Pr	**4,122.09**	**3**	**12**
Cr	**4,122.16**	**5**	**30**
U	**4,122.166**	**8**	**10**
Ti	**4,122.166**	**10**	**40**
Tb	**4,122.21**	—	**3**
Ce	**4,122.24**	**3**	**3**
Co	**4,122.27**	—	**10*h***
Pr	**4,122.31**	**2**	**8**
La	4,123.23	500	500

Co 4,530.96 *J* 8* *O* 1000

	λ	I			λ	I	
		J	O			J	O
Ce	**4,530.94**	—	**3**	**Ce**	**4,531.07**	—	**3**
Re	**4,530.89**	—	**20**	**Pr**	**4,531.09**	**2*W***	**10*W***
Ru	**4,530.854**	—	**60**	**Er**	**4,531.11**	—	**5**
Ta	**4,530.846**	**50**	**300**	**Gd**	**4,531.12**	—	**5**
Cu	**3,530.82**	**50**	**200**	**Fe**	**4,531.15**	—	**125**
Ce	**4,530.81**	—	**3**	**Ho**	**4,531.28**	**2*h***	**2**
V	**4,530.79**	**9**	**12**	**Ce**	**4,531.31**	—	**8**
Cr	**4,530.74**	**125**	**150**	**Sr**	**4,531.35**	—	**10**
Rh	**4,530.59**	**2**	**4**	**Pr**	**4,531.51**	—	**3**
La	**4,530.56**	**25**	**15**	**Ho**	**4,531.62**	**2**	**5**
W	**4,530.47**	**4**	**15**	**Ce**	**4,531.63**	—	**4**
Nd	**4,530.33**	—	**10**	**Fe**	**4,531.65**	**1**	**8**
Er	**4,530.31**	**2**	**2**	**Th**	**4,531.72**	**3**	**3**
Ho	**4,530.08**	**2**	**3**	**Ru**	**4,531.80**	—	**5**
Ta	**4,529.98**	**10*h***	**3**	**Gd**	**4,531.81**	**1**	**3**
Sm	**4,529.95**	—	**4**	**Tb**	**4,531.82**	—	**1*W***
Nd	**4,529.935**	—	**40**	**Pr**	**4,531.846**	—	**2**
Pr	**4,529.930**	**4*W***	**25*W***	**Er**	**4,531.849**	—	**2**
Re	**4,529.928**	—	**40*W***	**Ce**	**4,532.01**	—	**5**
Ce	**4,529.91**	—	**5**	**Ti**	**4,532.14**	—	**3**
Yb	**4,529.87**	—	**12**	**Tm**	**4,532.15**	**1**	**15**
Cr	**4,529.85**	**8**	**25**	**Er**	**4,532.16**	—	**3**
Mn	**4,529.79**	—	**50**	**Th**	**4,532.26**	**4**	**8**
Dy	**4,529.78**	**2**	**2**				
Nd	**4,529.764**	—	**3**	Co	4,533.99	8	500
Tb	**4,529.76**	—	**4**	Co	4,543.81	—	500*W*
W	**4,529.760**	**4**	**15**				
U	**4,529.71**	**3**	**12**				
Fe	**4,529.68**	**2**	**10**	*Co*	*4,549.66*	—	*600*
Os	**4,529.67**	**2**	**80**	*Ba*	*4,554.04*	*200*	*1000R*
				Ru	*4,554.51*	*200*	*1000R*
				Cs	*4,555.35*	*100*	*2000R*
Rh	4,528.72	60	500*r*				
Fe	4,528.62	200	600				
Sn	4,524.74	50	500*wh*				
Eu	4,522.58	—	500				
In	*4,511.32*	*4000R*	*5000R*				

Co 4,581.60 *J* 10* *O* 1000

	λ	J	O		λ	J	O
Sm	**4,581.584**	—	**8**	**Nb**	**4,581.62**	**50**	**30**
Pr	**4,581.584**	—	**10*W***	**Er**	**4,581.70**	—	**7**
Th	**4,581.581**	**6**	**6**	**U**	**4,581.72**	**18**	**8**
Fe	**4,581.52**	**2**	**60**	**Sm**	**4,581.74**	—	**40**
Dy	**4,581.46**	**2**	**2**	**Yt**	**4,581.77**	—	**3**

	λ	I J	I O
Tb	4,581.43	—	8
Ca	4,581.40	10	100
Yt	4,581.302	—	3
Gd	4,581.301	4	30
V	4,581.23	4	5
Th	4,581.22	3	3
La	4,581.21	5	15
Nd	4,581.20	—	3
Gd	4,581.09	—	5
Ce	4,581.088	1	4
Er	4,581.07	—	2
Cr	4,581.063	1	15
U	4,581.058	3	2
U	4,580.83	4	2
Pr	4,580.82	—	4*W*
*bh*La	4,580.8	—	3
Eu	4,580.76	—	6*W*
Yb	4,580.73	—	2
Ta	4,580.69	10	200*W*
Re	4,580.668	—	30
Pt	4,580.668	1	3
Ni	4,580.62	—	5
Fe	4,580.60	—	6
Pt	4,580.55	1	3
Nb	4,580.50	2	1
Pr	4,580.46	—	3
V	4,580.401	25*h*	30*h*
Tb	4,580.40	—	4
Co	4,580.14	3	300
Co	*4,565.59*	*12*	*800W*
Cs	*4,555.35*	*100*	*2000R*

	λ	I J	I O
Mn	4,581.83	—	125
Ir	4,581.92	—	20*h*
Cr	4,582.06	—	5
Yt	4,582.17	2	2
Nb	4,582.286	5	5
Zr	4,582.292	—	8
Mo	4,582.35	10	10
Yb	4,582.36	6	80
U	4,582.37	4	4
Gd	4,582.382	2	6
Ce	4,582.385	—	5
Cr	4,582.45	—	8
Mo	4,582.498	10	10
Ce	4,582.502	8	10
Gd	4,582.51	3	10
Tb	4,582.56	—	2
Pr	4,582.57	—	5*W*
Th	4,582.77	—	2
Mn	4,582.80	—	20
Fe	4,582.83	1	1
Tb	4,582.86	—	3
Cs	4,593.02	50*R*	1000*R*
Eu	4,594.02	200	500*R*
Co	4,594.63	—	400
Li	*4,602.86*	*—*	*800*
Sr	*4,607.33*	*50R*	*1000R*

Co 4,780.01 *J* 500 *O* 500

	λ	I J	I O
Tb	4,780.00	—	3
Ti	4,779.95	100*h*	10
Cr	4,779.939	—	10
Gd	4,779.939	—	2*h*
La	4,779.89	—	6
Er	4,779.74	—	2
N	4,779.71	(15)	—
Eu	4,779.69	—	2
U	4,779.63	3	8
Th	4,779.60	2	2

	λ	I J	I O
Yt	4,780.18	1	2
Dy	4,780.19	—	2
U	4,780.20	—	3
W	4,780.289	—	3
Tb	4,780.29	—	3
Br	4,780.31	(125)	—
Ne	4,780.34	(50)	—
W	4,780.52	1	10
Nd	4,780.54	—	2
La	4,780.56	—	4

	λ	I J	I O		λ	I J	I O
Nd	**4,779.44**	—	**30**	**Ce**	**4,780.73**	—	**6**
Dy	**4,779.44**	—	**2*h***	**Ne**	**4,780.88**	**(30)**	—
W	**4,779.42**	—	**4**	**Ta**	**4,780.93**	**200**	**50**
Sb	**4,779.40**	**8**	—	**Er**	**4,781.02**	**2**	**35**
Se	**4,779.35**	**40**	**80**	**Yt**	**4,781.03**	**5**	**10**
Sm	**4,779.22**	—	**10**	**Dy**	**4,781.04**	**2**	**3**
Pr	**4,779.20**	—	**15**	**Gd**	**4,781.06**	—	**25**
Xe	**4,779.18**	**(50)**	—	**Ru**	**4,781.11**	—	**7**
Mn	**4,779.15**	—	**20*W***	**N**	**4,781.17**	**(5)**	—
S	**4,779.11**	**(25)**	—	**Ne**	**4,781.24**	**(2)**	—
Yb	**4,778.98**	—	**2**	**Tm**	**4,781.30**	**10**	**5**
Cl	**4,778.93**	**(45)**	—	**Eu**	**4,781.318**	—	**15**
Tb	**4,778.81**	—	**6**	**Cl**	**4,781.32**	**(75)**	—
P	**4,778.80**	—	**(30)**	**Co**	**4,781.43**	**2*h***	**400**
U	**4,778.797**	—	**2**	**Nd**	**4,781.46**	—	**3**
Eu	**4,778.65**	—	**20**	**Tb**	**4,781.55**	—	**2**
Sm	**4,778.64**	—	**5**	**Sm**	**4,781.57**	—	**8**
***bh*Sc**	**4,778.5**	—	**3**				
Er	**4,778.50**	—	**3**				
Nd	**4,778.40**	—	**5**	Mn	4,783.42	60	400
				Br	4,785.50	(400)	—
Tb	**4,778.36**	—	**6**	Ni	4,786.54	2	300*W*
Pr	**4,778.30**	—	**35**	Ne	4,788.93	(300)	—
				Cr	4,789.38	100	300
Co	4,776.32	—	300	Co	4,792.86	5	600*W*
Co	4,771.11	—	500*W*	Os	4,793.99	6	300
Co	4,768.08	10	300	Cl	4,794.54	(250)	—
Kr	4,765.74	(1000)	—				
				Ar	*4,806.07*	*(500)*	—
Se	*4,763.65*	*(800)*	—				
Kr	*4,762.43*	*(300)*	—				
Ne	*4,752.73*	*(1000)*	—				

Co 4,813.48 *J* 6 *O* 1000

	λ	J	O		λ	J	O
Rh	**4,813.40**	—	**2**	**Eu**	**4,813.55**	—	**8**
Ru	**4,813.23**	—	**6**	**Tb**	**4,813.767**	—	**25**
Tb	**4,813.18**	—	**2**	**Gd**	**4,813.768**	—	**15**
Mo	**4,813.16**	**5**	**5**	**Os**	**4,813.80**	—	**18**
Ti	**4,812.91**	—	**2**	**Co**	**4,813.98**	**2**	**100**
Ru	**4,812.85**	—	**4**	**Mo**	**4,814.01**	**5*h***	**2**
Dy	**4,812.81**	**2**	**2**	**Cr**	**4,814.26**	—	**100**
Ta	**4,812.75**	**5**	**150**	**Sm**	**4,814.33**	—	**3**
Nd	**4,812.67**	—	**3**	**Pr**	**4,814.34**	**3**	**30**
Os	**4,812.62**	—	**2**	**Mo**	**4,814.47**	**8**	**10**
***bh*V**	**4,812.6**	—	**2**	**Tb**	**4,814.49**	—	**2**

	λ	I J	I O		λ	I J	I O
W	**4,812.59**	—	**4**	**Eu**	**4,814.52**	**2**	**2**
Mo	**4,812.48**	**3**	**3**	**Ni**	**4,814.62**	—	**5**
Ti	**4,812.25**	**2**	**18**	**Ru**	**4,814.72**	—	**4**
Ru	**4,812.20**	—	**5**	**Tb**	**4,814.73**	—	**3**
Ni	**4,812.00**	—	**10**	**U**	**4,814.84**	—	**2**
Zn	4,810.53	300*h*	400*W*	Co	4,840.27	150	700*W*

*P**

Cr $^{24}_{51.996}$

t_0 1,615°C t_1 2,482°C

I.	II.	III.	IV.	V.
6.761	16.7	–	–	73.0

λ	*I*		eV
	J	*O*	
II 2,835.63	400*R*	100	12.6
II 2,843.25	400*R*	125	12.6
II 2,849.84	150*R*	80	12.6
II 2,855.68	200	60	12.5
II 2,860.93	100	60	12.5
I 3,578.69	400*R*	500*R*	3.5
I 3,593.49	400*R*	500*R*	3.4
I 3,605.33	400*R*	500*R*	3.4
I 4,254.35	1000	5000*R*	2.9
I 4,274.80	800*R*	4000*R*	2.9
I 4,289.72	800*R*	3000*R*	2.9
I 5,204.52	100	400*R*	3.3
I 5,206.04	200	500*R*	3.3
I 5,208.44	100	500*R*	3.3

Cr 2,835.653 *J* 400*R* *O* 100

	λ	I J	I O		λ	I J	I O
Ce	**2,835.60**	**—**	**10**	**V**	**2,835.635**	**2**	**12**
U	**2,835.57**	**6**	**5**	**W**	**2,835.636**	**10**	**12**
Fe	**2,835.46**	**100**	**100**	**Ti**	**3,835.643**	**6**	**8**
Au	**2,835.43**	**8**	**—**	**Nd**	**2,835.65**	**—**	**5**
Ce	**2,835.39**	**—**	**2**	**Ir**	**2,835.66**	**—**	**12**
Kr	**2,835.35**	**(8*h*)**	**—**	**Ce**	**2,835.800**	**—**	**2**
V	**2,835.347**	**10**	**1**	**U**	**2,835.803**	**10**	**8**
Mo	**2,835.33**	**40**	**20**	**Au**	**2,835.89**	**10**	**—**
Ir	**2,835.20**	**—**	**2**	**Mo**	**2,835.91**	**—**	**15**
Lu	**2,835.25**	**10*hd***	**—**	**In**	**2,835.95**	**3**	**—**
Ne	**2,835.23**	**(15)**	**—**	**Fe**	**2,835.955**	**10**	**15**
Hf	**2,835.18**	**3*h***	**1**	**Mo**	**2,836.03**	**—**	**8**
Dy	**2,835.14**	**—**	**2**	**Ce**	**2,836.04**	**—**	**6**
Nb	**2,835.12**	**100**	**5*d***	**Th**	**2,836.05**	**8**	**10**
				Ir	**2,836.097**	**1**	**5**
Pt	2,834.71	5	80	**Ti**	**2,836.100**	**—**	**5**
Co	2,834.43	—	50	**Tb**	**2,836.11**	**20**	**—**
Rh	2,834.12	30	70	**Ru**	**2,836.15**	**—**	**20**
Re	2,834.06	—	100*r*				
Ta	2,833.64	40*w*	300*w*	Ta	2,836.62	2	80*r*
Ce	2,833.31	—	50*d*	Rh	2,836.69	—	60
Pb	2,833.07	80*R*	500*R*	C	2,836.71	200	—
Fe	2,832.44	200	300	Cd	2,836.91	80	200
Fe	2,831.56	500	1	In	2,836.92	80	80
Mn	2,830.79	—	50				
				Co	2,837.15	—	75*r*
Pt	2,830.39	600*r*	1000*R*	Zr	2,837.23	100	—
				Ce	2,837.29	—	50*s*
				Cu	2,837.55	250	—
Fe	*2,828.81*	*60*	*100*	Fe	2,838.12	150	150
Eu	*2,828.69*	*150*	*200 W*				
Ta	*2,827.18*	*10*	*200*	Os	2,838.63	100	100*R*
Rh	*2,826.67*	*50d*	*100*	Sn	2,839.99	300*R*	300*R*
Tl	*2,826.16*	*100R*	*200R*	Gd	2,840.24	60	50
				Fe	2,840.42	20	125
Fe	*2,825.69*	*60*	*70*	Ru	2,840.54	8	60
Fe	*2,825.56*	*150*	*150*				
Co	*2,825.24*	*200*	*5*				
Ta	*2,824.81*	*5h*	*60W*	*Ru*	*2,841.68*	*200*	*50*
				Ta	*2,842.81*	*50*	*200*

*N**

Cr 2,843.25 *J* 400*R* *O* 125

	λ	I J	I O		λ	I J	I O
Fe	**2,843.24**	**2**	**6**	**Tl**	**2,843.27**	**—**	**5**
Ru	**2,843.17**	**3**	**30**	**Fe**	**2,843.32**	**3**	**—**
Re	**4,843.001**	**—**	**30**	**Ar**	**2,843.37**	**(2)**	**—**
Rn	**2,843.00**	**(3)**	**—**	**Ho**	**2,843.39**	**10*h***	**—**
Yb	**2,843.00**	**5**	**—**	**Fe**	**2,843.48**	**3**	**—**

	λ	I J	I O		λ	I J	I O
Fe	**2,842.93**	**2**	**5**	**Ta**	**2,843.51**	**80**	**3**
Ce	**2,842.92**	—	**3**	**Zr**	**2,843.52**	**8**	**8**
Mo	**2,842.91**	—	**4**	**Fe**	**2,843.63**	**100**	**125**
U	**2,842.77**	**6*h***	**4**	**Nb**	**4,843.64**	**10**	**3**
Ce	**2,842.83**	—	**25*d***	**La**	**2,843.66**	**4**	**4**
Eu	**2,842.82**	**1*h***	**2**	**Br**	**2,843.69**	**(2)**	—
Ta	**2,842.815**	**50**	**200**	**Mo**	**2,843.73**	**8**	**1**
Th	**2,842.815**	**10**	**12**	**W**	**2,843.78**	**8**	**9**
Cr	**2,842.87**	**12**	—				
Ru	**2,842.75**	—	**20**	Fe	2,843.98	300	300
V	**2,842.69**	**12**	**1**	Ta	2,844.25	50	400*r*
				Os	2,844.40	25	50
				Ta	2,844.46	200	200
*P**				*N**			

Cr 2,849.84 *J* 150*R* *O* 80

	λ	J	O		λ	J	O
Ta	**2,849.82**	**50*W***	**15*Ws***	**U**	**2,849.98**	**8*h***	**5**
Tl	**2,849.80**	**(200)**	—	**Co**	**2,850.04**	—	**75**
Ir	**2,849.72**	**20*h***	**40*h***	**Fe**	**2,850.11**	—	**2**
J	**2,849.70**	**(4)**	—	**Hf**	**2,850.15**	**20**	**20**
Xe	**2,849.66**	**(4)**	—	**Ir**	**2,850.25**	—	**5**
Fe	**2,849.61**	**50**	—	**Re**	**2,850.28**	—	**2**
Ru	**2,849.565**	**18**	—	**Cr**	**2,850.29**	**5**	—
Nb	**2,849.557**	**100*w***	**2*w***	**Nb**	**2,850.38**	**5**	**1**
Ta	**2,849.55**	**1*h***	**5*h***	**W**	**2,850.39**	**7**	**9**
U	**2,849.48**	**15**	**18**				
W	**2,849.46**	**6**	**7**	Ta	2,850.49	100	200
Ag	**2,849.42**	**2*h***	—	*bh*B	2,850.6	—	50
Mo	**2,849.38**	**5**	**50**	Sn	2,850.62	100*wh*	80
Rh	**2,849.343**	—	**3**	Os	2,850·76	25	75
Yb	**2,849.34**	**4**	—	Re	2,850.977	—	40
Os	**2,849.30**	**4**	**8**	Ta	2,850.985	150	400
Ru	**2,849.289**	**100**	**3**	Ti	2,851.10	80	20
Cr	**2,849.287**	**30**	**35**	Sb	2,851.11	45	50
				Cr	2,851.36	80	20
				Fe	2,851.80	150	200
Hf	2,849.21	100	30	Ce	2,852.12	1	50*d*
Fe	2,848.72	30	60	Mg	2,852.129	100*R*	300*R*
Ru	2,848.58	3	50	Fe	2,852.13	80	150
Ta	2,848.525	50	300	Ta	2,852.35	100	5
Zr	2,848.523	—	100	Na	2,852.83	20	100*R*
Mo	2,848.23	200*h*	125	V	2,852.87	7*h*	60
Ta	2,848.05	15	150	Na	2,853.03	15	80*R*
Hg	2,847.83	100	15	Cr	2,853.22	100*R*	5
Hg	2,847.67	(300)	—	Mo	2,853.23	100*h*	25
V	2,847.57	150	15	Ru	2,854.07	35	60
Lu	2,847.51	125	40				
Ta	2,846.75	10*h*	150*hs*				

	λ	I J	I O		λ	I J	I O
V	2,846.57	20*h*	50	V	2,854.34	100*R*	20
Os	2,846.39	10	40	Pd	2,854.58	500*h*	4
Fe	2,845.59	7	125				
Fe	2,845.54	7	125				
Ta	2,845.35	10	150				
V	2,845.24	80	18				
Ta	2,844.76	30	150				
Ru	2,844.71	150	—				
Tm	2,844.67	15	40				
Zr	2,844.58	50	50				
Ta	2,844.46	200	200				
Os	2,844.40	25	50				

*P** *N**

Cr 2,855.676 *J* 200 *O* 60*

	λ	I J	I O		λ	I J	I O
Bi	**2,855.674**	**12**	—	**Tb**	**2,855.68**	**20**	**5**
Fe	**2,855.670**	**200**	**2**	**Mo**	**2,855.71**	**20**	—
U	**2,855.60**	**8*h***	**12*h***	**Ir**	**2,855.82**	**2*h***	**10**
Nb	**2,855.54**	**10**	**1**	**J**	**2,855.87**	**(12)**	—
Re	**2,855.53**	—	**15**	**Th**	**2,855.900**	**8**	**10**
W	**2,855.51**	**10**	—	**La**	**2,855.902**	**50*h***	**3**
Ti	**2,855.49**	**5*wh***	—	**U**	**2,855.96**	**2**	**4**
Er	**2,855.41**	**3**	**15**	**Mo**	**2,855.99**	**25**	**2**
W	**2,855.35**	**3**	**9**	**W**	**2,856.03**	**9**	**10*s***
Ru	**2,855.340**	**18**	—	**Ru**	**2,856.05**	**1**	**6**
Os	**2,855.337**	**8**	**25**	**Zr**	**2,856.06**	**4*h***	**2**
Ar	**2,855.33**	**(5)**	—	**Fe**	**2,856.14**	**40**	**1**
V	**2,855.29**	**35**	**1**	**Rh**	**2,856.16**	**30*h***	**60**
V	**2,855.22**	**1**	**50**	**U**	**2,856.18**	**2**	**3**
U	**2,855.19**	**2**	**3**	**U**	**2,856.22**	**2**	**8**
Nb	**2,855.08**	**4**	—	**Ti**	**2,856.24**	**100*wh***	—
Cr	**2,855.073**	**100**	**4**	**Yt**	**2,856.29**	**15**	**8**
Se	**2,855.07**	**(5)**	—				
Tb	**2,854.95**	**3*h***	**5**				

*P** *N**

Cr 2,860.93 *J* 100 *O* 60*

	λ	I J	I O		λ	I J	I O
J	**2,860.92**	**(12)**	—	**Os**	**2,860.96**	**25**	**100**
Cs	**2,860.85**	**(8)**	—	**Na**	**2,861.011**	**(2)**	**1**
Ti	**2,860.84**	**25*wh***	—	**Hf**	**2,861.012**	**90**	**40**
U	**2,860.80**	**12**	**15**	**Te**	**2,861.03**	**(25)**	—
Rh	**2,860.76**	**10*h***	**30**	**W**	**2,861.05**	**5**	**1**
Ar	**2,860.73**	**(5)**	—	**Nb**	**2,861.09**	**100**	**10**
Cl	**2,860.71**	**(5)**	—	**Ru**	**2,861.10**	**20**	—

	λ	I	
		J	O
W	**2,860.679**	**9**	**6**
Pt	**2,860.678**	**150*h***	**30**
Ir	**2,860.66**	**2**	**12**
Pb	**2,860.64**	**2**	—
Hf	**2,860.56**	**2**	**20**
In	**2,860.52**	**5**	—
U	**2,860.47**	**30**	**35**
As	**2,860.45**	**50**	**50*r***
Yb	**2,860.40**	**8**	**1**
Hf	**2,860.312**	**30**	**15**
Lu	**2,860.31**	**3**	—
Nb	2,859.04	50	1
Cr	2,858.91	80*Wh*	50
Ta	2,858.43	300	100
Fe	2,858.34	200	3
Te	2,858.29	(100)	2
Ti	2,857.81	70*wh*	—
Ru	2,857.78	60	4
Pd	2,857.73	100	—
Cr	2,857.40	80	20
Cr	2,856.77	60	20
Ru	2,856.55	50	—
Ti	2,856.24	100*wh*	—
P*			

	λ	I	
		J	O
U	**2,861.13**	**15**	**10**
Fe	**2,861.19**	**30**	**1**
W	**2,861.21**	**6**	**2**
Yb	**2,861.21**	**30**	**3**
Ti	**2,861.30**	**15**	**7**
Yb	**2,861.31**	**25**	**3**
Tb	**2,861.34**	**10**	**10**
Th	**2,861.37**	**8*d***	**8*d***
O	**2,861.38**	**(10*h*)**	—
V	**2,861.40**	**15**	—
Ru	**2,861.41**	**35**	**60**
W	**2,861.44**	**7*d***	**8**
Ho	**2,861.49**	**10**	—
Mo	**2,861.56**	**3*h***	—
Hf	2,861.70	125	50
Ti	2,861.99	100*Wh*	—
Fe	2,862.50	50	100
Eu	2,862.57	70	100*W*
Cr	2,962.571	300*R*	80
Ru	2,862.88	60	6
Rh	2,862.93	60	150
Ru	2,863.32	80	30
Sn	2,863.33	300*R*	300*R*
Fe	2,863.43	80	100
Ni	2,863.70	250	—
Mo	2,863.81	100*h*	30
Fe	2,863.86	100	125
Ni	2,864.15	300*wh*	—
Pb	2,864.26	60	—
Pb	2,864.51	60	—
Xe	2,864.73	(100)	—
Fe	2,864.97	50	—
Pt	2,865.05	80*h*	20
Cr	2,865.11	200*R*	60
U	2,865.679	50	30
In	2,865.684	(50)	—
Fe	2,866.63	80	125
Cr	2,866.74	125*R*	80
Ta	*2,867.41*	*150*	*5h*
Cr	*2,867.65*	*100R*	*80*
Nb	*2,868.52*	*300*	*15*
Te	*2,868.86*	*(100)*	—
V	*2,869.13*	*150r*	*25*
Tm	*2,869.22*	*300*	*100*
Ti	*2,870.04*	*100wh*	—
Cr	*2,870.44*	*300W*	*25*

	λ	I J	I O		λ	I J	I O
				Co	*2,871.24*	*100*	*—*
				Mo	*2,871.51*	*100h*	*100*
				Fe	*2,873.47*	*300*	*—*

Cr 3,578.687 *J* 400*R* *O* 500*R*

	λ	J	O		λ	J	O
V	**3,578.64**	**80*d***	**35*d***	**Ti**	**3,578.687**	**5**	**25**
Nb	**3,578.584**	**1**	**15**	**Tb**	**3,578.70**	**—**	**8**
Gd	**3,578.58**	**—**	**8**	**Hg**	**3,578.747**	**(40)**	**—**
Pr	**3,578.57**	**1**	**2**	**Pr**	**3,578.75**	**5**	**8**
Eu	**3,578.49**	**1**	**2**	**Ta**	**3,578.76**	**1*h***	**2**
Dy	**3,578.47**	**2**	**4**	**Dy**	**3,578.80**	**1**	**2**
Pr	**3,578.42**	**3**	**9**	**Ce**	**3,578.82**	**—**	**3**
Tb	**3,578.40**	**—**	**8**	**La**	**3,578.89**	**3*h***	**2*h***
Fe	**3,578.38**	**5**	**40**	**Co**	**3,578.90**	**2**	**8**
Gd	**3,578.34**	**5**	**5**	**Se**	**3,578.93**	**(12)**	**—**
Er	**3,578.33**	**—**	**3*w***	**Ce**	**3,578.95**	**—**	**3**
U	**3,578.327**	**1**	**8**	**Co**	**3,579.03**	**—**	**5**
Ti	**3,578.27**	**1**	**8**	**Mo**	**3,579.07**	**25**	**—**
Nb	**3,578.234**	**5**	**1**	**Ta**	**3,579.076**	**1**	**15**
Zr	**3,578.226**	**6**	**5**	**Eu**	**3,579.08**	**1*h***	**6*w***
Eu	**3,578.110**	**1**	**6**	**V**	**3,579.09**	**1**	**7**
Pr	**3,578.106**	**5**	**10**	**Ho**	**3,579.12**	**4**	**—**
Co	**3,578.07**	**—**	**18**	**Dy**	**3,579.125**	**1**	**3**
Co	**3,578.03**	**30**	**—**	**Re**	**3,579.13**	**—**	**50**
Tb	**3,578.00**	**—**	**8**	**Tb**	**3,579.20**	**50**	**50**
Dy	**3,577.99**	**50**	**150**	**Th**	**3,579.34**	**8**	**4**
Rb	**3,577.96**	**(15)**	**—**	**Eu**	**3,579.36**	**1*h***	**4**
U	**3,577.92**	**—**	**10**	**U**	**3,579.364**	**10**	**6**
Mn	**3,577.88**	**25**	**50**	**Dy**	**3,579.42**	**1**	**2**
V	**3,577.87**	**40**	**50**	**Er**	**3,579.44**	**—**	**15*d***
Pr	**3,577.85**	**—**	**6**	**Ta**	**3,579.448**	**35**	**—**
Ce	**3,577.80**	**2**	**2**	**Ca**	**3,579.450**	**2**	**—**
U	**3,577.78**	**1**	**4**	**Sm**	**3,579.50**	**2**	**4**
Sm	**3,577.77**	**3**	**20**	**Gd**	**3,579.55**	**5**	**5**
Fe	**3,577.75**	**—**	**2**	**Fe**	**3,579.56**	**1**	**2**
Nb	**3,577.72**	**15**	**10**	**Mn**	**3,579.66**	**—**	**3**
Co	**3,577.67**	**—**	**2*h***	**Sm**	**3,579.668**	**3**	**4**
V	**3,577.644**	**3**	**—**	**Ba**	**3,579.672**	**8**	**10**
W	**3,577.64**	**6*d***	**—**	**Ru**	**3,579.77**	**8**	**3**
Ir	**3,577.62**	**—**	**4**	**Fe**	**3,579.83**	**—**	**1**
Ba	**3,577.61**	**—**	**4**				
P	**3,577.60**	**(50*d*)**	**—**	Nb	3,580.27	300	100
Kr	**3,577.60**	**(4*h*)**	**—**	Fe	3,581.19	600*r*	1000*R*
Zr	**3,577.55**	**1**	**12**	Dy	3,585.07	100	300
J	**3,577.53**	**(5)**	**—**	Te	3,585.34	(350)	—

	λ	I J	I O
Tm	**3,577.51**	**8**	**5**
U	**3,577.487**	—	**3**
Os	**3,577.487**	—	**3**
Pr	**3,577.465**	**8**	**30**
Ce	3,577.458	12	300
Ar	3,576.62	(300)	—
Dy	3,576.25	—	300
Ar	3,572.29	(390)	—
Ni	3,571.87	40	1000*R*
Rh	3,570.18	150	400*r*
Fe	3,570.10	300	300
Co	3,569.38	100	400*R*
Ar	3,567.66	(300)	—
Ni	*3,566.37*	*100wh*	*2000R*
Fe	*3,565.38*	*300*	*400*
Al	3 586.55	(200)	—
Al	3,586.69	(200)	—
Al	3,586.80	(200*wh*)	—
Al	3,586.91	(500*h*)	—

*N**

Cr 3,593.49 *J* 400*R* *O* 500*R*

	λ	J	O
Hg	**3,593.48**	**(10)**	—
Gd	**3,593.43**	**15**	**15**
Sm	**3,593.402**	**3**	**2**
Re	**3,593.397**	—	**15*w***
Te	**3,593.34**	**(5)**	—
V	**3,593.334**	**300*R***	**30**
Fe	**3,593.329**	**2**	**7**
La	**3,593.29**	—	**2**
Ba	**3,593.28**	—	**3**
K	**3,593.22**	**(5)**	—
Sr	**3,593.21**	**2**	**3**
U	**3,593.20**	**1**	**10**
Dy	**3,593.15**	**4**	**6**
Ce	**3,593.134**	—	**8**
Eu	**3,593.13**	—	**2*w***
Ho	**3,593.13**	**6**	**6**
Zr	**3,593.129**	**1**	**7**
Pb	**3,593.12**	**30**	—
*bh***Zr**	**3,593.1**	—	**2**
Tb	**3,593.10**	**8**	**15**
Ti	**3,593.09**	**30**	**5**
Pr	**3,593.04**	**2*h***	**5**
Ru	**3,593.02**	**150**	**60**
W	**3,592.98**	**3**	**5**
Hg	**3,592.97**	**2*h***	—
Ho	**3,592.95**	—	**6**
Eu	**3,592.93**	**1*h***	**7**
Ne	**3,593.526**	**(500)**	—
Rh	**3,593.530**	**2**	**10**
Nd	**3,593.55**	**3**	**15**
W	**3,593.56**	**3**	**4**
N	**3,593.60**	**(10)**	—
Tl	**3,593.61**	**(10)**	—
Ne	**3,593.64**	**(250)**	—
Pr	**3,593.685**	**2**	**4**
U	**3,593.685**	**1**	**8**
Sm	**3,593.73**	**2**	**4**
Tb	**3,593.75**	**8**	**30**
Th	**3,593.88**	**5**	**5**
Rh	**3,593.89**	—	**3**
Er	**3,593.95**	—	**8**
Nb	**3,593.966**	**50**	**80**
W	**3,593.971**	**8**	**9**
Sm	**3,593.99**	—	**2**
Pt	**3,594.01**	—	**3**
Cu	**3,594.02**	**2**	**15**
Ce	**3,594.03**	—	**5**
Ce	**3,594.10**	**2**	**5**
U	**3,594.11**	**4**	**1**
Th	**3,594.116**	**2**	**5**
Ca	**3,594.12**	**2**	**3**
Er	**3,594.13**	—	**9**
Ir	**3,594.14**	**4**	**10**
Au	**3,594.15**	**6**	**25**

	λ	I J	I O		λ	I J	I O
Pb	**3,592.92**	**3**	—	**Ne**	**3,594.18**	**(12)**	—
Yt	**3,592.91**	**25**	**80**	**Tb**	**3,594.25**	**8**	**15**
Sm	**3,582.90**	—	**4**	**Cr**	**3,594.31**	**4**	—
Fe	**3,592.89**	—	**3**	**Ir**	**3,594.39**	**30**	**15**
W	**3,592.85**	**3*d***	**6*d***	**Hf**	**3,594.43**	**2**	—
U	**3,592.801**	**6**	**1**	**S**	**3,594.46**	**(35)**	—
Xe	**3,592.80**	**(2)**	—	**W**	**3,594.53**	**7**	**5**
Th	**3,592.78**	**6**	**2*d***	**Mo**	**3,594.55**	**4**	**3**
Gd	**3,592.70**	**70**	**50**	**Dy**	**3,594.57**	**2**	**4**
Fe	**3,592.69**	**2**	**12**	**vzduch**	**3,594.60**	**3**	—
Mo	**3,592.65**	**20**	—	**Fe**	**3,594.64**	**100**	**125**
Sm	**3,592.595**	**50**	**40**	Co	3,594.87	—	200*W*
Nd	**3,592.595**	**30**	**20**				
V	**3,592.53**	—	**40**				
Ta	**3,592.495**	**1*h***	**2**				
Fe	**3,592.486**	**1**	**3**				
Cs	**3,592.48**	**(4)**	—				
W	**3,592.42**	**35**	**9**				
Bi	**3,592.40**	**(5)**	—				
V	3,592.02	300*R*	50				
Nd	3,590.35	300*W*	400*W*				
V	3,589.76	600*R*	80				
Ar	3,588.44	(300)	—				

*P** *N**

Cr 3,605.33 *J* 400*R* *O* 500*R*

	λ	J	O		λ	J	O
U	**3,605.28**	—	**8**	**Eu**	**3,605.34**	**1*h***	**3**
Gd	**3,605.25**	**15**	**15**	**Co**	**3,605.36**	—	**60**
Fe	**3,605.211**	**2**	**12**	**Fe**	**3,605.458**	**150**	**300**
Ca	**3,605.21**	**2r**	**2**	**Yt**	**3,605.46**	**6*h***	**2*h***
Dy	**3,605.09**	**2**	**4**	**U**	**3,605.48**	**4**	**1**
Pr	**3,605.05**	**10**	**20**	**Cs**	**3,605.53**	**(4)**	—
Mo	**3,605.02**	**5**	**2**	**V**	**3,605.59**	**20*h***	**30**
Co	**3,605.01**	—	**15**	**Rn**	**3,605.60**	**(10)**	—
Sm	**3,604.99**	**1**	**3**	**Eu**	**3,605.61**	—	**4*w***
Ta	**3,604.98**	**1*h***	**7**	**Gd**	**3,605.62**	**15**	**20**
Ce	**3,604.93**	**1**	**6**	**Ru**	**3,605.64**	**9**	**2**
Er	**3,604.901**	**4**	**12**	**Th**	**3,605.66**	**6**	**8**
Tb	**3,604.90**	**15**	**3**	**Mn**	**3,605.691**	—	**10**
Gd	**3,604.88**	**12**	**50**	**Er**	**3,605.695**	—	**3**
Dy	**3,604.85**	**2**	**4*w***	**U**	**3,605.75**	**4**	**1**
Er	**3,604.715**	**2**	**9**	**Ce**	**3,605.78**	—	**5**
Sm	**3,604.712**	—	**2**	**Hg**	**3,605.80**	**(200)**	—
Ce	**3,604.70**	**1**	**3**	**U**	**2,605.82**	**8**	**8**
Mn	**3,604.69**	—	**12**	**Rh**	**3,605.86**	**30**	**25**
Eu	**3,604.684**	**1*h***	**7**	**Ar**	**3,605.89**	**(15)**	—

	λ	I: J	I: O
Th	3,604.680	1	5
Ce	3,604.65	—	3
Nb	3,604.64	10*w*	5*w*
Rb	3,604.60	(2)	—
Mo	3,604.56	5	5
Ce	3,604.52	1	2
*bh*La	3,604.5	—	5*h*
Os	3,604.475	100	15
Co	3,604.470	—	3
U	3,604.396	6	1
Re	3,604.396	—	15*w*
Fe	3,604.380	1	10
V	3,604.377	50*h*	—
Dy	3,604.36	1	3
Ir	3,604.31	—	5
Ti	3,604.284	6	20
Sm	3,604.276	20	6
Fe	3,604.27	—	5
U	3,604.22	6	1
Ce	3,604.20	1	8
W	3,604.16	3	—
Yt	3,600.73	300	100
Zr	3,605.90	—	3
Fe	3,605.91	1	3
Pr	3,605.96	3	10
Ti	3,605.06	1	12
W	3,606.07	10	12
Er	3,606.08	4	15*d*
Ho	3,606.09	4	—
U	3,606.116	6	1
Tb	3,606.12	15*d*	30
Dy	3,606.126	100	200
Ce	3,606.130	—	5
Ru	3,606.15	1	2
Th	3,606.20	5*h*	—
Eu	3,606.21	1*h*	2*h*
Nb	3,606.27	5*wh*	2
Nd	3,606.28	4	12
Mo	3,606.29	10	1
U	3,606.32	12	8
Sm	3,606.33	1	2
W	3,606.34	8	10
Nb	3,606.35	3*h*	—
Eu	3,606.38	—	5
Nd	3,606.40	2	12
La	3,606.43	3*h*	2
Yb	3,606.47	60	15
Ar	3,606.52	(1000)	—
Fe	3,608.86	400	500
Pd	3,609.55	700*R*	1000*R*
Ni	3,610.46	—	1000*r*
Cd	3,610.51	500	1000
Ni	3,612.74	50*h*	400
Cd	3,612.87	500	800
Fe	*3,618.77*	*400*	*400*
Ni	*3,619.39*	*150h*	*2000R*

*P**

Cr 4,254.35 *J* 1000 *O* 5000*R*

	λ	J	O
Er	4,254.32	1	7
W	4,254.29	3	4
Bi	4,254.15	10	10
Ga	4,254.13	2	—
Be	4,254.12	(5)	—
W	4,254.06	2	8
Gd	4,254.03	—	3
Ce	4,254.004	—	3
Ce	4,254.37	—	8
Nb	4,254.39	15	10
Pr	4,254.420	18	35
V	4,254.425	5	3
Mo	4,254.429	5	10
Ho	4,254.43	20	100
Th	4,254.46	6	8
Nb	4,254.69	10	10

	λ	I J	I O		λ	I J	I O
Tb	**4,254.00**	—	**5**	**Ce**	**4,254.70**	—	**20**
O	**4,253.98**	**(100*h*)**	—	**N**	**4,254.75**	—	**15**
Fe	**4,253.93**	**1**	**2**	**Yb**	**4,254.77**	**4**	**4**
Th	**4,253.875**	**4**	**6**	**Kr**	**4,254.85**	**(100*h*)**	—
Nd	**4,253.868**	**3**	**12**	**Ce**	**4,254.90**	—	**8**
U	**4,253.85**	**3**	**12**	**Fe**	**4,254.94**	**1**	**2**
Ce	**4,253.83**	—	**3**	**Ar**	**4,254.95**	**(10)**	—
Eu	**4,253.81**	**3**	**15**	**Mo**	**4,254.955**	**25**	**25**
Be	**4,253.76**	**(15)**	—	**Pr**	**4,255.00**	**1**	**4**
O	**4,253.74**	**(50*h*)**	—	**Re**	**4,255.11**	—	**2*h***
Sm	**4,253.72**	**2**	**5**	**Ce**	**4,255.18**	—	**2**
U	**4,253.70**	**2**	**8**	**Tb**	**4,255.24**	**1*h***	25
Nb	**4,253.69**	**40**	**25**	**Eu**	**4,255.27**	—	**15**
Dy	**4,253.65**	—	**3**	**Re**	**4,255.34**	—	**2*h***
Gd	**4,253.62**	**50**	**50**	**Ce**	**4,255.36**	—	**10**
Er	**4,253.582**	**2**	**9**	**U**	**4,255.39**	**1*h***	**2*h***
Mo	**4,253.576**	**5**	**8**	**Gd**	**4,255.44**	—	**10**
Zr	**4,253.57**	—	**20**	**Nb**	**4,255.442**	**50**	**30**
Th	**4,253.54**	**3**	**8**	**Fe**	**4,255.499**	**2**	**5**
Cl	**4,253.51**	**(75)**	—	**Cr**	**4,255.502**	**30**	**30**
Sr	**4,253.50**	—	**2**				
Gd	**4,253.37**	**4**	**50**	Ar	4,259.36	(1200)	—
Ce	**4,253.36**	**3**	**40*s***				
Cu	**4,253.34**	—	**7*wh***				
N	**4,253.28**	**(15)**	—				
Be	**4,253.05**	**(20)**	—				
Ar	4,251.18	(800)	—				
Sc	4,246.83	500	80				

*N**

Cr 4,274.80 *J* 800*R* *O* 4000*R*

	λ	I J	I O		λ	I J	I O
Zr	**4,274.77**	—	**9**	**Nb**	**4,274.89**	**5*h***	**5**
Nb	**4,274.69**	**3**	**5**	**Os**	**4,274.90**	**1**	**9**
Ne	**4,274.66**	**(50)**	—	**W**	**4,274.94**	**5**	**10**
Ti	**4,274.58**	**40**	**100**	**Pr**	**4,274.96**	—	**3**
W	**4,274.55**	**12**	**20**	**Tl**	**4,274.98**	**(100)**	—
Mo	**4,274.41**	**30**	—	**Dy**	**4,275.00**	—	**5**
Ti	**4,274.40**	—	**5**	**U**	**4,275.02**	**1**	**2**
Tb	**4,274.36**	—	**2*h***	**Co**	**4,275.07**	—	**3**
Re	**4,274.34**	—	**20*w***	**Nd**	**4,275.08**	**10**	**20**
Th	**4,274.33**	**4**	**8**	**Cu**	**4,275.13**	**30**	**80**
Pr	**4,274.27**	**1*d***	**8*d***	**W**	**4,275.15**	**2**	**6**
Tl	**4,274.24**	**(8)**	—	**Pr**	**4,275.17**	**2**	**12**
Yt	**4,274.18**	—	**2*h***	**Ar**	**4,275.19**	**(10)**	—
Gd	**4,274.17**	—	**100**	**F**	**4,275.21**	**(100*h*)**	—
Sm	**4,274.16**	**1**	**5**	**Tb**	**4,275.21**	**1*h***	**15**

	λ	I J	I O		λ	I J	I O
Mo	4,274.05	6	6	Pr	4,275.32	1	5
Dy	4,274.04	2	5	Tb	4,275.37	—	3
Sm	4,274.035	1	3	Dy	4,275.45	—	5
Th	4,274.032	8	10	O	4,275.47	(50*h*)	—
U	4,273.975	15	12	W	4,275.49	10	15
Kr	4,273.970	(1000)	—	Ne	4,275.560	(70)	—
Pt	4,273.909	—	2	Ce	4,275.561	4	25
Ca	4,273.907	2	—	La	4,275.64	500	40
Fe	4,273.87	2	10	Mo	4,275.68	12	12
Ce	4,273.79	—	2	U	4,275.73	8	12
Tb	4,273.74	—	3*W*	Kr	4,275.75	(2)	—
Nd	4,273.739	2	10	Nd	4,275.76	3	15
Hg	4,273.72	(2*h*)	—	Pr	4,275.82	4	18
Rb	4,273.70	2	—	Eu	4,275.91	—	4
W	4,273.69	5	10				
Ho	4,273.63	2	2				
Ar	4,272.17	(1200)	—				
Fe	4,271.76	700	1000				
C	4,267.27	500	—				
Ar	4,266.29	(1200)	—				

*P** *N**

Cr 4,289.72 *J* 800*R* *O* 3000*R*

	λ	I J	I O		λ	I J	I O
Se	4,289.65	(10)	—	Tb	4,289.73	—	30
U	4,289.56	3	4	Ne	4,289.80	(2)	—
Ce	4,289.45	4	25	U	4,289.88	12	12
Nb	4,289.44	20	10	Pr	4,289.89	4	15
Pr	4,289.42	3	12	Gd	4,289.90	100	40
Mo	4,289.41	15	20	Ti	4,289.92	2	15
Sm	4,289.365	2	15	Ce	4,289.94	25	50
Ca	4,289.364	20	35	Gd	4,290.07	—	6
Nd	4,289.363	3*h*	15	Mn	4,290.11	—	8
Dy	4,289.35	—	2	Ca	4,290.12	5*h*	—
Sb	4,289.30	(6)	—	W	4,290.14	3	8
W	4,289.29	5	1	Ho	4,290.17	2	2
Ar	4,289.09	(5)	—	Mo	4,290.18	25*h*	30*h*
Ti	4,289.07	50	125	Zr	4,290.21	20	40
La	4,289.01	—	4	Ti	4,290.23	60	35
Fe	4,288.96	1	5	Ce	4,290.29	—	2
U	4,288.84	2	20	Nb	4,290.35	2*h*	—
V	4,288.81	2	3	Fe	4,290.38	5	35
Rh	4,288.71	100	400	Ne	4,290.40	(100)	—
Th	4,288.68	1	5	Ce	4,290.43	—	2
Ce	4,288.67	—	30	Dy	4,290.44	—	5
K	4,288.65	(15)	—	Pr	4,290.51	—	4

	λ	I J	I O		λ	I J	I O
Mo	**4,288.64**	**100**	**80**	**Ru**	**4,290.53**	—	**6**
Eu	**4,288.60**	**1**	**10**	**Se**	**4,290.57**	**(20)**	—
				Ce	**4,290.59**	—	**4**
				Na	**4,290.60**	—	**3**
				Kr	**4,290.78**	**(4)**	—
				Kr	4,292.92	(600)	—
				Fe	4,298.04	400	100
				Fe	4,299.24	400	500
				Ar	4,300.10	(1200)	—
				Fe	4,307.91	800*R*	1000*R*

*P**

Cr 5,204.52 *J* 100 *O* 400*R*

5,206.04 *J* 200 *O* 500*R*

	λ	J	O		λ	J	O
W	**5,204.514**	—	**40**	**Fe**	**5,204.58**	—	**125**
Tm	**5,204.51**	**10**	**6**	**Ce**	**5,204.73**	—	**8***w*
Nd	**5,204.38**	—	**2**	**Hg**	**5,204.78**	**(40)**	—
U	**5,204.32**	**10**	**10**	**Nb**	**5,205.13**	**5**	**8**
Ce	**5,204.27**	—	**10**	**Ce**	**5,205.14**	—	**10**
J	**5,204.20**	**(50)**	—	**W**	**5,205.15**	**6**	—
La	**5,204.15**	**300**	**50**	**U**	**5,205.18**	**6***h*	**10**
Nb	**5,204.03**	**1**	**3**	**Sm**	**5,205.39**	**2**	**15**
Mo	**5,203.94**	**5**	**12**	**As**	**5,205.40**	**12**	—
Ne	**5,203.89**	**(150)**	—	**Ce**	**5,205.52**	—	**8**
Th	**5,203.86**	—	**5**	**J**	**5,205.54**	**(8)**	—
Sm	**5,203.67**	—	**3**	**Yt**	**5,205.72**	**80**	**50**
Rh	**5,203.33**	—	**8**	**Th**	**5,205.78**	**1**	**8**
Ce	**5,203.28**	—	**8**	**Ar**	**5,205.79**	**(10)**	—
W	**5,203.26**	—	**30**	**Ra**	**5,205.93**	**(250)**	—
Os	**5,203.23**	—	**12**	*bh***Mg**	**5,206.0**	—	**3**
Nb	**5,203.22**	**8**	**15**				
Si	**5,202.85**	**2**	—				
Tb	**5,202.77**	—	**15**	Cr	*5,206.04*	*200*	*500R*
Sm	**5,202.73**	—	**50**				
Os	**5,202.63**	—	**30***h*	**Ti**	**5,206.08**	**1**	**40**
				W	**5,206.19**	—	**30**
				Ta	**5,206.27**	—	**5***w*
Fe	5,202.34	10	300				
Eu	5,200.92	—	300	**Tl**	**5,206.29**	**(2)**	—
Sm	5,200.59	1	200	**Eu**	**5,206.43**	—	**40**
Yt	5,200.41	150	60	**Lu**	**5,206.47**	—	**10**
Kr	5,200.22	(60*whs*)	—	**Th**	**5,206.49**	—	**6**
				Er	**5,206.52**	—	**20**
Eu	5,199.85	—	500				
Fe	5,194.95	15	—	**Ta**	**5,206.55**	—	**2**
Ne	5,193.22	(150)	200				
Rh	5,193.14	3	200				
Ne	5,193.13	(150)	—				

*N**

	λ	I J	I O		λ	I J	I O
V	5,192.99	75*h*	100				
Ar	5,192.72	(60)	—				
Fe	5,192.36	50	400				
Xe	5,192.10	(50)	—				
Fe	5,191.47	35	400				
P	5,191.40	(100)	—				
Xe	5,191.37	(200)	—				
Ti	5,188.70	100	80				
Ne	5,188.61	(150)	—				
La	5,188.23	500	50				
Xe	5,188.11	(100)	—				
Ar	5,187.75	(800)	—				
Kr	5,186.99	(60*whs*)	—				
In	*5,184.44*	*(300)*	—				
Mg	*5,183.62*	*300*	*500wh*				
La	*5,183.42*	*400*	*300*				
Br	*5,182.36*	*(100)*	—				
Ge	*5,178.58*	*100*	—				
Co	*5,176.08*	—	*500r*				
Se	*5,175.98*	*(600)*	—				
In	*5,175.56*	*(150)*	—				
In	*5,175.42*	*(300)*	—				
In	*5,175.29*	*(400)*	—				
Mg	*5,172.70*	*100wh*	*200wh*				
Fe	*5,169.03*	*200h*	*2*				
Fe	*5,167.49*	*150*	*700*				

Cr 5,208.44 *J* 100 *O* 500*R*

	λ	J	O		λ	J	O
Kr	**5,208.32**	**(500)**	—	**Pt**	**5,208.59**	—	**2**
Ar	**5,208.04**	**(10)**	—	**Fe**	**5,208.60**	**8**	**200**
W	**5,207.974**	—	**5**	**Bi**	**5,208.80**	**70**	**5***wh*
Tb	**5,207.97**	—	**15**	**Sb**	**5,208.80**	**(8)**	—
Pr	**5,207.90**	**1**	**15**	**Ne**	**5,208.86**	**(70)**	—
Eu	**5,207.89**	—	**20***h*	**Ce**	**5,208.90**	—	**2**
Ti	**5,207.87**	—	**25**	**Pd**	**5,208.91**	—	**10**
*bh***F**	**5,207.8**	—	**2**	**Ag**	**5,209.07**	**1000***R*	**1500***R*
U	**5,207.79**	—	**4**	**W**	**5,209.10**	—	**10**
V	**5,207.68**	**8**	**8**	**Bi**	**5,209.29**	**600***h*	—
Sm	**5,207.64**	—	**3**	**Zr**	**5,209.30**	—	**5**
Ce	**5,207.34**	—	**8**	**Tb**	**5,209.34**	—	**15**
Ar	**5,207.17**	**(10)**	—	**Cs**	**5,209.44**	**(15)**	—
Sm	**5,207.15**	—	**25***d*	**Ru**	**5,209.50**	—	**7**
Cu	**5,207.13**	**20**	—	**Cs**	**5,209.62**	**(15)**	—
Rh	**5,206.95**	**1**	**6**	**Nd**	**5,209.90**	—	**2**
*bh***F**	**5,206.9**	—	**2**	**Sm**	**5,209.93**	—	**20**

	λ	I	
		J	O
Th	**5,206.66**	**1**	**4**
O	**5,206.61**	**(60)**	**—**
V	**5,206.608**	**25**	**25**
Ne	**5,206.565**	**(3)**	**—**
Pr	**5,206.562**	**2**	**30**
Ta	**5,206.55**	**—**	**2**
*P**			
Co	**5,210.06**	**—**	**100**
Er	**5,210.28**	**—**	**12**
Ti	5,210.39	35	200
Ar	5,210.49	(200)	—
Ne	5,210.57	(50)	—
Hg	5,210.79	(60	—
La	5,211.87	5	300*r*
Cc	5,212.71	—	300*w*
Ar	5,214.77	(200)	—
Eu	5,215.09	—	1000
Fe	5,215.18	5	200
Fe	5,216.278	10	300
Ar	5,216.28	(60)	—
Cl	5,217.92	(100)	—
Cu	5,218.20	—	700
Ar	5,221.27	(500)	—
Cl	5,221.34	(70)	—
Ne	5,222.35	(50)	—
Hg	5,222.81	(80)	—
Eu	5,223.48	—	700
Ti	5,226.55	50*h*	30
Fe	5,226.87	15	200
Cs	5,227.00	(200)	—
Fe	5,227.19	60	400
Se	*5,227.51*	*(600)*	—
Kr	*5,229.52*	*(60)*	—
Fe	*5,229.87*	*15h*	*200*
Fe	*5,232.94*	*150*	*800*
Br	*5,238.23*	*(100)*	—
Ti	*5,238.58*	*100*	*50*

Cs $^{55}_{132.905}$

t_0 28.5°C t_1 690°C

I.	II.	III.	IV.	V.
3.893	32.453	~35	~51	~58

λ	*I*		eV
	J	*O*	
I 3,876.39	—	300	3.17
I 3,888.65	10	150	3.16
I 4,555.35	100	2000*R*	2.7
I 4,593.18	50	1000*R*	2.7

Cs 3,876.39 *O* 300

	λ	*I*	
		J	*O*
Fe	**3,876.38**	—	**2**
Ce	**3,876.37**	—	**2**
Pr	**3,876.18**	**30**	**80**
Ce	**3,876.14**	**1**	**6**
U	**3,876.134**	**2**	**15**
Tb	**3,876.13**	—	**8**
V	**3,876.09**	**30**	**50**
Ru	**3,876.08**	**3**	**20**
Fe	**3,876.04**	**15**	**40**
Eu	**3,875.95**	—	**3***W*
V	**3,875.902**	**3**	**40**
Ce	**3,875.902**	**2***s*	**5**
Nd	**3,875.87**	**4**	**30**
Ca	**3,875.81**	—	**50**
Nb	**3,875.76**	**50**	**10**
Nd	**3,875.74**	**6**	**15**
Nb	**3,875.70**	—	**3***h*
W	**3,875.68**	**9**	**12**
Th	**3,875.65**	—	**3**
Sm	**3,875.54**	**10**	**5**
V	**3,875.43**	**2**	**10**
Nb	**3,875.42**	**10**	**5**
Fe	**3,875.383**	**1***h*	**2***h*
Th	**3,875.379**	**5**	**10**
Eu	**3,875.35**	—	**10***W*
U	**3,875.34**	**6**	**12**
Gd	**3,875.31**	—	**4**
Ce	**3,875.261**	**2**	**4**
Ti	**3,875.256**	**8**	**35***h*
Re	**3,875.25**	—	**40**
Tb	**3,875.21**	**8**	**20**
Ta	**3,875.21**	**2***h*	**5***h*
Cr	**3,875.209**	**1***h*	**2***h*
Sm	**3,875.17**	**10**	**50**
U	**3,875.16**	—	**8**
Dy	**3,875.15**	—	**3**
V	**3,875.075**	**50**	**70***r*
***A*u**	**3,875.072**	—	**3**
Ce	**3,875.04**	—	**6**
Th	**3,874.87**	**3**	**5**
Tb	3,874.18	200	200
Co	3,873.95	80	400*R*
Co	3,873.11	80	500*R*
Fe	3,872.50	300	300
Dy	3,872.12	150	300
La	3,871.63	15	200

	λ	*I*	
		J	*O*
Er	**3,876.45**	**1**	**15***d*
Tb	**3,876.47**	—	**3**
Lu	**3,876.55**	—	**3**
Ta	**3,876.56**	**5**	**10**
U	**3,876.59**	**4**	**6**
Lu	**3,876.63**	**100**	**50**
Ru	**3,876.65**	**1**	**12**
Tb	**3,876.67**	—	**8**
Fe	**3,876.671**	—	**1**
Lu	**3,876.69**	—	**15**
Nd	**3,876.73**	—	**3**
V	**3,876.74**	**2**	**25**
Os	**3,876.77**	**50**	**300**
Yt	**3,876.82**	**3**	**5**
Co	**3,876.83**	**40**	**300***w*
Re	**3,876.89**	—	**60**
U	**3,876.95**	**2**	**3**
Er	**3,876.95**	**1***w*	**15***d*
Nb	**3,876.96**	**5**	**5**
Ce	**3,876.97**	**3***s*	**15**
Hf	**3,877.10**	**30**	**4**
Ce	**3,877.12**	—	**4**
Sm	**3,877.18**	**5**	**30**
Pr	**3,877.22**	**80***w*	**125***w*
Eu	**3,877.27**	—	**5***w*
Os	**3,877.31**	**10**	**20**
Rh	**3,877.34**	**5**	**20**
U	**3,877.45**	**4***h*	**5**
Ag	**3,877.47**	**1**	**2**
Sm	**3,877.473**	**8**	**50**
Fe	**3,877.51**	—	**2**
Er	**3,877.52**	**1**	**15***d*
Ru	**3,877.53**	**5***h*	**3**
Nb	**3,877.558**	**20**	**50**
Ce	**3,877.559**	—	**3***s*
Tb	**3,877.56**	—	**6**
Ti	**3,877.592**	—	**3**
Zr	**3,877.595**	**1**	**10**
Eu	**3,877.88**	—	**4***w*
Fe	3,878.02	300	400
*N**			

	λ	I	
		J	O
Fe	3,867.22	100	150
Fe	3,865.53	400	600
Pr	3,865.46	125*r*	200*r*
Mo	3,864.11	500*R*	1000*R*
Co	*3,861.16*	*15*	*300R*
Fe	*3,859.91*	*600*	*1000r*
Ni	*3,858.30*	*70h*	*800r*
Fe	*3,856.37*	*300*	*500*
Fe	*3,849.97*	*400*	*500*
Tm	*3,848.02*	*250*	*400*

Cs 3,888.65 *J* 10* *O* 150

	λ	J	O
U	3,888.61	3	4
Fe	3,888.52	3*w*	4*w*
Dy	3,888.40	—	6
Ce	3,888.39	4	15
Rh	3,888.335	2	5
V	3,888.329	1	10
Pr	3,888.29	4	5
Bi	3,888.23	2	40
Tb	3,888.21	—	30
U	3,888.206	6	12
Mo	3,888.18	8	10
Ce	3,888.11	—	2
Er	3,888.09	1	18
V	3,888.08	4	15
Ti	3,888.02	—	15
Mo	3,887.96	4	3
Re	3,887.95	—	20
Bi	3,887.93	2	5
Tb	3,887.88	—	10
Nd	3,887.87	20	25
Eu	3,887.81	—	5*w*
Yt	3,887.78	3	3
Ru	3,887.77	8	15
Gd	3,887.74	—	5
U	3,887.70	10	20
Mo	3,887.672	3	3
Tb	3,887.67	3	25
Dy	3,887.54	2	12
Tb	3,887.49	—	4
Re	3,887.48	—	20
U	3,887.45	—	20
Ir	3,887.38	—	3
Ti	3,887.36	—	5
Eu	3,888.68	5*w*	8*w*
Fe	3,888.82	15	40
Mo	3,888.87	8	8
W	3,888.89	4	3
Gd	3,888.93	—	6
Ho	3,888.95	20	40
Dy	3,888.99	—	20
Ce	3,888.997	3	12
Zr	3,889.004	2*h*	2*h*
Er	3,889.02	1	9
Ca	3,889.14	—	8
Sm	3,889.15	4	4
Sm	3,889.216	7	5
Nd	3,889.218	15	10
Hf	3,889.232	1*h*	4
V	3,889.235	2	12
Mo	3,889.285	3	2
U	3,889.285	—	12
Ce	3,889.30	1	6
Ba	3,889.32	2	10
Hf	3,889.328	1	5
Pr	3,889.330	70	150
W	3,889.37	5	—
U	3,889.41	5	1
Pr	3,889.42	6	10
Zr	3,889.446	—	2
Mn	3,889.450	50	25
Au	3,889.47	8	3
Ce	3,889.48	—	4
Eu	3,889.51	5*W*	10*W*
Ce	3,889.55	—	3
Ir	3,889.58	3	2
Nb	3,889.63	5	3

	λ	I	
		J	O
Tm	**3,887.35**	**8**	**80**
Yb	**3,887.31**	**40**	**6**
Tb	**3,887.29**	**—**	**2**
Gd	**3,887.18**	**10**	**10**
Er	**3,887.16**	**1**	**2**
Cr	3,886.79	125	125
La	3,886.37	200	400
Fe	3,886.28	400	600
Fe	3,885.51	60	100
Pr	3,885.19	40*w*	100*w*
Co	3,884.61	—	100
Fe	3,884.36	35	80
Tm	3,883.43	30	150
Tm	3,883.13	10	100
Co	3,881.87	30	300*R*
Os	3,881.86	20	125
Pr	3,880.47	60	80
Pr	3,879.21	80	100
Fe	3,878.57	300	300*R*
Fe	3,878.02	300	400
Pr	3,877.22	80*w*	125*w*
P*			

	λ	I	
		J	O
Nd	**3,889.66**	**25**	**30**
Ni	**3,889.67**	**10*h***	**30**
Er	**3,889.80**	**1**	**15**
Tb	**3,889.85**	**3**	**10**
Fe	**3,889.92**	**—**	**2**
Nd	**3,889.93**	**20**	**30**
Mo	**3,889.953**	**4**	**2**
Ti	**3,889.954**	**5**	**25**
Re	**3,889.96**	**—**	**25**
Pr	**3,889.97**	**3**	**5**
Ce	**3,889.99**	**8**	**50**
U	**3,890.01**	**3**	**1**
Sm	**3,890.07**	**10**	**10**
Pr	**3,890.17**	**2**	**8**
V	3,890.18	30	100
Zr	3,890.32	6	150
Ho	3,891.02	40	200
Zr	3,891.38	5	100
Fe	3,891.93	70	100
Fe	3,893.39	8	100
Co	3,894.08	100	1000*R*
Pd	3,894.20	200*W*	200*W*
Gd	3,894.71	80	150*W*
Co	3,894.98	3	300*R*
Fe	3,895.66	300	400
Fe	3,897.89	60	100
Ce	3,898.27	6	80
Co	3,898.49	6	80*r*
Dy	3,898.54	—	100
Tb	3,899.19	100	200
Fe	3,899.71	300	600
Zr	3,900.52	—	100
Tm	3,900.79	50	80
Os	3,901.71	20	150
Gd	3,902.40	80	100
Cr	3,902.91	100	100
Fe	3,902.95	400	500
Mo	3,902.96	500*R*	1000*R*
Co	*3,906.29*	*—*	*150*
Fe	*3,906.48*	*200*	*300*
Eu	*3,907.11*	*500R*	*1000RW*
Cr	*3,908.75*	*150*	*200*
Co	*3,909.93*	*—*	*200 W*
Sc	*3,911.81*	*30*	*150*
Pr	*3,912.90*	*80*	*150*
Li	*3,915.00*	*—*	*200wh*

	λ	I	
		J	O
Ir	*3,915.38*	*50*	*150*
La	*3,915.04*	*400*	*400*
Gd	*3,916.59*	*100*	*150w*
Fe	*3,917.18*	*70*	*150*

Cs 4,555.35 *J* 100 *O* 2000*R*

	λ	J	O
W	**4,555.33**	**1**	**7**
Yt	**4,555.298**	**2**	**2**
Cr	**4,555.296**	—	**15**
Te	**4,555.27**	**(30)**	—
Tm	**4,555.26**	**50**	**25**
Dy	**4,555.24**	—	**4**
Nd	**4,555.14**	—	**15**
Zr	**4,555.13**	—	**15**
U	**4,555.095**	**40**	**20**
Cr	**4,555.092**	**50**	**15**
Ti	**4,555.08**	**2**	**12**
Th	**4,555.07**	—	**3**
Cr	**4,555.03**	**40*h***	—
Gd	**4,554.99**	**2**	**6**
Nd	**4,554.97**	**1**	**5**
Cr	**4,554.83**	**2**	**25**
Ne	**4,554.82**	**(40)**	—
P	**4,554.80**	**(100)**	—
Pr	**4,554.79**	**1**	**4**
Ir	**4,554.78**	—	**4**
W	**4,554.68**	—	**4**
Tm	**4,554.65**	—	**5**
Pt	**4,554.593**	**5**	**10**
Ca	**4,554.592**	**2**	—
Ne	**4,554.561**	**(5)**	—
Ce	**4,554.557**	—	**6**
Ru	**4,554.51**	**200**	**1000*R***
Pr	**4,554.50**	—	**2**
Yt	**4,554.46**	**6**	**3**
Sm	**4,554.44**	—	**60**
Ne	**4,554.41**	**(10)**	—
Ce	**4,554.33**	—	**2**
Ar	**4,554.32**	**(15)**	—
Dy	**4,554.23**	—	**2**
Ba	**4,554.04**	**200**	**1000*R***
Ce	**4,554.035**	—	**35*s***
Mo	**4,554.028**	**4**	**1**
Zr	**4,553.97**	**12**	**4**
Eu	**4,555.38**	**4**	**4**
Ne	**4,555.39**	**(30)**	—
Ce	**4,555.42**	—	**5**
Ti	**4,555.49**	**60**	**125**
Zr	**4,555.52**	**2**	**30**
Nb	**4,555.56**	**2*h***	**3*h***
Th	**4,555.61**	**10*h***	—
Ce	**4,555.62**	—	**2**
Er	**4,555.69**	—	**3**
Eu	**4,555.71**	—	**12*W***
Th	**4,555.81**	—	**3**
Fe	**4,555.89**	**12**	**12**
Cu	**4,555.92**	**70**	**2**
Xe	**4,555.94**	**(100*wh*)**	—
Pr	**4,555.98**	—	**3**
U	**4,556.01**	**4**	—
Re	**4,556.02**	—	**2**
Mo	**4,556.03**	**4**	**2**
Fe	**4,556.12**	**35**	**150**
Nd	**4,556.14**	**1**	**20**
Cr	**4,556.17**	**12**	**40**
W	**4,556.219**	**1**	**6**
Ce	**4,556.224**	—	**3**
Pr	**4,556.26**	—	**4**
U	**4,556.33**	**3**	**1**
Ta	**4,556.35**	**5**	**200**
Tb	**4,556.45**	—	**20*w***
Dy	**4,556.46**	**4**	**3**
Sm	**4,556.50**	—	**5**
Br	**4,556.55**	**(4)**	—
Kr	**4,556.61**	**(200*h*)**	—
Sm	**4,556.63**	—	**10**
Pr	**4,556.66**	—	**2**
Tm	**4,556.67**	**70**	**35**
Ne	**4,556.70**	**(2)**	—
Te	4,557.84	(300)	—
P	4,558.03	(100)	—
La	4,558.46	200	100
Cr	4,558.66	600*wh*	20

	λ	I	
		J	O
Yb	4,553.56	60	20
Ne	4,553.16	(50)	—
Ti	4,552.46	60	150
S	4,552.38	(200)	—
As	4,552.37	50	—
Te	4,552.20	(50)	—
Ti	4,549.63	200	100
S	4,549.55	(80)	—
Fe	4,549.47	100	100
As	4,549.23	125	—
P	4,548.40	(50)	—
Gd	4,548.01	50	50
Fe	4,547.85	100	200
Te	4,546.64	(50)	—
P	4,546.03	(70)	—
Te	4,545.97	(70)	—
Cr	4,545.96	125	200
Xe	4,545.23	(200*wh*)	—
Ar	4,545.08	(200)	—
Ti	4,544.69	60	150
Cr	4,544.62	70	100
Ne	4,544.50	(50)	—
Sm	4,543.94	50	100
As	4,543.76	200	—
U	4,543.63	80	50
Br	4,542.93	(250)	—
Gd	4,542.03	50	50
Xe	*4,540.89*	*(200h)*	—
Gd	*4,540.02*	*200*	*80*
As	*4,539.97*	*200*	—
Ne	*4,538.31*	*(300)*	—
Gd	*4,537.82*	*100*	*150*
Ne	*4,537.75*	*(1000)*	—
Ne	*4,537.68*	*(300)*	—
Ne	*4,536.31*	*(150)*	—
Pr	*4,535.92*	*100*	*125*
Cr	*4,535.72*	*100*	*125*
Ti	*4,533.97*	*150*	*30*
Ra	*4,533.11*	*(300)*	—
Xe	*4,532.49*	*(100)*	—
Co	*4,530.96*	*8*	*1000w*
P	*4,530.78*	*(150)*	—
Cr	*4,530.74*	*125*	*150*
Fe	*4,528.62*	*200*	*600*
La	4,559.29	150	100
Ti	4,563.77	200	100
Se	4,563.95	(200)	—
V	4,564.59	150	—
P	4,565.21	(100)	—
Co	4,565.59	12	800*W*
Ne	4,565.89	(60)	—
N*			

Cs 4,593.18 *J* 50 *O* 1000*R*

	λ	*I* *J*	*I* *O*
Ce	4,593.10	—	3
Tb	4,593.06	—	3
Ru	4,593.02	—	6
Sc	4,592.94	—	2
U	4,592.933	—	3
Er	4,592.93	—	3
Kr	4,592.80	(150*wh*)	—
Fe	4,592.65	50	200
W	4,592.57	5	15
U	4,592.56	3	1
Re	4,592.55	—	2*h*
Cr	4,592.54	1	25
Ni	4,592.53	2	200
Ru	4,592.52	—	100
Te	4,592.49	(30)	—
W	4,592.42	10	20
Tb	4,592.391	—	5
Eu	4,592.39	—	4*w*
Se	4,592.34	(8)	—
Ce	4,592.27	—	3
Mo	4,592.21	20	20
Pr	4,592.15	2	8
Cr	4,592.06	35*h*	3
Xe	4,592.05	(150*wh*)	—
Ru	4,592.01	—	6
Pr	4,591.92	—	4
Sb	4,591.88	20	—
Ba	4,591.829	3	15
Sm	4,591.826	—	100
Dy	4,591.78	3	3
Cr	4,591.39	125	200
V	4,591.22	25	30
S	4,591.05	(35)	—
As	4,591.02	30	—
J	4,590.95	(25)	—
O	4,590.94	(300)	—
Ti	4,589.95	100	40
Ar	4,589.93	(150)	—
O	4,589.89	(30)	—
P	4,589.78	(300*w*)	—
Ar	4,589.29	(80)	—
Cr	4,588.22	600*h*	10
Al	4,588.19	(30)	—
Mo	4,588.15	30	25
Ne	4,588.13	(30)	—
P	4,587.90	(300*w*)	—

	λ	*I* *J*	*I* *O*
Ru	4,593.21	—	7
Ne	4,593.24	(50)	—
Th	4,593.29	2	3
Br	4,593.30	(4)	—
Yb	4,593.37	—	4
Sm	4,593.41	—	2
Ar	4,593.44	(2)	—
Sm	4,593.53	50	50
Pr	4,593.57	2*d*	8*d*
Th	4,593.642	50*wh*	—
Mo	4,593.643	8	8
In	4,593.698	20*h*	—
Xe	4,593.70	(5)	—
Ce	4,593.72	—	2
Nb	4,593.78	15*h*	—
Cr	4,593.83	—	12
*bh*La	4,593.9	—	5
Pr	4,593.926	2	30
Ce	4,593.932	30	30
Nd	4,593.94	—	2*h*
Tb	4,593.95	—	4
Yt	4,594.00	—	3
Er	4,594.01	—	2
Eu	4,594.02	200	500*R*
V	4,594.108	25*wh*	30*wh*
Mn	4,594.108	—	12
Ce	4,594.13	—	4
Sb	4,594.21	(15)	—
U	4,594.29	8	6
Tb	4,594.31	—	2
Cr	4,594.40	—	10
Nd	4,594.45	—	10
Co	4,594.63	—	400
Mo	4,595.16	40	40
Ne	4,595.25	(50)	—
Sm	4,595.30	60	100
Cr	4,595.59	60	50
K	4,595.61	(40)	—
P	4,595.98	(30)	—
Ar	4,596.10	(1000)	—
O	4,596.13	(150)	—
Sb	4,596.90	(70)	—
Gd	4,596.99	25	25
Gd	4,597.92	40	25
Se	4,597.93	(25)	—
Yb	4,598.37	70	25

	λ	I	
		J	O
Cu	4,586.95	80*w*	250*w*
Sb	4,586.84	(25)	—
V	4,586.36	30*h*	40*h*
Al	4,585.82	(40)	—
Sn	4,585.64	25*wh*	—
Xe	4,585.48	(200*wh*)	—
Sm	4,584.83	50	60
Ru	4,584.44	80	150*R*
Fe	4,583.85	150	150
Kr	4,582.85	(300*h*)	—
Xe	4,582.75	(300)	—
Ne	4,582.45	(150)	—
Ne	4,582.03	(150)	—
P	4,581.76	(30*h*)	—
Nb	4,581 62	50	30
Co	4,581.60	10	1000*w*
Xe	4,580.70	(40*wh*)	—
V	4,580.40	25*h*	30*h*
Ne	4,580.35	(30)	—
La	4,580.057	100	80
Cr	4,580.056	125	300
J	4,579.99	(30)	—
Br	4,579.95	(25)	—
Ba	4,579.67	40	75
Nb	4,579.45	30*h*	5*h*
Ar	4,579.39	(80)	—
In	4,578.39	(60)	—
In	4,578.09	(50)	—
Rn	4,577.72	(250)	—
Sm	4,577.69	50	100
Kr	4,577.20	(800)	—
Xe	4,577.06	(100*wh*)	—
Br	4,575.75	(100)	—
Ne	4,575.06	(300)	—
La	4,574.87	300	300
Ne	4,573.56	(50)	—
Nb	4,573.07	50	30
Cl	4,572.13	(100)	—
Ti	4,571.98	300	150
Te	4,569.71	(70)	—
Cl	4,569.42	(50)	—
Ne	4,569.01	(70)	—
Sm	4,566.20	50	100
Ne	4,565.89	(60)	—

	λ	I	
		J	O
Kr	4,598.49	(50*h*)	—
Sb	4,599.09	(40)	—
Hf	4,599.44	25	10
J	4,599.81	(30)	—
Se	4,599.96	(70)	—
Cr	4,600.10	50	30
V	4,600.15	60*h*	1
Cr	4,600.75	150	150
U	4,601.13	25	18
Ta	4,601.42	100*wh*	60
N	4,601.49	(100)	—
P	4,601.96	(300*w*)	—
Te	4,602.37	(800)	—
As	4,602.73	200	—
Li	4,602.86	—	800
Fe	4,602.94	100	300
Xe	4,603.03	(300*h*)	—
Tm	4,603.42	25	35
U	4,603.66	40	25
Cs	4,603.75	(60)	—
Kr	4,604.02	(60*h*)	—
Sc	4,604.34	(300)	—
Rn	4,604.40	(200)	—
Sb	4,604.77	(30)	—
U	4,605.15	25	12
Hf	4,605.77	30	20
La	4,605.78	100	100
V	4,606.15	25	30
Nb	4,606.77	50	50
N	*4,607.17*	*(50)*	—
Sr	*4,607.33*	*50R*	*1000R*
As	*4,607.46*	*200*	—
Rn	*4,609.38*	*(250)*	—
O	*4,609.39*	*(60h)*	—
Ar	*4,609.60*	*(300)*	—
Ne	*4,609.91*	*(150)*	—
Kr	*4,610.65*	*(60h)*	—
Xe	*4,611.89*	*(700)*	—
In	*4,612.13*	*150*	—
Sb	*4,612.92*	*(50)*	—
Cr	*4,613.37*	*60*	*150*
La	*4,613.39*	*100*	*100*
Ne	*4,614.39*	*(100)*	—
Br	*4,614.60*	*(200)*	—
Xe	*4,615.06*	*(50h)*	—
Kr	*4,615.28*	*(500)*	—
Xe	*4,615.50*	*(100)*	—

	λ	I	
		J	O
Tm	4,615.93	300	200
Ne	4,615.98	(50)	—
Cr	4,616.14	200	300r
Cr	4,616.66	50	—
In	4,617.16	(200)	—
Ti	4,617.27	100	200
Xe	4,617.50	(50)	—
Ne	4,617.84	(70)	—
Sc	4,618.77	(100)	—
Cr	4,618.83	80h	6
Kr	4,619.15	(1000)	—
La	4,619.88	200	150
In	4,620.05	(80)	—
In	4,620.24	(200)	—

Cu $^{29}_{63.54}$

t_0 1,853°C

t_1 2,595°C

I.	II.	III.	IV.	V.
7.723	20.283	–	–	–

λ	*I*		eV
	J	*O*	
II 2,192.26	500	25	13.5
II 2,246.99	500	30	13.2
I 2,824.37	300	1000	5.78
I 3,247.54	2000*R*	5000*R*	3.8
I 3,273.96	1500*R*	3000*R*	3.8
I 4,539.70	80	100	7.88
I 4,586.95	80	250	7.80
I 5,105.54	–	500	3.8
I 5,153.23	–	600	6.2
I 5,218.20	–	700	6.2
I 5,782.13	–	1000	3.79

Cu 2,192.26 *J* 500 *O* 25*

	λ	*I*			λ	*I*	
		J	*O*			*J*	*O*
Ir	**2,192.22**	**25***w*	—	**Ni**	**2,192.36**	**8***h*	—
Si	**2,192.22**	**(10)**	—	**W**	**2,192.40**	**3**	**3**
Ni	**2,192.12**	**6***h*	—	**Nb**	**2,192.42**	**3**	**2**
W	**2,192.10**	**6**	—	**Co**	**2,192.49**	**25**	**2**
				Pt	**2,192.50**	**4**	**6**

Cu 2,246.995 *J* 500 *O* 30*

	λ	*J*	*O*		λ	*J*	*O*
Sb	**2,246.99**	**4**	—	**Pd**	**2,247.01**	**8**	—
Nb	**2,246.98**	**2**	—	**U**	**2,247.08**	**8**	—
W	**2,246.98**	**5**	—	**Mo**	**2,247.11**	**15**	—
Mo	**2,246.95**	**20**	—	**Ni**	**2,247.23**	**12**	**2**
Cd	**2,246.93**	**2**	—				
Fe	**2,246.91**	**20**	—				
Ir	**2,246.90**	**100**	**10**				
Pb	**2,246.89**	**100***R*	**30***R*				
Cd	**2,246.80**	**(2***d***)**	—				
Nb	**2,246.76**	**10**	—				
Ag	2,246.41	300*hs*	25				

Cu 2,824.369 *J* 300 *O* 1000

	λ	*J*	*O*		λ	*J*	*O*
Co	**2,824.36**	—	**2**	**Ag**	**2,824.370**	**200***w*	**150***wh*
Er	**2,824.32**	—	**3**	**Pt**	**2,824.41**	**2**	**2**
W	**2,824.30**	**6**	**1**	**V**	**2,824.44**	**15**	**2**
U	**2,824.28**	**30**	**25**	**Ir**	**2,824.45**	**15**	**20**
Re	**2,824.25**	—	**20**	**Cr**	**2,824.537**	**10**	—
Ho	**2,824.19**	**3**	**20**	**Eu**	**2,824.54**	—	**2***w*
Mo	**2,824.172**	**3**	—	**Ce**	**2,824.629**	—	**2**
Os	**2,824.166**	**4**	**20**	**U**	**2,824.630**	**6**	**10**
J	**2,824.15**	**(4)**	—	**Fe**	**2,824.67**	**2***d*	**2***d*
Cs	**2,824.12**	**(8)**	—	**Th**	**2,824.68**	**25***d*	—
Ta	**2,824.05**	—	**4**	**Ru**	**2,824.77**	—	**10**
Ce	**2,824.03**	—	**15**	**Ho**	**2,824.79**	**10h**	—
In	**2,823.99**	**3**	—	**Ta**	**2,824.81**	**5***h*	**60***W*
Ag	**2,823.95**	**3***h*	—	**Zr**	**2,824.82**	—	**4**
Nb	**2,823.88**	**10**	**1**	**J**	**2,824.84**	**(8)**	—
				U	**2,824.86**	**8**	**12**
Fe	2,823.28	300	200	**Ce**	**2,824.88**	—	**10**
Ru	2,822.55	150	30				
Eu	2,820.77	200*W*	200*W*				
Au	2,819.95	150	—	Co	2,825.24	200	5
				Fe	2,825.56	150	150

	λ	I			λ	I	
		J	O			J	O
Mo	*2,816.15*	*300h*	*200*	Eu	2,838.69	150	200*W*
Eu	*2,813.95*	*300wh*	*300w*	Ti	2,828.90	150*W*	—
Ra	*2,813.76*	*(400)*	—				
				Pt	*2,830.29*	*600r*	*1000R*
				Fe	*2,831.56*	*500*	*1*

Cu 3,247.540 *J* 2000*R* *O* 5000*R*

	λ	J	O		λ	J	O
Eu	3,247.53	5	50*W*	Mn	3,247.542	—	125
Er	3,247.52	2	18*d*	Sb	3,247.547	10	2*h*
Cs	3,247.50	(4)	—	Ag	3,247.55	15	15
Nb	3,247.47	100*w*	50*w*	Ar	3,247.55	(3)	—
Fe	3,247.39	2	—	Ce	3,247.552	—	15
Sm	3,247.37	—	4	In	3,247.51	15	—
Eu	3,247.30	1	4	Mo	3,247.62	20	30
Fe	3,247.28	10	20	Ir	3,247.67	1	3
Cr	3,247.27	1	20	U	3,247.71	5	4
Ce	3,247.25	—	3	Xe	3,246.74	(4*wh*)	—
Fe	3,247.213	10	10	Tb	3,247.79	—	8
U	3,247.209	1	2	Rn	3,246.80	(10)	—
Tb	3,247.18	3	15	Ce	3,247.90	—	3
Co	3,247.179	—	80	V	3,247.91	5	—
Fe	3,247.171	10	—	Os	3,248.00	5	1
Sm	3,247.17	10	1	Kr	3,248.03	(6*wh*)	—
Ce	3,247.12	—	3	Ce	3,248.06	—	5
La	3,247.04	2	8	Sm	3,248.14	4	10
U	3,247.03	—	8	Ne	3,248.15	(7)	—
Kr	3,247.00	(12*wh*)	—	Fe	3,248.21	150	200
Co	3,246.997	—	35	Ca	3,248.28	3	1*h*
Fe	3,246.96	70	100	U	3,248.34	3	5
Ta	3,246.90	—	35*h*	Dy	3,248.36	4	12
Sm	3,246.84	2	9				
Eu	3,246.780	2	4*wh*				
Nb	3,246.780	—	5	In	3,256.09	600*R*	1500*R*
Nb	3,246.69	5	—				
Pd	3,242.70	600R	2000*wh*				

Cu 3,273.962 *J* 1500*R* *O* 3000*R*

	λ	J	O		λ	J	O
Mo	3,273.961	—	20	Ce	3,273.964	—	5
Ca	3,273.958	4	2	Sb	3,273.97	4	—
Co	3,273.931	—	10	In	3,274.02	15	—
Ce	3,273.926	—	5	Ti	3,274.05	1	7
Nb	3,273.89	100*W*	20r	Ce	3,274.06	—	10

	λ	I	
		J	O
Th	**3,273.88**	**15**	**10**
Au	**3,273.68**	**2**	—
Hf	**3,273.65**	**10**	**20**
Ru	**3,273.621**	**5**	**2**
Sc	**3,273.619**	**12**	**35**
U	**3,273.60**	—	**5**
Mo	**3,273.58**	**25**	—
O	**3,273.54**	**(35)**	—
Ce	**3,273.52**	—	**4**
Nb	**3,273.51**	**20**	**1**
Fe	**3,273.50**	**6**	—
Sm	**3,273.48**	**8**	**30**
J	**3,273.45**	**(3)**	—
Os	**3,273.38**	**5**	**15**
Ar	**3,273.36**	**(5)**	—
Er	**3,273.325**	—	**7**
Sm	**3,273.316**	**3**	**6**
U	**3,273.22**	**2**	**2**
Nd	**3,273.17**	**2**	**8**
Ta	**3,273.13**	**3*h***	**70**
Tb	**3,273.12**	—	**15**
Er	**3,273.08**	**15**	**25**
Ru	**3,273.078**	**20**	**60**
Pt	**3,273.06**	**5**	—
Zr	**3,273.05**	**80**	**50**
V	**3,273.03**	**5**	**30**
In	3,256.09	600*R*	1500*R*

	λ	I	
		J	O
Ce	**3,274.11**	—	**8**
Mo	**3,274.20**	**5*h***	—
Na	**3,274.22**	**(40)**	**15**
Tb	**3,274.24**	—	**70**
Eu	**3,274.29**	**1*h***	**3**
Th	**3,274.40**	**3**	**6**
Ag	**3,274.41**	**5*h***	—
Fe	**3,274.45**	**60**	**80**
Ta	**3,274.46**	**1**	**35**
Ir	**3,274.56**	—	**4**
Pr	**3,274.58**	—	**8**
Ce	**3,274.62**	—	**8**
Mo	**3,274.63**	**30**	—
Be	**3,274.64**	**(50)**	—
Ca	**3,274.66**	—	**20**
Ru	**3,274.71**	**25**	**60**
Er	**3,274.75**	**1**	**10**
Rh	**3,274.78**	—	**2**
Nb	**3,274.79**	**10**	**1**
Ce	**3,274.86**	**8**	**35**
Eu	3,280.68	—	1000*R*
Ag	3,280.68	1000*R*	2000*R*
Yb	*3,289.37*	*1000R*	*500R*

Cu 4,539.695 *J* 80 *O* 100

	λ	J	O
Eu	**4,539.69**	**3**	**3*w***
Mo	**4,539.64**	**6**	**6**
Yt	**4,539.61**	—	**2**
Ce	**4,539.59**	—	**3**
Nd	**4,539.423**	—	**3**
Er	**4,539.416**	—	**2**
Tm	**4,539.39**	—	**10**
Yb	**4,539.294**	—	**2**
Eu	**4,539.29**	**2**	**12**
Pr	**4,539.288**	**3*w***	**30*d***
U	**4,539.28**	**3*h***	—
Nd	**4,539.26**	**2**	**10**
Mo	**4,539.21**	**3**	**4**
Ne	**4,539.17**	**(50)**	—
Dy	**4,539.16**	—	**3**

	λ	J	O
W	**4,539.696**	**1**	**4**
Ce	**4,539.75**	**10**	**20**
Hf	**4,539.76**	**6**	**3**
Cr	**4,539.79**	**25**	**40**
Os	**4,539.92**	**2**	**100**
As	**4,539.97**	**200**	—
Zr	**4,539.98**	—	**15**
V	**4,540.01**	**12**	**15**
Gd	**4,540.02**	**200**	**80**
Nb	**4,540.08**	**1**	**5**
Pr	**4,540.15**	**1**	**5**
Er	**4,540.18**	—	**5**
Sm	**4,540.19**	—	**40**
P	**4,540.20**	**(70)**	—
Cu	**4,540.21**	**2**	—

	λ	I J	I O
Ti	**4,539.10**	**4**	**15**
U	**4,539.08**	—	**5**
Ce	**4,539.07**	—	**8**
Cs	**4,538.94**	**(30)**	—
Mo	**4,538.88**	**4**	**5**
La	**4,538.87**	**8*h***	—
Br	**4,538.77**	**(15*h*)**	—
Dy	**4,538.76**	**2**	**5**
Tb	**4,538.73**	—	**5**
Ir	**4,538.67**	—	**25**
Ru	**4,538.66**	—	**10**
Sm	**4,538.55**	—	**10**
Mn	**4,538.46**	—	**40**
Ce	**4,538.42**	—	**2**
Mo	**4,538.41**	**5**	**5**
Ne	4,538.31	(300)	—
U	4,538.19	40	25
Sm	4,537.95	25	50
Gd	4,537.82	100	150
Ne	4,537.75	(1000)	—
Ne	4,537.68	(300)	—
Os	4,537.61	—	50
Te	4,537.07	(50)	—
Xe	4,536.92	(40*wh*)	—
Mo	4,536.80	80	40
Ne	4,536.31	(150)	—
Pr	4,535.92	100	125
Cr	4,535.72	100	125
Ti	4,535.57	50	80
Cr	4,535.15	30	50
Ti	4,534.78	40	100
Pr	4,534.15	80	150
Co	4,533.99	8	500
Ti	4,533.97	150	30
Ti	4,533.24	40	150
Hf	4,533.15	40	20
Ra	4,533.11	(300)	—
Ir	4,532.87	2	80
Xe	4,532.49	(100)	—
Sm	4,532.44	—	60
Fe	4,531.15	—	125
Co	4,530.96	8	1000*w*
Ru	4,530.854	—	60
Ta	4,530.846	50	300
Cu	4,530.82	50	200
P	4,530.78	(150)	—
Cr	4,530.74	125	120
Mn	4,529.79	—	50

	λ	I J	I O
W	**4,540.31**	**1**	**7**
Cu	**4,540.33**	**3**	—
Ne	**4,540.38**	**(50)**	—
Th	**4,540.42**	**8**	**8**
Ti	**4,540.48**	**1**	**7**
Cr	**4,540.50**	**40**	**40*d***
Tb	**4,540.57**	—	**3**
Ce	**4,540.626**	—	**4**
J	**4,540.63**	**(8)**	—
Eu	**4,540.66**	**4**	**4*W***
La	**4,540.71**	**25**	**4**
Cr	**4,540.72**	**40**	**40*d***
Mo	**4,540.75**	**25*d***	**25*d***
Rb	**4,540.77**	**10**	—
Co	**4,540.78**	—	**30**
Ti	**4,540.87**	**1**	**6**
Xe	**4,540.89**	**(200*h*)**	—
Hf	**4,540.93**	**2**	**50**
P	4,541.12	(70)	—
Nd	4,541.27	4	50
Gd	4,542.03	50	50
Sm	4,542.049	—	50
Nd	4,542.050	5	50
Mn	4,542.44	5	80
Nd	4,542.60	5	50
Br	4,542.93	(250)	—
U	4,543.63	80	50
As	4,543.76	200	—
Co	4,543.81	—	500*W*
Sm	4,543.94	50	100
Mn	4,544.41	5	60
Ne	4,544.50	(50)	—
Cr	4,544.62	70	100
Ti	4,544.69	60	150
Ar	4,545.08	(200)	—
Xe	4,545.23	(200*wh*)	—
Co	4,545.24	—	50
Ir	4,545.68	4	200
Cr	4,545.96	125	200
Te	4,545.97	(70)	—
P	4,546.03	(70)	—
Te	4,546.64	(50)	—
Ni	4,546.93	2	50
Ta	4,547.15	2	150
Fe	4,547.85	100	200
Gd	4,548.01	50	50
P	4,548.40	(50)	—
Ir	4,548.48	5	100

	λ	I J	I O
Br	4,529.77	(80)	—
Os	4,529.67	2	80
Ti	4,529.46	40	5
Tm	4,529.37	5	80
Rh	4,528.72	60	500r
Fe	4,528.62	200	600
J	4,528.10	(40)	—
Co	4,527.93	2h	100
Yt	4,527.79	40	25
Ta	4,527.49	5	150
Ce	4,527.35	25	50
Ti	4,527.31	50	100
Yt	4,527.237	50	40
Er	4,527.236	10	50
Ca	4,526.93	3wh	100
Eu	4,526.69	—	100
La	4,526.11	150	100
Br	4,525.62	(125)	—
La	4,525.30	100	100
Fe	4,525.15	50	100
S	4,524.95	(150)	—
Sn	4,524.74	50	500wh
Xe	4,524.68	(400)	—
Xe	4,524.21	(100)	—
Sm	4,523.91	50	100
Kr	4,523.14	(400h)	—
Ti	4,522.80	70	100
Re	4,522.72	—	100
Eu	4,522.60	200	200
Eu	4,522.58	—	500
Tm	4,522.57	300	200
La	4,522.37	400	200
Ar	4,522.32	(800)	—
Ta	4,521.09	10h	200
Gd	4,519.66	100	150
Sm	4,519.63	80	150
Lu	4,518.57	40	300
Ti	4,518.03	60	100
Ne	4,517.74	(100)	—
Co	4,517.11	6	300
Ru	4,516.89	—	100
Yb	4,515.15	100	45
Sm	4,515.10	—	100
Br	4,513.44	(100)	—

	λ	I J	I O
Pr	4,548.54	30	60
Mn	4,548.58	5	80
Os	4,548.66	5	100
Ti	4,548.77	25	125
As	4,549.23	125	—
Fe	4,549.47	100	100
S	4,549.55	(80)	—
Ti	4,549.63	200	100
Co	4,549.66	—	600
Kr	4,550.30	(40)	—
Os	4,550.41	10	150
Ir	4,550.77	—	80
Fe	4,550.79	—	50
Os	4,551.30	8	150
Ta	4,551.95	8	400
Te	4,552.20	(50)	—
Pr	4,552.26	—	60
As	4,552.37	50	—
S	4,552.38	(200)	—
Pt	4,552.42	10	60
Ti	4,552.46	50	150
Si	4,552.50	40	—
Ta	4,553.69	2	200
Ba	4,554.04	200	1000R
Ru	4,554.51	200	1000R
P	4,554 80	(100)	—
Cs	4,555 35	100	2000R
Ti	4,555.49	60	125
Xe	4,555.94	(100wh)	—
Fe	4,556.12	35	150
Ta	4,556.35	5	200
Kr	4,556.61	(200)h	—
Te	4,557.84	(300)	—
P	4,558.03	(100)	—
La	4,558.46	200	100
Cr	4,558.66	600wh	20
La	4,559.29	150	100
Pr	4,563.13	40	100
Ti	4,563.77	200	100
Se	4,563.95	(200)	—
V	4,564.59	150	—
P	4,565.21	(100)	—
Co	4,565.59	12	800W
Ta	4,565.85	15	200

Cu 4,586.95 *J* 80 *O* 250

	λ	*I*	
		J	*O*
Co	4,586.94	—	15
Lu	4,586.93	6	2
U	4,586.92	2	—
Er	4,586.848	—	3
W	4,586.846	5	30
Sb	4,586.84	(25)	—
Ca	4,586.82	2	—
Mo	4,586.79	15	15
Dy	4,586.64	2	4
Th	4,586.63	3	2
Nd	4,586.614	—	50
Ar	4,586.610	(10)	—
J	4,586.58	(8)	—
Mo	4,586.57	15	15
Pr	4,586.53	—	5*w*
Eu	4,586.44	—	30*w*
V	4,586.36	30*h*	40*h*
Tm	4,586.35	—	10
Hf	4,586.25	10	8
Dy	4,582.21	2	2
Pr	4,586.16	—	2
Cr	4,586.14	6	25
Mn	4,586.11	—	30
Mo	4,586.06	20	20
Re	4,586.054	—	2*w*
Ce	4,586.051	—	2
Ca	4,586.03	3*h*	2
Pr	4,585.94	—	4
Eu	4,585.89	—	2*w*
Ne	4,585.88	(10)	—
Ca	4,585.871	10	125
Tb	4,585.87	—	2
Al	4,585.82	(40)	—
Dy	4,585.72	3	—
Eu	4,585.67	5	8*w*
Sn	4,585.64	25*wh*	—
Ir	4,585.59	—	8
U	4,585.588	2	2
Xe	4,585.48	(200*wh*)	—
Sm	4,584.83	50	60
Ru	4,584.44	80	150*R*
Fe	4,583.85	150	150
Ta	4,583.17	10	150
Kr	4,582.85	(300*h*)	—
Xe	4,582.75	(300)	—
Ne	4,582.45	(150)	—

	λ	*I*	
		J	*O*
Nd	4,586.97	—	10
Gd	4,586.98	2	15
Te	4,586.98	(15)	—
Pr	4,586.982	—	6
Ru	4,587.10	—	10
La	4,587.130	—	3*h*
Re	4,587.135	—	15
Pr	4,587.135	—	8
Fe	4,587.14	2	12
Ar	4,587.21	(5)	—
U	4,587.27	—	6*h*
Mo	4,587.40	5	5
Ir	4,587.41	—	20
Ir	4,587.47	—	5
Tb	4,587.71	—	20
Cr	4,587.87	—	30
Au	4,587.89	15	—
P	4,587.90	(300*w*)	—
Ar	4,587.90	(2)	—
Dy	4,587.93	3	6
Pr	4,588.00	—	5*d*
Tm	4,588.06	—	5
Al	4,588.08	(2)	—
Ne	4,588.13	(30)	—
Er	4,588.14	—	3
Mo	4,588.147	30	23
Tb	4,588.15	—	8*d*
Al	4,588.19	(30)	—
Cr	4,588.22	600*h*	10
Th	4,588.24	4	4
Ar	4,589.29	(80)	—
P	4,589.78	(300*w*)	—
Ar	4,589.93	(150)	—
Ti	4,589.95	100	40
O	4,590.94	(300)	—
Cr	4,591.39	125	200
Xe	4,592.05	(150*wh*)	—
Ni	4,592.53	2	200
Fe	4,592.65	50	200
Kr	4,592.80	(150*wh*)	—
Cs	4,593.18	50*R*	1000*R*
Ne	4,593.24	(50)	—
Sm	4,593.53	50	50
Th	4,593.64	50*wh*	—
Eu	4,594.02	200	500*R*
Co	4,594.63	—	400

	λ	I J	I O
Ne	4,582.03	(150)	—
Mn	4,581.83	—	125
Nb	4,581.62	50	50
Co	4,581.60	10	1000*w*
Xe	4,580.70	(40*wh*)	—
Ta	4,580.69	10	200*W*
Co	4,580.14	3	300
La	4,580.057	100	80
Cr	4,580.056	125	300
Ba	4,579.67	40	75
Ar	4,579.39	(80)	—
In	4,578.39	(60)	—
In	4,578.09	(50)	—
Rn	4,577.72	(250)	—
Sm	4,577.69	50	100
Kr	4,577.20	(800)	—
Xe	4,577.05	(100*wh*)	—
Mo	4,576.50	40	40
Br	4,575.75	(100)	—
Ne	4,575.06	(300)	—
La	4,574.87	300	300
Ta	4,574.31	20	300
U	4,573.69	40	30
Ne	4,573.56	(50)	—
Ta	4,573.29	2*h*	200
Cl	*4,572.14*	*(100)*	—
Ti	*4,571.98*	*300*	*150*
Co	*4,570.02*	—	*300*
Co	*4,565.59*	*12*	*800 W*
Mo	4,595.16	40	40
Ne	4,595.25	(50)	—
Sm	4,595.30	60	100
Cr	4,595.59	60	50
K	4,595.61	(40)	—
Ar	4,596.10	(1000)	—
O	4,596.13	(150)	—
Sb	4,596.90	(70)	—
Co	4,596.905	—	400
Gd	4,597.92	40	25
Yb	4,598.37	70	25
Kr	4,598.49	(50*h*)	—
Sb	4,599.09	(40)	—
Se	4,599.96	(70)	—
Cr	4,600.10	50	20
V	4,600.15	60*h*	1
Ni	4,600.37	—	200
Cr	4,600.75	150	150
Ta	*4,601.42*	*100wh*	*60*
N	*4,601.49*	*(100)*	—
P	*4,601.96*	*(300w)*	—
Te	*4,602.37*	*(800)*	—
As	*4,602.73*	*200*	—
Li	*4,602.86*	—	*800*
Fe	*4,602.94*	*100*	*300*
Xe	*4,603.03*	*(300h)*	—
Se	*4,604.34*	*(300)*	—
Rn	*4,604.40*	*(200)*	—
Ni	*4,604.99*	*10h*	*300*
La	*4,605.78*	*100*	*100*
Sr	*4,607.33*	*50R*	*1000R*
As	*4,607.46*	*200*	—
Rn	*4,609.38*	*(250)*	—
Ar	*4,609.60*	*(300)*	—
Ne	*4,609.91*	*(150)*	—
Xe	*4,611.89*	*(700)*	—
In	*4,612.13*	*150*	—
La	*4,613.39*	*100*	*100*
Ne	*4,614.39*	*(100)*	—
Br	*4,614.60*	*(100)*	—

Cu 5,105.54 *O* 500

	λ	J	O
Zr	**5,105.53**	—	**5**
W	**5,105.48**	—	**15**
Nd	**5,105.35**	—	**10**
La	**5,106.23**	**10**	**100**
Fe	**5,106.44**	—	**25**
Ru	**5,106.55**	—	**10**

	λ	I J	I O
Nd	5,105.211	1	5
Ce	5,105.208	—	2
Re	5,105.16	—	20
V	5,105.14	40	40
Yb	5,105.06	5	—
Os	5,104.740	—	8
Ar	5,104.74	(20)	—
Ne	5,104.705	(35)	—
Mo	5,104.701	1	3
Re	5,104.63	—	50*w*
Sm	5,104.47	1	125
N	5,104.45	(15)	—
Yb	5,104.43	50	1
W	5,104.42	1	5
*bh*V	5,104.3	—	7
P	5,104.13	(13)	—
Cl	5,104.08	(25)	—
Sm	5,104.06	—	15*d*
Tb	5,104.02	—	10
Cl	5,103.85	(6)	—
Th	5,103.77	—	6
Cd	5,085.82	500	1000*wh*
Sm	5,106.62	3	10*d*
Nd	5,106.64	—	8
U	5,106.75	—	4
Sm	5,106.99	—	3
Ru	5,107.07	—	40
Ce	5,107.20	—	10
U	5,107.34	6	6
Fe	5,110.41	—	300
*N**			

Cu 5,153.23 *O* 600

	λ	I J	I O
Ru	5,153.20	—	7
Hf	5,153.12	1	3
Nb	5,153.03	2	8
*bh*F	5,152.8	—	2
Nb	5,152.63	10	100
Sm	5,152.58	—	3
U	5,152.38	—	2
Pr	5,152.21	—	15*w*
Ti	5,152.20	2	90
Os	5,152.01	—	7
Sm	5,151.95	—	4
Fe	5,151.91	—	70
*bh*F	5,151.9	—	2
Th	5,151.86	—	2
Nd	5,151.78	—	5
Th	5,151.62	—	4
Pr	5,151.36	—	3
Na	5,149.09	—	400
Co	5,146.74	—	400*w*
Rh	5,153.34	—	2
Ta	5,153.42	—	15
Nd	5,153.45	—	4
W	5,153.53	3	9
Na	5,153.64	—	600
W	5,153.87	1*h*	7
Co	5,154.05	—	200*W*
Ti	5,154.08	15*h*	10
Th	5,154.25	—	8
Sm	5,154.27	1	125*d*
Ce	5,154.39	—	10
Ne	5,154.42	(50)	—
W	5,154.45	—	8
Hf	5,154.64	—	2*d*
Cd	5,154.68	—	6*r*
P	5,154.84	(10)	—
W	5,154.89	—	6
Sm	5,155.02	1	125
Ru	5,155.14	—	125
Co	5,156.34	—	300*w*

	λ	I J	I O		λ	I J	I O
				Fe	5,162.29	—	300*h*
				Pd	5,163.84	8	300
				Fe	5,167.49	150	700
				Fe	*5,171.60*	*60h*	*300*

*P**

Cu 5,218.20 *O* 700

	λ	J	O		λ	J	O
W	**5,218.18**	**—**	**6**	**Re**	**5,218.23**	**—**	**2*h***
Ru	**5,218.11**	**—**	**10**	**Er**	**5,218.25**	**—**	**30**
Sm	**5,218.08**	**2**	**8*d***	**Sm**	**5,218.40**	**—**	**25**
Fe	**5,217.92**	**—**	**6**	**W**	**5,218.43**	**3**	**7**
La	**5,217.83**	**10*h***	**2*h***	**Ta**	**5,218.45**	**—**	**40**
Gd	**5,217.49**	**—**	**25**	**Nb**	**5,218.46**	**1**	**3**
Re	**5,217.45**	**—**	**2**	**Th**	**5,218.53**	**2**	**10**
Fe	**5,217.40**	**3**	**150**	**Ta**	**5,218.66**	**—**	**40**
Eu	**5,217.02**	**—**	**125**	**Co**	**5,219.03**	**—**	**10**
U	**5,216.95**	**—**	**8**	**Pr**	**5,219.05**	**2**	**50**
Pr	**5,216.76**	**—**	**8**	**Nb**	**5,219.09**	**10**	**100**
Th	**5,216.590**	**2**	**12**	**Th**	**5,219.11**	**—**	**8**
V	**5,216.588**	**40**	**40**	**Eu**	**5,219.21**	**—**	**12**
Ni	**5,216.49**	**—**	**10**	**Gd**	**5,219.40**	**—**	**25**
Sm	**5,216.43**	**—**	**5**	**Mo**	**5,219.405**	**20**	**25**
Ce	**5,216.38**	**—**	**8**	**Eu**	**5,219.42**	**—**	**4**
				Sc	**5,219.67**	**12**	**10**
				Ti	**5,219.71**	**2**	**60**
Fe	5,216.28	10	300	**Cu**	**5,220.07**	**—**	**100**
Eu	5,215.09	—	1000				
Ag	5,209.07	1000*R*	1500*R*	Eu	5,223.48	—	700
Cr	5,208.54	100	500*R*	Fe	5,227.19	60	400
Cr	5,206.04	200	500*R*	Fe	5,232.94	150	800
Cr	5,204.52	100	400*R*				

Cu 5,782,132 *O* 1000

	λ	J	O		λ	J	O
Cr	**5,782.131**	**—**	**2**	**Mg**	**5,782.14**	**—**	**2**
*bh***F**	**5,782.1**	**—**	**150**	**Sb**	**5,782.15**	**—**	**6**
U	**5,781.96**	**—**	**3**	**Tm**	**5,782.36**	**5**	**10**
Sm	**5,781.89**	**—**	**100**	**Ce**	**5,782.44**	**—**	**8**
Cr	**5,781.81**	**—**	**20**	**K**	**5,782.60**	**—**	**60**
Yt	**5,781.69**	**5**	**5**	**Y**	**5,782.61**	**—**	**30**
Eu	**5,781.36**	**—**	**4**	*bh***Yt**	**5,782.7**	**—**	**10**
Cr	**5,781.19**	**—**	**18**	**Ce**	**5,782.81**	**—**	**5**
Os	**5,780.81**	**—**	**50**	**U**	**5,782.823**	**—**	**2**
Ti	**5,780.78**	**20**	**20**	**Er**	**5,782.824**	**—**	**12**

	λ	I	
		J	O
Ta	**5,780.71**	—	**80**
Mo	**5,780.64**	—	**10*h***
U	**5,780.61**	—	**40**
*bh***F**	**5,780.5**	—	**50**
Si	**5,780.45**	—	**15**
Nb	**5,780.33**	**3**	**3**
Ce	**5,780.21**	—	**2**
Mn	**5,780.19**	—	**10**
Mo	**5,780.11**	—	**12**
Ta	**5,780.02**	—	**60**
Ho	**5,780.01**	—	**12**
Tb	**5,779.91**	—	**15**
Ba	5,777.66	100*R*	500*R*
Hg	5,769.59	200	600
Eu	5,765.20	—	2000

	λ	I	
		J	O
Re	**5,783.03**	—	**3**
Cr	**5,783.11**	—	**30*h***
U	**5,783.17**	—	**2**
Ta	**5,783.24**	—	**2**
Mo	**5,783.33**	—	**20*h***
*bh***F**	**5,783.4**	—	**30**
V	**5,783.50**	—	**30**
Sm	**5,783.53**	—	**8**
Nd	**5,783.68**	—	**3*h***
Eu	**5,783.71**	—	**150*s***
*bh***Zr**	**5,783.8**	—	**5**
Cd	**5,783.93**	—	**5**
Cr	**5,783.934**	—	**30*h***
Ce	**5,783.99**	—	**3**
Mo	**5,784.00**	—	**5**
Ba	**5,784.06**	**2**	**6**
V	**5,784.38**	**30**	**50**
Er	**5,784.63**	—	**12**
Ho	**5,784.64**	—	**8**
Eu	5,818.74	—	1000
Eu	5,831.05	—	2000*W*

F $^{9}_{18.9984}$

t_0 −223°C t_1 −187°C

I.	II.	III.	IV.	V.
17.42	34.979	62.647	87.142	114.22

λ	*I*	eV
II 3,501.42	(200)	28.66
II 3,503.10	(400)	28.66
II 3,505.61	(600)	28.66
II 3,847.09	(800)	25.12
II 3,849.99	(600)	25.11
II 3,851.67	(200)	25.11
II 4,246.16	(300)	31.56
I 6,239.64	(300)	14.68
I 6,856.02	(1000)	14.50
I 6,902.46	(500)	14.52

F 3,501.42 (200)

	λ	I	
		J	O
Ru	**3,501.354**	**3**	**30**
Zr	**3,501.347**	**1**	**15**
Nb	**3,501.341**	**30**	**3**
U	**3,501.34**	**3***d*	**5***d*
Cu	**3,501.338**	**1***h*	**5**
Sm	**3,501.23**	**1**	**2**
Ne	**3,501.22**	**(150)**	–
Os	**3,501.16**	**15**	**100**
Ba	**3,501.12**	**20**	**1000**
Ni	**3,500.852**	**80**	**500***wh*
Th	**3,500.851**	**3**	**5**
Tb	**3,500.84**	**15**	**70**
V	**3,500.82**	**25**	**35**
Nb	**3,500.74**	**304**	—
Rn	**3,500.60**	**(2)**	—
Fe	**3,500.57**	**20**	**50**
Th	**3,500.55**	**5**	**6**
Dy	**3,500.50**	**4**	**2**
Sm	**3,500.498**	**15**	**30**
Xe	**3,500.36**	**(15)**	—
V	**3,500.35**	**2**	**2**
Ru	3,498.94	200	500*R*
Sb	3,498.46	300*wh*	—
Fe	3,497.84	200	200
Mn	3,497.54	150	15
S	3,497.34	(100)	—
Fe	3,497.11	100	200
V	3,497.03	150	—
Zr	3,496.21	100	100
Mn	3,495.84	150	25
Cd	3,495.34	(100)	—
Hg	3,493.85	(100)	—
V	3,493.17	100	15
Ni	3,492.96	100*h*	1000*R*
Rb	3,492.76	(300)	—
Fe	*3,490.57*	*300*	*400*
Mn	*3,488.68*	*200*	*50*
Pt	*3,485.27*	*200R*	*150*
Mn	*3,482.91*	*250*	*50*
Ta	*3,480.52*	*200w*	*70*
Dy	**3,501.44**	**3**	**10**
Ce	**3,501.45**	**3**	**18***w*
Th	**3,501.46**	**10**	**10**
V	**3,501.48**	**20**	**25**
F	**3,501.487**	**(6)**	—
Zr	**3,501.494**	**3**	**12**
Se	**3,501.52**	**(50)**	—
Cu	**3,501.529**	**1***h*	**2***h*
Nd	**3,501.535**	**4**	**10***d*
F	**3,501.56**	**(15)**	—
U	**3,501.65**	**2**	**4**
Ag	**3,501.68**	**20***h*	**8***h*
Os	**3,501.69**	**5**	**30**
Co	**3,501.72**	**100**	**5**
Xe	**3,501.77**	**(10***h***)**	—
Dy	**3,501.87**	**1**	**4**
Pb	**3,501.90**	**(10)**	—
Ag	**3,501.94**	**2**	**5**
Yt	**3,501.96**	**2**	**3**
Mo	**3,501.963**	**30**	—
Ta	**3,501.97**	**1**	**3**
Nd	**3,502.01**	**4**	**8**
Ca	**3,502.02**	**3**	—
Dy	**3,502.09**	**4**	**6**
W	**3,502.23**	**10**	—
Co	**3,502.28**	**20**	2000*R*
Cr	**3,502.31**	**6**	**35**
He	**3,502.38**	**(2)**	—
Ru	**3,502.42**	**4**	**20**
Eu	**3,502.46**	**2**	**10***w*
Ta	**3,502.49**	**2***h*	**7**
Rh	3,502.52	150	1,000
*N**			

F 3,503.10 (400)

	λ	I J	I O		λ	I J	I O
Pr	3,503.06	4	15	Xe	3,503.15	(8)	—
U	3,503.041	1	2	V	3,503.176	4	10
W	3,503.038	7	8	Dy	3,503.179	3*h*	8
P	3,502.99	(70)	—	Nb	3,503.20	5	10
F	3,502.95	(60)	—	W	3,503.23	5	—
Ir	3,502.94	4	8	Kr	3,503.25	(50*wh*)	—
Ta	3,502.87	3	7	Er	3,503.37	1	10
F	3,502.86	(10)	—	Pr	3,503.43	2	10
Eu	3,502.81	5	15*d*	Os	3,503.45	8	15
Er	3,502.783	1	10	Fe	3,503.47	1*h*	1*h*
Th	3,502.783	8	8	Mo	3,503.50	3	3
I	3,502.75	(10)	—	W	3,503.558	5	9
Nd	3,502.70	4	8*d*	Al	3,503.56	2	—
Mo	3,502.67	20	—	Ne	3,503.61	(18)	—
Co	3,502.62	5	60	Th	3,503.62	8	8
Rn	3,502.60	(2)	—	Dy	3,503.66	2	4
Kr	3,502.55	(20)	—	Cs	3,503.67	(4)	—
Rh	3,502.52	152	1000	S	3,503.78	(8*h*)	—
				Ta	3,503.87	10*h*	70
				Kr	3,503.90	(15)	—
				U	3,504.01	1	10
				Th	3,504.039	3	4
				Tb	3,504.04	3	15
				Er	3,504.06	1	10*d*
				Nd	3,504.08	5	10*d*
				Ce	3,504.09	3	5

*P** *N**

F 3,505.61 (600)

	λ	I J	I O		λ	I J	I O
Cl	3,505.53	(4)	—	Nb	3,505.63	5	3
Gd	3,505.52	60	60	Zr	3,505.67	30	40
F	3,505.51	(20)	—	V	3,505.69	35	50
Th	3,505.50	6	6	F	3,505.76	(10)	—
Zr	3,505.48	30	30	Nb	3,505.81	2	5
Dy	3,505.46	2	70	Dy	3,505.83	2	20
U	3,505.45	6	4	Tb	3,505.90	8	30
Cl	3,505.44	(12)	—	Ti	3,505.901	5	2
Ho	3,505.42	4	—	Eu	3,505.97	1	10*w*
Rh	3,505.41	3	30	Nb	3,506.02	3	3
Mo	3,505.31	20	10	Zr	3,506.05	4	8
Eu	3,505.305	5*w*	25*w*	Ce	3,506.25	1	15
Nd	3,505.297	6	15	Ca	3,506.28	2*h*	—
Hf	3,505.23	50	20	Co	3,506.31	15	400*R*
Ta	3,505.17	2*h*	23	Ar	3,506.46	(30)	—

	λ	I J	I O
Pb	3,505.15	5	—
Tb	3,505.09	3	15
U	3,505.074	15	8
Er	3,505.074	8	12
Fe	3,505.06	10	10
Os	3,505.01	5h	—
Ta	3,504.89	2	70
Ti	3,504.89	150	20
Fe	3,504.86	5	10
Cs	3,504.85	(4)	—
Pr	3,504.78	2	5
Co	3,504.73	2	18
U	3,504.661	3	1
Os	3,504.659	20	300
W	3,504.65	6	9s
Ce	3,504.61	10wh	5
Dy	3,504.522	3	90
Er	3,504.518	4h	25
P	3,504.50	(15)	—
Pr	3,504.496	1	4
V	3,504.44	200	60
Zr	3,506.48	2	6
Yt	3,506.49	2	2
Fe	3,506.50	30	50
Xe	3,506.56	(8)	—
V	3,506.564	7	—
W	3,506.641	7	8
Ti	3,506.643	3	35
Kr	3,506.66	(3)	—
Rh	3,507.32	125	500
P	3,507.36	(100w)	—
Lu	3,507.39	150	100
V	3,509.04	150	2
Tb	3,509.17	200	200
Ru	3,509.20	100	10
Nb	3,510.26	200	15
Ti	3,510.84	125	40
Fe	3,513.82	300	400
Nb	3,515.42	300	30
Pd	*3,516.94*	*500R*	*1000R*
Nb	*3,517.67*	*200*	*2*
Tl	*3,519.24*	*1000R*	*2000R*
Ne	*3,520.47*	*(1000)*	—
Fe	*3,521.26*	*200*	*300*
Rb	*3,521.44*	*(200)*	—

*P**

F 3,847.09 (800)

	λ	I J	I O
U	3,847.06	6	5
Zr	3,847.01	4	10
Dy	3,846.99	2	10
Ce	3,846.989	3	2
Sm	3,846.986	3	9
Nd	3,846.97	20	20
Kr	3,846.83	(5h)	—
Ce	3,846.803	1	4
Fe	3,846.803	100	125
Nd	3,846.71	30d	30d
Ho	3,846.68	10	10
Ru	3,846.676	10	12
U	3,846.66	10	2
Sc	3,846.65	8	8
Ta	3,846.64	4h	5
Pr	3,846.60	30	70d
Ce	3,846.520	1	4
Yt	3,846.516	2	2
W	3,847.24	4	5
Mo	3,847.248	25	25
Nd	3,847.25	10	20
V	3,847.33	70h	100
N	3,847.38	(10)	—
Au	3,847.42	5	5
W	3,847.49	15	18
Sm	3,847.52	10	6
Ce	3,847.81	6	6
U	3,847.83	12h	8
Nd	3,847.85	50	60
Yt	3,847.87	4	8
O	3,847.89	(10)	—
Er	3,847.89	1	18d
Tm	3,848.02	250	400
Ta	3,848.05	5	30
U	3,848.07	6	6
Mg	3,848.09	3	2

	λ	I	
		J	O
Ti	3,846.45	3	15
Fe	3,846.415	75	50
Os	3,846.411	12	15
Eu	3,846.39	2	4
Sm	3,846.28	3	8
U	3,846.243	1	10
Th	3,846.242	8	10
W	3,846.21	20	20
Mo	3,846.18	5	5
Kr	3,846.12	(2)	—
Bi	3,846.03	100	—
Fe	3,846.001	1	10
Sm	3,845.997	4	10
La	3,845.997	50	40
Nd	3,845.99	25	30
Se	3,845.98	(12)	—
Kr	3,845.978	(15)	—
V	3,845.97	2	5
Ir	3,845.96	2h	4h
Mo	3,845.95	20	20
Ce	3,845.93	1	2
Nb	3,845.90	30	10
U	3,845.86	4	4
W	3,845.845	9	—
Ti	3,845.842	1	4
Cl	3,845.82	(30)	—
Bi	3,845.8	(100)	—
Co	3,845.47	100	500R
Mn	3,843.98	100	75
Cl	3,843.26	(100)	—
Fe	3,843.259	100	125
Fe	3,841.05	400	500
Fe	3,840.44	300	400
Mn	3,839.78	125	100
Mg	3,838.26	200	300
Mo	3,837.29	100w	—
Nd	3,836.54	100	80
Ar	3,834.68	(800)	—
Fe	3,834.22	400	400
Ta	3,833.74	200	40
Mg	*3,832.31*	*200*	*250*
Nb	*3,831.84*	*300*	*5*
Fe	*3,827.82*	*200*	*200*
F	*3,825.88*	*400*	*500*
Fe	*3,820.43*	*600*	*800*
Eu	*3,819.66*	*500wd*	*500wd*

	λ	I	
		J	O
Ce	3,848.10	1	8
Ir	3,848.16	2	10
Yt	3,848.19	4h	2
Eu	3,848.20	1	2w
Nd	3,848.23	10d	10d
Mg	3,848.24	10	10
Fe	3,848.298	2	5
Mo	3,848.301	20	25
Ti	3,848.31	2	10
Pr	3,848.33	2	3
Eu	3,848.40	1h	2
*N**			

F 3,849.99 (600)

	λ	I J	I O		λ	I J	I O
Fe	3,849.97	400	500	Cr	3,850.04	40*r*	40*r*
Os	3,849.94	20	125	V	3,850.16	2	15*W*
Xe	3,849.87	(25*wh*)	—	Nd	3,850.227	2	10
U	3,849.85	4	4	Sb	3,850.23	20	—
Mo	3,849,78	5	5	U	3,850.25	8*h*	1
Ta	3,849.758	4*h*	3*h*	Mg	3,850.40	5	8
V	3,849.756	3	—	V	3,850.405	10	1
Sm	3,849.752	3	8	Ru	3,850.43	10	50
Nb	3,849.746	2	2	Dy	3,850.45	1	4
U	3,849.71	10*h*	2	Ar	3,850.57	(400)	—
Se	3,849.60	(4)	—	Cl	3,850.58	(12)	—
Ni	3,849.58	15*h*	—	Gd	3,850.70	2	5
Cr	3,849.53	20	20	O	3,850.81	(5)	—
Hf	3,849.52	4	2*h*	Mo	3,850.819	25	—
Ta	3,849.42	2*h*	4*h*	Fe	3,850.820	75	200
Eu	3,849.40	3*w*	4*w*	I	3,850.82	(15)	—
Cr	3,849.36	30*h*	40*h*	Pr	3,850.825	15	50
U	3,849.34	3	2	S	3,850.93	(8)	—
V	3,849.32	25	60	U	3,850.935	4	1
Hg	3,849.27	(2)	—	Gd	3,850.98	6	10
Zr	3,849.25	4	10	O	3,851.04	(10)	—
Hf	3,849.18	15	15				
Ce	3,849.07	4	4	*N**			
La	3,849.013	150	200				
Sm	3,849.012	4	10				
Bi	3,849.01	2	—				
Rh	3,849.005	2	2				
Er	3,849.002	2	12				
Cr	3,848.98	50*d*	80*d*				
Ru	3,848.94	12	8				
Mg	3,848.92	2	—				
Sm	3,848.81	10	150*d*				
Mg	3,848.77	12	2				
Tb	3,848.75	200	100				
U	3,848.72	3	4				
U	3,848.62	8	10				
*P**							

F 3,851.667 (200)

	λ	I J	I O		λ	I J	I O
Pr	3,851.62	150*w*	200*w*	Tl	3,851.67	(6)	—
Eu	3,851.59	5*wh*	8*h*	Cl	3,851.69	(30)	—
W	3,851.57	3	6	U	3,851.73	2	15
O	3,851.47	(2)	—	Nd	3,851.75	15	8
Ta	3,851.44	2	1	Tb	3,851.68	8	5

	λ	I	
		J	O
Cl	**3,851.42**	**(75)**	—
Mo	**3,851.39**	**5**	**6**
Ce	**3,851.35**	**4**	**3**
U	**3,851.30**	**8**	**4**
Cl	**3,851.02**	**(100)**	—
Rb	**3,851.20**	**(20)**	—
V	**3,851.17**	**15**	**50**
W	**3,851.12**	**6**	—
O	**3,851.04**	**(10)**	—
*P**			

	λ	I	
		J	O
Sm	**3,851.88**	**8**	**10**
Ra	**3,851.90**	**(25)**	—
U	**3,851.96**	**2**	**8**
Mo	**3,851.99**	**15**	**10**
W	**3,852.00**	**2**	**5**
U	**3,852.09**	**8**	**4**
V	**3,852.096**	**10**	**20**
Tl	**3,852.10**	**(10)**	—
Ce	**3,852.11**	**3**	**6**
Ru	**3,852.14**	**10**	**12**
Cr	**3,852.22**	**12**	**60**
Nd	**3,852.38**	**50**	**60**
Ce	**3,852.387**	**25**	**8**
Sc	**3,852.392**	**15**	**15**
Gd	**3,852.50**	**8**	**100**
Ru	**3,852.56**	**5*h***	**4**
Fe	**3,852.57**	**100**	**150**
Nb	**3,852.61**	**8*h***	**1**
U	**3,852.70**	**6*h***	**6**
Pr	**3,852.80**	**50**	**100**
W	**3,852.83**	**5**	**5**
Ru	**3,852.84**	**3*h***	**5**
Nd	**3,852.90**	**8**	**10**
Ce	**3,852.93**	**2**	**2**
Th	**3,852.96**	**10**	**8**
U	**3,852.98**	**6*h***	**6**
Si	**3,853.01**	**5**	—
Ti	**3,853.05**	**4**	**18**
Zr	**3,853.06**	**2**	**8**
S	**3,853.09**	**(8)**	—
Nb	**3,853.095**	**3**	**1**
Ce	**3,853.16**	**3**	**25**
Pb	3,854.05	100	—
V	3,855.84	200	200
Fe	3,856.37	300	500
Fe	3,859.22	100	100
Fe	3,859.91	600	1000*r*
Cl	3,860.83	(150)	—
Cl	3,860.99	(100)	—
Bi	3,863.90	(100)	—
Mo	3,864.11	500*R*	1000*R*
Bi	3,864.20	150*h*	—
La	3,864.49	150	100
Nb	3,865.02	200*h*	—
Pr	3,865.46	125*r*	200*r*
Os	*3,865.47*	*200*	*125*

	λ	I			λ	I	
		J	O			J	O
				Fe	*3,865.53*	*400*	*600*
				Fe	*3,872.50*	*300*	*300*
				Tb	*3,874.18*	*200*	*200*
				Fe	*3,878.02*	*300*	*400*
				Fe	*3,878.57*	*300*	*300R*
				Nb	*3,879.35*	*300*	*5*

F 4,246.16 (300)

	λ	J	O		λ	J	O
Pr	**4,246.15**	**2**	**10**	**U**	**4,246.26**	**2**	**30**
Ca	**4,246.095**	**5**	**2*h***	**Nb**	**4,246.29**	**10**	**8**
Fe	**4,246.090**	**30**	**80**	**Th**	**4,246.34**	**1**	**3**
Mo	**4,246.02**	**30**	**30**	**Tm**	**4,246.38**	**4**	**20**
Ce	**4,245.98**	**2**	**8**	**Te**	**4,246.47**	**(15)**	—
U	**4,245.95**	**3**	**1**	**Gd**	**4,246.55**	**3**	**150**
Dy	**4,245.92**	**4**	**25**	**Tb**	**4,246.59**	**1*h***	**12**
Ce	**4,245.88**	**2**	**6**	**Mo**	**4,246.62**	**25**	—
Cd	**4,245.87**	**2**	—	**Ce**	**4,246.71**	**4**	**30**
Eu	**4,245.86**	**1**	**5**	**Sc**	**4,246.83**	**500**	**80**
Hf	**4,245.84**	**12**	**9**	**Nd**	**4,246.879**	**4**	**10**
Ti	**4,245.51**	**3**	**20**	**P**	**4,246.88**	**(150*w*)**	**70**
Th	**4,245.464**	**1**	**4**	**Eu**	**4,247.07**	**3**	**15**
Pr	**4,245.46**	**1**	**10**	**U**	**4,247.14**	**8**	**10**
Eu	**4,245.39**	**1**	**15*d***	**Dy**	**4,247.36**	**4**	**30**
Ho	**4,245.39**	**2**	**3**				
Xe	**4,245.38**	**(200*h*)**	—				
Ta	**4,245.35**	**15**	**30**	Fe	4,250.13	150	250
				Kr	4,250.58	(150)	—
Sb	**4,245.34**	**2*h***	—	Fe	4,250.79	250	400
In	**4,245.33**	**5**	—	Ar	4,251.18	(800)	—
Fe	**4,245.26**	**40**	**80**	Sm	4,251.79	200	200
Sm	**4,245.17**	**6**	**8**				
Hf	**4,245.16**	**2**	**10**	Cr	4,254.35	1000*R*	5000*R*
				Sm	4,256.40	150	150
Pr	**4,245.14**	**1*w***	**10*w***				
				Ar	*4,259.36*	*(1200)*	—
Rb	4,242.60	(150)		*Fe*	*4,260.48*	*300*	*400*
Gd	4,238.78	200	200	*Te*	*4,261.08*	*(300)*	—
La	4,238.38	300	500	*Ar*	*4,266.29*	*(1200)*	—
Xe	4,238.25	(200*h*)	—	*C*	*4,267.02*	*350*	—
Fe	4,235.94	200	300				
				C	*4,267.27*	*500*	—

F 6,239.64 (300)

	λ	J	O		λ	J	O
Tl	**6,239.46**	**15**	—	**Se**	**6,239.70**	**(2)**	—
Sc	**6,239.41**	**5**	**8**	**Ar**	**6,239.73**	**(4)**	—
J	**6,238.74**	**(15)**	—	**Sc**	**6,239.78**	**30**	**3*h***

	λ	I J	I O
Hf	**6,238.58**	**8**	**6**
Tl	**6,238.49**	**10**	**—**
Fe	**6,238.41**	**2**	**—**
Te	6,230.80	(300)	—
Lu	*6,221.87*	*1000*	*500*
Ne	*6,217.28*	*(1000)*	—
Ne	*6,213.88*	*(150)*	—
Ra	*6,200.30*	*(1000)*	—
Xe	*6,182.42*	*(300)*	—

	λ	I J	I O
V	**6,240.13**	**2**	**20**
Se	**6,240.37**	**(10)**	**—**
J	**6,240.93**	**(15)**	**—**
Kr	6,241.39	(10)	—
Xe	6,242.09	(8)	—
Se	6,242.21	(80)	—
Hg	6,242.24	(10)	—
Lu	6,242.34	200	40
N	6,242.52	(70)	—
Cl	6,242.66	(2)	—
Te	6,245.61	(150)	—
Ti	*6,258.70*	*250*	*300*
La	*6,262.30*	*150*	*125*
Se	*6,266.23*	*(200)*	—
Ne	*6,266.49*	*(1000)*	—
Xe	*6,270.82*	*(250)*	—
Se	*6,284.47*	*(300)*	—
S	*6,286.35*	*(300)*	—
S	*6,287.06*	*(1000)*	—
J	*6,294.08*	*(300)*	—

F 6,856.02 (1000)

	λ	I J	I O
Hf	**6,855.29**	**50**	**7**
Fe	**6,855.18**	**80*h***	**60**
Te	**6,854.70**	**(50)**	**—**
F	6,834.26	(300)	—

	λ	I J	I O
J	**6,856.82**	**(15)**	**—**
Hf	**6,857.03**	**10**	**5**
Fe	**6,857.25**	**5*wh***	**5*wh***
F	6,870.22	(150)	—

F 6,902.46 (500)

	λ	I J	I O
J	**6,902.13**	**(150)**	**—**
Xe	6,872.11	(700)	—

	λ	I J	I O
Fe	**6,902.80**	**5*h***	**—**
Nb	**6,902.89**	**10**	**60**
In	**6,902.98**	**5**	**—**
Ra	**6,903.10**	**(30)**	**—**
F	6,909.82	(150)	—
Ne	*6,929.47*	*(1000)*	—

Fe $^{26}_{55.847}$

t_0 1,535°C

t_1 2,735°C

I.	II.	III.	IV.	V.
7.86	16.240	30.6	—	—

λ	*I*		eV
	J	*O*	
II 2,382.04	100*R*	40*R*	13.1
II 2,395.62	100	50	13.1
I 2,483.27	50	500*R*	5.0
II 2,599.40	1000	1000	12.6
II 2,755.74	100	300	13.3
I 3,020.64	600*R*	1000*R*	4.1
I 3,581.19	600*R*	1000*R*	4.3
I 3,719.93	700	1000*R*	3.3
I 3,734.87	600	1000*R*	4.2
I 3,737.13	600	1000*R*	3.4
I 3,745.56	500	500	3.4
I 3,748.26	200	500	3.4
I 4,271.76	700	1000	4.4
I 4,307.91	800*R*	1000*R*	4.4
I 4,325.76	700	1000	4.5
I 4,383.55	800	1000	4.3
I 4,404.75	700	1000	4.4

Fe 2,382.04 *J* 100*R* *O* 40*R*

	λ	*I* *J*	*I* *O*		λ	*I* *J*	*I* *O*
*bh*C	2,382.00	—	30	Hg	2,382.06	(8)	—
Ru	2,381.99	150	50	Zn	2,382.22	—	4
Dy	2,381.97	—	8	Nb	2,382.246	—	2
Fe	2,381.83	—	3	Pd	2,382.25	2*h*	—
Ir	2,381.82	50	8	Co	2,382.33	4	2
W	2,381.79	10	—	W	2,382.34	10	3
J	2,381.79	(20)	—	Fe	2,382.35	—	3
Pd	2,381.78	(2)	—	Zr	2,382.36	2	1*h*
Co	2,381.75	12	4				
Er	2,381.72	5	—	Os	2,382.46	5	30
Lu	2,381.69	30*h*	—	V	2,382.47	100*w*	—
Ir	2,381.62	2*h*	30	Rh	2,382.89	5	50
				V	2,383.00	80	8
				Te	2,383.25	300	500
Rb	2,381.30	(100)	—				
As	2,381.18	4	75	Cr	2,383.33	—	20
Re	2,381.14	7	40	Rh	2,383.40	10	50
Pd	2,381.02	9	25	Pd	2,383.40	50*wh*	—
Hf	2,381.00	30	20	Re	2,383.48	5	25
				Sb	2,383.63	20	75
V	2,380.82	50	6				
Os	2,380.82	20	30	Pt	2,383.64	20	30
Co	2,380.48	10	20*d*	Mn	2,384.05	3	40
Rb	2,380.44	(125)	—	Zr	2,384.17	1*h*	25
Tl	2,380.34	60	20	Fe	2,384.39	5	20
				Os	2,384.62	5	30
Hf	2,380.30	60	30				
Ce	2,380.16	50	—	Rh	2,384.65	—	25
Os	2,379.84	8	25	Ir	2,384.81	80	4
Tl	2,379.69	200*R*	100*R*	Pd	2,385.01	50*wh*	—
La	2,379.42	60	—				
Os	2.379.39	9	40	*Te*	*2,385.76*	*(300)*	*600*
Ir	2,379.38	8	24	*V*	*2,385.82*	*100*	*—*
Os	2,378.74	5	25	*Ir*	*2,385.86*	*3*	*20*
Co	2,378.62	50*w*	25	*Rh*	*2,386.14*	*8*	*80*
				Ir	*2,386.89*	*15*	*50*
Al	*2,378.41*	*20*	*40*	*Os*	*2,387.29*	*15*	*40*
Os	*2,377.61*	*15*	*50*	*Ru*	*2,387.90*	*3*	*60*
Os	*2,377.03*	*30*	*50*	*Pb*	*2,388.77*	*18*	*40*
Ru	*2,375.63*	*80*	*50*				
Ru	*2,375.27*	*5*	*80*				

Fe 2,395.62 *J* 100 *O* 50

	λ	*J*	*O*		λ	*J*	*O*
Ni	2,395.61	4	10	Ag	2,395.63	5	4
Co	2,395.52	6	—	vzduch	2,395.66	3	—
W	2,395.47	—	8	W	3,395.71	8	5
V	2,395.43	5	5	Cr	2,395.79	2*h*	25
Co	2,395.42	2	6	Er	2,395.81	4	—

	λ	I	
		J	O
Fe	**2,395.41**	**5*h***	**6**
Os	**2,395.39**	**20**	**10**
Br	**2,395.34**	**(25)**	—
Nb	**2,395.32**	**2**	**15**
W	**2,395.30**	—	**6**
Tl	**2,395.27**	**2**	—
Sb	2,395.20	15	50
Ir	2,394.33	—	30
Os	2,394.29	10	45
Cr	2,394.01	50	—
Ru	2,393.97	—	50
Os	2,393.86	5	30
Hf	2,393.83	100	80
Pb	2,393.79	1000	2500
V	2,393.57	500	—
Hf	2,393.36	80	50
Ru	2,393.25	1	80
V	2,392.90	—	25
Cr	2,392.89	2*h*	40
Ru	2,392.42	6	80
Cr	2,392.37	2*h*	25
Lu	2,392.19	100	30
Ir	*2,391.18*	—	*50*
V	*2,289.70*	*100*	*5*
In	*2,389.54*	—	*50R*

	λ	I	
		J	O
Nb	**2,395.84**	**8**	**2*h***
Ir	**2,395.86**	**8**	**15**
W	**2,395.89**	—	**6**
Mo	**2,395.98**	**3**	—
Pt	2,396.17	18	25
Ta	2,396.30	—	80
Rh	2,396.55	200	2
Ru	2,396.71	80	60
Os	2,396.77	12	100
Co	2,396.774	—	90
Re	2,396.81	7	30
Re	2,397.36	8	30
V	2,397.78	—	30
Os	2,398.18	3	25
bhB	2,398.50	—	200
Ca	2,398.56	20	100*R*
Ir	2,398.75	150	10
Cr	2,399.06	2	50
V	*2,399.68*	*150*	—
Cu	*2,400.11*	*100*	*5*
Bi	*2,400.88*	*100*	*200R*
Pt	*2,401.87*	*30*	*200*
Pb	*2,401.95*	*40*	*50*
Ru	*2,402.72*	*150r*	*100*

Fe 2,483.27 *J* 50* *O* 500*R*

	λ	J	O
W	**2,483.23**	**4**	**1**
V	**2,483.07**	**150**	**20**
Rh	2,482.73	100	2
Fe	2,479.78	30	200*R*
C	*2,478.57*	*(400)*	*400*

	λ	J	O
Hf	**2,483.33**	**5**	**6**
Rh	**2,483.333**	**5**	**100*r***
Pt	**2,483.37**	**2**	**40**
Sn	**2,483.40**	**125**	**125**
Fe	**2,483.53**	—	—
V	**2,483.65**	—	**10**
Pt	2,487.17	20	600*r*
Fe	*2,488.15*	*100r*	*600r*

Fe 2,599.396 *J* 1000 *O* 1000

	λ	*I*			λ	*I*	
		J	*O*			*J*	*O*
Fe	**2,599.22**	**—**	**1**	**Ta**	**2,599.397**	**30**	**100**
Hf	**2,599.215**	**10**	**10**	**Ir**	**2,599.400**	**—**	**40**
Co	**2,599.207**	**—**	**5**	**Nb**	**2,499.52**	**5*wh***	**—**
Mo	**2,599.18**	**15**	**1**	**Fe**	**2,599.57**	**—**	**1000**
Yb	**2,599.15**	**50**	**2**	**W**	**2,599.642**	**5**	**1**
Fe	**2,599.131**	**—**	**1**	**Mo**	**2,599.643**	**1**	**20**
Os	**2,599.129**	**1**	**5**	**Ru**	**2,599.658**	**25**	**10**
Ta	**2,599.09**	**—**	**2**	**Fe**	**2,599.661**	**—**	**6**
Ir	**2,599.040**	**10**	**25**	**W**	**2,599.76**	**12**	**2**
Mn	**2,599.036**	**12**	**2**	**U**	**2,599.80**	**2**	**4**
Dy	**2,598.97**	**—**	**3**	**Re**	**2,599.86**	**—**	**80**
Fe	2,598.37	1000*h*	700	Ni	2,601.13	2000*h*	—
Mn	*2,593.73*	*1000R*	*200R*	*Mn*	*2,605.69*	*500R*	*100R*
Cu	*2,592.63*	*50*	*1000*	*Ni*	*2,606.39*	*600h*	*—*
Nb	*2,590.94*	*800*	*15*	*Fe*	*2,607.09*	*400*	*300*

Fe 2,755.737 *J* 100 *O* 300

	λ	*J*	*O*		λ	*J*	*O*
W	**2,755.69**	**6**	**—**	**Ir**	**2,755.739**	**—**	**3**
V	**2,755.66**	**—**	**15**	**Dy**	**2,755.76**	**—**	**6**
Mn	**2,755.65**	**—**	**3**	**Ne**	**2,755.82**	**(15)**	**—**
Er	**2,755.643**	**3**	**20**	**W**	**2,755.94**	**6**	**10**
Nb	**2,755.637**	**2**	**5**	**Ce**	**2,755.96**	**—**	**2**
J	**2,755.58**	**(12)**	**—**	**Yb**	**2,756.01**	**3**	**—**
Nb	**2,755.56**	**4**	**1**	**Mo**	**2,756.07**	**50*h***	**10**
Ce	**2,755.41**	**—**	**8**	**Ce**	**2,756.09**	**—**	**2**
Mo	**2,755.37**	**10**	**15**	**Ir**	**2,756.11**	**5**	**5**
Br	**2,755.31**	**(5)**	**—**	**Mo**	**2,756.259**	**—**	**15**
Nb	**2,755.29**	**10**	**5**	**Fe**	**2,756.264**	**100**	**300**
Cr	**2,755.27**	**2**	**50*d***				
W	**2,755.26**	**4**	**10**				
Ru	**2,755.23**	**30**	**—**	Cr	2,756.30	100	1
Re	**2,755.22**	**—**	**25**	Fe	2,756.33	100	300
Cs	**2,755.20**	**(20)**	**—**				
Fe	**2,755.18**	**—**	**15**				
U	**2,755.13**	**12**	**10**				
Pt	2,754.92	5*h*	200				
Nb	2,754.52	100	10*w*				
Cr	2,754.28	50	3				
Lu	2,754.17	125	40				
In	2,753.88	300*wh*	300*R*				

	λ	I (J)	I (O)
Hf	2,753.61	60	—
Ru	2,753.44	50	50
V	2,753.40	200R	50
Fe	2,753.29	150h	25
Nb	2,753.14	80	2
Nb	2,753.01	50h	5
Ti	2,752.881	60wh	—
Cr	2,752.877	40	300r
Hg	2,752.775	50	100R
Ru	2,752.766	150	50
Ta	2,752.49	300	300
Ta	2,752.29	8	150
Ru	2,752.10	60	—
Cr	2,751.87	125	20
Hf	2,751.81	80	25
Ti	2,751.70	200wh	—
Fe	2,751.12	70	—
Cr	2,750.73	150	30
Yb	2,750.48	150	20
Fe	*2,750.14*	*100*	*300h*
Cr	*2,748.98*	*200*	*35*
Ta	*2,748.78*	*50*	*400*
Cd	*2,748.58*	*200*	*5*
Cr	*2,748.29*	*5*	*300*
Ru	*2,747.97*	*100*	*50*
Rh	*2,747.63*	*100*	*1*
Fe	*2,746.98*	*300wh*	*200*
Ti	*2,746.71*	*150wh*	—
Fe	*2,746.48*	*300wh*	*150*
Cu	*2,745.45*	*150*	*8*
Ru	*2,745.25*	*150*	*12*
Zn	2,756.45	100	200
Ag	2,756.51	200	5
Cr	2,757.10	10	300r
Nb	2,757.26	50	3
Fe	2,757.32	60	100
Cr	2,757.72	150	35
Ta	2,758.31	40	200
Nb	2,758.78	100w	1
Ni	2,759.02	500wh	—
Fe	2,759.82	60	100
V	2,760.70	100h	25
Rh	2,761.26	50	1
Ta	*2,761.68*	*150*	*200*
Cr	*2,761.75*	*35*	*300r*
Fe	*2,761.81*	*200*	*50*
Cr	*2,762.59*	*100*	*40*
Pd	*2,763.09*	*30r*	*300r*
Rh	*2,764.83*	*125*	*15r*
Ru	*2,765.44*	*150*	*50*
V	*2,765.67*	*200h*	*50*
Cu	*2,766.37*	*25*	*500*
V	*2,766.45*	*100h*	*40*
Cr	*2,766.540*	*300r*	*40*
Rh	*2,766.54*	*150*	*5*
Ru	*2,766.55*	*100*	—

Fe 3,020.640 *J* 600*R* *O* 1000*R*

	λ	J	O
Co	**3,020.639**	—	**60**
Tb	**3,020.58**	—	**3**
U	**3,020.57**	**6**	**6**
Lu	**3,020.54**	**100**	—
Hf	**3,020.53**	**2**	**15**
Os	**3,020.50**	**3**	—
Fe	**3,020.49**	**300r**	**300r**
Zr	**3,020.47**	**30**	**50**
Cs	**3,020.30**	**(4)**	—
Tb	**3,020.29**	**15**	**8**
U	**3,020.25**	**6**	**8**
W	**3,020.21**	**6**	**7**
Dy	**3,020.65**	—	**10**
Nb	**3,020.666**	**8**	**5**
Cr	**3,020.673**	**100**	**200r**
Mo	**3,020.69**	**10**	**5**
Yb	**3,020.70**	**5**	—
Ru	**3,020.882**	**40**	**60**
Ce	**3,020.883**	—	**15**
Cs	**3,020.90**	**(4)**	—
U	**3,020.92**	**12d**	**8d**
Ce	**3,021.04**	—	**15**
Fe	**3,021.07**	**300r**	**700r**
J	**3,021.219**	**(18)**	—

	λ	I J	I O		λ	I J	I O
Ca	3,020.15	2h	2	U	3,021.22	15	10
Te	3,020.02	(15)	—	Sm	3,021.225	—	6
Ir	3,020.01	1	35				
				Fe	3,024.03	200	300
Te	3,017.51	(350)	—	Fe	3,025.64	100	100
				Fe	3,025.84	300r	400r
				Fe	3,026.46	200	200
Ni	*3,012.00*	*125W*	*800R*				
Fe	*3,009.57*	*400*	*500*				
In	*3,008.31*	*500W*	*—*	*Fe*	*3,030.15*	*300*	*300*
Fe	*3,008.14*	*400r*	*600r*	*Fe*	*3,031.21*	*150*	*150*
				Fe	*3,031.64*	*200*	*200*

Fe 3,581.19 *J* 600*R* *O* 1000*R*

	λ	J	O		λ	J	O
Mo	3,581.00	3	2	W	3,581.24	8	8
Re	3,580.97	—	40w	Ca	3,581.29	2	—
Sc	3,580.93	40	12	Cs	3,581.30	(4)	—
U	3,580.92	—	3	Ar	3,581.62	(15)	—
Sm	3,580.91	6	40	Fe	3,581.65	3	4
Ta	3,580.89	1h	3	La	3,581.68	20h	—
Ir	3,580.86	3	15	Sn	3,581.68	6	—
V	3,580.82	50	50	Os	3,581.79	8	5
Ce	3,580.78	2	10	Mo	3,581.80	5	10
Ho	3,580.75	4	—	Fe	3,581.81	2	3
Tb	3,580.63	—	8	U	3,581.838	15	6
Gd	3,580.629	—	5	Pr	3,581.840	7	10
Sm	3,580.59	—	2	Er	3,581.842	2	10h
Eu	3,580.57	1h	3	Co	3,581.87	—	3
Ce	3,580.56	—	2	W	3,581.887	9	4
Mo	3,580.54	10	5	Mo	3,581.891	15	10
Os	3,580.53	4	2	Gd	3,581.92	15	15
Er	3,580.48	5	12d	Th	3,582.01	6	5
Hf	3,580.45	3	2	U	3,582.02	15	6
P	3,580.35	(30)	—	Dy	3,582.03	10	25
Ti	3,580.291	5	15	Nb	3,582.06	5	3
Mo	3,580.29	10	—	Zr	3,582.08	3	3
Nb	3,580.273	300	100	Ce	3,582.12	—	2
Ta	3,580.267	—	7	Fe	3,582.20	30	30
Rh	3,580.262	2	10	W	3,582.24	6	7
Lu	3,580.26	—	3	Pr	3,582.25	2h	—
U	3,580.24	1	6				
Th	3,580.235	5	5				
Eu	3,580.232	1h	4	Ar	3,582.35	(50)	—
Re	3,580.13	—	80	Te	3,585.34	(350)	—
				Al	3,586.91	(500h)	—
U	3,580.12	2	—	Fe	3,586.99	150	200
Mn	3,580.102	—	2				
La	3,580.099	3h	2				

	λ	I J	I O		λ	I J	I O
Au	**3,580.08**	**15**	**20**	*Ar*	*3,588.44*	*(300)*	*—*
Dy	**3,580.04**	**30**	**80**	*V*	*3,589.76*	*600R*	*80*
				Nd	*3,590.35*	*300W*	*400W*
				V	*3,592.02*	*300R*	*50*
Cr	3,578.69	400*r*	500*r*	*Cr*	*3,593.48*	*400R*	*500R*
Ar	3,576.62	(300)	—				
				Ne	*3,593.53*	*(500)*	*—*
				Ni	*3,597.70*	*60h*	*1000r*
Ar	*3,572.29*	*(300)*	*—*				
Ni	*3,571.87*	*40h*	*1000h*				
Fe	*3,570.10*	*300*	*300*				
Ni	*3,566.37*	*100wh*	*2000R*				
Fe	*3,565.38*	*300*	*400*				
Fe	*3,558.52*	*300*	*400*				

Fe 3,719.935 *J* 700 *O* 1000*R*

	λ	J	O		λ	J	O
Ba	**3,719.93**	—	**2**	**Ce**	**3,719.95**	—	**2***w*
Ce	**3,719.80**	**5**	**15***s*	**Th**	**3,719.97**	**1***h*	**2**
Mo	**3,719.74**	**30**	—	**Os**	**3,720.13**	**40**	**80**
Tm	**3,719.72**	**4**	**10**	**Pr**	**3,720.22**	**6**	**15**
Sb	**3,719.70**	**8**	—	**Mo**	**3,720.25**	**40**	**10**
Mo	**3,719.692**	—	**3**	**Th**	**3,720.309**	**10**	**15**
U	**3,719.69**	**8**	**1***h*	**As**	**3,720.31**	**15**	—
Nb	**3,719.63**	**50***w*	—	**Tb**	**3,720.36**	—	**8**
Nd	**3,719.59**	**8**	**10**	**Ce**	**3,720.380**	**2**	**2**
Mo	**3,719.55**	**3**	**5**	**Ti**	**3,720.384**	**10**	**40**
Os	**3,719.52**	**12**	**40**	**U**	**3,720.39**	**10***h*	**6**
Gd	**3,719.46**	**40**	**40**	**Ca**	**3,720.40**	**3**	—
Sm	**3,719.451**	**10**	**50**	**Ar**	**3,720.43**	**(10)**	—
Tb	**3,719.45**	**8**	**30**	**Nb**	**3,720.46**	**100***h*	**5**
Th	**3,719.44**	**10**	**30**	**W**	**3,720.51**	**10**	**8**
Pr	**3,719.435**	**10**	**15**	**Nd**	**3,720.54**	**8**	**8**
Ce	**3,719.430**	—	**8**	**Sm**	**3,720.57**	**4**	**4**
Ta	**3,719.42**	—	**3**	**Ce**	**3,720.59**	—	**2**
W	**3,719.40**	**10**	**12**	**Sm**	**3,720.64**	—	**2**
Ru	**3,719.33**	**25**	**20**	**Eu**	**3,720.69**	—	**4***w*
Sm	**3,719.30**	—	**3**	**Pt**	**3,720.74**	—	**5**
Er	**3,719.295**	—	**4***d*	**Ho**	**3,720.74**	**4**	**6**
U	**3,719.293**	—	**12**	**La**	**3,720.75**	**3***h*	—
Hf	**3,719.28**	**30**	**15**	**Cu**	**3,720.77**	**1***h*	**10**
Eu	**3,719.17**	—	**30**	**Th**	**3,720.78**	—	**5**
Ce	**3,719.08**	—	**3**	**Xe**	**3,720.80**	**(20)**	—
Mo	**3,719.05**	**20**	—	**Ir**	**3,720.82**	—	**6**
Mn	**3,718.93**	**100**	**75**	**Pr**	**3,720.83**	**8**	**10**
V	**3,718.912**	**5**	**20**	**Ba**	**3,720.85**	—	**2**
Pd	**3,718.909**	**200**	**300**	**Sm**	**3,721.02**	**1**	**4***h*
Sm	**3,718.880**	**5**	**100**	**V**	**3,721.06**	**2**	—

	λ	I J	I O		λ	I J	I O
Pr	**3,718.877**	**3**	**3**	**W**	**3,721.12**	**12**	**—**
U	**3,718.844**	**4**	**—**	**Ba**	**3,721.18**	**—**	**2**
Zr	**3,718.843**	**9**	**9**				
Hf	**3,718.842**	**3**	**—**				
In	**3,718.836**	**(40)**	**—**	Fe	3,722.56	500	500
In	**3,718.71**	**(18)**	**—**	V	3,727.34	200	40
Li	**3,718.70**	**—**	**30**	Ar	3,729.29	(200)	—
Sm	**3,718.698**	**1**	**8**	Ti	3,729.81	150	500
				Fe	3,732.40	150	200
Kr	3,718.02	(300*h*)	—	*V*	*3,732.76*	*500R*	*70R*
Nb	3,717.07	100	8	*Fe*	*3,733.32*	*300*	*400*
V	3,715.47	400*R*	70				
Fe	3,709.25	400	600				
Fe	*3,705.57*	*500*	*700*				
Fe	*3,701.09*	*200*	*300*				

*N**

Fe 3,734.867 *J* 600 *O* 100*R*

	λ	I J	I O		λ	I J	I O
Co	**3,734.867**	**—**	**60**	**Ga**	**3,734.87**	**(4)**	**—**
Nd	**3,734.862**	**4**	**6**	**Ca**	**3,734.91**	**3**	**—**
Ce	**3,734.856**	**2**	**2**	**Tb**	**3,734.94**	**—**	**8**
Pb	**3,734.80**	**(10)**	**—**	**Ne**	**3,734.94**	**(40)**	**—**
Tb	**3,734.80**	**—**	**—**	**Re**	**3,735.00**	**—**	**30*W***
Ir	**3,734.77**	**30**	**100**	**Sn**	**3,735.02**	**5**	**—**
Nb	**3,734.74**	**5**	**2**	**Ru**	**3,735.021**	**—**	**4**
Al	**3,734.72**	**(2)**	**—**	**V**	**3,735.17**	**25**	**3**
Yb	**3,734.70**	**5**	**25**	**Nd**	**3,735.22**	**—**	**2**
U	**3,734.68**	**1**	**6**	**Rh**	**3,735.28**	**2**	**70**
Eu	**3,734.66**	**—**	**2*W***	**Re**	**3,735.329**	**—**	**40**
Th	**3,734.60**	**10**	**8**	**Fe**	**3,735.330**	**20**	**30**
Er	**3,734.59**	**—**	**4**	**Rn**	**3,735.40**	**(5)**	**—**
Al	**3,734.57**	**(2)**	**—**	**Ar**	**3,735.49**	**(5)**	**—**
Se	**3,734.50**	**(8)**	**—**	**Ir**	**3,735.50**	**—**	**8**
Er	**3,734.46**	**—**	**4**	**Th**	**3,735.53**	**—**	**8**
V	**3,734.43**	**5**	**10**	**Os**	**3,735.536**	**10**	**20**
Pr	**3,734.41**	**30**	**40**	**Nd**	**3,735.518**	**—**	**10**
H	**3,734.372**	**(8)**	**—**	**Ce**	**3,735.599**	**4**	**3**
Mo	**3,734.370**	**5**	**15**	**Nd**	**3,735.599**	**50**	**10**
J	**3,734.35**	**(18)**	**—**	**Eu**	**3,735.62**	**—**	**3*W***
Cs	**3,734.34**	**(10)**	**—**	**Mo**	**3,735.622**	**5**	**5**
Ir	**3,734.33**	**—**	**5**	**Ti**	**3,735.67**	**4**	**15**
V	**3,734.28**	**—**	**3**	**Ba**	**3,735.75**	**2**	**1**
Eu	**3,734.23**	**—**	**5*W***	**Pr**	**3,735.76**	**2**	**2**
Ce	**3,734.211**	**—**	**2**	**Ce**	**3,735.77**	**—**	**2**
Dy	**3,734.21**	**—**	**4**	**Kr**	**3,735.78**	**(40*h*)**	**—**
Co	**3,734.14**	**—**	**70**	**Br**	**3,735.83**	**(6)**	**—**

	λ	I	
		J	O
Tm	3,734.13	50	150
Ce	3,734.06	—	2
Pr	3,734.03	2*h*	4
Ca	3,733.92	2	—
Al	3,733.91	(5)	—
Mo	3,733.84	25	—
Hf	3,733.79	5	12
Ti	3,733.78	2	10
Ce	3,733.77	—	2
Er	3,733.76	—	2
U	3,733.752	1	3
Mo	3,733.749	1	4
Cl	3,733.73	(10)	—
Eu	3,733.66	2*w*	8*w*
Nb	3,733.62	10	10
La	3,735.85	10	20
Mo	3,735.909	3	5
Ce	3,735.912	—	2
Co	3,735.93	—	200*R*
O	3,735.94	(10)	—
Eu	3,735.95	—	5
Sm	3,735.97	8	50
V	3,736.02	20	3
U	3,736.03	—	3*h*
Eu	3,736.04	1	6
Ce	3,736.06	2	3

*P**

*N**

Fe 3,737.133 *J* 600 *O* 1000*R*

	λ	I	
		J	O
Rh	3,737.12	1	50
Ce	3,737.02	—	3
Th	3,736.95	5	5
Ce	3,736.91	1	2
Ca	3,736.901	50	12
Mn	3,736.899	25	25
Re	3,736.83	—	15
Ni	3,736.81	15	300
Ru	3,736.80	—	3
Ti	3,736.79	2	10
Ta	3,736.76	3	35
Ir	3,736.75	—	2
Ir	3,736.65	—	2
U	3,736.60	—	2
Zr	3,736.51	2	2
Pr	3,736.50	20	20
Ir	3,736.480	—	2
U	3,736.476	—	2
Ce	3,736.47	3	—
Nd	3,736.44	4	—
La	3,736.41	6	—
Ce	3,736.402	3	3
Mo	3,736.40	20	—
Mo	3,736.35	2	6
Nb	3,736.33	5*h*	3
Be	3,736.28	—	10
Eu	3,736.26	—	4*w*
W	3,736.21	10	5
Sm	3,737.134	3	10
Xe	3,737.20	(3*wh*)	—
U	3,737.25	10	8
Rh	3,737.27	10	50
Eu	3,737.37	—	2
Zr	3,737.39	2	3
Ru	3,737.40	12	12
V	3,737.43	2	2
Sm	3,737.47	2	2
Th	3,737.51	1	5
Ce	3,737.52	—	5
Cr	3,737.55	18	—
Ho	3,737.65	10*h*	6
Pr	3,737.67	7	10
Ce	3,737.736	2	10
Ru	3,737.741	5	6
W	3,737.84	5	4
Sm	3,737.87	—	2
Hf	3,737.88	25	15
Ar	3,737.89	(15)	—
Mo	3,737.91	20	20
Ce	3,737.96	—	4
V	3,737.99	5	50
Al	3,738.00	(10)	—
U	3,738.05	20	8
Nd	3,738.06	—	25
Eu	3,738.07	10	10
W	3,738.111	7	6

	λ	I			λ	I	
		J	O			J	O
Tm	**3,736.20**	**8**	**2**	**Zr**	**3,738.115**	**2**	**4**
Mo	**3,736.17**	**1**	**4**	**Er**	**3,738.17**	**2**	**8**
				Ta	**3,738.21**	**1**	**10**
*P**				**Ce**	**3,738.25**	**—**	**5**
				Sm	**3,738.26**	**—**	**5**
				Fe	**3,738.31**	**100**	**100**
				V	**3,738.35**	**4*h***	**—**
				Cr	**3,738.38**	**40**	**6**
				Nb	3,739.79	200	100
				Ti	3,741.64	200	30
				Kr	3,741.69	(200*h*)	—
				Ru	3,742.28	100	70
				Fe	3,743.36	150	200
				Ar	3,743.76	(100)	—
				*N**			

Fe 3,745.564 *J* 500 *O* 500

	λ	J	O		λ	J	O
W	**3,745.56**	**25**	**—**	**Sm**	**3,745.62**	**30**	**40**
Co	**3,745.50**	**—**	**300*R***	**Dy**	**3,745.64**	**—**	**4**
Mo	**3,745.48**	**20*d***	**30*hd***	**Xe**	**3,745.69**	**(4)**	**—**
Sm	**3,745.46**	**10**	**40**	**Sn**	**3,745.80**	**6**	**—**
Re	**3,745.44**	**—**	**40*W***	**V**	**3,745.80**	**600**	**35**
Ir	**3,745.43**	**—**	**3**	**Fe**	**3,745.90**	**100**	**150**
Xe	**3,745.38**	**(10)**	**—**	**Th**	**3,745.98**	**20**	**15**
Er	**3,745.11**	**1**	**4**	**Zr**	**3,745.99**	**15**	**—**
Tb	**3,745.07**	**3**	**15**	**Eu**	**3,746.03**	**—**	**10**
Er	**3,744.99**	**2**	**9**	**Sm**	**3,746.05**	**2*h***	**—**
Hf	**3,744.98**	**20**	**15**	**Er**	**3,746.06**	**1*W***	**12*W***
Ir	**3,744.95**	**—**	**5**	**Nd**	**3,746.13**	**—**	**2**
Mo	**3,744.94**	**8**	**5**	**Ru**	**3,746.22**	**4**	**4**
W	**3,744.91**	**3**	**5**	**Ce**	**3,746.25**	**—**	**3**
La	**3,744.85**	**2*h***	**1**	**Ta**	**3,746.36**	**5**	**35**
Pr	**3,744.82**	**—**	**3**	**Ce**	**3,746.37**	**2**	**8**
Gd	**3,744.805**	**12**	**25**	**Rb**	**3,746.38**	**(10)**	**—**
U	**3,744.805**	**2**	**—**	**Mo**	**3,746.41**	**40*w***	**—**
Nd	**3,744.804**	**4**	**8**	**U**	**3,746.413**	**12**	**3**
Kr	**3,744.80**	**(150*hs*)**	**—**	**Gd**	**3,746.42**	**10*h***	**5**
Th	**3,744.74**	**10**	**15**	**Ar**	**3,746.46**	**(5)**	**—**
Ce	**3,744.72**	**2**	**3**	**Os**	**3,746.47**	**20**	**100**
Ne	**3,744.66**	**(12)**	**—**	**Fe**	**3,746.48**	**1**	**3**
Ni	**3,744.56**	**—**	**6**	**Tb**	**3,746.54**	**8**	**8**
Eu	**3,744.55**	**5*wh***	**10*w***	**Ce**	**3,746.57**	**1**	**2**
Cr	**3,744.49**	**12**	**30**	**Mn**	**3,746.616**	**25**	**25**
K	**3,744.40**	**(20)**	**—**	**Er**	**3,746.62**	**—**	**2**

	λ	I J	I O		λ	I J	I O
Ru	**3,744.39**	**35**	**8**	**U**	**3,746.68**	**5**	**1**
Mo	**3,744.37**	**80**	**20**	**Eu**	**3,746.72**	**3**	**4**
				Hf	**3,746.80**	**8**	**20**
*P**				*N**			

Fe 3,748.26 *J* 200 *O* 500

	λ	I J	I O		λ	I J	I O
Rh	**3,748.22**	**100**	**200**	**Th**	**3,748.30**	**8**	**10**
Cu	**3,748.21**	**10**	**—**	**Ca**	**3,748.34**	**—**	**12**
Ho	**3,748.17**	**40**	**60**	**Cl**	**3,748.46**	**(15)**	**—**
Sm	**3,748.15**	**4**	**5**	**Mo**	**3,748.490**	**10**	**15**
Mo	**3,748.13**	**50**	**1**	**Fe**	**3,748.490**	**—**	**2*h***
Ti	**3,748.10**	**1**	**10**	**Pr**	**3,748.50**	**3**	**5**
J	**3,748.06**	**(5)**	**—**	**Sm**	**3,748.51**	**—**	**5**
Pr	**3,748.058**	**4**	**9**	**Ce**	**3,748.53**	**—**	**2**
Ce	**3,748.056**	**3*h***	**10**	**Nb**	**3,748.55**	**10**	**10**
Ti	**3,748.00**	**25**	**2**	**Ir**	**3,748.56**	**—**	**3**
Ld	**3,747.99**	**4**	**2**	**Cr**	**3,748.61**	**30**	**40**
V	**3,747.98**	**4**	**50**	**Sm**	**3,748.63**	**5**	**10**
Dy	**3,747.83**	**20**	**60**	**U**	**3,748.68**	**25**	**15**
Ti	**3,747.78**	**1**	**7**	**Pr**	**3,748.82**	**3**	**9**
Sm	**3,747.75**	**—**	**3**	**Fe**	**3,748.969**	**20**	**35**
Tb	**3,747.64**	**—**	**30**	**Th**	**3,748.970**	**5**	**8**
Sm	**3,747.62**	**25**	**25**	**Cr**	**3,748.998**	**125*R***	**125*R***
Yt	**3,747.554**	**15**	**12**	**Pr**	**3,749.005**	**—**	**2**
Er	**3,747.554**	**10**	**20**	**Ho**	**3,749.02**	**—**	**5**
Th	**3,747.547**	**30**	**30**	**Ni**	**3,749.04**	**5**	**50**
Ce	**3,747.546**	**—**	**6**	**Os**	**3,749.07**	**2**	**4**
Kr	**3,747.50**	**(2*wh*)**	**—**	**Pb**	**3,749.15**	**2**	**—**
Hf	**3,747.49**	**8**	**6**	**U**	**3,749.16**	**8**	**2**
Pr	**3,747.47**	**3**	**10**	**Ce**	**3,749.36**	**—**	**5**
W	**3,747.46**	**5**	**7**	**O**	**3,749.47**	**(125)**	**—**
Er	**3,747.44**	**—**	**2**	**Rn**	**3,749.48**	**(8)**	**—**
Sm	**3,747.35**	**1**	**8**	**Fe**	**3,749.487**	**700**	**1000*r***
Pr	**3,747.265**	**2**	**5**	**Ce**	**3,749.493**	**2*h***	**2**
Cr	**3,747.264**	**6**	**12**				
Ta	**3,747.25**	**—**	**35**				
				As	3,749.77	100	—
Eu	**3,747.24**	**1*wh***	**5*W***	V	3,751.23	100	4
Ir	**3,747.20**	**60**	**100**	Os	3,752.52	100	400*R*
Mo	**3,747.19**	**10**	**15**	Fe	3,753.61	100	150
Tb	**3,747.17**	**—**	**30**	Tb	3,755.24	100	50
V	**3,747.14**	**2**	**25**				
				Ti	3,757.69	100	30
U	**3,747.11**	**10**	**6**	Fe	3,758.23	700	700
Ce	**3,747.09**	**1**	**2**	La	3,759.08	150	400
				Ti	3,759.29	400*R*	100
				Fe	3,760.05	100	150
*P**							

	λ	I J	I O
Ti	*3,761.32*	*300r*	*100*
Fe	*3,763.79*	*400*	*500*
Fe	*3,767.19*	*400*	*500*
Ar	*3,770.37*	*(400)*	—

Fe 4,271.764 *J* 700 *O* 1000

	λ	I J	I O
Pr	4,271.764	15	18
Cs	4,271.74	(10)	—
Tm	4,271.72	2	20
Ca	4,271.56	3	2*h*
V	4,271.55	10	20
Sb	4,271.54	(10)	—
J	4,271.53	(5)	—
Ta	4,271.51	5*wh*	40
Ce	4,271.48	—	5
Ar	4,271.24	(2)	—
Fe	4,271.161	300	400
La	4,271.157	—	30
Th	4,271.10	3	5
Ce	4,271.07	—	2
Cr	4,271.06	12	30
W	4,270.91	3	5
U	4,270.89	6	6
Sm	4,270.83	15	15
Os	4,270.79	1	12
Gd	4,270.77	1*h*	4*h*
Th	4,270.73	—	3
Ce	4,270.72	—	25
Sm	4,270.70	3	5
Nb	4,270.69	50	30
U	4,270.65	4	5
C	4,267.27	500	—
C	4,267.02	350	—
Ar	4,266.29	(1200)	—
Fe	*4,260.48*	*300*	*400*
Ar	*4,259.36*	*(1200)*	—
Cr	*4,254.35*	*1000R*	*5000R*
Ar	*4,251.18*	*(800)*	—
Fe	*4,250.79*	*250*	*400*

	λ	I J	I O
Sm	4,271.86	2	5
Ca	4,271.87	7*h*	—
Er	4,271.97	1	4
Tl	4,272.00	(8)	—
Sm	4,272.01	3	5
Ce	4,272.03	—	2
Mo	4,272.06	15	15
Yb	4,272.12	—	6
Yt	4,272.14	—	3
Ar	4,272.17	(1200)	—
Tb	4,272.22	—	3
Pr	4,272.27	35	50
U	4,272.28	3	3
W	4,272.31	3	8
Ce	4,272.339	—	2
Eu	4,272.34	—	4
Ti	4,272.43	10	40
Er	4,272.44	1*h*	4
Bi	4,272.49	10*wh*	—
Pb	4,272.55	2	—
Pb	4,272.63	30	—
Rb	4,272.64	2	—
Eu	4,272.76	1	5
Nd	4,272.79	5	15
Hf	4,272.848	20	12
Ce	4,272.855	—	2
Cs	4,272.87	(10)	—
Th	4,272.88	—	4
Kr	4,273.97	(1000)	—
Cr	4,274.80	800*r*	4000*r*
La	4,275.64	500	40
Fe	*4,282.41*	*300*	*600*
Cr	*4,289.72*	*800r*	*3000r*
Fe	*4,294.13*	*400*	*700*

Fe 4,307.91 *J* 800*R* *O* 1000*R*

	λ	*I*			λ	*I*	
		J	*O*			*J*	*O*
Ti	4,307.90	100	100	Cs	4,307.94	(8)	—
Br	4,307.80	(10)	—	Se	4,307.95	(12)	—
Ir	4,307.799	—	8	Sr	4,308.10	—	20*h*
Nd	4,307.78	6	5	Nb	4,308.12	(10)	5
Rn	4,307.76	(400)	—	Bi	4,308.177	12	50
Ca	4,307.74	20	45	Ir	4,308.182	—	3
Pr	4,307.67	10	30	Gd	4,308.21	—	10
W	4,307.64	12	12	Th	4,308.26	2	3
Ru	4,307.59	50	20	Dy	4,308.34	—	7
Cr	4,307.49	1	35	W	4,308.50	4	8
Co	4,307.422	—	3	Ti	4,308.504	2	20
Cl	4,307.42	(75)	—	Bi	4,308.52	1	4
U	4,307.32	3	3	Dy	4,308.62	12	100
Ni	4,307.29	—	4	Er	4,308.63	3	30
Pr	4,307.24	2*w*	30	Ho	4,308.65	2	4
Tb	4,307.20	1*h*	12	Mo	4,308.651	6	6
Th	4,307.185	3	5	Tb	4,308.68	2	10
V	4,307.184	20	30	Nb	4,308.69	10	5
Al	4,307.16	(20)	—	Co	4,308.74	—	2
Ti	4,306.94	1	10	Re	4,308.78	—	4
W	4,306.87	15	20	Rh	4,308.83	2	4
Eu	4,306.86	—	2*w*	Os	4,308.88	2	18
Cd	4,306.82	3	8	Pr	4,308.89	5	25
Pr	4,306.81	—	4	W	4,308.95	4	10
Tl	4,306.80	(40)	—	Pr	4,309.00	2	8
Ta	4,306.80	5*h*	1	Sm	4,309.004	150	200
U	4,306.78	4	40*r*	Fe	4,309.04	10	40
Ar	4,300.10	(1200)	—	Fe	4,315.09	300	500
Fe	4,299.24	400	500	Kr	4,317.81	(500*wh*)	—
Fe	4,298.04	400	100	Kr	4,318.55	(400)	—
				Sm	4,318.93	300	300
Fe	*4,294.13*	*400*	*700*				
*P**				*N**			

Fe 4,325.765 *J* 700 *O* 1000

O	4,325.76	(20)	—	Nd	4,325.766	30	100
In	4,325.757	(5)	—	Te	4,325.77	(30)	—
Ba	4,325.73	(50)	—	Mo	4,325.82	20	25
Li	4,325.70	(3)	—	Tb	4,325.83	—	100
Gd	4,325.69	250	500*R*	C	4,325.88	10	—
Ca	4,325.66	4	—	In	4,325.885	(10)	—
Ni	4,325.61	—	70	U	4,325.90	10	10
Gd	4,325.57	—	5	In	4,326.02	(10)	—

	λ	I J	I O
Eu	4,325.53	—	30*w*
Tb	4,325.50	—	15
Zr	4,325.43	2	8
Ni	4,325.361	—	10
Pr	4,325.358	2*h*	20*h*
Ce	4,325.31	—	4
Yb	4,325.28	10	—
Mo	4,325.26	20	15
V	4,325.22	5	1
W	4,325.17	3	2
Ba	4,325.16	3	15
Sm	4,325.15	2*h*	3
Ho	4,325.14	2*h*	—
Dy	4,325.14	4	10
Ti	4,325.134	40	100
Ce	4,325.126	—	2
Cr	4,325.07	130	125
Ru	4,325.05	10	25
Sc	4,325.01	40	50
Pr	4,324.80	—	3
Ce	4,324.79	3	18
Ce	4,324.60	—	10
Kr	4,319.58	(1000)	—
*P**			
Sm	4,326.127	—	2
Eu	4,326.13	—	4*w*
Mo	4,326.14	40	50
La	4,326.18	—	5
In	4,326.23	(15)	—
Hf	4,326.24	—	6*h*
Os	4,326.25	12	30
Gd	4,326.29	—	3
Cs	4,326.31	(10)	—
Nb	4,326.327	5	30
In	4,326.335	(5)	—
Ti	4,326.35	25	60
Dy	4,326.39	—	2
Ho	4,326.39	1	2
Yb	4,326.40	—	20*h*
U	4,326.43	2	2
Eu	4,326.44	—	8*w*
Sr	4,326.445	—	8
Tb	4,326.48	4	150
U	4,326.59	2	2
Ba	4,326.74	(5)	—
Mo	4,326.743	50	50
Mn	4,326.756	30	80
Fe	4,326.760	4	10
In	4,326.761	(15)	—
In	4,326.81	(5)	—
Ru	4,326.825	—	20
Ce	4,326.826	—	15
Gd	4,327.10	100	500*R*
Xe	4,330.52	(500*wh*)	—
Ar	4,333.56	(1000)	—
La	4,333.73	500	800
Ar	4,335.34	(800)	—
Fe	4,337.05	150	400
Cr	*4,337.57*	*300*	*500*
Ra	*4,340.64*	*(1000)*	—
Ar	*4,345.17*	*(1000)*	—

Fe 4,383.547 *J* 800 *O* 1000

	λ	I J	I O
*bh*La	4,383.50	—	15
La	4,383.45	50	10
*bh*La	4,383.40	—	8
Ru	4,383.36	—	12
Rn	4,383.30	(35)	—
Ce	4,383.555	1	8
U	4,383.63	1	4
Ce	4,383.74	—	4
In	4,383.76	5*h*	—
Ar	4,383.79	(10)	—

	λ	I J	I O		λ	I J	I O
U	**4,383.27**	**1**	**10**	**Ce**	**4,383.88**	—	**2**
Dy	**4,383.18**	—	**3**	**Xe**	**4,383.91**	**(100)**	—
Eu	**4,383.16**	**20**	**100*W***	**Br**	**4,384.00**	**(20)**	—
Gd	**4,383.14**	**40**	**30**	**Tb**	**4,384.06**	—	**10**
Mn	**4,383.07**	—	**10**	**Ne**	**4,384.08**	**(5)**	—
Ce	**4,382.96**	—	**4**	**Pr**	**4,384.14**	**25*w***	**30*w***
B	**4,382.95**	**4**	—	**Mo**	**4,384.19**	**8**	**8**
Ca	**4,382.93**	**2**	**5**	**Sm**	**4,384.29**	**50**	**50**
Se	**4,382.87**	**(800)**	—	*bh***Ca**	**4,384.30**	—	**6**
Cr	**4,382.85**	**2**	**12**	**Dy**	**4,384.30**	**2**	**4**
Nb	**4,382.84**	**5**	**3**	**Cs**	**4,384.43**	**(25)**	—
U	**4,382.83**	—	**2**	**Ce**	**4,384.45**	—	**4**
Pr	**4,382.82**	**8*w***	**25*w***	**Nd**	**4,384.51**	**2**	**5**
Fe	**4,382.77**	**10**	**10**	**Ni**	**4,384.54**	**1**	**25**
Nd	**4,382.74**	**10**	**15**	**U**	**4,384.59**	**2**	**2**
Zr	**4,382.73**	—	**3**	**Ce**	**4,384.63**	—	**3**
Mn	**3,382.63**	—	**80*h***	**Mg**	**4,384.64**	—	**8**
*bh***C**	**4,382.50**	—	—	**Er**	**4,384.698**	**5**	**30**
Nb	**4,382.49**	**5*h***	**3*h***	**Fe**	**4,384.699**	**2**	**5**
Tb	**4,382.45**	—	**25**	**Tb**	**4,384.70**	—	**2**
Pr	**4,382.42**	**20**	**30**	**V**	**4,384.72**	**125*R***	**125*R***
Mo	**4,382.41**	**20**	**10**	**Ho**	**4,384.76**	**3*h***	**3**
U	**4,382.34**	**5**	**18**				
Kr	4,376.12	(800)	—				
Fe	4,375.93	200	500				
Rh	4,374.80	500	1000*W*				
Fe	*4,369.77*	*100*	*200*				
O	*4,368.30*	*(1000)*	—				

*N**

Fe 4,404.75 *J* 700 *O* 1000

	λ	J	O		λ	J	O
Pr	**4,404.71**	**4**	**25*w***	**Ce**	**4,404.77**	—	**2**
Ce	**4,404.57**	—	**3**	**Yt**	**4,404.82**	—	**2**
J	**4,404.56**	**(8)**	—	**Lu**	**4,404.86**	—	**5**
Mo	**4,404.55**	**8**	**15**	**Hg**	**4,404.86**	**(50)**	—
As	**4,404.53**	**15**	—	*bh***La**	**4,404.90**	—	**3**
W	**4,404.46**	**1**	**3**	**Ti**	**4,404.90**	**10**	**15**
Tb	**4,404.43**	—	**3**	**Ar**	**4,404.91**	**(2)**	—
Ti	**4,404.40**	**7**	**12**	**U**	**4,404.91**	**2**	—
Ce	**4,404.38**	—	**2**	**Co**	**4,404.96**	—	**5**
Kr	**4,404.33**	**(30*h*)**	—	**V**	**4,405.01**	**6**	**12**
Ti	**4,404.27**	**30**	**50**	**Mo**	**4,405.03**	**4**	**5**
Ce	**4,404.25**	—	**4**	**Pr**	**4,405.14**	**20**	**25**
Os	**4,404.21**	**1**	**18**	**Ce**	**4,405.15**	—	**2**

	λ	I J	I O		λ	I J	I O
Mo	**4,404.18**	**15*h***	**1**	**Ba**	**4,405.23**	**(20)**	—
U	**4,404.04**	**1*h***	**2*h***	**Cs**	**4,405.25**	**(35)**	—
W	**4,403.95**	**9**	**20**	**Eu**	**4,405.27**		**15*W***
***bh*Ca**	**4,403.90**	—	**6**	**Ce**	**4,405.30**	—	**4**
Cs	**4,403.85**	**(20)**	—	**Tb**	**4,405.41**	—	**15**
Ir	**4,403.78**	**10**	**300**	**Ce**	**4,405.47**	**2**	**18**
Nd	**4,403.77**	—	**5**	**Te**	**4,405.49**	**(15)**	—
Ta	**4,403.72**	**8**	—	**Ir**	**4,405.50**	—	**3**
Re	**4,403.68**	—	**12**	**Dy**	**4,405.58**	**2**	**3**
V	**4,403.67**	**15**	**20**	**Sm**	**4,405.64**	**8**	**8**
Pr	**4,403.60**	**40**	**100**	**Ti**	**4,405.68**	**6**	**20**
Dy	**4,403.56**	—	**3**	**U**	**4,405.73**	**2**	**6**
Ce	**4,403.55**	—	**6**	**Pr**	**4,404.85**	**100**	**100**
J	**4,403.55**	**(20)**	—	**Nd**	**4,405.91**	**2**	**8**
				U	**4,405.95**	**8**	**4**
Ni	4,401.55	30	1000*W*				
Ir	4,399.47	100	400	Fe	4,415.12	400	600
*P**				*Fe*	*4,422.57*	*125*	*300*
				Fe	*4,427.31*	*200*	*500*

Ga 31 69.72

t_0 29.75°C t_1 2,071°C

I.	II.	III.	IV.	V.
5.997	20.509	30.7	64.1	—

λ	*I*		eV
	J	*O*	
I 2,874.24	15*R*	10	4.3
I 2,943.64	10	20*R*	4.3
I 4,032.98	500*R*	1000*R*	3.1
I 4,172.06	1000*R*	2000*R*	3.1
I 6,396.61	20	—	4.98

Ga 2,874.24 *J* 15*R* *O* 10

	λ	I *J*	I *O*		λ	I *J*	I *O*
F	**2,874.22**	**(2)**	**—**	**La**	**2,874.28**	**3**	**5**
V	**2,874.21**	**20**	**7**	**Fe**	**2,874.30**	**—**	**2**
Zr	**2,874.173**	**2*wh***	**—**	**U**	**2,874.47**	**4**	**6**
Fe	**2,874.172**	**200**	**300**	**Ta**	**2,874.52**	**3**	**20**
Ta	**2,874.167**	**15**	**150**	**Ar**	**2,874.55**	**(5)**	**—**
Ce	**2,874.13**	**—**	**30*w***	**Ce**	**2,874.551**	**—**	**2**
U	**2,874.08**	**10**	**15**	**Cu**	**2,874.56**	**—**	**3*h***
Ru	**2,874.05**	**—**	**5**	**Nb**	**2,874.57**	**3**	**5**
Rh	**2,873.99**	**—**	**6**	**Os**	**2,874.63**	**4**	**10**
Rb	**2,873.93**	**(2)**	**—**	**U**	**2,874.79**	**2**	**3**
Cr	**2,873.82**	**40**	**20**	**F**	**2,874.80**	**(15)**	**—**
Er	**2,873.81**	**2**	**10**	**Ce**	**2,874.82**	**—**	**2**
Ir	**2,873.80**	**5**	**2**	**Er**	**2,874.83**	**1**	**12**
Ru	**2,873.73**	**12**	**—**	**Mo**	**2,874.85**	**60**	**2**
Kr	**2,873.72**	**(4*wh*)**	**—**	**Fe**	**2,874.88**	**20**	**60**
Th	**2,873.717**	**5**	**8**	**Ce**	**2,874.92**	**—**	**2**
Ag	**2,873.65**	**100*wh***	**3**				
Fe	**2,873.65**	**2**	**8**				
Hf	**2,287.650**	**—**	**10**	Ta	2,874.94	2	25
U	**2,873.640**	**6**	**3**	Os	2,874.95	15	60
				Ru	2,874.98	50	80
Mo	**2,873.637**	**1**	**10**	U	2,875.20	12	18
				W	2,875.21	4	10
Rh	2,873.62	10	60	J	2,875.25	(12)	—
Ta	2,873.56	50	150	Re	2,875.29	—	80
Ag	2,873.53	2	3	Cs	2,875.30	(8)	—
Fe	2,873.528	1	15	Fe	2,875.304	50	125
Cr	2,873.48	125	30	Fe	2,875.35	70	—
Fe	2,873.40	300	—	Nb	2,875.39	300	50*r*
W	2,873.38	10	4	Br	2,875.42	(7)	—
Ta	2,873.360	40*h*	200*w*	Ir	2,875.60	15	25
V	2,873.356	—	20	V	2,875.69	40	10
Ir	2,873.33	3	18	Ti	2,875.79	40*wh*	—
Pb	2,873.32	60	100*R*	Pt	2,875.85	80*h*	20
Ru	2,873.31	60	4	Yb	2,875.88	8	—
U	2,873.30	4	6	U	2,875.881	4	5
Hg	2,873.24	(20)	—	Zr	2,875.983	1	70
Cr	2,873.19	—	25	Ir	2,875.983	15	25
V	2,873.18	50	4	Cr	2,875.99	80*wh*	30
Tm	2,873.01	20	15	Fe	2,876.01	—	15
U	2,873.005	6	10	Nd	2,876.05	10	—
Pb	2,873.00	(20)	—	Eu	2,876.06	—	20
Ne	2,873.00	(10)	—	Ta	2,876.11	5	50*r*
J	2,872.89	(60)	—	Cr	2,876.24	80*Wh*	25
Mo	2,872.88	50	2	Hf	2,876.33	100	30
Er	2,872.84	1	6	Th	2,876.42	8	10
Nb	2,872.80	10	1	Ne	2,876.43	(18)	—
Re	2,872.67	—	10	U	2,876.433	2	8

	λ	I	
		J	O
Ne	2,872.66	(35)	—
Br	2,872.59	(25)	—
Mn	2,872.58	—	30
W	2,872.50	4	5
Os	2,872.40	8	50
Fe	2,872.38	20	2
Cs	2,872.35	(8)	—
Fe	2,872.34	50	150
Re	2,872.30	—	15
U	2,871.99	4*h*	6*h*
W	2,871.90	10	—
Mo	2,871.89	—	10
Re	2,871.81	—	50
Er	2,871.68	1	9
U	2,871.64	4	6
Ru	2,871.64	5	50
Ce	2,871.633	—	6
Cr	2,871.632	2	50
Ce	2,871.58	—	6
Mo	2,871.51	100*h*	100
Ru	2,871.47	30	—
Cr	2,871.45	80	—
Nd	2,871.44	—	5
Ta	2,871.42	50	200
F	2,871.40	(25)	—
Pd	2,871.37	10*wh*	—
W	2,871.367	8	10
Rh	2,871.35	10*h*	100
Na	2,871.27	(40)	—
Xe	2,871.24	(25*hs*)	—
Co	2,871.24	100	—
Mo	2,871.18	—	10
Fe	2,871.13	20	—
Ce	2,871.07	—	15
Fe	2,871.06	40	—
Tb	2,871.05	10	3
U	2,870.97	20	18
W	2,870.905	12	9
Mo	2,870.903	—	15
Na	2,870.89	6	6
Th	2,870.82	5	6
Ce	2,870.62	—	6
Fe	2,870.60	15	—
Ru	2,870.548	50	—
V	2,870.547	20*r*	50*r*
W	2,870.53	2	6
Pt	2,870.47	2	10
Cr	2,870.44	200*W*	25
U	2,870.42	2	8

	λ	I	
		J	O
F	2,876.49	(10)	—
Mo	2,876.54	2	15
Cr	2,876.66	20	—
Fe	2,876.71	2	5
Fe	2,876.80	100	—
Re	2,876.87	—	15
W	2,876.93	12	8
V	2,876.94	25	4
Nb	2,876.95	500*W*	40*W*
Rn	2,877.00	(7)	—
Nb	2,877.03	10*w*	3
U	2,877.05	12	6
Ru	2,877.09	1	5
Hf	2,877.16	—	12
Cs	2,877.29	(8)	—
Fe	2,877.30	125	200
Os	2,877.35	2	30
Ti	2,877.44	100	30
Pt	2,877.52	200*h*	40
P	2,877.53	(10)	—
U	2,877.57	4	6
N	2,877.66	(8*h*)	—
Ir	2,877.68	10	20
Ta	2,877.686	80*h*	15
V	2,877.688	100*R*	15
Cu	2,877.689	20	5
Eu	2,877.76	—	5
Pd	2,877.87	15*wh*	—
Sb	2,877.91	150	250*W*
Cr	2,877.98	100	30
V	2,878.02	10	2
Ru	2,878.04	15	—
W	2,878.08	8	4
Ta	2,878.20	15*h*	4
Tm	2,878.21	20	3
V	2,878.299	7	—
W	2,878.30	10	1
Tm	2,878.36	20	10
Mo	2,878.38	—	20
Os	2,878.40	12	40
Cr	2,878.449	80	20
Sm	2,878.449	25	4
Tb	2,878.52	10	—
Co	2,878.56	—	12
Ce	2,878.63	—	8
J	2,878.64	(400)	—
Rh	2,878.65	10	50
W	2,878.716	8	10
U	2,878.719	4	6

	λ	I J	I O		λ	I J	I O
Th	2,870.41	20	18	Nb	2,878.74	10	3
Nb	2,870.35	10*wh*	—	Fe	2,878.76	5	8
Ir	2,870.22	—	5	Eu	2,878.86	—	20
Mo	2,870.179	1	10	U	2,878.87	4	5
Cr	2,870.178	—	12	Er	2,878.91	2*h*	7
V	2,870.11	12	1	Ta	2,878.95	3*s*	50*r*
Yb	2,870.06	10	1	O	2,879.04	(7)	—
Cr	2,870.04	—	12*h*	Mo	2,879.05	100*h*	15
Ti	2,870.04	100*wh*	—	Ru	2,879.06	60	—
V	2,870.04	—	7*w*	W	2,879.107	10	10
Co	2,870.03	50*wh*	—	Hf	2,879.112	10	15
V	2,869.96	20	7	V	2,879.16	35	50
Ca	2,869.95	7	—	Th	2,879.20	5	10
Th	2,869.93	8	6	Fe	2,879.24	25	—
Rh	2,869.91	1*h*	8	Cs	2,879.25	(8)	—
Fe	2,869.83	5	10	Cr	2,879.27	12	60
Nb	2,869.828	10	1	Re	2,879.28	—	25
Hf	2,869.825	20	25	Nd	2,879.36	2*wh*	5*h*
Rb	2,869.82	(10)	—	Nb	2,879.364	10	2*d*
Zr	2,869.81	30	30	W	2,879.39	10	10
Te	2,869.72	(10)	—	Ir	2,879.41	5	20
Ir	2,869.70	2	15	Fe	2,879.43	2	7
W	2,869.61	10	7*s*	Mn	2,879.488	5	12
Mo	2,869.56	—	15	Nb	2,879.495	2	25
V	2,869.46	—	7	Ta	2,879.52	10*h*	50*h*
U	2,869.37	4	6	Th	2,879.53	4	6
Fe	2,869.31	70	300	U	2,879.59	4	6
Tm	2,869.22	300	100	Co	2,879.62	—	25
Mo	2,869.217	15	—	Mo	2,879.69	25	1
V	2,869.13	150*r*	25	N	2,879.73	(25*h*)	—
W	2,869.10	3	9	Ta	2,879.74	10	150
U	2,868.97	6	6	Ru	2,879.75	12	50
Ce	2,868.96	—	6	U	2,879.99	2	9
Fe	2,868.87	60	5	Ta	2,880.02	50	150
Te	2,868.86	(100)	—	V	2,880.03	150*r*	25
J	2,868.77	(30)	—	Ir	2,880.07	—	7
Ti	2,868.74	25	15	Ru	2,880.08	25	—
W	2,868.73	25	6	W	2,880.16	7	—
Rn	2,868.70	(70)	—	Ir	2,880.21	2*h*	10
Th	2,868.68	6	8	Ho	2,880.27	10	20
Ta	2,868.65	40	150	Co	2,880.29	50*wh*	—
Ru	2,868.54	—	5	Ti	2,880.293	20	—
Nb	2,868.525	300	15	Se	2,880.31	(25)	—
Al	2,868.52	(80)	—	Ce	2,880.36	—	6
Fe	2,868.45	40	80	U	2,880.49	15	12
Cl	2,868.41	(10)	—				
Cs	2,868.33	(8)	—	*Fe*	*2,880.58*	5	*15*
Mo	2,868.32	20	2	*Ce*	*2,880.637*	—	*18*

	λ	I	
		J	O
Ru	2,868.31	5	8
Rh	2,868.28	—	10
Cd	2,868.26	80	100
Ce	2,868.23	—	8
Fe	2,868.21	8	15
Ru	2,868.188	4	8
U	2,868.187	8	10
Re	2,868.15	—	5
Mo	2,868.11	20	3
V	2,868.10	30r	40
Tm	2,868.01	40	15
W	2,867.92	10	1
Mo	2,867.82	40h	—
Hf	2,867.70	2	10
Cr	2,867.65	100R	80
Ir	2,867.63	2	12
Fe	2,867.56	30	60
Ta	2,867.41	150	5h
Fe	2,867.31	30	60
Re	2,867.20	—	40
Cr	2,867.10	35	20
Ru	2,867.09	18	—
Mo	2,867.05	—	10
Yb	2,867.04	40	4
V	2,866.95	—	20
Pt	2,866.89	15	6
Ce	2,866.81	—	12
Cr	2,866.74	125R	80
Mo	2,866.693	30	30
Ir	2,866.689	2	20
Ru	2,866.64	25	60
Fe	2,866.63	80	125
V	2,866.59	6	25
In	2,866.46	(18)	—
V	2,866.42	—	35
Hf	2,866.373	12	50
W	2,866.373	10	10
Ru	2,866.25	30	—
P	2,866.14	(20)	—
Ru	2,866.08	30	—
W	2,866.06	10	15
Cr	2,865.89	15	1
Ar	2,865.85	(20)	—
In	2,865.684	(50)	—
Os	2,865.680	5	15
U	2,865.679	50	30
Cr	2,865.676	20	—

	λ	I	
		J	O
La	2,880.644	40	4
Rh	2,880.65	2wh	10
Nb	2,880.71	50	4w
Fe	2,880.76	50	15
Cd	2,880.77	125	200R
V	2,880.80	20	2
Fe	2,880.83	25	1
Cr	2,880.87	25	20
Ho	2,880.99	10	20
Ce	2,881.13	—	12
Na	2,881.140	(60)	8
Cr	2,881.141	2	25
Th	2,881.15	10	10
Ir	2,881.158	3	15
Cs	2,881.16	(20)	—
Cd	2,881.23	(30)	50R
Ta	2,881.232	5	30
Rh	2,881.25	—	20
Ru	2,881.28	3	30
Tb	2,881.31	—	10
Al	2,881.46	(30)	—
Gd	2,881.578	—	40
Ce	2,881.578	2	40
Si	2,881.578	400	500
Cr	2,881.93	30	1
J	2,882.01	(20)	—
Th	2,882.014	12	10
Ru	2,882.12	200	30
Zn	2,882.15	(25)	—
Co	2,882.22	—	30
Ta	2,882.33	80	3
Rh	2,882.37	10	80
Mo	2,882.38	25	1
V	2,882.50	200r	35
Ce	2,882.61	—	15
Ir	2,882.63	6	40
U	2,882.74	20	18
P	2,882.75	(18)	—
Mn	2,882.90	—	25
Nb	2,883.18	800R	100
Re	2,883.450	—	60
Au	2,883.45	20	15
Ru	2,883.59	5	30
Co	2,883.60	—	15
Fe	2,883.70	300	—
Fe	2,883.73	—	30
Rn	2,883.80	(25)	—
O	2,883.82	(15)	—

	λ	I	
		J	O
Mo	2,865.62	20	5
Nb	2,865.61	30	5
Ru	2,865.53	20	—
Cr	2,865.33	15	15
U	2,865.14	10	10
Cr	2,865.11	200R	60
Pt	2,865.05	80h	20
Fe	2,864.97	50	—
Xe	2,864.73	(100)	—
Mo	2,864.66	3	40
Re	2,864.57	—	10
V	2,864.52	35	6
Pb	2,864.51	60	—
Ta	2,864.50	30	125
Eu	2,864.42	10h	10
Rh	2,864.40	10	70
V	2,864.36	25r	40
Mo	2,864.31	2	40
U	2,864.28	12h	18h
Pb	2,864.26	60	—
Ni	2,864.15	300wh	—
W	2,863.88	12	10
Fe	2,863.86	100	125
Ir	2,863.84	—	15
Mo	2,863.81	100h	30
Tm	2,863.76	40	15
Bi	2,863.75	18	80w
Ni	2,863.70	250	—
Fe	2,863.43	80	100
Tm	2,863.35	—	10
Ce	2,863.34	—	12
Sn	2,863.33	300R	300R
Ru	2,863.32	80	30
Mo	2,863.12	20	—
V	2,863.05	7wh	20
La	2,862.97	15h	2
Rh	2,862.93	60	150
Ru	2,862.88	60	6
Mo	2,862.84	—	10
Ce	2,862.78	—	15
Co	2,862.61	—	50
Cr	2,862.571	300R	80
Eu	2,862.57	70	100W
Fe	2,862.50	50	100
U	2,862.41	10	15
V	2,862.38	—	12
Ti	2,862.32	40	20

	λ	I	
		J	O
P	2,883.90	(25)	—
Re	2,884.05	—	20
Ti	2,884.11	125	35
Al	2,884.20	(30)	—
Th	2,884.29	12	12
Cu	2,884.38	30	—
Os	2,884.41	5	20
Ru	2,884.507	5	20
As	2,884.51	25	—
Re	2,884.63	—	25
Fe	2,884.779	25h	—
V	2,884.785	200r	40
Th	2,885.04	15	12
Lu	2,885.14	3	40h
Tb	2,885.14	70	—
La	2,885.141	50	5
U	2,885.19	6	12
N	2,885.25	(50h)	—
Ce	2,885.29	—	10
Hf	2,885.47	15	12
Mo	2,885.74	25	10
Tb	2,885.90	10	10
Fe	2,885.928	70	—
Re	2,885.931	15	—
Na	2,886.25	(20)	—
Tb	2,886.28	10	15
Fe	2,886.32	15	50
Co	2,886.44	2	50
U	2,886.45	6h	15
Yt	2,886.457	6	15
Tm	2,886.459	5	30
Ru	2,886.54	50	60
Mo	2,886.61	1	30
Cs	2,886.67	(20)	—
Mn	2,886.68	6	15

	λ	I J	I O		λ	I J	I O
Cd	*2,862.31*	*10*	*15*				
V	*2,862.30*	*25*	*1*				
Ta	*2,862.02*	*10*	*40*				
Ti	*2,861.99*	*100wh*	—				
Ta	*2,861.98*	*5*	*30*				
Tm	*2,861.74*	*30*	*20*				

Ga 2,943.637 *J* 10 *O* 20*R*

	λ	J	O		λ	J	O
V	**2,943.636**	**4**	**2**	**Ce**	**2,943.67**	—	**8**
Ir	**2,943.63**	—	**3*h***	**Ir**	**2,943.72**	—	**3*h***
Sn	**2,943.62**	**2**	—	**Ta**	**2,943.77**	**5**	**10**
Fe	**2,943.57**	**6**	**12**	**Sm**	**2,943.79**	**2**	**8**
La	**2,943.55**	**6*h***	**2**	**V**	**2,943.83**	—	**7*h***
Nd	**2,943.50**	**5**	**3**	**Ir**	**2,943.87**	**5*d***	**4*d***
Sm	**2,943.49**	**3**	**8**	**Cd**	**2,943.89**	**(5)**	—
Co	**2,943.484**	—	**30**	**U**	**2,943.895**	**25**	**10**
Ru	**2,943.481**	—	**30**	**Mn**	**2,943.908**	**2**	**2**
Xe	**2,943.41**	**(2)**	—	**Ni**	**2,943.914**	**20**	**50*r***
U	**2,943.40**	**4**	**6**	**Ru**	**2,943.92**	**5**	**50**
Mo	**2,943.380**	**25**	**1**	**W**	**2,943.96**	**4**	**5**
Re	**2,943.380**	—	**15**	**Ce**	**2,943.987**	—	**6**
Tm	**2,943.36**	**15**	**6**	**Mo**	**2,943.989**	**3**	—
W	**2,943.33**	**6**	**7**	**Er**	**2,944.07**	—	**12**
Ir	**2,943.26**	—	**4**	**Cs**	**2,944.10**	**(2)**	—
Ce	**2,943.21**	—	**6**	**Ga**	**2,944.17**	**15*r***	**10**
Cl	**2,943.20**	**(4)**	—	**Ru**	**2,944.18**	—	**12**
V	**2,943.196**	**25*r***	**30**	**U**	**2,944.19**	**12**	**8**
U	**2,943.18**	**8**	**5**	**Cl**	**2,944.20**	**(4)**	—
Ir	**2,943.151**	**20**	**30**	**Zr**	**2,944.20**	**3**	**2*h***
Co	**2,943.15**	**100*wh***	—	**Mo**	**2,944.21**	**2**	**25**
Re	**2,943.14**	—	**60**				
Mn	**2,943.132**	**3**	**1**				
Ti	**2,943.13**	**60*wh***	—	Re	2,944.32	—	10
				Ce	2,944.35	—	18
O	**2,943.00**	**(5*h*)**	—	W	2,944.39	20	30
				Fe	2,944.40	600	70
				Ho	2,944.50	20	10
Ar	2,942.90	(100)	—				
O	2,942.88	(5*h*)	—	V	2,944.57	300*r*	50
Th	2,942.86	10	10*s*	U	2,944.64	8	2
U	2,942.853	6	6	Hf	2,944.71	1	20
Mo	2,942.850	1	10	Pt	2,944.75	2	15
				Mo	2,944.82	50*h*	2
Os	2,942.848	8	30				
Yb	2,942.82	5	1	Tl	2,945.04	25	50
Pt	2,942.75	3	20	Fe	2,945.05	30	100
Mn	2,942.74	1	10	Ru	2,945.101	50	6
Fe	2,942.63	5	10	He	2,945.104	(100)	—
				Er	2,945.28	—	15

	λ	I	
		J	O
W	2,942.61	10	2
V	2,942.35	10*h*	80*r*
Pr	2,942.30	10	—
J	2,942.27	(20)	—
W	2,942.26	10	2
Ru	2,942.25	100	30
Cs	2,942.21	(8)	—
Os	2,942.20	5	8
Te	2,942.16	(100*h*)	—
W	2,942.139	10	6
Ta	2,942.137	40	150
U	2,942.12	12	8
Mg	2,942.11	2*h*	20
Xe	2,942.10	(10*h*)	—
Ho	2,942.05	10	—
Yb	2,942.03	6	1
Ti	2,941.99	150	100
U	2,941.92	30	15
Cr	2,941.88	25	12
Tb	2,941.70	10	3
Re	2,941.56	—	15
J	2,941.55	(12)	—
Nb	2,941.54	300	50
V	2,941.49	150*r*	12
Se	2,941.48	(15)	—
V	2,941.37	300*r*	40
U	2,941.343	12	8
Fe	2,941.343	300	600
Mo	2,941.22	40	2
Ir	2,941.08	5	20
In	2,941.05	(80)	—
Mn	2,941.04	1	25
Mo	2,940.978	1	10
Cr	2,940.978	10	—
Cs	2,940.953	(8)	—
Te	2,940.95	(5)	—
Nb	2,940.85	5	—
Eu	2,940.82	—	10
Ce	2,940.78	—	15
Hf	2,940.77	12	60
Fe	2,940.59	80	200
Ir	2,940.54	10	15
Yb	2,940.51	25	3
Mn	2,940.48	—	40*wh*
Eu	2,940.67	—	15
Mn	2,940.39	—	40*Wh*
U	2,940.37	12	6
Ru	2,940.36	3	50
Cr	2,940.22	30	—
Zr	2,945.46	10*h*	4
Ti	2,945.47	100*wh*	—
P	2,945.52	(5)	—
Mo	2,945.66	2	20
Ru	2,945.67	300	60
Na	2,945.69	(20)	2
Fe	2,945.70	5	10
Tb	2,945.70	3	10
Ho	2,945.83	70*h*	3
Nb	2,945.88	100	2
U	2,945.89	12	8
Yb	2,945.90	60	10
Nd	2,945.91	5	—
Yt	2,945.95	100	2
Mo	2,946.01	40	20
Ce	2,946.06	—	10
Nb	2,946.12	20	4
Ta	2,946.26	2	10
Yb	2,946.29	15	3
Mo	2,946.42	6	10
W	2,946.43	12	—
V	2,946.53	5	10
Re	2,946.58	—	10
Er	2,946.61	5	8
Mo	2,946.69	25	3
Nd	2,946.72	10	—
Yb	2,946.75	8	1
Tm	2,946.84	30	10*d*
Cr	2,946.842	30	5
Nb	2,946.90	30	3
Ta	2,946.91	10	150
Ir	2,946.97	10	20*h*
W	2,946.98	18	20
Ru	2,946.99	12	60
Hg	2,947.08	(25)	—
Hf	2,947.13	15	15
Mo	2,947.28	25	2
Ne	2,947.297	(150)	—
Eu	2,947.300	—	30
Fe	2,947.36	20	30
Ir	2,947.37	2	12
W	2,947.38	10	12
Na	2,947.44	(40)	6
Ni	2,947.45	10	—
Cr	2,947.50	25	—
Fe	2,947.66	100	10
W	2,947.72	8	—
Ta	2,947.80	2	10
Fe	2,947.88	200	600*r*

	λ	I	
		J	O
Ta	2,940.21	50	150
W	2,940.20	18	4
Mo	2,940.10	40	2
Ta	2,940.06	40*w*	100
Tb	2,940.03	3*h*	10
Ru	2,939.94	3	30
Mn	2,939.90	—	12
W	2,939.74	12	2
Ru	2,939.69	5	12
Th	2,939.61	5	5
Ce	2,939.54	—	12
Nd	2,939.535	5	—
Yb	2,939.53	6	2
Fe	2,939.51	30	3
K	2,939.50	(5)	—
Cr	2,939.45	20	6
In	2,939.36	10	—
Mn	2,939.30	—	50
Ho	2,939.29	10	—
Ta	2,939.28	40*h*	200
Ir	2,939.27	15	20
Ru	2,939.13	—	12
Fe	2,939.08	20	80
Hg	2,939.03	(10)	—
W	2,938.85	9	1
Pt	2,938.81	2	15
Mo	2,938.77	5	2
Ir	2,938.76	—	10
In	2,938.71	10	—
Ti	2,938.70	100*wh*	—
V	2,938.67	2	12
Mo	2,938.59	10	1
Ta	2,938.56	—	10
Ag	2,938.55	200*wh*	200
Mg	2,938.54	—	25
Cs	2,938.50	(20)	—
W	2,938.50	6	8
Ir	2,938.47	12	18*h*
Ta	2,938.43	3	50
Mo	2,938.300	30	1
Bi	2,938.298	300*w*	300*w*
V	2,938.25	60	2
Nb	2,938.07	5	3
Mn	2,937.92	—	25*Wh*
Te	2,937.90	(15)	—
Fe	2,937.81	150	300
Hf	2,937.79	100	50
Na	2,937.72	(40)	6
Tm	2,948.01	20	15
V	2,948.07	70	2
U	2,948.09	12	10
P	2,948.14	(5*h*)	—
Tm	2,948.155	10	15
Cd	2,948.16	35	—
Os	2,948.23	5	12
Ti	2,948.25	30	100
Ce	2,948.38	—	10
Yt	2,948.40	5*h*	20
Fe	2,948.43	70	80
Pb	2,948.72	125	—
Fe	2,948.726	7	10
Tl	2,948.73	(30)	—
Zr	2,948.94	20	12
Fe	2,948.95	4	10
Ne	2,949.04	(10)	—
Th	2,949.07	8	10
Re	2,949.098	—	20*h*
J	2,949.105	(30)	—
Eu	2,949.16	5	5*d*
V	2,949.17	80	6
Ho	2,949.19	20	—
Mn	2,949.20	30	100
Re	2,949.26	—	15
Ne	2,949.32	(15)	—
Cr	2,949.44	18	—
Ru	2,949.500	12	80
Nb	2,949.502	5	—
Te	2,949.52	(100)	—
Os	2,949.53	10	30
Kr	2,949.54	(15*h*)	—
U	2,949.61	6	5
V	2,949.63	15*r*	30
Rb	2,949.68	(10)	—
Fe	2,949.70	5	10
Lu	2,949.73	1	20*h*
Ir	2,949.76	10	25
Os	2,949.81	5	8
Re	2,949.88	—	20
Ru	*2,949.96*	*20*	*4*
Kr	*2,950.21*	*(30h)*	—
Te	*2,950.21*	*(10)*	—
Fe	*2,950.24*	*300*	*700*
Yb	*2,950.32*	*10*	*2*
V	*2,950.35*	*100r*	*25*
W	*2,950.44*	*20*	—

	λ	I J	I O
Nb	2,937.71	20wh	—
V	2,937.69	10	20w
Mo	2,937.665	1	20
W	2,937.661	10	6d
Ru	2,937.55	—	20
Sm	2,937.49	5	12
Th	2,937.44	5	5
Ru	2,937.34	—	20
Nb	2,937.33	15	2
Ti	2,937.32	5	35
Yb	2,937.18	10	2
W	2,937.14	12	8
V	2,937.04	25	2
J	2,937.00	(20)	—
Cr	2,936.93	18	4
Fe	2,936.90	500r	700r
Mo	2,936.78	25	2
Ho	2,936.77	1000R	—
W	2,936.67	20	10
Nb	2,936.66	30	—
Ir	2,936.62	—	40
Mg	2,936.54	—	20
Re	2,936.50	—	25
Th	2,936.47	12	12
U	2,936.45	20	12
Zr	2,936.31	15	10
Th	2,936.19	10	10
Ti	2,936.17	100wh	—
Fe	2,936.024	10	5
Ru	2,936.016	—	20
Tm	2,936.00	300	80
V	2,935.87	15h	30
Xe	2,935.86	(30h)	—
W	2,935.75	10d	—
Mo	2,935.69	15	1
Mn	2,935.66	—	20
Ru	2,935.52	80	10
W	2,935.35	10	5
Nb	2,935.29	15	2
Mo	2,935.20	10	2
Ta	2,935.16	2	20
Cr	2,935.14	40	8
Yb	2,935.10	15	3
W	2,934.99	12	15
bhB	2,934.90	—	100
Ta	2,934.85	4	40
Pd	2,934.84	10h	—
La	2,950.49	50	3
Ru	2,950.54	2	30
Hf	2,950.68	12	15
Re	2,950.83	—	25
Nb	2,950.88	200	150
Na	2,951.23	(100)	40
Tm	2,951.26	150	30
Zr	2,951.479	15	15
Te	2,951.48	(15)	—
Lu	2,951.69	80	20
Cd	2,951.82	25	—
Ta	2,951.92	200	400w
V	2,952.07	150R	35
Ti	2,952.08	25wh	—
Ru	2,952.25	25	4
Se	2,952.28	(12)	—
W	2,952.29	30	12s
Na	2,952.39	(12)	—
Cr	2,952.46	20	—
Ru	2,952.50	2	60
Eu	2,952.68	150	200W
Ru	2,952.69	10	—
Ta	2,952.99	100h	30h
Ho	2,953.11	10	20
Yt	2,953.28	12h	3
Cr	2,953.36	50	4
Fe	2,953.49	50	100
Pr	2,953.54	10	10
Tm	2,953.59	30	5
Cr	2,953.71	25	6
Fe	2,953.78	80	5
Fe	2,953.94	150	400r
Mo	2,953.95	20	1
Ru	2,954.10	20	6
Hf	2,954.20	10	15
Kr	2,954.28	(12h)	—
V	2,954.33	20	30
Re	2,954.34	—	25
U	2,954.39	15	12
J	2,954.39	(12)	—
Au	2,954.39	50	—
Pd	2,954.394	60h	—
Ru	2,954.49	20	100
Fe	2,954.65	70	100
Co	2,954.74	100	2
Ti	2,954.76	150wh	—
U	2,954.77	15	10
Te	2,954.82	(10)	—

	λ	I	
		J	O
Tb	2,934.79	10	5
Os	2,934.64	5	30
Cr	2,934.49	—	25h
V	2,934.40	50h	10
Mo	2,934.30	50h	30
Ag	2,934.23	200h	10
Ru	2,934.18	—	30
J	2,934.09	(12)	—
Mn	2,934.02	—	25
Cr	2,933.97	40	2h
U	2,933.86	10	8
V	2,933.83	35	2
Ta	2,933.55	150	400
Ru	2,933.24	150	20
Ir	2,933.14	5	20
Mn	2,933.06	15	80
Ho	2,933.05	10	—
W	2,932.86	10	1
Ne	2,932.72	(75)	—
Cr	2,932.70	25	2
Ta	2,932.69	80w	400
Nb	2,932.66	80	1h
In	2,932.62	300	500
U	2,932.61	25	10
Ar	2,932.60	(20)	—
Tm	2,932.59	25	5
Th	2,932.52	25wh	—
V	2,932.32	80	12
Au	2,932.19	40	8
Mo	2,932.189	15	—
Nb	2,932.13	50	—
Rh	2,931.94	20	80
W	2,931.87	10	2
V	2,931.86	15	2
Sr	2,931.83	8	30
J	2,931.73	(20)	—
V	2,931.62	30	4
Fe	2,931.60	15	—
W	2,931.53	12	—
Ar	2,931.49	(20)	—
Nb	2,931.47	50	3
U	2,931.41	12	12
Os	2,931.28	10	40
Ti	2,931.26	150wh	—
Tm	2,955.06	10	4
Hg	2,955.13	(100)	—
Mo	2,955.16	25	1
Ru	2,955.36	1	50
Co	2,955.386	—	30
Ar	2,955.39	(40)	—
Rh	2,955.41	—	20
Te	2,955.55	(10)	—
V	2,955.58	60	1
Gd	2,955.60	25	2
U	2,955.65	10	12
Ne	2,955.73	(40)	—
Lu	2,955.78	60h	2
Zr	2,955.783	30	10
V	2,955.80	2	20
Mo	2,955.84	30	2
Ce	2,955.94	—	25
Yt	2,956.04	15h	7
Mo	2,956.05	25	10
U	2,956.06	60	10
Mn	2,956.10	1	20
Rb	2,956.12	(70)	—
Ti	2,956.13	25	125

Ga 4,032.982 *J* 500*R* *O* 1000*R*

	λ	I *J*	I *O*
Sm	4,032.977	8	20
Pr	4,032.974	10	15
Ar	4,032.97	(20)	—
Sc	4,032.89	(10)	—
V	4,032.86	1	2
Br	4,032.85	(20)	—
Dy	4,032.847	—	8
S	4,032.81	(125)	—
Ce	4,032.75	—	2
Tb	4,032.70	—	3
Ti	4,032.632	1	35
Fe	4,032.630	15	80
Tb	4,032.626	—	4
Ce	4,032.55	—	3
Th	4,032.54	8	10
Nb	4,032.524	50	30
Ru	4,032.521	5	10
Nd	4,032.51	—	2
Mo	4,032.50	8	8
Pr	4,032.49	12	20
Dy	4,032.480	12	20
Er	4,032.477	—	9
Zr	4,032.471	—	5
Fe	4,032.469	1	4
W	4,032.385	7	6
Sr	4,032.379	—	20
U	4,032.29	2*h*	2
Tb	4,032.28	—	30
Hf	4,032.27	2	5
Se	4,032.22	(8)	—
Ir	4,032.21	—	10
Ru	4,032.20	20	20
Cl	4,032.19	(4)	—
Re	4,032.15	—	10
J	4,032.092	(10)	—
Th	4,032.089	—	3
Fe	4,031.965	50	80
J	4,031.96	(10)	—
Ta	4,031.96	—	5
V	4,031.83	3	10
Nd	4,031.81	15	15
Mn	4,031.79	10	8
U	4,031.78	2	8
Pr	4,031.755	30	50
Ti	4,031.754	1	35
Ho	4,031.75	1*h*	4
La	4,031.692	300	400
Tb	4,033.04	5	125
In	4,033.066	—	4
Ta	4,033.069	10	100
Cr	4,033.072	2	15
Mn	4,033.073	20	400*r*
Sr	4,033.19	—	6
Nb	4,033.20	5	5
Pr	4,033.24	2	3
Cr	4,033.26	8	30
Re	4,033.31	—	40
Ce	4,033.38	—	2*h*
U	4,033.43	10	12
Gd	4,033.49	5	10
Nd	4,033.50	5	10
Sb	4,033.54	60	70
Zr	4,033.58	—	3
Mn	4,033.630	5	5
Mo	4,033.631	6	6
N	4,033.64	(2)	—
Dy	4,033.67	4	15
P	4,033.69	(15)	—
Eu	4,033.69	—	8*w*
U	4,033.73	12	12
Ir	4,033.76	25	100
Ce	4,033.79	—	6
Ar	4,033.83	(30)	—
Pr	4,033.86	35	50
Hf	4,033.88	8	5
Nd	4,033.90	4	10
Ti	2,033.906	3	40
W	4,033.913	10	—
Mo	4,033.999	3	3
U	4,034.002	4	4
S	4,034.01	(8)	—
Nd	4,034.012	2	4
Cr	4,034.05	—	20
Zr	4,034.09	2	5
Eu	4,034.11	3	2
Nd	4,034.15	2	10*d*
Pt	4,034.17	5	—
Sc	4,034.23	2*h*	8
Th	4,034.256	10	10
Ce	4,034.259	—	2
Pr	4,034.30	5	20
Gd	4,034.38	—	5
Mn	4,034.49	20	250*r*
Nb	4,034.52	5	10

	λ	I	
		J	O
Er	**4,031.690**	—	6*w*
W	**4,031.675**	**7**	**8**
Ce	**4,031.669**	—	**10***d*
Tb	**4,031.64**	**3**	**50**
Re	**4,031.64**	—	**2**
Al	**4,031.63**	**(2)**	—
Ir	**4,031.56**	—	**3**
Nd	**4,031.55**	**3**	**10**
Ar	**4,031.41**	**(2)**	—
Sc	**4,031.40**	**2**	**10**
Eu	**4,031.38**	**3**	**7**
Zr	**4,031.35**	**1**	**3**
Mn	4,030.75	20	500*r*
F	4,025.49	(300)	—
F	4,024.73	(500)	—
Cu	4,022.66	25	400
Co	4,020.90	—	500*w*

	λ	I	
		J	O
Ce	**4,034.57**	—	**2**
La	4,042.91	300	400
K	4,044.14	400	800
Ar	4,044.42	(1200)	—
Co	4,045.39	—	400
Fe	4,045.81	300	400
Hg	4,046.56	300	200
K	4,047.20	200	400
Ag	*4,055.26*	*500R*	*800R*
In	*4,056.94*	*(500)*	—
Pb	*4,057.82*	*300R*	*2000R*
Nb	*4,058.94*	*400w*	*1000w*

Ga 4,172.06 *J* 1000*R* *O* 2000*R*

	λ	J	O
Dy	**4,171.99**	—	**15**
Ce	**4,171.96**	—	**2**
Dy	**4,171.92**	**2**	**4**
Ti	**4,171.903**	**70**	**15**
Mo	**4,171.90**	**25**	—
Cu	**4,171.85**	**5**	—
Pr	**4,171.82**	**40**	**75**
Tb	**4,171.80**	—	**8**
Ce	**4,171.77**	**3**	**2**
Gd	**4,171.71**	**1**	**25**
Er	**4,171.708**	—	**15**
Fe	**4,171.70**	**2**	**8**
Cr	**4,171.67**	**8**	**70**
N	**4,171.63**	**(5*h*)**	—
U	**4,171.59**	**30**	**30**
Sc	**4,171.57**	—	**3**
J	**4,171.56**	**(8)**	—
Sm	**4,171.55**	**12**	**7**
Zr	**4,171.48**	—	**20**
Mo	**4,171.45**	**3**	**10**
Ce	**4,171.39**	—	**18**
Th	**4,171.35**	**12**	**10**
V	**4,171.30**	**7**	**15**
W	**4,171.18**	**12**	**25**
La	**4,171.13**	—	**8**
Mo	**4,171.07**	**10**	**15**

	λ	J	O
Fe	**4,172.13**	**50**	**80**
Ce	**4,172.16**	**1**	**18**
U	**4,172.18**	**3**	**3**
Yb	**4,172.23**	**2**	**2**
Ho	**4,172.23**	—	**2**
Pr	**4,172.27**	**40**	**75**
La	**4,172.32**	—	**8**
Kr	**4,172.51**	**(20*h*)**	—
Ir	**4,172.56**	**12**	**150**
Os	**4,172.57**	**3**	**60**
Tb	**4,172.59**	**2**	**12**
Ti	**4,172.61**	—	**5**
Ce	**4,172.62**	—	**2**
Fe	**4,172.65**	**1**	**3**
Fe	**4,172.750**	**10**	**60**
Mo	**4,172.751**	**2**	**3**
Cr	**4,172.77**	**15**	**35**
P	**4,172.79**	**(5)**	—
Eu	**4,172.80**	**3**	**12**
Tb	**4,172.82**	—	**10**
Kr	**4,172.83**	**(3)**	—
Ce	**4,172.88**	**2**	**2**
U	**4,172.97**	**15**	**10**
Zr	**4,173.13**	**1**	**3**
Sc	4,175.32	(800)	—

	λ	I J	I O
Sb	**4,171.045**	**3***wh*	—
Tb	**4,171.04**	**3**	**20**
Pr	**4,171.04**	**4**	**12**
Ce	**4,171.040**	**14**	**6**
Ti	**4,171.03**	**7**	**35**
Rn	4,166.43	(500)	—
Ar	4,164.18	(1000)	—
S	4,162.70	(600)	—
Ar	*4,158.59*	(*1200*)	—
N	*4,151.46*	(*1000*)	—
Xe	4,180.10	(500*h*)	—
Se	4,180.94	(800)	—
Ar	4,181.88	(1000)	—
Ar	*4,191.03*	(1200)	—

Ga 6,396.61 *J* 20

	λ	J	O
S	**6,396.54**	**(15)**	—
Te	**6,396.46**	**(50)**	—
Sb	**6,396.21**	**(2*h*)**	—
J	**6,395.26**	**(30)**	—
S	**6,395.07**	**(15)**	—
Hg	**6,394.94**	**(25)**	—
Kr	**6,394.28**	**(4*hs*)**	—
Fe	**6,393.60**	**80*h***	**100**
V	**6,393.27**	**2**	**4**
Se	6,391.96	(15)	—
Kr	6,391.14	(30)	—
La	6,390.48	100	70
Te	6,390.19	(15)	—
J	6,388.94	(30)	—
Sr	**6,388.24**	**10**	**35**
Sr	**6,386.50**	**10**	**35**
Hf	**6,386.23**	**20**	**15**
S	6,384.89	(300)	—
Ar	6,384.72	(100)	—
Hg	6,383.34	(15)	—
Ne	6,382.99	(1000)	—
Rh	6,380.45	(12)	—
V	6,380.11	20	—
N	6,379.63	(70)	—
Sc	6,378.82	15	8
J	6,378.80	(30)	—
Tl	6,378.32	(10)	—
Cu	6,377.84	20	—
Se	6,376.74	(15)	—
Xe	6,375.28	(80)	—
Ar	**6,396.63**	**(2)**	—
S	**6,397.30**	**(300)**	—
V	**6,397.35**	**2**	—
Sc	**6,397.69**	**(15)**	—
Xe	**6,397.99**	**(50)**	—
S	**6,398.05**	**(300)**	—
Cl	**6,398.64**	**(40)**	—
V	**6,398.75**	**2**	—
La	**6,399.05**	**200**	**15**
Ar	**6,399.23**	**(8)**	—
Cl	**6,399.41**	**(10)**	—
W	**6,399.74**	**1**	**5**
Se	6,399.99	(15)	—
Fe	6,400.02	150*h*	200
Ne	6,401.08	(100)	—
Se	6,401.17	(15)	—
Ne	6,402.25	(2000)	—
Se	6,404.53	(15)	—
Te	6,405.90	(18)	—
As	6,405.95	10	—
Fe	6,408.03	30*h*	50
Sr	6,408.47	20	50
J	6,409.54	(15)	—
Ne	6,409.75	(150)	—
Kr	6,409.84	(10*hs*)	—
Br	6,410.32	(30)	—
Se	6,410.99	(15)	—
Cu	6,411.18	10	—
J	6,411.29	(30)	—
Fe	6,411.66	80*h*	100

	λ	I	
		J	O
O	6,374.29	(70)	—
La	6,374.11	15	6
Kr	6,373.58	(30)	—
J	6,371.76	(100)	—
Si	6,371.09	30	2
Se	6,370.62	(30)	—
Ar	6,369.58	(30)	—
S	6,369.34	(50)	—
Se	6,368.13	(30)	—
J	6,367.34	(70)	—
Te	6,367.10	(70)	—
O	6,366.28	(50)	—
Ne	6,365.01	(100)	—
Ar	6,364.89	(20)	—
J	6,363.26	(15)	—
In	6,362.96	(40)	—
In	6,362.90	(40)	—
In	6,362.37	(12)	—
Zn	6,362.35	500	1000 *Wh*
In	6,263.13	(40)	—
Hg	6,362.10	(15)	—
In	6,361.74	(20)	—
In	6,361.49	(20)	—
Cd	6,359.93	50	10
Sb	6,359.50	10	—
J	*6,359.19*	*(60)*	—
N	*6,357.00*	*(30)*	—
Xe	*6,356.35*	*(300)*	—
Xe	*6,355.77*	*(20)*	—
Yb	*6,355.40*	*50h*	*2*
Nb	*6,354.72*	*40*	*5*
In	*6,354.32*	*(20)*	—
Sm	*6,353.44*	*30*	*2*
Xe	*6,353.29*	*(30h)*	—
Sm	*6,353.01*	*20*	*20*
Br	*6,352.94*	*(25)*	—
Ne	*6,351.87*	*(100)*	—
Br	*6,350.74*	*(200)*	—
Se	*6,348.86*	*(30)*	
F	*6,348.50*	*(200)*	—
J	*6,248.34*	*(50)*	—
Si	*6,347.01*	*50*	*2*
Kr	*6,346.66*	*(20)*	—
Te	*6,345.51*	*(30)*	—
Pb	*6,345.00*	*(25)*	—
Xe	*6,343.96*	*(200)*	—
Ba	*6,341.68*	*50*	*90*

	λ	I	
		J	O
Te	6,412.15	(70)	—
Xe	6,412.38	(10)	—
Sc	6,413.35	25	10
F	6,413.66	(150)	—
S	6,413.71	(500)	—
Ga	6,414.01	15	—
Cu	6,414.62	20	—
S	6,415.50	(10)	—
Kr	6,415.65	(20)	—
J	6,415.79	(20)	—
Ar	6,416.31	(100)	—
Kr	6,416.61	(60*hs*)	—
N	6,417.05	(10)	—
J	6,417.66	(15)	—
Se	6,417.69	(10)	—
Xe	6,418.41	(30)	—
Xe	6,418.58	(30)	—
Hg	6,418.95	(25)	—
Xe	6,418.98	(30*h*)	—
Se	6,419.27	(15)	—
Ga	6,419.40	(25)	—
Cs	6,419.54	(10)	—
Fe	6,419.98	15*h*	18*h*
Kr	6,420.18	(300)	—
N	6,420.47	(30)	—
Kr	6,421.03	(100)	—
Fe	6,421.35	40*h*	60
Rn	6,421.48	(10)	—
Ne	6,421.71	(100)	—
Se	6,422.90	(125)	—
N	6,422.93	(10)	—
Te	6,422.96	(70)	—
Cu	6,423.90	30	—
Mo	6,424.37	20	100
S	6,425.64	(15*h*)	—
K	6,427.69	(20)	—
J	6,428.54	(30)	—
Se	6,428.67	(15)	—
Xe	6,430.15	(20)	—
Nb	6,430.46	10	80
Fe	6,430.85	80	100
Tm	6,430.95	60	15
J	6,430.97	15*h*	—
Yb	*6,432.73*	*40*	*30*
Cl	*6,434.80*	*(25)*	—
Yt	*6,435.00*	*50*	*150*
N	*6,437.01*	*(30)*	—

	λ	I J	I O
J	6,341.09	(30)	—
N	6,340.67	(50)	—
J	6,339.97	(100h)	—
J	6,339.52	(300)	—
J	6,338.97	(100)	—
J	6,338.02	(100)	—
Ra	6,336.90	(500)	—
Fe	6,336.84	35h	60
In	6,336.57	(20)	—
Al	6,335.70	(25)	—
Fe	6,335.33	20h	50
Ne	6,334.43	(1000)	—
Xe	6,333.97	(40h)	—
J	6,333.58	(30)	—
Xe	6,331.50	(20)	—
Ne	6,330.90	(150)	—
J	6,330.45	(50)	—
Ne	6,328.17	(300)	—
Te	6,437.06	(1000)	—
Cd	6,438.47	1000	2000
Ra	6,438.90	(30)	—
Ca	6,439.07	50	150
J	6,440.22	(100)	—
N	6,440.95	(25)	—
Se	6,441.43	(20)	—
Cu	6,441.698	40	—
N	6,441.70	(70)	—
La	6,443.05	25h	8
J	6,443.26	30h	—
Se	6,444.25	(100)	—
J	6,444.51	(100)	—
Ne	6,444.72	(150)	—
Ra	6,446.20	(1000)	—
Fe	6,446.34	20	—
La	6,446.60	100	15
Ba	6,450.85	20	100
Sn	6,453.58	300wh	6
O	6,453.69	(100)	—
O	6,454.55	(150)	—
S	6,455.36	(70)	—
Hf	6,455.85	20	2
O	6,456.07	(500)	—
Kr	6,456.29	(200)	—
N	6,457.93	(25)	—
Rb	6,458.35	400	—
P	6,460.10	(30)	—
Tm	6,460.28	80	400
Sn	6,462.36	300wh	4
Ca	6,462.57	50	125
Lu	6,463.12	800	400
Yb	6,463.15	100	10
Cd	6,464.98	50	5

Ge $^{32}_{72.59}$

t_0 958.5°C t_1 2,700°C

I.	II.	III.	IV.	V.
8.126	15.93	34.216	45.7	93.43

λ	I		eV
	J	O	
2,651.18	20	40	4.8
I 2,651.57	20	30	4.7
2,709.63	20	30	4.6
2,754.59	20	30	—
I 3,039.06	1000	1000	4.9
I 3,124.82	80	200	4.85
I 3,269.49	300	300	4.7
I 4,226.58	50	200	4.9
II 4,814.80	200	—	12.41
II 5,131.70	100	—	—
II 5,893.46	100	—	9.84

Ge 2,651.180 *J* 20 *O* 40
2,651.575 *J* 20 *O* 30

	λ	I J	I O		λ	I J	I O
W	**2,651.44**	4*d*	9*d*	**Se**	**2,651.48**	**(10)**	—
Ce	**2,651.42**	—	**2**	**Ta**	**2,651.483**	—	**50**
Ru	**2,651.292**	**5**	**60**	**Pt**	**2,651.50**	**4**	—
Fe	**2,651.291**	**2**	—	**Ru**	**2,651.51**	—	**20**
Ta	**2,651.22**	**80**	—	**V**	**2,651.57**	**5*h***	—
Ge	***2,651.18***	***20***	***40***	***Ge***	***2,651.575***	***20***	***30***
Hf	**2,651.16**	**40**	**15**	**Fe**	**2,651.706**	**60**	**60**
Nb	**2,651.12**	**200**	**3**	**La**	**2,651.71**	**8**	**1**
Pd	**2,651.09**	**10*h***	—	**Yb**	**2,651.71**	**60**	**2**
W	**2,651.02**	**10**	**1**	**Ir**	**2,651.77**	**5**	—
Ne	**2,651.01**	**(50)**	—	**Nb**	**2,651.81**	**15**	**2**
Ce	**2,651.006**	**2**	**25**	**Cl**	**2,651.82**	**(5)**	—
Mn	**2,650.99**	**150**	—	**Ru**	**2,651.841**	**9**	**100**
Pt	**2,650.86**	**100**	**700**	**U**	**2,651.844**	**15*h***	**3**
Be	**2,650.78**	—	**25**	**Mn**	**2,651.86**	**12**	—
Br	**2,650.76**	**(2)**	—	**W**	**2,651.87**	**12**	**2**
Yb	**2,650.74**	**4**	**2**	**V**	**2,651.888**	**4**	**50**
Si	**2,650.73**	**(5)**	—	**Re**	**2,651.902**	—	**100**
W	**2,650.71**	3*h*	**2**	**Ar**	**2,651.95**	**(2)**	—
Be	**2,650.70**	—	**10**	**Ce**	**2,652.006**	—	**10**
Cs	**2,650.70**	**(20)**	—	**W**	**2,652.012**	**2**	**8**
Be	2,650.64	—	25	**Hg**	**2,652.04**	**60**	**100**
Be	2,650.61	15	20	Yb	2,652.23	60	2
Be	2,650.55	—	30	Ta	2,652.32	—	15
Fe	2,650.49	150*h*	—	Mo	2,652.35	10	1
Be	2,650.47	15	100	Mn	2,652.48	100	3
Ru	2,650.40	—	50	Al	2,652.49	60	150*R*
Pb	2,650.40	80	100	Fa	2,652.57	40	1
In	2,650.35	(18)	—	Sb	2,652.606	75	50
Ta	2,650.28	—	25	W	2,652.609	12	10
Tm	2,650.27	7	20	Rh	2,652.66	25	100
Co	2,650.270	25	50*w*	V	2,652.78	40*h*	—
Ru	2,650.25	10	—	U	2,652.83	10	12
Al	2,650.10	(30)	—	Re	2,652.91	—	30
Mo	2,650.00	20	1	V	2,652.92	—	18
Ru	2,649.99	—	20	Ru	2,653.08	10	—
Co	2,649.94	5	50*w*	Ta	2,653.27	15	200
Te	2,649.80	(15)	—	Mo	2,653.35	150	25
Yb	2,649.78	10	4	Mn	2,653.56	15	—
W	2,649.69	12	—	W	2,653.57	15	10
Re	2,649.58	—	20	Cr	2,653.59	35	12
Ru	2,649.51	3	30	Eu	2,653.60	15*h*	15
Pd	2,649.47	200	—				
Fe	2,649.464	70	—				
Mo	2,649.460	10	30*h*				

	λ	I	
		J	O
Se	2,649.42	(50)	—
V	2,649.36	100h	2
Os	2,649.33	6	25
Ti	2,649.30	—	20
Kr	2,649.27	(20)	—
Mo	2,649.25	25	10
Hf	2,649.15	20	10
U	2,649.07	15	15
Re	2,649.05	—	100
Mn	2,648.94	50	1
Mn	2,648.81	—	20w
Ru	2,648.78	150	30
Ni	2,648.72	80	—
Co	2,648.63	40w	5
Te	2,648.61	(10)	—
Ne	2,648.56	(25)	—
V	2,648.47	60	2
Ru	2,648.45	1	20
Ne	2,648.21	(15)	—
Cl	2,648.19	(10)	—
Kr	2,648.15	(20wh)	—
Mn	2,648.04	25	—
W	2,647.74	20	10
Os	2,647.73	6	25
V	2,647.71	10h	50
Mn	2,647.61	15	—
Fe	2,647.56	70	100
Nb	2,647.50	5	15h
Ta	2,647.47	10	200
Ne	2,647.42	(150)	—
Ru	2,647.31	5	50
Hf	2,647.292	125	40
Ar	2,647.29	(10)	—
Ba	2,647.289	40	10
Mo	2,647.25	—	20
Re	2,647.12	—	100
Ni	2,647.06	500wh	—
Os	2,646.892	5	15
Pt	2,646.886	100	1000h
Cl	2,646.88	(25)	—
Ta	2,646.77	50	50h
W	2,646.73	10	12
Ti	2,646.64	15	20
Tb	2,646.50	10	—
Mo	2,646.49	100	25
Yb	2,646.46	10	2
Re	2,646.38	—	20
Ta	2,646.37	2	125
Hg	2,653.68	40	80
Ru	2,653.697	1wh	20
Co	2,653.703	40	5
Ho	2,653.72	10	—
Re	2,653.738	—	15
Er	2,653.743	10	15
Yb	2,653.75	200	50
Ir	2,653.76	5h	15
Mo	2,653.79	20	1
V	2,653.83	15	20
Ru	2,653.95	80	—
Ta	2,654.01	—	15
Re	2,654.12	—	50
U	2,654.58	6h	15
In	2,654.65	(18)	—
In	2,654.76	(30)	—
Rh	2,654.77	25	—
J	2,654.85	(12)	—
Ti	2,654.93	1	20
Mo	2,655.03	8	50h
Hg	2,655.12	40	80
Re	2,655.18	—	25
Os	2,655.19	2	15
Ru	2,655.22	—	20
Ni	2,655.47	400wh	—
W	2,655.55	10	4
W	2,655.670	10	5
Ta	2,655.675	—	15
V	2,655.68	100h	10
Mn	2,655.79	10	5
Mo	2,655.81	20	—
Re	2,655.82	—	25W
J	2,655.84	(20)	—
Ni	2,655.907	500Wh	—
Mn	2,655.914	20	5
Mo	2,655.93	1	15
Nb	2,656.08	200	8
Fe	2,656.15	40	70
Mn	2,656.17	12	—
V	2,656.22	4	40
Ru	2,656.25	150	20
Kr	2,656.38	(15h)	—
Ru	*2,656.56*	—	*30*
Ta	*2,656.61*	*2*	*200R*
Ru	*2,656.69*	—	*30*
Fe	*2,656.80*	*25*	*50*
Br	*2,656.83*	*(25)*	—
Ag	*2,656.92*	*20h*	—

	λ	I J	I O
Nb	2,646.26	200wh	8
Ta	2,646.22	2	50
Ti	2,646.11	200wh	—
Ru	2,646.02	150	20
V	2,645.84	100	15
Ne	2,645.70	(35)	—
Ne	2,645.51	(50)	—
U	2,645.47	25	20
Fe	2,645.43	10	50
Pt	2,645.37	5	40
Ta	2,645.10	30h	80
Fe	2,645.08	20	—
Cs	2,644.70	(20)	—
Ru	2,644.61	100	8
Ta	2,644.60	50W	20W
V	2,644.355	100h	12
Mo	2,644.353	60	30
Yb	2,644.32	40	5
Ti	2,644.26	12	100
P	2,644.20	(25)	—
Ir	2,644.19	5	35
Os	2,644.11	10	75
Fe	2,644.00	150	150
Ta	2,643.89	—	50
Ra	2,643.73	(125)	—
Tm	2,643.58	30	—
J	2,643.31	(20)	—
W	2,643.29	20	2
Ru	2,643.13	30	5
Kr	2,643.06	(20h)	—
Ru	2,642.96	—	150
Ru	2,642.79	150	—
Rh	2,642.76	20	1
Re	2,642.760	—	125
Cs	2,642.63	(20)	—
Yb	2,642.55	80	3
Mo	2,642.41	30	—
V	2,642.27	40	10
Nb	2,642.24	300	5
V	2,642.21	40	10
Pd	2,642.17	100	—
Ti	2,642.15	150wh	—
Cr	2,642.12	3	35
Fe	2,642.01	20	—
Ru	2,641.98	—	20
Fe	2,641.65	60	100
Ru	2,641.62	100	—
Au	2,641.49	20	5
C	2,641.44	20	—
Mo	2,656.98	25	—
Ru	2,657.17	—	30
Ru	2,657.19	50	—
V	2,657.29	35	1
Ta	2,657.30	—	25
Rh	2,657.32	70	—
Hf	2,657.50	20	10
Pd	2,657.56	200	—
Lu	2,657.80	150	50
Hf	2,657.84	25	20
Fe	2,657.92	20	—
Nb	2,658.03	200	—
W	2,658.04	20	10
Mo	2,658.11	5	40
Pt	2,658.17	10	100
Ru	2,658.24	35	—
Fe	2,658.25	80	—
Rh	2,658.36	25	2
Cr	2,658.59	35	18
Os	2,658.60	10	50
Pt	2,658.70	5	40
Pd	2,658.72	300	20
Cl	2,658.74	(100)	—
Ta	2,658.86	50	25
Tb	2,658.91	500	—
V	2,658.97	40	10
Nb	2,659.05	30	3h
Mn	2,659.08	25	—
Rh	2,659.11	100	—
J	2,659.27	(20)	—
Pt	2,659.45	500R	2000R
V	2,659.606	40	9
Ru	2,659.615	15	80
Os	2,659.83	8	30
Nb	2,660.03	30	2
Tm	2,660.09	10	30
Al	2,660.39	60	150R
Cd	2,660.40	(5)	50h
Fe	2,660.402	15	40
Ag	2,660.46	150	30
Br	2,660.49	(25)	—
Mo	2,660.58	125	25
J	2,660.75	(20)	—
Mg	2,660.755	6	40
Mg	2,660.82	6	40
Na	2,661.00	(80)	—
Te	2,661.13	(30)	—
Pd	2,661.14	100	—
Ru	2,661.17	100	20

	λ	I				λ	I	
		J	O				J	O
Hf	*2,641.41*	*125*	*40*		*Ta*	*2,661.34*	*10*	*200*
J	*2,641.39*	*(40)*	—					

Ge 2,709.63 *J* 20 *O* 30

	λ	J	O		λ	J	O
Mn	2,709.61	12	—	Br	2,709.67	(35)	—
Nb	2,709.60	5	1	Yb	2,709.71	2	—
Al	2,709.58	(6)	—	Eu	2,709.73	—	8*w*
W	2,709.57	15	6	W	2,709.75	—	8
Mo	2,709.53	—	15	Mo	2,709.76	6	—
Rh	2,709.52	2	50	Zr	2,709.78	—	3
U	2,709.51	8	8	N	2,709.82	(50)	—
Ce	2,709.41	—	8	Pd	2,709.82	(5)	—
Fe	2,709.37	4	—	Os	2,709.86	5	20
Zr	2,705.33	—	12	La	2,709.92	3	1*h*
Cr	2,709.31	60	2	Mn	2,709.96	25	—
Ta	2,709.27	150	40	Eu	2,709.990	—	20
Mo	2,709.25	1	20	Fe	2,709.993	10	40
Tl	2,709.23	200*R*	400*R*	Hf	2,709.997	4	—
Pd	2,709.21	30*wh*	—	W	2,710.000	4	8
Ru	2,709.20	8	60	Nd	2,710.04	2	5
In	2,709.20	5	—	Ir	2,710.08	5	10
Fe	2,709.06	100	3	Ta	2,710.13	3	200
Co	2,709.05	30*wh*	—	V	2,710.16	60	6
Cl	2,709.03	(10)	—	Mo	2,710.19	20	10
Ra	2,708.96	(200)	—	Cr	2,710.23	—	50*h*
Ru	2,708.84	—	20	Ru	2,710.232	100	50
Co	2,708.82	2	30	Re	2,710.234	—	15*w*
Mn	2,708.81	12	—	In	2,710.26	200*Rh*	800*R*
Ni	2,708.791	500	—	Mn	2,710.33	40*h*	12
Cr	2,708.79	40	3	Cl	2,710.38	(10)	—
Ir	2,708.67	10*h*	—	P	2,710.40	(10)	—
Ru	2,708.65	—	20	Hg	2,710.45	(12)	—
W	2,708.58	15	10	Yb	2,710.54	10*d*	2
Fe	2,708.57	50	80	Fe	2,710.55	35	80
Mn	2,708.45	50*h*	15	Mn	2,710.62	25*h*	—
Ar	2,708.28	(40)	—	Tl	2,710.67	10	30*R*
Th	2,708.181	20	10	Ta	2,710.72	10*h*	1
Os	2,708.179	15	10	Ru	2,710.738	—	20
J	2,708.16	(30)	—	Mo	2,710.742	1	20
Ru	2,707.97	3	50	W	2,710.78	15	6*d*
Rn	2,707.90	(12)	—	Cr	2,710.92	70	1
V	2,707.86	150	70	Pt	2,710.922	15	1
Nb	2,707.83	80	3	Mo	2,710.93	25	1
				Ag	2,711.21	300*wh*	1*h*

	λ	I	
		J	O
Mn	2,707.53	50*h*	10
Co	2,707.50	100*wh*	—
Ru	2,707.47	—	30
Fe	2,707.45	6	20
Re	2,707.41	—	25
Ru	2,707.29	60	—
Rh	2,707.23	4	100
Cd	2,707.14	(30)	—
Fe	2,707.13	70	—
Ti	2,707.04	15	2
W	2,707.02	15	6
U	2,706.95	20*d*	15*d*
Ta	2,706.92	—	15
Ce	2,706.88	—	15
Cs	2,706.79	(20)	—
Co	2,706.74	100*wh*	—
Hf	2,706.73	50	10
Os	2,706.702	8	50
V	2,706.698	200*R*	60
W	2,706.696	20	6
Ta	2,706.69	1	50
Mn	2,706.63	12	—
Fe	2,706.581	150	150
W	2,706.580	10	12
Sn	2,706.51	150*R*	200*R*
Nb	2,706.39	15	2
Tb	2,706.28	10	10
V	2,706.17	400*R*	100
Mo	2,706.12	20*Wh*	20
Re	2,706.06	—	25
Fe	2,706.01	40	60
Pt	2,705.89	200*wh*	1000*wh*
Co	2,705.85	100*w*	15*w*
Mn	2,705.73	25*h*	25
Rh	2,705.63	300*wh*	100
W	2,705.57	12	1
Mn	2,705.56	20	4
Tl	2,705.55	(20)	—
Hg	2,705.36	(30)	—
Eu	2,705.26	—	50
Mo	2,705.24	1	20
V	2,705.22	50	25
Rh	2,704.96	40	3
Mo	2,704.93	50	1
Ru	2,704.81	25	—
Cr	2,704.75	2	15*r*

	λ	I	
		J	O
Mo	2,711.49	25	1
Zr	2,711.51	20	40
Mn	2,711.58	125*h*	2
Te	2,711.61	(50)	—
Fe	2 711.65	50	100
J	2,711.66	(20)	—
V	2,711.74	150*R*	50
Rb	2,711.82	(10)	—
Fe	2,711.84	100	4
Hf	2,711.99	10	10
Ag	2,712.06	100*h*	3
Ru	2,712.09	—	30
Hf	2,712.14	10	10
V	2,712.22	10*h*	6
J	2,712.23	(100)	—
Cr	2,712.31	70	30
Mo	2,712.35	40	1
Fe	2,712.39	100	2
Kr	2,712.40	(80*h*)	—
Ru	2,712.41	300	80
Zr	2,712.420	15	20
Hf	2,712.425	50	25
Re	2,712.48	—	30
Zn	2,712.49	8	300
Cd	2,712.57	20	75
W	2,712.69	20	1
Ir	2,712.74	10	40
Ru	2,712.88	—	30
Re	2,713.03	—	25
V	2,713.05	80	40
Ru	2,713.07	80	—
Pt	2,713.13	10	200
Re	2,713.16	—	25
Ru	2,713.19	2	60
Bi	2,713.30	10*h*	—
Rh	2,713.32	25	—
Mn	2,713.33	—	300*Wh*
U	2,713.49	15*wh*	—
Cu	2,713.50	300*w*	50
Mo	2,713.51	40	20
Ru	2,713.58	100	—
Re	2,713.67	—	15
Ru	2,713.737	2*h*	60
Br	2,713.74	(25)	—
*bh*B	2,713.8	—	200
Mn	2,713.84	15	—
Rb	2,713.92	(10)	—
In	2,713.93	125*wh*	200*R*

	λ	I J	I O
Ru	*2,704.57*	*100*	—
In	*2,704.48*	*(30)*	—
Ta	*2,704.31*	—	*50*
Nb	*2,704.26*	*50*	*2w*
Ru	*2,704.19*	*35*	*10*
Mn	*2,703.990*	*25*	*100wh*
Fe	*2,703.989*	*400*	*30*
Cr	*2,703.86*	*30*	*8*
Ru	*2,703.80*	*3*	*60*
Rh	*2,703.73*	*25*	*150*
Cr	*2,703.55*	*50*	—
Te	*2,703.54*	*(25)*	—
W	*2,703.46*	*20*	*2*
Cu	*2,703.18*	*200*	*10*
Ta	*2,703.06*	—	*60*
Ru	*2,702.83*	*8*	*80*
Ta	*2,702.80*	*30wh*	*40d*
Ba	*2,702.64*	*5h*	*50*
Nb	*2,702.52*	*80*	*8*
Pt	*2,702.40*	*300*	*1000*
Nb	*2,702.20*	*100*	*10*
V	*2,702.19*	*300R*	*80*
Co	*2,702.114*	*25wh*	—
W	*2,702.111*	*25*	*8*
Cr	*2,701.99*	*8*	*35*
Eu	*2,701.89*	*200*	*300W*
Mo	*2,701.87*	*30*	*2*
Lu	*2,701.71*	*150*	*40*
Mn	*2,701.70*	*40*	*150*
Mo	*2,701.42*	*100*	*20*
Ru	*2,701.34*	*8*	*60*
Cs	*2,701.18*	*(20)*	—
Eu	*2,701.12*	*60h*	*60h*
Ru	*2,700.99*	*50*	—
Cu	*2,700.96*	*400*	*20*
V	*2,700.94*	*500R*	*125*
Au	*2,700.89*	*25*	*20*
Nb	*2,700.88*	*20*	*2*
Ru	*2,700.67*	*1h*	*30*
Cr	*2,700.60*	*2*	*30*
Rh	*2,700.59*	*80*	*2*
Ru	*2,700.48*	—	*50*
Ga	*2,700.47*	*(70)*	—
Ru	*2,700.15*	*100*	—
Zr	*2,700.13*	*50*	*50*
Ru	*2,699.88*	*3*	*30*

	λ	I J	I O
Fe	2,714.06	3	20
Ir	2,714.10	10*h*	—
V	2,714.20	100	60
Pd	2,714.32	150	—
Rh	2,714.410	5	150
Fe	2,714.412	400	200
Co	2,714.42	200*W*	12
Os	*2,714.64*	*10*	*50r*
Ta	*2,714.67*	*8*	*200*
Fe	*2,714.87*	*15*	*40*
Pd	*2,714.90*	*200*	—
Rh	*2,715.04*	*2*	*50*
Rh	*2,715.31*	*500wh*	*50*
W	*2,715.338*	*20*	*8*
Nb	*2,715.344*	*100*	*2*
Re	*2,715.47*	—	*100*
V	*2,715.69*	*300R*	*50*
Re	*2,715.77*	—	*30*
Ru	*2,715.78*	—	*50*
Nb	*2,715.88*	*100*	*2*
Co	*2,715.99*	*75*	*75w*
Ru	*2,716.12*	*100*	—
Fe	*2,716.22*	*150*	*20*
Ti	*2,716.25*	*70*	*5*
Nb	*2,716.308*	*30*	*3*
W	*2,716.315*	*20*	*8*
Ru	*2,716.58*	*80*	—
Nb	*2,716.62*	*200*	*10*
Mn	*2,716.79*	*40*	*1*
Rh	*2,716.82*	*3*	*50*
Eu	*2,716.97*	*300*	*300*
Ru	*2,717,01*	—	*30*
Ta	*2,717.18*	—	*100*
Ti	*2,717.30*	*35*	*4*
Nb	*2,717.33*	*20*	*2*
Mo	*2,717.35*	*100*	*20*
Ru	*2,717.40*	*100*	*50*
Pb	*2,717.50*	*(20)*	—
Cr	*2,717.509*	*40*	*12*
Rh	*2,717.512*	*5*	*100*
Pt	*2,717.62*	*40*	*4*
Nb	*2,717.63*	*20*	*2*
Fe	*2,717.79*	*25*	*50*
Ru	*2,717.86*	*50*	—
Rh	*2,717.98*	*50*	—
W	*2,718.04*	*20*	*10*
Yb	*2,718.34*	*30*	*5*

	λ	I J	I O
Ta	2,718.38	—	80
Fe	2,718.43	60	80
Hf	2,718.51	20	10
Rh	2,718.54	20	150
Cu	2,718.77	300w	40
Ru	2,718.83	5h	30
Sb	2,718.89	50	50
W	2,718.90	20	25
Mn	2,719.00	20	—
Fe	2,719.02	300r	500r
Pt	2,719.04	100W	1000w
Mn	2,719.30	20h	—
W	2,719.33	20 Ws	15

Ge 2,754.59 *J* 20 *O* 30

	λ	J	O
U	**2,754.56**	**2**	**3**
Nb	**2,754.52**	**100**	**10w**
Fe	**2,754.43**	**20**	**70**
Mo	**2,754.29**	**20**	**20**
Cr	**2,754.28**	**50**	**3**
Ce	**2,754.23**	**—**	**5**
Zr	**2,754.21**	**1**	**3**
Ca	**2,754.18**	**3**	**1**
Lu	**2,754.17**	**125**	**40**
U	**2,754.15**	**35**	**20**
Cl	**2,754.10**	**(25)**	**—**
Rh	**2,754.09**	**25**	**1**
Nb	**2,754.07**	**3**	**3**
Cl	**2,754.04**	**(6)**	**—**
Fe	**2,754.037**	**35**	**90**
Ar	2,753.92	(10)	—
In	2,753.88	300wh	300R
Pt	2,753.86	4	100
Mo	2,753.82	10	1
Pt	2,753.76	—	15
Fe	2,753.69	25	70
Te	2,753.64	(10)	—
Hf	2,753.61	60	—
Ru	2,753.44	50	50
V	2,753.40	200R	50
bhB	2,753.4	—	100
W	2,753.32	12	1
Fe	2,753.29	150h	25
Tm	2,753.19	20	20
Nb	2,753.14	80	2
Ru	**2,754.61**	**1**	**50**
W	**2,754.71**	**5**	**—**
J	**2,754.72**	**(20)**	**—**
Cr	**2,754.90**	**—**	**10**
Fe	**2,754.907**	**18**	**—**
Ar	**2,754.91**	**(2)**	**—**
W	**2,754.918**	**9**	**12**
Pt	**2,754.920**	**5h**	**200**
Yb	**2,754.93**	**3**	**2**
Pd	**2,754.943**	**5**	**—**
Er	**2,754.944**	**1**	**15**
Fe	**2,754.95**	**—**	**25**
V	**2,755.07**	**20h**	**—**
U	**2,755.13**	**12**	**10**
Fe	2,755.18	—	15
Cs	2,755.20	(20)	—
Re	2,755.22	—	25
Ru	2,755.23	30	—
Cr	2,755.27	2	50d
Nb	2,755.29	10	5
Mo	2,755.37	10	15
J	2,755.58	(12)	—
Er	2,755.64	3	20
V	2,755.66	—	15
Fe	2,755.74	100	300
Ne	2,755.82	(15)	—
Mo	2,756.07	50h	10
Mo	2,756.259	—	15
Fe	2,756.264	100	300
Cr	2,756.30	100	1

	λ	I	
		J	O
Fe	2,753.10	—	25
W	2,753.051	10*d*	15
Re	2,753.046	—	40
Nb	2,753.01	50*h*	5
Ti	2,752.881	50*wh*	—
Cr	2,752.877	40	300*r*
La	2,752.858	10	1
Re	2,752.857	—	25
Hg	2,752.84	10*h*	40
Rh	2,752.837	2	50
Hg	2,752.775	50	100*R*
Ru	2,752.766	150	50
Ta	2,752.49	300	300
Ru	2,752.45	—	50
U	2,752.44	12	4
J	2,752.42	(12)	—
Cr	2,752.39	10	—
Mn	2,752.32	—	20*d*
Ta	2,752.29	8	150
Ru	2,752.27	—	30
W	2,752.24	20	2
Te	2,752.21	(10)	—
Zr	2,752.206	40	40
Th	2,752.172	12	15
Ce	2,752.166	—	15
Eu	2,752.16	—	20*w*
Fe	2,752.159	10	—
V	2,752.13	35*h*	1*h*
Ru	2,752.10	60	—
Fe	2,752.09	20	1
Co	2,752.07	1	40
J	2,751.94	(12)	—
U	2,751.93	20*h*	2*d*
Cr	2,751.87	125	20
Hf	2,751.812	80	25
Fe	2,751.811	5	15
Ti	2,751.70	200*wh*	—
Cr	2,751.60	—	30
Te	2,751.59	(10)	—
Mo	2,751.47	5	50
Yb	2,751.45	15	2
J	2,751.42	(20)	—
Fe	2,751.37	—	15
Fe	2,751.12	70	—
Ta	2,751.04	2*w*	20
Fe	2,750.88	20	60
Tm	2,750.77	30	10
W	2,750.76	12	—
Cr	2,750.73	150	30
Rn	2,756.30	(25)	—
Fe	2,756.33	100	300
Ru	2,756.39	20	—
Zn	2,756.45	100	200
Ag	2,756.51	200	5
V	2,756.57	15*h*	—
Ne	2,756.68	(10)	—
W	2,756.77	12*s*	1
Cd	2,756.79	—	50*h*
Rh	2,756.83	10	—
Hf	2,756.91	40	15
Cr	2,756.93	30*r*	—
Fe	2,757.02	30	10
Ru	2,757.07	—	30
Cr	2,757.10	10	300*r*
W	2,757.21	12	6
Nb	2,757.26	50	3
Fe	2,757.32	60	100
Ti	2,757.39	2	25
J	2,757.43	(12*h*)	—
Re	2,757.48	—	20
Nb	2,757.51	30	2
Pt	2,757.69	—	15
W	2,757.70	12	—
Cr	2,757.72	150	35
Os	2,757.808	8	25
Ru	2,757.808	—	50
Xe	2,757.86	(20h)	—
Fe	2,757.865	10	25
Re	2,758.00	—	60*w*
Ru	2,758.01	—	20
Ti	2,758.07	4	70
Ta	2,758.31	40	200
W	2,758.33	12	6
Ti	2,758.34	15*wh*	—
Mo	2,758.506	20	2
Fe	2,758.51	25*h*	2
V	2,758.512	12*h*	2
P	2,758.52	(20)	—
Co	2,758.54	—	30
Nb	2,758.61	15	10
Cr	2,758.62	30	—
Mo	2,758.63	10	10
Tb	2,758.67	10	—
Re	2,758.71	—	15
Nb	2,758.78	100*w*	1
Zr	2,758.813	30	30
V	2,758.814	10	4
Os	2,758.82	6	15
Zn	2,758.86	(10)	—

	λ	I J	I O
Fe	2,750.72	—	15
Nb	2,750.58	30	2
Ta	2,750.53	—	20
Yb	2,750.48	150	20
Ho	2,750.44	10*h*	—
Ta	2,750.41	10	3*h*
Ru	2,750.35	—	50
W	2,750.32	12	2
Er	2,750.17	3	15
Fe	2,750.144	100	300*h*
Co	2,750.141	—	15
Ti	2,750.140	2	30
Mo	2,750.03	50	2
U	2,749.96	12	10
Se	2,749.94	(15)	—
Ta	2,749.83	50	200
Cr	2,749.82	20	—
In	2,749.80	(30)	—
In	2,749.70	(50)	—
Ru	2,749.68	10	50
Fe	2,749.48	20	15
Fe	2,749.32	20	30
Fe	2,749.184	40	40
Os	2,749.180	6	15
Ti	*2,749.06*	—	*30*
Cr	*2,748.98*	*200*	*35*
Al	*2,748.86*	*(30h)*	—
Ta	*2,748.78*	*50*	*400*
In	*2,748.72*	*25wh*	—
Yb	*2,748.66*	*30*	*5*
Cd	*2,748.58*	*200*	*4*
Cr	*2,748.29*	*5*	*300*
Au	*2,748.26*	*80*	*40*
Cs	*2,748.18*	*(20)*	—
Ru	*2,747.97*	*100*	*50*
Ru	*2,747.69*	*25*	*8*
Rh	*2,747.63*	*100*	*1*
Pt	*2,747.61*	*2*	*150*
Fe	*2,747.56*	*5*	*30*
O	*2,747.50*	*(25)*	—
V	*2,747.47*	*60*	*6*
Re	*2,747.44*	—	*30*
Ta	*2,747.25*	*5*	*50*
Fe	*2,746.98*	*300wh*	*200*
Nb	*2,746.91*	*20*	*5*
Ni	*2,746.75*	*50*	*125*
W	*2,746.73*	*20*	*12*
Ti	2,758.90	10*wh*	—
U	2,758.96	10	10
Cr	2,758.978	40	1
Tm	2,758.98	30	10
Ni	2,759.02	500*wh*	—
Ru	2,759.14	20	—
Nb	2,759.16	10	—
J	2,759.25	(12)	—
Ho	2,759.30	10*h*	—
Fe	2,759.33	12	2
Cr	2,759.39	35	10
Tb	2,759.47	10	10
W	2,759.53	10	—
V	2,759.580	20*h*	2
Mo	2,759.582	1	20
Hg	2,759.71	15	20
Cr	2,759.73	15	1
Fe	2,759.82	60	100
Nb	2,759.97	10*h*	2
Cr	2,760.05	10	1
V	*2,760.12*	*35h*	*20*
Cr	*2,760.52*	*20*	—
Mo	*2,760.53*	*30*	*1*
Ni	*2,760.67*	*40*	—
V	*2,760.70*	*100h*	*25*
W	*2,760.74*	*20*	*8*
Mn	*2,760.93*	—	*80*
Rh	*2,761.26*	*50*	*1*
Ti	*2,761.29*	*35*	*10*
Co	*2,761.37*	*5*	*75*
Os	*2,761.42*	*10*	*50*
Mo	*2,761.53*	*20*	*40*
Ta	*2,761.55*	—	*80*
Ta	*2,761.68*	*150*	*200*
Cr	*2,761.75*	*35*	*300r*
Fe	*2,761.78*	—	*200*
Fe	*2,761.81*	*200*	*50*
Hg	*2,761.97*	*(40)*	—
Fe	*2,762.03*	*60*	*100*
Ta	*2,762.05*	*40*	*1*
Ru	*2,762.31*	*3*	*50*
Cr	*2,762.59*	*100*	*40*
Fe	*2,762.66*	*1*	*50*
U	*2,762.85*	*20*	*15*
Cr	*2,763.06*	*4*	*35*
Pd	*2,763.09 r*	*30*	*300r*
Fe	*2,763.11*	*70*	*100*
Ru	*2,763.14*	*5*	*30*

	λ	I	
		J	O
Ti	2,746.71	150wh	—
Ta	2,746.68	5	100
C	2,746.50	25	—
Fe	2,746.48	300wh	150
Mo	2,746.30	25	30
Cr	2,746.18	25	—
Nb	2,746.10	50h	—
Ru	2,746.07	8	50
Co	2,746.03	—	50
V	2,745.90	35	—
Zr	2,745.85	25	25
Ru	2,745.83	80	50
Nb	2,745.73	50	3
Te	2,745.57	(25)	—
Cu	2,745.45	150	8
Nb	2,745.30	30w	5w
Cu	2,745.28	30	30
Ru	2,745.25	150	12
Co	2,745.10	60	50
Ru	2,745.08	—	30
W	2,745.03	25	2
Cr	2,744.98	20	—
Ti	2,744.85	—	30
Ar	2,744.82	(20)	—
Cr	2,744.59	40	—
V	2,744.54	20	2
Fe	2,744.53	50	70
Ru	2,744.45	50	30
Mo	2,744.19	25	2
Tm	2,744.09	50	20
Fe	2,744.07	8	150
Ru	2,743.94	100	50
Ag	2,743.90	50	—
V	2,743.77	30	7
Mo	2,743.71	1	30
Cr	2,743.64	125	30

	λ	I	
		J	O
J	2,763.23	(30)	—
Mo	2,763.30	20	1
Re	2,763.31	—	40w
Ta	2,763.37	60d	25d
Ru	2,763.42	15	50
Cr	2,763.59	30	1
Mo	2,763.62	50h	25
He	2,763.800	(20)	—
Re	2,763.803	—	50
Cd	2,763.89	50	100h
Ru	2,763.90	1	30
Fe	2,763.91	25	1
Cr	2,763.97	30	—
Gd	2,764.08	30	25
Cd	2,764.11	25	50h
Co	2,764.19	—	100r
W	2,764.27	60	20
Fe	2,764.33	40	70
Cr	2,764.35	6	200r
Cs	2,764.42	(20)	—
Ru	2,764.72	1	50
Fe	2,764.78	20	—
Ti	2,764.82	70	15
Rh	2,764.83	125	15r
Ru	2,765.13	80	—
Ru	2,765.44	150	50
Cr	2,765.47	20	2

Ge 3,039.064 *J* 1000 *O* 1000

Mo	3,039.058	25	—
Ce	3,038.99	—	4
Nd	3,038.96	2	4
Mo	3,038.79	3	—
Ru	3,038.784	6	3
Fe	3,038.779	3	—
Ti	3,038.706	40	2
V	3,038.706	1	20

Bi	3,039.12	2	—
Sm	3,039.13	9	15
U	3,039.14	5	5
Nb	3,039.19	3	2
Ca	3,039.21	4	1h
Ce	3,039.25	—	4
Ir	3,039.260	2	25
U	3,039.263	12	15

	λ	I J	I O		λ	I J	I O
Ho	**3,038.69**	**6**	**4**	**Cs**	**3,039.31**	**(4)**	**—**
In	**3,038.67**	**3**	**—**	**W**	**3,039.311**	**12**	**10**
Se	**3,038.66**	**(60)**	**—**	**Fe**	**3,039.32**	**15**	**20**
Tb	**3,038.66**	**3**	**8**	**In**	**3,039.356**	**500*R***	**1000*R***
Mn	**3,038.602**	**—**	**3**	**Sm**	**3,039.358**	**2**	**6**
Th	**3,038.600**	**12**	**12**	**Nb**	**3,039.41**	**5**	**3**
Zr	**3,038.596**	**2**	**—**	**Se**	**3,039.50**	**(20)**	**—**
Yb	**3,038.54**	**2**	**—**	**U**	**3,039.401**	**3**	**10**
V	**3,038.52**	**45**	**—**	**Ce**	**3,039.51**	**—**	**8**
Mn	**3,038.50**	**4**	**4**	**Mn**	**3,039.55**	**6*h***	**1**
U	**3,038.49**	**6**	**3**	**Co**	**3,039.567**	**—**	**70**
Yt	**4,038.47**	**—**	**3**	**Ce**	**3,039.572**	**—**	**8**
J	**3,038.39**	**(18)**	**—**	**Cd**	**3,039.572**	**—**	**4**
Kr	**3,038.38**	**(3*wh*)**	**—**	**W**	**3,039.575**	**20**	**2**
				Ne	**3,039.65**	**(7)**	**—**
				Yb	**3,039.66**	**15**	**3**
Ni	3,037.93	100	800*R*	**Rb**	**3,039.68**	**(15)**	**—**
Fe	3,037.39	400*r*	700*R*				
				Nb	**3,039.682**	**10*h***	**3**
				Ru	**3,039.684**	**—**	**12**
				Eu	**3,039.70**	**—**	**10**
				Bi	**3,039.71**	**2*h***	**—**
				Ir	**3,039.714**	**—**	**2**
				V	**3,039.76**	**5**	**—**
				Fe	3,040.43	400	400
				Cr	3,040.85	200	500*R*
				Fe	*3,047.60*	*500r*	*800r*
				Ni	*3,050.82*	*—*	*1000R*

Ge 3,124.82 *J* 80 *O* 200

	λ	J	O		λ	J	O
W	**3,124.73**	**3**	**8**	**Ce**	**3,124.86**	**—**	**3**
Ru	**3,124.61**	**2**	**50**	**Eu**	**3,124.87**	**5**	**6*w***
Nd	**3,124.57**	**10**	**12**	**Fe**	**3,124.89**	**7**	**15**
Tb	**3,124.54**	**8**	**8**	**U**	**3,124.900**	**10**	**12**
W	**3,124.50**	**5**	**6**	**Tm**	**3,124.90**	**30**	**15**
U	**3,124.43**	**6**	**10**	**Ir**	**3,124.927**	**—**	**5**
Na	**3,124.41**	**(15)**	**2**	**Sm**	**3,124.927**	**3**	**9**
Rh	**3,124.40**	**—**	**5**	**Sc**	**3,124.935**	**1*h***	**2**
Cd	**3,124.40**	**(10)**	**—**	**Os**	**3,124.94**	**10*wh***	**150*h***
Th	**3,124.39**	**15**	**12**	**Ta**	**3,124.97**	**10**	**50**
Ru	**3,124.37**	**2**	**30**	**Cr**	**3,124.98**	**125**	**20**
P	**3,124.30**	**(15)**	**—**	**V**	**3,125.00**	**50**	**4**
Gd	**3,124.26**	**—**	**4**	**Eu**	**3,125.10**	**1*h***	**5**
Eu	**3,124.21**	**1*h***	**2*h***	**Th**	**3,125.16**	**8**	**10**

	λ	I	
		J	O
F	**3,124.19**	**(3*h*)**	**—**
Ru	**3,124.17**	**8**	**60**
U	**3,124.13**	**2**	**5**
Ce	**3,124.10**	**—**	**20**
Fe	**3,124.09**	**—**	**1**
Ir	**3,124.08**	**—**	**4**
Nd	**3,124.07**	**2*d***	**4**
Rh	3,122.70	2	150
V	3,122.89	300*r*	12
Au	3,122.78	5	500*h*
Cr	3,122.60	80	10
Ti	3,122.06	50	2
Mo	3,122.00	150	5
Xe	3,121.87	(150)	—
Rh	3,121.75	—	150
Co	3,121.41	6	150*r*
Re	3,121.37	—	100
V	3,121.14	200*r*	60
Fe	3 120.87	50	80
V	3,120.73	80	12
Re	3,120.43	80	100
Cr	3,120.37	150	40
Ti	3,119.80	150	4
As	3,119.60	50	100
Fe	3,119.49	80	100
Cr	3,118.65	200	35
V	3,118.38	200*R*	70
Os	3,118.33	20	150
Re	3,118.19	—	200
Ru	3,118.07	50	50
Pb	3,117.70	(100)	—
Ti	3,117.67	200	15
As	*3,116.59*	*150*	—
Fe	*3,116.59*	*150*	—
Hg	*3,116.24*	*(100)*	—
Bi	*3,115.42*	*500*	—
Fe	*3,114.29*	*80*	—
Ni	*3,114.12*	*50*	*300*
Pd	*3,114.04*	*500w*	*400w*
V	*3,113.57*	*100*	*7*
Eu	*3,111.43*	—	*200*
V	*3,110.71*	*300R*	*70*
Ti	*3,110.67*	*100*	*10*
Er	**3,125.17**	**2**	**15**
Zr	**3,125.19**	**2**	**3**
Ir	**3,125.25**	**—**	**2*h***
Fe	**3,125.27**	**2**	**5**
V	**3,125.28**	**200*R***	**80**
Cs	**3,125.30**	**(4)**	**—**
Ce	**3,125.358**	**—**	**2**
W	**3,125.358**	**9**	**10**
Yb	**3,125.44**	**2**	**1**
Cl	**3,125.44**	**(6)**	**—**
Th	**3,125.46**	**10**	**10**
Cr	**3,125.47**	**4**	**8**
Re	**3,125.52**	**—**	**30**
Ti	**3,125.55**	**—**	**2**
Pb	3,125.60	(50)	—
Fe	3,126.65	300	400
Hg	3,125.66	150	200
Fe	3,125.17	70	150
Ne	3,126.19	(150)	—
V	3,126.21	100*R*	60
Ru	3,126.61	50	12
Co	3,127.25	—	100
Nb	3,127.53	50	10*w*
Ta	3,127.76	100	18*w*
Ru	3,127.91	100	10
V	3,128.28	50	2
Ti	3,128.64	70*wh*	12
V	3,128.69	70	3
Cr	3,128.70	150	30
Yt	3,128.79	40	10
Re	3,128.95	—	100*W*
Ni	3,129.31	—	125
Fe	3,129.33	60	100
Na	3,129.37	(60)	35
Yt	3,129.93	50	8
V	3,130.27	200*r*	50
Be	3,130.42	200	200
Ta	3,130.58	35	100*W*
Eu	3,130.74	100	100*W*
Nb	3,130.79	100	100
Ti	3,130.80	100	25
Be	3,131.07	150	200
Os	3,131.11	30	125
Tm	3,131.26	500	400
Hg	3,131.55	300	400
Hg	3,131.83	100	200
Cr	3,132.06	125	25

	λ	I J	I O		λ	I J	I O
				Mo	*3,132.59*	*300R*	*1000R*
				Ta	*3,132.64*	*25*	*250w*
				Cd	*3,133.17*	*300*	*200*
				V	*3,133.33*	*200r*	*50*
				Tm	*3,133.89*	*200*	*200*
				Ni	*3,134.108*	*150*	*1000R*
				Fe	*3,134.111*	*125*	*200*
				Hf	*3,134.72*	*125*	*80*
				O	*3,134.79*	*(100)*	*—*
				Ru	*3,134.80*	*100*	*10*
				V	*3,134.93*	*150r*	*30*
				Fe	*3,135.36*	*100*	*1*
				Ru	*3,135.80*	*80*	*10*
				Ta	*3,135.89*	*100*	*35*
				V	*3,136.51*	*200*	*20*
				Pb	*3,137.83*	*100*	*—*
				Pt	*3,139.39*	*80*	*300*
				V	*3,139.74*	*150*	*15*

Ge 3,269.49 *J* 300 *O* 300

	λ	J	O		λ	J	O
Th	**3,269.47**	**10**	**10**	**Dy**	**3,269.53**	**1**	**10**
U	**3,269.46**	**2**	**2**	**W**	**3,269.63**	**12**	**10**
Eu	**3,269.414**	**1*h***	**2**	**Zr**	**3,269.657**	**1**	**12**
Er	**3,269.411**	**4**	**18**	**Eu**	**3,269.66**	**—**	**4*wh***
Sm	**3,269.39**	**15**	**—**	**Cr**	**3,269.76**	**35**	**—**
Fe	**3,269.235**	**6**	**20**	**Fe**	**3,269.77**	**4**	**—**
U	**3,269.229**	**1**	**2**	**U**	**3,269.78**	**6**	**10**
Os	**3,269.21**	**20**	**200**	**Ag**	**3,269.82**	**10**	**—**
Ta	**3,269.14**	**7**	**70*r***	**Ne**	**3,269.86**	**(7)**	**—**
Ce	**3,269.13**	**—**	**10**	**Os**	**3,269.89**	**2**	**8**
Dy	**3,269.12**	**3**	**20**	**Sc**	**3,269.90**	**12**	**30**
Nb	**3,269.117**	**10**	**2**	**Fe**	**3,269.96**	**3**	**5**
Ca	**3,269.10**	**2**	**10**	**Re**	**3,269.05**	**—**	**5*h***
Ar	**3,269.05**	**(5)**	**—**	**Ba**	**3,270.11**	**—**	**4**
Re	**3,269.04**	**—**	**30**	**V**	**3,270.116**	**5**	**3**
U	**3,269.01**	**2**	**3**	**U**	**3,270.124**	**25**	**20**
Tm	**3,269.00**	**150**	**40**	**Ce**	**3,270.13**	**—**	**12**
Mo	**3,268.98**	**25**	**—**	**Co**	**3,270.20**	**—**	**10**
Ni	**3,268.97**	**—**	**2**	**Te**	**3,270.22**	**(5)**	**—**
Ru	**3,268.94**	**—**	**4**	**Th**	**3,270.23**	**5**	**2*d***
W	**3,268.92**	**10*s***	**9**	**Ru**	**3,270.24**	**—**	**3**
Re	**3,268.90**	**—**	**30**	**W**	**3,270.26**	**4**	**9**
Co	**3,268.89**	**—**	**4**	**Mn**	**3,270.35**	**30**	**30**
Te	**3,268.84**	**(15)**	**—**				
Er	**3,268.80**	**1**	**12**	Fe	3,271.00	300	300
Ru	**3,268.79**	**60**	**4**	Rh	3,271.61	60	200

	λ	I	
		J	O
Mn	**3,268.72**	**30**	**30**
Eu	**3,268.66**	—	**6*W***
U	**3,268.65**	—	**2**
Ti	**3,268.61**	—	**4**
W	**3,268.58**	**9**	**8**
Os	3,267.94	30	400*R*
Sb	3,267.50	150*Wh*	150
Pd	3,267.35	200*h*	—
La	3,265.67	200	300
Fe	3,265.62	300	300
Fe	3,265.05	150	200
Hg	3,264.06	(200)	—
Nb	3,263.37	500	3
Rh	3,263.14	40	200
Sn	3,262.33	300*h*	400*h*
Os	3,262.29	50	500*R*
Tl	3,261.60	300*r*	70
Cd	3,261.06	300	300
Nb	3,260.56	300	15
Pd	*3,258.78*	*200h*	*300*
In	*3,258.56*	*300R*	*500R*
In	*3,256.09*	*600R*	*1500R*
Co	*3,254.21*	—	*300R*
Nb	*3,254.07*	*300*	*20*
Cd	*3,252.52*	*300*	*300*
Pd	*3,251.64*	*500*	*200*
Fe	*3,251.23*	*150*	*300*
Cu	3,273.96	1500*R*	3000*R*
Ta	3,274.95	35*W*	200
Os	3,275.20	15	200
V	3,276.12	200*R*	50
Fe	3,277.35	200	40
Eu	*3,280.682*	—	*1000R*
Ag	*3,280.683*	*1000R*	*2000R*
Zn	*3,282.33*	*300*	*500R*
Fe	*3,286.75*	*400*	*500*
Pd	*3,287.25*	*25*	*300w*

Ge 4,226.58 *J* 50* *O* 200

	λ	J	O
Mo	**4,226.55**	**5**	**10**
Os	**4,226.53**	**6**	**12**
Tb	**4,226.44**	—	**50**
Fe	**4,226.43**	**25**	**80**
W	**4,226.34**	**3**	**10**
Mo	**4,226.29**	**20**	**20**
Nb	**4,226.25**	**3**	**4**
Nb	**4,226.21**	**5**	**3*w***
Ir	**4,226.205**	—	**2**
Ce	**4,226.196**	—	**3**
Sm	**4,226.17**	**2**	**5**
Tb	**4,226.15**	—	**3**
B	**4,226.15**	**2**	**4**
U	**4,226.06**	**8**	**6**
Fe	**4,225.96**	**30**	**80**
U	**4,226.50**	**2**	**1**
V	**4,226.62**	**3**	**8**
Ir	**4,226.63**	—	**5**
Ru	**4,226.66**	—	**15**
Mo	**4,226.726**	**2**	**15**
Yt	**4,226.726**	**15**	**5**
Ca	**4,226.728**	**50*R***	**500**
Ce	**4,226.724**	**30**	**50*R***
Ir	**4,226.725**	—	**30**
Cr	**4,226.76**	**30**	**125**
W	**4,226.91**	**3**	**15**
Nd	**4,226.99**	**5**	**6**
Mo	**4,227.08**	**25**	**3**
Gd	**4,227.14**	**20**	**50**
U	**4,227.33**	**8**	**6**

	λ	I J	I O
Gd	4,225.85	50	150
U	4,225.75	3	1
Ce	4,225.746	8	6
Sc	4,225.58	—	4
Nd	4,225.56	2	3
Fe	4,224.18	80	200
Pr	4,222.98	40	125
Cr	4,222.73	15	100
Fe	4,222.22	200	200
P	4,222.15	(150*w*)	300
Re	4,221.08	—	100
Sm	4,220.65	100	100
Fe	4,219.36	200	250
Cr	4,217.63	70	150
Fe	4,217.555	100	200
La	4,217.554	100	200
Ru	4,217.27	20	100
Gd	4,217.19	100	100
Fe	4,216.19	100	200
Rb	*4,215.56*	*300*	*1000R*
Sr	*4,215.52*	*400W*	*200r*
Gd	*4,215.02*	*150*	*200*
Pd	*4,212.95*	*300W*	*500W*
Dy	*4,211.72*	*15*	*300*
Ag	*4,210.94*	*30h*	*200h*
Fe	*4,210.35*	*200*	*300*
Ce	4,227.41	—	10
Fe	4,227.43	250	300
Re	4,227.46	—	200*W*
Eu	4,227.58	—	8
Zr	4,227.76	8	150
La	4,230.95	50	150
Mo	4,232.59	100	125
Fe	4,233.17	100	100
Fe	4,233.61	150	250
Co	4,234.00	—	100*W*
Eu	4,235.60	—	400*r*
Fe	4,235.94	200	300
La	*4,238.38*	*300*	*500*
Gd	*4,238.78*	*200*	*200*
Fe	*4,238.82*	*100*	*200*
Cr	*4,240.70*	*30*	*200*
Tm	*4,242.15*	*100*	*500*
Fe	*4,247.43*	*100*	*200*

Ge 4,814.80 *J* 200

	λ	I J	I O
Eu	4,814.52	2	2
Mo	4,814.47	8	10
Pr	4,814.344	3	30
Ne	4,814.338	(50)	—
P	4,814.20	(30*wh*)	—
Mo	4,814.10	5*h*	2
Co	4,813.98	2	100
V	4,813.94	25	—
Co	4,813.48	6	1000*W*
Si	4,813.28	6	—
Kr	4,811.76	(300)	—
Ne	4,810.63	(100)	—
Zn	4,810.53	300*h*	400*w*
Ne	4,810.063	(150)	—
Cl	4,810.06	(200)	—
Lu	4,815.05	2	20
Ta	4,815.12	15	5
S	4,815.51	(800)	—
Sm	4,815.81	80	125
Co	4,815.89	1	4
W	4,816.11	1	10
Br	4,816.71	(300)	—
Rn	4,817.15	(100)	—
Ne	4,817.64	(300)	—
Xe	4,818.02	(100)	—
Ne	4,818.79	(150)	—
Cl	4,819.46	(200)	—
Fe	4,821.05	200*h*	200*h*
Ne	4,821.92	(300)	—
Ne	4,823.17	(100)	—

	λ	I	
		J	O
La	4,809.01	150	150
Xe	4,807.02	(500)	—
Ar	4,806.07	(500)	—
Ti	4,805.10	125	15
La	4,804.04	150	150
Gd	4,801.08	200	200
Cd	4,799.92	300*w*	300*w*
Hg	*4,797.01*	*(300)*	—
Cl	*4,794.54*	*(250)*	—
Fe	*4,791.25*	*200R*	*200*
Ne	*4,788.93*	*(300)*	—
Yb	*4,786.60*	*200*	*50*
Br	*4,785.50*	*(400)*	—
Lu	*4,785.42*	*200*	*100*
Xe	4,823.41	(150*h*)	—
La	4,824.07	150	150
Kr	4,825.18	(300)	—
Ra	4,825.91	(800)	—
Ne	4,827.34	(1000)	—
Ne	4,827.59	(300)	—
K	4,829.21	(100)	—
Xe	4,829.71	(400)	—
Mo	4,830.51	100	125
Te	*4,831.29*	*(800)*	—
Kr	*4,832.07*	*(800)*	—
Ne	*4,837.31*	*(500)*	—
Se	*4,840.63*	*(800)*	—
Xe	*4,843.29*	*(300)*	—
Xe	*4,844.33*	*(1000)*	—
Se	*4,844.96*	*(800)*	—

Ge 5,131.70 *J* 100

	λ	I	
		J	O
Th	**5,131.08**	**1**	**10**
Te	**5,130.99**	**(15)**	—
U	**5,130.85**	**3**	**3**
Nd	**5,130.60**	**2**	**40**
O	**5,130.53**	**(30)**	—
V	**5,130.52**	**2**	—
In	**5,130.36**	**(15)**	—
In	**5,129.94**	**(70)**	—
V	5,128.53	75*h*	75
Hg	5,128.45	(150)	—
Ar	5,127.78	(60)	—
Kr	5,125.73	(400*wh*)	—
Xe	5,125.70	(50)	—
Bi	5,124.30	100*wh*	—
La	5,122.99	200	150
Xe	5,122.42	(150)	—
Ne	5,122.34	(150)	—
Ne	5,122.26	(150)	—
In	5,121.34	(50)	—
In	5,121.10	(400)	—
In	5,120.96	(150)	—
In	5,120.85	(100)	—
J	5,119.28	(500)	—
Ar	5,118.20	(60)	—
In	5,117.41	(300)	—
Nd	**5,132.33**	**1**	**10**
Yb	**5,132.41**	**5*h***	**2**
C	**5,132.96**	**30**	—
Te	**5,133.23**	**(8)**	—
C	**5,133.29**	**15**	—
Nb	**5,133.34**	**3**	**10**
Pr	**5,133.42**	**1**	**60**
Yb	5,135.98	50	6
Fe	5,136.79	100	3
V	5,138.42	50*h*	50
V	5,139.53	50	50
P	5,141.49	(50)	—
Fe	5,141.75	100*h*	100
Se	5,142.14	(500)	—
Kr	5,143.05	(600*h*)	—
Bi	5,144.48	300*h*	2
Ne	5,144.94	(500)	—
Ne	5,145.01	(500)	—
C	5,145.16	70	—
Ar	5,145.36	(200)	—
O	5,146.06	(70)	—
Yb	5,147.03	50	3
V	5,148.72	60	60

	λ	I J	I O
In	5,117.37	(50)	—
In	5,116.75	(70)	—
Ne	5,116.50	(150)	—
In	5,115.91	(100)	—
La	5,114.57	200	150
Sb	5,113.86	70	—
Ne	5,113.67	(75)	—
In	*5,109.36*	*(300w)*	—
As	*5,107.80*	*150*	—
As	*5,105.80*	*150*	—
Cl	*5,103.04*	*(125)*	—
Cl	*5,099.30*	*(100)*	—
Ra	*5,097.56*	*(250)*	—
Se	*5,096.57*	*(350)*	—
Ar	*5,151.39*	*(200)*	—
Rb	*5,152.09*	*100*	—
La	*5,157.43*	*100*	*40*
J	*5,161.19*	*(300)*	—
Ar	*5,162.28*	*(500)*	—
Fe	*5,167.49*	*150*	*700*

Ge 5,893.46   *J* 100

	λ	I J	I O
Nb	**5,893.44**	**3**	**15**
J	**5,893.43**	**(8)**	—
Mo	**5,893.38**	**15**	**70*h***
Xe	**5,893.29**	**(150)**	—
La	**5,892.66**	**4**	**1**
Pr	**5,892.23**	**1**	**10**
C	**5,891.65**	**30**	—
Te	**5,891.43**	**(15)**	—
Se	**5,891.29**	**(8)**	—
S	**5,890.98**	**(8)**	—
C	5,889.97	60	—
Na	5,889.95	1000*R*	9000*R*
Rn	5,888.60	(80)	—
Ar	5,888.59	(300)	—
Mo	5,888.33	100	150
Ar	5,882.62	(100)	—
Ne	5,881.89	(1000)	—
La	5,880.65	50	30
Ti	5,880.31	125	60
Kr	5,879.90	(50)	—
He	5,875.62	(1000)	—
Xe	5,875.02	(100)	—
Ne	5,872.15	(75)	—
Kr	5,870.92	(3000)	—
Ne	5,868.42	(75)	—
Kr	5,866.75	(50)	—
Ti	5,866.46	400	300
J	**5,894.05**	**(60)**	—
Zn	**5,894.35**	**(30)**	**3**
Rn	**5,894.40**	**(30)**	—
Kr	**5,894.56**	**(8*wh*)**	—
Bi	**5,894.50**	**6**	—
Xe	**5,894.99**	**(100)**	—
Sb	**5,895.09**	**(150*wh*)**	—
Xe	**5,895.62**	**(24)**	—
Tm	**5,895.63**	**20**	**80**
Pb	**5,895.70**	**2**	**20*hl***
Na	**5,895.92**	**500*R***	**5000*R***
In	**5,896.02**	**5**	—
Te	**5,896.65**	**(25)**	—
Yb	5,897.22	100*h*	7
Yb	5,898.80	50*h*	3
Ti	5,899.32	150	150
Nb	5,900.62	200	200
Ne	5,902.46	(50)	—
In	5,903.04	(70)	—
In	5,903.14	(100)	—
In	5,903.24	(100)	—
In	5,903.37	(100)	—
In	5,903.47	(70)	—
In	5,903.63	(500)	—
In	5,903.75	(150)	—
Xe	5,905.13	(100)	—
Ne	5,906.43	(50)	—
Ar	5,912.08	(500)	—

	λ	I	
		J	O
Mo	*5,858.27*	*200*	*200*
Ba	*5,853.68*	*100h*	*300*
In	*5,853.43*	*(300)*	—
In	*5,853.11*	*(150)*	—
In	*5,852.83*	*(100)*	—
Ne	*5,852.49*	*(2000)*	—
Br	*5,852.10*	*(150)*	—
V	*5,849.30*	*100h*	*100*

	λ	I	
		J	O
Ne	5,913.63	(250)	—
In	5,914.68	(70)	—
In	5,914.83	(50)	—
In	5.915.45	(50)	—
In	5,915.63	(50)	—
In	5,915.97	(100)	—
Xe	5,917.44	(50)	—
In	5,918.65	(70)	—
In	5,918.78	(50)	—
Ne	5,918.91	(250)	—
Ti	*5,922.12*	*100*	*100*
Ar	*5,928.80*	*(200)*	—
N	*5,931.79*	*(150)*	—
Xe	*5,934.17*	*(100)*	—
N	*5,941.67*	*(200)*	—
Ne	*5,944.83*	*(500)*	—
Xe	*5,945.53*	*(200)*	—
Yb	*5,946.02*	*100*	*4*

H $^{1}_{1.00797}$

t_0 −259.1°C t_1 −252.5°C

I.	II.	III.	IV.	V.
13.595	—	—	—	—

λ	*I*	eV
3,797.91	(20)	13.45
3,835.40	(40)	13.42
3,889.06	(60)	13.38
3,970.07	(80)	13.31
4,101.74	(100)	13.21
4,340.47	(200)	13.01
4,861.33	(500)	12.74
6,562.72	(1000)	—
6,562.85	(2000)	12.09

H 3,797.910 *J* (20)

	λ	*I*			λ	*I*	
		J	*O*			*J*	*O*
Cs	3,797.908	(4)	—	Th	3,797.91	5	2
Nd	3,797.89	10*d*	20*d*	Hf	3,797.92	25	25
Cu	3,797.83	2*h*	—	Fe	3,797.95	2	4
U	3,797.77	1	10	Ru	3,798.05	40	30
Sm	3,797.73	5	25	Ce	3,798.08	2	2
Cr	3,797.72	20	100	Th	3,798.10	1	5
U	3,797.520	3	—	Nb	3,798.12	80	50
Fe	3,797.517	200	300	La	3,798.19	2	—
Th	3,797.516	10	8	Ho	3,798.25	4	—
Sn	3,797.42	3	—	Mo	3,798.252	1,000*R*	1,000*R*
Ta	3,797.40	3	6	U	3,798.26	8	2
Mo	3,797.30	5	5	Br	3,798.28	(6)	—
Rb	3,797.28	(2)	—	Ti	3,798.31	6	10
Pr	3,797.23	8	15	F	3,798.46	(6)	—
Rb	3,797.17	(2)	—	Ce	3,798.51	3	3
U	3,797.14	3	1	Fe	3,798.513	300	400
Cr	3,797.13	30	50	Ce	3,798.62	3	3
Sb	3,797.05	4	1	V	3,798.661	5	7
Mo	3,797.03	5	5	Hf	3,798.662	3	5
Bi	3,797.00	5	—	Tm	3,798.76	10	20
Th	3,796.952	5	10	Cl	3,798.80	(50)	—
W	3,796.947	4	5	U	3,798.84	1	15
Pr	3,796.92	20*d*	40*d*	Ru	3,798.90	100	70
Nd	3,796.89	10*d*	20*d*	W	3,798.92	10	7
Ti	3,796.885	15	12	Tb	3,798.95	8	8
Kr	3,796.884	(20)	—	In	3,799.05	(20)	—
Nb	3,796.85	15*w*	10*w*	Ho	3,799.06	4*h*	—
U	3,796.845	15	12	In	3,799.12	(10)	—
Dy	3,796.84	1*h*	2*h*	Pd	3,799.19	150	200*w*
Rb	3,796.82	(40)	—	U	3,799.202	12	6
Ho	3,796.73	40	20	In	3,799.204	(18)	—
Ar	3,796.60	(5)	—	S	3,799.21	(8)	—
Nb	3,796.59	5	5	Th	3,799.22	5	10
Mo	3,796.57	20	—	Nd	3,799.24	10*d*	10*d*
U	3,796.54	15	10	Mn	3,799.26	50	50
Zr	3,796.48	15	8	V	3,799.27	7	10
V	3,796.47	12	30				
Nb	3,796.44	15	15				
Gd	3,796.39	150	150*w*				
Ec	3,796.31	10*w*	5*w*	Rh	3,799.311	100	25
Xe	3,796.30	(40)	—	In	3,799.314	(10)	—
Ta	3,796.21	50	—	Ru	3,799.35	100	70
U	3,796.20	12	8	In	3,799.37	(25)	—
Nd	3,795.80	15	20	Ar	3,799.39	(15)	—
Pr	3,795.765	20*d*	30*d*	In	3,799.42	(10)	—
Tm	3,795.765	150	250	Sm	3,799.546	10	30
Th	3,795.75	10*w*	20*w*	Fe	3,799.549	300	400
				As	3,799.58	10	—

	λ	I J	I O
Os	3,795.67	12	40
Mo	3.795.59	20*d*	—
Nb	3,795.54	20*h*	10
Nd	3,795.45	10	6
Th	3,795.39	10	20
Pr	3,795.34	25*d*	25*d*
In	3,795.27	(10)	—
In	3.795.21	(50)	—
Ta	3,795.18	30	—
In	3,795.17	(18)	—
P	3,795.09	(30)	—
Fe	3,795.00	400	500
V	3,794.96	50*h*	50*h*
Ru	3,794.92	30	20
La	3,794.77	200	400
Os	3,794.66	20	40
Cr	3,794.61	30	50
O	3,794.48	(10)	—
Mo	3,794.43	10	5
Pr	3,794.38	25*d*	50*d*
V	3,794.36	25*h*	1
W	3,794.345	10	10
Th	3,794.341	10	10
Fe	3,794.340	50	80
Th	3,794.15	10	10
Sm	3,793.97	30	10
Tl	3,793.95	(25)	—
Os	3,793.91	300	125
Cr	3,793.879	30	50
Fe	3,793.877	10	25
Ir	3,793.792	10	30
Pr	3,793.789	30*d*	40*d*
Cl	3,793.75	(25)	—
Se	3,793.63	(25)	—
V	3,793.61	15	35
P	3,793.60	(30)	—
U	3,793.58	12	10
Tb	3,793.55	15	15
Hf	3,793.37	20	10
Cr	3,793.29	30	50
U	3,793.28	12	20
Rh	3,793.22	60	200
U	3,793.10	18	12
Bi	3,793.00	(25)	—
Bi	3.792.80	500*h*	—
Nb	3,792.79	20*h*	—
W	3,792.77	15	15
Nd	3,792.55	10*d*	20*d*

	λ	I J	I O
Pr	3,799.68	25	50
V	3,799.91	50	60
Ne	3,800.02	(18)	—
Nd	3,800.026	20	30*d*
Sc	3,800.029	12	5
Ir	3,800.12	100	150
K	3,800.14	(30)	—
Ru	3,800.26	40	12
Pr	3,800.30	50	100
Hf	3,800.39	12	20
Os	3,800.44	75	50
Kr	3,800.54	(30)	—
Eu	3,800.55	10*w*	10*w*
Mn	3,800.552	60	60
Sm	3,800.89	25	20
Xe	3,800.99	(10)	—
Sn	3,801.00	150*h*	200*h*
Nd	3,801.12	15	15
Ta	3,801.147	10	1
Nb	3,801.154	20	3
Gd	3,801.33	15	15
Eu	3,801.36	10*w*	9*w*
Nd	3,801.37	40	60
Xe	3,801.39	(30)	—
In	3,801.50	(50)	—
Eu	3,801.61	10*w*	7*w*
Fe	3,801.68	25	50
Mo	3,801.84	25	20
Mn	3,801.91	20	20
W	3,801.921	10	9
Rb	3,801.925	(20)	—
P	3,802.07	(100)	—
Fe	3,802.28	10	25
Pr	3,802.76	15	30*d*
Nb	3,802.92	50	50
Th	3,803.08	15	15
Eu	3,803.10	10*w*	9*w*
Pr	3,803.11	20	50*d*
O	3,803.13	(20)	—
Ar	3,803.19	(15)	—
Mo	3,803.41	20*w*	—
Nd	3,803.474	20	20
V	3,803.474	40	50
W	3,803.68	15*w*	—
Nb	3,803.88	20	30
Au	3,804.00	150	25
Fe	3,804.01	10	40
Nd	3,804.10	10*d*	15*d*

	λ	I J	I O
B	3,792.50	(500)	—
S	3,792.46	(35)	—
F	3,792.40	(10)	—
Gd	3,792.396	25	20
Fe	3,792.166	20	40
Cr	3,792.14	40	60
Mo	3,792.09	15	1
Ta	3,792.01	10	50
Nd	3,791.50	10	20
Mo	3,791.41	10*h*	—
Cr	3,791.38	40	80
Nb	3,791.21	80	80
Gd	3,791.13	25	25
Nd	3,790.84	10*d*	20*d*
La	3,790.82	300	400
Os	3,790.73	20	100
Gd	3,790.62	15	8
Ru	3,790.51	150	70
Cr	3,790.45	15	50
V	3,790.32	12	40
Cr	3,790.23	10	30
Mn	3,790.21	125	100
Nb	3,790.15	50*r*	200*r*
Os	3,790.14	30	80
Fe	3,790.09	100	200
Rn	3,790.07	(12)	—
Pr	3,789.99	20	25
W	3,789.77	12	—
Cr	3,789.72	10	50
U	3,789.60	12	4
Nb	3,789.49	10	10
Ti	3,789.29	15	50
Fe	3,789.18	50	80
Th	3,789.12	20	20
Os	3,789.11	12	30
Nd	3,788.97	10*d*	15*d*
Pr	3,788.93	25*d*	50*d*
Cr	3,788.86	10	60
Eu	3,788.76	10*w*	15*w*
Yt	3,788.70	30	30
Rh	3,788.47	25	50
Dy	3,788.45	40	100
Mo	3,788.26	15	15
Sm	3,788.13	10	25
P	3,788.06	(15*h*)	—
Fe	3,787.88	300	500
Gd	3,787.56	25	25
V	3,787.24	20	2
U	3,787.23	10	1

	λ	I J	I O
Eu	3,804.26	10*w*	10*w*
Mo	3,804.52	20	20
Kr	3,804.67	(30*h*)	—
Nb	3,804.74	50	20
Nd	3,804.77	10	10
Cr	3,804.80	30	100
Pr	3,804.85	50*w*	25*w*
Cs	3,805.10	(25)	—
Gd	3,805.11	10	10
Cl	3,805.25	(75)	—
Fe	3,805.34	300	400
Nd	3,805.36	30	50
Gd	3,805.53	10	10
Th	3,805.82	15	20
F	3,805.90	(15)	—
Rh	3,805.92	50	25
Mo	3,805.99	10	10
Hf	3,806.07	20	5
Nb	3,806.20	10	8
Fe	3,806.22	20	40
Dy	3,806.28	80	25
Zn	3,806.37	15	3
Hg	3,806.38	(200)	—
Ce	3,806.39	12	12
Pr	3,806.41	15	25*d*
Nd	3,806.54	20*d*	20
Nb	3,806.63	204	3*h*
Fe	3,806.70	150	200
Mn	3,806.72	20	50*h*
Rh	3,806.76	50	50
V	3,806.80	12	35
Cr	3,806.83	35	35
Tb	3,806.85	50	50
Mo	3,807.01	20*w*	1
Ni	3,807.14	40*h*	800*w*
Nd	3,807.23	20	15
V	3,807.50	50	80
Fe	3,807.537	100	150
Yb	3,807.54	15	3
Eu	3,807.59	15*w*	20*w*
Th	3,807.87	10	10
Cr	3,807.926	12	25
Nd	3,807.935	20	30
I	3,808.07	(40)	—
Ce	3,808.12	35	35
Th	3,808.15	10	10
Zr	3,808.20	25	30
V	3,808.52	30	50

	λ	I	
		J	O
Sm	3,787.20	35	100
Th	3,787.19	20*w*	40*w*
As	3,787.18	15	—
Fe	3,787.17	15	25
Nd	3,787.16	50	60*d*
Sc	3,787.154	10	30
Ta	3,787.146	20*w*	—
Nb	3,787.06	30	30
Th	3,786.884	15	15
Pr	3,786.882	15	30*d*
Eu	3,786.83	10*w*	5*w*
P	3,786.69	(15)	—
Fe	3,786.68	50	125
Se	3,786.61	(12)	—
U	3,786.57	10	5
Ar	3,786.40	(15)	—
W	3,786.38	10	10
Mo	3,786.36	125	2
Pb	3,786.24	10*h*	—
Dy	3,786.20	15	20
Er	3,786.187	12	20
Fe	3,786.176	60	100
Ru	3,786.05	100	70
Ti	3,786.04	40	40
Pb	3,786.00	40	—
Fe	3,785.95	80	125
Eu	3,785.80	10*wh*	10*wh*
Th	3,785.65	15	15
W	3,785.63	10	—
Pr	3,785.50	20	50
Eu	3,785.47	10*wh*	25*w*
Hf	3,785.46	15	20
Cs	3,785.42	(20)	—
U	3,785.35	15*h*	—
Nd	3,785.09	10	10*d*
Mo	3,785.03	10	8
Nb	3,784.88	10	1
Nd	3,784.85	20	20
Ta	3,784.254	50*w*	150
Nd	3,784.250	25	25
Eu	3,784.23	10*w*	10*w*
Pr	*3,783.86*	*30d*	*200d*
U	*3,783.840*	*25*	*20*
Nb	*3,783.839*	*20*	*15*
Nd	*3,783.78*	*20d*	*20d*
Ni	*3,783.53*	*40h*	*500*
Pr	*3,783.51*	*20d*	*30d*

	λ	I	
		J	O
Ar	3,808.61	(10)	—
Ru	3,808.68	30	50
Fe	3,808.73	70	100
Nd	3,808.77	12	20
La	3,808.78	10	—
Nd	3,809.05	20*d*	15
Pr	3,809.16	40	80
U	3,809.22	12	15
W	3,809.23	20	25
Ar	3,809.49	(25)	—
Cl	3,809.51	(40)	—
Mn	3,809.59	150	150
V	3,809.60	40	70
Sm	3,809.74	10	10
Xe	3,809.84	(30)	—
Nd	3,809.90	20	20
Pr	3,809.96	10*r*	20*r*
Cl	3,810.10	(30)	—
W	3,810.38	15	15
Sm	3,810.43	10	5
Nd	3,810.48	20	20*d*
Nb	3,810.49	50	30
Hf	3,810.61	15	15
Mn	3,810.69	15	15
Ho	3,810.70	40	20
Tm	3,810.73	10	30
Fe	3,810.76	25	70
W	3,810.80	10*s*	10*s*
Ag	3,810.86	10*h*	100*wh*
U	3,810.93	12	8
Nb	3,811.03	12	10
Xe	3,811.05	(20)	—
Nd	3,811.07	10	10
Bi	3,811.14	(150)	—
Nd	3,811.34	10*d*	20*d*
Pr	3,811.35	10	10*d*
Th	3,811.38	10	20
Ti	3,811.40	10	25
Mo	*3,811.84*	*25d*	—
Gd	*3,812.05*	*200h*	*200R*
Nd	*3,812.05*	*20d*	*20d*
Sm	*3,812.07*	*20*	*20*
Kr	*3,812.22*	*(20)*	—
Mo	*3,812.28*	*25*	—
Fe	*3,812.96*	*300*	*400*
Th	*3,813.06*	*20*	*15*
Ti	*3,813.39*	*20*	*4*
Fe	*3,813.89*	*25*	*50*

	λ	I J	I O
K	*3,783.19*	*(30)*	—
Mo	*3,783.18*	*40*	—
Kr	*3,783.13*	*(500h)*	—
Gd	*3,783.06*	*20*	*25*
Th	*3,783.02*	*20*	*20*
U	*3,782.84*	*30*	*25*
Yb	*3,782.56*	*25*	*5*
Gd	*3,782.28*	*50*	*25*
S	*3,782.26*	*(35)*	—
Os	*3,782.19*	*200*	*400R*
Mo	*3,782.07*	*100*	—
Pr	*3,781.64*	*50d*	*100d*
Mo	*3,781.59*	*20*	*25*
Eu	*3,781.39*	*30W*	*25W*
Nb	*3,781.38*	*200*	*5*
Ar	*3,781.36*	*(300)*	—
Nd	*3,781.31*	*20*	*20*
Cl	*3,781.23*	*(30)*	—
Ru	*3,781.18*	*40*	*50*
Nb	*3,781.01*	*20*	*20*
Nd	*3,780.92*	*20*	—
Ar	*3,780.84*	*(50)*	—
Sm	*3,780.76*	*20*	*20*
U	*3,780.72*	*20*	*15*
La	*3,780.67*	*50*	*50*
La	*3,780.51*	*20*	—
Mo	*3,779.77*	*30*	*25*
Nb	*3,779.58*	*20*	*2*
Fe	*3,779.45*	*70*	*100*
Pr	*3,778.75*	*20*	*40*
Fe	*3,778.51*	*25*	*60*
V	*3,778.36*	*35*	*3*
Nd	*3,778.14*	*40*	*40*
Sm	*3,778.13*	*100*	*40*
Kr	*3,778.09*	*(500h)*	—
Nb	*3,777.67*	*30*	*20*
Ru	*3,777.59*	*50*	*60*
Ne	*3,777.16*	*(75)*	—
Th	*3,777.12*	*20*	*40*
Os	*3,776.99*	*20*	*150*
Mn	*3,776.53*	*25*	*25*
Tb	*3,776.49*	*100*	*100*
Fe	*3,776.46*	*70*	*125*
Hg	*3,776.26*	*(30)*	—
Ti	*3,776.06*	*60*	*8*
Tl	*3,775.72*	*1000R*	*3000R*
Mo	*3,775.65*	*20*	*15*
Ni	*3,775.57*	*40h*	*500h*
Nd	*3,775.50*	*20*	*10*
Mo	*3,813.90*	*20*	*2*
Gd	*3,813.98*	*60*	*100w*
Ra	*3,814.42*	*(2000)*	—
Fe	*3,814.52*	*40*	*80*
Ti	*3,814.58*	*35*	*12*
Cr	*3,814.62*	*30*	*35*
Ru	*3,814.86*	*35*	*20*
Rh	*3,815.01*	*20*	*20*
V	*3,815.39*	*150h*	*1*
Nb	*3,815.51*	*30*	*20*
Bi	*3,815.80*	*(300)*	—
Fe	*3,815.84*	*700*	*700*
Au	*3,816.13*	*40*	*20*
Pr	*3,816.166*	*40*	*40*
Bi	*3,816.173*	*25h*	—
Yb	*3,816.20*	*35*	*12*
Co	*3,816.33*	*50r*	*60*
Fe	*3,816.34*	*20*	*25*
K	*3,816.55*	*(30)*	—
Mo	*3,816.61*	*20*	—
Mn	*3,816.75*	*50*	*60*
Dy	*3,816.77*	*50*	*100*
Ir	*3,817.24*	*30*	*15*
Ru	*3,817.27*	*60*	*50*
Nd	*3,817.37*	*25*	*30*
K	*3,817.54*	*(40)*	—
Nd	*3,817.67*	*40*	*40*
Cr	*3,817.84*	*20*	*30*
Rh	*3,818.19*	*25*	*50*
Pr	*3,818.28*	*100*	*125*
Yt	*3,818.34*	*50*	*30*
Cl	*3,818.40*	*(30)*	—
Ne	*3,818.44*	*(25)*	—
Cr	*3,818.48*	*20*	*50*
Nb	*3,818.86*	*300*	*8*
Ru	*3,819.03*	*30*	*50*
Hf	*3,819.38*	*20*	*15*
Cr	*3,819.56*	*40*	*60*
He	*3,819.61*	*(50)*	—
Eu	*3,819.66*	*500wd*	*500wd*
Sm	*3,819.68*	*20*	*10*
V	*3,819.96*	*35*	*60*
Cl	*3,820.25*	*(100)*	—
V	*3,820.30*	*20*	*25*
Fe	*3,820.429*	*600*	*800*
Nd	*3,820.431*	*40*	—
Th	*3,820.81*	*20*	*20*
Fe	*3,821.18*	*100*	*100*
J	*3,821.35*	*(25)*	—

	λ	I	
		J	O
P	*3,775.02*	*(30)*	—
Fe	*3,774.83*	*40*	*100*
Hg	*3,774.52*	*(30)*	—
Yt	*3,774.33*	*100*	*12*
Nd	*3,774.32*	*30*	*20*
Cl	*3,774.25*	*(25)*	—
Pr	*3,774.06*	*50*	*100*
Th	*3,773.76*	*20*	*20*
Cl	*3,773.68*	*(20)*	—
Kr	*3,773.42*	*(50)*	—
La	*3,773.12*	*20*	*2*
V	*3,772.97*	*40*	*2*
Pr	*3,772.85*	*20*	*80d*
Mo	*3,772.82*	*20*	*20*
U	*3,772.81*	*20*	*6*
Mo	*3,771.95*	*30*	*30*
Nb	*3,771.85*	*20*	*20*
Pr	*3,771.77*	*20d*	*40d*
Ti	*3,771.65*	*30*	*70*
Th	*3,771.38*	*20w*	*30w*
Kr	*3,771.34*	*(30h)*	—
Eu	*3,771.16*	*20W*	*10W*
V	*3,770.97*	*60*	*30*
Sm	*3,770.73*	*50*	*25*
Gd	*3,770.70*	*60*	*50*
Nb	*3,770.65*	*20h*	—
Ar	*3,770.37*	*(400)*	—
Eu	*3,770.23*	*25W*	*15W*
Fe	*3,769.99*	*30*	*80*
Rh	*3,769.97*	*30*	*25*
Nd	*3,769.64*	*20*	*100*
Ni	*3,769.455*	*50h*	*2*
Gd	*3,769.453*	*40*	*50*
V	*3,821.49*	*30*	*50*
Nd	*3,821.77*	*40w*	*50w*
Pr	*3,821.82*	*50*	*50*
Fe	*3,821.84*	*30*	*50*
V	*3,822.01*	*40*	*70*
N	*3,822.07*	*(35)*	—
Ru	*3,822.09*	*25*	*50*
Rh	*3,822.26*	*100*	*100*
V	*3,822.89*	*25*	*40*
Mo	*3,823.15*	*20h*	—
Pr	*3,823.18*	*25*	*125*
V	*3,823.21*	*20*	*35*
O	*3,823.47*	*(125)*	—
Mn	*3,823.51*	*75*	*75h*
Cr	*3,823.52*	*30*	*40*
Mn	*3,823.89*	*50*	*50h*
Th	*3,824.35*	*20w*	*20w*
Fe	*3,824.44*	*100*	*150*
Nd	*3,824.79*	*30*	*40*
Nb	*3,824.88*	*50*	*30*
Ru	*3,824.93*	*25*	*30*
O	*3,825.09*	*(20)*	—
Zr	*3,825.27*	*60*	*40*

H 3,835.40 J (40)

	λ	J	O
Ho	**3,835.35**	**6*h***	**2**
Th	**3,835.34**	**3**	**5**
Mo	**3,835.31**	**25**	**8**
Er	**3,835.26**	**2*d***	**25*d***
In	**3,835.18**	**15**	—
Nb	**3,835.17**	**20**	**20**
U	**3,835.14**	**8**	**4**
La	**3,835.07**	**15**	**12**
W	**3,835.049**	**20**	**15**
Ru	**3,835.048**	**6**	**50**
Gd	**3,835.00**	**10**	**8**
Co	**3,835.50**	**2**	**3**
V	**3,835.56**	**12**	**50**
Co	**3,835.69**	**4**	**10**
W	**3,835.70**	**5**	—
Sm	**3,835.73**	**15**	**15**
Ce	**3,835.75**	**2**	**4**
W	**3,835.88**	**5**	**7**
U	**3,835.92**	**3**	**4**
Zr	**3,835.96**	**5**	**25**
Ru	**3,835.99**	**1*h***	**6**
V	**3,836.05**	**15**	**10**

	λ	I	
		J	O
Mo	3,834.97	6	8
Pr	3,834.92	15	30
Eu	3,834.91	4*W*	4*W*
N	3,834.84	(5)	—
V	3,834.811	1	6
U	3,834.809	6	3
Rh	3,834.75	5	4
Cr	3,834.72	12	25
In	3,834.72	(18)	—
Br	3,834.69	(2)	—
In	3,834.686	(25)	—
Ar	3,834.68	(800)	—
In	3,834.65	(35)	—
Mo	3,834.64	6	8
Th	3,834.608	10*w*	15*w*
In	3,834.606	(40)	—
Sm	3,834.60	15	25
U	3,834.58	4	2
Pr	3,834.57	2	—
In	3,834.563	(40)	—
Ce	3,834.556	4	10
Dy	3,834.55	1	3
Sm	3,834.47	2	5
Mn	3,834.36	75	75*r*
N	3,834.24	(15)	—
Fe	3,834.225	400	400
V	3,834.224	12	20
Ce	3,834.220	12	6
Pr	3,834.20	4	—
V	3,834.15	2	5
W	3,834.04	10	7
Rh	3,833.89	50	25
Mn	3,833.86	75	75
Mo	3,833.75	25	80
Ta	3,833.74	200	40
Mo	3,833.47	25	—
Fe	3,833.31	60	100
Pr	3,833.04	20	20
Nd	3,833.03	30	60
Yt	3,832.89	80	30
Pb	3,832.83	50	—
Mg	3.832.31	200	250
Tl	3,832.30	(30)	—
Pd	3,832.29	(150)	(150)
Nb	3,831.84	300	5
Ru	3,831.79	50	60
U	3,831.46	25	25
Os	3,836.06	20	150
Cr	3,836.07	8	25
Ti	3,836.08	30	15
C	3,836.10	2*h*	—
Nd	3,836.108	15	40
Ce	3,836.112	6	15
Sm	3,836.12	4	20
Cu	3,836.15	2*wh*	—
Se	3,836.25	(20)	—
Fe	3,836.33	60	100
Nb	3,836.45	3	5
Lu	3,836.48	30	30
Ca	3,836.49	2*h*	2
Er	3,836.505	10	40
Th	3,836.505	50*w*	50*w*
Dy	3,836.51	40	100
Gd	3,836.514	60	100*wh*
Sm	3,836.515	25	50
Sc	3,836.519	25	25
U	3,836.520	15	6
Kr	3,836.54	(30*wh*)	—
Nd	3,836.541	100	80
Ti	3,836.598	3	10
Se	3,836.60	(8)	—
Ta	3,836.603	3	30
Ru	3,836.70	40	8
Nb	3,836.74	15	4*w*
Zr	3,836.76	20	15
Ti	3,836.77	5	18
P	3,837.15	(30)	—
Mo	3,837.141	20	—
Mo	3,837.29	100*w*	—
Kr	3,837.82	(30)	—
Tm	3,838.20	60	80
Mg	3,838.26	200	300
Nd	3,838.33	25	40
Cl	3,838.37	(20)	—
N	3,838.39	(25)	—
W	3,838.50	20	15
Nd	3,838.72	25	50
Nd	3,838.98	30	20
Fe	3,839.259	75	100
Hg	3,839.26	(50)	—
Ru	3,839.69	30	50
Mn	3,839.78	125	100
Nd	3,839.81	20	30
Os	3,840.30	20	150
Zn	3,840.34	(50)	3

	λ	I	
		J	O
Hf	3,831.13	25	25
Cr	3,831.032	25	40
Nd	3,831.030	30	60*d*
Pr	3,830.72	60	100
Nb	3,830.64	20*w*	1*h*
N	3,830.39	(150)	—
Cr	3,830.03	50	150*w*
Ne	3,829.77	(40)	—
Mn	3,829.68	60	60
Nd	3,829.63	30	40
Nd	3,829.41	40	50
Th	3,829.410	20*w*	40*w*
Au	3,829.38	20	25
Mg	3,829.35	150	100*w*
Mo	3,828.87	30	40
Nd	3,828.84	20	20
Rh	3,828.48	60	100
Sc	3,828.18	30	30
Nd	3,828.17	40	50
Fe	3,827.82	200	200
Cl	3,827.62	(150)	—
P	3,827.44	(150)	—
Rb	3,827.20	(40)	—
Pb	3,827.20	(20)	—
Mo	3,827.16	25	25
Os	3,827.14	20	50
Pd	3,827.13	25*wh*	4
V	3,826.97	25	1
Nd	3,926.91	25*w*	40*w*
Mo	3,826.69	25	30
Hg	3,826.61	(30)	—
Cr	3,826.425	20	40
Nd	3,826.416	20	10
Pr	3,826.29	50*d*	80*d*
Fe	3,825.88	400	500
Au	3,825.65	30	20

	λ	I	
		J	O
Fe	3,840.44	20	40
Ce	3,840.45	35	30
La	3,840.71	70	50
Fe	3,841.05	400	500
Mn	3,841.08	50	50
Cr	3,841.28	80	150
Bi	3,841.60	(25)	—
Pb	3,841.62	60	—
Se	3,841.93	(20*h*)	—
Nd	3,841.95	30	30
Th	3,841.96	20	20
Co	3,842.05	20	400*R*
In	3,842.17	(25)	—
In	3,842.22	(25)	—
In	3,842.27	(35)	—
Kr	3,842.28	(20*wh*)	—
Pr	3,842.36	40	80
Tb	3,842.49	50	40
Mg	3,842.58	25	—
As	3,842.82	50	—
Ce	3,842.988	25	18
Nd	3,842.988	40	30
Sc	3,843.00	20	25
Zr	3,843.02	40	40
Fe	3,843.259	100	125
Cl	3,843.26	(100)	—
Nd	3,843.51	20	—
Mn	3,843.98	100	75
V	3,844.44	50*h*	100
Kr	3,844.45	(50*wh*)	—
Hg	3,845.15	(30)	—
Fe	3,845.17	60	100
Cl	3,845.42	(50)	—
Zr	3,845.44	20	20
Co	3,845.47	100	500*R*
Cl	3,845.68	(75)	—
Bi	3,845.80	(100)	—
Cl	3,845.82	(30)	—
Nb	3,845.90	30	10
Mo	3,845.95	20	20
Nd	3,845.99	25	30
La	3,846.00	50	40
Bi	3,846.03	100	—
W	3,846.21	20	20
Fe	3,846.41	75	50
Pr	3,846.60	30	70*d*
Nd	3,846.71	30*d*	30*d*
Fe	3,846.80	100	125

	λ	I			λ	I	
		J	O			J	O
				Nd	3,846.97	20	20
				F	3,847.09	(800)	—
				Mo	3,847.25	25	25
				V	3,847.33	70*h*	100
				Nd	3,847.85	50	60
				Tm	3,848.02	250	400
				Mo	3,848.30	20	25
				Nd	3,848.52	20	10
				Tb	3,848.75	200	100
				Cr	3,848.98	50*d*	80*d*
				La	3,849.01	150	200
				Fe	*3,849.97*	*400*	*500*
				F	*3,849.99*	*(600)*	—
				Cr	*3,850.04*	*40r*	*40r*
				Ar	*3,850.57*	*(400)*	—
				Fe	*3,850.82*	*75*	*200*
				Cl	*3,851.02*	*(100)*	—
				Cl	*3,851.42*	*(75)*	—
				Pr	*3,851.62*	*150w*	*200w*
				F	*3,851.67*	*(200)*	—
				Nd	*3,852.38*	*50*	*60*
				Fe	*3,852.57*	*100*	*150*
				Pr	*3,852.80*	*50*	*100*
				Pb	*3,854.05*	*100*	—
				La	*3,854.91*	*40*	—
				V	*3,855.37*	*50r*	*50r*
				Nb	*3,855.50*	*50h*	—
				V	*3,855.84*	*200*	*200*
				Fe	*3,856.37*	*300*	*500*
				Mo	*3,857.20*	*60*	—
				Ni	*3,858.30*	*70h*	*800r*
				Nb	*3,858.95*	*50*	*20*
				Fe	*3,859.22*	*100*	*100*
				Fe	*3,859.91*	*600*	*1000r*
				Cl	*3,860.83*	*(150)*	—
				Cl	*3,860.99*	*(100)*	—
				Cl	*3,861.34*	*(50)*	—
				Fe	*3,861.342*	*50*	*80*
				Ru	*3,862.65*	*60*	*2*

H 3,889.06 *J* (60)

Er	**3,889.02**	**1**	**9**	**Sm**	**3,889.15**	**4**	**4**
Zr	**3,889.004**	**2*h***	**2*h***	**Sm**	**3,889.216**	**7**	**5**
Ce	**3,888.997**	**3**	**12**	**Nd**	**3,889.218**	**15**	**10**

	λ	I J	I O
Ho	3,888.95	20	40
W	3,888.89	4	3
Mo	3,888.87	8	8
Fe	3,888.82	15	40
Eu	3,888.68	5*w*	8*w*
Cs	3,888.65	10	150
He	3,888.645	(1000)	—
U	3,888.61	3	4
Br	3,888.52	(10)	—
Fe	3,888.517	3*w*	4*w*
Ce	3,888.39	4	15
Rh	3,888.335	2	5
V	3,888.329	1	10
Pr	3,888.29	4	5
Bi	3,888.23	2	40
U	3,888.21	6	12
Mo	3,888.18	8	10
Er	3,888.09	1	18
V	3,888.08	4	15
Mo	3,887.96	4	3
W	3,887.94	10	—
Bi	3,887.93	2	5
Nd	3,887.87	20	25
Yt	3,887.78	3	3
Ru	3,887.77	8	15
U	3,887.70	10	20
Mo	3,887.672	3	3
Tb	3,887.67	3	25
Dy	3,887.54	2	12
Yb	3,887.31	40	6
Tl	3,887.15	(30)	—
Mo	3,886.82	30	30
Cr	3,886.79	125	125
La	3,886.37	200	400
Fe	3,886.28	400	600
Nb	3,885.68	30	15
Fe	3,885.51	60	100
Nb	3,885.44	100	50
Sm	3,885.28	50	50
Cr	3,885.22	50	40
Pr	3,885.19	40*w*	100*w*
P	3,885.17	(115)	—
V	3,884.84	70	4
Fe	3,884.36	35	80
Tm	3,883.43	30	150
Cr	3,883.292	80	60
Fe	3,883.298	40	70
Nb	3,883.14	30	30
O	3,882.19	(35)	—
Hf	3,889.232	1*h*	4
V	3,889.235	2	12
Mo	3,889.28	3	2
Ce	3,889.30	1	6
Ba	3,889.32	2	10
Hf	3,889.328	1	5
Pr	3,889,330	70	150
W	3,889.37	5	—
U	3,889.41	5	1
Pr	3,889.42	6	10
Mn	3,889.45	50	25
Ta	3,889.46	40*W*	—
Au	3,889.47	8	3
Eu	3,889.51	5*W*	10*W*
Ir	3,889.58	3	2
Nb	3,889.63	5	3
Nd	3,889.66	25	30
Ni	3,889.67	10*h*	30
In	3,889.78	(100)	—
Er	3,889.80	1	15
Tb	3,889.85	3	10
Nd	3,889.93	20	30
Mo	3,889.953	4	2
Ti	3,889.954	5	25
Pr	3,889.97	3	5
Ce	3,889.99	8	50
U	3,890.01	3	1
Cu	3,890.073	3	—
Sm	3,890.074	10	10
Pr	3,890.17	2	8
V	3,890.18	30	100
Ru	3,890.20	8	30
Nd	3,890.22	8	12
Mg	3,890.24	8	3
Zr	3,890.32	5	150
U	3,890.36	30	35
Fe	3,890.39	2	4
W	3,890.42	8	10
Mo	3,890.45	3	2
As	3,890.46	5	—
Ta	3,890.50	10	2
Tm	3,890.52	10	40
Fe	3,890.84	30	60
Ho	3,891.02	40	200
Nb	3,891.30	100	50
Fe	3,891.93	70	100
Ru	3,892.21	40	50

	λ	I J	I O
Co	3,881.87	30	300R
Hf	3,880.82	30	20
Sm	3,880.75	30	40
Pr	3,880.47	60	80
Nb	3,879.35	300	5
Pr	3,879.21	80	100
V	3,878.71	100	35
Fe	3,878.57	300	300R
Fe	3,878.02	300	400
Se	3,877.28	(50)	—
Pr	3,877.22	80w	125w
Hf	3,877.10	30	4
Co	3,876.83	40	300w
Os	3,876.77	50	300
C	3,876.67	40	—
Lu	3,876.63	100	50
C	3,876.41	60	—
C	3,876.19	125	—
Pr	3,876.18	30	80
V	3,876.09	30	50
C	3,876.05	40	—
Nb	3,875.76	50	10
Kr	3,875.44	(150wh)	—
V	3,875.07	50	70r
Hg	3,874.98	(30h)	—
Tb	3,874.18	200	200
Co	*3,873.95*	*80*	*400R*
Fe	*3,873.76*	*80*	*125*
Co	*3,883.11*	*80*	*500R*
Fe	*3,872.50*	*300*	*300*
Dy	*3,872.12*	*150*	*300*
Fe	*3,871.75*	*60*	*100*
Fe	*3,869.56*	*80*	*100*
S	*3,867.56*	*(150)*	—
Fe	*3,867.22*	*100*	*150*
Fe	*3,865.53*	*400*	*600*
Os	*3,865.47*	*200*	*125*
Pr	*3,865.46*	*125r*	*200r*
Nb	*3,865.02*	*200h*	—
La	*3,864.49*	*150*	*100*
Bi	*3,864.20*	*150h*	—
Mo	*3,864.11*	*500R*	*1000R*
Bi	*3,863.90*	*(100)*	—
Ru	*3,862.65*	*60*	*2*

	λ	I J	I O
S	3,892.32	(35)	—
U	3,892.68	30	20
V	3,892.86	35	60
Nb	3,894.034	30	15
Cr	3,894.035	40	60
Co	3,894.08	100	1000R
Pd	3,894.20	200W	200W
Ar	3,894.66	(300)	—
Ta	3,894.67	40h	—
Gd	3,894.708	80	150W
Mn	3,894.708	40	40
Kr	3,894.71	(60wh)	—
P	3,895.02	(100)	—
Fe	3,895.66	300	400
Nb	3,895.90	30	10
V	3,896.16	40	50
Sm	3,896.97	50	50
J	3,897.26	(40)	—
K	3,897.87	(60)	—
Fe	3,897.89	60	100
Fe	3,898.01	50	80
Nb	3,898.28	200	3
Tb	3,899.19	100	200
Fe	3,899.71	300	500
Ar	3,899.86	(100)	—
Nd	3,900.23	30	30
Ti	3,900.54	50h	30
Al	3,900.68	(200)	—
Tm	3,900.79	50	80
Th	3,900.89	30	30
Nd	3,901.85	30	30
Cr	3,902.11	30	40
Gd	3,902.40	80	100
Pr	3,902.47	40	60
Cr	3,902.91	100	100
Fe	3,902.95	400	500
Mo	3,902.96	500R	1000R
Cr	3,903.16	30	35
Sm	3,903.41	60	60
Fe	3,903.90	80	100
P	*3,904.78*	*(100)*	—
Yb	*3,904.83*	*150*	*12*
Fe	*3,906.48*	*200*	*300*
Eu	*3,907.11*	*500R*	*1000R*
Gd	*3,907.12*	*100*	*100w*
Fe	*3,907.94*	*60*	*100*
Pr	*3,908.43*	*60*	*100*
Cr	*3,908.75*	*150*	*200*

	λ	I			λ	I	
		J	O			J	O
				O	3,911.95	(150)	—
				Kr	3,912.59	(70)	—
				Pr	3,912.90	80	150
				Ti	3,913.46	70	40
				P	3,914.26	(100)	—
				Br	3,914.28	(150)	—
				Hg	3,914.29	(100)	—
				V	3,914.33	70wh	25
				Nb	3,914.70	100	30
				Cr	3,915.84	80	125
				La	3,916.04	400	400
				Cr	3,916.24	60	100
				Gd	3,916.59	100	150w
				Fe	3,916.73	80	100
				Fe	3,917.18	70	150
				Hg	3,918.92	(200)	—
				C	3,918.89	80	—

H 3,970.074 *J* (80)

Pr	3,970.070	4	15	Ta	3,970.10	40	100
Pt	3,970.06	15	4	U	3,970.14	5	1
Hf	3,970.05	3	10	Tb	3,970.18	3	30
Ce	3,970.04	3	12	Fe	3,970.26	2	5
N	3,969.95	(2)	—	Fe	3,970.39	30	50
Th	3,969.83	5	5	Ce	3,970.42	1	6
Ru	3,969.79	4	8	Sm	3,970.53	10	15
Pr	3,969.755	3	12	U	3,970.59	2	10
Cr	3,969.748	90	200	Br	3,970.60	(10)	—
Os	3,969.671	100	100	Ce	3,970.64	3	15
Nd	3,969.666	4	20	Nb	3,970.65	10s	3
Fe	3,969.63	5	5	S	3,970.69	(5)	—
Th	3,969.53	5	5	W	3,970.80	12	12
Pr	3,969.51	3	8	Mo	3,970.96	5	5
Er	3,969.43	1	6	U	3,971.03	2	—
Th	3,969.34	2w	3w	Gd	3,971.08	20	20
Fe	3,969.26	400	600	Pr	3,971.16	60	100
W	3,969.20	10	12	Cl	3,971.18	(7)	—
Te	3,969.18	(10)	—	Cr	3,971.25	50	80
Ir	3,969.17	10	30	Pb	3,971.30	(30)	—
Nb	3,969.135	20h	—	Fe	3,971.33	125	200
In	3,969.13	15	—	Mo	3,971.37	3	2
Co	3,969.12	6	100w	Sm	3,971.39	30	50
Cr	3,969.06	50	80	U	3,971.40	4	—
U	3,969.02	8	5	F	3,971.63	(3)	—
Mo	3,969.01	3	3	Nd	3,971.669	4	10
Th	3,969.00	5	10	Rn	3,971.67	(80)	—

	λ	I	
		J	O
Nd	**3,968.88**	**4**	**20**
Mo	**3,968.75**	**50**	**8**
Br	**3,968.66**	**(8)**	**—**
W	**3,968.59**	**6**	**6**
Nb	**3,968.471**	**10**	**3**
Ce	**3,968.469**	**35*w***	**35**
Ca	**3,968.468**	**500*R***	**500*R***
Ru	**3,968.46**	**200**	**12**
Ar	3,968.36	(200)	—
Ag	3,968.22	60	100
V	3,968.09	50	25
Hg	3,968.03	(50)	—
Xe	3,967.54	(200)	—
Fe	3,967.42	100	125
Nb	3,967.37	50*h*	—
Fe	3,966.63	50	80
Pr	3,966.57	70*d*	100*d*
Pt	3,966.36	40	80
Fe	3,966.07	70	100
Sm	3,966.05	50	60
Pr	3,965.26	50	100
Ru	3,964.90	40	50
Pr	3,964.82	80*d*	125*d*
He	3,964.73	(50)	—
Nb	3,964.28	50	1
Ti	3,964.27	40	80
Pr	3,964.26	50	60
Cr	3,963.69	300	300
Gd	3,963.66	60	50
Os	3,963.63	50	500
Fe	3,963.11	50	125
Sm	3,963.00	40	50
Pr	3,962.44	50	60
Al	3,961.53	2000	3000
Mo	3,961.50	500	5
Sm	3,959.53	40	50
Rh	3,958.86	100	200
Pd	3,958.64	200	500*w*
Zr	3,958.22	150	500
Ti	3,958.21	100	150
Tm	3,958.10	40	200
Gd	3,957.68	200	300*w*
P	3,957.62	(100)	—
Nd	3,957.46	40*s*	60
Fe	3,956.68	150	150
Fe	3,956.46	100	100
Ti	3,956.34	50	100
Kr	3,954.78	(90*wh*)	—
Ce	**3,971.68**	**6**	**35**
Pr	**3,971.693**	**40**	**60**
Nb	**3,971.695**	**15*h***	**—**
Pr	3,972.16	80	125
Ta	3,973.18	400*w*	1
O	3,973.27	(125)	—
V	3,973.64	40	25
Gd	3,974.07	80	100
Xe	3,974.42	(40)	—
Mn	3,975.89	50	40
Ir	3,976.31	70	10
Nb	3,976.51	80*h*	—
Cr	3,976.66	300	300
Tb	3,976.82	200	150
Mn	3,977.08	100	50
Os	3,977.23	40	300
Fe	3,977.74	150	300
Rb	3,978.21	(40)	—
Ru	3,978.44	70	60
Cr	3,978.68	40	80
Ce	3,978.89	50	50
Sm	3,979.19	50	50
Hf	3,979.40	40	6
Ru	3,979.42	60	60
Ta	3,981.01	40	2
Cr	3,981.23	50	100
Rn	3,981.68	(150)	—
Ti	3,981.76	70	100
Fe	3,981.77	100	150
Tb	3,981.88	200	80
Dy	3,981.94	100	150
Pr	3,982.06	100	125
Yt	3,982.59	100	60
Gd	3,983.03	40	40
Sm	3,983.14	60	100
Cr	3,983.91	60	200
Fe	3,983.96	125	200
Hg	3,983.98	(400)	—
Cr	3,984.34	60	80
Ru	3,984.86	70	60
Mn	3,985.24	100	75
Fe	3,985.39	40	125
V	3,985.79	40	1
Gd	*3,987.22*	*100*	*100*
Yb	*3,987.99*	*500R*	*1000R*
La	*3,988.52*	*800*	*1000*
Zn	*3,989.23*	*(100)*	*—*
Pr	*3,989.72*	*125*	*200*

	λ	I	
		J	O
O	3,954.60	(40)	—
O	3,954.38	(100)	—
Pr	3,953.52	100	150
Fe	3,951.17	125	150
Xe	3,950.92	(125)	—
Yt	3,950.36	100	60
Fe	3,949.96	100	150
Pr	3,949.44	100	150
La	3,949.11	800	1000
Ar	3,948.98	(2000)	—
Fe	3,948.78	100	150
Hg	3,948.29	(100)	—
Ar	3,947.50	(1000)	—
O	3,947.33	(300)	—
Ru	3,945.57	100	50
Gd	3,945.54	150	200w
Hg	3,945.09	(100)	—
Dy	3,944.69	150	300
Al	3,944.03	1000	2000
Hg	3,942.59	(100)	—
Hg	3,942.24	(100)	—
Mo	3,941.48	150	5
Fe	3,940.88	80	150
Rb	3,940.57	(200)	—
J	3,940.24	(500)	—
Tb	3,939.60	200	200
Nb	3,938.55	100h	—
Ti	3,989.76	100	150
Kr	3,994.83	(100)	—
N	3,994.99	(300)	—
La	3,995.85	300	600
Gd	3,996.32	100	100
Dy	3,996.70	80	200
Fe	3,997.40	150	300
Kr	3,997.95	(100wh)	—
Fe	3,998.05	100	150
Ti	3,998.64	100	150
Dy	4,000.45	300	400
Cr	4,001.44	80	200

H 4,101.735 J (100)

	λ	J	O
Fe	4,101.68	2	5
Ce	4,101.55	3	2
Nd	4,101.46	6	8
Dy	4,101.43	2	5
Ce	4,101.36	2	1
U	4,101.32	2h	2
Sm	4,101.31	2h	15d
Fe	4,101.27	10	40
Ru	4,101.23	2h	7
Cr	4,101.16	2	30
Ho	4,101.09	40	40
Te	4,101.07	(50)	—
La	4,101.01	2h	2
V	4,101.00	7	5
Hf	4,100.93	1	10
Nb	4,100.92	200w	300w
Ru	4,101.745	60	20
Ce	4,101.772	6	33
In	4,101.773	1000R	2000R
W	4,101.85	8	7
U	4,101.90	1	18
Dy	4,101.95	2	8
Mo	4,102.15	25	30
V	4,102.16	15	30
N	4,102.18	(5)	—
U	4,102.21	4	1
Ru	4,102.28	10	15
Ce	4,102.36	2	18
Yt	4,102.38	30	150
Ho	4,102.40	3	—
Tb	4,102.52	2	25w
Br	4,102.53	(10)	—

	λ	I J	I O		λ	I J	I O
Tb	**4,100.90**	**2**	**50*d***	**Nd**	**4,102.56**	**5**	**10**
W	**4,100.895**	**6**	**7**	**W**	**4,102.70**	**30**	**35**
Ce	**4,100.889**	**1**	**8**	**Ti**	**4,102.71**	**1**	**3**
Th	**4,100.83**	**18**	**18**	**Eu**	**4,102.72**	**3**	**10**
Pr	**4,100.75**	**50**	**200**	**Ce**	**4,102.724**	**1**	**2**
Fe	**4,100.74**	**30**	**80**	**O**	**4,103.01**	**(50)**	**—**
				F	**4,103.08**	**(150)**	**—**
				Dy	**4,103.31**	**50**	**50**
N	4,099.94	(150)	—				
La	4,099.54	100	100				
Gd	4,098.91	100	100	F	4,103.52	(300)	—
Xe	4,098.89	(50*h*)	—	F	4,103.72	(50)	—
Ne	4,098.77	(50)	—	Mo	4,103.84	400	400
				Hg	4,103.87	(50)	—
Kr	4,098.72	(250)	—	F	4,103.87	(50)	—
Se	4,097.91	(60)	—				
Ru	4,097.79	125	25	Ar	4,103.91	(200)	—
O	4,097.24	(70)	—	O	4,104.73	(50)	—
O	4,092.94	(80)	—	O	4,105.00	(125)	—
				Fe	4,107.49	100	120
Sc	4,091.95	(70)	—	Se	4,108.83	(800)	—
Ac	4,088.37	100	—				
Kr	4,088.33	(500)	—	F	4,109.17	(100)	—
Gd	4,087.71	100	80	P	4,109.19	(70)	—
Pd	4,087.34	100	500	Kr	4,109.23	(100*hs*)	—
				Xe	4,109.71	(60)	—
La	4,086.71	500	500	Fe	4,109.81	100	120
				N	4,109.98	(1000)	—
				V	4,111.78	100 *wR*	100 *wR*
				Ru	4,112.74	200	125
				Fe	4,118.55	700	200
				O	*4,119.22*	*(300)*	—
				Nb	*4,119.28*	*200*	*2*

H 4,340.47 **J** (200)

	λ	I J	I O		λ	I J	I O
U	**4,340.45**	**8**	**5**	**Tl**	**4,340.53**	**(20)**	**—**
Ne	**4,340.42**	**(2)**	**—**	**Ce**	**4,340.56**	**1**	**12**
O	**4,340.29**	**(10*h*)**	**—**	**Eu**	**4,340.59**	**1**	**8*w***
Ne	**4,340.26**	**(2)**	**—**	**Bi**	**4,340.59**	**40*h***	**—**
Cr	**4,340.13**	**30**	**80**	**Tb**	**4,340.63**	**2**	**40**
K	**4,339.98**	**(20)**	**—**	**Ra**	**4,340.64**	**(1000)**	**—**
Eu	**4,339.94**	**1**	**3*w***	**U**	**4,340.70**	**2**	**1**
Sm	**4,339.93**	**3**	**8**	**La**	**4,340.73**	**3**	**50**
Mo	**4,339.82**	**15**	**15**	**Mo**	**4,340.75**	**20**	**20**
Bi	**4,339.80**	**(12*w*)**	**—**	**V**	**4,341.01**	**30**	**60**
Ne	**4,339.78**	**(5)**	**—**	**Th**	**4,341.03**	**6**	**6**
Cr	**4,339.72**	**150**	**150**	**Ca**	**4,341.11**	**1**	**5**
Pr	**4,339.68**	**2**	**20**	**Zr**	**4,341.133**	**4**	**50**

	λ	I	
		J	O
Dy	4,339.68	8	15
Se	4,339.59	(200)	—
Zr	4,339.555	1	3
Tl	4,339.55	(8)	—
W	4,339.452	3	8
Cr	4,339.450	300	300r
Sm	4,339.35	2	2
Ti	4,337.92	125	70
Cr	4,337.57	300	500
Fe	4,337.05	150	400
As	4,336.85	100	—
Ar	4,335.34	(800)	—
Sm	4,334.15	200	300
Pr	4,333.91	100	150
La	4,333.73	500	800
Ar	4,333.56	(1000)	—
Ar	4,331.25	(200)	—
Xe	4,330.52	(500wh)	—
Se	4,330.28	(200)	—
Gd	4,329.58	100	100
Sm	*4,329.02*	*300*	*300*
Se	*4,328.70*	*(200)*	—
Fe	*4,325.76*	*700*	*1000*
Gd	*4,325.69*	*250*	*500R*
Kr	*4,319.58*	*(1000)*	—
Sm	*4,318.93*	*300*	*300*
Kr	*4,318.55*	*(400)*	—

	λ	I	
		J	O
W	4,341.134	2	4
Gd	4,341.29	125	200
Kr	4,341.33	(8wh)	—
Pr	4,341.375	1	10
Ti	3,341.375	40	12
Ne	4,341.42	(15)	—
Mo	4,341.42	25	25
Gd	4,342.19	200	200
Cl	4,343.62	(100)	—
Cr	4,344.51	300	400r
Ar	4,345.17	(1000)	—
O	4,345.57	(125)	—
Sm	4,345.85	100	100
Gd	4,347.32	100	100
Ar	4,348.11	(500h)	—
O	4,349.43	(300)	—
Rn	4,349.60	(5000)	—
Sm	4,350.46	150	150
Cr	4,351.05	150	100
O	4,351.27	(125)	—
Kr	4,351.36	(100)	—
Cr	*4,351.77*	*300*	*300*
Sm	*4,352.10*	*100*	*125*
As	*4,352.25*	*200*	—
Kr	*4,355.48*	*(3000)*	—
N	*4,358.27*	*(250)*	—
Hg	*4,358.35*	*500*	*3000w*
Kr	*4,362.64*	*(500)*	—

H 4,861.33 *J* (500)

Ta	4,861.04	1	3
U	4,861.01	10	10
O	4,860.93	(20)	—
La	4,860.91	100	100
Mo	4,860.75	8	10
Mo	4,860.56	4	5
N	4,860.35	(5)	—
Mo	4,860.05	20	25
Br	4,860.04	(12)	—
Yt	4,859.85	5	50
U	4,859.750	8	8
Fe	4,859.748	40	150

Hf	4,861.49	2	3
Er	4,861.59	1	5
Kr	4,861.84	(2h)	—
Cr	4,861.842	8	125
Mn	4,862.05	5	40
Kr	4,862.10	(2h)	—
J	4,862.13	(25)	—
Ra	4,862.27	(4)	—
J	4,862.31	(700)	—
Xe	4,862.54	(400h)	—
Gd	4,862.608	2	100
V	4,862.609	12	15
P	4,862.83	(15)	—

	λ	I J	I O
Kr	4,846.60	(700)	—
Sc	4,844.96	(800)	—
Xe	*4,844.33*	*(1000)*	—
Se	*4,840.63*	*(800)*	—
Ne	*4,837.31*	*(500)*	—
Kr	*4,832.07*	*(800)*	—
Te	*4,831.29*	*(800)*	—

	λ	I J	I O
Te	4,864.10	(800)	—
Te	4,866.22	(800)	—
Ne	*4,884.91*	*(1000)*	—
Ne	*4,892.09*	*(500)*	—

H 6,562.72 *J* (1000)

6,562.85 *J* (2000)

	λ	J	O
H	**6,562.72**	**(1000)**	—
J	**6,560.87**	**(30)**	—
Rb	**6,560.84**	**150**	—
Xe	**6,560.65**	**(4*h*)**	—
He	**6,560.13**	**(100)**	—
Xe	**6,559.97**	**(25)**	—
Br	**6,559.81**	**(150)**	—
Ne	*6,506.53*	*(1000)*	—
Ra	*6,487.32*	*(1000)*	—

	λ	J	O
H	**6,562.85**	**(2000)**	—
Hf	**6,562.86**	**10**	**2**
Xe	**6,563.19**	**(15)**	—
W	**6,563.22**	**1**	**2**
Sn	**6,563.24**	**50*wh***	—
Co	**6,563.42**	**5**	**200*wh***
Te	**6,563.95**	**(50*h*)**	—
Cu	**3,564.50**	**10**	—
Pr	**6,564.63**	**1**	**10**
J	**6,564.80**	**(30)**	—
S	**6,565.04**	**(15)**	—
Kr	**6,565.32**	**(6*h*)**	—
Hf	**6,565.76**	**3**	—
J.	6,565.48	(400)	—
C	6,578.03	500	—
Ra	*6,593.34*	*(500)*	—
Ne	*6,598.95*	*(1000)*	—

He $^{2}_{4.0026}$

t_0 −272.2°C t_1 −268.9°C

I.	II.	III.	IV.	V.
24.581	54.405	—	—	—

λ	I	eV
I 3,187.74	(200)	23.70
II 3,203.14	(100)	52.23
I 3,888.65	(1000)	23.00
II 4,685.75	(300)	51.00
I 5,875.62	(1000)	23.06

He 3,187.74 J (200)

	I	I	
		J	O
Tl	3,187.74	(50)	—
V	3,187.71	100R	35
Fe	3,187.680	2	6
Dy	3,187.676	10	25
Kr	3,187.61	(4)	—
Bi	3,187.6	2h	—
Ne	3,187.60	(4)	—
Mo	3,187.59	50	1
U	3,187.51	1	3
Nb	3,187.49	10	12
Cl	3,187.42	(5)	—
Tm	3,187.41	25	10
Th	3,187.408	2	4
Ho	3,187.37	4	—
Os	3,187.33	12	80
Fe	3,187.29	60	—
Tb	3,187.25	15	15
Sm	3,187.21	8	15
Fe	3,187.16	1	5
W	3,187.13	20	—
In	3,187.03	12	—
Sm	3,187.01	8	15
Th	3,187.002	8	10
Eu	3,187.00	2	6
Os	3,186.98	15	100
V	3,186.86	10	1
Fe	3,186.74	300	20
V	3,185.40	400R	500R
Fe	3,184.90	150	200
Nb	3,184.22	150	5
V	3,183.98	400R	500R
V	3,183.41	100R	200R
Cr	3,183.32	150	6
Fe	3,180.75	100	100
Cr	3,180.70	150	30
Nb	3,180.29	200	5
Fe	3,180.23	300	300
Ca	3,179.33	400w	100
Fe	*3,177.53*	*300*	*5*
Ru	*3,177.05*	*200*	*60*
Fe	*3,175.45*	*200*	*200*
Sn	*3,175.02*	*400hr*	*500h*
Tm	*3,172.82*	*200*	*200*

	λ	I	
		J	O
W	3,187.76	7	8
Sm	3,187.77	8	25
Er	3,187.78	12	25
Ag	3,187.83	8h	1
Rh	3,187.88	4	2
Rn	3,187.9	(10)	—
Cr	3,188.01	60h	150h
Fe	3,188.026	10	4
Tb	3,188.03	8	15
Pt	3,188.08	5	1
Mo	3,188.09	3	5
Th	3,188.19	4d	4d
Ru	3,188.338	50	60
U	3,188.339	10	10
Te	3,188.37	(10)	—
Co	3,188.373	2h	100
Mo	3,188.40	3	5
Ta	3,188.46	1	2
V	3,188.51	100R	35
Fe	3,188.57	100	150
Fe	3,188.82	100	150
Nb	3,189.28	300r	10w
V	3,190.68	150R	50
Ti	3,190.87	200r	40
Hg	3,191.03	(100)	—
Nb	3,191.10	300w	100w
Nb	3,191.43	200	3
Fe	3,191.66	150	200
J	3,193.95	(100)	—
Nb	3,194.27	150w	2w
Nb	3,194.98	300	30
Ru	3,195.15	100	—
*N**			

He 3,203.14 J (100)

	λ	I J	I O
Nb	3,203.14	5wh	—
W	3,203.054	8	9
Cl	3,203.05	(20)	—
Fe	3,202.951	2	3
Tb	3,202.95	3	8
Ar	3,202.85	(5)	—
Os	3,202.83	10	40
F	3,202.74	(200)	—
U	3,202.73	4h	4
Tb	3,202.70	3	8
Fe	3,202.65	1	2
Dy	3,202.56	1	4
Fe	3,202.560	20	40
Kr	3,202.54	(15h)	—
Ti	3,202.538	200	25
Th	3,202.52	4	5
Cr	3,202.51	15	—
U	3,202.47	4h	4
V	3,202.38	20r	100r
Cr	3,201.260	50	1
Ru	3,201.258	100	2
Fe	3,200.47	150	150
Kr	3,200.40	(50h)	—
Ar	3,200.39	(100)	—
Ti	3,199.91	150	200
Fe	3,199.52	200	300
Ta	3,199.22	70	—
Ta	3,198.94	70	—
Rb	3,198.77	(60)	—
Nb	3,198.22	50	1
Lu	3,198.12	80	40
Fe	3,196.93	300	500
Fe	3,196.08	150	10
Mo	3,195.96	50	12
Yt	3,195.61	50	30
Ru	3,195.15	100	—
Nb	3,194.98	300	30
Fe	3,194.42	70	100
Nb	*3,194.27*	*150w*	*2w*
*P**			
Ca	3,203.22	3	1
U	3,203.223	6	8
Th	3,203.23	6	10
Yt	3,203.32	50	30
Os	3,203.327	5	5
Ho	3,203.33	6	—
W	3,203.34	20	4
Nb	3,203.35	50	8
Er	3,203.36	12	20
U	3,203.41	6	8
W	3,203.42	12	5
Ti	3,203.43	15	10
Nd	3,203.46	4	10
Cr	3,203.51	12	—
Ar	3,203.66	(10)	—
Hf	3,203.67	10	5
U	3,203.726	4	8
Ta	3,203.735	2h	3
Ti	3,203.83	6	40
Si	3,203.87	3	—
Th	3,203.88	6	10
Er	3,203.95	1	6
Pt	3,204.04	100	250
Nb	3,204.97	150	10
Fe	3,205.40	200	300
Nb	3,206.34	300	—
Fe	3,207.09	50	80
V	3,208.35	100	10
Fe	3,208.47	80	100
Nb	3,208.58	100w	3w
Mo	3,208.83	60	150r
Cr	3,209.18	125	40
Fe	3,209.30	125	200
Fe	3,210.24	100	150
Pd	3,210.448	60h	—
Fe	3,210.451	50	5
Tm	3,210.57	50	40
Fe	3,210.83	100	150
Fe	3,211.68	50	80
Mn	*3,212.88*	*100*	*100*
Fe	*3,213.31*	*300*	*50*
Fe	*3,214.04*	*200*	*400*
V	*3,214.75*	*100*	*20*
Nb	*3,215.59*	*200*	*50*

	λ	I J	I O		λ	I J	I O
				Fe	*3,215.94*	*150*	*300*
				Cr	*3,216.56*	*125*	*3*
				Nb	*3,217.02*	*200w*	*2w*
				Ti	*3,217.06*	*150*	*40*
				Fe	*3,217.38*	*125*	*200*
				Ti	*3,218.27*	*150*	*15*
				Te	*3,218.44*	*(100)*	*—*
				P	*3,219.30*	*(100w)*	*—*
				Fe	*3,219.58*	*125*	*200*

He 3,888.65 *J* (1000)

	λ	J	O		λ	J	O
U	**3,888.61**	**3**	**4**	**Cs**	**3,888.65**	**10**	**150**
Br	**3,888.52**	**(10)**	**—**	**Eu**	**3,888.68**	**5***w*	**8***w*
Fe	**3,888.517**	**3***w*	**4***w*	**Fe**	**3,888.82**	**15**	**40**
Ce	**3,888.39**	**4**	**15**	**Mo**	**3,888.87**	**8**	**8**
Rh	**3,888.335**	**2**	**5**	**W**	**3,888.89**	**4**	**3**
V	**3,888.329**	**1**	**10**	**Ho**	**3,888.95**	**20**	**40**
Pr	**3,888.29**	**4**	**5**	**Ce**	**3,888.997**	**3**	**12**
Bi	**3,888.23**	**2**	**40**	**Zr**	**3,889.004**	**2***h*	**2***h*
U	**3,888.21**	**6**	**12**	**Er**	**3,889.02**	**1**	**9**
Mo	**3,888.18**	**8**	**10**	**H**	**3,889.05**	**(60)**	**—**
Er	**3,888.09**	**1**	**18**	**Sm**	**3,889.15**	**4**	**4**
V	**3,888.08**	**4**	**15**	**Sm**	**3,889.216**	**7**	**5**
Mo	**3,887.96**	**6**	**3**	**Nd**	**3,889.218**	**15**	**10**
W	**3,887.94**	**10**	**—**	**Hf**	**3,889.232**	**1***h*	**4**
Bi	**3,887.93**	**2**	**5**	**V**	**3,889.235**	**2**	**12**
Nd	**3,887.87**	**20**	**25**	**Mo**	**3,889.28**	**3**	**2**
Yt	**3,887.78**	**3**	**3**	**Ce**	**3,889.30**	**1**	**6**
Ru	**3,887.77**	**8**	**15**	**Ba**	**3,889.32**	**2**	**10**
U	**3,887.70**	**10**	**20**	**Hf**	**3,889.328**	**1**	**5**
Mo	**3,887.672**	**3**	**3**	**Pr**	**3,889.330**	**70**	**150**
Tb	**3,887.67**	**3**	**25**	**W**	**3,889.37**	**5**	**—**
Dy	**3,887.54**	**2**	**12**	**U**	**3,889.41**	**5**	**1**
Kr	**3,887.54**	(5*wh*)	**—**	**Pr**	**3,889.42**	**6**	**10**
Tm	**3,887.35**	**8**	**80**	**Mn**	**3,889.45**	**50**	**25**
Nb	**3,887.32**	**5**	**—**	**Ta**	**3,889.46**	**40***W*	**—**
Yb	**3,887.31**	**40**	**6**	**Au**	**3,889.47**	**8**	**3**
U	**3,887.20**	**3**	**—**	**Eu**	**3,889.51**	**5***W*	**10***W*
Gd	**3,887.18**	**10**	**10**	**Ir**	**3,889.58**	**3**	**2**
Er	**3,887.16**	**1**	**2**	**Nb**	**3,889.63**	**5**	**3**
Tl	**3,887.15**	**(30)**	**—**	**Nd**	**3,889.66**	**25**	**30**
				Ni	**3,889.67**	**10***h*	**30**
Fe	3,886.28	400	600	**In**	**3,889.78**	**(100)**	**—**
				Er	**3,889.80**	**1**	**15**
				Tb	**3,889.85**	**3**	**10**
				Nd	**3,889.93**	**20**	**30**

	λ	I J	I O		λ	I J	I O
				Mo	3,889.953	4	2
				Ti	3,889.954	5	25
				Pr	3,889.97	3	5
				Ce	3,889.99	8	50
				U	3,890.01	3	1
				Cu	3,890.073	3	—
				Sm	3,890.074	10	10
				Mo	3,902.96	500*R*	1000*R*

He 4,685.75 *J* (300)

	λ	I J	I O		λ	I J	I O
N	**4,685.74**	**(10)**	—	**Mo**	**4,685.81**	**12**	**12**
U	**4,685.72**	**18**	**10**	**Ho**	**4,685.83**	**2**	**3**
U	**4,685.533**	**3**	**1**	**Nb**	**4,685.93**	**2**	**2**
Nb	**4,685.527**	**1**	**2**	**Mo**	**3,686.09**	**6**	**8**
Se	**4,685.45**	**(12)**	—	**Cr**	**4,686.19**	**1**	**20**
Hg	**4,685.30**	**(5)**	—	**Ni**	**4,686.22**	**1**	**200**
Ta	**4,685.27**	**2**	**80**	**Pr**	**4,686.28**	**1**	**4**
Ca	**4,685.26**	**1**	**25**	**Kr**	**4,686.30**	**(8*wh*)**	—
In	**4,685.22**	**(100)**	—	**Gd**	**4,686.40**	**5*h***	**10**
Zr	**4,685.19**	**3**	**2**	**Th**	**4,686.59**	**1**	**2**
Nb	**4,685.13**	**20**	**15**	**V**	**4,686.920**	**12**	**15**
In	**4,685.04**	**(10)**	—	**Ti**	**4,686.921**	**2**	**8**
Pr	**4,684.94**	**10*w***	**125**	**Te**	**4,686.95**	**(300)**	—
In	**4,684.93**	**(15)**	—	**Kr**	**4,687.28**	**(10*h*)**	—
Pb	**4,684.90**	**(5)**	—				
Ta	**4,684.87**	**2**	**100**	La	4,692.50	300	200
In	**4,684.76**	**(25*h*)**	—	Kr	4,694.44	(300*h*)	—
U	**4,684.64**	**8**	**8**	Xe	4,697.02	(300)	—
In	**4,684.59**	**(25)**	—	Xe	4,698.01	(150*h*)	—
Ti	**3,684.48**	**1**	**7**	La	4,699.63	200*r*	200*r*
In	**4,684.449**	**(20)**	—				
V	**4,684.447**	**7**	**8**	*Ar*	*4,702.32*	*(1200)*	—
Mo	**4,684.34**	**4**	**5**	*La*	*4,703.28*	*300r*	*200r*
In	**4,684.22**	**(35)**	—	*Ne*	*4,704.39*	*(1500)*	—
Ra	**4,682.28**	**(800)**	—	*O*	*4,705.32*	*(300)*	—
				Ne	*4,708.85*	*(1200)*	—
In	4,682.00	250*w*	—	*Ne*	*4,710.06*	*(1000)*	—
In	4,681.11	(200)	—	*Ne*	*4,712.06*	*(1000)*	—
Rn	4,680.83	(500)	—				
Kr	4,680.41	(500)	—				
Zn	4,680.14	200*h*	300*w*				
Ne	4,679.13	(150)	—				
Br	4,678.69	(200)	—				
Ne	4,678.22	(300)	—				
Cd	4,678.16	200*W*	200*W*				

	λ	I J	I O		λ	I J	I O
La	4,671.83	150	100				
Xe	4,671.23	(2000)	—				
Sc	*4,670.40*	*300wh*	*100*				
La	*4,668.91*	*300r*	*200r*				
Te	*4,664.34*	*(800)*	—				
Kr	*4,658.87*	*(2000)*	—				
Ne	*4,656.39*	*(300)*	—				
La	*4,655.50*	*300*	*150*				

He 5,875.62 *J* (1000)

	λ	I J	I O		λ	I J	I O
Nb	**5,875.26**	**3*wh***	**5**	**He**	**5,875.87**	**(10)**	—
J	**5,875.13**	**(13)**	—	**Nb**	**5,876.31**	**1**	**10**
Xe	**5,875.02**	**(100)**	—	**Pb**	**5,876.70**	**(40)**	—
Nb	**5,874.700**	**5**	**30**	**Rn**	**5,877.56**	**(10)**	—
Yb	**5,874.70**	**30*h***	**1**	**Nb**	**5,877.79**	**5**	**5**
Te	**5,874.60**	**(3*s*)**	—				
La	**5,874.00**	**6**	**2**	Ne	5,881.89	(1000)	—
Fe	**5,873.22**	**2**	**8**	Na	5,889.95	1000*R*	9000*R*
				Na	5,895.92	500*R*	5000*R*
Kr	5,870.92	(3000)	—				
Ne	5,852.49	(2000)	—	*In*	*5,903.63*	*(500)*	—

Hf 72 178.49

t_0 1,700°C t_1 >3,200°C

I.	II.	III.	IV.	V.
~5.5	—	—	—	—

λ	I		eV
	J	O	
II 2,513.03	70	25	—
II 2,516.88	100	35	—
II 2,641.41	125	40	—
II 2,820.22	100	40	—
I 2,916.48	15	50	—
I 2,940.77	12	60	—
I 3,072.88	18	80	4.04
II 3,134.72	125	80	4.33
II 5,040.82	150	100	3.94
II 5,298.06	100	80	4.23
II 5,311.60	150	100	4.12
II 6,644.60	200	100	3.64

Hf 2,513.03 J 70 O 25

	λ	I J	I O
W	**2,512.93**	**12**	**10**
Os	**2,512.87**	**3**	**40**
Ru	**2,512.81**	**2**	**80**
Th	**2,512.74**	**15*d***	—
Hf	**2,512.69**	**50**	**25**
Fe	2,512.52	40	—
Yb	2,512.06	50	10
Ta	2,512.04	800	5
C	2,512.03	400	—
Ru	2,511.99	40	20
Nb	2,511.97	50*h*	—
Fe	2,511.76	100	25
C	2,511.71	60	—
Ta	2,511.69	100*d*	2*d*
He	2,511.22	(50)	—
Nb	2,511.00	100	3
Ni	2,510.87	250*h*	50*h*
Fe	2,510.83	50	300*R*
Rh	2,510.65	200*wh*	—
Fe	2,509.12	50	1
C	*2,509.01*	*200*	
Ru	*2,507.01*	*80*	*60*
Si	*2,506.90*	*200*	*300*
Co	*2,506.46*	*200h*	*50w*
Cu	*2,506.27*	*500r*	—
V	*2,576.22*	*150*	*10*
Fe	*2,506.09*	*70*	*2*
Rh	*2,505.10*	*200*	*2*
Ta	**2,513.10**	**1**	**10**
Al	**2,513.15**	**(15)**	—
J	**2,513.21**	**(20)**	—
Os	**2,513.25**	**8**	**50**
Ru	**2,513.32**	**80**	**50**
U	**2,513.329**	**20**	**1**
Fe	**2,513.330**	**4*wh***	**6**
Rh	**2,513.36**	**6**	**8**
Pt	2,513.88	50	6
Si	2,514.33	200	300
B	2,514.39	50	—
Pd	2,514.48	200	—
Hf	2,515.16	40	1
Hf	2,515.48	30	20
V	2,516.118	100	25
Si	2,516.123	500	500
*N**			

Hf 2,516.88 J [100] O 35*

	λ	I J	I O
Yb	**2,516.83**	**20**	**2**
Ar	**2,516.81**	**(20)**	—
Mn	**2,516.73**	**12**	—
Ru	**2,516.70**	**8**	**6**
Cr	**2,516.92**	**2**	**35**
U	**2,516.97**	**2**	**4**
W	**2,517.01**	**4**	—
Ca	**2,517.03**	**6**	—

	λ	I J	I O
Cr	**2,516.65**	**6*h***	—
W	**2,516.58**	**3**	**12**
Se	**2,216.57**	**(15)**	—
Fe	**2,516.570**	**1**	**10**
P*			

	λ	I J	I O
Fe	**2,517.120**	**60**	**10**
U	**2,517.122**	**2**	**8**
Ti	**2,517.124**	**2**	**3**
V	**2,517.14**	**25**	**35*r***
Fe	**2,517.21**	**3**	**1**
W	**2,517.24**	**4**	—
Cd	**2,517.25**	**(3)**	—
Ru	2,517.32	80	60
O	2,517.41	(50)	—
Rh	2,517.52	150	—
Fe	2,518.10	50	200*R*
Ru	2,518.40	50	3
Fe	2,519.05	70	—
Si	2,519.207	300	300
Ru	2,519.208	80	20
La	2,519.215	50	—
V	2,519.62	50	125*r*
Tm	2,519.80	50	1
Co	2,519.82	200	40
Rh	2,520.53	1000*wh*	10
Co	*2,521.36*	*150*	*75R*
Nb	*2,521.40*	*100*	*3*
V	*2,523.95*	*100*	*10*
Si	*2,524.12*	*400*	*400*

Hf 2,641.41 *J* 125 *O* 40

	λ	I J	I O
J	**2,641.39**	**(40)**	—
Ba	**2,641.375**	**5**	**5**
Cr	**2,641.368**	**2**	—
Eu	**2,641.26**	**60**	**100*w***
Os	**2,641.174**	—	**10**
Re	**2,641.170**	—	**20**
Mo	**2,641.15**	**20**	—
Fe	**2,641.126**	**20**	—
U	**2,641.126**	**2*h***	**3**
Ti	**2,641.10**	**20**	**150**
W	**2,641.08**	**8**	—
Nb	**2,641.06**	**30**	**2**
Re	**2,641.02**	—	**5**
Cs	**2,641.00**	**(2)**	—
Mo	**2,640.99**	**40*wh***	**40*h***

	λ	I J	I O
C	**2,641.44**	**20**	—
Ce	**2,641.456**	—	**8**
Ru	**2,641.463**	—	**12**
Au	**2,641.49**	**20**	**5**
Ce	**2,641.493**	—	**5**
Th	**2,641.494**	**10**	**15**
O	**2,641.53**	**(10*h*)**	—
U	**2,641.55**	**5**	**10**
Ru	**2,641.62**	**100**	—
Fe	**2,641.649**	**60**	**100**
Pd	**2,641.650**	**2**	—
Rh	**2,641.65**	**10**	**1**
Cr	**2,641.80**	**8**	—
Ru	2,641.98	—	20

	λ	I J	I O
Tm	2,640.77	8	40
Ru	2,640.33	5	60
Pd	2,640.18	70	—
Ru	2,639.87	—	50
Mn	2,639.83	80h	12
Ir	2,639.71	15	100h
Fe	2,639.55	100	1
Cd	2,639.50	15	75
Cr	2,639.43	—	30
Pt	2,639.35	50	500
Rh	2,639.25	100	2
Ru	2,639.12	5	60
Zr	2,639.09	15	20
Eu	2,638.764	200	300
Mo	2,638.758	125	30
Rh	2,638.74	100	2
Hf	2,638.71	100	40
Ti	2,638.70	100wh	—
Ru	2,638.51	4	60
Mo	2,638.30	5	25
Mn	2,638.17	80h	25
Yb	2,638.09	60	1
Fe	2,637.64	200	2
V	2,637.22	15	40h
Os	2,637.13	30	150
Pd	2,637.07	100	—
Re	2,637.01	—	20
Hf	*2,637.00*	*1*	*10*
Ta	*2,636.90*	*3*	*100*
Ta	*2,636.673*	*1*	*70*
Ru	*2,636.670*	—	*60*
Mo	*2,636.670*	*150*	*25*
Re	*2,636.64*	—	*125*
Fe	*2,636.48*	*20*	*50*
Pd	*2,635.94*	*300*	*50*
Ta	*2,635.93*	—	*50h*
Fe	*2,635.81*	*200*	*300*
Hf	*2,635.79*	*20*	*12*
Te	*2,635.55*	*(350)*	—
Dy	*2,634.81*	*20*	*40*
Re	*2,633.62*	—	*40*
Ru	*2,633.46*	*1h*	*50*
Nb	*2.633.26*	*200*	—
Ni	*2,632.89*	*2000wh*	—

	λ	I J	I O
Hf	2,642.08	—	5
Cr	2,642.12	3	35
Ti	2,642.15	150wh	—
Pd	2,642.17	100	—
Nb	2,642.24	300	5
Yb	2,642.55	80	3
Hf	2,642.75	1	15
Re	2,642.76	—	125
Ru	2,642.79	150	—
Ru	2,642.96	—	150
V	2,643.16	10	25
Ra	2,643.73	(125)	—
Mo	2,643.81	—	20
Ta	2,643.89	—	50
Fe	2,644.00	150	150
Os	2,644.11	10	75
Ir	2,644.19	5	35
Ti	2,644.26	12	100
Mo	2,644.353	60	30
V	2,644.355	100h	12
Ta	2,644.60	50W	20W
Ru	2,644.61	100	8
Ta	2,645.10	30h	80
Pr	2,645.37	5	40
Fe	2,645.43	10	50
U	2,645.47	25	20
Re	2,645.80	—	20
V	*2,645.84*	*100*	*15*
Ru	*2,646.02*	*150*	*20*
Ti	*2,646.11*	*200wh*	—
Ta	*2,646.22*	*2*	*50*
Nb	*2,646.26*	*200wh*	*8*
Ta	*2,646.37*	*2*	*125*
Ta	*2,646.77*	*50*	*50h*
Pt	*2,646.89*	*100*	*1000h*
Ni	*2,647.06*	*500wh*	—
Re	*2,647.12*	—	*100*
Hf	*2,647.29*	*125*	*40*
Ru	*2,647.31*	*5*	*50*
Ne	*2,647.42*	*(150)*	—
Ta	*2,647.47*	*10*	*200*
Fe	*2,647.56*	*70*	*100*
V	*2,647.71*	*10h*	*50*
Ru	*2,648.78*	*150*	*30*
Re	*2,649.05*	—	*100*
Pd	*2,649.47*	*200*	—
Co	*2,649.94*	*5*	*50w*

Hf 2,820.224 *J* 100 *O* 40

	λ	*I*	
		J	*O*
Fe	2,820.22	3	4
Er	2,820.19	3	18
Os	2,820.18	5	12
Se	2,820.10	(15)	—
Xe	2,820.06	(5*h*)	—
Co	2,820.01	—	50
Mo	2,820.003	8	—
Hg	2,820.0	20*h*	10*h*
Ti	2,820.00	70*wh*	—
Re	2,819.955	—	150*W*
Au	2,819.95	150	—
Fe	2,819.90	3*h*	—
Nb	2,819.894	10*wh*	—
Tb	2,819.89	20	3
Cd	2,819.89	(3)	—
U	2,819.83	25	6
Er	2,819.81	2	6
Rn	2,819.8	(12)	—
Ho	2,819.74	10*h*	—
Hf	2,819.738	1	20
La	2,819.73	2	—
Mn	2,819.728	—	10
Nd	2,819.64	—	5*h*
Ta	2,819.37	5	100
Hf	2,818.94	—	15
Mn	2,818.92	—	20
Ru	2,818.81	—	30
Mh	2,818.77	—	25
Yb	2,818.75	80	—
Co	2,818.60	—	30
Tm	2,818.48	20	30
Ru	2,818.361	12	50
Cr	2,818.359	80	8
Mo	2,818.30	1	25
Pt	2,818.25	4	70
Mn	2,817.97	1	50
Ti	2,817.87	200	10
Hf	2,817.67	1	18
Fe	2,817.51	60	100
Ta	2,817.503	10	80
V	2,817.500	50	18
Ti	2,817.40	—	20
Ta	2,817.10	100	80*d*
Ru	2,817.09	4	50
Re	2,816.96	—	30
Nb	2,816.68	50	2

	λ	*I*	
		J	*O*
U	2,820.266	10	6
Cs	2,820.268	(2)	—
Ce	2,820.32	—	2
Th	2,820.34	6	10
Ti	2,820.36	15	8
Hf	2,820.42	5	10
U	2,820.51	2	2
Os	2,820.55	5	10
Al	2,820.632	(3)	—
Ir	2,830.632	—	4
Mo	2,820.633	—	8
J	2,820.65	(12)	—
Te	2,820.75	(10)	—
Ce	2,820.74	—	2
Eu	2,820.77	200*W*	200*W*
Fe	2,820.81	15	20
Cr	2,820.82	1	20
U	2,821.12	35	20
Ru	2,821.18	—	30
Lu	2,821.23	50*h*	2
Os	2,821.25	6	20
Ni	2,821.29	125	125
Ti	2,821.42	70*wh*	—
Ru	2,821.424	100	5
Mn	2,821.45	—	40
Co	2,821.74	10*h*	30*h*
Cr	2,822.01	80	10
Ru	2,822.03	5	50
Re	2,822.12	—	20
Se	2,822.13	20	50
Pt	2,822.27	60*h*	10
Cr	2,822.37	100	20
V	2,822.44	70*h*	4
Ru	2,822.55	150	30
Hf	2,822.68	90	30
Au	2,822.72	80	—
Mo	2,822.86	6	20
Ru	2,823.18	80	20
Pb	2,823.189	40	150*R*
Re	2,823.193	—	25
Fe	2,823.28	300	200
Rh	2,823.37	—	25
Os	2,824.17	4	20
Ho	2,824.19	3	20
Re	2,824.25	—	20

	λ	I	
		J	O
Kr	2,816.46	(60)	—
Re	2,816.33	—	40
Eu	2,816.18	50	50w
Mo	2,816.15	300h	200
Hf	2,816.07	6	10
Mo	2,815.91	—	20
Hf	2,815.81	—	10
Os	2,815.78	4	40
Co	2,815.56	—	50r
Ag	2,815.54	80wh	3
Fe	2,815.51	25	40
Ta	2,815.12	4	100
Mn	2,815.02	75	25
Ta	2,815.01	15	150
Co	2,814.98	—	25
Ce	2,814.95	—	20
Hg	2,814.93	(200)	—
Zr	2,814.90	1	70
Ru	2,814.87	—	30
Ce	2,814.81	—	40d
Ta	2,814.80	5	125
Hf	2,814.76	35	15
Re	*2,814.68*	—	*50*
Hf	*2,814.47*	*40*	*25*
Ta	*2,814.31*	*50*	*50r*
Os	*2,814.20*	*25*	*50*
Eu	*2,813.95*	*300wh*	*300w*
Hf	*2,813.86*	*30*	*25*
Ra	*2,813.76*	*(400)*	—
Ru	*2,813.71*	*125*	*50*
Sn	*2,813.58*	*50*	*50*
Fe	*2,813.29*	*400*	*400*
Eu	*2,811.75*	—	*50*
Ta	*2,811.72*	*150*	*1*
Co	*2,811.52*	—	*50w*
Co	*2,811.13*	—	*50*
Ta	*2,810.92*	*40W*	*200W*
Nb	*2,810.81*	*100*	—
Ru	*2,810.55*	*200*	*50*
Ti	*2,810.30*	*150*	*6*
V	*2,810.27*	*50*	*50*
Fe	*2,810.26*	*15*	*40*
Ru	*2,810.03*	*12*	*50*
bhB	*2,809.9*	—	*60*
Fe	*2,809.81*	*100h*	*1*
Gd	*2,809.72*	*80*	*60*
U	2,824.28	30	25
Cu	2,824.369	300	1000
Ag	2,824.370	200w	150wh
Ir	2,824.45	15	20
Ta	2,824.81	5h	60w
Ru	2,825.06	60	—
Co	2,825.15	—	75w
Ni	2,825.236	125	—
Co	2,825.242	200	6
Re	2,825.461	—	20
Ru	2,825.463	80	—
Zr	2,825.558	30	30
Fe	2,825.560	150	150
Mo	2,825.67	1	25
Fe	2,825.69	60	70
Tl	*2,826.16*	*100R*	*200R*
Ta	*2,826.18*	*5*	*60*
Ru	*2,826.22*	*80*	—
Mo	*2,826.54*	*5*	*40*
Rh	*2,826.67*	*50d*	*100*
Ru	*2,826.68*	*100*	—
Cr	*2,826.75*	*3*	*70*
Co	*2,826.80*	—	*50W*
Ta	*2,827.18*	*10*	*200*
Rh	*2,827.31*	—	*50*
Re	*2,827.53*	—	*30*
Ta	*2,827.55*	*100d*	*3d*
Fe	*3,827.89*	*50*	*70*
Tm	*2,827.92*	*100*	*50*
Hf	*2,828.149*	*1*	*3*
Ti	*2,828.150*	*200h*	*2*
Ta	*2,828.58*	*100*	*75*
Eu	*2,828.69*	*150*	*200W*
Mn	*2,828.76*	—	*50wh*
Fe	*2,828.81*	*60*	*100*
Ti	*2,828.9*	*150wh*	—
Ru	*2,829.16*	*8*	*50*
Os	*2,829.27*	*6*	*40*
Hf	*2,829.32*	*30*	*15*
Pt	*2,830.29*	*600r*	*1000R*
Mn	*2,830.79*	—	*50*

	λ	I J	I O
Bi	2,809.62	100	200 W
Hf	2,809.61	2	3

Hf 2,916.48 J 15 O 50

	λ	J	O
U	2,916.46	10	8
Yb	2,916.43	10	1
W	2,916.38	4	4
Ir	2,916.36	2	25
Ta	2,916.32	1	5
U	2,916.30	2h	5h
Tb	2,916.26	10	5
Ru	2,916.255	25	100
Zr	2,916.250	—	6
Cr	2,916.16	15	18
Fe	2,916.15	6	2h
W	2,916.11	4	5
Mo	2,916.10	—	20
Ta	2,916.04	—	2h
V	2,916.02	—	9
Zr	2,915.99	20	25
V	2,915.86	30	1
Ta	2,915.49	40	150
Rh	2,915.42	40	80
Ta	2,915.34	50	150w
V	2,914.93	50r	60
Mn	2,914.60	—	150wh
Fe	2,914.31	25	50
Ru	2,914.30	—	50
Ta	2,914.12	30	200
Os	2,913.84	8	30
Cr	2,913.73	5	60r
Sn	2,913.542	125wh	100wh
Pt	2,913.542	25	300
Pt	2,913.25	—	25
Ru	2,913.17	3	50
Rh	2,912.62	20	50
Ru	2,912.43	3	30
Os	2,912.33	50	50
Pt	2,912.26	25	300
Fe	2,912.16	150	150
Ti	2,912.08	15	35
Mo	2,911.91	50h	30
Er	2,911.42	15	30
Lu	2,911.39	300	100
Nb	2,916.30	3	3
Tm	2,916.52	40	15
Zr	2,916.63	4	4
Ce	2,916.68	—	10
U	2,916.71	8	6
Nd	2,916.73	—	2
W	2,916.79	6	1
Ta	2,916.86	3h	2
Ce	2,917.03	—	2
Nb	2,917.05	100r	10w
U	2,917.08	3	2
Os	2,917.26	20	40
Th	2,917.39	6	25d
Hf	2,917.49	8h	10
Ru	2,917.77	2	60
Fe	2,918.03	100	125
Tm	2,918.27	50	25
Tl	2,918.32	200R	400R
Fe	2,918.36	25	40
Hf	2,918.58	8	30
Ce	2,918.66	—	30s
Pt	2,919.34	40	150
Re	2,919.41	—	25
Co	2,919.55	—	30
Hf	2,919.59	80	40
Ru	2,919.61	12	80
Os	2,919.79	15	100
Fe	2,919.85	36	80
Ru	2,920.26	—	30
Fe	2,920.69	80	150
Ru	2,920.96	30	30
Pt	2,921.38	6	100
Tl	2,921.52	100R	200R
Pd	2,922.49	25	200
Fe	2,922.62	25	50
Fe	2,923.29	35	50
V	2,923.62	150r	50r
Fe	2,923.85	70	100

	λ	I J	I O
Cr	2,911.14	8	40
V	2,911.06	200r	30
Cr	2,910.90	8	60r
V	2,910.39	150r	35
Rh	*2,910.17*	*12*	*50*
Hf	*2,909.91*	*20*	*30*
Fe	*2,909.50*	*35*	*70*
Os	*2,909.06*	*400*	*500R*
Cr	*2,909.05*	*12*	*60r*
Ta	*2,908.91*	*10*	*150*
Fe	*2,908.86*	*40*	*80*
V	*2,908.82*	*400R*	*70r*
Fe	*2,907.52*	*80*	*100*
Mn	*2,907.22*	—	*50*
Rh	*2,907.21*	*30*	*100*
Eu	*2,906.68*	*300*	*300W*
Fe	*2,906.42*	*25*	*60*
Pt	*2,905.90*	*15*	*100*
Ru	*2,905.65*	*12*	*50*
Re	*2,905.58*	—	*50*
Cr	*2,905.49*	*8*	*60r*
Ta	*2,905.24*	*100*	*80*
Hf	*2,904.75*	*6*	*30*
Ta	*2,904.07*	*40*	*300w*

	λ	I J	I O
Rh	*2,924.024*	—	*100*
V	*2,924.025*	*300R*	*70r*
Hf	*2,924.61*	*2*	*25*
V	*2,924.64*	*200r*	*60*
Eu	*2,925.03*	*100*	*150*
Ta	*2,925.19*	*5*	*100*
Ta	*2,925.26*	*40ws*	*100W*
Fe	*2,925.36*	*50*	*70*
Hg	*2,925.41*	*50*	*60*
Mn	*2,925.57*	*1Wh*	*150Wh*
Ta	*2,925.66*	*4*	*100*
Hf	*2,926.32*	*2*	*10*
Ta	*2,926.46*	*10*	*100*
Fe	*2,926.59*	*400*	*150*
Tm	*2,926.75*	*60*	*80*
Re	*2,927.40*	—	*125w*
Ru	*2,927.54*	*200*	*50*
Co	*2,927.67*	*1*	*50*
Nb	*2,927.81*	*800r*	*200*
Co	*2,928.81*	*1h*	*50*
Fe	*2,929.008*	*100*	*150*
Hf	*2,929.009*	—	*15*

Hf 2,940.77 *J* 12* *O* 60

	λ	I J	I O
Ce	**2,940.66**	—	**4**
Fe	**2,940.59**	**80**	**200**
Ir	**2,940.54**	**10**	**15**
Yb	**2,940.51**	**25**	**3**
Mn	**2,940.48**	—	**40Wh**
Eu	**2,940.46**	—	**15**
Ir	**2,940.43**	—	**5**
Mn	**2,940.39**	—	**40Wh**
U	**2,940.37**	**12**	**6**
Ru	**2,940.36**	**3**	**50**
Cr	**2,940.34**	—	**8hs**
Ta	**2,940.21**	**50**	**150**
W	**2,940.20**	**18**	**4**
Mo	**2,940.10**	**40**	**2**
Ta	2,940.06	40w	100
Ru	2,939.94	3	30
Mn	2,939.30	—	50
Ta	2,939.28	40h	200

	λ	I J	I O
Ce	**2,940.78**	—	**15**
Eu	**2,940.82**	—	**10**
Ce	**2,940.88**	—	**5**
W	**2,940.95**	**3**	**5**
Mo	**2,940.98**	**1**	**10**
Mn	**2,940.04**	**1**	**25**
Dy	**2,941.06**	**1**	**2**
Ir	**2,941.08**	**5**	**20**
Pt	**2,941.09**	—	**2**
V	**2,941.11**	—	**2**
Er	**2,941.18**	**1**	**6**
Ce	**2,941.20**	—	**5**
Mo	**2,941.22**	**40**	**2**
W	**2,941.24**	**4**	**8**
Fe	**2,941.343**	**300**	**600**
U	**2,941.343**	**12**	**8**
V	**2,941.37**	**300r**	**40**
V	**2,941.49**	**150r**	**12**

	λ	I	
		J	O
Fe	2,939.08	20	80
Ag	2,938.55	200*wh*	200
Ta	2,938.43	3	50
Bi	2,938.30	300*w*	300*w*
Fe	2,937.81	150	300
Hf	2,937.79	100	50
Ti	2,937.32	5	35
Fe	2,936.90	500*r*	700*r*
Ir	2,936.62	—	40
Tm	2,936.00	300	80
V	2,935.87	15*h*	30
Hf	2,935.37	—	5
*bh*B	2,934.9	—	100
Ta	2,934.85	4	40
Os	2,934.64	5	30
Ta	2,933.55	150	400
Mn	*2,933.06*	*15*	*80*
Ta	*2,932.69*	*80w*	*400*
In	*2,932.62*	*300*	*500*
Rh	*2,931.94*	*20*	*80*
Hf	*2,929.90*	*3*	*5*
Pt	*2,929.79*	*200w*	*800R*
Hf	*2,929.63*	*50*	*30*
Co	*2,929.51*	—	*75*
Rh	*2,929.11*	*10*	*100*
Hf	*2,929.01*	—	*15*
Fe	*2,929.01*	*100*	*150*
Nb	2,941.54	300	50
Ti	2,941.99	150	100
Ta	2,942.14	40	150
Ru	2,942.25	100	30
V	2,942.35	20*h*	80*r*
Os	2,942.85	8	30
Re	2,943.14	—	60
Ir	2,943.15	20	30
V	2,943.20	25*r*	30
Ru	2,943.481	—	30
Co	2,943.484	—	30
Ni	2,943.91	20	50*r*
Ru	2,943.92	5	50
W	2,944.39	20	30
Fe	2,944.40	600	70
V	2,944.57	300*r*	50
Hf	2,944.71	1	20
Tl	2,945.04	25	50
Fe	2,945.05	30	100
Ru	2,945.67	300	60
Ta	2,946.91	10	150
Ru	2,946.99	12	60
Hf	2,947.13	15	15
Fe	2,947.88	200	600*r*
Ti	*2,948.25*	*30*	*100*
Fe	*2,948.43*	*70*	*80*
Mn	*2,949.20*	*30*	*100*
Ru	*2,949.50*	*12*	*80*
Fe	*2,950.24*	*300*	*700*
Hf	*2,950.68*	*12*	*15*
Nb	*2,950.88*	*200*	*150*
Hf	*2,951.90*	—	*8*
Ta	*2,951.92*	*200*	*400w*
Ru	*2,952.50*	*2*	*60*
Eu	*2,952.68*	*150*	*200w*

Hf 3,072.88 *J* 18 *O* 80

U	**3,072.78**	**20**	**20**
W	**3,072.73**	**12**	**3**
V	**3,072.71**	**40*r***	**70*r***
Co	**3,072.66**	—	**20**
Sm	**3,072.65**	**2**	**2**
Tb	**3,072.60**	**15**	**3**
Gd	**3,072.57**	**25**	**20**
Er	**3,072.52**	**6**	**15**
Ce	**3,072.89**	**1**	**20**
Re	**3,072.96**	—	**40**
Ti	**3,072.97**	**200*r***	**35**
Tm	**3,073.085**	**150**	**60**
Nd	**3,073.085**	**2**	**4**
Mn	**3,073.13**	**20**	**75**
U	**3,073.17**	**3**	**6**
Ca	**3,073.18**	**2**	**1**

	λ	I	
		J	O
Nb	**3,072.51**	**15**	**2**
Re	**3,072.453**	**—**	**5**
U	**2,072.448**	**3**	**6**
Nb	**3,072.41**	**2**	**5**
Ce	**3,072.39**	**—**	**20**
Ta	**3,072.35**	**1*h***	**2*s***
Co	**3,072.344**	**100**	**200*R***
U	**3,072.336**	**6**	**6**
Ru	**3,072.335**	**40**	**5**
Fe	**3,072.335**	**7**	**8**
Yt	**3,072.334**	**8**	**8**
Nb	**3,072.30**	**1**	**2**
Nb	**3,072.18**	**5**	**1**
Zn	3,072.06	125	200
Co	3,071.96	2	80
Pt	3,071.94	15	60
Ba	3,071.59	50*R*	100*R*
Re	3,071.16	—	50
Mo	3,070.89	3	40
Hf	3,070.48	—	5
Mn	3,070.27	25	100
Re	3,069.945	—	125
Os	3,069.936	15	125
Hf	3,069.68	—	3
Ta	3,069.24	70	150
Hf	3,069.18	—	15
Ir	3,068.89	20	40
Gd	3,068.65	50	50
Ru	3,068.26	8	60
Fe	3,068.18	150	150
Bi	3,067.72	2000*wh*	3000*R*
Hf	3,067.41	10	30
Re	3,067.39	—	60
Rh	3,067.30	1	80
Fe	3,067.24	300	300
Ge	3,067.01	40	60
Fe	3,066.48	40	60
V	3,066.37	125*r*	400*r*
Os	3,066.12	10	50
Pd	3,066.10	2	150
Mn	3,066.02	—	75
Pt	*3,064.71*	*300R*	*2000R*
Hf	*3,064.68*	*30*	*10*
Ni	*3,064.62*	*50*	*200r*
Co	*3,064.37*	*—*	*100*
Mo	*3,064.28*	*10*	*80*
Nb	**3,073.236**	**15**	**2**
Fe	**3,073.24**	**—**	**2**
W	**3,073.280**	**10**	**12**
Ir	**3,073.281**	**—**	**10**
Ce	**3,073.336**	**—**	**10**
Ru	**3,073.336**	**5**	**50**
Er	**3,073.35**	**4**	**12**
Mo	**3,073.38**	**10**	**12**
Ta	**3,073.39**	**5**	**18**
Tm	**3,073.50***	**60**	**25**
U	**3,073.501**	**5**	**8**
Ru	**3,073.51**	**80**	**10**
Co	**3,073.52**	**2**	**60**
Ce	**3,073.538**	**—**	**2**
Dy	**3,073.542**	**8**	**30**
Cu	3,073.80	20	70
V	3,073.82	20*r*	60
Fe	3,073.98	25	40
Os	3,074.08	20	125
Hf	3,074.10	1	25
Fe	3,074.15	25	40
Mo	3,074.37	15	60
F	3,074.44	25	40
Hf	3,074.79	4	30
Os	3,074.96	20	125
Ti	3,075.22	300*R*	40
Hf	3,075.30	—	15
As	3,075.32	35	60
Fe	3,075.72	400	400
Zn	3,075.90	50	150
Tb	3,076.04	3	30
Hf	3,076.70	—	4
Ru	3,076.78	3	50
Hf	3,076.87	8	5
Os	3,077.06	12	100
Ta	3,077.24	50*w*	150*w*
Os	3,077.44	8	80
Lu	3,077.60	200	100
Fe	3,077.64	25	60
Os	3,077.72	30	100
Fe	3,078.02	80	100
Os	3,078.11	15	125
Ta	3,078.23	5	50
Os	3,078.38	15	125
Fe	3,078.43	50	80
Ti	3,078.64	500*R*	60
Co	3,079.40	2	80

	λ	I J	I O		λ	I J	I O
Hf	*3,063.78*	*1*	*25*	Os	3,079.56	10	40
Cu	*3,063.41*	*50*	*300*	Mn	3,079.63	40	125
Os	*3,062.19*	*30*	*100*	Mo	3,079.88	3	40
Mn	*3,062.12*	*20*	*75*	Ta	3,079.95	5*h*	50*w*
Co	*3,061.82*	*125*	*200R*	*Mo*	*3,080.41*	*6*	*60*
Mo	*3,061.59*	*3*	*50*	*Hf*	*3,080.66*	*100*	*30*
V	*3,060.46*	*100r*	*150r*	*Ni*	*3,080.75*	*60*	*200*
Os	*3,060.30*	*30*	*100*	*Cd*	*3,080.83*	*100*	*150*
Ta	*3,060.29*	*35W*	*125*	*Hf*	*3,080.84*	*5*	*25*
Co	*3,060.05*	*1*	*150*	*Lu*	*3,081.47*	*8*	*80*
Fe	*3,059.09*	*400*	*600r*	*Gd*	*3,082.00*	*60*	*100*
Os	*3,058.66*	*500*	*500R*	*V*	*3,082.11*	*2h*	*80r*
				Al	*3,082.15*	*800*	*800*
				Re	*3,082.43*	—	*100w*
				Hf	*3,083.65*	—	*6*
				Fe	*3,083.74*	*500*	*500*
				Rh	*3,083.96*	*2*	*150*
				Mo	*3,085.61*	*25*	*125*
				Co	*3,086.40*	*2*	*80*
				Co	*3,086.78*	—	*200R*

Hf 3,134.72 *J* 125 *O* 80

	λ	J	O		λ	J	O
Eu	**3,134.70**	**1**	**3*W***	**O**	**3,134.79**	**(100)**	—
U	**3,134.69**	**2**	**5**	**Cs**	**3,134.8**	**(4)**	—
Ti	**3,134.65**	—	**2**	**Ru**	**3,134.80**	**100**	**10**
Ce	**3,134.60**	—	**4**	**Nd**	**3,134.897**	**30**	**40**
Ca	**3,134.56**	**2**	**1**	**Rb**	**3,134.90**	**(10)**	—
Th	**3,134.43**	**12**	**10**	**Cr**	**3,134.92**	—	**10**
Se	**3,134.42**	**(70)**	—	**V**	**3,134.93**	**150*r***	**30**
Fe	**3,134.405**	**1*h***	**3*h***	**Mg**	**3,135.0**	**2*h***	**4**
Ho	**3,134.40**	**4*h***	**6**	**Ag**	**3,135.02**	**8*h***	—
Nb	**3,134.34**	**15**	**2**	**Gd**	**3,135.04**	**2**	**5**
O	**3,134.32**	**(10)**	—	**Re**	**3,135.068**	—	**15**
Cr	**3,134.31**	**50**	**3**	**Ti**	**3,135.069**	—	**2**
Tb	**3,134.26**	**8**	**8**	**Kr**	**3,135.10**	**(8)**	—
Sm	**3,134.19**	**1**	**6**	**Yt**	**3,135.17**	**18**	**10**
Fe	**3,134.111**	**125**	**200**	**Ce**	**3,135.18**	—	**15**
Ni	**3,134.108**	**150**	**1000*R***	**V**	**3,135.19**	—	**8**
Re	**3,134.02**	—	**30**	**Ir**	**3,135.23**	—	**2*h***
U	**3,133.92**	**5**	**6**	**Cr**	**3,135.34**	**25**	**1**
				Tb	**3,135.35**	**8**	**15**
				Pr	**3,135.351**	**2**	**10**
Tm	3,133.89	200	200	**Fe**	**3,135.36**	**100**	**1**
Hf	3,133.50	5	15	**Dy**	**3,135.37**	**50**	**100**
V	3,133.33	200*r*	50				

	λ	I J	I O
Ir	3,133.32	2*h*	40
Cd	3,133.17	300	200
Hf	3,133.10	3	5
Ru	3,132.88	5	60
Ta	3,132.64	25	250*w*
Mo	3,132.594	300*R*	1000*R*
V	3,132.594	20	80*r*
Fe	3,132.51	40	70
Co	3,132.22	—	40
Cr	3,132.06	125	25
Hg	3,131.83	100	200
Hf	3,131.81	10	40
Hg	3,131.55	300	400
Tm	3,131.26	500	400
Os	3,131.11	30	125
Be	3,131.07	150	200
Ti	3,130.80	100	25
Rh	3,130.790	2	60
Nb	3,130.786	100	100
Eu	3,130.74	100	100*W*
Ta	3,130.58	35	100*W*
Be	3,130.42	200	200
V	3,130.27	200*r*	50
Ta	3,129.95	8	50
Ru	3,129.84	4	60
Ru	3,129.60	1	50
Hf	3,129.58	—	15
Ta	3,129.55	7	50
Co	3,129.48	2	40
Na	3,129.37	(60)	35
Fe	3,129.33	60	100
Ni	3,129.31	—	125
Os	3,129.23	15	60
Re	3,128.95	—	100*W*
Hf	3,128.75	—	20
Cu	3,128.701	15	70
Cr	3,128.699	150	30
V	3,128.69	70	3
Ti	3,128.64	70*wh*	12
Dy	3,128.41	10	40
V	3,128.28	60	2
Ru	3,127.91	100	10
Ta	3,127.76	100	18*w*
Ce	3,127.53	—	40
Co	3,127.25	—	100
Hf	*3,126.29*	*5*	*18*
Ne	*3,126.19*	*(150)*	—

	λ	I J	I O
Nb	**3,135.40**	**10**	**2**
Fe	**3,135.45**	**3**	**10**
Rh	**3,135.47**	—	**5**
Ru	3,135.80	80	10
Ta	3,135.89	100	35
Fe	3,136.50	40	60
V	3,136.51	200	20
Ru	3,136.55	6	60
Co	3,136.73	—	60
Co	3,137.33	—	150*r*
Co	3,137.45	—	50
Hf	3,137.51	10	30
Rh	3,137.71	—	100
Co	3,137.75	2	60
Pb	3,137.83	100	—
V	3,138.06	70	—
Pt	3,139.39	80	300
Hf	3,139.65	20	25
V	3,139.74	150	15
Fe	3,139.91	40	70
Co	3,139.94	10	150*r*
Cu	3,140.312	12	50
Os	3,140.314	12	60
Fe	3,140.39	80	100
Ir	3,140.41	1	50*r*
Ru	3,140.48	—	50
Dy	3,140.64	20	40
Hf	3,140.76	25	25
Os	3,140.94	12	50
Ru	3,140.97	6	60
Se	3,141.13	(100)	—
Dy	3,141.13	20	50
V	3,141.48	100	5
Pt	3,141.66	5*h*	150
Fe	*3,142.45*	*100*	*125*
La	*3,142.76*	*50*	*150*
Pd	*3,142.81*	*100*	*300*
Fe	*3,142.88*	*70*	*80*
Ti	*3,143.76*	*125*	*18*
Fe	*3,143.99*	*150*	*200*
Fe	*3,144.49*	*100*	*150*
Hf	*3,145.32*	*20*	*50*
Nb	*3,145.40*	*100*	*10*
Ni	*3,145.72*	*3*	*200*
Cu	*3,146.82*	*20*	*100*
Co	*3,147.06*	—	*150R*

	λ	I J	I O		λ	I J	I O
Fe	*3,126.17*	*70*	*150*	*Cr*	*3,147.23*	*150*	*25*
Cu	*3,126.11*	*20*	*80*	*Ti*	*3,148.04*	*150*	*25*
Ru	*3,125.96*	*12*	*70*	*Mn*	*3,148.18*	*40*	*150*
Hg	*3,125.66*	*150*	*200*	*Hf*	*3,148.414*	*1*	*20*
Fe	*3,125.65*	*300*	*400*	*Fe*	*3,148.414*	*40*	*100*
V	*3,125.28*	*200R*	*80*				
Cr	*3,124.98*	*125*	*20*				
Os	*3,124.94*	*10wh*	*150h*				
Ge	*3,124.82*	80	*200*				
Hf	*3,123.95*	—	*8*				
Rh	*3,123.70*	*2*	*150*				
V	*3,122.89*	*300r*	*12*				
Au	*3,122.78*	*5*	*500h*				
Hf	*3,122.55*	*1*	*8*				
Mo	*3,122.00*	*150*	*5*				
Xe	*3,121.87*	*(150)*	—				
Rh	*3,121.75*	—	*150*				
Co	*3,121.41*	*6*	*150r*				
Re	*3,121.37*	—	*100*				
V	*3,121.14*	*200r*	*60*				
Fe	*3,120,87*	*50*	*80*				
Fe	*3,120.43*	*80*	*100*				
Cr	*3,120.37*	*150*	*40*				
Hf	*3,119.98*	*1*	*25*				

Hf 5,040.82 *J* 150 *O* 100

	λ	J	O		λ	J	O
Ru	**5,040.744**	—	**10**	**Ce**	**5,040.85**	—	**30**
P	**5,040.74**	**(70)**	—	**Si**	**5,041.03**	**4***wh*	—
Ti	**5,040.62**	**40**	**40**	**Ni**	**5,041.077**	—	**30**
Tb	**5,050.56**	—	**10**	**Fe**	**5,041.081**	—	**125**
Tl	**5,040.55**	**(3)**	—	**Ar**	**5,041.23**	**(10)**	—
Ar	**5,040.51**	**(10)**	—	**Cu**	**5,041.32**	**10**	—
W	**5,040.36**	**1***h*	**35**	**Ra**	**5,041.56**	**(35)**	—
Ru	**5,040.35**	—	**7**	**Ca**	**5,041.62**	—	**30**
Kr	**5,040.34**	**(7)**	—	**C**	**5,041.66**	**(30)**	—
Nd	**5,040.19**	—	**10**	**Fe**	**5,041.759**	—	**300**
*bh***B**	**5,040.10**	—	**12**	**W**	**5,041.762**	—	**7**
Ti	**5,039.953**	**25**	**125**	**Ta**	**5,041.87**	—	**4**
Sm	**5,039.948**	—	**25**	**Sm**	**5,041.93**	—	**5**
Ce	**5,039.93**	—	**8**	**Er**	**5,042.038**	—	**60**
Ce	**5,039.75**	—	**8**	**Tb**	**5,042.043**	—	**25**
Ru	**5,039.63**	—	**6**	**Ce**	**5,042.09**	—	**10**
Sm	**5,039.45**	—	**2**	**Ni**	**5,042.19**	—	**80**
Fe	**5,039.261**	**2***h*	**100**	**Pb**	**5,042.5**	**(200)**	—
Ni	**5,039.259**	—	**20**				
Th	**5,039.23**	—	**8**				

	λ	I	
		J	O
Os	**5,039.12**	—	**50**
Sm	**5,039.11**	—	**2**
Nb	5,039.04	30	200
Ni	5,038.60	—	50
Ti	5,038.40	20	100
Ne	5,037.75	(500)	—
Ta	5,037.37	—	60
Ti	5,036.47	25	125
Sm	5,036.21	—	50
Ni	5,035.96	—	70*w*
Ti	5,035.91	30	125
Ni	5,035.37	5	300*w*
Hf	5,034.90	—	3
Tm	5,034.21	100	100
Kr	5,033.85	(100*wh*)	—
Eu	5,033.54	2	60
Ne	5,031.35	(350)	—
Se	5,031.02	200*h*	50
Fe	5,030.78	125	1
Eu	5,029.48	—	500*w*
Sm	5,028.44	—	200
Xe	5,028.28	(200)	—
Fe	5,028.14	—	100
Fe	5,027.21	—	60
Fe	5,027.14	—	60
Pr	5,026.97	1	80
Nb	5,026.36	8	50
Hf	5,025.91	2	2
N	5,025.66	(100)	—
Ti	5,025.58	8	100
Ti	5,024.84	15	100
Fe	*5,013.48*	*300*	*10*
Hf	*5,023.08*	—	*4*
Eu	*5,022.90*	—	*125*
Ti	*5,022.87*	*18*	*100*
Kr	*5,022.40*	*(200)*	—
Fe	*5,022.25*	—	*150*
Hf	*5,021.75*	—	*4*
Hf	*5,021.11*	*2*	*4*
Ti	*5,020.03*	*80*	*100*
Hf	*5,018.20*	*2*	*20*
Ni	*5,017.59*	*1*	*100w*
Ti	*5,016.17*	*15*	*100*
Gd	*5,015.06*	—	*100w*
Fe	*5,014.96*	—	*500*
V	*5,014.62*	*125*	*125*

	λ	I	
		J	O
Ta	5,043.32	—	60
Cs	5,043.80	(80)	—
Pt	5,044.04	1	60
Sm	5,044.28	—	150
Sb	5,044.56	(100)	—
Xe	5,044.92	(100)	—
N	5,045.10	(200)	—
Kr	5,046.31	(80*wh*)	—
La	5,046.88	—	80
Hf	5,047.45	5	15
Fe	5,048.45	—	50
Ar	5,048.81	(500)	—
Ni	5,048.85	2*h*	80
Fe	5,049.82	1	400
La	5,050.57	4	80
Hf	5,051.32	1	3
Ni	5,051.53	—	50
Fe	5,051.64	—	200
Cr	5,051.90	—	50
C	5,052.12	(100)	—
Sm	5,052.75	2	150
Ti	5,052.87	3	50
W	5,053.30	10	60
Ar	5,054.18	(300)	—
Br	5,054.65	(200)	—
La	5,056.46	—	80
Ar	5,056.53	(200)	—
Hf	5,057.03	30	20
Ru	5,057.29	—	60
Ru	5,057.33	—	100
Sm	5,057.75	1	100
Nb	5,058.01	10	50
Hf	*5,058.18*	*10*	*8*
Ar	*5,060.08*	*(500)*	—
Ar	*5,062.07*	*(200)*	—
Ti	*5,064.65*	*35*	*150*
Se	*5,068.65*	*(250)*	—
Fe	*5,068.79*	*200*	*400*
Sm	*5,069.44*	*1*	*150*
Hf	*5,069.80*	—	*2*
Sm	*5,071.19*	—	*100*
Hf	*5,071.23*	*8*	*6*
Ar	*5,073.08*	*(200)*	—
Yb	*5,074.34*	*5*	*200*

	λ	I J	I O
Ti	*5,014.24*	*30*	*100*
Eu	*5,013.14*	*1*	*125*
Fe	*5,012.07*	—	*300*
Ar	*5,009.35*	*(200)*	—
N	*5,007.32*	*(150)*	—
Ti	*5,007.21*	*40*	*200*

Hf 5,298.06 *J* 100 *O* 80

	λ	I J	I O
Cr	**5,297.98**	**1*h***	**5*h***
Dy	**5,297.83**	—	**3**
Th	**5,297.74**	—	**8**
Cd	**5,297.64**	—	**3**
Tb	**5,297.60**	—	**10**
U	**5,297.45**	—	**8**
Cr	**5,297.36**	**2*h***	**5*h***
Ti	**5,297.26**	**1**	**70**
Mn	**5,296.97**	**(40)**	—
Sm	**5,296.94**	**2**	**6**
*bh***F**	**5,296.8**	—	**100**
Zr	**5,296.79**	—	**9**
Cr	**5,296.69**	**15**	**15**
Ce	**5,296.60**	—	**15**
Tb	**5,296.55**	—	**10**
J	**5,296.52**	**(150)**	—
Nb	**5,296.34**	**2**	**5**
Ar	**5,296.32**	**(5)**	—
P	**5,296.09**	**(300*w*)**	—
Ti	5,295.79	1	50
Pd	5,295.63	10	200
Ta	5,295.01	—	40
Hf	5,294.87	2	12
Eu	5,294.60	—	300
Eu	5,293.68	—	50*w*
Nd	5,293.17	4	60
*bh*F	5,292.9	—	150
Hf	5,292.78	—	2
Pr	5,292.63	2	50
Cu	5,292.52	—	50
Xe	5,292.22	(800)	—
Rh	5,292.14	1	80
Pr	5,292.10	—	60*w*
Eu	5,291.25	—	200
bhF	5,291.0	—	200
La	5,290.84	100	60
Hf	5,289.98	10	3
Mo	**5,298.060**	**3**	**15**
Pr	**5,298.11**	—	**25**
Nb	**5,298.12**	**2**	**3**
Eu	**5,298.16**	—	**20**
Ne	**5,298.19**	**(150)**	—
Cr	**5,298.27**	**25**	**15*R***
Ce	**5,298.29**	—	**10**
*bh***Zr**	**5,298.3**	—	**2**
Sm	**5,298.35**	**2**	**10**
Bi	**5,298.36**	**8**	**20**
Ti	**3,298.44**	**1**	**40**
Gd	**5,298.59**	—	**15**
*bh***F**	**5,298.6**	—	**100**
Er	**5,298.64**	—	**12**
Fe	**5,298.780**	—	**12**
Os	**5,298.781**	—	**20**
U	**5,298.81**	—	**2**
Mn	**5,298.85**	—	**4**
Nd	**5,298.88**	—	**3**
Nd	**5,298.98**	**2*h***	**3**
O	**5,298.00**	**(70)**	—
Sm	**5,299.19**	—	**2**
Zr	**5,299.20**	—	**2**
Mn	**5,299.28**	**(50)**	—
U	**5,299.47**	**3**	**6**
Zr	**5,299.51**	—	**2**
Hg	**5,299.53**	**(10)**	—
J	**5,299.79**	**(10)**	—
Kr	**5,299.79**	**(2)**	—
Hf	**5,299.85**	**10**	**8**
Yb	**5,299.85**	—	**3**
Ti	**5,300.02**	—	**8**
Yb	5,300.94	60	6
Pt	5,301.02	10	150
Co	5,301.06	—	700*w*
La	5,301.98	200	300*r*
Fe	5,302.31	—	300

	λ	I J	I O
Eu	5,289.25	—	125
Fe	5,287.92	20	100
Eu	5,287.23	—	125
Cr	5,287.19	—	40
Hf	5,286.09	—	4
Ar	5,286.08	(60)	—
Al	5,285.85	(50)	—
Eu	5,285.73	—	40
Pr	5,285.63	1	40
Eu	5,285.46	—	40
Fe	5,284.09	70*h*	—
Ru	5,284.08	—	100
Al	5,283.77	(100)	—
Fe	5,283.63	40	400
Co	5,283.49	—	125*w*
Ti	5,283.45	2	50
Ra	5,283.28	(250)	—
Sm	5,282.91	—	100
Eu	5,282.82	—	1000
Fe	5,281.80	20	300
Mo	5,280.86	25	50
Co	5,280.65	—	500*w*
Ar	5,280.40	(60)	—
Al	5,280.21	(50)	—
Ne	5,280.07	(50)	—
Ta	2,279.82	—	60*w*
Yb	5,279.55	100	15
Re	5,278.24	—	100
Yb	*5,277.07*	*6*	*200*
Kr	*5,276.50*	*(100h)*	*—*
Hf	*5,276.39*	*3*	*1*
Nb	*5,276.195*	*50*	*200*
Co	*5,276.192*	*—*	*400w*
Re	*5,275.53*	*—*	*500w*
Hf	*5,275.04*	*1*	*7*
Fe	*5,273.17*	*4*	*80*
Eu	*5,272.48*	*—*	*400*
Eu	*5,271.95*	*—*	*2000*
Nb	*5,271.53*	*50*	*200*
Sm	*5,271.40*	*—*	*150*
Se	*5,271.22*	*(150)*	*—*
La	*5,271.20*	*20*	*100*
Re	*5,270.98*	*—*	*200W*
Fe	*5,270.36*	*80*	*400*
Fe	*5,269.54*	*200*	*800*
Co	*5,268.51*	*—*	*500w*
Fe	*5,266.58*	*40*	*500*

	λ	I J	I O
Mn	5,302.32	(60)	—
La	5,302.62	150	50
*bh*F	5,302.7	—	100
Sm	5,302.91	—	50
P	5,303.21	(50)	—
La	5,303.56	125	100
Eu	5,303.87	—	300
Hf	*5,304.19*	*—*	*2*
bhF	*5,304.4*	*—*	*100*
Ne	*5,304.76*	*(70)*	*—*
Ru	*5,304.86*	*—*	*60*
Se	*5,305.35*	*(500)*	*—*
Tm	*5,307.11*	*20*	*100*
Yb	*5,307.12*	*—*	*40*
Fe	*5,307.36*	*—*	*125*
Hf	*5,307.82*	*—*	*4*
Kr	*5,308.66*	*(200)*	*—*
bhF	*5,308.7*	*—*	*80*
In	*5,309.03*	*(70)*	*—*
Ru	*5,309.267*	*—*	*125*
Xe	*5,309.27*	*(150)*	*—*
In	*5,309.40*	*(70)*	*—*
Ar	*5,309.52*	*(200)*	*—*
Hf	*5,309.68*	*1*	*6*
*N**			

	λ	I J	I O		λ	I J	I O
Co	*5,266.49*	—	*500w*				
Eu	*5,266.40*	—	*1000*				
Co	*5,266.30*	—	*100*				
Hf	*5,264.95*	*80*	*50*				
Cr	*5,264.15*	*20*	*100r*				
Fe	*5,263.33*	—	*300*				
Xe	*5,261.95*	*(200)*	—				
Hf	*5,260.54*	*40*	*30*				
Xe	*5,260.44*	*(300)*	—				
Pr	*5,259.74*	*3*	*125*				
Hf	*5,258.75*	—	*2*				

Hf 5,311.60 *J* 150 *O* 100

	λ	J	O		λ	J	O
Re	**5,311.55**	—	**4*h***	**Zr**	**5,311.77**	*2*	*2*
Nd	**5,311.46**	**1**	**15**	**Gd**	**5,311.86**	—	*10*
Zr	**5,311.40**	—	**10**	**Dy**	**5,311.88**	—	*3h*
W	**5,311.37**	—	**7**	**U**	**5,311.881**	*18*	*18*
Pr	**5,311.12**	**1**	**10**	**Er**	**5,311.89**	—	*8*
Te	**5,311.07**	**(35)**	—	**V**	**5,311.92**	*5*	—
Zn	**5,311.02**	—	**7**	**Th**	**5,311.98**	—	**8**
Zrť	**5,310.99**	—	**2**	**Eu**	**5,312.17**	—	**4*h***
Ta	**5,310.96**	—	**2**	**Sm**	**5,312.21**	—	**100**
Al	**5,310.76**	**(10)**	—	**Al**	**5,312.32**	**(35)**	—
N	**5,310.52**	**(3)**	—	**Pr**	**5,312.33**	**1**	**5**
U	**5,310.48**	**4**	**4**	**Th**	**5,312.51**	—	**6**
*bh***F**	**5,310.3**	—	**80**	**Pd**	**5,312.57**	—	**15**
Th	**5,310.260**	**1**	**12*d***	**Dy**	**5,312.62**	—	**3**
Kr	**5,310.26**	**(4*h*)**	—	**Co**	**5,312.66**	—	**400*w***
Zn	**5,310.24**	—	**7**	**U**	**5,312.73**	**5**	**5**
K	**5,310.21**	**(21)**	—	**Mo**	**5,312.76**	**2**	**10**
Co	**5,310.20**	—	**20**	**Cr**	**5,312.878**	—	**40**
Sm	**5,310.19**	—	**2**	**Th**	**5,312.883**	—	**6**
U	**5,310.04**	**8**	**10**	**Ir**	**5,313.01**	—	**2**
Er	**5,310.03**	—	**12**	**W**	**5,313.08**	**1**	**4**
Eu	**5,310.02**	—	**4*h***	**Ti**	**5,313.26**	—	**7**
In	**5,309.83**	**(100)**	—	**U**	**5,313.27**	**4**	**3**
Hf	**5,309.68**	**1**	**6**	**Pr**	**5,313.39**	—	**3**
				Xe	5,313.87	(500)	—
				bhF	5,314.7	—	50
				Al	5,316.07	(70)	—
				P	5,316.07	(150*w*)	—
				*bh*F	5,316.2	—	80
				Fe	5,316.61	150	—
				Co	5,316.78	—	300*w*
				Nb	5,318.60	12	100

*P**

	λ	I	
		J	O
Nd	5,319.82	2	60
Ra	5,320.29	(250)	—
Sm	5,320.60	—	100
Sm	5,321.87	4	50d
Fe	5,324.18	70	400
Hf	5,324.26	30	20
Co	5,325.28	—	300w
Ne	5,326.40	(75)	—
Re	5,327.46	—	100
Fe	5,328.05	100	400
Fe	5,328.53	35	150
N	5,328.70	(70)	—
O	5,328.98	(100)	—
O	5,329.59	(150)	—
O	5,330.66	(600)	—
Ne	5,330.78	(600)	—
Co	5,331.47	80	500w
As	5,331.54	200	—
Co	5,332.67	—	200w
Kr	5,333.41	(500h)	—
Co	5,333.65	—	100
Yb	5,335.16	400	150
Ru	5,335.93	—	100
J	5,338.19	(300)	—
Xe	5,339.38	(500)	—
Co	5,339.53	—	100w
Fe	5,339.94	30	200
Fe	5,341.05	15	200
Ta	5,341.05	80	150w
Mn	5,341.06	100	200
Ne	5,341.09	(1000)	—
Co	5,341.33	—	300w
Co	5,342.71	—	800w
Ne	5,343.28	(600)	—
Co	5,343.39	—	600w
Nb	5,344.17	200	400
P	5,344.72	(150w)	—
J	5,345.15	(300)	—
Cr	5,345.81	25	300R
Hf	5,346.30	40	10
Yb	5,347.20	200	40
Ar	5,347.41	(200)	—
Cr	5,348.32	15	150R
Hf	5,348.40	15	10
Ne	5,349.21	(150)	—
Tl	5,350.46	2000R	5000R

	λ	I J	I O
Nb	5,350.74	50	150
Eu	5,351.67	—	150

Hf 6,644.60 *J* 200 *O* 100

	λ	I J	I O
La	6,644.41	—	40
Yb	6,644.07	5*h*	2
Gd	6,644.04	—	8
Sm	6,643.88	—	4
Ar	6,643.79	(100)	—
Ba	6,643.77	—	4*h*
Ir	6,643.67	—	3
Co	6,643.65	—	2*h*
Ni	6,643.64	—	300*w*
Sr	6,643.540	—	100
Yb	6,643.54	—	50
Dy	6,643.41	—	4
Cr	6,643.02	—	5
La	6,642.79	25	10
Gd	6,642.75	—	5
Tb	6,642.27	—	4
*bh*F	6,642.0	—	50
U	6,641.74	—	2
Sm	6,641.57	—	3
S	6,641.50	(15)	—
Cu	6,641.41	10	—
Ce	6,641.15	—	3
O	6,640.90	(70)	—
Ne	6,640.80	(5)	—
Eu	6,640.8	—	2
Sm	6,637.17	—	60*d*
Hf	6,635.38	4	2
Gd	6,634.35	—	150
Fe	6,633.77	25*Wh*	60*h*
*bh*F	6,632.7	—	300
Co	6,632.44	—	150
Sm	6,632.28	—	100
Br	6,631.64	(200)	—
Sm	6,630.61	—	50*d*
Cr	6,630.01	—	50
*bh*F	6,629.40	—	200
Sm	6,628.88	—	50*d*
*bh*F	6,626.1	—	200
Co	6,623.79	—	70*W*
*bh*F	6,622.9	—	200
Ta	6,621.30	—	200
Th	6,644.66	—	3
N	6,644.96	(500)	—
Rh	6,645.01	—	2
Re	6,645.02	—	25
*bh*Sc	6,645.1	—	6
Eu	6,645.15	—	1000
La	6,645.16	—	8
Gd	6,645.18	—	150
*bh*Zr	6,645.3	—	2
Co	6,645.33	—	4*h*
*bh*F	6,645.4	—	80
Tb	6,645.41	—	4
Ra	6,645.95	(15)	—
Sm	6,646.22	—	10*d*
N	6,646.52	(15)	—
Cs	6,646.56	(15)	—
Hg	6,646.7	(10)	—
Gd	6,646.84	—	10
Hf	6,647.06	100	30
Pr	6,647.12	—	5
Ce	6,647.38	—	2
Sc	6,647.43	(15)	—
Sb	6,647.44	(60)	—
Ni	6,647.80	—	5
Kr	6,647.94	(3)	—
Sm	6,648.12	—	3
Sb	6,648.13	—	3
Eu	6,648.3	—	2
Pt	6,648.31	—	10
Te	6,648.52	(100)	—
*bh*F	6,648.7	—	100
Sm	6,649.02	—	50
Mo	6,650.37	6	80
La	6,650.80	—	100
Sm	6,651.61	—	80*d*
Ne	6,652.09	(150)	—
*bh*F	6,652.1	—	100
Re	6,652.40	—	80*W*
Ba	6,654.05	—	50
*bh*F	6,655.6	—	100
Sm	6,656.19	—	100
Tm	6,657.73	10	70

	λ	I J	I O
Xe	6,620.02	(100)	—
bhF	6,619.8	—	150
J	6,619.69	(200)	—
Mo	6,619.13	15	300
Sm	6,617.61	—	50d
Sr	6,617.26	—	150
bhF	6,616.6	—	100
La	6,616.58	—	125
Hf	6,616.14	4	2
Ta	6,611.95	—	300
Lu	6,611.71	150	100
Hf	6,609.12	12h	25
Re	6,605.19	—	100W
Tm	6,604.97	60	300
Sm	6,604.56	—	200d
Sm	6,601.83	—	150d
Hf	6,599.76	4	1
Ti	6,599.11	—	100
Ne	6,598.95	(1000)	—
Xe	6,597.25	(200)	—
Co	6,595.90	—	150
Ba	6,595.32	300	1000
Xe	6,595.01	(400)	—
Eu	6,593.82	—	400
Ra	6,593.34	(500)	—
Fe	6,592.92	80	150
Sm	6,589.73	—	400d
Sm	6,588.92	—	100
Hf	6,587.23	10	5
Cs	6,586.51	(5)	500
Sm	6,585.21	—	150d
Hf	6,584.53	40	4
C	6,582.85	200	—
Hf	6,581.15	2	2
La	6,578.51	—	100
C	6,578.03	500	—
Ta	6,574.84	—	200
Sm	6,574.38	—	100
Sm	6,570.67	—	200d
Sm	6,569.31	—	500d
Eu	6,567.87	—	600
Hf	6,567.39	60	6

	λ	I J	I O
Pb	6,660.0	(500)	—
Ar	6,660.64	(100)	—
Nb	6 660.84	80	300
Cr	6,661.08	—	100
J	6,661.16	(100)	—
La	6,661.40	—	70
J	6,662.14	(100)	—
Ru	6,663.14	—	100
Fe	6,663.45	25h	70
Ar	6,664.02	(100)	—
Ne	6,666.89	(100)	—
Sm	6,667.22	—	50d
Yb	6,667.85	20	1000
Xe	6,668.92	(150)	—
Cr	6,669.26	—	80
Hf	6,669.31	3	1
Hf	6,671.29	3	1
Sm	6,671.48	—	50
Ta	6,673.73	—	200
Ba	6,675.27	100	500
Ta	6,675.53	—	400
Nb	6,677.33	50	200
Fe	6,677.99	150	250
He	6,678.15	(100)	—
Ne	6,678.28	(500)	—
Co	6,678.81	—	125
Sm	6,679.1	—	80
Sm	6,679.24	—	80
Gd	6,681.22	—	100
Sm	6,681.53	—	60d
Ru	6,690.00	—	300
Hf	6,691.67	3	2
Sm	6,693.55	—	100d
Ba	6,693.88	100	600
Eu	6,693.97	—	500
Xe	6,694.32	(200)	—
Nb	6,701.20	15	100
Li	6,707.84	200	3000R
Mo	6,707.85	—	300W
Co	6,707.86	—	200Wh
Hf	6,709.41	6	1
La	6,709.50	—	150
Hf	6,713.5	20	10
Tl	6,713.69	40	100
Hf	6,716.0	5	1
Ra	6,719.32	(500)	—
Hf	6,719.40	50	2
O	6,721.21	(300)	—

Hg 80 200.59

t_0 −38.9°C t_1 357°C

I.	II.	III.	IV.	V.
10.434	18.752	34.5	∼72	∼82

λ	*I*		eV
	J	*O*	
I 2,536.52	1000*R*	2000*R*	4.9
I 3,125.66	150	200	8.85
I 3,131.55	300	400	8,82
I 3,131.83	100	200	8.85
I 3,650.15	500	200	8.9
I 3,663.28	400	500	8.8
I 4,046.56	300	200	7.7
I 4,358.35	500	3000	7.7
I 5,460.74	(2000)	—	7.7
I 5,769.59	200	600	8.85
I 5,790.65	(1000)	—	8.85

Hg 2,536.52 *J* 1,000*R* *O* 2,000*R*

	λ	*I* *J*	*I* *O*		λ	*I* *J*	*I* *O*
Co	**2,536.493**	**2**	**1**	**Th**	**2,536.558**	**10**	**5**
Pt	**2,536.487**	**10**	**100**	**Bi**	**2,536.56**	**2*h***	**5*h***
U	**2,536.24**	**2**	**4**	**U**	**2,536.600**	**2**	**3**
Ta	**2,536.23**	—	**100*W***	**W**	**2,536.605**	**12**	**1**
Fe	**2,536.224**	—	**3**	**Ir**	**2,536.66**	—	**3**
Ru	**2,536.216**	**6**	**12**	**In**	**2,536.669**	**(10)**	—
Ir	**2,536.13**	**1**	**2**	**Ta**	**2,536.67**	—	**2*h***
				Fe	**2,536.673**	**5**	**1**
				Rh	**2,536.706**	**5**	**15**
Fe	2,535.60	—	1,000	**Pr**	**2,536.71**	**8**	—
				Tb	**2,536.75**	**10**	**3**
Hg	*2,531.69*	*(3)*	*3h*	**La**	**2,536.76**	**3*d***	—
Hg	*2,530.04*	*(2)*	—	**U**	**2,536.79**	**2**	**4**
Hg	*2,529.53*	*(1)*	*2*	**Co**	**2,536.80**	**2**	—
				Fe	**2,536.82**	**4**	**10**
				Mn	**2,536.83**	**5**	—
				Mo	**2,536.85**	**4**	**25**
				Hg	2,539.01	(3)	3*h*
				Hg	*2,543.37*	*(3)*	—

Hg 3,125.663 *J* 150 *O* 200

	λ	*J*	*O*		λ	*J*	*O*
Ti	**3,125.656**	—	**5**	**Th**	**3,125.71**	**3**	**8**
Fe	**3,125.654**	**300**	**400**	**La**	**3,125.72**	**3**	—
Er	**3,125.65**	**2**	**10**	**Ce**	**3,125.76**	—	**20**
Pr	**3,125.6**	**(50)**	—	**Nb**	**3,125.89**	**15**	**1**
Ti	**3,125.55**	—	**2**	**Te**	**3,125.91**	**(10)**	—
Re	**3,125.52**	—	**30**	**Zr**	**3,125.918**	**5**	**8**
Cr	**3,125.47**	**4**	**8**	**J**	**3,125.92**	**(10)**	—
Th	**3,125.46**	**10**	**10**	**Cl**	**3,125.96**	**(5)**	—
Cl	**3,125.44**	**(6)**	—	**Ru**	**3,125.963**	**12**	**70**
Yb	**3,125.44**	**2**	**1**	**Sm**	**3,126.00**	—	**5**
W	**3,125.358**	**9**	**10**	**Tm**	**3,126.01**	—	**25**
Ce	**3,125.358**	—	**2**	**Kr**	**3,126.02**	**(6*h*)**	—
Cs	**3,125.3**	**(4)**	—	**Mo**	**3,126.03**	**10**	—
V	**3,125.28**	**200*R***	**80**	**Yb**	**3,126.06**	**25*h***	**1**
Fe	**3,125.27**	**2**	**5**	**Cu**	**3,126.11**	**20**	**80**
Ir	**3,125.25**	—	**2*h***	**Fe**	**3,126.173**	**70**	**150**
Zr	**3,125.19**	**2**	**3**	**U**	**3,126.174**	**20**	**12**
Er	**3,125.17**	**2**	**15**	**Yt**	**3,126.176**	**6**	**6**
Th	**3,125.16**	**8**	**10**	**Dy**	**3,126.18**	**1*h***	**15**
Eu	**3,125.10**	**1*h***	**5**	**Er**	**3,126.18**	—	**4**

	λ	I	
		J	O
V	**3,125.00**	**50**	**4**
Cr	**3,124.98**	**125**	**20**
Ta	**3,124.97**	**20**	**50**
Os	**3,124.94**	**10wh**	**150h**
Sc	**3,124.935**	**1h**	**2**
Sm	**3,124.927**	**3**	**9**
Ir	**3,124.927**	—	**4**
Ge	3,124.82	80	200
Rh	3,123.70	2	150
V	3,122.89	300r	12
Au	3,122.78	5	500h
Cr	3,122.60	80	10
Ti	3,122.06	40	2
Mo	3,122.00	150	5
Xe	3,121.87	(150)	—
Rh	3,121.75	—	150
Co	3,121.41	6	150r
Re	3,121.37	—	100
V	3,121.14	200r	60
Fe	3,120.87	50	80
V	3,120.73	80	12
Fe	3,120.43	80	100
Cr	3,120.37	150	40
Ti	3,119.80	150	4
As	3,119.60	50	100
Fe	3,119.49	80	100
Cr	3,118.65	200	35
V	3,118.38	200R	70
Os	3,118.33	20	150
Re	3,118.19	—	200
Pb	*3,117.7*	*(100)*	—
Ti	*3,117.67*	*200*	*15*
As	*3,116.63*	*150*	—
Fe	*3,116.59*	*150*	—
Hg	*3,116.24*	*(100)*	—
Bi	*3,115.42*	*500*	—
Ni	*3,114.12*	*50*	*300*
Pd	*3,114.04*	*500w*	*400w*
V	*3,113.57*	*100*	*7*
Eu	*3,111.43*	—	*200*
V	*3,110.71*	*300R*	*70*
Ti	*3,110.67*	*100*	*10*

	λ	I	
		J	O
Ne	**3,126.19**	**(150)**	—
V	**3,126.21**	**100R**	**60**
Hf	**3,126.29**	**5**	**18**
Ru	3,126.61	50	12
Co	3,127.25	—	100
Nb	3,127.53	50	10w
Ta	3,127.76	100	18w
Ru	3,127.91	100	10
V	3,128.28	60	2
Ti	3,128.64	70wh	12
V	3,128.96	70	3
Cr	3,128.70	150	30
Re	3,128.95	—	100W
Ni	3,129.31	—	125
Fe	3,129.33	60	100
Na	3,129.37	(60)	35
Yt	3,129.93	50	8
V	3,130.27	200r	50
Be	3,130.42	200	200
Ta	3,130.58	35	100W
Eu	3,130.74	100	100W
Nb	3,130.79	100	100
Ti	3,130.80	100	25
*N**			

Hg 3,131.55 *J* 300 *O* 400

3,131.83 *J* 100 *O* 200

	λ	*I*	
		J	*O*
Eu	**3,131.75**	**5**	**20***W*
Fe	**3,131.72**	**35**	—
Ce	**3,131.68**	—	**4**
U	**3,131.62**	**2**	**1**
Hg	***3,131.55***	***300***	***400***
U	**3,131.53**	**2**	**4**
Ce	**3,131.50**	—	**2***d*
Os	**3,131.48**	**10**	**20**
Fe	**3,131.45**	—	**2***h*
Tb	**3,131.35**	**3**	**8**
U	**3,131.32**	**6**	**8**
Tm	**3,131.26**	**500**	**400**
Ta	**3,131.22**	**25***h*	**5***h*
Cr	**3,131.21**	**6**	**20**
Mo	**3,131.19**	**5**	—
Os	**3,131.115**	**30**	**125**
Zr	**3,131.110**	—	**7**
Be	**3,131.072**	**150**	**200**
Er	**3,131.07**	—	**3**
Th	**3,131.070**	**10**	**12**
Ho	**3,130.99**	**6**	**6**
Ce	**3,130.92**	—	**5**
Ce	**3,130.87**	**2**	**30**
Gd	**3,130.81**	**1**	**3**
Ar	**3,130.80**	**(20)**	—
Ti	**3,130.800**	**100**	**25**
*P**			

	λ	*I*	
		J	*O*
Sr	**3,131.75**	**10**	—
Hf	**3,131.81**	**10**	**40**
Co	**3,131.830**	—	**8**
Hg	***3,131.833***	***100***	***200***
U	**3,131.99**	**6**	**8**
Nb	**3,132.01**	**5**	**1**
Re	**3,132.02**	—	**2**
Er	**3,132.03**	**2**	**6**
Ce	**3,132.04**	—	**15**
Cr	**3,132.06**	**125**	**25**
Zr	**3,132.07**	**2**	**10**
Dy	**3,132.12**	**2**	**5**
Nb	**3,132.137**	**2**	**1**
La	**3,132.14**	**3***h*	**2**
Eu	**3,132.16**	—	**15***w*
U	**3,132.18**	**1**	**2**
Co	**3,132.216**	—	**40**
Ne	**3,132.22**	**(4)**	—
Mn	**3,132.28**	—	**15**
Er	**3,132.51**	**3**	**12**
Pd	**3,132.513**	**15***h*	—
Fe	**3,132.514**	**40**	**70**
Te	**3,132.58**	**(10)**	—
Se	**3,132.58**	**(20)**	—
Mo	**3,132.59**	**300***R*	**1000***R*
Ta	**3,132.64**	**25**	**250***w*
Cd	3,133.17	300	200
V	3,133.33	200*r*	50
Tm	3,133.89	200	200
Ni	3,134.108	150	1000*R*
Fe	3,134.111	125	200
Cr	3,134.31	50	3
Se	3,134.42	(70)	—
Hf	3,134.72	125	80
O	3,134.79	(100)	—
Ru	3,134.80	100	10
V	3,134.93	150*r*	30
Fe	3,135.36	100	1
Dy	3,135.37	50	100
Hg	3,135.76	—	10
Ru	3,135.80	80	10
Ta	3,135.89	100	35

	λ	J	O
V	3,136.51	200	20
Cr	3,136.80	50	20
Co	3,137.33	—	150*r*
Ta	3,137.44	50	3
Rh	3,137.71	—	100
Pb	3,137.83	100	—
V	3,138.06	70	—
In	3,138.56	(50)	—
In	3,138.64	(50)	—
Pt	*3,139.39*	*80*	*300*
V	*3,139.74*	*150*	*15*
Se	*3,141.13*	*(100)*	—
V	*3,141.48*	*100*	*5*
Fe	*3,142.45*	*100*	*125*
V	*3,142.47*	*100r*	*15*
Pd	*3,142.81*	*100*	*300*
Ti	*3,143.76*	*125*	*18*
Fe	*3,143.99*	*150*	*200*
Hg	*3,144.48*	—	*10*
Fe	*3,144.49*	*100*	*150*
Nb	*3,145.40*	*100*	*10*
Ni	*3,145.72*	*3*	*200*

Hg 3,650.15 *J* 500 *O* 200

	λ	J	O		λ	J	O
Ce	3,650.121	1	10	Sm	3,650.168	5	25
Xe	3,650.12	(3)	—	La	3,650.174	60	100
Cl	3,650.10	(4)	—	Pr	3,650.18	6	30
Mo	3,650.05	25	3	N	3,650.19	(70)	—
Fe	3,650.03	30	70	Fe	3,650.28	50	70
Dy	3,650.00	2	5	Ru	3,650.32	12	3
Cr	3,649.86	4	15	U	3,650.339	—	3
Nb	3,649.85	20	20	Cr	3,650.344	40	—
Ar	3,649.83	(800)	—	Os	3,650.38	4	15
Eu	3,649.82	2	5	Tb	3,650.40	100	50
Th	3,649.74	3	8	Er	3,650.414	2	15
U	3,649.729	8	1	Nd	3,650.415	6	12
Ce	3,649.726	1	10	Zr	3,650.47	—	3
Cd	3,649.60	15	20	Th	3,650.51	1*d*	2*d*
Ra	3,649.55	(1000)	—	Nb	3,650.52	3	2
Ho	3,649.52	4*h*	2	Hf	3,650.53	2	2
La	3,649.509	8	40	Fe	3,650.53	—	3
Fe	3,649.508	100	100	Mo	3,650.58	2	3
Re	3,649.507	—	5*h*	Ca	3,650.62	3	—
Sm	3,649.506	30	100	U	3,650.68	2	3
Mo	3,649.47	5	5	Nd	3,650.69	6	15

	λ	I	
		J	O
Gd	**3,649.44**	**5**	**5**
Tb	**3,649.41**	**3**	**8**
U	**3,649.410**	**4**	—
Pr	**3,649.40**	**3**	**10**
Co	**3,649.35**	**4**	**200**
Fe	**3,649.30**	**25**	**60**
Th	**3,649.25**	**10**	**10**
Al	**3,649.22**	**(2)**	—
Al	**3,649.18**	**(5)**	—
Hf	**3,649.10**	**5**	**20**
Au	**3,649.09**	**5*h***	**3**
W	**3,649.02**	**7**	**6**
Sm	**3,649.01**	**1**	**2**
Pb	**3,649.0**	**(20)**	—
Cr	**3,649.00**	**20**	**40**
Os	3,648.81	10	100
Fe	3,647.84	400	500
Lu	3,647.77	5	100
Co	3,647.67	8	100
Gd	3,646.20	150	200*w*
Dy	3,645.42	100	300
La	3,645.41	60	100
Ca	3,644.41	15	200
Hg	3,644.32	(40)	—
Re	3,642.99	—	100
Ti	3,642.67	125	300
Ta	3,642.06	18	125
V	3,641.10	30*wh*	100*h*
Fe	3,640.39	200	300
Os	3,640.33	40	200
Pb	3,639.58	50*h*	300
Rh	3,639.51	70	125
Co	3,639.44	20	200
Pt	3,638.79	10	250
Hg	3,638.34	(100)	—
Zr	*3,636.45*	*30*	*200*
Ti	*3,635.46*	*100*	*200*
Pd	*3,634.69*	*1000R*	*2000R*
Fe	*3,631.46*	*300*	*500*
Hg	*3,630.65*	*(100)*	—
Pt	*3,628.11*	*20*	*300W*
Co	*3,627.81*	—	*200*
Zr	**3,650.72**	**2*h***	**2**
Au	**3,650.75**	**10**	**5**
Th	**3,650.77**	**10**	**10**
Nb	**3,650.81**	**15**	**15**
Cu	**3,650.85**	**1*h***	**4**
Ce	**3,650.88**	**3**	**12**
Ar	**3,650.90**	**(5)**	—
Tb	**3,650.93**	**8**	**8**
Gd	**3,650.97**	**30**	**25**
Sm	**3,650.98**	**10**	**24**
Zr	**3,650.99**	—	**2**
W	**3,651.00**	**12**	**10**
Kr	**3,651.02**	**(25*h*)**	—
Pr	**3,651.04**	**2**	**9**
Al	**3,651.06**	**(50)**	—
Cs	**3,651.07**	**(4)**	—
Al	**3,651.09**	**(18)**	—
Fe	**3,651.10**	**3**	**10**
Mo	**3,651.11**	**50**	**2**
Gd	**3,651.12**	**15**	**10**
Nb	**3,651.19**	**400**	**10**
U	**3,651.24**	—	**2**
Co	**3,651.26**	—	**20**
Nb	**3,651.30**	—	**2**
Fe	3,651.47	200	300
Co	3,652.54	—	200*r*
Tl	3,652.95	50	150
Ti	3,653.50	200	500
Cr	3,653.91	25	100
Kr	3,653.97	(250*h*)	—
Os	3,654.49	15	100
Ti	3,654.59	40	100
Gd	3,654.64	200	200*W*
Hg	3,654.83	(200)	—
Gd	3,656.16	200	200*W*
Os	3,656.90	30	150
Rh	3,657.99	200*W*	500*W*
Ti	3,658.10	60	150
Tb	3,658.88	100	100
Fe	3,659.52	80	125
Nb	3,659.61	500	15
Sm	3,661.35	50	100

Hg 3,663.28 *J* 400 *O* 500

	λ	*I*	
		J	*O*
Fe	**3,663.27**	**3**	**8**
Cr	**3,663.21**	**20**	**35**
Th	**3,663.20**	**5**	**4**
Nb	**3,663.18**	**3*h***	**5**
W	**3,663.15**	**6**	**7**
Tb	**3,663.12**	**15**	**50**
Ta	**3,663.10**	—	**18*d***
Pt	**3,663.095**	**2**	**50**
Hg	**3,663.09**	**3**	**5*d***
Nd	**3,663.033**	**2**	**6**
Re	**3,663.029**	—	**5**
Mo	**3,662.990**	**10**	**8**
Ce	**3,662.990**	**1**	**8*s***
Ho	**3,662.98**	**4**	**6**
Eu	**3,662.94**	**5*w***	**15*w***
Nb	**3,662.92**	**2**	**2**
Sm	**3,662.90**	**10**	**25**
Hg	**3,662.88**	**400**	**50**
Er	**3,662.87**	—	**10**
Fe	**3,662.85**	**5**	**30**
Cr	**3,662.84**	**8**	**25**
Rb	**3,662.78**	**(15)**	—
Ce	**3,662.73**	—	**2**
Sm	**3,662.69**	**10**	**25**
U	**3,662.66**	**2**	**15*r***
Tb	**3,662.64**	—	**8**
Ba	**3,662.53**	**5**	**10**
Eu	**3,662.52**	—	**3*w***
Ce	**3,662.49**	—	**8**
Pr	**3,662.46**	**1**	**3**
Cr	**3,662.38**	**4*h***	**6**
Ta	**3,662.34**	**15**	**15**
U	**3,662.331**	**10**	**10**
Eu	**3,662.33**	—	**8**
Er	**3,662.275**	**1**	**15**
Ho	**3,662.27**	**10**	**20**
Gd	**3,662.268**	**200**	**200*w***
Ce	**3,662.265**	—	**2**
Nd	**3,662.263**	**30**	**30**
Sm	**3,662.254**	**50**	**50**
Tb	**3,662.25**	—	**8**
Dy	**3,662.24**	—	**6**
Ti	**3,662.237**	**100**	**40**
Th	**3,662.19**	**4**	**4**
Co	**3,662.16**	**25**	**100**
Mo	**3,662.15**	**8**	**6**
Zr	**3,662.14**	**5**	**5**

	λ	*I*	
		J	*O*
Mo	**3,663.30**	**8**	**8**
U	**3,663.35**	**1**	**3**
W	**3,663.36**	**9**	**8**
Ru	**3,663.37**	**60**	**5**
Nb	**3,663.439**	**3**	**3**
Kr	**3,663.44**	**(20)**	—
Eu	**3,663.44**	**4*w***	**4*w***
Fe	**3,663.46**	**7**	**25**
V	**3,663.59**	**1*wh***	**150**
Zr	**3,663.651**	**10**	**100**
Sm	**3,663.654**	**4**	**3**
Mo	**3,663.66**	**4**	**4**
Ce	**3,663.70**	**1**	**10*s***
Th	**3,663.71**	**10**	**10**
Nb	**3,663.75**	**8*h***	—
Ar	**3,663.76**	**(5)**	—
W	**3,663.82**	**6**	**7**
Ta	**3,663.84**	—	**15**
Rb	**3,663.86**	**(15)**	—
Mo	**3,663.88**	**3**	**3**
Xe	**3,663.93**	**(3*h*)**	—
U	**3,663.934**	—	**10**
Fe	**3,663.95**	**2**	**4**
Mo	**3,663.96**	**3**	**3**
Sm	**3,664.01**	—	**15**
Ni	**3,664.09**	**30**	**300**
Ne	**3,664.11**	**(250)**	—
P	**3,664.19**	**(100*w*)**	—
Sc	**3,664.254**	**2**	**4**
U	**3,664.255**	**6**	**1**
Ir	**3,664.27**	—	**3**
Tb	**3,664.28**	—	**8**
Mo	**3,664.30**	**5**	**5**
Ce	**3,664.437**	—	**2**
Er	**3,664.440**	**20*h***	**40**
U	**3,664.53**	**2**	**2**
Gd	3,664.62	200	200*w*
Hg	3,666.55	—	9
V	3,669.41	300	20*W*
Fe	3,669.52	150	200
Fe	3,670.07	200	200
Ar	3,670.64	(300)	—
Os	3,670.89	20	200
Hg	3,673.02	(5)	—
Yb	3,675.09	200	50
Ar	3,675.22	(300)	—

	λ	I J	I O
Re	3,662.12	—	10
La	3,662.07	40	60
Nb	3,662.05	3	5
Er	3,662.04	1	9
*P**			

	λ	I J	I O
Fe	*3,679.91*	*300*	*500*
Hg	*3,684.91*	*(18)*	—
Ti	*3,685.19*	*700R*	*150*
Eu	*3,688.44*	*500W*	*1000W*

Hg 4,046.56 *J* 300 *O* 200

	λ	I J	I O
Ir	4,046.54	—	8
U	4,046.52	5	5
Sc	4,046.49	—	10
Pt	4,046.45	20	—
U	4,046.40	5	3
Ce	4,046.34	10	30
Nb	4,046.27	1	3
V	4,046.26	15	1
Sm	4,046.15	10	12
Zr	4,046.08	—	3
Dy	4,045.98	12	150
Ce	4,045.973	—	5
Tb	4,045.97	1	25
Ar	4,045.966	(150)	—
Ho	4,045.95	2	10
Er	4,045.89	—	2
Ag	4,045.82	2	10
Fe	4,045.81	300	400
Ru	4,045.76	—	25
Ne	4,045.66	(2)	—
Ho	4,045.43	80	200
Co	4,045.39	—	400
P	4,044.49	(150*w*)	—
Ar	4,044.42	(1200)	—
K	4,044.14	400	800
Hg	4,044.10	10	5
La	4,042.91	300	400
Os	4,041.82	6	100
Mn	4,041.36	50	100
Ho	4,040.84	30	150
Hg	4,040.40	(20)	—
Cr	4,039.10	40	100
Gd	4,037.91	30	100
Gd	4,037.34	30	100
Co	4,035.55	3	150
Mn	*4,034.49*	*20*	*250r*

	λ	I J	I O
Pr	4,046.63	3	8
W	4,046.701	12	10
Nd	4,046.702	2	8
Ce	4,046.73	—	2
Cr	4,046.760	3	30
Ni	4,046.761	—	2
Gd	4,046.84	10	10
Ce	4,046.85	—	3
Mo	4,046.89	20	3
Er	4,046.96	—	8
U	4,047.05	8	6
Gd	4,047.09	5	6
Pr	4,047.10	12	20
Nd	4,047.158	10	12
Tb	4,047.16	—	9
Te	4,047.18	(15)	—
Cs	4,047.184	(20)	—
K	4,047.201	200	400
Mo	4,047.204	3	4
Ce	4,047.27	2	18
Fe	4,047.31	—	3
Ir	4,047.33	—	4
Ca	4,047.35	3	2
Sm	4,047.36	6	8
Yb	4,047.39	4	—
Ce	4,047.392	—	4
Mo	4,047.40	4	4
Rb	4,047.4	(4)	—
Gd	4,047.85	50	150
Gd	4,049.90	60	100
Ru	4,051.40	200	125
Gd	4,053.30	80	100
Gd	4,053.65	40	100
Ho	4,053.92	200	400
Zr	4,055.03	5	100
Ag	4,055.26	500*R*	800*R*
In	4,056.94	(500)	—

	λ	I J	I O		λ	I J	I O
Mn	*4,033.07*	*20*	*400r*	Kr	4,057.01	(300*hs*)	—
Ga	*4,032.98*	*500R*	*1000R*	Pb	4,057.82	300*R*	2000*R*
La	*4,031.69*	*300*	*400*	Nb	4,058.94	400*w*	1000*w*
Mn	*4,030.75*	*20*	*500r*	Hg	4,060.89	(10)	—
				Cu	4,062.70	20	500*w*
				Fe	*4,063.60*	*300*	*400*

Hg 4,358.35 *J* 500 *O* 3000

	λ	J	O		λ	J	O
Pt	**4,358.34**	—	**2**	**Ca**	**4,358.42**	**5**	—
Th	**4,358.33**	—	**3**	**Tb**	**4,358.43**	—	**3**
Mo	**4,358.32**	**40**	—	**Dy**	**4,358.46**	**4**	**25**
Ir	**4,358.28**	—	**8**	**In**	**4,358.50**	**(2)**	—
N	**4,358.27**	**(250)**	—	**Fe**	**4,358.505**	**20**	**70**
Er	**4,358.172**	—	**4**	**Mo**	**4,358.55**	**10*w***	**20*w***
Nd	**4,358.169**	**20**	**50**	**Th**	**4,358.56**	**4**	**4**
Os	**4,358.14**	**1**	**9**	**Pd**	**4,358.60**	—	**25**
Ta	**4,358.03**	**10*h***	**3**	**Sc**	**4,358.64**	—	**10**
Os	**4,357.98**	—	**12**	**Ta**	**4,358.654**	—	**10**
Re	**4,357.97**	—	**15**	**U**	**4,358.654**	**3**	**2**
In	**4,357.92**	**(2)**	—	**Re**	**4,358.69**	—	**80**
Ne	**4,357.918**	**(5)**	—	**Nd**	**4,358.70**	**8**	**15**
La	**4,357.917**	**2*h***	**5**	**Yt**	**4,358.73**	**50**	**60**
Ce	**4,357.91**	**1**	**12**	**Ho**	**4,358.74**	—	**3**
Sm	**4,357.89**	**2*h***	**3**	**Zr**	**4,358.742**	—	**10**
Mo	**4,357.87**	**2**	**6**	**Tb**	**4,358.77**	—	**2**
Gd	**4,357.80**	—	**2**	**Ne**	**4,358.816**	**(2)**	—
Eu	**4,357.76**	**2**	**7**	**Tl**	**4,358.82**	**(5)**	—
Yt	**4,357.73**	—	**10**	**Th**	**4,358.83**	**2**	**3**
U	**4,357.63**	**3**	**3**	**Cs**	**4,359.02**	**(10)**	—
Th	**4,357.59**	**4**	**8**	**Tb**	**4,359.05**	—	**2**
Fe	**4,357.574**	**3**	**2**	**Ce**	**4,359.07**	**2**	**15**
Nd	**4,357.572**	—	**3**	**Sc**	**4,359.08**	—	**12**
Cr	**4,357.52**	**4**	**12**	**Ac**	**4,359.09**	**30**	—
Pr	**4,357.50**	**5**	**25**	**Pr**	**4,359.11**	**25**	**70**
Tm	**4,357.49**	**2**	**6**	**Gd**	**4,359.16**	—	**20**
Tb	**4,357.47**	—	**5**	**Nd**	**4,359.24**	**3**	**15**
V	**4,357.45**	**6**	**7**	**Re**	**4,359.31**	—	**12**
Er	**4,357.39**	—	**2**	**Tb**	**4,359.34**	—	**2*d***
Mo	**4,357.335**	**15**	**5**	**Ce**	**4,359.37**	—	**2**
Se	**4,357.33**	**(20)**	—	**Th**	**4,359.38**	**2**	**4**
Ne	**4,357.30**	**(2)**	—	**Co**	**4,359.43**	**1**	**15**
Nd	**4,357.22**	—	**8**	**U**	**4,359.47**	**3**	**2**
Co	**4,357.17**	—	**10**	**Ba**	**4,359.55**	**3**	**15**
				Kr	4,362.64	(500)	—

	λ	I J	I O		λ	I J	I O
Kr	4,355.48	(3000)	—	Te	4,364.02	(400)	—
Cr	4,351.77	300	300	O	4,368.30	(1000)	—
Rn	4,349.60	(5000)	—				
O	4,349.43	(300)	—				
Ar	4,348.11	(500h)	—	*Rh*	*4,374.80*	*500*	*1000W*
				Kr	*4,376.12*	*(800)*	—
Hg	4,347.50	50	200	*Hg*	*4,376.19*	*(59)*	—
				Hg	*4,382.16*	(10)	—
Ar	*4,345.17*	*(1000)*	—				
Hg	*4,343.63*	*5*	*20*				
Ra	*4,340.64*	*(1000)*	—				
Ar	*4,335.34*	*(800)*	—				

Hg 5,460.74 J (2000)

	λ	J	O		λ	J	O
Br	**5,460.70**	**(10)**	—	**P**	**5,460.85**	**(100)**	—
Mo	**5,460.53**	**15h**	**20**	**Nb**	**5,460.93**	**1**	**3**
Xe	**5,460.39**	**(200)**	—	**Th**	**5,461.74**	**1**	**12**
Xe	**5,460.04**	**(15)**	—	**Tm**	**5,461.96**	**25**	**5**
Ar	**5,459.61**	**(20)**	—	**Th**	**5,462.61**	**2**	**12**
Kr	**5,459.47**	**(4)**	—	**N**	**5,462.62**	**(30)**	—
U	**5,459.28**	**2**	**8**	**Kr**	**5,462.65**	**(2)**	—
Kr	**5,458.80**	**(7)**	—				
La	**5,458.69**	**50**	**5**				
U	**5,458.56**	**2**	**2**	J	5,464.61	(900)	—
				Lu	5,476.69	1000	500
Se	**5,458.52**	**(35)**	—				
				Hg	*5,499.8*	(5)	—
Hg	5,457.8	(8)	—				
S	5,453.88	(750)	—				
Hg	*5,425.25*	*(200)*	—				

Hg 5,769.59 J 200 O 600

	λ	J	O		λ	J	O
Eu	**5,769.56**	—	**2h**	**Mo**	**5,769.748**	—	**12**
La	**5,769.35**	—	**70**	**Gd**	**5,769.754**	—	**8**
Pr	**5,769.15**	—	**3**	**Pr**	**5,769.78**	—	**3**
Te	**5,769.1**	**(6)**	—	**Nd**	**5,769.87**	—	**10**
La	**5,769.07**	**60**	**30**	**Er**	**5,769.93**	—	**20**
Se	**5,769.06**	**(15)**	—	**Ce**	**5,769.95**	—	**6**
Ir	**5,768.91**	—	**3**	**La**	**5,770.01**	—	**25**
Ce	**5,768.89**	—	**15**	**Ne**	**5,770.31**	**(50)**	—
Sm	**5,768.78**	—	**2**	**Ce**	**5,770.43**	—	**8**
Ho	**5,768.41**	—	**8**	**Co**	**5,770.44**	—	**3**

	λ	I J	I O
Th	5,768.18	—	6
Sm	5,768.10	—	3
Ru	5,767.92	—	7
Ta	5,767.91	—	100*w*
Pb	5,767.9	(40)	—
Th	5,767.79	—	4
U	5,767.72	—	2
Sm	5,767.63	—	2
Eu	5,767.62	—	10
U	5,767.46	—	3
N	5,767.43	(30)	—
Nd	5,767.33	—	5
Yb	5,767.23	10*h*	2
Tb	5,767.22	—	10
Hf	5,767.18	30	15
Ru	5,767.16	—	5
Sm	5,767.10	—	2
Sr	5,767.05	—	4
Eu	5,765.20	—	2000
Ne	5,764.42	(700)	—
J	5,764.33	(100)	—
Xe	5,758.65	(150)	—
Te	5,755.87	(250)	—
Mo	5,751.40	100	125
Xe	5,751.03	(200)	—
Ne	5,748.30	(500)	—
Ar	*5,739.52*	*(500)*	—
Xe	*5,726.91*	*(200)*	—
Rb	*5,724.45*	—	*600*
Ne	*5,719.22*	*(500)*	—

	λ	I J	I O
Nd	5,770.50	—	20
V	5,770.55	2	18
Te	5,770.92	(35)	—
Tb	5,770.94	—	10
Mo	5,771.05	—	15
U	5,771.078	—	4
Nb	5,771.08	2	10
Kr	5,771.41	(100)	—
Nd	5,771.48	—	2
Yb	5,771.67	50	30
Sm	5,771.74	—	3
*bh*F	5,771.9	—	100
Ta	5,771.93	—	7
W	5,771.99	—	12
W	5,772.00	—	5
Zn	5,772.10	—	4
Ar	5,772.12	(100)	—
Xe	5,776.39	(150)	—
Re	5,776.84	—	300*w*
Ba	5,777.66	100*R*	500*R*
Cu	5,782.13	—	1000
Hg	5,789.66	(500)	—
Hg	5,790.65	(1000)	—
Ne	*5,804.45*	*(500)*	—
Ne	*5,811.42*	*(300)*	—
Ra	*5,813.63*	*(500)*	—
Eu	*5,818.74*	—	*1000*
Ne	*5,820.15*	*(500)*	—

Hg 5,790.65 J (1000)

	λ	I J	I O
Cl	5,790.50	(25)	—
Ar	5,790.39	(5)	—
Se	5,790.03	(15)	—
Nb	5,789.79	5*h*	5
Hg	5,789.66	(500)	—
Sb	5,789.52	(2)	—
Ar	5,789.48	(20)	—
Te	5,789.22	(9)	—
Kr	5,788.24	(7)	—
Hg	5,769.59	200	600
Ne	*5,764.42*	*(700)*	—

	λ	I J	I O
Nb	5,790.80	5*h*	—
Fe	5,791.04	2*h*	6*h*
Mo	5,791.87	60	100
Xe	5,791.88	(3*h*)	—
Ne	5,804.45	(500)	—
Ra	5,813.63	(500)	—
Hg	*5,838.77*	*(5)*	—

In $^{49}_{114.82}$

t_0 155°C

t_1 1,450°C

I.	II.	III.	IV.	V.
5.785	18.867	28.030	58.037	—

λ	*I*		eV
	J	*O*	
I 2,560.23	50*R*	150*R*	4.81
I 2,710.27	200*R*	800*R*	4.81
I 2,753.88	300	300*R*	5.01
I 3,039.36	500*R*	1,000*R*	4.1
I 3,256.09	600*R*	1,500*R*	4.1
I 3,258.56	300*R*	500*R*	4.1
I 4,101.77	1,000*R*	2,000*R*	3.02
I 4,511.32	4,000*R*	5,000*R*	3.02

In 2,560.23 *J* 50*R* *O* 150*R*

	λ	I	
		J	*O*
Sc	**2,560.23**	**30**	**10**
Rh	**2,560.21**	—	**2*r***
Dy	**2,560.20**	—	**7**
V	**2,560.15**	**10**	—
Tb	**2,560.12**	**10**	**2**
W	**2,560.119**	**8**	**15**
Nb	**2,560.11**	**10**	**1**
Co	**2,560.09**	**60*wd***	**1*d***
U	**2,560.02**	**2*h***	**3**
Fe	**2,559.93**	**15**	**2**
Rh	**2,559.90**	**100**	**5**
Re	**2,559.88**	—	**15**
Ru	**2,559.81**	**8**	**3**
Cr	**2,559.80**	**8**	—
Fe	2,559.77	30	3
J	2,559.72	(40)	—
Te	2,559.71	(25)	—
Mn	2,559.66	40*d*	—
Ta	2,559.43	2	100
Mn	2,559.412	50	2
Co	2,559.407	60*wh*	10
J	2,559.28	(25)	—
Hf	2,559.19	40	20
Mo	2,558.88	30*h*	5*h*
Mn	2,558.59	80	—
Zn	2,557.96	300	10
Nb	2,557.94	100*h*	—
Rh	2,557.92	50	1
Ni	2,557.87	80	—
Al	2,557.71	(40)	—
Ta	2,557.709	100	50
Mn	2,557.54	50	1
Fe	2,557.50	50	1
Co	2,557.35	30	2
Rh	2,557.20	100	1
Ru	2,557.13	50	5
Nb	2,556.94	200	5
Br	2,556.93	(25)	—
Mn	2,556.89	40	3
Co	2,556.76	150	50*w*
Ru	2,556.70	30	—
Mn	2,556.57	80*h*	10
Re	2,556.511	—	100
Ta	2,556.510	30	25
Al	2,556.01	(30)	—
Ti	2,555.99	80	15

	λ	I	
		J	*O*
Ru	**2,560.26**	**5**	**60**
Fe	**2,560.27**	**80**	**10**
U	**2,560.28**	**2**	**6**
Ni	**2,560.300**	**500*h***	—
Ir	**2,560.300**	**2*h***	**8**
Cs	**2,560.37**	**(8)**	—
La	**2,560.374**	**50**	—
W	**2,560.49**	—	**4**
Ir	**2,560.54**	**2*h***	**10**
Fe	**2,560.56**	—	**15**
Yb	**2,560.57**	**5**	—
Nb	**2,560.73**	**15**	**2**
Ta	2,560.68	—	70
Hf	2,560.74	25	—
Pd	2,561.02	200	—
J	2,561.49	(150)	—
Tm	2,561.65	30	60
Se	2,561.69	(25)	—
Rh	2,561.92	50	—
Mo	2,562.08	30	1
Ta	2,562.097	—	100
Fe	2,562.097	25	2
Pb	2,562.28	100	—
Nb	2,562.41	100	4
J	2,562.47	(30)	—
Fe	2,562.53	150	50
Li	2,562.54	15	150
V	2,562.76	25	1
W	2,563.162	30	8
Os	2,563.164	25	8
Fe	2,563.40	25*h*	5
Fe	2,563.47	125	70
Lu	2,563.52	80*h*	1
Hf	2,563.61	35	20
Mn	2,563.65	50*wh*	25
Ta	2,563.70	—	80
Ru	2,563.89	35	—
Co	2,564.04	100*wh*	15*w*
Nb	2,564.07	30	1
Eu	2,564.18	60	20
Mo	2,564.34	40	3
J	2,564.40	(70)	—
Zn	2,564.45	(25)	—
Te	2,564.58	(150)	—

	λ	I J	I O
V	2,555.91	80	2
Nb	2,555.63	80	2
Mo	2,555.42	50	3
Rh	2,555.36	60	100
Yb	2,555.31	50h	—
Ni	2,555.11	1000h	—
Ta	2,554.91	50h	50h
Ta	2,554.62	100	50
In	2,554.48	(50)	—
In	2,554.40	(50)	—
Mn	2.553.26	50	—
Tm	2,552.49	50	—
Pt	2,552.25	20	150
Cd	2,552.18	(100)	5h
In	2,552.0	2	—
Ru	2,551.98	150	10
Mn	2,551.88	100d	1
Pd	2,551.85	100h	—

	λ	I J	I O
In	2,564.72	(10)	—
In	2,565.13	(40)	—
Mn	2,565.22	80	—
Ni	2.565.37	150wh	—
In	2,565.40	(5)	
Pd	2,565.51	200	2
Ru	2,565.70	50	4
Ni	2,566.08	600h	—
Ru	2,566.23	50	—
J	2,566.26	(300)	—
Fe	2,566.91	150	60
Yb	2,567.63	150	5
Zr	2,567.64	100	100
Al	2,567.99	80R	200R
Fe	2,568.40	80	—
Rh	2,568.83	100	2
Zr	2,568.87	200	100

In 2,710.265 J 200R O 800R

	λ	J	O
Re	2,710.234	—	15w
Ru	2,710.232	100	50
Cr	2,710.23	—	50h
Mo	2,710.19	20	10
V	2,710.16	60	6
Ta	2,710.13	3	200
Ir	2,710.08	5	10
Nd	2,710.04	2	5
W	2,710.000	4	8
Hf	2,709.997	4	—
Fe	2,709.993	10	40
Eu	2,709.990	—	20
Mn	2,709.96	25	—
La	2,709.92	3	1h
Os	2,709.86	5	20
Pd	2,709.82	(5)	—
N	2,709.82	(50)	—
Zr	2,709.78	—	3
Mo	2,709.76	6	—
Ta	2,709.27	150	40
Tl	2,709.23	200R	400R
Fe	2,709.06	100	3
Ra	2,708.96	(200)	—
Ni	2,708.79	500	—

	λ	J	O
Kr	2,710.27	(3)	—
W	3,710.32	7	—
Mn	2,710.33	40h	12
Cl	2,710.38	(10)	—
P	2,710.40	(10)	—
Hg	2,710.45	(12)	—
Cs	2,710.5	(2)	—
Yb	2,710.54	10d	2
Fe	2,710.55	35	80
Ca	2,710.60	4	—
U	2,710.602	6	—
Mn	2,710.62	25h	—
Hf	2,710.66	3	—
Tl	2,710.67	10	30R
La	2,710.68	—	3
Rn	2,710.7	(3)	—
Ta	2,710.72	10h	1
Ru	2,710.738	—	20
Mo	2,710.742	1	20
Ag	2,711.21	300wh	1h
Mn	2,711.58	125R	2
V	2,711.74	150R	50
Fe	2,711.84	100	4
Ag	2,712.06	200h	3

	λ	I J	I O
V	2,707.86	150	70
Co	2,707.50	100wh	—
Co	2,706.74	100wh	—
V	2,706.70	200R	60
Fe	2,706.58	150	150
Sn	2,706.51	150R	200R
V	2,706.17	400R	100
Pt	2,705.89	200wh	1000wh
Co	2,705.85	100w	15w
Rh	2,705.63	300wh	100
In	2,704.48	(30)	—
Fe	2,703.99	400	30
Cu	2,703.18	200	10
Pt	2,702.40	200	1000
V	2,702.19	300R	80
Eu	2,701.89	200	300W
Cu	2,700.96	400	20
V	2,700.94	500R	125
J	2,712.23	(100)	—
Fe	2,712.39	100	2
Ru	2,712.41	300	80
Cu	2,713.50	300w	50
Ru	2,713.58	100	—
In	2,713.93	125wh	200R
V	2,714.20	100	60
Pd	2,714.32	150	—
Fe	2,714.41	400	200
Co	2,714.42	200W	12
Pd	2,714.90	200	—
Rh	2,715.31	500wh	50
V	2,715.69	300R	50
Nb	2,716.62	200	10
Eu	2,716.97	300	300
Cu	2,718.77	300w	40
Fe	2,719.02	300r	500r
Pt	2,719.04	100W	1000w

In 2,753.88 *J* 300 *O* 300*R*

	λ	I J	I O
Pt	2,753.86	4	100
Mo	2,753.82	10	1
Fe	2,753.81	—	5
Cd	2,753.80	2	—
Pt	2,753.76	—	15
Nb	2,753.74	5	2
Os	2,753.72	4	8
Fe	2,753.692	25	70
Re	2,753.688	—	2
Yb	2,753.67	2h	—
Te	2,753.64	(10)	—
Hf	2,753.61	60	—
Nb	2,753.55	3	2
Ce	2,753.54	—	2
Ru	2,753.44	50	50
V	2,753.40	200R	50
bhB	2,753.4	—	100
Co	2,753.34	4h	—
Fe	2,753.29	150h	25
Cr	2,752.88	40	300r
In	2,752.83	(5)	—
Ru	2,752.77	150	50
Ta	2,752.49	300	300
Ta	2,752.29	150	8
Ti	2,751.70	200wh	—
Cr	2,750.73	150	30
Ar	2,753.92	(10)	—
Mo	2,753.922	5	—
Fe	2,754.037	35	90
Cl	2,754.04	(6)	—
Nb	2,754.07	3	3
Rh	2,754.09	25	1
Cl	2,754.10	(25)	—
U	2,754.15	35	20
Lu	2,754.17	125	40
Ca	2,754.18	3	1
Zr	2,754.21	1	3
Ce	2,754.23	—	5
Cr	2,754.28	50	3
Mo	2,754.29	20	20
Fe	2,754.43	20	70
Pt	2,754.92	5h	200
Fe	2,755.74	100	300
Fe	2,756.26	100	300
Fe	2,756.33	100	300
Zn	2,756.45	100	200
Ag	2,756.51	200	5
Cr	2,757.10	10	300r
Cr	2,757.72	150	35
Ta	2,758.31	40	200
Ni	2,759.02	500wh	—

	λ	I J	I O		λ	I J	I O
Yb	2,750.48	150	20	*Cr*	*2,761.75*	*35*	*300r*
Fe	2,750.14	100	300*h*	*Pd*	*2,763.09*	*30r*	*300r*
Ta	2,749.83	50	200				
In	2,749.80	(30)	—				
In	2,749.70	(50)	—				
Cr	2,748.98	200	35				
Ta	2,748.78	50	400				
In	2,748.72	25*wh*	—				
Cd	2,748.58	200	5				
Cr	*2,748.29*	*5*	*300*				
In	*2,747.9*	*2*	—				
Fe	*2,746.98*	*300wh*	*200*				
Fe	*2,746.48*	*300wh*	*150*				
Ni	*2,742.99*	*500*	—				

In 3,039.356 *J* 500*R* *O* 1000*R*

	λ	I J	I O		λ	I J	I O
Fe	**3,039.32**	**15**	**20**	**Sm**	**3,039.358**	**2**	**6**
W	**3,039.311**	**12**	**10**	**Nb**	**3,039.41**	**5**	**3**
Cs	**3,039.31**	**(4)**	—	**Se**	**3,039.50**	**(20)**	—
U	**3,039.263**	**12**	**15**	**U**	**3,039.501**	**3**	**10**
Ir	**3,039.260**	**2**	**25**	**Ce**	**3,039.51**	—	**8**
Ce	**3,039.25**	—	**4**	**Mn**	**3,039.55**	**6*h***	**1**
Ca	**3,039.21**	**4**	**1*h***	**Co**	**3,039.567**	—	**70**
N*b*	**3,039.19**	**3**	**2**	**Ce**	**3,039.572**	—	**8**
U	**3,039.14**	**5**	**5**	**Cd**	**3,039.572**	—	**4**
Sm	**3,039.13**	**9**	**15**	**W**	**3,039.575**	**20**	**2**
Bi	**3,039.12**	**2**	—	**Ne**	**3,039.65**	**(7)**	—
Ge	**3,039.064**	**1000**	**1000**	**Yb**	**3,039.66**	**15**	**3**
Mo	**3,039.058**	**25**	—	**Rb**	**3,039.68**	**(15)**	—
Ce	**3,038.99**	—	**4**	**Nb**	**3,039.682**	**10*h***	**3**
Nd	**3,038.96**	**2**	**4**	**Ru**	**3,039.684**	—	**12**
Mo	**3,038.79**	**3**	—	**Eu**	**3,039.70**	—	**10**
Ru	**3,038.784**	**6**	**3**	**Bi**	**3,039.71**	**2*h***	—
Fe	**3,038.779**	**3**	—	**Ir**	**3,039.714**	—	**2**
Ti	**3,038.706**	**40**	**2**	**V**	**3,039.76**	**5**	—
V	**3,038.706**	**1**	**20**	**Ir**	**3,039.77**	—	**2**
Ho	**3,038.69**	**6**	**4**	**Cr**	**3,039.78**	**35**	**80**
In	**3,038.67**	**3**	—	**Nb**	**3,039.81**	**300**	**5**
				Mo	**3,039.82**	**2**	**20**
Ni	*3,037.93*	*100*	*800R*	**Eu**	**3,039.89**	**5**	**15**
Fe	*3,737.39*	*400r*	*700R*	**Sc**	**3,039.92**	**20*h***	**4**
In	*3,027.11*	*12*	—	**U**	**3,039.93**	**10**	**5**
Nb	*3,032.77*	*300*	*3*	**Ru**	**3,039.96**	**2**	**30**

	λ	I J	I O		λ	I J	I O
				Sb	3,040.67	(400*wh*)	—
				Cr	3,040.85	200	500*R*
				Pt	3,042.64	250*R*	200*R*
				In	3,043.40	3	—
				In	*3,047.00*	*6h*	—
				Fe	*3,047.60*	*500r*	*800r*
				In	*3,047.93*	*3*	—
				Ni	*3,050.82*	—	*1000R*

In 3,256.09 *J* 600*R* *O* 1500*R*

	λ	J	O		λ	J	O
W	**3,255.96**	**8**	**9**	**Nb**	**3,256.13**	**2*wh***	—
Pt	**3,255.92**	**30**	**3**	**Mn**	**3,256.14**	**50**	**75**
Fe	**3,255.89**	**100**	**20**	**Mo**	**3,256.21**	**25**	**40**
Sm	**3,255.84**	**2**	**6**	**W**	**3,256.230**	**7**	**8**
Ca	**3,255.81**	**3**	**1*h***	**Ce**	**3,256.232**	—	**6**
Re	**3,255.80**	—	**4**	**Dy**	**3,256.25**	**5**	**25**
Er	**3,255.79**	**1**	**10**	**Ce**	**3,256.251**	—	**12**
Ta	**3,255.69**	**2*wh***	**8*h***	**Ho**	**3,256.27**	**4**	—
Sc	**3,255.68**	**8**	**15**	**Th**	**3,256.273**	**15**	**10**
V	**3,255.65**	**5**	**25**	**Re**	**3,256.29**	—	**8**
U	**3,255.626**	—	**3**	**Ru**	**3,256.33**	**3**	**50**
Nd	**3,255.625**	**4**	**8**	**Er**	**3,256.35**	**2**	**10**
Sm	**3,255.624**	**3**	**8**	**Pt**	**3,256.43**	—	**2**
Th	**3,255.51**	**10**	**10**	**U**	**3,256.458**	**1**	**2**
Ne	**3,255.39**	**(4)**	—	**V**	**3,256.46**	**1**	**8**
Cs	**3,255.35**	**(10)**	—	**J**	**3,256.463**	**(18)**	—
Ge	**3,255.34**	**100*wh***	—	**La**	**3,256.60**	—	**3**
Hf	**3,255.28**	**30**	**20**	**Kr**	**3,256.67**	**(4)**	—
Os	**3,255.270**	**3**	**10*h***	**Ce**	**3,256.68**	—	**20**
Nb	**3,255.269**	**20**	**2*h***	**Fe**	**3,256.698**	**7**	**20**
Mo	**3,255.25**	**40**	—	**Ag**	**3,256.70**	**2*h***	—
Tb	**3,255.22**	**3**	**15**	**Mo**	**3,256.73**	**2**	**4**
Ce	**3,255.21**	—	**8**	**Nb**	**3,256.74**	**4**	—
Ir	**3,255.20**	—	**2**	**Ta**	**3,256.77**	**1**	**100**
				V	**3,256.777**	—	**8**
Fe	3,254.36	150	200	**Ir**	**3,256.784**	—	**4**
Lu	3,254.31	150	50	**Te**	**3,256.81**	**(35)**	—
Co	3,254.21	—	300*R*	**Tb**	**3,256.83**	—	**8**
Nb	3,254.07	300	20	**Nd**	**3,256.90**	**2**	**10**
Ti	3,252.91	200*r*	60	**Os**	**3,256.92**	**12**	**80**
Cd	3,252.52	300	300	*bh***B**	**3,257.0**	—	**100**
Ti	3,251.91	150	50	**Nb**	**3,257.01**	**3*h***	**4**
Pd	3,251.64	500	200				
Te	3,251.37	(150)	—				
Fe	3,251.23	150	300	*N**			

	λ	I	
		J	O
In	3,250.38	10	—
La	3,249.35	80	300
Ti	3,248.60	200*r*	25
Fe	3,248.21	150	200
In	3,247.61	15	—
Cu	3,247.54	2000*R*	5000*R*
La	*3,245.12*	*300*	*400*
Pd	*3,242.70*	*600R*	*2000wh*
Ti	*3,241.99*	*300R*	*60*
Sb	*3,241.28*	*(350wh)*	—
Fe	*3,239.44*	*300*	*400*
Ti	*3,239.04*	*300R*	*60*

In 3,258.56 *J* 3000*R* *O* 500*R*

	λ	I	
		J	O
Er	**3,258.48**	**2**	**10**
Ho	**3,258.47**	**4*h***	—
Si	**3,258.45**	**2*h***	—
Mn	**3,258.41**	**40**	**75**
Tb	**3,258.38**	**3**	**8**
Pr	**3,258.31**	—	**4**
He	**3,258.27**	**(5)**	—
Sm	**3,258.25**	**3**	**9**
Ta	**3,258.24**	**2*h***	**10**
W	**3,258.14**	**3**	**5**
Th	**3,258.11**	**2*h***	**2*h***
U	**3,258.10**	**2**	**3**
Re	**3,258.07**	—	**4*w***
Ho	**3,258.06**	**4**	—
Tm	**3,258.04**	**60**	**125**
Ru	**3,258.040**	**8**	**50**
Co	**3,258.02**	—	**60**
Na	**3,257.96**	**(60)**	**35**
Th	**3,257.935**	**8**	**8**
J	**3,257.93**	**(5)**	—
Fe	**3,257.894**	**3**	—
V	**3,257.889**	**40**	**6**
Ir	**3,257.888**	—	**2**
S	**3,257.83**	**(10)**	—
Cr	**3,257.822**	**30**	**40**
Ta	**3,257.822**	**2**	**25**
Ce	**3,257.81**	—	**6**
W	**3,257.80**	**10**	—
U	**3,257.77**	**3**	**1**
U	**3,257.71**	—	**6**
Tm	**3,258.62**	**5**	**10**
Eu	**3,258.67**	**2*h***	**2*h***
Mo	**3,258.69**	**25**	**1**
Cr	**2,258.77**	**50**	—
Fe	**3,258.773**	**150**	—
Pd	**3,258.78**	**200*h***	**300**
Ho	**3,258.8**	**4*h***	—
K	**3,258.81**	**(10)**	—
Re	**3,258.85**	—	**100**
Ce	**3,258.87**	**1**	**25**
Pr	**3,258.94**	**1**	**10**
U	**3,258.95**	**1**	**3**
Ru	**3,258.97**	**60**	**10**
Fe	**3,259.048**	**200**	**1**
Er	**3,259.048**	**6*d***	**18*d***
Th	**3,259.06**	**1**	**6**
Yb	**3,259.10**	**8**	**2**
Nb	**3,259.14**	**3**	**3**
Mo	**3,259.16**	**4**	**5**
Ho	**3,259.17**	**6**	—
Cl	**3,259.18**	**(8)**	—
Nd	**3,259.23**	**4**	**10**
Th	**3,259.25**	**6**	**5**
Xe	**3,259.36**	**(6)**	—
Tb	**3,259.38**	**3**	**15**
Ti	**3,259.42**	—	**2**
W	**3,259.44**	**9**	**9**
Nb	3,260.56	300	15
Cd	3,261.06	300	300

	λ	I J	I O		λ	I J	I O
Er	**3,257.69**	—	**2**	Tl	3,261.60	300*r*	70
				Os	3,262.29	50	500*R*
				Sn	3,262.33	300*h*	400*h*
*P**							
				Nb	3,263.37	500	3
				In	3,264.00	(10)	—
				In	3,264.04	(5)	—
				Hg	3,264.06	(200)	—
				Fe	3,265.05	150	200
				Fe	3,265.62	300	300
				La	3,265.67	200	300
				Pd	3,267.35	200*h*	—
				Sb	3,267.50	150*Wh*	150
				Ge	*3,269.49*	*300*	*300*
				Fe	*3,271.00*	*300*	*300*
				Co	*3,273.96*	*1500R*	*3000R*
				In	*3,274.02*	*15*	—

In 4,101.77 *J* 1000*R* *O* 2000*R*

	λ	J	O		λ	J	O
Ce	**4,101.77**	**6**	**35**	**W**	**4,101.85**	**8**	**7**
Ru	**4,101.74**	**60**	**20**	**U**	**4,101.90**	**1**	**18**
H	**4,101.73**	**(100)**	—	**Dy**	**4,101.95**	**2**	**8**
Nd	**4,101.682**	—	**8**	**Ce**	**4,102.07**	—	**2**
Fe	**4,101.679**	**2**	**5**	**Mo**	**4,102.15**	**25**	**30**
Ir	**4,101.67**	—	**3**	**Re**	**4,102.159**	—	**4*h***
Zn	**4,101.66**	—	**5**	**V**	**4,102.159**	**15**	**30**
Tb	**4,101.65**	—	**20*w***	**N**	**4,102.18**	**(5)**	—
Ce	**4,101.55**	**3**	**2**	**U**	**4,102.21**	**4**	**1**
Nd	**4,101.46**	**6**	**8**	**Zr**	**4,102.281**	—	**10**
Eu	**4,101.44**	—	**2**	**Ru**	**4,102.285**	**10**	**15**
Dy	**4,101.43**	**2**	**5**	*bh***Sr**	**4,102.3**	—	**2**
Ce	**4,101.36**	**2**	**1**	**Ce**	**4,102.36**	**2**	**18**
U	**4,101.32**	**2*h***	**2**	**Yt**	**4,102.38**	**30**	**150**
Sm	**4,101.31**	**2*h***	**15*d***	**Ho**	**4,102.40**	**3**	—
Fe	**4,101.27**	**10**	**40**	**Tb**	**4,102.52**	**2**	**25*w***
Ru	**4,101.23**	**2*h***	**7**	**Br**	**4,102.53**	**(10)**	—
Cr	**4,101.16**	**2**	**30**	**Nd**	**4,102.56**	**5**	**10**
Er	**4,101.093**	—	**3**	**W**	**4,102.70**	**30**	**35**
Ho	**4,101.09**	**40**	**40**	**Ti**	**4,102.71**	**1**	**3**
Te	**4,101.07**	**(50)**	—	**Eu**	**4,102.72**	**3**	**10**
La	**4,101.01**	**2*h***	**2**	**Ce**	**4,102.724**	**1**	**2**
V	**4,101.00**	**7**	**5**				
Hf	**4,100.93**	**1**	**10**	Ho	4,103.84	400	400
Nb	**4,100.92**	**200*w***	**300*w***	Se	4,108.83	(800)	—
				In	4,109.34	(5*h*)	—
Tb	**4,100.90**	**2**	**50*d***	N	4,109.98	(1000)	—

	λ	I J	I O
W	4,100.895	6	7
Ce	4,100.889	1	8
Th	4,100.83	18	18
Pr	4,100.75	50	200
In	4,115.48	(2)	—

In 4,511.32 *J* 4000*R* *O* 5000*R*

	λ	J	O
Sm	4,511.31	—	40
Sn	4,511.30	—	200
Ne	4,511.29	(5)	—
Nd	4,511.290	—	25
Pt	4,511.26	1*h*	2
Mn	4,511.24	—	2
Ru	4,511.20	—	25
Ti	4,511.170	10	40
Zr	4,511.170	—	5
U	4,511.16	8	4
Pr	4,511.091	—	3
Nb	4,511.089	15*h*	5
Ta	4,510.98	50*W*	200*W*
Ce	4,510.921	—	6
Er	4,510.917	—	2*w*
Al	4,510.84	6*h*	—
Ho	4,510.81	1	2
Ce	4,510.76	—	3
Ar	4,510.73	(1000)	—
Th	4,510.53	20	30
Gd	4,510.39	10*h*	10*h*
U	4,510.32	30	20
Mn	4,510.21	(6)	—
Rh	4,510.2	(10)	—
Ne	4,510.170	(15)	—
Ce	4,510.166	—	4
Pr	4,510.16	125	200
Ru	4,510.097	—	25
Os	4,510.096	—	2
Ce	4,510.08	—	2*h*
In	4,500.95	(50)	—
In	4,500.77	(30)	—
In	*4,487.36*	*10*	—
Cd	4,511.34	—	5
Pr	4,511.35	—	4
Ne	4,511.37	(50)	—
V	4,511.34	2	2
Pr	4,511.45	—	5
Ta	4,511.50	40*W*	300
Ne	4,511.51	(20)	—
Tb	4,511.52	—	40
Eu	4,511.53	3	3*w*
Ce	4,511.63	—	10
Er	4,511.71	1*h*	4*w*
U	4,511.75	8*h*	5*h*
Pr	4,511.81	—	3
Nd	4,511.82	10	50
Sm	4,511.83	100	100
Cr	4,511.90	100	80
Ir	4,512.03	—	3
Nb	4,512.13	3	2
Eu	4,512.14	—	2
Mo	4,512.15	25	25
U	4,512.18	4	2
Er	4,512.20	—	3
Pr	4,512.27	—	8
Ca	4,512.28	—	10
Nd	4,512.29	—	3
Sm	4,512.30	—	2
U	4,512.39	3	2
Th	4,512.49	4	4
In	4,517.42	10	—
In	*4,530.05*	*10*	—
In	*4,535.50*	*5*	—

Ir 77 192.2

t_0 2,350°C t_1 4,800°C

I.	II.	III.	IV.	V.
9.2	—	—	—	—

λ	I		eV
	J	O	
I 2,543.97	100	200	>4.9
I 2,639.71	15	100	>4.7
I 2,664.79	50	200	>4.6
I 2,694.23	50	150	>4.6
I 2,849.72	20	40	4.3
I 3,133.32	2	40	>4,0
I 3,220.78	30	100	4.20
I 3,513.64	100	100	3.52
I 4,268.10	15	200	3.79
4,311.50	10	300	—
4,399.47	100	400	—
4,426.27	10	400	—

Ir 2,543.971 *J* 100 *O* 200

	λ	*I*	
		J	*O*
Rh	**2,543.94**	**100*r***	**15**
Cs	**2,543.92**	**(20)**	**—**
Fe	**2,543.920**	**20**	**40**
Na	**2,543.87**	**—**	**12*R***
Sb	**2,543.84**	**25**	**—**
Re	**2,543.83**	**—**	**20**
Dy	**2,543.82**	**—**	**10**
Na	**2,543.817**	**—**	**6*R***
Os	**2,543.80**	**1**	**10**
V	**2,543.73**	**9**	**12**
Te	**2,543.72**	**(10)**	**—**
Gd	**2,543.71**	**—**	**2**
Ru	**2,543.68**	**—**	**20**
Re	**2,543.67**	**—**	**20*r***
Re	**2,543.65**	**—**	**700**
Zr	**2,543.64**	**2**	**2**
Mo	**2,543.61**	**15**	**3**
Mn	2,543.45	100	4
Fe	2,543.38	50	5
Ru	2,543.25	150	50
Mn	2,542.92	100	1
Rh	2,542.16	50	1
Zr	2,542.10	50	100
Ir	2,542.02	10	35
Co	2,541.94	300*h*	40
Ir	2,541.47	2	12
Nb	2,541.42	80	3
Mn	2,541.11	80	—
Fe	2,540.98	10	100*R*
Ir	2,540.66	—	4
Nb	2,540.61	150	2
Ir	2,540.40	2	10
Ru	2,540.30	100	10
Tb	2,540.12	50	3
Ru	*2,539.72*	*100*	*12*
Ir	*2,539.61*	*5*	*2*
Pt	*2,539.20*	*20*	*400*
Ni	*2,539.10*	*250w*	*—*
Ir	*2,538.88*	*—*	*8*
Mo	*2,538.455*	*125*	*30*
Ir	*2,538.453*	*—*	*3*
Pd	*2,537.97*	*100*	*—*
Te	*2,537.80*	*(300)*	*—*
Ir	*2,537.67*	*3*	*15*

	λ	*I*	
		J	*O*
Mn	**2,543.973**	**3*h***	**—**
Cl	**2,543.98**	**(10)**	**—**
Nb	**2,543.981**	**1**	**4**
W	**2,544.00**	**5*d***	**—**
U	**2,544.04**	**2**	**4**
W	**2,544.17**	**2**	**8**
Au	**2,544.19**	**8**	**30**
Re	**2,544.21**	**—**	**25**
Ru	**2,544.222**	**6**	**60**
Rh	**2,544.223**	**4**	**8**
Au	**2,544.25**	**10**	**—**
Co	**2,544.253**	**100**	**50*r***
Ta	**2,544.267**	**—**	**2**
Ce	**2,544.269**	**—**	**3**
Mn	**2,544.29**	**3**	**—**
Cr	**2,544.32**	**2**	**—**
U	**2,544.36**	**2**	**8**
Ir	2,544.56	5	—
Fe	2,544.71	5	100
Cu	2,544.802	700*R*	—
Nb	2,544.803	300	5
Pd	2,544.83	200	—
Rh	2,545.35	150	8
Ir	2,545.54	2	10
Al	2,545.60	(50)	—
Nb	2,545.64	150	1
Ir	2,545.78	—	2
Ni	2,545.90	900*h*	20
Fe	2,545.98	30	100*R*
Ir	2,546.03	20	100*R*
Sn	2,546.55	100	100
Co	2,546.74	50	1
Ni	2,547.19	100	—
Ru	2,547.67	80	5
Ir	2,547.69	5	15
Mn	*2,548.74*	*150*	*8*
Ru	*2,549.18*	*150*	*5*
V	*2,549.28*	*150*	*20*
Ni	*2,549.56*	*150*	*—*
Ir	*2,549.69*	*—*	*5*
Ru	*2,549.79*	*100*	*3*
Ir	*2,550.21*	*—*	*2*
Pd	*2,550.66*	*150*	*—*
Ir	*2,550.76*	*—*	*2*
Ir	*2,550.90*	*—*	*2*

	λ	I J	I O
Ir	*2,537.22*	*10*	*35*
Pd	*2,537.17*	*100*	—
Rh	*2,537.04*	*100*	*15*
Ir	*2,536.66*	—	*3*
Hg	*2,536.52*	*1000R*	*2000R*
Nb	*2,551.38*	*100*	*5*
Ir	*2,551.399*	*4*	*20*
Hf	*2,551.40*	*125d*	*25d*

Ir 2,639.71 *J* 15* *O* 100

	λ	I J	I O
Ru	**2,639.71**	**35**	**10**
U	**2,639.58**	**2**	**3**
Fe	**2,639.55**	**100**	**1**
Th	**2,639.51**	**4**	**8**
Cd	**2,639.50**	**15**	**75**
Yb	**2,639.44**	**15**	**5**
Cr	**2,639.43**	—	**30**
Ir	**2,639.42**	**5**	**15**
W	**2,639.40**	**6*d***	**1**
U	**2,639.38**	**2*h***	**2**
Ce	**2,639.37**	—	**3**
Pt	**2,639.35**	**50**	**500**
Ru	**2,639.12**	**5**	**60**
Ir	**2,638.97**	—	**5**
Eu	**2,638.76**	**200**	**300**
Ru	**2,638.51**	**4**	**60**
Ir	**2,637.31**	—	**3**
Os	2,637.13	30	150
Ta	2,636.90	3	100
Ir	2,636.88	—	3
Ta	2,636.673	1	70
Ru	2,636.670	—	60
Re	2,636.64	—	125
Fe	2,636.48	20	50
Pd	2,635.94	300	50
Ta	2,635.93	—	50*h*
Fe	2,635.81	200	300
Ir	*2,635.27*	*3*	*15*
Ir	*2,634.25*	*5*	*15*
Ir	*2,634.17*	—	*10*
Fe	*2,632.24*	*60*	*100*
Fe	*2,631.32*	*60*	*150*
Fe	*2,631.05*	*125*	*200*
Ce	**2,639.77**	—	**2**
Mn	**2,639.83**	**80*h***	**12**
U	**2,639.84**	**2**	**10**
Ru	**2,639.87**	—	**50**
Nb	**2,639.886**	**10**	**2**
U	**2,639.89**	**2**	**8**
Th	**2,639.895**	**2**	**5**
Hg	**2,639.93**	**10**	**5*wh***
Zr	**2,640.15**	—	**3**
Ru	**2,640.33**	**5**	**60**
Ir	**2,640.38**	**1**	**5**
Ti	2,641.10	20	150
Eu	2,641.26	60	100*w*
Fe	2,641.65	60	100
Re	2,642.76	—	125
Ru	2,642.96	—	150
Ta	2,643.89	—	50
Fe	2,644.00	150	150
Os	2,644.11	10	75
Ir	*2,644.19*	*5*	*35*
Ti	*2,644.26*	*12*	*100*
Ir	*2,645.41*	*5*	*10*
Ta	*2,646.37*	*2*	*125*
Pt	*2,646.89*	*100*	*1000h*
Re	*2,647.12*	—	*100*
Ta	*2,647.47*	*10*	*200*
Fe	*2,647.56*	*70*	*100*

Ir 2,664.79 *J* 50 *O* 200

	λ	*I*	
		J	*O*
Ru	**2,664.76**	**5**	**60**
La	**2,664.75**	**3**	**1**
Ce	**2,664.74**	—	**2**
Fe	**2,664.66**	**300**	**20**
Pt	**2,664.64**	—	**30**
Rh	**2,664.49**	**30**	**1**
Ir	**2,664.45**	**10*h***	—
Kr	**2,664.37**	**(4)**	—
W	**2,664.32**	**20**	**10**
Ta	2,664.23	100*h*	10*h*
Mn	2,664.03	25	—
Re	2,664.63	—	150
Co	2,663.53	60*w*	15*w*
Ir	2,663.31	5	10
V	2,663.25	100	12
Pb	2,663.17	40	300*wh*
Ru	2,662.86	40	—
In	2,662.68	(30)	—
Ir	2,662.63	10	40
In	2,662.58	(30)	—
Mn	2,662.54	50	—
Ru	2,662.16	30	—
Te	2,662.11	(25)	—
Fe	2,662.06	40	70
Se	2,662.05	(25)	—
Ir	2,661.98	15	150*h*
Hf	2,661.87	40	25
Ru	2,661.61	150	80
V	2,661.42	80	100
Ta	2,661.34	10	200
Ir	2,661.26	4	2
Ru	2,661.17	100	20
Pd	2,661.14	100	—
Te	2,661.13	(30)	—
Na	2,661.00	(80)	—
Ir	2,660.99	—	5
Mo	2,660.58	125	25
Br	2,660.49	(25)	—
Ag	2,660.46	150	30
Al	2,660.39	60	150*R*
Ir	2,660.08	—	3
Nb	2,660.03	30	2
Ir	2,659.95	—	2
Pt	*2,659.45*	*500R*	*2000R*
Rh	*2,659.11*	*100*	—
Re	**2,664.81**	—	**25**
Sn	**2,664.93**	**(5)**	—
W	**2,664.96**	**6**	**12**
Ce	**2,665.00**	—	**2**
Yb	**2,665.03**	**60**	**10**
Er	**2,665.04**	**5**	**20**
Ga	**2,665.05**	**(40)**	—
Ir	**2,665.067**	—	**8**
Mn	**2,665.068**	—	**10**
Mo	**2,665.10**	**2**	**25**
Au	**2,665.15**	**5**	—
Zr	**2,665.177**	**3**	**1**
Mn	**2,665.185**	**15**	—
Ni	**2,665.249**	**125**	—
Nb	**2,665.251**	**300**	**3**
Ce	**2,665.271**	—	**5**
V	**2,665.275**	**3**	**2**
J	2,665.31	(30)	—
Ta	2,665.60	40*h*	80*d*
Hf	2,665.97	35	20
Yb	2,666.08	150	5
Ir	2,666.41	2	10
Ir	2,666.58	—	4
Nb	2,666.59	50	5
Fe	2,666.63	80	5
Mn	2,666.77	50*h*	8
Yb	2,666.97	150	5
Mn	2,667.00	50*h*	5
Cl	2,667.36	(40)	—
Ru	2,667.39	150	10
Ir	2,667.46	2	10
Nb	2,667.76	30	3
Ru	2,667.79	80	—
Eu	2,668.33	400	300*w*
Ta	2,668.619	100	1
In	2,668.623	(30)	—
In	2,668.68	(50)	—
Ir	2,668.99	5	50
Al	2,669.17	100	3
W	2,669.30	30*d*	15
Ru	2,669.42	100	—
Ir	2,669.46	5	10
Fe	2,669.50	25	50
Co	*2,669.910*	*100wh*	—

	λ	I J	I O
Tb	*2,658.91*	*500*	—
Ta	*2,658.86*	*50*	*25*
Cl	*2,658.74*	*(100)*	—
Pd	*2,658.72*	*300*	*20*
Fe	*2,658.25*	*80*	—
Nb	*2,658.03*	*200*	—
Lu	*2,657.90*	*150*	*50*
Ir	*2,657.71*	*5*	*10*
Pd	*2,657.56*	*100*	—
Ir	*2,657.50*	*3*	*3*
Rh	*2,657.32*	*70*	—
Ru	*2,657.19*	*50*	—
Ir	*2,656.81*	*3*	*15*
Ta	*2,656.61*	*2*	*200R*
Ru	*2,656.25*	*150*	*20*
Nb	*2,656.08*	*200*	*8*
Ni	*2,655.91*	*500wh*	—
V	*2,655.68*	*100h*	*10*
Ni	*2,655.47*	*400wh*	—

	λ	I J	I O
Ir	*2,669.913*	*10*	*60*
V	*2,670.23*	*70*	*2*
Ni	*2,670.33*	*80*	—
Zn	*2,670.53*	*4*	*200*
Mn	*2,671.80*	*50*	—
Na	*2,671.829*	*(60)*	*12*
Mo	*2,671.834*	*100*	*1*
Ir	*2,671.84*	*10*	*50*
Nb	*2,671.92*	*200*	*20*
V	*2,672.00*	*300R*	*50*
Cl	*2,672.19*	*(50)*	—
Mn	*2,672.59*	*125h*	*15*
Yb	*2,672.65*	*80*	*20*
Ir	*2,672.80*	—	*4*
Mo	*2,672.84*	*100*	*15*
Ru	*2,673.01*	*50*	*8*
V	*2,673.23*	*60*	*10*
Mo	*2,673.27*	*100*	*1*
Mn	*2,673.37*	*50*	—
Eu	*2,673.41*	*50*	*125*
Nb	*2,673.57*	*500*	*10*
Ir	*2,673.61*	*10*	*40*
Ru	*2,674.19*	*50*	—
Rh	*2,674.44*	*200w*	*1*
Pt	*2,674.57*	*10*	*200*

Ir 2,694.233 *J* 50 *O* 150

	λ	J	O
Br	**2,694.23**	**(5)**	—
U	**2,694.22**	**4**	**10**
La	**2,694.21**	**3**	**1***h*
Yt	**2,694.20**	—	**8**
V	**2,694.10**	**1**	**4**
Mn	**2,694.09**	**50**	**8**
J	**2,694.07**	**(30)**	—
Zr	**2,694.06**	**15**	**15**
V	**2,693.92**	**2***h*	**10**
Rb	**2,693.91**	**(2)**	—
Pd	**2,693.882**	**15***wh*	—
In	**2,693.882**	**(30)**	—
Fe	**2,693.86**	**30**	—
Cr	2,693.52	40	1
Ir	2,693.49	—	10
Ru	2,693.29	2*h*	80
Mo	2,693.18	25	1
Co	2,693.12	25	—

	λ	J	O
Pt	**2,694.24**	—	**2**
Rh	**2,694.308**	**4**	**5**
Nb	**2,694.315**	**5**	**1***h*
Au	**2,694.37**	**3**	—
W	**2,694.38**	**18**	**9**
Co	**2,694.40**	—	**25**
V	**2,694.47**	**20**	**1**
Os	**2,694.52**	**3**	**10**
Ir	**2,694.53**	—	**2***h*
Fe	**2,694.54**	**35**	**100**
Te	**2,694.55**	**(15)**	—
W	**2,694.59**	**18**	—
V	**2,694.65**	**10***h*	—
Co	**2,694.68**	**200***w*	**25**
V	2,694.74	70	2
Mo	2,695.22	40	5
Mn	2,695.36	50	100*R*
La	2,695.46	35	3

	λ	I J	I O
Co	2,693.01	25	—
Ir	2,692.88	—	10
Mn	2,692.66	—	150
Mo	2,692.61	40	2
Fe	2,692.60	300	—
Ta	2,692.39	1	100
Ir	2,692.34	2	15
Sb	2,692.25	40	40
Pt	2,692.24	40	4
Ir	2,692.19	—	5
Ru	2,692.06	200	8
Eu	2,692.02	200	200
Nb	2,691.77	100	10
Fe	2,691.73	35	—
Ta	2,691.31	—	150
Ga	2,691.29	(25)	—
Rh	2,691.12	60	1
Ir	2,691.06	—	15
Cr	2,691.041	125	35
U	2,691.038	30	15
Yb	2,690.99	30	—
Mn	2,690.98	25W	—
V	2,690.79	300R	70
Ni	2,690.64	250h	1
Ir	2,690.58	10h	—
V	2,690.24	200R	50
Fe	2,690.07	30	30
V	2,689.88	150R	50
Fe	2,689.83	40	40
Rh	2,689.62	100	1
Cu	*2,689.30*	*300*	—
Fe	*2,689.21*	*150*	*150*
J	*2,688.99*	*(100)*	—
V	*2,688.71*	*100R*	*35*
Pd	*2,688.55*	*200*	—
Mn	*2,688.25*	*100h*	*3*
Ru	*2,688.11*	*100*	*8*
Cl	*2,688.04*	*(150)*	—
Mo	*2,687.99*	*100*	*30*
V	*2,687.96*	*500R*	*150*
Pd	*2,687.66*	*150*	—
Ru	*2,687.50*	*100*	*12*
Mn	*2,687.41*	*80r*	*25*
Cr	*2,687.09*	*60*	*30*
Ru	*2,687.07*	*50*	—
Nb	*2,686.39*	*300*	*2*
Mn	*2,685.94*	*100w*	*12h*
Eu	*2,685.65*	*100*	*150*
Ir	2,695.47	5	5
U	2,695.49	30	12
Fe	2,695.53	30	40
Te	2,695.55	(50)	—
Kr	2,695.70	(30h)	—
Ir	2,695.93	2	8
Fe	2,696.00	50	80
Fe	2,696.28	50	90
Ni	2,696.49	50	2
Bi	2,696.61	100	100
Ta	2,696.81	1	125
Mo	2,696.83	40	1
V	2,696.99	2	70
Fe	2,697.02	25	50
Co	2,697.04	60	—
Nb	2,697.06	500	10
Ru	2,697.07	30	4
Fe	2,697.46	50	3
Cr	2,697.50	35	1
W	2,697.71	25	15
V	2,697.74	50R	100
Cr	2,697.91	35	3
U	2,698.06	50	20
Ta	2,698.30	3	150
J	2,698.32	(50)	—
V	2,698.38	300	30
Cr	2,698.41	35	12
Pt	2,698.43	50	500
Ti	2,698.52	200h	—
Pd	2,698.55	200	—
Ir	2,698.61	1	5
Nb	2,698.86	200	5
Fe	2,699.11	60	100
Zr	*2,700.13*	*50*	*50*
Ru	*2,700.15*	*100*	—
Ga	*2,700.47*	*(70)*	—
Rh	*2,700.59*	*80*	*2*
V	*2,700.94*	*500R*	*125*
Cu	*2,700.96*	*400*	*20*
Ru	*2,700.99*	*50*	—
Ir	*2,701.11*	*1*	*4*
Eu	*2,701.12*	*60h*	*60h*
Mo	*2,701.42*	*100*	*20*
Mn	*2,701.70*	*40h*	*150*
Lu	*2,701.71*	*150*	*40*
Eu	*2,701.89*	*200*	*300w*
V	*2,702.19*	*300R*	*80*

	λ	I J	I O
Nb	*2,702.20*	*100*	*10*
Pt	*2,702.40*	*300*	*1000*
Nb	*2,702.52*	*80*	*8*

Ir 2,849.72 *J* 20* *O* 40

	λ	I J	I O
Nb	**2,849.56**	**100***w*	**2***w*
Ta	**2,849.55**	**1***h*	**5***h*
U	**2,849.48**	**15**	**18**
W	**2,849.46**	**6**	**7**
Mo	**2,849.38**	**5**	**50**
Rh	**2,849.34**	—	**3**
Os	**2,849.30**	**4**	**8**
Ru	**2,849.289**	**100**	**3**
Cr	**2,849.287**	**30**	**35**
Hf	**2,849.21**	**100**	**30**
Ce	**2,849.20**	—	**2**
Fe	**2,849.19**	—	**2**
V	**2,849.17**	—	**25**
V	2,848.77	2	20
Fe	2,848.72	30	60
Ru	2,848.58	2	50
Ta	2,848.525	50	300
Zr	2,848.523	—	100
Ir	2,848.44	—	2
Mg	2,848.42	—	20
Os	2,848.25	15	30
Mo	2,848.23	200*h*	125
Ta	2,848.05	15	150
Lu	2,847.51	125	40
Re	2,846.98	—	30*w*
Fe	2,846.82	12	20
Ta	2,846.75	10*h*	150*hs*
Ir	2,846.65	—	10
V	2,846.57	20*h*	50
Os	2,846.39	10	40
Cr	2,846.02	4	25
Ta	2,845.84	10*h*	30
Hf	2,845.83	5	25
Fe	2,845.59	7	125
Fe	2,845.54	7	125
Ta	2,845.35	10	150
Lu	2,845.13	2	30*h*
Ir	2,844.85	—	5
Ta	2,844.76	30	150
Ta	**2,849.82**	**50***w*	**15***ws*
Cr	**2,849.84**	**150***r*	**80**
U	**2,849.98**	**8***h*	**5**
Co	**2,850.04**	—	**75**
Fe	**2,850.11**	—	**2**
Hf	**2,850.15**	**20**	**20**
Ir	**2,850.25**	—	**5**
Ta	2,850.49	100	200
*bh*B	2,850.6	—	50
Sn	2,850.62	100*wh*	80
Os	2,850.76	25	75
Mo	2,850.79	—	20
Co	2,850.95	—	30
Hf	2,850.96	5	25
Re	2,850.977	—	40
Ta	2,850.985	150	400
Ir	2,851.03	—	2
Ti	2,851.10	80	20
Sb	2,851.11	45	50
Hf	2,851.21	50	25
Cr	2,851.36	80	20
Ir	2,851.41	—	8
Mg	2,851.65	—	25
V	2,851.75	4	30
Fe	2,851.80	150	200
Hf	2,852.01	50	20
Ce	2,852.12	1	50*d*
Mg	2,852.129	100*R*	300*R*
Fe	2,852.13	80	150
Ir	2,852.131	—	20
Ir	2,852.48	—	2
Na	2,852.83	20	100*R*
Re	2,852.86	—	30*w*
V	2,852.87	7*h*	60
Na	2,853.03	15	80*R*
Mo	2,853.23	100*h*	25
Ir	2,853.31	—	8
Ru	2,854.07	35	60
V	2,854.34	100*R*	20

	λ	I			λ	I	
		J	OI			J	O
Tm	2,844.67	15	40	Ce	2,864.67	—	30s
Zr	2,844.58	50	50	Ce	2,854.88	—	25s
Ta	2,844.46	200	200	V	2,855.22	1	50
Os	2,844.40	25	50	Cr	2,855.68	200wh	60
Mo	2,844.39	5	30	Ir	2,855.82	2h	10
Ta	2,844.25	50	400r	Ir	2,855.93	—	10
Fe	2,843.98	300	300	Rh	2,856.16	20h	60
Fe	2,843.63	100	125	Cr	2,856.77	60	20
Cr	2,843.25	400r	125	Ir	2,856.94	2h	10
Ta	2,842.81	50	200	Ta	2,857.28	5	60
Ir	2,842.28	2	12	V	2,857.94	7h	50
Ti	2,841.94	125	40	Ta	2,858.43	300	100
Ir	2,841.69	2	7	Mn	2,858.66	—	50
Ru	2,841.68	200	50	Fe	2,858.90	30	100
Ru	2,840.54	8	60	Cr	2,858.91	80wh	50
Fe	2,840.42	20	125	V	2,858.98	20	40
Re	2,840.35	—	40	Ir	2,859.02	—	4
Gd	2,840.24	60	50	Co	2,859.66	—	40
Ir	2,840.22	10	15	Eu	2,859.67	—	40
Sn	2,839.99	300R	300R	V	2,859.97	10	50
Ir	2,839.24	—	12	Ru	2,860.02	12	60
Ir	2,839.16	15	25	As	2,860.45	50	50r
				Ir	2,860.66	2	12

Ir 3,133.32 *J* 2* *O* 40

	λ	J	O		λ	J	O
Eu	3,133.24	—	10	Ce	3,133.327	—	20
Zr	3,133.231	—	5	V	3,133.328	200r	50
Fe	3,133.226	4	5	Ce	3,133.41	—	5
Cd	3,133.17	300	200	U	3,133.42	6	8
Tb	3,133.15	—	8	Zr	3,133.47	12	6
Ti	3,133.13	—	4	Hf	3,133.50	5	15
Hf	3,133.10	3	5	Ce	3,133.53	—	2
Sc	3,133.096	10	2	Ta	3,133.55	3	15
Gd	3,133.092	1	3	Nd	3,133.60	10	15
Ir	3,133.086	1	20	Th	3,133.62	10	10
Nb	3,133.08	4	3	Ru	3,133.70	—	12
Dy	3,132.98	2	6	W	3,133.72	2h	6
Ru	3,132.88	5	60	Gd	3,133.86	25	25
Ce	3,132.86	—	3	W	3,133.889	10	10
Cr	3,132.82	2	20	Tm	3,133.89	200	200
Mn	3,132.79	—	8	U	3,133.92	5	6
Er	3,132.775	5	15	Re	3,134.02	—	30
Ag	3,132.775	2h	1h				
Nb	3,132.76	15	1				

	λ	I	
		J	O
Ti	**3,132.71**	**—**	**2**
Fe	**3,132.68**	**3**	**5**
Ta	**3,132.643**	**25**	**250*w***
Ce	**3,132.640**	**—**	**15**
Yb	**3,132.60**	**6**	**15**
Mo	**3,132.594**	**300*R***	**1000*R***
V	**3,132.594**	**20**	**80*r***
Ce	**3,132.590**	**—**	**25**
Fe	3,132.51	40	70
Co	3,132.22	—	40
Cr	3,132.06	125	25
Hg	3,131.83	100	200
Hf	3,131.81	10	40
Eu	3,131.75	5	20*w*
Hg	3,131.55	300	400
Os	3,131.48	10	20
Tm	3,131.26	500	400
Cr	3,131.21	6	20
Os	3,131.11	30	125
Be	3,131.07	150	200
Ce	3,130.87	2	30
Ti	3,130.80	100	25
Rh	3,130.790	2	60
Nb	3,130.786	100	100
Eu	3,130.74	130	100*w*
Ir	3,130.578	—	3
Ta	3,130.578	35	100*w*
Be	3,130.42	200	200
Ce	3,130.33	1	30
Ir	3,130.28	—	4
V	3,130.27	200*r*	50
Ag	3,130.01	15*h*	25*h*
Os	3,130.00	9	30
Ta	3,129.95	8	50
Ru	3,129.84	4	60
Ru	3,129.60	1	50
Ta	3,129.55	7	50
Co	3,129.48	2	40
Na	3,129.37	(60)	35
Fe	3,129.33	60	100
Ni	3,129.31	—	125
Ir	3,129.24	1*h*	2*h*
Os	3,129.23	15	60
Co	3,129.00	—	25
Re	3,128.95	—	100*W*
Hf	3,128.75	1	20
Cu	3,128.701	15	70

	λ	I	
		J	O
Ni	3,134.108	150	1000*R*
Fe	3,134.111	125	200
Hf	3,134.72	125	80
Nd	3,134.90	30	40
V	3,134.93	150*r*	30
Dy	3,135.37	50	100
Ta	3,135.89	100	35
Ir	3,136.10	—	2*h*
Ir	3,136.48	—	3
Fe	3,136.50	40	60
V	3,136.51	200	20
Ru	3,136.55	6	60
Cr	3,136.68	50	20
Ce	3,136.72	—	25
Co	3,136.73	—	60
Ir	3,136.90	—	3
Ir	3,137.27	—	2*h*
Co	3,137.33	—	150*r*
Co	3,137.45	—	50
Hf	3,137.51	10	30
Os	3,137.52	10	25
Ce	3,137.60	—	25
Rh	3,137.71	—	100
Ir	3,137.72	3	8
Co	3,137.75	2	60
Ir	3,139.17	—	3
Ru	3,139.27	—	20
Pt	3,139.39	80	300
U	3,139.56	25	25
Ir	3,139.59	—	5
Tb	3,139.64	15	30
Hf	3,139.65	20	25
Re	3,139.80	—	25*w*
Fe	3,139.91	40	70
Co	3,139.94	10	150*r*
Cu	3,140.312	12	50
Os	3,140.314	12	60
Fe	3,140.39	80	100
Ir	3,140.41	1	50*r*
Ru	3,140.48	—	50
Dy	3,140.64	20	40
Co	3,140.72	3	20
Ir	3,140.74	—	2
Hf	3,140.76	25	25
Os	*3,140.94*	*12*	*50*
Ru	*3,140.97*	*6*	*60*
Dy	*3,141.13*	*20*	*50*
Pt	*3,141.66*	*5h*	*150*

	λ	I J	I O
Cr	3,128.699	150	30
Os	3,128.44	5	20
Dy	3,128.41	10	40
Ir	3,128.39	1	20
Ce	3,127.53	—	40
Co	3,127.25	—	100
Ce	3,127.10	—	20
Co	3,126.72	—	70
Co	3,126.49	—	20
V	3,126.21	100R	60
Fe	3,126.17	70	150
Cu	3,126.11	20	80
Tm	3,126.01	—	25
Ru	3,125.96	12	70
Hg	*3,125.66*	*150*	*200*
Fe	*3,125.65*	*300*	*400*
V	*3,125.28*	*200R*	*80*
Ir	*3,125.25*	—	*2h*
Ta	*3,124.97*	*20*	*50*
Os	*3,124.94*	*10wh*	*150h*
Ir	*3,124.93*	—	*4*
Ge	*3,124.82*	*80*	*200*
Ru	*3,124.61*	*2*	*50*
Ru	*3,124.17*	*8*	*60*
Ir	*3,124.08*	—	*4*
Ir	*3,123.92*	—	*2*
Rh	*3,123.70*	*2*	*150*
Ir	*3,123.22*	—	*2*
Au	*3,122.78*	*5*	*500h*
Fe	*3,122.30*	*20*	*70*
Ir	*3,121.78*	*1*	*35*
Rh	*3,121.75*	—	*150*
Pr	*3,121.571*	*4*	*50*
Co	*3,121.57*	*3*	*60r*
Co	*3,121.41*	*6*	*150r*
Re	*3,121.37*	—	*100*
V	*3,121.14*	*200r*	*60*
Fe	*3,120.87*	*50*	*80*
Ir	*3,120.76*	*2*	*50*
Fe	*3,120.43*	*80*	*100*
Cr	*3,120.37*	*150*	*40*
Mn	*3,120.34*	—	*50*
Ir	*3,119.67*	*1*	*5*
As	*3,119.60*	*50*	*100*
Fe	*3,119.49*	*80*	*100*
Ir	*3,118.84*	*2*	*2*
Ru	*3,118.68*	*3*	*50*

	λ	I J	I O
Ir	*3,141.81*	—	*2h*
Cu	*3,142.44*	*15*	*60*
Fe	*3,142.45*	*100*	*125*
Ir	*3,142.51*	—	*2h*
Mn	*3,142.67*	—	*50*
La	*3,142.76*	*50*	*150*
Pd	*3,142.81*	*100*	*300*
Fe	*3,142.88*	*77*	*80*
Ir	*3,143.01*	—	*2*
Fe	*3,143.24*	*30*	*60*
Ir	*3,143.80*	—	*3*
Fe	*3,143.99*	*150*	*200*
Ru	*3,144.26*	*8*	*60*
Cr	*3,144.41*	*12*	*50*
Fe	*3,144.49*	*100*	*150*
Gd	*3,145.01*	*30*	*50*
Fe	*3,145.06*	*25*	*40*
Ir	*3,145.07*	*2*	*20*
Hf	*3,145.32*	*20*	*50*
Ni	*3,145.72*	*3*	*200*
Os	*3,145.96*	*10*	*60*
Cu	*3,146.82*	*20*	*100*
Tb	*3,147.04*	*3*	*50*
Co	*3,147.06*	—	*150R*
Ru	*3,147.21*	*3*	*50*
Cr	*3,147.23*	*150*	*25*
Ta	*3,147.37*	*50*	*70*
Fe	*3,147.79*	*15*	*40*
Ta	*3,148.03*	*7*	*50*
Mn	*3,148.18*	*40*	*150*
Fe	*3,148.41*	*40*	*100*

	λ	I J	I O		λ	I J	I O
Lu	*3,118.43*	*5*	*40*				
V	*3,118.38*	*200R*	*70*				
Os	*3,118.33*	*20*	*150*				

Ir 3,220.790 *J* 30* *O* 100

	λ	I J	I O		λ	I J	I O
Pt	**3,220.778**	—	**2**	**Rh**	**3,220.78**	**2**	**4**
Er	**3,220.73**	**5**	**25**	**Mo**	**3,220.85**	**8**	**8**
Hf	**3,220.61**	**35**	**25**	**Ce**	**3,220.87**	—	**30**
Pb	**3,220.54**	**5**	**50*h***	**Nb**	**3,220.93**	**10**	**10**
Nd	**3,220.53**	**4*h***	**2*h***	**Ce**	**3,221.06**	—	**2**
Nd	**3,220.49**	**5**	**3*h***	**Tm**	**3,221.08**	—	**7**
Dy	**3,220.46**	**2**	**10**	**Nb**	**3,221.12**	**5**	**4**
Ce	**3,220.40**	—	**12*s***	**Ti**	**3,221.15**	—	**4**
Th	**3,220.30**	**10**	**12**	**Ce**	**3,221.17**	**8**	**50**
Ti	**3,220.28**	—	**3*h***	**Ru**	**3,221.19**	**3**	**4**
Os	**3,220.19**	**10**	**30**	**W**	**3,221.21**	**10*d***	**12*d***
Tb	**3,220.17**	—	**15**	**Yb**	**3,221.22**	**3**	**2**
Ru	**3,220.07**	**3**	**4**	**Ni**	**3,221.27**	—	**35**
W	**3,220.06**	**7**	**8**	**Ir**	**3,221.28**	**1*h***	**10**
Tb	**3,219.95**	**50**	**50**	**Tb**	**3,221.29**	**3**	**8**
Ce	**3,219.948**	—	**2**	**Th**	**3,221.293**	**40*h***	**15**
Re	**3,219.92**	—	**2**	**Ta**	**3,221.31**	**15*s***	**70**
Sm	**3,219.86**	**2**	**10**	**Ti**	**3,221.381**	**6**	**25**
Ni	**3,219.811**	—	**4**	**Os**	**3,221.383**	**5**	**15**
				U	**3,221.41**	**2**	**6**
Fe	3,219.810	80	100	**Ce**	**3.221.47**	—	**2**
Fe	3,219.58	125	200	**Dy**	**3,221.50**	**2**	**12**
Ir	3,219.51	2	35	**W**	**3,221.62**	**6**	**7**
Co	3,219.15	—	60	**Ba**	**3,221.63**	—	**2**
Pd	3,218.97	8	300	**Ni**	**3,221.65**	**4**	**200**
Ce	3,218.94	8	50	Fe	3,222.07	100	200
Tb	3,218.93	50	50	Ir	3,223.01	—	2
Cr	3,218.69	2*wh*	80*wh*	Ru	3,223.27	35	60
Sm	3,218.60	25	100	Ir	3,223.51	—	3
Ir	3,218.46	1	20	Ta	3,223.83	50*W*	200*W*
Rh	3,218.28	—	60	Os	3,223.86	8	100
Rh	3,217.88	20	60	Ir	3,223.89	1*h*	3
K	3,217.50	25	50*R*	Co	3,224.64	—	60
Fe	3,217.38	125	200	Mn	3,224.76	40	75
K	3,217.02	20*h*	100*R*	Ni	3,225.02	6	300
Mn	3,216.95	75	75	Nb	3,225.48	800*wr*	150*w*
Ta	3,216.92	18*w*	100*w*	Fe	3,225.79	150	300
Ir	3,216.80	—	3*h*	Ca	3,225.90	10	80
Fe	3,215.94	150	300	Ru	3,226.37	12	50
Nb	3,215.59	200	50	Ir	3,226.71	1	20
Sm	3,215.24	15	50	Ni	3,226.984	—	100

	λ	I	
		J	O
Fe	3,214.40	50	100
Rh	3,214.32	20	70
Fe	3,214.04	200	400
Eu	3,213.75	20h	100
Ir	3,213.45	2	10
Fe	3,213.314	300	50
Os	3,213.312	40	50
Re	3,212.94	—	50
Mn	3,212.88	100	100
Eu	3,212.81	20	200
V	3,212.43	50	70
Ir	3,212.22	—	5
Ir	3,212.12	15	25
Sm	3,211.75	15	100
Fe	3,210.83	100	150
Fe	3,210.24	100	150
Fe	3,209.30	125	200
Mo	3,208.83	60	150r
Fe	3,208.47	80	100
Ir	3,208.15	—	10
Ta	3,207.85	15	70
Ir	3,207.09	—	5
Fe	3,205.40	200	300
Ir	3,205.09	5	20
Re	3,204.20	—	300
Pt	3,204.04	100	250
Co	3,226.985	—	80r
Os	3,227.28	12	125
Ta	3,227.32	10	70
Fe	3,227.75	300	200
Mn	3,228.09	100	100
Fe	3,228.25	80	100
Ru	3,228.53	150	50
Fe	3,228.90	40	80
Fe	3,229.12	50	80
Os	3,229.21	5	125
Ta	3,229.24	70w	300w
Tl	3,229.75	800	2000
Fe	3,230.21	80	100
Pt	3,230.29	6	100
Sm	3,230.54	30	100
Ir	3,230.76	1	20
Ta	3,230.85	18w	200
Fe	3,230.97	200	300
Os	3,231.42	12	150
Ir	3,232.00	1	20
Os	3,232.05	20	500R
Sb	3,232.50	250wh	150
Os	3,232.54	10	150
Li	3,232.61	500	1000R
Ni	3,232.96	35	300R
Fe	3,233.05	60	100
Fe	3,233.97	150	300
Os	3,234.20	12	150
Ir	3,234.51	—	3
Ti	3,234.52	500r	100
Fe	3,234.61	125	200
Ni	3,234.65	15	300
Os	3,234.73	10	100
Fe	3,236.22	200	300
Sm	3,236.63	40	100
Tm	3,236.80	80	100
Co	3,237.03	—	100

Ir 3,513.64 *J* 100 *O* 100

	λ	J	O
Ta	3,513.61	1h	35
Ni	3,513.484	—	3
Co	3,513.480	25	300R
Ce	3,513.47	—	4
U	3,513.37	10	3
Gd	3,513.66	—	4
U	3,513.68	1	10
Cl	3,513.69	(12)	—
Mo	3,513.71	1	3
Th	3,513.76	3	3

	λ	I	
		J	O
Eu	3,513.34	—	10*w*
Ce	3,513.283	—	6
Hf	3,513.276	1	10
Pr	3,513.27	—	5
Cl	3,513.22	(35)	—
Th	3,513.219	3	3
As	3,513.13	5	—
Rh	3,513.102	3	50
Tb	3,513.10	8	30
Dy	3,513.10	—	4
Fe	3,513.06	2	10
Sm	3,513.048	4	15*d*
Ce	3,513.048	2	1
Cr	3,513.044	8	—
U	3,514.039	6	6
Tm	3,513.038	20	10
Os	3,512.99	30	—
Fe	3,512.96	1	2
W	3,512.93	7	5
La	3,512.92	15	50
Sm	3,512.909	3	6*d*
Nd	3,512.909	10	8
U	3,512.887	1	8
Yt	3,512.886	3	8
Ru	3,512.88	—	8
Ta	3,512.83	—	15
Er	3,512.77	—	2
Nb	3,512.76	2	3
Th	3,512.75	6	5
Dy	3,512.71	—	12
J	3,512.70	(18)	—
Er	3,512.69	1	9
Zr	3,512.68	3*h*	5
U	3,512.67	5	1
Cr	3,512.65	—	8*h*
Sm	3,512.650	3	15
Co	3,512.64	100	400*R*
W	3,512.62	5	7
Tb	3,512.60	3	15
Ce	3,512.57	—	5
Dy	3,512.56	—	10
Re	3,512.29	—	50
Ir	3,512.20	5	12
Cu	3,512.12	30	50
Ir	3,511.89	7	20
Cr	3,511.84	50	20
Rh	3,511.78	3	50
Rb	3,511.19	(60)	—

	λ	I	
		J	O
Ce	3,513.79	—	6
Tl	3,513.80	(3)	—
Fe	3,513.820	300	400
Os	3,513.822	3	10
Dy	3,513.84	—	4
Ce	3,513.856	—	8
Tb	3,513.86	—	30
V	3,513.877	30*h*	—
Rb	3,513.88	(10)	—
Ni	3,513.93	40*h*	200
Cs	3,514.02	(6)	—
Nb	3,514.04	50	—
La	3,514.06	3	8
Dy	3,514.10	2	4
Ru	3,514.14	—	4
W	3,514.16	10	—
Tb	3,514.18	—	8
Co	3,514.21	20	—
Eu	3,514.22	—	15
Zr	3,514.32	—	2
Ce	3,514.33	—	6
Ar	3,514.39	(125)	—
Ta	3,514.40	1*h*	5
V	3,514.42	40	1
Dy	3,514.43	—	6
Ru	3,514.488	40	70
Eu	3,514.49	5	18
Th	3,514.53	8	8
Xe	3,514.58	(4*wh*)	—
Rn	3,514.60	(12)	—
U	3,514.61	5	18
Fe	3,514.63	2	7
Zr	3,514.64	2	3
Pt	3,514.71	2	2
Ni	3,515.05	50*h*	1000*R*
Ne	3,515.19	(150)	—
Nb	3,515.42	300	20
Ir	3,515.95	15	35
P	3,516.15	(70)	—
Re	3,516.65	—	60
Pd	3,516.94	500*R*	1000*R*
Dy	3,517.27	4	70
Re	3,517.33	—	50
Nb	3,517.67	200	2
Co	3,518.35	100	200*w*
P	3,518.60	(50*h*)	—
Ni	3,518.63	8	90
Os	3,518.72	30	200

	λ	I	
		J	O
Nb	3,511.16	1	50
Ta	3,511.04	35w	100
Bi	3,510.85	30	200wh
Ti	3,510.84	125	40
Ne	3,510.72	(50)	—
Ir	3,510.64	10	20
Ni	3,510.34	50h	900R
Nb	3,510.26	200	15
Tb	3,510.10	8	50
Co	3,509.84	40	400R
Ru	3,509.72	2	50
Ir	3,509.24	2	8
Ru	3,509.20	100	10
Tb	3,509.17	200	200
V	3,509.04	150	2
Ir	3,508.72	—	2
Ir	3,508.58	1	6
Yb	3,507.83	60	12
Ni	3,507.69	12	100
V	3,507.54	50	—
Ir	3,507.61	2	1
Tb	3,507.45	8	50
Lu	3,507.38	150	100
P	3,507.36	(100w)	—
Rh	3,507.32	125	500
Ir	3,507.22	—	2
Dy	3,506.82	—	80
Fe	3,506.50	30	50
Co	3,506.31	15	400R
V	3,505.69	35	50
F	3,505.61	(600)	—
Gd	3,505.52	60	60
Dy	3,505.46	2	70
Hf	3,505.23	50	20
Ta	3,504.98	2	70
Ti	3,504.89	150	20
Ir	3,504.87	—	2
Os	3,504.66	20	300
Dy	3,504.52	3	90
V	3,504.44	200	60
Ir	3,504.12	—	5
Ta	3,503.87	10h	70
Kr	3,503.25	(50wh)	—
F	3,503.09	(400)	—
Re	3,503.06	—	80
P	3,502.99	(70)	—
F	3,502.95	(60)	—
Ir	3,502.94	4	8

	λ	I	
		J	O
Ir	3,518.91	—	2h
Tl	3,519.24	1000R	2000R
Zr	3,519.60	10	100
Ru	3,519.63	30	70
Tb	3,519.76	15	50
Ni	3,519.77	30	500h
V	3,520.02	50	5
Co	3,520.08	—	100W
Ru	3,520.13	40	60
Yb	3,520.29	70	10
Ne	3,520.472	(1000)	—
Sb	3,520.474	(125)	—
Eu	3,521.09	4h	50
Fe	3,521.26	200	300
Rb	3,521.44	(200)	—
Co	3,521.57	25	200r
V	3,521.839	80	20
Fe	3,521.841	20	50
Ir	3,522.03	50	—
Fe	3,522.28	30	50
Nb	3,522.36	50W	—
Co	3,523.43	25	300r
Ni	3,523.44	—	100
Os	3,523.64	30	150
Tb	3,523.66	50	30
Fe	3,524.07	40	50
Hg	3,524.19	(100)	—
Fe	3,524.24	50	60
Ni	*3,524.54*	*100wh*	*1000R*
Ir	*3,526.76*	*2*	*20*
Co	*3,526.85*	*25*	*300R*
Fe	*3,527.80*	*80*	*100*
Ni	*3,527.98*	*15*	*200*
Rh	*3,528.02*	*150*	*1000w*
Os	*3,528.60*	*50*	*400R*
Co	*3,529.03*	—	*200R*
Tl	*3,529.43*	*800*	*1000*
Co	*3,529.81*	*30*	*1000R*
Fe	*3,529.82*	*80*	*125*
Os	*3,530.06*	*20*	*100*
Ir	*3,530.74*	*2*	*10*
V	*3,530.77*	*100*	*40*
Rb	*3,531.60*	*(100)*	—
Dy	*3,531.71*	*100*	*100*
Hg	*3,532.63*	*(200)*	—
Os	*3,532.80*	*20*	*100*
Na	*3,533.01*	*(200)*	*50*

	λ	I J	I O		λ	I J	I O
Ni	*3,502.59*	*—*	*100*	*Co*	*3,533.36*	*—*	*200w*
Rh	*3,502.52*	*150*	*1000*				
Co	*3,502.28*	*20*	*2000R*	*Dy*	*3,534.96*	*—*	*125*
Co	*3,501.72*	*100*	*5*	*Ir*	*3,534.98*	*—*	*10*
F	*3,501.42*	*(200)*	*—*	*Nb*	*3,535.30*	*500*	*300*
Ne	*3,501.22*	*(150)*	*—*				
Os	*3,501.16*	*15*	*100*				
Ba	*3,501.12*	*20*	*1000*				
Ni	*3,500.85*	*80*	*500wh*				
Ir	*3,498.95*	*—*	*25*				
Ru	*3,498.94*	*200*	*500R*				
Ir	*3,498.74*	*—*	*15*				
Rh	*3,498.73*	*60*	*500*				
Sb	*3,498.46*	*300wh*	*—*				
Fe	*3,497.84*	*200*	*200*				
Mn	*3,497.54*	*150*	*15*				
S	*3,497.34*	*(100)*	*—*				
Fe	*3,497.11*	*100*	*200*				
V	*3,497.03*	*150*	*—*				
Co	*3,496.68*	*4*	*150R*				
Zr	*3,496.21*	*100*	*100*				
Mn	*3,495.84*	*150*	*25*				
Co	*3,495.69*	*25*	*1000R*				
Cd	*3,495.34*	*(100)*	*—*				
Fe	*3,495.29*	*60*	*100*				
Dy	*3,494.50*	*5*	*100*				
Mg	*3,493.85*	*(100)*	*—*				
V	*3,493.17*	*100*	*15*				
Ni	*3,492.96*	*100h*	*1000R*				
Rb	*3,492.76*	*(300)*	*—*				

Ir 4,268.10 *J* 15* *O* 200

	λ	J	O		λ	J	O
Mo	**4,268.07**	**20**	**20**	**Ta**	**4,268.25**	**15*h***	**50**
W	**4,268.05**	**5**	**10**	**Nd**	**4,268.28**	—	**3*d***
Co	**4,268.03**	—	**3*h***	**Ce**	**4,268.30**	—	**10**
Zr	**4,268.02**	**1**	**40**	**Dy**	**4,268.31**	—	**8**
U	**4,267.93**	**4**	**15**	**Pd**	**4,268.33**	—	**10*h***
Dy	**4,267.92**	—	**7**	**Co**	**4,268.44**	**2**	**2**
Ce	**4,267.852**	—	**2**	**Tm**	**4,268.56**	—	**3**
Eu	**4,267.85**	—	**10**	**V**	**4,268.64**	**20**	**40**
Fe	**4,267.83**	**60**	**125**	**Nb**	**4,268.67**	**10**	**5**
Pr	**4,267.78**	—	**12**	**Gd**	**4,268.74**	**40**	**40**
La	**4,267.74**	—	**8**	**Re**	**4,268.75**	—	**2*h***
Ce	**4,267.67**	—	**2**	**Fe**	**4,268.758**	**10**	**30**
U	**4,267.303**	**10**	**12**	**Ir**	**4,268.761**	—	**3**
Ta	**4,267.298**	**20**	**1**	**Cr**	**4,268.79**	**3**	**30**

	λ	I J	I O
Ce	**4,267.22**	—	**10**
Gd	**4,267.02**	**40**	**40**
Yb	**4,266.99**	**8*h***	**4**
Fe	**4,266.97**	**30**	**70**
Yt	**4,266.89**	—	**2**
Ir	4,266.04	—	10
Mn	4,265.92	50	100
Ir	4,265.30	2	60
La	4,263.58	150	150
Cr	4,263.14	80	125
Ti	4,263.13	35	125
Sm	4,262.68	150	200
Gd	4,262.09	10	150
Ir	4,261.89	—	10
Ti	4,261.60	8	70
Cr	4,261.35	50	125
Ir	4,261.26	—	25
Ir	4,260.90	—	10
Os	4,260.85	200	200
Fe	4,260.48	300	400
Ir	4,260.03	—	25
Ir	4,259.11	10	200
Mn	4,257.66	40	100
Re	4,257.59	—	125*w*
Ir	4,257.37	—	30
Cr	*4,254.35*	*1000R*	*5000R*
Sm	*4,251.79*	*200*	*200*
Gd	*4,251.74*	*10*	*300*
Fe	*4,250.79*	*250*	*400*
Fe	*4,250.13*	*150*	*250*
Fe	*4,247.43*	*100*	*200*

	λ	I J	I O
U	**4,268.85**	**10**	**12**
Ti	**4,268.93**	—	**8**
Ir	**4,268.94**	**2**	**10**
U	**4,269.08**	**1**	**3**
Pr	**4,269.10**	**4**	**12**
La	4,269.49	150	150
Ir	4,269.71	—	8
Fe	4,271.16	300	400
Fe	4,271.76	700	1000
Li	4,273.28	100*h*	200*h*
Gd	4,274.17	—	100
Ti	4,274.58	40	100
Cr	4,274.80	800*r*	4000*R*
Tb	4,278.51	100	200
Gd	*4,280.50*	*100*	*200*
Sm	*4,280.78*	*200*	*200*
Fe	*4,282.41*	*300*	*600*
Sm	*4,285.48*	*200*	*200*
Ir	*4,286.62*	*3*	*200*
La	*4,286.97*	*300*	*400*
Rh	*4,288.71*	*100*	*400*
Cr	*4,289.72*	*800r*	*3000R*

Ir 4,311.50 *J* 10* *O* 300

	λ	J	O
Os	**4,311.40**	**9**	**150**
Nb	**4,311.39**	**5**	**5**
Eu	**4,311.305**	**1**	**4*W***
Tb	**4,311.30**	—	**4**
Nb	**4,311.26**	**100**	**30**
Nd	**4,311.25**	**1**	**15**
Pr	**4,311.102**	**10*w***	**50**
W	**4,311.10**	**3**	**8**
Ag	**4,311.07**	**25**	**5**
Ho	**4,311.04**	**1**	**2**

	λ	J	O
Pr	**4,311.54**	**1**	**15**
Tb	**4,311.57**	**1*h***	**10**
Ce	**4,311.59**	**3**	**25**
Ti	**4,311.653**	**7**	**25**
Mo	**4,311.653**	**40**	**2**
Nb	**4,311.70**	**3**	**3**
La	**4,311.74**	**5*h***	**15**
Th	**4,311.80**	**1**	**6**
Pr	**4,311.92**	**2**	**10**
Dy	**4,311.93**	—	**4**

	λ	I J	I O
Ce	4,311.038	—	2
Eu	4,311.03	2	10
Mo	4,311.02	30	1
Gd	4,310.99	—	100
Tb	4,310.98	1	12
Ce	4,311.70	3	30*s*
Tb	4,310.68	—	3
Ir	4,310.59	8	150
Eu	4,310.58	2	7
Nd	4,310.51	6	20
Tb	4,310.45	1	20
Re	4,310.44	—	5
Er	4,310.43	—	2
Mo	4,310.390	12	15
Ce	4,310.388	—	10
U	4,310.386	8	8
Ti	4,310.37	1	10
Ir	4,309.77	—	3
Sm	4,309.00	150	200
Fe	4,307.91	800*R*	1000*R*
Ir	4,307.80	—	8
Gd	4,306.35	80	200
Ir	4,305.95	2	25
Ti	4,305.92	150	300
Pr	4,305.76	90	150
Cr	4,305.45	20	150
Ir	4,305.20	—	15
Ir	4,301.60	10	200
Ti	4,301.09	50	150
Ir	4,300.64	2	10
Fe	*4,299.24*	*400*	*500*
Fe	*4,294.13*	*400*	*700*
Nd	4,312.07	1	5
Tb	4,312.10	2	12
W	4,312.348	3	10
Pr	4,312.355	1	8
Yb	4,312.37	—	7
Dy	4,312.43	—	2
Cb	4,312.45	5	5
Cr	4,312.47	1	30
Ru	4,312.48	—	6
Mn	4,312.55	20	100
Ce	4,312.56	—	10
U	4,312.63	4	6
Gd	4,313.85	80	200
Fe	4,315.09	300	500
Gd	4,316.06	60	150
Ir	4,316.30	—	12
Tb	4,318.85	30	150
Sm	4,318.93	300	300
La	4,322.50	150	150
Gd	*4,325.69*	*250*	*500R*
Fe	*4,325.76*	*700*	*1000*
Gd	*4,327.10*	*100*	*500R*
Sm	*4,329.02*	*300*	*300*
Ir	*4,329.90*	*2*	*30*
La	*4,333.73*	*500*	*800*

Ir 4,399.47 *J* 100 *O* 400

	λ	I J	I O
Zr	4,399.44	—	3
V	4,399.42	4	7
Kr	4,399.39	(15*hs*)	—
Th	4,399.38	1	3
Pr	4,399.33	20	40
Eu	4,399.31	1	12*w*
Mo	4,399.22	4	4
Ce	4,399.20	6	35
Tb	4,399.18	—	4
Se	4,399.17	(15)	—
Cs	4,399.49	(20)	—
Ce	4,399.54	—	4
Nd	4,399.58	3	10
Ru	4,399.59	—	20
Eu	4,399.60	—	2*w*
*bh*Sr	4,399.6	—	6
Ni	4,399.61	5	10
U	4,399.63	6	15
Ca	4,399.64	10	—
Br	4,399.72	(10)	—

	λ	I	
		J	O
Cl	4,399.14	(15)	—
Th	4,399.10	8	8
J	4,399.09	(20)	—
Yb	4,398.95	1	5
Ce	4,398.79	3	20
Ni	4,398.625	3	2
Hg	4,398.62	(300)	—
Tb	4,398.61	—	2
J	4,398.59	(8)	—
Ce	4,398.54	—	4
Mo	4,398.49	12	6
Te	4,398.45	(70)	—
Ta	4,398.449	10	40
Ti	4,398.31	10	3
W	4,398.27	1*h*	5
Pr	4,398.26	9	25
Yt	4,398.01	100	150
Ne	4,397.94	(100)	—
Pr	4,396.12	50	80
Te	4,396.00	(100)	—
O	4,395.95	(80)	—
Xe	4,395.77	(200*wh*)	—
Ne	4,395.56	(50)	—
Ti	4,395.03	150	50
Xe	4,393.20	(200*wh*)	—
Ir	4,392.59	4	100
Gd	4,392.07	100	100
Ne	4,391.94	(150)	—
Gd	4,390.95	100	100
Sm	4,390.86	150	150
Ru	4,390.43	80	150*R*
Gd	4,389.99	80	30
V	4,389.97	60*R*	80*R*
Te	4,389.92	(50)	—
Fe	4,388.41	50	125
Eu	4,387.88	—	200
Kr	*4,386.54*	*(300h)*	—
Ac	*4,386.37*	*100*	—
P	*4,385.33*	*(100)*	—
Cr	*4,384.98*	*200*	*150*
V	*4,384.72*	*125R*	*125R*
Xe	*4,383.91*	*(100)*	—
Fe	*4,383.55*	*800*	*1000*
Se	*4,382.87*	*(800)*	—
Mo	*4,381.64*	*150*	*150*
Kr	*4,381.52*	*(100h)*	—

	λ	I	
		J	O
Dy	4,399.73	—	3
Ti	4,399.77	100	40
Cr	4,399.82	3	20
Re	4,399.84	—	12*w*
Sm	4,399.86	15	20
Se	4,399.94	(12)	—
Kr	4,399.97	(200)	—
*bh*La	4,400.0	—	2
Pr	4,400.03	20	30
Ar	4,400.09	(30)	—
Tb	4,400.10	—	3
Dy	4,400.10	2	4
*bh*La	4,400.1	—	4
Ce	4,400.15	—	6
Gd	4,400.18	5	10
Sm	4,400.19	2	2
W	4,400.21	6	12
Zr	4,400.24	—	8
Pr	4,400.25	10	25
Fe	4,400.351	1	20
Nb	4,400.354	10	5
Sc	4,400.355	30	150
Th	4,400.39	3	6
Tb	4,400.51	—	2
U	4,400.53	2	1
Ce	4,400.54	2	10
V	4,400.57	40	60
Os	4,400.579	1	18
Ti	4,400.582	2	25
Mo	4,400.66	3	3
Kr	4,400.87	(100*h*)	—
P	4,400.99	(50)	—
Se	4,401.02	(100)	—
Ir	4,401.25	—	12
Ni	4,401.55	30	1000*W*
Gd	4,401.85	100	200
Te	4,401.89	(100)	—
Hg	4,402.06	(50)	—
Gd	4,403.14	100	100
Sm	4,403.36	50	50
Ir	4,403.78	10	300
Fe	4,404.75	700	1000
Hg	4,404.86	(50)	—
Pr	4,405.85	100	100
Se	4,406.58	(70)	—
Gd	4,406.67	200	70
Ir	4,406.76	2	25
Xe	4,406.88	(100*wh*)	—
Fe	4,407.72	50	100

	λ	I	
		J	O
Gd	*4,380.64*	*125*	*100*
Ne	*4,379.50*	*(100)*	—
V	*4,379.24*	*200R*	*200R*
Sm	*4,378.23*	*100*	*100*
Mo	*4,377.76*	*200*	*5*
Ir	*4,377.01*	*4*	*100*
Kr	*4,376.12*	*(800)*	—
Fe	*4,375.93*	*200*	*500*
Gd	4,408.26	150	100
Fe	4,408.42	60	125
Pr	4,408.84	100	125
J	4,408.96	(250)	—
Ne	4,409.30	(150)	—
Sm	4,409.34	100	100
Ru	4,410.03	80	150
Kr	4,410.37	(50)	—
Te	4,410.95	(50)	—
Ti	4,411.08	100	7
Gd	4,411.16	50	100
Ir	4,411.18	2	40
Ac	*4,413.17*	*100*	—
P	*4,414.28*	*(100)*	—
Cd	*4,414.63*	*200*	—
Xe	*4,414.84*	*(150)*	—
O	*4,414.89*	*(300)*	—
Fe	*4,415.12*	*400*	*600*
O	*4,416.97*	*(150)*	—
Gd	*4,419.04*	*200*	*200*
Os	*4,420.47*	*100*	*400R*
Sm	*4,420.53*	*200*	*200*
Sm	*4,421.14*	*150*	*150*
Ir	*4,421.96*	*2*	*20*
Ne	*4,422.52*	*(300)*	—
Fe	*4,422.57*	*125*	*300*
Kr	*4,422.70*	*(100hs)*	—

Ir 4,426.27 *J* 10* *O* 400

Hf	4,426.18	8	3*h*
Gd	4,426.15	2	50
U	4,426.14	1	3
Nd	4,426.10	—	5
Ce	4,426.08	—	5
Ti	4,426.05	25	80
Ru	4,426.01	—	10
V	4,426.00	15*h*	25*h*
Th	4,425.99	4	6
Sm	4,425.98	7	9
Tm	4,425.96	—	3
Ce	4,425.92	—	4
W	4,425.91	6	15
U	4,425.87	2	4
Pr	4,425.86	—	2
Tb	4,426.30	—	5
Ce	4,426.301	—	4
Pr	4,426.32	—	3
Tm	4,426.34	—	4
Eu	4,426.42	2	4*w*
Pr	4,426.51	—	3
Mo	4,426.67	30	30
U	4,426.676	15	18
Nb	4,426.685	8	8
Er	4,426.77	—	12
Eu	4,426.80	—	10*W*
Nd	4,426.82	2	12
Dy	4,426.87	2	4
U	4,426.94	1	20
Pr	4,426.96	—	2

	λ	I J	I O
Ti	**4,425.83**	**1**	**10**
Dy	**4,425.82**	—	**3**
*bh*Ca	**4,425.8**	—	**3**
Re	**4,425.77**	—	**3**
Ir	**4,425.76**	—	**10**
V	**4,425.71**	**7**	**9**
Ce	**4,425.61**	—	**6**
Pr	**4,425.50**	—	**3**
Ca	**4,425.44**	**20**	**100**
U	**4,425.41**	—	**8**
Ce	**4,425.33**	—	**3*s***
Tb	**4,425.25**	—	**2*h***
Pr	**4,425.22**	—	**2**
U	**4,425.20**	**5**	**2**
Cr	**4,425.13**	**1**	**15**
Ce	**4,425.12**	—	**2**
Sm	4,424.34	300	300

	λ	I J	I O
Ce	**4,427.07**	**3**	**20**
Ti	**4,427.10**	**60**	**125**
Zr	**4,427.24**	—	**10**
Fe	**4,427.312**	**200**	**500**
V	**4,427.312**	**15**	**20**
W	**4,427.38**	**3**	**10**
Tb	**4,427.39**	—	**4**
Yb	**4,427.44**	—	**5**
Pr	4,429.24	125	200
La	4,429.90	300	200
Fe	4,430.62	8	200
Sm	4,433.88	200	200
Sm	4,434.32	200	200
Eu	4,435.53	—	2000
Eu	4,435.60	100	400*R*
Fe	4,442.34	200	500
Pt	4,442.55	25	800
Ir	4,450.18	3	60

*P**

J 53 126.9044

t_0 112.9°C t_1 183.0°C

I.	II.	III.	IV.	V.
10.44	19.010	—	—	—

λ	I	eV
2,062.38	(900)	16.4
II 2,566.26	(300)	—
II 2,582.81	(400)	—
II 2,878.64	(400)	15.19
3,055.37	(350)	—
3,078.77	(350)	—
3,931.01	(400)	—
3,940.24	(500)	—
II 4,452.88	(700)	14.76
I 4,862.31	(700)	9.47
I 5,119.29	(500)	9.34
II 5,161.19	(300)	22.6
5,464.61	(900)	22.7

J 2,062.38 (900)

	λ	I			λ	I	
		J	O			J	O
Ni	**2,062.341**	**2**	**15**	**Cu**	**2,062.49**	**25**	—
Cr	**2,062.34**	**20**	—	**Tl**	**2,062.5**	**2*d***	—
Ir	**2,062.28**	**5**	**2**	**Ir**	**2,062.54**	**2**	**2*h***
Os	**2,062.16**	**3**	**12**	**Pd**	**2,062.56**	**10**	—
				Se	*2,062.79*	*(800)*	—

J 2,566.258 (300)

	λ	I			λ	I	
		J	O			J	O
Ru	**2,566.23**	**50**	—	**Mo**	**2,566.259**	**20**	**10**
Fe	**2,566.21**	**40**	**2**	**B**	**2,566.26**	**15**	—
Ni	**2,566.081**	**600*h***	—	**Fe**	**2,566.29**	**10**	**3**
La	**2,566.078**	**10*h***	—	**W**	**2,566.31**	**12**	—
Nb	**2,566.07**	**30**	**2**	**B**	**2,566.40**	**2**	—
Rh	**2,566.04**	**4**	**5**	**Fe**	**2,566.41**	**15**	**1**
V	**2,566.03**	**12**	**1**	**Os**	**2,566.49**	**4**	**25**
Gd	**2,566.01**	**3**	—	**Br**	**2,566.52**	**(4)**	—
Tm	**2,565.98**	**2**	**5**	**J**	**2,566.53**	**(10)**	—
Mn	**2,565.95**	**10**	**15*h***	**Cr**	**2,566.55**	**3**	**10**
U	**2,565.91**	**4**	—	**Ru**	**2,566.587**	**25**	**30**
Cd	**2,565.88**	**(10)**	**3**	**Th**	**2,566.589**	**10*h***	**15*s***
Ru	**2,565.81**	**5**	**12**	**V**	**2,566.60**	**20**	—
W	**2,565.79**	**8**	**3**	**Fe**	**2,566.62**	**15**	**3**
				Hf	**2,566.63**	**3*h***	—
Pd	2,565.51	200	2	**U**	**2,566.66**	**2**	**4**
Ni	2,565.37	150*wh*	—				
Te	2,564.58	(150)	—				
J	2,564.40	(70)	—	Fe	2,566.91	150	60
Fe	2,562.53	150	50	Yb	2,567.63	150	5
				Zr	2,568.87	200	100
J	2,562.47	(30)	—	Pd	2,569.55	150	20
J	*2,561.49*	*(150)*	—	*Zr*	*2,571.39*	*400R*	*300R*
Ni	*2,560.30*	*500h*	—				
J	*2,559.72*	*(40)*	—				
Zn	*2,557.96*	*300*	*10*				

J 2,582.81 (400)

	λ	J	O		λ	J	O
Ru	**2,582.63**	**9**	**4**	**Ru**	**2,582.818**	**20**	**3**
Fe	**2,582.58**	**80**	**25**	**W**	**2,582.825**	**8**	—
La	**2,582.56**	**6**	—	**Rh**	**2,582.83**	**10**	**2**
Hf	**2,582.54**	**35**	**25**	**C**	**2,582.88**	**(4)**	—
W	**2,582.52**	**12**	—	**Hf**	**2,582.92**	**2*h***	**1*h***

	λ	I J	I O
Cs	2,582.5	(2)	—
Zn	2,582.49	40	300
Co	2,582.37	3	—
Co	2,582.24	500wh	50w
J	2,581.60	(20)	—
J	2,580.60	(20)	—
Yb	2,579.58	200	5
Mn	*2,576.10*	*2000R*	*300R*
J	*2,575.47*	*(20)*	—

	λ	I J	I O
La	2,582.960	8	—
Mn	2,582.965	50	—
O	2,582.99	(5)	—
V	2,583.01	50	—
Pd	2,583.03	6	—
Ru	2,583.04	40	30
Fe	2,583.05	3	—
Mn	2,583.09	1	5
Nb	2,583.11	2	8
Pt	2,583.16	10	—
Co	2,583.18	40w	—
Nb	2,583.220	2	12
Ti	2,583.225	1	15
Pd	2,583.85	200	—
Nb	2,583.986	800wh	10
Ni	2,583.99	200	—
Ta	2,584.03	200	80w
J	2,584.76	(20)	—
J	*2,588.68*	*(40)*	—
Nb	*2,590.94*	*800*	*15*

J 2,878.643 (400)

	λ	I J	I O
Fe	2,878.64	2	3
Tb	2,878.52	10	—
Xe	2,878.48	(2wh)	—
Sm	2,878.449	25	4
Cr	2,878.449	80	20
Os	2,878.40	12	40
Tm	2,878.36	20	10
W	2,878.30	10	1
V	2,878.299	7	—
U	2,878.25	2	5
Tm	2,878.21	20	3
Ta	2,878.20	15h	4
W	2,878.08	8	4
Ru	2,878.04	15	—
V	2,878.02	10	2
Pt	2,877.52	200h	40
Nb	2,876.95	500W	40W
Nb	2,875.39	300	50r
J	2,875.25	(12)	—
Fe	2,874.17	200	300
Fe	2,873.40	300	—

	λ	I J	I O
Rh	2,878.65	10	50
Dy	2,878.70	1h	2
W	2,878.716	8	10
U	2,878.719	4	6
Nb	2,878.74	10	3
Ar	2,878.76	(2)	—
Fe	2,878.761	5	8
U	2,878.87	4	5
Er	2,878.91	2h	7
Ta	2,878.95	3s	40r
O	2,879.04	(7)	—
Mo	2,879.05	100h	15
Ru	2,879.06	60	—
W	2,879.107	100	10
Hf	2,879.112	10	15
Yb	2,879.16	5	—
V	2,879.163	35	50
Th	2,879.20	5	10
Nb	2,879.22	5h	—
Fe	2,879.24	25	—
Cs	2,879.25	(8)	—

	λ	I			λ	I	
		J	O			J	O
J	2,872.89	(60)	—	Si	2,881.58	400	500
				J	2,882.01	(20)	—
				Ru	2,882.12	200	30
J	*2,868.77*	*(30)*	—	V	2,882.50	200*r*	35
				Nb	2,883.18	800*R*	100
				Fe	2,883.70	300	—
				V	2,884.78	200*r*	40
				J	*2,887.36*	*(20)*	—
				J	*2,889.54*	*(20)*	—

J 3,055.370 (350)

Fe	**3,055.368**	**2**	—	**W**	**3,055.40**	**10**	**9**
Mo	**3,055.32**	**5**	**50**	**Hf**	**3,055.44**	**15**	**15**
Kr	**3,055.31**	**(3)**	—	**Cr**	**3,055.47**	**20**	—
Pd	**3,055.29**	**3*h***	—	**Nb**	**3,055.522**	**100**	**2**
Fe	**3,055.26**	**150**	**200**	**Sm**	**3,055.524**	**1**	**3**
Ce	**3,055.24**	**3**	**18**	**U**	**3,055.59**	**2**	**2**
Yt	**3,055.22**	**50**	**8**	**Rh**	**3,055.65**	**2**	**2*h***
Os	**3,055.21**	**15**	**80**	**Fe**	**3,055.71**	**6**	**10**
Yb	**3,055.15**	**3**	**1**	**U**	**3,055.88**	**2**	**4**
U	**3,055.09**	**2**	**5**	**V**	**3,055.94**	**50*w***	—
Os	**3,054.97**	**10**	**50**	**Kr**	**3,056.01**	**(30*wh*)**	—
Ru	**3,054.94**	**12**	**70**	**Cs**	**3,056.04**	**(6)**	—
Eu	**3,054.93**	**3**	**400*w***	**Ru**	**3,056.058**	**1**	**30**
Ho	**3,054.88**	**4*h***	—	**Tm**	**3,056.06**	**100**	**40*d***
Zr	**3,054.83**	**25**	**15**	**Pt**	**3,056.07**	**5**	**1**
Ta	**3,054.80**	**1**	**3**	**Th**	**3,056.10**	**4**	**10**
Mo	**3,054.76**	**3**	—				
U	**3,054.73**	**6**	**10**				
Al	**3,054.70**	**10**	**20**	Ru	3,056.86	150	12
Ne	**3,054.69**	**(18)**	—	Ne	3,057.39	(250)	—
				Fe	3,057.45	400	400
				Lu	3,057.90	150*h*	3
Rn	3,054.3	(250)	—	Os	3,058.66	500	500*R*
Cr	3,053.88	150	3*r*				
Pd	3,052.15	150*h*	—	Fe	3,059.09	400	600*r*
Tm	3,050.73	150	50	Pd	3,059.43	150*w*	—
Cr	3,050.14	150	10	Fe	3,062.23	400	2
Nb	3,049.52	150	—				
				Bi	*3,067.72*	*2000wh*	*3000hR*
				J	*3,068.98*	*(18)*	—
Fe	*3,047.60*	*500r*	*800r* t				
Te	*3,047.00*	*(350)*	—				
Sb	*3,040.67*	*(400wh)*	—				
Fe	*3,040.43*	*400*	*400*				

J 3,078.77 (350)

	λ	I			λ	I	
		J	O			J	O
Fe	**3,078.70**	**15*h***	**4**	**Th**	**3,078.83**	**25*h***	**10**
Yt	**3,078.65**	**8**	**6**	**Tb**	**3,078.86**	**80**	**30**
Ti	**3,078.64**	**500*R***	**60**	**Er**	**3,078.873**	**2**	**8*s***
Ir	**3,078.58**	**5**	**5**	**Ne**	**3,078.875**	**(75)**	—
Rb	**3,078.50**	**(2)**	—	**V**	**3,078.95**	**15**	—
Fe	**3,078.434**	**50**	**80**	**Ca**	**3,079.10**	**3**	**1**
U	**3,078.433**	**2**	**4**	**Ba**	**3,079.12**	**5*w***	—
Os	**3,078.38**	**15**	**125**	**Ne**	**3,079.17**	**(75)**	—
Dy	**3,078.35**	**1**	**12**	**W**	**3,079.22**	**5**	**8**
Na	**3,078.31**	**(60)**	**12**	**U**	**3,079.26**	**8**	**8**
In	**3,078.27**	**10**	—	**Cr**	**3,079.33**	**25**	—
Er	**3,078.233**	**1**	**5**	**Dy**	**3,079.34**	**5**	**15**
Ta	**3,078.232**	**5**	**50**	**Ncl**	**3,079.36**	**2**	**8**
Os	**3,078.11**	**15**	**125**	**Ta**	**3,079.37**	**2*h***	**3**
Cs	**3,078.09**	**(6)**	—	**Co**	**3,079.40**	**2**	**80**
Fe	**3,078.02**	**80**	**100**	**U**	**3,079.47**	**2**	**4**
J	3,077.92	(10)	—	J	3,081.66	(100)	—
J	3,077.65	(18)	—	Al	3,082.15	800	800
Lu	3,077.60	200	100	J	3,082.87	(25)	—
Fe	3,077.17	300	1	Fe	3,083.74	500	500
Ne	3,076.97	(150)	—				
Fe	3,075.72	400	400	*Ti*	*3,088.02*	*500R*	*70*
Ti	3,075.22	300*R*	40	*J*	*3,088.19*	*(35)*	—
J	3,073.82	(10)	—	*J*	*3,091.63*	*(25)*	—
Ti	3,072.97	200*r*	35	*Al*	*3,092.71*	*1000*	*1000*
				V	*3,093.11*	*400R*	*100R*
*P**							

J 3,931.01 (400)

	λ	J	O		λ	J	O
U	**3,930.98**	**35**	**12**	**Ce**	**3,931.09**	**8**	**35**
W	**3,930.97**	**12**	**10**	**In**	**3,031,11**	**3**	—
Pr	**3,930.96**	**3**	**9**	**Fe**	**3,931.12**	**15**	**35**
Ta	**3,930.94**	**3**	**5**	**Sm**	**3,931.16**	**2**	**4**
U	**3,930.81**	**3*h***	—	**U**	**3,931.20**	**6**	**5**
Ce	**3,930.806**	**3**	**12**	**Nd**	**3,931.23**	**2*h***	**6**
Tb	**3,930.76**	**3**	**8**	**Ar**	**3,931.24**	**(15)**	—
Yt	**3,930.67**	**25**	**20**	**As**	**3,931.28**	**15**	—
Pr	**3,930.62**	**3**	**9**	**V**	**3,931.34**	**20**	**25**
Nd	**3,930.61**	**30**	**20**	**Ce**	**3,931.37**	**4**	**15**
Eu	**3,930.50**	**400*R***	**1000*R***	**Hf**	**3,931.38**	**3**	**10**
W	**3,930.48**	**10**	**8**	**Mo**	**3,931.40**	**5**	**5**
La	**3,930.47**	**3**	—	**J**	**3,931.42**	**(3)**	—
U	**3,930.43**	**10**	**6**	**Te**	**3,931.43**	**(5)**	—

	λ	I	
		J	O
Sn	**3,930.37**	**(6)**	**—**
Fe	**3,930.30**	**400**	**600**
W	**3,930.25**	**10**	**12**
Ta	**3,930.23**	**2**	**1**
Mo	**3,930.20**	**5**	**5**
C	**3,930.2**	**(6)**	**—**
Yt	**3,930.11**	**3**	**2*h***
U	**3,930.07**	**4**	**—**
V	**3,930.023**	**20**	**50**
Nb	**3,930.022**	**10*h***	**—**
Os	**3,930.00**	**12**	**80**
Ce	**3,929.96**	**1**	**8**
Nd	**3,929.95**	**6**	**10**
Pr	**3,929.88**	**6**	**8**
Ti	**3,929.87**	**35**	**70**
In	**3,929.80**	**(5)**	**—**
V	**3,929.73**	**35**	**2**
U	**3,929.72**	**4**	**2**
Th	**3,929.67**	**20**	**30**
Mn	**3,929.65**	**25**	**12**
Mo	**3,929.584**	**25**	**—**
Tm	**3,929.584**	**50**	**70**
Br	**3,929.56**	**(15)**	**—**
Hf	**3,929.535**	**1*h***	**3**
In	**3,929.53**	**(10)**	**—**
Zr	**3,929.530**	**6**	**100**
La	3,929.22	300	400
Fe	3,927.92	300	500
Tb	3,925.45	200	150
S	3,923.48	(200)	—
Fe	3,922.91	400	600
La	3,921.53	200	400
C	3,920.68	200	—
Fe	3,920.26	300	500
Kr	3,920.14	(200*h*)	—
Hg	3,918.92	(200)	—
La	3,916.04	400	400
J	*3,915.22*	*(18)*	—
J	*3,912.48*	*(25)*	—
Eu	*3,907.11*	*500R*	*1000RW*
Mo	*3,902.96*	*500R*	*1000R*
Fe	*3,902.95*	*400*	*500*
Nb	**3,931.46**	**3**	**4**
U	**3,931.49**	**6**	**25**
Os	**3,931.52**	**12**	**40**
Er	**3,931.53**	**4**	**20**
Se	**3,931.56**	**(5)**	**—**
Se	**3,931.72**	**(8)**	**—**
Hf	**3,931.75**	**2**	**2**
Ru	**3,931.76**	**70**	**50**
Sb	**3,931.78**	**5**	**2**
Nb	**3,931.79**	**20*h***	**1**
Rn	**3,931.82**	**(250)**	**—**
Ce	**3,931.83**	**3**	**18**
Bi	**3,931.9**	**(10)**	**—**
W	**3,931.93**	**2**	**2**
S	**3,931.94**	**(15)**	**—**
Ti	**3,932.02**	**30**	**20**
U	**3,932.03**	**50**	**35**
Pr	**3,932.13**	**5**	**25**
Ce	**3,932.14**	**2**	**12**
Th	**3,932.23**	**10**	**15**
Er	**3,932.25**	**5**	**20**
S	**3,932.30**	**(10)**	**—**
Hf	**3,932.40**	**10**	**3**
Sm	3,933.59	200*h*	200
Fe	3,933.60	200	200
Ca	3,933.67	600*R*	600*R*
Ru	3,933.68	200	5
Nb	3,936.02	200	5
J	3,937.22	(12)	—
J	3,937.91	(25)	—
Tb	3,939.60	200	200
*N**			

J 3,940.24 (500)

	λ	I	
		J	O
Pr	**3,940.15**	**15**	**80**
Eu	**3,939,98**	**1**	**2*W***
W	**3,939.91**	**5**	**5**
La	**3,939.85**	**3*h***	**2**
Nd	**3,939.83**	**4**	**20**
U	**3,939.76**	**10**	**6**
Dy	**3,939.70**	**2**	**3**
Br	**3,939.69**	**(15)**	—
Ba	**3,939.67**	**(5)**	—
Sm	**3,939.64**	**2**	**3**
Tb	**3,939.60**	**200**	**200**
Se	**3,939.60**	**(8)**	—
Sb	**3,939.57**	**3**	—
Os	**3,939.566**	**12**	**50**
Nd	**3,939.55**	**3**	**20**
Ce	**3,939.52**	**1**	**18**
Mo	**3,939.49**	**4**	**4**
U	**3,939.45**	**1*h***	**6**
W	**3,939.44**	**10**	**8**
V	**3,939.33**	**15**	**40**
Mo	**3,939.14**	**4**	**4**
U	**3,939.11**	**12**	**6**
Al	**3,939.07**	**(2)**	—
F	**3,939.03**	**(30)**	—
Gd	**3,938.98**	**40**	**20**
Fe	**3,938.97**	**4**	**4**
Xe	**3,938.92**	**(10)**	—
Pr	**3,938.91**	**1**	**4**
V	**3,938.89**	**1**	**15**
Kr	**3,938.88**	**(20*wh*)**	—
Nd	**3,938.87**	**25**	**30**
Ho	**3,938.84**	**2*h***	**2**
U	**3,940.26**	**6**	**4**
Ce	**3,940.34**	**6**	**35**
Ho	**3,940.37**	**4*h***	**6**
Pr	**3,940.38**	**3*d***	**10*d***
Hg	**3,940.40**	**(2)**	—
U	**3,940.49**	**15**	**10**
Ho	**3,940.55**	**4*h***	**12**
Rb	**3,940.57**	**(200)**	—
V	**3,940.59**	**4**	**15**
Fe	**3,940.88**	**80**	**150**
Kr	**3,940.92**	**(5*wh*)**	—
Ce	**3,940.97**	**2**	**10**
Br	**3,941.05**	**(6)**	—
U	**3,941.09**	**6**	**3**
Cr	**3,941.15**	**8**	**15**
Tb	**3,941.16**	**8**	**15**
V	**3,941.25**	**12**	**30**
Nb	**3,941.27**	**15**	**10**
Fe	**3,941.28**	**10**	**60**
In	**3,941.33**	**3**	—
Sc	**3,941.45**	**(20)**	—
U	**3,941.46**	**6**	**5**
Mo	**3,941.48**	**150**	**5**
Cr	**3,941.49**	**60**	**200*r***
Pr	**3,941.507**	**3**	**10**
Nd	**3,941.512**	**30**	**60**
F	**3,941.52**	**(6)**	—
Eu	**3,941.56**	**4**	**5*W***
Zr	**3,941.62**	**14**	**20**
Ru	**3,941.65**	**8**	**12**
Rn	**3,941.72**	**(25)**	—
Th	**3,941.73**	**14**	**8*h***
Mo	**3,941.76**	**2**	**6**
J	3,942.46	(15)	—
Al	3,944.03	1000	1000
O	3,947.33	(300)	—
Ar	3,947.50	(1000)	—
Ar	3,948.98	(2000)	—
La	3,949.11	800	1000
J	3,949.90	(20)	—
Mo	*3,961.50*	*500*	*5*
Al	*3,961.53*	*2000*	*3000*
J	*3,965.53*	*(15)*	—
Ca	*3,968.47*	*500R*	*500R*

*P**

J 4,452.88 (700)

	λ	I J	I O
Ir	4,452.81	2	15
Mo	4,452.74	15	—
Sm	4,452.71	200	200
V	4,452.701	6	10
Hf	4,452.70	10	—
Ne	4,452.56	(2)	—
Mo	4,452.58	8	12
P	4,452.44	(150)	—
O	4,452.41	(70)	—
La	4,452.15	5	15
V	4,452.01	15	20
U	4,451.98	5*h*	1
Nd	4,451.978	20	50
Mo	4,451.97	30	—
Pr	4,451.95	20	80
Mg	4,451.64	(7)	—
J	4,451.21	(8)	—
J	4,451.08	(8)	—
Kr	*4,436.81*	*(600)*	—
J	*4,434.23*	*(20)*	—
J	*4,431.73*	*(20)*	—
J	*4,428.22*	*(35)*	—

	λ	I J	I O
Sm	4,452.95	6	8
Ne	4,452.98	(15)	—
Hf	4,453.00	2	8*h*
Mn	4,453.005	20	50
V	4,453.12	3	7
Kr	4,453.21	(50*whs*)	—
U	4,453.22	1*h*	10
Ne	4,453.25	(5)	—
Ti	4,453.321	70	150
Ne	4,453.324	(2)	—
V	4,453.35	20*h*	—
Ti	4,453.71	40	80
Pb	4,453.80	(3)	—
La	4,453.85	3	3
Mo	4,453.87	5	4
Kr	4,453.92	(600)	—
Gd	4,453.93	10	10
Tm	4,454.04	1	20
J	4,456.61	(10)	—
J	4,458.47	(35*h*)	—
Xe	4,462.19	(500*wh*)	—
Kr	4,463.69	(800)	—
J	4,464.32	(30)	—
J	*4,473.44*	*(80)*	—
Kr	*4,475.00*	*(800hs)*	—
J	*4,476.05*	*(60)*	—
J	*4,478.64*	*(15)*	—

J 4,862.31 (700)

	λ	I J	I O
Ra	4,862.27	(4)	—
J	4,862.13	(25)	—
Kr	4,862.1	(2*h*)	—
Mn	4,862.05	5	40
Cr	4,861.842	8	125
Kr	4,861.84	(2*h*)	—
Er	4,861.59	1	5
Hf	4,861.49	2	3
H	4,861.33	(500)	—
Ta	4,861.04	1	3
U	4,861.01	10	10
O	4,860.93	(20)	—
La	4,860.91	100	100
Mo	4,860.75	8	10

	λ	I J	I O
Xe	4,862.54	(400*h*)	—
Gd	4,862.608	2	100
V	4,862.609	12	15
P	4,862.83	(15)	—
Ne	4,863.08	(100)	—
Th	4,863.18	10	20
Hf	4,863.27	3	20
Te	4,864.10	(800)	—
J	4,864.51	(25)	—
Te	4,866.22	(800)	—
J	*4,882.18*	*(15)*	—

	λ	I	
		J	O
J	4,850.46	(25)	–
J	4,850.26	(25)	–
Kr	4,846.60	(700)	–
Se	*4,844.96*	*(800)*	–
Xe	*4,844.33*	*(1000)*	–
Se	*4,840.63*	*(800)*	–
J	*4,835.37*	*(15)*	–
J	*4,835.18*	*(25)*	–
Kr	*4,832.07*	*(800)*	–
Te	*4,831.29*	*(800)*	–
J	*4,884.82*	*(25)*	–
Ne	*4,884.91*	*(1000)*	–
J	*4,891.54*	*(15)*	–

J 5,119.29 (500)

	λ	I	
		J	O
Yt	**5,119.12**	**20**	**7**
Ar	**5,118.20**	**(60)**	–
Bi	**5,118.2**	**25**	–
Nb	**5,118.06**	**1**	**3**
J	**5,118.01**	**(8)**	–
Se	**5,117.77**	**(25)**	–
Xe	**5,117.76**	**(3*h*)**	–
Yb	**5,117.75**	**3*h***	–
In	5,117.41	(300)	–
J	5,114.40	(25)	–
In	5,109.36	(300*w*)	–
Cd	*5,085.82*	*500*	*1000w*
Rn	**5,119.3**	**(20)**	–
C	**5,119.55**	**15**	–
P	**5,120.12**	**(5)**	–
Nb	**5,120.30**	**10**	**50**
Ti	**5,120.42**	**4**	**100**
Ne	**5,120.51**	**(25)**	–
In	**5,120.53**	**(30)**	–
Hg	**5,120.55**	**(20)**	–
Cu	**5,120.74**	**20**	–
In	**5,120.85**	**(100)**	–
Sn	**5,120.95**	**2**	–
In	**5,120.96**	**(150)**	–
In	**5,121.10**	**(400)**	–
*N**			

J 5,161.19 (300)

	λ	I	
		J	O
Th	**5,160.69**	**2**	**12**
Nb	**5,160.333**	**15**	**200**
U	**5,160.326**	**20**	**18**
Ba	**5,159.92**	**10**	**50*h***
O	**5,159.88**	**(40)**	–
Ar	**5,159.69**	**(10)**	–
V	**5,159.35**	**40**	**40**
J	5,156.45	(25)	–
J	5,155.06	(15)	–
Ar	5,151.39	(200)	–
Ar	5,145.36	(200)	–
Ne	5,145.01	(500)	–
As	**5,161.25**	**30**	–
Pr	**5,161.74**	**1**	**40**
P	**5,161.97**	**(30)**	–
Hg	**5,162.15**	**(5)**	–
P	**5,162.27**	**(30)**	–
Ar	**5,162.28**	**(500)**	–
Cl	**5,162.34**	**(10)**	–
Xe	**5,162.71**	**(10)**	–
N	**5,162.78**	**(5)**	–
Ar	**5,162.80**	**(2)**	–
Sm	**5,162.86**	**3**	**15**
Fe	5,167.49	150	700

	λ	I J	I O
Ne	5,144.94	(500)	—
Bi	5,144.48	300h	2
Kr	5,143.05	(600h)	—
Se	*5,142.14*	*(500)*	—
Kr	*5,125.73*	*(400wh)*	—

	λ	I J	I O
Fe	5,169.03	200h	2
In	5,175.29	(400)	—
In	5,175.42	(300)	—
In	5,175.56	(150)	—
Se	5,175.98	(600)	—
J	5,178.13	(25)	—
La	*5,183.42*	*400*	*300*
Mg	*5,183.62*	*300*	*500wh*
In	*5,184.44*	*(300)*	—
J	*5,185.21*	*(30)*	—
Ar	*5,187.75*	*(800)*	—
La	*5,188.23*	*500*	*50*
J	*5,198.88*	*(25)*	—

J 5,464.61 (900)

	λ	I J	I O
La	**5,464.38**	**30**	**25**
Sb	**5,464.08**	**(100)**	—
Sn	**5,463.6**	**2**	**2h**
Hf	**5,463.38**	**10**	**10**
Kr	**5,462.65**	**(2)**	—
N	**5,462.62**	**(30)**	—
Th	**5,462.61**	**2**	**12**
Hg	**5,460.74**	**(2000)**	—
J	**5,457.06**	**(15)**	—
S	**5,453.88**	**(750)**	—
Ar	**5,451.65**	**(500)**	—
J	*5,438.00*	*(35)*	—
J	*5,435.84*	*(125)*	—
J	*5,435.48*	*(35)*	—
J	*5,427.10*	*(50)*	—

	λ	I J	I O
Te	**5,465.17**	**(15)**	—
Ag	**5,465.49**	**500R**	**1000R**
Mo	**5,465.57**	**5**	**20**
U	**5,465.69**	**8**	**12**
Br	**5,466.23**	**(150)**	—
Yt	**5,466.47**	**20**	**150**
Fe	**5,466.95**	**5**	—
J	5,470.53	(25)	—
Xe	5,472.51	(500)	—
Ar	5,473.44	(500)	—
S	5,473.63	(750)	—
Lu	5,476.69	1000	500
J	5,479.60	(15)	—
J	*5,491.57*	*(100)*	—
J	*5,493.50*	*(20)*	—
J	*5,494.07*	*(15)*	—
Ar	*5,495.89*	*(1000)*	—
J	*5,496.60*	*(30)*	—
J	*5,496.92*	*(900)*	—
J	*5,497.66*	*(15)*	—
J	*5,504.72*	*(60)*	—

$K^{19}_{39.102}$

t_0 62.3°C t_1 774°C

I.	II.	III.	IV.	V.
4.340	31.811	45.7	—	—

λ	I		eV
	J	O	
2,240.89	40	—	—
I 3,102.03	—	50*R*	3.99
I 3,217.02	20	100*R*	3.86
I 3,217.50	25	50*R*	3.86
I 3,446.39	100*R*	150*R*	3.60
I 3,447.39	79*R*	100*R*	3.60
I 4,044.14	400	800	3.06
I 4,047 20	200	400	3 06
II 4,388.13	40	—	23.45
II 4,608.43	40	—	—
I 5,339.67	—	40	3.94
I 5,782.60	—	60	3.76
I 7,664.91	400	9000*R*	1.62
I 7,698.98	200	5000*R*	1.61

K 2,240.89 *J* 40

	λ	*I* *J*	*I* *O*		λ	*I* *J*	*I* *O*
W	2,240.85	5	4	Kr	2,240.89	(2)	—
Nb	2,240.65	15	10	Ir	2,240.96	34	5
Ir	2,240.64	15	—	Cd	2,240.99	(2)	—
Fe	2,240.63	34	20	Nb	2,241.02	6	—
V	2,240.62	100	—	W	2,241.08	15	10
Ir	2,240.42	30	20	Hf	2,241.1	2	—
Ag	2,240.39	20*h*	—	Se	2,241.12	(5)	—
J	2,240.14	(20)	—	Mo	2,241.17	4	—
Yb	2,240.09	25	5	V	2,241.53	200	—
Rh	2,240.00	50	—	Cr	2,241.84	30	—
Cd	2,239.86	30	80	Rh	2,242.05	25	25
Ta	2,239.48	35	20	Mo	2,242.21	25	—
J	2,238.81	(20)	—	Ce	2,242.33	60	—
Ir	2,238.29	80	—	Nb	2,242.58	50*h*	5
Ar	*2,238.2*	*(60)*	—	Cu	2,242.61	50*h*	25
Rh	*2,237.71*	*100*	—	Ir	2,232.68	300	50
Cr	*3,237.58*	*40*	—	Au	2,242.71	30	—
Ir	*2,237.09*	*100*	—	Al	2,243.05	(30)	—
Lu	*2,236.17*	*150h*	—	Yt	2,243.06	35*W*	25
Cr	*2,235.93*	*50*	—	Hf	2,243.18	20	15
Ar	*2,235.77*	*(40)*	—	Cr	2,243.31	25	—
				K	2,243.4	(5)	—
				Fe	*2,245.50*	*50*	—
				Ir	*2,245.76*	*150*	*10*
				Sn	*2,246.05*	*100R*	*100R*
				K	*2,246.32*	*(5)*	—

K 3,102.03 *O* 50*R*

	λ	*J*	*O*		λ	*J*	*O*
Sm	3,101.93	8	10	Yb	3,102.07	8	1
Gd	3,101.925	3	5	Ir	3,102.144	—	4
Nb	3,101.917	10	2	Fe	3,102.144	2	2
Mo	3,101.915	—	5	Hf	3,102.147	1	6
Dy	3,101.91	2	4	Sc	3,102.149	4	3
Ni	3,101.88	150	400*R*	Dy	3,102.19	14	4
U	3,101.86	4	6	W	3,102.22	12	2
Ce	3,101.79	—	20	K	3,102.25	—	20*R*
Ta	3,101.72	2	7	V	3,102.299	300*R*	70
Re	3,101.703	—	3	Sm	3,102.299	4	8
U	3,101.699	8	8	Ir	3,102.355	—	4
Th	3,101.69	8	10	Hf	3,102.358	3*h*	15
Mn	3,101.56	50	50	Ce	3,102.358	—	12
Ni	3,101.55	150	1000*R*	Ca	3,102.359	4	24
Os	3,101.528	20	125	Fe	3,102.361	2	2

	λ	I	
		J	O
Ti	3,101.526	12	8
Hf	3,101.40	90	60
Ce	3,101.39	—	10
Yb	3,101.36	12	2
Mo	3,101.34	10	80
Pr	3,101.27	2	40
Mo	3,101.087	2	40
Ru	3,100.84	50	70
Fe	3,100.666	100	100
Re	3,100.666	—	100
Ti	3,100.666	15	30
Gd	3,100.51	80	100
Fe	3,100.304	100	100
Mn	3,100.302	60	60
Ir	3,100.29	2	30
Fe	3,099.97	40	40
Cu	3,099.934	10	60
Mo	3,099.932	3	25
Fe	3,099.90	60	60
Co	3,099.67	—	50
Ru	3,099.28	60	70
Os	3,099.26	8	40
Ni	3,099.11	50	200
Tm	3,098.59	60	80
Co	3,098.20	5	100*r*
Fe	3,098.19	60	70
Eu	3,098.17	5	25*d*
Eu	3,097.46	5	200*w*
Ru	3,097.23	—	30
Ni	3,097.12	50	200
Mn	3,097.06	40	75*w*
Mg	3,096.90	25	150
Fe	3,096.84	20	30
Hf	3,096.76	1	25
Co	3,096.70	3	60
Ru	3,096.57	60	70
Cr	3,096.53	—	35
Ce	3,096.50	—	25
Co	3,096.40	3	60
Er	3,095.88	8	25
Cr	3,095.86	3	125
Re	3,095.81	—	30*w*
Co	3,095.72	2	60
Mo	3,095.70	—	25
Ta	3,095.39	18*w*	70*w*
Re	3,095.06	—	40
Fe	3,094.90	15	30

	λ	I	
		J	O
U	3,102.39	20*d*	12*d*
Ru	3,102.40	—	20
Co	3,102.41	4	60
Ce	3,102.43	—	10
Ti	3,102.52	—	5
Rh	3,102.53	—	10
Gd	3,102.557	25	25
Ce	3,102.563	—	15*h*
U	3,102.61	5	6
Fe	3,102.64	5	6
Ag	3,102.66	—	3
Th	3,102.67	12	15
Er	3,102.69	2	15
Os	3,102.72	8	20
Fe	3,102.87	20	30
Tb	3,102.97	15	30
Dy	3,103.246	1	30
Ta	3,103.251	15	70
Co	3,103.74	2	80
Dy	3,103.84	10	30
Co	3,103.98	—	60*r*
Ru	3,104.46	—	30
La	3,104.59	50	200
Re	3,104.65	—	30
Os	3,104.98	15*d*	200*d*
Ru	3,105.28	40	50
Ru	3,105.41	1	50
Ni	3,105.47	35	200
Co	3,105.92	—	30
Os	3,105.99	20	150
Ti	3,106.23	150	25
Ru	3,106.84	3	50
Co	3,107.04	3	70
Os	3,107.38	10	40
Ce	3,107.47	—	25
Ni	3,107.714	—	25
Ru	3,107.715	5	60
Os	3,107.86	8	30
Os	3,107.98	8	25
Ru	3,108.43	—	30
Ca	3,108.58	3	30
Re	3,108.81	—	125
Os	3,108.98	15	125
Hf	3,109.12	100	50
Cr	3,109.34	12	30
Os	3,109.38	20	125
La	3,109.44	2	25
Co	3,109.51	14	60

	λ	I	
		J	O
V	3,094.69	—	40
Mo	3,094.66	25	150
Ru	3,094.39	3	50
Nb	3,094.18	1000	100
Cu	3,093.99	50	150
Ta	3,093.87	15	50
Fe	3,093.81	40	50
Re	3,093.65	—	60
Os	3,093.59	15	125
Fe	3,093.36	40	70
V	3,093.11	400R	100R
Mg	3,092.99	20	125
Al	3,092.84	18	50R
Fe	3,092.78	30	50
Na	3,092.73	(200)	50
V	3,092.72	50r	100r
Al	3,092.71	1000	1000
Ta	3,092.44	15	50
Ru	3,091.87	5	50
Fe	3,091.58	200	300
Mg	3,091.08	10	80
Os	3,090.49	15	80
Os	3,090.30	12	100
Co	3,090.25	1	80
Ru	3,090.23	6	50
Os	3,090.08	15	100
Ru	3,089.80	5	60
Co	3,089.59	—	100r
Ru	3,089.14	12	60
Re	3,088.77	—	60
bhB	3,088.6	—	100
Os	3,088.27	12	60
Ir	3,088.04	2	50
Ti	3,088.02	500R	70
Co	3,087.81	—	60
Os	3,087.75	10	50

	λ	I	
		J	O
Co	3,110.02	2	60
Ru	3,110.55	6	60
V	3,110.71	300R	70
Co	3,110.82	2	60
Re	3,110.86	—	100
Os	3,111.09	20	100
Eu	3,111.43	—	200
Ru	3,111.91	5	50
Ru	3,112.68	3	50
Ru	3,113.40	—	50
Co	3,113.48	4	100
Ta	3,113.90	35w	50
Pd	3,114.04	500w	400w
Ni	3,114.12	50	300
Os	3,114.81	12	50
Rh	3,114.91	2	100
Mn	3,115.46	25	50
Ta	3,115.86	18w	50
Cu	3,116.35	12	50
Os	3,116.47	15	50
Fe	3,116.63	—	150

K 3,217.017 *J* 20* *O* 100*R*
3,217.50 *J* 25* *O* 50*R*

	λ	I J	I O		λ	I J	I O
Ce	3,217.13	—	12	Yb	3,217.17	2	2
La	3,217.12	10*h*	2	Nb	3,217.29	5	10
Nd	3,217.116	—	2*d*	Hf	3,217.30	15	30
V	3,217.11	80*h*	30	Fe	3,217.38	125	200
Ti	3,217.060	150	40	Cr	3,217.40	20	30
Os	3,217.057	10	20	Th	3,217.46	10	10
K	*3,217.017*	*20h*	*100R*	*K*	*3,217.50*	*25*	*50R*
Nb	3,217.02	200*w*	2*w*	Ce	3,217.52	—	20
Co	3,217.00	—	2	Nd	3,217.53	—	4
Mn	3,216.95	75	75	Sm	3,217.58	2*h*	3
Ta	3,216.92	18*w*	100*w*	Th	3,217.73	5	8
Ti	3,216.90	—	2	Er	3,217.79	—	4
Sm	3,216.84	10	40	Nb	3,217.80	1	3
Ni	3,216.82	—	6	Ni	3,217.83	5	10
Bi	3,216.80	2*h*	8*h*	Nb	3,217.86	5	5
Ir	3,216.8	—	3*h*	Ce	3,217.87	—	3
Mo	3,216.78	5	5	Rh	3,217.88	20	60
Ce	3,216.72	—	12	Ti	3,217.94	1	12
Yt	3,216.68	70	40	Ta	3,217.98	1*h*	2*r*
Tm	3,216.64	10	5	Os	3,218.02	4	20
Dy	3,216.63	15	25	Sm	3,218.04	5	15
Th	3,216.625	20*h*	3	Hf	3,218.20	8	8
Ta	3,216.620	—	3	Pr	3,218.24	—	8
Tb	3,216.61	3	8	Ti	3,218.27	150	15
Ce	3,216.587	—	3	Rh	3,218.28	—	60
Zr	3,216.586	—	2	Th	3,218.31	4	8
Cr	3,216.56	125	3	Yb	3,218.32	10	2
Ru	3,216.52	6	12	U	3,218.34	6	8
Ce	3,216.39	—	3				
Ce	3,216.225	—	3	Ce	3,218.38	2	30
				Sm	3,218.60	25	100
W	3,216.22	10*s*	6	Cr	3,218.69	2*wh*	80*wh*
Ti	3,216.203	—	8	Tb	3,218.93	50	50
U	3,216.201	3	5	Ce	3,218.94	8	50
Nb	3,216.19	10	2				
Ce	3,216.18	—	3	Pd	3,218.97	8	300
				Os	3,219.13	8	25
				Co	3,219.15	—	60
Fe	3,215.94	150	300	Sm	3,219.42	10	25
Nb	3,215.59	200	50	Ir	3,219.51	2	35
Sm	3,215.24	15	50				
Dy	3,215.19	10	40	Pr	3,219.55	5	20
Mo	3,215.07	20	25	Fe	3,219.58	125	200
				Fe	3,219.81	80	100
Rh	3,214.87	—	30	Tb	3,219.95	50	50

	λ	I	
		J	O
Fe	3,214.40	50	100
Rh	3,214.32	20	70
Ti	3,214.24	6	30
Sm	3,214.12	6	30
Re	3,214.11	—	25
Fe	3,214.04	200	400
Eu	3,213.75	20*h*	100
Fe	3,213.314	300	50
Os	3,213.312	40	50
Ti	3,213.14	25	25
Re	3,212.94	—	50
Mn	3,112.88	100	100
Eu	3,212.81	20	200
Os	3,212.72	10	40
V	3,212.43	50	70
Na	3,212.19	(60)	35
Ir	3,212.12	15	25
Tm	3,212.01	40	50
Fe	3,211.99	50	70
Re	3,211.76	—	40
Sm	3,211.75	15	100
Fe	3,211.68	50	80
Fe	3,211.49	40	80
Cr	3,211.31	12	35
Fe	3,210.83	100	150
Tm	3,210.82	10	40
Tm	3,210.57	50	40
Eu	3,210.566	—	80
Fe	3,210.24	100	150
Co	3,210.23	2	80
Ca	3,209.93	2	30
Fe	3,209.30	125	200
Cr	3,209.18	125	40
Mo	3,208.83	60	150*r*
Fe	*3,208.47*	*80*	*100*
Cu	*3,208.23*	*15*	*25*
Ta	*3,207.85*	*15*	*70*
V	*3,207.41*	*20*	*80r*
Sm	*3,207.18*	*9*	*50*
Fe	*3,207.09*	*50*	*80*
Mn	*3,206.91*	*—*	*60*
Ta	*3,206.39*	*15*	*70*
Fe	*3,205.40*	*200*	*300*
Au	*3,204.74*	*30*	*50*
Re	*3,204.20*	*—*	*300*
Pt	*3,204.04*	*100*	*250*
V	*3,202.38*	*20r*	*100r*
Ce	*3,201.71*	*10*	*50*

	λ	I	
		J	O
Os	3,220.19	10	30
Pb	3,220.54	5	50*h*
Hf	3,220.61	35	25
Er	3,220.73	5	25
Ir	3,220.78	30	100
Ce	3,220.87	—	30
Ce	3,221.17	8	50
Ni	3,221.27	—	35
Ta	3,221.31	15*s*	70
Ti	3,221.38	6	25
Ni	3,221.65	4	300
Fe	3,222.07	100	200
Ru	3,223.27	35	60
Ta	3,223.83	50*w*	200*w*
Os	3,223.86	8	100
Co	3,224.64	—	60
Cu	3,224.66	10	25
Mn	3,224.76	40	75
Ni	3,225.02	6	300
Nb	3,225.48	800*wr*	150*w*
Ce	3,225.67	3	25
Fe	*3,225.79*	*150*	*300*
Ca	*3,225.90*	*10*	*80*
Ru	*3,226.37*	*12*	*50*
Ni	*3,226.984*	*—*	*100*
Co	*3,226.985*	*—*	*80r*
Os	*3,227.28*	*12*	*125*
Ta	*3,227.32*	*10*	*70*
Fe	*3,227.75*	*300*	*200*
Mn	*3,228.09*	*100*	*100*
Fe	*3,228.25*	*80*	*100*
Ru	*3,228.53*	*150*	*50*
Fe	*3,228.90*	*40*	*80*
Fe	*3,229.12*	*50*	*80*
Os	*3,229.21*	*5*	*125*
Ta	*3,229.24*	*70w*	*300w*
Tl	*3,229.75*	*800*	*2000*
Fe	*3,230.21*	*80*	*100*
Pt	*3,230.29*	*6*	*100*
Sm	*3,230.54*	*30*	*100*
Mn	*3,230.72*	*75*	*75*
Ta	*3,230.85*	*18w*	*200*
Fe	*3,230.97*	*200*	*300*
Os	*3,231.42*	*12*	*150*
Os	*3,232.05*	*20*	*500R*
Sb	*3,232.50*	*250wh*	*150*
Os	*3,232.54*	*10*	*150*

	λ	I J	I O		λ	I J	I O
Pt	*3,200.71*	*40*	*100*	*Li*	*3,232.61*	*500*	*1000R*
Fe	*3,200.47*	*150*	*150*	*Ru*	*3,232.75*	*4*	*50*
Re	*3,200.04*	—	*50w*	*Co*	*3,232.87*	*25*	*60*
				Ni	*3,232.96*	*35*	*300R*
				Fe	*3,233.05*	*60*	*100*
				Mo	*3,233.14*	*30*	*50*
				Mn	*3,233.968*	—	*75wh*
				Fe	*3,233.971*	*150*	*300*
				Os	*3,234.20*	*12*	*150*

K 3,446.39 *J* 100*R* *O* 150*R*
3,447.39 *J* 75*R* *O* 100*R*

	λ	J	O		λ	J	O
Nd	3,446.88	—	12	Y	3,446.88	50	10
Er	3,446.87	1	15	W	3,446.90	7	7
Ce	3,446.86	—	2	Ta	3,446.91	150*W*	2*W*
Fe	3,446.79	—	1	Nb	3,446.93	15	—
U	3,446.76	—	4	Dy	3,447.00	8	50
Ce	3,446.721	1	15	Cr	3,447.01	25	35
Ir	3,446.64	2	8	Mo	3,447.12	20	25*r*
Zr	3,446.61	1	15	W	3,447.13	3	8
Ti	3,446.603	—	4	Hg	3,447.22	—	2
Rn	3,446.6	(2)	—	Tm	3,447.27	3	7
W	3,446.60	5	—	Ce	3,447.273	—	3
Re	3,446.58	—	2*w*	Ga	3,447.280	(2)	—
Th	3,446.55	3	5	Dy	3,447.28	2	10
Kr	3,446.51	(50*wh*)	—	Co	3,447.281	—	10
Ru	3,446.49	—	50	Fe	3,447.281	60	100
Ga	3,446.46	(3)	—	Ta	3,447.29	3	55
Tb	*3,446.40*	—	*50*	Tb	3,447.32	—	8
K	*3,446.39*	*100R*	*150R*	*Zr*	*3,447.36*	*3*	*150w*
Co	3,446.388	60	3	*K*	*3,447.39*	*75R*	*100R*
Eu	3,446.37	3	5	Cr	3,447.43	35	35
Er	3,446.36	1	8	Er	3,447.52	—	10*d*
Xe	3,446.34	(12*h*)	—	Ce	3,447.53	—	2
Ir	3,446.30	4	30	He	3,447.59	(15)	—
Ni	3,446.26	50*h*	1000*R*	Nd	3,447.62	—	6
Ce	3,446.21	—	15	Th	3,447.63	2	3
Nd	3,446.18	4	8	Pt	3,447.67	5	—
Co	3,446.09	—	60*h*	Ne	3,447.70	(150)	—
Mo	3,446.08	40*h*	1	W	3,447.73	4	6
Ru	3,446.07	6	50	Rh	3,447.74	5	50
Ru	3,445.94	—	10	Cr	3,447.76	30	35
Ta	3,445.91	3	7	Pt	3,447.78	15	1
Nd	3,445.89	—	2	Dy	3,447.78	4	10

	λ	I J	I O
U	**3,445.87**	**1**	**3*h***
Eu	**3,445.84**	**2**	**3**
Mo	**3,445.808**	**—**	**3**
V	**3,445.807**	**12**	**15**
Fe	**3,445.77**	**3**	**10**
Th	**3,445.75**	**8**	**8**
W	**3,445.72**	**7**	**10**
U	**3,445.71**	**3**	**3*h***
Nb	**3,445.68**	**80**	**50**
Sm	**3,445.621**	**10*R***	**150*R***
Cr	**3,445.618**	**80**	**100**
Dy	**3,445.58**	**8**	**80**
Er	**3,445.575**	**8**	**20**
Ti	**3,445.566**	**—**	**3**
Ho	**3,445.56**	**6**	**6**
Os	**3,445.55**	**15**	**80**
Ta	**3,445.51**	**1*h***	**2*h***
Mo	**3,445.47**	**25**	**2**
Er	**3,445.46**	**—**	**6**
Hg	**3,445.43**	**2**	**—**
W	**3,445.40**	**5**	**7**
Fe	3,445.15	150	300
Os	3,444.46	12	50
Ti	3,444.31	150	60
Nb	3,444.28	50	1*h*
Se	3,444.27	(35)	—
Fe	3,443.88	200	400
Co	3,443.64	100	500*R*
Ti	3,443.39	35	3
Co	3,442.93	15	400*R*
Fe	3,442.36	15	50
Mn	3,441.99	75	75
Tm	3,441.51	80	150
Dy	3,441.45	5	50
Cr	3,441.44	90	80
Pd	3,441.40	2*h*	800*h*
Fe	3,440.99	200	300
Fe	3,440.61	300	500
Nb	3,440.59	80	15
Rh	3,440.53	100	2
Ta	3,440.24	50	18
Ru	3,440.20	30	100
K	3,440.05	(40)	—
Gd	3,439.99	50	70
Nb	3,439.92	50	4
Gd	3,439.78	50	50
Rb	3,439.34	(200)	—

	λ	I J	I O
Sm	**3,447.783**	**1**	**2**
Co	**3,447.80**	**—**	**5*h***
Pr	**3,447.84**	**—**	**8**
Eu	**3,447.87**	**—**	**2**
O	**3,447.92**	**(18)**	**—**
W	**3,447.95**	**7**	**8*d***
Ce	**3,447.97**	**—**	**2**
Ce	**3,448.02**	**—**	**2**
U	**3,448.05**	**—**	**6**
Er	**3,448.07**	**2**	**18**
Dy	**3,448.08**	**2**	**2**
S	**3,448.09**	**(8)**	**—**
Eu	**3,448.16**	**2**	**4**
Cr	**3,448.19**	**1**	**12**
Pr	**3,448.20**	**4**	**15**
W	**3,448.21**	**12**	**10**
Nb	**3,448.22**	**50**	**3*h***
Ti	**3,448.25**	**—**	**6**
Nd	**3,448.27**	**2*h***	**8*d***
Ce	**3,448.29**	**—**	**12**
Ca	**3,448.31**	**2**	**—**
Co	**3,448.36**	**2**	**10**
Ru	3,448.95	20	70
Ir	3,448.97	10	60
Co	3,449.17	125	500*R*
Os	3,449.20	20	100
Re	3,449.37	—	**100*r***
Co	3,449.44	125	500*R*
Rh	3,450.29	10	100
Fe	3,450.330	80	150
Cu	3,450.332	30	150
Gd	3,450.38	100	100
Nb	3,450.762	50	2
Ne	3,450.765	(50)	—
V	3,451.04	60	—
Rh	3,451.15	2	50
Gd	3,451.24	40	50
Pd	3,451.35	400*h*	—
B	3,451.41	100	—
Hg	3,451.69	(200)	—
Re	3,451.81	—	100
Fe	3,451.92	60	100
La	3,452.18	40	50
Fe	3,452.28	8	150
Nb	3,452.35	200	5
Ti	3,452.47	100	12
Ni	3,452.89	50	600*R*

	λ	I	
		J	O
Gd	3,439.21	35	60
Ta	3,439.00	70W	—
Yb	3,438.84	100	20
Yb	3,438.72	80	4
Co	3,438.71	—	80W
Nb	3,438.42	50wh	1
Ru	3,438.37	35	70
In	3,438.34	(50)	—
Zr	3,438.23	200	250
Re	3,437.72	—	100
Co	3,437.69	—	150Wh
Ta	3,437.37	300wh	7
Ni	3,437.28	40	600R
Fe	3,437.05	15	80
Nb	3,436.96	50	20r
Ru	3,436.74	150	300R
Rh	*3,434.89*	*200r*	*1000r*
Ni	*3,433.56*	*50wh*	*800R*
Pd	*3,433.45*	*500h*	*1000h*
Cr	*3,433.31*	*150*	*30*
Co	*3,433.04*	*150*	*1000R*
Nb	*3,432.70*	*100*	*10*
Co	*3,431.57*	*40*	*500R*
Bi	*3,431.23*	*(150)*	—
Bi	*3,430.83*	*(200)*	—
Tm	*3,429.97*	*100*	*100*
Yb	*3,428.46*	*80*	*25*
Ru	*3,428.31*	*100*	*100*
Co	*3,428.23*	*2*	*100W*
Nb	*3,426.57*	*200*	*5*

	λ	I	
		J	O
Ru	3,452.90	6	60
V	3,453.08	60	—
La	3,453.17	40	50
Cr	3,453.33	35	35
Co	3,453.50	200	3000R
Tm	3,453.66	80	150
Nb	3,453.97	100	—
Tb	3,454.06	30	80
Yb	3,454.07	250	40
Ne	3,454.19	(75)	—
Dy	3,454.33	10	100
Ce	3,454.47	40	10
Nb	3,454.71	50	3
Nb	3,454.91	80	3
Cr	3,454.99	100	—
Os	3,455.03	15	50
Rh	3,455.22	12	300
Co	3,455.23	10	2000R
Bi	3,455.27	100h	—
Rh	3,455.42	2	50
Pb	3,455.49	70	—
Cr	3,455.60	35	50
Ho	3,456.00	60	60
Ti	3,456.39	125	25
Dy	3,456.57	30	50
Ru	3,456.62	8	60
Tb	3,457.03	8	50
Rh	3,457.07	4	100
V	3,457.15	150	2
Cr	*3,457.63*	*125*	*4*
Rh	*3,457.93*	*10*	*125*
Yb	*3,458.28*	*100*	*12*
Os	*3,458.38*	*12*	*200*
Ni	*3,458.47*	*50h*	*800R*
Os	*3,459.02*	*10*	*100*
Mn	*3,460.33*	*500*	*60*
Re	*3,460.47*	—	*1000W*
Ne	*3,460.52*	*(75)*	—
Pd	*3,460.77*	*600h*	*300r*
Dy	*3,460.97*	*3*	*100*
Ar	*3,461.08*	*(300)*	—
Co	*3,461.18*	*3*	*100wh*
Ti	*3,461.50*	*125*	*80*
Rb	*3,461.57*	*(200)*	—
Ni	*3,461.65*	*50h*	*800R*
Rh	*3,462.04*	*150*	*1000*
Tm	*3,462.20*	*200*	*250*
Co	*3,462.80*	*80*	*1000R*

	λ	I J	I O
Gd	*3,463.98*	*125*	*100*
Ne	*3,464.34*	*(75)*	—
Yb	*3,464.37*	*50r*	*200R*
Sr	*3,464.46*	*200*	*200*
Re	*3,464.72*	—	*100*
Co	*3,465.80*	*25*	*2000R*
Fe	*3,465.86*	*400*	*500*
Cd	*3,466.20*	*500*	*1000*
Ne	*3,466.58*	*(150)*	—
Gd	*3,467.28*	*100*	*100*

K 4,044.14 *J* 400 *O* 800

	λ	J	O
Sm	**4,044.11**	—	**4**
Nb	**4,044.105**	**10**	**5**
Hg	**4,044.10**	**10**	**5**
Cl	**4,044.09**	**(4)**	—
Ce	**4,044.06**	—	**3**
U	**4,044.04**	**4**	**1**
Gd	**4,044.03**	—	**3*h***
Eu	**4,043.97**	—	**20**
Ce	**4,043.95**	—	**3**
Fe	**4,043.90**	**7**	**25**
J	**4,043.88**	**(20)**	—
Sc	**4,043.80**	**4**	**12**
Ti	**4,043.77**	—	**20**
Cu	**4,043.751**	**10**	—
Ce	**4,043.747**	—	**3**
Mo	**4,043.74**	**8**	**8**
Gd	**4,043.71**	**5**	**4**
Cr	**4,043.70**	**2**	**30**
Tb	**4,043.66**	**1*w***	**8*w***
Nd	**4,043.60**	**5**	**15**
Zr	**4,043.58**	—	**25**
N	**4,043.54**	**(10*h*)**	—
Cu	**4,043.50**	**25**	—
Ce	**4,043.47**	—	**4*s***
Cs	**4,043.42**	**(20)**	—
Th	**4,043.40**	**1**	**8**
Sm	**4,043.36**	**3**	**5**
Mo	**4,043.27**	**2**	**3**
La	**4,042.91**	**300**	**400**
K	**4,042.59**	**(30)**	—
K	4,039.69	(15)	—
Xe	4,037.59	(100)	—
Mn	4,033.07	20	400*r*
W	**4,044.29**	**12**	**15**
Ce	**4,044.33**	**1**	**4*h***
Nd	**4,044.35**	—	**3**
Hf	**4,044.39**	**4**	**10**
U	**4,044.416**	**25**	**18**
Ar	**4,044.418**	**(1200)**	—
Ca	**4,044.419**	**3**	**5*d***
Tm	**4,044.47**	—	**15**
P	**4,044.49**	**(150*w*)**	—
Zr	**4,044.56**	**2**	**25**
Cl	**4,044.58**	**(10)**	—
Fe	**4,044.61**	**35**	**70**
Xe	**4,044.64**	**(3*wh*)**	—
Kr	**4,044.67**	**(80)**	—
Nb	**4,044.71**	**3**	**5**
N	**4,044.75**	**(2)**	—
Pr	**4,044.818**	**35**	**50**
U	**4,044.824**	**2**	**10**
Ir	**4,044.89**	**6**	**2**
Xe	**4,044.90**	**(4*wh*)**	—
Sm	**4,044.95**	**10**	**10**
Gd	**4,045.01**	**5**	**20**
Sm	**4,045.05**	**6**	**10**
*N**			

	λ	I J	I O		λ	I J	I O
Ga	4,032.98	500R	1000R				
La	4,031.69	300	400				
Mn	4,030.75	20	500r				
S	4,028.79	(200)	—				

K 4,047.201 *J* 200 *O* 400

	λ	J	O		λ	J	O
Cs	**4,047.184**	**(20)**	**—**	**Mo**	**4,047.204**	**3**	**4**
Te	**4,047.18**	**(15)**	**—**	**Ce**	**4,047.27**	**2**	**18**
Tb	**4,047.16**	**—**	**9**	**Fe**	**4,047.31**	**—**	**3**
Nd	**4,047.158**	**10**	**12**	**Ir**	**4,047.33**	**—**	**4**
Pr	**4,047.10**	**12**	**20**	**Ca**	**4,047.35**	**3**	**2**
Gd	**4,047.09**	**5**	**6**	**Sm**	**4,047.36**	**6**	**8**
U	**4,047.05**	**8**	**6**	**Yb**	**4,047.39**	**4**	**—**
Er	**4,046.96**	**—**	**8**	**Ce**	**4,047.392**	**—**	**4**
Mo	**4,046.89**	**20**	**3**	**Mo**	**4,047.398**	**4**	**4**
Ce	**4,046.85**	**—**	**3**	**Rb**	**4,047.4**	**(4)**	**—**
Gd	**4,046.84**	**10**	**10**	**Ho**	**4,047.50**	**3**	**3**
Ni	**4,046.761**	**—**	**2**	**Ar**	**4,047.51**	**(2)**	**—**
Cr	**4,046.760**	**3**	**30**	**Mo**	**4,047.56**	**4**	**4**
Ce	**4,046.73**	**—**	**2**	**U**	**4,047.61**	**3**	**18**
Nd	**4,046.702**	**2**	**8**	**Ce**	**4,047.62**	**—**	**3**
W	**4,046.701**	**12**	**10**	**Yt**	**4,047.63**	**10**	**50**
Pr	**4,046.63**	**3**	**8**	**Ce**	**4,047.70**	**—**	**3**
Hg	**4,046.56**	**300**	**200**	**Dy**	**4,047.740**	**2**	**4**
Ir	**4,046.54**	**—**	**8**	**Eu**	**4,047.74**	**—**	**2**
U	**4,046.52**	**5**	**5**	**Se**	**4,047.77**	**(8)**	**—**
Sc	**4,046.49**	**—**	**10**	**Sc**	**4,047.79**	**10**	**25**
Pt	**4,046.45**	**20**	**—**	**Gd**	**4,047.846**	**50**	**150**
U	**4,046.40**	**5**	**3**	**Er**	**4,047.850**	**—**	**5**
Ce	**4,046.34**	**10**	**30**	**Cl**	**4,047.87**	**(6)**	**—**
Nb	**4,046.27**	**1**	**3**	**Ce**	**4,047.875**	**—**	**5**
Ar	**4,045.97**	**(150)**	**—**	**W**	**4,047.93**	**8**	**9**
Fe	**4,045.81**	**300**	**400**	**Hf**	**4,047.96**	**25**	**8**
				Os	**4,048.05**	**2**	**20**
Ho	4,045.43	80	200	**Th**	**4,048.057**	**8**	**8**
Co	4,045.39	—	400	**U**	**4,048.065**	**2**	**8**
				Ru	4,051.40	200	125
				Ho	4,053.92	200	400
				Ru	4,054.05	100	40
				Ag	4,055.26	500R	800R
				In	4,056.94	(500)	—
				Kr	4,057.01	(300hs)	—
				Pb	4,057.82	300R	2000R
				Nb	4,058.94	400w	1000w
				Cu	4,062.70	20	500w

	λ	I J	I O		λ	I J	I O
				Fe	*4,063.60*	*300*	*400*
				Kr	*4,065.11*	*(300)*	*—*

K 4,388.13 *J* 40

Ti	**4,388.07**	**5**	**25**	**Mo**	**4,388.27**	**20**	**6**
Ce	**4,388.01**	**3**	**8**	**Nb**	**4,388.358**	**15**	**10**
Co	**4,387.929**	**1**	**3**	**Er**	**4,388.36**	**1**	**8**
He	**4,387.928**	**(30)**	**—**	**Fe**	**4,388.411**	**50**	**125**
Fe	**4,387.90**	**35**	**150**	**Th**	**4,388.413**	**3**	**8**
Nb	**4,387.74**	**5**	**3**	**J**	**4,388.51**	**(15*h*)**	**—**
U	**4,387.59**	**1**	**4**	**Ti**	**4,388.55**	**1**	**10**
Cl	**4,387.53**	**(6)**	**—**	**Pd**	**4,388.62**	**2**	**8**
Cr	**4,387.50**	**15**	**15*d***	**Pr**	**4,388.73**	**1**	**10**
Pr	**4,387.44**	**1**	**5**	**Cs**	**4,388.76**	**(10)**	**—**
Cr	**4,387.38**	**2**	**10*d***	**Kr**	**4,388.90**	**(3*h*)**	**—**
U	**4,387.31**	**4**	**15**	**Sm**	**4,388.987**	**8**	**10**
Mo	**4,387.30**	**2**	**3**	**Gd**	**4,388.991**	**4**	**4**
V	**4,387.21**	**12**	**15**	**Eu**	**4,389.07**	**2**	**5**
Th	**4,387.10**	**3**	**4**	**Fe**	**4,389.25**	**2**	**35**
Ce	**4,387.06**	**1**	**5**	**Cl**	**4,389.32**	**(6)**	**—**
Ti	4,386.85	800	8	Kr	4,389.72	(20*h*)	—
Pb	4,386.58	(20)	—	Cl	4,389.76	(25)	—
Kr	4,386.54	(300*h*)	—	Gd	4,389.88	40*h*	40
Ac	4,386.37	100	—	Te	4,389.92	(50)	—
Xe	4,385.77	(70)	—	V	4,389.97	60*R*	80*R*
Nd	4,385.66	20	40	Gd	4,389.99	80	30
Ru	4,385.65	50	125	Ru	4,390.43	80	150*R*
Mo	4,385.56	20	—	Sm	4,390.86	150	150
Ru	4,385.39	40	125	Fe	4,390.954	35	100
P	4,385.33	(100)	—	Gd	4,390.955	100	100
Kr	4,385.27	(50*wh*)	—	Ti	4,391.03	25	6
Te	4,385.08	(50)	—	Th	4,391.11	40	50
Cr	4,384.98	200	150	Gd	4,391.44	25	15
Xe	4,384.93	(30)	—	Br	4,391.61	(25)	—
V	4,384.72	125*R*	125*R*	Cr	4,391.75	35	50
Cs	4,384.43	(25)	—	S	4,391.84	(30)	—
Sm	4,384.29	50	50	Ne	4,391.94	(150)	—
Pr	4,384.14	25*w*	30*w*	Gd	4,392.07	100	100
Br	4,384.00	(20)	—	Yb	4,392.83	20	3
Xe	4,383.91	(100)	—	Xe	4,393.20	(200*wh*)	—
Fe	4,383.55	800	1000	Br	4,393.56	(25)	—
La	4,383.45	50	10	Ti	4,395.03	150	50
Rn	4,383.30	(35)	—	V	4,395.23	40*R*	60*R*
Eu	4,383.16	20	100*W*	Ne	4,395.56	(50)	—

	λ	I	
		J	O
Gd	4,383.14	40	30
Se	4,382.87	(800)	—
Pr	4,382.42	20	30
Mo	4,382.41	20	10
Th	4,381.86	30	30
Mn	4,381.70	20	80
Mo	4,381.64	150	150
Kr	4,381.52	(100wh)	—
Ne	4,381.22	(30)	—
Nb	4,381.13	20	3
Cr	4,381.11	25	30
Rb	4,380.7	(20)	—
Gd	4,380.64	125	100
Pr	4,380.32	20	50
Mo	4,380.29	25	30
Rh	4,379.92	25	60
Ar	4,379.74	(80)	—
Ne	4,379.50	(100)	—
Bi	4,379.4	20	25
V	4,379.24	200R	200R
Sm	4,378.23	100	100
Cu	4,378.20	30w	200w
La	4,378.10	30	40
Nb	4,377.96	30	10
Mo	4,377.76	200	5
Kr	4,377.71	(40h)	—
Te	4,377.10	(70)	—
Cr	4,376.80	20	20
Hg	4,376.19	(50)	—
J	4,376.16	(20)	—
Kr	4,376.12	(800)	—
Fe	*4,375.93*	*200*	*500*
Sm	*4,374.97*	*200*	*200*
Yt	*4,374.93*	*150*	*150*
Rh	*4,374.80*	*500*	*1000W*
C	*4,374.28*	*40*	—
Se	*4,374.24*	*(40)*	—
Cr	*4,374.16*	*60*	*50*
Gd	*4,373.84*	*80*	*200*
Xe	*4,373.78*	*(50wh)*	—
Sm	*4,373.46*	*50*	*50*
Se	*4,373.31*	*(40)*	—
Cr	*4,373.25*	*50*	*50*
Te	*4,373.00*	*(50)*	—
Cl	*4,372.91*	*(80)*	—
In	*4,372.87*	*(80)*	—
Ru	*4,372.21*	*100*	*125*
Xe	4,395.77	(200wh)	—
Ti	4,395.84	30	10
Ag	4,395.93	30	10
O	4,395.95	(80)	—
Te	4,396.00	(100)	—
Pr	4,396.12	50	80
Br	4,396.40	(20)	—
Mo	4,396.66	25	25
Mo	4,397.29	30	30
Ne	4,397.94	(100)	—
Yt	4,398.01	100	150
Te	4,398.45	(70)	—
Hg	4,398.62	(300)	—
J	4,399.09	(20)	—
Pr	4,399.33	20	40
Ir	4,399.47	100	400
Cs	4,399.49	(20)	—
Ti	4,399.77	100	40
Kr	4,399.97	(200)	—
Pr	4,400.03	20	30
Ar	4,400.09	(30)	—
V	*4,400.57*	*40*	*60*
Kr	*4,400.87*	*(100h)*	—
P	*4,400.99*	*(50)*	—
Ar	*4,401.02*	*(40)*	—
Se	*4,401.02*	*(100)*	—
Gd	*4,401.85*	*100*	*200*
Te	*4,401.89*	*(100)*	—
Hg	*4,402.06*	*(50)*	—
Gd	*4,403.14*	*100*	*100*
Sm	*4,403.36*	*50*	*50*
Pr	*4,403.60*	*40*	*100*
Fe	*4,404.75*	*700*	*1000*
Hg	*4,404.86*	*(50)*	—
Pr	*4,405.85*	*100*	*100*
Sc	*4,406.58*	*(70)*	—
Gd	*4,406.67*	*200*	*70*
Xe	*4,406.88*	*(100wh)*	—
Fe	*4,407.72*	*50*	*100*
Gd	*4,408.26*	*150*	*100*
Fe	*4,408.42*	*60*	*125*
Pr	*4,408.84*	*100*	*125*
Kr	*4,408.89*	*(40hs)*	—
J	*4,408.96*	*(250)*	—
Sm	*4,409.34*	*100*	*100*
Ru	*4,410.03*	*80*	*150*
Kr	*4,410.37*	*(50)*	—

	λ	I J	I O
Pr	4,371.61	40w	125
As	4,371.38	50	—
Ar	4,371.36	(80)	—
Cr	4,371.28	150	200
Hf	4,370.97	40	30
Yb	4,370.81	40	15
Gd	4,369.775	150	250
Fe	4,369.774	100	200
Ne	4,369.77	(70)	—
Kr	4,369.69	(200)	—
O	4,369.28	(50)	—
Xe	4,369.20	(100wh)	—
Pr	4,368.33	90	125
O	4,368.30	(1000)	—
Sm	4,368.03	60	60
Nb	4,367.97	50h	2
Fe	4,367.91	70	60
Fe	4,367.58	50	100
O	4,366.91	(100)	—
Br	4,365.60	(200)	—
La	4,364.67	50	50
Te	4,410.95	(50)	—
Ti	4,411.08	100	7
Gd	4,411.16	50	100
C	4,411.20	40	—
C	4,411.52	40	—
Te	4,411.78	(50)	—

K 4,608.43 J 40

	λ	I J	I O
Rh	4,608.119	5	15
Mo	4,608.116	5	5
Hf	4,608.09	4	25
Gd	4,608.02	3	8
Fe	4,607.65	5	50
Au	4,607.50	15r	—
As	4,607.46	200	—
Au	4,607.34	15	30
Sr	4,607.33	50R	1000R
V	4,607.23	3	4
N	4,607.17	(50)	—
Nb	4,606.77	50	50
V	4,606.15	25	30
La	4,605.78	100	100
Hf	4,605.77	30	20
U	4,605.15	25	12
Sb	4,604.77	(30)	—
Rn	4,604.40	(200)	—
Se	4,604.34	(300)	—
Mo	4,604.23	20	—
Kr	4,604.02	(60h)	—
Cu	4,608.46	2	—
Ca	4,608.54	2	—
Nb	4,608.58	3h	2
Ho	4,608.67	1	3
Mo	4,608.71	10	10
U	4,608.81	5	3
W	4,608.84	1wh	5
Dy	4,609.06	2h	2
Ho	4,609.32	2h	4
Ne	4,609.36	(30)	—
Ti	4,609.37	2h	15
Th	4,609.277	4	6
Rn	4,609.38	(250)	—
O	4,609.39	(60h)	—
Rb	4,609.5	(15)	—
Sc	4,609.53	5	10
N	4,609.60	(30)	—
Ar	4,609.60	(300)	—
V	4,609.65	9	10
Al	4,609.7	(4)	—
Hg	4,609.72	(10)	—
Kr	4,609.72	(20hs)	—
U	4,609.86	20	15

	λ	I	
		J	O
Cs	4,603.75	(60)	—
U	4,603.66	40	25
Tm	4,603.42	25	35
Xe	4,603.03	(300*h*)	—
Fe	4,602.94	100	300
As	4,602.73	200	—
Te	4,602.37	(800)	—
P	4,601.96	(300*w*)	—
N	4,601.49	(100)	—
Ta	4,601.42	100*wh*	60
Br	4,601.36	(20)	—
U	4,601.13	25	18
Cl	4,601.00	(20)	—
Cr	4,600.75	150	150
V	4,600.15	60*h*	1
Cr	4,600.10	50	20
Se	4,599.96	(70)	—
J	4,599.81	(30)	—
Hf	4,599.44	25	10
Mo	4,599.16	20	25
Sb	4,599.09	(40)	—
Ar	4,598.77	(20)	—
Kr	4,598.49	(50*h*)	—
Yb	4,598.37	70	25
Se	4,597.93	(25)	—
Gd	4,597.92	40	25
Mo	4,597.88	20	15
Hg	4,597.72	(20)	—
Gd	4,596.99	25	25
Sb	4,596.90	(70)	—
O	4,596.13	(150)	—
Ar	4,596.10	(1000)	—
P	4,595.98	(30)	—
K	4,595.61	(40)	—
Cr	4,595.59	60	50
Sm	4,595.30	60	100
Ne	4,595.25	(50)	—
Mo	4,595.16	40	40
Sb	4,594.93	(20)	—
V	4,594.11	25*wh*	30*wh*
Eu	*4,594.02*	*200*	*500R*
Th	*4,593.64*	*50wh*	—
Sm	*4,593.53*	*50*	*50*
Ne	*4,593.24*	*(50)*	—
Cs	*4,593.18*	*50R*	*1000R*
Kr	*4,592.80*	*(150wh)*	—
Fe	*4,592.65*	*50*	*200*
Xe	*4,592.05*	*(150wh)*	—

	λ	I	
		J	O
Th	**4,609.876**	**3**	—
Mo	4,609.879	40	40
Cr	4,609.89	3	15
Ne	4,609.91	(150)	—
Kr	4,610.65	(60*h*)	—
J	4,611.22	(20)	—
Fe	4,611.29	25	200
U	4,611.44	25	12
Xe	4,611.89	(700)	—
In	4,612.13	150	—
P	4,612.83	(50)	—
Sb	4,612.92	(50)	—
Tm	4,613.19	20	15
Cr	4,613.37	60	150
La	4,613.39	100	100
Hf	4,613.74	25*wh*	12
P	4,613.8	(30)	—
Nd	4,613.84	25*h*	4
N	4,613.89	(30)	—
Ne	4,614.39	(100)	—
Gd	4,614.50	25	50
Br	4,614.60	(100)	—
Xe	4,615.06	(50*h*)	—
Kr	4,615.28	(500)	—
Xe	4,615.50	(100)	—
Sm	4,615.69	40	50
Tm	4,615.93	300	200
Ne	4,615.98	(50)	—
Cr	4,616.14	200	300*r*
In	4,616.176	(20)	—
Br	4,616.18	(20)	—
Cr	4,616.66	50	—
In	4,617.16	(200)	—
Ti	4,617.27	100	200
Xe	4,617.50	(50)	—
Ne	4,617.84	(70)	—
U	4,618.39	20	5
Se	4,618.77	(100)	—
Cr	4,618.83	80*h*	6
C	4,618.85	25*h*	—
Kr	4,619.15	(1000)	—
Cr	4,619.55	30	50
La	4,619.88	200	150
In	4,620.05	(80)	—
In	4,620.24	(200)	—
Mo	4,621.37	25	30
N	4,621.40	(50)	—
J	4,621.89	(35)	—

	λ	I J	I O
Cr	4,591.39	125	200
O	4,590.94	(300)	—
Ti	4,589.95	100	40
Ar	4,589.93	(150)	—
P	4,589.78	(300w)	—
Ar	4,589.29	(80)	—
Cr	4,588.22	600h	10
P	4,587.90	(300w)	—
Cu	4,586.95	80w	250w
Al	4,585.82	(40)	—
Xe	4,585.48	(200wh)	
Sm	4,584.83	50	60
Ru	4,584.44	80	150R
Fe	4,583.85	150	150
Kr	4,582.85	(300h)	—
Xe	4,582.75	(300)	—
Ne	4,582.45	(150)	—
Ne	4,582.03	(150)	—
Nb	4,581.62	50	30
Cr	4,621.96	40	50
Rb	4,622.45	50	—
Hf	4,622.699	60	20
P	4,622.70	(50h)	—
Br	4,622.75	(200)	—
Ti	4,623.09	40	125
Se	4,623.77	(150)	—
Xe	4,624.28	(1000)	—
Rn	4,625.48	(500)	—
Cr	4,626.19	125	100
Mo	4,626.47	80	100
P	4,626.60	(70)	—
U	4,627.08	60	30
In	4,627.38	(150)	—
Mo	4,627.47	80	80
As	4,627.80	200	—
Ne	4,628.31	(150)	—
Ar	4,628.44	(1000)	—
P	4,628.70	(50)	—
Pr	4,628.75	50w	200
As	4,630.14	200	—
N	4,630.55	(300)	—
Te	4,630.57	(50)	—
J	4,632.32	(50)	—
Kr	4,633.88	(800)	—
Cr	4,634.09	80h	5

K 5,339.67 O 40

	λ	J	O
Rh	5,339.65	—	2
Co	5,339.53	—	100w
Re	5,339.413	—	10
Sc	5,339.408	—	4
bhF	5,339.3	—	20
W	5,339.28	—	5
Mo	5,339.08	1	4
Re	5,338.63	—	10
V	5,338.61	6	4
Zr	5,338.43	—	2
Ti	5,338.33	—	8
Eu	5,338.27	—	4h
Nd	5,338.01	—	2
Tb	5,337.90	—	10
Gd	5,337.55	—	20
Os	5,336.23	—	20
Fe	5,339.94	30	200
Sm	5,340.312	—	3
Dy	5,340.314	—	6
Cr	5,340.44	—	50
La	5,340.67	100	80
bhF	5,340.7	—	20
Ir	5,340.74	—	5
Nb	5,340.80	5	8
Fe	5,341.03	15	200
Sc	5,341.04	—	5
Ta	5,341.05	80	150w
Mn	5,341.06	100	200
Sm	5,341.29	—	80
Co	5,341.33	—	300w
Ti	5,341.50	—	8
Eu	5,341.88	—	20

	λ	I J	I O
Co	5,336.17	—	50
Ta	5,336.13	—	40
Ru	5,335.93	—	100
Yb	5,335.16	400	150
Nb	5,334.87	10	50
Co	5,334.84	—	70
Ru	5,334.70	—	60
*bh*F	5,334.5	—	30
Er	5,334.20	—	20
Re	5,333.85	—	30
Co	5,333.65	—	100
Gd	5,333.25	—	20
*bh*F	5,333.1	—	30
Ru	5,332.93	—	40
Co	5,332.67	—	200*w*
Re	5,331.90	—	80
Co	5,331.47	80	500*w*
Ce	5,330.58	—	25
Sr	5,329.82	2	40
Rh	5,329.74	2	30
Fe	5,328.53	35	150
Ta	5,328.38	—	40
*bh*F	5,328.3	—	30
Fe	5,328.05	100	400
Re	5,327.46	—	100
Gd	5,327.33	—	20
Mo	5,327.06	12	20
*bh*F	5,326.9	—	30
Co	5,325.95	—	25
Co	5,325.28	—	300*w*
Sm	5,324.89	3	25*d*
Hf	5,324.26	30	20
Fe	5,324.18	70	400
K	5,323.23	—	40
Pr	5,322.78	3	30
*bh*F	5,322.2	—	30
Fe	5,322.05	—	30
Sm	5,321.87	4	50*d*
Re	5,321.26	—	40
*bh*F	5,320.7	—	30
Sm	5,320.60	—	100
Nd	5,319.82	2	60
Tb	*5,319.23*	—	*40*
Nb	*5,318.60*	*12*	*100*
Re	*5,317.28*	—	*40w*
Co	*5,316.78*	—	*300w*
bhF	*5,316.2*	—	*80*
Ta	5,342.25	—	80
Co	5,342.71	—	800*w*
K	5,342.97	—	30
Co	5,343.39	—	600*w*
Er	5,343.93	—	30
Nb	5,343.17	200	400
Er	3,343.51	—	30
*bh*F	5,345.6	—	20
Yb	5,345.67	100	20
Cr	5,345.81	25	300*R*
Tm	5,346.48	25	40
*bh*F	5,347.0	—	20
Yb	5,347.20	200	40
Co	5,347.49	—	80
Gd	5,347.83	—	20
Er	5,348.05	—	20
Sm	5,348.09	—	20
Cr	5,348.32	15	150*R*
Gd	5,348.69	—	20
Sm	5,348.75	—	20
W	5,348.95	—	30
Co	5,349.087	—	80
Ta	5,349.093	—	80
Lu	5,349.12	2	25
Sm	5,349.14	—	25
Sc	5,349.29	—	30
Rh	5,349.31	—	20
Ta	5,349.57	—	30
Mn	5,349.88	—	20
Eu	5,350.40	—	60*h*
Gd	5,350.41	—	25
Tl	5,350.46	2000*R*	5000*R*
Nb	5,350.74	50	150
Ti	5,351.08	60	50
Yb	5,351.32	3	50
Eu	5,351.67	—	150
W	5,351.90	—	20
Co	5,352.05	—	500*w*
Pr	5,352.40	2	80
Eu	5,352.82	—	80
Yb	5,352.96	250	100
Gd	5,353.28	—	25
Fe	5,353.39	2	60
V	5,353.412	50	50
Ni	5,353.415	—	40
Co	5,353.48	—	500*w*
Ce	5,353.53	30	50
*bh*C	5,354.1	—	100
Rh	5,354.40	5	300

	λ	I	
		J	O
Rh	5,314.79	1	40
bhF	5,314.7	—	50
Cr	5,312.88	—	40
Co	5,312.66	—	400w
Sm	5,312.21	—	100
Hf	5,311.60	150	100
bhF	5,310.3	—	80
Ru	5,309.27	—	125
bhF	5,308.7	—	80
Fe	5,307.36	—	125
Yb	5,307.12	—	40
Tm	5,307.11	20	100
Ru	5,304.86	—	60
bhF	5,304.4	—	100
Eu	5,303.87	—	300
La	5,303.56	135	100
Sm	5,302.91	—	50
bhF	5,302,7	—	100
La	5,302.62	150	50
Fe	5,302.31	—	300
La	5,301.98	200	300r
Co	5,301.06	—	700w
Pt	5,301.02	10	150

	λ	I	
		J	O
Ta	5,354.68	—	80r
Mo	5,354.879	15	25
Tb	5,354.88	—	40
Eu	5,355.08	—	200
Sc	5,356.10	—	40
Rh	5,356.47	1	30
Mo	5,356.48	12	25
Eu	5,356.72	—	40
Eu	5,375.61	—	1000
La	5,357.87	—	40
Co	5,358.92	—	40w
Co	5,359.18	—	300w
K	5,359.21	—	40
Mo	5,360.56	70h	100h
Eu	5,360.81	—	150
Eu	5,361.59	—	300
Ru	5,361.77	—	100
Rb	5,362.60	—	50
Co	5,362.77	—	500w
Mo	5,364.28	25h	70h
Sm	5,364.36	—	50
Fe	5,364.88	10h	200h
Fe	5,365.40	—	40
Fe	5,367.46	15h	200h
Sm	5,368.36	—	80
Pt	5,368.99	1	50
Sm	5,369.16	—	40
Co	5,369.58	—	500w
Tb	5,369.72	—	40
Re	5,369.81	—	40
Fe	5,369.96	20h	150h
Fe	5,371.49	—	700
Tb	5,375.97	—	40
Os	5,376.79	—	50
Eu	5,376.91	—	200
Re	5,377.04	—	300W
Mn	5,377.63	—	40
Rh	5,379.06	3	100

K 5,782.60 *O* 60

	λ	J	O
Ce	5,782.44	—	8
Tm	5,782.36	5	10
Sb	5,782.15	—	6
Mg	5,782.14	—	2
Cu	5,782.132	—	1000
V	5,782.61	—	30
bhYt	5,782.7	—	10
Ce	5,782.81	—	5
U	5,782.823	—	2
Er	5,782.824	—	12

	λ	I	
		J	O
Cr	5,782.131	—	2
*bh*F	5,782.1	—	150
U	5,781.96	—	3
Sm	5,781.89	—	100
Cr	5,781.81	—	20
Yt	5,781.69	5	5
Eu	5,781.36	—	4
Cr	5,781.19	—	18
Os	5,780.81	—	50
Ti	5,780.78	20	20
Ta	5,780.71	—	80
Mo	5,780.64	—	10*h*
U	5,780.61	—	40
bhF	5,780.5	—	50
Si	5,780.45	—	15
Nb	5,780.33	3	3
Ce	5,780.21	—	2
Mn	5,780.19	—	10
Mo	5,780.11	—	12
Ta	5,780.02	—	60
*bh*F	5,779.4	—	200
Pr	5,779.29	—	50
Sm	5,779.25	—	50
*bh*Zr	5,778.5	—	60
Sm	5,778.33	—	50
Ba	5,777.66	100*R*	500*R*
*bh*F	5,777.6	—	100
Re	5,776.84	—	300*w*
Ta	5,776.77	—	80
V	5,776.68	25	50
Nb	5,776.07	3	30
Lu	5,775.40	5	50
*bh*F	5,774.8	—	100
Ti	5,774.05	50	70*W*
Sm	5,773.77	—	100
Ce	5,773.12	—	30
V	5,772.42	25	50
Si	5,772.26	—	30
*bh*F	5,771.9	—	100
Yb	5,771.67	50	30
Hg	5,769.59	200	600
La	5,769.35	—	70
La	5,769.07	60	30
Ta	5,767.91	—	100*w*
Ta	5,766.56	—	80*w*
Ti	5,766.35	50	70*W*
Eu	5,765.20	—	2000
Tm	5,764.29	—	50

	λ	I	
		J	O
Re	5,783.03	—	3
Cr	5,783.11	—	30*h*
U	5,783.17	—	2
Ta	5,783.24	—	2
Mo	5,783.33	—	20*h*
*bh*F	5,783.4	—	30
V	5,783.50	—	30
Sm	5,783.53	—	8
Nd	5,783.68	—	3*h*
Eu	5,783.71	—	150*s*
*bh*Zr	5,783.8	—	5
Cd	5,783.93	—	5
Cr	5,783.934	—	30*h*
Ce	5,783.99	—	3
Mo	5,784.00	—	5
Ba	5,784.06	2	6
V	5,784.38	30	50
Er	5,784.63	—	12
Ho	5,784.64	—	8
Sm	5,784.72	—	3
W	5,784.75	—	3
*bh*F	5,784.8	—	100
Ce	5,784.85	—	12
Th	5,784.89	—	8
Nd	5,784.96	—	20
Cr	5,785.00	—	20
Tb	5,785.18	—	40
Ti	5,785,98	60	100*W*
V	5,786.16	—	75
*bh*F	5,786.4	—	30
Sm	5,786.99	—	200
Nb	5,787.54	15	80
*bh*F	5,787.6	—	100
Cr	5,787.99	—	50*wh*
Nd	5,788.22	—	30
Sm	5,788.39	—	30
La	5,789.25	—	125
*bh*F	5,789.3	—	30
*bh*F	5,790.3	—	100
Cr	5,791.00	—	40*wh*
La	5,791.34	—	200
Mo	5,791.85	60	100
Eu	5,792.72	—	30
Rh	5,792.77	—	40
*bh*F	5,793.1	—	100
C	5,793.51	—	30
*bh*F	5,795.9	—	100

	λ	I	
		J	O
Pt	5,763.57	—	30
Fe	5,763.01	35	80
Er	5,762.79	—	30
Ti	5,762.27	50	70*h*
La	5,761.84	—	60
Ni	5,760.85	—	50
Nb	5,760.34	30	30
Sm	5,759.50	—	60
Er	5,757.62	—	30
Ni	*5,754.67*	—	*150w*
Re	*5,752.95*	—	*200w*
Mo	*5,751.40*	*100*	*125*
bhZr	*5,748.1*		*100*
Tb	*5,747.58*	—	*60*
Ta	*5,746.71*	—	*60*
La	*5,744.41*	—	*80*
V	*5,743.45*	*20*	*60*
Sm	*5,741.19*	—	*60*
Sm	*5,740.88*	—	*80*
La	*5,740.66*	*1*	*100*
Ti	*5,739.51*	*80*	*70*
Eu	*5,739.00*	—	*300*
V	*5,737.06*	*100*	*100*
Lu	*5,736.55*	*15*	*150*
Sm	*5,732.95*	—	*100*

	λ	I	
		J	O
La	5,797.59	150	80
Zr	5,797.74	—	50
U	5,798.55	1	35
*bh*F	5,798.90	—	80
V	5,799.9	2	40
Eu	5.800.27	—	200
Ba	5,800.28	20	100
Sm	5,800.50	—	80
Lu	5,800.59	2	30
Os	5,800.60	—	50
*bh*F	5,801.8		80
K	5,801.96	20	50*h*
Mo	5,802.67		30
Sm	5,802.82	—	80
Tb	5,803.15	—	40
Nd	5,804.02	—	100
Ti	5.804.26	50	100*h*
*bh*F	5,804.7	—	80
Ni	5,805.23	—	50
Ba	5,805.69	—	70
La	5,805.78	120	60
Rh	5,806.91	1	100
V	5,807.14	40	75
*bh*F	5,807.7	—	80
Ta	*5,811.10*	—	*100*
K	*5,812.52*	—	*30*
Sm	*5,814.87*	—	*60*
V	*5,817.53*	—	*100h*
Eu	*5,818.74*	—	*1000*
Pr	*5;823.72*	*1*	*60w*
Ba	*5,826.29*	—	*150wh*
V	*5,830.72*	*80*	*100*
Sm	*5,831.01*	—	*80*
Eu	*5,831.05*	—	*2000W*
Rh	*5,831.58*	*1*	*80*
K	*5,832.09*	—	*50*

K 7,664.91 *J* 400 *O* 9000*R*

Cd	**7,664.74**	**3**	**—**
Cu	**7,664.70**	**70**	**5**
Xe	**7,664.56**	**(30)**	**—**
La	**7,664.34**	**3**	**8**
Fe	**7,664.30**	**—**	**15**
Xe	**7,664.02**	**(10)**	**—**
Kr	**7,663.75**	**(2)**	**—**
Sc	**7,665.72**	**—**	**5**
***bh*Ti**	**7,666.4**	**—**	**6**
Xe	**7,666.61**	**(10)**	**—**
Kr	7,685.25	(1000)	—
Kr	7,694.54	(1000)	—

*N**

	λ	I J	I O		λ	I J	I O
Nd	**7,663.52**	—	**2**				
Ti	**7,663.47**	—	**12**				
Xe	7,642.02	(500)	—				
Ar	7,635.10	(500)	—				

K 7,698.98 *J* 200 *O* 5000*R*

	λ	I J	I O		λ	I J	I O
Nd	**7,698.94**	—	**2*h***	**Ta**	**7,699.14**	—	**5**
Yt	**7,698.00**	—	**4**	**Yb**	**7,699.49**	—	**2000**
Sc	**7,697.73**	—	**20**	**J**	**7,700.20**	**(8)**	—
				Ar	7,723.76	(200)	—
				Ar	7,724.21	(200)	—

*P**

Kr $^{36}_{83.80}$

t_0 −156.7°C t_1 −152.0°C

I.	II.	III.	IV.	V.
13.996	26.5	36.94	68	—

λ	I	eV
II 2,464.77	(100)	22.8
II 2,712.40	(80)	23.2
II 2,833.00	(100)	22.79
II 3,607.88	(100)	22.97
II 3,631.87	(200)	21.89
II 3,653.97	(250)	21.92
II 3,718.02	(300)	23.83
II 3,920.14	(200)	21.93
I 4,319.58	(1000)	12.78
II 4,355.48	(3000)	21.92
II 4,658.87	(2000)	19.30
I 5,570.29	(2000)	12.14
I 5,870.92	(3000)	12.14

Kr 2,464.77 (100)

	λ	I			λ	I	
		J	O			J	O
Ru	2,464.769	6	—	Br	2,464.81	(15)	—
Ru	2,464.70	4	50	Ir	2,464.901	1	5
J	2,464.68	(100)	—	Fe	2,464.903	50	4
Nb	2,464.65	15	1	Tm	2,464.94	6	6
Ho	2,464.63	10	—	J	2,464.95	(12)	—
W	2,464.626	20	3	Ti	2,464.98	1	10
Os	2,464.50	1	4	Ru	2,464.99	4	—
Yb	2,464.49	2	5	Hf	4,465.06	15	10
Nb	2,464.43	1	5	Os	2,465.16	40	25
Co	2,464.195	150	40	Fe	2,465.20	50	—
Hf	2,464.19	100	30	V	2,465.28	100	2
Fe	2,464.01	80	5	Fe	2,465.91	100	7
Rh	2,463.44	150	2	Rh	2,466.15	100	1
Fe	2,463.28	60	6	Mn	2,466.21	50	—
Fe	2,462.64	50	200*R*	Co	2,467.06	80	2
Ag	2,462.24	80*h*	5	Rh	2,467.23	50	—
Nb	2,462.05	100	1	Pt	2,467.44	100	800*R*
Fe	2,461.86	70	15	Pt	2,467.59	100	5
Fe	2,461.28	50	5	Te	2,467.72	(50)	—
J	2,461.13	(60)	—	Cu	2,468.58	70	5
Rh	2,461.04	200*wh*	80				
				Pd	*2,469.25*	*150*	—
Nb	*2,460.40*	*200wh*	—	*Te*	*2,469.62*	*(300)*	—
Rh	*2,458.90*	*300*	*50*	*Cd*	*2,469.84*	*(500)*	—
Nb	*2,457.00*	*200*	*1*	*Pd*	*2,470.01*	*150*	—
				Pd	*2,471.15*	*150*	—
				Rh	*2,471.77*	*100*	*3*
				Pd	*2,472.51*	*150*	—

Kr 2,712.40 (70)

	λ	J	O		λ	J	O
Fe	2,712.39	100	2	Ru	2,712.41	300	80
Mo	2,712.35	40	1	Zr	2,712.420	15	20
Cr	2,712.31	70	30	Hf	2,712.425	50	25
J	2,712.23	(100)	—	Zn	2,712.49	8	300
V	2,712.22	10*h*	6	Cd	2,712.57	20	75
Hf	2,712.14	10	10	U	2,712.58	2	—
Er	2,712.120	1	7	Yb	2,712.65	5	2
U	2,712.116	2	3	Er	2,712.66	2	4
Rn	2,712.1	(7)	—	W	2,712.69	20	1
U	2,712.061	6	8	Ir	2,712.74	10	40
Ag	2,712.06	200*h*	3	V	2,712.81	4*wh*	9*h*
Hf	2,711.99	10	10				

	λ	I	
		J	O
Fe	2,711.84	100	4
V	2,711.74	150R	50
Fe	2,711.65	50	100
Te	2,711.61	(50)	—
Mn	2,711.58	125h	2
Ag	2,711.21	300wh	1h
Cr	2,810.92	70	1
Mn	2,710.33	40h	12
In	2,710.26	200rh	800R
Ru	2,710.16	100	50
V	2,710.16	60	6
N	2,709.82	(50)	—
Cr	2,709.31	60	2
Ta	2,709.27	150	40
Tl	2,709.23	200R	400R
Fe	2,709.06	100	3
Ra	2,708.96	(200)	—
Ni	2,708.791	500	—
Cr	2,708.79	40	3
Fe	2,708.57	50	80
Mn	2,708.45	50h	15
Ar	2,708.28	(40)	—
V	2,707.86	150	70
Nb	2,707.83	80	3
Mn	2,707.53	50h	10
Co	2,707.50	100wh	—
Co	*2,706.74*	*100wh*	—
V	*2,706.70*	*200R*	*60*
Fe	*2,706.58*	*150*	*150*
Sn	*2,706.51*	*150R*	*200R*
V	*2,706.17*	*400R*	*100*
Pt	*2,705.89*	*200wh*	*1000wh*
Co	*2,705.85*	*100w*	*15w*
Rh	*2,705.63*	*300wh*	*100*
Ru	*2,704.57*	*100*	—
Fe	*2,703.99*	*400*	*30*
Cu	*2,703.18*	*200*	*10*

	λ	I	
		J	O
V	2,713.05	80	40
Ru	2,713.07	80	—
Cu	2,713.50	300w	50
Mo	2,713.51	40	20
Ru	2,713.58	100	—
In	2,713.93	125wh	200R
V	2,714.20	100	60
Pd	2,714.32	150	—
Fe	2,714.41	400	200
Co	2,714.42	200W	12
Pd	2,714.90	200	—
Rh	2,715.31	500wh	50
Nb	2,715.34	100	2
V	2,715.69	300R	50
Nb	2,715.88	100	2
Co	2,715.99	75	75w
Ru	2,716.12	100	—
Kr	2,716.16	(10h)	—
Fe	2,716.22	150	20
Ti	2,716.25	70	5
Ru	2,716.58	80	—
Nb	2,716.62	200	10
Mn	2,716.79	40	1
Eu	2,716.97	300	300
Mo	2,717.35	100	20
Ru	2,717.40	100	50
Cr	2,717.51	40	12
Cu	*2,718.77*	*300w*	*40*
Fe	*2,719.02*	*300r*	*500r*
Pt	*2,719.04*	*100w*	*1000w*
Tm	*2,721.19*	*100*	*60*
Cu	*2,721.67*	*150*	—
Nb	*2,721.98*	*200*	*10*

Kr 2,833.00 (100)

	λ	J	O
W	**2,832.952**	**3**	**10**
Zn	**2,832.95**	**(25)**	—
Ne	**2,832.921**	**(8)**	—
J	**2,832.917**	**(20)**	—
Nb	**2,832.79**	**5h**	—
Cl	**2,833.03**	**(4)**	—
In	**2,833.06**	**3**	—
Zr	**2,833.061**	**1**	**2**
Pb	**2,833.07**	**80R**	**500R**
Fe	**2,833.10**	**5h**	—

	λ	I J	I O
Mo	**2,832.66**	**5**	**—**
U	**2,832.64**	**2**	**2**
La	**2,832.55**	**5**	**4**
W	**2,832.48**	**9**	**10**
Cr	**2,832.46**	**125**	**2**
Xe	**2,832.46**	**(24)**	**—**
Fe	2,832.44	200	300
Kr	2,832.39	(2)	—
Ti	2,832.16	100	25
U	2,832.06	50	35
Ru	2,831.84	50	10
Ho	2,831.60	70	—
Fe	2,831.56	500	1
Rn	2,830.6	70	—
Cr	2,830.47	80*h*	15
V	2,830.40	60	10
Pt	2,830.29	600*r*	1000*r*
Nb	2,829.75	50	3
Ti	2,828.9	150*wh*	—
Fe	2,828.81	60	100
Eu	2,828.69	150	200*w*
Fe	2,828.63	80	—
Ta	2,828.58	100	75
Ti	2,828.15	200*h*	2
Tm	2,827.92	100	50
Fe	2,827.89	50	70
Ta	*2,827.55*	*100d*	*3d*
Ru	*2,826.68*	*100*	*—*
Tl	*2,826.16*	*100R*	*200R*
Fe	*2,825.56*	*150*	*150*
Co	*2,825.242*	*200*	*5*
Ni	*2,825.236*	*150*	*—*
Ag	*2,824.370*	*200w*	*150w*
Cu	*2,824.369*	*300*	*1000*
Fe	*2,823.28*	*300*	*200*
Ru	*2,822.55*	*150*	*30*
Cr	*2,822.37*	*100*	*20*

	λ	I J	I O
K	**2,833.14**	**(2)**	**2*h***
Ir	**2,833.236**	**20**	**7**
U	**2,833.244**	**4**	**8**
Eu	**2,833.25**	**5*wh***	**10*w***
Hf	**2,833.28**	**4**	**25**
Nb	**2,833.30**	**10**	**1**
Tl	**2,833.31**	**(25)**	**—**
Th	**2,833.34**	**8**	**8**
Tb	**2,833.37**	**20**	**—**
Cr	**2,833.39**	**3**	**3**
Fe	**2,833.40**	**8**	**10**
W	**2,833.63**	**12**	**15**
Ru	2,833.78	80	—
Cd	2,834.19	(100)	—
Cr	2,834.26	125	—
Co	2,834.94	75	2
Nb	2,835.12	100	5*d*
Fe	2,835.46	100	100
Cr	2,835.63	400*r*	100
V	2,836.52	80	20
Ti	2,836.64	100*wh*	—
C	2,836.71	200	—
Cd	2,836.91	80	200
In	2,836.92	80	80
Cu	2,837.55	250	—
Au	2,838.03	80	—
Fe	2,838.12	150	150
Ta	2,838.24	150	2
Rn	2,838.5	(70)	—
Os	*2,838.63*	*100*	*100R*
Kr	*2,838.79*	*(20)*	*—*
Ti	*2,839.80*	*100wh*	*—*
Pd	*2,839.89*	*100*	*—*
Sn	*2,839.99*	*300R*	*300R*
Cr	*2,840.02*	*125*	*25*
Pd	*2,841.03*	*100*	*—*
Ru	*2,841.12*	*125*	*—*
Nb	*2,841.15*	*100*	*10*
Ru	*2,841.68*	*200*	*50*
Ti	*2,841.94*	*125*	*40*
Rn	*2,842.1*	*(150)*	*—*
Ni	*2,842.42*	*150*	*—*
Nb	*2,842.65*	*100*	*10*
Cr	*2,843.25*	*400r*	*125*
Fe	*2,843.63*	*100*	*125*
Fe	*2,843.98*	*300*	*300*

Kr 3,607.88 (100)

	λ	I	
		J	O
Pr	3,607.82	3	5
U	3,607.78	3	1
Nd	3,607.72	6	10
Ce	3,607.62	8	15
Hg	3,607.60	(18)	—
Au	3,607.54	20	—
Mn	3,607.537	40	75
U	3,607.52	2	—
J	3,607.51	(18)	—
Mo	3,607.412	4	4
Xe	3,607.41	(5)	—
Ta	3,607.406	35	70
Th	3,607.39	4	3
Zr	3,607.38	9	8
Tm	3,607.36	10	8
Nb	3,607.33	5	2
F	3,607.32	(6)	—
Pr	3,607.22	2	10
Ti	3,607.13	5	25
W	3,607.066	12	10
Gd	3,607.066	20	15
Nb	3,607.01	10	1
U	3,606.95	1	3
Hg	3,606.92	(30)	—
Mo	3,606.91	30	—
Nb	3,606.81	1	3
F	3,606.80	(10)	—
Ti	3,606.79	4	12
V	3,606.69	70	80
Fe	3,606.68	150	200
Ar	3,606.52	(1000)	—
Yb	3,606.47	60	15
Dy	3,606.13	100	200
Hg	3,605.80	(200)	—
Fe	3,605.46	150	300
Cr	3,605.33	400*R*	500*R*
Os	3,604.47	100	15
V	3,604.38	50*h*	—
Hg	3,604.09	(50)	—
Cr	3,603.74	50	15
Fe	3,603.21	80	150
Eu	3,603.20	50	100*w*
F	3,602.85	(60*d*)	—
Yt	3,601.92	60	18
Yt	3,600.73	300	100
Tb	3,600.44	50	8
Ne	3,600.17	(75)	—
Pt	3,607.89	5	—
Th	3,607.91	2	1
U	3,608.00	4	2
Nb	3,608.01	5	5
Dy	3,608.06	2	4
Rh	3,008.09	3	10
U	3,608.146	1	2
Fe	3,608.150	25	15
La	3,608.16	6	2
Cs	3,608.28	(10)	—
Nb	3,608.31	3	3
Ir	3,608.354	2	6
U	3,608.354	2*h*	3
Mo	3,608.369	15	15
W	3,608.37	10	—
Th	3,608.38	1*d*	4
Cr	3,608.40	8	12
Pr	3,608.47	3	10
Mn	3,608.49	40	60
U	3,608.69	4*h*	—
Eu	3,608.72	1*h*	10
Ru	3,608.73	8	2
Gd	3,608.76	125	100
Tm	3,608.77	20	100
Ta	3,608.78	1*d*	15*r*
Fe	3,608.86	400	500
K	3,608.87	(10)	—
F	3,608.89	(6)	—
U	3,608.96	10	18
Ne	3,609.18	50	—
Sm	3,609.48	100	60
Pd	3,609.55	700*R*	1000*R*
Ti	3,610.156	70	100
Fe	3,610.162	90	100
Cd	3,610.51	500	1000
Yt	3,611.05	60	40
Yb	3,611.31	50	12
Fe	3,612.07	50	80
Al	3,612.467	80*h*	—
Rh	3,612.470	50	200
Ni	3,612.74	50*h*	400
Cd	3,612.87	500	800
Sc	3,613.84	70	40
Cd	3,614.45	100	60
Ir	3,614.77	80	40
Kr	3,615.47	(20)	—

	λ	I J	I O
Kr	3,599.90	(40h)	—
Ru	3,599.76	100	12
Kr	3,599.21	(25h)	—
Ni	3,597.70	50h	1000R
Sb	3,597.51	200wh	2
Rh	3,597.15	100	200
Sb	3,596.96	(100)	—
Ru	*3,596.18*	*100*	*30*
Ti	*3,596.05*	*125*	*50*
Dy	*3,595.05*	*100*	*200*
Fe	*3,594.64*	*100*	*125*
Ne	*3,593.64*	*(250)*	—
Ne	*3,593.53*	*(500)*	—
Cr	*3,593.49*	*400R*	*500R*
V	*3,593.33*	*300R*	*30*
Ru	*3,593.02*	*150*	*60*
Gd	*3,592.70*	*70*	*50*
V	*3,592.02*	*300R*	*50*
Dy	*3,591.42*	*100*	*200*
Nd	*3,590.35*	*300W*	*400W*
V	*3,589.76*	*600R*	*80*
Kr	*3,589.65*	*(70wh)*	—
Nb	*3,589.36*	*100*	*100*
Ru	*3,589.21*	*100*	*60*
Ar	*3,588.44*	*(300)*	—
Al	*3,587.06*	*(100)*	—
Fe	*3,586.99*	*150*	*200*
Al	*3,586.91*	*(500h)*	—
Al	*3,586.80*	*(200wh)*	—
Al	*3,586.69*	*(200)*	—
Al	*3,586.55*	*(200)*	—
Dy	*3,585.78*	*100*	*150*
Te	*3,585.34*	*(350)*	—
Fe	*3,585.32*	*100*	*150*
Dy	*3,585.07*	*100*	*300*
Gd	*3,584.96*	*100*	*100*
S	3,616.92	(60)	—
P	3,617.09	(100w)	—
Fe	3,617.79	80	125
Hg	3,618.53	(50)	—
Fe	3,618.77	400	400
V	3,618.93	100	—
Mn	3,619.28	50	75
Ni	*3,619.39*	*150h*	*2000R*
Nb	*3,619.51*	*200*	*5*
Nb	*3,619.73*	*300*	*3*
Yb	*3,619.81*	*100*	*30*
*N**			

Kr 3,631.87 (200)

	λ	I J	I O
Ho	**3,631.75**	**4h**	**8**
Ir	**3,631.714**	**2**	**10**
Ru	**3,631.711**	**8**	**2**
Cr	**3,631.69**	**60**	**10**
Ca	**3,631.51**	**2**	—
J	**3,631.88**	**(2)**	—
Zn	**3,631.93**	**(1)**	**15**
W	**3,631.95**	**10**	**15**
Fe	**3,632.04**	**50**	**50**
Ce	**3,632.11**	**2**	**10**

	λ	I	
		J	O
Fe	3,631.46	300	500
P	3,631.40	(50)	—
Co	3,631.39	25	50*W*
Se	3,631.38	(25)	—
Na	3,631.27	(100)	12
Ce	3,631.19	3	50
Sm	3,631.14	15	40
Fe	3,631.10	10	25
Nd	3,631.02	8	10
Pr	3,630.97	20	50
W	3,630.95	7	—
Hf	3,630.87	6	15
Sm	3,630.85	3	3
W	3,630.82	8	9
Ca	3,630.75	9	150
Sc	3,630.74	70	50
U	3,630.73	20	8
Hg	3,630.65	(100)	—
Kr	3,628.16	(10)	—
V	3,625.61	125	—
Ti	3,624.82	125	60
Kr	3,623.61	(30*h*)	—
Fe	3,622.00	100	125
Fe	3,621.46	100	125
Rn	3,621.0	(250)	—
*P**			

	λ	I	
		I	O
V	3,632.12	70	—
U	3,632.17	3*hd*	—
Eu	3,632.175	10	30
Hg	3,632.38	(10)	—
J	3,632.48	(2)	—
Kr	3,632.49	(4)	—
Fe	3,632.556	25	30
Cu	3,632.558	3	25
Au	3,632.56	2*wh*	—
Th	3,632.63	6	6
Ar	3,632.68	(300)	—
Hf	3,632.69	2	5
W	3,632.708	8	9
U	3,632.713	2	3
Dy	3,632.73	4	10
Ne	3,632.75	(4)	—
Er	3,632.78	1	5
Cr	3,632.84	35	80
Fe	3,632.98	8	12
Nb	3,632.999	3	5
Dy	3,633.00	3	10
Yt	3,633.12	100	50
Ar	3,634.46	(300)	—
Pd	3,634.69	1000*R*	2000*R*
Rn	3,634.8	(25)	—
Ru	3,634.93	100	50
Ti	3,635.46	100	200
Kr	3,637.48	(20*h*)	—
Hg	3,638.34	(100)	—
Fe	3,640.39	200	300
F	3,640.89	(100)	—
Ti	3,641.33	150	60
Ti	3,642.67	125	300
Ar	3,643.09	(100)	—
Fe	*3,647.84*	*400*	*500*
Kr	*3,648.61*	*(40h)*	—
Ra	*3,649.55*	*(1000)*	—
Ar	*3,649.83*	*(800)*	—
Hg	*3,650.15*	*500*	*200*
Kr	*3,651.02*	*(25h)*	—
Nb	*3,651.19*	*400*	*10*
Fe	*3,651.47*	*200*	*300*
*N**			

Kr 3,653.97 (250)

	λ	I			λ	I	
		J	O			J	O
Au	3,653.93	2	5	Fe	3,653.98	2*h*	4
Sr	3,653.928	3	15	Pt	3,653.99	1*h*	2
Cr	3,653.91	25	100	U	3,654.13	1	5
Mo	3,653.90	2	3	Nd	3,654.16	10	10
Tb	3,653.87	8	15	Dy	3,654.17	1	4
Ta	3,653.83	1	3	W	3,654.20	10	20
Fe	3,653.76	10	25	Nb	3,654.23	5*h*	—
Os	3,653.72	10	30	U	3,654.29	4	3
Ce	3,653.67	8	18	Cu	3,654.30	2*h*	10*wh*
Pr	3,653.65	1	4	Pr	3,654.34	2	8
Nb	3,653.615	5	10	Bi	3,654.38	5*s*	7
Tm	3,653.61	20	30	Ru	3,654.40	40	3
Yt	3,653.606	2	5	Nb	3,654.423	10	10
Th	3,653.59	2	3	Pd	3,654.425	4	—
Mo	3,653.55	4	4*d*	Ho	3,654.45	6	6
W	3,653.52	1	5	Th	3,654.47	1	4
Ti	3,653.50	200	500	Os	3,654.49	15	100
Ca	3,653.496	3	—	S	3,654.51	(8)	—
Au	3,653.491	5	3	Th	3,654.58	4	4
Sm	3,653.48	2	15	Mo	3,654.58	20	25
Ta	3,653.39	1	3	Ti	3,654.59	40	100
P	3,653.38	(100*w*)	—	Gd	3,654.64	200	200*w*
W	3,653.34	12	1	Au	3,654.67	3	3
Sr	3,653.27	8	30	W	3,654.71	12	—
U	3,653.21	1	10	Hg	3,654.83	(200)	—
Os	3,653.20	5*W*	10*W*	Sm	3,654.85	1	9
Ir	3,653.19	50	15	Rh	3,654.87	10	40
Nd	3,653.15	4	10	Dy	3,654.878	1	5
Sm	3,653.113	2	10	Tb	3,654.88	30	70
Ce	3,653.108	5	15	Se	3,654.89	(3)	—
Se	3,653.05	(25)	—	Ce	3,654.97	3	15
Tb	3,652.97	8	15	Al	3,654.98	(18)	—
Tl	3,652.95	50	150	Fe	3,654.99	1	3
Er	3,652.88	4	20	Al	3,655.00	(100)	—
W	3,652.76	4	3	Nd	3,655.03	1	5
Fe	3,652.73	—	6	Mo	3,655.08	20	8
				U	3,655.11	4	8
				Pr	3,655.12	2	9
*P**				Sm	3,655.18	3	7
				Sb	3,655.25	4	1
				Gd	3,656.16	200	200*W*
				Rh	3,657.99	200*W*	500*W*
				Tb	3,658.88	100	100
				Ar	3,659.50	(100)	—
				Nb	3,659.61	500	15
				Ti	3,569.76	150	50

	λ	I J	I O
Kr	3,661.00	(15)	—
V	3,661.38	150	10
Gd	3,662.27	200	200*w*
Hg	3,662.88	400	50
Hg	3,663.28	400	500
Kr	3,663.44	(20)	—
Ne	3,664.11	(250)	—
Gd	3,664.62	200	200*w*
Kr	3,665.33	(80)	—
Kr	*3,668.74*	*(10)*	—
Kr	*3,669.01*	*(150h)*	—
V	*3,669.41*	*300*	*20W*
Ar	*3,670.64*	*(300)*	—
Ar	*3,675.22*	*(300)*	—

Kr 3,718.02 (300)

	λ	I J	I O
Pr	**3,718.016**	**2**	**3*h***
La	**3,717.98**	**3**	**3**
Cl	**3,717.94**	**(15)**	—
Os	**3,717.93**	**12**	**10**
Tm	**3,717.92**	**10**	**100**
Pr	**3,717.84**	**3**	**8**
Th	**3,717.83**	**10**	**20**
W	**3,717.80**	**12**	—
Hf	**3,717.800**	**8**	**20**
Eu	**3,717.69**	**2**	**18*w***
Ru	**3,717.68**	**5**	**6**
P	**3,717.62**	**(70)**	—
Nb	**3,717.54**	**8*h***	**10**
Gd	**3,717.49**	**50**	**100*w***
Ce	**3,717.48**	**1**	**8**
Tb	**3,717.47**	**3**	**8**
U	**3,717.42**	**8**	**10**
Ti	**3,717.40**	**50**	**80**
Sr	**3,717.30**	**5**	**1**
Ti	**3,717.26**	**1**	**6**
Rn	**3,717.2**	**(10)**	—
Xe	**3,717.20**	**(10)**	—
Ar	**3,717.17**	**(10)**	—
Sc	**3,717.096**	**4**	**4**
W	**3,717.095**	**10*d***	**12*d***
Nb	**3,717.07**	**1000**	**8**
Ca	**3,717.03**	**3**	—
U	**3,717.01**	**4**	**1**
Ru	**3,717.00**	**25**	**30**
U	**3,718.106**	**12**	**8**
Yt	**3,718.108**	**4**	**12**
V	**3,718.16**	**70**	**5**
Th	**3,718.17**	**10**	**15**
Ce	**3,718.19**	**5**	**15**
Ar	**3,718.21**	**(15)**	—
In	**3,718.22**	**(18)**	—
W	**3,718.32**	**6**	—
In	**3,718.33**	**(25)**	—
Os	**3,718.339**	**10**	**30**
Pb	**3,718.34**	**2**	—
Ce	**3,718.38**	**5**	**15**
In	**3,718.39**	**(25)**	—
Ru	**3,718.40**	**2**	**8**
Fe	**3,718.41**	**50**	**80**
Tb	**3,718.44**	**3**	**15**
Ta	**3,718.45**	**1**	**3**
Mo	**3,718.48**	**5**	**5**
Nb	**3,718.52**	**3**	**2**
Ru	**3,718.611**	**9**	**2**
U	**3,718.614**	**8**	**4**
Kr	**3,718.63**	**(200*h*)**	—
In	**3,718.634**	**(25)**	—
Th	**3,718.66**	**8**	**20*h***
Sm	**3,718.70**	**1**	**8**
In	**3,718.71**	**(18)**	—
In	**3,718.836**	**(40)**	—
Hf	**3,718.842**	**3**	—
Zr	**3,718.843**	**9**	**9**

	λ	I J	I O		λ	I J	I O
Pr	**3,716.98**	**2*h***	**10 *h***	**U**	**3,718.844**	**4**	**—**
Mo	**3,716.97**	**3*h***	**—**	**Pr**	**3,718.877**	**3**	**3**
Dy	**3,716.93**	**5**	**5**	**Sm**	**3,718.880**	**5**	**100**
Ce	**3,716.930**	**2**	**10**	**Pd**	**3,718.909**	**200**	**300**
Yt	**3,716.91**	**5*h***	**2**	**V**	**3,718.912**	**5**	**20**
Eu	**3,716.91**	**2**	**10*w***	**Mn**	**3,718.93**	**100**	**75**
Mo	**3,716.87**	**25**	**4**	**Mo**	**3,719.05**	**20**	**—**
U	**3,716.78**	**8**	**6**	**Hf**	**3,719.28**	**30**	**15**
V	3,715.47	400*R*	70	Fe	3,719.93	700	1000*R*
Kr	3,715.04	(12*h*)	—	Kr	3,721.35	(150)	—
Ne	3,713.08	(250)	—	Fe	3,722.56	400	500
Gd	3,712.71	250	200*W*	Fe	3,724.38	150	200
Yt	3,710.29	150	80	Ru	3,726.93	150	100
Fe	3,709.25	400	600	V	3,727.34	200	40
P	3,706.05	(150*w*)	—	Fe	3,727.62	150	200
Fe	3,705.57	500	700	Ru	3,728.03	150	100
				Ar	3,729.29	(200)	—
				Ti	3,720.81	150	500
Yb	*3,694.20*	*1000R*	*500R*				
Fe	*3,694.01*	*300*	*400*				
				Kr	*3,732.61*	*(15h)*	—
				V	*3,732.76*	*500R*	*70R*
				Fe	*3,733.32*	*300*	*400*
				Fe	*3,734.87*	*600*	*1000r*
				Kr	*3,735.78*	*(40h)*	—
				Fe	*3,737.13*	*600*	*1000r*
				Kr	*3,741.69*	*(200h)*	—
				Kr	*3,744.80*	*(150hs)*	—

Kr 3,920.14 (200)

	λ	I J	I O		λ	I J	I O
Sm	**3,920.09**	**2**	**3**	**Nb**	**3,920.20**	**100**	**30**
Mo	**3,920.08**	**3**	**3**	**Th**	**3,920.259**	**5**	**8**
W	**3,920.04**	**5**	**5**	**Fe**	**3,920.260**	**300**	**500**
J	**3,920.00**	**(5)**	**—**	**U**	**3,920.35**	**2**	**—**
V	**3,919.99**	**7**	**25**	**V**	**3,920.49**	**15**	**35**
Nd	**3,919.92**	**6*d***	**20*d***	**Pr**	**3,920.52**	**10**	**30**
Ti	**3,919.82**	**2**	**20**	**U**	**3,920.53**	**2**	**6**
Ce	**3,919.81**	**2**	**45**	**In**	**3,920.54**	**18**	**—**
Nb	**3,919.720**	**100**	**2**	**Cu**	**3,920.64**	**2**	**—**
U	**3,919.719**	**5**	**5**	**Br**	**3,920.65**	**(15)**	**—**
Co	**3,919.63**	**1**	**4**	**C**	**3,920.677**	**200**	**—**
Pr	**3,919.62**	**15**	**35**	**S**	**3,920.68**	**(8)**	**—**
Tb	**3,919.54**	**15**	**40**	**Tb**	**3,920.72**	**3**	**10**
Br	**3,919.51**	**(15)**	**—**	**Nb**	**3,920.76**	**50*h***	**1**
Mo	**3,919.453**	**20**	**—**	**Fe**	**3,920.84**	**2**	**7**

	λ	I	
		J	O
Ho	3,919.45	4	6
U	3,919.34	4	3
O	3,919.28	(35)	—
Nb	3,919.165	5	5
Cr	3,919.159	125	300r
Fe	3,919.07	7	15
N	3,919.003	(35)	—
Th	3,919.003	1	10
Nb	3,919.000	5	5
C	3,918.98	80	—
Os	3,918.97	10	30
Hg	3,918.92	(200)	—
Nd	3,918.90	4	20
Pr	3,918.86	30	100
Fe	3,918.65	40	60
Kr	3,917.64	(50*wh*)	—
Gd	3,916.59	100	150*w*
La	3,916.04	400	400
Nb	3,914.70	100	30
Hg	3,914.29	(100)	—
Br	3,914.28	(150)	—
P	3,914.26	(100)	—
Kr	3,912.59	(70)	—
O	3,911.95	(150)	—
Cr	3,908.75	150	200
Gd	3,907.12	100	100*W*
Eu	3,907.11	500*R*	1000*RW*
Fe	3,906.48	200	300
Kr	3,906.25	(150*h*)	—
Mo	*3,902.96*	*500R*	*1000R*
Fe	*3,902.95*	*400*	*500*
Kr	*3,901.15*	*(10h)*	—
Al	*3,900.68*	*(200)*	—
Fe	*3,899.71*	*300*	*500*
Nb	*3,898.28*	*200*	*3*
Fe	*3,895.66*	*300*	*400*
Kr	*3,894.71*	*(60wh)*	—
Ar	*3,894.66*	*(300)*	—
Pd	*3,894.20*	*200W*	*200W*
Os	3,920.87	12	20
Ru	3,920.91	20	20
Mo	3,920.92	5	5
Nd	3,920.96	15	40
Cr	3,921.02	40	150
U	3,921.24	6	8
Cu	3,921.27	1*h*	40
Nb	3,921.35	10*h*	—
Pr	3,921.41	2	10
Ti	3,921.42	6	40
La	3,921.53	200	400
Mo	3,921.54	20	—
U	3,921.55	8	8
Kr	3,921.68	(6*h*)	—
Fe	3,922.91	400	600
Ru	3,923.47	100	60
S	3,923.48	(200)	—
Tb	3,925.45	200	150
Pr	3,925.46	100	125
Hg	3,925.65	(100)	—
Ru	3,925.92	100	60
Fe	3,927.92	300	500
Ar	3,928.62	(125)	—
La	3,929.22	300	400
Kr	3,929.26	(20*h*)	—
Fe	3,930.30	400	600
Eu	3,930.50	400*R*	1000*R*
J	3,931.01	(400)	—
Rn	3,931.82	(250)	—
Sm	3,933.59	200*h*	200
Fe	3,933.60	200	200
Ca	3,933.67	600*R*	600*R*
Ru	3,933.68	200	5
Nb	*3,936.02*	*200*	*5*
Kr	*3,938.88*	*(20wh)*	—
Tb	*3,939.60*	*200*	*200*
J	*3,940.24*	*(500)*	—
Rb	*3,940.57*	*(200)*	—
Kr	*3,942.93*	*(20wh)*	—
Al	*3,944.03*	*1000*	*2000*
O	*3,947.33*	*(300)*	—
Ar	*3,947.50*	*(1000)*	—
Ar	*3,948.98*	*(2000)*	—
La	*3,949.11*	*800*	*1000*

Kr 4,319.58 (1000)

	λ	I J	I O
Ca	4,319.54	2	2h
Sm	4,319.53	15	50
U	4,319.52	4	5
Mo	4,319.512	4h	5h
Hf	4,319.51	10wh	2
Os	4,319.339	1	9
Nd	4,319.336	2	8
Te	4,319.26	(50)	—
Ca	4,319.13	3	—
Sr	4,319.12	20	50
Kr	4,319.12	(4)	—
Th	4,319.11	4	5
U	4,319.06	5	1
Pr	4,319.00	8	25
Sm	4,318.93	300	500
C	4,318.92	10	—
Se	4,318.91	(8)	—
Tb	4,318.85	30	150
Ne	4,318.83	(5)	—
Ta	4,318.81	5	15
V	4,318.70	2	5
S	4,318.68	(40)	—
Ca	4,318.65	20	60
Ti	4,318.639	50	100
Ce	4,318.636	1	12
Se	4,318.61	(10)	—
W	4,318.57	3	9
Kr	4,318.5525	(400)	—
Mo	4,318.550	6	5
Kr	4,317.81	(500wh)	—
Fe	4,315.09	300	500
Fe	*4,307.91*	*800R*	*1000R*
Kr	*4,302.45*	*(10)*	—
Kr	*4,301.53*	*(40)*	—
Kr	*4,300.49*	*(200)*	—
Ar	*4,300.10*	*(1200)*	—

	λ	I J	I O
Cr	4,319.64	20	100
O	4,319.65	(150)	—
Ce	4,319.68	3wh	6
Fe	4,319.717	1	1
Pr	4,319.723	2h	8
U	4,319.78	8	8
Ru	4,319.87	43	20
Th	4,320.13	10	12
Eu	4,320.16	1	12
Pr	4,320.17	2h	20
Sb	4,320.27	3	—
V	4,320.273	4	6
Co	4,320.387	1	2
Se	4,320.39	(100)	—
U	4,320.40	2	5
W	4,320.50	1h	3
Gd	4,320.53	20	60
Cr	4,320.59	6	125
Th	4,320.60	5	8
Hf	4,320.67	20	15
Kr	4,322.98	(150wh)	—
Fe	4,325.76	700	1000
Xe	4,330.52	(500wh)	—
Kr	4,331.24	(80wh)	—
Kr	4,333.34	(50wh)	—
Ar	4,333.56	(1000)	—
Ar	4,335.34	(800)	—
Ra	*4,340.64*	*(1000)*	—

Kr 4,355.48 (3000)

	λ	I J	I O
Nd	4,355.35	4	12
Th	4,355.33	8	12
Ti	4,355.31	1	10
Mo	4,355.21	20	—
Pr	4,355.19	5	25

	λ	I J	I O
U	4,355.63	10	6
B	4,355.68	6	—
U	4,355.74	20	10
Pr	4,355.75	1	5
V	4,355.94	20	25

	λ	I J	I O
W	**4,355.16**	**9**	**15**
Se	**4,355.15**	**(40)**	—
Ta	**4,355.14**	**10*h***	**80**
Eu	**4,355.10**	**20**	**150**
V	**4,354.98**	**15**	**20**
Pr	**4,354.91**	**30**	**80**
W	**4,354.90**	**2**	**6**
La	**4,354.80**	**3**	**40**
Eu	**4,354.79**	**2**	**100*w***
W	**4,354.72**	**2**	**6**
Mo	**4,354.69**	**5**	**5**
Se	**4,354.64**	**(2)**	—
Sc	**4,354.61**	**10**	**60**
U	**4,354.548**	**8**	**10**
Os	**4,354.46**	**1**	**9**
La	**4,354.40**	**100**	**80**
U	**4,354.36**	**6**	**4**
Kr	4,351.36	(100)	—
Kr	4,351.02	(40*wh*)	—
Rn	4,349.60	(5000)	—
Ar	4,345.17	(1000)	—

	λ	I J	I O
Nd	**4,356.02**	**15**	**30**
Mo	**4,356.07**	**30**	—
Cr	**4,356.29**	**1**	**8**
Hf	**4,356.33**	**4**	**30**
U	**4,356.55**	**2**	**5**
Mn	**4,356.63**	**5**	**20**
Kr	4,362.64	(500)	—
Kr	*4,369.69*	*(200)*	—
Kr	*4,371.25*	*(20h)*	—
Kr	*4,376.12*	*(800)*	—
Kr	*4,377.71*	*(40h)*	—

Kr 4,658.87 (2000)

	λ	I J	I O
U	**4,658.81**	**2**	**1**
Mo	**4,658.54**	**5**	**4**
Yt	**4,758.32**	**15**	**8**
W	**4,658.26**	**8**	**1**
Nb	**4,658.18**	**3**	**2**
P	**4,658.11**	**(100)**	—
Pr	**4,658.09**	**1**	**20**
Lu	**4,658.02**	**15**	**100**
Pt	**4,657.96**	**4**	**9**
Ar	**4,657.94**	**(150)**	—
Sb	**4,657.91**	**30**	—
Rb	**4,657.9**	**(8)**	—
U	**4,657.55**	**3**	**2**
Dy	**4,657.48**	**2**	**2**
Mo	**4,657.478**	**10**	**10**
W	**4,657.44**	**12**	**50**
Te	4,654.38	(800)	—
Kr	4,650.17	(30)	—
Kr	*4,636.14*	*(20)*	—
Kr	*4,633.88*	*(800)*	—

	λ	I J	I O
Yt	**4,658.89**	**4**	**5**
Lu	**4,659.03**	**1**	**10**
Hf	**4,659.21**	**8**	**8**
K	**4,659.317**	**(40)**	—
Rb	**4,659.320**	**(8)**	—
U	**4,659.35**	**3**	**1**
Ce	**4,659.40**	**6**	**6**
Dy	**4,659.50**	**2**	**2**
Gd	**4,659.847**	**10**	**10**
U	**4,659.850**	**2**	**1**
W	**4,659.87**	**70**	**200**
Se	**4,659.89**	**(20)**	—
N	**4,660.05**	**(5)**	—
Hg	**4,660.28**	**(200)**	—
Cu	**4,660.29**	**2**	—
Te	4,664.34	(800)	—
Xe	4,671.23	(2000)	—
Kr	4,671.61	(10)	—
Kr	*4,680.41*	*(500)*	—
Kr	*4,687.28*	*(10h)*	—

Kr 5,570.29 (2000)

	λ	I			λ	I	
		J	O			J	O
Nd	**5,569.96**	**2**	**4**	**Mo**	**5,570.45**	**100**	**200**
Fe	**5,569.62**	**15**	**300**	**Ca**	**5,570.49**	**2**	—
Mo	**5,569.48**	**10**	**15**	**U**	**5,570.68**	**15**	**15**
Te	**5,569.38**	**(15)**	—	**Hg**	**5,571.0**	**(4)**	—
Cl	**5,569.17**	**(4)**	—	**Th**	**5,571.20**	**1**	**8**
Cl	**5,568.81**	**(15)**	—	**Nb**	**5,571.44**	**3*h***	**5**
Kr	**5,568.65**	**(100)**	—	**Nb**	**5,572.00**	**2*wh***	—
Mo	**5,568.62**	**15**	**30**	**Xe**	**5,572.19**	**(50)**	—
Yb	**5,568.12**	**1**	**20**	**Th**	**5,572.48**	**1**	**6**
J	**5,568.10**	**(2)**	—	**Ar**	**5,572.548**	**(500)**	—
				Yb	**5,572.55**	**6**	**1**
Sb	5,568.09	200*wh*	6				
Kr	5,562.23	(500)	—				
Kr	5,552.99	(100*whs*	—	Kr	5,575.56	(10)	—
				Kr	5,580.39	(80)	—

Kr 5,870.92 (3000)

	λ	I			λ	I	
		J	O			J	O
Ar	**5,870.26**	**(2)**	—	**Ne**	**5,870.97**	**(3)**	—
S	**5,869.08**	**(6)**	—	**Br**	**5,871.61**	**(15)**	—
Nb	**5,868.90**	**1**	**3**	**Hg**	**5,871.73**	**(40)**	—
Pr	**5,868.83**	**1**	**10**	**Te**	**5,871.80**	**(15)**	—
Ne	**5,868.42**	**(75)**	—	**Hg**	**5,872.03**	**(10)**	—
Yb	**5,868.40**	**6**	—	**Ne**	**5,872.15**	**(75)**	—
				Ne	**5,872.83**	**(35)**	—
				Fe	**5,873.22**	**2**	**8**
Kr	5,866.75	(50)	—				
Kr	5,860.75	(10*wh*)	—				
Ne	5,852.49	(2000)	—	He	5,875.62	(1000)	—
				Kr	5,879.90	(50)	—
				Ne	5,881.89	(1000)	—
Kr	*5,832.86*	(*100*)	—	Na	5,889.95	1000*R*	9000*R*
Kr	*5,827.07*	(*20*)	—				
Kr	*5,824.50*	(*40*)	—				
Kr	*5,820.10*	(*10*)	—	*Kr*	*5,911.72*	(*10wh*)	—

La 57 138.91

t_0 826°C t_1 1,800°C

I.	II.	III.	IV.	V.
5.614	11.43	19.17	—	—

λ	I		eV
	J	O	
III 3,171.67	1000	2	5.59
II 3,337.49	300	800	4.12
II 3,344.56	200	300	3.94
II 3,916.05	400	400	3.40
II 3,949.11	800	1000	3.54
II 4,077.34	400	600	3.27
II 4,123.23	500	500	3.32
I 5,455.15	1	200	2.40
I 5,930.65	—	250	2.1
I 6,249.93	—	300	2.49

La 3,171.668 *J* 1000 *O* 2*

	λ	*I*			λ	*I*	
		J	*O*			*J*	*O*
Yt	**3,171.667**	**2*h***	**6**	**Ho**	**3,171.71**	**15**	**10**
Th	**3,171.665**	**1*d***	**5*d***	**V**	**3,171.74**	**30**	—
Fe	**3,171.663**	**10**	**30**	**Ca**	**3,171.75**	**2**	**1*h***
Os	**3,171.62**	**5**	**10**	**Nb**	**3,171.79**	**25*h***	—
Er	**3,171.52**	**2**	**6**	**K**	**3,171.81**	**(5)**	—
Sb	**3,171.45**	**(6)**	—	**P**	**3,171.84**	**(50)**	—
Nb	**3,171.43**	**10*w***	**5**	**Tb**	**3,171.90**	**3**	**8**
Zn	**3,171.40**	**(3)**	—	**Eu**	**3,171.94**	**1**	**10**
Mo	**3,171.37**	**4**	**5**	**Mo**	**3,172.03**	**25**	**2**
Lu	**3,171.36**	**5**	**40**	**U**	**3,172.06**	**3**	**6**
Fe	**3,171.35**	**80**	**100**	**Fe**	**3,172.07**	**100**	**100**
Th	**3,171.28**	**1**	**8**	**Cr**	**3,172.08**	**200**	**2**
Tb	**3,171.19**	**3**	**8**	**Th**	**3,172.10**	**3*d***	**5*d***
Yb	**3,171.188**	**4**	**2**	**Ar**	**3,172.18**	**(5)**	—
Nb	**3,171.17**	**3**	**2**	**Zn**	**3,172.18**	**(12)**	—
U	**3,171.05**	**2**	**3**	**V**	**3,172.23**	**25**	—
Eu	**3,170.97**	**10**	**6**	**Ag**	**3,172.24**	**4*h***	**2*h***
Ti	**3,170.92**	**1**	**8**	**Nb**	**3,172.26**	**3*h***	—
U	**3,170.85**	**8**	**10**	**Nb**	**3,172.30**	**2**	**2**
				Pr	**3,172.31**	**5**	**25**
				Mo	**3,172.37**	**20**	**5**
				Th	**3,172.50**	**5**	**8**
				Nb	**3,172.51**	**3**	**3**
				Sn	3,175.02	400*hr*	500*h*

La 3,337.488 *J* 300 *O* 800

	λ	*J*	*O*		λ	*J*	*O*
U	**3,337.39**	**3**	**3**	**Th**	**3,337.488**	**1**	**3**
Re	**3,337.254**	—	**5*d***	**Ta**	**3,337.498**	**1**	**35**
Er	**3,337.25**	**3**	**15**	***bh*Sr**	**3,337.5**	—	**8**
Cr	**3,337.22**	—	**10**	**Cl**	**3,337.5**	**(2)**	—
Ho	**3,337.20**	**12**	**12**	**Ce**	**3,337.502**	—	**20**
Co	**3,337.172**	**2**	**60**	**W**	**3,337.505**	**3**	**5**
Yb	**3,337.17**	**30**	**25**	**Eu**	**3,337.58**	**2**	**2*h***
Zn	**3,337.16**	**(5)**	—	**Fe**	**3,337.666**	**100**	**125**
Th	**3,337.16**	**5**	**6**	**Tb**	**3,337.67**	—	**15**
Sb	**3,337.15**	**10*h***	**2**	**K**	**3,337.67**	**(2)**	—
Os	**3,337.14**	**3**	**2**	**W**	**3,337.68**	**8**	**10**
Dy	**3,337.12**	**2**	**2**	**U**	**3,337.79**	**10**	**12**
U	**3,337.04**	—	**10**	**Er**	**3,337.79**	**3**	**20**
Ni	**3,337.01**	—	**6**	**Ta**	**3,337.80**	**18*s***	**100**
Ti	**3,337.00**	**6**	—	**Tm**	**3,337.82**	**15**	**5**
Cr	**3,336.98**	**8**	**18**	**Ru**	**3,337.823**	**8**	**60**
Ti	**3,336.97**	—	**4**	**Cu**	**3,337.84**	**50**	**70**

	λ	I J	I O
W	**3,336.85**	**5**	**6**
Dy	**3,336.83**	**2**	**2**
V	**3,336.82**	**1**	**6**
Th	**3,336.77**	**2**	**2**
Er	**3,336.76**	**1**	**15**
Ce	**3,336.74**	—	**10**
U	**3,336.685**	**6**	**10**
Mg	**3,336.680**	**60**	**125**
Ru	**3,336.64**	**4**	**50**
Mo	**3,336.571**	—	**2**
W	**3,336.571**	**9**	**8**
Ce	**3,336.55**	—	**10**
Mo	**3,336.51**	**25*d***	**25*d***
Os	**3,336.15**	**50**	**200*R***
Fe	3,335.77	100	125
Ti	3,335.19	150	60
Ne	3,334.87	(250)	—
Cr	3,334.69	—	150*wh*
Co	3,334.14	—	250*R*
Ta	3,331.01	200*w*	18
Ti	3,329.46	200*r*	80
Rh	*3,323.09*	*200*	*1000*
Ti	*3,322.94*	*300R*	*80*
Be	*3,321.34*	*30*	*1000r*
Ar	*3,319.34*	*(300)*	—

	λ	I J	I O
V	**3,337.846**	**150**	**2**
Yt	**3,337.849**	**2**	**2**
Ti	**3,337.853**	**60**	**12**
Th	**3,337.869**	**15**	**12**
Ce	**3,337.872**	—	**10**
Zr	**3,337.92**	**2**	**2**
U	**3,337.93**	**4**	**4**
Er	**3,338.02**	**1*h***	**12**
Tb	**3,338.03**	**15**	**30**
Mo	**3,338.11**	**20**	**1**
Re	**3,338.17**	—	**150**
Pt	**3,338.18**	—	**2**
W	**3,338.24**	**10**	—
Ir	**3,338.37**	**2**	**15**
Th	**3,338.40**	**8**	**8**
Zr	**3,338.408**	**12**	**15**
In	**3,338.415**	**(5)**	—
Hg	**3,338.42**	**(5)**	—

*N**

La 3,344.560 *J* 200 *O* 300

	λ	I J	I O
Ce	**3,344.552**	—	**4*w***
Yt	**3,344.55**	**2*h***	**3**
Ru	**3,344.53**	**6**	**60**
Ca	**3,344.513**	**7**	**100**
Cr	**3,344.51**	**2**	**20**
Tb	**3,344.50**	**3**	**8**
Dy	**3,344.49**	**2**	**3**
Ho	**3,344.46**	**6**	**4**
W	**3,344.44**	**7**	**8**
Ne	**3,344.43**	**(18)**	—
Er	**3,344.36**	**1**	**12**
Sm	**3,344.349**	**10**	**40**
Re	**3,344.347**	—	**150**
Th	**3,344.34**	**2**	**2**
Ce	**3,344.33**	—	**8**

	λ	I J	I O
U	**3,344.561**	—	**4**
Pr	**3,344.564**	**1**	**8**
Ti	**3,344.63**	—	**2**
Ag	**3,344.66**	**2*h***	—
Rb	**3,344.73**	**(15)**	—
Mo	**3,344.746**	**40**	**50**
Er	**3,344.75**	**1**	**8**
Pr	**3,344.758**	—	**2**
Ce	**3,344.761**	**8**	**50**
Zr	**3,344.786**	**15**	**15**
Ar	**3,344.79**	**(3)**	—
Ru	**3,344.80**	—	**5**
U	**3,344.87**	**6**	**6**
Th	**3,344.88**	**8**	**8**
W	**3,344.90**	**12**	**3**

	λ	I J	I O
U	3,344.32	1	4
Nb	3,344.245	10	—
Co	3,344.245	—	4
Ir	3,344.22	1	10
Rh	3,344.20	20	100
Sm	3,344.172	—	4
Er	3,344.17	—	4
Rn	3,344.1	(2)	—
U	3,344.08	1	2
Cs	3,344.00	(10)	—
Nb	3,343.96	15	3
Pr	3,343.93	1	9
Pt	3,343.90	80	100
Ce	3,343.86	6	50
W	3,343.85	4	7
Th	3,343.815	3	3
Zr	3,343.813	15	20
Ti	3,343.77	70	60
Cu	3,343.743	10	—
Cr	3,343.74	—	30h
Mn	3,343.73	—	30
Mo	3,343.72	5	10
Zr	3,343.711	—	8
Nb	3,343.709	20	15
Er	3,343.70	2	10w
Fe	3,343.68	1	2
Sm	3,343.64	5	25
Co	3,342.73	—	150W
Cr	3,342.59	125	30
Re	3,342.26	—	200
La	3,342.22	5	80
Ti	3,341.87	300R	100
Hg	3,341.48	100	100
Fe	3,340.57	100	125
Ti	3,340.34	100	80
Cr	3,339.80	150	25
Co	3,339.78	—	150w
Rh	3,338.54	50	200
P*			

	λ	I J	I O
Ti	3,344.93	—	8
Xe	3,344.97	(2h)	—
Zn	3,345.020	300	800
Sm	3,345.02	3h	1
Nd	3,345.088	2	12
W	3,345.089	8	9
Ta	3,345.11	3h	3
Cr	3,345.15	2	15
Ce	3,345.23	—	3
Eu	3,345.31	2	3
Ru	3,345.317	5	60
K	3,345.32	(30)	—
Mn	3,345.35	—	15
Cr	3,345.37	1	18
Dy	3,345.37	2	3
Ce	3,345.44	—	20
Be	3,345.45	—	2
Er	3,345.46	1	6
Ne	3,345.49	(7)	—
Zn	3,345.57	100	500
V	3,345.90	125	—
Zn	3,345.93	50	150
Cr	3,346.74	80r	150R
Cr	3,347.84	125	35
Fe	3,347.93	100	150
Ti	3,349.03	800R	125
Nb	3,349.06	100	80
Nb	3,349.35	100	5
Ti	3,349.41	400R	100
Gd	3,350.48	180	150
Rb	3,350.89	—	150
Sr	3,351.25	15	300
Sn	3,352.43	100h	—
Cl	3,353.39	(125)	—
V	3,353.77	100	2
Cr	3,358.50	200	40
Cr	3,360.29	200	50
Ti	3,361.21	600R	100
V	3,361.51	200	—
Ni	3,361.56	20	500W
Tm	3,362.61	200	250

La 3,916.05 *J* 400 *O* 400

	λ	I J	I O
Eu	3,915.99	—	20
Nd	3,915.95	20	25
Zr	3,915.94	15	25
U	3,915.884	30	20
Ti	3,915.876	1	15
Cr	3,915.84	80	125
Er	3,915.69	1	3
Mo	3,915.66	1	3
Ce	3,915.63	—	3
Dy	3,915.60	—	80
U	3,915.57	—	6
Ce	3,915.52	3	10
Co	3,915.510	1	2
Cr	3,915.507	10	15
Pr	3,915.47	20	40
W	3,915.46	7	—
Tb	3,915.45	3	30
Mo	3,915.44	50	6
Ir	3,915.38	50	150
V	3,915.36	—	2
Eu	3,915.242	—	5*w*
W	3,915.239	5	—
Re	3,915.23	—	2
Ce	3,915.225	—	3
U	3,915.219	1	15
J	3,915.216	(18)	—
Th	3,915.21	1*h*	15*h*
Br	3,915.18	(3)	—
Nd	3,915.13	12	20
V	3,915.12	3	15
W	3,915.08	7	—
Mo	3,915.01	1	3
Li	3,915.0	—	200*wh*
U	3,914.96	1*h*	2*h*
Ce	3,914.95	2	18
Nd	3,914.94	—	2
Ir	3,914.91	3	20
Dy	3,914.88	—	50
Pr	3,914.872	3	6
Er	3,914.867	1	15
Ru	3,914.85	15	20
Eu	3,914.84	—	4*w*
Pr	3,914.764	8	10
Ar	3,914.76	(25)	—
Ti	3,914.74	2	18
U	3,914.732	8	12
Tb	3,914.73	—	2
U	3,916.08	—	2
Ti	3,916.10	1	10
Ce	3,916.14	4	20
Cr	3,916.24	60	100
Hg	3,916.25	(18)	—
U	3,916.35	2	8
Sm	3,916.36	5*h*	6*h*
W	3,916.40	4	5
Th	3,916.413	3	10
V	3,916.413	40	15
Mo	3,916.43	5	5
Pr	3,916.46	2	6
Tm	3,916.47	8	80
Se	3,916.49	(8)	—
U	3,916.527	3	12
Ce	3,916.530	—	2
Gd	3,916.59	100	150*w*
Mn	3,916.61	—	12
Tb	3,916.64	3	4
Zr	3,916.645	—	10
Gd	3,916.67	—	2
Ce	3,916.68	—	6
Cl	3,916.70	(20)	—
Th	3,916.730	10	15
Fe	3,916.733	80	100
Pr	3,916.80	4	4
Eu	3,916.84	—	10*W*
Ce	3,916.89	2*h*	12
Kr	3,916.90	(3*h*)	—
Mo	3,916.92	5	8
Tb	3,916.94	—	4
Cr	3,916.98	8	30
Se	3,917.06	(20)	—
U	3,917.064	2	1
Co	3,917.11	10	80
V	3,917.14	—	2
Fe	3,917.18	70	150
Rn	3,917.20	(25)	—
Pr	3,917.23	10	25
Ce	3,917.25	—	6
Th	3,917.26	1	8
Re	3,917.27	—	100*w*
W	3,917.28	6	—
Eu	3,917.282	—	20*W*
Dy	3,917.30	—	5
Tb	3,917.32	3	6
Dy	3,917.37	—	5

	λ	I J	I O
Ba	**3,914.73**	**(15)**	**—**
Au	**3,914.72**	**3**	**15**
Nb	**3,914.70**	**100**	**30**
Tb	**3,914.59**	**—**	**5**
La	3,910.81	*5h*	10
Co	3,909.93	—	200*W*
Cr	3,908.75	150	200
Eu	3,907.11	500*R*	1000*RW*
Fe	3,906.48	200	300
Mo	3,902.96	500*R*	1000*R*
Fe	3,902.95	400	500
La	3,902.58	—	20
Fe	*3,899.71*	*300*	*500*
La	*3,898.60*	*5*	*30*
Fe	*3,895.66*	*300*	*400*
La	*3,895.65*	*—*	*10*
Co	*3,894.08*	*100*	*1000R*
He	*3,888.65*	*(1000)*	*—*
La	*3,886.37*	*200*	*400*
Fe	*3,886.28*	*400*	*600*
U	**3,917.39**	**3**	**3**
Sm	**3,917.44**	**15**	**20**
Hf	**3,917.45**	**15**	**5**
Mo	**3,917.54**	**10**	**15**
Hg	3,918.92	(200)	—
Cr	3,919.16	125	300*r*
Kr	3,920.14	(200*h*)	—
Fe	3,920.26	300	500
C	3,920.68	200	—
La	3,921.53	200	400
Fe	3,922.91	400	600
S	3,923.48	(200)	—
Tb	3,925.45	200	150
La	3,927.56	—	40
Fe	3,927.92	300	500
La	3,929.22	300	400
Fe	3,930.30	400	600
Eu	3,930.50	400*R*	1000*R*
J	3,931.01	(400)	—
Ca	*3,833.67*	*600R*	*600R*
Co	*3,935.97*	*15*	*400R*
J	*3,940.24*	*(500)*	*—*
Al	*3,944.03*	*1000*	*2000*

La 3,949.11 *J* 800 *O* 1000

	λ	I J	I O
U	**3,948.99**	**—**	**8**
Ar	**3,948.98**	**(2000)**	**—**
Th	**3,948.97**	**30**	**30**
Ce	**3,948.95**	**—**	**4**
Ca	**3,948.90**	**15**	**40**
Cr	**3,948.85**	**2**	**25**
Se	**3,948.80**	**(25)**	**—**
Fe	**3,948.779**	**100**	**150**
Eu	**3,948.779**	**—**	**4**
Nd	**3,948.778**	**8**	**—**
As	**3,948.74**	**50**	**—**
Xe	**3,948.72**	**(10)**	**—**
Ti	**3,948.67**	**40**	**80*h***
Mo	**3,948.65**	**10**	**6**
Pr	**3,948.62**	**2**	**9**
Eu	**3,948.61**	**—**	**3**
Ir	**3,948.51**	**—**	**4**
Ce	**3,948.45**	**—**	**2**
Pt	**3,948.39**	**5**	**60**
Ce	**3,949.116**	**—**	**6**
Eu	**3,949.123**	**1**	**25*w***
Fe	**3,949.15**	**1**	**4**
Gd	**3,949.208**	**20**	**10**
W	**3,949.213**	**3**	**—**
Tm	**3,949.27**	**5**	**50**
U	**3,949.31**	**6**	**2**
Nb	**3,949.33**	**2*h***	**3**
Ce	**3,949.38**	**3**	**20**
Tb	**3,949.39**	**—**	**4**
Ru	**3,949.42**	**5**	**10**
Pr	**3,949.438**	**100**	**150**
Ag	**3,949.44**	**8**	**3**
Nb	**3,949.45**	**50**	**1**
Nd	**3,949.46**	**—**	**2**
Tb	**3,949.509**	**3**	**10**
Ba	**3,949.51**	**(20)**	**—**
U	**3,949.52**	**8**	**2**
Ca	**3,949.57**	**2**	**—**

	λ	I	
		J	O
Tb	3,948.35	15	20
Nd	3,948.325	10	20
Ir	3,948.317	3	10
Hg	3,948.29	(100)	—
W	3,948.20	5	—
Xe	3,948.16	(60)	—
C	3,948.15	2	—
Sm	3,948.108	50	50
Fe	3,948.107	50	125
Er	3,948.07	—	8
Th	3,948.04	1	3
W	3,947.98	9	10
Ce	3,947.97	3	20
U	3,947.96	8	2
Te	3,947.93	(10)	—
Sm	3,947.83	8	15
Ti	3,947.77	35	70
Ce	3,947.71	—	2
Kr	3,947.66	(5*h*)	—
Th	3,947.65	1	3
Pr	3,947.63	60*d*	125*d*
O	3,947.61	(18)	—
Nd	3,947.608	—	5
Ar	3,947.50	(1000)	—

	λ	I	
		J	O
Cr	3,949.58	1	5
Eu	3,949.59	—	50*W*
Cr	3,949.62	—	6
Os	3,949.78	10	50
Ce	3,949.82	—	10
Eu	3,949.837	—	5
Sm	3,949.845	2	2
Tb	3,949.87	—	6
J	3,949.90	(20)	—
Ca	3,949.92	2*r*	—
Nb	3,949.93	3	4
Fe	3,949.957	100	150
Cl	3,949.96	(10)	—
Ru	3,950.00	8	10
W	3,950.12	12	9
Tb	3,950.126	—	4
U	3,950.132	6	3
Ru	3,950.21	15	12
V	3,950.23	10	20
Mo	3,950.26	4	4
Zr	3,950.29	—	2*h*
Er	3,950.35	10*h*	30
Yt	3,950.36	100	60
Th	3,950.39	30	30
Dy	3,950.40	—	50
Ru	3,950.41	10	10
Tb	3,950.416	—	20*d*
Nd	3,950.417	10	20
Ce	3,950.424	3	10
U	3,950.48	—	8
Ho	3,950.50	—	8
La	3,953.68	—	30
Pd	3,958.64	200	500*w*
Al	3,961.53	2000	3000
Zr	3,961.59	8	500
Os	3,963.63	50	500
Eu	*3,971.99*	—	*1000Rwh*

*P**

La 4,077.34 *J* 400 *O* 600

	λ	J	O
Yb	4,077.27	100	30
Ti	4,077.153	2	18
Nd	4,077.150	4	10
Dy	4,077.35	—	4
Yt	4,077.37	40	50
Co	4,077.41	2*h*	100*wh*

	λ	I	
		J	O
Cr	4,077.089	10	35
Nb	4,077.088	5	3
As	4,077.08	10	—
W	4,077.06	5	6
Zr	4,077.05	1	3
Ar	4,076.96	(10)	—
Eu	4,076.95	—	10
Sm	4,076.85	5	10
Ir	4,076.81	—	2*h*
Fe	4,076.80	1	8
Ru	4,076.73	25	60
U	4,076.72	10	8
La	4,076.71	5	15
Ar	4,076.64	(20)	—
Fe	4,076.637	50	80
Sm	4,076.63	15	25
Co	4,076.57	1	3*h*
Ir	4,076.55	—	25
Zr	4,076.53	—	10
Mo	4,076.51	5	5
Fe	4,076.50	1	2
Ti	4,076.375	1	15
Ir	4,076.370	2	2
Yt	4,076.35	8	30
Au	4,076.33	25*w*	4
Ce	4,076.24	1	12
U	4,076.225	3	2
Fe	4,076.222	1	2
Pr	4,076.21	1	10
Mo	4,076.19	25	25
Co	4,076.13	—	70
Nb	4,076.09	3	4
Cr	4,076.06	15	30
Er	4,076.01	—	2
Dy	4,076.01	—	2
C	4,076.00	80	—
Ru	4,075.98	25	5
Eu	4,075.96	—	7
U	4,075.94	1*h*	6
Fe	4,075.937	5	5
Mo	4,075.937	5	10
Tb	4,075.90	—	4
O	4,075.87	(800)	—
Ce	4,075.85	—	4*s*
Sm	4,075.83	40	40
Ce	4,075.79	2	10
P	4,075.74	(15)	—
Ce	4,075.714	2	15
Th	4,075.713	1	5
Ce	4,077.47	4	18
Rh	4,077.57	4	5
Pb	4,077.61	2	—
Nd	4,077.62	3	8
Cr	4,077.677	10	30
Mo	4,077.682	10	8
Pr	4,077.69	2	4
Sr	4,077.71	500*W*	400*r*
Cu	4,077.716	—	5
Sn	4,077.72	3	2
Ta	4,077.721	2*h*	4
U	4,077.79	6	15
Hg	4,077.81	150	150
Er	4,077.970	18*s*	20*s*
Tb	4,077.97	2	25
Dy	4,077.974	100	150*r*
V	4,077.977	—	2
Pr	4,077.98	2	10
Ho	4,078.00	3	3
Mo	4,078.07	4	4
Re	4,078.124	—	10
W	4,078.124	6	7
Zn	4,078.14	(5)	—
Pr	4,078.16	1	5
Eu	4,078.23	—	3
Tb	4,078.26	—	3
Zr	4,078.31	—	10
Ce	4,078.32	4	15
Nb	4,078.35	3	4
Fe	4,078.36	40	80
Mo	4,078.381	3	5
Sb	4,078.385	4	—
Gd	4,078.46	10	15
Tb	4,078.47	1	5
Ti	4,078.474	50	125
Ce	4,078.51	1	5
Sc	4,078.58	10	10
Ce	4,078.61	—	5
Gd	4,078.709	10	20
V	4,078.709	—	2
Tb	4,078.78	—	5
Xe	4,078.82	(100)	—
U	4,078.83	8	3
O	4,078.85	(70)	—
Ga	4,078.90	5	—
Ce	4,079.02	1*h*	15
La	4,079.18	3	25

	λ	I J	I O
U	4,075.66	6	5
V	4,075.65	2	—
Ir	4,075.63	—	10
Eu	4,075.62	—	2
Te	4,073.57	(300)	—
In	4,072.40	200wh	—
O	4,072.16	(300)	—
Fe	4,071.74	200	300
Se	4,070.16	(500)	—
La	4,067.39	80	150
La	4,065.58	2	30
Kr	4,065.11	(300)	—
La	4,064.78	3	40
Fe	4,063.60	300	400
Cu	4,062.70	20	500w
La	4,060.32	5	80
Nb	4,058.94	400w	1000w
Pb	4,057.82	300R	2000R
In	4,056.94	(500)	—
Ag	4,055.26	500R	800R
La	4,050.08	60	60
K	4,044.14	400	800
La	4,042.91	300	400
Nb	4,079.73	200w	500w
Ru	4,080.60	300	125
Co	4,086.31	15	400
La	4,086.71	500	500
Pd	4,087.34	100	500
Kr	4,088.33	(500)	—
La	4,089.61	5	40
Co	4,092.39	15	600W
Tm	4,094.18	30	300
La	4,099.54	100	100
In	4,101.77	1000R	2000R
Ho	4,103.84	400	400
La	4,104.88	2	40
Se	4,108.83	(800)	—
La	4,109.48	5	15
La	4,109.80	10	15
N	4,109.98	(1000)	—
Co	4,110.53	—	600

La 4,123.23 J 500 O 500

	λ	J	O
V	4,123.19	3	6
Ta	4,123.17	4w	50r
Ti	4,123.14	—	10
U	4,123.14	—	3
Na	4,123.07	(15)	10
Ru	4,123.06	35	25
W	4,123.055	8	7
Er	4,123.053	—	8
Gd	4,123.010	10	8
Sm	4,123.006	2	5
Nd	4,123.005	8	15
Pr	4,122.98	2	8
Th	4,122.97	5	3
Eu	4,122.94	1	3
Yb	4,122.87	20	10
Ce	4,122.86	—	3
Nb	4,122.81	10	5
In	4,122.791	(30h)	—
Er	4,123.24	—	5
Mn	4,123.28	5	12
Cu	4,123.29	1h	30w
Ti	4,123.306	6	25
U	4,123.310	2h	5
Cr	4,123.39	15	35
Ce	4,123.49	5	20
Re	4,123.50	—	2
Hf	4,123.53	10	1
Mn	4,123.54	5	15
V	4,123.566	12r	30r
Th	4,123.569	5	5
Ti	4,123.572	10	40
Gd	4,123.59	—	3
Pd	4,123.62	2	10
Mo	4,123.65	25	5
Ta	4,123.657	1	5
U	4,123.66	6	5

	λ	I J	I O		λ	I J	I O
Ru	4,122.787	—	6	As	4,123.71	5	—
Eu	4,122.786	—	3	Fe	4,123.74	20	80
Re	4,122.76	—	10	Ni	4,123.778	—	2
Fe	4,122.51	30	70	Tb	4,123.78	—	10
Sm	4,122.50	10	5	Nb	4,123.810	125	200
Lu	4,122.49	2	15	Ru	4,123.813	10	20
Tb	4,122.47	—	8	Ce	4,123.87	6	25
Mo	4,122.39	50	15	Nd	4,123.88	20	40
U	4,122.35	4	15	Sm	4,123.95	20	10
Pr	4,122.31	2	8	U	4,123.96	1	20*r*
Co	4,122.27	—	10*h*	Cl	4,124.00	(12)	—
Ce	4,122.24	3	3	Eu	4,124.02	—	2
				Pr	4,124.06	2	5
Co	4,121.32	25	1000*R*	Hg	4,124.071	(30)	—
O	4,119.22	(300)	—	V	4,124.072	5	8
Co	4,118.77		1000*R*	N	4,124.10	(2)	—
Pt	4,118.69	10	400	Nd	4,124.11	—	2
Pr	4,118.48	50*d*	250*d*				
La	4,117.68	8	20	Rh	4,128.87	150	300
La	4,113.28	20	10	Fe	4,132.06	200	300
				Li	4,132.29	—	400*wh*
				La	4,132.50	10*h*	—
				La	4,133.33	25	15
				La	4,137.02	—	15
				La	*4,141.47*	*200*	*200*

*P**

La 5,455.15 *J* 1* *O* 200

	λ	J	O		λ	J	O
Yb	5,455.08	6	25	Sm	5,455.27		5*d*
Ru	5,454.82	—	100	Gd	5,455.28	—	10
V	5,454.81	4	6	Fe	5,455.43		50
Co	5,454.55	—	300*w*	Dy	5,455.45		4
Ir	5,454.50	—	20	Se	5,455.58	(15)	—
Er	5,454.27	—	30	Fe	5,455.61	30	300
N	5,454.26	(15)	—	Er	5,455.62	—	8
Yb	5,454.02	—	5	Nd	5,455.81	—	30
Ce	5,453.95	—	12	Ar	5,456.01	(5)	—
Nd	5,453.88	—	3	Ru	5,456.13	—	40
Tb	5,453.68	—	10	Nb	5,456.19	1	3*h*
Ti	5,453.65	25	12	Re	5,456.31	—	2*h*
Lu	5,453.57	1	8	Sm	5,456.39	1	5
Gd	5,453.46	—	8	Ce	5,456.41	—	15
U	5,453.44	1	5	Mo	5,456.46	12	20
Os	5,453.39	—	18	*bh*Zr	5,456.5	—	40
Pr	5,453.26	—	4	Tb	5,456.53	—	10

	λ	I	
		J	O
Ni	**5,453.25**	**—**	**3**
Mo	**5,453.03**	**6**	**15**
Sm	**5,453.02**	**—**	**100**
Eu	**5,452.96**	**—**	**1000*s***
Eu	5,451.53	—	1000*s*
Fe	5,446.92	35	300
Fe	5,445.04	—	150
Co	5,444.57	—	400*w*
Eu	5,443.56	—	125
Fe	5,434.53	35	300
*bh*La	5,433.0	—	30
*bh*La	5,431.1	—	20
La	5,429.86	—	10
Fe	5,429.70	40	500
Eu	5,426.93	—	200
Fe	5,424.08	20	400
Fe	*5,415.21*	*20*	*500*

	λ	I	
		J	O
W	**5,456.59**	**—**	**18**
Ta	**5,456.59**	**—**	**5*h***
***bh*La**	**5,456.6**	**—**	**10**
Er	**5,456.62**	**—**	**30**
Tb	**5,456.98**	**—**	**15**
Pr	**5,457.06**	**—**	**5**
Ce	**5,457.21**	**—**	**12**
Os	**5,457.30**	**—**	**25**
Mn	**5,457.47**	**—**	**25**
Zr	**5,457.59**	**—**	**2**
*bh*La	5,458.7	—	20
Fe	5,463.28	—	100
La	5,464.38	30	25
Ag	5,465.49	500*R*	1000*R*
Yt	5,466.47	20	150
Co	5,469.30	—	125
Ag	5,471.55	100	500*h*
Eu	5,472.33	—	1000*s*
Fe	5,473.92	—	100
Fe	5,474.92	—	100
Lu	5,476.69	1000	500
Ni	5,476.91	8	400*w*
La	*5,482.27*	*50*	*25*
Co	*5,483.34*	*—*	*500w*
bhLa	*5,484.3*	*—*	*10*
Eu	*5,488.65*	*—*	*500*
La	*5,493.45*	*20*	*15*

La 5,930.65 *O* 250

Ta	**5,930.62**	**—**	**15*s***
Ru	**5,930.31**	**—**	**4**
Gd	**5,930.28**	**—**	**6**
Fe	**5,930.19**	**10*h***	**30**
Ce	**5,929.83**	**—**	**8**
Ce	**5,929.50**	**—**	**5**
Th	**5,929.49**	**—**	**5**
Hf	**5,929.35**	**5**	**3**
U	**5,929.33**	**—**	**5**
Sm	**5,929.20**	**—**	**3**
Nd	**5,928.92**	**—**	**4**
Mo	**5,928.88**	**—**	**100*h***
Ce	**5,928.73**	**—**	**2**
W	**5,928.58**	**—**	**6**
La	**5,928.50**	**—**	**10**

Sm	**5,930.89**	**—**	**2**
Ta	**5,931.05**	**—**	**40*s***
***bh*Yt**	**5,931.1**	**—**	**6**
Sc	**5,931.23**	**—**	**4*h***
Nd	**5,931.42**	**—**	**3**
Sm	**5,931.53**	**—**	**3**
Ta	**5,931.68**	**—**	**40**
Fe	**5,931.71**	**—**	**3**
Ce	**5,931.72**	**—**	**4**
Dy	**5,931.88**	**—**	**2**
Ti	**5,932.13**	**—**	**80**
Nd	**5,932.140**	**—**	**3**
Sm	**5,932.142**	**—**	**10**
Ce	**5,932.16**	**—**	**6**
Ru	**5,932.38**	**—**	**15**

	λ	I J	I O
Sm	**5,928.36**	—	**2**
Ce	**5,928.34**	—	**25**
Nb	**5,928.20**	**2**	**5**
*bh*Sc	**5,928.1**	—	**15**
Nd	**5,927.96**	—	**2**
V	5,924.57		250*W*
Ho	5,921.76		200
*bh*La	5,920.8	—	50
La	5,917.63	—	25
Eu	5,915.76	—	200
Co	5,915.54	—	200*w*
U	5,915.40	—	125
Nb	5,900.62	200	200
*bh*La	5,896.7		80
Na	5,895.92	500*R*	5000*R*
La	5,894.85	—	25
*bh*La	5,893.6	—	60
Na	*5,889.95*	*1000R*	*9000R*
La	*5,880.65*	*50*	*30*
La	*5,874.74*	—	*20*
Sm	**5,932.42**	—	**35*d***
U	**5,932.44**	—	**4**
Sm	**5,932.89**	—	**50**
Tb	**5,932.95**	—	**10**
Fe	**5,933.08**	—	**4**
*bh*Yt	**5,933.2**	—	**2**
Ho	5,933.71	—	200
La	5,935.29	—	20
Co	5,935.39	—	150
La	5,936.22	20	15
Ho	5,948.03	—	200
Ti	5,953.17	250	150
La	*5,960.59*	—	*10*
Eu	*5,966.07*	—	*1000*
Eu	*5,967.16*	—	*2000s*
Eu	*5,972.78*	—	*300*
bhYt	*5,973.0*	—	*600*
La	*5,973.53*	*120*	*15*

La 6,249.93 *O* 300

	λ	I J	I O
Ta	**6,249.79**	—	**100**
Co	**6,249.51**	—	**125**
Hf	**6,248.95**	**100**	**80**
Lu	**6,248.80**	—	**10**
Sr	**6,248.52**	**2**	**4*h***
Nd	**6,248.28**	—	**15**
Sm	**6,248.11**	—	**10**
Th	**6,247.990**	—	**2**
Yb	**6,247.99**	—	**3**
*bh*F	**6,247.8**	—	**10**
Fe	**6,247.56**	**8**	**1**
V	**6,247.55**	**1**	**2**
Ta	**6,247.32**	—	**2**
Co	**6,247.28**	—	**8**
Yb	**6,246.97**	**60**	**40**
Dy	**6,246.86**	—	**2**
Nd	**6,246.83**	—	**5**
Sc	**6,249.96**	**10**	**10**
*bh*F	**6,250.3**	—	**10**
Nd	**6,250.43**	—	**10**
Eu	**6,250.46**	—	**70**
Dy	**6,250.65**	—	**2**
*bh*Pb	**6,250.8**	—	**5**
*bh*Yt	**6,251.0**	—	**5**
Pr	**6,251.02**	—	**3*w***
Yt	**6,251.05**	**2**	**3**
W	**6,251.50**	—	**2**
Sm	**6,251.73**	—	**5**
Nb	**6,251.76**	**10**	**30**
U	**6,251.819**	—	**8**
V	**6,251.823**	**8**	**70**
Re	**6,252.06**	—	**2**
*bh*F	**6,252.2**	—	**10**
Sm	**6,252.24**	—	**2**
Fe	**6,252.56**	**25*h***	**60**
Re	**6,253.04**	—	**2**

	λ	I	
		J	O
Sm	6,246.76	—	80
Li	6,240.1	—	300
La	6,238.59	—	25
La	6,234.85	—	25
La	6,233.50	—	15
Co	6,230.97	—	200*w*
Lu	*6,221.87*	*1000*	*500*
Rb	*6,206.31*	*100*	*800*
Eu	*6,195.06*	—	*300*
Ni	*6,191.19*	*1*	*500*
Eu	*6,188.10*	—	*500 W*
Ni	6,256.36	10	600*w*
Ta	6,256.68	—	300
Ti	6,258.10	100	200
Ti	6,258.70	250	300
Ti	6,261.10	100	300
Eu	6,262.28	—	600
La	6,262.30	150	125
La	6,266.03	—	80
Sm	6,267.28	—	150
Ta	6,268.70	—	200
Au	*6,278.18*	*20*	*700*
Co	*6,282.63*	—	*300w*
Eu	*6,288.34*	—	*300*
La	*6,293.56*	—	*30*
La	*6,296.10*	*150*	*50*
Rb	*6,298.33*	*150*	*1000*
Rb	*6,299.22*	*50*	*300*
Eu	*6.299.76*	—	*500 W*
Eu	*6,303.39*	—	*700*
La	*6,310.93*	*100*	*10*
Ni	*6,314.67*	—	*300*

Li $^{3}_{6.939}$

t_0 186°C t_1 1,372°C

I.	II.	III.	IV.	V.
5.390	75.622	122.427	—	—

λ	I		eV
	J	*O*	
I 2,562.54	15	150	4.83
I 2,741.31	—	200	4.52
I 3,232.61	500	1000*R*	3.83
I 3,915.0	—	200	5.02
I 4,132.29	—	400	4.85
I 4,273.28	100	200*R*	4.75
I 4,602.86	—	800	4.54
I 4,971.99	—	500	4.34
I 6,103.64	300	2000*R*	3.87
I 6,707.84	200	3000*R*	1.90

Li 2,562.54 *J* 15* *O* 150

	λ	*I*			λ	*I*	
		J	*O*			*J*	*O*
Fe	2,562.53	150	50	Zn	2,562.61	—	10*h*
Ce	2,562.42	—	5	Fe	2,562.63	—	2
Nb	2,562.41	100	4	Os	2,562.66	1	12
Fe	2,562.23	—	15	V	2,562.76	25	1
Mg	2,562.22	—	3	Ru	2,562.840	1	3
Ru	2,562.17	—	5	U	2,562.841	4	5
Co	2,562.15	—	10*r*	U	2,562.94	6	15
V	2,562.13	4	50				
Ni	2,562.12	—	15	Ta	2,563.70	—	80
Fe	2,562.10	25	2				
Ta	2,562.10	—	100	*Al*	*2,567.99*	*80R*	*200R*
				Zr	*2,571.39*	*400R*	*300R*
In	2,560.23	50*Rh*	150*R*				
Ta	2,559.43	2	100				

Li 2,741.31 *O* 200

	λ	*J*	*O*		λ	*J*	*O*
Nb	2,741.15	10	4	Mo	2,741.32	20	2
Fe	2,741.11	3	10	Ta	2,741.38	2	5
Cr	2,741.07	30	35	Ce	2,741.52	—	3
U	2,741.06	6	6	Zr	2,741.55	8	10
Mo	2,740.96	—	8	V	2,741.567	10	2
Ti	2,740.87	—	6	Ce	2,741.575	—	3
U	2,740.86	8	12	Fe	2,741.58	1	4
				Mo	2,741.618	25	5
				U	2,741.621	2	5
Ta	2,740.21	10*h*	100	Yb	2,741.72	15	2
Fe	2,739.55	300*h*	200				
Pt	2,738.48	5	100	U	2,741.75	6	18
Fe	2,737.31	150	300*r*				
Rh	2,736.76	3	100				
				Fe	2,744.07	8	150
Cr	2,736.47	50	300*r*				
				Fe	*2,746.98*	*300wh*	*200*
Ta	*2,736.25*	*8s*	*300s*	*Cr*	*2,748.29*	*5*	*300*
Pt	*2,733.96*	*200h*	*1000h*	*Ta*	*2,748.78*	*50*	*400*
Fe	*2,733.58*	*200*	*300*	*Ta*	*2,749.83*	*50*	*200*
Cr	*2,731.91*	*30*	*300r*	*Fe*	*2,750.14*	*100*	*300h*

Li 3,232.61 *J* 500 *O* 1000*R*

	λ	*J*	*O*		λ	*J*	*O*
Cl	3,232.58	(4)	—	Sm	3,232.62	3	10
Os	3,232.54	10	150	Te	3,232.63	(10)	—
Rh	3,222.504	—	6	Dy	3,232.652	4	15
Sb	3,232.499	250*wh*	150	W	3,232.652	8	9
Sm	3,232.497	—	4	Ce	3,232.66	—	3

	λ	I J	I O		λ	I J	I O
W	3,232.49	8	9	Ru	3,232.75	4	50
Ne	3,232.38	(7)	—	J	3,232.78	(5)	—
Cd	3,232.36	2	—	Fe	3,232.791	50	—
Pb	3,232.35	—	30	Ti	3,232.791	—	8
Pd	3,232.32	—	2	Nb	3,232.793	2*h*	—
Eu	3,232.31	1	4*w*	Kr	3,232.80	(2)	—
Th	3,232.308	—	8	Co	3,232.874	25	60
Ce	3,232.29	—	15	Ce	3,232.875	—	3
Ti	3,232.280	100	30	Ni	3,232.96	35	300*R*
Ta	3,232.279	1	25	Ca	3,233.02	4	—
Se	3,232.27	(8)	—	Fe	3,233.05	60	100
W	3,232.23	2	3	Zr	3,233.11	—	6
U	3,232.16	12	12	Mo	3,233.140	30	50
Kr	3,232.15	(2*h*)	—	W	3,233.142	6	2
W	3,232.13	9	6	Ag	3,232.15	10	—
Th	3,232.12	25*h*	3	Ni	3,233.17	—	4
Os	3,232.05	20	500*R*	V	3,233.19	3	40
Yt	3,232.027	3	3	Ce	3,233.21	—	2
Er	3,232.026	3	18	Cr	3,233.23	4	30
Ir	3,232.001	1	20	W	3,233.25	9	2
Tb	3,232.00	3	8	Ag	3,233.25	2	5
Yb	3,231.98	6	2	Rh	3,233.32	5	2
Ce	3,231.980	—	3	Th	3,233.33	8	8
V	3,231.95	100	8	Ho	3,233.36	2*h*	6
Sm	3,231.94	8	20	Dy	3,233.422	—	2
Eu	3,231.86	—	3*w*	Pt	3,233.422	10	40
Ce	3,231.81	—	2	Ce	3,233.44	2	30
Tl	3,229.75	800	2000	Ti	3,234.52	500*r*	100
Fe	3,227.75	300	200	Ti	3,236.57	300*r*	70
Nb	3,225.48	800*wr*	150*w*	Ti	3,239.04	300*R*	60
				Fe	3,239.44	300	400
				Pd	*3,242.70*	*600R*	*2000wh*
				Cu	*3,247.54*	*2000R*	*5000R*

Li 3,915.0 *O* 200

U	3,914.96	1*h*	2*h*	Mo	3,915.01	1	3
Ce	3,914.95	2	18	V	3,915.12	3	15
Nd	3,914.94	—	2	Nd	3,915.13	12	20
Ir	3,914.91	3	20	Th	3,915.21	1*h*	15*h*
Dy	3,914.88	—	50	U	3,915.219	1	15
Pr	3,914.872	3	6	Ce	3,915.225	—	3
Er	3,914.867	1	15	Re	3,915.23	—	2
Ru	3,914.85	15	20	Eu	3,915.24	—	5*w*
Eu	3,914.84	—	4*w*	V	3,915.36	—	2
Pr	3,914.76	8	10	Ir	3,915.38	50	150

	λ	I	
		J	O
Ti	**3,914.74**	**2**	**18**
U	**3,914.732**	**8**	**12**
Tb	**3,914.73**	**—**	**2**
Au	**3,914.72**	**3**	**15**
Nb	**3,914.70**	**100**	**30**
Tb	**3,914.59**	**—**	**5**
Th	**3,914.47**	**1**	**2**
Ce	**3,914.42**	**—**	**2**
Sm	**3,914.39**	**—**	**6**
Zr	**3,914.34**	**8**	**70**
Cr	**3,914.335**	**2**	**3**
Ti	**3,914.335**	**10**	**50**
V	**3,914.330**	**70*wh***	**25**
Fe	**3,914.28**	**3**	**15**
Sb	**3,914.27**	**2*h***	**3**
U	**3,914.268**	**18**	**10**
Au	**3,914.19**	**4**	**10**
Ce	**3,914.17**	**1**	**6**
Eu	**3,914.15**	**—**	**5*w***
Sm	**3,914.07**	**3**	**3**
Ce	**3,913.99**	**2**	**10**
Ho	**3,913.96**	**—**	**3**
Dy	**3,913.95**	**—**	**8**
Re	**3,913.92**	**—**	**30**
Th	**3,913.83**	**8**	**10**
Gd	**3,913.791**	**10**	**10**
Tb	**3,913.79**	**—**	**2**
Eu	**3,913.74**	**1**	**5*w***
Dy	**3,913.72**	**—**	**2**
Nd	**3,913.69**	**2*h***	**8**
Mo	**3,913.68**	**3**	**4**
Sm	**3,913.64**	**3**	**10**
Fe	**3,913.635**	**25**	**100**
Dy	**3,913.628**	**—**	**6**
Pr	**3,913.561**	**30**	**80**
V	**3,913.56**	**1**	**2**
Nd	**3,913.55**	**—**	**4**
Mo	**3,913.53**	**4**	**4**
Rh	**3,913.51**	**2**	**4**
U	**3,913.50**	**2**	**4**
Pr	3,912.90	80	150
Sc	3,911.81	30	150
Co	3,909.93	—	200*W*
Cr	3,908.75	150	200
Pr	3,980.43	60	100
Pr	3,908.03	50	100
Fe	3,907.94	60	100
Sc	3,907.48	25	125
Mo	**3,915.44**	**50**	**6**
Tb	**3,915.45**	**3**	**30**
Pr	**3,915.47**	**20**	**40**
Cr	**3,915.507**	**10**	**15**
Co	**3,915.510**	**1**	**2**
Ce	**3,915.52**	**3**	**10**
U	**3,915.57**	**—**	**6**
Dy	**3,915.60**	**—**	**80**
Ce	**3,915.63**	**—**	**3**
Mo	**3,915.66**	**1**	**3**
Er	**3,915.69**	**1**	**3**
Cr	**3,915.84**	**80**	**125**
Ti	**3,915.876**	**1**	**15**
U	**3,915.884**	**30**	**20**
Zr	**3,915.94**	**15**	**25**
Nd	**3,915.95**	**20**	**25**
Eu	**3,915.99**	**—**	**20**
La	**3,916.04**	**400**	**400**
U	**3,916.08**	**—**	**2**
Ti	**3,916.10**	**1**	**10**
Ce	**3,916.14**	**4**	**20**
Cr	**3,916.24**	**60**	**100**
U	**3,916.35**	**2**	**8**
Sm	**3,916.36**	**5*h***	**6*h***
W	**3,916.40**	**4**	**5**
Th	**3,916.413**	**3**	**10**
V	**3,916.413**	**40**	**15**
Mo	**3,916.43**	**5**	**5**
Pr	**3,916.46**	**2**	**6**
Tm	**3,916.47**	**8**	**80**
Gd	3,916.59	100	150*w*
Fe	3,916.73	80	100
Fe	3,917.18	70	150
Re	3,917.27	—	100*w*
Pr	3,918.86	30	100
Cr	3,919.16	125	300*r*
Fe	3,920.26	300	500
Cr	3,921.02	40	150
La	3,921.53	200	400
Zr	3,921.79	4	100
Co	3,922.75	—	100
Ta	3,922.78	15	100
Fe	3,922.914	400	600
Ta	3,922.915	10*h*	100
Pt	3,922.96	20*r*	100
Pr	3,924.14	15	100
Tb	3,925.45	200	150
Pr	3,925.46	100	125

	λ	I	
		J	O
Gd	3,907.12	100	100 *W*
Eu	3,907.11	500 *R*	1000 *RW*
Fe	3,906.48	200	300
Co	3,906.29	—	150
Fe	3,903.90	80	100
Mo	3,902.96	500 *R*	1000 *R*
Fe	3,902.95	400	500
Cr	3,902.91	100	100
Gd	3,902.40	80	100
Os	3,901.71	20	150
Zr	3,900.52	—	100
Fe	*3,899.71*	*300*	*500*
Tb	*3,899.19*	*100*	*200*
Fe	*3,895.66*	*300*	*400*
Co	*3,894.98*	*3*	*300 R*
Pd	*3,894.20*	*200 W*	*200 W*
Co	*3,894.08*	*100*	*1000 R*
Ho	*3,891.02*	*40*	*200*
La	*3,886.37*	*200*	*400*
Fe	*3,886.28*	*400*	*600*

	λ	I	
		J	O
Fe	3,927.92	300	500
Cr	3,928.64	40	150
La	3,929.22	300	400
Zr	3,929.53	6	100
Re	3,929.84	—	100
Fe	*3,930.30*	*400*	*600*
Eu	*3,930.50*	*400 R*	*1000 R*
Dy	*3,931.54*	*—*	*200*
Sm	*3,933.59*	*200h*	*200*
Fe	*3,933.60*	*200*	*200*
Ca	*3,933.67*	*600 R*	*600 R*
Ir	*3,934.84*	*50*	*200*
Co	*3,935.97*	*15*	*400 R*
Tb	*3,939.60*	*200*	*200*
Cr	*3,941.49*	*60*	*200r*
Co	*3,941.73*	*—*	*200wh*
Al	*3,944.03*	*1000*	*2000*
Dy	*3,944.69*	*150*	*300*

Li 4,132.29 *O* 400

	λ	J	O
Re	4,132.282	—	20
Gd	4,132.281	20	25
Eu	4,132.24	1*h*	4
Mo	4,132.230	20	20
Pr	4,132.230	8	15
Tb	4,132.22	—	3
W	4,132.21	8	6
Co	4,132.15	—	15*h*
Yb	4,132.14	—	3*h*
Fe	4,132.06	200	300
V	4,132.02	10	12
Mo	4,131.92	10	20
Ce	4,131.85	2	5
Pr	4,131.82	4	12
Lu	4,131.79	—	10
U	4,131.78	6	6*h*
La	4,131.74	3	3
Nb	4,131.53	10	2
Er	4,131.50	—	9
Gd	4,131.48	50	50
Tb	4,131.45	2	10
Th	4,131.433	8	8
Mn	4,131.430	—	10

	λ	J	O
Ce	4,132.31	2*s*	8
Ba	4,132.43	4	10
Ce	4,132.45	—	4
Tb	4,132.47	—	5
Nd	4,132.55	5	4
Ce	4,132.64	2*h*	8
Tm	4,132.69	15	12
Mo	4,132.750	10	12
Th	4,132.751	12	10
Tb	4,132.83	—	9
Dy	4,132.85	2	9
V	4,132.88	—	2
Mo	4,133.00	—	15
Sc	4,133.01	3	8
Ce	4,133.05	—	3
Eu	4,133.11	1*h*	4
Ir	4,133.15	—	10
Sm	4,133.17	4	4
U	4,133.20	8	8
Re	4,133.42	—	200
Rh	4,135.27	150	300
Os	4,135.78	50	200

	λ	I J	I O		λ	I J	I O
Cr	**4,131.36**	**20**	**30**	La	4,141.74	200	200
U	**4,131.353**	**6**	**6**	Pr	4,143.14	50	200
				Fe	4,143.87	250	400
Ce	**4,131.347**	**—**	**2**				
Nd	**4,131.31**	**4**	**8**				
Gd	**4,130.38**	**10**	**200**				
Ta	4,129.38	40	200				
Rh	4,128.87	150	300				
Lu	4,124.73	10	200				
Nb	4,123.81	125	200				
La	4,123.23	500	500				
Co	4,121.32	25	1000*R*				
Co	*4,118.77*	*—*	*1000R*				
Pt	*4,118.69*	*10*	*400*				

Li 4,273.28 *J* 100 *O* 200*R*

	λ	I J	I O		λ	I J	I O
Gd	**4,273.25**	**—**	**4**	**Ti**	**4,273.30**	**1**	**20**
Pr	**4,273.21**	**1**	**3**	**Fe**	**4,273.32**	**2*h***	**3*h***
Na	**4,273.2**	**—**	**3**	**Th**	**4,273.36**	**15**	**20**
U	**4,273.19**	**—**	**8**	**Te**	**4,273.40**	**(70)**	**—**
Tb	**4,273.18**	**1**	**8**	**Rh**	**4,273.43**	**10**	**25**
Dy	**4,273.14**	**4**	**4**	**Ce**	**4,273.444**	**2**	**20**
Mo	**4,273.07**	**10**	**20**	**U**	**4,273.445**	**5**	**8**
Ce	**4,273.01**	**—**	**2**	**Kr**	**4,273.48**	**(4)**	**—**
Nb	**4,272.97**	**4**	**2**	**Zr**	**4,273.517**	**3**	**9**
U	**4,272.94**	**1**	**3**	**Rb**	**4,273.524**	**2**	**—**
Pr	**4,272.93**	**1**	**5**	**Eu**	**4,273.61**	**—**	**2**
Cr	**4,272.91**	**30**	**40**	**Ho**	**4,273.63**	**2**	**2**
Th	**4,272.88**	**—**	**4**	**W**	**4,273.69**	**5**	**10**
Cs	**4,272.87**	**(10)**	**—**	**Rb**	**4,272.70**	**2**	**—**
Ce	**4,272.855**	**—**	**2**	**Hg**	**4,273.72**	**(2*h*)**	**—**
Hf	**4,272.848**	**20**	**12**	**Nd**	**4,273.739**	**2**	**10**
Nd	**4,272.79**	**5**	**15**	**Tb**	**4,273.74**	**—**	**3*W***
Eu	**4,272.76**	**1**	**5**	**Ce**	**4,273.79**	**—**	**2**
Rb	**4,272.64**	**2**	**—**	**Fe**	**4,273.87**	**2**	**10**
Pb	**4,272.63**	**30**	**—**	**Ca**	**4,273.907**	**2**	**—**
Pb	**4,272.55**	**2**	**—**	**Pt**	**4,273.909**	**—**	**2**
Bi	**4,272.49**	**10*wh***	**—**	**Kr**	**4,273.9700**	**(1000)**	**—**
Er	**4,272.44**	**1*h***	**4**	**U**	**4,273.975**	**15**	**12**
Ti	**4,272.43**	**10**	**40**	**Th**	**4,274.032**	**8**	**10**
Eu	**4,272.34**	**—**	*4*	**Sm**	**4,274.035**	**1**	**3**
Ce	**4,272.339**	**—**	**2**	**Dy**	**4,274.04**	**2**	**5**
W	**4,272.31**	**3**	**8**	**Mo**	**4,274.05**	**6**	**6**
U	**4,272.28**	**3**	**3**	**Sm**	**4,274.16**	**1**	**5**

	λ	I	
		J	O
Pr	**4,272.27**	**35**	**50**
Tb	**4,272.22**	**—**	**3**
Ar	**4,272.17**	**(1200)**	**—**
Fe	4,271.76	700	1000
Fe	4,271.16	300	400
Nb	4,270.69	50	30
Ne	4,270.23	(50)	—
Ne	4,269.72	(70)	—
La	4,269.49	150	150
Kr	4,268.81	(100wh)	—
Kr	4,268.57	(60wh)	—
Ir	4,268.10	15	200
Ne	4,268.01	(70)	—
Ba	4,267.95	(80)	—
Fe	4,267.83	60	125
S	4,267.80	(60)	—
Nd	4,267.49	200wh	—
C	4,267.27	500	—
C	4,267.02	350	—
Ar	4,266.53	(200)	—
Ar	4,266.29	(1200)	—
Mn	4,265.92	50	100
Cs	4,264.67	(50)	—
La	4,263.58	150	150
Cr	4,263.14	80	125
Ti	4,263.13	35	125
Sm	4,262.68	150	200
Gd	4,262.09	10	150
Te	*4,261.08*	*(300)*	*—*
Os	*4,260.85*	*200*	*200*
Fe	*4,260.48*	*300*	*400*
Ar	*4,259.36*	*(1200)*	*—*
Ir	*4,259.11*	*10*	*200*
Sm	*4,256.40*	*150*	*150*
Kr	*4,254.85*	*(100h)*	*—*
Cr	*4,254.35*	*1000R*	*5000R*
O	*4,253.98*	*(100h)*	*—*
Sm	*4,251.79*	*200*	*200*
Gd	*4,251.74*	*10*	*300*
Ar	*4,251.18*	*(800)*	*—*
Fe	*4,250.79*	*250*	*400*

	λ	I	
		J	O
Gd	**4,274.17**	**—**	**100**
Yt	**4,274.18**	**—**	**2h**
Tl	**4,274.24**	**(8)**	**—**
Pr	**4,274.27**	**1d**	**8d**
Th	**4,274.33**	**4**	**8**
Re	**4,274.34**	**—**	**20w**
Tb	**4,274.36**	**—**	**2h**
Ti	**4,274.40**	**—**	**5**
Ti	4,274.58	40	100
Ne	4,274.66	(50)	—
Cr	4,274.80	800r	4000R
Tl	4,274.98	(100)	—
F	4,275.21	(100h)	—
O	4,275.47	(50h)	—
Ne	4,275.56	(70)	—
La	4,275.64	500	40
Te	4,276.68	(50)	—
Cs	4,277.10	(50)	—
Ar	4,277.55	(80)	—
Tb	4,278.51	100	200
Mo	4,279.02	100	8
Sm	4,279.67	100	100
Se	4,280.36	(150)	—
Cr	4,280.40	50	80
Gd	4,280.50	100	200
Sm	4,280.78	200	200
Mn	4,281.10	50	100
Se	4,282.10	(100)	—
Fe	4,282.41	300	600
Kr	4,282.97	(100)	—
Sm	*4,285.48*	*200*	*200*
Ir	*4,286.62*	*3*	*200*
La	*4,286.97*	*300*	*400*
Mo	*4,288.64*	*100*	*80*
Rh	*4,288.71*	*100*	*400*
Cr	*4,289.72*	*800r*	*3000R*
Gd	*4,289.90*	*100*	*40*
Ne	*4,290.40*	*(100)*	*—*
Br	*4,291.40*	*(150)*	*—*
Kr	*4,292.92*	*(600)*	*—*
Mo	*4,293.21*	*100*	*125*
Fe	*4,294.13*	*400*	*700*

Li 4,602.863 *O* 800

	λ	I: J	I: O		λ	I: J	I: O
Nb	4,602.860	5	2	Th	4,602.88	3	5
Ru	4,602.81	—	15	Gd	4,602.944	2	10
Ce	4,602.75	—	4	Fe	4,602.944	100	300
Hf	4,602.71	3	6	V	4,602.946	6	7
*bh*La	4,602.7		3	Tb	4,602.95	—	8
Eu	4,602.63	—	15*w*	Sm	4,603.12	—	5
Lu	4,602.60	—	3*h*	Tm	4,603.21	—	10
Zr	4,602.57	—	12	Tm	4,603.42	25	35
Pr	4,602.56	—	10	Pr	4,603.48	—	4
Tb	4,602.50	—	4	Mo	4,603.56	6*h*	6
Nd	4,602.24	—	10	U	4,603.66	40	25
Ta	4,602.19	2	100	*bh*Yt	4,603.7	—	5
Er	4,602.06	1	2	Er	4,603.73	1	3
La	4,602.05	—	20	Nb	4,603.80	3	2
Sm	4,602.02	—	2	Pr	4,603.81	2	10
Fe	4,602.01	2	20	Nd	4,603.82	—	25
U	4,601.79	4	4	Ir	4,603.84	—	2
Ru	4,601.76	—	20	Sm	4,603.99	—	2*h*
Rh	4,601.64	2	3	Tb	4,604.10	—	10
Ce	4,601.57	—	2	Sm	4,604.18	—	60
				Pr	4,604.20	—	3
Co	4,596.90	—	400	Ce	4,604.21	—	3
Co	4,594.63		400				
Eu	4,594.02	200	500*R*				
Cs	4,593.18	50*R*	1000*R*	Sr	4,607.33	50*R*	1000*R*
Co	*4,581.60*	*10*	*1000w*				

Li 4,971.99 *O* 500

	λ	I: J	I: O		λ	I: J	I: O
Co	4,971.959	—	150	U	4,972.10	6	8
Pd	4,971.958	—	10	Sm	4,972.17	—	80
Ce	4,971.94	—	8	Th	4,972.18	1	8
Nb	4,971.93	5	20	Ce	4,972.24	—	8
Dy	4,971.79	—	2*h*	Pr	4,972.48	—	2
Sr	4,971.67	—	2	W	4,972.57	—	15
Ce	4,971.66	—	2	Th	4,972.61	—	3
Pr	4,971.54	—	2	Gd	4,972.62	—	25
Ce	4,971.47	—	12	Na	4,972.8	—	3
Tb	4,971.41	—	4	Re	4,972.85	—	3
Ni	4,971.35	—	100	Tb	4,973.02	—	4
Co	4,971.05	—	6	Ti	4,973.05	2	35
Tb	4,970.99	—	5	Fe	4,973.11	—	100
Nd	4,970.922	—	5	Nb	4,973.14	5	20
Pr	4,970.918	—	5	Mo	4,973.36	5	25

	λ	I J	I O		λ	I J	I O
Tm	**4,970.86**	**3**	**5**	**Th**	**4,973.39**	—	**10**
Mo	**4,970.77**	**1**	**4**	**Dy**	**4,973.57**	—	**2**
Ce	**4,970.67**	—	**12**	**Sc**	**4,973.67**	**5**	**6**
W	**4,970.65**	—	**8**	**Sm**	**4,973.73**	—	**40**
Fe	**4,970.50**	—	**20**	**Gd**	**4,973.85**	—	**25**
Ir	**4,970.48**	**2**	**6**	Ni	4,980.16	1	500*W*
La	**4,970.39**	**150**	**125**	Ti	4,981.73	125	300
Mo	**4,970.37**	**2**	**3**	Ni	4,984.13	1	500*W*
Fe	4,966.10	1	300	Co	4,988.04	—	500*R*
Fe	4,957.61	150	300				

Li 6,103.64 *J* 300 *O* 2000*R*

	λ	J	O		λ	J	O
Ar	**6,103.56**	**(8)**	—	**Dy**	**6,103.67**	—	**2**
Nb	**6,103.49**	**1*h***	**6**	**Sm**	**6,103.72**	—	**4**
Sm	**6,103.374**	—	**30**	**Xe**	**6,103.88**	**(3)**	—
Dy	**6,103.370**	—	**2**	**Sm**	**6,103.95**	—	**2**
Fe	**6,103.33**	**40**	—	**Nd**	**6,104.11**	—	**3**
Fe	**6,103.18**	—	**8*h***	**Sm**	**6,104.20**	—	**3**
Ce	**6,102.75**	—	**3**	**Tb**	**6,104.27**	—	**6**
Co	**6,102.74**	—	**10**	**Sm**	**6,104.39**	—	**6**
Ca	**6,102.721**	**50**	**80**	**Th**	**6,104.57**	**2**	**12**
Rh	**6,102,721**	—	**100**	**Ar**	**6,104.60**	**(6)**	—
Cr	**6,102.71**	—	**10**	**Sm**	**6,104.82**	—	**30*d***
Th	**6,102.59**	—	**4**	**Er**	**6,105.20**	—	**6**
Zn	**6,102.54**	**(20)**	**6**	**Nb**	**6,105.30**	**1*h***	**10**
S	**6,102.26**	**(50)**	—	**Sm**	**6,105.36**	—	**3**
Fe	**6,102.18**	**20*h***	**15**	**Co**	**6,105.47**	—	**10*h***
Sm	**6,101.96**	—	**30**	**Nd**	**6,105.518**	—	**5**
Se	**6,101.96**	**(200)**	—	**Co**	**6,105.519**	—	**4**
*bh***Sc**	**6,101.9**	—	**30**	**Pr**	**6,105.55**	—	**6*w***
Mo	**6,101.87**	**4**	**40**	**Ar**	**6,105.64**	**(60)**	—
U	**6,101.79**	—	**6**	**Cu**	**6,105.97**	**5**	—
Nd	**6,101.75**	—	**10**	**Tb**	**6,106.04**	—	**6**
Rn	**6,101.72**	**(4)**	—	**Nd**	**6,106.13**	—	**2**
Au	**6,101.65**	—	**5**	**Gd**	**6,106.18**	—	**15**
Ta	**6,101.58**	—	**150**	**Yb**	**6,106.22**	**4**	—
Xe	**6,101.43**	**(200)**	—	**Dy**	**6,106.23**	—	**2**
Sm	**6,101.42**	—	**4**	**O**	**6,106.40**	**(30)**	—
Ar	**6,101.16**	**(6)**	—	**Zr**	**6,106.47**	**2**	**8**
Se	**6,100.81**	**(50)**	—	**Ta**	**6,106.51**	—	**3**
Co	**6,100.78**	—	**4*h***	**Te**	**6,106.55**	**(15)**	—
La	**6,100.38**	**15**	**30**	As	6,110.30	150	—
Hg	**6,100.36**	**(25)**	—	As	6,110.66	150	—
				Sb	6,129.98	150*h*	10*h*

	λ	I J	I O		λ	I J	I O
Xe	6,097.59	(600)	—	*Ba*	*6,141.72*	*2000wh*	*2000wh*
Ne	6,096.16	(300)	—	*Ne*	*6,143.06*	*(1000)*	—
Xe	6,093.56	(150)	—	*O*	*6,156.78*	*(300)*	—
J	6,082.46	(1000)	—	*O*	*6,158.20*	*(1000)*	—
Pb	6,081.5	(200)	—	*Ne*	*6,163.59*	*(1000)*	—
Pb	6,075.8	(200)	—				
Ne	*6,074.34*	*(1000)*	—				
Se	*6,055.96*	*(1000)*	—				
Xe	*6,051.15*	*(700)*	—				

Li 6,707.84 *J* 200 *O* 3000*R*

	λ	I J	I O		λ	I J	I O
Ru	**6,707.52**	—	**5**	**Mo**	**6,707.85**	—	**300** *W*
Sm	**6,707.45**	—	**50***d*	**Co**	**6,707.86**	—	**200** *Wh*
Sm	**6,707.1**	—	**2**	**V**	**6,708.17**	—	**2**
Sm	**6,706.85**	—	**5**	**W**	**6,708.18**	—	**20**
Tb	**6,707.79**	—	**4**	**F**	**6,708.28**	**(40)**	—
Ta	**6,706.46**	—	**5**	**Hf**	**6,708.33**	**3**	—
				Te	**6,708.34**	**(30)**	—
				N	**6,708.81**	**(50)**	—
Se	6,699.56	(125)	—				
Ar	6,698.85	(100)	—				
Xe	6,694.32	(200)	—	Ar	6,719.20	(100)	—
Ba	6,693.88	100	600	Ra	6,719.32	(500)	—
				O	6,721.21	(300)	—
				N	*6,723.12*	*(500)*	—
				Mo	*6,728.04*	*(200)*	—

Mg $^{12}_{24\cdot312}$

t_0 651°C t_1 1,107°C

I.	II.	III.	IV.	V.
7.645	15.032	80.119	109.533	–

λ	*I*		eV
	J	*O*	
II 2,795.53	300	150	12.0
II 2,802.69	300	150	12.0
I 2,852.13	100*R*	300*R*	4.34
I 3,829.35	150	100	5.94
I 3,832.31	200	250	5.94
I 3,838.26	200	300	5.94
I 5,167.34	50	100	5.11
I 5,172.70	100	200	5.11
I 5,183.62	300	500	5.11

Mg 2,795.53 *J* 300 *O* 150

	λ	*I*	
		J	*O*
Au	2,795.53	15	—
Ce	2,795.52	8	30*s*
Ar	2,795.45	(2)	—
Ce	2,795.38	—	2
Ru	2,795.35	8*h*	30
Cu	2,795.331	2	—
Ar	2,795.33	(2)	—
U	2,795.23	12	18
Ra	2,795.21	(125)	—
Nb	2,795.14	15	1
Zr	2,795.13	—	5
Ne	2,795.10	(35)	—
Yb	2,795.07	3	—
Fe	2,795.01	35	50
Mn	2,794.817	5	1000*R*
Co	2,794.816	15	100*R*
Fe	2,793.89	150	8
Pt	2,793.27	5	100
Te	3,793.24	(300)	—
Ta	2,791.67	10	100
Ta	2,791.37	150	25
Rh	2,791.16	1*h*	100
Ta	2,790.71	10	150
Ta	*2,788.30*	*3*	*150*
Fe	*2,788.10*	*150*	*150*
Ta	*2,787.69*	*40*	*400R*

	λ	*I*	
		J	*O*
Ag	2,795.53	10	10
Fe	2,795.54	60	90
W	2,795.55	10	—
Co	2,795.60	10*wh*	—
Yb	2,795.63	8	—
Rh	2,795.70	—	15
Fe	2,795.77	3*h*	—
Kr	2,795.81	(80*h*)	—
Cr	2,795.818	3	35
Co	2,795.819	—	15
Fe	2,795.85	10	15*W*
Nb	2,795.86	3	4
Ce	2,795.94	—	2
Ne	2,795.96	(8)	—
Re	2,796.08	—	10
Ta	2,796.34	80	400
Ta	2,796.56	—	150
Rh	2,796.63	1	100
Os	2,796.73	15	100
Gd	2,796.94	80	70
Nb	2,797.69	200	10
Ta	2,797.76	100*d*	100*d*
Fe	2,797.77	80	150
Zr	2,798.270	—	100
Mn	2,798.271	80	800*R*
Ta	2,798.40	—	150
Ni	2,798.65	—	125
Bi	2,798.68	25	200

*N**

Mg 2,802.69 *J* 300 *O* 150

	λ	*J*	*O*
Cu	2,802.68	2	10
Mn	2,802.65	—	2*d*
U	2,802.56	30	15
Bi	2,802.55	3	—
Te	2,802.53	(10)	—
Er	2,802.528	1	12
Ti	2,802.50	15	100
Ta	2,802.49	2	15
Pd	2,802.47	10	—
Bi	2,802.44	2	—
Mn	2,802.42	—	12
Mo	2,802.35	25	15
Ce	2,802.28	—	2

	λ	*J*	*O*
Ce	2,802.699	5*h*	18
Dy	2,802.70	—	2
Ta	2,802.702	4	10
Co	2,802.706	200*h*	100
Bi	2,802.707	10	—
Nb	2,802.715	5	2
Hg	2,802.76	15	5
V	2,802.797	25*h*	15
Mn	2,802.800	—	12
Ru	2,802.81	150	50
Re	2,802.84	—	5*w*
Te	2,802.86	(15)	—
Eu	2,802.86	—	150*w*

	λ	I J	I O		λ	I J	I O
Ni	**2,802.274**	**15**	**50**	**Er**	**2,802.87**	**1**	**6**
Br	**2,802.27**	**(3)**	**—**	**W**	**2,802.95**	**10*d***	**12*d***
Re	**2,802.251**	**—**	**10*h***	**Xe**	**2,803.02**	**(3)**	**—**
Fe	**2,802.25**	**—**	**3**	**Ce**	**2,803.04**	**—**	**8**
S	**2,802.23**	**(8)**	**—**	**Tm**	**2,803.11**	**10**	**7**
Au	**2,802.19**	**200**	**—**	**Fe**	**2,803.12**	**15**	**35**
Mn	**2,802.17**	**—**	**5**	**Ce**	**2,803.121**	**—**	**2**
Ru	**2,802.162**	**40**	**30**	**Mo**	**2,803.13**	**—**	**3**
U	**2,802.157**	**4**	**6**	**Ni**	**2,803.15**	**—**	**5**
				Fe	**2,803.17**	**—**	**15**
				In	**2,803.18**	**2**	**—**
Ta	2,802.07	80	300	**Kr**	**2,803.20**	**(20*h*)**	**—**
Pb	2,802.00	100*h*	250*Rh*				
Mn	2,801.064	60	600*R*	**Re**	**2,803.236**	**—**	**8*h***
Zn	2,801.056	20	100	**Pt**	**2,803.239**	**5**	**400**
Zn	2,800.87	300	400				
Cr	2,800.77	150	12	Co	2,803.77	12	100
Ti	2,800.61	150*wh*	—	Os	2,804.07	20	80
Ta	2,800.57	40*h*	150*W*	Fe	2,804.52	200	300
				Ti	2,805.01	200*wh*	—
				Ni	2,805.67	200*h*	—
				Ta	2,806.30	50	300
				Ta	2,806.58	50	200
				Os	2,806.91	1	100*w*
				Fe	2,806.98	200	200
				Bi	*2,809.62*	*100*	*200W*
				Ta	*2,810.92*	*40W*	*200W*
				Fe	*2,813.29*	*400*	*400*

*P**

Mg 2,852.129 *J* 100*R* *O* 300*R*

	λ	J	O		λ	J	O
Dy	**2,852.129**	**—**	**5**	**Fe**	**2,852.13**	**80**	**150**
Ce	**2,852.12**	**1**	**50*d***	**Ir**	**2,852.131**	**—**	**20**
W	**2,852.10**	**18*h***	**1*d***	**Mo**	**2,852.131**	**10**	**10*h***
Hf	**2,852.01**	**50**	**20**	**Ce**	**2,852.24**	**—**	**3**
Nb	**2,851.98**	**5**	**4**	**Fe**	**2,852.35**	**—**	**2**
Zr	**2,851.97**	**20**	**12**	**Ta**	**2,852.355**	**100**	**5**
Fe	**2,851.80**	**150**	**200**	**Xe**	**2,852.39**	**(2*h*)**	**—**
V	**2,851.75**	**4**	**30**	**Re**	**2,852.40**	**—**	**10**
Mg	**2,851.65**	**—**	**25**	**Cs**	**2,852.41**	**(2)**	**—**
Ce	**2,851.60**	**—**	**3**	**Hg**	**2,852.42**	**(8)**	**—**
Fe	**2,851.50**	**2**	**5**	**U**	**2,852.47**	**4**	**6**
				Ir	**2,852.48**	**—**	**2**
				Th	**2,852.50**	**6**	**2*h***
Cr	2,851.36	80	20	**Ag**	**2,852.53**	**5*wh***	**1**
Hf	2,851.21	50	25	**V**	**2,852.536**	**35**	**6**
Yb	2,851.12	50	10				

	λ	I J	I O
Ti	2,851.10	80	20
Ta	2,850.98	150	400
Sn	2,850.62	100wh	80
Ta	2,850.49	100	200
Cr	2,849.84	150r	80
Ta	2,849.82	50W	15 Ws
Tl	2,949.80	(200)	—
Fe	2,849.61	50	—
Nb	2,849.56	100w	2w
Ru	2,849.29	100	3
Hf	2,849.21	100	30
V	2,849.05	50	7
Ta	2,848.52	50	300
Co	2,848.37	60h	—
Mo	2,848.23	200h	125
Ta	2,848.054	15	150
Fe	2,848.046	70	—
Hg	2,847.83	100	15
Hg	2,847.67	(300)	—
V	2,847.57	150	15
Lu	2,847.51	125	40
Ta	2,846.750	10h	150hs
Mg	2,846.75	4	18
Nb	2,846.28	50	10
Ti	2,846.09	70wh	—
Ru	*2,844.71*	*150*	*—*
Ta	*2,844.46*	*200*	*200*
Ta	*2,844.25*	*50*	*400r*
Fe	*2,843.98*	*300*	*300*
Fe	*2,843.63*	*100*	*125*
Cr	*2,843.25*	*400r*	*125*
Ta	*2,842.81*	*50*	*200*
Nb	*2,842.65*	*100*	*10*
Ni	*2,842.42*	*150*	*—*
Rn	*2,842.1*	*(150)*	*—*
Ti	*2,841.94*	*125*	*40*
Ru	*2,841.68*	*200*	*50*
Nb	*2,841.15*	*100*	*10*
Ru	*2,841.12*	*125*	*—*
Pd	*2,841.03*	*100*	*—*
Cr	*2,840.02*	*125*	*25*
Sn	*2,839.99*	*300R*	*300R*
Pd	*2,839.89*	*100*	*—*
Ti	*2,839.80*	*100wh*	*—*
Au	**2,852.54**	**5**	**—**
Eu	**2,852.56**	**—**	**5W**
U	**2,852.75**	**15h**	**15**
Na	2,852.83	20	100R
Cr	2,853.22	100R	5
Mo	2,853.23	100h	25
V	2,854.34	100R	20
Pd	2,854.58	500h	4
Ru	2,854.72	60	2
Cr	2,855.07	100	4
Fe	2,855.67	200	2
Cr	2,855.68	200Wh	60
La	2,855.90	50h	3
Ti	2,856.24	100wh	—
Ru	2,856.55	50	—
Cr	2,856.77	60	20
Cr	2,857.40	80	20
Pd	2,857.73	100	—
Ru	2,857.78	60	4
Ti	2,857.81	70wh	—
Te	2,858.29	(100)	2
Fe	2,858.34	200	3
Ta	*2,858.43*	*300*	*100*
Pt	*2,860.68*	*150h*	*30*
Cr	*2,860.93*	*100*	*60*
Nb	*2,861.09*	*100*	*10*
Hf	*2,861.70*	*125*	*50*
Ti	*2,861.99*	*100wh*	*—*
Cr	*2,862.57*	*300R*	*80*
Sn	*2,863.33*	*300R*	*300R*
Ni	*2,863.70*	*250*	*—*
Mo	*2,863.81*	*100h*	*30*
Fe	*2,863.86*	*100*	*125*
Ni	*2,864.15*	*300wh*	*—*

Mg 3,829.35 *J* 150 *O* 100

	λ	*I*			λ	*I*	
		J	*O*			*J*	*O*
Ru	**3,829.332**	**3**	**8**	**Au**	**3,829.38**	**20**	**25**
S	**3,829.33**	**(5)**	—	**U**	**3,829.39**	**5**	**6**
Ce	**3,829.28**	—	**2**	**Ti**	**3,829.406**	**12**	**30**
Cl	**3,829.27**	**(15)**	—	**Sm**	**3,829.406**	—	**15**
Nb	**3,829.21**	**10*h***	—	**Ce**	**3,829.408**	**2**	**3**
Pb	**3,829.2**	**(2)**	—	**Th**	**3,829.410**	**20*w***	**40*w***
Sm	**3,829.16**	**6**	**5**	**Nd**	**3,829.41**	**40**	**50**
Nd	**3,829.15**	**15**	**10**	**Gd**	**3,829.411**	—	**25*h***
W	**3,829.127**	**10**	**12**	**Pr**	**3,829.435**	**10**	**10**
Fe	**3,829.126**	**2**	**4**	**Eu**	**3,829.435**	**10**	**5**
Zr	**3,829.11**	**4*h***	**5*h***	**Fe**	**3,829.46**	**8**	**15**
Lu	**3,829.07**	—	**10*h***	**Ru**	**3,829.48**	**5**	**4**
U	**3,829.03**	**10**	**8**	**Er**	**3,829.52**	**2**	**12**
Ta	**3,828.95**	**10**	**25**	**V**	**3,829.53**	**4**	—
Eu	**3,828.93**	**8*wh***	**6*wh***	**Nd**	**3,829.63**	**30**	**40**
Ba	**3,828.92**	—	**5**	**Pr**	**3,829.65**	**10**	**10**
Mo	**3,828.87**	**30**	**40**	**Nb**	**3,829.659**	**3*h***	**3*h***
Nd	**3,828.842**	**20**	**20**	**V**	**3,829.661**	**6**	**2**
V	**3,828.836**	**3**	**15**	**Hf**	**3,829.67**	**1**	**3**
U	**3,828.81**	**5**	**2**	**Ir**	**3,829.676**	—	**4*h***
As	**3,828.74**	**10**	—	**Mn**	**3,829.680**	**60**	**69**
Ru	**3,828.71**	**8**	**30**	**Ce**	**3,829.69**	**2**	**4**
Ce	**3,828.60**	—	**4**	**Ti**	**3,829.73**	**2**	**7**
V	**3,828.56**	**5**	**20**	**Br**	**3,829.76**	**(3)**	—
Br	**3,828.55**	**(12)**	—	**Ne**	**3,829.77**	**(40)**	—
U	**3,828.51**	—	**6**	**Xe**	**3,829.77**	**(5*h*)**	—
Fe	**3,828.50**	**1**	**2**	**Fe**	**3,829.77**	**2**	**8**
Br	**3,828.49**	**(6)**	—	**Mo**	**3,829.79**	**5**	**5**
Rh	**3,828.48**	**60**	**100**	**N**	**3,829.80**	**(10)**	—
Ir	**3,828.46**	—	**30**	**U**	**3,829.803**	—	**15**
Sm	**3,828.41**	**2**	**8**	**Re**	**3,829.81**	—	**25**
Ce	**3,828.389**	**2**	**3**	**Ce**	**3,829.82**	—	**2**
Th	**3,828.388**	**5**	**5**	**Mo**	**3,829.91**	**3**	**4**
Re	**3,828.33**	—	**30**	**Ce**	**3,829.942**	—	**3**
Nb	**3,828.24**	**15**	**5**	**Mn**	**3,829.945**	—	**8**
Ti	**3,828.193**	**15**	**35**	**Nd**	**3,830.000**	**8**	**20**
Dy	**3,828.19**	**3**	**10**	**Nb**	**3,830.002**	**5**	**5**
Er	**3,828.18**	**3*w***	**15*w***	**Hf**	**3,830.02**	**6**	**10**
Sc	**3,828.179**	**30**	**30**	**Ce**	**3,830.030**	**2**	**4**
Nd	**3,828.17**	**40**	**50**	**Cr**	**3,830.032**	**50**	**150*w***
Th	**3,828.14**	**8**	**8**	**Ir**	**3,830.051**	—	**2**
U	**3,828.06**	**8**	**3**	**Mo**	**3,830.055**	**6**	**5**
Sm	**3,828.05**	**8**	—	**Er**	**3,830.067**	—	**6**
Ti	**3,828.02**	**1**	**6**	**Th**	**3,830.069**	**3**	**4**
Nd	**3,827.99**	**8**	**6**	**W**	**3,830.13**	**8**	**7**
Ce	**3,827.98**	—	**3**	*bh***B**	**3,830.2**	—	**50**
				U	**3,830.21**	**6**	**5**

	λ	I J	I O
Fe	3,827.82	200	200
Cl	3,827.62	(150)	—
P	3,827.44	(150)	—
Os	3,827.14	20	50
Pr	3,826.29	50*d*	80*d*
Fe	3,825.88	400	500
Fe	3,824.44	100	150
Mn	3,823.89	50	50*h*
Mn	3,823.51	75	75*h*
O	3,823.47	(125)	—
Pr	3,823.18	25	125
Rh	3,822.26	100	100
Ru	3,822.09	25	50
V	3,822.01	40	70
Fe	3,821.84	30	50
Pr	3,821.82	50	50
Nd	3,821.77	40*w*	50*w*
V	3,821.49	30	50
Fe	3,821.18	100	100
Fe	3,820.43	600	800
Cl	3,820.25	(100)	—
V	3,819.96	35	60
Eu	3,819.66	500*wd*	500*wd*
Cr	3,819.56	40	60
Ru	3,819.03	30	50
Nb	3,818.86	300	8
Cr	3,818.48	20	50
Pr	3,818.28	100	125
Rh	3,818.19	25	50
Fe	3,817.65	15	50
Tm	3,817.40	15	60
Ru	3,817.27	60	50
Co	3,816.87	5	70
Dy	3,816.77	50	100
Mn	3,816.75	50	60
Co	3,816.47	—	60
Co	3,816.33	50*r*	60
Fe	3,815.84	700	700
Ce	3,815.83	5	50
Bi	3,815.8	(300)	—
V	3,815.39	150*h*	1
Fe	3,814.52	40	80
Ra	3,814.42	(2000)	—
Gd	3,813.98	60	100*w*
Fe	3,812.96	300	400
Co	3,812.47	—	100*w*
Gd	3,812.02	200*h*	200*R*
Bi	3,811.14	(150)	—
Ag	3,810.86	10*h*	100*wh*

	λ	I J	I O
V	**3,830.27**	**4**	**40**
Tb	**3,830.29**	**8**	**30**
Sm	**3,830.30**	**10**	**50**
Ir	**3,830.34**	**2*h***	**2**
Pr	**3,830.36**	**7**	**10**
N	**3,830.39**	**(150)**	—
Ar	**3,830.43**	**(10)**	—
O	**3,830.45**	**(18)**	—
Nd	**3,830.48**	**10**	**10**
Ho	**3,830.49**	**6**	**2**
Er	**3,830.51**	**1*h***	**8*w***
Ir	**3,830.52**	—	**2*h***
Ce	**3,830.55**	—	**4**
U	**3,830.61**	**5**	**3*h***
Nb	**3,830.64**	**20*w***	**1*h***
Ir	**3,830.65**	—	**4**
Pr	**3,830.719**	**60**	**100**
W	**3,830.722**	**9**	**8**
*N**			

	λ	I J	I O
Mn	3,809.59	150	150
Fe	3,808.73	70	100
Co	3,808.11	7	200*w*
Fe	3,807.54	100	150
Ni	3,807.14	40*h*	800*W*
Fe	3,806.70	150	200
Hg	3,806.38	(200)	—
Fe	*3,805.34*	*300*	*400*
Cr	*3,804.80*	*30*	*100*
Au	*3,804.00*	*150*	*25*

Mg 3,832.31 *J* 200 *O* 250

	λ	I J	I O
Tl	3,832.30	(30)	—
Pd	3,832.29	150	150
Ta	3,832.27	3*h*	5*h*
W	3,832.24	6	4
Ce	3,832.23	—	5
U	3,832.20	—	6
Ir	3,832.179	—	15
Os	3,832.176	10	30
C	3,832.12	6*h*	—
Mo	3,832.11	8	10
Ce	3,832.05	2	3
Ce	3,831.93	2	3
Ho	3,831.9	6*h*	—
U	3,831.86	3*h*	15
Tb	3,831.85	—	15
Nb	3,831.84	300	5
V	3,831.83	2	15
Ru	3,831.79	50	60
Ce	3,831.78	4*h*	3
Mo	3,831.76	3	4
Th	3,831.74	8	5
Ni	3,831.69	10	300
Ce	3,831.67	1	3
Dy	3,831.64	—	10
Ce	3,831.55	—	4
Sm	3,831.51	15	6
Zr	3,831.49	—	2
U	3,831.46	25	25
S	3,831.41	(10)	—
Pr	3,831.39	2	6
Zr	3,831.30	1	5*h*
Nb	3,831.20	2	2
V	3,832.31	2	—
Ce	3,832.34	—	2
Cr	3,832.35	2	18
Mo	3,832.37	10*h*	—
Re	3,832.41	—	20
Eu	3,832.43	4*W*	6*W*
Th	3,832.431	5	8
V	3,832.432	—	2
Er	3,832.44	1	12
Hg	3,832.46	(18)	—
U	3,832.56	1	5
W	3,832.64	5	4
Pr	3,832.65	8	10
Ce	3,832.66	4	2
Tb	3,832.68	—	8
Ce	3,832.75	—	5
Th	3,832.79	5	3
Sm	3,832.80	1	6
Pb	3,832.83	50	—
V	3,832.835	5	25
W	3,832.86	7	8
Ni	3,832.87	—	25
Ce	3,832.88	2	3
Yt	3,832.89	80	30
Co	3,832.899	—	5
Dy	3,832.90	—	10
Pb	3,832.904	5	—
U	3,833.02	15	20*r*
Nd	3,833.03	30	60
Th	3,833.037	10*w*	20*w*
Er	3,833.041	—	4
Pr	3,833.042	20	20

	λ	I J	I O		λ	I J	I O
Eu	**3,831.17**	**6***Wh*	**4***W*	**Sm**	**3,833.05**	**3**	**15**
Kr	**3,831.17**	**(2***wh***)**	—	**Sc**	**3,833.06**	**8**	**10**
Au	**3,831.14**	**5**	**8**	**O**	**3,833.10**	**(10)**	—
Hf	**3,831.13**	**25**	**25**	**Gd**	**3,833.15**	—	**2***h*
Ce	**3,831.08**	**2**	**10**	**Ti**	**3,833.19**	**3**	**12**
Mo	**3,831.07**	**4**	**4**	**V**	**3,833.23**	**2**	**10**
Dy	**3,831.04**	**3**	**5**	**Nb**	**3,833.26**	**10**	**5***w*
Ir	**3,831.033**	—	**10**	**Ir**	**3,833.28**	—	**2**
Cr	**3,831.032**	**25**	**40**	**Fe**	**3,833.31**	**60**	**100**
V	**3,831.032**	**5**	**1**	**J**	**3,833.40**	**(7)**	—
Nd	**3,831.030**	**30**	**60***d*	**Tb**	**3,833.40**	**3**	**8**
W	**3,830.99**	**7**	**6**	**Mo**	**3,833.47**	**25**	—
S	**3,830.94**	**(8)**	—	**He**	**3,833.57**	**(4)**	—
				Pr	**3,833.604**	**8**	**10**
				Nd	**3,833.604**	**12**	**20**
				Ce	**3,833.63**	**4**	**4**
				Hf	**3,833.67**	**6**	**10**
				Ti	**3,833.68**	**1**	**5**
				Ta	3,833.74	200	40
				Mn	3,833.86	75	75
				Fe	3,834.22	400	400
				Ar	3,834.68	(800)	—
				Os	3,836.06	20	150
				Fe	3,836.33	60	100
				Dy	3,836.51	40	100
				Gd	3,836.514	60	100*wh*
				Nd	3,836.54	100	80

*P** *N**

Mg 3,838.26 *J* 200 *O* 300

	λ	I J	I O		λ	I J	I O
Mn	**3,838.25**	—	**10**	**Zr**	**3,838.28**	**4**	**10**
Eu	**3,838.24**	—	**4**	**Ca**	**3,838.32**	**2**	**2**
J	**3,838.24**	**(10)**	—	**Nd**	**3,838.33**	**25**	**40**
Tm	**3,838.20**	**60**	**80**	**Er**	**3,838.339**	—	**10**
Li	**3,838.15**	—	**5**	**Pr**	**3,838.341**	**3**	**5**
U	**3,838.150**	**10**	**8**	**Eu**	**3,838.36**	—	**2***w*
He	**3,838.09**	**(2)**	—	**Hf**	**3,838.37**	**3**	**2**
Ru	**3,838.07**	**5***wh*	**12**	**Cl**	**3,838.37**	**(20)**	—
Fe	**3,838.04**	**1**	**1**	**N**	**3,838.39**	**(25)**	—
W	**3,837.92**	**6**	—	**W**	**3,838.50**	**20**	**15**
Rb	**3,837.910**	**(2)**	—	**Ce**	**3,838.54**	**3**	**35**
Nd	**3,837.909**	**2**	**10**	**Os**	**3,838.59**	—	**3**
Mo	**3,837.882**	**3**	**3**	**Dy**	**3,838.67**	—	**10**
Eu	**3,837.88**	**5***W*	**6***W*	**Nd**	**3,838.72**	**25**	**50**
Th	**3,837.880**	**8**	**10**	**Ru**	**3,838.73**	**10**	**10**
Dy	**3,837.86**	**1**	**2**	**Sm**	**3,838.93**	**15**	**10**

	λ	I	
		J	O
V	3,837.85	—	5
Tb	3,837.83	—	8
U	3,837.828	8*h*	12
Kr	3,837.82	(30)	—
Ir	3,837.72	4	15
Kr	3,837.70	(30)	—
Si	3,837.65	3	—
Er	3,837.631	1	15
V	3,837.631	—	10
Ho	3,837.60	6*h*	2
Ce	3,837.534	—	2
Ir	3,837.527	—	8
U	3,837.51	2*h*	2
Tl	3,837.49	(6)	—
Ho	3,837.45	6*h*	15
Cs	3,837.449	(4)	—
Eu	3,837.43	3*Wh*	3*W*
Er	3,837.42	1	10
Mo	3,837.29	100*w*	—
U	3,837.27	—	15
W	3,837.23	10	—
Ce	3,837.21	—	8
Tb	3,837.18	—	5
Fe	3,837.142	6	25
Mo	3,837.141	20	—
P	3,837.14	(30)	—
Nb	3,837.08	10	5
Ru	3,836.97	—	4
W	3,836.96	7	9
Gd	3,836.92	18	15
Mo	3,836.90	1	4
*P**			
Nd	3,838.98	30	20
U	3,838.996	10	6
V	3,839.002	10	60
Ta	3,839.03	5	30
Zr	3,839.130	10	10
Ir	3,839.134	—	6
Ce	3,839.15	3	4
Tb	3,839.18	—	8
Sm	3,839.190	3	25
S	3,839.19	(8)	—
W	3,839.257	6	7
Fe	3,839.259	75	100
Hg	3,839.26	(50)	—
Ir	3,839.32	—	4
Kr	3,839.37	(4*wh*)	—
V	3,839.38	5	30
Ta	3,839.42	—	2
Yb	3,839.45	15	2
Mo	3,839.47	5	5
Ce	3,839.497	1	6
Nd	3,839.497	12	10
U	3,839.53	6	—
Tb	3,839.62	—	15
Fe	3,839.63	4	3
U	3,839.632	2	30
Gd	3,839.64	—	25
Mn	3,839.78	125	100
Os	3,840.30	20	150
Fe	3,840.44	300	400
Fe	3,841.05	400	500
Cr	3,841.28	80	150
Co	3,842.05	20	400*R*
Fe	3,843.259	100	125
Cl	3,843.26	(100)	—
Mn	3,843.98	100	75
Mg	3,844.97	10	2
Co	3,845.47	100	500*R*
Bi	3,845.8	(100)	—
Bi	3,846.03	100	—
Fe	3,846.80	100	125
F	3,847.09	(800)	—
Tm	3,848.02	250	400
Mg	3,848.24	10	10
bhB	3,848.7	—	200
Tb	3,848.75	200	100
Mg	3,848.77	12	2
Sm	3,848.81	10	150*d*

	λ	I J	I O		λ	I J	I O
				La	3,849.01	150	200
				Os	3,849.94	20	125
				Fe	3,849.97	400	500
				F	3,849.99	(600)	—
				Ar	3,850.57	(400)	—
				Fe	3,850.82	75	200
				Cl	3,851.02	(100)	—
				Pr	3,851.62	150*w*	200*w*
				F	3,851.67	(200)	—
				V	*3,855.84*	*200*	*200*
				Fe	*3,856.37*	*300*	*500*
				Ni	*3,858.30*	*70h*	*800r*
				Fe	*3,859.91*	*600*	*1000r*
				Co	*3,861.16*	*15*	*300R*
				Mo	*3,864.11*	*500R*	*1000R*
				Nb	*3,865.02*	*200h*	—
				Os	*3,865.47*	*200*	*125*
				Fe	*3,865.53*	*400*	*600*

Mg 5,167.34 *J* 50 *O* 100

	λ	J	O		λ	J	O
La	**5,167.28**	**10**	**3**	**Hf**	**5,167.42**	—	**3**
Sm	**5,167.27**	—	**3**	**Fe**	**5,167.49**	**150**	**700**
Eu	**5,167.19**	—	**4**	**Mo**	**5,167.76**	**20**	**25**
Th	**5,166.90**	**2***h*	**3***W*	**La**	**5,167.79**	—	**20**
Dy	**5,166.84**	—	**4**	**Nd**	**5,167.92**	—	**5**
Kr	**5,155.80**	**(80)**	—	**Kr**	**5,168.06**	**(4)**	—
Ta	**5,166.79**	—	**30**	**Eu**	**5,168.23**	—	**4**
Eu	**5,166.72**	—	**125**	**N**	**5,168.24**	**(5)**	—
Hf	**5,166.38**	—	**2**	**Pr**	**5,168.31**	—	**3**
Sb	**5,166.32**	**(30***wh***)**	—	**Sm**	**5,168.34**	—	**6**
Fe	**5,166.30**	—	**125**	**Ni**	**5,168.66**	—	**70**
Cr	**5,166.23**	**2**	**80**	**Fe**	**5,168.90**	—	**80**
Nd	**5,166.09**	—	**4**	**La**	**5,168.971**	—	**3**
Co	**5,166.06**	—	**10**	**Os**	**5,168.975**	—	**8**
Sm	**5,166.05**	—	**125***d*	**Fe**	**5,169.03**	**200***h*	**2**
Zr	**5,165.96**	—	**7**	**Tb**	**5,169.12**	—	**10**
*bh***F**	**5,165.9**	—	**5**	**Sm**	**5,169.60**	—	**50***d*
Ar	**5,165.82**	**(20)**	—				
Hg	**5,165.8**	**(5)**	—				
Sr	**5,165.46**	—	**15**				
Fe	5,165.42	—	50				
Fe	5,164.61	—	70*h*				
Nb	5,164.38	20	150				
Pd	5,163.84	8	300				
Pb	5,163.8	(25)	—				

	λ	I	
		J	O
La	5,163.61	40	25
*bh*Mg	5,162.5	—	2
Fe	5,162.29	—	300*h*
Ar	5,162.28	(500)	—
P	5,162.27	(30)	—
P	5,161.97	(30)	—
Ta	5,161.81	—	80*w*
As	5,161.25	30	—
J	5,161.19	(300)	—
Nb	5,160.33	15	200
Eu	5,160.07	—	200
Ba	5,159.92	10	50*h*
O	5,159.88	(40)	—
V	5,159.35	40	40
Ne	5,158.89	(50)	—
La	5,158.693	—	50
Rh	5,158.69	1	80
La	5,157.43	100	40
La	5,156.74	40	40
P	5,156.72	(50)	—
Ne	5,156.66	(50)	—
Ta	5,156.56	—	80*W*
J	5,156.45	(25)	—
Co	5,156.34	—	300*w*
Sr	5,156.07	18	80
Pb	5,155.8	(25)	—
Ni	5,155.76	1	80
Rh	5,155.54	1	150
Ni	5,155.140	—	50
Ru	5,155.136	—	125
Sm	5,155.02	1	125
Ne	5,154.42	(50)	—
Sm	5,154.27	1	125*d*
Co	5,154.05	—	200*W*
Na	5,153.64	—	600
Cu	5,153.23	—	600
Nb	5,152.63	10	100
Ti	5,152.200	2	90
P	5,152.20	(50)	—
Tl	5,152.14	50*w*	—
Rb	5,152.09	100	—
Ne	5,151.96	(75)	—
Fe	5,151.91	—	70
Ar	5,151.39	(200)	—
C	5,151.08	30	—
Fe	5,150.84	—	150
Ne	5,150.08	(35)	—
Co	5,149.79	—	100
Os	5,149.74	—	80

	λ	I			λ	I	
		J	O			J	O
Na	5,149.09	—	400				
V	5,148.72	60	60				
Yb	5,147.03	50	3				
Co	5,146.74	—	400w				
Ni	5,146.48	1	150				
O	5,146.06	(70)	—				
Ti	5,145.47	4	100				
La	5,145.42	10	100				
Ar	5,145.36	(200)	—				
C	5,145.16	70	—				
Ne	5,145.01	(500)	—				
Ne	5,144.94	(500)	—				
Bi	5,144.48	300h	2				
Kr	*5,143.05*	*(600h)*	—				
Fe	*5,142.94*	—	*125*				
Ni	*5,142.77*	—	*100*				
Fe	*5,142.54*	—	*100h*				
Se	*5,142.14*	*(500)*	—				
Fe	*5,141.75*	*100h*	*100*				
P	*5,141.49*	*(50)*	—				
V	*5,139.53*	*50*	*50*				
Fe	*5,139.48*	*40*	*200*				
Fe	*5,139.26*	—	*125*				
V	*5,138.42*	*50h*	*50*				
Fe	*5,137.39*	—	*200h*				
Ni	*5,137.07*	*1*	*150*				
Fe	*5,136.79*	*100*	*3*				
Ru	*5,136.55*	—	*125*				
Yb	*5,135.98*	*50*	*6*				
Lu	*5,135.09*	*20*	*200*				
Nb	*5,134.75*	*15*	*200*				
Fe	*5,133.68*	*1h*	*200h*				
Eu	*5,133.48*	—	*150*				
Ge	*5,131.7*	*100*	—				
Fe	*5,131.47*	—	*125*				
In	*5,129.94*	*(70)*	—				
Pr	*5,129.52*	—	*100*				

Mg 5,172.70 *J* 100 *O* 200

	λ	J	O		λ	J	O
Co	**5,172.69**	—	**10**	**Re**	**5,172.72**	—	**2**
Sb	**5,172.46**	**(15)**	—	**Sm**	**5,172.74**	**1**	**80**
Re	**5,172.38**	—	**2**	**Er**	**5,172.76**	—	**12**
Kr	**5,172.36**	**(2)**	—	**La**	**5,172.92**	**20**	**5**
N	**5,172.32**	**(5)**	—	**Mo**	**5,172.94**	**25h**	**70h**

	λ	I J	I O
V	**5,172.09**	**18*h***	**18*h***
*bh***F**	**5,171.8**	—	**2**
Os	**5,171.72**	—	**8**
Ta	**5,171.63**	—	**20**
Fe	**5,171.60**	**60*h***	**300**
Se	**5,171.54**	**(18)**	—
N	**5,171.46**	**(5)**	—
Mo	**5,171.25**	**4**	**12**
*bh***Sc**	**5,171.1**	—	**4**
Mo	**5,171.08**	**6*h***	**30*h***
Ru	**5,171.03**	—	**150**
*P**			

	λ	I J	I O
Te	**5,172.99**	**(15)**	—
Yb	**5,173.13**	**15*h***	**1**
Cl	**5,173.15**	**(25)**	—
F	**5,173.16**	**(15)**	—
N	**5,173.37**	**(15)**	—
Sm	**5,173.63**	—	**3**
Ce	**5,173.70**	—	**3**
Ti	**5,173.75**	**20**	**125**
La	**5,173.85**	**25*h***	**20**
Pr	**5,173.90**	**4**	**100**
Eu	**5,173.99**	—	**4**
Mo	**5,174.18**	**25*h***	**70*h***
Th	**5,174.198**	—	**5**
Nb	**5,174.205**	**2**	**3**
U	**5,174.34**	—	**8**
N	**5,174.46**	**(5)**	—
V	**5,174.53**	**10*h***	**10*h***
Ce	**5,174.54**	—	**25**
In	**5,175.29**	**(400)**	—
Sm	**5,175.418**	**1**	**60**
In	**5,175.422**	**(300)**	—
In	5,175.56	(150)	—
Rh	5,175.97	1	200
Se	5,175.98	(600)	—
Co	5,176.08	—	500*r*
P	5,176.38	(70)	—
Sb	5,176.55	(50)	—
Ni	5,176.56	2	70
V	5,176.77	50	60
La	5,177.31	30	150
*bh*Mg	5,177.4	—	2
Cr	5,177.43	—	50
Ar	5,177.53	(40)	—
Sm	5,178.06	—	100*d*
J	5,178.13	(25)	—
Ge	5,178.58	100	—
Xe	5,178.82	(50)	—
Re	5,178.91	—	100*W*
N	5,179.50	(70)	—
Nb	5,180.31	15	150

Mg 5,183.62 *J* 300 *O* 500

	λ	I			λ	I	
		J	O			J	O
Co	**5,183.610**	—	**35**	**Zr**	**5,183.70**	—	**6**
Eu	**5,183.606**	—	**4**	**Ti**	**5,183.72**	—	**8**
La	**5,183.42**	**400**	**300**	**Nb**	**5,183.82**	**1**	**5*h***
Cu	**5,183.36**	**20**	—	**Pr**	**5,183.85**	—	**5**
Nb	**5,183.33**	**1**	**5*h***	**Br**	**5,183.88**	**(10)**	—
N	**5,183.21**	**(15)**	—	**La**	**5,183.92**	—	**25**
Ce	**5,183.20**	—	**10**	**W**	**5,183.97**	—	**20**
Tl	**5,183.10**	**(10)**	—	**Th**	**5,183.99**	**1**	**6**
Se	**5,183.01**	**(15)**	—	**Ru**	**5,184.03**	—	**4**
Nd	**5,182.60**	—	**8**	**Yb**	**5,184.18**	**30**	**8**
Th	**5,182.52**	**2**	**5**	**Rh**	**5,184.19**	**1**	**100**
Br	**5,182.36**	**(100)**	—	**Eu**	**5,184.27**	—	**2**
As	**5,182.32**	**30**	—	**Fe**	**5,184.29**	—	**20**
Os	**5,182.28**	—	**5**	**In**	**5,184.44**	**(300)**	—
Ru	**5,182.22**	—	**5**	**Xe**	**5,184.48**	**(40)**	—
Si	**5,182.13**	**3*wh***	—	**Ni**	**5,184.58**	—	**50**
Zn	**5,181.99**	**2**	**200**	**U**	**5,184.587**	**15**	**12**
Tl	**5,181.95**	**(10)**	—	**Cr**	**5,184.590**	**1**	**60**
Ce	**5,181.93**	—	**10**	**In**	**5,184.66**	**(70)**	—
Hf	**5,181.86**	**10**	**25**	**Th**	**5,184.73**	—	**3**
N	**5,181.80**	**(15)**	—	**N**	**5,184.97**	**(15)**	—
Re	**5,181.76**	—	**25**	*bh***Zr**	**5,185.0**	—	**30**
Ce	**5,181.75**	—	**5**	**Rh**	**5,185.02**	—	**8**
				Dy	**5,184.15**	—	**3**
*P**				**J**	**5,185.21**	**(30)**	—
				Tm	**5,185.24**	—	**5**
				*bh***F**	**5,185.6**	—	**5**
				Ti	**5,185.90**	**35**	**8**
				Kr	5,186.99	(60*whs*)	—
				Ar	5,187.75	(800)	—
				Xe	5,188.11	(100)	—
				La	5,188.23	500	50
				Ne	5,188.61	(150)	—
				Ti	5,188.70	100	80
				Xe	5,191.37	(200)	—
				P	5,191.40	(100)	—
				Fe	5,191.47	35	400
				Xe	5,192.10	(50)	—
				Co	5,192.35	—	100*w*
				Fe	5,192.36	50	400
				Ar	5,192.72	(60)	—
				Ti	5,192.97	25	150
				V	5,192.99	75*h*	100
				Nb	5,193.07	20	100
				Ne	5,193.13	(150)	—
				Rh	5,193.14	3	200

	λ	I J	I O
Ne	5,193.22	(150)	—
Fe	5,194.95	15	200
Ru	5,195.02	—	100
Fe	5,195.48	—	100*h*
Eu	5,199.85	—	500
Kr	5,200.22	(60*whs*)	—
Yt	5,200.41	150	60
Sm	5,200.59	1	200
Eu	5,200.92	—	300
Fe	5,202.34	10	300
Ne	5,203.89	(150)	—
La	5,204.15	300	50
J	5,204.20	(50)	—
Cr	5,204.52	100	400R
Fe	5,204.58	—	125
Yt	5,205.72	80	50
Ra	5,205.93	(250)	—
Cr	5,206.04	200	500R
O	5,206.61	(60)	—
Kr	5,208.32	(500)	—
Cr	5,208.44	100	500R
Fe	5,208.60	8	200
Bi	5,208.8	70	5wh
Ne	5,208.86	(70)	—
Ag	5,209.07	1000R	1500R
Bi	5,209.29	600	—
Co	5,210.06	—	100
Ti	5,210.39	35	200
Ar	5,210.49	(200)	—
La	5,211.87	5	300r
Co	5,212.71	—	300w
Ar	5,214.77	(200)	—
Eu	5,215.09	—	1000
Fe	5,215.18	5	200
Fe	5,216.28	10	300
Cl	5,217.92	(100)	—
Cu	5,218.20	—	700
Ar	5,221.27	(500)	—

Mn $^{25}_{54.9381}$

t_0 1,260°C t_1 2,151°C

I.	II.	III.	IV.	V.
7.429	15.636	—	—	76.0

λ	I		eV
	J	O	
II 2,576.10	2000*R*	300*R*	12.2
II 2,593.73	1000*R*	200*R*	12.2
II 2,605.69	500*R*	100*R*	12.2
2,794.82	15	1000*R*	—
2,798.27	80	800*R*	—
2,801.06	60	600*R*	—
I 4,030.75	20	500*R*	3.08
I 4,033.07	20	400*R*	3.08
I 4,034.49	20	250*R*	3.08
I 4,783.42	60	400	4.89
I 4,823.52	80	400	4.89
I 5,341.07	100	200	4.44

Mn 2,576.10 *J* 2000*R* *O* 300*R*

	λ	*I* *J*	*I* *O*		λ	*I* *J*	*I* *O*
Co	**2,576.10**	**—**	**30**	**Zr**	**2,576.10**	**—**	**8**
Ru	**2,576.08**	**50**	**—**	**W**	**2,576.16**	**20**	**2**
Nb	**2,575.96**	**10**	**1**	**Br**	**2,576.17**	**(15)**	**—**
Ce	**2,575.93**	**—**	**6**	**Rh**	**2,576.23**	**2**	**2**
W	**2,575.90**	**1**	**9**	**Hg**	**2,576.30**	**15**	**20**
Cr	**2,575.80**	**4**	**—**	**Re**	**2,576.32**	**—**	**15**
Rh	**2,575.75**	**4**	**2**	**Th**	**2,576.34**	**3**	**6**
Ag	**2,575.744**	**3*h***	**10*h***	**W**	**2,576.36**	**15**	**2**
Fe	**2,575.744**	**10**	**80**	**Pd**	**2,576.40**	**100**	**—**
Ir	**2,575.743**	**2**	**10**	**Ta**	**2,576.43**	**3*h***	**—**
Te	**2,575.70**	**(10)**	**—**	**V**	**2,576.48**	**50**	**1**
Mn	2,575.51	1	150	Pb	2,576.55	(100)	—
Al	2,575.10	80*R*	200*R*	Mn	2,578.35	10*h*	—
Mn	2,572.76	50	200	Mn	2,578.91	25*hw*	—
Mn	2,572.43	10	—	Mn	2,579.67	1	125
Mn	2,571.89	12	—	Mn	2,580.18	—	10
Zr	*2,571.39*	*400R*	*300R*	*Mn*	*2,581.65*	*15*	*—*
Mn	*2,570.94*	*80*	*—*	*Zn*	*2,582.49*	*40*	*300*
Mn	*2,570.09*	*10*	*—*	*Mn*	*2,582.96*	*50*	*—*
Mn	*2,569.32*	*10*	*—*	*Mn*	*2,583.28*	*—*	*15*
Mn	*2,568.72*	*15*	*—*	*Mn*	*2,584.11*	*5w*	*15*
Mn	*2,568.52*	*12*	*—*	*Mn*	*2,584.31*	*15*	*150w*
Mn	*2,568.31*	*10*	*—*	*Mn*	*2,584.53*	*12*	*12*

Mn 2,593.73 *J* 1000*R* *O* 200*R*

	λ	*I* *J*	*I* *O*		λ	*I* *J*	*I* *O*
Fe	**2,593.73**	**70**	**15**	**Br**	**2,593.76**	**(20)**	**—**
Si	**2,593.71**	**(2)**	**—**	**Nb**	**2,593.764**	**100**	**2*h***
Mo	**2,593.705**	**40**	**20**	**Ce**	**2,593.80**	**—**	**2**
Ru	**2,593.700**	**—**	**20**	**U**	**2,593.82**	**4**	**1*h***
Ta	**2,593.66**	**100**	**80*d***	**Na**	**2,593.83**	**—**	**20*R***
Ti	**2,593.642**	**2**	**20**	**Na**	**2,593.93**	**—**	**15*R***
Ru	**2,593.636**	**—**	**10**	**Mo**	**2,594.00**	**—**	**15**
Rh	**2,593.62**	**15**	**—**	**Fe**	**2,594.038**	**2**	**20**
U	**2,593.57**	**6**	**18**	**Bi**	**2,594.039**	**4**	**12*h***
Fe	**2,593.52**	**—**	**25**	**Os**	**2,594.14**	**3**	**10**
J	**2,593.47**	**(150)**	**—**	**Fe**	**2,594.15**	**2**	**20**
Hg	**2,593.41**	**3**	**5**	**Co**	**2,594.16**	**—**	**10*w***
W	**2,593.382**	**4**	**12**	**Mn**	**2,594.72**	**12**	**1**
Mo	**2,593.378**	**20**	**3**	**Mn**	**2,595.65**	**25**	**—**
Br	**2,593.34**	**(2)**	**—**	**Mn**	**2,595.76**	**25**	**200**
Mn	**2,593.28**	**2**	**—**	**Pt**	**2,596.00**	**20**	**200**
Pd	**2,593.27**	**100**	**3**				

	λ	I: J	I: O
Ta	2,593.08	1	150
Mn	2,592.94	3	150
Cu	2,592.63	50	1000
Mn	2,592.30	2	12
Ir	2,592.06	20	100
Cr	2,591.85	12	100*r*
Mn	2,591.42	12	2
Mn	2,591.23	12	—
Nb	2,590.94	800	15
Mn	*2,589.71*	*50h*	*10*
Mn	*2,588.96*	*80*	—
Mn	*2,587.28*	*12*	—
Mn	*2,586.57*	*10*	—
Mn	*2,585.48*	*10wh*	—

Mn 2,605.69 *J* 500*R* *O* 100*R*

	λ	J	O		λ	J	O
Pd	**2,605.678**	**2**	—	**Cl**	**2,605.70**	**(2)**	—
Co	**2,605.677**	**200**	**30**	**U**	**2,605.74**	**2**	**4**
Cl	**2,605.67**	**(5)**	—	**Ta**	**2,605.81**	—	**3**
Fe	**2,605.65**	**10**	**80**	**Ru**	**2,605.86**	**3**	**50**
Cr	**2,605.61**	**4**	—	**Te**	**2,605.89**	**(10)**	—
J	**2,605.55**	**(20)**	—	**W**	**2,605.92**	**2**	**6**
Xe	**2,605.54**	**(25)**	—	**Fe**	**2,605.90**	**20**	—
W	**2,605.51**	**2**	**12**	**Mo**	**2,605.93**	**10**	**5**
W	**2,605.47**	**4**	—	**W**	**2,605.96**	**8**	**2**
P	**2,605.45**	**(20)**	—	**Tm**	**2,606.02**	**10**	**20**
Tb	**2,605.42**	**10**	—	**P**	**2,606.02**	**(10)**	—
Fe	**2,605.41**	**40**	—	**Cr**	**2,606.07**	**2**	—
Cs	**2,605.40**	**(20)**	—	**Ce**	**2,606.09**	—	**4**
Ce	**2,605.38**	—	**4**	**Co**	**2,606.12**	—	**40**
Ru	**2,605.349**	**4**	**50**				
Ni	**2,605.347**	**250***wh*	—	Ag	2,606.16	200*wh*	10
Ta	**2,605.32**	—	**10**	Mn	2,606.25	5	—
Fe	**2,605.30**	**50**	—	Ni	2,606.39	600*h*	—
Ti	**2,605.15**	**12**	**100**	Mn	2,606.58	8	—
				Fe	2,606.82	30	200
Mn	2,604.36	10	—	Tm	2,607.05	30	50
Re	2,603.87	—	100	Fe	2,607.09	400	300
Mn	2,603.72	50	5	Mo	2,607.37	5	50*h*
Ta	2,603.57	300	5	Ir	2,608.25	10	50
Pt	2,603.14	20	300	Mn	2,608.43	20	—
Mn	2,602.72	80	—	Zn	2,608.56	50	200
Mn	2,602.13	1	12	Fe	2,608.58	10	100
Mn	2,601.97	10	—	Ta	2,608.63	4	125
Mn	2,601.83	10	—	Zn	2,608.64	100	300
In	2,601.76	15*wh*	50*R*	Mn	2,608.81	20	—

	λ	I J	I O
Ni	2,601.13	2000h	—
Mn	2,600.59	10	—
Mn	2,600.27	10	1
Mn	2,599.87	1	10
Fe	2,599.57	—	1000
Ta	2,599.397	30	100
Fe	2,599.396	1000h	1000
Mn	2,599.04	12	2
Mn	2,598.90	100wd	5
Fe	2,598.37	1000h	700
Mn	2,598.17	—	12
Sb	2,598.06	100	200
Mn	2,597.51	10	—
Tl	2,608.99	10	80R
Ta	2,609.00	1	80
Ru	2,609.06	12	80
Mn	2,609.55	15	—
Ni	2,610.09	900h	—
Mn	2,610.20	100h	15
Mn	2,610.85	10	—
Ta	2,611.34	—	100
Fe	2,611.87	500	500
Ru	2,612.07	30	100
Mn	2,612.85	2	15
Mn	2,613.59	2	12
Fe	2,613.82	400	400
Ta	2,614.17	—	200wh
Pb	2,614.18	80	200r

Mn 2,794.82 *J* 15* *O* 1000*R*

	λ	I J	I O
Co	2,794.82	15	100R
Fe	2,794.70	30	50
Tm	2,794.60	20	60
Mo	2,794.57	3	3
Ir	2,794.55	—	3
Yb	2,794.43	5	1
U	2,794.42	4	3
Mn	2,791.08	—	15
Mn	2,790.36	—	20
Mn	2,789.35	—	12h
Mn	2,789.20	—	15
Mn	2,787.82	—	12
Fe	2,795.01	35	50
Zr	2,795.13	—	5
Nb	2,795.14	15	1
U	2,795.23	12	18
Ru	2,795.35	8h	30
Mg	2,795.53	300	150
Ta	2,796.34	80	400

*N**

Mn 2,798.27 *J* 80* *O* 800*R*

	λ	I J	I O
Zr	2,798.27	—	100
Yb	2,798.22	10	4
U	2,798.21	4hd	1hd
Ir	2,798.18	5	15
Re	2,798.11	—	20
Mg	2,798.06	80	30
Mo	2,798.01	30	15
Mo	2,797.93	2	15
Pt	2,797.81	20	2
V	2,797.79	70h	12
Ta	2,798.40	—	150
W	2,798.45	1	8
La	2,798.55	40h	2
Ni	2,798.65	—	125
Co	2,798.65	—	4
Th	2,798.670	4	6
Cr	2,798.672	20	10
Bi	2,798.68	25	200
V	2,798.76	80h	25
Dy	2,798.86	—	2

	λ	I J	I Ú		λ	I J	I O
Fe	**2,797.77**	**80**	**150**	**Ce**	**2,798.89**	**—**	**2**
Ta	**2,797.76**	**100*d***	**100*d***	**Nb**	**2,798.91**	**15**	**2**
Th	**2,797.74**	**10**	**10**				
Ce	**2,797.72**	**—**	**4**				
Ru	**2,797.71**	**4*h***	**12**				

*P** *N**

Mn 2,801.06 *J* 60 *O* 600*R*

	λ	J	Ú		λ	J	O
Zn	**2,801.06**	**20**	**100**	**Zn**	**2,801.17**	**—**	**5**
W	**2,801.05**	**15**	**6**	**W**	**2,801.17**	**8**	**10**
Zn	**2,800.87**	**300**	**400**	**Ce**	**2,801.20**	**—**	**2**
Ir	**2,800.82**	**5**	**18**	**Hg**	**2,801.22**	**—**	**2**
Cr	**2,800.77**	**150**	**12**	**Sc**	**2,801.31**	**5*h***	**6**
Mo	**2,800.73**	**15**	**2**	**U**	**2,801.34**	**2**	**3**
Ru	**2,800.695**	**—**	**8**	**Dy**	**2,801.41**	**1*h***	**3**
Re	**2,800.695**	**—**	**15*h***	**W**	**2,801.43**	**6**	**4**
Ta	**2,800.572**	**40*h***	**150*W***	**Mo**	**2,801.47**	**3**	**20**
Th	**2,800.572**	**8**	**6**	**Mo**	**2,801.55**	**1**	**20**
Ce	**2,800.56**	**—**	**2**				
Tb	**2,800.51**	**40**	**10**	Pb	2,802.00	100*h*	250*Rh*
Fe	**2,800.46**	**10**	**50**	Ta	2,802.07	80	300
				Mn	2,802.80	—	12
				Pt	2,803.24	5	400
				Mn	2,803.62	—	12
				Mn	2,804.09	—	15
				Fe	2,804.52	200	300
				Mn	2,804.92	—	10
				Mn	2,806.14	—	25
				Ta	2,806.30	50	300
				Mn	*2,806.79*	—	*12*
				Mn	*2,809.11*	—	*25*

*P**

Mn 4,030.75 *J* 20* **O** 500*R*

	λ	J	Ú		λ	J	O
Cr	4,030.68	30	40	U	4,030.758	6	5
Ta	4,030.67	1*d*	10	Zr	4,030.759	—	20
Eu	4,030.66	—	5*w*	Yt	4,030.83	2	2
Sc	4,030.657	2	10	Ce	4,030.853	—	2
Ti	4,030.51	18	80	Th	4,030.855	8	10
Fe	4,030.49	60	120	Gd	4,030.88	—	8
Nd	4,030.47	15	20	Mo	4,030.91	5	3
Sm	4,030.42	3	10	Ru	4,031.00	12	15
Sr	4,030.38	—	40	Dy	4,031.08	—	7
Ce	4,030.34	4	18	Pr	4,031.09	8	12

	λ	I	
		J	O
Ca	**4,030.3**	**2*h***	**10**
Th	**4,030.29**	**5**	**8**
Eu	**4,030.20**	**—**	**10*w***
Fe	**4,030.19**	**4**	**20**
Ce	**4,030.15**	**—**	**5**
Ru	**4,030.14**	**—**	**7**
Zr	**4,030.04**	**2**	**35**
Eu	**4,030.00**	**—**	**5**
Tb	**4,029.97**	**—**	**3**
W	**4,029.95**	**7**	**6**
Mo	**4,029.941**	**30**	**1**
Ta	**4,029.94**	**5**	**50**
U	**4,029.92**	**4**	**6**
Nd	**4,029.91**	**3**	**10**
Yt	**4,029.84**	**—**	**5**
U	**4,029.80**	**2**	**1**
Ce	**4,029.75**	**—**	**4**
Pr	**4,029.73**	**12**	**15**
Zr	**4,029.68**	**15**	**40**
Re	**4,029.639**	**—**	**80**
Fe	**4,029.636**	**25**	**80**
W	**4,029.61**	**7**	**6**
Eu	**4,029.58**	**—**	**7**
Mo	**4,029.51**	**3**	**3**
Dy	**4,029.41**	**2**	**5**
Th	**4,029.31**	**5**	**8**
Ce	**4,029.26**	**1*h***	**5**
Tb	**4,029.22**	**—**	**2**
Hf	**4,029.17**	**12**	**10**
Ir	**4,029.14**	**—**	**6**
Co	4,027.04	4	200
Mn	4,026.43	40	50
Fe	4,024.74	30	120
Co	4,023.40	—	200
Cu	4,022.66	25	400
Fe	4,021.87	100	200
Co	4,020.90	—	500*w*
Mn	4,020.09	5	10
Mn	4,018.10	60	80
Fe	4,014.53	100	200
Co	*4,013.94*	*—*	*300*
Mn	*4,011.90*	*10*	*12*
Mn	*4,011.53*	*15*	*15*
Mn	*4,008.02*	*5*	*15*
Mn	*4,007.04*	*5*	*10*
Fe	*4,005.25*	*200*	*250*
Mn	*4,003.25*	*5h*	*20*

	λ	I	
		J	O
Th	**4,031.10**	**5**	**5**
Cr	**4,031.13**	**6**	**30**
V	**4,031.22**	**3**	**10**
Fe	**4,031.24**	**—**	**2**
U	**4,031.31**	**8**	**8**
Th	**4,031.33**	**5**	**5**
Ce	**4,031.34**	**8**	**40**
Zr	**4,031.35**	**1**	**3**
Eu	**4,031.38**	**3**	**7**
Sc	**4,031.40**	**2**	**10**
Nd	**4,031.54**	**3**	**10**
Ir	**4,031.56**	**—**	**3**
Re	**4,031.64**	**—**	**2**
Tb	**4,031.64**	**3**	**50**
Ce	**4,031.669**	**—**	**10*d***
W	**4,031.675**	**7**	**8**
Er	**4,031.690**	**—**	**6*w***
La	**4,031.692**	**300**	**400**
Ho	**4,031.75**	**1*h***	**4**
Ti	**4,031.754**	**1**	**35**
Pr	**4,031.755**	**30**	**50**
U	**4,031.78**	**2**	**8**
Mn	**4,031.79**	**10**	**8**
Nd	**4,031.81**	**15**	**15**
V	**4,031.83**	**3**	**10**
Ta	**4,031.96**	**—**	**5**
Fe	**4,031.965**	**50**	**80**
Th	**4,032.09**	**—**	**3**
Re	**4,032.15**	**—**	**10**
Ru	**4,032.20**	**20**	**20**
Ir	**4,032.21**	**—**	**10**
Hf	**4,032.27**	**2**	**5**
Tb	**4,032.28**	**—**	**30**
U	**4,032.29**	**2*h***	**2**
Sr	**4,032.38**	**—**	**20**
*N**			

	λ	I J	I O		λ	I J	I O
Mn	*4,002.16*	*5*	*15*				
Mn	*4,001.91*	*10*	*15*				
Mn	*4,001.18*	*5*	*12*				
Dy	*4,000.45*	*300*	*400*				

Mn 4,033.07 *J* 20* *O* 400*R*

	λ	J	O		λ	J	O
Cr	4,033.072	2	15	Sr	4,033.19	—	6
Ta	4,033.069	10	100	Nb	4,033.20	5	5
In	4,033.066	—	4	Pr	4,033.24	2	3
Tb	4,033.04	5	125	Cr	4,033.26	8	30
Ga	4,032.982	500*R*	1000*R*	Re	4,033.31	—	40
Sm	4,032.977	8	20	Ce	4,033.38	—	2*h*
Pr	4,032.97	10	15	U	4,033.43	10	12
V	4,032.86	1	2	Gd	4,033.49	5	10
Dy	4,032.85	—	8	Nd	4,033.50	5	10
Ce	4,032.75	—	2	Sb	4,033.54	60	70
Tb	4,032.70	—	3	Zr	4,033.58	—	3
Ti	4,032.632	1	35	Mn	4,033.630	5	5
Fe	4,032.630	15	80	Mo	4,033.631	6	6
Tb	4,032.626	—	4	Dy	4,033.67	4	15
Ce	4,032.55	—	3	Eu	4,033.69	—	8*w*
Th	4,032.54	8	10	U	4,033.73	12	12
Nb	4,032.524	50	30	Ir	4,033.76	25	100
Ru	4,032.521	5	10	Ce	4,033.79	—	6
Nd	4,032.51	—	2	Pr	4,033.86	35	50
Mo	4,032.50	8	8	Hf	4,033.88	8	5
Pr	4,032.49	12	20	Nd	4,033.90	4	10
Dy	4,032.480	12	20	Ti	4,033.91	3	40
Er	4,032.477	—	9	Mo	4,033.999	3	3
Zr	4,032.471	—	5	U	4,034.002	4	4
Fe	4,032.469	1	4				
W	4,032.385	7	6				
Sr	4,032.379	—	20				

*P** *N**

Mn 4,034.49 *J* 20* *O* 250*R*

	λ	J	O		λ	J	O
Gd	4,034.38	—	5	Nb	4,034.52	5	10
Pr	4,034.30	5	20	Ce	4,034.57	—	2
Ce	4,034.259	—	2	Tm	4,034.74	10	10
Th	4,034.256	10	10	Co	4,034.86	—	2
Sc	4,034.23	2*h*	8	Th	4,034.89	5	8
Nd	4,034.15	2	10*d*	Ti	4,034.91	2	25
Eu	4,034.11	3	2	Cr	4,035.00	—	8

	λ	I J	I O
Zr	**4,034.09**	**2**	**5**
Cr	**4,034.05**	**—**	**20**
Nd	**4,034.01**	**2**	**4**
*P**			

	λ	I J	I O
Pr	**4,035.07**	**1**	**2**
Nb	**4,035.098**	**3**	**4**
Sm	**4,035.101**	**3**	**50**
Nd	**4,035.17**	**—**	**2**
Cr	**4,035.24**	**—**	**8**
Ir	**4,035.33**	**—**	**6**
W	**4,035.35**	**9**	**10**
Nd	**4,035.399**	**2**	**8*d***
Gd	**4,035.403**	**5**	**8**
Pr	**4,035.42**	**1**	**10**
Co	**4,035.55**	**3**	**150**
V	**4,035.63**	**80**	**40**
Mo	**4,035.66**	**25**	**3**
Mn	**4,035.73**	**60**	**50**
Pr	**4,035.79**	**1**	**3**
Ti	**4,035.83**	**5**	**50**
Zr	**4,035.893**	**2**	**40**
Ta	**4,035.893**	**5*h***	**10**
V	**4,035.90**	**—**	**3**
Nb	**4,035.93**	**5**	**3**
Ce	**4,035.99**	**—**	**2**
Nd	**4,036.00**	**3**	**10**
Pr	**4,036.05**	**1**	**15**
Th	**4,036.06**	**3**	**10**
Ce	**4,036.09**	**—**	**5**
Ho	**4,036.10**	**—**	**2**
Eu	**4,036.11**	**—**	**50*W***
Er	**4,036.12**	**—**	**3**
Gd	4,037.34	30	100
Mn	4,038.73	15	15
Ho	4,040.84	30	150
Mn	4,041.36	50	100
La	4,042.91	300	400
K	4,044.14	400	800
Mn	4,045.13	—	15
Mn	4,045.21	15	15
Co	4,045.39	—	400
Ho	4,045.43	80	200
Fe	4,045.81	300	400
Hg	4,046.56	300	200
K	4,047.20	200	400
Gd	4,047.85	50	150
Mn	4,048.75	60	60
Mn	*4,051.73*	*20*	*15*
Mn	*4,052.47*	*20*	*20*
Ho	*4,053.92*	*200*	*400*
Mn	*4,055.21*	*5*	*10*

	λ	I			λ	I	
		I	*O*			*J*	*O*
				Ag	*4,055.26*	*500R*	*800R*
				Mn	*4,055.54*	*80*	*80*
				Pb	*4,057.82*	*300R*	*2000R*
				Mn	*4,057.95*	*20*	*80*
				Mn	*4,058.93*	*60*	*80*
				Nb	*4,058.94*	*400w*	*1000w*
				Mn	*4,059.39*	*15*	*20*
				Mn	*4,061.74*	*30*	*80*
				Cu	*4,062.70*	*20*	*500w*
				Mn	*4,063.53*	*60*	*100*
				Fe	*4,063.60*	*300*	*400*
				Mn	*4,065.08*	*20*	*20*
				Mn	*4,066.22*	*5*	*12*

Mn 4,783.42 *J* 60* *O* 400

Pr	**4,783.35**	**10*w***	**125**	**Gd**	**4,783.56**	—	**20**
Ti	**4,783.31**	—	**2**	**Tm**	**4,783.63**	**1**	**5**
Ru	**4,783.29**	—	**10**	**W**	**4,783.74**	—	**5**
Sm	**4,783.10**	—	**150**	**Nd**	**4,783.83**	—	**20**
Cr	**4,783.08**	—	**25*h***	**Eu**	**4,784.02**	—	**12**
Mo	**4,782.94**	**40**	**40**	**Tb**	**4,784.11**	—	**2**
Pr	**4,782.84**	—	**3**	**Ru**	**4,784.27**	—	**25**
Th	**4,782.76**	**3**	**4**	**Nb**	**4,784.28**	**1**	**3**
Hf	**4,782.74**	**5**	**40**	**Sr**	**4,784.32**	—	**30**
Ru	**4,782.64**	—	**10**	**Mo**	**4,784.41**	**5**	**5**
Zr	**4,782.60**	—	**3**	**Sm**	**4,784.43**	—	**6**
Tb	**4,782.58**	—	**2**	**V**	**4,784.47**	**10**	**12**
Ce	**4,782.22**	—	**10**	**Yb**	**4,784.53**	—	**2**
Pr	**4,782.10**	—	**3**	**Pr**	**4,784.63**	—	**3**
Gd	**4,781.93**	**50**	**200**	**Gd**	**4,784.64**	**50**	**100**
Tb	**4,781.92**	—	**2**	**Er**	**4,784.68**	—	**2**
Yb	**4,781.88**	**5**	**50**	**Ce**	**4,784.78**	—	**10**
Dy	**4,781.88**	—	**3**	**U**	**4,784.88**	—	**2**
				Dy	**4,784.918**	**2**	**4**
				Zr	**4,784.919**	—	**40**
Co	4,781.43	2*h*	400				
Co	4,780.01	500	500*w*				
Mn	4,779.15	—	20*W*	Ni	4,786.54	2	300*W*
Co	4,776.32	—	300	Pd	4,788.17	4*h*	200*h*
Mn	4,774.10	—	50	Cr	4,789.38	100	300
				Fe	4,791.25	200*R*	200
Co	4,771.11	—	500*W*	Re	4,791.42	—	200*w*
Co	4,768.08	10	300				
				Cr	4,792.51	40	200
				Au	4,792.60	60	200*W*
Mn	*4,766.43*	*30*	*80*	Co	4,792.86	5	600*W*
Mn	*4,765.86*	*25*	*60*	Os	4,793.99	6	300
				Mn	4,796.71	—	15

	λ	I J	I O		λ	I J	I O
Mn	*4,762.38*	*40*	*100*	Mn	4,797.70	—	25
Mn	*4,761.53*	*15*	*60*	Gd	4,798.92	300*w*	300*w*
Mn	*4,755.72*	*5*	*10*				
Mn	*4,754.04*	*60*	*400*	*Mn*	*4,807.18*	—	*12*

*N**

Mn 4,823.52 *J* 80* *O* 400

	λ	J	O		λ	J	O
Os	**4,823.43**	—	**8*h***	**Dy**	**4,823.73**	—	**2**
Yt	**4,823.31**	**10**	**15**	**Cr**	**4,823.92**	—	**25**
Er	**4,823.30**	**15**	**15**	**La**	**4,824.07**	**150**	**150**
Th	**4,823.18**	**2**	**3**	**Cr**	**4,824.12**	**35**	**4**
Gd	**4,823.08**	—	**5**	**Nd**	**4,824.18**	—	**10**
Pr	**4,822.98**	**10*w***	**125**	**Pt**	**4,824.22**	—	**2**
Mo	**4,822.93**	**5**	**6**	**Eu**	**4,824.26**	**2**	**2**
Sm	**4,822.86**	—	**5**	**Zr**	**4,824.29**	—	**20**
Ru	**4,822.57**	—	**10**	**Ru**	**4,824.357**	—	**5**
Ce	**4,822.54**	—	**25**	**Tb**	**4,824.364**	—	**3**
Mo	**4,822.42**	**12**	**15**	**Er**	**4,824.56**	—	**3**
Th	**4,822.16**	—	**2**	**Tb**	**4,824.59**	—	**2**
Yt	**4,822.13**	—	**8**	**Pr**	**4,824.65**	—	**5**
Er	**4,822.12**	**2**	**10**	**Sm**	**4,824.66**	—	**10**
Dy	**4,822.00**	—	**2**	**U**	**4,824.67**	**1**	**3**
Pr	**4,821.95**	—	**15**	**Dy**	**4,824.97**	**2*h***	**5**
Gd	**4,821.71**	**80**	**150**				
Fe	**4,821.05**	**200*h***	**200*h***				
Mn	**4,815.11**	—	**15*h***	Mn	4,825.59	5	20
Co	**4,813.48**	**6**	**1000 *W***	Mn	4,826.90	5	10
				Ni	4,829.03	2*h*	300*w*
Zn	**4,810.53**	**300*h***	**400*w***	Cr	4,829.38	40	200
Mn	**4,808.72**	—	**20**	Sm	4,829.58	—	200
				Ni	4,831.18	2	200
				Sr	4,832.07	8	200
				Mn	4,838.24	—	50
				Mn	*4,840.15*	—	*50*
				Co	*4,840.27*	*150*	*700w*
				Mn	*4,843.19*	—	*15*
				Mn	*4,844.31*	*5*	*80*

*P**

Mn 5,341.07 *J* 100 *O* 200

	λ	J	O		λ	J	O
Ta	**5,341.05**	**80**	**150*w***	**Ne**	**5,341.09**	**(1000)**	—
Sc	**5,341.040**	—	**5**	**Sm**	**5,341.29**	—	**80**
Cl	**5,341.04**	**(2)**	—	**Co**	**5,341.33**	—	**300*w***
Fe	**5,341.03**	**15**	**200**	**Ti**	**5,341.50**	—	**8**
Nb	**5,340.80**	**5**	**8**	**J**	**5,341.58**	**(2)**	—

	λ	I: J	I: O
Ir	5,340.74	—	5
*bh*F	5,340.7	—	20
La	5,340.67	100	80
Cr	5,340.44	—	50
Dy	5,340.314	—	6
Sm	5,340.312	—	3
N	5,340.15	(5)	—
Fe	5,339.94	30	200
K	5,339.67	—	40
Rh	5,339.65	—	2
Co	5,339.53	—	100*w*
Re	5,339.413	—	10
Sc	5,339.408	—	4
Xe	5,339.38	(500)	—
*bh*F	5,339.3	—	20
W	5,339.28	—	5
Kr	5,339.13	(20)	—
Mo	5,339.08	1	4
			8
J	5,338.19	(300)	—
Ru	5,335.93	—	100
Yb	5,335.16	400	150
Br	5,335.11	(70)	—
Co	5,333.65	—	100
Kr	5,333.41	(500*h*)	—
Ne	5,333.32	(50)	—
Co	5,332.67	—	200*w*
Br	5,332.04	(100)	—
As	5,331.54	200	—
Co	5,331.47	80	500*w*
Ne	5,330.78	(600)	—
O	5,330.66	(500)	—
O	5,329.59	(150)	—
O	5,328.98	(100)	—
N	5,328.70	(70)	—
Fe	5,328.53	35	150
Fe	5,328.05	100	400
Re	5,327.46	—	100
Ne	5,326.40	(75)	—
Co	5,325.28	—	300*w*
Fe	5,324.18	70	400
Kr	5,322.77	(60*h*)	—
Ra	*5,320.29*	*(250)*	—
Co	*5,316.78*	—	*300w*
Fe	*5,316.61*	*150*	—
P	*5,316.07*	*(150w)*	—
Xe	*5,313.87*	*(500)*	—

	λ	I: J	I: O
Ar	5,341.78	(10)	—
Gd	5,341.82	—	8
Eu	5,341.88	—	20
Pr	5,341.92	—	5
Ta	5,342.25	—	80
Tm	5,342.38	—	3
Hg	5,342.40	(12)	—
Pr	5,342.54	—	3
Co	5,342.71	—	800*w*
Sm	5,342.77	—	3
Sc	5,342.96	—	5
K	5,342.97	—	30
Gd	5,343.02	—	10
Ne	5,343.28	(600)	—
Co	5,343.39	—	600*w*
Nb	5,344.17	200	400
Mn	5,344.47	—	12
P	5,344.72	(150*w*)	—
J	5,345.15	(300)	—
Br	5,345.43	(80)	—
Yb	5,345.67	100	20
Cr	5,345.807	25	300*R*
P	5,345.81	(50)	—
Yb	5,345.83	50	10
Kr	5,346.76	(60*h*)	—
Yb	5,347.20	200	40
Ar	5,347.41	(200)	—
Mn	5,348.07	—	10
Cr	5,348.32	15	150*R*
Ne	5,349.21	(150)	—
Mn	5,349.88	—	20
Tl	5,350.46	2000*R*	5000*R*
Nb	5,350.74	50	150
Ti	5,351.08	60	50
Eu	5,351.67	—	150
Co	5,352.05	—	500*w*
Yb	5,352.96	250	100
V	5,353.41	50	50
Co	5,353.48	—	500*w*
*bh*C	5,354.1	—	100
Sb	5,354.24	(200)	—
Rh	5,354.40	5	300
Eu	5,355.08	—	200
Ne	5,355.18	(150)	—
Ne	5,355.42	(150)	—
N	5,356.77	(50)	—
Eu	5,357.61	—	1000
Cs	5,358.53	(500)	—

	λ	I	
		J	O
Co	*5,312.66*	—	*400w*
Hf	*5,311.60*	*150*	*100*
In	*5,309.83*	*(100)*	—
Ar	*5,309.52*	*(200)*	—
Xe	*5,309.27*	*(150)*	—
Kr	*5,308.66*	*(200)*	—
Se	*5,305.35*	*(500)*	—
Eu	*5,303.87*	—	*300*
La	*5,303.56*	*125*	*100*
La	*5,302.62*	*150*	*50*
Mn	*5,302.32*	*(60)*	—
Fe	*5,302.31*	—	*300*
La	*5,301.98*	*200*	*300r*
Co	*5,301.06*	—	*700w*

	λ	I	
		J	O
Yb	5,358.64	100	1*s*
Co	5,359.18	—	300*w*
Ne	5,360.01	(150)	—
Mo	5,360.56	70*h*	100*h*
Eu	5,360.81	—	150
Eu	*5,361.59*	—	*300*
Co	*5,362.77*	—	*500w*
Fe	*5,364.88*	*10h*	*200h*
Se	*5,365.47*	*(125)*	—
Fe	*5,367.46*	*15h*	*200h*
Xe	*5,368.07*	*(100)*	—
Co	*5,369.58*	—	*500w*
Se	*5,369.91*	*(175)*	—
Fe	*5,371.49*	—	*700*
Xe	*5,372.39*	*(200)*	—
Ar	*5,373.49*	*(500)*	—
Se	*5,374.14*	*(150)*	—
Eu	*5,376.91*	—	*200*
Re	*5,377.04*	—	*300 W*
La	*5,377.09*	*200*	*30*
Mn	*5,377.63*	—	*40*
C	*5,380.24*	*(300)*	—
La	*5,381.06*	*100*	*50*

Mo $^{42}_{95.94}$

t_0 2,622°C t_1 4,804°C

I.	II.	III.	IV.	V.
7.383	—	—	—	61.12

λ	*I*		eV
	J	*O*	
II 2,816.15	300	200	11.9
II 2,848.23	200	125	11.8
II 2,871.51	100	100	11.7
II 2,890.99	50	30	11.7
II 2,909.12	40	25	11.6
I 3,170.35	25*R*	1000*R*	3.91
I 3,798.25	1000*R*	1000*R*	3.26
I 3,864.11	500*R*	1000*R*	3.20
I 3,902.96	500*R*	1000*R*	3.17
I 5,506.49	100	200*R*	3.58

Mo 2,816.15 *J* 300 *O* 200

	λ	*I*	
		J	*O*
Th	**2,816.08**	**5**	**10**
Ba	**2,816.07**	**30**	—
Hf	**2,816.069**	**6**	**10**
Ce	**2,816.05**	—	**2**
U	**2,815.98**	**6**	**8**
V	**2,815.97**	—	**12**
Mo	**2,815.91**	—	**20**
Hf	**2,815.81**	—	**10**
Os	**2,815.78**	**4**	**40**
U	**2,815.76**	**8**	**6**
Re	**2,815.64**	—	**10**
Mn	**2,815.60**	—	**12*h***
Mo	**2,815.54**	—	**10**
Ta	2,815.12	4	100
Ta	2,815.01	15	150
Hg	2,814.93	(200)	—
Ta	2,814.80	5	125
Mo	2,814.67	20	1
Eu	2,813.95	300*wh*	300*w*
Ra	2,813.76	(400)	—
Ru	2,813.71	125	50
Fe	2,813.29	400	400
Mo	2,812.58	30	2
Ta	2,811.72	150	1
Mo	2,811.50	—	20
Mo	2,811.15	10	—
Ta	2,810.92	40*W*	200*W*
Ru	2,810.55	200	50
Mo	2,810.43	10	10
Ti	2,810.30	150	6
Mo	*2,809.95*	—	*20*
Bi	*2,809.62*	*100*	*200W*
Mo	*2,808.37*	*1*	*25*
Mo	*2,807.75*	*80h*	*60*
Mo	*2,807.35*	—	*20*
Fe	*2,806.98*	*200*	*200*
Ta	*2,806.58*	*50*	*200*
Ta	*2,806.30*	*50*	*300*
Mo	*2,806.18*	*15*	—
Al	**2,816.179**	**(15)**	—
Eu	**2,816.18**	**50**	**50*w***
Mn	**2,816.32**	**2**	**1**
Re	**2,816.329**	—	**40**
Ca	**2,816.33**	**3**	—
Yb	**2,816.33**	**2**	—
Ho	**2,816.38**	**10**	—
Dy	**2,816.39**	**2**	**5**
U	**2,816.42**	**6*h***	**3**
Kr	**2,816.46**	**(60)**	—
O	**2,816.52**	**(25)**	—
Tm	**2,816.56**	**10**	**2**
Fe	**2,816.66**	**4**	**5**
W	**2,816.67**	**1**	**5**
Nb	**2,816.68**	**50**	**2**
Cr	**2,816.70**	—	**12**
Mo	2,816.94	—	5
Ta	2,817.10	100	80*d*
Mo	2,817.44	25	8
Mo	2,817.50	25	15
Fe	2,817.51	60	100
Ti	2,817.87	200	10
Mo	2,818.30	1	25
Ta	2,819.37	5	100
Au	2,819.95	150	—
Re	2,819.955	—	150*W*
Eu	2,820.77	200*W*	200*W*
Ni	2,821.29	125	125
Mo	*2,821.83*	*25*	*1*
Mo	*2,822.03*	*20*	*15*
Mo	*2,822.43*	*4*	*15*
Ru	*2,822.55*	*150*	*30*
Mo	*2,822.86*	*6*	*20*
Pb	*2,823.19*	*40*	*150R*
Fe	*2,823.28*	*300*	*200*
Cu	*2,824.369*	*300*	*1000*
Ag	*2,824.370*	*200w*	*150wh*
Co	*2,825.24*	*200*	*5*
Mo	*2,825.67*	*1*	*25*
Mo	*2,825.99*	*4*	*15*
Tl	*2,826.16*	*100R*	*200R*
Mo	*2,826.54*	*5*	*40*
Ta	*2,827.18*	*10*	*200*

Mo 2,848.23 *J* 200 *O* 125

	λ	*I* *J*	*I* *O*
Zr	**2,848.19**	**12**	**12**
U	**2,848.17**	**2**	**5**
Fe	**2,848.12**	**10**	—
Ce	**2,848.08**	—	**2**
Ta	**2,848.054**	**15**	**150**
U	**2,848.051**	**8**	**8**
Fe	**2,848.046**	**70**	—
W	**2,848.03**	**12**	**15**
Nb	**2,848.02**	**8**	**1**
Hg	**2,847.83**	**100**	**15**
W	**2,847.82**	**12**	**9**
Ar	**2,847.81**	**(5)**	—
Re	**2,847.75**	—	**3**
U	**2,847.720**	**4*h***	**5**
J	**2,847.718**	**(15)**	—
Ce	**2,847.69**	—	**4**
Mo	**2,847.68**	**6**	—
Hg	2,847.67	(300)	—
V	2,847.57	150	15
Lu	2,847.51	125	40
Ta	2,846.75	10*h*	150*hs*
Mo	2,846.62	10	1
Mo	2,845.65	10	—
Fe	2,845.59	7	125
Fe	2,845.54	7	125
Ta	2,845.35	10	150
Mo	2,844.81	20	—
Ta	2,844.76	30	150
Ru	2,844.71	150	—
Ta	2,844.46	200	200
Mo	2,844.39	5	30
Ta	2,844.25	50	400*r*
Fe	2,843.98	300	300
Fe	2,843.63	100	125
Cr	2,843.25	400*r*	125
Ta	2,842.81	50	200
Mo	*2,842.46*	*30*	*3*
Mo	*2,842.37*	—	*10*
Mo	*2,842.15*	*40*	*2*
Mo	*2,841.77*	*15*	—
Ru	*2,841.68*	*200*	*50*
Fe	*2,840.42*	*20*	*125*
Sn	*2,839.99*	*300R*	*300R*
Mo	*2,839.58*	*1*	*25*

	λ	*I* *J*	*I* *O*
Os	**2,848.25**	**15**	**30**
Nb	**2,848.29**	**10**	**2**
Fe	**2,848.33**	**5*h***	—
La	**2,848.34**	**6**	**2**
Er	**2,848.368**	**4**	**10**
Co	**2,848.37**	**60*h***	—
Mg	**2,848.37**	**3**	—
Mo	**2,848.38**	—	**5**
Cr	**2,848.40**	**30**	**6**
Nd	**2,848.41**	**5**	—
Mg	**2,848.42**	—	**20**
Yb	**2,848.44**	**15**	**2**
Ir	**2,848.441**	—	**2**
Ta	**2,848.522**	**5**	**10**
Zr	**2,848.523**	—	**100**
Cu	**2,848.525**	**2**	—
Ta	**2,848.525**	**50**	**300**
Ru	**2,848.58**	**3**	**50**
U	**2,848.61**	**2**	**4**
Cu	**2,848.71**	**2**	—
Fe	**2,848.72**	**30**	**60**
V	**2,848.77**	**2**	**20**
Hf	2,849.21	100	30
Ru	2,849.29	100	3
Mo	2,849.38	5	50
Nb	2,849.56	100*w*	2*w*
Tl	2,849.80	(200)	—
Cr	2,849.84	150*r*	80
Co	2,850.04	—	75
Ta	2,850.49	100	200
Sn	2,850.62	100*wh*	80
Os	2,850.76	25	75
Mo	2,850.79	—	20
Mo	2,850.90	—	10
Ta	2,850.98	150	400
Mo	2,851.18	—	15
Fe	2,851.80	150	200
Mg	2,852.129	100*R*	300*R*
Fe	2,852.13	80	150
Mo	2,852.131	10	10*h*
Ta	2,852.35	100	5
Na	2,852.83	20	100*R*
V	2,852.87	7*h*	60
Na	2,853.03	15	80*R*
Cr	2,853.22	100*R*	5
Mo	2,853.23	100*h*	25

	λ	I J	I O
Mo	*2,839.16*	*25*	—
Fe	*2,838.12*	*150*	*150*
Mo	*2,837.90*	—	*15*
Cu	*2,837.55*	*250*	—

	λ	I J	I O
Mo	2,853.58	25	1
Mo	*2,854.11*	*10*	—
Pd	*2,854.58*	*500h*	*4*
Fe	*2,855.67*	*200*	*2*
Cr	*2,855.68*	*200 Wh*	*60*
Mo	*2,855.71*	*20*	—
Mo	*2,855.99*	*25*	*2*
Fe	*2,858.34*	*200*	*3*
Ta	*2,858.43*	*300*	*100*
Mo	*2,858.99*	*15*	—

Mo 2,871.51 *J* 100 *O* 100

	λ	J	O
Ru	**2,871.47**	**30**	—
Cr	**2,871.45**	**80**	—
Nd	**2,871.44**	—	**5**
Ta	**2,871.42**	**50**	**200**
F	**2,871.40**	**(25)**	—
Pd	**2,871.37**	**10*wh***	—
W	**2,871.367**	**8**	**10**
Rh	**2,871.35**	**10*h***	**100**
Cs	**2,871.32**	**(2)**	—
Na	**2,871.27**	**(40)**	**6**
Xe	**2,871.24**	**(25*hs*)**	—
Co	**2,871.24**	**100**	—
Ru	**2,871.19**	—	**4**
Mo	**2,871.18**	—	**10**
Fe	**2,871.13**	**20**	—
Ce	**2,871.07**	—	**15**
Fe	**2,871.06**	**40**	—
Tb	**2,871.05**	**10**	**3**
Ca	**2,870.98**	**2**	—
Mo	**2,870.90**	—	**15**
Ru	2,870.548	50	—
V	2,870.547	20*r*	50*r*
Cr	2,870.44	300*W*	25
Mo	2,870.18	1	10
Ti	2,870.04	100*wh*	—
Co	2,870.03	50*wh*	—
Mo	2,869.56	—	15
Fe	2,869.31	70	300
Tm	2,869.22	300	100
Mo	2,869.217	15	—
V	2,869.13	150*r*	25
Fe	2,868.87	60	5

	λ	J	O
Eu	**2,871.57**	**1*h***	**4**
Ce	**2,871.58**	—	**6**
Cr	**2,871.632**	**2**	**50**
Ce	**2,871.635**	—	**6**
Ru	**2,871.638**	**5**	**50**
U	**2,871.644**	**4**	**6**
Er	**2,871.68**	**1**	**9**
Fe	**2,871.73**	**1**	**3**
Re	**2,871.81**	—	**50**
Mo	**2,871.89**	—	**10**
W	**2,871.90**	**10**	—
U	**2,871.99**	**4*h***	**6*h***
Se	**2,872.08**	**(3)**	—
U	**2,872.110**	**4**	**3**
Ir	**2,872.111**	**3*h***	**2**
Te	**2,872.18**	**(5)**	—
Fe	2,872.34	50	150
Os	2,872.40	8	50
Mo	2,872.88	50	2
J	2,872.89	(60)	—
V	2,873.18	50	4
Ru	2,873.31	60	4
Pb	2,873.32	60	100*R*
Ta	2,873.36	40*h*	200*W*
Fe	2,873.40	300	—
Cr	2,873.48	125	30
Ta	2,873.56	50	150
Rh	2,873.62	10	60
Mo	2,873.64	1	10
Ag	2,873.65	100*wh*	3
Ta	2,874.167	15	150
Fe	2,874.172	200	300

	λ	I	
		J	O
Te	2,868.86	(100)	—
Rn	2,868.70	(70)	—
Ta	2,868.65	40	150
Nb	2,868.525	300	15
Al	2,868.52	(80)	—
Fe	2,868.45	40	80
Mo	2,868.32	20	2
Cd	2,868.26	80	100
Mo	2,868.11	20	3
Cr	2,867.65	100*R*	80
Fe	2,867.56	30	60
Ta	2,867.41	150	5*h*
Fe	2,867.31	30	60
Mo	2,867.05	—	10
Cr	2,866.74	125*R*	80
Mo	2,866.69	30	30
Ru	2,866.64	25	60
Fe	2,866.63	80	125
Hf	2,866.37	12	50
In	2,865.684	(50)	—
U	2,865.679	50	30
Mo	2,865.62	20	5
Cr	*2,865.11*	*200R*	*60*
Xe	*2,864.73*	*(100)*	—
Mo	*2,864.66*	*3*	*40*
Ta	*2,864.50*	*30*	*125*
Mo	*2,864.31*	*2*	*40*
Ni	*2,864.15*	*300wh*	—
Fe	*2,863.86*	*100*	*125*
Mo	*2,863.81*	*100h*	*30*
Ni	*2,863.70*	*250*	—
Fe	*2,863.43*	*80*	*100*
Sn	*2,863.33*	*300R*	*300R*
Mo	*2,863.12*	*20*	—
Rh	*2,862.93*	*60*	*150*
Mo	*2,862.84*	—	*10*
Cr	*2,862.571*	*300R*	*80*
Eu	*2,862.57*	*70*	*100W*
Fe	*2,862.50*	*50*	*100*
Ti	*2,861.99*	*100wh*	—
Mo	*2,861.86*	*10*	—
Hf	*2,861.70*	*125*	*50*
Nb	*2,861.09*	*100*	*10*
Os	*2,860.96*	*25*	*100*
Cr	*2,860.93*	*100*	*60*
Pt	*2,860.68*	*150h*	*30*

	λ	I	
		J	O
Mo	2,874.85	60	2
Fe	2,874.88	20	60
Os	2,874.95	15	50
Ru	2,874.98	50	80
Re	2,875.29	—	80
Fe	2,875.30	50	125
Fe	2,875.35	70	—
Nb	2,875.39	300	50*r*
Pt	2,875.85	80*h*	20
Zr	2,875.98	1	70
Cr	2,875.99	80*wh*	30
Ta	2,876.11	5	50*r*
Cr	2,876.24	80*Wh*	25
Hf	2,876.33	100	30
Mo	2,876.54	2	15
Fe	2,876.80	100	—
Nb	2,876.95	500*W*	40*W*
Fe	2,877.30	125	300
Ti	2,877.44	100	30
Pt	2,877.52	200*h*	40
Ta	2,877.686	80*h*	15
V	2,877.688	100*R*	15

Mo 2,890.99 *J* 50 *O* 30

	λ	*I*	
		J	*O*
W	2,890.99	8	2
Tm	2,890.93	15	60
Th	2,890.893	4	6
Ru	2,890.886	—	2
Os	2,890.85	4	10
Nd	2,890.84	2*h*	5
Dy	2,890.75	1	2
Cr	2,890.73	—	18
Ce	2,890.69	—	2*W*
W	2,890.66	12	2
Ti	2,896.61	50*wh*	—
Nb	2,890.56	5	1
V	2,890.55	12	1
Ir	2,890.53	—	8
Ta	2,890.52	2	1
Co	2,890.48	10	—
Dy	2,890.44	—	2
U	2,890.43	4	6
Fe	2,890.42	1	4
Ce	2,890.41	—	2
Pt	2,890.37	25	5*r*
In	2,890.161	(40)	—
Cr	2,890.16	—	20
Ta	2,890.06	2	15
Fe	2,889.99	8	15
Fe	2,889.88	15	20
Rh	2,889.839	30	70
Mo	2,889.837	—	12
U	2,889.63	50	30
V	2,889.621	150*r*	40
Hf	2,889.619	10	30
Cr	2,889.48	25	2
Re	2,889.46	—	30*w*
Ta	2,889.38	5	40
Cr	2,889.26	30	60*r*
Cr	2,889.20	20	15
Rh	2,889.11	1*h*	80
Ti	2,888.93	25	15
Nb	2,888.83	100	10
Cr	2,888.74	40	3
Mo	2,888.69	10	—
Ti	2,888.63	70*wh*	—
Mo	2,888.53	—	10
Cr	2,888.38	—	15
V	2,888.25	125*r*	20
Pt	2,888.20	—	50

	λ	*I*	
		J	*O*
Dy	2,891.026	1	3
Hf	2,891.03	—	9
Ta	2,891.04	30*h*	150*W*
Ti	2,891.066	50	20
U	2,891.074	10	8
Cr	2,891.10	40	10
Ru	2,891.14	—	5
Sb	2,891.21	(12)	—
Th	2,891.25	10	10
Mo	2,891.27	20	15
U	2,891.29	2	3
Mn	2,891.32	3	3
Yb	2,891.38	100	50*h*
Er	2,891.387	12	20
Ho	2,891.39	10	—
Fe	2,891.40	2	4
Tb	2,891.41	500	3
Nb	2,891.41	10*wh*	—
Cr	2,891.415	15	30
W	2,891.46	10	1
Re	2,891.48	—	25
Sb	2,891.51	20*wh*	—
Ar	2,891.61	(40)	—
V	2,891.64	200*r*	40
Fe	2,891.71	10	15
Ta	2,891.84	100	500*W*
Cr	2,891.88	35	—
Re	2,891.882	—	40
Fe	2,891.91	10	25
Ta	2,892.00	10	80*h*
Mo	2,892.03	15	—
Ce	2,892.14	—	15
*bh*B	2,892.20	—	200
Rh	2,892.22	—	30
Co	2,892.25	—	25
Zr	2,892.26	—	20
V	2,892.44	150*r*	30
Fe	2,892.48	40	100
Eu	2,892.54	30*h*	40*w*
Ru	2,892.560	6	20
Mo	2,892.565	—	5
Re	2,892.64	—	25
Mn	2,892.658	—	20
V	2,892.659	150*r*	30
Rn	2,892.7	(150)	—
Mo	2,892.81	30	25
N	2,892.86	(25*h*)	—

	λ	I	
		J	O
Mo	2,888.15	40	1
Fe	2,888.09	80	—
Ru	2,887.99	4	30
Eu	2,887.88	30	30
Th	2,887.82	18	18
Fe	2,887.81	60	80
Cr	2,887.77	35	—
Re	2,887.67	—	125
Mo	2,887.62	—	20*d*
Hf	2,887.54	2	15
Re	2,887.31	—	25
U	2,887.25	25	25
Rn	2.887.20	(125)	—
Hf	2,887.13	3	25
Cr	2,887.00	18	100
Mo	2,886.97	25	1
Re	2,886.94	—	20
Mn	2,886.68	6	15
Mo	2,886.61	1	30
Ru	2,886.54	50	60
Tm	2,886.459	5	30
Yt	2,886.457	6	15
U	2,886.45	6*h*	15
Co	2,886.44	2	50
Fe	2,886.32	15	50
Tb	2,886.28	10	15
Re	2,885.931	—	15
Fe	2,885.928	70	—
Mo	2,885.74	25	10
N	2,885.25	(50*h*)	—
La	2,885.141	50	5
Tb	2,885.14	70	—
Lu	2,885.14	3	40*h*
Mo	2,884.79	6	—
V	2,884.785	200*r*	40
Fe	2,884.779	25*h*	—
Re	2,884.63	—	25
Mo	2,884.59	—	5
Ti	2,884.11	125	35
Mo	2,883.96	10	—
Fe	*2,883.73*	—	*30*
Fe	*2,883.70*	*300*	—
Ru	*2,883.59*	*5*	*30*
Re	*2,883.45*	—	*60*
Mo	*2,883.30*	*6*	—
Nb	*2,883.18*	*800R*	*100*
Ir	*2,882.63*	*6*	*40*
Mo	*2,882.54*	—	*6*

	λ	I	
		J	O
Eu	2,893.03	—	40
Nb	2,893.069	100*W*	—
La	2,893.071	60	6
Pd	2,893.09	100	—
Pt	2,893.22	5	25
Mo	2,893.23	1	10
Cr	2,893.25	10	80*r*
Te	2,893.27	(30)	—
V	2,893.32	300*r*	50
Au	2,893.42	30	—
Br	2,893.44	(35)	—
Hg	2,893.59	50	40
Fe	2,893.76	8	15
Eu	2,893.85	100	150
Pt	2,893.86	25	500
Fe	2,893.88	20	25
Na	2,893.95	(60)	8
Cr	2,894.17	2	40
Cr	2,894.25	25	—
Re	2,894.33	—	20
Mo	2,894.45	80*h*	50
Tm	2,894.47	40	20
Fe	2,894.50	150	150
U	2,894.51	15	15
Fe	2,894.78	80	—
Lu	2,894.84	200	60
Mo	2,894.85	8	—
Ho	2,894.99	10	20
Fe	2,895.03	70	125
Os	2,895.06	8	25
Ta	2,895.10	15	125
Fe	2,895.21	80	—
Xe	2,895.22	(80*h*)	—
P	2,895.32	(25*w*)	—
Tl	2,895.41	15*s*	30*s*
Co	2,895.48	—	20
Te	2,895.49	(300*h*)	—
Re	2,895.66	—	25
Sc	2,895.88	(25)	—
Re	2,896.02	—	125*W*
Os	2,896.06	8	40
V	2,896.21	150*r*	35
Ta	2,896.438	50	2*h*
Mo	2,896.444	20	1
W	2,896.45	25	15
Cr	2,896.46	30	6
Ag	2,896.49	150*wh*	2
Ru	2,896.53	4	30

	λ	I J	I O
V	*2 882.50*	*200r*	*35*
Mo	*2,882.38*	*25*	*1*
Rh	*2,882.37*	*10*	*80*
Ta	*2,882.33*	*80*	*3*
Co	*2,882.22*	—	*30*
Ru	*2,882.12*	*200*	*30*
Si	*2,881.578*	*400*	*500*
Ce	*2,881.578*	*2*	*40*
Gd	*2,881.578*	—	*40*
Mo	*2,881.37*	*10*	—
Ru	*2,881.28*	*3*	*30*
Ta	*2,881.232*	*5*	*30*
Cd	*2,881.23*	*(30)*	*50R*
Na	*2,881.14*	*(60)*	*8*
Cd	*2,880.77*	*125*	*200R*
Fe	*2,880.76*	*50*	*15*
Nb	*2,880.71*	*50*	*4w*
Co	*2,880.29*	*50wh*	—
V	*2,880.03*	*150r*	*25*
Ta	*2,880.02*	*50*	*150*
Ru	*2,879.75*	*12*	*50*
Ta	*2,879.74*	*10*	*150*
Mo	*2,879.69*	*25*	*1*
Ta	*2,879.52*	*10h*	*50s*
Cr	*2,879.27*	*12*	*60*
V	*2,879.16*	*35*	*50*
Ru	*2,879.06*	*60*	—
Mo	*2,879.05*	*100h*	*15*
Ta	*2,878.95*	*3s*	*40r*
Rh	*2,878.65*	*10*	*50*
J	*2,878.64*	*(400)*	—
Cr	*2,878.45*	*80*	*20*
Os	*2,878.40*	*12*	*40*
Mo	*2,878.38*	—	*20*
Cr	*2,877.98*	*100*	*30*
Sb	*2,877.91*	*150*	*250w*
V	*2,877.688*	*100R*	*15*
Ta	*2,877.686*	*80h*	*15*
Pt	*2,877.52*	*200h*	*40*
Ti	*2,877.44*	*100*	*30*
Ce	2,896.73	—	20*s*
Cr	2,896.75	30	60*r*
*N**			

Mo 2,909.12 *J* 40 *O* 25

	λ	I J	I O
Os	**2,909.06**	**400**	**500*R***
Cr	**2,909.05**	**12**	**60*r***
Eu	**2,909.01**	—	**40**
Nd	**2,908.98**	**5*h***	—
W	**2,909.12**	**8**	**8**
Yb	**2,909.19**	**7**	**2**
Ru	**2,909.22**	**2**	**12**
U	**2,909.25**	**15**	**6**

	λ	I	
		J	O
Nb	**2,908.979**	**5**	**1**
Ta	**2,908.91**	**10**	**150**
Nb	**2,908.881**	**20**	**2**
Mn	**2,908.879**	—	**10**
U	**2,908.878**	**2**	**2**
Ru	**2,908.877**	**8**	—
Fe	**2,908.859**	**40**	**80**
Hf	**2,908.858**	**5**	**3**
V	**2,908.82**	**400*R***	**70*r***
Cd	**2,908.74**	—	**5**
Tm	**2,908.69**	**10**	**5**
Kr	**2,908.62**	**(5)**	—
Nd	**2,908.60**	**5**	—
Er	**2,908.53**	—	**6**
W	**2,908.493**	**10**	—
Tb	2,908.49	20	3
V	2,908.44	20	2
Ce	2,908.42	—	30*s*
Re	2,908.34	—	20
U	2,908.27	30	12
Nb	2,908.24	200	20*r*
Mo	2,908.16	8	5
Ti	2,908.14	25*wh*	—
Mn	2,907.99	—	20
Pt	2,907.90	4	15
Fe	2,907.86	20	2
Mo	2,907.78	—	6
Fe	2,907.52	80	100
V	2,907.47	150*h*	40
Ni	2,907.46	—	40
Ir	2,907.23	10	25
Mn	2,907.22	—	50
Rh	2,907.21	30	100
Xe	2,907.18	(40*h*)	—
Mo	2,907.12	30	10
Re	2,907.10	—	15
Au	2,907.06	25	—
U	2,906.91	15*h*	18*r*
U	2,906.80	50	15
Ti	2,906.68	100*wh*	—
Eu	2,906.676	300	300*W*
V	2,906.46	150*h*	40
Fe	2,906.42	25	60
Yb	2,906.34	40	—
Ru	2,906.31	5	30
Cl	2,906.25	(20)	—
Fe	**2,909.31**	**2**	**4**
Dy	**2,909.32**	—	**2**
Hg	**2,909.36**	**(25)**	—
Ho	**2,909.42**	**10**	**40**
Ti	**2,909.46**	**4*h***	—
Yb	**2,909.48**	**8**	**2**
Fe	**2,909.50**	**35**	**70**
Ir	**2,909.56**	**5**	**18**
Er	**2,909.58**	**1**	**7**
Fe	**2,909.59**	—	**2**
Ce	**2,909.61**	—	**5**
W	**2,909.628**	**3**	**4**
La	**2,909.631**	—	**2**
Os	**2,909.67**	**5**	**12**
Ru	**2,909.740**	**150**	—
U	**2,909.742**	**4**	**4**
Re	2,909.82	—	40
Hf	2,909.91	20	30
V	2,910.02	150*r*	35
Re	2,910.08	—	15
Rh	2,910.17	12	50
Er	2,910.36	6	20
V	2,910.39	150*r*	35
Nb	2,910.59	100	10
Cr	2,910.65	50	—
Cd	2,910.80	(30)	—
Cr	2,910.90	8	60*r*
Mo	2,910.93	10	—
V	2,911.06	200*r*	30
Cr	2,911.14	8	40
Sm	2,911.27	1*h*	15
Lu	2,911.39	300	100
Er	2,911.42	15	30
Ne	2,911.46	(25)	—
Yb	2,911.52	40	5
Cr	2,911.68	40	—
Nb	2,911.74	100	8
Mo	2,911.76	—	5
Mo	2,911.91	50*h*	30
Ti	2,912.08	15	35
Fe	2,912.16	150	150
Pt	2,912.26	25	300
Os	2,912.33	50	50
Ru	2,912.43	3	30
V	2,912.50	20*h*	1
Sm	2,912.56	—	15
Rh	2,912.62	20	50

	λ	I	
		J	O
V	2,906.13	25*h*	30
Fe	2,906.12	40	—
Mo	2,906.06	1	15
Re	2,906.02	—	30
Pt	2,905.902	15	100
Au	2,905.90	30	10
Mo	2,905.833	15	—
Ru	2,905.828	5	20
Ta	2,905.74	3	40
Os	2,905.73	8	40
Ti	2,905.66	2	30
Ru	2,905.65	12	50
Ir	2,905.64	4	20
V	2,905.60	20	4
Re	2,905.58	—	50
Cr	2,905.49	8	60*r*
Re	2,905.40	—	20
Gd	2,905.31	20	20
Mo	2,905.27	4	30
Ta	2,905.24	100	80
V	2,904.99	25	2
Na	2,904.91	(80)	20
Ir	2,904.80	10	25
Hf	2,904.75	6	30
Gd	2,904.71	30	—
Cr	2,904.68	2	20
Er	2,904.47	5	25
Ta	2,904.43	3	15
Hf	2,904.41	6	30
Mo	2,904.33	10	—
Fe	2,904.16	8	15
V	2,904.13	6*h*	20
Ta	2,904.07	40	300*w*
V	2,903.70	1	15
Zr	2,903.64	20*wh*	1
Tb	2,903.21	3*h*	15
Co	2,903.19	1	25
Tm	2,903.08	25	15
V	2,903.078	150*r*	35
Mo	2,903.07	100*h*	20
Lu	2,903.05	1	20
Ta	*2,902.63*	*2*	*30*
Mo	*2,902.62*	—	*5*
Re	*2,902.48*	—	*125W*
Zn	*2,902.26*	*(50)*	—
Mo	*2,902.24*	—	*5*
Mn	*2,902.20*	—	*50*
Lu	2,912.70	—	15*h*
Re	2,913.16	—	20
Ne	2,913.168	(150)	—
Ru	2,913.170	3	50
Pt	2,913.25	—	25
Ta	2,913.32	2	15*W*
Ti	2,913.33	50*wh*	—
Mo	2,913.52	1	20
Au	2,913.54	50	—
Pt	2,913.542	25	300
Sn	2,913.542	125*wh*	100*wh*
Ni	2,913.59	20	—
Cr	2,913.73	5	60*r*
Mo	2,913.81	50*w*	1
Os	2,913.84	8	30
Tm	2,913.96	—	15
U	2,913.963	4	14
Ru	2,913.97	50	—
Rh	2,913.99	—	20
Ni	2,914.01	—	20
Ta	2,914.12	30	200
Yb	2,914.21	60	10
U	2,914.25	25	18
Ru	2,914.299	—	50
V	2,914.301	35	7
Fe	2,914.306	25	50
Mo	2,914.314	10	—
Mo	2,914.43	5	1
Mn	2,914.60	—	150*Wh*
Cd	2,914.69	(45)	—
Tb	2,914.79	10*h*	15
V	2,914.93	50*r*	60
Ta	2,914.94	3	20
Cr	2,915.23	25	1
Mo	2,915.26	—	6
Yb	2,915.27	40	10
V	2,915.33	20	12
Ta	2,915.34	50	150*w*
Mo	2,915.38	1	10
Rh	*2,915.42*	*40*	*80*
Ta	*2,915.49*	*40*	*150*
Mg	*2,915.52*	*12*	*20*
Zr	*2,915.99*	*20*	*25*
Mo	*2,916.100*	—	*20*
Ti	*2,916.10*	*50wh*	—
Ru	*2,916.25*	*25*	*100*
Ir	*2,916.36*	*2*	*25*
Hf	*2,916.48*	*15*	*50*

	λ	I	
		J	O
Ag	2,902.07	200wh	5
Ta	2,902.05	200	1000w
Ir	2,901.95	15	25
Fe	2,901.91	40	125
Mo	2,901.79	15	—
Fe	2,901.38	80	100
Ta	2,901.05	3	100
Mo	2,900.79	40	2
Ta	2,900.75	100	3h
Nb	2,900.67	100wh	—
Mn	2,900.55	—	50
Ru	2,900.42	50	—
Ta	2,900.36	40	200
Lu	2,900.30	150	50
Rh	2,899.95	30	70
Co	2,899.82	1	25
Li	2,899.66	60	—
Ir	2,899.63	5	25
V	2,899.60	4h	30
Cr	2,899.48	40	2
Fe	2,899.41	100	125
Nb	2,899.24	500	20
Cr	2,899.21	25	50
V	2,899.20	7h	30
Ta	2,899.04	15	200
As	2,898.71	40	25r
Hf	2,898.709	50	25
Mo	2,898.65	6	20
Cr	2,898.54	40	12
Mo	2,898.48	15	—
Ta	2,898.42	5	30
Fe	2,898.35	30	100
Hf	2,898.26	12	50
Ru	2,898.22	60	—
Bi	2,897.97	500WR	500WR
Pt	2,897.87	15	400
Nb	2,897.81	150	15
Ru	2,897.71	60	6
Mo	2,897.63	25	20
Mo	2,897.42	20	—
Fe	2,897.26	200	—
Cr	2,896.75	30	60r

	λ	I	
		J	O
Tm	2,916.52	40	15
Nb	2,917.05	100r	10w
Mo	2,917.15	20	—
Os	2,917.26	20	40
Th	2,917.39	6	25h
Na	2,917.52	(40)	8
Ru	2,917.77	2	60
Fe	2,918.03	100	125
Tm	2,918.27	50	25
Tl	2,918.32	200R	400R
Fe	2,918.36	25	40
Ru	2,918.50	50	—
Hf	2,918.58	8	30
Ce	2,918.66	—	30s
Mo	2,918.83	25	15
Ta	2,918.96	50	2
Na	2,919.05	(40)	6
Mo	2,919.20	1	8
Pt	2,919.34	40	150
Yb	2,919.35	90	15
Mo	2,919.38	1	8
Re	2,919.41	—	25
Co	2,919.55	—	30
Hf	2,919.50	80	40
Ru	2,919.61	12	80
Os	2,919.79	15	100
Fe	2,919.85	35	80
Te	2,919.96	(50)	—
V	2,919,99	70r	10
Ag	2,920.04	100wh	—
Ru	2,920.257	—	3
Mo	2,920.263	10	1
V	2,920.38	125r	20
Fe	2,920.69	80	150
Ru	2,920.96	30	30
Pt	2,921.38	6	100
Tl	2,921.52	100R	200R

Mo 3,170.35 *J* 25*R** *O* 1000*R*

	λ	I J	I O		λ	I J	I O
Fe	**3,170.35**	**50**	**10**	**Eu**	**3,170.38**	**1**	**15**
Ta	**3,170.29**	**35**	**250*w***	**Th**	**3,170.43**	**4**	**10**
Sm	**3,170.203**	**5**	**15**	**Ce**	**3,170.53**	—	**2**
W	**3,170.204**	**9**	**15**	**U**	**3,170.54**	**3**	**3**
Nb	**3,170.16**	**3**	**2**	**Ag**	**3,170.58**	**3**	**5**
Ru	**3,170.093**	—	**30**	**Ni**	**3,170.71**	—	**4**
U	**3,170.09**	**2**	**2**	**Dy**	**3,170.75**	**3**	**10**
Ce	**3,170.07**	—	**12**	**U**	**3,170.85**	**8**	**10**
Nd	**3,170.01**	—	**10**	**Dy**	**3,170.92**	—	**5**
U	**3,169.99**	**3**	**6**	**Ti**	**3,170.925**	**1**	**8**
Dy	**3,169.98**	**50**	**100**	**Eu**	**3,170.97**	**10**	**6**
W	**3,169.93**	**15**	**10**	**Re**	**3,170.99**	—	**4*w***
Tm	**3,169.89**	**40**	**15**	**U**	**3,171.05**	**2**	**3**
Sm	**3,169.87**	**8**	**25**	**Nb**	**3,171.17**	**3**	**2**
Ca	**3,169.85**	**2*h***	**10**	**Yb**	**3,171.188**	**4**	**2**
Tb	**3,169.84**	**8**	**30**	**Tb**	**3,171.19**	**3**	**8**
Co	**3,169.77**	—	**100**				
Cu	**3,169.68**	**20**	**50**				
Fe	**3,169.615**	**2**	**2**	Sn	3,175.02	**400*hr***	**500*h***
Sm	**3,169.613**	—	**3**				
Cr	**3,169.58**	**2**	**25**	*Mo*	*3,183.03*	—	*10*
Dy	**3,169.55**	**1*h***	**5**	*V*	*3,183.98*	*400R*	*500R*
				Mo	*3,185.10*	*8*	*20*
				V	*3,185.40*	*400R*	*500R*
Mo	3,164.53	10	10				
Mo	3,163.90	20	10				
Mo	*3,159.34*	*5*	*20*				
Mo	*3,158.16*	*30r*	*300R*				

Mo 3,798.25 *J* 1000*R* *O* 1000*R*

	λ	I J	I O		λ	I J	I O
Ho	**3,798.25**	**4**	—	**U**	**3,798.25**	**8**	**2**
Ce	**3,798.24**	—	**2**	**Er**	**3,798.26**	—	**5**
La	**3,798.19**	**2**	—	**Ir**	**3,798.27**	—	**8**
Yb	**3,798.17**	—	**4**	**Br**	**3,798.28**	**(6)**	—
Nb	**3,798.12**	**80**	**50**	**Ti**	**3,798.31**	**6**	**10**
Th	**3,798.10**	**1**	**5**	**Yb**	**3,798.44**	—	**4**
Ce	**3,798.08**	**2**	**2**	**F**	**3,798.46**	**(6)**	—
Ir	**3,798.06**	—	**6**	**Ce**	**3,798.51**	**3**	**3**
Ru	**3,798.05**	**40**	**30**	**Fe**	**3,798.51**	**300**	**400**
Fe	**3,797.95**	**2**	**4**	**Tm**	**3,798.55**	—	**15**
Tb	**3,797.93**	—	**15**	**Tb**	**3,798.59**	—	**15**
Ir	**3,797.924**	—	**2**	**Ce**	**3,798.62**	**3**	**3**
Hf	**3,797.923**	**25**	**25**	**Er**	**3,798.65**	—	**4**
Th	**3,797.91**	**5**	**2**	**Dy**	**3,798.65**	—	**2**

	λ	I	
		J	O
H	3,797.910	(20)	—
Cs	3,797.908	(4)	—
Nd	3,797.89	10*d*	20*d*
Cu	3,797.83	2*h*	—
U	3,797.77	1	10
Dy	3,797.76	—	4
Sm	3,797.73	5	25
Cr	3,797.72	20	100
Re	3,797.59	—	40
U	3,797.520	3	—
Fe	3,797.517	200	300
Th	3,797.516	10	8
Sn	3,797.42	3	—
Ta	3,797.40	3	6
Ce	3,797.32	—	2
Mo	3,797.30	5	5
Sm	3,797.278	—	15
Rb	3,797.276	(2)	—
Pr	3,797.23	8	15
Rb	3,797.17	(2)	—
U	3,797.14	3	1
Cr	3,797.13	30	50
Er	3,797.06	—	2
Sb	3,797.05	4	1
Mo	3,797.03	5	5
Bi	3,797.00	5	—
Tb	3,796.98	—	15
Th	3,796.952	5	10
W	3,796.947	4	5
Pr	3,796.92	20*d*	40*d*
Nd	3,796.89	10*d*	20*d*
Ti	3,796.885	15	12
Kr	3,796.884	(20)	—
Mo	3,796.57	20	—
Mo	3,796.04	5	10
Mo	3,795.59	20*d*	—
Fe	3,795.00	400	500
La	3,794.77	200	400
Mo	3,794.43	10	5
Bi	3,792.80	500*h*	—
B	3,792.50	(500)	—
Mo	3,792.09	15	1
Mo	3,791.41	10*h*	—
La	3,790.82	300	400
Mo	3,788.26	15	15
Fe	3,787.88	300	500
Mo	3,786.36	125	2

	λ	I	
		J	O
V	3,798.661	5	7
Hf	3,798.662	3	5
Ir	3,798.666	—	5
Eu	3,798.71	—	2*wh*
Tm	3,798.76	10	20
Cl	3,798.80	(50)	—
U	3,798.84	1	15
Ru	3,798.90	100	70
W	3,798.92	10	7
Tb	3,798.95	8	8
Zn	3,799.00	—	5
Eu	3,799.01	—	100*wh*
Ce	3,799.04	—	4
In	3,799.05	(10)	—
Ho	3,799.06	4*h*	—
Ce	3,799.10	—	3
In	3,799.12	(10)	—
Tb	3,799.15	—	8
Pd	3,799.19	150	200*w*
U	3,799.202	12	6
In	3,799.204	(18)	—
S	3,799.21	(8)	—
Th	3,799.22	5	10
Nd	3,799.24	10*d*	10*d*
Mn	3,799.26	50	50
V	3,799.27	7	10
Rh	3,799.311	100	25
In	3,799.314	(10)	—
Ru	3,799.35	100	70*r*
Pr	3,799.36	3	6
In	3,799.37	(25)	—
Ar	3,799.39	(15)	—
In	3,799.42	(10)	—
Nb	3,799.49	2	5
Eu	3,799.50	—	18*wd*
Hf	3,799.52	4	—
Sm	3,799.546	10	30
Ce	3,799.547	4	3
Fe	3,799.549	300	400
U	3,799.552	4	1
As	3,799.58	10	—
Th	3,799.66	2	2
Mo	3,801.84	25	20
Mo	3,803.41	20*w*	—
Mo	3,804.52	20	20
Fe	3,805.34	300	400
Mo	3,805.99	10	10

	λ	I J	I O		λ	I J	I O
Mo	3,785.03	10	8	Mo	3,807.01	20*w*	1
				Ni	3,807.14	40*h*	800*W*
				Mo	3,811.84	25*d*	—
Ni	*3,783.53*	*40h*	*500*				
Mo	*3,783.18*	*40*	*—*				
Mo	*3,782.07*	*100*	*—*	*Mo*	*3,812.28*	*25*	*—*
Mo	*3,781.59*	*20*	*25*	*Mo*	*3,813.90*	*20*	*2*
Mo	*3,779.77*	*30*	*25*	*Fe*	*3,815.84*	*700*	*700*
				Mo	*3,816.61*	*20*	*—*
Tl	*3,775.72*	*1000R*	*3000*	*Mo*	*3,817.97*	*10*	*5*
Mo	*3,775.65*	*20*	*15*				
Mo	*3,772.82*	*20*	*20*	*Mo*	*3,818.66*	*15*	*8*
Mo	*3,771.95*	*30*	*30*	*Fe*	*3,820.43*	*600*	*800*
				Mo	*3,822.98*	*15*	*20*
				Mo	*3,823.15*	*20h*	*—*

Mo 3,864.11 *J* 500*R* *O* 1000*R*

	λ	J	O		λ	J	O
Eu	**3,864.10**	**10*w***	**7*w***	**Br**	**3,864.12**	**(2)**	**—**
Pr	**3,864.06**	**5**	**3**	**Cu**	**3,864.121**	**2*wh***	**—**
Sm	**3,864.05**	**2**	**10**	**Bi**	**3,864.2**	**150*h***	**—**
Eu	**3,864.04**	**—**	**2*w***	**Cs**	**3,864.25**	**(6)**	**—**
Bi	**3,863.9**	**(100)**	**—**	**Ar**	**3,864.26**	**(10)**	**—**
Zr	**3,863.874**	**6**	**20**	**V**	**3,864.300**	**—**	**30**
V	**3,863.866**	**15**	**25**	**U**	**3,864.305**	**10**	**8**
Nb	**3,863.78**	**—**	**2**	**Fe**	**3,864.306**	**—**	**2**
Re	**3,863.750**	**—**	**15**	**Zr**	**3,864.33**	**20**	**50**
Ce	**3,863.746**	**—**	**3**	**W**	**3,864.34**	**10**	**12**
Fe	**3,863.74**	**30**	**60**	**Nb**	**3,864.36**	**5**	**3**
Br	**3,863.71**	**(3)**	**—**	**Cs**	**3,864.37**	**(4)**	**—**
Eu	**3,863.65**	**—**	**3**	**O**	**3,864.42**	**(18)**	**—**
Co	**3,863.61**	**—**	**30**	**Ce**	**3,864.46**	**3**	**2**
Ce	**3,863.59**	**1**	**2**	**U**	**3,864.48**	**10**	**8**
O	**3,863.49**	**(5)**	**—**	**La**	**3,864.49**	**150**	**100**
W	**3,863.47**	**7**	**8**	**Ti**	**3,864.50**	**6**	**15**
Yb	**3,863.46**	**5**	**1**	**Pr**	**3,864.53**	**4**	**2**
Er	**3,863.46**	**3**	**5**	**Cl**	**3,864.60**	**(15)**	**—**
Fe	**3,863.413**	**1*h***	**1*h***	**O**	**3,864.66**	**(5)**	**—**
Nd	**3,863.409**	**20**	**20**	**Ti**	**3,864.70**	**—**	**4**
Sm	**3,863.407**	**8**	**15*d***	**U**	**3,864.72**	**4**	**4**
V	**3,863.404**	**1**	**2**	**Hf**	**3,864.75**	**20**	**2**
U	**3,863.403**	**20**	**5**	**Re**	**3,864.76**	**—**	**3**
Th	**3,863.39**	**20**	**20**	**Fe**	**3,864.77**	**—**	**2*h***
Nb	**3,863.38**	**10**	**15**	**Eu**	**3,864.78**	**—**	**2*W***
Zr	**3,863.33**	**—**	**2**	**Er**	**3,864.81**	**1**	**7**
Nd	**3,863.327**	**4**	**10**	**Ru**	**3,864.857**	**—**	**5**
Dy	**3,863.20**	**—**	**5**	**V**	**3,864.862**	**50*r***	**100*r***
Re	**3,863.15**	**—**	**10**	**Ce**	**3,864.87**	**—**	**2**

	λ	I	
		J	O
La	3,863.11	2	2
Er	3,863.08	1	3
U	3,863.08	1	10
Ni	3,863.07	5	4
Gd	3,863.06	10	8
Nb	3,863.05	20	3
Ce	3,862.95	2	4
Eu	3,862.94	—	5wh
Nb	3,862.93	5	10
Ti	3,862.83	4	30h
Ce	3,862.79	2	4
Cu	3,862.76	—	10
Dy	3,862.66	1	7
Ru	3,862.65	60	2
Cr	3,862.55	20	25
Fe	3,859.91	600	1000r
Mo	3,858.84	15	—
Ni	3,858.30	70h	800r
Mo	3,857.20	60	—
Fe	3,856.37	300	500
Mo	3,851.99	15	10
Mo	3,850.82	25	—
Ar	3,850.57	(400)	—
F	3,849.99	(600)	—
Fe	3,849.97	400	500
Mo	*3,848.30*	*20*	*25*
Mo	*3,847.25*	*25*	*25*
F	*3,847.09*	*(800)*	—
Mo	*3,845.95*	*20*	*20*
Mo	*3,842.58*	*25*	—
Mo	*3,837.29*	*100w*	—
Mo	*3,837.14*	*20*	—
Mo	*3,835.31*	*25*	*8*
Ar	*3,834.68*	*(800)*	—

	λ	I	
		J	O
Nd	3,864.90	—	4
Nb	3,865.02	200h	—
Th	3,865.036	3	8
Nb	3,865.039	—	10d
Os	3,865.045	5	10
Ce	3,865.10	—	2
Be	3,865.146	—	5
Pr	3,865.147	2	—
U	3,865.148	8	5h
Mo	3,865.153	20	—
Sm	3,865.24	—	10
Eu	3,865.29	2	2
W	3,865.32	8	7
Ce	3,865.39	2	5
Ru	3,865.40	4	10
Be	3,865.43	—	30
Dy	3,865.45	2	10
Pr	3,865.458	125r	200r
Sr	3,865.46	—	50
Os	3,865.47	200	125
Be	3,865.52	—	10
Fe	3,865.53	400	600
Eu	3,865.56	—	30W
Cr	3,865.60	3	5
Ir	3,865.64	20	25
Mo	3,866.79	10	—
Mo	3,869.08	30	25
Mo	3,871.45	50	—
Mo	3,871.88	25	—
Fe	3,872.50	300	300
Co	3,873.11	80	500R
Mo	3,873.26	10	—
Co	3,873.95	80	400R
Fe	3,878.02	300	400
Fe	3,878.57	300	300R
Nb	*3,879.35*	*300*	*5*
Mo	*3,879.52*	*8*	*10*
Mo	*3,882.95*	*15*	*5*
Mo	*3,883.70*	*20*	—
Fe	*3,886.28*	*400*	*600*
La	*3,886.37*	*200*	*400*
Mo	*3,886.82*	*30*	*30*
Mo	*3,888.18*	*8*	*10*
He	*3,888.65*	*(1000)*	—
Mo	*3,891.88*	*20d*	*2*
Mo	*3,892.29*	*15*	*5*
Co	*3,894.08*	*100*	*1000R*

Mo 3,902.96 *J* 500*R* *O* 1000*R*

	λ	*I* *J*	*I* *O*		λ	*I* *J*	*I* *O*
Fe	3,902.95	400	500	Th	3,903.09	5	15
Cr	3,902.91	100	100	Tb	3,903.11	—	3
Ce	3,902.89	1	8	Ce	3,903.12	2	2
Ir	3,902.85	2	8	Cu	3,903.163	5	—
Cl	3,902.84	(9)	—	Cr	3,903.164	30	35
Re	3,902.821	—	3	Eu	3,903.24	—	10*W*
Ru	3,902.816	—	5	V	3,903.262	3	3
Er	3,902.77	1	10	U	3,903.262	1	10
Ce	3,902.74	—	2	W	3,903.30	7	6
Gd	3,902.72	—	25	Dy	3,903.33	—	8
W	3,902.68	4	5	Ce	3,903.34	3	15
Ir	3,902.66	8	8	Sm	3,903.41	60	60
Re	3,902.583	—	6	Nd	3,903.51	2	10
La	3,902.576	—	20	Ce	3,903.53	—	3
U	3,902.561	18	18	Eu	3,903.63	—	10
V	3,902.558	2	6	Hg	3,903.64	—	3
Ce	3,902.509	—	3	Pt	3,903.72	1	2
Ir	3,902.506	15	10	Zr	3,903.77	2	1
Pr	3,902.47	40	60	F	3,903.82	(10)	—
Th	3,902.46	5	8	Fe	3,903.90	80	100
Nd	3,902.45	8	10	Er	3,903.906	1*h*	5
Ce	3,902.44	1	2	Pr	3,903.908	8	10
Gd	3,902.40	80	100	Eu	3,903.92	—	10*W*
Dy	3,902.39	—	5	Ce	3,903.93	3	10
Tb	3,902.35	—	10	W	3,903.98	10	10
Sm	3,902.32	—	10	Er	3,903.99	—	5
Ce	3,902.27	—	2	U	3,904.01	8	8
U	3,902.26	2	1	Mg	3,904.03	12	2
V	3,902.25	5	20	Co	3,904.05	—	8
Ho	3,902.24	6	8	Th	3,904.09	20	20
Ce	3,902.13	—	2	Dy	3,904.14	—	20
In	3,902.123	(10)	—	Ce	3,904.16	—	4
Th	3,902.122	10	10	Nb	3,904.182	20	10
Cr	3,902.11	30	40	Sm	3,904.185	1	3
In	3,902.08	(18)	—	Tb	3,904.186	—	4
Ru	3,902.07	—	4	V	3,904.218	2	10
In	3,902.02	(18)	—	Rh	3,904.224	2	3
S	3,901.99	(20)	—	Gd	3,904.29	10	10
Tb	3,901.98	3	15	U	3,904.299	15	8
F	3,901.955	(15)	—	Mn	3,904.305	5	5
B	3,901.95	2	3	Ce	3,904.34	3	12
Hg	3,901.90	5	15	Pt	3,904.39	1	5
F	3,901.852	(3)	—	C	3,904.40	(6*h*)	—
Nd	3,901.850	30	30	V	3,904.40	—	2
W	3,901.83	10	10	Ho	3,904.45	2	4
U	3,901.79	2	1	Mo	3,906.48	10	5
Mo	3,901.77	20	15				

	λ	I			λ	I	
		J	O			J	O
Os	3,901.71	20	150	Eu	3,907.11	500R	1000RW
V	3,901.69	—	5	Mo	3,908.61	30	—
Eu	3,901.681	—	5w	Mo	3,911.09	20	20
				Mo	3,915.44	50	8
Tb	3,901.68	3	6	La	3,916.04	400	400
Pr	3,901.679	3	10				
Th	3,901.67	—	5	Mo	3,917.54	10	15
Er	3,901.66	—	2	Mo	3,917.78	5	15
U	3,901.55	—	15				
Se	3,901.53	(5)	—	*Mo*	*3,919.45*	*20*	—
Tl	3,901.53	(12)	—	*Mo*	*3,921.54*	*20*	—
Zr	3,901.51	—	8	*Mo*	*3,922.32*	*10*	*10*
				Mo	*3,923.75*	*20*	*10*
Fe	*3,899.71*	*300*	*500*	*Mo*	*3,925.65*	*20*	—
Mo	*3,896.38*	*10*	*8*				
Fe	*3,895.66*	*300*	*400*	*Mo*	*3,925.83*	*30*	*1*
Ar	*3,894.66*	*(300)*	—	*Mo*	*3,926.95*	*20*	—
				Mo	*3,927.61*	*10*	—
				Mo	*3,929.58*	*25*	—
				Eu	*3,930.50*	*400R*	*1000R*
*P**				*Ca*	*3,933.67*	*600R*	*600R*

Mo 5,506.49 *J* 100 *O* 200*R*

	λ	J	O		λ	J	O
Ce	5,506.45	—	6	Dy	5,506.50	—	3
Ar	5,506.11	(500)	—	In	5,506.71	(30)	—
Ce	5,506.09	—	6	Br	5,506.78	(300)	—
La	5,506.00	—	50	Fe	5,506.782	10	150
Fe	5,505.88	—	9h	In	5,506.82	(15)	—
Mn	5,505.87	—	40	S	5,507.01	(25)	—
V	5,505.86	10	10	In	5,507.11	(30)	—
Er	5,505.67	—	12	P	5,507.13	(70)	—
Ta	5,505.65	—	25	Cs	5,507.17	(15)	—
Re	5,505.61	—	2h	Nd	5,507.19	—	2
Yb	5,505.501	2	40	La	5,507.33	—	5
Ra	5,505.50	(25)	—	In	5,507.335	(70)	—
Sm	5,505.184	—	4	Ne	5,507.34	(25)	—
Ar	5,505.18	(10)	—	Xe	5,507.46	(10)	—
Gd	5,505.08	—	8	Ar	5,507.63	(10)	—
Nd	5,504.98	—	10	Nd	5,507.66	—	8
Eu	5,504.91	—	2h	V	5,507.75	60	60
V	5,504.87	15	15	Zr	5,507.87	—	2
J	5,504.72	(60)	—	Cr	5,508.21	—	6
Rh	5,504.65	—	10	Mo	5,508.24	—	20
Nb	5,504.58	3h	30w	Br	5,508.38	(5)	—
Ho	5,504.51	—	12	Nd	5,508.40	—	15
				bhLa	5,508.40	—	10
				W	5,508.49	—	4
La	5,503.81	—	100	V	5,508.62	5	5

	λ	I	
		J	O
Ra	5,501.98	(250)	—
Mo	5,501.87	10	20
Mo	5,501.54	15	20
Fe	5,501.47	—	150
La	5,501.34	50	200
Kr	5,500.71	(50)	—
P	5,499.71	(150)	—
Kr	5,499.54	(50)	—
Mo	5,498.49	10	20
As	5,497.98	200	—
In	5,497.64	(50)	—
In	5,497.55	(70)	—
Fe	5,497.52	5	150
J	5,496.92	(900)	—
Ar	5,495.87	(1000)	—
N	5,495.70	(70)	—
Eu	5,495.17	—	125
Br	5,495.06	(150)	—
Ne	5,494.41	(50)	—
Mo	5,493.80	12*h*	20*h*
U	5,492.97	50	60
W	5,492.32	50	50
Mo	5,492.17	8	15
J	5,491.57	(100)	—
Kr	5,490.94	(50)	—
Mo	5,490.28	10	20
Ar	5,490.13	(60)	—
Co	5,489.65	—	150*w*
Br	5,488.79	(70)	—
Eu	5,488.65	—	500
Te	5,488.07	(50)	—
Co	*5,483.96*	—	*150w*
Co	*5,483.34*	—	*500w*
Ra	*5,482.13*	*(100)*	—
Ni	*5,476.91*	*8*	*400w*
Lu	*5,476.69*	*1000*	*500*
Mo	*5,475.90*	*12*	*20*
S	*5,473.63*	*(750)*	—
Ar	*5,473.44*	*(500)*	—
Mo	*5,473.37*	*25*	*50*
Xe	*5,472.61*	*(500)*	—
Eu	*5,472.33*	—	*1000s*
Ag	*5,471.55*	*100*	*500h*
Kr	*5,468.17*	*(200hs)*	—
Br	*5,466.23*	*(150)*	—
Mo	*5,465.57*	*5*	*20*
Ag	*5,465.49*	*500R*	*1000R*
J	*5,464.61*	*(900)*	—
Sb	*5,464.08*	*(100)*	—
W	**5,508.63**	—	**8**
La	**5,510.34**	—	**200**
Eu	**5,510.55**	—	**300*s***
Ru	5,510.71	—	100
In	5,510.88	(70)	—
Mo	5,511.49	4	15
Ti	5,512.53	12	125
O	5,512.70	(70*h*)	—
In	5,512.82	(150)	—
In	5,512.92	(100)	—
In	5,513.00	(70)	—
In	5,513.06	(70)	—
In	5,513.10	(50)	—
Sm	5,516.14	—	200
Ta	5,518.91	—	100*W*
Ba	5,519.11	60	200*wh*
In	5,519.36	(500)	—
Mo	5,520.04	12	20
Mo	5,520.64	12	20
Se	5,522.42	(750)	—
Rb	5,522.79	100	—
Kr	5,522.94	(60)	—
Ce	5,522.99	—	100
In	5,523.00	(50)	—
In	5,523.287	(50)	—
Co	5,523.292	—	300*w*
Os	5,523.53	—	100
In	5,523.61	(50)	—
In	5,523.86	(70)	—
In	5,523.91	(100)	—
Hf	5,524.35	50	40
Ar	5,524.93	(300)	—
Mo	5,526.52	20	25
Sc	5,526.81	300*wh*	100
Mo	5,526.97	15	25
Yt	5,527.54	15	100
Co	*5,530.77*	—	*500*
Xe	*5,531.07*	*(300)*	—
Mo	*5,533.05*	*100*	*200*
Ba	*5,535.55*	*200R*	*1000R*
La	*5,535.67*	*100*	*50*
Yb	*5,539.06*	*5*	*200*
Mo	*5,539.41*	*12*	*25*
Mo	*5,541.65*	*5*	*12*
Mo	*5,543.12*	*15*	*20*
Mo	*5,544.49*	*12*	*20*
Eu	*5,547.45*	—	*1000*

N 7
14.0067

t_0 −209.9°C

t_1 −195.8°C

I.	II.	III.	IV.	V
14.545	29.606	47.609	77.4	97.87

λ	*I*	eV
II 3,006.86	(50)	24.53
II 3,995.00	(300)	21.60
I 4,099.94	(150)	13.69
I 4,109.98	(1000)	13.70
I 4,935.03	(250)	13.19
II 5,666.64	(300)	20.62
II 5,676.02	(100)	20.62
II 5,679.56	(500)	20.66

N 3,006.86 (50)

	λ	I J	I O
Ca	**3,006.86**	**7**	**25**
Rn	**3,006.80**	**(300)**	**—**
W	**3,006.66**	**5**	**6**
Ru	**3,006.59**	**15**	**70**
Hg	**3,006.57**	**(50)**	**—**
Ta	**3,006.56**	**3h**	**3w**
V	**3,006.50**	**50**	**—**
Te	**3,006.35**	**(50)**	**—**
W	**3,006.31**	**12**	**9**
Cl	3,006.05	(20)	—
V	3,005.81	50	—
Nb	3,005.77	50	1
Yb	3,005.76	100	10
Fe	3,005.31	40	70
Ru	3,005.13	60	—
Cr	3,005.06	125	300r
Hg	3,004.47	(30)	—
Mo	3,004.46	40	5
As	3,003.93	50	—
Cr	3,003.92	150	1
Ir	3,003.632	30	60
Ni	3,003.629	80	500R
V	3,003.46	70	8
Fe	3,003.03	100	200
Pd	3,002.65	60	100r
Fe	3,002.65	150	20
Yb	3,002.61	150	15
Ni	3,002.49	100	1000R
Pt	3,002.27	30	200
Nb	3,002.21	30	5W
V	3,001.75	30	1
Ne	3,001.65	(25)	—
V	3,001.20	200r	20
Pt	3,001.17	50w	3
Fe	3,000.95	300r	800r
Cr	3,000.89	125	150r
Fe	3,000.45	80	100
Mo	3,000.23	30	25
Ru	*2,999.81*	*50*	*8*
Pd	*2,999.55*	*100h*	*—*
Fe	*2,999.51*	*300*	*500*
Ru	*2,998.89*	*100*	*50*
Cr	*2,998.79*	*70*	*200r*
Pt	*2,997.97*	*200r*	*1000r*
Fe	*2,997.30*	*60*	*—*

	λ	I J	I O
Yb	**3,006.86**	**3**	**1**
O	**3,006.90**	**(12h)**	**—**
Th	**3,006.93**	**15**	**12**
Cl	**3,006.98**	**(20)**	**—**
J	**3,007.01**	**(10)**	**—**
O	**3,007.08**	**(10)**	**—**
Tb	**3,007.11**	**3**	**8**
Fe	**3,007.14**	**80**	**100**
O	**3,007.28**	**(2)**	**—**
Fe	**3,007.281**	**60**	**80**
V	**3,007.283**	**50**	**2**
La	**3,007.32**	**4**	**4**
W	**3,007.40**	**3**	**7**
Na	**3,007.44**	**(20)**	**—**
Mn	3,007.65	40	40
Fe	3,008.14	400r	600r
In	3,008.31	500W	—
Ti	3,008.32	25	1
Nb	3,008.41	50	—
V	3,008.50	50	—
V	3,008.61	70	3
Fe	3,009.09	60	80
Sn	3,009.15	200h	300h
Fe	3,009.57	400	500
Gd	3,010.14	100	100
Zr	3,010.28	25	12
Ru	3,010.51	45	10
Cr	3,010.64	40	—
Rn	3,010.8	(100)	—
Cu	3,010.839	30	250
Ta	3,010.844	70	5
Sb	3,011.07	70	—
Ta	3,011.12	25	100W
Mn	3,011.16	25	25
Mn	3,011.38	25	25
Fe	3,011.48	125	125
Ni	3,012.00	125W	800R
V	3,012.01	50	2
Te	3,012.05	(25)	—
Ne	3,012.13	(50)	—
Fe	3,012.45	30	50
Ta	3,012.54	100	125
Hf	3,012.90	100	80
Ne	3,012.95	(50)	—
Cr	3,013.03	40	80
V	3,013.10	70	10

	λ	I	
		J	O
Te	*2,997.05*	*(50)*	—
Cr	*2,996.58*	*125*	*300r*
Fe	*2,996.39*	*50*	*90*
V	*2,996.00*	*70*	*5*
Ru	*2,995.99*	*50*	*4*
Ti	*2,995.75*	*70wh*	*10h*
Cr	*2,995.10*	*75*	*200r*
Au	*2,994.99*	*100*	—
Yb	*2,994.80*	*80*	*10*
Nb	*2,994.73*	*300*	*100*
V	*2,994.54*	*50*	*2h*
Fe	*2,994.43*	*600r*	*1000R*
Cr	*2,994.07*	*50*	*150*
Nb	*2,993.96*	*50*	—
J	*2,993.86*	*(70)*	—
Bi	*2,993.34*	*100wh*	*200wh*

	λ	I	
		J	O
Cr	*3,013.71*	*150*	*200r*
Mo	*3,014.16*	*50*	*5*
Cr	*3,014.76*	*100*	*300r*
V	*3,014.82*	*100*	*10*
Cr	*3,014.91*	*100*	*300r*
Cr	*3,015.19*	*80*	*200r*
Tm	*3,015.29*	*100*	*125*
Na	*3,015.40*	*(60)*	—
Ru	*3,015.41*	*50*	*8*
Cr	*3,015.51*	*150*	*1*
Fe	*3,015.91*	*50*	*70*
Fe	*3,016.18*	*150*	*200*
V	*3,016.78*	*80*	*15*
Ti	*3,017.19*	*200*	*15*
Ru	*3,017.24*	*50*	*100*
Ne	*3,017.35*	*(50)*	—
Xe	*3,017.43*	*(50h)*	—
Te	*3,017.51*	*(350)*	—
Cr	*3,017.57*	*200*	*300r*
Fe	*3,017.63*	*150*	*150*
Ru	*3,017.81*	*60*	*10*
Os	*3,018.04*	*50*	*300R*
Cr	*3,018.496*	*125*	*200r*
Pd	*3,018.498*	*50h*	—
Mo	*3,018.55*	*100*	*1*
Cr	*3,018.82*	*60*	*200r*
Fe	*3,018.98*	*150*	*150*

N 3,995.00 (300)

U	**3,994.980**	**20*wh***	**8**
J	**3,994.979**	**(35)**	—
Os	**3,994.93**	**5**	**30**
Sb	**3,994.90**	**10**	—
Pr	**3,994.834**	**25**	**300**
Kr	**3,994.83**	**(100)**	—
Kr	**3,994.82**	**(3)**	—
Ar	**3,994.81**	**(10)**	—
W	**3,994.762**	**2**	**4**
Rn	**3,994.76**	**(10)**	—
U	**3,994.72**	**2**	**2**
Nd	**3,994.68**	**40**	**80**
Mo	**3,994.63**	**3**	**3**
Th	**3,994.55**	**10**	**30**
Sn	**3,994.51**	**(2)**	—
La	**3,994.47**	**5**	—
Nb	**3,994.46**	**2*h***	—

K	**3,995.10**	**30**	—
In	**3,995.16**	**18*Wh***	—
Cl	**3,995.18**	**(2)**	—
Co	**3,995.31**	**20**	**1000*R***
Mo	**3,995.48**	**2**	**3**
Sm	**3,995.59**	**3**	**8**
Rh	**3,995.61**	**10**	**15**
Ba	**3,995.66**	**5**	**18**
Rn	**3,995.7**	**(3)**	—
La	**3,995.75**	**300**	**600**
U	**3,995.77**	**3**	**4**
Pr	**3,995.85**	**2**	**4**
Al	**3,995.86**	**(30)**	—
U	**3,995.97**	**8**	**10**
Ru	**3,995.98**	**30**	**30**
Fe	**3,995.989**	**20**	**60**
Eu	**3,995.994**	**2**	**3**

	λ	I J	I O
U	3,994.29	6	6
Gd	3,994.18	6	8
Fe	3,994.12	10	25
Nd	3,994.10	2	20
Pr	3,994.01	3	20
Cr	3,993.97	20	60
Mo	3,993.934	5	5
Eu	3,993.928	2	8
W	3,993.90	6	7
Cs	3,993.86	(4)	—
Ce	3,993.82	6	50
U	3,993.81	4	12
Th	3,993.72	5	8
Ho	3,993.71	4	10
Tb	3,993.55	8	30*d*
Ru	3,993.531	5	10
S	3,993.526	(50)	—
Ba	3,993.40	50r	100*R*
La	3,988.52	800	1000
Yb	3,987.99	500*R*	1000*R*
Hg	3,983.98	(400)	—
Tb	3,981.88	200	80
Rn	3,981.68	(150)	—
Cr	*3,976.66*	*300*	*300*
Ta	*3,973.18*	*400 W*	*1*
Fe	*3,969.26*	*400*	*600*
Ca	*3,968.47*	*500 R*	*500 R*
Cr	*3,963.69*	*300*	*300*
Xe	3,996.05	(2)	—
Th	3,996.066	10	15
Al	3,996.075	(2)	—
Rh	3,996.15	10	25
Al	3,996.16	(18)	—
Ta	3,996.17	30*h*	100
Al	3,996.18	(2)	—
Mo	3,996.25	2	3
Al	3,996.323	(2)	—
Gd	3,996.325	100	100
Al	3,996.38	(10)	—
Ir	3,996.45	2	8
Ru	3,996.51	4	10
Tm	3,996.52	40	200
Sc	3,996.61	10	40
Fe	3,997.40	150	300
Dy	4,000.45	300	400
Fe	4,005.25	200	250
Se	4,007.90	(150)	—
Se	*4,011.88*	*(200)*	—
Se	*4,012.96*	*(150)*	—
F	*4,024.73*	*(500)*	—
F	*4,025.49*	*(300)*	—

N 4,099.94 (150)

	λ	I J	I O
U	4,099.93	5	5
Dy	4,099.89	2	6
Pr	4,099.88	1	4
V	4,099.80	12	25
Ce	4,099.75	2	12
Eu	4,099.72	5	12*d*
Kr	4,099.71	(3)	—
La	4,099.54	100	100
Ar	4,099.47	(5)	—
Tb	4,099.46	1	25*w*
S	4,099.44	(8)	—
Pd	4,099.27	4*h*	—
U	4,099.269	1	20*r*
Ti	4,099.17	8	25
Tb	4,099.14	1	9
He	4,100.00	(2)	—
Ir	4,100.15	3*h*	100
Fe	4,100.17	1	10
Pr	4,100.22	10	15
Nd	4,100.24	8	10
Gd	4,100.26	5	15
Os	4,100.300	3	60
Ne	4,100.30	(5*h*)	—
Mo	4,100.32	20	—
Sc	4,100.33	1	5
Xe	4,100.34	(10)	—
Th	4,100.35	5	8
Ru	4,100.37	10	12
Nb	4,100.40	20	15
Er	4,100.56	1	15

	λ	I (J)	I (O)
Zr	4,099.08	1	3
Nb	4,099.07	5	5
Ca	4,099.029	1	2h
W	4,099.026	6	7
Cr	4,099.02	8	30
Ce	4,098.98	1	15
Th	4,098.939	10	10
Nd	4,098.936	8	8
Gd	4,098.91	100	100
Xe	4,098.89	(50h)	—
Ne	4,098.77	(50)	—
Mo	4,098.743	20	20
Cl	4,098.74	(8)	—
La	4,098.73	4	—
Kr	4,098.72	(250)	—
Pr	4,098.65	3	8
Gd	4,098.61	6	25h
Tb	4,098.60	2	2
Ca	4,098.53	3	15
Nd	4,098.51	5	10
W	4,098.49	3	4
Pr	4,098.41	12	20
Cl	4,098.40	(12)	—
O	4,098.24	(5h)	—
Ru	4,097.79	125	25
O	4,097.24	(70)	—
O	4,092.94	(80)	—
Se	4,091.95	(70)	—
Ac	4,088.37	100	—
Kr	4,088.33	(500)	—
Gd	4,087.71	100	80
Pd	4,087.34	100	500
La	4,086.71	500	500
Ta	4,085.80	80W	5
Gd	4,085.65	80	100
Fe	4,085.32	70	100
O	4,085.12	(70)	—
Fe	4,084.50	80	120
Ru	*4,080.60*	*300*	*125*
P	*4,080.04*	*(150)*	*—*
Nb	*4,079.73*	*200w*	*500w*
Hg	*4,077.81*	*150*	*150*
Sr	*4,077.71*	*500W*	*400r*
La	*4,077.34*	*400*	*600*
O	*4,075.87*	*(800)*	*—*
Te	*4,073.57*	*(300)*	*—*
In	*4,072.40*	*200wh*	*—*

	λ	I (J)	I (O)
Fe	4,100.74	30	80
Pr	4,100.75	50	200
Th	4,100.83	18	18
Ce	4,100.889	1	8
W	4,100.895	6	7
Tb	4,100.90	2	50d
Nb	4,100.92	200w	300w
Hf	4,100.93	1	10
V	4,101.00	7	5
La	4,101.01	2h	2
Te	4,101.07	(50)	—
Ho	4,101.09	40	40
Cr	4,101.16	2	30
Ru	4,101.23	2h	7
Fe	4,101.27	10	40
Sm	4,101.31	2h	15d
U	4,101.32	2h	2
Ce	4,101.36	2	1
Dy	4,101.43	2	5
Nd	4,101.46	6	8
Ce	4,101.55	3	2
H	4,101.73	(100)	—
In	4,101.77	1000R	2000R
F	4,103.08	(150)	—
F	4,103.52	(300)	—
Ho	4,103.84	400	400
Ar	4,103.91	(200)	—
O	4,105.00	(125)	—
Fe	4,107.49	100	120
*N**			

	λ	I J	I O		λ	I J	I O
O	*4,072.16*	*(300)*	—				
Ar	*4,072.01*	*(150)*	—				
Fe	*4,071.74*	*200*	*300*				
J	*4,070.75*	*(150)*	—				
Se	*4,070.16*	*(500)*	—				
Kr	*4,065.11*	*(300)*	—				

N 4,109.98 (1000)

	λ	J	O		λ	J	O
Ca	**4,109.98**	**3**	**2**	**Br**	**4,110.00**	**(10)**	—
Ce	**4,109.90**	**3*h***	—	**Zr**	**4,110.05**	**1**	**3**
Fe	**4,109.81**	**100**	**120**	**Pr**	**4,110.11**	**1**	**3**
La	**4,109.80**	**10**	**15**	**Kr**	**4,110.16**	**(5*wh*)**	—
V	**4,109.79**	**20*r***	**40*r***	**Sm**	**4,110.18**	**5**	**5**
W	**4,109.76**	**20**	**20**	**Mo**	**4,110.287**	**5**	**3**
Xe	**4,109.71**	**(60)**	—	**Nb**	**4,110.295**	**10**	**1**
Ru	**4,109.65**	**6**	**20**	**Ce**	**4,110.38**	**10*h***	**35**
Cr	**4,109.58**	**10**	**40**	**Xe**	**4,110.41**	**(15)**	—
La	**4,109.48**	**5**	**15**	**U**	**4,110.44**	**1*h***	**6**
Nd	**4,109.45**	**30**	**30**	**Pr**	**4,110.47**	**1**	**5**
Pr	**4,109.41**	**2**	**15**	**Nd**	**4,110.472**	**10**	**10**
Sm	**4,109.407**	**10**	**10**	**Sn**	**4,110.51**	**(2)**	—
In	**4,109.34**	**(5*h*)**	—	**W**	**4,110.57**	**6**	**7**
Kr	**4,109.23**	**(100*hs*)**	—	**Gd**	**4,110.60**	**15**	**10**
P	**4,109.19**	**(70)**	—	**Th**	**4,110.64**	**5**	**5**
F	**4,109.17**	**(100)**	—	**Mo**	**4,110.70**	**5**	**6**
Pr	**4,109.09**	**1**	**5**	**V**	**4,110.76**	**2**	**2**
Nd	**4,109.073**	**15**	**15**	**Pb**	**4,110.77**	**(5)**	—
Fe	**4,109.072**	**2**	**12**	**O**	**4,110.79**	**(40)**	—
V	**4,109.04**	**2**	**3**	**Yt**	**4,110.80**	**2*h***	**7**
Se	**4,108.83**	**(800)**	—	**U**	**4,110.830**	**8**	**8**
Nb	**4,108.70**	**1*h***	**2**	**Nb**	**4,110.832**	**5*h***	—
Mo	**4,108.69**	**8*h***	—	**Th**	**4,110.87**	**8**	**8**
Ho	**4,108.63**	**40**	**100**	**Mn**	**4,110.90**	**40**	**80*r***
Ca	**4,108.56**	**3**	**6**	**Pr**	**4,110.92**	**2**	**10**
W	**4,108.53**	**9**	**8**	**Os**	**4,111.03**	**1**	**6**
U	**4,108.475**	**4**	**4**	**Eu**	**4,111.07**	**1**	**10**
Ir	**4,108.467**	**2**	**4**	**Tl**	**4,111.2**	**(2)**	—
Kr	**4,108.43**	**(3)**	—	**Sb**	**4,111.20**	**10**	—
Th	**4,108.426**	**15**	**15**	**Dy**	**4,111.346**	**12**	**30**
Gd	**4,108.42**	**2**	**4**	**Er**	**4,111.351**	**1**	**15**
Cr	**4,108.40**	**4**	**30**	**Ho**	**4,111.37**	**2*h***	—
Pr	**4,108.34**	**3**	**15**	**Ce**	**4,111.39**	**5**	**35**
Sm	**4,108.32**	**4**	**5**	**Gd**	**4,111.44**	**15**	**15**
J	**4,108.28**	**(20)**	—	**U**	**4,111.53**	**2**	**2**
				S	**4,111.56**	**(30)**	—
*P**				**Sn**	**4,111.59**	**(2)**	—

	λ	I J	I O
V	4,111.78	100*WR*	100*WR*
Ru	4,112.74	200	125
Pr	4,113.89	70*w*	30*w*
Rn	4,114.56	(80)	—
Xe	4,116.11	(80)	—
Fe	4,118.55	100	200
O	4,119.22	(300)	—
Nb	4,119.28	200	2
La	4,123.23	500	500
Nb	4,123.81	125	200
Se	4,126.57	(150)	—
Rh	*4,128.87*	*150*	*300*
Se	*4,129.15*	*(200)*	—
Fe	*4,132.06*	*200*	*300*
Cl	*4,132.48*	*(200)*	—
Se	*4,132.76*	*(200)*	—

N 4,935.03 (250)

	λ	I J	I O
La	**4,934.82**	**100**	**150**
Kr	**4,934.48**	**(4*h*)**	—
Hf	**4,934.45**	**50**	**40**
Sc	**4,934.24**	**8**	—
Mn	**4,934.15**	**5**	**25**
Th	**4,934.088**	**2**	**4**
Ba	**4,934.086**	**400*h***	**400*h***
Th	**4,933.85**	**1**	**8**
Mo	**4,933.73**	**4**	**12**
U	**4,933.66**	**8**	**8**
Fe	**4,933.63**	**70**	**2**
Ta	**4,933.53**	**1**	**5*s***
Mo	**4,933.46**	**2**	**4**
Br	4,928.79	(150)	—
Xe	4,923.15	(500)	—
La	4,921.78	400	500
Xe	4,921.48	(500)	—
La	4,920.97	400	500
Fe	4,920.50	125	500
Xe	4,919.66	(125)	—
Xe	*4,916.51*	*(500)*	—

	λ	I J	I O
Yb	**4,935.50**	**10**	**200**
P	**4,935.55**	**15**	—
Sc	**4,935.73**	**1**	**5**
Ni	**4,935.83**	**1**	**150**
Cr	**4,936.33**	**5**	**200**
Gd	**4,936.34**	**2**	**30**
Cl	**4,936.99**	**(25)**	—
P	4,943.41	(150)	—
Kr	4,945.59	(300)	—
Ne	*4,955.38*	*(150)*	—
Ne	*4,957.03*	*(1000)*	—

N 5,666.64 (300)

	λ	I J	I O		λ	I J	I O
Ag	**5,666.34**	**2**	**5**	**Nb**	**5,666.86**	**2**	**3**
Te	**5,666.26**	**(50)**	—	**N**	**5,667.04**	**(5)**	—
Ar	**5,665.82**	**(5)**	—	**Fe**	**5,667.52**	**3*h***	**5*h***
Nb	**5,665.63**	**30**	**100**	**Xe**	**5,667.56**	**(300)**	—
Th	**5,665.62**	**2**	**10**	**Te**	**5,667.86**	**(15)**	—
Lu	**5,664.89**	**20**	**2**	**V**	**5,668.36**	**50**	**75**
S	**5,664.73**	**(15)**	—	**Sc**	**5,669.03**	**15**	**12**
Nb	**5,664.71**	**30*r***	**100*r***				
Cu	**5,664.47**	**3**	—				
Yt	5,662.92	400	20				
Ra	5,660.81	(1000)	—				
S	5,659.93	(600)	—				
Xe	5,659.38	(150)	—				
Ar	5,659.13	(500)	—				
Ne	5,656.66	(500)	—				
As	5,651.53	200	—				
Ar	5,650.70	(1500)	—				
Te	5,649.30	(250)	—				
Ar	5,648.66	(200)	—				
V	5,646.11	150	150				
Ti	5,644.14	200	150				
S	*5,640.37*	*(500)*	—				
S	*5,639.98*	*(500)*	—				
In	*5,636.75*	*(300)*	—				
Se	*5,623.13*	*(300)*	—	*N**			

N 5,676.02 (100)
5,679.56 (500)

	λ	I J	I O		λ	I J	I O
Nb	**5,677.47**	**3**	**5**				
Hg	**5,677.17**	**(300)**	—	**J**	**5,678.06**	**(80)**	—
P	**5,676.89**	**(20)**	—	**Fe**	**5,679.02**	**4**	**5**
N	*5,676.02*	*(100)*	—	*N*	*5,679.56*	*(500)*	—
Hg	**5,675.86**	**(80)**	—	**Te**	**5,679.71**	**(25)**	—
Tm	**5,675.83**	**100**	**100**				
Ti	**5,675.44**	**125**	**90**	**J**	**5,679.97**	**(2)**	—
Yt	**5,675.26**	**1**	**5**	**Kr**	**5,681.89**	**(400)**	—
Xe	**5,675.15**	**(5)**	—	**Pr**	**5,681.896**	**1**	**7**
Kr	**5,674.52**	**(30*hs*)**	—	**Ar**	**5,681.900**	**(500)**	—

	λ	I	
		J	O
Pr	**5,674.13**	**1**	**5***w*
J	**5,673.66**	**(15)**	—
*P**			
Ar	5,689.64	(200)	—
Ne	5,689.82	(150)	—
Ar	5,689.91	(200)	—
Kr	5,690.35	(200*whs*)	—
V	5,698.52	300	300
Te	*5,708.07*	*(250)*	—
Ne	*5,719.22*	*(500)*	—

Na $^{11}_{22.9898}$

t_0 97.5°C

t_1 892°C

I.	II.	III.	IV.	V.
5.138	47.292	71.650	—	—

λ	*I*		eV
	J	*O*	
I 3,302.32	300*R*	600*R*	3.75
I 3,302.99	150*R*	300*R*	3.75
I 5,682.66	—	80	4.29
I 5,688.22	—	300	4.29
I 5,889.95	1000*R*	9000*R*	2.11
I 5,895.92	500*R*	5000*R*	2.10

Na 3,302.32 *J* 300*R* *O* 600*R*
3,302.99 *J* 150*R* *O* 300*R*

	λ	I *J*	I *O*		λ	I *J*	I *O*
Zn	3,302.59	300	800	Nb	3,302.62	10	1
J	3,302.56	(10)	—	Tb	3,302.64	—	8
Bi	3,302.55	—	150	Pr	3,302.66	2	15
Kr	3,302.54	(10)	—	Zr	3,302.67	6	10
B	3,302.51	10	—	Mo	3,302.72	25	—
U	3,302.492	1	3	Ta	3,302.76	1*h*	50
J	3,302.489	(5)	—	U	3,302.82	4	6
Dy	3,302.47	2	4	Fe	3,302.86	5	1
Tm	3,302.45	80	125	Cr	3,302.88	2	30
Yb	3,302.44	5	7	Ce	3,302.91	—	10*W*
J	3,302.43	(2)	—	Ca	3,302.94	6	4
Ta	3,302.33	1*h*	3	Zn	3,302.941	300*R*	700*R*
Na	*3,302.32*	*300R*	*600R*	*Na*	*3,302.99*	*150R*	*300R*
Kr	3,302.28	(4*h*)	—	J	3,303.08	(25)	—
U	3,302.26	2*h*	1*h*	Ir	3,303.09	—	3
Re	3,302.23	—	30	La	3,303.11	150	400
Cr	3,302.19	1*h*	50*h*	Mo	3,303.113	2	4
Nb	3,302.18	10	5	Re	3,303.21	—	30
Yt	3,302.17	2	5	Ce	3,303.22	—	10
Pd	3,302.13	200*h*	1,000*wh*	Mn	3,303.28	—	40
Pt	3,302.12	—	2*h*	Nb	3,303.32	30	1
Ti	3,302.10	20	8	W	3,303.33	3	7
Sm	3,302.09	4	8	Mo	3,303.34	5	25
Dy	3,302.02	—	2	U	3,303.37	3*h*	4
Eu	3,301.95	2	25	Fe	3,303.47	5	5
Er	3,301.93	2	15	Th	3,303.49	10	5
Ru	3,301.91	8	30	Cu	3,303.52	2	—
Ce	3,301.905	—	10	Mo	3,303.54	20	—
Pr	3,301.898	—	10	Fe	3,303.57	10	70
Ta	3,301.895	3	25	U	3,303.597	12	10
Ar	3,301.87	(10)	—	Dy	3,303.60	—	2
Pt	3,301.861	250*W*	300	Sm	3,303.605	—	5
Al	3,301.86	2	—	Ir	3,303.63	—	4
W	3,301.85	10	1	W	3,303.66	—	6
Ir	3,301.76	—	2	Ce	3,303.67	—	5
U	3,301.754	3	10	Cs	3,303.72	(4)	—
Kr	3,301.75	(5*h*)	—	Re	3,303.75	—	40
Sr	3,301.73	10	100	Ce	3,303.77	—	8
Dy	3,301.72	—	2	U	3,303.84	—	6
Mo	3,301.71	4	4	Co	3,303.88	—	60*r*
Er	3,301.68	—	3	F	3,303.89	(20)	—
Sm	3,301.67	8	15				
Ce	3,301.652	—	3				

	λ	I	
		J	O
Th	3,301.651	1h	6
V	3,301.651	80	—
U	3,301.650	1	6
K	3,301.60	(10)	—
Re	3,301.600	—	50
Ru	3,301.59	40	70
Eu	3,301.57	1	15w
O	3,301.56	(10)	—
Os	3,301.559	50	500R
Ag	3,301.55	8h	—
Ca	3,301.52	6	4
Tm	3,301.51	10	5
Nb	3,301.49	100	1
Te	3,301.48	(5)	—
Fe	3,301.43	—	1
F	3,301.41	(6)	—
Na	3,301.35	(5)	—
Fe	3,298.13	150	200
Cr	3,295.43	200	10
Nb	3,294.36	100	2
Ru	3,294.11	200	60
Tb	3,293.07	100	50
Fe	3,292.59	150	300
Mo	3,292.31	300	10
Cr	3,291.76	200	10
Yb	3,289.85	—	1000
Yb	3,289.37	1000R	500R
Ti	3,287.65	200	40
Pd	3,287.25	25	300w
Fe	3,286.75	400	500
Na	3,285.75	(100)	40
V	3,304.47	125	—
Nb	3,305.61	100	1
Yb	3,305.73	125	30
Fe	3,305.97	300	400
Zr	3,306.28	80	80
Fe	3,306.35	150	200
Rn	3,306.7	(100)	—
Te	3,306.99	(150)	—
Sr	3,307.53	10s	200
V	3,308.48	80	—
Ti	3,308.81	100	35
P	3,308.85	(100w)	—
Fe	3,310.34	80	100
Cr	3,310.65	200	2
Os	3,310.91	30	200
Ta	3,311.16	70w	300w
Cr	3,311.93	125	6
Cr	3,312.18	125	5
Fe	3,314.74	200	200
Ni	3,315.66	20	400R
Na	3,318.03	(20)	6
Ar	3,319.34	(300)	—
Ni	3,320.26	15	400w
Be	3,321.34	30	1000r
V	3,321.54	150	3

Na 5,682.66 *O* 80

	λ	J	O
Er	5,682.53	—	8
Cr	5,682.48	—	6wh
Sm	5,682.42	—	3
Ni	5,682.20	—	60
Ho	5,682.12	—	30
Pr	5,681.90	1	7
Ho	5,681.41	—	20
Cr	5,681.20	—	3wh
bhZr	5,681.1	—	8
bhLa	5,681.1	—	20wh
Ru	5,681.09	—	4
Ce	5,682.78	—	3
Mo	5,682.89	—	20
V	5,683.22	2	50
U	5,683.33	—	2
Gd	5,683.34	—	3h
Cr	5,683.51	—	3
Yb	5,683.60	6	2
Ce	5,683.77	—	2
Gd	5,684.11	—	2
Sc	5,684.19	—	10
Eu	5,684.27	—	125

	λ	I	
		J	O
Eu	**5,681.05**	—	**40**
Re	**5,680.95**	—	**2*h***
Zr	**5,680.90**	—	**50**
Os	**5,680.88**	—	**20**
Gd	**5,680.86**	—	**6**
Sm	**5,680.85**	—	**4**
Pd	**5,680.80**	—	**2**
***bh*F**	**5,680.40**	—	**2**
U	**5,680.38**	—	**2**
Ce	**5,680.27**	—	**2**
Ba	5,680.20	—	60
Ti	5,679.94	—	50
Tm	5,675.83	100	100
Na	5,675.70	—	150*wh*
Ti	5,676.44	125	90
Ho	5,674.70	—	200
Eu	5,673.84	—	200
Se	5,671.80	—	300*w*
Nb	5,671.02	10	200
V	5,670.85	70	150
Na	5,670.18	—	100*Wh*
Pd	5,670.07	—	100
Ce	5,669.97	—	50
Nd	5,669.77	—	40
V	5,668.36	50	75
Re	5,667.90	—	100
Nb	5,665.63	30	100
Ta	5,664.90	—	60
Nb	5,664.71	30*r*	100*r*
Zr	5,664.51	—	50
Ti	5,662.91	—	40
Fe	5,662.52	50	50
Ti	5,662.16	100	100
Sm	5,659.86	—	60
Fe	5,658.83	80	100
V	*5,657.44*	*60*	*150*
Ce	*5,655.13*	—	*40*
Rb	*5,653.74*	—	*200*
Mo	*5,650.13*	*50*	*90*
Ti	*5,648.58*	*60*	*80*
La	*5,648.25*	—	*150*
Rb	*5,648.10*	—	*400*
Co	*5,647.22*	—	*600w*
V	*5,646.11*	*150*	*150*
Ta	*5,645.91*	—	*80*
Eu	*5,645.80*	—	*1000*

	λ	I	
		J	O
Re	**5,684.32**	—	**2**
Si	**5,684.52**	—	**30**
Pt	**5,684.72**	—	**6**
Tm	**5,684.75**	**80**	**40**
Er	**5,684.76**	—	**8**
Tb	**5,685.76**	—	**40**
*N**			

	λ	I J	I O		λ	I J	I O
Ti	*5,644.14*	*200*	*150*				
Nb	*5,642.11*	*20*	*80*				
Ho	*5,640.62*	—	*100*				
Sm	*5,637.27*	—	*80*				
Ru	*5,636.23*	—	*100*				

Na 5,688.22 *O* 300

	λ	I J	I O		λ	I J	I O
Sm	**5,688.20**	—	**2**	**Ta**	**5,688.25**	—	**100***w*
V	**5,687.76**	**30**	**15**	**Eu**	**5,688.34**	—	**2***h*
Sm	**5,687.66**	—	**2**	**Pr**	**5,688.46**	**1**	**8**
Mo	**5,687.63**	—	**3**	**Yb**	**5,688.47**	—	**15**
Pd	**5,687.48**	—	**3**	**Ce**	**5,688.48**	—	**3**
*bh***Zr**	**5,687.40**	—	**20**	**Nd**	**5,688.52**	—	**150**
Ir	**5,687.36**	—	**3**	**Co**	**5,688.59**	—	**10**
Er	**5,687.35**	—	**8**	**Na**	**5,688.61**	—	**10**
Pr	**5,687.19**	—	**4***h*	**Tb**	**5,688.76**	—	**10**
Tb	**5,687.12**	—	**15**	**Mo**	**5,689.14**	**40**	**80**
Sm	**5,686.98**	—	**5**	**Sm**	**5,689.26**	—	**12**
Sc	**5,686.83**	—	**200**	**Ti**	**5,689.47**	**80**	**80**
Sm	**5,686.79**	—	**5**	**Nd**	**5,689.51**	—	**10**
Sm	**5,686.73**	**20**	**50**	**Mo**	**5,689.52**	—	**12**
Gd	**5,686.67**	—	**4**	**Re**	**5,689.74**	—	**2***h*
Nd	**5,686.64**	—	**10**	**Yb**	**5,689.92**	—	**10**
Yb	**5,686.55**	**10***h*	**5**	**Pd**	**5,690.14**	—	**10**
Fe	**5,686.52**	**8**	**10**	**Nd**	**5,690.25**	—	**8**
Pr	**5,686.50**	—	**5***h*	**Sm**	**5,690.36**	—	**2**
Tb	**5,686.48**	—	**15**	**Si**	**5,690.47**	—	**25**
Rh	**5,686.38**	**1**	**100**	Ho	5,691.47	—	200
Ce	**5,685.86**	—	**5**	Ni	5,695.00	—	40
				Pd	5,695.09	—	50
				La	5,696.19	—	50
				Sm	5,696.24	3	40
				Sm	5,696.73	—	40
				Ce	5,696.99	—	40
				V	5,698.52	300	300
				Ru	5,699.05	125	—
				Ce	5,699.23	—	40
				Ta	5,699.24	—	80*w*
				Sc	5,700.23	—	400*R*
				Cu	5,700.24	—	350
				Fe	5,701.56	25	50
				Ti	5,702.67	40	60
				V	5,703.56	60	200
				Ta	5,704.31	—	40
				Mo	5,705.72	40	40
*P**				Nd	5,706.21	—	40

	λ	J	O
Ta	5,706.28	—	50
Ho	5,706.88	—	50
V	5,706.98	—	200
Pr	5,707.61	—	100*w*
Nd	5,708.28	—	60
Si	5,708.44	2*h*	40
Fe	5,709.38	—	100*h*
Ni	5,709.56	1	100*w*
In	5,709.75	—	50*wh*
Sc	5,711.75	—	100
Ti	5,711.88	40	50
Ni	5,711.90	—	50
bhZr	*5,718.1*	—	*150*
Yb	*5,720.01*	*8*	*300*
Os	*5,721.93*	—	*80*
Mo	*5,722.73*	*60*	*80*
Rb	*5,724.45*	—	*600*
V	*5,727.03*	*150*	*150*
bhC	*5,730.0*	—	*150*
Eu	*5,730.89*	—	*300*
V	*5,731.25*	*100*	*250*
Sm	*5,732.95*	—	*100*
Lu	*5,736.55*	*15*	*150*

Na 5,889.95 *J* 1000*R* *O* 9000*R*

	λ	J	O
S	5,889.75	(5)	—
Se	5,889.74	(15)	—
Sm	5,889.69	—	20
Xe	5,889.12	(20)	—
Tb	5,889.06	—	10
Hg	5,888.94	(20)	—
Te	5,888.89	(8)	—
Cr	5,888.78	—	3
Ti	5,888.67	—	15
Rn	5,888.6	(80)	—
Ar	5,888.59	(300)	—
Ta	5,888.49	—	5
Mo	5,888.33	100	150
Th	5,888.28	—	8
Sm	5,888.25	—	2
Th	5,888.15	—	10
Cr	5,888.01	—	20
Nd	5,887.91	—	25
C	5,889.97	60	—
Mo	5,889.98	—	50*h*
Cr	5,889.99	—	12
Hg	5,890.16	(40)	—
In	5,890.26	10	—
W	5,890.33	—	7
Hf	5,890.45	—	8
Co	5,890.48	—	7
Nd	5,890.50	—	5
Sm	5,890.63	—	4
S	5,890.98	(8)	—
Tb	5,891.12	—	10
Eu	5,891.27	—	200
Se	5,891.29	(8)	—
Ru	5,891.30	—	4
Sm	5,891.42	—	15
Te	5,891.43	(15)	—
Nd	5,891.53	—	20
Mo	5,891.56	—	25*h*

	λ	I J	I O
Sc	5,887.76	—	4*wh*
Kr	5,887.68	(3)	—
Ho	5,887.58	—	12
*bh*Sc	5,887.4	—	20
Ir	5,887.36	—	10
Sm	5,887.27	—	3
Ne	5,881.89	(1000)	—
He	5,875.62	(1000)	—
Kr	5,870.92	(3000)	—
Ti	5,866.46	400	300
Ne	*5,852.49*	*(2000)*	—

	λ	I J	I O
W	5,891.61	—	12
C	5,891.65	30	—
Sm	5,891.91	—	2
Pr	5,892.23	1	10
Mo	5,892.29	—	20
Sm	5,892.40	—	5*d*
Ta	5,892.45	—	2
Ho	5,892.56	—	50
U	5,892.63	—	4

*N**

Na 5,895.92 *J* 500*R* *O* 5000*R*

	λ	I J	I O
Pb	5,895.70	2	20*h*
Tm	5,895.63	20	80
Xe	5,895.62	(2*h*)	—
Nd	5,895.58	—	2
Fe	5,895.50	—	4
Eu	5,895.29	—	25
Ta	5,895.20	—	2*h*
Sm	5,895.15	—	6
Sb	5,895.09	(150*wh*)	—
Xe	5,894.99	(100)	—
La	5,894.85	—	25
Sm	5,894.72	—	15*d*
Sc	5,894.63	—	5
Bi	5,894.6	6	—
Kr	5,894.56	(8*wh*)	—
Dy	5,894.47	—	2
Rn	5,894.4	(30)	—
Zn	5,894.35	(30)	3
Pr	5,894.29	—	15*w*
Ir	5,894.06	—	20
J	5,894.05	(60)	—
*bh*Yt	5,893.9	—	10
Mo	5,893.74	—	5
*bh*La	5,893.6	—	60
Th	5,893.50	—	8
Ge	5,893.46	100	—
Nb	5,893.44	3	15
J	5,893.43	(8)	—
Mo	5,893.38	15	70*h*
Xe	5,893.29	(150)	—

*P**

	λ	I J	I O
In	5,896.02	5	—
Sm	5,896.28	—	5
Yb	5,896.61	—	5
Hf	5,896.61	—	2
Te	5,896.65	(25)	—
*bh*La	5,896.7	—	80
Sm	5,896.87	—	3
Yb	5,897.22	100*h*	7
Sm	5,897.38	—	100
Kr	5,897.47	(2*wh*)	—
V	5,897.54	30	—
Gd	5,897.59	—	7
Mo	5,897.86	—	5
Ta	5,897.93	—	2
Cu	5,897.99	25	—
Ne	5,898.41	(20)	—
Xe	5,898.56	(8)	—
Mo	5,898.785	—	8
U	5,898.785	—	8
In	5,903.63	(500)	—
Ar	5,912.08	(500)	—
Ne	5,913.63	(250)	—
Ne	5,918.91	(250)	—
Ne	*5,944.83*	*(500)*	—

Nb 41 92.906

t_0 2,500°C t_1 3,700°C

I.	II.	III.	IV.	V.
6.77	—	24.332	—	~50

λ	I		eV
	J	*O*	
II 3,094.18	1000	100	4.52
II 3,130.79	100	100	4.40
II 3,194.98	300	30	4.21
II 3,225.48	800*R*	150	4.14
I 3,580.27	300	100	3.59
I 4,058.94	400	1000	3.18
I 4,079.73	200	500	3.12
I 4,100.92	200	300	3.07
I 4,123.81	125	200	3.02
5,344.17	200	400	—

Nb 3,094.183 *J* 1000 *O* 100

	λ	*I*	
		J	*O*
Ne	3,094.08	(12)	—
Os	3,094.07	8	30
W	3,094.03	10	8
Ir	3,094.01	10	20
Cu	3,093.99	50	150
Ce	3,093.950	—	4
Cr	3,093.947	25	—
Ru	3,093.90	100	30
Nd	3,093.883	—	6
Fe	3,093.883	30	40
Yb	3,093.879	15	3
Ta	3,093.869	15	50
U	3,093.868	4	3
Dy	3,093.82	1	10
Ti	3,093.813	—	5
Fe	3,093.806	40	50
V	3,093.79	—	30
Yt	3,093.77	5*d*	9*d*
Er	3,093.75	1	6
Cd	3,093.74	2	3
Mo	3,093.68	—	10
Re	3,093.65	—	60
Ce	3,093.61	—	18
Os	3,093.59	15	125
Bi	3,093.58	8	10*w*
W	3,093.51	10	12
Cr	3,093.49	100	1
Rh	3,093.48	3	2
Ca	3,093.46	3	2
Ru	3,093.45	—	4
Yb	3,093.44	2	—
Fe	3,093.36	40	70
V	3,093.11	400*R*	100*R*
Mg	3,092.99	20	125
Al	3,092.84	18	50*R*
Fe	3,092.78	20	50
Na	3,092.73	(200)	50
V	3,092.72	50*r*	100*r*
Al	3,092.71	1000	1000
Ta	3,092.44	15	50
Ru	3,091.87	5	50
Fe	3,091.58	200	300
Mg	3,091.08	10	80
Os	3,090.49	15	80
Os	3,090.30	12	100
Co	3,090.25	1	80
Eu	3,094.184	—	10*w*
V	3,094.20	125*r*	20
Fe	3,094.33	2*h*	2*h*
Nb	3,094.36	—	2
Ir	3,094.37	—	5
Ru	3,094.39	3	50
U	3,094.46	4	3
Xe	3,094.53	(12*h*)	—
Ru	3,094.56	50	6
Ir	3,094.617	—	3
Fe	3,094.622	4	3
Sc	3,094.625	1	4
Eu	3,094.63	1*h*	3
Mo	3,094.66	25	150
Sn	3,094.69	(6)	—
V	3,094.692	—	40
Th	3,094.74	6	10
La	3,094.76	3	2
Zr	3,094.80	—	5
Cs	3,094.82	(4)	—
U	3,094.83	10	10
Fe	3,094.90	15	30
Yb	3,094.92	2	1
Cr	3,094.93	15	—
Ta	3,095.39	18*w*	70*w*
Co	3,095.72	2	60
Cr	3,095.86	3	125
Co	3,096.40	3	60
Ru	3,096.57	60	70
Co	3,096.70	3	60
Mg	3,096.90	25	150
Mn	3,097.06	40	75*w*
Ni	3,097.118	50	200
Nb	3,097.122	100*w*	3*w*
Eu	3,097.46	5	200*w*
Fe	3,098.19	60	70
Co	3,098.20	5	100*r*
Tm	3,098.59	60	80
Ni	3,099.11	50	200
Ru	3,099.28	60	70
Co	3,099.67	—	50
Fe	3,099.90	60	60
Cu	3,099.93	10	60
Nb	3,100.25	20	—
Mn	3,100.302	60	60
Fe	3,100.304	100	100

	λ	I J	I O
Ru	3,090.23	6	50
Os	3,090.08	15	100
Ru	3,089.80	5	60
Co	3,089.59	—	100r
Ru	3,089.14	12	60
Re	3,088.77	—	60
bhB	3,088.6	—	100
Os	3,088.27	12	60
Ir	3,088.04	2	50
Ti	3,088.02	500R	70
Co	3,087.81	—	60
Os	3,087.75	10	50
Co	3,086.78	—	200R
Mo	3,085.61	25	125
Rh	3,083.96	2	150
Fe	3,083.74	500	500
Co	3,082.62	50	150R
Re	3,082.43	—	100w
Al	3,082.15	800	800
Gd	3,082.00	60	100
Cd	3,080.83	100	150
Ni	3,080.75	60	200
Mn	3,079.63	40	125
Os	3,078.38	15	125
Os	3,078.11	15	125
Fe	3,078.02	80	100
Os	3,077.72	30	100
Gd	3,100.51	80	100
Re	3,100.666	—	100
Fe	3,100.666	100	100
Ru	3,100.84	50	70
Mo	3,101.34	10	80
Hf	3,101.40	90	60
Os	3,101.53	20	125
Ni	3,101.55	150	1000R
Mn	3,101.56	50	50
Ni	3,101.88	150	400R
La	3,104.59	50	200
Os	3,104.98	15d	200d
Ni	3,105.47	35	200
Os	3,105.99	20	150
Re	3,108.81	—	125
Os	3,108.98	15	125
Os	3,109.38	20	125

Nb 3,130.786 J 100 O 100

	λ	J	O
Eu	3,130.74	100	100W
U	3,130.73	6	10
S	3,130.71	(15)	—
Cs	3,130.7	(4)	—
Ir	3,130.578	—	3
Ta	3,130.578	35	100W
Ba	3,130.570	3	2
Fe	3,130.567	4	4
Cr	3,130.565	12	1
U	3,130.564	15	—
Ce	3,130.52	—	5
Si	3,130.48	5	—
W	3,130.46	8	10
Be	3,130.42	200	200
Xe	3,130.40	(2wh)	—
P	3,130.38	(30)	—
Rh	3,130.790	2	60
Ti	3,130.800	100	25
Ar	3,130.80	(20)	—
Gd	3,130.81	1	3
Ce	3,130.87	2	30
Ce	3,130.92	—	5
Ho	3,130.99	6	6
Th	3,131.070	10	12
Er	3,131.07	—	3
Be	3,131.072	150	200
Zr	3,131.110	—	7
Os	3,131.115	30	125
Mo	3,131.19	5	—
Cr	3,131.21	6	20
Ta	3,131.22	25h	5h
Tm	3,131.26	500	400

	λ	I J	I O
Ho	3,130.38	4*h*	—
Ti	3,130.376	1	2
U	3,130.37	—	3*h*
Ce	3,130.33	1	30
Ta	3,130.29	1	15
Ir	3,130.285	—	4
Fe	3,130.278	4	5
V	3,130.27	200*r*	50
La	3,130.25	4	3
Sm	3,130.23	—	5
Ce	3,130.20	—	15
Ti	3,130.17	—	2
Dy	3,130.16	2*h*	4
W	3,130.15	4	5
Se	3,130.12	(8)	—
Zr	3,130.063	—	3
Mo	3,130.060	20	—
Ta	3,129.95	8	50
Yt	3,129.93	50	8
Ru	3,129.84	4	60
Ru	3,129.60	1	50
Ta	3,129.55	7	50
Na	3,129.37	(60)	35
Fe	3,129.33	60	100
Ni	3,129.31	—	125
O	3,129.23	15	60
Re	3,128.95	—	100*W*
Cu	3,128.701	15	70
Cr	3,128.699	150	30
V	3,128.69	70	3
Ti	3,128.64	70*wh*	12
V	3,128.28	60	2
Ru	3,127.91	100	10
Ta	3,127.76	100	18*w*
Nb	3,127.53	50	10*w*
Co	3,127.25	—	100
Co	3,126.72	—	70
Ru	3,126.61	50	12
V	3,126.26	100*R*	60
Ne	3,126.19	(150)	—
Fe	3,126.17	70	150
Cu	3,126.11	20	80
Ru	3,125.96	12	70
Hg	3,125.66	150	200
Fe	3,125.65	300	400
Pb	3,125.60	(50)	—
V	3,125.28	200*R*	80
V	3,125.00	50	4
U	3,131.32	6	8
Tb	3,131.35	3	8
Fe	3,131.45	—	2*h*
Os	3,131.48	10	20
Ce	3,131.50	—	2*d*
U	3,131.53	2	4
Hg	3,131.55	300	400
Hg	3,131.83	100	200
Cr	3,132.06	125	25
Fe	3,132.51	40	70
V	3,132.594	20	80*r*
Mo	3,132.594	300*R*	1000*R*
Ta	3,132.64	25	250*w*
Ru	3,132.88	5	60
Cd	3,133.17	300	200
V	3,133.33	200*r*	50
Tm	3,133.89	200	200
Ni	3,134.108	150	1000*R*
Fe	3,134.111	125	200
Hf	3,134.72	125	80
O	3,134.79	(100)	—
Ru	3,134.80	100	10
V	3,134.93	150*r*	30
Fe	3,135.36	100	1
Dy	3,135.37	50	100
Ru	3,135.80	80	10
Ta	3,135.89	100	35
Fe	3,136.50	40	60
V	3,136.51	200	20
Ru	3,136.55	6	60
Cr	3,136.68	50	20
Co	3,136.73	—	60
Co	3,137.33	—	150*r*
Ta	3,137.44	50	3
Co	3,137.45	—	50
Rh	3,137.71	—	100
Co	3,137.75	3	60
Pb	3,137.83	100	—
V	3,138.06	70	—
Pt	*3,139.39*	*80*	*300*
V	*3,139.74*	*150*	*15*
Fe	*3,139.91*	*40*	*70*
Co	*3,139.94*	*10*	*150r*
Cu	*3,140.312*	*12*	*50*
Os	*3,140.314*	*12*	*60*
Fe	*3,140.39*	*80*	*100*

	λ	I J	I O		λ	I J	I O
Cr	3,124.98	125	20	*Se*	*3,141.13*	*(100)*	—
Ta	3,124.97	20	50	*V*	*3,141.48*	*100*	*5*
Os	3,124.94	10*wh*	150*h*	*Pt*	*3,141.66*	*5h*	*150*
Ge	3,124.82	80	200				
Ru	3,124.61	2	50	*Fe*	*3,142.45*	*100*	*125*
Ru	3,124.17	8	60	*V*	*3,142.47*	*100r*	*15*
Rh	3,123.70	2	150	*La*	*3,142.76*	*50*	*150*
				Pd	*3,142.81*	*100*	*300*
				Fe	*3,142.88*	*70*	*80*
V	*3,122.89*	*300r*	*12*	*Ti*	*3,143.76*	*125*	*18*
Au	*3,122.78*	*5*	*500h*	*Fe*	*3,143.99*	*150*	*200*
Mo	*3,122.00*	*150*	*5*	*Fe*	*3,144.49*	*100*	*150*
Xe	*3,121.87*	*(150)*	—	*Nb*	*3,145.40*	*100*	*10*
Rh	*3,121.75*	—	*150*	*Ni*	*3,145.72*	*3*	*200*
Co	*3,121.41*	*6*	*150r*				
Re	*3,121.37*	—	*100*				
V	*3,121.14*	*200r*	*20*				
Fe	*3,120.43*	*80*	*100*				
Cr	*3,120.37*	*150*	*40*				
As	*3,119.60*	*50*	*100*				
Fe	*3,119.49*	*80*	*100*				
Cr	*3,118.65*	*200*	*35*				
V	*3,118.38*	*200R*	*70*				
Os	*3,118.33*	*20*	*150*				
Re	*3,118.19*	—	*200*				
Pb	*3,117.7*	*(100)*	—				
Ti	*3,117.67*	*200*	*15*				
Fe	*3,116.633*	—	*150*				
As	*3,116.63*	*150*	—				
Fe	*3,116.59*	*150*	—				
Hg	*3,116.24*	*(100)*	—				

Nb 3,194.98 *J* 300 *O* 30*

	λ	J	O		λ	J	O
Mo	**3,194.87**	**3**	**3**	**W**	**3,195.08**	**5**	**9**
U	**3,194.85**	**2*h***	—	**Ar**	**3,195.12**	**(5)**	—
Ta	**3,194.84**	**40**	**1*h***	**Ru**	**3,195.15**	**100**	—
Ti	**3,194.76**	**40*wh***	—	**Ba**	**3,195.21**	**4**	—
Ca	**3,194.75**	**2**	**1**	**Fe**	**3,195.231**	**3**	**6**
Ru	**3,194.74**	**1**	**4**	**Mo**	**3,195.235**	**20**	**1**
Au	**3,194.71**	**20**	**25**	**U**	**3,195.25**	**1**	**2**
Tb	**3,194.71**	**8**	**15**	**Th**	**3,195.31**	**6**	**8**
La	**3,194.70**	**2**	—	**Tm**	**3,195.34**	**15**	**30**
Os	**3,194.69**	**10**	**50**	**Os**	**3,195.38**	**12**	**100**
In	**3,194.63**	**(25)**	—	**V**	**3,195.48**	**25*h***	—
Ne	**3,194.61**	**(12)**	—	**Kr**	**3,195.50**	**(2)**	—
Tm	**3,194.6**	**5**	**10**	**Cs**	**3,195.50**	**(4)**	—
Fe	**3,194.597**	**20**	**60**	**U**	**3,195.53**	**3**	**3**

	λ	I J	I O
Ti	3,194.56	8	2
Rh	3,194.55	10h	50
Hf	3,194.47	3	6
In	3,194.43	(10)	—
Fe	3,194.42	70	100
Te	3,194.41	(10)	—
V	3,194.374	1	8
Eu	3,194.37	5h	2
Cd	3,194.36	2	—
In	3,194.360	(10)	—
In	3,194.28	(5)	—
Nb	3,194.27	150w	2w
Ti	3,194.26	7	2
Ar	3,194.25	(10)	—
Yb	3,194.24	3	1
Os	3,194.23	15	125
Hf	3,194.19	40	40
Fe	3,191.66	150	200
Nb	3,191.43	200	3
Nb	3,191.10	300w	100w
Ti	3,190.87	200r	40
V	3,190.68	150R	50
Nb	3,189.28	300r	10w
He	3,187.74	(200)	—
Fe	3,186.74	200	20
V	*3,185.40*	*400R*	*500R*
Nb	*3,184.22*	*150*	*5*
V	*3,183.98*	*400R*	*500R*
Nb	*3,181.40*	*20*	*5*
Nb	*3,180.29*	*200*	*5*
Fe	*3,180.23*	*300*	*300*
Ca	*3,179.33*	*400w*	*100*
Ho	3,195.56	4	—
Yb	3,195.58	6	2
Tb	3,195.60	15	15
Hf	3,195.614	8	6
Yt	3,195.615	50	30
Th	3,195.69	1	8
Ti	3,195.72	20	5
Li	3,195.80	(3)	—
Mo	3,195.96	50	12
Os	3,195.97	10	30
Fe	3,196.08	150	10
Fe	3,196.93	300	500
Nb	3,198.22	50	1
Fe	3,199.52	200	300
Ti	3,199.91	150	200
Fe	3,200.47	150	150
Ti	3,202.54	200	25
F	3,202.74	(200)	—
Nb	3,203.35	50	8
Nb	*3,206.34*	*300*	*10*
Nb	*3,207.33*	*30*	*5*
Nb	*3,208.58*	*100w*	*3w*

Nb 3,225.479 *J* 800*R* *O* 150

	λ	I J	I O
Gd	3,225.44	1	2
Tb	3,225.43	3	8
U	3,225.412	—	8
Ce	3,225.408	—	2
Th	3,225.359	15	12
Cr	3,225.356	20	—
B	3,225.29	—	5
Pr	3,225.20	—	3
Ce	3,225.485	—	2
Zr	3,225.49	2h	5r
Ar	3,225.58	(20)	—
Th	3,225.62	5	5
W	3,225.63	8	10
Ce	3,225.67	3	25
Fe	3,225.79	150	300
Yb	3,225.85	15	3

	λ	I J	I O
Nb	**3,225.19**	**1**	**3**
Sm	**3,225.18**	**2**	**8**
Yt	**3,225.17**	**12h**	**8**
W	**3,225.16**	**4**	**1**
Dy	**3,225.08**	**1**	**10**
Xe	**3,225.08**	**(10)**	—
Hf	**3,225.03**	**3**	**4**
Co	**3,225.023**	—	**3**
Ni	**3,225.020**	**6**	**300**
U	**3,225.016**	**1**	**5**
Pr	**3,225.00**	**1**	**10**
Fe	**3,224.93**	**1**	**3**
Ce	**3,224.83**	—	**10**
Ne	**3,224.82**	**(12)**	
Mn	**3,224.76**	**40**	**75**
Cu	**3,224.56**	**10**	**25**
Ru	**3,224.65**	—	**4**
Co	**3,224.64**	—	**60**
Os	3,223.86	8	100
Ta	3,223.83	50W	200W
Nb	3,223.32	100	10
Nb	3,222.070	30	5
Fe	3,222.069	100	200
Ni	3,221.65	4	300
Ta	3,221.31	15s	70
Ir	3,220.78	30	100
Fe	3,219.81	80	100
Fe	3,219.58	125	200
Pd	3,218.97	8	300
Cr	3,218.69	2wh	80wh
Sm	3,218.60	25	100
Fe	3,217.38	125	200
Nb	3,217.02	200w	2w
K	3,217.017	20h	100R
Mn	3,216.95	75	75
Fe	*3,215.94*	*150*	*300*
Nb	*3,215.59*	*200*	*50*
Fe	*3,214.04*	*200*	*400*
Eu	*3,212.81*	*20*	*200*
Fe	*3,210.83*	*100*	*150*
Fe	*3,210.24*	*100*	*150*
Fe	*3,209.30*	*125*	*200*
Mo	*3,208.83*	*60*	*150r*

	λ	I J	I O
Th	**3,225.86**	**2**	**5**
Ca	**3,225.90**	**10**	**80**
Sm	**3,225.92**	**4**	**10**
Dy	**3,225.96**	**3**	**18**
Na	**3,225.98**	**(20)**	**2**
Ar	**3,226.00**	**(3)**	—
La	**3,226.01**	**2**	**2**
Fe	**3,226.02**	**1**	**2**
Mn	**3,226.03**	**20**	**40**
Ce	**3,226.04**	—	**8**
V	**3,226.11**	**5**	**25**
Th	**3,226.12**	**8**	**10**
Ti	**3,226.128**	**7**	**25**
Ca	**3,226.13**	—	**8**
U	**3,226.17**	**10**	**10**
Ti	**3,226.24**	—	**2**
Ta	**3,226.31**	**1**	**2**
Gd	**3,226.33**	**3**	**3**
Ni	3,226.984	—	100
Co	3,226.985	—	80r
Os	3,227.28	12	125
Ta	3,227.32	10	70
Fe	3,227.75	300	200
Mn	3,228.09	100	100
Fe	3,228.25	80	100
Fe	3,228.90	40	80
Fe	3,229.12	50	80
Os	3,229.21	5	125
Ta	3,229.24	70w	300w
Nb	3,229.56	50	5
Tl	3,229.75	800	2000
Fe	3,230.21	80	100
Pt	3,230.29	6	100
Sm	3,230.54	30	100
Mn	3,230.72	75	75
Ta	3,230.85	18w	200
Fe	3,230.96	200	300
Os	3,231.42	12	150
Os	3,232.05	20	500R
Sb	3,232.50	250wh	150
Os	3,232.54	10	150
Li	3,232.61	500	1000R
Ni	3,232.96	35	300R
Fe	3,233.05	60	100
Mn	3,233.968	—	75wh
Fe	3,233.971	150	300

	λ	I J	I O		λ	I J	I O
				Cr	*3,234.06*	*150*	*10*
				Os	*3,234.20*	*12*	*150*
				Ti	*3,234.52*	*500r*	*100*
				Fe	*3,234.61*	*125*	*200*
				Ni	*3,234.65*	*15*	*300*
				Fe	*3,236.22*	*200*	*300*
				Nb	*3,236.40*	*200*	*10*
				Nb	*3,238.02*	*200*	*20*
				Fe	*3,239.44*	*300*	*400*
				Ta	*3,239.99*	*18w*	*200*
				Tm	*3,241.53*	*125*	*150*
				Pd	*3,242.70*	*600R*	*2000wh*

Nb 3,580.273 *J* 300 *O* 100

	λ	J	O		λ	J	O
Ta	**3,580.267**	**—**	**7**	**Mo**	**3,580.29**	**10**	**—**
Rh	**3,580.262**	**2**	**10**	**Ti**	**3,580.291**	**5**	**15**
Lu	**3,580.26**	**—**	**3**	**P**	**3,580.35**	**(30)**	**—**
U	**3,580.24**	**1**	**6**	**Hf**	**3,580.45**	**3**	**2**
Th	**3,580.235**	**5**	**5**	**Er**	**3,580.48**	**5**	**12*d***
Eu	**3,580.232**	**1*h***	**4**	**Os**	**3,580.53**	**4**	**2**
Re	**3,580.13**	**—**	**80**	**Mo**	**3,580.54**	**10**	**5**
U	**3,580.12**	**2**	**—**	**Ce**	**3,580.56**	**—**	**2**
Mn	**3,580.102**	**—**	**2**	**Eu**	**3,580.57**	**1*h***	**3**
La	**3,580.099**	**3*h***	**2**	**Sm**	**3,580.59**	**—**	**2**
Au	**3,580.08**	**15**	**20**	**Gd**	**3,580.629**	**—**	**5**
Dy	**3,580.04**	**30**	**80**	**Tb**	**3,580.63**	**—**	**8**
Ta	**3,580.02**	**—**	**10**	**Ho**	**3,580.75**	**4**	**—**
Tb	**3,579.98**	**3**	**8**	**Ce**	**3,580.78**	**2**	**10**
Pr	**3,579.95**	**3**	**7**	**V**	**3,580.82**	**50**	**50**
Sm	**3,579.948**	**1**	**2**	**Ir**	**3,580.86**	**3**	**15**
Zr	**3,579.92**	**5*h***	**6*h***	**Ta**	**3,580.89**	**1*h***	**3**
Hf	**3,579.90**	**10**	**15**	**Sm**	**3,580.91**	**6**	**40**
Os	**3,579.87**	**5**	**2**	**U**	**3,580.92**	**—**	**3**
Fe	**3,579.83**	**—**	**1**	**Sc**	**3,580.93**	**40**	**12**
Ru	**3,579.77**	**8**	**3**	**Re**	**3,580.97**	**—**	**40*w***
Ba	**3,579.672**	**8**	**10**	**Mo**	**3,581.00**	**3**	**2**
Sm	**3,579.668**	**3**	**4**	**Fe**	**3,581.19**	**600*r***	**1000*R***
Mn	**3,579.66**	**—**	**3**	**W**	**3,581.24**	**8**	**8**
Fe	**3,579.56**	**1**	**2**	**Ca**	**3,581.29**	**2**	**—**
Gd	**3,579.55**	**5**	**5**	**Cs**	**3,581.30**	**(4)**	**—**
Sm	**3,579.50**	**2**	**4**	**Ar**	**3,581.62**	**(15)**	**—**
Ca	**3,579.450**	**2**	**—**				
Ta	**3,579.448**	**35**	**—**				
Er	**3,579.44**	**—**	**15*d***	Re	3,583.02	—	100*w*
				Rh	3,583.10	125	200
Dy	**3,579.42**	**1**	**2**	Fe	3,583.34	15	50
U	**3,579.364**	**10**	**6**	V	3,583.70	30	60

	λ	I	
		J	O
Eu	**3,579.36**	**1*h***	**4**
Th	**3,579.34**	**8**	**4**
Tb	**3,579.20**	**50**	**50**
Re	**3,579.13**	**—**	**50**
Cr	3,578.69	400*r*	500*R*
Nb	3,578.58	1	15
Dy	3,577.99	50	150
Mn	3,577.88	25	50
V	3,577.87	40	50
Nb	3,577.72	15	10
Ce	3,577.46	12	300
Dy	3,576.87	50	200
Fe	3,576.76	40	80
Ar	3,576.62	(300)	—
Dy	3,576.25	—	300
Fe	3,575.98	25	80
Nb	3,575.85	80	50
Zr	3,575.79	5	100
Co	3,575.36	25	200*r*
Co	3,574.96	25	200
Dy	3,574.16	100	200
Cr	3,573.64	15	60
Fe	3,573.40	20	50
Pb	3,572.73	20	200
Zr	3,572.47	80	60
Ar	3,572.29	(300)	—
Fe	3,572.00	80	100
Ni	3,571.87	40*h*	1000*R*
Fe	3,570.26	15	50
Rh	3,570.18	150	400*r*
Fe	3,570.10	300	300
Os	3,569.77	30	100
Nb	3,569.47	15	20
Co	3,569.38	100	400*R*
Dy	3,568.99	—	100
Fe	3,568.98	35	50
Nb	*3,568.51*	*50*	*10*
Nb	*3,567.99*	*50*	*2*
Lu	*3,567.84*	*7*	*100*
Ar	*3,567.66*	*(300)*	*—*
Nb	*3,567.10*	*30w*	*5*
Ni	*3,566.37*	*100wh*	*2000R*
Fe	*3,565.38*	*300*	*400*
Co	*3,564.95*	*—*	*150w*
Nb	*3,564.08*	*20*	*2*
Nb	*3,563.50*	*30*	*30*

	λ	I	
		J	O
Ta	3,584.21	7*h*	50
Dy	3,584.43	20	50
Fe	3,584.66	60	100
Gd	3,584.96	100	100
Nb	3,584.97	50	30
Dy	3,585.07	100	300
Co	3,585.16	—	60
Fe	3,585.32	100	150
Te	3,585.34	(350)	—
Fe	3,585.71	80	125
Dy	3,585.78	100	150
Fe	3,586.11	80	80
Mn	3,586.54	40	50*h*
Al	3,586.55	(200)	—
Al	3,586.69	(200)	—
Nb	3,586.76	20	2
Al	3,586.80	(200*wh*)	—
Al	3,586.91	(500*h*)	—
Fe	3,586.99	150	200
Rb	3,587.08	40	200
Co	3,587.19	50*h*	200*r*
Os	3,587.31	15	60
Ar	3,588.44	(300)	—
Nb	3,589.107	30	50
Fe	3,589.107	30	70
Ru	3,589.21	100	60
Nb	3,589.36	100	100
Fe	3,589.46	30	50
V	3,589.76	600*R*	80
Nd	3,590.35	300*W*	400*W*
Ce	3,590.60	1	50
Nb	3,591.20	50	2
Dy	3,591.42	100	200
Rb	3,591.59	20	80
V	*3,592.02*	*300R*	*50*
V	*3,593.33*	*300R*	*30*
Cr	*3,593.49*	*400R*	*500R*
Ne	*3,593.53*	*(500)*	*—*
Ne	*3,593.64*	*(250)*	*—*
Nb	*3,593.97*	*50*	*80*
Fe	*3,594.64*	*100*	*125*
Co	*3,594.87*	*—*	*200W*
Dy	*3,595.05*	*100*	*200*
Bi	*3,596.11*	*50*	*150wh*
Rh	*3,596.19*	*50*	*200*
Rh	*3,597.15*	*100*	*200*
Ni	*3,597.70*	*50h*	*1000r*

	λ	I J	I O		λ	I J	I O
Dy	3,563.15	100	200	*Os*	*3,598.11*	*30*	*300*
Tb	3,561.74	200	200	*Nb*	*3,599.28*	*15*	*15*
Co	3,560.89	25	200				
Os	3,560.85	100	150*R*	*Yt*	*3,600.73*	*300*	*100*
Ce	3,560.80	2	300	*Zr*	*3,601.19*	*15*	*400*
				Co	*3,602.08*	*35*	*200*
				Ni	*3,602.28*	*15*	*150*
Os	*3,559.79*	*50*	*150*	*Nb*	*3,602.56*	*30*	*30*
Nb	*3,559.60*	*100*	*2*				
Gd	*3,558.519*	*50*	*100r*	*Eu*	*3,603.20*	*50*	*100w*
Fe	*3,558.518*	*300*	*400*	*Fe*	*3,603.21*	*80*	*150*

Nb 4,058.94 *J* 400 *O* 1000

	λ	J	O		λ	J	O
Mn	4,058.930	60	80	Mg	4,058.96	—	2
Ca	4,058.930	—	3*d*	Zr	4,058.98	—	8
Tm	4,058.92	6	20	U	4,059.02	2	1
Ru	4,058.88	—	10	Eu	4,059.03	—	4
Sm	4,058.87	20	30	Ir	4,059.23	4	30
Tb	4,058.81	—	3*W*	W	4,059.25	4	5
Pr	4,058.78	15	25	Th	4,059.26	5	8
Cr	4,058.77	50	80	P	4,059.27	(100)	—
Fe	4,058.76	10	40	Ce	4,059.32	—	8*h*
Zr	4,058.62	1	9	Gd	4,059.35	3	10*w*
Mo	4,058.61	10	—	Ce	4,059.367	1*h*	3
Co	4,058.60	—	100	Pr	4,059.37	3	4
Si	4,058.49	3	—	Eu	4,059.38	—	25
Ta	4,058.46	5	10	Mn	4,059.39	15	20
Eu	4,058.45	—	2	Tb	4,059.40	1	3
Tb	4,058.44	2	5	Ru	4,059.43	—	7
Dy	4,058.25	—	2*h*	Nb	4,059.506	2	5
Ce	4,058.24	—	18	Er	4,059.509	—	7
Gd	4,058.231	60	100	Mo	4,059.61	10	10
Fe	4,058.229	25	80	Fe	4,059.72	8	15
Se	4,058.20	(20)	—	Er	4,059.78	2	18
Pr	4,058.19	8	6	Ca	4,059.87	3	2*h*
Co	4,058.190	—	100	Gd	4,059.881	20	50
U	4,058.16	4	10	Th	4,059.883	8	8
Ti	4,058.144	6	50	Nd	4,059.96	12	20
Ta	4,058.136	1	2	Eu	4,060.02	—	3
La	4,058.08	—	4	Zr	4,060.08	—	10
Tb	4,058.02	—	2	U	4,060.10	8	4
*bh*Sr	4,058.0	—	3	Ce	4,060.169	1	6
U	4,057.955	6	1	J	4,060.175	(5)	—
Mn	4,057.950	20	80	Hf	4,060.20	4	10
In	4,057.87	10	80	U	4,060.232	1	3
V	4,057.825	2	10	W	4,060.234	9	4
Th	4,057.823	3	2	Ti	4,060.26	25	60
Pb	4,057.820	300*R*	2000*R*	Ho	4,060.30	2	3

	λ	I	
		J	O
Er	**4,057.819**	—	**30**
Zn	**4,057.71**	—	**80**
Tb	**4,057.68**	—	**2**
Sm	**4,057.65**	**3**	**10**
Mg	**4,057.63**	—	**10*w***
Ti	**4,057.62**	**6**	**40**
Mo	**4,057.58**	**4**	**10**
Ce	**4,057.56**	—	**2**
Ho	**4,057.55**	**2*h***	**2**
Xe	**4,057.46**	**(100*wh*)**	—
W	**4,057.45**	**7**	**6**
Mo	**4,057.44**	**4**	**4**
Hf	**4,057.43**	**2**	**3**
Dy	**4,057.40**	**4**	**4**
P	**4,057.39**	**(50)**	—
Ni	**4,057.347**	—	**2**
Fe	**4,057.346**	**3**	**20**
Th	**4,057.34**	**8**	**10**
Ce	**4,057.30**	—	**2**
Co	4,057.20	—	100
In	4,057.07	(100)	—
Kr	4,057.01	(300*hs*)	—
In	4,056.94	(500)	—
Ag	4,055.26	500*R*	800*R*
Ru	4,054.05	100	40
Ho	4,053.92	200	400
Ru	4,051.40	200	125
Gd	4,047.85	50	150
K	4,047.20	200	400
Hg	4,046.56	300	200
Dy	4,045.98	12	150
Ar	4,045.97	(150)	—
Fe	4,045.81	300	400
Ho	4,045.43	80	200
Co	4,045.39	—	400
P	4,044.49	(150*w*)	—
Ar	4,044.42	(1200)	—
K	4,044.14	400	800
La	4,042.91	300	400
Nb	*4,039.53*	*50*	*30*
Nb	*4,037.66*	*20*	—
Mn	*4,033.07*	*20*	*400r*
Ga	*4,032.98*	*500R*	*1000R*
Nb	*4,032.52*	*50*	*30*
La	*4,031.69*	*300*	*400*
Mn	*4,030.75*	*20*	*500r*
S	*4,028.69*	*(200)*	—
F	*4,025.49*	*(300)*	—
F	*4,024.73*	*(500)*	—

	λ	I	
		J	O
Eu	**4,060.301**	—	**2**
Nb	**4,060.31**	**5**	**5**
La	**4,060.32**	**5**	**80**
Tb	**4,060.38**	**1**	**20**
Ce	**4,060.47**	**2**	**10**
Nd	**4,060.56**	**5**	**10**
Zr	**4,060.579**	—	**10**
Dy	**4,060.58**	**2**	**8**
O	**4,060.60**	**(30*h*)**	—
Nb	4,060.79	10*W*	10
Fe	4,062.44	100	120
Cu	4,062.70	20	500*w*
*N**			

Nb 4,079.73 *J* 200 *O* 500

	λ	*I*	
		J	*O*
Ti	**4,079.72**	**7**	**40**
Ce	**4,079.67**	—	**15**
Th	**4,079.61**	**3**	**5**
Ar	**4,079.60**	**(20)**	—
Dy	**4,079.59**	—	**8**
Mn	**4,079.42**	**40**	**50**
Re	**4,079.363**	—	**20**
Ne	**4,079.359**	**(2)**	—
Mo	**4,079.34**	**4**	**4**
Ru	**4,079.277**	**5**	**12**
Ce	**4,079.277**	—	**6**
Dy	**4,079.27**	—	**6**
W	**4,079.26**	**3**	**6**
Mn	**4,079.24**	**40**	**50**
Bi	**4,079.21**	**(40*w*)**	**2*h***
Ta	**4,079.19**	**4**	**10**
La	**4,079.18**	**3**	**25**
Ir	**4,079.17**	—	**3**
Tb	**4,079.15**	—	**4**
Nb	**4,079.13**	**3*w***	**1*h***
Ce	**4,079.02**	**1*h***	**15**
Ga	**4,078.90**	**5**	—
O	**4,078.85**	**(70)**	—
U	**4,078.83**	**8**	**3**
Xe	**4,078.82**	**(100)**	—
Tb	**4,078.78**	—	**5**
V	**4,078.709**	—	**2**
Gd	**4,078.709**	**10**	**20**
Ce	**4,078.61**	—	**5**
Sc	**4,078.58**	**10**	**10**
Ce	**4,078.51**	**1**	**5**
Ti	**4,078.474**	**50**	**125**
Tb	**4,078.47**	**1**	**5**
Gd	**4,078.465**	**10**	**15**
Sb	**4,078.385**	**4**	—
Mo	**4,078.381**	**3**	**5**
Fe	**4,078.36**	**40**	**80**
Nb	**4,078.35**	**3**	**4**
Ce	**4,078.32**	**4**	**15**
Zr	**4,078.31**	—	**10**
Tb	**4,078.26**	—	**3**
Eu	**4,078.23**	—	**3**
Pr	**4,078.16**	**1**	**5**
Zn	**4,078.14**	**(5)**	—
W	**4,078.124**	**6**	**7**
Re	**4,078.124**	—	**10**
Mo	**4,078.07**	**4**	**4**
W	**4,079.78**	**3**	**4**
Pr	**4,079.79**	**35**	**50**
Sm	**4,079.83**	**1**	**20**
Fe	**4,079.84**	**40**	**80**
U	**4,079.85**	**2**	**1**
Cl	**4,079.88**	**(15)**	—
Ir	**4,079.897**	—	**25**
J	**4,079.901**	**(5)**	—
Ce	**4,080.02**	—	**5**
P	**4,080.04**	**(150)**	—
*bh***Sr**	**4,080.1**	—	**2**
Ne	**4,080.15**	**(50)**	—
Cr	**4,080.221**	**1**	**15**
Fe	**4,080.221**	**10**	**60**
Nd	**4,080.23**	**10**	**20**
Dy	**4,080.37**	—	**3**
Ce	**4,080.436**	**2**	**8**
Hf	**4,080.442**	**15**	**15**
Ir	**4,080.479**	—	**5**
Ne	**4,080.48**	**(15*h*)**	—
Eu	**4,080.51**	—	**2**
Gd	**4,080.53**	**10**	**10**
Rh	**4,080.54**	**2**	**2**
Sm	**4,080.547**	**10**	**15**
Cu	**4,080.553**	—	**30*w***
Ce	**4,080.553**	**1**	**6**
Ru	**4,080.60**	**300**	**125**
U	**4,080.61**	**20**	**12**
Ar	**4,080.67**	**(10)**	—
Th	**4,080.71**	**3**	**8**
Eu	**4,080.77**	—	**7**
Gd	**4,080.78**	—	**5*h***
Fe	**4,080.89**	**1**	**5**
Yt	**4,080.926**	**4**	**5**
Ba	**4,080.93**	**3**	**2**
Pr	**4,081.02**	**25**	**50**
Eu	**4,081.04**	**1*h***	**5**
Mo	**4,081.08**	**30**	—
Hg	**4,081.20**	**(10)**	—
Zr	**4,081.21**	**7**	**150**
Ce	**4,081.222**	**8**	**40**
Yt	**4,081.223**	**3**	**6**
Tb	**4,081.23**	**1**	**30**
Er	**4,081.24**	**1**	**10*s***
U	**4,081.26**	**4**	**10**
W	**4,081.30**	**25**	**2**
Ba	**4,081.33**	—	**2**

	λ	I J	I O		λ	I J	I O
Dy	4,077.97	100	150*r*	**Th**	**4,081.39**	**3**	**8**
Hg	4,077.81	150	150	**Ir**	**4,081.397**	**2**	**15**
Sr	4,077.71	500*W*	400*r*	**J**	**4,081.40**	**(3*h*)**	—
La	4,077.34	400	600				
Yb	4,077.27	100	30	**Re**	**4,081.43**	—	**30**
				Mo	**4,081.44**	**50**	**50**
O	4,075.87	(800)	—				
Te	4,073.57	(300)	—				
In	4,072.40	200*wh*	—				
O	4,072.16	(300)	—				
Ar	4,072.01	(150)	—				
Fe	4,071.74	200	300				
J	4,070.75	(150)	—				
Se	4,070.16	(500)	—				
O	4,069.90	(125)	—				
Co	4,068.54	100	150				
Fe	4,067.98	100	150				
La	4,067.39	80	150				
Os	4,066.69	100	100				
Kr	4,065.11	(300)	—				
V	4,065.08	100	2				
Fe	4,063.60	300	400				
Pr	4,062.82	50	150				
*P**				*N**			

Nb 4,100.92 *J* 200 *O* 300

	λ	I J	I O		λ	I J	I O
Tb	**4,100.90**	**2**	**50*d***	**Hf**	**4,100.93**	**1**	**10**
W	**4,100.895**	**6**	**7**	**V**	**4,101.00**	**7**	**5**
Ce	**4,100.889**	**1**	**8**	**La**	**4,101.01**	**2*h***	**2**
Th	**4,100.83**	**18**	**18**	**Te**	**4,101.07**	**(50)**	—
Pr	**4,100.75**	**50**	**200**	**Ho**	**4,101.09**	**40**	**40**
Fe	**4,100.74**	**30**	**80**	**Er**	**4,101.093**	—	**3**
Eu	**4,100.71**	—	**2**	**Cr**	**4,101.16**	**2**	**30**
Er	**4,100.56**	**1**	**15**	**Ru**	**4,101.23**	**2*h***	**7**
Nb	**4,100.40**	**20**	**15**	**Fe**	**4,101.27**	**10**	**40**
Ru	**4,100.37**	**10**	**12**	**Sm**	**4,101.31**	**2*h***	**15*d***
Th	**4,100.35**	**5**	**8**	**U**	**4,101.32**	**2*h***	**2**
Xe	**4,100.34**	**(10)**	—	**Ce**	**4,101.36**	**2**	**1**
Sc	**4,100.33**	**1**	**5**	**Dy**	**4,101.43**	**2**	**5**
Mo	**4,100.32**	**20**	—	**Eu**	**4,101.44**	—	**2**
Ne	**4,100.30**	**(5*h*)**	—	**Nd**	**4,101.46**	**6**	**8**
Os	**4,100.300**	**3**	**60**	**Ce**	**4,101.55**	**3**	**2**

	λ	I J	I O		λ	I J	I O
Gd	**4,100.26**	**5**	**15**	**Tb**	**4,101.65**	**—**	**20***w*
Eu	**4,100.25**	**—**	**2**	**Zn**	**4,101.66**	**—**	**5**
Nd	**4,100.24**	**8**	**10**	**Ir**	**4,101.67**	**—**	**3**
Pr	**4,100.22**	**10**	**15**	**Fe**	**4,101.679**	**2**	**5**
Ho	**4,100.20**	**—**	**4**	**Nd**	**4,101.682**	**—**	**8**
Fe	**4,100.17**	**1**	**10**	**H**	**4,101.73**	**(100)**	**—**
Ir	**4,100.15**	**3***h*	**100**	**Ru**	**4,101.74**	**60**	**20**
He	**4,100.00**	**(2)**	**—**	**Ce**	**4,101.772**	**6**	**35**
Sm	**4,099.95**	**—**	**4**	**In**	**4,101.773**	**1000***R*	**2000***R*
N	**4,099.94**	**(150)**	**—**	**W**	**4,101.85**	**8**	**7**
U	**4,099.93**	**5**	**5**	**U**	**4,101.90**	**1**	**18**
Dy	**4,099.89**	**2**	**6**	**Dy**	**4,101.95**	**2**	**8**
Pr	**4,099.88**	**2**	**4**	**Ce**	**4,102.07**	**—**	**2**
V	**4,099.80**	**12**	**25**	**Mo**	**4,102.15**	**25**	**30**
Ce	**4,099.75**	**2**	**12**	**Re**	**4,102.159**	**—**	**4***h*
Eu	**4,099.72**	**5**	**12***d*	**V**	**4,102.159**	**15**	**30**
Kr	**4,099.71**	**(3)**	**—**	**N**	**4,102.18**	**(5)**	**—**
La	**4,099.54**	**100**	**100**	**U**	**4,102.21**	**4**	**1**
Ir	**4,099.48**	**—**	**4**	**Zr**	**4,102.281**	**—**	**10**
Ar	**4,099.47**	**(5)**	**—**	**Ru**	**4,102.285**	**10**	**15**
Tb	**4,099.46**	**1**	**25***w*	*bh***Sr**	**4,102.3**	**—**	**2**
S	**4,099.44**	**(8)**	**—**	**Ce**	**4,102.36**	**2**	**18**
Ce	**4,099.39**	**—**	**6**	**Yt**	**4,102.38**	**30**	**150**
Zr	**4,099.31**	**—**	**10**	**Ho**	**4,102.40**	**3**	**—**
Pd	**4,099.27**	**4***h*	**—**	**Tb**	**4,102.52**	**2**	**25***w*
U	**4,099.269**	**1**	**20***r*	**Br**	**4,102.53**	**(10)**	**—**
Dy	**4,099.21**	**—**	**4**	**Nd**	**4,102.56**	**5**	**10**
Gd	4,098.91	100	100	Mn	4,102.96	20	100
Kr	4,098.72	(250)	—	F	4,103.08	(150)	—
Ru	4,097.79	125	25	F	4,103.52	(300)	—
Tm	4,094,18	30	300	Ho	4,103.84	400	400
Co	4,092.39	15	600*W*	Ar	4,103.91	(200)	—
Ac	4,088.37	100	—	Fe	4,104.13	25	100
Kr	4,088.33	(500)	—	Nb	4,104.17	30	2
Gd	4,087.71	100	80	O	4,105.00	(125)	—
Pd	4,087.34	100	500	Tm	4,105.84	30	300
La	4,086.71	500	500	Sm	4,106.60	5	100
Co	4,086.31	15	400				
*P**				*N**			

Nb 4,123.810 *J* 125 *O* 200

	λ	J	O		λ	J	O
Tb	**4,123.78**	**—**	**10**	**Ru**	**4,123.813**	**10**	**20**
Ni	**4,123.778**	**—**	**2**	**Ce**	**4,123.87**	**6**	**25**

	λ	I	
		J	O
Fe	4,123.74	20	80
As	4,123.71	5	—
U	4,123.66	6	5
Ta	4,123.657	1	5
Mo	4,123.65	25	5
Pd	4,123.62	2	10
Gd	4,123.59	—	3
Ti	4,123.572	10	40
Th	4,123.569	5	5
V	4,123.566	12*r*	30*r*
Mn	4,123.54	5	15
Hf	4,123.53	10	1
Re	4,123.50	—	2
Ce	4,123.49	5	20
Cr	4,123.39	15	35
U	4,123.310	2*h*	5
Ti	4,123.306	6	25
Cu	4,123.29	1*h*	30*w*
Mn	4,123.28	5	12
Er	4,123.24	—	5
La	4,123.23	500	500
V	4,123.19	3	6
Ta	4,123.17	4*w*	50*r*
Ti	4,123.143	—	10
U	4,123.14	—	3
Na	4,123.07	(15)	10
Ru	4,123.06	35	25
W	4,123.055	8	7
Er	4,123.053	—	8
Gd	4,123.010	10	8
Sm	4,123.06	2	5
Nd	4,123.00	8	15
Pr	4,122.98	2	8
Th	4,122.97	5	3
Eu	4,122.94	1	3
Yb	4,122.87	20	10
Ce	4,122.86	—	3
Nb	4,122.81	10	5
In	4,122.791	(30*h*)	—
Ru	4,122.787	—	6
Eu	4,122.786	—	3
Re	4,122.76	—	10
Fe	4,122.51	30	70
Sm	4,122.50	10	5
Lu	4,122.49	2	15
Tb	4,122.47	—	8
Mo	4,122.39	50	15
U	4,122.35	4	15

	λ	I	
		J	O
Nd	4,123.88	20	40
Sm	4,123.95	20	10
U	4,123.96	1	20*r*
Cl	4,124.00	(12)	—
Eu	4,124.02	—	2
Pr	4,124.06	2	5
Hg	4,124.071	(30)	—
V	4,124.072	5	8
N	4,124.10	(2)	—
Nd	4,124.11	—	2
Tb	4,124.27	—	3*h*
Pr	4,124.35	2	5
Ir	4,124.540	—	4
Mo	4,124.545	25	30
Eu	4,124.55	2	8
Os	4,124.60	9	30
Dy	4,124.63	8	15
Th	4,124.65	3	3
*bh*Sr	4,124.7	—	2
U	4,124.72	25	30
Lu	4,124.73	10	200
Ir	4,124.74	—	8
Ce	4,124.78	5	25
Re	4,124.79	—	4
Er	4,124.80	—	2
Rh	4,124.90	3	5
Eu	4,124.910	2	8
Yt	4,124.910	18	7
Hg	4,124.91	(5)	—
Nd	4,125.05	4	4*d*
Pr	4,125.06	5	12
Hf	4,125.10	6	2
U	4,125.13	1	15
W	4,125.18	12	9
Tb	4,125.21	1*w*	10*w*
Sm	4,125.24	1	5
Nb	4,125.25	20	10
Tb	4,125.42	—	2
Ce	4,125.43	2	6
Nd	4,125.50	1*h*	8
Fe	4,126.19	60	80
Cr	4,126.52	50	100
Se	4,126.57	(150)	—
Ho	4,127.16	60	150
P	4,127.49	(70)	—
Fe	4,127.61	80	100
Yt	4,128.30	30	150

	λ	I J	I O
Pr	**4,122.31**	**2**	**8**
Co	**4,122.27**	**—**	**10*h***
Ce	**4,122.24**	**3**	**3**
Tb	**4,122.21**	**—**	**3**
Ti	**4,122.166**	**10**	**40**
U	**4,122.166**	**8**	**10**
Cr	**4,122.16**	**5**	**30**
Ce	**4,122.14**	**2**	**—**
Hg	**4,122.12**	**(20)**	**—**
Fe	4,121.81	40	100
Rh	4,121.68	50	150
Bi	4,121.53	50	125*wh*
Co	4,121.32	25	1000*R*
N	4,119.72	20	1*h*
Rh	4,119.68	25	1000
Nb	4,119.28	200	2
O	4,119.22	(300)	—
Co	4,118.77	—	1000*R*
Pt	4,118.69	10	400
Fe	4,118.55	100	200
Pr	4,118.48	50*d*	250*d*
Xe	4,116.11	(80)	—
Ir	4,115.78	30	100
Rn	4,114.56	(80)	—
Pr	4,113.89	70*w*	30*w*
Ru	4,112.74	200	125
Os	4,112.02	9	150
V	4,111.78	100*WR*	100*WR*
Co	4,110.53	—	600
N	4,109.98	(1000)	—
Fe	4,109.81	100	120
Kr	4,109.23	(100*hs*)	—
P	4,109.19	(70)	—
F	4,109.17	(100)	—
Se	4,108.83	(800)	—
Ho	4,108.63	40	100
Fe	4,107.49	100	120
Rh	4,128.87	150	300
Se	4,129.15	(200)	—
Ta	4,129.38	40	200
Nb	4,129.43	20	15
Eu	4,129.74	50*R*	150*R*
Nb	4,129.93	30	15
Gd	4,130.38	10	200
Ne	4,131.05	(70)	—
Ar	4,131.73	(80)	—
Fe	4,132.06	200	300
Li	4,132.29	—	400*wh*
Cl	4,132.48	(200)	—
Se	4,132.76	(200)	—
O	4,132.82	(100)	—
Re	4,133.42	—	200
Fe	4,134.68	100	150
Rh	4,135.27	150	300
Os	4,135.78	50	200
Se	4,136.28	(100)	—
Re	4,136.45	—	150*w*
Fe	4,137.00	80	100
Nb	4,137.09	60	100
Os	4,137.84	3	100
Kr	4,139.11	(100*wh*)	—
Nb	4,139.71	50	50
Hg	4,140.38	(200)	—
Pd	4,140.83	—	100*r*
La	*4,141.74*	*200*	*200*
S	*4,142.29*	*(150)*	*—*
Pr	*4,143.14*	*50*	*200*
Fe	*4,143.42*	*100*	*200*
Mo	*4,143.55*	*100*	*100*
Fe	*4,143.87*	*250*	*400*
Ru	*4,144.16*	*200*	*150*
S	*4,145.10*	*(250)*	*—*
Kr	*4,145.12*	*(250)*	*—*
Ru	*4,145.74*	*150*	*125*
Fe	*4,147.67*	*100*	*200*
Nb	*4,150.12*	*20*	*15*
N	*4,151.46*	*(1000)*	*—*
La	*4,151.95*	*300*	*200*
Nb	*4,152.58*	*300*	*100*
S	*4,153.10*	*(600)*	*—*
O	*4,153.31*	*(200)*	*—*
Ar	*4,158.59*	*(1200)*	*—*

Nb 5,344.17 *J* 200 *O* 400

	λ	*I* *J*	*I* *O*
Er	**5,343.93**	—	**30**
Pr	**5,343.86**	—	**8**
Eu	**5,343.76**	—	**2**
Nd	**5,343.65**	—	**3**
Zr	**5,343.60**	—	**2*h***
Th	**5,343.585**	—	**12**
Nb	**5,343.580**	**1*h***	**5*h***
Fe	**5,343.47**	—	**12*h***
Co	**5,343.39**	—	**600*w***
Ne	**5,343.28**	**(600)**	—
Gd	**5,343.02**	—	**10**
K	**5,342.97**	—	**30**
Sc	**5,342.96**	—	**5**
Sm	**5,342.77**	—	**3**
Co	**5,342.71**	—	**800*w***
Pr	**5,342.54**	—	**3**
Hg	**5,342.40**	**(12)**	—
Tm	**5,342.38**	—	**3**
Ta	**5,342.25**	—	**80**
Pr	**5,341.92**	—	**5**
Co	5,341.33	—	300*w*
Ne	5,341.09	(1000)	—
Mn	5,341.06	100	200
Fe	5,341.03	15	200
La	5,340.67	100	80
Fe	5,339.94	30	200
Xe	5,339.38	(500)	—
J	5,338.19	(300)	—
Yb	5,335.16	400	150
Nb	5,334.87	10	50
Kr	5,333.41	(500*h*)	—
Co	5,332.67	—	200*w*
Br	5,332.04	(100)	—
As	5,331.54	200	—
Co	5,331.47	80	500*w*
Ne	5,330.78	(600)	—
O	5,330.66	(500)	—
O	5,329.59	(150)	—
O	5,328.98	(100)	—
Fe	5,328.05	100	400
Co	5,325.28	—	300*w*
Fe	5,324.18	70	400
Ra	*5,320.29*	*(250)*	—
Nb	*5,318.60*	*12*	*100*
Co	*5,316.78*	—	*300w*
Ar	**5,344.28**	**(5)**	—
Mn	**5,344.47**	—	**12**
Er	**5,344.51**	—	**30**
Co	**5,344.57**	—	**10**
P	**5,344.72**	**(150*w*)**	—
Cr	**5,344.76**	—	**15**
Er	**5,344.93**	—	**8**
Ce	**5,345.097**	—	**5**
Os	**5,345.102**	—	**5**
Pd	**5,345.105**	—	**10**
Gd	**5,345.14**	—	**10**
J	**5,345.15**	**(300)**	—
Re	**5,345.22**	—	**2**
Th	**5,345.31**	**1**	**8**
Br	**5,345.43**	**(80)**	—
*bh***F**	**5,345.6**	—	**20**
S	**5,345.66**	**(25)**	—
Yb	**5,345.67**	**100**	**20**
Gd	**5,345.69**	—	**10**
Nd	**5,345.71**	—	**6**
Cr	**5,345.807**	**25**	**300*R***
P	**5,345.81**	**(50)**	—
Ar	**5,345.81**	**(20)**	—
Yb	**5,345.83**	**50**	**10**
Er	**5,346.02**	—	**8**
Os	**5,346.03**	—	**8**
Tb	**5,346.14**	—	**10**
Yb	5,347.20	200	40
Ar	5,347.41	(200)	—
Ne	5,349.21	(150)	—
Tl	5,350.46	2000*R*	5000*R*
Nb	5,350.74	50	150
Co	5,352.05	—	500*w*
Yb	5,352.96	250	100
Co	5,353.48	—	500*w*
Sb	5,354.24	(200)	—
Rh	5,354.40	5	300
Eu	5,355.08	—	200
Ne	5,355.18	(150)	—
Ne	5,355.42	(150)	—
Eu	5,357.61	—	1000
Cs	5,358.53	(500)	—
Yb	5,358.64	100	15
Co	5,359.18	—	300*w*
Ne	5,360.01	(150)	—
Eu	5,361.59	—	300

	λ	I J	I O
Xe	*5,313.87*	*(500)*	—
Co	*5,312.66*	—	*400w*
Ar	*5,309.52*	*(200)*	—
Kr	*5,308.66*	*(200)*	—
Se	*5,305.35*	*(500)*	—
Co	5,362.77	—	500*w*
Co	*5,369.58*	—	*500w*
Fe	*5,371.49*	—	*700*
Xe	*5,372.39*	(200)	—
Ar	*5.373.49*	(500)	—
La	*5,377.09*	200	30
C	*5,380.24*	(300)	—
Fe	*5,383.37*	40	*400h*

Ne $^{10}_{20.183}$

t_0 −248.7°C t_1 −246.0°C

I.	II.	III.	IV.	V.
21.559	40.958	63.427	96.897	126.43

λ	*I*	eV
I 3,369.81	(500)	20.29
I 3,369.91	(700)	20.29
I 3,520.47	(1000)	20.36
I 4,537.75	(1000)	21.10
I 4,884.92	(1000)	20.65
I 4,884.92	(1000)	21.10
I 4,957.03	(1000)	21.10
I 5,400.56	(2000)	18.95
I 5,852.49	(2000)	18.96
I 6,402.25	(2000)	18.56

Ne 2,647.42 (150)

	λ	I J	I O
U	**2,647.363**	**2**	**3**
La	**2,647.36**	**4**	**—**
Ru	**2,647.31**	**5**	**50**
Hf	**2,647.292**	**125**	**40**
Ar	**2,647.29**	**(10)**	**—**
Ba	**2,647.289**	**40**	**10**
Yb	**2,647.25**	**4**	**—**
W	**2,647.10**	**8**	**10**
Ni	**2,647.06**	**500*wh***	**—**
U	**2,647.02**	**2*h***	**3**
Pt	2,646.89	100	1,000*h*
Mo	2,646.49	100	25
Nb	2,646.26	200*wh*	8
Ti	2,646.11	200*wh*	
Ru	2,646.02	150	20
V	2,645.84	100	15
Ne	2,645.70	(35)	—
Ne	2,645.51	(50)	—
Ru	2,644.61	100	8
V	2,644.35	100*h*	12
Fe	2,644.00	150	150
Ra	2,643.73	(125)	—
Ru	*2,642.79*	*150*	*—*
Nb	*2,642.24*	*300*	*5*
Ti	*2,642.15*	*150wh*	*—*
Ne	*2,639.97*	*(15)*	*—*
Eu	*2,638.76*	*200*	*300*
Yb	**2,647.47**	**7**	**2**
Ta	**2,647.472**	**10**	**200**
Nb	**2,647.50**	**5**	**15*h***
U	**2,647.53**	**2**	**6**
Fe	**2,647.56**	**70**	**100**
Mn	**2,647.61**	**15**	**—**
V	**2,647.71**	**10*h***	**50**
Os	**2,647.73**	**6**	**25**
W	**2,647.74**	**20**	**10**
Ne	**2,647.76**	**(8)**	**—**
Ne	2,648.21	(15)	—
Ne	2,648.56	(25)	—
Ni	2,648.72	80	—
Ru	2,648.78	150	30
V	2,649.36	100*h*	2
Fe	2,649.46	70	—
Pd	2,649.47	200	—
Pb	2,650.4	80	100
Fe	2,650.49	150*h*	—
Pt	2,650.86	100	700
Mn	2,650.99	150	—
Ne	2,651.01	(50)	—
Nb	2,651.12	200	3
Ta	2,651.22	80	—
Mo	*2,653.35*	*150*	*25*
Yb	*2,653.75*	*200*	*50*
Ni	*2,655.47*	*400wh*	*—*
Ni	*2,655.91*	*500wh*	*—*
Nb	*2,656.08*	*200*	*8*

Ne 2,974.71 (250)

	λ	I J	I O
Yt	**2,974.59**	**10**	**12**
Nb	**2,974.55**	**1**	**2**
Er	**2,974.48**	**1**	**4**
U	**2,974.46**	**1**	**3**
W	**2,974.39**	**12**	**8**
Ru	**2,974.33**	**2**	**30**
Tm	**2,974.29**	**30**	**10**
Na	**2,974.24**	**(5)**	**—**
Hg	**2,974.12**	**(10)**	**—**
Ir	**2,974.101**	**5**	**10**
Nb	**2,974.098**	**200**	**5**
Nb	**2,974.72**	**10**	**—**
Ga	**2,974.77**	**(8)**	**—**
Fe	**2,974.78**	**6**	**10**
Ru	**2,974.859**	**5**	**—**
Xe	**2,974.86**	**(20*wh*)**	**—**
U	**2,974.91**	**4**	**5**
Ti	**2,974.93**	**5**	**15**
Ir	**2,974.95**	**10**	**25**
Na	**2,974.99**	**(60)**	**15**
V	**2,975.06**	**2**	**15**
W	**2,975.08**	**10**	**—**

	λ	I	
		J	O
Hf	**2,974.09**	**6**	**3**
U	**2,974.08**	**2**	**2**
Kr	**2,974.04**	**(25*h*)**	—
Fe	2,973.24	400*R*	500*R*
Fe	2,973.13	400*R*	500*R*
Yb	2,970.56	150	150
Fe	2,970.10	200	400
Te	2,967.21	(300)	—
Fe	*2,966.90*	*600r*	*1000R*
Ta	*2,965.13*	*800*	*40*
Cu	*2,961.16*	*300*	*350*
Cs	**2,975.13**	**(2)**	—
Hg	**2,975.19**	**(15)**	—
U	**2,975.22**	**6**	**8**
Os	**2,975.34**	**8**	**15**
Ne	**2,975.52**	**(35)**	—
Ta	2,976.26	150	2*wh*
Ru	2,976.59	200	60
Nb	2,977.68	300	1
Ta	2,978.18	150	—
Ru	2,978.64	150	50
Xe	2,979.32	(200)	—
Ne	2,979.81	(50)	—
Cd	2,980.63	500	1000*R*
Ne	2,980.64	(40)	—
Pd	2,980.65	200*R*	—
Sb	2,980.96	(125*hd*)	—
Fe	2,981.45	200	300
Ne	*2,982.66*	*(250)*	—
Fe	*2,983.57*	*400r*	*1000R*
Fe	*2,984.83*	*400*	*200r*
Fe	*2,985.55*	*300*	*80*

Ne 3,369.81 (500)
3,369.91 (700)

	λ	J	O
Nb	**3,369.83**	**3**	**10**
W	**3,369.82**	**4**	**1**
Ne	**3,369.81**	**(500)**	—
W	**3,369.69**	**4**	**5**
Rh	**3,369.68**	**5**	**25**
Ag	**3,369.673**	**5*h***	—
Ru	**3,369.669**	**2**	**30**
Dy	**3,369.64**	**2**	**2**
Tm	**3,369.64**	**20**	**10**
Er	**3,369.620**	**1**	**12**
Cd	**3,369.616**	**2**	**2**
Nd	**3,369.58**	**2*d***	**4*d***
Ni	**3,369.57**	**100**	**500*R***
Fe	**3,369.55**	**200**	**300**
Sm	**3,369.45**	**5**	**10**
Ru	**3,369.282**	**60**	**12**
Ta	**3,369.280**	**5**	**18**
Dy	**3,369.27**	**2**	**2**
Zr	**3,369.26**	**5**	**7**
Ta	**3,369.84**	**1**	**3**
Ne	**3,369.91**	**(700)**	—
Mo	**3,369.94**	**20**	**5**
U	**3,369.96**	**1*h***	**6**
Ru	**3,370.05**	**10**	**3**
U	**3,370.13**	**4**	**6**
Tb	**3,370.14**	**3**	**15**
Nb	**3,370.16**	**80**	—
Os	**3,370.20**	**10**	**50**
Cr	**3,370.23**	**1*h***	**30**
W	**3,370.24**	**10*s***	**5**
Pr	**3,370.30**	**5*d***	**10*d***
Co	**3,370.33**	**2**	**80**
Th	**3,370.39**	**2*d***	**2*d***
Ti	**3,370.44**	**15**	**80**
U	**3,370.45**	**2*h***	**2**
Cu	**3,370.454**	**15**	—
C	**3,370.50**	**(20)**	—
W	**3,370.519**	**8*s***	**9**

	λ	I J	I O		λ	I J	I O
Mo	**3,369.25**	**15**	**20**	**Mo**	**3,370.520**	**25**	**2**
Ti	**3,369.21**	**25**	**6**	**Eu**	**3,370.52**	**3*d***	**3*d***
Nb	**3,369.16**	**50**	**5**	**Sm**	**3,370.586**	**3**	**9**
Tl	**3,369.15**	**(40)**	—	**Os**	**3,370.588**	**30**	**300*R***
Th	**3,369.10**	**2**	**1**	**Er**	**3,370.59**	**3**	**15**
Eu	**3,369.052**	**5**	**40**	**Nb**	**3,370.61**	**30**	**3**
Nd	**3,369.050**	**8**	**10**	**Ir**	**3,370.63**	**2*h***	**8**
Sm	**3,369.04**	**3**	**8**	**In**	**3,370.64**	**3**	—
Pr	**3,368.96**	**1**	**5**	**Hf**	**3,370.69**	**3*h***	**3**
Er	**3,368.953**	**1**	**10**	**W**	**3,370.70**	**4**	**5**
Sc	**3,368.946**	**20**	**50**	**Rn**	**3,370.70**	**(2)**	—
Nd	**3,368.88**	**2**	**6**	**Fe**	**3,370.786**	**200**	**300**
U	**3,368.80**	**6**	**5**	**Th**	**3,370.788**	**1**	**4**
				Ho	**3,370.86**	**8**	**6**
				Dy	**3,370.86**	**4**	**7**
Ne	3,367.20	(25)	—	**Cd**	**3,370.91**	**2**	—
Ti	3,361.21	600*R*	100				
				Ti	3,372.80	400*R*	80
Ne	*3,355.05*	*(40)*	—	Pd	3,373.00	500*wh*	800*r*
				Ar	3,373.48	(300)	—
				Ne	3,375.65	(50)	—
				Ag	*3,382.89*	*700R*	*1000R*
				Ne	*3,388.46*	*(25)*	—

Ne 3,417.90 (500)

	λ	J	O		λ	J	O
Nb	**3,417.86**	**5**	**5**	**Ne**	**3,418.01**	**(50)**	—
Fe	**3,417.84**	**100**	**150**	**Cs**	**3,418.11**	**(6)**	—
Co	**3,417.79**	**2**	**30**	**Sm**	**3,418.13**	**6**	**20**
Th	**3,417.73**	**4**	**3**	**Dy**	**3,418.14**	**5**	**15**
Tb	**3,417.72**	**3**	**8**	**Fe**	**3,418.18**	**1*h***	**4**
Ca	**3,417.72**	**4**	**2**	**Ta**	**3,418.31**	**2*h***	**8**
Ne	**3,417.71**	**(18)**	—	**Mo**	**3,418.34**	**1**	**4**
Ar	**3,417.68**	**(3)**	—	**Sr**	**3,418.36**	**2**	**1**
Er	**3,417.64**	**4**	**10**	**Xe**	**3,418.37**	**(2)**	—
Nd	**3,417.53**	**6**	**8**	**Yb**	**3,418.39**	**5**	**3**
Mo	**3,417.51**	**4**	**5**	**U**	**3,418.39**	**2**	**10**
Ce	**3,417.45**	**5**	**30**	**Pr**	**3,418.47**	**3**	**30**
Eu	**3,417.42**	**2**	**4**	**Sm**	**3,418.510**	**10**	**50**
Cd	**3,417.40**	**15**	**10**	**Ar**	**3,418.51**	**(3)**	—
Ru	**3,417.35**	**70**	**1**	**Fe**	**3,418.512**	**100**	**150**
Hf	**3,417.34**	**2**	**10**	**V**	**3,418.515**	**20**	**25**
U	**3,417.30**	**1**	**3**	**Mo**	**3,418.52**	**12**	**10**
Er	**3,417.29**	**3**	**10**	**Sc**	**3,418.53**	**2*h***	**5**
F	**3,417.21**	**(10)**	—	**Er**	**3,418.731**	**2**	**9**
Nb	**3,417.18**	**10**	—	**Gd**	**3,418.735**	**25**	**50**

	λ	I			λ	I	
		J	O			J	O
Dy	3,417.14	5	15	Th	3,418.78	8	6
Ga	3,417.12	5	—	Eu	3,418.81	2	8
Th	3,417.12	3	3	U	3,418.85	3	3
V	3,417.06	25	35	Ce	3,418.930	2	20
Ta	3,417.03	15	3	Th	3,418.931	3*h*	2*d*
F	3,417.02	(20)	—	Tb	3,418.95	8	15
Mo	3,417.00	6	—	Mo	3,418.96	10*h*	8
Ti	3,416.957	50	7	J	3,419.04	(5)	—
Gd	3,416.957	30	50	Zr	3,419.11	4	6
Yb	3,416.89	15	7	Th	3,419.15	10	8
Nc	3,416.87	(12)	—				
				Pd	3,421.24	1000*R*	2000*R*
				Ne	3,423.91	(50)	—
Fe	3,413.13	300	400	Tm	3,425.08	300	200
Fe	3,407.46	400	400	Nb	3,425.42	300	30*r*
Ne	*3,406.88*	*(18)*	—	*Ne*	*3,428.76*	*(18)*	—
Pd	*3,404.58*	*1000R*	*2000R*	*Pd*	*3,433.45*	*500h*	*1000h*
Cd	*3,403.65*	*500h*	*800*				

Ne 3,520.472 (1000)

	λ	J	O		λ	J	O
Yb	3,520.29	70	10	Sb	3,520.474	(125)	—
Ti	3,520.25	18	10	Ta	3,520.49	18	25*wh*
Mo	3,520.20	20	1	Ce	3,520.52	2	30
Ho	3,520.16	10*h*	8	V	3,520.54	20	1
Eu	3,520.14	4	4	Tl	3,520.60	3	—
Ru	3,520.13	40	60	La	3,520.711	4	4
Ce	3,530.10	1	2	Nb	3,520.715	5	3
Sb	3,530.07	2*h*	4	Th	3,520.72	2	5
Nb	3,520.05	20	20	Tb	3,520.79	8	15
Er	3,520.034	4	12	U	3,520.793	10	6
Cu	3,520.031	10	30	Fe	3,520.85	4	10
V	3,520.02	50	5	Zr	3,520.87	4	9
Ar	3,520.00	(15)	—	Ce	3,520.98	1	5
Os	3,519.998	20	30	Th	3,521.06	1*d*	3*d*
U	3,519.96	12	6	Te	3,521.07	(15)	—
Ho	3,519.92	10*h*	10	Eu	3,521.09	4*h*	30
Co	3,519.82	25*h*	2*h*				
Nd	3,519.77	2	10*d*	Ce	3,521.128	1	2
Dy	3,519.770	1*h*	2	Dy	3,521.13	6	15
Ni	3,519.766	30	500*h*	Nb	3,521.14	10*h*	2
				Ag	3,521.16	2	5
Tb	3,519.76	15	50				
Th	3,519.69	6	3*d*	Mo	3,521.17	10*w*	3
Ho	3,519.65	4*h*	—	Fe	3,521.26	200	300
Nb	3,519.649	20	5	Ar	3,521.27	(10)	—
Ru	3,519.63	30	70	Mo	3,521.41	3	8
				Rb	3,521.44	(200)	—

	λ	I J	I O
Zr	**3,519.60**	**10**	**100**
Rh	**3,519.54**	**2**	**40**
Nb	**3,519.33**	**3**	**2**
Tl	3,519.24	1000*R*	2000*R*
Pd	3,516.94	500*R*	1000*R*
Ne	3,515.19	(150)	—
Ne	3,510.72	(50)	—
F	*3,505.61*	*(600)*	—
Ne	*3,503.61*	*(18)*	—
F	*3,503.09*	*(400)*	—

	λ	I J	I O
U	**3,521.48**	**8*h***	**3**
Tl	3,529.43	800	1000

Ne 3,593.526 (500)

	λ	I J	I O
Cr	**3,593.49**	**400*R***	**500*R***
Hg	**3,593.48**	**(10)**	**—**
Gd	**3,593.43**	**15**	**15**
Sm	**3,593.40**	**3**	**2**
Te	**3,593.34**	**(5)**	**—**
V	**3,593.334**	**300*R***	**30**
Fe	**3,593.329**	**2**	**7**
K	**3,593.22**	**(5)**	**—**
Sr	**3,593.21**	**2**	**3**
U	**3,593.20**	**1**	**10**
Dy	**3,593.15**	**4**	**6**
Ho	**3,593.13**	**6**	**6**
Zr	**3,593.129**	**1**	**7**
Pb	**3,593.12**	**30**	**—**
Tb	**3,593.10**	**8**	**15**
Ti	**3,593.09**	**30**	**5**
Pr	**3,593.04**	**2*h***	**5**
Ru	**3,593.02**	**150**	**60**
W	**3,592.98**	**3**	**5**
Hg	**3,592.97**	**2*h***	**—**
Eu	**3,592.93**	**1*h***	**7**
Pb	**3,592.92**	**3**	**—**
Yt	**3,592.91**	**25**	**80**
W	**3,592.85**	**3*d***	**6*d***
U	**3,592.801**	**6**	**1**
Xe	**3,592.80**	**(2)**	**—**
Th	**3,592.78**	**6**	**2*d***
Gd	**3,592.70**	**70**	**50**
Fe	**3,592.69**	**2**	**12**
Mo	**3,592.65**	**20**	**—**
Sm	**3,592.595**	**50**	**40**
Nd	**3,592.595**	**30**	**20**

	λ	I J	I O
Rh	**3,593.530**	**2**	**10**
Nb	**3,593.55**	**3**	**15**
W	**3,593.56**	**3**	**4**
N	**3,593.60**	**(10)**	**—**
Tl	**3,593.61**	**(10)**	**—**
Ne	**3,593.64**	**(250)**	**—**
Pr	**3,593.685**	**2**	**4**
U	**3,593.685**	**1**	**8**
Sm	**3,593.73**	**2**	**4**
Tb	**3,593.75**	**8**	**30**
Th	**3,593.88**	**5**	**5**
Nb	**3,593.966**	**50**	**80**
W	**3,593.971**	**8**	**9**
Cu	**3,594.02**	**2**	**15**
Ce	**3,594.10**	**2**	**5**
U	**3,594.11**	**4**	**1**
Th	**3,594.116**	**2**	**5**
Ca	**3,594.12**	**2**	**3**
Ir	**3,594.14**	**4**	**10**
Au	**3,594.15**	**6**	**25**
Ne	**3,594.18**	**(12)**	**—**
Tb	**3,594.25**	**8**	**15**
Cr	**3,594.31**	**4**	**—**
Ir	**3,594.39**	**30**	**15**
Hf	**3,594.43**	**2**	**—**
S	**3,594.46**	**(35)**	**—**
W	**3,594.53**	**7**	**5**
Mo	**3,594.55**	**4**	**3**
Dy	**3,594.57**	**2**	**4**
vzduch	**3,594.6**	**3**	**—**
Fe	**3,594.64**	**100**	**125**

	λ	I	
		J	O
Ta	**3,592.495**	**1*h***	**2**
Fe	**3,592.486**	**1**	**3**
Cs	**3,592.48**	**(4)**	—
W	**3,592.42**	**35**	**9**
Bi	**3,592.40**	**(5)**	—
Os	**3,592.32**	**12**	**20**
U	**3,592.30**	**4**	**4**
V	3,582.02	300*R*	50
Nd	3,590.35	300*W*	400*W*
V	3,589.76	600*R*	80
Ar	3,588.44	(300)	—
Al	3,586.91	(500*h*)	—
Al	3,586.80	(200*wh*)	—
Al	3,586.69	(200)	—
Al	3,586.55	(200)	—
Te	3,585.34	(350)	—
Fe	*3,581.19*	*600r*	*1000R*
Yt	3,600.73	300	100
Cr	3,605.33	400*R*	500*R*
Ar	*3,606.52*	*(1000)*	—
Ne	*3,609.18*	*(50)*	—
Pd	*3,609.55*	*700R*	*1000R*
Cd	*3,610.51*	*500*	*1000*
Cd	*3,612.87*	*500*	*800*

Ne 4,537.75 (1000)

	λ	J	O
Ne	**4,537.68**	**(300)**	—
Ar	**4,537.67**	**(10)**	—
V	**4,537.66**	**20**	**25**
Nb	**4,537.59**	**5*h***	**5**
Mo	**4,537.32**	**3**	**1**
Ti	**4,537.23**	**3**	**10**
Yt	**4,537.16**	**2*h***	**3**
U	**4,537.12**	**4**	**2**
Te	**4,537.07**	**(50)**	—
Th	**4,537.070**	**8**	**20**
Hg	**4,537.01**	**(10)**	—
Gd	**4,536.98**	**2**	**20**
Xe	**4,536.92**	**(40*wh*)**	—
U	**4,536.82**	**3**	**1**
Mo	**4,536.80**	**80**	**40**
Cl	**4,536.78**	**(20)**	—
W	**4,536.66**	**6**	**15**
U	**4,536.61**	**2**	**8**
Ne	4,536.31	(150)	—
Ne	4,535.47	(30)	—
Ne	4,529.48	(30)	—
Ne	4,526.18	(50)	—
Ne	4,525.76	(70)	—
Gd	**4,537.82**	**100**	**150**
Sm	**4,537.95**	**25**	**50**
Kr	**4,538.06**	**(3)**	—
Eu	**4,538.07**	**20*W***	—
U	**4,538.19**	**40**	**25**
Ne	**4,538.31**	**(300)**	—
Mo	**4,538.41**	**5**	**5**
Dy	**4,538.76**	**2**	**5**
Br	**4,538.77**	**(15)**	—
La	**4,538.87**	**8*h***	—
Mo	**4,538.88**	**4**	**5**
Cs	**4,538.94**	**(30)**	—
Ne	4,539.17	(50)	—
Ne	4,540.38	(50)	—
Ne	*4,544.50*	*(50)*	—
Ne	*4,552.60*	*(30)*	—
Ne	*4,553.16*	*(50)*	—
Ne	*4,554.82*	*(40)*	—
Ne	*4,555.39*	*(30)*	—

	λ	I J	I O
Xe	*4,524.68*	*(400)*	—
Kr	*4,523.14*	*(400h)*	—
Ne	*4,522.66*	*(50)*	—
Ar	*4,522.32*	*(800)*	—
Ne	*4,517.74*	*(100)*	—
Ne	*4,516.94*	*(50)*	—
Ne	*4,515.41*	*(30)*	—
Ne	*4,514.89*	*(70)*	—

Ne 4,884.92 (1000)
4,884.92 (1000)

	λ	I J	I O
J	**4,884.82**	**(25)**	—
Mo	**4,884.33**	**4**	**10**
Xe	**4,884.15**	**(50*wh*)**	—
J	**4,884.13**	**(15)**	—
V	**4,884.06**	**6*h***	—
Eu	**4,884.05**	**2**	**15**
Ta	**4,883.95**	**3**	**150**
Ar	**4,883.86**	**(5)**	—
U	**4,883.78**	**8**	**10**
Yt	**4,883.69**	**300**	**20**
P	**4,883.65**	**(30)**	—
Xe	**4,883.53**	**(300*h*)**	—
Si	**4,883.51**	**2**	**4**
V	**4,883.42**	**. 20*h***	**1**
Ne	**4,883.40**	**(15)**	—
Ar	**4,883.27**	**(30)**	—
Hf	**4,883.26**	**4**	—
Ar	4,879.90	(300)	—
Ne	*4,867.01*	*(70)*	—
Ne	*4,866.48*	*(80)*	—
Te	*4,866.22*	*(800)*	—
Ne	*4,865.50*	*(100)*	—
Ne	*4,864.35*	*(30)*	—
Te	*4,864.10*	*(800)*	—
Ne	*4,863.08*	*(100)*	—
J	*4,862.31*	*(100)*	—
H	*4,861.33*	*(500)*	—
Ne	*4,852.65*	*(100)*	—
Mo	**4,884.95**	**5**	**5**
Ne	**4,885.084**	**(100)**	—
Ti	**4,885.085**	**25**	**150**
Xe	**4,885.19**	**(2*h*)**	—
Te	**4,885.22**	**(100)**	—
Mo	**4,885.32**	**4**	**5**
Rb	**4,885.627**	**10**	—
S	**4,885.63**	**(30)**	—
Mo	**4,885.64**	**3**	**5**
V	**4,885.65**	**6**	**7**
Hf	**4,885.74**	**4**	**2**
Nb	**4,885.77**	**1**	**3**
Pr	**4,886.04**	**1**	**20**
Ar	**4,886.29**	**(30)**	—
Yt	**4,886.291**	**8**	**4**
N	**4,886.30**	**(5)**	—
U	**4,886.33**	**6**	**5**
Mo	**4,886.47**	**15**	**25**
Ne	4,892.09	(500)	—
Ne	4,897.92	(70)	—
Ne	4,899.01	(50)	—

Ne 4,957.03 (1000)

	λ	I	
		J	O
U	**4,956.79**	**2**	—
Ar	**4,956.75**	**(100)**	—
Nb	**4,956.59**	**2**	**3**
Mo	**4,956.58**	**5**	**15**
Cu	**4,955.96**	**2**	—
O	**4,955.78**	**(30)**	—
Ne	**4,955.38**	**(150)**	—
Ne	4,944.99	(100)	—
Ne	*4,939.04*	*(100)*	—
Ne	*4,928.23*	*(70)*	—
Ne	**4,957.12**	**(150)**	—
Ba	**4,957.16**	**(50)**	—
Tm	**4,957.19**	**5**	**50**
Fe	**4,957.31**	**20**	**100**
Dy	**4,957.36**	**3**	**20**
Nb	**4,957.39**	**3**	**8**
Mo	**4,957.54**	**25**	**60**
Fe	**4,957.61**	**150**	**300**
V	**4,957.64**	**2**	**3**
Rn	**4,957.8**	**(10)**	—
J	**4,957.86**	**(8)**	—
Ne	4,973.54	(100)	—
Ne	*4,974.76*	*(50)*	—

Ne 5,400.56 (2000)

	λ	I	
		J	O
Mo	**5,400.47**	**15**	**20**
Xe	**5,400.45**	**(4*h*)**	—
Ra	**5,400.23**	**(500)**	—
Se	**5,399.88**	**(8)**	—
Ra	**5,399.80**	**(250)**	—
Yb	**5,399.74**	**15**	**1**
Ar	**5,399.01**	**(20)**	—
Mo	**5,398.73**	**1**	**5**
Ne	5,383.26	(25)	—
Ne	*5,374.97*	*(50)*	—
Ne	*5,372.31*	*(75)*	—
Ne	*5,366.22*	*(25)*	—
Ne	*5,362.25*	*(25)*	—
Ne	*5,360.44*	*(35)*	—
Ne	*5,360.01*	*(150)*	—
Cr	**5,400.61**	**1**	**30**
U	**5,400.95**	**6**	**10**
Se	**5,401.01**	**(75)**	—
Mg	**5,401.05**	**5**	**2**
V	**5,401.93**	**100**	**100**
In	**5,402.44**	**(30)**	—
Ta	**5,402.51**	**30**	**80**
Ne	5,412.65	(250)	—
Ne	5,418.55	(150)	—
Xe	5,419.15	(1000)	—
Ne	5,420.15	(50)	—
Ne	*5,433.65*	*(250)*	—

Ne 5,852.49 (2000)

	λ	I	
		J	O
Br	**5,852.10**	**(150)**	—
Cu	**5,851.93**	**2**	—
Hg	**5,851.89**	**(8)**	—
In	**5,852.83**	**(100)**	—
Kr	**5,852.86**	**(5)**	—
In	**5,853.11**	**(150)**	—

	λ	I	
		J	O
Mo	**5,851.52**	**20*h***	**40*h***
Te	**5,851.09**	**(75)**	—
V	**5,850.32**	**20**	**40**
Xe	**5,849.85**	**(3*h*)**	—
Ne	5,828.91	(75)	—
Ne	*5,820.15*	*(500)*	—
Ne	*5,816.64*	*(50)*	—
Ne	*5,811.42*	*(300)*	—
Ne	*5,804.45*	*(500)*	—
Ne	*5,804.10*	*(75)*	—

	λ	I	
		J	O
In	**5,853.43**	**(300)**	—
Al	**5,853.62**	**(35)**	—
Ba	**5,853.68**	**100*h***	**300**
Bi	**5,853.9**	**10**	**3**
Kr	**5,854.04**	**(4*wh*)**	—
N	**5,854.16**	**(15)**	—
Yb	**5,854.52**	**1**	**30**
In	**5,854.58**	**5**	—
Ne	5,868.42	(75)	—
Kr	5,870.92	(3000)	—
Ne	5,872.15	(75)	—
He	5,875.62	(1000)	—
Ne	*5,881.89*	*(1000)*	—
Na	*5,889.95*	*1000R*	*9000R*
Ne	*5,898.41*	*(20)*	—
Ne	*5,906.43*	*(50)*	—

Ne 6,402.25 (2000)

	λ	I	
		J	O
W	**6,402.07**	**1**	**5**
Yt	**6,402.00**	**7**	**12**
Tm	**6,401.45**	**5**	**40**
Se	**6,401.17**	**(15)**	—
Ne	**6,401.08**	**(100)**	—
Mo	**6,401.07**	**6**	**20**
We	6,382.99	(1000)	—

	λ	I	
		J	O
Ar	**6,403.10**	**(2)**	—
Sc	**6,403.15**	**2**	**2*h***
S	**6,403.58**	**(2)**	—
Cu	**6,403.70**	**5**	—
Ne	6,409.75	(150)	—
Ne	*6,421.71*	*(100)*	—

Ni 28 58.71

t_0 1,452°C

t_1 2,732°C

I.	II.	III.	IV.	V.
7.633	18.2	—	—	—

λ	*I*		eV
	J	*O*	
II 2,253.86	300	100	6.81
II 2,264.46	400	150	6.72
II 2,270.21	400	100	6.61
II 2,287.08	500	100	7.28
I 3,002.49	100	1000*R*	4.16
I 3,050.82	—	1000*R*	4.09
I 3,414.77	50	1000*R*	3.65
I 3,461.65	50	800*R*	3.60
I 3,492.96	100	1000*R*	3.65
I 3,515.05	50	1000*R*	3.63
I 3,524.54	100	1000*R*	3.54
I 3,619.39	150	2000*R*	3.85
I 4,401.55	30	1000	6.00
I 4,714.42	8	1000	6.00
I 4,786.54	2	300	6.00
I 5,035.37	5	300	6.09

Ni 2,253.86 *J* 300 *O* 100

	λ	I J	I O		λ	I J	I O
Nb	**2,253.80**	—	**3**	**Mg**	**2,253.87**	—	**8**
Co	**2,253.78**	—	**10**	**W**	**2,253.91**	**10**	—
Pd	**2,253.74**	**(2)**	—	**Pb**	**2,253.95**	**5**	**40**
Ni	**2,253.66**	**10**	—	**Hf**	**2,254.01**	**80**	**60**
Pd	**2,253.65**	**25**	**2**	**Re**	**2,254.05**	—	**15**
Ru	**2,253.64**	—	**50**	**Rh**	**2,254.07**	**25**	**3**
Zn	**2,253.59**	**(3)**	—	**Fe**	**2,254.08**	**10*h***	—
Ni	**2,253.55**	—	**10**				
				Ru	2,255.52	3	80
Ni	2,251.48	3	10	Os	2,255.85	2	125
				Ni	2,256.15	10	—
				Ni	*2,257.89*	*12*	—
				Ni	*2,258.15*	*2*	*15*
				Ni	*2,259.57*	—	*300w*

Ni 2,264.46 *J* 400 *O* 150

	λ	I J	I O		λ	I J	I O
Co	**2,264.42**	**2**	**10**	**Nb**	**2,264.55**	**8**	**2**
Fe	**2,264.39**	**5**	**35**	**Fe**	**2,264.59**	**20**	—
Mo	**2,264.35**	**10**	**4**	**Ir**	**2,264.61**	**15**	**30**
U	**2,264.30**	**2*h***	**5**	**Lu**	**2,264.63**	**2**	—
Pd	**2,264.28**	**25**	—	**Pd**	**2,264.68**	**(3)**	—
Xe	**2,264.20**	**(3)**	—	**Mo**	**2,264.74**	**18**	—
Mo	**2,264.20**	**5**	—				
W	**2,264.18**	**12**	**3**				
Ir	**2,264.17**	**3**	**12**	Cd	2,265.02	300	25*d*
Rh	**2,264.14**	**25**	**50**	Zn	2,266.00	(250)	—
				Ni	2,266.35	2	20
				Hf	2,266.83	80	60
Al	2,263.45	25	60*R*				
Rh	2,263.43	200	5				
Ni	2,261.42	3	12				
*P**				*N**			

Ni 2,270.21 *J* 400 *O* 100

	λ	I J	I O		λ	I J	I O
Pd	**2,270.209**	**40*wh***	—	**W**	**2,270.24**	**20**	**12**
Nb	**2,270.18**	**20**	**6**	**Mo**	**2,270.33**	**5**	—
Os	**2,270.17**	**15**	**60**	**Rh**	**2,270.34**	**20**	—
Pd	**2,270.11**	**(40)**	—	**Fe**	**2,270.35**	**4**	**2**
Sb	**2,270.08**	**15**	**25**				

	λ	I J	I O
Ir	2,270.02	3	20
Co	2,269.98	3	—
Hg	2,269.92	(10)	—
Sn	2,268.91	100R	100R
P*			
Ni	2,271.95	3	10
Ru	2,272.09	3	100
Nb	*2,274.13*	*300*	*12*
Ni	*2,274.66*	—	*12*
Ni	*2,274.73*	*15*	—
Re	*2,275.25*	*300r*	*300r*
Ni	*2,275.69*	*15*	*2*

Ni 2,287.08 *J* 500 *O* 100

	λ	I J	I O
Si	2,287.06	(10)	—
Rb	2,286.97	(5d)	—
Nb	2,286.89	3h	2
Ni	2,286.80	—	2h
Pd	2,286.78	(8)	—
Sn	2,286.68	40	60
Co	2,286.16	300	40
Ru	2,285.38	1	80
Zr	2,285.23	—	100
Os	*2,282.26*	*125*	*100*
W	2,287.19	3	—
Ar	2,287.19	(20)	—
Fe	2,287.248	6	20
Cr	2,287.25	2h	—
Ni	2,287.32	—	5
Ni	2,287.65	20	2
Cd	2,288.02	300R	1500R
As	2,288.12	5	250R
Ni	2,288.39	2	12
Ni	2,289.98	20	20r
Rh	2,290.03	500	25
Pt	*2,292.39*	*100*	*400*

Ni 3,002.491 *J* 100 *O* 1000R

	λ	I J	I O
Mn	3,002.489	—	12
Ir	3,002.485	—	6
Kr	3,002.48	(2h)	—
Cl	3,002.45	(8)	—
Tb	3,002.45	—	3
V	3,002.442	1	10
U	3,002.441	—	3
Th	3,002.39	12	15
Dy	3,002.38	1h	10
Ce	3,002.376	—	12s
Si	3,002.37	3	—
Fe	3,002.33	5	—
W	3,002.28	15	1
Pt	3,002.27	30	200
Ir	3,002.25	10	50
Mo	3,002.213	3	40
Yb	3,002.61	150	15
U	3,002.64	—	2
Fe	3,002.649	150	20
V	3,002.65	—	10
Pd	3,002.652	60	100r
Er	3,002.66	3	15
Ti	3,002.73	2	10
Pb	3,002.74	10	—
Mo	3,002.74	—	10
Ce	3,002.75	1	20
W	3,002.82	6	7
Gd	3,002.87	20	15
Cs	3,002.88	(6)	—
Ar	3,003.00	(3)	—
Fe	3,003.03	100	200
V	3,003.07	—	4

	λ	I J	I O
Nb	**3,002.212**	**30**	**5*W***
Ce	**3,002.14**	**1**	**20**
Ru	**3,002.07**	—	**6**
Yb	**3,002.03**	**15**	—
W	**3,001.98**	**9**	**10**
V	**3,001.90**	**3**	**18*h***
Hf	**3,001.85**	**1**	**5**
Nb	**3,001.85**	**15*wh***	—
Al	**3,001.82**	**(10)**	—
V	3,001.20	200*r*	20
Pt	3,001.17	50*W*	3
Fe	3,000.95	300*r*	800*R*
Cr	3,000.89	125	150*r*
Ru	2,999.81	50	8
Pd	2,999.55	100*h*	—
Fe	2,999.51	300	500
Ru	2,998.89	100	50
Cr	2,998.79	70	200*r*
Pt	2,997.97	200*r*	1000*R*
Te	2,997.05	(50)	—
Cr	2,996.58	125	300*r*
Fe	2,996.39	50	90
V	2,996.00	70	5
Ru	2,995.99	50	4
Ti	2,995.75	70*wh*	10*h*
Au	*2,994.99*	*100*	—
Nb	*2,994.73*	*300*	*100*
Ni	*2,994.46*	*10*	*125R*
Fe	*2,994.43*	*600r*	*1000R*
Bi	*2,993.34*	*100wh*	*200wh*
Ni	*2,992.59*	*10*	*80R*
Ne	*2,992.44*	*(150)*	—
Nb	*2,991.95*	*100*	*1*
Ru	*2,991.62*	*100*	*50*
Ni	*2,991.09*	—	*15*
Fe	*2,990.39*	*100*	*150*
Nb	*2,990.26*	*200*	*5*
Bi	*2,989.03*	*100wh*	*250wh*
Ru	*2,988.95*	*100*	*250*

	λ	I J	I O
Mo	**3,003.17**	—	**10**
V	3,003.46	70	8
Ni	3,003.63	80	500*R*
Cr	3,003.92	150	1
As	3,003.93	50	—
Cr	3,005.06	125	300*r*
Ru	3,005.13	60	—
Yb	3,005.76	100	10
Nb	3,005.77	50	1
V	3,005.81	50	—
Te	3,006.35	(50)	—
V	3,006.50	50	—
Hg	3,006.57	(50)	—
Rn	3,006.8	(300)	—
N	3,006.86	(50)	—
Fe	3,007.14	80	100
Fe	3,007.281	60	80
V	3,007.283	50	2
Fe	3,008.14	400*r*	600*r*
In	3,008.31	500*W*	—
Nb	3,008.41	50	—
V	3,008.50	50	—
V	3,008.61	70	3
Fe	3,009.09	60	80
Sn	3,009.15	200*h*	300*h*
Fe	*3,009.57*	*400*	*500*
Gd	*3,010.14*	*100*	*100*
Rn	*3,010.8*	*(100)*	—
Fe	*3,011.48*	*125*	*125*
Ni	*3,012.00*	*125W*	*800R*
Ta	*3,012.54*	*100*	*125*
Hf	*3,012.90*	*100*	*80*
Cr	*3,013.71*	*150*	*200r*
Cr	*3,014.75*	*100*	*300r*
V	*3,014.82*	*100*	*10*
Cr	*3,014.91*	*100*	*300r*
Tm	*3,015.29*	*100*	*125*
Cr	*3,015.51*	*150*	*1*

Ni 3,050.819 *O* 1000*R*

	λ	I J	I O
Sm	**3,050.819**	**8**	**5**
Rh	**3,050.818**	—	**2**
Hf	**3,050.76**	**10**	**50**
Er	**3,050.83**	**1*h***	**5**
V	**3,050.89**	—	**30**
Co	**3,050.93**	—	**60**

	λ	I J	I O		λ	I J	I O
Tm	**3,050.73**	**150**	**50**	**Th**	**3,050.99**	**10**	**10**
V	**3,050.730**	**50**	**1**	**Sm**	**3,051.00**	**2**	**6**
Ce	**3,050.59**	—	**12**	**Ir**	**3,051.09**	—	**5**
U	**3,050.502**	**6**	**4**	**Nd**	**3,051.11**	**2**	**6**
Co	**3,050.500**	—	**60**	**Tb**	**3,051.12**	**8**	**15**
Hg	**3,050.46**	—	**2*h***	**U**	**3,051.14**	**10*r***	**15*r***
V	**3,050.40**	**1**	**30**	**Ir**	**3,051.15**	**1**	**5**
Os	**3,050.39**	**50**	**100**	**Ce**	**3,051.16**	—	**6**
Pr	**3,050.35**	**4**	**2**	**Os**	**3,051.17**	**15**	**80**
Zr	**3,050.32**	—	**3**	**Ce**	**3,051.287**	—	**2**
Ce	**3,050.30**	—	**8**	**W**	**3,051.291**	**30**	**10**
Mo	**3,050.21**	**1**	**10**	**U**	**3,051.30**	**6*d***	**4*d***
U	**3,050.20**	**8**	**12**	**Nb**	**3,051.34**	**10**	**1**
Ru	**3,050.19**	—	**5**	**U**	**3,051.42**	**5**	**5**
Cr	**3,050.14**	**150**	**10**				
Ta	**3,050.10**	**5**	**35**				
Al	**3,050.08**	**10**	**18**	Ni	3,054.32	100	400*R*
				Ni	3,057.64	125	400*R*
Fe	3,047.60	500*r*	800*r*				
Ni	3,045.01	10	200	*Os*	*3,058.66*	*500*	*500R*
				Fe	*3,059.09*	*400*	*600r*
				Ni	*3,064.62*	*50*	*200r*
In	*3,039.36*	*500R*	*1000R*	*Pt*	*3,064.71*	*300R*	*2000R*
Ge	*3,039.06*	*1000*	*1000*				
Ni	*3,037.93*	*100*	*800R*				

Ni 3,414.765 *J* 50* *O* 1000*R*

	λ	I J	I O		λ	I J	I O
Co	**3,414.74**	—	**200*W***	**Ce**	**3,414.766**	—	**5**
Zr	**3,414.66**	**15**	**20**	**Pr**	**3,414.767**	—	**7**
Ru	**3,414.64**	**5**	**50**	**Eu**	**3,414.773**	—	**40**
U	**3,414.63**	**1*d***	**3*d***	**Er**	**3,414.79**	**1**	**8**
Ce	**3,414.60**	—	**5**	**Dy**	**3,414.83**	**5**	**35**
Ag	**3,414.55**	—	**4**	**Ho**	**3,414.92**	**30**	**30**
Th	**3,414.51**	**8**	**6**	**Zr**	**3,414.950**	—	**2**
Yt	**3,414.49**	**2**	**3**	**Sm**	**3,414.953**	**2**	**10**
U	**3,414.36**	**1*d***	**8*d***	**Ce**	**3,415.07**	—	**8**
Ce	**3,414.31**	—	**8**	**Pr**	**3,415.08**	**3**	**4**
Cr	**3,414.305**	—	**8*wh***	**Tb**	**3,415.12**	—	**8**
Dy	**3,414.30**	**2**	**2**	**Th**	**3,415.13**	**1**	**2**
Nd	**3,414.298**	**2**	**10**	**Os**	**3,415.22**	**8**	**5**
Ru	**3,414.28**	—	**12**	**Ir**	**3,415.24**	**2*h***	**10**
Os	**3,414.24**	**8**	**10**	**Ta**	**3,415.270**	**35**	**3**
V	**3,414.20**	**40**	**20**	**Mo**	**3,415.273**	**5**	**5**
Ce	**3,414.17**	—	**8*s***	**Ce**	**3,415.31**	—	**3**
Ta	**3,414.14**	**100*W***	**18*W***	**U**	**3,415.32**	—	**3**
Nb	**3,414.07**	**8**	**10**	**Dy**	**3,415.35**	—	**4**
Eu	**3,414.02**	**2**	**4*h***	**Ca**	**3,415.38**	**3**	**2**

	λ	I J	I O
Ru	**3,413.98**	**—**	**3**
Ni	**3,413.94**	**10**	**300**
Sm	**3,413.89**	**2**	**6**
U	**3,413.81**	**10**	**4**
Dy	**3,413.79**	**9**	**40**
Ni	3,413.48	15	500
Co	3,412.63	40	1000*R*
Co	3,412.34	100	1000*R*
Ni	3,409.58	—	300
Co	3,409.18	125	1000*R*
Co	3,405.12	150	2000*R*
Pd	*3,404.58*	*1000R*	*2000R*
Cd	*3,403.65*	*500h*	*800*
Ni	*3,401.17*	*1*	*40*
Rh	*3,396.85*	*500*	*1000w*
W	**3,415.42**	**4**	**3**
Tb	**3,415.43**	**—**	**8**
Co	**3,415.530**	**—**	**20**
Fe	**3,415.534**	**20**	**60**
W	**3,415.55**	**6**	**7**
Nd	**3,415.553**	**6**	**15**
Cr	**3,415.57**	**—**	**8***wh*
U	**3,415.613**	**2**	**1**
Ce	**3,415.614**	**—**	**10**
Mo	**3,415.63**	**5**	**10**
Pr	**3,415.71**	**4**	**25**
Ir	**3,415.74**	**1***h*	**7**
Co	3,417.16	—	400*R*
Ni	3,420.74	3	30
Pd	3,421.24	1000*R*	2000*R*
Ni	3,423.71	25	600*R*
Co	*3,433.04*	*150*	*1000R*
Pd	*3,433.45*	*500h*	*1000h*
Ni	*3,433.56*	*50wh*	*800R*

Ni 3,461.652 *J* 50* *O* 800*R*

	λ	J	O
Nb	**3,461.61**	**2**	**5***h*
Ir	**3,461.58**	**—**	**4**
Ti	**3,461.50**	**125**	**80**
Ce	**3,461.42**	**—**	**2**
Sm	**3,461.405**	**4**	**5**
Er	**3,461.40**	**1**	**10**
Eu	**3,461.38**	**2**	**25**
W	**3,461.36**	**6**	**7**
Ce	**3,461.34**	**—**	**15***s*
Dy	**3,461.31**	**1**	**3**
Pr	**3,461.24**	**—**	**4**
Th	**3,461.218**	**4**	**10**
Ce	**3,461.217**	**—**	**4**
La	**3,461.184**	**2**	**10**
Co	**3,461.176**	**3**	**100***wh*
Tm	**3,461.17**	**5**	**15**
Sm	**3,461.14**	**3**	**8**
Zr	**3,461.09**	**1**	**20**
Pr	**3,461.06**	**2**	**10**
Th	**3,461.02**	**4**	**10**
Yt	**3,461.01**	**12**	**7**
U	**3,461.003**	**3**	**3**
Tb	**3,461.00**	**8**	**15**
W	**3,461.655**	**3**	**3**
Eu	**3,461.66**	**—**	**3**
Yb	**3,461.71**	**3**	**1**
Ce	**3,461.79**	**—**	**10**
W	**3,461.81**	**5**	**9**
Pr	**3,461.87**	**2**	**6**
Ru	**3,461.92**	**—**	**30**
Gd	**3,461.956**	**5**	**5**
Ho	**3,461.96**	**20**	**20**
Rh	**3,462.040**	**150**	**1000**
Ru	**3,462.040**	**—**	**5**
Pr	**3,462.10**	**—**	**6**
Sc	**3,462.18**	**2***h*	**4**
Os	**3,462.19**	**10**	**20**
Tm	**3,462.20**	**200**	**250**
Eu	**3,462.21**	**1***h*	**5**
Ce	**3,462.23**	**—**	**3**
La	**3,462.32**	**2***h*	**2**
Pr	**3,462.36**	**—**	**3**
Fe	**3,462.361**	**3**	**10**
Ce	**3,462.43**	**—**	**8**
Na	**3,462.49**	**(15)**	**2**
Tb	**3,462.51**	**3**	**8**

	λ	I J	I O
Ce	**3,460.999**	—	**12**
Dy	**3,460.971**	**3**	**100**
Er	**3,460.968**	**7**	**20**
Ho	**3,460.95**	**4**	**6**
W	**3,460.784**	**3**	**3**
Mo	**3,460.784**	**25**	**25**
Ce	**3,460.783**	—	**2**
U	**3,460.781**	—	**2**
Pd	**3,460.77**	**600*h***	**300*r***
Co	**3,460.72**	—	**18**
Sc	**3,460.70**	**2*h***	**6**
Pr	**3,460.66**	—	**3**
Dy	**3,460.64**	**1**	**4**
Sm	**3,460.63**	**1**	**3*w***
Ce	**3,460.581**	—	**2**
Nd	**3,460.581**	**6**	**25**
Re	3,460.47	—	1000*W*
Ni	3,458.47	50*h*	800*R*
Co	3,455.23	10	2000*R*
Co	3,453.50	200	3000*R*
Ni	3,452.89	50	600*R*
Ni	3,446.26	50*h*	1000*R*
Ni	3,442.04	—	15
Pd	3,441.40	2*h*	800*h*
Er	**3,462.58**	—	**10**
Ca	**3,462.62**	**4**	**2**
Hf	**3,462.64**	**12**	**15**
Nb	**3,462.65**	**3**	**5**
Sm	**3,462.69**	**2**	**5**
Cr	**3,462.73**	**8**	**2**
Co	3,462.80	80	1000*R*
Co	3,465.80	25	2000*R*
Fe	3,465.86	400	500
Cd	3,466.20	500	1000
Ni	3,467.50	15	300
Cd	3,467.66	400	800
Ni	3,469.49	20	300
Rh	3,470.66	125	500
Ni	3,472.54	40	800*R*
Co	*3,474.02*	*100*	*3000R*
Rh	*3,474.78*	*125*	*700*

Ni 3,492.956 *J* 100 *O* 1000*R*

	λ	I J	I O
Au	**3,492.95**	**4**	—
Sm	**3,492.89**	—	**3**
Mo	**3,492.82**	**4**	**3**
U	**3,492.80**	**10**	**2**
Sm	**3,492.775**	**3**	**3**
Hg	**3,492.77**	**(50)**	—
Rb	**3,492.76**	**(300)**	—
Pr	**3,492.73**	**1**	**4**
Th	**3,492.68**	**2**	**2**
Sm	**3,492.62**	**1**	**8**
Tm	**3,492.59**	**10**	**20**
Tb	**3,492.56**	**8**	**15**
Ce	**3,492.559**	—	**3**
Er	**3,492.54**	**2*d***	**25*d***
Dy	**3,492.52**	**1**	**2**
Ti	**3,492.50**	**35*wh***	—
Ce	**3,492.49**	—	**5**
U	**3,492.33**	**1**	**8**
Mn	**3,492.960**	—	**10**
Tb	**3,492.96**	—	**15**
Eu	**3,492.97**	—	**2**
Ce	**3,492.98**	—	**3**
Tb	**3,492.99**	—	**15**
W	**3,493.036**	**4**	**5**
Kr	**3,493.04**	**(8*wh*)**	—
Ho	**3,493.10**	**10**	**10**
Ce	**3,493.11**	**1**	**12**
Pr	**3,493.16**	**2**	**10**
V	**3,493.17**	**100**	**15**
W	**3,493.19**	**3**	**6**
F	**3,493.21**	**(3)**	—
Ru	**3,493.22**	**1**	**20**
Ar	**3,493.25**	**(20)**	—
Dy	**3,493.26**	—	**3**
Tb	**3,493.27**	—	**8**
Ti	**3,493.28**	**1**	**15**

	λ	I J	I O
W	3,492.26	—	4
Ce	3,492.25	—	12
U	3,492.21	8	2
Ru	3,492.10	—	6
Ir	3,492.060	—	2
W	3,492.058	12	2d
bhZr	3,492.0	—	30
Co	3,491.99	—	10
vzduch	3,491.98	6	—
Gd	3,491.97	25	50
Pr	3,491.94	2	10
Lu	3,491.92	3h	—
Th	3,491.90	—	3
Nb	3,491.89	15	1
Ar	3,491.54	(50)	—
Nb	3,491.03	50	30
Fe	3,490.57	300	400
P	3,490.44	(70)	—
Nb	3,489.09	50	5
P	3,488.77	(70)	—
Mn	3,488.68	200	50
V	3,485.92	70	8
Ni	3,485.89	30	150
Fe	3,485.34	50	100
Pt	3,485.27	200R	150
P	3,485.00	(50)	—
Nb	3,484.05	100	10
Ti	3,483.80	70wh	—
Ni	3,483.77	30	500R
Nb	3,482.95	100	2
Mn	3,482.91	250	50
Gd	*3,481.36*	*150*	*150*
Ta	*3,480.52*	*200ws*	*70*
Nb	*3,479.56*	*200*	*5*
Rh	*3,478.91*	*100*	*500*
Yb	*3,478.84*	*300*	*40*
V	*3,477.52*	*100*	—
Ti	*3,477.18*	*100*	*60*
Fe	*3,476.70*	*200*	*300*
Fe	*3,475.45*	*300*	*400*
Rh	*3,474.78*	*125*	*700*
Mn	*3,474.13*	*400*	*12*
Co	*3,474.02*	*100*	*3000R*
Sb	*3,473.91*	*300wh*	*3*
Ne	*3,472.57*	*(500)*	—
Ni	*3,472.54*	*40*	*800R*

	λ	I J	I O
Fe	3,493.29	—	1
U	3,493.33	15	6
Mo	3,493.34	10	6
U	3,493.307	5	2
Eu	3,493.407	—	7
Sm	3,493.408	1	2
Ta	3,493.46	—	15W
Nb	3,493.473	3	3
Fe	3,493.474	80	40
Th	3,493.53	15	30
Sm	3,493.597	4	5
Nd	3,493.600	6	6d
Zr	3,493.68	—	6
Fe	3,493.70	1	3
Er	3,493.721	—	5
Ce	3,493.724	—	18
Hg	3,493.85	(100)	—
Tb	3,493.90	3	8
Ce	3,493.94	—	8
U	3,494.00	2h	12
Gd	3,494.42	60	70
Fe	3,495.29	60	100
Cd	3,495.34	(100)	—
Co	3,495.69	25	1000R
Mn	3,495.84	150	25
Zr	3,496.21	100	100
Ni	3,496.35	—	15
V	3,497.03	150	—
Fe	3,497.11	100	200
S	3,497.34	(100)	—
Mn	3,497.54	150	15
Fe	3,497.84	200	200
Ne	3,498.06	(75)	—
Sb	3,498.46	300wh	—
Nb	3,498.63	50	30
Dy	3,498.67	50	50
Rh	3,498.73	60	500
Ru	3,498.94	200	500R
V	3,499.82	50	3
Nb	3,499.95	50	5
Ni	3,500.85	80	500wh
Ba	3,501.12	20	1000
Ne	3,501.22	(150)	—
F	3,501.42	(200)	t
Se	3,501.52	(50)	—
Co	3,501.72	100	5
Co	3,502.28	20	2000R

	λ	I J	I O
Lu	*3,472.48*	*150*	*50*
Rh	3,502.52	150	1000
Ni	3,502.59	—	100
F	3,502.95	(60)	—
P	3,502.99	(70)	—
F	3,503.09	(400)	—
Kr	3,503.25	(50*wh*)	—
V	*3,504.44*	*200*	*60*
Ti	*3,504.89*	*150*	*20*
F	*3,505.61*	*(600)*	—
Rh	*3,507.32*	*125*	*500*
P	*3,507.36*	*(100w)*	
Lu	*3,507.39*	*150*	*100*
V	*3,509.04*	*150*	*2*
Tb	*3,509.17*	*200*	*200*
Ru	*3,509.20*	*100*	*10*
Nb	*3,510.26*	*200*	*15*
Ni	*3,510.34*	*50h*	*900h*
Ti	*3,510.84*	*125*	*40*
Co	*3,512.64*	*100*	*400R*
Ir	*3,513.64*	*100*	*100h*
Fe	*3,513.82*	*300*	*400*

Ni 3,515.054 *J* 50* *O* 1000*R*

	λ	J	O
Dy	3,515.053	1	3
Tb	3,515.04	8	30
Re	3,515.00	—	2
Ce	3,514.966	—	2
Th	3,514.966	4	4
Er	3,514.89	5	15
La	3,514.87	2	3
Ce	3,514.802	1	2
*bh*Ca	3,514.8	—	4
Dy	3,514.80	—	2
Mo	3,514.78	3	4
Ru	3,514.77	4	12
Pt	3,514.71	2	2
Zr	3,514.64	2	3
Fe	3,514.63	2	7
U	3,514.61	5	18
Th	3,514.53	8	8
Eu	3,514.49	5	18
Ru	3,514.488	40	70
Dy	3,514.43	—	6
V	3,514.42	40	1
W	3,515.07	9	1
Zn	3,515.11	—	2*h*
U	3,515.23	10	3
Zr	3,515.24	—	3
Ce	3,515.28	—	4
Ce	3,515.39	—	2
Nb	3,515.42	300	20
Tb	3,515.44	3	15
U	3,515.45	—	4
Ce	3,515.546	—	2
Be	3,515.549	—	30
Er	3,515.57	5*w*	25*w*
Ho	3,515.58	40	40
Ce	3,515.637	—	8
Dy	3,515.64	—	6
U	3,515.676	1*h*	6
Ru	3,515.678	8*r*	10
Th	3,515.71	4	5
Re	3,515.75	—	3
Ce	3,515.78	1	8
*bh*Zr	3,515.8	—	2

	λ	I J	I O
Ta	3,514.40	1h	5
Ce	3,514.33	—	6
Zr	3,514.32	—	2
Eu	3,514.22	—	15
Tb	3,514.18	—	8
Ru	3,514.14	—	4
Dy	3,514.10	2	4
La	3,514.06	3	8
Ni	3,513.93	40h	200
Fe	3,513.82	300	400
Ni	3,510.34	50h	900R
Ni	3,507.69	12	100
Rh	3,507.32	125	500
Ni	*3,502.59*	—	*100*
Rh	*3,502.52*	*150*	*1000*
Co	*3,502.28*	*20*	*1000R*
Ba	*3,501.12*	*20*	*1000*
Ni	*3,500.85*	*80*	*500wh*
Co	*3,495.69*	*25*	*1000R*
Ni	*3,492.96*	*100h*	*1000R*

	λ	I J	I O
Ru	3,515.89	8	10
Ce	3,515.94	—	5
Ir	3,515.95	15	35
W	3,515.96	6d	7d
Er	3,516.00	1	10
Tb	3,516.14	3	15
Dy	3,516.15	—	6
Pd	3,516.94	500R	1000R
Ni	3,518.63	8	90
Tl	3,519.24	1000R	2000R
Ni	3,519.77	30	500R
Ni	3,523.44	—	100
Ni	3,524.54	100wh	1000R
Ni	*3,527.98*	*15*	*200*
Rh	*3,528.02*	*150*	*1000w*
Ni	*3,528.89*	—	*15*
Tl	*3,529.43*	*800*	*1000*
Co	*3,529.81*	*30*	*1000R*
Ni	*3,530.59*	—	*30*

Ni 3,524.541 J 100 O 1000R

	λ	J	O
Mn	3,524.540	—	15
Sm	3,524.538	—	10
Zr	3,524.538	2h	9
Pr	3,524.47	2	3
Ru	3,524.45	6	—
Ir	3,524.37	—	2h
Eu	3,524.34	1h	3w
J	3,524.28	(3d)	—
Hg	3,524.27	(15)	—
W	3,524.25	8	9
Zr	3,524.242	—	12
Fe	3,524.241	50	60
Ti	3,524.240	—	2
Cu	3,524.239	10	40
Mo	3,524.239	3	8
Gd	3,524.20	15	25
Hg	3,524.19	(100)	—
Ru	3,524.15	—	6
Ce	3,524.073	—	8
Cd	3,524.072	8	—
Fe	3,524.071	40	50

	λ	J	O
Dy	3,524.61	4	6
Mo	3,524.65	50h	5
W	3,524.68	5	7
V	3,524.71	60	10
Ti	3,524.87	5	—
Ru	3,524.90	2	12
Er	3,524.920	8	20
Dy	3,524.92	—	6
Nb	3,524.94	5	2
Mo	3,524.981	5	5
Ba	3,524.985	5	20
Sm	3,525.06	—	2
Co	3,525.09	—	2h
Tb	3,525.13	8	8
U	3,525.141	8	5
Th	3,525.142	1	2d
Ti	3,525.16	1	10
Nb	3,525.23	30w	15w
Os	3,525.29	5	4
Br	3,525.34	(3)	—
bhSr	3,525.4	—	8
Pr	3,525.49	2	4

	λ	I J	I O
Dy	**3,524.03**	**20**	**15**
Ce	**3,524.01**	—	**8**
Er	**3,523.98**	**8**	**25**
Ho	**3,523.97**	**4**	—
Mo	**3,523.89**	**10**	—
Ce	**3,523.73**	—	**5**
Co	**3,523.70**	**3**	**15**
Eu	**3,523.67**	**1*h***	**2*w***
Tb	**3,523.66**	**50**	**30**
Os	**3,523.636**	**30**	**150**
Zr	**3,523.636**	**1**	**5**
Nd	**3,523.62**	**8**	**15**
Ce	**3,523.61**	**2**	**2**
Cr	**3,523.59**	**1**	**10**
U	**3,523.565**	**15**	**4**
Th	**3,523.557**	**1*wh***	**5*wh***
W	**3,523.55**	**5**	—
Sr	**3,523.51**	**2**	**1**
Eu	**3,523.50**	—	**15*w***
Ni	3,523.44	—	100
Nb	3,522.36	50*W*	—
Ir	3,522.03	50	—
V	3,521.84	80	20
Rb	3,521.44	(200)	—
Fe	3,521.26	200	300
Sb	3,520.474	(125)	—
Ne	3,520.472	(1000)	—
Yb	3,520.29	70	10
V	3,520.02	50	5
Ni	3,519.77	30	500*h*
Tl	3,519.24	1000*R*	2000*R*
Ni	3,518.63	8	90
P	3,518.60	(50*h*)	—
Co	3,518.35	100	200*W*
Nb	3,517.67	200	2
Pd	3,516.94	500*R*	1000*R*
P	3,516.15	(70)	—
Nb	3,515.42	300	20
Ne	3,515.19	(150)	—
Ni	3,515.05	50*h*	1000*R*
Ar	3,514.39	(125)	—
Nb	3,514.04	50	—
Ni	3,513.93	40*h*	200

*P**

	λ	I J	I O
Sm	**3,525.50**	**2**	**15**
Tb	**3,525.61**	**8**	**50**
Fe	3,526.04	50	80
Fe	3,526.68	50	80
Fe	3,527.80	80	100
Ni	3,527.98	15	200
Rh	3,528.02	150	1000*w*
Nb	3,528.48	50	3
Os	3,528.60	50	400*R*
Ni	3,528.89	—	15
Tl	3,529.43	800	1000
Co	3,529.81	30	1000*R*
Fe	3,529.82	80	125
P	3,530.24	(70)	—
Ni	3,530.59	—	30
V	3,530.77	100	40
Rb	3,531.60	(100)	—
Tb	3,531.70	50	15
Dy	3,531.71	100	100
Hg	3,532.63	(200)	—
Fe	3,533.010	75	50
Na	3,533.100	(100)	50
Fe	3,533.20	50	50
Nb	3,535.30	500	300
Kr	3,535.35	(50*h*)	—
Ti	*3,535.41*	*125*	*15*
Fe	*3,536.56*	*200*	*300*
V	*3,538.24*	*100*	*10*
Nb	*3,540.96*	*500*	*15*
Fe	*3,541.09*	*200*	*200*
Rb	*3,541.22*	*(100)*	—
F	*3,541.76*	*(100)*	—
Fe	*3,542.08*	*100*	*150*
V	*3,545.20*	*300R*	*40*
Ar	*3,545.58*	*(300)*	—
Gd	*3,545.79*	*125*	*125*
Ar	*3,545.84*	*(125)*	—

Ni 3,619.392 *J* 150 *O* 2000*R*

	λ	*I*	
		J	*O*
Ce	3,619.391	—	8
Th	3,619.390	1	2
Fe	3,619.390	1	12
Bi	3,619.37	—	5
Sm	3,619.29	2	2
Mn	3,619.28	50	75
W	3,619.27	10	10
Ru	3,619.202	8	2
Nb	3,619.200	1	2
Ir	3,619.16	3	30
U	3,619.13	2	5
Pr	3,619.10	3	9
Nd	3,618.969	15	20
Ce	3,618.968	—	2
V	3,618.93	100	—
Er	3,618.92	6	10*w*
Nb	3,618.90	5	5
Cl	3,618.88	(3)	—
Ce	3,618.78	—	2
Fe	3,618.77	400	400
Th	3,618.74	—	2
Se	3,618.73	(35)	—
Rn	3,618.70	(8)	—
Ce	3,618.58	3	10
Hg	3,618.53	(50)	—
Sm	3,618.523	—	4
Dy	3,618.522	10	80
U	3,618.49	10	2
W	3,618.45	12	—
Nb	3,618.44	5	10
K	3,618.43	(20)	—
Ce	3,618.42	—	3
Fe	3,618.39	4	8
Th	3,618.38	1	5
Mo	3,618.35	20	—
Fe	3,618.30	1	4
Fe	3,617.79	80	125
P	3,617.09	(100*w*)	—
Zr	3,614.77	80	40
Cd	3,614.45	100	60
Cd	3,612.87	500	800
Ni	3,612.74	50*h*	400
Al	3,612.47	80*h*	—
Cd	3,610.51	500	1000
Ni	3,610.46	—	1000*r*
Fe	3,610.16	90	100

	λ	*I*	
		J	*O*
Zr	3,619.395	—	3
*bh*La	3,619.4	—	4*h*
Ho	3,619.43	6	6
Os	3,619.431	25	60
Cr	3,619.460	8	30
Ti	3,619.464	2	2
Dy	3,619.47	1	3
Tl	3,619.49	(6)	—
Nb	3,619.51	200	5
Pr	3,619.59	2	8
Th	3,619.71	4	4
Nb	3,619.727	300	3
Tb	3,619.73	8	15
Ag	3,619.77	7*h*	—
Fe	3,619.772	1	3
U	3,619.806	3	—
Yb	3,619.809	100	30
Ce	3,619.92	1	8
Dy	3,619.96	5	25
Hf	3,619.966	—	3
Tm	3,619.97	20	20
Er	3,619.98	1	9
Ti	3,620.02	—	10
Hf	3,620.04	—	8
Sn	3,620.08	(2)	—
U	3,620.085	2*h*	15
Sm	3,620.09	3	3
*bh*La	3,620.1	—	6*h*
Dy	3,620.176	20	80
Er	3,620.177	3	50
Fe	3,620.228	1	4
Au	3,620.23	10	—
Os	3,620.24	12	15
Ru	3,620.28	5	2
Tb	3,620.29	—	15
Lu	3,620.31	—	4
Ce	3,620.33	—	5
Cu	3,620.35	5	30
Th	3,620.37	6	6
Pr	3,620.426	2	5
Co	3,620.429	—	5
Rh	3,620.456	10	20
Gd	3,620.461	25	25
Fe	3,620.471	2	15
V	3,620.472	50	10
Sn	3,620.54	(6)	—

	λ	I	
		J	O
Pd	3,609.55	700R	1000R
Sm	3,609.48	100	60
Ni	3,609.31	15	200
Fe	3,608.86	400	500
Gd	3,608.76	125	100
Kr	3,607.88	(100wh)	—
Ni	*3,606.85*	*—*	*100r*
Fe	*3,606.68*	*150*	*200*
Ar	*3,606.52*	*(1000)*	*—*
Hg	*3,605.80*	*(200)*	*—*
Fe	*3,605.46*	*150*	*300*
Cr	*3,605.33*	*400R*	*500R*
Ni	*3,602.28*	*15*	*150*
Yt	*3,600.73*	*300*	*100*
Ni	*3,597.70*	*50h*	*1000r*
Sb	*3,597.51*	*200wh*	*2*
Rn	3,621.00	(250)	—
V	3,621.21	80	15
Fe	3,621.46	100	125
Fe	3,622.00	100	125
Fe	3,623.19	80	100
Ni	3,624.73	15	150
Ti	3,624.82	125	60
V	3,625.61	125	4
Hg	3,630.65	(100)	—
Fe	*3,631.46*	*300*	*500*
Kr	*3,631.87*	*(200h)*	*—*
Ar	*3,632.68*	*(300)*	*—*
Ar	*3,634.46*	*(300)*	*—*
Pd	*3,634.69*	*1000R*	*2000R*
Rn	*3,634.8*	*(250)*	*—*
Ni	*3,634.94*	*10*	*50*
Fe	*3,640.39*	*200*	*300*
Ti	*3,641.33*	*150*	*60*

Ni 4,401.55 J 30 O 1000

	λ	J	O
Tb	4,401.54	—	20
Ce	4,401.51	—	5
Fe	4,401.45	2	7
Rb	4,401.4	(30)	—
Fe	4,401.30	15	60
Ir	4,401.25	—	12
Ho	4,401.24	3	3
Nb	4,401.172	10	—
Sm	4,401.156	30	50
Ce	4,401.15	—	2
Yt	4,401.14	—	3
Se	4,401.02	(100)	—
Ar	4,401.02	(40)	—
P	4,400.99	(50)	—
Pb	4,400.89	10	—
Ce	4,400.879	2	10
Kr	4,400.87	(100h)	—
Ni	4,400.870	3	15
W	4,400.86	—	5
Pr	4,400.84	—	3
Nb	4,400.831	3	2
Nb	4,400.828	20	50
Gd	4,400.76	8	15
Mo	4,400.66	3	3
Ti	4,400.582	2	25
Th	4,401.67	3	3
Er	4,401.848	—	12
Gd	4,401.853	100	200
Te	4,401.89	(100)	—
Th	4,401.99	1	3
Ce	4,402.00	—	4
Nb	4,402.05	8	1
Hg	4,402.06	(50)	—
Pr	4,402.16	—	2
Yb	4,402.30	20	6
U	4,402.303	6	6
Ne	4,402.37	(2)	—
Ce	4,402.41	—	2
U	4,402.44	8	8
Nd	4,402.47	1	10
Mo	4,402.49	15	15
Ta	4,402.50	20h	100
Cl	4,402.52	(2)	—
Ba	4,402.55	10	80
Pr	4,402.58	—	4
Re	4,402.60	—	30
La	4,402.65	—	5
Co	4,402.68	—	5h
Os	4,402.74	3	50
Mo	4,402.90	20	20

	λ	I	
		J	O
Os	**4,400.579**	**1**	**18**
V	**4,400.57**	**40**	**60**
Ce	**4,400.54**	**2**	**10**
U	**4,400.53**	**2**	**1**
Tb	**4,400.51**	**—**	**2**
Th	**4,400.39**	**3**	**6**
Sc	**4,400.355**	**30**	**150**
Nb	**4,400.354**	**10**	**5**
Fe	**4,400.351**	**1**	**20**
Ar	4,400.09	(30)	—
Pr	4,400.03	20	30
Kr	4,399.97	(200)	—
Sm	4,399.86	15	20
Ti	4,399.77	100	40
Cs	4,399.49	(20)	—
Ir	4,399.47	100	400
Kr	4,399.39	(15*hs*)	—
Pr	4,399.33	20	40
Se	4,399.17	(15)	—
Cl	4,399.14	(15)	—
J	4,399.09	(20)	—
Hg	4,398.62	(300)	—
Te	4,398.45	(70)	—
Yt	4,398.01	100	150
Ne	4,397.94	(100)	—
Mo	4,397.29	30	30
Cs	3,396.91	(15)	—
Mo	4,396.66	25	25
Br	4,396.40	(20)	—
Pr	4,396.12	50	80
Te	4,396.00	(100)	—
O	4,395.95	(80)	—
Ag	4,395.93	30	10
Ti	4,395.84	30	10
Xe	4,395.77	(200*wh*)	—
Ne	4,395.56	(50)	—
V	4,395.23	40*R*	60*R*
Ti	4,395.03	150	50
Pr	4,395.00	15	25
Ne	4,394.77	(15)	—
Ne	4,394.37	(15)	—
Mo	4,394.32	15	20
Ti	4,394.06	15	5
Br	4,393.56	(25)	—
Xe	4,393.20	(200*wh*)	—
Yb	4,392.83	20	3
Mo	4,392.12	15	15
V	4,392.074	15	25
Gd	4,403.14	100	100
Mo	4,403.34	20	—
Sm	4,403.36	50	50
Cr	4,403.50	25	15*d*
J	4,403.55	(20)	—
Pr	4,403.60	40	100
V	4,403.67	15	20
Cs	4,403.85	(20)	—
Mo	4,404.18	15*h*	1
Ti	4,404.27	30	50
Kr	4,404.33	(30*h*)	—
As	4,404.53	15	—
Fe	4,404.75	700	1000
Hg	4,404.86	(50)	—
Pr	4,405.14	20	25
Ba	4,406.23	(20)	—
Cs	4,405.25	(35)	—
Te	4,405.49	(14)	—
Pr	4,405.85	100	100
V	4,406.15	15*W*	20*W*
Se	4,406.58	(70)	—
V	4,406.64	30	40
Gd	4,406.67	200	70
Xe	4,406.88	(100*wh*)	—
Mo	4,407.44	30	—
Fe	4,407.72	50	100
Be	4,407.91	(35)	20
Gd	4,408.26	150	100
Fe	4,408.42	60	125
V	4,408.51	20*R*	30*h*
Pr	4,408.84	100	125
Kr	4,408.89	(40*hs*)	—
J	4,408.96	(250)	—
Ne	4,409.30	(140)	—
J	4,409.31	(15)	—
Sm	4,409.34	100	100
Ne	4,409.62	(20)	—
Ru	4,410.03	80	150
C	4,410.06	30	—
Cs	4,410.208	(20)	—
Nb	4,410.213	30	15
Kr	4,410.37	(50)	—
Ni	4,410.52	4	25
Te	4,410.95	(50)	—
Nd	4,411.05	20	50
Ti	4,411.08	100	7
Gd	4,411.16	50	100
C	4,411.20	40	—
C	4,411.52	40	—

	λ	I	
		J	O
Gd	4,392.071	100	100
Ne	4,391.94	(150)	—
S	4,391.84	(30)	—
Cr	4,391.75	35	50
Ce	4,391.66	15	40
Br	4,391.61	(25)	—
Mo	4,391.53	15	15
Gd	4,391.44	25	15
Th	4,391.11	40	50
Ti	4,391.03	25	6
Gd	4,390.955	100	100
Fe	4,390.954	35	100
Sm	4,390.86	150	150
Nd	4,390.66	15	20
Ru	4,390.43	80	150*R*
J	4,390.34	(15*h*)	—
Gd	4,389.99	80	30
V	4,389.97	60*R*	80*R*
Te	4,389.92	(50)	—
Gd	4,389.88	40*h*	40
Cl	4,389.76	(25)	—
Kr	4,389.72	(20*h*)	—
Fe	*4,388.41*	*50*	*125*
K	*4,388.13*	*(40)*	—
He	*4,387.93*	*(30)*	—
Fe	*4,387.90*	*35*	*150*
Ti	*4,386.85*	*80*	*8*
Kr	*4,386.54*	*(300h)*	—
Ac	*4,386.37*	*100*	—
Xe	*4,385.77*	*(70)*	—
Ru	*4,385.65*	*50*	*125*
Ru	*4,385.39*	*40*	*125*
P	*4,385.33*	*(100)*	—
Kr	*4,385.27*	*(50wh)*	—
Te	*4,385.08*	*(50)*	—
Cr	*4,384.98*	*200*	*150*
Xe	*4,384.93*	*(30)*	—
V	*4,384.72*	*125R*	*125R*
Ni	*4,384.54*	*1*	*25*
Sm	*4,384.29*	*50*	*50*
Xe	*4,383.91*	*(100)*	—
Fe	*4,383.55*	*800*	*1000*
La	*4,383.45*	*50*	*10*
Rn	*4,383.30*	*(35)*	—
Gd	*4,383.14*	*40*	*30*
Se	*4,382.87*	*(800)*	—
Th	*4,381.86*	*30*	*30*

	λ	I	
		J	O
Mo	4,411.70	25	25
Mn	4,401.88	20	100
P	4,411.94	(15*h*)	—
V	4,412.14	20	25
Pr	4,412.15	15	50
Mn	4,412.22	25	—
Nd	4,412.26	15	40
Ne	4,412.28	(20)	—
J	4,412.37	(25)	—
Ne	4,412.54	(15)	—
Mo	4,412.77	20	30
Se	4,413.16	(20)	—
Ac	4,413.17	100	—
Ne	4,413.20	(50)	—
Ne	4,413.56	(15)	—
As	*4,413.64*	*50*	—
Pr	*4,413.76*	*40*	*90*
Gd	*4,414.16*	*60*	*100*
P	*4,414.28*	*(100)*	—
P	*4,414.60*	*(70)*	—
Cd	*4,414.63*	*200*	—
Gd	*4,414.73*	*50*	*100*
Xe	*4,414.84*	*(150)*	—
Mn	*4,414.88*	*60*	*150*
O	*4,414.89*	*(300)*	—
Fe	*4,415.12*	*400*	*600*
Xe	*4,416.07*	*(80wh)*	—
Ne	*4,416.82*	*(50)*	—
O	*4,416.97*	*(160)*	—
Kr	*4,417.24*	*(40)*	—
P	*4,417.30*	*(30)*	—
Hf	*4,417.35*	*50*	*25*
Sm	*4,417.58*	*80*	*80*
Ti	*4,417.72*	*80*	*40*
Kr	*4,418.76*	*(50)*	—
Gd	*4,419.04*	*200*	*200*
Pr	*4,419.06*	*30*	*80*
Pr	*4,419.67*	*50*	*100*
Os	*4,420.47*	*100*	*400R*
Sm	*4,420.53*	*200*	*200*
P	*4,420.64*	*(70)*	—
Ar	*4,420.90*	*(40)*	—
Sm	*4,421.14*	*150*	*150*
Te	*4,421.14*	*(70)*	—
Pr	*4,421.23*	*35w*	*100*
Ne	*4,421.38*	*(30)*	—
Ne	*4,421.56*	*(50)*	—
Ti	*4,421.95*	*35*	*6*

	λ	I J	I O		λ	I J	I O
Mo	*4,381.64*	*150*	*150*	*Gd*	*4,422.41*	*40*	*100*
Kr	*4,381.52*	*(100h)*	—	*Ne*	*4,422.52*	*(300)*	—
Ne	*4,381.22*	*(30)*	—				
Gd	*4,380.64*	*125*	*100*	*Fe*	*4,422.57*	*125*	*300*
Ar	*4,379.74*	*(80)*	—	*Yt*	*4,422.59*	*60*	*60*
				Kr	*4,422.70*	*(100hs)*	—
Ne	*4,379.50*	*(100)*	—	*P*	*4,423.55*	*(30)*	—
V	*4,379.24*	*200R*	*200R*	*Mo*	*4,423.62*	*40*	*40*
Sm	*4,378.23*	*100*	*100*				
Cu	*4,378.20*	*30w*	*200w*	*J*	*4,423.76*	*(80)*	—
La	*4,378.10*	*30*	*40*	*P*	*4,423.9*	*(30)*	—
				Ar	*4,423.99*	*(80)*	—
Nb	*4,377.96*	*30*	*10*	*Cr*	*4,424.28*	*35*	*25*
Mo	*4,377.76*	*200*	*5*	*Sm*	*4,424.342*	*300*	*300*
Kr	*4,377.71*	*(40h)*	—				
				Nd	*4,424.343*	*50*	*50*
				Pr	*4,424.59*	*35*	*90*
				Ne	*4,424.80*	*(300)*	—
				Kr	*4,425.19*	*(100)*	—
				Hg	*4,425.22*	*(30)*	—
				Ne	*4,425.40*	*(150)*	—

Ni 4,714.42 *J* 8* *O* 1000

Sc	**4,714.33**	**5**	**7**	**Mo**	**4,714.51**	**15**	**15**
Nd	**4,714.23**	—	**4**	**W**	**4,714.52**	—	**7**
Ru	**4,714.21**	—	**5**	**U**	**4,714.59**	—	**4**
Tm	**4,714.20**	**15**	**10**	**Sm**	**4,714.63**	—	**50**
La	**4,714.152**	**2**	**5**	**Pr**	**4,714.85**	—	**4**
Pr	**4,714.146**	—	**30**	**Tb**	**4,714.89**	—	**2**
V	**4,714.12**	**20**	**25**	**Hf**	**4,714.99**	**4**	**3**
Fe	**4,714.07**	**50**	**50**	**Er**	**4,714.995**	**1**	**4**
Ce	**4,714.00**	**3**	**10**	**Ce**	**4,715.07**	—	**6**
Cr	**4,713.93**	**2**	**15**	**Pr**	**4,715.23**	—	**4**
Sr	**4,713.96**	—	**3**	**Sm**	**4,715.27**	—	**100**
U	**4,713.94**	**5**	**4**	**Ti**	**4,715.30**	**2**	**40**
W	**4,713.86**	—	**6**	**V**	**4,715.43**	**3**	**4**
Er	**4,713.77**	—	**4**	**Th**	**4,715.44**	**2**	**3**
Tm	**4,713.69**	—	**10**	**Gd**	**4,715.51**	—	**8**
Eu	**4,713.61**	**2**	**400**	**Pr**	**4,715.588**	—	**6**
Nb	**4,713.50**	**15**	**15**	**Nd**	**4,715.589**	**1**	**30**
Hf	**4,713.48**	**2**	**6**	**Ru**	**4,715.61**	—	**4**
V	**4,713.45**	**4*h***	**5*h***	**U**	**4,715.67**	**5**	**10**
Zr	**4,713.43**	—	**5**	**Ni**	**4,715.78**	**2**	**200**
Tm	**4,713.32**	—	**10**	**Nb**	**4,715.83**	**50*h***	**3**
Pr	**4,713.10**	—	**30**				
Sm	**4,713.07**	—	**100**				
Er	**4,713.060**	—	**2**	Bi	4,722.55	100	1000
Nd	**4,713.059**	**2**	**30**				

	λ	I: J	I: O
Nb	**4,713.05**	**3**	**2**
Ni	4,712.07	—	30
Ni	4,703.81	—	200
Ni	4,701.54	—	150
Ni	4,701.34	—	100
Ni	*4,698.41*	—	*30*
Ni	*4,686.22*	*1*	*200*
Ni	*4,731.81*	*2*	*100*
Ni	*4,732.46*	—	*100*
Ni	*4,740.16*	—	*15*

Ni 4,786.541 *J* 2* *O* 300

	λ	I: J	I: O
Ce	**4,786.540**	—	**10**
V	**4,786.51**	**25*h***	**30**
Mo	**4,786.457**	**25**	**25**
Nd	**4,786.457**	—	**2**
Ni	**4,786.29**	—	**25**
Dy	**4,786.24**	—	**2**
Tb	**4,786.19**	—	**5**
Sm	**4,786.15**	—	**4**
Nd	**4,786.11**	—	**10**
W	**4,785.96**	—	**2**
U	**4,785.91**	—	**3**
Sm	**4,785.87**	—	**100**
Nb	**4,785.70**	**2**	**3**
Pr	**4,785.62**	—	**10*w***
Lu	**4,785.42**	**200**	**100**
Yb	**4,785.34**	—	**2**
Dy	**4,785.31**	**2**	**3**
Mo	**4,785.12**	**30**	**30**
Co	**4,785.07**	—	**50**
Zr	**4,784.919**	—	**40**
Dy	**4,784.918**	**2**	**4**
Mn	4,783.42	60	400
Sm	4,783.10	—	150
Gd	4,781.93	50	200
Co	4,781.43	2*h*	400
Co	4,780.01	500	500*w*
Co	4,776.32	—	300
Ni	4,773.41	—	15
Co	4,771.11	—	500*w*
Co	*4,768.08*	*10*	*300*
Ni	*4,762.63*	*1h*	*150*
Ni	*4,756.52*	*3*	*250*
Yt	**4,786.58**	**25**	**15**
Pr	**4,786.59**	—	**6**
Yb	**4,786.598**	**200**	**50**
Er	**4,786.603**	**2**	**2**
Ta	**4,786.64**	—	**20**
Tb	**4,786.78**	—	**35**
Gd	**4,786.806**	**2**	**40**
Fe	**4,786.810**	—	**150**
Yt	**4,786.88**	**15**	**15**
K	**4,786.89**	—	**2**
Dy	**4,786.93**	**4**	**8**
Tb	**4,787.16**	—	**3**
***bh*Sr**	**4,787.2**	—	**2**
W	**4,787.21**	—	**5**
Pr	**4,787.23**	—	**3**
Ru	**4,787.34**	—	**5**
Nd	**4,787.43**	—	**8**
Pr	**4,787.57**	—	**5**
Eu	**4,787.62**	—	**2**
Mo	**4,787.63**	**8**	**10**
Nb	**4,787.73**	—	**3*h***
Ir	**4,787.748**	—	**5**
Cr	**4,787.752**	—	**12**
W	**4,787.94**	**1**	**15**
Tm	**4,787.97**	—	**5**
Eu	**4,787.99**	—	**2**
Pd	**4,788.17**	**4*h***	**200*h***
Mo	4,788.18	15	15
Cr	4,789.38	100	300
Fe	4,791.25	200*R*	200
Re	4,791.42	—	200*w*
Sm	4,791.58	—	150
Gd	4,791.60	—	150

	λ	I J	I O
Cr	*4,756.11*	*100*	*300*
Ni	*4,754.77*	—	*100*
Mn	*4,754.04*	*60*	*400*
Cr	4,792.51	40	200
Au	4,792.60	60	200*W*
Co	4,792.86	5	60*W*
Os	4,793.99	6	300
Cd	4,799.92	300*w*	300*w*
Cr	4,801.03	70	200
Gd	4,801.08	200	200
Ni	*4,807.00*	*1*	*150W*
Ni	*4,808.86*	—	*25*
Zn	*4,810.53*	*300h*	*400w*
Co	*4,813.48*	*6*	*1000W*
Ni	*4,817.85*	—	*15*

Ni 5,035.37 *J* 5* *O* 300

	λ	I J	I O
Ir	**5,035.07**	—	**2*h***
Tb	**5,035.03**	—	**15**
Hf	**5,034.90**	—	**3**
Sm	**5,034.76**	—	**4**
Ru	**5,034.64**	—	**6**
Zr	**5,034.44**	—	**3**
Pr	**5,034.41**	—	**30**
Hf	**5,034.33**	**8**	**4**
*bh*F	**5,034.3**	—	**2**
Er	**5,034.25**	—	**6**
Tm	**5,034.21**	**100**	**100**
Mo	**5,034.175**	**1**	**4**
Os	**5,034.172**	—	**12**
Tb	**5,034.07**	—	**10**
Ce	**5,033.81**	—	**20**
Eu	**5,033.54**	**2**	**60**
Eu	5,029.48	—	500*w*
Sm	5,028.44	—	200
Fe	5,022.25	—	150
Ni	5,018.29	—	70*w*
Ni	*5,017.59*	*1*	*100w*
Fe	*5,014.96*	—	*500*
Ni	*5,014.24*	—	*25*
Ni	*5,012.46*	*2h*	*70*
Fe	*5,012.07*	—	*300*
Ni	*5,010.96*	—	*50*
Ni	*5,010.04*	—	*25*
Fe	*5,006.13*	*5*	*300*
Ni	*5,003.75*	—	*20*
Fe	*5,001.87*	*40*	*300*
Ni	*5,000.33*	—	*150w*
Eu	**5,035.43**	—	**20*w***
U	**5,035.53**	**2**	**2**
W	**5,035.74**	—	**6**
Ti	**5,036.91**	**30**	**125**
Er	**5,035.93**	—	**12**
Ni	**5,035.96**	—	**70*w***
Nb	**5,035.99**	**5**	**10**
Sm	**5,036.21**	—	**50**
*bh*F	**5,036.4**	—	**2**
Ti	**5,036.47**	**25**	**125**
U	**5,036.54**	—	**2**
Ce	**5,036.62**	—	**10**
Mo	**5,037.18**	**2**	**5**
Ni	5,038.60	—	50
Ni	5,039.26	—	20
Ni	5,041.08	—	30
Fe	5,041.76	—	300
Ni	5,042.19	—	80
Sm	5,044.28	—	150
Ni	5,048.08	—	20
Ni	5,048.85	2*h*	80
Fe	5,049.82	1	400
Ni	5,051.53	—	50
Fe	5,051.64	—	200
Sm	5,052.75	2	150
Ni	*5,058.03*	—	*20*
Fe	*5,068.79*	*200*	*400*

O 8 15.9994

t_0 −218.7°C

t_1 −183.12°C

I.	II.	III.	IV.	V.
13.615	35.082	55.118	77.28	113.7

λ	I	eV
I 3,823.47	(125)	15.77
II 3,911.95	(150)	28.83
I 3,947.33	(300)	12.28
II 3,973.27	(125)	26.56
II 4,072.16	(300)	28.69
II 4,075.87	(800)	28.80
I 4,368.30	(1000)	12.36
II 4,414.89	(300)	26.24
I 5,330.66	(500)	13.06
I 7,771.93	(1000)	10.73
I 7,774.14	(300)	10.73
I 7,775.43	(100)	10.73

O 3,823.47 (125)

	λ	I J	I O		λ	I J	I O
U	3,823.45	3	5	Ce	3,823.49	1	2
Zr	3,823.41	5	15	Sm	3,823.510	15	15
Ir	3,823.38	3	20	Hf	3,823.511	1*h*	2*h*
Eu	3,823.35	5	3	Mn	3,823.513	75	75*h*
Th	3,823.33	8	5	Cr	3,823.52	30	40
Ar	3,823.29	(3)	—	Th	3,823.589	10	5
Nd	3,823.25	6	4	Ta	3,823.595	10	15
V	3,823.21	20	35	Xe	3,823.74	(10)	—
Pr	3,823.18	25	125	V	3,823.75	2	10
Mo	3,823.15	20*h*	—	Ca	3,823.760	2	2
Nb	3,823.13	3	2	Sc	3,823.761	10	15
Tb	3,823.12	3	15	Sm	3,823.79	10	15
Ce	3,823.09	4*s*	4*s*	Nd	3,823.80	10*d*	20*d*
Th	3,823.08	10*w*	20*w*	Mn	3,823.89	50	50*h*
W	3,823.062	10	—	V	3,823.990	15	35
Eu	3,823.06	4*w*	4*w*	Nb	3,825.995	10*wh*	—
U	3,823.04	10*h*	4	Pr	3,824.07	2	2
Er	3,823.04	1	9	Fe	3,824.078	3	5
Mo	3,822.98	15	20	Nd	3,824.08	4*d*	6*d*
Nb	3,822.95	3	2	Ce	3,824.12	1	2
V	3,822.89	25	40	W	3,824.15	8	7
Ce	3,822.87	1	2	Mo	3,824.168	5	5
Th	3,822.86	3	5	Sm	3,824.169	8	5
V	3,822.70	2	2	Fe	3,824.30	7	7
U	3,822.55	12	12	Th	3,824.35	20*w*	20*w*
Nd	3,822.47	10	4	W	3,824.39	15	12
Zr	3,822.414	1	10	Ce	3,824.419	3	3
U	3,822.410	4	4	O	3,824.425	(10)	—
Sm	3,822.317	4	15	Fe	3,824.44	100	150
Pr	3,822.316	2	2	Ir	3,824.52	8	25
Er	3,822.31	2*w*	18*d*	Sm	3,824.53	8	5
Nd	3,822.29	10*d*	20*d*	U	3,824.72	10	5
Rh	3,822.26	100	100	Th	3,824.762	10	20*d*
W	3,822.21	8	—	Er	3,824.764	1	8*w*
Th	3,822.15	10	10	Mo	3,824.779	4	4
Cr	3,822.09	6	10	Yt	3,824.78	5*h*	1
				Nd	3,824.79	30	40
				Sm	3,824.80	1	10
N	3,822.07	(35)	—				
O	3,821.64	(10)	—				
Fe	3,821.18	100	100	O	3,825.09	(20)	—
Fe	3,820.43	600	800	O	3,825.25	(18)	—
Cl	3,820.25	(100)	—	Fe	3,825.88	400	500
				P	3,827.44	(150)	—
Eu	3,819.66	500*wd*	500*wd*	Cl	3,827.62	(150)	—
Nb	3,818.86	300	8				
Pr	3,818.28	100	125	Fe	3,827.82	200	200
Fe	3,815.84	700	700	Mg	3,829.35	150	100*w*
Bi	3,815.80	(300)	—				

	λ	I			λ	I	
		J	O			J	O
V	3,815.39	150*h*	1	N	3,830.39	(150)	—
Ra	3,814.42	(2000)	—	O	3,830.45	(18)	—
Fe	3,812.96	300	400	Nb	3,831.84	300	5
Gd	3,812.02	200*h*	200*R*				
Bi	3,811.14	(150)	—	Pd	3,832.29	150	150
				Mg	3,832.31	200	250
Mn	*3,809.59*	*150*	*150*	Yt	3,832.89	80	30
Fe	*3,806.70*	*150*	*200*	O	3,833.10	(10)	—
Hg	*3,806.38*	*(200)*	—	Ta	3,833.74	200	40
Fe	*3,805.34*	*300*	*400*				
Au	*3,804.00*	*150*	*25*	Mn	3,833.86	75	75
				Fe	3,834.22	400	400
Sn	*3,801.00*	*150h*	*200h*	Mn	3,834.36	75	75*r*
Fe	*3,799.55*	*300*	*400*	Ar	3,834.68	(800)	—
Pd	*3,799.19*	*150*	*200w*	Nd	3,836.54	100	80
Fe	*3,798.51*	*300*	*400*				
Mo	*3,798.25*	*1000R*	*1000R*	Mo	3,837.29	100*w*	—
Fe	*3,797.52*	*200*	*300*	*Mg*	*3,838.26*	*200*	*300*
Gd	*3,796.39*	*150*	*150w*	*Mn*	*3,839.78*	*125*	*100*
				Fe	*3,840.44*	*300*	*400*
				Fe	*3,841.05*	*400*	*500*
				O	*3,842.82*	*(10)*	—
				F	*3,847.09*	*(800)*	—
				O	*3,847.89*	*(10)*	—
				Tm	*3,848.02*	*250*	*400*
				Tb	*3,848.75*	*200*	*100*
				La	*3,849.01*	*150*	*200*
				Fe	*3,849.97*	*400*	*500*
				F	*3,849.99*	*(600)*	—
				Ar	*3,850.57*	*(400)*	—

O 3,911.95 (150)

Mo	**3,911.94**	**5**	**5**	**Pr**	**3,911.99**	**6**	**20*d***
Th	**3,911.914**	**3**	**8**	**Fe**	**3,912.05**	**5*wh***	**5*wh***
Nd	**3,911.914**	**6**	**12**	**O**	**3,912.09**	**(5)**	—
Os	**3,911.812**	**5**	**30**	**Ru**	**3.912.11**	**8**	**10**
Sc	**3,911.810**	**30**	**150**	**Ta**	**3,912.13**	**3*s***	**10**
Pr	**3,911.80**	**2**	**8**	**Ce**	**3,912.19**	**3**	**25**
Dy	**3,911.68**	**1**	**5**	**V**	**3,912.21**	**20**	**50**
U	**3,911.67**	**18**	**18**	**Nd**	**3,912.228**	**20**	**20**
Ar	**3,911.58**	**(10)**	—	**U**	**3,912.234**	**1**	**2**
Mn	**3,911.42**	**15**	**15**	**Pr**	**3,912.27**	**3**	**10**
Th	**3,911.308**	**2**	**5**	**Th**	**3,912.28**	**15**	**15**
Pr	**3,911.307**	**3**	**5**	**U**	**3,912.41**	**8**	**6**
Ce	**3,911.30**	**1**	**15**	**Eu**	**3,912.43**	**1**	**5**
W	**3,911.29**	**5**	**6**	**Ta**	**3,912.435**	**10*h***	**15**
U	**3,911.23**	**3**	**3**	**Ce**	**3,912.436**	**5**	**50**

	λ	I J	I O
Ti	**3,911.19**	**5**	**40**
Nd	**3,911.17**	**25**	**25**
Ru	**3,911.14**	**2**	**6**
Mo	**3,911.092**	**20**	**20**
Se	**3,911.09**	**(2)**	**—**
Th	**3,911.003**	**1**	**5**
Fe	**3,911.001**	**2**	**6**
U	**3,910.98**	**3**	**10**
Sm	**3,910.92**	**2**	**1**
Pt	**3,910.91**	**2**	**5**
Fe	**3,910.84**	**10**	**30**
La	**3,910.81**	**5*h***	**10**
V	**3,910.79**	**20**	**35**
Ce	**3,910.70**	**2*h***	**12**
U	**3,910.50**	**3*h***	**1**
Cr	3,908.75	150	200
O	3,907.44	(18)	—
Gd	3,907.12	100	100*W*
Eu	3,907.11	500*R*	1000*RW*
Fe	3,906.48	200	300
Kr	3,906.25	(150*h*)	—
Yb	3,904.83	150	12
P	3,904.78	(100)	—
Fe	3,903.90	80	100
Mo	3,902.96	500*R*	1000*R*
Fe	3,902.95	400	500
Cr	3,902.91	100	100
Gd	3,902.40	80	100
Al	3,900.68	(200)	—
Ar	3,899.86	(100)	—
Fe	3,899.71	300	500
Tb	3,899.19	100	200
Nb	3,898.28	200	3
Fe	*3,895.66*	*300*	*400*
Ar	*3,894.66*	*(300)*	*—*
Pd	*3,894.20*	*200W*	*200W*
He	*3,888.65*	*(1000)*	*—*
La	*3,886.37*	*200*	*400*
Fe	*3,886.28*	*400*	*600*
P	*3,885.17*	*(150)*	*—*
O	*3,882.19*	*(35)*	*—*
Ho	**3,912.44**	**2**	**4**
Fe	**3,912.45**	**1*h***	**2*h***
J	**3,912.48**	**(25)**	**—**
Kr	**3,912.59**	**(70)**	**—**
Pr	**3,912.61**	**5**	**10**
U	**3,912.74**	**4**	**2**
Mn	**3,912.75**	**5**	**5**
Tb	**3,912.78**	**8**	**5**
In	**3,912.81**	**15**	**—**
W	**3,912.82**	**6**	**7**
Rh	**3,912.83**	**1**	**2**
Dy	**3,912.85**	**1**	**5**
Kr	**3,912.88**	**(5*h*)**	**—**
V	**3,912.886**	**15**	**40**
Bi	**3,912.895**	**2*h***	**2*h***
Pr	**3,912.90**	**80**	**150**
Sm	**3,912.97**	**5**	**20**
Th	**3,913.012**	**15**	**15**
Nb	**3,913.013**	**10**	**5**
Nb	**3,913.14**	**5**	**5**
Fe	**3,913.21**	**1*h***	**2*h***
Eu	**3,913.24**	**1**	**4**
U	**3,913.25**	**3**	**8**
W	**3,913.26**	**5**	**—**
Mo	**3,913.362**	**8**	**8**
Sm	**3,913.365**	**8**	**5**
Cs	**3,913.37**	**(2)**	**—**
Ti	3,913.46	70	40
P	3,914.26	(100)	—
Br	3,914.28	(150)	—
Hg	3,914.29	(100)	—
Nb	3,914.70	100	30
Cr	3,915.84	80	125
La	3,916.04	400	400
Gd	3,916.59	100	150*w*
Fe	3,916.73	80	100
Hg	3,918.92	(200)	—
C	3,918.98	80	—
Cr	3,919.16	125	300*r*
O	3,919.28	(35)	—
Nb	3,919.72	100	2
Kr	3,920.14	(200*h*)	—
Nb	3,920.20	100	30
Fe	3,920.26	300	500
C	3,920.68	200	—
La	3,921.53	200	400
Fe	3,922.91	400	600

	λ	I J	I O
Ru	3,923.47	100	60
S	3,923.48	(200)	—
Tb	3,925.45	200	150
Pr	3,925.46	100	125
Hg	3,925.65	(100)	—
Ru	3,925.92	100	60
Fe	*3,927.92*	*300*	*500*
La	*3,929.22*	*300*	*400*
Fe	*3,930.30*	*400*	*600*
Eu	*3,930.50*	*400R*	*1000R*
J	*3,931.01*	*(400)*	—
Rn	*3,931.82*	*(250)*	—
Sm	*3,933.59*	*200h*	*200*
Fe	*3,933.60*	*200*	*200*
Ca	*3,933.67*	*600R*	*600R*
Ru	*3,933.68*	*200*	*5*
Nb	*3,936.02*	*200*	*5*
Tb	*3,939.60*	*200*	*200*
J	*3,940.24*	*(500)*	—
Rb	*3,940.57*	*(200)*	—
Mo	*3,941.48*	*150*	*5*

O 3,947.33 (300)

	λ	I J	I O
Mo	3,947.17	10	10
Co	3,947.13	2	20
U	3,947.08	2	1
Fe	3,947.00	20	50
Br	3,946.98	(3)	—
S	3,946.98	(5)	—
Pr	3,946.95	5	8
Mo	3,946.89	3	4
U	3,946.88	3h	1
Tb	3,946.87	30	150
Nd	3,946.81	8	20
Ce	3,946.681	1	20
U	3,946.677	10	15
Br	3,946.66	(2)	—
Sm	3,946.51	30	40
Al	3,946.41	(2)	—
Ru	3,946.314	3	8
W	3,946.311	8	9
Ir	3,946.27	15	50
U	3,946.25	5	6
Pr	3,946.224	1	8
Yt	3,946.217	3	2
Th	3,947.34	3	8
Ar	3,947.50	(1000)	—
U	3,947.509	10	10
O	3,947.51	(50)	—
Nb	3,947.528	10	10
Fe	3,947.532	20	70
O	3,947.61	(18)	—
Pr	3,947.63	60d	125d
Th	3,947.65	1	3
Kr	3,947.66	(5h)	—
Ti	3,947.77	35	70
Sm	3,947.83	8	15
Te	3,947.93	(10)	—
U	3,947.96	8	2
Ce	3,947.97	3	20
W	3,947.98	9	10
Th	3,948.04	1	3
Fe	3,948.107	50	125
Sm	3,948.108	50	50
C	3,948.15	2	—
Xe	3,948.16	(60)	—
W	3,948.20	5	—

	λ	I J	I O
Th	**3,946.15**	**20**	**20**
Ar	**3,946.10**	**(25)**	—
Hf	**3,946.012**	**4*h***	—
Br	**3,946.01**	**(3)**	—
W	**3,945.98**	**5**	**1**
Cr	**3,945.97**	**7**	**50**
As	**3,945.93**	**10**	—
Th	**3,945.82**	**8**	**10**
Ru	3,945.57	100	50
Gd	3,945.54	150	200 *W*
Hg	3,945.09	(100)	—
O	3,945.04	(20)	—
Dy	3,944.69	150	300
Al	3,944.03	1000	2000
Mn	3,942.85	75	75
Hg	3,942.59	(100)	—
Hg	3,942.24	(100)	—

*P**

	λ	I J	I O
Hg	**3,945.09**	**(100)**	—
O	**3,945.04**	**(20)**	—
Nd	**3,948.325**	**10**	**20**
Tb	**3,948.35**	**15**	**20**
Pt	**3,948.39**	**5**	**60**
Pr	**3,948.62**	**2**	**9**
Mo	**3,948.65**	**10**	**6**
Ti	**3,948.67**	**40**	**80*h***
Xe	**3,948.72**	**(10)**	—
As	**3,948.74**	**50**	—
Nd	**3,948.778**	**8**	—
Fe	**3,948.779**	**100**	**150**
Se	**3,948.80**	**(25)**	—

*N**

O 3,973.266 (125)

	λ	I J	I O
Er	**3,973.26**	**1*wh***	**3**
U	**3,973.234**	**8**	**6**
Th	**3,973.230**	**3**	**8**
Ta	**3,973.18**	**400 *W***	**1**
Co	**3,973.15**	**6**	**150*w***
Pr	**3,973.12**	**3**	**4**
Sm	**3,973.11**	**2**	**5**
P	**3,973.10**	**(15)**	—
Mo	**3,972.95**	**20**	**4**
J	**3,972.82**	**(15)**	—
W	**3,972.81**	**3**	**4**
Gd	**3,972.71**	**10**	**20**
Rn	**3,972.69**	**(5)**	—
Cr	**3,972.688**	**12**	**60**
F	**3,972.67**	**(10)**	—
Th	**3,972.65**	**3**	**5**
Ho	**3,972.60**	**2*h***	**2**
Xe	**3,972.58**	**(25*wh*)**	—
W	**3,972.56**	**6**	—
K	**3,972.55**	**(30)**	—
Co	**3,972.53**	**6**	**100**
Nb	**3,972.52**	**15**	**20**
F	**3,972.41**	**(2)**	—
Nd	**3,972.39**	**8**	**20**
U	**3,972.34**	**10**	**4**

	λ	I J	I O
Nd	**3,973.269**	**25**	**40**
W	**3,973.28**	**4*h***	**4**
V	**3,973.36**	**3**	**30**
Yt	**3,973.46**	**4**	**5**
Co	**3,973.561**	**5**	**15**
Ni	**3,973.562**	**10**	**800**
Er	**3,973.58**	**2**	**18**
Ho	**3,973.59**	**4**	**1**
Nb	**3,973.62**	**10**	**5**
V	**3,973.64**	**40**	**25**
Nd	**3,973.650**	**20**	**30**
Fe	**3,973.654**	**10**	**40**
Ca	**3,973.71**	**15**	**200**
J	**3,973.75**	**(5)**	—
Mo	**3,973.77**	**25**	**25**
C	**3,973.84**	**2*h***	—
Nb	**3,973.87**	**1*h***	**2*d***
Pr	**3,973.90**	**3**	**10**
Mo	**3,973.93**	**3**	**5**
U	**3,973.94**	**4**	**15**
Ce	**3,974.00**	**5**	**4**
Gd	**3,974.07**	**80**	**100**
Th	**3,974.23**	**5**	**5**
Cs	**3,974.24**	**(6)**	—
Tb	**3,974.27**	**8**	**15**

	λ	I	
		J	O
U	**3,972.22**	**4**	**10**
Gd	**3,972.18**	**4**	**4**
Ni	**3,972.17**	**6**	**100**
Pr	**3,972.164**	**80**	**125**
Th	**3,972.159**	**8**	**15**
Ce	**3,972.07**	**4**	**25**
Tb	**3,972.051**	**8**	**12**
Eu	**3,972.047**	**5**	**5**
F	**3,972.047**	**(6)**	—
U	**3,972.004**	**3**	**3**
W	**3,972.000**	**5**	**6**
V	**3,971.95**	**2**	**5**
Nb	**3,971.92**	**5**	**15**
U	**3,971.86**	**6**	**6**
Nb	**3,971.85**	**5**	**5**
Gd	**3,971.76**	**20**	**25**
Nb	**3,971.695**	**15*h***	—
Pr	**3,971.693**	**40**	**60**
Ce	**3,971.68**	**6**	**35**
Rn	**3,971.67**	**(80)**	—
Nd	**3,971.69**	**4**	**10**
Fe	3,971.33	125	200
H	3,970.07	(80)	—
Cr	3,969.75	90	200
Os	3,969.67	100	100
Fe	3,969.26	400	600
Ca	3,968.47	500*R*	500*R*
Ru	3,968.46	200	12
Ar	3,968.36	(200)	—
Xe	3,967.54	(200)	—
Fe	3,967.42	100	125
Pr	3,966.57	70*d*	100*d*
Fe	3,966.07	70	100
Pr	3,964.82	80*d*	125*d*
Cr	3,963.69	300	300
Al	3,961.53	2000	3000
Mo	3,961.50	500	5
Rh	3,958.86	100	200
Pd	3,958.64	200	500*w*
Zr	3,958.22	150	500
Ti	3,958.21	100	150
Gd	3,957.68	200	300*W*
P	3,957.62	(100)	—
Fe	*3,956.68*	*150*	*150*
Fe	*3,951.17*	*125*	*150*
Xe	*3,950.92*	*(125)*	—

	λ	I	
		J	O
Pr	**3,974.37**	**8**	**30**
Fe	**3,974.40**	**1**	**10**
Xe	**3,974.42**	**(40)**	—
Ar	**3,974.48**	**(10)**	—
Nd	**3,974.484**	**20**	**20**
Ru	**3,974.50**	**8**	**20**
Sm	**3,974.66**	**4**	**15**
Nd	**3,974.663**	**2**	**20**
Ho	**3,974.70**	**4**	**6**
Er	**3,974.72**	**3**	**15**
Co	**3,974.73**	**10**	**100**
Ar	**3,974.76**	**(15)**	—
F	**3,974.79**	**(20)**	—
Gd	**3,974.81**	**10**	**10**
Pr	**3,974.858**	**10**	**15**
Mo	**3,974.859**	**20**	—
Nb	3,976.51	80*h*	—
Cr	3,976.66	300	300
Tb	3,976.82	200	150
Mn	3,977.08	100	50
Fe	3,977.74	150	300
Ru	3,978.44	70	60
Rn	3,981.68	(150)	—
Ti	3,981.76	70	100
Fe	3,981.77	100	150
Tb	3,981.88	200	80
Dy	3,981.94	100	150
Pr	3,982.06	100	125
Yt	3,982.59	100	60
O	3,982.72	(20)	—
Fe	3,983.96	125	200
Hg	3,983.98	(400)	—
Ru	3,984.86	70	60
Mn	3,985.24	100	75
Mn	3,986.83	75	40
Gd	3,987.22	100	100
Yb	3,987.99	500*R*	1000*R*
La	3,988.52	800	1000
Zn	3,989.23	(100)	—
Pr	*3,989.72*	*125*	*200*
N	*3,994.99*	*(300)*	—
La	*3,995.75*	*300*	*600*
Fe	*3,997.40*	*150*	*300*
Dy	*4,000.45*	*300*	*400*
Fe	*4,005.25*	*200*	*250*
Tb	*4,005.55*	*125*	*100d*

	λ	I J	I O		λ	I J	I O
La	*3,949.11*	*800*	*1000*				
Ar	*3,948.98*	*(2000)*	—				

*P** *N**

O 4,072.16 (300)

	λ	J	O		λ	J	O
V	**4,072.14**	**2**	**15**	**Er**	**4,072.38**	**3*wh***	**10*d***
P	**4,072.13**	**(30)**	—	**Ir**	**4,072.39**	**3**	**10**
Se	**4,072.11**	**(20)**	—	**Ar**	**4,072.40**	**(40)**	—
Xe	**4,072.10**	**(4*h*)**	—	**In**	**4,072.40**	**200*wh***	—
Nb	**4,072.07**	**15**	—	**Fe**	**4,072.51**	**1**	**2**
U	**4,072.03**	**4**	**2**	**Pr**	**4,072.52**	**2**	**10**
Ar	**4,072.01**	**(150)**	—	**Dy**	**4,072.65**	**2**	**7**
J	**4,072.01**	**(2)**	—	**Cr**	**4,072.68**	**1**	**12**
Pr	**4,071.99**	**1**	**5**	**Zr**	**4,072.70**	**3**	**100**
W	**4,071.93**	**7**	**8**	**U**	**4,072.83**	**2*h***	**8**
Rb	**4,071.9**	**(12)**	—	**Ce**	**4,072.92**	**2**	**20**
Ce	**4,071.81**	**5**	**30**	**Ta**	**4,073.00**	**15*W***	**2**
Sn	**4,071.79**	**(4)**	—	**N**	**4,073.04**	**(5*h*)**	—
Fe	**4,071.74**	**200**	**300**	**Nb**	**4,073.09**	**8**	**2**
As	**4,071.64**	**10**	—	**Dy**	**4,073.11**	**15**	**80**
Os	**4,071.56**	**3**	**30**	**Ru**	**4,073.116**	**25**	**6**
V	**4,071.54**	**2**	**15**	**Er**	**4,073.120**	**1**	**20**
Ti	**4,071.47**	**1**	**8**	**Ho**	**4,073.13**	**2**	**3**
Ru	**4,071.40**	**20**	**12**	**W**	**4,073.15**	**8**	**9**
Pr	**4,071.38**	**1**	**4**	**Tb**	**4,073.16**	**1**	**10**
Ca	**4,071.30**	**2**	—	**Gd**	**4,073.21**	**80**	**80**
O	**4,071.24**	**(5)**	—	**Fe**	**4,073.45**	**8**	**8**
Hf	**4,071.22**	**4**	**2**	**Ce**	**4,073.48**	**8**	**50**
Ti	**4,071.21**	**1**	**10**	**Xe**	**4,073.50**	**(8)**	—
Eu	**4,071.20**	**2*h***	**6**	**Te**	**4,073.47**	**(300)**	—
U	**4,071.11**	**25**	**15**	**Os**	**4,073.62**	**3**	**6**
Zr	**4,071.09**	**2**	**3**	**Ce**	**4,073.73**	**3**	**30**
Ce	**4,071.08**	**2**	**20**	**Tb**	**4,073.75**	**2**	**9**
Os	**4,071.01**	**4**	**12**	**Gd**	**4,073.771**	**50**	**100**
Cr	**4,071.00**	**4**	**15**	**Fe**	**4,073.775**	**20**	**80**
Sm	**4,070.98**	**3**	**5**	**W**	**4,073.87**	**3**	**4**
Nb	**4,070.96**	**5**	**3**				
As	**4,070.92**	**10**	—				
Nd	**4,070.90**	**1**	**4*d***				
U	**4,070.98**	**1**	**6**				
Os	**4,070.85**	**4**	**20**				
Fe	**4,070.779**	**20**	**50**				
U	**4,070.777**	**4**	**2**				
V	**4,070.77**	**5**	**10**				
J	**4,070.75**	**(150)**	—				

*N**

	λ	I	
		J	O
Ir	**4,070.68**	**10**	**30**
W	**4,070.61**	**12**	**15**
Tb	**4,070.57**	**1**	**12*w***
Se	4,070.16	(500)	—
O	4,069.90	(125)	—
O	4,069.63	(60)	—
Kr	4,065.11	(300)	—
Fe	4,063.60	300	400
O	4,062.94	(30*h*)	—
O	4,061.00	(15*h*)	—
O	4,060.60	(30*h*)	—
Nb	4,058.94	400*w*	1000
Pb	4,057.82	300*R*	2000
Kr	4,057.01	(300*hs*)	—
In	4,056.94	(500)	—
Ag	4,055.26	500*R*	80
Hg	*4,046.56*	*300*	*200*
Fe	*4,045.81*	*300*	*400*
Ar	*4,044.42*	*(1200)*	—
K	*4,044.14*	*400*	*800*
La	*4,042.91*	*300*	*400*

O 4,075.87 (800)

	λ	J	O
Sm	**4,075.83**	**40**	**40**
Ce	**4,075.79**	**2**	**10**
P	**4,075.74**	**(15)**	—
Ce	**4,075.714**	**2**	**15**
Tb	**4,075.713**	**1**	**5**
U	**4,075.66**	**6**	**5**
V	**4,075.65**	**2**	—
J	**4,075.56**	**(25)**	—
Nd	**4,075.559**	**2**	**5*d***
Mo	**4,075.54**	**20**	**20**
Br	**4,075.51**	**(12)**	—
Pr	**4,075.33**	**1**	**5**
Nd	**4,075.27**	**4**	**12**
Mn	**4,075.254**	**5**	**25**
Mo	**4,075.246**	**25**	**25**
Tb	**4,075.23**	**1**	**8**
Nd	**4,075.12**	**10**	**15**
U	**4,075.08**	**2**	**2**
Sc	**4,074.98**	**3**	**10**
Zr	**4,074.93**	**1**	**12**
C	**4,074.89**	**40**	—
Mo	**4,075.937**	**5**	**10**
Fe	**4,075.937**	**5**	**5**
U	**4,075.94**	**1*h***	**6**
Ru	**4,075.98**	**25**	**2**
C	**4,076.00**	**80**	—
Cr	**4,076.06**	**15**	**30**
Nb	**4,076.09**	**3**	**4**
Mo	**4,076.19**	**25**	**25**
Pr	**4,076.21**	**1**	**10**
Fe	**4,076.222**	**1**	**2**
U	**4,076.225**	**3**	**2**
Ce	**4,076.24**	**1**	**12**
Au	**4,075.33**	**25*w***	**4**
Yt	**4,076.35**	**8**	**30**
Ir	**4,076.370**	**2**	**2**
Ti	**4,076.375**	**1**	**15**
Fe	**4,076.50**	**1**	**2**
Mo	**4,076.51**	**5**	**5**
Co	**4,076.57**	**1**	**3*h***
Sm	**4,076.63**	**15**	**25**
Fe	**4,076.637**	**50**	**80**

	λ	I J	I O
Cr	**4,074.86**	**10**	**25**
Pr	**4,074.84**	**1**	**5**
Fe	**4,074.79**	**40**	**80**
Yb	**4,074.69**	**2**	—
Os	**4,074.68**	**6**	**80**
Pr	**4,074.65**	**1**	**3**
C	**4,074.53**	**50**	—
Br	**4,074.49**	**(4)**	—
U	**4,074.488**	**10**	**10**
Mo	**4,074.44**	**20**	—
Nd	**4,074.42**	**3**	**8**
W	**4,074.364**	**45**	**50**
Ti	**4,074.362**	**1**	**15**
U	**4,074.24**	**2**	**6**
Ar	**4,076.64**	**(20)**	—
La	**4,076.71**	**5**	**15**
U	**4,076.72**	**10**	**8**
Ru	**4,076.73**	**25**	**60**
Fe	**4,076.80**	**1**	**8**
Sm	**4,076.85**	**5**	**10**
Ar	**4,076.96**	**(10)**	—
Zr	**4,077.05**	**1**	**3**
W	**4,077.06**	**5**	**6**
As	**4,077.08**	**10**	—
Nb	**4,077.088**	**5**	**3**
Cr	**4,077.089**	**10**	**35**
Nd	**4,077.150**	**4**	**10**
Ti	**4,077.153**	**2**	**18**
Yb	**4,077.27**	**100**	**30**
La	**4,077.34**	**400**	**600**
Yt	**4,077.37**	**40**	**50**
Co	**4,077.41**	**2*h***	**100*wh***
Ce	**4,077.47**	**4**	**18**
Rh	**4,077.57**	**4**	**5**
Sr	4,077.71	500*W*	400*r*
Hg	4,077.81	150	150
O	4,078.85	(70)	—
Nb	4,079.73	200*w*	500*w*
P	4,080.04	(150)	—
Ru	4,080.60	300	125
O	4,083.91	(40*h*)	—
O	4,084.66	(30)	—
O	4,085.12	(70)	—
La	4,086.71	500	500
O	4,087.14	(40*h*)	—
Kr	4,088.33	(500)	—
O	4,089.27	(60*h*)	—
O	*4,092.94*	*(80)*	—
O	*4,096.53*	*(30)*	—
O	*4,097.24*	*(70)*	—
In	*4,101.77*	*1000R*	*2000R*
O	*4,103.01*	*(50)*	—
F	*4,103.52*	*(300)*	—
Ho	*4,103.84*	*400*	*400*
O	*4,104.73*	*(50)*	—
O	*4,105.00*	*(125)*	—
Se	*4,108.83*	*(800)*	—
N	*4,109.98*	*(1000)*	—
O	*4,110.79*	*(40)*	—

O 4,368.30 (1000)

	λ	I J	I O
Cr	**4,368.25**	**6**	**15**
U	**4,368.24**	**3**	**3**
Ce	**4,368.23**	**1**	**8**
C	**4,368.14**	**(30*h*)**	—
Te	**4,368.09**	**(30)**	—
V	**4,368.04**	**8**	**12**
Sm	**4,368.03**	**60**	**0**
Nb	**4,367.97**	**50*h***	**2**
Fe	**4,367.906**	**70**	**60**
Hf	**4,367.905**	**20**	**25**
Tm	**4,367.90**	**10**	**20**
Ar	**4,367.87**	**(10)**	—
Cs	**4,367.66**	**(10)**	—
Ti	**4,367.658**	**25**	**8**
Fe	**4,367.58**	**50**	**100**
Ce	**4,367.56**	**1**	**6**
Nb	**4,367.39**	**10*h***	**1**
Sb	**4,367.38**	**3**	—
Pr	**4,367.23**	**5*w***	**25*w***
Nd	**4,367.16**	**2**	**10**
As	**4,367.10**	**5**	—
O	4,366.91	(100)	—
Kr	4,362.64	(500)	—
Kr	*4,355.48*	*(3000)*	—
O	*4,351.27*	*(125)*	—
Rn	*4,349.60*	*(5000)*	—
O	*4,349.43*	*(300)*	—
O	*4,347.43*	*(70)*	—
Ar	*4,345.17*	*(1000)*	—

	λ	I J	I O
Pr	**4,368.33**	**90**	**125**
Ar	**4,368.36**	**(5)**	—
Eu	**4,368.42**	**2**	**8*w***
Nb	**4,368.43**	**30**	**15**
V	**4,368.60**	**9**	**15**
Nd	**4,368.63**	**15**	**50**
W	**4,368.767**	**2**	**10**
Cs	**4,368.77**	**(10)**	—
Mo	**4,368.772**	**4**	**5**
Mo	**4,368.04**	**25**	**40**
W	**4,369.05**	**2**	**5**
V	**4,369.06**	**5**	**9**
Gd	**4,369.17**	**20**	**50**
Xe	**4,369.20**	**(100*wh*)**	—
Ce	**4,369.24**	**1**	**10**
Br	**4,369.25**	**(2)**	—
O	**4,369.28**	**(50)**	—
Th	**4,369.32**	**5**	**10**
Ta	**4,369.35**	**10**	**15**
Eu	**4,369.47**	**2**	**20*w***
Cl	**4,369.52**	**(12)**	—
Rh	4,374.80	500	1000*W*
Kr	4,376.12	(800)	—
O	4,379.55	(15*h*)	—
Se	*4,382.87*	*(800)*	—
Fe	*4,383.55*	*800*	*1000*

O 4,414.888 (300)

	λ	I J	I O
Nb	**4,414.881**	**3*w***	**4*w***
Mn	**4,414.879**	**60**	**150**
Xe	**4,414.84**	**(150)**	—
U	**4,414.74**	**3*h***	**6**
Gd	**4,414.73**	**50**	**100**
Eu	**4,414.65**	**2**	**6**
Cd	**4,414.63**	**200**	—
P	**4,414.60**	**(70)**	—
Zr	**4,414.544**	**3**	**4**
V	**4,414.541**	**5**	**10**

	λ	I J	I O
Zr	**4,414.893**	**2**	**3**
V	**4,415.06**	**4**	**7**
W	**4,415.07**	**6**	**15**
Fe	**4,415.12**	**400**	**600**
Ne	**4,415.14**	**(5)**	—
U	**4,415.24**	**12**	**12**
Yt	**4,415.39**	**2**	**3**
Sb	**4,415.40**	**4**	—
Gd	**4,415.43**	**1**	**3*h***
Sc	**4,415.56**	**25**	**100**

	λ	I J	I O
Nd	**4,414.43**	**3**	**12**
Pr	**4,414.40**	**4**	**20**
Mo	**4,414.35**	**4**	**5**
P	**4,414.28**	**(100)**	—
Pt	**4,414.25**	**1**	**2**
Gd	**4,414.16**	**60**	**100**
Cr	**4,413.87**	**15**	**25**
W	**4,413.86**	**1*h***	**5**
Ce	**4,413.80**	**1**	**20**
Nd	**4,413.78**	**1**	**5**
Pr	**4,413.76**	**40**	**90**
Ca	**4,413.74**	**3**	—
Ba	**4,413.68**	**3**	**10**
Ne	4,409.30	(150)	—
J	4,408.96	(250)	—
Gd	4,408.26	150	100
Gd	4,406.67	200	70
Fe	4,404.75	700	1000
Hg	*4,398.62*	*(300)*	—

	λ	I J	I O
Mo	**4,415.67**	**3**	**2**
In	**4,415.69**	**15*h***	—
Cd	**4,415.70**	**20**	**1**
W	**4,415.71**	**3**	**10**
Ta	**4,415.44**	**10**	**40**
Se	**4,415.85**	**(10)**	—
U	**4,415.86**	**6**	**8**
Nd	**4,415.98**	**4**	**12**
Sm	**4,416.00**	**3**	**5**
Xe	**4,416.07**	**(80*wh*)**	—
Th	**4,416.24**	**8**	**15**
O	4,416.97	(150)	—
Gd	4,419.04	200	200
Sm	4,420.53	200	200
Sm	4,421.14	150	150
Ne	4,422.52	(300)	—
Sm	4,424.34	300	300
Ne	4,424.80	(300)	—
Ne	4,425.40	(150)	—
Ar	4,426.01	(300)	—
La	*4,429.90*	*300*	*200*
Kr	*4,431.67*	*(500)*	—
Kr	*4,436.81*	*(600)*	—

O 5,330.66 (500)

	λ	I J	I O
Br	**5,330.57**	**(15)**	—
J	**5,330.16**	**(2)**	—
Sr	**5,329.82**	**2**	**40**
Rh	**5,329.74**	**2**	**30**
Cr	**5,329.72**	**2**	**5**
O	**5,329.59**	**(150)**	—
U	**5,329.22**	**1**	**10**
Kr	**5,329.15**	**(4*h*)**	—
O	**5,328.98**	**(100)**	—
V	**5,328.82**	**12**	**18*W***
Cu	**5,328.81**	**3**	—
N	**5,328.70**	**(70)**	—
Xe	**5,328.69**	**(2)**	—
Fe	**5,328.53**	**35**	**150**
Ra	5,320.29	(250)	—
Xe	5,313.87	(500)	—

	λ	I J	I O
Ne	**5,330.78**	**(600)**	—
Kr	**5,331.08**	**(2)**	—
Nb	**5,331.19**	**2*h***	—
Co	**5,331.47**	**80**	**500*w***
Pr	**5,331.48**	**1**	**6**
As	**5,331.54**	**200**	—
U	**5,331.85**	**2**	**2**
Br	**5,332.04**	**(100)**	—
Sn	**5,332.36**	**(20)**	—
V	**5,332.66**	**5**	—
Kr	5,333.41	(500*h*)	—
Yb	5,335.16	400	150
J	5,338.19	(300)	—
Xe	5,339.38	(500)	—
Ne	5,341.09	(1000)	—
Ne	5,343.28	(600)	—
J	5,345.15	(300)	—

	λ	I J	I O
Se	*5,305.35*	*(500)*	—
O	*5,299.00*	*(70)*	—
Xe	*5,292.22*	*(800)*	—
Tl	5,350.46	2000*R*	5000*R*
Cs	*5,358.53*	*(500)*	—

O 7,771.93 (1000)
7,774.14 (300)
7,775.43 (100)

	λ	I J	I O
Kr	**7,772.40**	**(5*h*)**	—
Xe	**7,772.12**	**(20*h*)**	—
Sb	**7,772**	**15**	—
O	***7,771.93***	***(1000)***	—
Cl	**7,771.13**	**(2)**	—
F	7,754.70	(60)	—
Kr	*7,735.69*	*(200h)*	—
O	**7,774.14**	**(300)**	—
Xe	**7,774.18**	**(4)**	—
O	***7,775.43***	***(100)***	—
Kr	**7,776.28**	**(40)**	—
In	**7,776.57**	**(2)**	—
In	**7,776.75**	**(5)**	—
In	**7,776.96**	**(10)**	—
Xe	**7,777.1**	**(10)**	—
Kr	7,781.97	(100*h*)	—
Xe	7,783.66	(50)	—
Xe	7,787.04	(50)	—
Xe	*7,802.65*	*(100)*	—
Rn	*7,809.82*	*(100)*	—

Os 76 190.2

t_0 2,700°C t_1 >5,300°C

I.	II.	III.	IV.	V.
~8.7	–	–	–	–

λ	*I*		eV
	J	*O*	
I 2,909.06	400	500*R*	4.2
I 3,058.66	500	500*R*	4.05
I 3,262.29	50	500*R*	4.32
I 3,267.95	30	400*R*	3.80
I 3,301.56	50	500*R*	3.76
I 4,260.85	200	200	2.91
I 4,420.47	100	400*R*	2.80

Os 2,909.06 *J* 400 *O* 500*R*

	λ	*I* *J*	*I* *O*		λ	*I* *J*	*I* *O*
Cr	**2,909.05**	**12**	**60*r***	**Mo**	**2,909.116**	**40*h***	**25**
Eu	**2,909.01**	—	**40**	**W**	**2,909.123**	**8**	**8**
Nd	**2,908.98**	**5*h***	—	**Yb**	**2,909.19**	**7**	**2**
Nb	**2,908.979**	**5**	**1**	**Ru**	**2,909.22**	**2**	**12**
Ta	**2,908.91**	**10**	**150**	**U**	**2,909.25**	**15**	**6**
Nb	**2,908.881**	**20**	**2**	**Fe**	**2,909.31**	**2**	**4**
Mn	**2,908.879**	—	**10**	**Dy**	**2,909.32**	—	**2**
U	**2,908.878**	**2**	**2**	**Hg**	**2,909.36**	**(25)**	—
Ru	**2,908.877**	**8**	—	**Ho**	**2,909.42**	**10**	**40**
Fe	**2,908.859**	**40**	**80**	**Ti**	**2,909.46**	**4*h***	—
Hf	**2,908.858**	**5**	**3**	**Yb**	**2,909.48**	**8**	**2**
V	**2,908.82**	**400*R***	**70*r***	**Fe**	**2,909.50**	**35**	**70**
Cd	**2,908.74**	—	**5**	**Ir**	**2,909.56**	**5**	**18**
Tm	**2,908.69**	**10**	**5**	**Er**	**2,909.58**	**1**	**7**
Kr	**2,908.62**	**(5)**	—	**Fe**	**2,909.59**	—	**2**
Nd	**2,908.60**	**5**	—	**Ce**	**2,909.61**	—	**5**
Er	**2,908.53**	—	**6**	**W**	**2,909.628**	**3**	**4**
W	**2,908.493**	**10**	—	**La**	**2,909.631**	—	**2**
Tb	**2,908.49**	**20**	**3**	**Os**	**2,909.67**	**5**	**12**
V	**2,908.44**	**20**	**2**				
				V	2,911.06	200*r*	30
Nb	2,908.24	200	20*r*	Lu	2,911.39	300	100
Eu	2,906.68	300	300*W*	Pt	2,912.26	25	300
Os	2,905.97	8	12	Os	2,912.33	50	50
Os	2,905.73	8	40	Pt	2,913.54	25	300
Ta	2,904.07	40	300*w*				
				Os	2,913.84	8	30
Ta	*2,902.05*	*200*	*1,000w*				
Nb	*2,899.24*	*500*	*20*	*Os*	*2,917.26*	*20*	*40*
Bi	*2,897.97*	*500WR*	*500WR*	*Os*	*2,919.79*	*15*	*100*

Os 3,058.66 *J* 500 *O* 500*R*

	λ	*I* *J*	*I* *O*		λ	*I* *J*	*I* *O*
Ru	**3,058.655**	**3**	**30**	**Re**	**3,058.786**	—	**50**
Ir	**3,058.653**	—	**5**	**Ru**	**3,058.786**	**3**	**30**
Ta	**3,058.64**	**3**	**50**	**Pr**	**3,058.97**	**5**	—
Cs	**3,058.6**	**(6)**	—	**U**	**3,058.98**	**5*d***	**8*d***
Mo	**3,058.597**	—	**20**	**Tm**	**3,058.99**	**30**	**10**
Ce	**3,058.55**	—	**12**	**Eu**	**3,059.00**	—	**20**
Fe	**3,058.49**	**3**	**3**	**S**	**3,059.07**	**(8)**	—
Th	**3,058.43**	**10**	**10**	**Fe**	**3,059.09**	**400**	**600*r***
Cr	**3,058.34**	**30**	—	**J**	**3,059.11**	**(5)**	—
Ga	**3,058.15**	**4*w***	—	**Ne**	**3,059.16**	**(7)**	—
Th	**3,058.142**	**3**	**8**	**Ru**	**3,059.17**	—	**50**
F	**3,058.14**	**(30)**	—	**Cd**	**3,059.22**	**2**	—
Ce	**3,058.11**	—	**5**	**U**	**3,059.27**	**3**	**4**

	λ	I J	I O
As	3,058.10	15	—
Ti	3,058.09	70	12
Cl	3,058.00	(40)	—
Ir	3,057.95	—	3
U	3,057.91	20d	20d
Ni	3,057.64	125	400R
Fe	3,057.45	400	400
Ne	3,057.39	(250)	—
Os	3,056.90	8	20
J	3,055.37	(350)	—
Os	3,055.21	15	40
Os	3,054.97	10	50
Eu	3,054.93	3	400 w
Ni	3,054.32	100	400R
Rn	3,054.3	(250)	—
Os	3,052.42	12	50
Os	3,051.17	15	80
Ni	*3,050.82*	—	*1000R*
Os	*3,050.39*	*50*	*100*
Os	*3,049.46*	*15*	*80*
Os	*3,049.04*	*8*	*40*
Fe	*3,047.60*	*500r*	*800r*
Os	*3,045.32*	*12*	*30*
Os	*3,044.41*	*10*	*50*
Os	*3,044.07*	*8*	*30*
Os	*3,043.64*	*12*	*60*

	λ	I J	I O
Nb	3,059.30	10	2
Rh	3,059.36	—	4
Os	3,060.30	30	100
Os	3,062.19	30	100
Os	3,062.47	8	20
Os	3,062.69	8	20
Cu	3,063.41	50	300
Pt	3,064.71	300R	2000R
Os	3,066.12	10	50
Os	*3,066.60*	*8*	*20*
Os	*3,066.83*	*10*	*30*
Bi	*3,067.72*	*2000wh*	*3000hR*
Os	*3,069.12*	*10*	*30*
Os	*3,069.94*	*15*	*125*

Os 3,262.29 J 50* O 500R

	λ	I J	I O
Fe	3,262.28	25	50
Ba	3,262.275	3	3
Dy	3,262.27	—	2
Sm	3,262.26	4	10
Mo	3,262.19	20	1
Ce	3,262.14	—	5
V	3,262.06	3	10
Dy	3,262.02	1	2
Fe	3,262.013	15	30
Ir	3,262.01	2	20
Ba	3,261.96	—	40
Nb	3,261.88	3	10
Mo	3,261.84	—	2
Tb	3,261.74	3	15
U	3,261.72	10	15

	λ	I J	I O
Sn	3,262.33	300h	400h
Pb	3,262.353	5h	20h
Mo	3,262.353	1	8
Nd	3,262.36	2h	10d
Pr	3,262.39	—	2
Tb	3,262.40	—	2
Co	3,262.43	—	2
Hf	3,262.47	1	10
Eu	3,262.49	2h	5
Sm	3,262.58	1	3
Eu	3,262.61	—	15
Mo	3,262.63	4	15
Th	3,262.67	15	12
Tb	3,262.68	3	8
Ir	3,262.72	—	3

	λ	I J	I O		λ	I J	I O
Nb	3,261.695	50	1	Os	3,262.75	20	100
Pt	3,261.692	2	8	Re	3,262.76	—	25
Tm	3,261.66	100	30	Er	3,262.80	4	15
Ce	3,261.63	2h	2	Ir	3,262.93	—	4
Tl	3,261.60	300r	70	Tb	3,262.97	3	30
Eu	3,261.59	—	2h	Ta	3,263.00	1	20
Re	3,261.55	—	50	Eu	3,263.02	—	2W
Th	3,261.54	10	8	Th	3,263.03	8	8
Ce	3,261.52	—	2	Fe	3,263.07	—	1
Yb	3,261.51	18	5	Ce	3,263.071	—	10
				W	3,263.10	7	8
Cd	3,261.06	300	300	Z	3,263.11	8	25
Os	3,260.57	5	15	Rh	3,263.14	40	200
Os	3,260.30	10	60	Co	3,263.21	—	30
Pd	3,258.78	200h	300				
In	3,258.56	300R	500R				
Os	3,256.92	12	80				
In	3,256.09	600R	1500R				
Os	3,254.91	12	60				
Fe	3,254.36	150	200				
Co	3,254.21	—	300R				
Os	*3,252.01*	*15*	*50*				
Cu	*3,247.54*	*2000R*	*5000R*				
La	*3,245.12*	*300*	*400*				

*N**

Os 3,267.95 *J* 30* *O* 400*R*

	λ	J	O		λ	J	O
Yb	3,267.89	—	6	Bi	3,267.97	—	2h
Mn	3,267.79	40	40	Ce	3,268.02	—	2
Dy	3,267.76	—	2	Ni	3,268.06	—	5
V	3,267.70	80R	30	Re	3,268.08	—	40
Nb	3,267.68	10	1	Tb	3,268.10	8	30
Gd	3,267.64	—	2	W	3,268.12	3	5
Mo	3,267.639	30	4	Mo	3,268.19	2	6
U	3,267.638	2	8	Ru	3,268.21	12	60
Ta	3,267.56	—	5h	Pr	3,268.23	1	10
Sm	3,267.51	—	2	Fe	3,268.24	100	125
Sb	3,267.50	150Wh	150	Cu	3,268.28	10	15
Rh	3,267.48	15	2	Gd	3,268.34	2	4
Ti	3,267.41	—	2	Pt	3,268.42	4	15
Tm	3,267.41	40	50	U	3,268.46	—	3
Zr	3,267.36	—	3	Rh	3,268.47	5	25
La	3,267.31	3	2	Re	3,268.48	—	40
Nd	3,267.25	4	10	Tb	3,268.52	3	15
U	3,267.244	6	8	W	3,268.58	9	8
Ce	3,267.237	—	12	Ti	3,268.61	—	4
Os	3,267.20	15	20	U	3,268.65	—	2

	λ	I J	I O
Hf	**3,267.17**	**1**	**10**
Er	**3,267.15**	**4**	**4**
Ir	**3,267.10**	**—**	**2**
Au	**3,267.08**	**8**	**10**
Ti	**3,267.07**	**—**	**10**
Nb	**3,267.05**	**5**	**5**
Eu	**3,267.01**	**2**	**4*wh***
La	3,265.67	200	300
Fe	3,265.62	300	300
Fe	3,265.05	150	200
Os	3,264.69	10	100*h*

*P**

	λ	I J	I O
Eu	**3,268.66**	**—**	**6*W***
Mn	**3,268.72**	**30**	**30**
Ru	**3,268.79**	**60**	**4**
Er	**3,268.80**	**1**	**12**
Os	3,269.21	20	200
Ge	3,269.49	300	300
Os	3,270.85	4	15*h*
Fe	3,271.00	300	300
Rh	3,271.61	60	200
Os	3,271.98	8	15
Os	3,272.10	10	40
Os	3,272.48	1	20
Os	3,273.38	5	15
Cu	3,273.96	1500*R*	2000*R*
Ta	3,274.95	35*W*	200
Os	3,275.20	15	200
Os	*3,277.97*	*8*	*80*
Eu	*3,280.682*	*—*	*1000R*
Ag	*3,280.683*	*1000R*	*2000R*
Zn	*3,282.33*	*300*	*500R*

*N**

Os 3,301.56 *J* 50* *O* 500*R*

	λ	J	O
Ca	**3,301.52**	**6**	**4**
Tm	**3,301.51**	**10**	**5**
Nb	**3,301.49**	**100**	**1**
Fe	**3,301.43**	**—**	**1**
Th	**3,301.35**	**5*d***	**6*d***
W	**3,301.28**	**8**	**2**
Eu	**3,301.24**	**1**	**4*w***
Ce	**3,301.23**	**—**	**3**
Fe	**3,301.22**	**7**	**15**
U	**3,301.09**	**1**	**8**
Er	**3,301.08**	**—**	**10**
Sm	**3,300.970**	**10**	**20**
Re	**3,300.970**	**—**	**25**
Ce	**3,300.955**	**—**	**8**
Mn	**3,300.948**	**—**	**3**
Dy	**3,300.93**	**2**	**4**
Nd	**3,300.91**	**2*d***	**8*d***
W	**3,300.820**	**12**	**15**
Zr	**3,300.819**	**—**	**2**
Cr	**3,300.80**	**—**	**10**
Eu	**3,300.75**	**1*h***	**4**
Mo	**3,300.69**	**6**	**6**

	λ	J	O
Eu	**3,301.57**	**1**	**15*w***
Ru	**3,301.59**	**40**	**70**
Re	**3,301.60**	**—**	**50**
U	**3,301.65**	**1**	**6**
Th	**3,301.651**	**1*h***	**6**
Ce	**3,301.652**	**—**	**3**
Sm	**3,301.67**	**8**	**15**
Er	**3,301.68**	**—**	**3**
Mo	**3,301.71**	**4**	**4**
Dy	**3,301.72**	**—**	**2**
Sr	**3,301.73**	**10**	**100**
U	**3,301.75**	**3**	**10**
Ir	**3,301.76**	**—**	**2**
W	**3,301.85**	**10**	**1**
Pt	**3,301.86**	**250*W***	**300**
Ta	**3,301.895**	**3**	**25**
Pr	**3,301.898**	**—**	**10**
Ce	**3,301.90**	**—**	**10**
Ru	**3,301.91**	**8**	**30**
Er	**3,301.93**	**2**	**15**
Eu	**3,301.95**	**2**	**25**
Dy	**3,302.02**	**—**	**2**

	λ	I J	I O
U	3,300.68	8*h*	15*h*
Fe	3,292.59	150	300
Os	*3,290.26*	*20*	*200*
Yb	*3,289.85*	*—*	*1000*
Yb	*3,289.37*	*1000R*	*500R*
Os	*3,288.84*	*15*	*30*
Fe	*3,286.75*	*400*	*500*
Os	*3,286.67*	*5*	*30*

	λ	I J	I O
Sm	3,302.09	4	8
Ti	3,302.10	20	8
Pt	3,302.12	—	2*h*
Pd	3,302.13	200*h*	1000*wh*
Yt	3,302.17	2	5
Nb	3,302.18	10	5
Cr	3,302.19	1*h*	30*h*
Re	3,302.23	—	30
U	3,302.26	2*h*	1*h*
Na	3,302.32	300*R*	600*R*
Ta	3,302.33	1*h*	3
Yb	3,302.44	5	7
Tm	3,302.45	80	125
Dy	3,302.47	2	4
Zn	3,302.59	300	800
Zn	3,302.94	300*R*	700*R*
Na	3,302.99	150*R*	300*R*
La	3,303.11	150	400
Os	3,305.38	10	20
Fe	3,305.97	300	400
Os	3,306.23	12	80
Os	*3,310.91*	*30*	*200*
Os	*3,315.42*	*15*	*50*
Os	*3,315.69*	*15*	*40*
Os	*3,316.69*	*15*	*30*

*P**

Os 4,260.85 *J* 200 *O* 200

	λ	J	O
Kr	4,260.85	(5*h*)	—
Ge	4,260.80	10*h*	20
Tb	4,260.78	—	8*h*
V	4,260.755	3	—
Ti	4,260.751	1	12
Ce	4,260.72	—	2
Mo	4,260.65	20	20
Br	4,260.58	(4)	—
Sb	4,260.55	(10)	—
Fe	4,260.48	300	400
Mo	4,260.36	20	20
Ce	4,260.345	—	2
Th	4,260.342	1	4
W	4,260.29	5	10
J	4,260.14	(15)	—
Fe	4,260.13	3	7

	λ	J	O
Ir	4,260.90	—	10
Re	4,260.91	—	10
Mo	4,260.977	15	15
Eu	4,260.980	—	12
Hf	4,260.98	—	20
Pr	4,261.04	1	4
Ce	4,261.07	—	3
Te	4,261.08	(300)	—
Ce	4,261.16	1	18
Eu	4,261.17	2	4
V	4,261.208	3	9
Zr	4,261.210	—	7
Cl	4,261.22	(20)	—
Ir	4,261.258	—	25
Re	4,261.16	—	5
Th	4,261.28	2	5

	λ	I	
		J	O
Gd	4,260.11	10	30
Pr	4,260.06	1	3
Ir	4,260.03	—	25
Sb	4,260.01	2	—
Ru	4,260.004	—	12
Fe	4,260.003	5	15
Tb	4,259.96	—	5
W	4,259.941	5	12
Pt	4,259.938	—	5
Au	4,259.89	15	—
Re	4,259.88	—	3
Ce	4,259.75	1	15
Ar	4,259.36	(1200)	—
Ir	4,259.11	10	200
Mn	4,257.66	40	100
Re	4,257.59	—	125w
Sm	4,256.40	150	150
Kr	4,254.85	(100h)	—
Ho	4,254.43	20	100
Cr	4,254.35	1000R	5000R
O	4,253.98	(100h)	—
Co	4,252.31	—	150
Sm	4,251.79	200	200
Gd	4,251.74	10	300
Ar	4,251.18	(800)	—
Fe	4,250.79	250	400
Mo	4,250.69	125	5
Kr	4,250.58	(150)	—
Fe	4,250.13	150	250
La	4,249.99	50	100
Fe	4,247.43	100	200
Sc	4,246.83	500	80
F	4,246.16	(300h)	—
Xe	4,245.38	(200h)	—
Tm	4,242.15	100	500
Cr	4,240.70	30	200
Fe	4,238.82	100	200
Gd	4,238.78	200	200
La	4,238.38	300	500
Mn	4,261.30	5	20
Cr	4,261.35	50	125
Zr	4,261.42	—	6
Mo	4,261.44	20	20
Ce	4,261.48	—	4
U	4,261.50	8	8
Ti	4,261.60	8	70
Cr	4,261.61	8	35
Nb	4,261.71	8	5
Ga	4,261.78	2	—
Eu	4,261.794	—	4
Pr	4,261.796	2	15
bhCa	4,261.8	—	2
Tb	4,261.83	—	8
Nd	4,261.84	10	20
Hg	4,261.88	(70)	—
Ir	4,261.89	—	10
Cr	4,261.91	30	—
Ga	4,261.93	4	—
Gd	4,262.09	10	150
Sm	4,262.68	150	200
Ti	4,263.13	35	125
Cr	4,263.14	80	125
La	4,263.58	150	150
Os	4,264.75	3	20
Mn	4,265.92	50	100
Ar	4,266.29	(1200)	—
Ar	4,266.53	(200)	—
C	4,267.02	350	—
C	4,267.27	500	—
Nd	4,267.49	200wh	—
Fe	4,267.83	60	125
Ir	4,268.10	15	200
Kr	4,268.81	(100wh)	—
La	4,269.49	150	150
Os	4,269.61	3	30
Os	4,270.79	1	12
Fe	4,271.16	300	400
Fe	4,271.76	700	1000
Ar	4,272.17	(1200)	—
Li	4,273.28	100h	200r
Kr	4,273.97	(1000)	—
Cr	4,274.80	800r	4000R
La	4,275.64	500	40
Tb	4,278.51	100	200
Gd	4,280.50	100	200
Sm	4,280.78	200	200
Fe	4,282.41	300	600

Os 4,420.47 *J* 100 *O* 400*R*

	λ	*I*	
		J	*O*
W	**4,420.47**	**10**	**30**
Zr	**4,420.46**	**1**	**20**
Nb	**4,420.45**	**3**	**3**
Ce	**4,420.42**	—	**3**
U	**4,420.41**	**3**	**12**
Ir	**4,420.35**	—	**10**
Pr	**4,420.30**	—	**4**
Tb	**4,420.20**	—	**12**
Pr	**4,420.02**	—	**3**
Na	**4,419.94**	—	**3**
V	**4,419.935**	**20**	**30**
Ce	**4,419.927**	—	**3**
Nb	**4,419.83**	**5**	**3**
Mn	**4,419.78**	**20**	**100**
Mo	**4,419.72**	**2**	**3**
Ce	**4,419.675**	—	**3**
Pr	**4,419.667**	**50**	**100**
Eu	**4,419.657**	**2**	**3***W*
Er	**4,419.61**	**4**	**25**
Ho	**4,419.60**	**2**	**3**
Ta	**4,419.55**	**5**	**10**
Fe	**4,419.54**	**1**	**8**
Nb	**4,419.444**	**20**	**10**
Sm	**4,419.338**	**20**	**40**
Ce	**4,419.30**	**1**	**18**
W	**4,419.26**	**2**	**10**
Gd	4,419.04	200	200
Kr	4,418.76	(50)	—
Ti	4,417.72	80	40
Sm	4,417.58	80	80
Hf	4,417.35	50	25
O	4,416.97	(150)	—
Ne	4,416.82	(50)	—
Xe	4,416.07	(80*wh*)	—
Fe	4,415.12	400	600
O	4,414.89	(300)	—
Mn	4,414.88	60	150
Xe	4.414.84	(150)	—
Gd	4,414.73	50	100
Cd	4,414.63	200	—
P	4,414.60	(70)	—
P	4,414.28	(100)	—
Gd	4,414.16	60	100
As	4,413.64	50	—
Ne	4,413.20	(50)	—
Ac	4,413.17	100	—

	λ	*I*	
		J	*O*
Tb	**4,420.52**	—	**4**
Sm	**4,420.529**	**200**	**200**
Er	**4,420.53**	—	**15**
Ho	**4,420.54**	**2**	**2**
Au	**4,420.63**	**4**	—
Nb	**4,420.635**	**15**	**15**
Gd	**4,420.64**	—	**10**
P	**4,420.64**	**(70)**	—
U	**4,420.641**	**3***h*	**2***h*
Sc	**4,420.66**	**2**	**20**
Mo	**4,420.74**	**5**	**6**
Rb	**4,420.80**	**(2)**	—
Ru	**4,420.84**	—	**40**
*bh***Sr**	**4,420.9**	—	**5**
Ar	**4,420.90**	**(40)**	—
Cr	**4,420.94**	—	**8**
Lu	**4,420.96**	—	**15**
W	**4,421.01**	**3**	**12**
Pd	**4,421.04**	—	**6**
As	**4,421.06**	**10**	—
Ce	**4,421.130**	—	**8**
Sm	**4,421.136**	**150**	**150**
Te	**4,421.14**	**(70)**	—
Tb	**4,421.14**	—	**2**
Pr	**4,421.13**	**35***w*	**100**
Gd	**4,421.24**	**8**	**100**
Ce	**4,421.32**	—	**4**
Co	**4,421.34**	—	**10**
Ne	**4,421.38**	**(30)**	—
Ru	**4,421.46**	—	**60**
Ce	**4,421.473**	—	**3**
Ti	**4,421.474**	—	**10**
Th	**4,421.557**	**2**	**8**
Ne	**4,421.559**	**(50)**	—
V	**4,421.57**	**30***h*	**30***h*
Nb	**4,421.67**	**10***h*	**1***h*
Ne	4,422.52	(300)	—
Fe	4,422.57	125	300
Yt	4,422.59	60	60
Kr	4,422.70	(100*hs*)	—
J	4,423.76	(80)	—
Ar	4,423.99	(80)	—
Sm	4,424.342	300	300
Nd	4,424.343	50	50
Ne	4,424.80	(300)	—
Kr	4,425.19	(100)	—

	λ	I	
		J	O
Te	4,411.78	(50)	—
Gd	4,411.16	50	100
Ti	4,411.08	100	7
Te	4,410.95	(50)	—
Kr	4,410.37	(50)	—
Ru	4,410.03	80	150
Sm	4,409.34	100	100
Ne	4,409.30	(150)	—
J	4,408.96	(250)	—
Pr	4,408.84	100	125
Gd	*4,408.26*	*150*	*100*
Xe	*4,406.88*	*(100wh)*	—
Gd	*4,406.67*	*200*	*70*
Pr	*4,405.85*	*100*	*100*
Fe	*4,404.75*	*700*	*1000*
Gd	*4,403.14*	*100*	*100*
Os	*4,402.74*	*3*	*50*
Te	*4,401.89*	*(100)*	—
Gd	*4,401.85*	*100*	*200*
Ni	*4,401.55*	*30*	*1000W*
Se	*4,401.02*	*(100)*	—
Kr	*4,400.87*	*(100h)*	—
Os	*4,400.58*	*1*	*18*
Kr	*4,399.97*	*(200)*	—
Ti	*4,399.77*	*100*	*40*
Ir	*4,399.47*	*100*	*400*
Hg	*4,398.62*	*(300)*	—
Yt	*4,398.01*	*100*	*150*
Ne	*4,397.94*	*(100)*	—
Os	*4,397.26*	*3*	*18*

	λ	I	
		J	O
Ne	4,425.40	(150)	—
Ar	4,426.01	(300)	—
Ir	4,425.27	10	400w
Ti	4,427.10	60	125
Fe	4,427.31	200	500
As	4,427.38	200	—
La	4,427.57	50	30
P	4,428.15	(70)	—
Ne	4,428.54	(100)	—
Pr	4,429.24	125	200
La	4,429.90	300	200
Ar	4,430.18	(100)	—
Fe	4,430.62	8	200
Ne	4,430.90	(50)	—
Ar	4,431.02	(80)	—
Kr	4,431.67	(500)	—
As	4,431.73	200	—
Se	4,432.33	(60)	—
S	4,432.41	(50)	—
Os	*4,432.41*	*1*	*30*
Mo	*4,433.50*	*125*	*4*
Sm	*4,433.88*	*200*	*200*
Sm	*4,434.32*	*200*	*200*
Rn	*4,435.05*	*(200)*	—
Eu	*4,435.53*	—	*2000*
Eu	*4,435.60*	*100*	*400R*
Gd	*4,436.22*	*100*	*30*
Ra	*4,436.27*	*(200)*	—
Kr	*4,436.81*	*(600)*	—
Gd	*4,438.27*	*100*	*40*
Os	*4,439.64*	*2*	*30*
Fe	*4,442.34*	*200*	*400*
Pt	*4,442.55*	*25*	*800*
Fe	*4,443.20*	*100*	*200*
Ti	*4,443.80*	*125*	*80*
Sm	*4,444.26*	*100*	*100*

P 15 30.9738

t_0 44.1°C t_1 280°C

I.	II.	III.	IV.	V.
10.977	19.653	30.157	51.356	65.01

λ	*I*		eV
	J	*O*	
I 2,534.01	(20)	50	7.2
I 2,535.65	(30)	100	7.2
I 2,553.28	(20)	80	7.1
I 2,554.93	(20)	60	7.1
III 4,222.15	(150)	300	17.54
II 5,296.09	(300)	—	13.14

P 2,534.01 (20) *O* 50

	λ	*I*	
		J	*O*
Ir	**2,534.006**	**—**	**2**
Ru	**2,534.001**	**80**	**4**
W	**2,533.98**	**1**	**10**
Pd	**2,533.97**	**5*h***	**—**
V	**2,533.96**	**10**	**1**
Nb	**2,533.92**	**12*h***	**1**
Cd	**2,533.01**	**(1)**	**2**
Co	**2,533.81**	**60**	**5*r***
V	**2,533.805**	**—**	**10**
U	**2,533.804**	**—**	**4**
Fe	**2,533.803**	**—**	**12**
Au	**2,533.69**	**10*h***	**—**
W	**2,533.633**	**5**	**10**
Fe	**2,533.627**	**50**	**8**
Zr	**2,533.626**	**2**	**—**
J	**2,533.619**	**(100)**	**—**
Ho	2,533.61	10	—
Mo	2,533.56	20	—
Cs	2,533.44	(20)	—
V	2,533.36	15	1
Mn	2,533.32	10	—
Re	2,533.31	—	25
Ru	2,533.24	—	50
Nb	2,533.19	80	2
La	2,533.14	15	—
Ir	2,533.13	20	100
Te	2,533.05	(10)	—
Ta	2,533.00	—	30
W	2,532.95	12	2
Mn	2,532.76	20	—
Bi	2,532.57	—	25*wh*
Zr	2,532.46	20	30
Si	2,532.38	40	30
Mo	2,532.31	30	8
Co	2,532.17	75	—
Ta	2,532.12	100	80
Al	2,532.10	(15)	—
Ni	2,532.08	—	30
Os	2,231.98	20	8
Tl	2,531.6	100	—
Na	2,531.55	(60)	6
Ta	2,531.52	—	25
Nb	2,531.25	80	1
Ti	2,531.25	125	30
Hf	2,531.19	50	25
Sn	2,531.11	15*W*	40

	λ	*I*	
		J	*O*
Rh	**2,534.07**	**4**	**4**
vzduch	**2,534.10**	**6**	**—**
Re	**2,534.101**	**—**	**10**
W	**2,534.146**	**15**	**2**
Mn	**2,534.15**	**25*d***	**—**
Ta	**2,534.16**	**—**	**3**
Os	**2,534.17**	**2**	**8**
Ce	**2,534.18**	**—**	**2**
Mn	**2,534.21**	**25**	**—**
J	**2,534.24**	**(12*h*)**	**—**
V	**2,534.26**	**12**	**2**
Hf	**2,534.33**	**5*h***	**—**
Cr	**2,534.34**	**10**	**8**
Fe	2,534.42	50	7
Nb	2,534.44	40	1
J	2,534.45	(12)	—
Ir	2,534.46	10	100
Ta	2,534.47	—	25
V	2,534.52	80	10
Rh	2,534.57	100*w*	2*w*
Pd	2,534.60	100	—
Ti	2,534.62	80	25
Ar	2,534.74	(40)	—
Hg	2,534.77	30	30
Re	2,534.80	—	100*w*
W	2,534.82	15	2
Ga	2,534.83	(20)	—
Ru	2,534.94	40	3
Ta	2,534.97	—	30
Ag	2,535.07	(10)	—
Ru	2,535.22	30	—

*N**

	λ	I	
		J	O
W	2,530.99	15	9
J	2,530.98	(60)	—
Nb	2,530.97	100	2
Tl	2,530.88	(80)	—
Tl	2,530.82	(50)	—
Te	2,530.70	(30)	—
W	2,530.70	12	8
Fe	2,530.69	70	25
Tl	2,530.67	(100)	—
Cr	2,530.45	1	35
Ir	2,530.41	10	—
Mo	2,530.34	25	3
O	2,530.30	(40)	—
V	2,530.18	70*R*	100
Co	2,530.13	300	40*w*
Fe	2,530.11	30	2
Cr	*2,529.95*	*20*	—
Mo	*2,529.84*	*20*	*25*
Fe	*2,529.83*	*2*	*50R*
Fe	*2,529.55*	*100*	*15*
Pt	*2,529.41*	—	*80*
Cu	*2,529.30*	*600*	—
Fe	*2,529.133*	*5*	*80R*
Mn	*2,529.132*	*5d*	*80*
Fe	*2,529.08*	*70*	—
Co	*2,528.97*	—	*50R*
Ru	*2,528.88*	*2*	*60*
Mo	*2,528.87*	*25*	*1*
V	*2,528.84*	*150R*	*25*
Co	*2,528.61*	*200*	*4*
Sb	*2,528.53*	*200*	*300R*
Si	*2,528.52*	*500*	*400*
Ba	*2,528.51*	*50r*	—
V	*2,528.47*	*150R*	*50*
Tb	*2,528.23*	*20*	—
Co	*2,528.18*	*40*	—
Ru	*2,528.04*	*60*	—
Se	*2,527.98*	*(25)*	—
Nb	*2,527.92*	*50*	*2*
V	*2,527.903*	*300R*	*35*
Ru	*2,527.897*	*40*	*2*
J	*2,527.72*	*(20)*	—
Fe	*2,527.69*	*30*	*1*
Mn	*2,527.44*	*12*	*150*
Fe	*2,527.43*	*50*	*200R*
Rh	*2,527.29*	*25*	*1*
Mo	*2,527.14*	*25*	*1*

	λ	I J	I O		λ	I J	I O
Fe	*2,527.10*	*60*	*1*				
O	*2,526.91*	*(25)*	*—*				
J	*2,526.88*	*(40)*	*—*				
Ru	*2,526.83*	*20*	*50*				
Cu	*2,526.73*	*50*	*—*				
Ta	*2,526.66*	*—*	*50*				
Pb	*2,526.62*	*(100)*	*—*				
Cu	*2,526.59*	*200*	*—*				
Ta	*2,526.45*	*—*	*150*				
Ta	*2,526.35*	*—*	*100*				
Fe	*2,526.29*	*60*	*10*				
V	*2,526.21*	*150R*	*150*				

P 2,535.65 (30) *O* 100

	λ	J	O		λ	J	O
Mn	**2,535.64**	**80**	**—**	**Th**	**2,535.870**	**2**	**5**
Fe	**2,535.604**	**—**	**1000**	**Ti**	**2,535.871**	**60**	**20**
Ta	**2,535.598**	**—**	**50*h***	**U**	**2,535.90**	**2**	**5**
Ru	**2,535.592**	**100**	**—**	**Pd**	**2,535.94**	**(3*h*)**	**—**
U	**2,535.58**	**8**	**—**	**Ta**	**2,535.96**	**—**	**4**
W	**2,535.576**	**10**	**—**	**Co**	**2,535.964**	**40**	**10*r***
J	**2,535.57**	**(30)**	**—**	**Pt**	**2,535.967**	**5**	**25**
Fe	**2,535.48**	**20**	**—**	**Ni**	**2,535.967**	**—**	**25**
Fe	**2,535.36**	**3**	**1**	**W**	**2,535.988**	**12**	**4**
W	**2,535.33**	**1**	**8**	**Yb**	**2,536.01**	**6**	**—**
Ag	**2,535.31**	**25**	**10**	**Ar**	**2,536.04**	**(20)**	**—**
Ar	**2,535.28**	**(2)**	**—**				
				Mn	2,536.08	12	—
				Ta	2,536.23	—	100*W*
				Pt	2,536.49	10	100
				Hg	2,536.52	1000*R*	2000*R*
				Th	2,536.56	10	5
				W	2,536.60	12	1
				In	2,536.67	(10)	—
				Tb	2,536.75	10	3
				Mo	2,536.85	4	25
				Lu	2,536.95	20	10
				Rh	2,537.04	100	15
				W	2,537.14	15	1
				Pd	2,537.17	100	—
				Ir	2,537.22	10	35
				Hf	2,537.33	15	20
				Co	2,537.46	10	—
				V	2,537.62	25	—
				Yb	2,537.64	15	4
				Rh	2,537.73	50	4
				Te	2,537.80	(300)	—

	λ	I	
		J	O
Ta	2,537.94	15	8
Pd	2,537.97	100	—
Os	2,538.00	15	—
Mn	2,538.05	25	—
Fe	2,538.20	35	2
Cr	2,538.29	25	—
Pr	2,538.38	10	—
Mo	2,538.45	125	30
Fe	2,538.50	10	2
K	2,538.60	(10)	—
Yb	2,538.67	20	10
Fe	2,538.81	30	15
Fe	2,539.00	20	10
Ni	2,539.10	250*w*	—
V	2,539.19	10*h*	—
Pt	2,539.20	20	400
Nb	2,539.23	10	1
W	2,539.31	15	6
Pd	2,539.36	50*wh*	—
Mn	2,539.40	20*Wh*	—
Tb	2,539.42	10	—
Mo	2,539.44	20	1
Se	2,539.54	(10)	—
Zr	*2,539.65*	—	*50*
Ru	*2,539.72*	*100*	*12*
W	*2,539.90*	*20*	*2*
Fe	*2,539.98*	*20h*	*3*
Tb	*2,540.12*	*50*	*3*
Al	*2,540.12*	*(30)*	—
Ru	*2,540.30*	*100*	*10*
Nb	*2,540.61*	*150*	*2*
Co	*2,540.65*	*40*	*6*
Fe	*2,540.67*	*30*	*6*
Fe	*2,540.98*	*10*	*100R*
Mn	*2,541.11*	*80*	—
Ru	*2,541.28*	—	*50*
Cr	*2,541.35*	*3*	*60*
Nb	*2,541.42*	*80*	*3*
Br	*2,541.45*	*(40)*	—
Co	*2,541.94*	*300h*	*40*
Zr	*2,542.10*	*50*	*100*
Rh	*2,542.16*	*50*	*1*
V	*2,542.44*	*20wh*	*1*
Os	*2,542.51*	*8*	*50*
Mo	*2,542.67*	*25*	*20*
Fe	*2,542.73*	*40*	*1*
Mn	*2,542.92*	*100*	*1*
V	*2,542.92*	*35*	*2*

	λ	I J	I O		λ	I J	I O
				Ru	*2,543.25*	*150*	*50*
				Fe	*2,543.38*	*50*	*5*
				Mn	*2,543.45*	*100*	*4*

P 2,553.28 (20) *O* 80

	λ	J	O		λ	J	O
Mn	**2,553.26**	**50**	—	**Ru**	**2,553.31**	**5**	**6**
Fe	**2,553.185**	**20**	**10**	**In**	**2,553.33**	**(5)**	—
Ta	**2,553.181**	**10*h***	**10*h***	**Ce**	**2,553.34**	—	**3**
W	**2,553.16**	**15**	**12**	**Co**	**2,553.37**	—	**10*r***
Cr	**2,553.06**	**1**	**20**	**Ni**	**2,553.38**	—	**20**
Zr	**2,553.05**	**2**	**2**	**La**	**2,553.40**	**3*h***	—
V	**2,553.02**	**15**	**7**	**Ag**	**2,553.41**	**10**	**2**
Co	**2,553.00**	—	**40*r***	**Ce**	**2,553.46**	—	**2**
Cd	**2,552.99**	**(10)**	—	**Nb**	**2,553.49**	**30**	**1**
Mn	**2,552.987**	**3**	—	**Fe**	**2,553.50**	—	**2**
Tl	**2,552.98**	—	**10*R***	**Re**	**2,553.556**	—	**4**
V	**2,552.96**	**30**	**10**	**Cd**	**2,553.56**	**(2)**	**25**
Cd	**2,552.91**	**(10)**	—	**In**	**2,553.56**	**(40)**	—
Mo	**2,552.873**	**2**	**20**	**Na**	**2,553.586**	**(2)**	—
Ga	**2,552.87**	**(5)**	—	**Ce**	**2,553.588**	—	**2**
Hg	**2,552.87**	**(25)**	—	**W**	**2,553.60**	**3**	**10**
				V	**2,553.67**	**20**	**10**
				Mo	**2,553.70**	—	**15**
Yb	2,552.70	30	8				
V	2,552.65	10	75*R*				
Tl	2,552.53	1	80*R*				
Tm	2,552.49	50	—				
Co	2,552.38	20	—				
Sc	2,552.359	40	25				
W	2,552.356	10	1				
Pt	2,552.25	20	150				
Cd	2,552.18	(100)	5*h*				
Yb	2,552.15	40	10				
Al	2,552.12	(40)	—				
Re	2,552.03	—	80				
Tb	2,552.01	10	—				
Ru	2,551.98	150	10				
Mn	2,551.88	100*d*	1				
Pd	2,551.848	100*h*	—				
Ta	2,551.732	4*h*	100				
V	2,551.729	25	—				
Ru	2,551.726	—	30				
Cr	2,551.59	18	1				
Ru	2,551.55	12	—				
Pr	2,551.52	15*s*	—				
W	2,551.45	12	—	*N**			

	λ	I J	I O		λ	I J	I O
J	2,551.43	(40)	—				
*bh*B	2,551.4	—	150				
Hf	2,551.40	125*d*	25*d*				
Nb	2,551.38	100	5				
Ta	2,551.19	—	150				
Ta	2,551.07	—	150				
Ni	2,551.03	80	—				
Pd	2,550.99	35	—				
Mo	2,550.85	3*h*	30*h*				
Rh	2,550.75	20	1				
Mo	2,550.75	15	—				
Zr	2,550.74	50	100				
Fe	2,550.68	30	3				
Pd	2,550.66	150	—				
Al	2,550.23	(15)	—				
Co	2,550.02	60	—				
K	2,550.02	(20)	—				
Fe	2,550.020	40	15				
Co	2,549.88	10*h*	—				
Cl	2,549.85	(50)	—				
Ru	2,549.79	100	3				
Rh	2,549.66	20	1				
V	2,549.656	10	—				
Fe	2,549.61	2	70*R*				
Ru	2,549.58	3	50				
Ni	2,549.56	150	—				
Th	2,549.51	15	—				
Pt	2,549.46	10	80				
Fe	2,549.46	15	1				
Fe	2,549.39	10	1				
Ta	2,549.38	—	100				
V	2,549.28	150	20				
Se	2,549.19	(50)	—				
Ru	2,549.18	150	5				
Th	2,549.10	12	—				
W	2,549.09	12	8				
Fe	2,549.08	30	3				
Tb	2,549.06	10	—				
Mn	*2,548.74*	*150*	*8*				
V	*2,548.69*	*80*	*10*				
Ce	*2,548.68*	*20h*	*20*				
Nb	*2,548.63*	*80*	*2*				
Co	*2,548.34*	*75*	*20*				
V	*2,548.23*	*25*	*4*				
V	*2,548.202*	*20wh*	—				
Hf	*2,548.20*	*20*	*15*				
Se	*2,547.98*	*(60)*	—				

	λ	I J	I O
Ru	2,547.67	80	5
Mo	2,547.57	20	15
Mo	2,547.35	20	1
Fe	2,547.33	35	1
Ni	2,547.19	100	—
Cl	2,546.94	(20)	—
Ta	2,546.80	—	80
Co	2,546.74	50	1
Fe	2,546.66	30	1
Sn	2,546.55	100	100
La	2,546.40	20*h*	—
Co	2,546.16	20	—
Ir	2,546.03	20	100*h*
Fe	2,545.98	30	100*R*
Ni	2,545.90	900*h*	20
Nb	2,545.64	150	1
Al	2,545.60	(50)	—
Re	2,645.49	—	60
Rh	2,545.35	150	8
Fe	2,545.22	25	4
Sc	2,545.20	20	15
Mn	2,545.16	25	—
Th	2,545.10	20	—
Co	2,545.04	30	3
Pd	2,544.83	200	—
Nb	2,544.803	300	5
Cu	2,544.812	700*R*	—
Ar	2,544.72	(40)	—
Fe	2,544.71	5	100

P 2,554.93 (20) *O* 60

	λ	J	O
Ta	2,554.91	50*h*	50*h*
W	2,554.862	10	15
V	2,554.862	2	15
Cs	2,554.8	(2)	—
Nb	2,554.79	20	1
Eu	2,554.78	1	10
S	2,554.76	(8)	—
Th	2,554.73	15*d*	—
Ru	2,554.69	—	8
W	2,554.67	15	4
Re	2,554.63	—	5
Ta	2,554.622	100	50
Mg	2,554.62	—	4
Sb	2,554.617	10	30
Se	2,554.58	(15)	—
Re	2,554.931	—	15
Ta	2,555.05	—	50
Fe	2,555.07	20	20
W	2,555.09	15	10
Ni	2,555.11	1000*h*	—
Th	2,555.20	15	—
W	2,555.205	—	10
U	2,555.21	4	4
Fe	2,555.22	—	10
Os	2,555.27	3	10
Ga	2,555.28	(6)	—
Yb	2,555.31	50*h*	—
Nb	2,555.32	30	1
Hf	2,555.33	2*h*	—
Ir	2,555.35	5	25

	λ	I	
		J	O
Fe	**2,554.52**	—	**2**
Mn	**2,554.513**	**20**	—
Cd	**2,554.51**	**(1)**	**3**
In	2,554.48	(50)	—
Ir	2,554.399	15*h*	15
In	2,554.399	(50)	—
V	2,554.22	50*wh*	—
Re	2,554.16	—	30*R*
Mn	2,553.981	10	—
J	2,553.98	(20)	—
Pd	2,553.74	(20)	—
Fe	2,553.73	15	—
*P**			

	λ	I	
		J	O
Rh	**2,555.36**	**60**	**100**
Mo	2,555.42	50	3
Fe	2,555.44	35	2
Nb	2,555.63	80	2
Sc	2,555.80	20	10
Ru	2,555.86	18	20
V	2,555.91	80	2
Ti	2,555.99	80	15
Ru	2,556.00	—	30
Al	2,556.01	(30)	—
Ru	2,556.05	18	—
F	2,556.10	(15)	—
U	2,556.19	12	15
Ru	2,556.31	—	50
Ta	2,556.510	30	25
Re	2,556.511	—	100
Mn	2,556.57	80*h*	10
Tb	2,556.61	20	—
Ru	2,556.70	30	—
Mo	2,556.75	20	2
Co	2,556.76	150	50*w*
Al	2,556.78	(15)	—
Mn	2,556.89	40	3
Br	2,556.93	(25)	—
Nb	2,556.94	200	5
Fe	2,557.08	10	—
Ru	2,557.13	50	5
Cr	2,557.15	2	35
Rh	2,557.20	100	1
J	2,557.28	(20)	—
Co	2,557.35	30	2
Mo	2,557.39	10	1
Fe	2,557.50	50	1
Mn	2,557.54	50	1
Ru	2,557.697	5*wh*	30
Ta	2,557.709	100	50
Al	2,557.71	(40)	—
Ni	2,557.87	80	—
Rh	2,557.92	50	1
Nb	2,557.94	100*k*	—
Zn	2,557.96	300	10
Ru	2,558.02	10	4
Sn	2,558.048	6	30
Re	2,558.049	—	60
Po	2,558.1	(20)	—
Mn	2,558.29	12	—
Fe	2,558.48	20	10

	λ	I J	I O
Ru	2,558.53	1	50
Mn	2,558.59	80	—
Mn	2,558.86	12	—
Mo	2,558.88	30h	5h
V	2,558.90	12	10
Re	2,559.08	—	30
Mo	2,559.15	10h	1
Eu	2,559.18	20	10
Hf	2,559.19	40	20
Tb	2,559.20	10	—
Ti	2,559.20	(15)	—
Fe	2,559.24	20	—
U	2,559.25	10h	3
J	2,559.28	(25)	—
Co	2,559.407	60wh	10
Mn	2,559.412	50	2
Ta	2,559.43	2	100
Mn	2,559.66	40d	—
Mo	2,559.69	20	—
Te	2,559.71	(25)	—
J	2,559.72	(40)	—
Fe	2,559.77	30	3
Rh	2,559.90	100	5
Co	2,560.09	60wd	1d
Sc	2,560.227	30	10
In	2,560.228	50Rh	150R
Ru	2,560.26	5	60
Fe	2,560.27	80	10
Ni	2,560.30	500h	—
La	2,560.37	50	—
Ta	2,560.68	—	70
Hf	2,560.74	25	—
Pd	2,561.02	200	—
J	2,561.49	(150)	—
Tm	2,561.65	30	60
Se	2,561.69	(25)	—
La	2,561.85	20	—
Rh	2,561.92	50	—
Mo	2,562.08	30	1
Ta	2,562.097	—	100
Fe	2,562.097	25	2
Ar	2,562.12	(20)	—
Pb	2,562.28	100	—
Nb	2,562.41	100	4
Fe	2,562.53	150	50
Li	2,562.54	15	150
V	2,562.76	25	1
W	2,563.162	30	8

	λ	I J	I O		λ	I J	I O
				Os	*2,563.164*	*25*	*8*
				Sc	*2,563.18*	*20*	*10*
				Fe	*2,563.40*	*25h*	*5*
				Fe	*2,563.47*	*125*	*70*
				Lu	*2,563.52*	*80h*	*1*
				Hf	*2,563.61*	*35*	*20*
				Mn	*2,563.65*	*50wh*	*25*
				Ta	*2,563.70*	—	*80*

P 4,222.15 (150) *O* 300

	λ	J	O		λ	J	O
U	**4,222.15**	**3*h***	**1**	**Kr**	**4,222.20**	**(20*h*)**	—
Nd	**4,222.06**	**2**	**5**	**Fe**	**4,222.221**	**200**	**200**
W	**4,222.05**	**8**	**15**	**Dy**	**4,222.221**	—	**10**
Tb	**4,222.02**	—	**3**	**Ho**	**4,222.26**	**1*h***	**3**
Tm	**4,221.95**	**2**	**2**	**Eu**	**4,222.31**	—	**5**
***bh*Ca**	**4,221.9**	—	**5**	**U**	**4,222.37**	**8**	**18**
Sc	**4,221.88**	—	**10**	**Br**	**4,222.40**	**(4)**	—
Sm	**4,221.86**	**2**	**4**	**Mo**	**4,222.41**	**15**	**20**
U	**4,221.80**	**8**	**12**	**Ce**	**4,222.60**	**18**	**80**
Hf	**4,221.76**	—	**2**	**Tm**	**4,222.67**	**4**	**10**
Ce	**4,221.73**	—	**2**	**Ar**	**4,222.67**	**(20)**	—
W	**4,221.72**	**1**	**6**	**Tb**	**4,222.71**	—	**9**
Nd	**4,221.71**	**2**	**8**	**Cr**	**4,222.73**	**15**	**100**
Ni	**4,221.70**	—	**5**	**U**	**4,222.74**	—	**6**
Ce	**4,221.63**	—	**4**	**O**	**4,222.78**	**(50)**	—
Se	**4,221.58**	**(20)**	—	**Na**	**4,222.8**	—	**3**
Cr	**4,221.57**	**35**	**80**	**Ce**	**4,222.88**	—	**3**
Ce	**4,221.49**	—	**2**	**Tb**	**4,222.91**	—	**6*W***
Pr	**4,221.45**	**1**	**4**	**U**	**4,222.94**	**2**	**1**
Tb	**4,221.38**	—	**3**	**Mo**	**4,222.96**	**10**	**15**
As	**4,221.23**	**5**	—	**K**	**4,222.97**	**(40)**	—
Ce	**4,221.17**	**1**	**12**	**Pr**	**4,222.98**	**40**	**125**
Nd	**4,221.14**	**8**	**15**	**Gd**	**4,222.98**	—	**10**
Cs	**4,221.12**	**(15)**	—	**Xe**	**4,223.00**	**(200*h*)**	—
Dy	**4,221.10**	**8**	**60**	**N**	**4,223.04**	**(25)**	—
				Tl	**4,223.05**	**(25)**	—
J	4,220.96	(80)	—	**Sm**	**4,223.06**	**2**	**3**
Sm	4,220.65	100	100	**Lu**	**4,223.09**	**4**	**1**
Te	4,220.42	(100)	—	**Ce**	**4,223.15**	—	**3**
Ne	4,219.76	(100)	—	**Ir**	**4,223.16**	**2*h***	**15**
Fe	4,219.36	200	250				
Cr	4,217.63	70	150	Br	4,223.88	(80)	—
Fe	4,217.555	100	200	Fe	4,224.18	80	200
La	4,217.554	100	200	Gd	4,225.85	50	150
Gd	4,217.19	100	100	Ge	4,226.57	50	200
Fe	4,216.19	100	200	Ca	4,226.73	50*R*	500*R*

	λ	I J	I O
Xe	4,215.60	(100)	—
Rb	4,215.56	300	1000*R*
Sr	4,215.52	400*W*	300*r*
Gd	4,215.024	150	200
Se	4,215.02	(150)	—
Nb	4,214.67	100	—
Xe	4,213.72	(200*h*)	—
Pd	4,212.95	300*W*	500*W*
Ag	4,212.68	20*h*	150*h*
Se	4,212.58	(200)	—
Ru	4,212.06	80	125
Gd	4,212.02	50	150
Os	4,211.85	50	150
Se	4,211.83	(200)	—
Dy	4,211.72	15	200
Rh	*4,211.14*	*200*	*15*
Fe	*4,210.35*	*200*	*300*
Xe	*4,208.48*	*(200h)*	—
Rn	*4,203.23*	*(200)*	—
Fe	*4,202.03*	*300*	*400*
Rb	*4,201.85*	*500*	*2000R*
Hg	4,227.29	(100)	—
Fe	4,227.43	250	300
Re	4,227.46	—	200*W*
Zr	4,227.76	8	150
Nb	4,229.15	100	50
La	4,230.95	50	150
Mo	4,232.59	100	125
Fe	*4,233.61*	*150*	*250*
Eu	*4,235.60*	—	*400r*
Fe	*4,235.95*	*200*	*300*
Xe	*4,238.25*	*(200h)*	—
La	*4,238.38*	*300*	*500*
Gd	*4,238.78*	*200*	*200*
Rb	*4,242.6*	*(150)*	

P 5,296.09 (300)

	λ	I J	I O
Ti	**5,295.79**	**1**	**50**
Pd	**5,295.63**	**10**	**200**
Mo	**5,295.47**	**12**	**20**
Mn	**5,295.29**	**(30)**	—
Hf	**5,294.87**	**2**	**12**
Ca	**5,294.48**	**1**	**2*h***
Mn	**5,294.22**	**(20)**	—
V	**5,294.04**	**18*wh***	**18*wh***
Xe	5,292.22	(800)	—
Ra	5,283.28	(250)	—
Xe	*5,260.44*	*(300)*	—
Ar	**5,296.32**	**(5)**	—
Nb	**5,296.34**	**2**	**5**
J	**5,296.52**	**(150)**	—
Cr	**5,296.69**	**15**	**15*rh***
Sm	**5,296.94**	**2**	**6**
Mn	**5,296.97**	**(40)**	—
Ti	**5,297.26**	**2**	**70**
Cr	**5,297.36**	**2*h***	**5*h***
Cr	**5,297.98**	**1*h***	**5*h***
Hf	**5,298.06**	**100**	**80**
Mo	**5,298.06**	**3**	**15**
Ne	5,298.19	3	5
La	5,301.98	200	300*r*
La	5,302.62	150	50
La	5,303.56	125	100
Se	5,305.35	(500)	—
Kr	5,308.66	(200)	—
Xe	5,309.27	(150)	—
Ar	5,309.52	(200)	—
Hf	5,311.60	150	100
Xe	5,313.87	(500)	—

	λ	I	
		J	O
P	5,316.07	(150w)	—
O	5,330.66	(500)	—
Ne	5,330.78	(600)	—
Kr	5,333.41	(500h)	—
Yb	5,335.16	400	150

Pb $^{82}_{207.19}$

t_0 327,5°C t_1 1,744°C

I.	II.	III.	IV.	V.
7.415	15.04	32.1	38.9	69.7

λ	I		eV
	J	O	
I 2,169.99	1000*R*	1000*R*	5.67
II 2,203.51	5000*R*	50	7.25
I 2,614.18	80	200*R*	5.68
I 2,833.07	80	500*R*	4.4
I 3,639.58	50	300	4.38
I 3,683.47	50	300	4.34
I 4,057.82	300*R*	2000*R*	4.38

Pb 2,169.99 *J* 1000*R* *O* 1000*R*

	λ	I J	I O		λ	I J	I O
Fe	**2,169.95**	—	**8**	**Er**	**2,170.00**	—	**8**
W	**2,169.94**	**12**	**10**	**Ir**	**2,170.05**	**3**	**10**
Ra	**2,169.90**	**(125)**	—	**Ni**	**2,170.06**	**4**	—
Nb	**2,169.89**	**5**	—	**V**	**2,170.07**	**2*wh***	—
V	**2,169.85**	—	**3**	**Hf**	**2,170.22**	**30**	**20**
Os	**2,169.81**	**1**	**5**	**Sb**	**2,170.23**	**8**	**2**
Yb	**2,169.77**	**5**	—				
Ir	**2,169.76**	**3**	—				
Pt	2,165.17	24	1,000*R*				

Pb 2,203.51 *J* 5000*R* *O* 50*R**

	λ	I J	I O		λ	I J	I O
Pd	**2,203.48**	**25**	—	**Ni**	**2,203.52**	**8*h***	—
Co	**2,203.43**	**3**	—	**Rh**	**2,203.55**	**50*w***	**15**
				Nd	**2,203.56**	**30**	—
				Nb	**2,203.63**	**40**	**15**
				V	**2,203.660**	—	**2**
				Sb	**2,203.66**	**2**	**4**
				Ag	**2,203.662**	**20*h***	—
				Ir	**2,203.68**	**25**	—
				N	**2,203.72**	**(3)**	—

Pb 2,614.178 *J* 80 *O* 200*R*

	λ	I J	I O		λ	I J	I O
Sn	**2,614.198**	**5**	**5**	**Fe**	**2,614.180**	**3**	—
Ta	**2,614.173**	—	**200*wh***	**Ir**	**2,614.20**	**2**	**10**
Co	**2,614.13**	—	**30**	**Ne**	**2,614.26**	**(5)**	—
K	**2,614.1**	**(5)**	—	**Ce**	**2,614.29**	—	**2**
Ru	**2,614.07**	**3**	**60**	**Hf**	**2,614.29**	**6**	**6**
Mn	**2,614.04**	**5**	—	**Zn**	**2,614.30**	**(10)**	—
U	**2,613.95**	**4**	**10**	**In**	**2,614.31**	**2**	—
Nb	**2,613.93**	**30**	**1**	**Nb**	**2,614.31**	**20**	**2**
Ce	**2,613.90**	**1**	**15**	**Co**	**2,614.36**	**60*w***	**6**
Nb	**2,613.85**	**30*h***	**3**	**W**	**2,614.38**	**6**	—
Fe	**2,613.823**	**400**	**400**	**V**	**2,614.40**	**35**	**1**
W	**2,613.823**	**9**	**15**	**Cu**	**2,614.41**	**5*w***	—
Re	**2,613.76**	—	**25**	**W**	**2,614.44**	**5*s***	**6**
				Fe	**2,614.49**	**5**	**40**
				Er	**2,614.55**	**2**	**3**
Hf	2,613.60	80	20	**Re**	**2,614.56**	—	**60**
Pd	2,613.43	100	—	**Ce**	**2,614.57**	—	**5**
Lu	2,613.40	100	30	**Ru**	**2,614.586**	**4**	**50**
Ta	2,612.61	40	50				

	λ	I J	I O
Ru	2,612.51	80	—
Sb	2,612.30	60	50
Ru	2,612.07	30	100
Fe	2,611.87	500	500
Na	2,611.81	(80)	—
Ni	2,611.65	125	—
Ru	2,611.51	80	3
Ta	2,611.34	—	100
Fe	2,611.07	80	20
V	2,610.64	40h	—
La	2,610.33	150	10
Mn	2,610.20	100h	15
Ni	2,610.09	900h	—
Ru	2,610.08	60	—
Fe	2,609.87	40	—
Pd	2,609.86	200	—
Rh	*2,609.17*	*200*	*3*
Zn	*2,608.64*	*100*	*300*
Zn	*2,608.56*	*50*	*200*
Ta	*2,607.84*	*150*	*20h*
Fe	*2,607.09*	*400*	*300*
Hf	*2,607.03*	*80*	*30*
Fe	*2,606.82*	*30*	*200*
Fe	*2,606.50*	*80*	—
Ni	*2,606.39*	*600h*	—
Ag	*2,606.16*	*200wh*	*10*
Mn	*2,605.69*	*500R*	*100R*
Co	*2,605.68*	*200*	*30*

	λ	I J	I O
Hf	**2,614.59**	**3h**	—
Ag	**2,614.59**	**300wh**	—
Pt	**2,614.61**	—	**10**
Cs	**2,614.62**	**(8)**	—
Ru	2,615.09	100	60
Ni	2,615.19	900h	—
V	2,615.40	50wh	
Lu	2,615.42	250	100
V	2,616.25	70	4
Mn	2,616.51	60	—
Pt	2,616.75	60	10
Ru	2,617.07	50	—
Fe	2,617.62	400	300
Fe	2,618.02	60	150
Mn	2,618.14	100h	50
Cu	2,618.37	100	500w
Fe	*2,619.08*	*150*	*5*
Lu	*2,619.26*	*100*	*30*
Pt	*2,619.57*	*5*	*300*
Nb	*2,620.45*	*200*	*3*
Fe	*2,620.69*	*80*	*3*
Yb	*2,621.12*	*100*	*2*
Fe	*2,621.67*	*400*	*200*
Hf	*2,622.74*	*80*	*30*
Mn	*2,622.90*	*15*	*200*

Pb 2,833.07 *J* 80 *O* 500*R*

	λ	J	O
Er	2,833.061	—	25
Zr	2,833.061	1	2
In	2,833.06	3	—
Eu	2,833.056	—	3
Ce	2,833.04	—	3
Cl	2,833.03	(4)	—
Kr	2,833.00	(100)	—
W	2,832.952	3	10
Zn	2,832.95	(25)	—
Ce	2,832.93	—	2
Ne	2,832.921	(8)	—
J	2,832.917	(20)	—
Nb	2,832.79	5h	—
Ir	2,832.774	—	5
Rh	2,832.769	—	5

	λ	J	O
Fe	2,833.10	5h	—
K	2,833.14	(2)	2h
Ir	2,833.236	20	7
U	2,833.244	4	8
Eu	2,833.25	5wh	10wh
Hf	2,833.28	4	25
Nb	2,833.304	10	1
Ce	2,833.309	—	50d
Tl	2,833.31	(25)	—
Th	2,833.34	8	8
Tb	2,833.37	20	—
Cr	2,833.39	3	—
Fe	2,833.40	8	10
Ce	2,833.58	—	2
W	2,833.63	12	15

	λ	I	
		J	O
Ce	**2,832.75**	**—**	**2**
Mo	**2,832.66**	**5**	**—**
U	**2,832.64**	**2**	**2**
Ru	**2,832.62**	**—**	**20**
Ce	**2,832.57**	**—**	**2**
La	**2,832.55**	**5**	**4**
Cr	2,832.46	125	2
Fe	2,832.44	200	300
Tu	2,832.16	100	25
U	2,832.06	50	35
Ru	2,831.84	50	10
Ho	2,831.60	70	—
Fe	2,831.562	500	1
Tm	2,831.56	40	10
Yb	2,830.98	40	2
Rn	2,830.60	(70)	—
Cr	2,830.47	80h	15
V	2,830.40	60	10
Pt	2,830.29	600r	1000R
Nb	2,829.75	50	3
He	2,829.07	(40)	—
Ti	2,828.9	150wh	—
Fe	2,828.81	60	100
Eu	2,828.69	150	200W
Fe	2,828.63	80	—
Ta	2,828.58	100	75
Ti	2,828.15	200h	2
Tm	2,827.92	100	50
Fe	2,827.89	50	70
Mo	2,827.74	40	8
Ta	*2,827.55*	*100d*	*3d*
Ti	*2,827.21*	*80wh*	—
Ru	*2,826.68*	*100*	—
Ru	*2,826.22*	*80*	—
Tl	*2,826.16*	*100R*	*200R*
Fe	*2.825.56*	*150*	*150*
Ru	*2,825.46*	*80*	—
Co	*2,825.242*	*200*	*5*
Ni	*2,825.236*	*125*	—
Ag	*2,824.370*	*200w*	*150wh*
Cu	*2,824.369*	*300*	*1000*
Fe	*2,823.28*	*300*	*200*
Ru	*2,823.18*	*80*	*20*
Au	*2,822.72*	*80*	—
Hf	*2,822.68*	*90*	*30*
Ru	*2,822.55*	*150*	*30*
Cr	*2,822.37*	*100*	*20*
Ta	2,833.64	40w	300w
Ru	2,833.78	80	—
Cd	2,834.19	(100)	—
Cr	2,834.26	125	—
Lu	2,834.35	40h	5
Mo	2,834.39	40	20
V	2,834.53	40h	—
Co	2,834.94	75	2
Nb	2,835.12	100	5d
Mo	2,835.33	40	20
Fe	2,835.46	100	100
Cr	2,835.63	400r	100
V	2,836.52	80	20
Ti	2,836.64	100wh	—
C	2,836.71	200	—
Cd	2,836.91	80	200
In	2,836.919	80	80
J	2,836.922	(40)	—
Pd	2,837.12	(40)	—
Cu	2,837.55	250	—
C	2,837.60	40	—
Au	2,838.03	80	—
Fe	2,838.12	150	150
Ta	2,838.24	150	2
Rn	2,838.5	(70)	—
Os	*2,838.63*	*100*	*100R*
Cr	*2,838.79*	*80*	*3*
Ti	*2,839.80*	*100wh*	—
Pd	*2,839.89*	*100*	—
Sn	*2,839.99*	*300R*	*300R*
Cr	*2,840.02*	*125*	*25*
Pd	*2,841.03*	*100*	—
Ru	*2,841.12*	*125*	—
Nb	*2,841.15*	*100*	*10*
Ru	*2,841.68*	*200*	*50*
Na	*2,841.72*	*(80)*	*20*
Ti	*2,841.94*	*125*	*40*
Rn	*2,842.1*	*(150)*	—
Ni	*2,842.42*	*150*	—
Nb	*2,842.65*	*100*	*10*
Cr	*2,843.25*	*400r*	*125*
Ta	*2,843.51*	*80*	*3*
Fe	*2,843.63*	*100*	*125*
Fe	*2,843.98*	*300*	*300*

Pb 3,639.58 *J* 50* *O* 300

	λ	I J	I O
Zr	**3,639.574**	—	**2**
Ce	**3,639.570**	—	**5**
Mo	**3,639.55**	**3**	**4**
Zn	**3,639.53**	**(5)**	**20**
Rh	**3,639.51**	**70**	**125**
U	**3,639.49**	—	**25**
Tb	**3,639.48**	—	**8**
Th	**3,639.45**	**10**	**10**
Co	**3,639.44**	**20**	**200**
Sm	**3,639.41**	**5**	**6**
Ir	**3,639.34**	—	**3**
Nb	**3,639.33**	**20**	**15**
Ce	**3,639.31**	—	**5**
Yt	**3,639.29**	**2**	**3**
Sm	**3,639.26**	**2**	**10**
La	**3,639.25**	**2**	**2**
Nd	**3,639.17**	**1**	**4*d***
Ir	**3,639.16**	—	**3**
Re	**3,639.15**	—	**10*h***
Nb	**3,639.055**	**20**	**2**
Gd	**3,639.053**	**15**	**10**
V	**3,639.02**	**60**	**70**
Er	**3,639.014**	—	**6**
Mo	**3,639.007**	**4**	**3**
Tb	**3,638.95**	—	**15**
Dy	**3,638.89**	**1*h***	**2**
Pr	**3,638.86**	—	**3**
W	**3,638.81**	**5**	**1**
Pt	**3,638.793**	**10**	**250**
Nb	**3,638.786**	**10**	**10**
Sm	**3,638.75**	**8**	**40**
Zr	**3,638.72**	—	**5**
O	**3,638.70**	**(10)**	—
Er	**3,638.68**	**1**	**15**
U	**3,638.65**	**8**	**5**
Th	**3,638.64**	**2**	**6**
Pr	**3,638.58**	**1**	**8**
Mo	**3,638.56**	**3**	**3**
Tb	**3,638.46**	**50**	**80**
Mo	**3,638.42**	**3**	**3**
Zr	3,636.45	30	200
Ti	3,635.46	100	200
Pd	3,634.69	1000*R*	2000*R*
Fe	3,631.46	300	500
Ca	3,630.75	9	150
Pt	3,628.11	20	300*W*
Ni	*3,619.39*	*150h*	*2000h*
Fe	*3,618.77*	*400*	*400*
Pr	**3,639.77**	**2**	**10**
Cr	**3,639.80**	**25**	**60**
Tb	**3,639.82**	**8**	**30**
Ir	**3,639.85**	—	**2**
Dy	**3,639.86**	**2**	**4**
Ce	**3,639.87**	—	**3**
Er	**3,639.91**	—	**8**
U	**3,640.03**	**4**	**1**
V	**3,640.05**	**2**	**5**
Mn	**3,640.10**	**2**	**2**
W	**3,640.13**	**8**	**9**
Gd	**3,640.19**	**8**	**10**
Nd	**3,640.23**	**10**	**16**
Dy	**3,640.24**	**6**	**15**
Er	**3,640.25**	**1**	**20**
Ti	**3,640.31**	—	**3**
Os	**3,640.33**	**40**	**200**
Cr	**3,640.388**	**5**	**30**
Fe	**3,640.390**	**200**	**300**
Ba	**3,640.391**	**4**	**8**
Mo	**3,640.62**	**5**	**8**
Nb	**3,640.636**	**10**	**8**
Ru	**3,640.64**	**12**	**3**
Ce	**3,640.69**	—	**10**
Ti	3,642.67	125	300
Ca	3,644.41	15	200
Dy	3,645.42	100	300
Gd	3,646.20	150	200*w*
Fe	3,647.84	400	500
Co	3,649.35	4	200
Hg	3,650.15	500	200
Fe	*3,651.47*	*200*	*300*
Ti	*3,653.50*	*200*	*500*
Rh	*3,657.99*	*200 W*	*500 W*

*N**

Pb 3,683.471 *J* 50* *O* 300

	λ	*I*			λ	*I*	
		J	*O*			*J*	*O*
Zr	**3,683.47**	—	**2**	**Zn**	**3,683.471**	**(15)**	**20**
Ag	**3,683.45**	**2**	**4**	**Er**	**3,683.472**	—	**25**
Ce	**3,683.393**	—	**2**	**Mn**	**3,683.474**	—	**12**
W	**3,683.392**	**7**	**8**	**Sb**	**3,683.48**	**2**	**3**
Th	**3,683.33**	**4**	**5**	**Ir**	**3,683.52**	—	**5**
W	**3,683.31**	**7**	**8**	**U**	**3,683.59**	**8*d***	**2*d***
Eu	**3,683.27**	**1*h***	**18*w***	**Ru**	**3,683.592**	—	**3**
Tb	**3,683.26**	—	**15**	**Fe**	**3,683.62**	**1**	**3**
Tm	**3,683.20**	**20**	**10**	**Pr**	**3,683.846**	**2**	**4**
Pr	**3,683.196**	**1**	**3**	**Eu**	**3,683.85**	**1*h***	**8*W***
V	**3,683.13**	**60**	**100**	**W**	**3,683.941**	**9**	**10**
Fe	**3,683.058**	**100**	**200**	**Th**	**3,683.944**	**3**	**3**
Ta	**3,683.058**	**1**	**18**	**Nb**	**3,683.97**	**3*h***	**2*h***
Co	**3,683.05**	—	**200*R***	**Er**	**3,684.01**	—	**6**
Pt	**3,682.98**	**2**	**8**	**Fe**	**3,684.11**	**200**	**300**
Au	**3,682.86**	**10*h***	**20**	**Sm**	**3,684.120**	**5**	**20**
Gd	**3,682.74**	**20**	**10**	**Gd**	**3,684.124**	**150**	**200W**
Er	**3,682.71**	**4**	**18**	**Nd**	**3,684.13**	—	**3**
Zr	**3,682.651**	**2*h***	**1**	**Mo**	**3,684.22**	**25*d***	**1**
Ho	**3,682.65**	**4**	**6**	**Ce**	**3,684.24**	—	**6**
Ce	**3,682.647**	**1**	**8**	**Nb**	**3,684.25**	**5**	**1**
Sm	**3,682.54**	—	**8*d***	**Er**	**3,684.28**	**3**	**15**
Dy	**3,682.52**	**2*wh***	**5*h***	**Nd**	**3,684.29**	**6**	**10**
Th	**3,682.49**	**1**	**4**	**Ta**	**3,684.31**	—	**2**
Ag	**3,682.47**	**4**	**50**	**Lu**	**3,684.32**	**1**	**15**
U	**3,682.46**	—	**8**	**Mo**	**3,684.327**	**1**	**5**
Eu	**3,682.44**	**1**	**12**	**Ir**	**3,684.327**	**5**	**6**
				V	**3,684.33**	**10**	**40**
*bh***Zr**	**3,682.4**	—	**30**	**Co**	**3,684.48**	—	**200*W***
Th	**3,682.36**	**4**	**4**	**Mn**	**3,684.52**	**5**	**5**
Tb	**3,682.26**	**30**	**50**				
				Sm	**3,684.54**	**6**	**6**
Hf	**3,682.24**	**30**	**25**	**Os**	**3,684.55**	**8**	**15**
				U	**3,684.62**	**5**	**5**
				Nd	**3,684.64**	**1**	**8**
Fe	3,682.21	300	400	**W**	**3,684.66**	**9**	**10**
Fe	3,679.91	300	500				
*bh*B	3,679.1	—	200	**Cu**	**3,684.67**	**4**	**12**
V	3,676.68	150*h*	300				
Fe	3,676.31	100	200				
				Ti	3,685.19	700*R*	150
Ni	3,674.15	50*r*	200	Fe	3,686.00	125	150
V	3,673.40	80*h*	150	Gd	3,686.34	200	150*W*
Ti	3,671.67	70	150	Fe	3,687.46	300	400
Gd	3,671.22	100	150*w*	Gd	3,687.76	200	200*w*
Os	3,670.89	20	200				
				V	3,688.07	200*R*	200
				Eu	3,688.44	500*W*	1000*W*
Ni	*3,664.09*	*30*	*300*	Os	3,689.06	30	200
Hg	*3,663.28*	*400*	*500*	Fe	3,689.46	150	200
				V	3,690.28	125	200

*P**

	λ	I J	I O		λ	I J	I O
				Pd	3,690.34	1000*w*	300*h*
				V	3,692.22	150*R*	200*R*
				Rh	3,692.36	150*wd*	500*hd*
				Fe	3,694.01	300	400
				Yb	3,694.20	1000*R*	500*R*
				Fe	3,695.05	150	200
				V	3,695.86	100*r*	150
				Fe	*3,701.09*	*200*	*300*
				Co	*3,704.06*	*35*	*300r*
				Fe	*3,705.57*	*500*	*700*

Pb 4,057.82 *J* 300*R* *O* 2000*R*

	λ	J	O		λ	J	O
Er	**4,057.819**	—	**30**	**Th**	**4,057.823**	**3**	**2**
Zn	**4,057.71**	—	**80**	**V**	**4,057.825**	**2**	**10**
Tb	**4,057.68**	—	**2**	**In**	**4,057.87**	**10**	**80**
Sm	**4,057.65**	**3**	**10**	**Mn**	**4,057.950**	**20**	**80**
Mg	**4,057.63**	—	**10***w*	**U**	**4,057.955**	**6**	**1**
Ti	**4,057.62**	**6**	**40**	*bh***Sr**	**4,058.0**	—	**3**
Mo	**4,057.58**	**4**	**10**	**Tb**	**4,058.02**	—	**2**
Ce	**4,057.56**	—	**2**	**La**	**4,058.08**	—	**4**
Ho	**4,057.55**	**2***h*	**2**	**Ta**	**4,058.136**	**1**	**2**
Xe	**4,057.46**	**(1000***wh***)**	—	**Ti**	**4,058.14**	**6**	**50**
W	**4,057.45**	**7**	**6**	**U**	**4,058.16**	**4**	**10**
Mo	**4,057.44**	**4**	**4**	**Co**	**4,058.190**	—	**100**
Hf	**4,057.43**	**2**	**3**	**Pr**	**4,058.19**	**8**	**6**
Dy	**4,057.40**	**4**	**4**	**Se**	**4,058.20**	**(20)**	—
P	**4,057.39**	**(50)**	—	**Fe**	**4,058.229**	**25**	**80**
Ni	**4,057.347**	—	**2**	**Gd**	**4,058.23**	**60**	**100**
Fe	**4,057.346**	**3**	**20**	**Ce**	**4,058.24**	—	**18**
Th	**4,057.337**	**8**	**10**	**Dy**	**4,058.25**	—	**2***h*
Ce	**4,057.30**	—	**2**	**Tb**	**4,058.44**	**1**	**5**
Co	**4,057.20**	—	**100**	**Eu**	**4,058.45**	—	**2**
In	**4,057.19**	**(10)**	—	**Ta**	**4,058.46**	**5**	**10**
Ca	**4,057.10**	**3**	**2**	**Si**	**4,058.49**	**3**	—
Cl	**4,057.08**	**(10)**	—	**Co**	**4,058.60**	—	**100**
V	**4,057.074**	**10**	**20**	**Mo**	**4,058.61**	**10**	—
In	**4,057.07**	**(100)**	—	**Zr**	**4,058.62**	**1**	**9**
Tb	**4,057.06**	**1**	**4**	**Fe**	**4,058.76**	**10**	**40**
Kr	**4,057.01**	**(300***hs***)**	—	**Cr**	**4,058.77**	**50**	**80**
Co	**4,056.98**	**2**	**20***h*	**Pr**	**4,058.78**	**15**	**25**
Nb	**4,056.941**	**5**	**3**	**Tb**	**4,058.81**	—	**3***W*
In	**4,056.936**	**(500)**	—	**Sm**	**4,058.87**	**20**	**30**
Ce	**4,056.90**	**2**	**15**	**Ru**	**4,058.88**	—	**10**
Nd	**4,056.84**	**8**	**8**	**Tm**	**4,058.92**	**6**	**20**
Al	**4,056.8**	**(2)**	—	**Ca**	**4,058.930**	—	**3***d*

	λ	I J	I O
Cr	4,056.79	3	15
In	4,056.78	(30)	—
In	4,056.75	(50)	—
U	4,056.74	2	1
Cu	4,056.70	—	8wh
Sr	4,056.67	4	4
In	4,056.59	(5)	—
Sc	4,056.58	2	5
Kr	4,056.57	(3)	—
Pr	4,056.54	60	100
Zr	4,056.51	—	6
Ir	4,056.47	2	12
W	4,056.46	5	2
Rh	4,056.342	2	3
Ce	4,056.338	—	4
J	4,056.321	(15)	—
Mo	4,056.318	15	15
U	4,056.29	—	5
V	4,056.26	3	1
Ce	4,056.25	—	3
Ti	4,056.21	2	3
Yb	4,056.18	10	2
Pr	4,056.13	2	5
Ag	4,055.26	500R	800R
Ho	4,053.92	200	400
Ru	4,051.40	200	125
K	4,047.20	200	400
Hg	4,046.56	300	200
Ar	4,045.97	(150)	—
Fe	4,045.81	300	400
P	4,044.49	(150w)	—
Ar	4,044.42	(1200)	—
K	4,044.14	400	800
La	4,042.91	300	400
Gq	*4,032.98*	*500R*	*1000R*
La	*4,031.69*	*300*	*400*
F	*4,025.49*	*(300)*	—
F	*4,024.73*	*(500)*	—

	λ	I J	I O
Mn	4,058.930	60	80
Nb	4,058.94	400w	1000w
Mg	4,058.96	—	2
Zr	4,058.98	—	8
U	4,059.02	2	1
Eu	4,059.03	—	4
Ir	4,059.23	4	30
W	4,059.25	4	5
Th	4,059.26	5	8
P	4,059.27	(100)	—
Ce	4,059.32	—	8h
Gd	4,059.35	3	10w
Ce	4,059.367	1h	3
Pr	4,059.37	3	4
Eu	4,059.38	—	25
Mn	4,059.39	15	20
Tb	4,059.40	1	3
Ru	4,059.43	—	7
Nb	4,059.51	2	5
Er	4,059.51	—	7
Pb	4,062.14	20	20
Fe	4,063.60	300	400
Kr	4,065.11	(300)	—
Se	4,070.16	(500)	—
J	4,070.75	(150)	—
Fe	4,071.74	200	300
A	4,072.01	(150)	—
O	4,072.16	(300)	—
In	4,072.40	200wh	—
Te	4,073.57	(300)	—
O	*4,075.87*	*(800)*	—
La	*4,077.34*	*400*	*600*
Sr	*4,077.71*	*500W*	*400r*
Ru	*4,080.60*	*300*	*125*
La	*4,086.71*	*500*	*500*
Kr	*4,088.33*	*(500)*	—

Pd $^{46}_{106.4}$

t_0 1,553°C t_1 2,200°C

I.	II.	III.	IV.	V.
8.334	19.9	—	—	—

λ	*I*		eV
	J	*O*	
I 3,404.58	1000*R*	2000*R*	4.46
I 3,421.24	1000*R*	2000*R*	4.58
I 3,516.94	500*R*	1000*R*	4.49
I 3,609.55	700*R*	1000*R*	4.40
I 3,634.70	1000	2000*R*	4.23
I 4,212.95	300	500	4.40

Pd 3,404.58 *J* 1000*R* *O* 2000*R*

	λ	*I*			λ	*I*	
		J	*O*			*J*	*O*
La	**3,404.52**	**2**	**9**	**Th**	**3,404.65**	**4**	**4**
In	**3,404.45**	**(10)**	—	**Tb**	**3,404.71**	**8**	**3**
Ce	**3,404.43**	—	**12**	**Re**	**3,404.72**	—	**100**
V	**3,404.42**	**50*h***	—	**Fe**	**3,404.75**	—	**2**
Fe	**3,404.36**	**50**	**100**	**Nd**	**3,404.763**	**6**	**4**
Mo	**3,404.34**	**25**	**20**	**Sm**	**3,404.767**	**2**	**2**
P	**3,404.33**	**(50)**	—	**Ne**	**3,404.77**	**(12)**	—
Fe	**3,404.304**	**25**	**25**	**W**	**3,404.80**	**6**	**8**
In	**3,404.297**	**(18)**	—	**Zr**	**3,404.83**	**35**	**40**
K	**3,404.24**	**(30)**	—	**Mo**	**3,404.86**	**4**	**6**
Tb	**3,404.24**	**3**	**15**	**Ce**	**3,404.91**	**2**	**18**
W	**3,404.22**	**7**	**8**	**Au**	**3,404.92**	**3*h***	—
Ta	**3,404.16**	**1**	**5**	**U**	**3,404.93**	**2**	**3**
W	**3,404.14**	**4**	—	**V**	**3,404.96**	**1**	**8**
In	**3,404.131**	**(18)**	—	**Ti**	**3,404.97**	**2*d***	—
Er	**3,404.13**	—	**8**	**Dy**	**3,404.99**	**4**	**4**
Ce	**3,404.130**	—	**18**	**Ag**	**3,405.03**	**2*h***	**3**
Yb	**3,404.10**	**30**	**9**	**Ti**	**3,405.09**	**2**	**20**
Sb	**3,403.91**	**10*h***	**2**	**Co**	**3,405.12**	**150**	**2,000*R***
U	**3,403.89**	**2**	**2**	**V**	**3,405.160**	**15**	**30**
Ce	**3,403.85**	—	**8**	**Kr**	**3,405.16**	**(80*wh*)**	—
W	**3,403.79**	**3**	**4**	**Mo**	**3,405.20**	**5**	**8**
Ru	**3,403.77**	—	**8**	**Cr**	**3,405.22**	**1**	**12**
Nb	**3,403.75**	**4**	**5**	**Ru**	**3,405.277**	—	**3**
Ce	**3,403.73**	—	**3**	**W**	**3,405.277**	**6**	**7**
In	**3,403.70**	**18**	—	**Bi**	**3,405.33**	**10**	**40**
Zr	**3,403.684**	**15**	**15**	**Nb**	**3,405.41**	**50**	**80**
Er	**3,403.678**	**1**	**9**	**Ce**	**3,405.44**	—	**6**
Tb	**3,403.66**	—	**8**	**Th**	**3,405.56**	**4**	**3**
Cd	**3,403.65**	**500*h***	**800**	**Fe**	**3,405.58**	—	**1**
Ce	**3,403.60**	—	**15**				
Cr	**3,403.59**	**3**	**35**	Co	3,409.18	125	1000*R*
				Co	3,412.34	100	1000*R*
				Co	3,412.63	40	1000*R*
Rh	3,396.85	500	1000*w*	Ni	3,414.76	50*wh*	1000*R*

Pd 3,421.24 *J* 1000*R* *O* 2000*R*

	λ	*J*	*O*		λ	*J*	*O*
Cr	**3,421.212**	**200**	**50**	**Mo**	**3,421.25**	**6**	**6**
Ta	**3,421.212**	**3*wh***	**1**	**Sm**	**3,421.30**	—	**2**
Ni	**3,421.20**	—	**4**	**Dy**	**3,421.32**	**5**	**7**
Th	**3,421.19**	**10**	**10**	**Ni**	**3,421.34**	**8**	**30**
Nb	**3,421.162**	**50*w***	**10*w***	**Co**	**3,421.35**	—	**3**
Rb	**3,421.16**	**(20)**	—	**U**	**3,421.38**	**2*d***	**8*d***
W	**3,421.13**	**20**	**3*d***	**Zr**	**3,421.42**	—	**3**
Pr	**3,421.11**	**1**	**10**	**W**	**3,421.43**	**7**	—

	λ	I J	I O		λ	I J	I O
Ce	3,421.07	—	6	Hf	3,421.45	4	3
Er	3,421.06	1	9	Ba	3,421.48	2*h*	2
Ba	3,421.01	2*h*	3	Ce	3,421.54	—	5
Ce	3,420.95	—	5	Re	3,421.58	—	15
Ca	3,420.82	2*h*	—	Co	3,421.63	2	20
Dy	3,420.81	2	5	Ho	3,421.64	20	20
Mn	3,420.795	—	8	Cr	3,421.64	6	2*h*
Co	3,420.792	2	80	Ar	3,421.64	(10)	—
Re	3,420.76	—	40	Eu	3,421.67	5	5
Ni	3,420.74	3	30	Os	3,421.687	10	30
Xe	3,420.73	(25)	—	U	3,421.69	8	8
V	3,420.71	10	—	Ce	3,421.71	—	3
Nb	3,420.63	50	5	Cr	3,421.72	—	12
O	3,420.63	(7*h*)	—	Pt	3,421.72	—	2*h*
La	3,420.54	2	—	Ir	3,421.76	—	6
Ce	3,420.53	—	8	Ta	3,421.80	18	—
Sm	3,420.51	2	3	Ce	3,422.00	—	5
Ir	3,420.49	2	20	Sm	3,422.06	—	5
Co	3,420.48	2	3	Dy	3,422.07	2	3
J	3,420.42	(10)	—	Fe	3,422.14	—	2
W	3,420.36	6	7	Sm	3,422.19	—	3
Yb	3,420.35	—	9	Ce	3,422.21	—	2
Tb	3,420.34	15	50				
Ba	3,420.338	2*h*	8*r*	Pd	3,433.45	500*h*	1000*h*
Si	3,420.32	2	—	Pd	3,441.40	2*h*	800*h*

*P**

Pd 3,516.94 *J* 500*R* *O* 1000*R*

	λ	I J	I O		λ	I J	I O
W	3,516.92	7	—	Dy	3,516.96	—	4
Nd	3,516.90	—	10	Ce	3,516.97	3*h*	1
Nb	3,516.86	10	2	Er	3,516.99	2	20*W*
U	3,516.85	15	6	Yb	3,517.01	—	25*w*
Ti	3,516.84	1	12	Sm	3,517.03	3*h*	—
Th	3,516.82	5	5	Ce	3,517.04	3*h*	20
W	3,516.78	3	7	U	3,517.05	8	4
Er	3,516.71	—	5	Nb	3,517.106	8	1
Co	3,516.67	—	2	La	3,517.11	25*h*	—
Re	3,516.65	—	60	Os	3,517.16	8	8
Tb	3,516.64	—	15	Ga	3,517.21	3	—
Os	3,516.63	20	20	Dy	3,517.27	4	70
Th	3,516.57	1	2	Os	3,517.29	10	15
Fe	3,516.56	4	30	V	3,517.30	30*WR*	20
Rb	3,516.53	(5)	—	Re	3,517.326	—	50
Er	3,516.52	—	8	He	3,517.327	(2)	—
J	3,516.51	(5)	—	Kr	3,517.37	(5*hs*)	—
Dy	3,516.48	—	4	Ce	3,517.38	6	40
Fe	3,516.420	15	40	Ru	3,517.41	—	10

	λ	I J	I O
Co	3,516.418	—	2
Th	3,516.36	5	5
F	3,516.32	(3)	—
Ni	3,516.23	—	5wh
W	3,516.22	8	—
Ce	3,516.21	—	2
Nb	3,516.20	5	3
Ru	3,516.19	—	3
Re	3,516.154	—	2
P	3,516.15	(70)	—
Dy	3,516.15	—	6
Tb	3,516.14	3	15
Cs	3,516.03	(4)	—
J	3,516.02	(10)	—
V	3,516.01	10	—
Er	3,516.00	1	10
W	3,515.96	6d	7d
Ir	3,515.95	15	35
Ce	3,515.94	—	5
Ru	3,515.89	8	10
Nb	3,515.42	300	20
Ni	3,515.05	50h	1000R
Fe	3,513.82	300	400
Ni	3,510.34	50h	900R
Pd	3,507.95	30h	—
Rh	3,507.32	125	500
F	*3,505.61*	*(600)*	—
Rh	*3,502.52*	*150*	*1000*
Co	*3,502.28*	*20*	*2000R*
Ba	*3,501.12*	*20*	*1000*
Co	*3,495.69*	*25*	*1000R*

	λ	I J	I O
Co	3,517.446	10	2
Ta	3,517.448	1h	3
U	3,517.457	5	1
W	3,517.46	4d	—
Zr	3,517.47	—	5
W	3,517.50	8	10
Mo	3,517.56	20	2
In	3,517.57	(5)	—
Dy	3,517.58	—	4
Tm	3,517.60	—	7
U	3,517.64	12	1
Nb	3,517.67	200	2
Nb	3,517.76	1	5
Tb	3,517.81	—	8
Zr	3,517.85	—	2
Er	3,517.89	1	8
Xe	3,517.90	(2)	—
Ar	3,517.90	(3)	—
Ce	3,517.91	1	5
Tl	3,519.24	1000R	2000R
Ni	3,519.77	30	500h
Ne	3,520.47	(1000)	—
Ni	3,524.54	100wh	1000R
Rh	*3,528.02*	*150*	*1000w*
Tl	*3,529.43*	*800*	*1000*
Co	*3,529.81*	*30*	*1000R*
Nb	*3,535.30*	*500*	*300*

Pd 3,609.548 *J* 700*R* *O* 1000*R*

	λ	J	O
Tm	3,609.54	25	15
Nd	3,609.495	30	25
Mo	3,609.493	10	5
Sm	3,609.484	100	60
Cr	3,609.479	12	20
Th	3,609.45	10	12
Er	3,609.44	1h	6d
Nb	3,609.360	5	1
Ta	3,609.357	1	8
Ni	3,609.314	15	200

	λ	J	O
Tb	3,609.55	—	15
Ti	3,609.59	2	12
Zr	3,609.64	—	3
U	3,609.68	12	15
Ce	3,609.69	10	40
Cl	3,609.74	(2)	—
Co	3,609.76	3	5
Ir	3,609.77	25	30
Nd	3,609.79	10	15
Tb	3,609.88	—	8

	λ	I			λ	I	
		J	O			J	O
Cu	3,609.307	5	25	Ce	3,609.89	—	2
Ca	3,609.30	2*h*	3	Ta	3,609.93	18	1*h*
V	3,609.29	30*W*	30*W*	Eu	3,609.94	—	3*w*
Dy	3,609.25	1	3	Nb	3,610.00	5	3
La	3,609.23	3	2	Cl	3,610.02	(4)	—
Th	3,609.228	4	5	Th	3,610.04	1	3
Ce	3,609.21	—	2	Cr	3,610.05	8	20
Ne	3,609.179	(50)	—	Ti	3,610.156	70	100
Ta	3,609.177	1	8	Fe	3,610.162	90	100
Os	3,609.15	12	20	La	3,610.24	5	7
Ru	3,609.11	4	3	Ce	3,610.26	—	5
N	3,609.09	(5)	—	Mn	3,610.30	40	60
Tb	3,609.06	—	8	Xe	3,610.32	(15)	—
U	3,608.96	10	18	Th	3,610.40	4	8
F	3,608.89	(6)	—	Ce	3,610.45	—	2
K	3,608.87	(10)	—	Ni	3,610.46	—	1000r
Fe	3,608.86	400	500	U	3,610.487	8	4
Ce	3,608.85	—	2	Re	3,610.49	—	40
Ta	3,608.78	1*d*	15*r*	Se	3,610.50	(35)	—
Tm	3,608.77	20	100	In	3,610.508	18	—
Gd	3,608.76	125	100	Cd	3,610.51	500	1000
Ru	3,608.73	8	2	Eu	3,610.57	2*h*	—
Eu	3,608.72	1*h*	10	Ta	3,610.60	15	—
U	3,608.69	4*h*	—	Mo	3,610.62	5	3
Mn	3,608.49	40	60	Pr	3,610.68	2	3
Pr	3,608.47	3	10	U	3,610.69	12*d*	3*d*
Cr	3,608.40	8	12	Fe	3,610.70	3	10
Ar	3,606.52	(1000)	—	Cd	3,612.87	500	800
Cr	3,605.33	400*R*	500*R*	Fe	3,618.77	400	400
				Ni	3,619.39	150*h*	2000*R*
Ni	*3,597.70*	*50h*	*1000r*				

Pd 3,634.70 *J* 1000 *O* 2000*R*

	λ	J	O		λ	J	O
Fe	3,634.69	2	20	Co	3,634.71	10	70
Ho	3,634.68	4*h*	6	Tb	3,634.75	—	8
Er	3,634.679	—	8	Cs	3,634.75	(6)	—
Nb	3,634.60	—	2	Gd	3,634.76	20	20
Th	3,634.58	4	8	Rn	3,634.8	(250)	—
U	3,634.56	1	8	Ar	3,634.83	(5)	—
Yb	3,634.55	—	2	Nd	3,634.87	15	30
Pr	3,634.47	4	20	Sm	3,634.91	2	20
Ar	3,634.46	(300)	—	Ru	3,634.93	100	50
Nb	3,634.44	15	5	Ni	3,634.94	10	50
Ce	3,634.43	—	2	Cr	3,634.99	12	25

	λ	I J	I O		λ	I J	I O
Kr	3,634.42	(3*wh*)	—	U	3,635.01	2	1
He	3,634.37	(2)	—	Nd	3,635.11	6	6
Fe	3,634.33	5	15	Au	3,635.13	10	3
Nd	3,634.28	30	25	Mo	3,635.14	10	100*h*
Sm	3,634.271	25	100	Tb	3,635.17	—	8
Dy	3,634.27	3*h*	3	Fe	3,635.196	2	12
S	3,634.25	(35*h*)	—	Ti	3,635.20	5	10
He	3,634.23	(15)	—	Th	3,635.25	5	4
Th	3,634.21	6	5	Dy	3,635.26	6	10
Zr	3,634.15	2	25	Cr	3,635.281	8	25
Cr	3,634.01	6	—	Pr	3,635.284	7	25
Ru	3,633.92	5	2	U	3,635.30	—	10
V	3,633.91	8	35	Sc	3,635.31	2	3
Ce	3,633.865	—	2	Nb	3,635.32	8	5
Ir	3,633.862	2	3	Yt	3,635.33	12	2*h*
Fe	3,633.83	3	7	Ho	3,635.35	4	—
Ta	3,633.79	10*h*	35	Th	3,635.37	10	15
Dy	3,633.77	1	4	U	3,635.40	5	6
Nb	3,633.71	3	8	Tb	3,635.42	8	20
Th	3,633.70	2	1*d*	Mo	3,635.429	5	25
Ne	3,633.665	(75)	—	W	3,635.434	—	2
Tb	3,633.66	—	8	V	3,635.463	5	40
Br	3,633.64	(6)	—	Nb	3,635.463	5	5
W	3,633.57	3	4	Ti	3,635.463	100	200
				Ir	3,635.49	4*h*	35
				Ru	3,635.52	8	2
				U	3,635.58	8	4
				Mo	3,635.61	3	4
				Eu	3,635.65	—	7*W*
				Ar	3,635.67	(3)	—
				Mn	3,635.68	5	10
				Ce	3,635.78	—	2
				Nb	3,635.85	8	3
				Ra	3,649.55	(1000)	—

*P**

Pd 4,212.95 *J* 300 *O* 500

	λ	J	O		λ	J	O
Sm	4,212.93	2	—	In	4,212.97	(15)	—
C	4,212.90	(5)	—	U	4,212.99	5	5
W	4,212.89	2	8	Yt	4,213.027	4	5
Tm	4,212.86	4	20	Sm	4,213.031	—	2
Nd	4,212.75	1	2	Ce	4,213.04	2	15
U	4,212.73	4*h*	5	Nd	4,213.06	—	2
Pr	4,212.70	2	5	Th	4,213.07	5*w*	10*w*
Ag	4,212.68	20	150*h*	In	4,213.10	(50)	—
Si	4,212.66	2	—	Cs	4,213.13	(30)	—
Cr	4,212.660	8	80	Ho	4,213.15	1	3

	λ	I J	I O
Zr	4,212.62	—	15
Se	4,212.58	(200)	—
Tb	4,212.54	—	5
Nb	4,212.534	3	2
Hg	4,212.53	(50)	—
Sc	4,212.49	—	4
U	4,212.48	3	5
Na	4,212.40	—	3
Ce	4,212.39	2	1
C	4,212.36	(5)	—
Sc	4,212.34	—	3
U	4,212.26	10	12
Tb	4,212.23	—	3*w*
Hg	4,212.22	(30)	—
Zr	4,212.16	—	3
Ru	4,212.06	80	125
Nb	4,212.04	3	4
Gd	4,212.02	50	150
Ce	4,211.91	2	1
Se	4,211.83	(200)	—
Rh	4,211.14	200	15
Fe	4,210.35	200	300
Xe	4,208.48	(200*h*)	—
Tm	4,203.73	25	250
Rn	4,203.23	(200)	—
Fe	*4,202.03*	*300*	*400*
Rb	*4,201.85*	*500*	*2000R*
Ar	*4,200.67*	*(1200)*	—
Ru	*4,199.90*	*300*	*150*
Ar	*4,198.32*	*(1200)*	—
Ca	4,213.17	5	—
Cr	4,213.179	8	60
Dy	4,213.182	8	5
Nd	4,213.21	—	10
Tm	4,213.26	—	3
Pr	4,213.26	2	4
Ir	4,213.28	2	3
Tb	4,213.49	2*h*	8*d*
Eu	4,213.54	2	3
Yt	4,213.55	2	4
Pr	4,213.57	10	18
In	4,213.58	(10)	—
Fe	4,213.65	60	100
Xe	4,213.72	(200*h*)	—
In	4,213.73	(15)	—
Os	4,213.859	3	30
Zr	4,213.865	3	40
U	4,213.88	4	20
Eu	4,213.91	—	2
Sm	4,213.92	3	10
Pr	4,213.96	3	12
Se	4,215.02	(150)	—
Gd	4,215.024	150	200
Sr	4,215.52	400*W*	300*r*
Rb	4,215.56	300	1000*R*
Fe	4,219.36	200	250
P	4,222.15	(150*w*)	300
Fe	4,222.22	200	200
Xe	4,223.00	(200*h*)	—
Ca	*4,226.73*	*50R*	*500R*

Pt $^{78}_{195.09}$

t_0 1,773.5°C

t_1 4,300°C

I.	II.	III.	IV.	V.
8.96	~19.3	–	–	–

λ	I		eV
	J	*O*	
I 2,659.45	500*R*	2000*R*	4.6
I 2,830.30	600*R*	1000*R*	4.4
I 2,929.79	200	800*R*	4.2
I 2,997.97	200*R*	1000*R*	4.23
I 3,064.71	300*R*	2000*R*	4.05
I 4,442.55	25	800	—
I 5,301.02	10	150	6.90

Pt 2,659.45 *J* 500*R* *O* 2000*R*

	λ	*J*	*O*		λ	*J*	*O*
Au	**2,659.43**	**5**	—	**Rh**	**2,659.47**	**3**	**3**
Eu	**2,659.42**	—	**4**	**Kr**	**2,659.60**	**(2*wh*)**	—
Ta	**2,659.41**	**10**	**20**	*bh***C**	**2,659.6**	—	**20**
Cd	**2,659.29**	**(5)**	—	**V**	**2,659.606**	**40**	**9**
Yb	**2,659.28**	**5**	**2**	**Ru**	**2,659.61**	**12**	**80**
J	**2,659.27**	**(20)**	—	**Ta**	**2,659.65**	—	**15**
Fe	**2,659.24**	**2**	**8**	**Hf**	**2,659.69**	**3*h***	—
Mn	**2,659.21**	**10*d***	—	**W**	**2,659.70**	**12**	—
W	**2,659.19**	**10**	—	**Ce**	**2,659.72**	—	**2**
Ir	**2,659.14**	**1**	**4**	**Cr**	**2,659.75**	**2**	—
Rh	**2,659.11**	**100**	—	**Re**	**2,659.79**	—	**15**
Mn	**2,659.08**	**25**	—	**Os**	**2,659.83**	**8**	**30**
Nb	**2,659.05**	**30**	**3*h***	**Th**	**2,659.86**	**1**	**3**
U	**2,659.025**	**2**	**8**	**Ga**	**2,659.87**	**12**	**5**
Re	**2,659.023**	—	**25**	**Ir**	**2,659.95**	—	**2**
Rh	**2,659.01**	**2**	**2**				
V	**2,658.97**	**40**	**10**	Pt	2,664.64	—	30
Tb	2,658.91	500	—				
Pd	2,658.72	300	20				
Pt	2,658.70	5	40				
Pt	2,658.17	10	100				
Ni	2,655.91	500*Wh*	—				
Ni	2,655.47	400*wh*	—				
Pt	*2,650.86*	*100*	*700*				

Pt 2,830.30 *J* 600*R* *O* 1000*R*

	λ	*J*	*O*		λ	*J*	*O*
W	**2,830.29**	**3**	**10**	**U**	**2,830.31**	**2**	**6**
Au	**2,830.27**	**2**	—	**Ce**	**2,830.34**	—	**2**
Ga	**2,830.19**	**2**	—	**V**	**2,830.40**	**60**	**10**
Ir	**2,830.17**	—	**12**	**Er**	**2,830.42**	**2**	**7**
W	**2,830.10**	**20**	**4**	**Kr**	**2,830.43**	**(3*h*)**	—
Re	**2,830.09**	—	**3*h***	**Ce**	**2,830.431**	—	**3**
U	**2,830.07**	**4*h***	**8**	**Th**	**2,830.44**	**8**	**10**
Ag	**2,830.06**	**1*h***	**2**	**As**	**2,830.45**	**25**	—
Ti	**2,830.04**	—	**8**	**Cr**	**2,830.47**	**80*h***	**15**
Ta	**2,830.02**	**1**	**20**	**Ir**	**2,830.51**	**5**	**8**
Mo	**2,829.942**	**5**	**25**	**Nb**	**2,830.57**	**30**	**1**
Th	**2,829.936**	**6**	**6**	**Rn**	**2,830.6**	**(70)**	—
Re	**2,829.88**	—	**3*h***	**Ru**	**2,830.70**	—	**20**
Na	**2,829.85**	**(5)**	**2**	**J**	**2,830.73**	**(15)**	—
W	**2,829.825**	**10**	**15**	**Ir**	**2,830.74**	—	**4**

	λ	I J	I O		λ	I J	I O
U	**2,829.820**	**4**	**6**	**Ho**	**2,830.79**	**10*hd***	—
Zr	**2,829.81**	—	**9**	**Mn**	**2,830.794**	—	**50**
Mo	**2,829.79**	—	**15**	**Re**	**2,830.83**	—	**2*h***
Nb	**2,829.75**	**50**	**3**	**Nb**	**2,830.84**	**5**	—
Cu	2,824.37	300	1000	Fe	2,831.56	500	1
Pt	2,822.27	60*h*	10	Pb	2,833.07	80*R*	500*R*
				Pt	2,834.71	5	80
				Cr	2,835.63	400*r*	100

Pt 2,929.79 *J* 200 *O* 800*R*

	λ	I J	I O		λ	I J	I O
U	**2,929.78**	**2**	**2**	**Pd**	**2,929.85**	**2*h***	—
Ir	**2,929.75**	**3*h***	—	**La**	**2,929.88**	**7**	**2**
Er	**2,929.73**	—	**12**	**Hf**	**2,929.90**	**3**	**5**
Ca	**2,929.71**	**4*h***	—	**Ru**	**2,929.93**	—	**12**
Xe	**2,929.66**	**(2*wh*)**	—	**W**	**2,929.99**	**3**	**5**
Hf	**2,929.63**	**50**	**30**	**Mo**	**2,930.06**	**20**	**1**
Fe	**2,929.62**	**10**	**50**	**Pr**	**2,930.08**	**5**	**3**
Re	**2,929.54**	—	**30**	**V**	**2,930.13**	**35**	**4**
Co	**2,929.509**	—	**75**	**Yt**	**2,930.14**	**20**	**8**
Os	**2,929.507**	**3**	**30**	**W**	**2,930.15**	**6**	**10**
Mo	**2,929.50**	**5**	—	**Ca**	**2,930.16**	**2*h***	—
Cr	**2,929.445**	**40**	—	**Te**	**2,930.17**	**(10)**	—
Ru	**2,929.442**	—	**20**	**Ir**	**2,930.176**	**5**	**12**
Ca	**2,929.44**	**2*h***	—	**J**	**2,930.18**	**(12)**	—
Ag	**2,929.35**	**40**	**20**	**Os**	**2,930.19**	**4**	**10**
Ne	**2,929.31**	**(15)**	—	**Mn**	**2,930.25**	—	**25**
Cd	**2,929.28**	**50**	—	**Nb**	**2,930.26**	**20*wh***	—
Fe	**2,929.24**	**1**	**3**	**Mo**	**2,930.400**	—	**5**
				Kr	**2,930.40**	**(2*wh*)**	—
Fe	2,929.01	100	150				
Ho	2,928.79	100	—	Co	2,930.43	150*wh*	—
Mg	2,928.75	100	25	Pt	2,930.79	3*h*	15
Nb	2,927.81	800*r*	200	V	2,930.81	150*r*	30
Ru	2,927.54	200	50	Ti	2,931.26	150*wh*	—
				In	2,932.62	300	500
Fe	2,926.59	400	150				
Eu	2,925.03	100	150	Ta	2,932.69	80*w*	400
V	2,924.64	200*r*	60	Ta	2,933.55	150	400
V	2,924.02	300*R*	70*r*	Ag	2,934.23	200*h*	10
V	2,923.62	150*r*	50*r*	Tm	2,936.00	300	80
Pt	*2,921.38*	*6*	*100*	*Ho*	*2,936.77*	*1000R*	—
Pt	*2,919.34*	*40*	*150*	*Fe*	*2,936.90*	*500r*	*700r*
Tl	*2,918.32*	*200R*	*400R*	*Bi*	*2,938.30*	*300w*	*300w*
				Ag	*2,938.55*	*200wh*	*200*
				Pt	*2,938.81*	*2*	*15*

	λ	I J	I O		λ	I J	I O
				Fe	*2,941.34*	*300*	*600*
				V	*2,941.37*	*300r*	*40*
				Nb	*2,941.54*	*300*	*50*

Pt 2,997.967 *J* 200*R* *O* 1000*R*

	λ	J	O		λ	J	O
V	**2,997.95**	**35**	—	**Ir**	**2,997.968**	—	**7**
Ga	**2,997.93**	**(5)**	—	**Ce**	**2,998.01**	—	**3**
W	**2,997.79**	**10**	**12**	**Yb**	**2,998.02**	**9**	**1**
O	**2,997.74**	**(7*h*)**	—	**Er**	**2,998.06**	**1**	**10**
Ce	**2,997.71**	—	**2**	**Cr**	**2,998.12**	**2**	**12**
Re	**2,997.70**	—	**15**	**Eu**	**2,998.14**	—	**4**
Mo	**2,997.67**	**3**	**2**	**Mo**	**2,998.15**	—	**10**
Os	**2,997.65**	**8**	**40**	**Al**	**2,998.17**	**(8)**	—
Ru	**2,997.61**	**5**	**30**	**Cs**	**2,998.20**	**(2)**	—
W	**2,997.60**	**12**	**8**	**Nb**	**2,998.23**	**2**	**2**
Pd	**2,997.491**	**2*wh***	—	**Sm**	**2,998.26**	—	**4**
Nb	**2,997.486**	**10**	**1**	**S**	**2,998.28**	**(25)**	—
Ce	**2,997.47**	—	**6**	**W**	**2,998.29**	**2**	**6**
Ru	**2,997.43**	**5**	**30**	**Ru**	**2,998.35**	**8**	**80**
Mo	**2,997.413**	—	**20**	**U**	**2,998.36**	**2**	**6**
Ir	**2,997.408**	**2**	**25**	**Yb**	**2,998.38**	**3**	—
Cu	**2,997.36**	**30**	**300**	**Cu**	**2,998.384**	**2*h***	**2*s***
U	**2,997.354**	**2**	**2**	**Zr**	**2,998.48**	**2*h***	**1**
Mo	**2,997.346**	**25**	**3**	**U**	**2,998.52**	—	**2**
Ca	**2,997.31**	**5**	**25**	**Re**	**2,998.55**	—	**8**
Fe	**2,997.30**	**60**	—	**V**	**2,998.63**	—	**2**
Cr	2,996.58	125	300*r*	Fe	2,999.51	300	500
Au	2,994.99	100	—	Pd	2,999.55	100*h*	—
Nb	2,994.73	300	100	Cr	3,000.89	125	150*r*
Fe	2,994.43	600*r*	1000*R*	Fe	3,000.95	300*r*	800*R*
Bi	2,993.34	100*wh*	200*wh*	Pt	3,001.17	50*W*	3
Ne	2,992.44	(150)	—	V	3,001.20	200*r*	20
Nb	2,991.95	100	1	Pt	3,002.27	30	200
Ru	2,991.62	100	50	Ni	3,002.49	100	1000*R*
				Yb	3,002.61	150	15
				Fe	3,002.65	150	20
Nb	*2,990.26*	*200*	*5*	Fe	3,003.03	100	200
Fe	*2,987.29*	*200*	*300*	Ni	3,003.63	80	500*R*
Fe	*2,985.55*	*300*	*80*	Cr	3,003.92	150	1
Fe	*2,984.83*	*400*	*200r*				
				Rn	*3,006.8*	*(300)*	—
				Fe	*3,008.14*	*400r*	*600r*
				In	*3,008.31*	*500W*	—
				Sn	*3,009.15*	*200h*	*300h*
				Fe	*3,009.57*	*400*	*500*

Pt 3,064.71 *J* 300*R* *O* 2000*R*

	λ	*I*			λ	*I*	
		J	*O*			*J*	*O*
Hf	**3,064.68**	**30**	**10**	**Ir**	**3,064.79**	—	**5**
Zr	**3,064.63**	**3**	**5**	**Ru**	**3,064.838**	**60**	**70**
Ni	**3,064.62**	**50**	**200*r***	**Er**	**3,064.84**	—	**7**
Rn	**3,064.60**	**(40)**	—	**U**	**3,064.908**	**2**	**1**
Re	**3,064.60**	—	**20**	**Yb**	**3,064.91**	**2**	**2**
U	**3,064.59**	**3**	**4**	**W**	**3,064.94**	**7**	**10*s***
Mo	**3,064.55**	—	**15**	**Cd**	**3,064.95**	**15*h***	—
Nb	**3,064.53**	**200**	**5*w***	**Er**	**3,064.97**	**1**	**3**
Ir	**3,064.51**	**1*h***	**20**	**Sm**	**3,065.01**	**2*h***	**4**
Ga	**3,064.45**	**2**	—	**Mo**	**3,065.04**	**10**	**30**
Hf	**3,064.38**	**2**	—	**Yb**	**3,065.05**	**30**	**4**
Na	**3,064.372**	**(20)**	**2**	**Cr**	**3,065.07**	**50**	**20**
Co	**3,064.370**	—	**100**	**Sc**	**3,065.11**	**25**	**12*d***
Al	**3,064.30**	**20**	**20**	**Dy**	**3,065.14**	**1*h***	**4**
Mo	**3,064.28**	**10**	**80**	**U**	**3,065.198**	**4**	**6**
Fe	**3,064.22**	**4**	**4**	**Sm**	**3,065.206**	**2**	**3**
Ho	**3,064.19**	**4**	—	**Zr**	**3,065.21**	**2**	**5**
U	**3,064.18**	**3**	**6**	**Rb**	**3,065.25**	**(2)**	—
Tb	**3,064.09**	**8**	**15**	**Nb**	**3,065.26**	**200**	**10**
Dy	**3,064.04**	—	**8**	**Re**	**3,065.28**	—	**5**
Ce	**3,064.02**	—	**15**	**Pd**	**3,065.306**	**100**	**10**
				Fe	**3,065.315**	**60**	—
Ne	3,063.69	(150)	—				
Fe	3,062.23	400	2	Fe	3,067.24	300	300
Pt	3,059.64	5	25	Bi	3,067.72	2000*wh*	3000*hR*
Pd	3,059.43	150*w*	—	Fe	3,068.18	150	150
Fe	3,059.09	400	600*r*	Pt	3,071.94	15	60
Os	3,058.66	500	500*R*				
				Ti	*3,075.22*	*300R*	*40*
				Fe	*3,075.72*	*400*	*400*
Fe	*3,057.45*	*400*	*400*	*Fe*	*3,077.17*	*300*	*1*
J	*3,055.37*	*(350)*	—				

Pt 4,442.55 *J* 25 *O* 800

	λ	*J*	*O*		λ	*J*	*O*
Nd	**4,442.48**	**6**	**6**	**La**	**4,442.67**	—	**6**
Sm	**4,442.47**	**12**	**12**	**Tm**	**4,442.74**	—	**8**
Ce	**4,442.46**	—	**6**	*bh***La**	**4,442.9**	—	**12**
Ni	**4,442.44**	—	**5**	**Pr**	**4,442.91**	—	**2**
Fe	**4,442.34**	**200**	**400**	**Zr**	**4,442.998**	**6**	**8**
Sm	**4,442.271**	**10**	**40**	*bh***La**	**4,443.0**	—	**25**
Cr	**4,442.268**	**3**	**15**	**Nb**	**4,443.01**	**5**	**1**
Mo	**4,442.20**	**30**	**40**	**Pd**	**4,443.039**	**2*h***	**5**
Ir	**4,442.14**	—	**2*wh***	**Ti**	**4,443.042**	—	**8*h***
*bh***Sr**	**4,442.1**	—	**1**	**Mo**	**4,443.069**	**20**	**25**

	λ	I J	I O		λ	I J	I O
Ce	**4,442.03**	—	**2**	**Hf**	**4,443.07**	**20**	**15**
Co	**4,441.95**	—	**5**	**Th**	**4,443.10**	**5**	**10**
Ce	**4,441.86**	—	**3**	**Ti**	**4,443.19**	—	**5*h***
W	**4,441.811**	**10**	**20**	**Fe**	**4,443.20**	**100**	**200**
Nb	**4,441.808**	**5**	**3**	**Sm**	**4,443.27**	**1**	**3**
Sm	**4,441.80**	**20**	**30**	**Ce**	**4,443.296**	—	**3**
V	**4,441.683**	**30**	**40*h***	**Nb**	**4,443.297**	**1**	**2**
Ta	**4,441.678**	**2*h***	**100**	**V**	**4,443.34**	**10**	**12**
Ce	**4,441.60**	—	**5**	**Nd**	**4,443.40**	**3**	**10**
Nb	**4,441.59**	**2*h***	**1**	**V**	**4,443.647**	**1*h***	**4**
Eu	**4,441.49**	—	**12*w***	**Yt**	**4,443.655**	—	**4**
Tb	**4,441.48**	—	**4**	**Cr**	**4,443.71**	**2**	**15**
Ni	**4,441.45**	—	**5**	**Ce**	**4,443.74**	**2**	**18**
Pr	**4,441.41**	—	**3**				
				Pt	4,445.55	2	20
Pt	4,437.28	2*h*	25				
Eu	4,435.60	100	400*R*				
Eu	4,435.53	—	2000				

Pt 5,301.02 *J* 10* *O* 150

	λ	I J	I O		λ	I J	I O
Yb	**5,300.94**	**60**	**6**	**Co**	**5,301.06**	—	**700*W***
Re	**5,300.77**	—	**3*h***	**Er**	**5,301.26**	—	**12**
Cr	**5,300.75**	**4**	**25**	**Th**	**5,301.41**	**2**	**12**
Er	**5,300.59**	—	**8**	**Dy**	**5,301.58**	—	**10**
U	**5,300.587**	**1**	**8**	**Gd**	**5,301.68**	—	**25**
Nd	**5,300.577**	—	**2**	**Sc**	**5,301.94**	**2**	**2**
Tb	**5,300.18**	—	**10**	**Zr**	**5,301.97**	—	**6**
Zr	**5,300.12**	—	**2**	**La**	**5,301.98**	**200**	**300*r***
Ti	**5,300.02**	—	**8**	**V**	**5,302.17**	**5**	**8**
Yb	**5,299.85**	—	**3**	**Nd**	**5,302.28**	**1**	**12**
Hf	**5,299.85**	**10**	**8**	**Er**	**5,302.30**	—	**30**
Zr	**5,299.51**	—	**2**	**Fe**	**5,302.31**	—	**300**
U	**5,299.47**	**3**	**6**	**Os**	**5,302.58**	—	**15**
Zr	**5,299.20**	—	**2**	**Nd**	**5,302.61**	—	**2*h***
Sm	**5,299.19**	—	**2**	**La**	**5,302.62**	**150**	**50**
				Sm	**5,302.66**	—	**3**
*bh*F	5,298.6	—	100	***bh*F**	**5,302.7**	—	**100**
Hf	5,298.06	100	80	**Eu**	**5,302.72**	—	**30**
*bh*F	5,296.8	—	100	**Gd**	**5,302.77**	—	**25**
Pd	5,295.63	10	200	**Ba**	**5,302.81**	**5**	**20**
Eu	5,294.60	—	300				
				Sm	**5,302.91**	—	**50**
*bh*F	5,292.9	—	150				
Rh	5,292.14	1	80				
Eu	5,291.25	—	200	La	5,303.56	125	100
*bh*F	5,291.0	—	200	Eu	5,303.87	—	300
Eu	5,289.25	—	125	*bh*F	5,304.4	—	100
				Tm	5,307.11	20	100

	λ	I	
		J	O
Fe	5,287.92	20	100
Eu	5,287.23	—	125
Ru	5,284.08	—	100
Fe	5,283.63	40	400
Co	5,283.49	—	125w
Sm	5,282.91	—	100
Eu	5,282.82	—	1000
Fe	5,281.80	20	300
Co	*5,280.65*	—	*500w*
Yb	*5,277.07*	*6*	*200*
Nb	*5,276.195*	*50*	*200*
Co	*5,276.192*	—	*400w*
Re	*5,275.53*	—	*500W*
Eu	*5,272.48*	—	*400*
Eu	*5,271.95*	—	*2000*
Nb	*5,271.53*	*50*	*200*
Sm	*5,271.40*	—	*150*
Re	*5,270.98*	—	*200W*
Fe	*5,270.36*	*80*	*400*
Fe	*5,269.54*	*200*	*800*
Co	*5,268.51*	—	*500w*
Fe	*5,266.58*	*40*	*500*
Co	*5,266.49*	—	*500w*
Eu	*5,266.40*	—	*1000*
Fe	*5,263.33*	—	*300*

	λ	I	
		J	O
Fe	5,307.36	—	125
*bh*F	5,308.7	—	80
Ru	5,309.27	—	125
*bh*F	5,310.3	—	80
Hf	5,311.60	150	100
Sm	5,312.21	—	100
Co	5,312.66	—	400w
*bh*F	5,316.2	—	80
Co	5,316.78	—	300w
Nb	5,318.60	12	100
Sm	5,320.60	—	100
Fe	*5,324.18*	*70*	*400*
Co	*5,325.28*	—	*300w*
Re	*5,327.46*	—	*100*
Fe	*5,328.05*	*100*	*400*
Fe	*5,328.53*	*35*	*150*
Co	*5,331.47*	*80*	*500w*
Co	*5,332.67*	—	*200w*
Yb	*5,335.16*	*400*	*150*
Fe	*5,339.94*	*30*	*200*
Fe	*5,341.03*	*15*	*200*

Ra $^{88}_{226.05}$

t_0 960°C t_1 1,140°C

I.	II.	III.	IV.	V.
5.278	10.145	—	—	—

λ	*I*	eV
II 2,708.96	(200)	—
II 2,813.76	(400)	—
II 3,649.55	(1000)	—
II 3,814.42	(2000)	3.25
II 4,682.28	(800)	2.65
I 4,825.91	(800)	2.57
I 5,660.81	(1000)	—

Ra 2,708.96 (200)

	λ	I J	I O
W	2,708.92	9	10
Yb	2,708.84	3	1
Co	2,708.82	2	30
Mn	2,708.81	12	—
W	2,708.794	2	9
Ni	2,708.791	500	—
Cr	2,708.79	40	3
Ir	2,708.767	10h	—
Br	2,708.66	(3)	—
W	2,708.58	15	10
Fe	2,708.57	50	80
V	2,707.86	150	70
Co	2,707.50	100wh	—
Co	2,706.74	100wh	—
V	2,706.70	200R	60
Fe	2,706.58	150	150
Sn	2,706.51	150R	200R
V	2,706.17	400R	100
Pt	2,705.89	200wh	1000wh
Co	2,705.85	100w	15w
Rh	2,705.63	300wh	100
Ru	2,704.57	100	—
Fe	*2,703.99*	*400*	*30*
Cu	*2,703.18*	*200*	*10*
Pt	*2,702.40*	*300*	*1000*
V	*2,702.19*	*300R*	*80*
Eu	*2,701.89*	*200*	*300W*
Cu	*2,700.96*	*400*	*20*
V	*2,700.94*	*500R*	*125*

	λ	I J	I O
B	2,709.00	2	—
Cs	2,709.0	(2)	—
U	2,709.03	4w	3w
Cl	2,709.03	(10)	—
Co	2,709.05	30wh	—
Fe	2,709.06	100	3
Ba	2,709.07	2	—
Zr	2,709.07	3	—
In	2,709.2	5	—
Ru	2,709.204	8	60
Pd	2,709.21	30wh	—
Tl	2,709.23	200R	400R
Mo	2,709.25	1	20
Ta	2,709.27	150	40
Cr	2,709.31	60	2
Fe	2,709.37	4	—
Ru	2,710.23	100	50
In	2,710.26	200Rh	800R
Ag	2,711.21	300wh	1h
Mn	2,711.58	125h	2
V	2,711.74	150R	50
Fe	2,711.84	100	4
Ag	2,712.06	200h	3
J	2,712.23	(100)	—
Fe	2,712.39	100	2
Ru	2,712.41	300	80
Cu	2,713.50	300w	50
Ru	2,713.58	100	—
Fe	*2,714.41*	*400*	*200*
Co	*2,714.42*	*200W*	*12*
Pd	*2,714.90*	*200*	—
Rh	*2,715.31*	*500wh*	*50*
V	*2,715.69*	*300R*	*50*
Nb	*2,716.62*	*200*	*10*
Eu	*2,716.97*	*300*	*300*
Cu	*2,718.77*	*300w*	*40*

Ra 2,813.76 (400)

	λ	I J	I O
La	2,813.72	5	2h
Ru	2,813.71	125	50
Yt	2,813.65	20h	7d
Fe	2,813.613	60	5
Si	2,813.61	(5)	—

	λ	I J	I O
Ho	2,813.77	10h	—
U	2,813.79	2	3
Os	2,813.84	4	10
Hf	2,813.86	30	25
Ca	2,813.88	4	—

	λ	I J	I O		λ	I J	I O
Sn	**2,813.58**	**50**	**50**	**Eu**	**2,813.95**	**300*wh***	**300*w***
U	**2,813.55**	**6**	**4**	**Te**	**2,813.96**	**(5)**	—
Mn	**2,813.47**	**1**	**30**	**U**	**2,813.98**	**4*h***	—
Cd	**2,813.41**	**(10)**	—	**Pt**	**2,814.001**	**15**	**4**
Rn	**2,813.40**	**(3)**	—	**Pd**	**2,814.004**	**10**	—
Ru	**2,813.30**	**75**	—	**Mo**	**2,814.05**	**6**	**2**
Fe	**2,813.29**	**400**	**400**	**Os**	**2,814.20**	**25**	**50**
				Cr	**2,814.22**	**5**	—
				Yb	**2,814.24**	**2**	—
Ru	2,810.55	200	50	**Ta**	**2,814.31**	**50**	**50*r***
				Hg	2,914.93	(200)	—
				Mo	2,816.15	300*h*	200
				Ti	2,817.87	200	10

Ra 3,649.55 (1000)

	λ	I J	I O		λ	I J	I O
Ho	**3,649.52**	**4*h***	**2**	**Cd**	**3,649.60**	**15**	**20**
La	**3,649.509**	**8**	**40**	**Ce**	**3,649.726**	**1**	**10**
Fe	**3,649.508**	**100**	**100**	**U**	**3,649.729**	**8**	**1**
Sm	**3,649.506**	**30**	**100**	**Th**	**3,649.74**	**3**	**8**
Mo	**3,649.47**	**5**	**5**	**Eu**	**3,649.82**	**2**	**5**
Gd	**3,649.44**	**5**	**5**	**Ar**	**3,649.83**	**(800)**	—
Tb	**3,649.41**	**3**	**8**	**Nb**	**3,649.85**	**20**	**20**
U	**3,649.410**	**4**	—	**Cr**	**3,649.86**	**4**	**15**
Pr	**3,649.40**	**3**	**10**	**Dy**	**3,650.00**	**2**	**5**
Co	**3,649.35**	**4**	**200**	**Fe**	**3,650.03**	**30**	**70**
Fe	**3,649.30**	**25**	**60**	**Mo**	**3,650.05**	**25**	**3**
Th	**3,649.25**	**10**	**10**	**Cl**	**3,650.10**	**(4)**	—
Al	**3,649.22**	**(2)**	—	**Xe**	**3,650.12**	**(3)**	—
Al	**3,649.18**	**(5)**	—	**Ce**	**3,650.121**	**1**	**10**
Hf	**3,649.10**	**5**	**20**	**Hg**	**3,650.15**	**500**	**200**
Au	**3,649.09**	**5*h***	**3**	**Sm**	**3,650.168**	**5**	**25**
W	**3,649.02**	**7**	**6**	**La**	**3,650.174**	**60**	**100**
Sm	**3,649.01**	**1**	**2**	**Pr**	**3,650.18**	**6**	**30**
Pb	**3,649.00**	**(20)**	—	**N**	**3,650.19**	**(70)**	—
Cr	**3,648.997**	**20**	**40**	**Fe**	**3,650.28**	**50**	**70**
Er	**3,648.98**	**1**	**4**	**Ru**	**3,650.32**	**12**	**3**
V	**3,648.97**	**50**	**80**	**Cr**	**3,650.34**	**40**	—
U	**3,648.93**	**4**	**4**	**Os**	**3,650.38**	**4**	**15**
Ti	**3,648.86**	**10**	**3**	**Tb**	**3,650.40**	**100**	**50**
Dy	**3,648.807**	**30**	**50**	**Er**	**3,650.414**	**2**	**15**
Os	**3,648.806**	**10**	**100**	**Nd**	**3,650.415**	**6**	**12**
Th	**3,648.64**	**2**	**4**	**Th**	**3,650.51**	**1*d***	**2*d***
Kr	**3,648.61**	**(40*h*)**	—	**Nb**	**3,650.52**	**3**	**2**
Mo	**3,648.607**	**10**	**10**	**Hf**	**3,650.53**	**2**	**2**
U	**3,648.54**	**2**	**5**	**Mo**	**3,650.58**	**2**	**3**

	λ	I J	I O		λ	I J	I O
Cr	3,648.53	8	10	Ca	3,650.62	3	—
Th	3,648.642	4	8	U	3,650.68	2	3
Dy	3,648.40	1	3	Nd	3,650.69	6	15
Pd	3,634.69	1000*R*	2000*R*	Nb	3,659.61	500	15

Ra 3,814.42 (2000)

	λ	I J	I O		λ	I J	I O
Os	3,814.26	10	30	W	3,814.43	6	2
Yb	3,814.23	10	3	Mo	3,814.49	4	4
Mo	3,814.10	3	4	Ce	3,814.517	3	4
U	3,814.07	15	25	Fe	3,814.523	40	80
La	3,814.02	2	2	Dy	3,814.57	3	4*h*
Cr	3,814.017	4	2	Ti	3,814.58	35	12
Gd	3,813.98	60	100*w*	Th	3,814.59	15	15
Tb	3,813.97	3	3	Ce	3,814.622	2	2
Mo	3,813.90	20	2	Cr	3,814.622	30	35
Fe	3,813.89	25	50	Sm	3,814.63	5	—
Se	3,813.85	(2)	—	F	3,814.65	(10)	—
U	3,813.79	15	20	Rn	3,814.70	(12)	—
Fe	3,813.632	15	35	Nd	3,814.72	8	6
Sm	3,813.632	4	10	Gd	3,814.75	10	8
Nb	3,813.47	5*h*	3	Fe	3,814.78	5	10
Ti	3,813.39	20	4	Ru	3,814.857	35	20
V	3,813.35	1	3	Ti	3,814.864	4	30
Er	3,813.32	1*d*	10*d*	Eu	3,814.87	5*wh*	5*wh*
Ti	3,813.27	3	12	Th	3,814.88	5	10
Ho	3,813.24	8*h*	15	Nd	3,814.89	6*d*	10*d*
U	3,813.22	8	6	Ce	3,814.93	3	2
Ru	3,813.18	3	4	Zr	3,814.96	5	5
Th	3,813.064	20	15	Rh	3,815.01	20	20
Fe	3,813.059	2	2	Mo	3,815.05	4	5
				Th	3,815.07	8	8
Th	3,815.07	8	8	U	3,815.151	6	1
				W	3,815.155	5	—
				Nd	3,815.34	4	6
				Mo	3,815.36	8	—
				V	3,815.39	150*h*	1
				Cr	3,815.43	12	35
				Eu	3,815.50	15*wh*	20*wh*
				Nb	3,815.511	30	20
				V	3,815.514	1	30
				Ir	3,815.52	3	25
				Br	3,815.68	(15)	—
				W	3,815.78	5	12
				Fe	3,815.84	700	700

Ra 4,682.28 (800)

	λ	I J	I O		λ	I J	I O
Mo	**4,682.24**	**5**	**4**	**Ar**	**4,682.29**	**(10)**	**—**
Ne	**4,682.15**	**(20)**	**—**	**Yt**	**4,682.32**	**100**	**60**
U	**4,682.13**	**2**	**4**	**U**	**4,682.60**	**2**	**—**
La	**4,682.12**	**3**	**3**	**Nb**	**4,682.66**	**3**	**2**
Dy	**4,682.03**	**2**	**3**	**Hf**	**4,682.67**	**8**	**2**
In	**4,682.00**	**250*W***	**—**	**V**	**4,682.75**	**2*h***	**3*h***
Cu	**4,681.99**	**20**	**—**	**Ne**	**4,682.91**	**(10)**	**—**
Mo	**4,681.933**	**3**	**3**	**Rh**	**4,682.96**	**2**	**3**
Ne	**4,681.930**	**(20)**	**—**	**Nb**	**4,682.98**	**2**	**2**
Tm	**4,681.92**	**2**	**50**	**U**	**4,683.05**	**8**	**2**
Ti	**4,681.916**	**100**	**200**	**Si**	**4,683.10**	**4**	**—**
Ta	**4,681.87**	**50**	**200**	**Ta**	**4,683.10**	**15**	**—**
Mo	**4,681.63**	**5**	**4**	**Ne**	**4,683.24**	**(5)**	**—**
Eu	**4,681.51**	**1**	**10**	**Gd**	**4,683.34**	**50**	**100**
S	**4,681.32**	**(5)**	**—**	**W**	**4,683.54**	**10**	**20**
Th	**4,681.21**	**2**	**2**	**Kr**	**4,683.68**	**(5)**	**—**
Ne	**4,681.20**	**(50)**	**—**	**Mo**	**4,683.72**	**5**	**5**
W	**4,681.19**	**1**	**10**				
U	**4,681.18**	**4**	**2**				
In	**4,681.11**	**(200)**	**—**	Ra	4,699.28	(40)	—
Eu	**4,681.06**	**1**	**15**	Ar	4,702.32	(1200)	—
Te	**4,681.06**	**(15*h*)**	**—**	Ne	4,704.39	(1500)	—
Mo	**4,681.05**	**8**	**6**	Ne	4,708.85	(1200)	—
V	**4,680.89**	**3**	**4**	Ne	4,710.06	(1000)	—
Cr	**4,680.87**	**8**	**60**				
Rn	**4,680.83**	**(500)**	**—**				
Kr	4,680.41	(500)	—				
Xe	4,671.23	(2000)	—				
Te	*4,664.34*	*(800)*	—				
Kr	*4,658.87*	*(2000)*	—				
Te	*4,654.38*	*(800)*	—				

Ra 4,825.91 (800)

	λ	I J	I O		λ	I J	I O
Eu	**4,825.65**	**2**	**5**	**Ar**	**4,825.97**	**(2)**	**—**
Hg	**4,825.62**	**(70)**	**—**	**Dy**	**4,826.56**	**2**	**4**
Mn	**4,825.59**	**5**	**20**	**S**	**4,826.77**	**(3)**	**—**
Ne	**4,825.53**	**(50)**	**—**	**La**	**4,826.89**	**30**	**15**
Nd	**4,825.48**	**8**	**100**	**Mn**	**4,826.90**	**5**	**10**
Ti	**4,825.46**	**1**	**10**	**Tm**	**4,826.99**	**25**	**10**
Cs	**4,825.42**	**(10)**	**—**	**Hg**	**4,827.1**	**(15)**	**—**
Kr	**4,825.18**	**(300)**	**—**	**Dy**	**4,827.12**	**2**	**2**
Dy	**4,824.97**	**2*h***	**5**	**Te**	**4,827.14**	**(50)**	**—**
U	**4,824.67**	**1**	**3**	**Ne**	**4,827.34**	**(1000)**	**—**

	λ	I	
		J	O
S	4,815.51	(800)	—

	λ	I	
		J	O
V	**4,827.45**	**15**	**20**
Xe	4,829.71	(400)	—
Te	4,831.29	(800)	—
Kr	4,832.07	(800)	—
Ne	4,837.31	(500)	—
Se	4,840.63	(800)	—
Xe	*4,844.33*	*(1000)*	—
Se	*4,844.96*	*(800)*	—
Kr	*4,846.60*	*(700)*	—
Ra	*4,856.57*	*(100)*	—

Ra 5,660.81 (1000)

	λ	J	O
Sb	**5,660.78**	**7**	**3**
Pb	**5,660.1**	**(4)**	—
S	**5,659.93**	**(600)**	—
Pr	**5,659.84**	**1**	**5**
Nd	**5,659.78**	**1**	**15**
Xe	**5,659.38**	**(150)**	—
Ar	**5,659.13**	**(500)**	—
Hf	**5,658.83**	**3**	**1**
Fe	**5,658.826**	**80**	**100**
Fe	**5,658.54**	**2**	**30**
Ne	5,656.66	(500)	—
Ar	5,650.70	(1500)	—
S	5,640.37	(500)	—
S	5,639.98	(500)	—
Ra	*5,620.47*	*(15)*	—
Ra	*5,616.66*	*(250)*	—

	λ	J	O
Ra	**5,661.73**	**(50)**	—
Ar	**5,662.00**	**(5)**	—
Ti	**5,662.16**	**100**	**100**
C	**5,662.51**	**50**	—
Fe	**5,662.52**	**50**	**50**
Ne	**5,662.55**	**(75)**	—
Kr	**5,662.67**	**(3)**	—
Yt	**5,662.92**	**400**	**20**
N	5,679.56	(500)	—
Kr	5,681.89	(400)	—
Ar	5,681.90	(500)	—

Rb $^{37}_{85.47}$

t_0 38.5°C

t_1 679°C

I.	II.	III.	IV.	V.
4.176	27.499	47	80	—

λ	*I*		eV
	J	*O*	
I 3,350.89	—	150	3.71
II 3,461.57	200	—	4.14
3,492.77	300	—	—
I 3,587.08	40	200	3.47
I 4,201.85	500	2000*R*	2.95
I 4,215.56	300	1000*R*	2.94
I 7,800.23	—	9000*R*	1.59
I 7,947.60	—	5000*R*	1.56

Rb 3,350.89 *O* 150

	λ	*I* *J*	*I* *O*
Sm	**3,350.88**	**4**	**9**
Ce	**3,350.75**	—	**10**
Nb	**3,350.69**	**7*h***	**10**
Ce	**3,350.68**	—	**20**
Dy	**3,350.66**	**2**	**3**
Sm	**3,350.65**	—	**8**
W	**3,350.61**	**10**	**4**
Ag	**3,350.56**	—	**2*h***
Ru	**3,350.549**	—	**12**
Ti	**3,350.548**	**1**	**10**
U	**3,350.52**	**2*h***	**8**
Gd	**3,350.482**	**180**	**150**
Ho	**3,350.46**	**8**	**6**
Eu	**3,350.43**	—	**30**
Mn	**3,350.40**	—	**4**
Co	**3,350.38**	—	**4*h***
Ca	**3,350.36**	**3**	**15**
Th	**3,350.33**	**2*d***	**2**
Mo	**3,350.30**	**6**	**5**
Fe	**3,350.285**	**4**	**8**
U	**3,350.28**	**1*hd***	**4*d***
Pr	**3,350.27**	**3**	**10**
Er	**3,350.255**	**2**	**10**
Ru	**3,350.251**	**2**	**10**
Ca	**3,350.21**	**10**	**100**
Ru	**3,350.11**	—	**12**
Gd	**3,350.10**	**3*h***	**3**
Er	**3,350.06**	**6**	**18**
Tm	**3,349.99**	—	**20**
Ce	**3,349.97**	**1**	**30**
Re	**3,349.91**	—	**20**
Ti	3,349.41	400*R*	100
Cr	3,349.07	40	125
Nb	3,349.06	100	80
Ti	3,349.03	800*R*	125
Rb	3,348.72	—	100
Fe	3,347.93	100	150
Co	3,346.93	1	100
Cr	3,346.74	80*r*	150*R*
Re	3,346.20	—	100
Zn	3,345.93	50	150
Zn	3,345.57	100	500
Zn	3,345.02	300	800
La	3,344.56	200*wh*	300
Ca	3,344.51	7	100
Re	3,344.35	—	150

	λ	*I* *J*	*I* *O*
U	**3,350.90**	**2**	**3**
Ce	**3,350.93**	—	**3**
Dy	**3,350.95**	**2**	**2**
Ta	**3,350.96**	**18*W***	**25*W***
Eu	**3,351.04**	**2**	**2**
Ce	**3,351.08**	—	**3**
Mo	**3,351.12**	**5**	**3**
Co	**3,351.15**	—	**4**
W	**3,351.18**	**3**	**5**
Eu	**3,351.19**	—	**15*W***
Zr	**3,351.228**	—	**4**
Th	**3,351.230**	**15**	**10**
Sr	**3,351.246**	**15**	**300**
U	**3,351.251**	**3**	**8**
Pr	**3,351.252**	**1**	**5**
Sm	**3,351.28**	**2**	**5**
Er	**3,351.30**	**2**	**10*w***
Hg	**3,351.30**	**20*h***	**10**
Mn	**3,351.43**	—	**4**
Ta	**3,351.510**	**15*r***	**18*r***
Mo	**3,351.512**	**8**	**5**
Ce	**3,351.520**	—	**12**
Fe	**3,351.521**	**60**	**70**
Co	**3,351.54**	**2**	**35**
Eu	**3,351.56**	**2**	**12**
Cr	**3,351.596**	**8**	**35**
Th	**3,351.599**	**4**	**4**
U	**3,351.66**	**2**	**4**
Mn	**3,351.662**	—	**8**
Ti	**3,351.67**	**5**	**1**
Sm	**3,361.69**	—	**3**
Os	**3,351.73**	**12**	**50**
Fe	**3,351.74**	**60**	**80**
Ce	**3,351.75**	—	**3**
Nd	**3,351.80**	**1*h***	**8*d***
U	**3,351.83**	**2**	**5**
Ta	**3,351.87**	**1*h***	**2*wh***
La	**3,351.89**	**2**	**2**
Co	3,354.38	—	200*R*
Ti	3,354.63	20	100
Fe	3,355.23	100	100
V	3,356.35	60	125
Co	3,356.47	2	150*W*
Os	3,357.97	15	100
Nb	3,358.42	100	100
Gd	3,358.63	100	100

	λ	I J	I O
Rh	3,344.20	20	100
Pt	3,343.90	80	100
Co	3,342.73	—	150W
Re	3,342.26	—	200
Nb	3,341.97	50	100r
Fe	3,341.90	80	100
Ti	3,341.87	300R	100
Mg	3,341.48	100	100
Co	*3,339.78*	—	*150w*
Rh	*3,338.54*	*50*	*200*
Re	*3,338.17*	—	*150*
La	*3,337.49*	*300wh*	*800*
Os	*3,336.15*	*50*	*200R*
Cr	*3,334.69*	—	*150wh*
Fe	*3,334.22*	*100h*	*150h*
Co	*3,334.14*	—	*250R*
Co	3,359.29	2	100
Lu	3,359.56	15	150
Rh	3,359.89	50	100
Ni	*3,361.56*	*20*	*500W*
Gd	*3,362.24*	*180*	*150*
Tm	*3,362.61*	*200*	*250*
Ni	*3,365.77*	*12*	*400w*
Ni	*3,366.17*	*12*	*400W*
Co	*3,367.11*	*30*	*300R*
Rh	*3,368.37*	*40*	*300*
Fe	*3,369.55*	*200*	*300*
Ni	*3,369.57*	*100*	*500R*
Os	*3,370.59*	*30*	*300R*
Fe	*3,370.79*	*200*	*300*

Rb 3,461.57 *J* 200

	λ	J	O
Ti	**3,461.50**	**125**	**80**
Sm	**3,461.405**	**4**	**5**
Er	**3,461.40**	**1**	**10**
Eu	**3,461.38**	**2**	**25**
W	**3,461.364**	**6**	**7**
Ho	**3,461.36**	**4**	—
Dy	**3,461.31**	**1**	**3**
Xe	**3,461.26**	**(50*h*)**	—
Th	**3,461.22**	**4**	**10**
La	**3,461.184**	**2**	**10**
Co	**3,461.176**	**3**	**100*wh***
Tm	**3,461.17**	**5**	**15**
Sm	**3,461.14**	**3**	**8**
Zr	**3,461.09**	**1**	**20**
Ar	**3,461.08**	**(300)**	—
Pr	**3,461.06**	**2**	**10**
Th	**3,461.02**	**4**	**10**
Yt	**3,461.01**	**12**	**7**
J	**3,461.007**	**(25)**	—
U	**3,461.003**	**3**	**3**
Tb	**3,461.00**	**8**	**15**
Dy	**3,460.971**	**3**	**100**
Er	**3,460.968**	**7**	**20**
Ho	**3,460.95**	**4**	**6**
Kr	**3,460.90**	**(2)**	—
W	**3,460.784**	**3**	**3**
Mo	**3,460.784**	**25**	**25**
V	**3,461.58**	**3**	—
Nb	**3,461.61**	**2**	**5*h***
Ni	**3,461.652**	**50*h***	**800*R***
W	**3,461.655**	**3**	**3**
Yb	**3,461.71**	**3**	**1**
W	**3,461.81**	**5**	**9**
Pr	**3,461.87**	**2**	**6**
Gd	**3,461.956**	**5**	**5**
Ho	**3,461.96**	**20**	**20**
Rh	**3,462.04**	**150**	**1000**
Mo	**3,462.06**	**25**	—
Sc	**3,462.18**	**2*h***	**4**
Os	**3,462.19**	**10**	**20**
Tm	**3,462.20**	**200**	**250**
Eu	**3,462.21**	**1*h***	**5**
La	**3,462.32**	**2*h***	**2**
Fe	**3,462.36**	**3**	**10**
Na	**3,462.49**	**(15)**	**2**
Tb	**3,462.51**	**3**	**8**
Ca	**3,462.62**	**4**	**2**
Hf	**3,462.64**	**12**	**15**
Nb	**3,462.65**	**3**	**5**
Gd	3,463.98	125	100
Sr	3,464.46	200	200
Fe	3,465.86	400	500
Cd	3,466.20	500	1000

	λ	I J	I O
Pd	**3,460.77**	**600h**	**300r**
Sc	**3,460.70**	**2h**	**6**
Dy	**3,460.64**	**1**	**4**
Sm	**3,460.63**	**1**	**3w**
Nd	**3,460.58**	**6**	**25**
Ne	**3,460.52**	**(75)**	**—**
Mn	3,460.33	500	60
Yb	3,458.28	100	12
Cr	3,457.63	125	4
V	3,457.15	150	2
Ti	3,456.39	125	25
Bi	3,455.27	100h	—
Cr	3,454.99	100	—
Yb	3,454.07	250	40
Nb	3,453.97	100	—
Co	3,453.50	200	3000R
Ti	3,452.47	100	12
Nb	3,452.35	200	5
Hg	3,451.69	(200)	—
B	3,451.41	100	—
Pd	3,451.35	400h	—
Fe	*3,443.88*	*200*	*400*
Fe	*3,440.99*	*200*	*300*
Fe	*3,440.61*	*300*	*500*
Ne	3,466.58	(150)	—
Gd	3,467.28	100	100
Cd	3,467.66	400	800
Gd	3,468.99	150	150
V	3,469.52	100	2
Nb	3,470.25	100	3h
Rh	3,470.66	125	500
O	3,470.77	(100)	—
Ne	*3,472.57*	*(500)*	*—*
Sb	*3,473.91*	*300wh*	*3*
Mn	*3,474.13*	*400*	*12*
Rb	*3,474.46*	*(20)*	*—*
Fe	*3,475.45*	*300*	*400*
Fe	*3,476.70*	*200*	*300*
Yb	*3,478.84*	*300*	*40*
Nb	*3,479.56*	*200*	*5*
Ta	*3,480.52*	*200ws*	*70*

Rb 3,492.77 *J* 300

	λ	J	O
Pr	**3,492.73**	**1**	**4**
Th	**3,492.68**	**2**	**2**
Sm	**3,492.62**	**1**	**8**
Tm	**3,492.59**	**10**	**20**
Tb	**3,492.56**	**8**	**15**
Er	**3,492.54**	**2d**	**25d**
Dy	**3,492.52**	**1**	**2**
Ti	**3,492.50**	**35wh**	**—**
U	**3,492.33**	**1**	**8**
U	**3,492.21**	**8**	**2**
W	**3,492.06**	**12**	**2d**
vzduch	**3,491.98**	**6**	**—**
Gd	**3,491.967**	**25**	**50**
Pr	**3,491.94**	**2**	**10**
Lu	**3,491.92**	**3h**	**—**
Nb	**3,491.89**	**15**	**1**
Mo	**3,491.87**	**3**	**3**
W	**3,491.83**	**4**	**9**
Hg	**3,492.77**	**(50)**	**—**
Sm	**3,492.775**	**3**	**3**
U	**3,492.80**	**10**	**2**
Mo	**3,492.82**	**4**	**3**
Au	**3,492.95**	**4**	**—**
Ni	**3,492.96**	**100h**	**1000R**
W	**3,493.036**	**4**	**5**
Kr	**3,493.04**	**(8wh)**	**—**
Ho	**3,493.10**	**10**	**10**
Ce	**3,493.11**	**1**	**12**
Pr	**3,493.16**	**2**	**10**
V	**3,493.17**	**100**	**15**
W	**3,493.19**	**3**	**6**
F	**3,493.21**	**(3)**	**—**
Ru	**3,493.22**	**1**	**20**
Ar	**3,493.25**	**(20)**	**—**
Ti	**3,493.28**	**1**	**15**
U	**3,493.33**	**15**	**6**

	λ	I J	I O
Al	3,491.80	2	—
Mo	3,491.77	3	3
Gd	3,491.74	8	4
Fe	3,490.57	300	400
Mn	3,488.68	200	50
Pt	3,485.27	200R	150
Mn	3,482.91	250	50

	λ	I J	I O
Mo	3,493.34	10	6
U	3,493.407	5	2
Sm	3,493.408	1	2
Nb	3,493.473	3	3
Fe	3,493.474	80	40
Th	3,493.53	15	30
Sm	3,493.597	4	5
Nd	3,493.60	6	6d
Fe	3,493.70	1	3
Hg	3,493.85	(100)	—
Mn	3,495.84	150	25
Mn	3,497.54	150	15
Fe	3,497.84	200	200
Sb	3,498.46	300wh	—
Ru	3,498.94	200	500R
Ne	3,501.22	(150)	—
F	3,501.42	(200)	—
Rh	3,502.52	150	1000
F	3,503.09	(400)	—
F	*3,505.61*	*(600)*	—
Rb	*3,511.19*	*(60)*	—
Fe	*3,513.82*	*300*	*400*

P*

Rb 3,587.08 J 40 O 200

	λ	I J	I O
Ce	3,587.079	2	3
Fe	3,586.99	150	200
Ag	3,586.98	—	6h
bhSr	3,586.9	—	8
Nb	3,586.87	—	2
Mo	3,586.86	6	6
Nb	3,586.76	20	2
Ce	3,586.753	4	10
Fe	3,586.747	2	4
Au	3,586.74	15	20
Nb	3,586.69	20	1
Er	3,586.64	—	8d
Mn	3,586.543	40	50h
Sm	3,586.54	3	3
Pr	3,586.53	4	10
Ba	3,586.52	3	5
Os	3,586.51	15	30
Sm	3,586.36	—	40r
U	3,586.33	—	3
Zr	3,586.294	3	20

	λ	I J	I O
Ti	3,587.13	25	12
Pr	3,587.14	8	8
Eu	3,587.15	1h	3
Co	3,587.190	50h	200r
Gd	3,587.191	15	25
Ru	3,587.20	70	5
Ce	3,587.22	1	5
Fe	3,587.24	2	5
Er	3,587.26	—	3
Os	3,587.31	15	60
Pr	3,587.34	5	10
Pt	3,587.40	—	4h
Nb	3,587.41	4	2
Fe	3,587.422	5	10
Ag	3,587.424	—	3h
Tb	3,587.44	15	15
Sm	3,587.47	3	15
Nd	3,587.50	20	8
Ce	3,587.64	2	10
Ce	3,587.68	—	2

	λ	I J	I O
Ta	3,586.291	3	18
Pr	3,586.25	1	2
Dy	3,586.12	2*h*	4
Fe	3,586.114	80	80
V	3,586.113	—	6
Tm	3,586.07	—	6
Dy	3,585.78	100	150
Fe	3,585.71	80	125
Fe	3,585.32	100	150
Dy	3,585.07	100	300
Gd	3,584.96	100	100
Fe	3,584.66	60	100
Rh	3,583.10	125	200
Re	3,583.02	—	100*w*
Fe	3,581.19	600*r*	1000*R*
Nb	3,580.27	300	100
Cr	3,578.69	400*r*	500*R*
Dy	3,577.99	50	150
Ce	3,577.46	12	300
Dy	3,576.87	50	200
Dy	3,576.25	—	300
Zr	3,575.79	5	100
Co	*3,575.36*	*25*	*200r*
Co	*3,574.96*	*25*	*200*
Dy	*3,574.16*	*100*	*200*
Pb	*3,572.73*	*20*	*200*
Ni	*3,571.87*	*40h*	*1000R*
Rh	*3,570.18*	*150*	*400r*
Fe	*3,570.10*	*300*	*300*
Co	*3,569.38*	*100*	*400R*
Ni	*3,566.37*	*100wh*	*2000R*
Fe	*3,565.38*	*300*	*400*
Er	3,587.748	3	10
Yt	3,587.753	2	15
Fe	3,587.758	25	50
Tb	3,587.76	—	30
U	3,587.78	—	2
Pr	3,587.86	6	8
Os	3,587.87	10	2
Ni	3,587.93	12	200
Nb	3,587.96	—	2
Zr	3,587.98	10	10
Fe	3,588.09	—	4
Eu	3,588.11	1	2
Ce	3,588.130	1	6
Ba	3,588.132	3	5
V	3,588.14	20	1
Ir	3,588.23	—	5
Nb	3,589.36	100	100
Nd	3,590.35	300*W*	400*W*
Dy	3,591.42	100	200
Rb	3,591.59	20	80
Cr	3,593.49	400*R*	500*R*
Fe	3,594.64	100	125
Co	3,594.87	—	200*W*
Dy	3,595.05	100	200
Bi	3,596.11	50	150*wh*
Rh	3,596.19	50	200
Rh	3,597.15	100	200
Ni	3,597.70	50*h*	1000*r*
Os	3,598.11	30	300
Zr	*3,601.19*	*15*	*400*
Co	*3,602.08*	*35*	*200*
Cr	*3,605.33*	*400R*	*500R*
Fe	*3,605.46*	*150*	*300*
Dy	*3,606.13*	*100*	*200*
Fe	*3,606.68*	*150*	*200*
Fe	*3,608.86*	*400*	*500*
Ni	*3,609.31*	*15*	*200*
Pd	*3,609.55*	*700R*	*1000R*

Rb 4,201.85 *J* 500 *O* 2000*R*

	λ	J	O
Ca	4,201.77	2	—
Mn	4,201.76	20	40
Ni	4,201.72	—	30
Th	4,201.85	10	8
Ta	4,201.97	—	5*h*
Ar	4,201.99	(20)	—

	λ	I J	I O
U	4,201.63	2	1
Er	4,201.60	—	8*d*
Ar	4,201.58	(2)	—
Pr	4,201.53	12*w*	30*w*
Nb	4,201.52	10	10
La	4,201.50	12*h*	2
Zr	4,201.46	3	50
Os	4,201.45	3	30
Kr	4,201.42	(30*wh*)	—
U	4,201.416	8	8
Dy	4,201.37	4	8
Br	4,201.35	(5)	—
Dy	4,201.318	—	30*w*
Mo	4,201.318	5	5
Ce	4,201.30	20	4
Xe	4,201.25	(8*wh*)	—
Ce	4,201.24	3	8
Pt	4,201.22	2	2
Pr	4,201.184	10	15
Mo	4,201.178	3	4
Tm	4,201.14	5	4
U	4,201.13	4	4
Dy	4,201.01	—	5
Tb	4,200.99	4	40
Fe	4,200.93	20	80
Eu	4,200.81	—	2
Ar	4,200.67	(1200)	—
Ru	4,199.90	300	150
Ar	4,198.32	(1200)	—
Ar	*4,191.03*	*(1200)*	—
Ar	*4,190.71*	*(600)*	—
O	*4,189.79*	*(500)*	—
Ar	*4,181.88*	*(1000)*	—
Se	*4,180.94*	*(800)*	—

	λ	I J	I O
Eu	4,202.03	—	5
Fe	4,202.031	300	400
Os	4,202.06	4	100
Ni	4,202.15	—	5
Mo	4,202.22	5	5
P	4,202.24	(30*h*)	—
Dy	4,202.25	4	20
V	4,202.34	15	6
Al	4,202.4	(8)	—
Pr	4,202.41	2	3
Ir	4,202.43	—	8
Br	4,202.50	(25)	—
Gd	4,202.51	—	15
Ir	4,202.516	—	8
Sr	4,202.522	—	6
U	4,202.676	2	6
Eu	4,202.68	—	15
Pr	4,202.70	3	8
Ce	4,202.71	—	2
Fe	4,202.76	4	10
Br	4,202.88	(4)	—

*N**

Rb 4,215.56 *J* 300 *O* 1000*R*

	λ	I J	I O
Tm	4,215.53	—	10
Sr	4,215.52	400*W*	300*r*
U	4,215.51	2	3
Re	4,215.506	—	20
Fe	4,215.42	15	60
Na	4,215.40	—	3
W	4,215.38	5	12
Zr	4,215.31	1	4

	λ	I J	I O
Xe	4,215.60	(100)	—
Cl	4,215.64	(6)	—
Zr	4,215.75	3	—
In	4,215.90	5	—
N	4,215.92	(5)	—
Tb	4,215.95	—	3
Er	4,215.96	—	9
Fe	4,215.97	1	2

	λ	I J	I O		λ	I J	I O
Dy	**4,215.17**	**8**	**50**	**Pr**	**4,215.98**	**3**	**4**
Os	**4,215.15**	**1**	**8**	**U**	**4,215.99**	**3*h***	**5**
Pr	**4,215.14**	**3**	**8**	*bh*C	**4,216.0**	—	—
Tb	**4,215.13**	**1*h***	**30*d***	**Ba**	**4,216.04**	**(25)**	—
Gd	**4,215.024**	**150**	**200**	**W**	**4,216.06**	**5**	**1**
Se	**4,215.02**	**(150)**	—	**Sc**	**4,216.10**	—	**3**
U	**4,215.01**	**4**	**4**	**Fe**	**4,216.19**	**100**	**200**
W	**4,214.96**	**1**	**5**	**Nb**	**4,216.23**	**10**	**1**
Co	**4,214.87**	**1**	**2**	**Cr**	**4,216.36**	**25**	**50**
Nb	**4,214.82**	**10**	**5**	**La**	**4,216.55**	—	**10**
Nb	**4,214.733**	—	**40**	**P**	**4,216.56**	**(15)**	—
N	**4,214.73**	**(25)**	—	**Ir**	**4,216.58**	—	**8**
Ce	**4,214.70**	—	**4**	**U**	**4,216.61**	**1**	**8**
Xe	**4,214.69**	**(3)**	—				
Nb	**4,214.67**	**100**	—				
Nd	**4,214.60**	**10**	**10**	Fe	4,219.36	200	250
Ru	**4,214.56**	—	**7**	P	4,222.15	(150*w*)	300
				Fe	4,222.22	200	200
Th	**4,214.55**	**3**	—	Xe	4,223.00	(200*h*)	—
Ir	**4,214.52**	—	**5**				
Xe	4,213.72	(200*h*)	—				
Pd	4,212.95	300*W*	500*W*				
Se	4,312.58	(200)	—				
Se	4,211.83	(200)	—				
Rh	4,211.15	200	15				
Fe	4,210.35	200	300				
Xe	4,208.48	(200*h*)	—				

*P**

Rb 7,800.23 *O* 9000*R*

Si	**7,800.00**	—	**4*Wh***	**Sc**	**7,800.44**	—	**40**
Re	**7,799.58**	—	**3**	**Zr**	**7,800.74**	—	**2**
Zr	**7,799.51**	—	**2**	**Re**	**7,801.05**	—	**4**
Ta	**7,799.51**	—	**3**	**Si**	**7,801.30**	—	**3*h***
Zn	**7,799.36**	—	**10**	**Hf**	**7,801.53**	**6**	**2**
Sm	**7,798.47**	—	**10**	**Sm**	**7,801.54**	—	**150**

Rb 7,947.60 *O* 5000*R*

	λ	I			λ	I	
		J	O			J	O
V	**7,947.38**	—	**8**	**Mn**	**7,948.10**	—	**2**
Sm	**7,947.00**	—	**15*d***	**Sm**	**7,948.12**	—	**100*d***
Sm	**7,946.15**	—	**10**	**Ru**	**7,948.15**	—	**15**
Fe	**7,945.88**	**2**	**30**	**La**	**7,948.30**	—	**5**
				*bh***Ti**	**7,948.6**	—	**8**
Rb	*7,925.54*	—	*70*	**Mo**	**7,949.03**	—	**2**
Rb	*7,925.26*	—	*100*	**Ti**	**7,949.17**	—	**50**

Re $^{75}_{186.2}$

t_0 3,167°C

t_1 – –

I.	II.	III.	IV.	V.
7.87	–	–	–	–

λ	*I*		eV
	J	*O*	
2,275.25	300*R*	300*R*	–
I 3,424.604	–	300	–
I 3,451.81	–	100	3.59
I 3,460.47	–	1000	3.58
I 3,464.72	–	100	3.57
I 4,889.17	–	2000	2.53
I 5,275.53	–	500	2.35

Re 2,275.25 *J* 300*R* *O* 300*R*

	λ	I J	I O		λ	I J	I O
V	2,275.23	15	—	Ag	2,275.26	50*wh*	—
Nb	2,275.22	20	—	Er	2,275.30	2	7
Fe	3,275.19	—	10	Zr	2,275.39	2*h*	—
U	2,275.12	2	—	Co	2,275.41	7	—
Mo	2,275.00	35	3	Mo	2,275.47	15	—
Ru	2,274.99	5	6	Ca	2,275.471	5	40
Xe	2,274.97	(2*h*)	—	Sr	2,275.48	(3)	2
				Cr	2,275.48	8	—
				W	2,275.50	5	—
Re	2,274.65	7	10	Re	2,276.65	3	40
Nb	2,274.13	300	12	Rh	2,276.96	150	20
Re	2,272.66	20	3	Hf	2,277.16	150	150
Re	*2,271.33*	*6*	*25*	*Re*	*2,278.57*	—	*15*
Ni	*2,270.21*	*400*	*100*	*Re*	*2,278.76*	*7*	*20*

Re 3,424.604 *O* 300

	λ	I J	I O		λ	I J	I O
Gd	3,424.601	—	30	Zr	3,424.63	—	2
Mo	3,424.60	5	8	Ir	3,424.69	2	20
Eu	3,424.596	2	3	Mo	3,424.758	3	5
U	3,424.56	15	20	Sm	3,424.765	2	5
Co	3,424.51	2	80	U	3,424.81	12	8
Ta	3,424.450	7*r*	35*r*	Zr	3,424.82	9	15
W	3,424.449	15	2	W	3,424.83	—	4
Er	3,424.42	1	9	Fe	3,425.01	40	70
Rh	3,424.38	5	30	Eu	3,425.02	4	50
Ce	3,424.37	—	3	W	3,425.03	4	5
Tb	3,424.35	3	8	Dy	3,425.06	5	40
Fe	3,424.29	150	200	V	3,425.07	20	25
Yt	3,424.14	2	3	Tm	3,425.08	300	200
U	3,424.13	2	1	U	3,425.12	3	6
Nd	3,424.07	2	2	Th	3,425.19	6	5
Ce	3,424.02	—	8*w*	Nd	3,425.21	4	8
Th	3,423.99	3	6	Sm	3,425.25	1	3
Tb	3,423.96	8	8	V	3,425.28	4	8
Gd	3,423.93	20	30	Ce	3,425.34	—	10
V	3,423.86	25	25	Ho	3,425.35	40	40
Ce	3,423.85	2	20*w*	Ir	3,425.37	—	3
W	3,423.84	3	6	U	3,425.41	3	4
Co	3,423.83	20	15	Tb	3,425.42	—	8
Dy	3,423.82	4	3	Nd	3,425.424	300	30*r*
Nb	3,423.76	5	15	Mo	3,425.48	8	8
Ni	3,423.71	25	600*R*	W	3,425.55	5	6
Ir	3,423.70	—	5	Fe	3,425.58	4	2

	λ	I J	I O
Nd	**3,423.63**	—	**6**
Ce	**3,423.62**	—	**3**
Re	3,421.58	—	15
Pd	3,421.24	1000*R*	2000*R*
Re	3,420.76	—	40
Re	3,419.40	—	80
Re	3,419.24	—	15
Fe	3,418.51	100	150
Fe	3,417.84	100	150
Re	3,417.80	—	40*r*
Co	3,417.16	—	400*r*
Ni	3,414.76	50*wh*	1000*R*
Co	3,414.74	—	200*W*
Ni	*3,413.94*	*10*	*300*
Re	*3,413.74*	—	*25*
Ni	*3,413.48*	*15*	*500*
Fe	*3,413.13*	*300*	*400*
Co	*3,412.63*	*40*	*1000R*
Co	*3,412.34*	*100*	*1000R*
Rh	*3,412.27*	*60*	*300*
Re	*3,409.83*	—	*30*
Ni	*3,409.58*	—	*300*
Co	*3,409.18*	*125*	*1000R*
Re	*3,408.68*	—	*100*
Fe	*3,407.46*	*400*	*400*
Re	*3,405.89*	—	*150*
Co	*3,405.12*	*150*	*2000R*
Re	*3,404.72*	—	*100*
Pd	*3,404.58*	*1000R*	*2000R*

	λ	I J	I O
Dy	**3,425.60**	—	**2**
Tm	3,425.63	50	100
Nb	3,425.85	30	50*r*
Re	3,426.19	—	30
Fe	3,426.39	20	80
Fe	3,426.64	60	80
Fe	3,427.12	50	50
Re	3,427.62	—	50
Os	3,427.67	15	80
Pt	3,427.93	6	50
Fe	3,428.20	50	50
Co	3,428.23	2	100*W*
Ru	3,428.31	100	100
Dy	3,429.44	5	50
Ru	3,429.54	5	60
Tm	3,429.97	100	100

*N**

Re 3,451.81 *O* 100

	λ	J	O
Mo	**3,451.749**	**20**	**10**
W	**3,451.747**	**7**	**9**
Th	**3,451.70**	**5**	**8**
Dy	**3,451.70**	**2**	**3**
Tb	**3,451.70**	—	**8**
Nb	**3,451.64**	**30**	**1**
Ce	**3,451.63**	—	**3**
Fe	**3,451.62**	**4**	**15**
Ce	**3,451.56**	—	**5**
Sm	**3,451.52**	**2*h***	**3**

	λ	J	O
Ce	**3,451.90**	—	**2**
Fe	**3,451.918**	**50**	**100**
W	**3,451.919**	**7**	**6**
U	**3,452.01**	—	**4**
Ce	**3,452.06**	—	**2**
La	**3,452.18**	**40**	**50**
Ce	**3,452.21**	—	**4**
Eu	**3,452.24**	—	**10**
Er	**3,452.26**	**1**	**8**
Fe	**3,452.277**	**8**	**150**

	λ	I	
		J	O
Pr	3,451.48	3	25
Mn	3,451.479	—	2
Ce	3,451.33	—	3
Gd	3,451.241	40	50
Er	3,451.239	1	10
Fe	3,451.23	1	1
U	3,451.21	5	5
Rh	3,451.149	2	50
W	3,451.148	10	4
La	3,451.12	4*h*	2
Ce	3,450.98	—	2
Yt	3,450.951	—	3
Th	3,450.949	8	8
Er	3,450.946	—	20
Ce	3,450.92	—	3
Cr	3,450.83	4	25
Th	3,450.81	2	3
Ce	3,450.80	—	2
Nb	3,450.76	50	2
W	3,450.75	6	4
Ti	3,450.735	—	4
Nd	3,450.73	2	8
Gd	3,450.38	100	100
Cu	3,450.332	30	150
Fe	3,450.33	80	150
Rh	3,450.29	10	100
Re	3,450.08	—	30
Co	3,449.44	125	500*R*
Re	3,449.37	—	100*r*
Os	3,449.20	20	100
Co	3,449.17	125	500*R*
Rh	3,447.74	5	50
K	3,447.70	75*R*	100*R*
Zr	3,447.36	3	150*w*
Fe	3,447.28	60	100
Dy	3,447.00	8	50
K	3,446.72	100*R*	150*R*
Ru	3,446.49	—	50
Tb	3,446.40	—	50
Ni	3,446.26	50*h*	1000*R*
Co	3,446.09	—	60*h*
Ru	3,446.07	6	50
Nb	3,445.68	80	50
Sm	3,445.621	10*R*	150*R*
Cr	3,445.618	80	100
Dy	3,445.58	8	80
Os	3,445.55	15	80

	λ	I	
		J	O
Ce	3,452.279	—	2
B	3,452.280	30	5
Sr	3,452.30	2	1
Hf	3,452.31	—	3
Co	3,452.315	5	8
Nb	3,452.35	200	5
Tb	3,452.37	3	15
Nb	3,452.37	—	15*W*
Yb	3,452.39	—	12
Ti	3,452.47	100	12
W	3,452.51	12	4
Ce	3,452.54	—	5
Mo	3,452.60	8	10
W	3,452.622	4	7
Ce	3,452.623	—	5
Nb	3,452.65	5	15
Th	3,452.68	10	10
Sm	3,452.77	1	4
Ce	3,452.81	—	5
Ni	3,452.89	50	600*R*
Ru	3,452.90	6	60
La	3,453.17	40	50
Co	3,453.50	200	3000*R*
Tm	3,453.66	80	150
Tb	3,454.06	30	80
Dy	3,454.33	10	100
Rh	3,455.22	12	300
Co	3,455.23	10	2000*R*
Rh	3,455.42	2	50
Cr	3,455.60	35	50
Ho	3,456.00	60	60
Dy	3,456.57	30	50
Ru	3,456.62	8	60
Tb	3,457.03	8	50
Rh	3,457.07	4	100
Cu	3,457.85	15	50
Rh	3,457.93	10	125
Co	3,458.03	—	60*w*
Fe	3,458.31	25	60
Os	3,458.38	12	200
Ni	3,458.47	50*h*	800*R*
Re	3,458.88	—	25
Os	3,459.02	10	100

	λ	I J	I O
Fe	3,445.15	150	300
Os	3,444.46	12	50
Ti	3,444.31	150	60
Fe	3,443.88	200	400
Co	3,443.64	100	500*R*
Re	3,442.96	—	20
Co	3,442.93	15	400*R*
Fe	3,442.36	15	50
Mn	3,441.99	75	75
Tm	3,441.51	80	150
Dy	3,441.45	5	50
Cr	3,441.44	90	80
Pd	3,441.40	2*h*	800*h*
Re	3,441.26	—	40
Fe	*3,440.99*	*200*	*300*
Fe	*3,440.61*	*300*	*500*
Ru	*3,440.20*	*30*	*100*
Zr	*3,438.23*	*200*	*250*
Re	*3,437.72*	—	*100*
Co	*3,437.69*	—	*150 Wh*
Ni	*3,437.28*	*40*	*600R*
Ru	*3,436.74*	*150*	*300R*
Rh	*3,434.89*	*200r*	*1000r*
Ni	*3,433.56*	*50wh*	*800R*
Pd	*3,433.45*	*500h*	*1000h*
Co	*3,433.04*	*150*	*1000R*
Co	*3,431.57*	*40*	*500R*

Re 3,460.47 *O* 1000

	λ	I J	I O
Cr	**3,460.43**	**30**	**40**
Dy	**3,460.40**	**2**	**20**
Tb	**3,460.38**	**8**	**15**
U	**3,460.35**	**5**	**3**
Mn	**3,460.33**	**500**	**60**
La	**3,460.31**	**3**	**2**
Eu	**3,460.29**	**1*h***	**15**
Yb	**3,460.27**	**5**	**30**
Mo	**3,460.23**	**5**	**5**
Ce	**3,460.16**	—	**6**
Nd	**3,460.13**	**2*h***	**6**
U	**3,460.06**	—	**5**
Dy	**3,460.05**	**1**	**5**
Mn	**3,460.02**	**5**	**5**
Zr	**3,459.93**	**2**	**20**
Ir	**3,460.54**	—	**3**
Nd	**3,460.581**	**6**	**25**
Ce	**3,460.581**	—	**2**
Sm	**3,460.63**	**1**	**3*w***
Dy	**3,460.64**	**1**	**4**
Pr	**3,460.66**	—	**3**
Sc	**3,460.70**	**2*h***	**6**
Co	**3,460.72**	—	**18**
Pd	**3,460.77**	**600*h***	**300*r***
U	**3,460.781**	—	**2**
Ce	**3,460.783**	—	**2**
Mo	**3,460.784**	**25**	**25**
W	**3,460.784**	**3**	**3**
Ho	**3,460.95**	**4**	**6**
Er	**3,460.968**	**7**	**20**

	λ	I J	I O		λ	I J	I O
Mo	3,459.923	8	8	Dy	3,460.971	3	100
Fe	3,459.918	50	80	Ce	3,460.999	—	12
U	3,459.917	—	8	Tb	3,461.00	8	15
Tb	3,459.87	—	8	U	3,461.003	3	3
Ce	3,459.83	1	12	Yt	3,461.01	12	7
Fe	3,459.74	1	2	Th	3,461.02	4	10
U	3,459.71	4	3	Pr	3,461.06	2	10
Nb	3,459.70	20	30	Zr	3,461.09	1	20
Th	3,459.64	5	5	Sm	3,461.14	3	8
Ru	3,459.569	—	30	Tm	3,461.17	5	15
Sb	3,459.568	3*h*	2	Co	3,461.176	3	100*wh*
W	3,459.52	8	9	La	3,461.184	2	10
Th	3,459.50	4	4	Ce	3,461.217	—	4
Ce	3,459.435	—	2	Th	3,461.218	4	10
Ti	3,459.431	1	20	Pr	3,461.24	—	4
Fe	3,459.429	3	10	Dy	3,461.31	1	3
Cu	3,459.428	2*h*	25	Ce	3,461.34	—	15*s*
Sm	3,459.40	2	10	W	3,461.36	6	7
Ce	3,459.38	15*wh*	2*h*	Eu	3,461.38	2	25
				Er	3,461.40	1	10
				Sm	3,461.405	4	5
				Ce	3,461.42	—	2
				Ti	3,461.50	125	80

*P** *N**

Re 3,464.722 *O* 100

	λ	I J	I O		λ	I J	I O
Tb	3,464.63	3	8	Ce	3,464.722	—	2
Er	3,464.54	3	10	Cr	3,464.84	3	30
Fe	3,464.49	1*h*	1*h*	Ce	3,464.86	—	12
U	3,464.47	—	4	U	3,464.87	—	2
Dy	3,464.46	1	2	Os	3,464.88	5	15
Sr	3,464.457	200	200	Fe	3,464.92	—	2
Sm	3,464.43	1	3	Nd	3,464.93	2*h*	10
Th	3,464.42	2*d*	2*d*	U	3,464.95	4	4
Yb	3,464.37	50*r*	200*r*	Ce	3,464.99	—	12
Eu	3,464.36	—	15*d*	Th	3,465.02	5	5
Ba	3,464.23	—	5	Er	3,465.12	—	8*d*
Ce	3,464.21	1*w*	8	Mn	3,465.19	—	2
Yb	3,464.19	—	12	Ir	3,465.22	1	12
Ce	3,464.16	—	10	V	3,465.249	25	2
Gd	3,464.14	3	4	Cr	3,465.250	30	35
Sm	3,464.08	10	30	Ru	3,465.29	2*h*	5
Cr	3,464.00	2	1	Dy	3,465.30	—	2
Gd	3,463.98	125	100	Ce	3,465.302	—	8
Pt	3,463.94	1*h*	2	Cu	3,465.40	2*h*	10
Ta	3,463.915	10	3	W	3,465.41	4	4

	λ	I	
		J	O
U	**3,463.91**	**—**	**4**
Dy	**3,463.88**	**2**	**12**
Nb	**3,463.81**	**50**	**30**
Ta	**3,463.77**	**—**	**50**
Ce	**3,463.76**	**1**	**15*s***
Th	**3,463.72**	**12**	**10**
Nb	**3,463.68**	**2**	**5**
Ru	3,463.14	4	60
Co	3,462.80	80	1000*R*
Tm	3,462.20	200	250
Rh	3,462.04	150	1000
Ni	3,461.65	50*h*	800*R*
*P**			

	λ	I	
		J	O
Ce	**3,465.42**	**—**	**8**
Os	**3,465.43**	**12**	**60**
Nd	**3,465.44**	**—**	**6*wh***
Sm	**3,465.46**	**—**	**5**
Ce	**3,465.50**	**—**	**5**
Ti	**3,465.56**	**60**	**6**
Cr	**3,465.58**	**8**	**30**
Zr	**3,465.63**	**—**	**3**
Mo	**3,465.66**	**6**	**6**
Pr	**3,465.76**	**4**	**9**
Th	**3,465.77**	**10**	**10**
Co	**3,465.80**	**25**	**2000*R***
Fe	3,465.86	400	500
Re	3,465.98	—	20
Cd	3,466.20	500	1000
Cr	3,467.02	20	50
Ru	3,467.05	3	50
Gd	3,467.28	100	100
Ni	3,467.50	15	300
Cd	3,467.66	400	800
Cr	3,467.71	30	50
Re	3,467.95	—	100*w*
Tb	3,468.03	15	50
Dy	3,468.43	3	50
Gd	3,468.99	150	150
Ni	3,469.49	20	300
Cr	3,469.59	15	50
Rh	3,469.62	10	100
Rh	3,470.66	125	500
Co	3,471.38	25*wh*	80
Ru	3,472.23	9	60
Rh	3,472.25	8	100
Lu	3,472.48	150	50
Ni	3,472.54	40	800*R*
Re	3,472.72	—	25
Tb	3,472.82	15	50
Gd	3,473.23	40	50
Dy	3,473.70	40	80
Ru	3,473.75	35	70
Co	3,474.02	100	3000*R*
Re	3,474.21	—	25
Nb	3,474.67	4	100
Rh	3,474.78	125	700
Sr	3,474.89	50	80
Fe	3,475.45	300	400
Co	*3,476.36*	—	*100R*

	λ	I J	I O		λ	I J	I O
				Re	3,476.44	—	30
				Fe	3,476.70	200	300
				Cs	3,476.88	—	100
				Re	3,477.14	—	15
				Os	3,478.53	15	100
				Rh	3,478.91	100	500
				Re	3,480.38	—	50
				Re	3,480.85	—	50
				Pd	3,481.15	2h	500r
				Gd	3,481.36	150	150
				Re	3,482.24	—	40
				Co	3,483.41	10	300R
				Ni	3,483.77	30	500R
				Pt	3,485.27	200R	150
				Fe	3,485.34	50	100
				Co	3,485.37	3	100wh
				Ni	3,485.89	30	150

Re 4,889.17 *O* 2000

	λ	J	O		λ	J	O
Nd	4,889.11	—	8	Gd	4,889.20	—	60
Fe	4,889.01	150h	2wh	Mo	4,889.22	5	25
Er	4,888.87	—	3	Dy	4,889.32	3	5
Fe	4,888.65	1	2	Nb	4,889.56	2	3
Ru	4,888.61	—	6	Ce	4,889.59	—	20
Cr	4,888.53	—	100	Pr	4,889.66	—	5
Ir	4,888.51	—	2wh	Ru	4,889.83	—	4
W	4,888.39	—	20	Hf	4,889.89	1h	2h
Ag	4,888.28	20	9	Dy	4,890.10	2	8
Lu	4,888.14	—	2h	Pr	4,890.26	—	15
Dy	4,888.09	2	5	W	4,890.29	—	15
Rh	4,887.92	—	2	Sm	4,890.329	—	3
Cr	4,887.71	—	15	Sc	4,890.335	1h	3
La	4,887.62	—	2	Th	4,890.44	—	3
				Nd	4,890.70	—	30
				Nb	4,890.75	8	15
				Sm	4,890.76	—	2
				Fe	4,890.77	15	100
				Re	4,908.57	—	20
				Re	4,915.02	—	30

Re 5,275.53 *O* 500

	λ	I *J*	I *O*
Dy	**5,275.31**	—	**5**
*bh***V**	**5,275.3**	—	**10**
Mo	**3,275.22**	**1**	**4**
Cr	**5,275.17**	—	**10**
Hf	**5,275.04**	**1**	**7**
Tb	**5,275.03**	—	**10**
Ta	**5,275.02**	—	**7**
W	**5,274.81**	—	**8**
Eu	**5,274.42**	—	**4***h*
Ce	**5,274.24**	**3**	**50**
Sm	**5,274.13**	—	**10**
Dy	**5,274.08**	—	**3**
Er	**5,273.86**	—	**12**
Ir	**5,273.78**	—	**5**
U	**5,273.74**	**4**	**3**
Eu	5,272.48	—	400
Eu	5,271.95	—	2000
Re	5,270.98	—	200 *W*
Fe	5,270.36	80	400
Fe	5,269.54	200	800
Co	5,268.51	—	500*w*
Fe	5,266.58	40	500
Co	5,266.49	—	500*w*
Eu	5,266.40	—	1000
Fe	5,263.33	—	300
Co	5,257.62	—	400*w*
Re	*5,248.85*	—	*25*
Co	*5,247.93*	—	*500w*
Re	*5,244.34*	—	*20*
Re	*5,236.67*	—	*20w*

	λ	I *J*	I *O*
W	**5,275.55**	—	**20**
Eu	**5,275.66**	—	**30**
V	**5,275.68**	**8***h*	**8***h*
Cr	**5,275.69**	**1***wh*	**4***h*
U	**5,275.92**	—	**12**
Fe	**5,275.99**	**15**	**2**
Er	**5,276.03**	—	**20**
Cr	**5,276.03**	**3***h*	**5***h*
Co	**5,276.192**	—	**400***w*
Nb	**5,276.195**	**50**	**200**
Mo	**5,276.27**	**10**	**20**
Hf	**5,276.39**	**3**	**1**
Th	**5,276.41**	—	**4**
La	**5,276.43**	—	**20**
Ag	**5,276.47**	—	**6**
Gd	**5,276.54**	—	**20**
U	**5,276.70**	**2**	**2**
Nd	**5,276.88**	**1**	**10**
Er	**5,276.90**	—	**8**
Ce	**5,276.97**	—	**5**
Yb	**5,277.07**	**6**	**200**
Eu	**5,277.10**	—	**2**
Pr	**5,277.32**	**1**	**5***W*
Mo	**5,277.36**	**2**	**10**
Dy	**5,277.40**	—	**3**
Zr	**5,277.41**	—	**10**
Th	**5,277.50**	**3**	**15***d*
Re	5,278.24	—	100
Co	5,280.65	—	500*w*
Fe	5,281.80	20	300
Eu	5,282.82	—	1000
Fe	5,283.63	40	400
Eu	5,294.60	—	300
Co	*5,301.06*	—	*700w*
Re	*5,305.56*	—	*25*

Rh 45 102.905

t_0 1,985°C t_1 >2,500°C

I.	II.	III.	IV.	V.
~7.7	—	—	—	—

λ	I		eV
	J	O	
I 3,283.57	—	150 *R*	4.10
I 3,323.09	200*R*	1000*R*	3.92
I 3,396.85	500*R*	1000*R*	3.64
I 3,434.89	200*R*	1000*R*	3.60
I 3,657.99	200	500	3.57
I 3,692.36	150	500	3.35
I 4,128.87	150	300	3.97
I 4,374.80	500	1000	—
I 4,528.73	60	500*R*	—

Rh 3,283.57 *O* 150*R*

	λ	I J	I O		λ	I J	I O
W	**3,283.56**	**8**	**9**	**Ir**	**3,283.59**	—	**20***Rh*
Fe	**3,283.543**	**3**	**7**	**Ce**	**3,283.68**	**1**	**25**
U	**3,283.537**	—	**3**	**Eu**	**3,283.71**	**5**	**2**
Nb	**3,283.463**	**100**	**2**	**Sm**	**3,283.74**	—	**5***h*
Co	**3,283.462**	—	**80**	**U**	**3,283.75**	**2***d*	**2***d*
Fe	**3,283.43**	**2**	**4**	**Co**	**3,283.78**	—	**60***w*
Tm	**3,283.40**	**40**	**40**	**Ta**	**3,283.807**	**2***wh*	**3**
Yb	**3,283.39**	—	**12**	**Tb**	**3,283.81**	**3**	**8**
Hf	**3,283.38**	**6**	**10**	**Sm**	**3,283.88**	**2***h*	**5***h*
Mo	**3,283.36**	**1**	**5**	**Ce**	**3,283.89**	—	**3**
Ce	**3,283.35**	**1**	**20***s*	**Ti**	**3,283.950**	—	**6**
Co	**3,283.33**	—	**7**	**La**	**3,283.95**	**4**	**3**
Pt	**3,283.312**	**3**	**8**	**Ce**	**3,284.22**	—	**20**
V	**3,283.311**	**10**	**35**	**Ni**	**3,284.34**	—	**2**
Yt	**3,283.21**	**4**	**3**	**V**	**3,284.36**	**5**	**25**
Pt	**3,283.209**	—	**8**	**U**	**3,284.368**	**2**	**6**
Ce	**3,283.17**	—	**3**	**Dy**	**3,284.37**	**1**	**2**
Pr	**3,283.12**	**2**	**15**	**Eu**	**3,284.39**	—	**4**
U	**3,283.104**	**4**	**6**	**Ce**	**3,284.42**	—	**8**
Tb	**3,283.10**	**8**	**30**				
Th	**3,282.97**	**12**	**8**	Fe	3,284.59	125	200
Mo	**3,282.91**	**30**	**1**	Fe	3,286.75	400	500
Fe	**3,282.892**	**80**	**80**	Ni	3,286.95	1	100
Be	**3,282.890**	—	**8**	Pd	3,287.25	25	300*w*
Zr	**3,282.834**	**10**	**10**	Rh	3,289.14	50	150
Ni	**3,282.827**	**1**	**25**	Yb	3,289.37	1000*R*	500*R*
Dy	**3,282.79**	**1**	**2**	Rh	3,289.64	5	50*r*
Nd	**3,282.78**	**2**	**8**	Yb	3,289.85	—	1000
Zr	**3,282.73**	—	**10**	Pt	3,290.22	10	150
Cu	**3,282.72**	**15***w*	**25**	Os	3,290.26	20	200
Ni	**3,282.70**	—	**100**	Fe	3,290.99	80	125
				Tm	3,291.00	80	125
				Fe	3,292.02	125	150
Zn	3,282.33	300	500*R*	Fe	3,292.59	150	300
Ag	3,280.683	1000*R*	2000*R*	Rh	3,294.28	25	60
Eu	3,280.682	—	1000*R*				
Rh	3,280.55	10	30*R*				
Fe	3,280.26	150	150	*Ta*	*3,295.33*	*20w*	*125w*
				Rh	*3,296.72*	*10*	*40*
Fe	3,278.73	60	100	*Fe*	*3,298.13*	*150*	*200*
Os	3,277.97	8	80	*Rh*	*3,300.46*	*20*	*100*
Fe	3,276.47	50	100	*Os*	*3,301.56*	*50*	*500R*
Os	3,275.20	15	200				
Ta	3,274.95	35*W*	200	*Pt*	*3,301.86*	*250W*	*300*
Fe	3,274.45	60	80				
Cu	*3,273.96*	*1500R*	*3000R*				
Rh	*3,271.61*	*60*	*200*				

	λ	I J	I O		λ	I J	I O
Fe	*3,271.00*	*300*	*300*				
Ge	*3,269.49*	*300*	*300*				
Os	*3,269.21*	*20*	*200*				
Rh	*3,268.47*	*5*	*25*				
Os	*3,267.94*	*30*	*400R*				
Sb	*3,267.50*	*150wh*	*150*				
La	*3,265.67*	*200*	*300*				
Fe	*3,265.62*	*300*	*300*				

Rh 3,323.09 *J* 200*R* *O* 1000*R*

	λ	J	O		λ	J	O
Fe	**3,323.07**	**100**	—	**Ce**	**3,323.10**	—	**3**
Te	**3,323.06**	**(15)**	—	**U**	**3,323.12**	**1***h*	**2**
Ba	**3,323.06**	—	**2**	**Os**	**3,323.17**	**3***h*	**3***h*
Zr	**3,322.99**	**10**	**10**	**Au**	**3,323.19**	**5**	**10**
Ti	**3,322.937**	**300***R*	**80**	**Er**	**3,323.20**	**4**	**25**
Nd	**3,322.936**	**2**	**25**	**B**	**3,323.20**	**5**	—
U	**3,322.93**	**1**	**4***d*	**Ho**	**3,323.21**	**4**	—
Ce	**3,322.92**	**1**	**2**	**Tm**	**3,323.22**	**40**	**20**
Ba	**3,322.874**	—	**30***r*	**Cr**	**3,323.25**	—	**25**
Ir	**3,322.87**	**15**	**20**	**Dy**	**3,323.27**	—	**4**
Nb	**3,322.81**	**2**	**3**	**Ce**	**3,323.29**	—	**10**
Cs	**3,322.8**	**(4)**	—	**B**	**3,323.34**	**10**	—
Eu	**3,322.74**	—	**3**	**Rb**	**3,323.34**	**(10)**	—
Cr	**3,322.70**	**20**	**1**	**Hf**	**3,323.36**	**15***h*	**12**
W	**3,322.675**	**5**	—	**Tb**	**3,323.38**	**3**	**15**
U	**3,322.670**	**2**	**4**	**W**	**3,323.41**	**8**	**5**
Ce	**3,322.63**	—	**10**	**Ce**	**3,323.45**	—	**2**
Ir	**3,322.60**	**18**	**30**	**J**	**3,323.50**	**(3)**	—
Re	**3,322.478**	—	**150**	**Cr**	**3,323.53**	**12**	—
Fe	**3,322.477**	**100**	**150**	**B**	**3,323.61**	**10**	—
Th	**3,322.47**	**2**	**2**	**Ti**	**3,323.66**	—	**2***h*
Ar	**3,322.44**	**(5)**	—	**Fe**	**3,323.74**	**150**	**150**
Ni	**3,322.31**	**10**	**400**	**Ne**	**3,323.75**	**(40)**	—
Rb	**3,322.3**	**(15)**	—	**Sm**	**3,323.77**	**8**	**15**
Ca	**3,322.30**	**7**	**1**	**Pt**	**3,323.79**	**10**	**150**
Tl	**3,322.30**	**(25)**	—	**Ti**	**3,323.80**	—	**5**
Zr	**3,322.28**	—	**8**	**Ar**	**3,323.81**	**(30)**	—
Pr	**3,322.27**	**1**	**7**	**Tb**	**3,323.89**	**3**	**15**
Eu	**3,322.26**	**2**	**20**	**Nb**	**3,323.894**	**50**	**1**
W	**3,322.25**	**12***d*	**10**	**Ti**	**3,323.90**	—	**3**
Sr	**3,322.231**	**8**	**100**	**Ca**	**3,323.93**	**5**	—
Ru	**3,322.226**	—	**8**	**Ru**	**3,323.94**	—	**5**
Re	**3,322.20**	—	**25***W*	**Mo**	**3,323.949**	**25**	**40**
Co	**3,322.199**	—	**100***W*	**Dy**	**3,323.95**	—	**2**
Ce	**3,322.175**	—	**2**	**Ce**	**3,323.98**	—	**8**
Mo	**3,322.170**	**30**	**4**				

	λ	I J	I O		λ	I J	I O
Ti	3,321.70	125	15	Ar	3,325.50	(100)	—
V	3,321.54	150	3	Ti	3,326.76	125	15
Be	3,321.34	30	1000*r*	Zr	3,326.80	100	100
Nb	3,320.81	100	3	Fe	3,328.87	100	150
Ar	3,319.34	(300)	—	Cl	3,329.12	(150)	—
Ti	3,318.02	125	60	Mo	3,329.21	100	2
Cl	3,315.44	(100)	—	Te	3,329.25	(100)	—
Ti	3,315.32	100	12	Ti	3,329.46	200*r*	80
Fe	3,314.74	200	200	Sn	3,330.59	100*h*	100*h*
Cr	3,314.56	100	10	Ta	3,331.01	200*w*	18
				Rh	3,331.09	10	50
Cr	*3,310.65*	*200*	*2*	Rh	3,331.24	10	40
Fe	*3,305.97*	*300*	*400*	Ti	3,332.11	125	40
Rh	*3,305.17*	*10*	*40*				
				Ne	*3,334.87*	*(250)*	—
				La	*3,337.49*	*300wh*	*800*
				Rh	*3,338.54*	*50*	*200*

Rh 3,396.85 *J* 500*R* *O* 1000*R*

	λ	J	O		λ	J	O
Er	**3,396.83**	**2**	**12*s***	**Fe**	**3,396.98**	**25**	**125**
Mo	**3,396.825**	**4**	**5**	**Yt**	**3,397.05**	**5**	**7**
Ru	**3,396.825**	—	**20**	**Lu**	**3,397.07**	**20*r***	**50**
Lu	**3,396.82**	**1**	**30**	**Ce**	**3,397.08**	—	**12**
Pd	**3,396.78**	—	**10**	**Th**	**3,397.11**	**3**	**2**
Th	**3,396.73**	**1**	**5**	**Cs**	**3,397.19**	**(6)**	—
Ce	**3,396.72**	—	**15**	**U**	**3,397.20**	—	**12**
Zr	**3,396.66**	**5**	**8**	**Re**	**3,397.210**	—	**15**
Pr	**3,396.58**	**4**	**4**	**Tb**	**3,397.21**	—	**15**
Eu	**3,396.58**	**10**	**100**	**Fe**	**3,397.213**	—	**1**
V	**3,396.51**	**12**	**15**	**Bi**	**3,397.213**	**50**	**100*wh***
Co	**3,396.46**	—	**3**	**Ni**	**3,397.25**	—	**2**
Th	**3,396.40**	**2*d***	**2*d***	**Hg**	**3,397.26**	**3**	**20**
Fe	**3,396.38**	—	**1**	**Nb**	**3,397.32**	**50**	—
Nb	**3,396.37**	**150**	**2**	**Ho**	**3,397.34**	**4*h***	—
Zr	**3,396.33**	**10**	**12**	**Dy**	**3,397.39**	**2**	**2**
Co	**3,396.324**	**3**	**30**	**Ta**	**3,397.42**	**1**	**3**
Yb	**3,396.32**	**20**	**4**	**Nd**	**3,397.45**	**2**	**2**
Ho	**3,396.315**	**10**	—	**W**	**3,397.46**	**4**	**4**
Ni	**3,396.184**	—	**8**	**Tm**	**3,397.50**	**50**	**100**
Sm	**3,396.183**	**15**	**35**	**Nb**	**3,397.517**	**15**	—
Nd	**3,396.178**	**8**	**10**	**Th**	**3,397.519**	**4**	**8**
Dy	**3,396.17**	**10**	**20**	**Pr**	**3,397.56**	**3**	**3**
Ho	**3,396.13**	**4**	—	**Fe**	**3,397.564**	**2**	**1**
Er	**3,396.05**	**3**	**12*d***	**V**	**3,397.58**	**25**	**3**
Fe	**3,395.942**	—	**3**	**Hf**	**3,397.597**	**3**	**20**
Hf	**3,395.94**	—	**10**	**Tb**	**3,397.60**	—	**15**

	λ	I J	I O
Nb	**3,395.93**	**5**	**10**
Sb	**3,395.91**	**4*wh***	—
S	**3,395.87**	**(15)**	—
Rh	3,394.89	5	15
Ar	3,393.75	(250)	—
Ni	3,392.99	—	600*R*
Zr	3,391.97	400	400
Rh	*3,385.78*	*10*	*30*
Ag	*3,382.89R*	*700R*	*1000R*
Rh	*3,377.71*	*5*	*25*
Rh	*3,377.14*	*10*	*40*
Fe	**3,397.64**	**2**	**10**
Zr	**3,397.65**	—	**8**
Re	**3,397.68**	—	**6**
Mo	**3,397.69**	**20**	**20**
Os	**3,397.758**	**10**	**12**
La	**3,397.76**	**8**	**5**
Sm	**3,397.763**	**3**	**9**
V	**3,397.84**	**20**	**30**
Rh	3,399.70	60	500
Cd	3,403.65	500*h*	800
Pd	3,404.58	1000*R*	2000*R*
Co	3,405.12	150	2000*R*
Rh	3,406.55	8	50
Fe	*3,407.46*	*400*	*400*
Co	*3,409.18*	*125*	*1000R*
Rh	*3,412.27*	*60*	*300*
Co	*3,412.34*	*100*	*1000R*
Co	*3,412.63*	*40*	*1000R*
Ni	*3,414.76*	*50wh*	*1000R*
*N**			

Rh 3,434.893 *J* 200*R* *O* 1000*R*

	λ	I J	I O
Hf	**3,434.89**	—	**5**
Os	**3,434.889**	**3**	**20**
U	**3,434.80**	**1*h***	**2*h***
Mo	**3,434.79**	**12**	**50**
Th	**3,434.762**	**8**	**8**
Ir	**3,434.757**	—	**10**
Pr	**3,434.757**	**2**	**20**
Ho	**3,434.75**	**4*h***	—
Eu	**3,434.74**	—	**3**
Hg	**3,434.73**	**15**	—
Rb	**3,434.72**	**(40)**	—
Ag	**3,434.65**	—	**2*h***
Er	**3,434.63**	**2**	**12**
U	**3,434.614**	**1*h***	**12**
Yb	**3,434.61**	**20**	**5**
Tb	**3,434.54**	**3**	**15**
Ta	**3,434.500**	**18**	**35**
Mo	**3,434.496**	**4**	**5**
Pr	**3,434.49**	—	**3**
Dy	**3,434.373**	**30**	**80**
Er	**3,434.367**	**8**	**25**
Ce	**3,434.29**	—	**3**
Tb	**3,434.92**	—	**30**
Eu	**3,435.06**	**3**	**12**
J	**3,435.07**	**(7)**	—
Ru	**3,435.19**	**20**	**60**
U	**3,435.200**	—	**10**
Eu	**3,435.205**	**2**	**35**
Ce	**3,435.21**	—	**20**
W	**3,435.24**	**5**	**6**
Os	**3,435.256**	**10**	**20**
Sm	**3,435.256**	**3**	**5**
Dy	**3,435.27**	**2**	**4**
V	**3,435.37**	**20**	—
Mo	**3,435.40**	**60*h***	—
Rn	**3,435.4**	**(18)**	—
Ti	**3,435.432**	**1**	**10**
Nd	**3,435.434**	**4**	**20**
Mo	**3,435.45**	—	**6**
Cr	**3,435.488**	**1**	**8**
Ni	**3,435.489**	—	**3**
U	**3,435.53**	—	**15*d***
Tb	**3,435.53**	**3**	**8**
Sc	**3,435.55**	**4**	**12**

	λ	I J	I O
U	3,434.283	—	6
Kr	3,434.28	(2)	—
Sr	3,434.28	—	2
Rb	3,434.26	(60)	—
Hg	3,434.16	2	—
U	3,434.15	10	6
Kr	3,434.14	(8)	—
Ce	3,434.13	—	5
Cr	3,434.11	25	30
Mo	3,434.04	4	4
V	3,434.03	20	1
Fe	3,434.02	—	2*h*
Th	3,434.00	12	12
Cu	3,433.97	—	5*h*
Nb	3,433.95	10	—
Zr	3,433.91	6	8
Ni	3,433.56	50*wh*	800*R*
Pd	3,433.45	500*h*	1000*h*
Cr	3,433.31	150	30
Co	3,433.04	150	1000*R*
Nb	3,432.70	100	10
Co	3,431.57	40	500*R*
Bi	3,431.23	(150)	—
Bi	3,430.83	(200)	—
Tm	3,429.97	100	100
Ru	3,428.31	100	100
Re	3,427.62	—	50
Nb	3,426.57	200	5
Re	3,426.19	—	30
Nb	3,425.42	300	30*r*
Tm	3,425.08	300	200
Rh	*3,424.38*	*5*	*30*
Rh	*3,422.29*	—	*12*
Pd	*3,421.24*	*1000R*	*2000R*
Cr	*3,421.21*	*200*	*50*
Ne	*3,417.90*	*(500)*	—

	λ	I J	I O
Nb	3,435.59	10*wh*	—
Ho	3,435.61	4*h*	—
Eu	3,435.65	—	4*h*
Cr	3,435.679	12	20
Ce	3,435.682	—	3
W	3,435.71	12	9
Eu	3,435.73	9	12
Co	3,435.75	—	8
Cr	3,435.82	12	30
Ru	3,436.74	150	300*R*
Ni	3,437.28	40	600*R*
Ta	3,437.37	300*wh*	7
Zr	3,438.23	200	250
Yb	3,438.84	100	20
Rb	3,439.34	(200)	—
Rh	3,440.53	100	2
Fe	3,440.61	300	500
Fe	3,440.99	200	300
Pd	3,441.40	2*h*	800*h*
Co	3,442.93	15	400*R*
Co	3,443.64	100	500*R*
Fe	3,443.88	200	400
Ti	3,444.31	150	60
Ni	*3,446.26*	*50h*	*1000R*
Rh	*3,447.74*	*5*	*50*
Rh	*3,448.57*	*2*	*25*
Rh	*3,450.29*	*10*	*100*
Rh	*3,451.15*	*2*	*50*
Pd	*3,451.35*	*400h*	—
Hg	*3,451.69*	*(200)*	—
Nb	*3,452.35*	*200*	*5*
Co	*3,453.50*	*200*	*3000R*
Yb	*3,454.07*	*250*	*40*

Rh 3,657.99 *J* 200 *O* 500

	λ	I J	I O
Co	3,657.91	5	18
Fe	3,657.904	4	20
Mn	3,657.902	10	12
Nb	3,657.897	5	4
W	3,657.88	15	5
U	3,654.82	—	10
Sm	3,657.81	3*h*	2

	λ	I J	I O
Nd	3,657.991	2*h*	4*h*
Fe	3,658.02	1	3
Th	3,658.07	10	10
Ce	3,658.09	—	5
Ti	3,658.100	60	150
Nd	3,658.104	—	2
Mo	3,658.15	5	5

	λ	I J	I O
Nb	3,657.75	—	3*d*
Xe	3,657.74	(3)	—
Ce	3,657.68	—	5
Pr	3,657.625	1	10
Ir	3,657.616	2	5
Eu	3,657.60	—	2*W*
W	3,657.59	25	10
Ru	3,657.55	50	—
Th	3,657.54	2	3
Ta	3,657.495	2	15
V	3,657.492	5	20
Fe	3,657.43	—	2
Pr	3,657.42	4	25
Eu	3,657.41	2*h*	2*h*
Ce	3,657.38	—	2
Mo	3,657.35	30	30
U	3,657.322	15	6
Sm	3,657.316	1	15
Ta	3,657.27	1*h*	18
Rn	3,657.20	(10)	—
Ca	3,657.174	2	2
Ru	3,657.173	4	2
Fe	3,657.14	7	20
Nb	3,657.11	5	5
J	3,657.06	(15)	—
Co	3,656.97	6	60
Cl	3,656.95	(15)	—
U	3,656.93	8	8
Os	3,656.90	30	150
Ta	3,656.89	2*h*	15
Zr	3,656.88	—	10
Dy	3,656.85	2	4
Cu	3,656.79	3	4
As	3,656.78	5	—
Tb	3,656.77	3	8
Ce	3,656.75	—	10*s*
Gd	3,656.16	200	200*W*
Al	3,655.00	(100)	—
Rh	3,654.87	10	40
Hg	3,654.83	(200)	—
Gd	3,654.64	(200)	200*W*
Kr	3,653.97	(250*h*)	—
Ti	3,653.50	200	500
P	3,653.38	(100*W*)	—
Fe	3,651.47	200	300
Nb	3,651.19	400	10

	λ	I J	I O
U	3,658.16	2	—
Cr	3,658.168	20	—
Th	3,658.173	6	8
Gd	3,658.19	—	2
Pr	3,658.21	2	15
Tb	3,658.22	3	15
Ce	3,658.258	—	10
V	3,658.264	10	1
Zr	3,658.30	—	2
Cl	3,658,33	(4)	—
Mo	3,658.333	25	4
Pr	3,658.35	1	7
W	3,658.37	3	6
La	3,658.41	3	3
Ce	3,658.42	—	2
Xe	3,658.44	(3*h*)	—
Se	3,658.46	(5)	—
Mn	3,658.52	5	5
Ho	3,658.55	4	—
Fe	3,658.552	1	5
Nb	3,658.60	5*h*	5*h*
Dy	3,658.64	1*h*	3
U	3,658.68	—	6
Re	3,658.74	—	4
Ce	3,658.77	—	6
Ta	3,658.78	3	35
Eu	3,658.80	—	2*W*
Th	3,658.81	1	5
Er	3,658.84	1*W*	10*W*
Tb	3,658.88	100	100
Mo	3,658.92	20	—
U	3,659.01	8	2
Pr	3,659.038	8	10
Hf	3,659.045	8	3
Eu	3,659.13	2*h*	2*h*
W	3,659.15	5	3
U	3,659.16	1	15
Ce	3,659.23	6	20
Ar	3,659.50	(100)	—
Nb	3,659.61	500	15
Ti	3,659.76	150	50
Ru	3,661.35	100	60
V	3,661.38	150	10
Ti	3,662.24	100	40
Gd	3,662.27	200	200*W*
Hg	3,662.88	400	50

	λ	I J	I O		λ	I J	I O
Tb	3,650.40	100	50	Hg	3,663.28	400	500
Hg	3,650.15	500	200	Ni	3,664.09	30	300
Ar	3,649.83	(800)	—				
Ra	3,649.55	(1000)	—	Ne	3,664.11	(250)	—
				P	3,664.19	(100*w*)	—
Fe	3,649.51	100	100	Yt	3,664.61	100	100
Fe	3,647.84	400	500	Gd	3,664.62	200	200*w*
Gd	3,646.20	150	200*w*	Rh	3,666.21	30	70
Fe	*3,640.39*	*200*	*300*				
Rh	*3,639.51*	*70*	*125*				
Rn	*3,634.8*	*(250)*	*—*				
Pd	*3,634.69*	*1000R*	*2000R*				
Ar	*3,634.46*	*(300)*	*—*				

*N**

Rh 3,692.36 *J* 150 *O* 500

	λ	J	O		λ	J	O
La	**3,692.31**	**2**	**—**	**Ru**	**3,692.37**	**—**	**6**
U	**3,692.310**	**2**	**—**	**O**	**3,692.44**	**(50)**	**—**
Pr	**3,692.29**	**1**	**4**	**Yt**	**3,692.53**	**7**	**7**
J	**3,692.26**	**(5)**	**—**	**Ce**	**3,692.55**	**—**	**2**
V	**3,692.225**	**150*R***	**200*R***	**Th**	**3,692.57**	**2**	**10**
Ce	**3,692.222**	**—**	**3**	**Zr**	**3,692.63**	**1**	**2**
Sm	**3,692.221**	**40**	**90**	**Mo**	**3,692.64**	**150**	**3**
Eu	**3,692.22**	**—**	**4*wh***	**Ho**	**3,692.65**	**15**	**10**
Ir	**3,692.20**	**5**	**2**	**Fe**	**3,692.652**	**1**	**5**
Nb	**3,692.18**	**5**	**—**	**Er**	**3,692.652**	**12**	**20**
Ti	**3,692.134**	**1**	**12**	**Eu**	**3,692.66**	**—**	**5*W***
Re	**3,692.126**	**—**	**15**	**Ir**	**3,692.69**	**4**	**15**
Mo	**3,692.081**	**3**	**5**	**W**	**3,692.72**	**3**	**7**
Th	**3,692.080**	**4**	**8**	**U**	**3,692.75**	**—**	**10**
U	**3,692.05**	**4**	**4**	**Sm**	**3,692.76**	**6**	**20**
Tb	**3,692.02**	**8**	**15**	**Nd**	**3,692.77**	**8**	**12**
Eu	**3,691.98**	**1**	**3*W***	**Mn**	**3,692.81**	**50**	**50**
U	**3,691.92**	**8**	**6**	**Sm**	**3,692.90**	**2**	**10**
Th	**3,691.881**	**2**	**8**	**W**	**3,692.91**	**7**	**—**
W	**3,691.88**	**20*d***	**4*d***	**U**	**3,692.91**	**10**	**5**
Ta	**3,691.879**	**15**	**—**	**Tb**	**3,692.95**	**8**	**30**
Yb	**3,691.69**	**—**	**10**	**Fe**	**3,693.03**	**7**	**15**
Th	**3,691.62**	**1**	**6**	**Ta**	**3,693.05**	**3*h***	**35*r***
Sm	**3,691.57**	**3**	**25*w***	**Cr**	**3,693.09**	**5**	**10**
W	**3,691.551**	**9**	**5**	**Co**	**3,693.11**	**15**	**80**
Eu	**3,691.55**	**—**	**20*d***	**Nd**	**3,693.13**	**8**	**20**
H	**3,691.55**	**(2)**	**—**	**Mo**	**3,693.23**	**—**	**3**
Re	**3,691.50**	**—**	**100W**	**U**	**3,693.32**	**6**	**4**
Pr	**3,691.48**	**3**	**9**	**Pr**	**3,693.363**	**8**	**30**
U	**3,691.47**	**3**	**3**	**Co**	**3,693.364**	**—**	**18**

	λ	I J	I O
Ho	**3,691.45**	**4**	**6**
Th	**3,691.41**	**1**	**5**
Re	**3,691.38**	—	**12**
Fe	**3,691.335**	—	**2**
Rh	**3,691.333**	**2**	**4**
Ho	**3,691.26**	—	**6**
Nb	**3,691.18**	**15**	**5**
Dy	**3,691.16**	**1**	**3**
Fe	**3,691.152**	—	**4**
Tb	**3,691.15**	**50**	**50**
Ar	3,690.90	(300)	—
Rh	3,690.70	50	125
Pd	3,690.34	1000*w*	300*h*
V	3,690.28	125	200
Ru	3,690.03	100	5
Fe	3,689.46	150	200
Eu	3,688.44	500*W*	1000*W*
Mo	3,688.31	150	4
J	3,688.21	(125)	—
V	3,688.07	200*R*	200
Nb	3,687.97	300*w*	20*w*
Gd	3,687.76	200	200
Fe	3,687.46	300	400
Gd	3,686.34	200	150*W*
V	3,686.26	100	100
Kr	3,686.15	(80*wh*)	—
Fe	3,686.00	125	150
Ne	3,685.74	(75)	—
Ti	3,685.19	700*R*	150
Gd	3,684.12	150	200*W*
Fe	3,684.11	200	300
Pb	3,683.47	50	300
Fe	3,683.06	100	200
Ne	3,682.24	(75)	—
Fe	3,682.21	300	400
Rh	3,681.04	25	20
Kr	3,680.37	(100*wh*)	—
Fe	3,679.91	300	500
V	*3,676.68*	*150h*	*300*
Tb	*3,676.35*	*200*	*100*
Ar	*3,675.22*	*(300)*	—
Yb	*3,675.09*	*200*	*50*
Ar	*3,670.64*	*(300)*	—
Fe	*3,670.07*	*200*	*200*
Fe	*3,669.52*	*150*	*200*
V	*3,669.41*	*300*	*20W*
Kr	*3,669.01*	*(150h)*	—
Nb	**3,693.368**	**10**	**8**
Mo	**3,693.37**	**3**	**5**
Ce	**3,693.42**	**1**	**10**
Co	**3,693.48**	**10**	**35**
Xe	**3,693.49**	**(40)**	—
Pr	**3,693.493**	**1**	**2**
Tb	**3,693.56**	—	**30**
Ru	**3,693.59**	**5**	**20**
Sm	3,694.00	150	100
Fe	3,694.01	300	400
Ne	3,694.197	(250)	—
Yb	3,694.203	1000*R*	500*R*
Fe	3,695.05	150	200
Rh	3,695.52	10	15
V	3,695.86	100*r*	150
Nb	3,695.90	200	4
Gd	3,697.79	200*w*	200*w*
Zr	3,698.17	80	50
Rh	3,698.26	10	15
Rh	3,698.59	15	20
Gd	3,699.75	250	200*w*
Tm	3,700.26	80	150
V	3,700.34	100	10
Rh	3,700.91	150*d*	150*d*
Fe	3,701.09	200	300
Tm	3,701.36	80	150
Mo	3,702.55	150	10
Tb	3,702.85	200	50
V	3,703.58	100*R*	200*R*
Tb	3,703.92	100	70
Co	3,704.06	35	300*r*
Fe	3,704.46	100	125
V	3,704.70	150*R*	200*R*
Fe	*3,705.57*	*500*	*700*
P	*3,706.05*	*(150W)*	—
Fe	*3,709.25*	*400*	*600*
Yt	*3,710.29*	*150*	*80*
Gd	*3,712.71*	*250*	*200W*
Rh	*3,713.02*	*100r*	*100*
Ne	*3,713.08*	*(250)*	—
V	*3,715.47*	*400*	*70*
Nb	*3,717.07*	*1000*	*8*

Rh 4,128.87 *J* 150 *O* 300

	λ	I J	I O		λ	I J	I O
V	**4,128.86**	**5**	**9**	**U**	**4,128.89**	—	**4**
Mo	**4,128.83**	**40**	**40**	**Ce**	**4,128.90**	**1*h***	**3**
U	**4,128.76**	**2**	**2**	**Os**	**4,128.96**	**3**	**60**
Fe	**4,128.73**	**2*h***	**2**	**Ce**	**4,129.10**	—	**3**
Tb	**4,128.72**	—	**4**	**Ca**	**4,129.12**	**2**	**2*h***
Nd	**4,128.70**	**10**	**10**	**Dy**	**4,129.13**	**2**	**10**
J	**4,128.69**	**(35)**	—	**J**	**4,129.13**	**(2)**	—
Ar	**4,128.65**	**(20)**	—	**Pr**	**4,129.148**	**15**	**20**
Ce	**4,128.48**	—	**3**	**Se**	**4,129.15**	**(200)**	—
Tb	**4,128.46**	—	**3**	**Ti**	**4,129.17**	**7**	**15**
Ir	**4,128.43**	**2**	**10**	**Ce**	**4,129.18**	**1**	**5**
Cr	**4,128.40**	**6**	**10**	**Fe**	**4,129.22**	**1**	**5**
Pd	**4,128.37**	—	**5*h***	**Sm**	**4,129.225**	**12**	**5**
Ce	**4,128.36**	**4**	**10*s***	**O**	**4,129.34**	**(15)**	—
U	**4,128.34**	**20**	**18**	**Cr**	**4,129.37**	—	**10**
Ho	**4,128.32**	**3**	—	**Ta**	**4,129.38**	**40**	**200**
Yt	**4,128.30**	**30**	**150**	**Tb**	**4,129.42**	**1*W***	**10**
Mo	**4,128.28**	**25**	**50**	**Dy**	**4,129.425**	**8**	**20**
Dy	**4,128.24**	**4**	**20**	**Nb**	**4,129.426**	**20**	**15**
Pb	**4,128.21**	**5**	—	**Er**	**4,129.433**	**2**	**10**
Mn	**4,128.14**	**15*s***	**15**	**Ho**	**4,129.44**	**1**	**3**
Nd	**4,128.13**	**4**	**8**	**U**	**4,129.46**	**5**	**6**
Sm	**4,128.12**	**2**	**5**	**Tm**	**4,129.54**	**2**	**1**
Si	**4,128.11**	**20*W***	—	**Eu**	**4,129.62**	—	**15**
Eu	**4,128.10**	—	**2**	**U**	**4,129.66**	**4**	**4**
Re	**4,128.09**	—	**6**	**Mo**	**4,129.69**	**4**	**8**
Ca	**4,128.08**	**4**	**2**	**Ar**	**4,129.70**	**(10)**	—
Ne	**4,128.072**	**(30)**	—	**Eu**	**4,129.74**	**50*R***	**150*R***
V	**4,128.071**	**20*r***	**30*r***				
Ce	**4,128.068**	**3**	**10**	Gd	4,130.38	10	200
Zr	**4,127.96**	**2**	**4**	Ar	4,131.73	(80)	—
U	**4,127.925**	**3**	**6**	Fe	4,132.06	200	300
Ir	**4,127.917**	**2**	**30**	Li	4,132.29	—	400*wh*
Ta	**4,127.88**	**8**	**15**	Cl	4,132.48	(200)	—
				Se	4,132.76	(200)	—
Fe	4,127.61	80	100	O	4,132.82	(100)	—
Ho	4,127.16	60	150	Re	4,133.42	—	200
Se	4,126.57	(150)	—	Fe	4,134.68	(100)	150
Lu	4,124.73	10	200	Rh	4,135.27	150	300
Nb	4,123.81	125	200				
				Os	4,135.78	50	200
La	4,123.23	500	500	Se	4,136.28	(100)	—
Rh	4,121.68	50	150	Re	4,136.45	—	150*W*
Co	4,121.32	25	1000*R*	Fe	4,137.00	80	100
Rh	4,119.68	25	100				
Nb	4,119.28	200	2				
				Hg	*4,140.38*	*(200)*	—
O	4,119.22	(300)	—	*La*	*4,141.74*	*200*	*200*
				S	*4,142.29*	*(150)*	—

	λ	I	
		J	O
Co	*4,118.77*	—	*1000R*
Pt	*4,118.69*	*10*	*400*
Rh	*4,116.33*	*10*	*30*
Ru	*4,112.74*	*200*	*125*
Co	*4,110.53*	—	*600*
N	*4,109.98*	*(1000)*	—

	λ	I	
		J	O
Fe	*4,143.87*	*250*	*400*
Ru	*4,144.16*	*200*	*150*
S	*4,145.10*	*(250)*	—
Kr	*4,145.12*	*(250)*	—
Ru	*4,145.74*	*150*	*125*

Rh 4,374.80 *J* 500 *O* 1000

	λ	J	O
Dy	**4,374.80**	**4**	**12**
Th	**4,374.79**	**10**	**15**
Nb	**4,374.78**	**5**	**3**
Ce	**4,374.76**	—	**3**
Ca	**4,374.61**	**2*h***	**10**
Se	**4,374.45**	**25**	**100**
Tb	**4,374.43**	—	**3**
Co	**4,374.429**	—	**2*wh***
Pr	**4,374.41**	**15**	**50**
C	**4,374.28**	**40**	—
Gd	**4,374.25**	**20**	**20**
Er	**4,374.241**	—	**7**
Tb	**4,374.24**	—	**8**
Dy	**4,374.24**	**4**	**12**
Se	**4,374.24**	**(40)**	—
Ta	**4,374.21**	**15**	**15**
Cr	**4,374.16**	**60**	**50**
Th	**4,374.13**	**1**	**4**
Ce	**4,374.08**	—	**3**
Th	**4,373.91**	**5**	**10**
Gd	**4,373.84**	**80**	**200**
Er	**4,373.829**	—	**5**
V	**4,373.827**	**12**	**20**
Ce	**4,373.82**	**4**	**40**
Pr	**4,373.81**	**10**	**40**
Xe	**4,373.78**	**(50*wh*)**	—
Sm	**4,373.73**	**2**	**2**
Ir	**4,373.72**	—	**5**
Nd	**4,373.657**	—	**2*h***
Cr	**4,373.656**	**12**	**15**
Co	**4,373.63**	—	**15*wh***
Tb	**4,373.59**	—	**3**
Rh	4,373.04	10	60
O	4,368.30	(1000)	—
Te	4,364.02	(400)	—

	λ	J	O
Ti	**4,374.82**	**35**	**7**
Tb	**4,374.83**	—	**6**
Ar	**4,374.87**	**(5)**	—
K	**4,374.870**	**(2)**	—
Mo	**4,374.89**	**5**	**8**
Lu	**4,374.90**	—	**5**
Nd	**4,374.923**	**5**	**20**
Co	**4,374.925**	**3**	**10**
Er	**4,374.925**	**25*wh***	**40*wh***
Yt	**4,374.93**	**150**	**150**
Mn	**4,374.95**	**20**	**150**
Sm	**4,374.97**	**200**	**200**
Gd	**4,374.99**	—	**25**
Ne	**4,375.00**	**(2)**	—
Mo	**4,375.01**	**8**	**8**
Nd	**4,375.04**	**5**	**30**
In	**4,375.08**	**(2*h*)**	—
*bh***Zr**	**4,375.1**	—	**8**
Eu	**4,375.12**	**2**	**4**
Ta	**4,375.14**	**5**	**15**
Sc	**4,375.170**	—	**3**
Ce	**4,375.174**	—	**12**
V	**4,375.30**	**12**	**20**
Dy	**4,375.33**	**4**	**10**
Tb	**4,375.33**	—	**6**
Cr	**4,375.333**	**30**	**25**
Ti	**4,375.42**	**1**	**10**
Co	**4,375.54**	—	**5**
B	**4,375.60**	**2**	—
Tb	**4,375.61**	—	**3**
Yt	**4,375.62**	—	**4**
Ce	**4,375.69**	—	**2**
*bh***La**	**4,375.7**	—	**15**
Pr	**4,375.73**	**2**	**10**
*bh***La**	**4,375.8**	—	**30**
Ce	**4,375.92**	**5**	**40**
Fe	**4,375.93**	**200**	**500**
Ar	**4,375.96**	**(20)**	—

	λ	I J	I O
Kr	*4,362.64*	*(500)*	—
Hg	*4,368.35*	*500*	*3000w*
Kr	*4,355.48*	*(3000)*	—

	λ	I J	I O
Kr	4,376.12	(800)	—
Rh	4,379.92	25	60
Se	4,382.87	(800)	—
Fe	4,383.55	800	1000
Kr	4,386.54	(300h)	—

Rh 4,528.73 *J* 60* *O* 500*R*

	λ	J	O
Ce	4,528.70	—	2
Fe	4,528.619	200	600
Mo	4,528.618	10	15
Eu	4,528.49	—	8
Ce	4,528.472	15	30
V	4,528.468	10*wh*	2
Ir	4,528.42	—	2
U	4,528.17	3	1
Yt	4,528.10	—	2
Pr	4,528.07	—	2
V	4,527.99	4*w*	10
U	4,527.97	5	1
Ce	4,527.96	—	3
Co	4,527.93	2*h*	100
Tm	4,527.88	—	15
Mo	4,527.87	4*h*	4*h*
Yt	4,527.79	40	25
Er	4,527.783	9	40
Dy	4,527.78	4	8
Eu	4,527.74	—	2
Th	4,527.73	2	3
U	4,527.69	2	3
Nb	4,527.65	30	1
Ta	4,527.49	5	150
Cr	4,527.47	6	15*d*
Ti	4,527.45	—	3
Sn	4,524.74	50	500*wh*
Eu	4,522.58	—	500
Lu	4,518.57	40	300
Co	4,517.11	6	300
In	*4,511.32*	*4000R*	*5000R*
Rh	*4,503.78*	*10*	*30*

	λ	J	O
Pr	4,528.910	—	3
La	4,528.915	—	2
Re	4,528.94	—	2
Ru	4,529.05	—	10
Nd	4,529.15	—	2
Ce	4,529.28	—	3
V	4,529.29	3	9
Tm	4,529.37	5	80
Er	4,529.38	—	4
Mo	4,529.40	20	25
Nb	4,529.42	3	2
Ti	4,529.46	40	5
Th	4,529.49	2	3
Fe	4,529.56	—	6
V	4,529.59	8	15
Os	4,529.67	2	80
Fe	4,529.68	2	10
U	4,529.71	3	12
W	4,529.760	4	15
Tb	4,529.76	—	4
Nd	4,529.76	—	3
Dy	4,529.78	2	2
Mn	4,529.79	—	50
Cr	4,529.85	8	25
Yb	4,529.87	—	12
Ce	4,529.91	—	5
Re	4,529.928	—	40*W*
Pr	4,529.930	4*W*	25*W*
Nd	4,529.935	—	40
Sm	4,529.95	—	4
Ta	4,529.98	10*h*	3
Ta	4,530.85	50	300
Co	4,530.96	8	1000*W*
Co	4,533.99	8	500
Co	*4,543.81*	—	*500W*
Co	*4,549.66*	—	*600*
Rh	*4,551.64*	*10*	*25*
Ba	*4,554.04*	*200*	*1000R*
Ru	*4,554.51*	*200*	*1000R*

Rn $^{86}_{222}$

t_0 −71°C

t_1 −61.8°C

I.	II.	III.	IV.	V.
10.746	—	—	—	—

λ	*I*	eV
3,006.8	(300)	—
3,054.3	(250)	—
3,621.0	(250)	—
3,634.8	(250)	—
I 4,349.60	(5000)	—
4,625.48	(500)	—
4,680.83	(500)	—
I 7,055.42	(400)	8.4
7,450.00	(600)	8.5

Rn 3,006.8 (300)

	λ	I J	I O		λ	I J	I O
W	**3,006.66**	**5**	**6**	**Ca**	**3,006.858**	**5**	**25**
Ru	**3,006.59**	**15**	**70**	**N**	**3,006.86**	**(50)**	**—**
Hg	**3,006.57**	**(50)**	**—**	**Yb**	**3,006.86**	**3**	**1**
Ta	**3,006.56**	**3*h***	**3*W***	**O**	**3,006.90**	**(12*h*)**	**—**
V	**3,006.50**	**50**	**—**	**Th**	**3,006.93**	**15**	**12**
Te	**3,006.35**	**(50)**	**—**	**Cl**	**3,006.98**	**(20)**	**—**
W	**3,006.31**	**12**	**9**	**J**	**3,007.01**	**(10)**	**—**
Sm	**3,006.15**	**3**	**6**	**O**	**3,007.08**	**(10)**	**—**
				Tb	**3,007.11**	**3**	**8**
				Fe	**3,007.14**	**80**	**100**
Cr	3,003.92	150	1				
Fe	3,002.65	150	20	**O**	**3,007.28**	**(2)**	**—**
Yb	3,002.61	150	15	**Fe**	**3,007.281**	**60**	**80**
V	3,001.20	200*r*	20	**V**	**3,007.283**	**50**	**2**
Fe	3,000.95	300*r*	800*R*	**La**	**3,007.32**	**4**	**4**
				W	**3,007.40**	**3**	**7**
Fe	*2,999.51*	*300*	*500*	**Na**	**3,007.44**	**(20)**	**—**
Nb	*2,994.73*	*300*	*100*				
Rn	*2,994.5*	*(20)*	*—*				
Fe	*2,994.43*	*600r*	*1,000R*	Fe	3,008.14	400*r*	600*r*
				In	3,008.31	500*W*	—
				Sn	3,009.15	200*h*	300*h*
				Fe	3,009.57	400	500
				Rn	3,010.8	(100)	—
				Rn	*3,014.5*	*(18)*	*—*
				Te	*3,017.51*	*(350)*	*—*
				Rn	*3,019.2*	*(18)*	*—*

Rn 3,054.3 (250)

	λ	I J	I O		λ	I J	I O
Se	**3,054.27**	**(20)**	**—**	**U**	**3,054.31**	**4**	**2**
V	**3,054.24**	**5**	**—**	**Ni**	**3,054.32**	**100**	**400*R***
Cs	**3,054.20**	**(4)**	**—**	**Mn**	**3,054.36**	**40**	**75**
U	**3,054.09**	**4**	**6**	**U**	**3,054.39**	**4**	**3**
Tm	**3,054.05**	**60**	**30**	**Er**	**3,054.42**	**2**	**10**
La	**3,054.02**	**6**	**2**	**Yt**	**3,054.42**	**2**	**3**
W	**3,054.01**	**8**	**9**	**Hf**	**3,054.52**	**15**	**15**
Ho	**3,053.99**	**4*h***	**6**	**Cs**	**3,054.56**	**(4)**	**—**
V	**3,053.89**	**60*r***	**10**	**Ne**	**3,054.69**	**(18)**	**—**
U	**3,053.887**	**2**	**1**	**Al**	**3,054.70**	**10**	**20**
Cr	**3,053.88**	**150**	**3*r***	**U**	**3,054.73**	**6**	**10**
Er	**3,053.78**	**1**	**6**	**Mo**	**3,054.76**	**3**	**—**
Cl	**3,053.74**	**(10)**	**—**	**Ta**	**3,054.80**	**1**	**3**
Tm	**3,053.71**	**30**	**7**	**Zr**	**3,054.83**	**25**	**15**
Bi	**3,053.7**	**(60)**	**—**	**Ho**	**3,054.88**	**4*h***	**—**
Cr	**3,053.67**	**5**	**—**	**Eu**	**3,054.93**	**3**	**400*W***

	λ	I			λ	I	
		J	O			J	O
Na	3,053.66	(60)	8	Ru	3,054.94	12	70
Nb	3,053.64	30	1	Os	3,054.97	10	50
Ir	3,053.60	1h	12				
Gd	3,053.58	3	5				
Tb	3,053.55	30	8	Fe	3,055.26	150	200
				J	3,055.37	(350)	—
Cs	3,053.50	(4)	—	Ru	3,056.86	150	12
				Ta	3,057.12	125	25W
				Ne	3,057.39	(250)	—
Pd	3,052.15	150h	—				
Tm	3,050.73	150	50	Fe	3,057.45	400	400
Cr	3,050.14	150	10	Ni	3,057.64	125	400R
Rn	3,050.10	(15)	—	Lu	3,057.90	150h	3
Nb	3,049.52	150	—	Os	3,058.66	500	500R
				Fe	3,059.09	400	600r
V	3,048.21	125r	10				
Fe	3,047.60	500r	800r	Pd	3,059.43	150w	—
Te	3,047.00	(350)	—	Rn	3,059.7	(20)	—
Rn	*3,045.20*	*(60)*	—	*Fe*	*3,062.23*	*400*	*2*
Pt	*3,042.64*	*250R*	*200R*	*Rn*	*3,064.6*	*(40)*	—
Sb	*3,040.67*	*(400wh)*	—	*Pt*	*3,064.71*	*300R*	*2000R*
Fe	*3,040.43*	*400*	*400*	*Fe*	*3,067.24*	*300*	*300*
Nb	*3,039.81*	*300*	*5*	*Bi*	*3,067.72*	*2000wh*	*3000hR*
In	*3,039.36*	*500R*	*1000R*	*Rn*	*3,068.9*	*(100)*	—

Rn 3,621.0 (250)

	λ	J	O		λ	J	O
Er	3,620.945	15	25	Yb	3,621.00	5	—
Yt	3,620.941	12	2h	Nb	3,621.027	15	10
Sm	3,620.938	3	4d	U	3,621.032	5	2
Br	3,620.93	(6)	—	Pr	3,621.09	2h	20h
Ho	3,620.91	4	—	Th	3,621.12	10	10
Eu	3,620.89	5	7	Co	3,621.179	50h	15
Pb	3,620.85	5	—	Pr	3,621.181	1	9
W	3,620.83	8	—	V	3,621.21	80	15
Nd	3,620.74	4	10	Sm	3,621.22	10	60
P	3,620.65	(15)	—	Cu	3,621.24	5	20
Nb	3,620.59	5h	—	Fe	3,621.27	1	—
Sm	3,620.58	2	4	Tb	3,621.39	3	8
Dy	3,620.56	1	3	Fe	3,621.46	100	125
U	3,620.557	2	1	Mo	3,621.62	3	4
Sn	3,620.54	(6)	—	Pt	3,621.66	1	4
V	3,620.472	50	10	Fe	3,621.71	5	6
Fe	3,620.471	2	15	La	3,621.77	8	5
Gd	3,620.461	25	25	U	3,621.78	2	1
Rh	3,620.456	10	20	Yt	3,621.86	1	2
Pr	3,620.43	2	5	Eu	3,621.90	10	8
Th	3,620.37	6	6	Xe	3,621.98	(2h)	—
Cu	3,620.35	5	30	Fe	3,622.00	100	125

	λ	I			λ	I	
		J	O			J	O
Ru	**3,620.28**	**5**	**2**	**Tb**	**3,622.11**	**3**	**30**
Os	**3,620.24**	**12**	**15**	**Ce**	**3,622.14**	**5**	**15**
Au	**3,620.23**	**10**	**—**	**Ar**	**3,622.15**	**(15)**	**—**
Fe	**3,620.228**	**1**	**4**				
Er	**3,620.177**	**3**	**50**	Ti	3,624.82	125	60
				V	3,625.61	125	4
Dy	**3,620.176**	**20**	**80**	Rn	3,626.5	25	—
Sm	**3,620.09**	**3**	**3**	Fe	3,631.46	300	500
U	**3,620.085**	**2*h***	**15**	Kr	3,631.87	(200*h*)	—
Sn	**3,620.08**	**(2)**	**—**				
Er	**3,619.98**	**1**	**9**	*Ar*	*3,632.68*	*(300)*	—
Tm	**3,619.97**	**20**	**20**				
Dy	**3,619.96**	**5**	**25**				
Ce	**3,619.92**	**1**	**8**				
Nb	3,619.73	300	3				
Nb	3,619.51	200	5				
Ni	3,619.39	150*h*	2000*R*				
Fe	3,618.77	400	400				
Rn	3,615.0	(30)	—				
Cd	3,612.87	500	800				
Rn	3,612.61	(20)	—				
Cd	3,610.51	500	1000				
Pd	*3,609.55*	*700R*	*1000R*				
Fe	*3,608.86*	*400*	*500*				
Ar	*3,606.52*	*(1000)*	—				
Cr	*3,605.33*	*400R*	*500R*				
Yt	*3,600.73*	*300*	*100*				

*N**

Rn 3,634.8 (250)

	λ	J	O		λ	J	O
Gd	**3,634.76**	**20**	**20**	**Ar**	**3,634.83**	**5**	**—**
Cs	**3,634.75**	**(6)**	**—**	**Nd**	**3,634.87**	**15**	**30**
Co	**3,634.71**	**10**	**70**	**Sm**	**3,634.91**	**2**	**20**
Pd	**3,634.695**	**1000R**	**2000*R***	**Ru**	**3,634.93**	**100**	**50**
Fe	**3,634.687**	**2**	**20**	**Ni**	**3,634.94**	**10**	**50**
Ho	**3,634.68**	**4*h***	**6**	**Cr**	**3,634.99**	**12**	**25**
Th	**3,634.58**	**4**	**8**	**U**	**3,635.01**	**2**	**1**
U	**3,634.56**	**1**	**8**	**Nd**	**3,635.11**	**6**	**6**
Pr	**3,634.47**	**4**	**20**	**Au**	**3,635.13**	**10**	**3**
Ar	**3,634.46**	**(300)**	**—**	**Mo**	**3,635.14**	**10**	**100*h***
Nb	**3,634.44**	**15**	**5**	**Fe**	**3,635.196**	**2**	**12**
Kr	**3,634.42**	**(3*wh*)**	**—**	**Ti**	**3,635.20**	**5**	**10**
He	**3,634.37**	**(2)**	**—**	**Th**	**3,635.25**	**5**	**4**
Fe	**3,634.33**	**5**	**15**	**Dy**	**3,635.26**	**6**	**10**
Nd	**3,634.28**	**30**	**25**	**Cr**	**3,635.281**	**8**	**25**

	λ	I J	I O		λ	I J	I O
Sm	**3,634.271**	**25**	**100**	**Pr**	**3,635.284**	**7**	**25**
Dy	**3,634.27**	**3*h***	**3**	**Sc**	**3,635.31**	**2**	**3**
S	**3,634.25**	**(35*h*)**	—	**Nb**	**3,635.32**	**8**	**5**
He	**3,634.23**	**(15)**	—	**Yt**	**3,635.33**	**12**	**2*h***
Th	**3,634.21**	**6**	**5**	**Ho**	**3,635.35**	**4**	—
Zr	**3,634.15**	**2**	**25**	**Th**	**3,635.37**	**10**	**15**
Cr	**3,634.01**	**6**	—	**U**	**3,635.40**	**5**	**6**
Ru	**3,633.92**	**5**	**2**	**Tb**	**3,635.42**	**8**	**30**
V	**3,633.91**	**8**	**35**	**Mo**	**3,635.43**	**5**	**25**
Fe	**3,633.83**	**3**	**7**	**V**	**3,635.463**	**5**	**40**
Ta	**3,633.79**	**10*h***	**35**	**Nb**	**3,635.463**	**5**	**5**
Dy	**3,633.77**	**1**	**4**	**Ti**	**3,635.463**	**100**	**200**
Nb	**3,633.71**	**3**	**8**	**Ir**	**3,635.49**	**4*h***	**35**
Th	**3,633.70**	**2**	**1*d***	**Ru**	**3,635.52**	**8**	**2**
Ne	**3,633.66**	**(75)**	—	**U**	**3,635.58**	**8**	**4**
				Mo	**3,635.61**	**3**	**4**
				Ar	**3,635.67**	**(3)**	—
				Mn	**3,635.68**	**5**	**10**
				Nb	**3,635.85**	**8**	**3**
				Eu	**3,635.87**	**4**	—
				V	**3,635.874**	**25*h***	**50**
				Cu	**3,635.92**	**7**	**50**
				Th	**3,635.94**	**4**	**8**
				Ca	**3,635.95**	**2*h***	**4**
				Fe	3,640.39	200	300
				Ti	3,641.33	150	60
				Ti	3,642.67	125	300
				Gd	3,646.20	150	200 *W*
				Fe	*3,647.84*	*400*	*500*
				Ra	*3,649.55*	*(1000)*	—
				Ar	*3,649.83*	*(800)*	—
				Hg	*3,650.15*	*500*	*200*
				Nb	*3,651.19*	*400*	*10*
				Kr	*3,653.97*	*(250h)*	—

*P**

Rn 4,349.60 (5000)

	λ	I J	I O		λ	I J	I O
Kr	**4,349.55**	**(2)**	—	**Hf**	**4,349.74**	**3**	**8**
Eu	**4,349.48**	**2**	**2**	**Ce**	**4,349.79**	**5**	**40**
O	**4,349.43**	**(300)**	—	**V**	**4,349.97**	**3**	**1**
S	**4,349.41**	**(4)**	—	**Nd**	**4,350.20**	**5**	**10**
Ce	**4,349.390**	**6*wh***	**3*W***	**Mo**	**4,350.34**	**40**	**50**
U	**4,349.388**	**5**	**3**	**Ba**	**4,350.37**	**20**	**40**
Te	**4,349.29**	**(30)**	—	**Pr**	**4,350.40**	**25**	**70**
Mo	**4,349.22**	**4**	**6**	**Sm**	**4,350.46**	**150**	**150**

	λ	I J	I O		λ	I J	I O
Rh	**4,349.17**	**2**	**4**	**Hf**	**4,350.51**	**40**	**20**
Nd	**4,349.10**	**2**	**10**	**Mo**	**4,350.73**	**15**	**40**
Dy	**4,349.09**	**2**	**2**				
Nb	**4,349.03**	**10**	**10**	Kr	4,355.48	(3000)	—
Fe	**4,348.94**	**2**	**8**				
Nb	**4,348.65**	**20**	**15**	*Rn*	*4,371.53*	*(30)*	—
Nd	**4,348.53**	**15**	**25**				
Rn	4,335.78	(35)	—				

Rn 4,625.48 (500)

	λ	I J	I O		λ	I J	I O
Ar	**4,625.46**	**(10)**	—	**Er**	**4,625.56**	**2**	**3**
Se	**4,625.37**	**(8)**	—	**In**	**4,625.74**	**5**	—
Pr	**4,625.30**	**1**	**4**	**Cr**	**4,625.92**	**6**	**8**
W	**4,625.17**	**1**	**5**	**U**	**4,625.93**	**2**	—
Fe	**4,625.05**	**12**	**100**	**Cr**	**4,626.19**	**125**	**100**
Th	**4,625.02**	**3**	**4**	**Er**	**4,626.28**	**1** *W*	**2**
Ce	**4,624.90**	**10**	**8**	**Tm**	**4,626.31**	**20**	**50**
Er	**4,624.78**	**1**	**3**	**Hf**	**4,626.42**	**6**	**15**
U	**4,624.71**	**8**	**6**	**Mo**	**4,626.47**	**80**	**100**
Cr	**4,624.57**	**6**	**15**	**V**	**4,626.48**	**20**	**25**
Gd	**4,624.43**	**8**	**15**	**Mn**	**4,626.54**	**15**	**80**
Dy	**4,624.42**	**2**	**3**	**Tm**	**4,626.55**	**3**	**50**
V	**4,624.41**	**15**	**20**	**P**	**4,626.60**	**(70)**	—
Xe	**4,624.28**	**(1000)**	—	**Ar**	**4,626.78**	**(30)**	—
Mo	**4,624.24**	**25**	**25**				
Th	**4,624.14**	**2**	**3**	Ar	4,628.44	(1000)	—
S	**4,624.11**	**(20)**	—	N	4,630.55	(300)	—
Dy	**4,624.10**	**2**	**3**	Kr	4,633.88	(800)	—
Kr	4,619.15	(1000)	—	*Rn*	*4,644.18*	*(300)*	—
Tm	4,615.93	300	200	*Se*	*4,648.44*	*(800)*	—
Kr	4,615.28	(500)	—				
Xe	4,611.89	(700)	—				
Rn	*4,609.38*	*(250)*	—				
Rn	*4,604.40*	*(200)*	—				
Te	*4,602.37*	*(800)*	—				

Rn 4,680.83 (500)

	λ	I J	I O		λ	I J	I O
Nd	**4,680.73**	**1**	**50**	**Cr**	**4,680.87**	**8**	**60**
Th	**4,680.65**	**3**	**4**	**V**	**4,680.89**	**3**	**4**
Cr	**4,680.54**	**25**	**50**	**Mo**	**4,681.05**	**8**	**6**

	λ	I			λ	I	
		J	O			J	O
W	**4,680.52**	**40**	**150**	**Te**	**4,681.06**	**(15*h*)**	**—**
Sc	**4,680.48**	**4*h***	**5**	**Eu**	**4,681.06**	**1**	**15**
Kr	**4,680.41**	**(500)**	**—**	**In**	**4,681.11**	**(200)**	**—**
Ne	**4,680.36**	**(100)**	**—**	**U**	**4,681.18**	**4**	**2**
Zn	**4,680.14**	**200*h***	**300*w***	**W**	**4,681.19**	**1**	**10**
Gd	**4,680.06**	**25**	**50**	**Ne**	**4,681.20**	**(50)**	**—**
V	**4,679.77**	**3**	**4**	**Th**	**4,681.21**	**2**	**2**
Xe	**4,679.45**	**(3*h*)**	**—**	**S**	**4,681.32**	**(5)**	**—**
				Eu	**4,681.51**	**1**	**10**
				Mo	**4,681.63**	**5**	**4**
Ne	4,678.22	(300)	—	**Ta**	**4,681.87**	**50**	**200**
Xe	4,671.23	(2000)	—	**Ti**	**4,681.92**	**100**	**200**
La	4,668.91	300*r*	200*r*				
J	4,666.52	(250)	—	**Tm**	**4,681.92**	**2**	**50**
				Ne	**4,681.93**	**(20)**	**—**
				Mo	**4,681.933**	**3**	**3**
Te	*4,664.34*	*(800)*	—	**Cu**	**4,681.99**	**20**	**—**
Kr	*4,658.87*	*(2000)*	—	**In**	**4,682.00**	**250*W***	**—**
Te	*4,654.38*	*(800)*	—				
				Dy	**4,682.03**	**2**	**3**
				La	**4,682.12**	**3**	**3**
				U	**4,682.13**	**2**	**4**
				Ne	**4,682.15**	**(20)**	**—**
				Mo	**4,682.24**	**5**	**4**
				Ra	**4,682.28**	**(800)**	**—**
				Ar	**4,682.29**	**(10)**	**—**
				He	4,685.75	(300)	—
				Te	4,686.95	(300)	—
				La	4,692.50	300	200
				Rn	*4,701.70*	*(50)*	—
				Ar	*4,702.32*	*(1200)*	—
				Ne	*4,704.39*	*(1500)*	—
				Ne	*4,708.85*	*(1200)*	—
				Ne	*4,710.06*	*(1000)*	—

Rn 7,055.42 (400)

	λ	J	O		λ	J	O
Ar	**7,055.01**	**(4)**	**—**	**Al**	**7,056.56**	**(20)**	**—**
Yt	**7,054.28**	**3**	**6**				
				Ne	7,059.11	(200)	—
Ne	7,032.41	(1000)	—	Se	7,062.06	(1000)	—
				Ar	7,067.22	(400)	—

Rn 7,450.00 (600)

	λ	I			λ	I	
		J	O			J	O
Al	**7,449.42**	**(18)**	—	**Yt**	**7,450.30**	**10**	**1**
Fe	**7,449.34**	**6**	—	**S**	**7,450.33**	**(15)**	—
S	**7,449.09**	**(5)**	—	**Xe**	**7,451.00**	**(25)**	—
Ne	7,438.90	(300)	—	Rn	7,470.89	(15)	—
				Rn	7,483.13	(15)	—

Ru $^{44}_{101.07}$

t_0 2,450°C t_1 2,700°C

I.	II.	III.	IV.	V.
~7.5	—	—	—	—

λ	I		eV
	J	O	
2,678.76	300	100	11.0
2,712.41	300	80	11.0
I 3,436.74	150	300*R*	3.75
I 3,498.94	200	500*R*	3.54
I 3,596.18	100	30	3.70
I 4,554.51	200	1000*R*	3.53
I 4,584.45	80	150*R*	3.70

Ru 2,678.76 *J* 300 *O* 100

	λ	I	
		J	O
V	2,678.675	—	12
Mo	2,678.666	—	20
Nb	2,678.656	10	3
Zr	2,678.63	100	80
Ce	2,678.62	—	2
V	2,678.57	150*R*	30
W	2,678.52	7	10
Hf	2,678.43	12	10
U	2,678.42	2	3
Eu	2,678.28	100	150
Ru	2,678.18	—	10
V	2,677.80	300*R*	70
Cd	2,677.64	25	100
Ru	2,677.32	—	12
Cr	2,677.16	300*r*	35
Pt	2,677.15	200*w*	800*w*
Ru	2,676.97	—	12
Cl	2,676.95	(150)	—
Ru	2,676.76	18	—
Ru	2,676.35	3	50
Rh	2,676.25	100	1
Ru	2,676.19	100	8
Au	2,675.95	100	250*R*
Ta	2,675.90	200	150
Ne	2,675.64	(150)	—
Ru	2,675.52	50	—
*bh*B	2,675.3	—	60
Ne	2,675.24	(150)	—
Ru	2,675.18	30	—
Pt	2,674.57	10	200
Rh	2,674.44	200*w*	1
Re	2,674.34	—	100
Ru	2,674.19	50	—
Ru	2,673.60	3	50
Nb	2,673.57	500	10
Ru	2,673.48	3	50
Eu	2,673.41	50	125
Ru	2,673.01	50	8
Ru	2,672.36	40	8
Ru	2,672.21	25	—
V	2,672.00	300R	50
Zn	2,670.53	4	200
Ru	2,670.49	30	8
Ru	2,669.42	100	—

	λ	I	
		J	O
Cr	2,678.79	80	10
Ta	2,678.80	20	3
U	2,678.86	4	6
V	2,678.87	—	15
W	2,678.88	10	20
Cs	2,678.92	20	—
Th	2,678.94	—	5
Ce	2,678.95	—	2
Nb	2,679.01	2	10
Fe	2,679.062	200	200
Ir	2,679.063	3	15
Re	2,679.08	—	25
Pd	2,679.12	15*wh*	—
Pt	2,679.13	50	6
Mn	2,679.16	40	—
Bi	2,679.2	2	—
Ni	2,679.24	500*wh*	—
V	2,679.32	300*R*	70
Ru	2,679.43	20	—
Co	2,679.756	—	75*W*
Ru	2,679.762	1*h*	12
Ti	2,679.93	12	100
Na	2,680.33	10	60*R*
Ru	2,680.54	50	5
V	2,682.87	200*R*	50
V	2,683.09	150*R*	35
Mo	2,683.23	150	20
Rh	2,683.563	200	1
Re	2,683.56	—	50
Ru	2,683.68	—	20
Ru	2,684.09	2h	20
Zn	2,684.16	6	300
Ta	2,684.28	2	150
Ni	2,684.41	600wh	—
Ru	2,684.60	30	10
Fe	2,684.75	400	3
Ru	2,685.16	35	10
Eu	2,685.65	100	150
Ru	2,686.29	12	80
Nb	2,686.39	300	2
Ru	2,686.89	30	—
Ru	2,687.07	50	—
Ru	2,687.14	—	12
Ru	2,687.50	100	12

	λ	I J	I O		λ	I J	I O
				V	*2,687.96*	*500R*	*150*
				Ru	*2,688.11*	*100*	*8*
				Re	*2,688.53*	*—*	*100*

Ru 2,712.41 *J* 300 *O* 80

	λ	J	O		λ	J	O
Kr	**2,712.40**	**(80*h*)**	**—**	**Zr**	**2,712.420**	**15**	**20**
Fe	**2,712.39**	**100**	**2**	**Hf**	**2,712.425**	**50**	**25**
Ce	**2,712.37**	**—**	**2**	**Re**	**2,712.48**	**—**	**30**
Mo	**2,712.35**	**40**	**1**	**Zn**	**2,712.49**	**8**	**300**
Cr	**2,712.31**	**70**	**30**	**Cd**	**2,712.57**	**20**	**75**
J	**2,712.23**	**(100)**	**—**	**U**	**2,712.582**	**2**	**—**
V	**2,712.22**	**10*h***	**6**	**W**	**2,712.585**	**—**	**5**
Hf	**2,712.14**	**10**	**10**	**Yb**	**2,712.65**	**5**	**2**
Er	**2,712.120**	**1**	**7**	**Er**	**2,712.66**	**2**	**4**
U	**2,712.116**	**2**	**3**	**Mo**	**2,712.689**	**—**	**3**
*bh***C**	**2,712.1**	**—**	**12**	**W**	**2,712.695**	**20**	**1**
Rn	**2,712.1**	**(7)**	**—**	**Ir**	**2,712.74**	**10**	**40**
Ru	**2,712.09**	**—**	**30**	**V**	**2,712.81**	**4*wh***	**9*h***
U	**2,712.061**	**6**	**8**	**Ru**	**2,712.88**	**—**	**30**
Ag	**2,712.06**	**200*h***	**3**				
Hf	**2,711.99**	**10**	**10**	V	2,713.05	80	40
Ce	**2,711.93**	**—**	**2**	Ru	2,713.07	80	—
Mo	**2,711.92**	**—**	**10**	Pt	2,713.13	10	200
				Ru	2,713.19	2	60
				Mn	2,713.33	—	300*wh*
V	2,711.74	150*R*	50	Cu	2,713.50	300*w*	50
Fe	2,711.65	50	100	Ru	2,713.58	100	—
Zr	2,711.51	20	40	Ru	2,713.74	2*h*	60
Ag	2,711.21	300*wh*	1*h*	*bh*B	2,713.8	—	200
Ru	2,710.74	—	20	In	2,713.93	125*wh*	200*R*
Fe	2,710.55	35	80	V	2,714.20	100	60
In	2,710.26	200*Rh*	800*R*	Pd	2,714.32	150	—
Ru	2,710.232	100	50	Rh	2,714.410	5	150
Cr	2,710.23	—	50*h*	Fe	2,714.412	400	200
Ta	2,710.13	3	200	Co	2,714.42	200*W*	12
Fe	2,709.99	10	40	Os	2,714.64	10	50*r*
Rh	2,709.52	2	50	Ta	2,714.67	8	200
Ta	2,709.27	150	40	Fe	2,714.87	15	40
Tl	2,709.23	200*R*	400*R*	Pd	2,714.90	200	—
Ru	2,709.20	8	60	Rh	2,715.04	2	50
Ra	2,708.96	(200)	—	Ru	2,715.24	5*h*	20
Ru	2,708.84	—	20	Rh	2,715.31	500*wh*	50
Ni	2,708.79	500	—	Re	2,715.47	—	100
Ru	2,708.65	—	20	Ru	2,715.50	2*h*	12
Fe	2,708.57	50	80	V	2,715.69	300*R*	50
Ru	2,707.97	3	50	Ru	2,715.78	—	50
V	2,707.86	150	70				

	λ	I J	I O
Ru	*2.707.47*	—	*30*
Ru	*2,707.29*	*60*	—
Rh	*2,707.23*	*4*	*100*
Fe	*2,706.58*	*150*	*150*
Sn	*2,706.51*	*150R*	*200R*
V	*2,706.17*	*400R*	*100*
Pt	*2,705.89*	*200wh*	*1000wh*
Rh	*2,705.63*	*300wh*	*100*
Ru	*2,705.33*	—	*12*
Ru	*2,704.81*	*25*	—
Ru	*2,704.57*	*100*	—
Ru	*2,704.19*	*35*	*10*
Mn	*2,703.990*	*25*	*100wh*
Fe	*2,703.989*	*400*	*30*
Ru	*2,703.80*	*3*	*60*
Rh	*2,703.73*	*25*	*150*
Co	2,715.99	75	75*w*
Ru	2,716.12	100	—
Fe	2,716.22	150	20
Ru	2,716.58	80	—
Nb	2,716.62	200	10
Rh	2,716.82	3	50
Eu	2,716.97	300	300
Ru	2,717.01	—	30
Ta	2,717.18	—	100
Ru	*2,717.40*	*100*	*50*
Rh	*2,717.51*	*5*	*100*
Ru	*2,717.86*	*50*	—
Ta	*2,718.38*	—	*80*
Fe	*2,718.43*	*60*	*80*
Rh	*2,718.54*	*20*	*150*
Cu	*2,718.77*	*300w*	*40*
Ru	*2,718.83*	*5h*	*30*
Fe	*2,719.02*	*300r*	*500r*
Pt	*2,719.04*	*100w*	*1000w*
Ru	*2,719.51*	*30*	*100*
Ru	*2,719.72*	*50*	*12*
Rh	*2,720.14*	*6*	*100*
Ta	*2,720.76*	*1*	*150*
Fe	*2,720.90*	—	*700r*
Ru	*2,721.56*	*5*	*60*

Ru 3,436.74 *J* 150 *O* 300*R*

	λ	J	O
Ce	**3,436.73**	—	**3**
Sm	**3,436.70**	**2**	**3**
Th	**3,436.69**	**5**	**3*d***
Ga	**3,436.66**	**(2)**	—
Pr	**3,436.63**	—	**15**
Cu	**3,436.54**	—	**7**
Pr	**3,436.46**	**3*h***	—
Yb	**3,436.46**	**30**	**5**
V	**3,436.39**	**3**	—
Er	**3,436.34**	**1**	**10**
Ru	**3,436.33**	**8**	**12**
Ho	**3,436.32**	**4*h***	—
Ce	**3,436.30**	—	**15**
Ce	**3,436.20**	—	**5**
U	**3,436.19**	—	**6**
Cr	**3,436.187**	**50**	**50**
Sm	**3,436.14**	—	**3**
Fe	**3,436.11**	**15**	**5**
U	**3,436.78**	**15**	**12**
Nb	**3,436.83**	**20**	**2**
J	**3,436.94**	**(10)**	—
Dy	**3,436.95**	—	**3**
Ce	**3,436.959**	—	**5**
Co	**3,436.960**	—	**10**
Nb	**3,436.962**	**50**	**20*r***
Tb	**3,436.97**	—	**15**
Ir	**3,437.01**	**15**	**20**
Th	**3,437.03**	**2**	**5**
Ho	**3,437.05**	**4*h***	—
Fe	**3,437.051**	**15**	**80**
Ta	**3,437.07**	—	**2*r***
U	**3,437.09**	**5*d***	**3*d***
Sm	**3,437.11**	**2**	**15**
Zr	**3,437.14**	**10**	**15**
N	**3,437.16**	**(35)**	—
Mo	**3,437.216**	**25**	**25**

	λ	I J	I O
Eu	3,436.07	—	5
Ta	3,436.00	18W	70W
Th	3,435.98	10	12
Dy	3,435.91	4	3
Cr	3,435.82	12	30
Co	3,435.75	—	8
Eu	3,435.73	9	12
Ru	3,435.19	20	60
Rh	3,434.89	200r	1000r
Ni	3,433.56	50wh	800R
Pd	3,433.45	500h	1000h
Cr	3,433.13	150	30
Ru	3,433.26	25	60
Co	3,433.04	150	1000R
Ru	3,432.74	40	70
Nb	3,432.70	100	10
Ru	3,432.21	12	50
Co	3,431.57	40	500R
Bi	3,431.23	(150)	—
Bi	3,430.83	(200)	—
Ru	3,430.77	45	70
Tm	3,429.97	100	100
Ru	3,429.54	25	60
Ru	3,428.63	12	30
Yb	3,428.46	80	25
Ru	3,428.31	100	100
Nb	*3,426.57*	*200*	*5*
Ru	*3,425.96*	*4*	*30*
Nb	*3,425.42*	*300*	*30r*
Tm	*3,425.08*	*300*	*200*
Re	*3,424.60*	—	*300W*
Fe	*3,424.29*	*150*	*200*
Ni	*3,423.71*	*25*	*600R*
Pd	*3,421.24*	*1000R*	*2000R*
Cr	*3,421.21*	*200*	*50*
Ru	*3,420.08*	*8*	*60*
Ru	*3,419.25*	*2*	*30*
Ne	*3,417.90*	*(500)*	—
Ru	*3,417.35*	*70*	*1*
Co	*3,417.16*	—	*400R*
W	3,437.220	10	3
Ni	3,437.28	40	600R
Th	3,437.31	2	6
Ce	3,437.32	—	12
Pb	3,437.36	10	—
Ta	3,437.37	300wh	7
Sb	3,437.49	8	2h
Ir	3,437.50	3	30
Pr	3,437.60	—	6
Sm	3,437.61	2h	3
Tm	3,437.64	15	10
Er	3,437.64	1	10
U	3,437.67	—	6
Co	3,437.69	—	150Wh
Re	3,437.72	—	100
Xe	3,437.73	(2wh)	—
Zr	3,438.23	200	250
Ru	3,438.37	35	70
Yb	3,438.72	80	4
Yb	3,438.84	100	20
Rb	3,439.34	(200)	—
Ru	3,439,68	—	30
Ru	3,440.20	30	100
Rh	3,440.53	100	2
Nb	3,440.59	80	15
Fe	3,440.61	300	500
Fe	3,440.99	200	300
Pd	3,441.40	2h	800h
Cr	3,441.44	90	80
Tm	3,441.51	80	150
Mn	3,441.99	75	75
Co	3,442.93	15	400R
Ru	3,443.15	—	30
Co	3,443.64	100	500R
Fe	3,443.88	200	400
Ti	3,444.31	150	60
Fe	3,445.15	150	300
Ru	3,445.30	—	12
Cr	3,445.618	80	100
Sm	3,445.621	10R	150R
Nb	3,445.68	80	50
Ru	3,446.07	6	50
Ni	3,446.26	50h	1000R
Ru	3,446.49	—	50
K	3,446.72	100R	150R
Ta	*3,446.91*	*150W*	*2W*

	λ	J	O		λ	J	O
				Ne	*3,447.70*	*(150)*	—
				Ru	*3,448.95*	*20*	*70*
				Co	*3,449.17*	*125*	*500R*
				Co	*3,449.44*	*125*	*500R*
				Pd	*3,451.35*	*400h*	—
				Hg	*3,451.69*	*(200)*	—
				Nb	*3,452.35*	*200*	*5*
				Ni	*3,452.89*	*50*	*600R*
				Ru	*3,452.90*	*6*	*60*
				Co	*3,453.50*	*200*	*3000R*
				Yb	*3,454.07*	*250*	*40*
				Rh	*3,455.22*	*12*	*300*
				Co	*3,455.23*	*10*	*2000R*
				Ru	*3,455.39*	—	*20*
				Ru	*3,455.73*	—	*12*
				Ru	*3,456.62*	*8*	*60*

Ru 3,498.942 *J* 200 *O* 500*R*

	λ	J	O		λ	J	O
Dy	**3,498.939**	—	**15**	**Yt**	**3,498.943**	**6**	**8**
Ce	**3,498.924**	—	**2**	**Ir**	**3,498.951**	—	**25**
Mo	**3,498.923**	**1**	**3**	**Hf**	**3,498.98**	—	**6**
Kr	**3,498.92**	**(2*wh*)**	—	**Th**	**3,498.99**	**8**	**8**
Sc	**3,498.91**	**2**	**8**	**Mo**	**3,499.071**	**20**	—
Ho	**3,498.87**	**10**	**8**	**U**	**3,499.071**	**8**	**6**
Ce	**3,498.83**	—	**2**	**Ho**	**3,499.08**	**10**	**10**
Fe	**3,498.76**	—	**1*h***	**Pr**	**3,499.09**	**3**	**40**
Ir	**3,498.74**	—	**15**	**Ti**	**3,499.099**	**10**	**25**
Tb	**3,498.73**	—	**8**	**Er**	**3,499.104**	**15**	**18**
Rh	**3,498.730**	**60**	**500**	**Ir**	**3,499.11**	—	**5**
Er	**3,498.71**	**3**	**10**	**Mo**	**3,499.20**	—	**3**
Ce	**3,498.68**	—	**15**	**Os**	**3,499.27**	**2**	**3**
Dy	**3,498.67**	**50**	**50**	**Ce**	**3,499.30**	—	**4**
He	**3,498.64**	**(3)**	—	**U**	**3,499.33**	**15**	**6**
Nb	**3,498.63**	**50**	**30**	**Tb**	**3,499.34**	—	**15**
Th	**3,498.62**	**3**	**6**	**Ir**	**3,499.37**	—	**2**
U	**3,498.60**	**4**	**2**	**Ce**	**3,499.39**	—	**4**
Ce	**3,498.56**	—	**8**	**Ce**	**3,499.47**	—	**2**
Os	**3,498.54**	**15**	**80**	**Ar**	**3,499.49**	**(10)**	—
Kr	**3,498.50**	**(4*wh*)**	—	**Os**	**3,499.53**	**3**	**10**
Sb	**3,498.46**	**300*wh***	—	**Pr**	**3,499.57**	**4**	**10**
Nb	**3,498.43**	**1**	**2**	**Zr**	**3,499.58**	**9**	**10**
U	**3,498.38**	**3**	**8**	**Dy**	**3,499.61**	—	**4**
V	**3,498.20**	**7**	**12**	**Si**	**3,499.61**	**2**	—
Pt	**3,498.17**	**2**	**1**	**Ce**	**3,499.62**	—	**5**
W	**3,498.15**	**10**	—	**W**	**3,499.63**	**9**	**9**
Os	**3,498.07**	**10**	**15**	**Sr**	**3,499.67**	—	**50**

	λ	I	
		J	O
Ne	3,498.064	(75)	—
Cu	3,498.063	5	20
Tb	3,498.06	—	15
J	3,498.03	(25)	—
Th	3,498.02	8	8
Ru	3,497.94	5	30
Zr	3,497.92	10	—
Fe	3,497.84	200	200
Mn	3,497.54	150	15
S	3,497.34	(100)	—
Fe	3,497.11	100	200
V	3,497.03	150	—
Zr	3,496.21	100	100
Ru	3,496.13	—	12
Ru	3,495.97	10	60
Mn	3,495.84	150	25
Co	3,495.69	25	1000R
Cd	3,495.34	(100)	—
Ru	3,494.25	8	50
Hg	3,493.85	(100)	—
Ru	3,493.22	1	20
V	3,493.17	100	15
Ni	3,492.96	100h	1000R
Rh	3,492.76	(300)	—
Fe	3,490.57	300	400
Ru	3,489.75	—	12
Mn	3,488.68	200	50
Ru	3,486.79	—	30
Ru	3,486.21	—	20
Pt	3,485.27	200R	150
Ni	3,483.77	30	500R
Ru	3,483.29	10	60
Ru	3,483.16	8	50
Mn	3,482.91	250	50
Ru	3,482.34	3	30
Ru	3,481.30	35	70
Pd	3,481.15	2h	500r
Ta	3,480.52	200ws	70
Nb	3,479.56	200	5
Rh	3,478.91	100	500
Yb	3,478.84	300	40

	λ	I	
		J	O
Ir	3,499.76	—	3
Ce	3,499.78	—	3
Dy	3,499.82	—	5
V	3,499.824	50	3
Sm	3,499.825	5	8
Fe	3,499.88	2h	2h
Nb	3,499.95	50	5
Er	3,499.965	—	9
Tm	3,499.965	—	10
Nd	3,499.97	4	8
Mo	3,499.98	25	—
Ce	3,499.994	1	15
Th	3,499.996	8	8
Cd	3,500.00	15	25
Tb	3,500.02	3	8
Ni	3,500.85	80	500wh
Ba	3,501.12	20	1000
Ne	3,501.22	(150)	—
Ru	3,501.35	3	30
F	3,501.42	(200)	—
Co	3,501.72	100	5
Co	3,502.28	20	2000R
Ru	3,502.42	4	20
Rh	3,502.52	150	1000
F	3,503.09	(400)	—
V	3,504.44	200	60
Os	3,504.66	20	300
Ti	3,504.89	150	20
F	3,505.61	(600)	—
Co	3,506.31	15	400R
Rh	3,507.32	125	500
P	3,507.36	(100w)	—
Lu	3,507.39	150	100
V	3,509.04	150	2
Tb	3,509.17	200	200
Ru	3,509.20	100	10
Co	3,509.84	40	400R
Nb	3,510.26	200	15
Ni	3,510.34	50h	900R
Fe	3,513.82	300	400
Ru	3,514.49	40	70
Ru	3,514.77	4	12
Ni	3,515.05	50h	1000R
Nb	3,515.42	300	20
Pd	3,516.94	500R	1000R
Nb	3,517.67	200	2

	λ	I J	I O
Ru	*3,518.98*	*3*	*30*
Tl	*3,519.24*	*1000R*	*2000R*
Ru	*3,519.63*	*30*	*70*
Ni	*3,519.77*	*30*	*500h*
Ru	*3,520.13*	*40*	*60*
Ne	*3,520.47*	*(1000)*	*—*

Ru 3,596.179 *J* 100 *O* 30*

	λ	I J	I O
Ce	**3,596.11**	**2**	**12**
Ho	**3,596.11**	**4*h***	**—**
Bi	**3,596.110**	**50**	**150*wh***
Dy	**3,596.07**	**20**	**50**
Ti	**3,596.05**	**125**	**50**
Rh	**3,596.02**	**5**	**25**
S	**3,595.99**	**(50)**	**—**
U	**3,595.94**	**3**	**1**
F	**3,595.92**	**(15)**	**—**
Rb	**3,595.91**	**(2)**	**—**
Fe	**3,595.86**	**1**	**4**
Os	**3,595.79**	**5**	**1**
Mo	**3,595.71**	**4**	**3**
Ta	**3,595.64**	**5**	**70**
Th	**3,595.62**	**1**	**3**
Mo	**3,595.55**	**5**	**8**
U	**3,595.53**	**3*h***	**2**
Er	**3,595.47**	**1**	**7*d***
Mo	**3,595.40**	**10**	**—**
W	**3,595.39**	**6**	**8**
Th	**3,595.32**	**3*d***	**3*d***
Fe	**3,595.31**	**7**	**20**
Mn	**3,595.12**	**25**	**50**
W	**3,595.07**	**5**	**—**
Ne	3,593.64	(250)	—
Ne	3,593.53	(500)	—
Cr	3,593.49	400*R*	500*R*
V	3,593.33	300*R*	30
Ru	3,593.02	150	60
Gd	3,592.70	70	50
Sm	3,592.59	50	40
V	3,592.02	300*R*	50
Dy	3,591.42	100	200
Nb	3,591.20	50	2
Nd	3,590.35	300*W*	400*W*
V	3,589.76	600*R*	80
Kr	3,589.65	(70*wh*)	—
W	**3,596.18**	**15**	**—**
Pr	**3,596.193**	**8**	**25**
Rh	**3,596.194**	**50**	**200**
Fe	**3,596.20**	**5**	**15**
Eu	**3,596.23**	**1*h***	**3*W***
Mo	**3,596.35**	**30**	**—**
Tb	**3,596.38**	**15**	**50**
W	**3,596.43**	**4**	**5**
Dy	**3,596.49**	**1**	**3**
Ti	**3,596.55**	**2**	**3**
Ba	**3,596.57**	**4**	**1**
W	**3,596.61**	**4**	**5**
Sm	**3,596.64**	**2**	**4**
La	**3,596.65**	**3**	**2**
Pd	**3,596.66**	**1**	**5**
Ce	**3,596.72**	**1**	**10**
U	**3,596.76**	**8**	**2**
Kr	**3,596.86**	**(2*h*)**	**—**
Eu	**3,596.86**	**2*h***	**3*W***
Ta	**3,596.861**	**1**	**7**
U	**3,596.88**	**8**	**6**
Sb	**3,596.96**	**(100)**	**—**
Rn	**3,597.0**	**(5)**	**—**
Fe	**3,597.061**	**10**	**40**
Nb	**3,597.25**	**10**	**8**
W	**3,597.26**	**9**	**10**
Sb	3,597.51	200*wh*	2
Ni	3,597.70	50*h*	1000*r*
Ru	3,599.76	100	12
Ne	3,600.17	(75)	—
Tb	3,600.44	50	8
Yt	3,600.73	300	100
Yt	3,601.92	60	18
F	3,602.85	(60*d*)	—
Eu	3,603.20	50	100*w*
Fe	3,603.21	80	150

	λ	I J	I O
Nb	3,589.36	100	100
Ru	3,589.21	100	60
Bi	3,588.5	(60)	—
Ar	3,588.44	(300)	—
Al	3,587.44	(80)	—
Ru	3,587.20	70	5
Co	3,587.19	50*h*	200*r*
Al	3,587.06	(100)	—
Fe	3,586.99	150	200
Al	3,586.91	(500*h*)	—
Al	3,586.80	(200*wh*)	—
Al	3,586.69	(200)	—
Al	3,586.55	(200)	—
Fe	3,586.11	80	80
Dy	3,585.78	100	150
Fe	3,585.71	80	125
Te	3,585.34	350	—
Fe	3,585.32	100	150
Dy	3,585.07	100	300
Tb	3,585.03	50	15
Nb	3,584.97	50	30
Gd	3,584.96	100	100
Rh	*3,583.10*	*125*	*200*
Fe	*3,581.19*	*600r*	*1000R*
Nb	*3,580.27*	*300*	*100*
Cr	*3,578.69*	*400r*	*500R*
Ar	*3,576.62*	*(300)*	—
Ru	*3,574.58*	*15*	*4*
Dy	*3,574.16*	*100*	*200*
Ir	*3,573.72*	*100*	*8*
Cr	3,603.74	50	15
Hg	3,604.09	(50)	—
V	3,604.38	50*h*	—
Os	3,604.47	100	15
Cr	3,605.33	400*R*	500*R*
Fe	3,605.46	150	300
Hg	3,605.80	(200)	—
Dy	3,606.13	100	200
Yb	3,606.47	60	15
Ar	3,606.52	1000	—
Fe	3,606.68	150	200
V	3,606.69	70	80
Kr	*3,607.88*	*(100wh)*	—
Gd	*3,608.76*	*125*	*100*
Fe	*3,608.86*	*300*	*500*
Sm	*3,609.48*	*100*	*60*
Pd	*3,609.55*	*700R*	*1000R*
Cd	*3,610.51*	*500*	*1000*
Cd	*3,612.87*	*500*	*800*
Cd	*3,614.45*	*100*	*60*
P	*3,617.09*	*(100w)*	—
Fe	*3,618.77*	*400*	*400*
V	*3,618.93*	*100*	—

Ru 4,554.51 *J* 200 *O* 1000*R*

	λ	I J	I O
Pr	**4,554.50**	—	**2**
Yt	**4,554.46**	**6**	**3**
Sm	**4,554.44**	—	**60**
Ne	**4,554.41**	**(10)**	—
Ce	**4,554.33**	—	**2**
Ar	**4,554.32**	**(15)**	—
Dy	**4,554.23**	—	**2**
Ba	**4,554.04**	**200**	**1000*R***
Ce	**4,554.035**	—	**35*S***
Mo	**4,554.028**	**4**	**1**
Zr	**4,553.97**	**12**	**4**
Cr	**4,553.95**	**3**	**20**
U	**4,553.86**	**1**	**4**
Ce	**4,554.557**	—	**6**
Ne	**4,554.56**	**(5)**	—
Ca	**4,554.590**	**2**	—
Pt	**4,554.593**	**5**	**10**
Tm	**4,554.65**	—	**5**
W	**4,554.68**	—	**4**
Ir	**4,554.78**	—	**4**
Pr	**4,554.79**	**1**	**4**
P	**4,554.80**	**(100)**	—
Ne	**4,554.82**	**(40)**	—
Cr	**4,554.83**	**2**	**25**
Nd	**4,554.97**	**1**	**5**
Gd	**4,554.99**	**2**	**6**

	λ	I J	I O
Th	4,553.85	2	3
Nb	4,553.84	8	5
Mo	4,553.80	20	20
Hf	4,553.78	—	10
Ce	4,553.75	—	2
bhPb	4,553.7	—	6
Ta	4,553.69	2	200
W	4,553.66	1	6
Yb	4,553.56	60	20
Mo	4,553.503	25	—
Pr	4,553.498	—	4
Ce	4,553.42	—	2
Ti	4,553.41	—	6
Co	4,553.33	—	25
Mo	4,553.32	4	12
Mn	4,553.31	—	12h
Eu	4,553.3	—	3W
Ca	4,553.27	4h	—
Pr	4,553.26	—	5h
Hf	4,553.24	—	5h
Mo	4,553.22	6	12
Ni	4,553.17	—	15r
Ne	4,553.16	(50)	—
Er	4,553.13	—	5
S	4,552.38	(200)	—
Co	4,549.66	—	600
Ti	4,549.63	200	100
Fe	4,549.47	100	100
As	4,549.23	125	—
Ru	4,547.853	—	20
Fe	4,547.851	100	200
Ru	4,547.33	—	25
Ru	4,546.93	—	15
Cr	4,545.96	125	200
Xe	4,545.23	(200wh)	—
Ar	4,545.08	(200)	—
Co	4,543.81	—	500W
As	4,543.76	200	—
Ru	4,543.69	—	15
Br	4,542.93	(250)	—
Xe	4,540.89	(200h)	—
Gd	*4,540.02*	*200*	*80*
As	*4,539.97*	*200*	—
Ne	*4,538.31*	*300*	—
Ne	*4,537.75*	*(1000)*	—
Ne	*4,537.68*	*(300)*	—
Cr	4,555.03	40h	—
Th	4,555.07	—	3
Ti	4,555.08	2	12
Cr	4,555.092	50	15
U	4,555.095	40	20
Zr	4,555.13	—	15
Nd	4,555.14	—	15
Dy	4,555.24	—	4
Tm	4,555.26	50	25
Te	4,555.27	(30)	—
Cr	4,555.296	—	15
Yt	4,555.298	2	2
W	4,555.327	1	7
Cs	4,555.35	100	2000R
Eu	4,555.38	4	4
Ne	4,555.39	(30)	—
Ce	4,555.42	—	5
Ti	4,555.49	60	125
Zr	4,555.52	2	30
Nb	4,555.56	2h	3h
Th	4,555.61	10h	—
Ce	4,555.62	—	2
Er	4,555.69	—	3
Eu	4,555.71	—	12W
Th	4,555.81	—	3
Fe	4,555.89	12	12
Xe	4,555.94	(100wh)	—
Kr	4,556.61	(200h)	—

	λ	I			λ	I	
		J	O			J	O
Ra	*4,533.11*	*(300)*	*—*				
Co	*4,530.96*	*8*	*1000w*				
Ru	*4,530.85*	*—*	*60*				
Fe	*4,528.62*	*200*	*600*				

Ru 4,584.445 *J* 80 *O* 150*R*

	λ	J	O		λ	J	O
Pr	**4,584.43**	**—**	**2**	**Mo**	**4,584.450**	**3**	**3**
Th	**4,584.38**	**4**	**5**	**Ir**	**4,584.75**	**—**	**3**
Gd	**4,584.282**	**2**	**3**	**Dy**	**4,584.78**	**2**	**3**
Cl	**4,584.28**	**(20)**	**—**	**Fe**	**4,584.82**	**1**	**8**
U	**4,584.26**	**3**	**—**	**Sm**	**4,584.83**	**50**	**60**
W	**4,584.25**	**—**	**3**	**Tb**	**4,584.84**	**—**	**12**
Zr	**4,584.24**	**—**	**5**	**U**	**4,584.847**	**15**	**10**
Ce	**4,584.18**	**—**	**2*h***	**Nb**	**4,584.848**	**3*h***	**3**
Nb	**4,584.10**	**3*h***	**—**	**Er**	**4,584.85**	**—**	**8**
Cr	**4,584.09**	**1**	**20**	**Mn**	**4,584.92**	**—**	**20**
Nd	**4,584.04**	**2**	**12**	**Cr**	**4,584.93**	**1**	**10**
Tb	**4,583.95**	**—**	**4**	**Os**	**4,584.94**	**—**	**2**
Cr	**4,583.90**	**—**	**10**	**Ar**	**4,584.96**	**(10)**	**—**
Co	**4,583.88**	**—**	**2**	**Cl**	**4,585.03**	**(15)**	**—**
Fe	**4,583.85**	**150**	**150**	**Cr**	**4,585.09**	**2**	**10**
V	**4,583.78**	**150**	**150**	***bh*La**	**4,585.10**	**—**	**4**
Th	**4,583.70**	**—**	**2**	**Ir**	**4,585.13**	**—**	**2**
Dy	**4,583.64**	**—**	**2**	**Xe**	**4,585.48**	**(200*wh*)**	**—**
Nb	**4,583.49**	**3*wh***	**2**	***bh*Sr**	**4,585.50**	**—**	**3**
Ti	**4,583.44**	**10**	**5**	**Tb**	**4,585.51**	**—**	**2*h***
U	**4,583.27**	**10**	**8**	**U**	**4,585.588**	**2**	**2**
Ta	**4,583.17**	**10**	**150**	**Ir**	**4,385.59**	**—**	**8**
				Sn	**4,585.64**	**25*wh***	**—**
				Eu	**4,585.67**	**5**	**8*w***
Kr	4,582.85	(300*h*)	—	**Dy**	**4,585.72**	**—**	**3**
Xe	4,582.75	(300)	—				
Ne	4,582.45	(150)	—	**Al**	**4,585.82**	**(40)**	**—**
Yb	4,582.36	6	80	**Tb**	**4,585.87**	**—**	**2**
Ne	4,582.03	(150)	—	**Ca**	**4,585.871**	**10**	**125**
				Ne	**4,585.88**	**(10)**	**—**
Mn	4,581.83	—	125	**Eu**	**4,585.89**	**—**	**2*w***
Nb	4,581.62	50	30				
Co	4,581.60	10	1000*w*				
Ca	4,581.40	10	100	Cu	4,586.95	80*w*	250*w*
Xe	4,580.70	(40*wh*)	—	P	4,587.90	(300*w*)	—
				Cr	4,588.22	600*h*	10
Ta	4,580.69	10	200*W*	Co	4,588.70	1	100
Co	4,580.14	3	300	Ar	4,589.29	(80)	—
Ru	4,580.07	—	25				
La	4,580.057	100	80	P	4,589.78	(300*w*)	—
Cr	4,580.056	125	300	Ar	4,589.93	(150)	—
				Ti	4,589.95	100	40
Ba	4,579.67	40	75	O	4,590.94	(300)	—
Ar	4,579.39	(80)	—	Ru	4,591.10	—	60

	λ	I	
		J	O
Ca	4,578.56	5	80
In	4,578.39	(60)	—
In	4,578.09	(50)	—
Rn	4,577.72	(250)	—
Sm	4,577.69	50	100
Kr	4,577.20	(800)	—
Xe	4,577.06	(100wh)	—
Mo	4,576.50	40	40
Yb	4,576.20	10	90
Br	4,575.75	(100)	—
Ne	4,575.06	(300)	—
La	4,574.87	300	300
Ta	4,574.31	20	300
U	4,573.69	40	30
Ne	4,573.56	(50)	—
Ta	4,573.29	2h	200
Be	4,573.1	40	—
Nb	4,573.07	50	30
Cl	4,572.13	(100)	—
Ti	4,571.98	300	150
Cr	4,571.68	40	50
Co	*4,570.02*	—	*300*
Ta	*4,565.85*	*15*	*200*
Co	*4,565.59*	*12*	*800W*
P	*4,565.21*	*(100)*	—
V	*4,564.59*	*150*	—
Se	*4,563.95*	*(200)*	—
Ti	*4,563.77*	*200*	*100*
Ru	*4,562.60*	—	*15*
Ru	*4,559.98*	—	*20*
La	*4,559.29*	*150*	*100*
Cr	*4,558.66*	*600wh*	*20*
La	*4,558.46*	*200*	*100*
P	*4,558.03*	*(100)*	—
Te	*4,557.84*	*(300)*	—

	λ	I	
		J	O
Cr	4,591.39	125	100
Ru	4,591.56	—	20
Sm	4,591.83	—	100
Xe	4,592.05	(1500wh)	—
Ru	4,592.52	—	100
Ni	4,592.53	2	200
Fe	4,592.65	50	200
Kr	4,592.80	(150wh)	—
Cs	4,593.18	50R	1000R
Ne	4,593.24	(50)	—
Sm	4,593.53	50	50
Th	4,593.64	50wh	—
Eu	4,594.02	200	500R
Co	4,594.63	—	400
Os	4,595.04	4	80
Mo	4,595.16	40	40
Ne	4,595.25	(50)	—
Sm	4,595.30	60	100
Cr	4,595.59	60	50
K	4,595.61	(40)	—
Ar	4,596.10	(1000)	—
O	4,596.13	(150)	—
Ru	4,596.71	—	20
Sb	4,596.90	(70)	—
Co	4,596.905	—	400
Os	4,597.16	4	100
Gd	4,597.92	40	25
Yb	4,598.37	70	25
Ru	*4,599.08*	—	*100*
Ni	*4,600.37*	—	*200*
Cr	*4,600.75*	*150*	*150*
Ta	*4,601.42*	*100wh*	*60*
N	*4,601.49*	*(100)*	—
Ru	*4,601.76*	—	*20*
P	*4,601.96*	*(300w)*	—
Te	*4,602.37*	*(800)*	—
As	*4,602.73*	*200*	—
Ru	*4,602.81*	—	*15*
Li	*4,602.86*	—	*800*
Fe	*4,602.94*	*100*	*300*
Xe	*4,603.03*	*(300h)*	—
Se	*4,604.34*	*(300)*	—
Rn	*4,604.40*	*(200)*	—
Ta	*4,604.85*	—	*200W*
Ni	*4,604.99*	*10h*	*300*
Mn	*4,605.36*	*15*	*150*
Ru	*4,605.66*	—	*15*
La	*4,605.78*	*100*	*100*

	λ	I J	I O

	λ	I J	I O
Sr	*4,607.33*	*50R*	*1000R*
As	*4,607.46*	*200*	—
Rn	*4,609.38*	*(250)*	—
Ar	*4,609.60*	*(300)*	—
Ne	*4,609.91*	*(150)*	—
Fe	*4,611.29*	*25*	*200*
Xe	*4,611.89*	*(700)*	—

S 16 32.064

t_0 112.8°C t_1 444.6°C

I.	II.	III.	IV.	V.
10.357	23.405	35.048	47.294	62.2

λ	I		eV
	J	O	
IV 3,097.46	—	50	26.49
III 3,497.34	(100)	—	—
II 4,153.10	(600)	—	18.88
I 4,694.13	(50)	—	9.16
I 4,695.45	(30)	—	9.16
I 9,212.91	(200)	—	7.87
I 9,228.11	(200)	—	7.86
I 9,237.49	(200)	—	7.86

S 3,097.46 *O* 50

	λ	*I*	
		J	*O*
Eu	3,097.456	5	200*W*
Mo	3,097.38	—	10
Ir	3,097.35	—	2*h*
Th	3,097.27	8	10
Ru	3,097.23	—	30
Mo	3,097.20	—	20
Ti	3,097.19	150	20
Nb	3,097.122	100*W*	3*W*
Ni	3,097.118	50	200
Mn	3,097.06	40	75*w*
Tm	3,096.97	50	15
Ir	3,096.941	—	2*h*
Pr	3,096.94	—	2
Mg	3,096.899	25	150
Ca	3,096.898	2*h*	1
Ce	3,096.883	—	20
U	3,096.88	5*d*	5*d*
Sm	3,096.877	4	10
Tb	3,096.86	8	3
Fe	3,096.84	20	30
Nd	3,096.83	2	2
Ir	3,096.82	—	5
Sc	3,096.80	3	7*d*
Hf	3,096.76	1	25
Ce	3,096.73	—	2
Co	3,096.70	3	60
Ru	3,096.57	60	70
Cr	3,096.53	—	35
Ce	3,096.50	—	25
Co	3,096.40	3	60
Er	3,095.88	8	25
Cr	3,095.86	3	125
Re	3,095.81	—	30*w*
Co	3,095.72	2	60
Mo	3,095.70	—	25
Ta	3,095.39	18*w*	70*w*
Re	3,095.06	—	40
Fe	3,094.90	15	30
V	3,094.69	—	40
Mo	3,094.66	25	150
Ru	3,094.39	3	50
Nb	3,094.18	1000	100
Os	3,094.07	8	30
Cu	3,093.99	50	150
Ru	3,093.90	100	30
Fe	3,093.88	30	40
Fe	3,097.51	1*h*	1*h*
Ru	3,097.60	18	2
Ti	3,097.63	5	1
Mo	3,097.69	30	10
Pr	3,097.78	2	15
Fe	3,097.815	10*h*	20*h*
Ir	3,097.815	10	12
Ru	3,097.88	15	6
Th	3,097.96	50*h*	1
Ag	3,097.98	—	5
Sm	3,098.00	2	1
Tm	3,098.01	20	5
U	3,098.011	10	10
Dy	3,098.05	—	4
Ce	3,098.15	—	5
Eu	3,098.17	5	25*d*
Fe	3,098.19	60	70
Co	3,098.20	5	100*r*
Tm	3,098.59	60	80
Ni	3,099.11	50	200
Os	3,099.26	8	40
Ru	3,099.28	60	70
Co	3,099.67	—	50
Fe	3,099.90	60	60
Mo	3,099.932	3	25
Cu	3,099.934	10	60
Fe	3,099.97	40	40
Ir	3,100.29	2	30
Mn	3,100.302	60	60
Fe	3,100.304	100	100
Gd	3,100.51	80	100
Ti	3,100.666	15	30
Re	3,100.666	—	100
Fe	3,100.666	100	100
Ru	3,100.84	50	70
Mo	3,100.87	2	40
Pr	3,101.27	2	40
Mo	3,101.34	10	80
Hf	3,101.40	90	60
Os	3,101.53	20	125
Ni	3,101.55	150	1000*R*
Mn	3,101.56	50	50
Ni	3,101.88	150	400*R*
K	3,102.03	—	50*R*
V	3,102.30	300*R*	70
Co	3,102.41	4	60

	λ	I J	I O
Ta	3,093.87	15	50
Fe	3,093.81	40	50
V	3,093.79	—	30
Re	3,093.65	—	60
Os	3,093.59	15	125
Fe	3,093.36	40	70
Tm	3,093.12	60	30
V	3,093.11	400R	100R
Mg	3,092.99	20	125
Al	3,092.84	18	50R
Fe	3,092.78	30	50
Na	3,092.73	(200)	50
V	3,092.72	50r	100r
Al	3,092.71	1000	1000
Ta	3,092.44	15	50
Mo	3,092.07	100	30
Ru	3,091.87	5	50
Fe	3,091.58	200	300
Os	3,091.25	15	40
Mg	3,091.08	10	80
Ru	3,090.90	1	30
Os	3,090.49	15	80
Os	3,090.30	12	100
Co	3,090.25	1	80
Ru	3,090.23	6	50
Fe	3,090.21	15	30
Os	3,090.08	15	100
Ru	*3,089.80*	*5*	*60*
Co	*3,089.59*	—	*1000r*
Ru	*3,089.14*	*12*	*60*
Re	*3,088.77*	—	*60*
bhB	*3,088.6*	—	*100*
Os	*3,088.27*	*12*	*60*
Ir	*3,088.04*	*2*	*50*
Ti	*3,088.02*	*500R*	*70*
Co	*3,087.81*	—	*60*
Os	*3,087.75*	*10*	*50*
Co	*3,086.78*	—	*200R*
Co	*3,086.40*	*2*	*80*
Os	*3,086.27*	*10*	*50*
Ru	*3,086.07*	*6*	*60*
Mo	*3,085.61*	*25*	*125*
Ta	*3,085.53*	*18*	*70*
Os	*3,084.60*	*10*	*60*
Rh	*3,083.96*	*2*	*150*
Fe	*3,083.74*	*500*	*500*
V	*3,083.54*	—	*50*
Ru	*3,083.15*	*3*	*50*
Co	*3,082.62*	*50*	*150R*

	λ	I J	I O
Gd	3,102.56	25	25
Fe	3,102.87	20	30
Tb	3,102.97	15	30
Dy	3,103.246	1	30
Ta	3,103.251	15	70
Co	3,103.74	2	80
Dy	3,103.84	10	30
Co	3,103.98	—	60r
Ru	3,104.46	—	30
La	3,104.59	50	200
Re	3,104.65	—	30
Os	*3,104.98*	*15d*	*200d*
Ru	*3,105.28*	*40*	*50*
Ru	*3,105.41*	*1*	*50*
Ni	*3,105.47*	*35*	*200*
Os	*3,105.99*	*20*	*150*
Ru	*3,106.84*	*3*	*50*
Co	*3,107.04*	*3*	*70*
Ru	*3,107.71*	*5*	*60*
Re	*3,108.81*	—	*125*
Os	*3,108.98*	*15*	*125*
Hf	*3,109.12*	*100*	*50*
Os	*3,109.38*	*20*	*125*
Co	*3,109.51*	*1h*	*60*
Co	*3,110.02*	*2*	*60*
Ru	*3,110.55*	*6*	*60*
V	*3,110.71*	*300R*	*70*
Co	*3,110.82*	*2*	*60*
Re	*3,110.86*	—	*100*
Os	*3,111.09*	*20*	*100*
Eu	*3,111.43*	—	*200*
Ru	*3,111.91*	*5*	*50*

S 3,497.34 (100)

	λ	I J	I O
U	3,497.265	4	6
Th	3,497.264	5	6
Nd	3,497.25	6	10
Hf	3,497.16	4	10
Dy	3,497.111	3	3
Fe	3,497.108	100	200
U	3,497.07	3	3
Er	3,497.03	1h	3
V	3,497.030	150	—
Th	3,497.02	5	6
U	3,496.91	4	1
Er	3,496.86	20	25
Th	3,496.812	3	5
Mn	3,496.807	30	10
Dy	3,496.72	2	3
Mo	3,496.70	3	3
Co	3,496.68	4	150R
U	3,496.41	15	8
O	3,496.32	(5)	—
Nb	3,496.28	1w	3d
Zr	3,496.21	100	100
Mn	3,495.84	150	25
Cd	3,495.34	(100)	—
Fe	3,495.29	60	100
Gd	3,494.42	60	70
Hg	3,493.85	(100)	—
Fe	3,493.47	80	40
V	3,493.17	100	15
Ni	3,492.96	100h	1000R
Hg	3,492.77	(50)	—
Rb	3,492.76	(300)	—
Ar	3,491.54	(50)	—
Nb	3,491.03	50	30
Fe	3,490.57	300	400
P	3,490.44	(70)	—
Nb	3,489.09	50	5
P	3,488.77	(70)	—
Mn	3,488.68	200	50
Pt	*3,485.27*	*200R*	*150*
Nb	*3,484.05*	*100*	*10*
Nb	*3,482.95*	*100*	*2*
Mn	*3,482.91*	*250*	*50*
Gd	*3,481.36*	*150*	*150*
Ta	*3,480.52*	*200ws*	*70*
Nb	*3,479.56*	*200*	*5*

	λ	I J	I O
J	3,497.38	(7)	—
V	3,497.39	12	—
Kr	3,497.45	(3h)	—
Hf	3,497.49	10	20
Mn	3,497.54	150	15
U	3,497.62	12	6
Th	3,497.70	3	3
Nb	3,497.81	15	30
Dy	3,497.841	3	30
Fe	3,497.843	200	200
Ta	3,497.85	5	70
Zr	3,497.92	10	—
Ru	3,497.94	5	30
Th	3,498.02	8	8
J	3,498.03	(25)	—
Cu	3,498.063	5	20
Ne	3,498.064	(75)	—
Os	3,498.07	10	15
W	3,498.15	10	—
Pt	3,498.17	2	1
V	3,498.20	7	12
U	3,498.38	3	8
Nb	3,498.43	1	2
Nb	3,498.46	300wh	—
Nb	3,498.63	50	30
Dy	3,498.67	50	50
Rh	3,498.73	60	500
Ru	3,498.94	200	500R
V	3,499.82	50	3
Nb	3,499.95	50	5
Ni	3,500.85	80	500wh
Ne	3,501.22	(150)	—
F	3,501.42	(200)	—
Se	3,501.52	(50)	—
Co	3,501.72	100	5
Rh	3,502.52	150	1000
F	3,502.95	(60)	—
P	3,502.99	(70)	—
F	3,503.09	(400)	—
Kr	3,503.25	(50wh)	—
V	3,504.44	200	60
Ti	3,504.89	150	20
Hf	3,505.23	50	20
Gd	3,505.52	60	60
F	3,505.61	(600)	—
Rh	3,507.32	125	500

	λ	I	
		J	O
Rh	*3,478.91*	*100*	*500*
Yb	*3,478.84*	*300*	*40*
V	*3,477.52*	*100*	—
Ti	*3,477.18*	*100*	*60*
Fe	*3,476.70*	*200*	*300*
P	3,507.36	(100*w*)	—
Lu	3,507.39	150	100
V	3,507.54	50	—
Yb	3,507.83	60	12
V	*3,509.04*	*150*	*2*
Tb	*3,509.17*	*200*	*200*
Ru	*3,509.20*	*100*	*10*
Nb	*3,510.26*	*200*	*15*
Ti	*3,510.84*	*125*	*40*
Co	*3,512.64*	*100*	*400R*
Ir	*3,513.64*	*100*	*100h*
Fe	*3,513.82*	*300*	*400*
Ar	*3,514.39*	*(125)*	—
Ne	*3,515.19*	*(150)*	—
Nb	*3,515.42*	*300*	*20*
Pd	*3,616.94*	*500R*	*1000R*
Nb	*3,517.67*	*200*	*2*
Co	*3,518.35*	*100*	*200W*

S 4,153.10 (600)

	λ	J	O
Cr	**4,153.07**	**6**	**40**
Pb	**4,152.93**	**5**	—
Ce	**4,152.928**	**1**	**4**
La	**4,152.775**	**50**	**40**
Cr	**4,152.775**	**12**	**50**
Ho	**4,152.75**	**40**	**30**
V	**4,152.66**	**4**	**9**
Nb	**4,152.58**	**300**	**100**
Ho	**4,152.54**	**30**	**30**
Ar	**4,152.54**	**(20)**	—
Dy	**4,152.43**	**4**	**15*r***
P	**4,152.4**	**(15)**	—
Sc	**4,152.35**	**4**	**10**
Se	**4,152.34**	**(80)**	—
Pr	**4,152.34**	**4**	**5**
Eu	**4,152.25**	**1*h***	**2**
Sm	**4,152.21**	**15**	**8**
Fe	**4,152.17**	**5**	**70**
Sm	**4,152.07**	**3**	**3**
La	4,151.95	300	200
N	4,151.46	(1000)	—
S	4,146.94	(30)	—
S	4,145.10	(250)	—
Ce	**4,153.13**	**2**	**12**
Mo	**4,153.17**	**2**	**10**
O	**4,153.31**	**(200)**	—
Sm	**4,153.326**	**10**	**5**
V	**4,153.328**	**3**	**12**
Fe	**4,153.40**	**1*h***	**1**
Eu	**4,153.44**	**2**	**5**
U	**4,153.48**	**8**	**8**
Co	**4,153.62**	**2*h***	—
Ca	**4,153.65**	**6**	—
Nd	**4,153.73**	**4**	**3**
U	**4,153.76**	**2**	**2*h***
Cr	**4,153.82**	**30**	**50**
Fe	**4,153.91**	**100**	**120**
Ce	**4,153.92**	**3**	**12**
Se	**4,153.93**	**(40)**	—
U	**4,153.98**	**5**	**15**
Pr	**4,154.05**	**2**	**10**
Lu	**4,154.08**	**3**	**40**
Te	**4,154.14**	**(15)**	—
Ar	4,158.59	(1200)	—
S	4,152.39	(18)	—
S	4,162.70	(600)	—

	λ	I J	I O		λ	I J	I O
S	*4,142.29*	*(150)*	—	*Ar*	*4,164.18*	(1000)	—
				S	*4,168.41*	*(50)*	—
				Ga	*4,172.06*	*1000*	*2000R*

S 4,694.13 (50)
4,695.45 (30)

	λ	J	O		λ	J	O
Kr	4,694.84	(4)	—	Th	4,694.92	2	3
W	4,694.66	1	12	Sb	4,694.95	(6)	—
N	4,694.55	(10*h*)	—	U	4,695.12	2	1
Nb	4,694.51	8	5	Cr	4,695.15	4	50
Kr	4,694.44	(200*h*)	—	U	4,695.23	6	8
Gd	4,694.34	8	50	Eu	4,695.34	1	10*W*
Tm	4,694.27	3	10	Ne	4,695.36	(20)	—
S	*4,694.13*	*(50)*	—	*S*	*4,695.45*	*(30)*	—
Th	4,694.09	10	15	Nb	4,695.47	8	5
Cr	4,693.95	20	50	Gd	4,695.49	10	25
Mo	4,693.93	25	20	Cs	4,695.61	(10)	—
W	4,693.73	12	50	U	4,695.64	3	1
Ti	4,693.68	3	25	Kr	4,695.66	(50*h*)	—
Dy	4,693.67	2	3	Pr	4,695.77	2	60
Nd	4,693.63	1	20	Mo	4,695.86	6	4
Ta	4,693.35	3	150	U	4,696.08	1	2
Mo	4,693.342	3	2	Hg	4,696.25	(5)	—
Xe	4,693.34	(10*h*)	—	S	4,696.25	(15)	—
Br	4,693.27	(40)	—	Bi	4,696.3	7	—
Co	4,693.21	25	500	O	4,696.32	(30)	—
Dy	4,692.74	2	3	Te	4,696.35	(50)	—
Mo	4,692.69	4	2	Br	4,696.43	(2)	—
				Mo	4,696.51	10	8
La	4,692.50	300	200	Gd	4,696.61	2	25
Ba	4,691.62	40	100	Yt	4,696.80	3	5
Ne	4,691.58	(15)	—				
O	4,691.37	(15)	—				
Ti	4,691.34	25	125	Xe	4,697.02	(300)	—
				Xe	4,698.01	(150*h*)	—
Kr	4,691.28	(100)	—	Sc	4,698.28	15	5
La	4,691.18	25	10	P	4,698.56	(30)	—
Tm	4,691.10	20	35	Ti	4,698.76	20	100
Xe	4,690.97	(100)	—				
Mo	4,690.86	25	20	O	4,698.99	(30)	—
				O	4,699.21	(100)	—
Cr	4,689.37	35	80	Ra	4,699.28	(40)	—
U	4,689.07	40	30	Nb	4,699.55	15*h*	—
La	4,688.66	20	6	La	4,699.63	200*r*	200*r*
Mo	4,688.22	30	25				
Ne	4,687.67	(100)	—	Kr	4,699.69	(30*wh*)	—

	λ	I J	I O
J	4,687.43	(15)	—
Te	4,686.95	(300)	—
He	4,685.75	(300)	—
U	4,685.72	18	10
In	4,685.22	(100)	—
Nb	4,685.13	20	15
In	4,684.93	(15)	—
In	4,684.76	(25*h*)	—
In	4,684.59	(25)	—
In	4,684.45	(20)	—
In	4,684.32	(35)	—
Yb	4,683.82	20	7
Ne	4,683.76	(30)	—
Gd	4,683.34	50	100
Ta	4,683.10	15	—
Yt	4,682.32	100	60
Ra	4,682.28	(800)	—
Ne	4,682.15	(20)	—
In	4,682.00	250*W*	—
Cu	4,681.99	20	—
Ne	4,681.93	(20)	—
Ti	4,681.92	100	200
Ta	4,681.87	50	200
Ne	4,681.20	(50)	—
In	4,681.11	(200)	—
Te	4,681.06	(15*h*)	—
Rn	4,680.83	(500)	—
Cr	4,680.54	25	50
W	4,680.52	40	150
Kr	4,680.41	(500)	—
Ne	4,680.36	(100)	—
Zn	4,680.14	200*h*	300*w*
Gd	4,680.06	25	50
Ne	*4,679.13*	*(150)*	—
P	*4,678.94*	*(100)*	—
Fe	*4,678.85*	*100*	*150*
Br	*4,678.69*	*(200)*	—
Ne	*4,678.60*	*(50)*	—
Ne	*4,678.22*	*(300)*	—
In	*4,678.17*	*30*	—
Gd	*4,678.16*	*200 W*	*200 W*
J	*4,676.94*	*(80)*	—
Sm	*4,676.91*	*50*	*100*
Xe	*4,676.46*	*(100wh)*	—
O	*4,676.25*	*(125)*	—
P	*4,675.78*	*(70)*	—
J	*4,675.53*	*(50)*	—
Rh	*4,675.03*	*50*	*100*

	λ	I J	I O
Hf	4,699.71	25	20
Mo	4,700.49	25	25
P	4,700.79	(50)	—
O	4,701.16	(20)	—
Rn	4,701.70	(50)	—
Cs	4,701.79	(25)	—
U	4,702.05	18	8
Ar	4,702.316	(200)	—
Gd	4,702.323.	100	50
U	4,702.52	20	10
Ne	4,702.53	(150)	—
Gd	4,703.136	20	50
O	4,703.14	(30)	—
La	4,703.28	300*r*	200*r*
Ne	4,704.39	(1500)	—
Cu	4,704.60	50	200
Hg	4,704.63	(200)	—
Br	4,704.86	(250)	—
O	4,705.32	(300)	—
Bi	4,705.35	50	—
Mo	4,706.06	25	25
Nb	4,706.14	50	50
Te	4,706.53	(70)	—
Mo	4,707.25	125	125
As	4,707.82	200	—
Gd	4,707.89	30	30
Cr	4,708.04	150	200
Mo	4,708.22	30	30
Nb	4,708.29	30	50
Ti	4,708.66	20	2
Ne	4,708.85	(1200)	—
Ba	4,708.94	(80)	—
Fe	4,708.96	50	50
Sc	4,709.34	15*h*	15
Ru	4,709.48	80	150
Ar	4,709.50	(30)	—
Mn	4,709.71	15	150
Gd	4,709.78	25	100
O	4,710.00	(60)	—
Ne	4,710.06	(1000)	—
Ne	*4,710.48*	*(30)*	—
Te	*4,711.16*	*(70)*	—
Sb	*4,711.26*	*(100)*	—
Gd	*4,711.98*	*30*	*30*
Ne	*4,712.06*	*(1000)*	—
Gd	*4,712.819*	*40*	*25*
Yb	*4,712.82*	*30h*	—

	λ	I J	I O		λ	I J	I O
Yt	4,674.85	100	80	La	4,712.93	150	100r
Cu	4,674.76	30W	200	Ne	4,713.13	(100)	—
Sm	4,674.59	40	80	He	4,713.14	(40)	—
O	4,673.71	(30)	—				
Be	4,673.46	(100)	—	Fe	4,714.07	50	50
				Ne	4,714.34	(70)	—
O	4,672.75	(30)	—	Ne	4,715.13	(30)	—
As	4,672.70	50	—	Xe	4,715.18	(80)	—
Xe	4,672.20	(50h)	—	Ne	4,715.25	(30)	—
Be	4,672.2	100	—				
Nb	4,672.09	100	150	Ne	4,715.34	(1500)	—
				Nb	4,715.83	50h	3
Mo	4,671.90	30	30	S	4,716.23	(600)	—
La	4,671.83	150	100	La	4,716.44	200	100
U	4,671.41	30	20	Te	4,716.81	(30)	—
Xe	4,671.23	(2000)	—				
Ne	4,670.88	(70)	—	Ne	4,717.61	(70)	—
				Mo	4,717.92	50	50
V	4,670.49	40r	60R	Cr	4,718.43	150	200
Sc	4,670.40	300wh	100	Hf	4,719.10	40	30
Te	4,670.11	(30)	—	Br	4,719.77	(80)	—
Sm	4,669.65	40	50				
Sm	4,669.39	35	40	La	4,719.95	300	200r
				P	4,720.26	(30)	—
S	4,669.14	(35)	—	Co	4,721.41	100h	8
Ne	4,669.02	(50)	—	Ne	4,721.54	(70)	—
La	4,668.91	300r	200r	Rn	4,721.76	(150)	—
Na	4,668.60	100	200				
S	4,668.58	(50)	—	Zn	4,722.16	300h	400w
				Bi	4,722.55	100	1000
Xe	4,668.49	(50)	—	U	4,722.73	50	40
Ag	4,668.48	70	200	Gd	4,723.73	200	100
Se	4,667.80	(70)	—	P	4,724.25	(75)	—
Ne	4,667.36	(200)	—				
U	4,666.86	40	25				
Ne	4,666.65	(50)	—				
J	4,666.52	(250)	—				
Te	4,665.33	(70)	—				
Se	4,664.98	(150)	—				

S 9,212.91 *J* 200

	λ	J	O		λ	J	O
Ne	**9,212.9**	**(2)**	—	**In**	**9,212.95**	**(4)**	—
In	**9,212.69**	**(3)**	—	**In**	**9,213.28**	**(6)**	—
In	**9,212.47**	**(2)**	—	**In**	**9,213.66**	**(8)**	—
Xe	**9,212.38**	**(25)**	—	**Se**	**9,213.30**	**(2)**	—
Se	**9,211.03**	**(2*h*)**	—				
Ne	9,201.76	(600)	—				
Ar	9,194.68	(150)	—				

S 9,228.11 *J* (200)
9,237.49 *J* (200)

	λ	I				λ	I	
		J	O				J	O
H	*9,229.7*	(*4*)	—		*Kr*	***9,233.18***	(***12***)	—
S	*9,228·11*	(*200*)	—		*S*	***9,237.49***	(***200***)	—
Ne	**9,226.67**	**(200)**	—		**Kr**	**9,238.48**	**(125)**	—
Xe	**9,226.39**	(7*h*)	—					
Ar	**9,224.50**	**(1000)**	—					
Ne	**9,221.59**	**(200)**	—					
Ne	**9,220.05**	**(400)**	—					

*P**

Sb 51 121.75

t_0 630.5°C t_1 1,440°C

I.	II.	III.	IV.	V.
8.64	18.6	24.825	44.147	55.69

λ	*I*		eV
	J	*O*	
I 2,068.38	3	300*R*	5.98
I 2,175.89	40	300	5.69
I 2,528.54	200	300*R*	6.12
I 2,877.92	150	250	5.36
I 3,232.50	250	150	6.12
I 3,267.50	150	150	5.82
I 5,568.09	200	6	7.92
II 6,004.6	200	—	11.46

Sb 2,068.38 *J* 3* *O* 300*R*

	λ	*I* *J*	*I* *O*		λ	*I* *J*	*I* *O*
Ni	2,068.35	—	2	Os	2,068.38	10	10
Rh	2,068.32	—	8	Hf	2,068.58	—	5
Re	2,068.30	—	4*wh*	Ni	2,068.61	—	10
Ir	2,068.25	2	10				
Pt	2,068.16	25	2				
Sb	2,063.44	12	30				

Sb 2,175.89 *J* 40* *O* 300

	λ	*J*	*O*		λ	*J*	*O*
Nb	2,175.84	25	5	Fe	2,176.03	—	3
V	2,175.835	5	2				
Re	2,175.83	5	25				

Sb 2,528.54 *J* 200 *O* 300*R*

	λ	*J*	*O*		λ	*J*	*O*
Si	2,528.52	500	400	Ta	2,528.59	1	4
Ba	2,528.51	50*r*	—	Lu	2,528.59	5*h*	—
Dy	2,528.49	—	4	Co	2,528.61	200	4
Xe	2,528.49	(3)	—	Mn	2,528.70	2	12
V	2,528.47	150*R*	50	Ar	2,528.71	(5)	—
Ir	2,528.39	10*h*	—	Ru	2,528.715	—	30
Th	2,528.36	5	—	U	2,528.77	2	5
Ar	2,528.33	(10)	—	Cs	2,528.8	(8)	—
Yb	2,528.33	3	—	V	2,528.84	150*R*	25
Ce	2,528.29	1	15	Mo	2,528.866	25	1
Cr	2,528.24	—	30	Ru	2,528.878	2	60
Tb	2,528.23	20	—	Fe	2,528.879	2	2
Co	2,528.18	40	—	Cl	2,528.88	(8)	—
Fe	2,528.17	—	6	W	2,528.91	15	3
In	2,528.17	5	—				
W	2,528.16	6	—	Cu	2,529.30	600	—
				Fe	2,529.55	100	15
				Co	2,530.13	300	40*W*
V	2,527.90	300*R*	35	Tl	2,530.67	(100)	—
Mn	2,527.44	12	150	Nb	2,530.97	100	2
Fe	2,527.43	50	200*R*				
Pb	2,526.62	(100)	—	Ti	2,531.25	125	30
Cu	2,526.59	200	—	Tl	2,531.6	100	—
				Ta	2,532.12	100	80
Ta	2,526.45	—	150				
V	2,526.21	150*R*	150				
Ti	2,525.60	125	35	*Fe*	*2,535.60*	—	*1000*
Ni	2,525.39	300*Wh*	—	*Hg*	*2,536.52*	*1000R*	*2000R*
Co	2,524.96	700	50*W*				
Si	*2,524.12*	*400*	*400*				
Fe	*2,522.85*	*50*	*300R*				

Sb 2,877.92 *J* 150 *O* 250

	λ	*I*	
		J	*O*
Ca	**2,877.91**	**4**	**1*h***
Eu	**2,877.89**	**1**	**2**
Dy	**2,877.88**	**1*h***	**2**
Pd	**2,877.87**	**15*wh***	—
Nb	**2,877.85**	**4**	**2**
Ru	**2,877.84**	**1**	**3**
U	**2,877.83**	**2**	**2**
Eu	**2,877.76**	—	**5**
Cu	**2,877.689**	**20**	**5**
V	**2,877.688**	**100*R***	**15**
Ta	**2,877.686**	**80*h***	**15**
Ir	**2,877.678**	**10**	**20**
N	**2,877.66**	**8*h***	—
Nb	**2,877.62**	**5*W***	—
U	**2,877.57**	**4**	**6**
Zr	**2,877.55**	**4*h***	**4**
P	**2,877.53**	**(10)**	—
Pt	**2,877.52**	**200*h***	**40**
Ti	**2,877.44**	**100**	**30**
Os	**2,877.35**	**2**	**30**
Fe	**2,877.30**	**125**	**200**
Cs	**2,877.29**	**(8)**	—
Nb	2,876.95	500*W*	40*W*
Fe	2,876.80	100	—
Hf	2,876.33	100	30
Cr	2,876.24	80*wh*	25
Cr	2,875.99	80*wh*	30
Pt	2,875.85	80*h*	20
Nb	2,875.39	300	50*r*
Fe	2,874.172	200	300
Ta	2,874.167	15	150
Ag	2,873.65	100*wh*	3
Ta	2,873.56	50	150
Cr	2,873.48	125	30
Fe	2,873.40	300	—
Ta	2,873.36	40*h*	200*W*
Fe	2,872.34	50	150
Mo	2,871.51	100*h*	100
Cr	*2,870.44*	*300W*	*25*
Fe	*2,869.31*	*70*	*300*
Tm	*2,869.22*	*300*	*100*
V	*2,869.13*	*150r*	*25*
Nb	*2,868.52*	*300*	*15*
Ta	*2,867.41*	*150*	*5h*
Cr	**2,877.98**	**100**	**30**
Ce	**2,878.018**	—	**2**
V	**2,878.022**	**10**	**2**
Ru	**2,878.04**	**15**	—
W	**2,878.08**	**8**	**4**
Nb	**2,878.16**	—	**2**
Ta	**2,878.20**	**15*h***	**4**
Tm	**2,878.21**	**20**	**3**
U	**2,878.24**	**2**	**5**
V	**2,878.299**	**7**	—
W	**2,878.30**	**10**	**1**
Tm	**2,878.36**	**20**	**10**
Mo	**2,878.38**	—	**20**
Os	**2,878.40**	**12**	**40**
Cr	**2,878.45**	**80**	**20**
Sm	**2,878.45**	**25**	**4**
Xe	**2,878.48**	**2*wh***	—
Ir	**2,878.51**	—	**4**
Tb	**2,878.52**	**10**	—
J	2,878.64	(400)	—
Mo	2,879.05	100*h*	15
Ta	2,879.74	10	150
Ta	2,880.02	50	150
V	2,880.03	150*r*	25
Cd	2,880.77	125	200*R*
Si	2,881.58	400	500
Ru	2,882.12	200	30
Ta	2,882.33	80	3
V	2,882.50	200*r*	35
Nb	2,883.18	800*R*	100
Fe	2,883.70	300	—
Ti	2,884.11	125	35
V	*2,884.78*	*200r*	*40*
V	*2,889.62*	*150r*	*40*

Sb 3,232.50 *J* 250 *O* 150

	λ	*I*			λ	*I*	
		J	*O*			*J*	*O*
Sm	**3,232.497**	—	**4**	**Rh**	**3,232.504**	—	**6**
W	**3,232.49**	**8**	**9**	**Os**	**3,232.54**	**10**	**150**
Ne	**3,232.38**	**(7)**	—	**Cl**	**3,232.58**	**(4)**	—
Cd	**3,232.36**	**2**	—	**Li**	**3,232.51**	**500**	**1000*R***
Pb	**3,232.35**	—	**30**	**Sm**	**3,232.62**	**3**	**10**
Pd	**3,232.32**	—	**2**	**Te**	**3,232.63**	**10**	—
Eu	**3,232.31**	**1**	**4*W***	**Dy**	**3,232.652**	**4**	**15**
Th	**3,232.308**	—	**8**	**W**	**3,232.652**	**8**	**9**
Ce	**3,232.29**	—	**15**	**Ce**	**3,232.66**	—	**3**
Ti	**3,232.28**	**100**	**30**	**Ru**	**3,232.75**	**4**	**50**
Ta	**3,232.279**	**1**	**25**	**J**	**2,232.78**	**(5)**	—
Se	**3,232.27**	**(8)**	—	**Fe**	**3,232.791**	**50**	—
W	**3,232.23**	**2**	**3**	**Ti**	**3,232.791**	—	**8**
U	**3,232.16**	**12**	**12**	**Nb**	**3,232.793**	**2*h***	—
Kr	**3,232.15**	**(2*h*)**	—	**Kr**	**3,232.80**	**(2)**	—
W	**3,232.13**	**9**	**6**	**Co**	**3,232.874**	**25**	**60**
Th	**3,232.12**	**24*h***	**3**	**Ce**	**3,232.875**	—	**3**
Os	**3,232.05**	**20**	**500**	**Ni**	**3,232.96**	**35**	**300*R***
Yt	**3,232.027**	**3**	**3**	**Ca**	**3,233.02**	**4**	—
Er	**3,232.026**	**3**	**18**	**Fe**	**3,233.05**	**60**	**100**
Ir	**3,232.001**	**1**	**20**	**Zr**	**3,233.11**	—	**6**
Tb	**3,232.00**	**3**	**8**	**Mo**	**3,233.140**	**30**	**50**
Yb	**3,231.98**	**6**	**2**	**W**	**3,233.142**	**6**	**2**
Ce	**3,231.980**	—	**3**	**Ag**	**3,233.15**	**10**	—
V	**3,231.95**	**100**	**8**	**Ni**	**3,233.17**	—	**4**
Sm	**3,231.94**	**8**	**20**	**V**	**3,233.19**	**3**	**40**
Eu	**3,231.86**	—	**3*W***	**Ce**	**3,233.21**	—	**2**
Ce	**3,231.81**	—	**2**	**Cr**	**3,233.23**	**4**	**30**
Fe	**3,231.71**	**30**	—	**W**	**3,233.24**	**9**	**2**
Cl	**3,231.70**	**(5)**	—	**Ag**	**3,233.25**	**2**	**5**
Zr	**3,231.69**	**10**	**10**	**Rh**	**3,233.32**	**5**	**2**
Ta	**3,231.66**	**1*h***	**18*h***	**Th**	**3,233.33**	**8**	**8**
Os	3,231.42	12	150	Mn	3,233.968	—	75*wh*
Fe	3,230.97	200	300	Fe	3,233.971	150	300
Ta	3,230.85	18*w*	200	Cr	3,234.06	150	10
Mn	3,230.72	75	75	Os	3,234.20	12	150
Sm	3,230.54	30	100	Ti	3,234.52	500*r*	100
Pt	3,230.29	6	100	Fe	3,234.61	125	200
Fe	3,230.21	80	100	Ni	3,234.65	15	300
Tl	3,229.75	800	2000	Os	3,234.73	10	100
Ta	3,229.24	70*w*	300*w*	Tm	3,235.45	40	80
Os	3,229.21	5	125	Fe	3,236.22	200	300
Fe	3,229.12	50	80	Nb	3,236.40	200	10
Fe	3,228.90	40	80	Ti	3,236.57	300*r*	70
Ru	3,228.53	150	50	Sm	3,236.63	40	100
Fe	3,228.25	80	100	Mn	3,236.78	75	75

	λ	I J	I O		λ	I J	I O
Mn	3,228.09	100	100	Tm	3,236.80	80	100
Fe	3,227.75	300	200	Co	3,237.03	—	100
Os	3,227.28	12	125	Nb	3,238.02	200	20
Co	3,226.985	—	80*r*	Ru	3,238.53	45	100
Ni	3,226.984	—	100	Os	3,238.63	20	100
Ca	3,225.90	10	80	Ti	3,239.04	300*R*	60
Fe	3,225.79	150	300	Fe	3,239.44	300	400
Nb	3,225.48	800*wr*	150*w*	Sm	3,239.64	25	100
Ni	3,225.02	6	300	Ta	3,239.99	18*w*	200
Mn	3,224.76	40	75	Tm	3,240.23	80	100
Ti	3,224.24	150	15				
Ta	*3,223.83*	*50 W*	*200 W*	*Sb*	*3,241.28*	*(350 wh)*	—
Fe	*3,222.07*	*100*	*200*	*Tm*	*3,241.53*	*125*	*150*
Ni	*3,221.65*	*4*	*300*	*Ti*	*3,241.99*	*300 R*	*60*
Fe	*3,219.58*	*125*	*200*	*Pd*	*3,242.70*	*600 R*	*2000 wh*
Pd	*3,218.97*	*8*	*300*	*Ni*	*3,243.06*	*15*	*400 R*
Fe	*3,217.38*	*125*	*200*	*Fe*	*3,244.19*	*200*	*300*
Fe	*3,215.94*	*150*	*300*	*La*	*3,245.12*	*300*	*400*
				Fe	*3,245.98*	*150*	*200*
				Cu	*3,247.54*	*2000 R*	*5000 R*
				Fe	*3,248.21*	*150*	*200*
				Ni	*3,248.46*	*2*	*150*
				La	*3,249.35*	*80*	*300*

Sb 3,267.50 *J* 150 *O* 150

	λ	I J	I O		λ	I J	I O
Rh	**3,267.48**	**15**	**2**	**Sm**	**3,267.51**	—	**2**
W	**3,267.45**	**12**	—	**Ta**	**3,267.56**	—	**5*h***
Ti	**3,267.41**	—	**2**	**U**	**3,267.638**	**2**	**8**
Tm	**3,267.41**	**40**	**50**	**Mo**	**3,267.639**	**30**	**4**
Zr	**3,267.36**	—	**3**	**Gd**	**3,267.64**	—	**2**
Ag	**3,267.35**	**12**	—	**Nb**	**3,267.68**	**10**	**1**
Pd	**3,267.35**	**200*h***	—	**V**	**3,267.70**	**80*R***	**30**
Xe	**3,267.34**	**(4)**	—	**Dy**	**3,267.76**	—	**2**
La	**3,267.31**	**3**	**2**	**Mn**	**3,267.79**	**40**	**40**
Nd	**3,267.25**	**4**	**10**	**Yb**	**3,267.89**	—	**6**
U	**3,267.244**	**6**	**8**	**Os**	**3,267.94**	**30**	**400*R***
Ce	**3,267.237**	—	**12**	**Bi**	**3,267.97**	—	**2*h***
Os	**3,267.20**	**15**	**20**	**Ce**	**3,263.02**	—	**2**
Hf	**3,267.17**	**1**	**10**	**Se**	**3,268.05**	**(20)**	—
Er	**3,267.15**	**4**	**4**	**Ni**	**3,268.06**	—	**5**
Cs	**3,267.13**	**(30)**	—	**Re**	**3,268.08**	—	**40**
Ir	**3,267.10**	—	**2**	**Tb**	**3,268.10**	**8**	**30**
Au	**3,267.08**	**8**	**10**	**W**	**3,268.12**	**3**	**5**
Ti	**3,267.07**	—	**10**	**Mo**	**3,268.19**	**2**	**6**
Xe	**3,267.05**	**(3)**	—	**Ru**	**3,268.21**	**12**	**60**

	λ	I J	I O
Nb	3,267.050	5	5
Fe	3,267.04	4	—
Eu	3,267,01	2	4wh
Hf	3,267.009	—	8
Th	3,267.00	12	10
Si	3,266.95	2	—
U	3,266.940	1	12
Fe	3,266.938	15	—
Ta	3,266.890	1h	35r
Mo	3,266.887	40	—
Ce	3,266.86	—	3
Re	3,266.85	—	30
J	3,266.84	(5)	—
W	3,266.76	6	7
Os	3,266.75	5	10
Gd	3,266.73	—	3
La	3,265.67	200	300
Fe	3,265.62	300	300
Fe	3,265.05	150	200
Mn	3,264.71	50	75
Os	3,264.69	10	100h
Hg	3,264.06	(200)	—
Nb	3,263.37	500	3
Rh	3,263.14	40	200
Os	3,262.75	20	100
Sn	3,262.33	300h	400h
Os	3,262.29	50	500R
Tm	3,261.66	100	30
Ti	3,261.60	300r	70
Cd	3,261.06	300	300
Nb	3,260.56	300	15
Ru	3,260.35	50	100
Mn	3,260.23	50	75
Ta	3,260.18	18W	125
Fe	3,259.99	100	150
Re	3,259.55	—	100
Fe	3,259.05	200	1
Re	3,258.85	—	100
Pd	3,258.78	200h	300
Fe	3,258.77	150	—
In	3,258.56	300R	500R
Mn	3,258.41	40	75
In	3,256.09	600R	1500R
Fe	3,254.36	150	200
Lu	3,254.31	150	50
Co	3,254.21	—	300R
Pr	3,268.23	1	10
Fe	3,268.24	100	125
Cu	3,268.28	10	15
Cs	3,268.31	(10)	—
Rb	3,268.33	(2)	—
Gd	3,268.34	2	4
Pt	3,268.42	4	15
Tm	3,269.00	150	40
Os	3,269.21	20	200
Gd	3,269.49	300	300
Fe	3,271.00	300	300
Ni	3,271.12	1	125
Rh	3,271.61	60	200
Ti	3,271.65	125	35
Ti	3,272.08	100	25
Zr	3,273.05	80	50
Nb	3,273.89	100W	20r
Cu	3,273.96	1500R	3000R
Fe	3,274.45	60	80
Ta	3,274.95	35W	200
Os	3,275.20	15	200
Yb	3,275.81	100	12
V	3,276.12	200R	50
Fe	3,276.47	50	100
Fe	3,277.35	200	40
Ti	3,278.92	150	40
Fe	3,280.26	150	150
Eu	3,280.682	—	1000R
Ag	3,280.683	1000R	2000R
Ti	3,282.329	150	30
Zn	3,282.333	300	500R
Rh	3,283.57	—	150
Fe	3,284.59	125	200

	λ	I J	I O		λ	I J	I O
Nb	*3,254.07*	*300*	*20*				
Ti	*3,252.91*	*200r*	*60*				
Cd	*3,252.52*	*300*	*300*				
Ti	*3,251.91*	*150*	*50*				
Pd	*3,251.64*	*500*	*200*				
Te	*3,251.37*	*(150)*	—				
Fe	*3,251.23*	*150*	*300*				
La	*3,249.35*	*80*	*300*				

Sb 5,568.09 *J* 200 *O* 6*

	λ	I J	I O		λ	I J	I O
U	**5,567.96**	**2**	**1**	**J**	**5,568.10**	**(2)**	—
Xe	**5,567.77**	**(2*h*)**	—	**Yb**	**5,568.12**	**1**	**20**
Yt	**5,567.75**	**2**	**5**	**Mo**	**5,568.62**	**15**	**30**
N	**5,567.63**	**(5)**	—	**Kr**	**5,568.65**	**(100)**	—
U	**5,567.34**	**1**	**2**	**Cl**	**5,568.81**	**(15)**	—
Sb	**5,567.00**	**40**	—	**Cl**	**5,569.17**	**(4)**	—
La	**5,566.94**	**20**	**8**	**Te**	**5,569.38**	**(15)**	—
Se	**5,566.93**	**(500)**	—	**Mo**	**5,569.48**	**10**	**15**
Pr	**5,566.91**	**1**	**3**	**Fe**	**5,569.62**	**15**	**300**
Cs	**5,566.7**	**(40)**	—	**Nd**	**5,569.96**	**2**	**4**
Mo	**5,566.66**	**3**	**3**	**Kr**	**5,570.29**	**(2000)**	—
Xe	**5,566.61**	**(100)**	—				
Xe	**5,566.22**	**(5*h*)**	—				
Ar	**5,565.96**	**(5)**	—	Mo	5,570.45	(100)	200
				Ar	5,572.55	(300)	—
S	5,564.93	150	—	Te	5,576.40	(100)	—
N	5,564.37	(200)	—	In	5,576.75	(300)	—
Cs	5,563.02	(125)	—	In	5,576.91	(150)	—
Ne	5,562.77	(500)	—				
Ne	5,562.44	(150)	—	In	5,577.04	(100)	—
				Rn	5,582.4	(200)	—
Kr	5,562.23	(500)	—	Yb	5,588.47	100	30
N	5,560.37	(200)	—	Ar	5,588.49	(500)	—
Ar	5,559.62	(200)	—	Br	5,589.93	(250)	—
Ar	5,558.70	(500)	—				
As	5,558.31	200	—	*Se*	*5,591.16*	*(500)*	—
Na	5,556.04	(100)	—	*Al*	*5,593.23*	*(200)*	—
Ra	5,555.85	(500)	—	*Ar*	*5,597.46*	*(500)*	—
O	5,554.94	(100*h*)	—	*S*	*5,606.10*	*(700)*	—
Ra	5,553.57	(250)	—	*Ar*	*5,606.73*	*(500)*	—
Kr	5,552.99	(100*wh*)	—				
Bi	5,552.35	100	500*wh*				
Ba	*5,535.55*	*200R*	*1000R*				
Xe	*5,531.07*	*(300)*	—				
Se	*5,526.81*	*300wh*	*100*				
Ar	*5,524.93*	*(300)*	—				
Se	*5,522.42*	*(750)*	—				

Sb 6,004.6 *J* 200

	λ	*I*	
		J	*O*
Gd	**6,004.57**	**8**	**35**
Lu	**6,004.52**	**40**	**400**
Te	**6,004.40**	**(50)**	—
Fe	**6,003.03**	**15**	**30**
V	**6,002.63**	**2**	**20**
Sc	**6,002.34**	**5**	**3**
V	**6,002.31**	**2**	**10**
Kr	**6,002.19**	**(3)**	—
Pb	**6,001.884**	**3*h***	**40*h***
Al	**6,001.88**	**(50)**	—
Al	**6,001.76**	**(60)**	—
Ne	6,000.95	(100)	—
Kr	5,992.22	(200)	—
Yb	5,991.50	150	50
Ne	5,987.91	(150)	—
Sb	5,980.98	15	—
Ti	5,978.56	150	125
Xe	5,976.46	(800)	—
Ne	*5,975.53*	*(600)*	—
Te	*5,974.70*	*(250)*	—
Ne	*5,974.63*	*(500)*	—
Ti	*5,965.84*	*200*	*150*
Ne	*5,965.47*	*(500)*	—
Te	**6,005.0**	**(5*w*)**	—
Sb	**6,005.00**	**12*h***	**8*h***
J	**6,005.09**	**(15)**	—
Sb	**6,005.21**	**(200)**	—
Ar	**6,005.74**	**(4)**	—
Sb	**6,006.10**	**(10)**	**4*h***
Pr	**6,006.35**	**1**	**12**
Hf	**6,006.39**	**2**	**2**
Al	**6,006.42**	**(30)**	—
Yb	**6,007.42**	**80*h***	**4**
N	6,008.48	(800)	—
Xe	6,008.92	(100)	—
Te	6,014.49	(100)	—
As	6,022.81	150	—
J	6,024.13	(300)	—
Ne	6,030.00	(1000)	—
Mo	6,030.66	125	300
Xe	*6,036.20*	*(500)*	—
Xe	*6,051.15*	*(700)*	—
Sb	*6,053.41*	*(15)*	—
Se	*6,055.96*	*(1000)*	—

Sc 21 44.956

t_0 1,200°C t_1 2,400°C

I.	II.	III.	IV.	V.
6.56	12.9	24.753	73.913	91.8

λ	*I*		eV
	J	*O*	
II 3,613.84	70	40	3.45
II 3,630.74	70	50	3.42
II 3,642.79	50	60	3.40
I 3,907.48	25	125	3.17
I 3,911.81	30	150	3.19
I 4,020.40	20	50	3.08
I 4,023.69	25	100	3.10
II 4,670.40	300	100	4.01
I 5,700.23	—	400*R*	3.60

Sc 3,613.84 *J* 70 *O* 40

	λ	*I*			λ	*I*	
		J	*O*			*J*	*O*
Bi	3,613.82	30	—	Sm	3,613.90	—	2
J	3,613.81	(10*h*)	—	Rh	3,613.94	—	2
Mg	3,613.80	—	4	U	3,613.997	2	—
*bh*La	3,613.8	—	3*h*	Au	3,614.00	20	—
W	3,613.79	30	10	Th	3,614.01	8	6
Th	3,613.78	3	5	Ce	3,614.03	1	5
Cu	3,613.761	7	60	Dy	3,614.083	10	30
Eu	3,613.76	1*wh*	2*h*	Eu	3,614.084	2*d*	2*d*
Ti	3,613.756	—	12	Fe	3,614.12	2	10
U	3,613.75	—	4	*bh*La	3,614.2	—	3*h*
Zr	3,613.702	—	9	Ti	3,614.21	4	35
Ce	3,613.701	5	18	Cu	3,614.22	6	50
Pr	3,613.70	2*wh*	6*dh*	Ce	3,614.23	—	3
Tb	3,613.68	3	15	Mo	3,614.253	30	50*d*
Cr	3,613.67	8	10	W	3,614.255	2	5
Mo	3,613.642	4	3	Ce	3,614.36	—	3
He	3,613.641	(30)	—	*bh*La	3,614.4	—	4*h*
Fe	3,613.609	—	3	Gd	3,614.41	—	3
Hg	3,613.607	(40)	—	Cd	3,614.450	100	60
Sm	3,613.59	2	3	Ir	3,614.454	2*h*	2
W	3,613.55	6	—	Rh	3,614.51	2	4
Sm	3,613.52	2	2	Fe	3,614.56	6	15
Nb	3,613.452	3	5	Tb	3,614.63	8	30
Fe	3,613.452	2	10	Er	3,614.64	1	12
Zr	3,613.451	1	7	Mo	3,614.69	20	1
Ti	3,613.435	—	8	Eu	3,614.70	—	3
U	3,613.403	3*h*	4	Dy	3,614.707	5	15
Gd	3,613.400	6	8	Fe	3,614.715	1*h*	6
Mo	3,613.37	5	5	Sm	3,614.74	1	4
Ho	3,613.33	8	6	Zr	3,614.774	80	40
Os	3,613.329	12	30	Rh	3,614.775	10	15
Nd	3,613.244	4	10	*bh*La	3,614.8	—	5*h*
Nb	3,613.239	5	2	W	3,614.801	2	6
Cr	3,613.18	8	5	Eu	3,614.86	2	—
U	3,613.152	2	3	Fe	3,614.87	2	—
Fe	3,613.148	2	6	Cs	3,614.99	(4)	—
Zr	3,613.10	40	40	Pr	3,615.16	4	40
Er	3,613.096	—	7	Sm	3,615.24	2	40
La	3,613.08	3	12	Nb	3,615.50	30	30
Dy	3,613.06	4	6	Cr	3,615.64	10	30
Xe	3,613.06	(8)	—	Tb	3,615.66	15	50
Tb	3,613.06	3	15	Nd	3,615.82	10	20
S	3,613.03	(12)	—	Eu	3,616.15	10	40
Nb	3,613.01	2	2	Fe	3,616.572	7	30
Fe	3,612.94	4	20	Er	3,616.573	20	30
Te	3,612.90	(5)	—	Os	3,616.574	20	150
Cd	3,612.875	500	800				

	λ	I	
		J	O
Er	**3,612.87**	**2*h***	**10*w***
Th	**3,612.869**	**1**	**5**
Cl	**3,612.86**	**(10)**	—
In	**3,612.86**	**15**	—
Ce	**3,612.84**	—	**3**
Ti	**3,612.824**	—	**4**
Ta	**3,612.82**	—	**3*W***
Dy	**3,612.79**	**10**	**25**
Yt	**3,612.78**	—	**2**
Ce	**3,612.77**	**3**	—
Ni	**3,612.74**	**50*h***	**400**
Pr	**3,612.71**	**2**	**10**
U	**3,612.67**	**8**	**6**
Cr	3,612.61	25	35
Rh	3,612.470	50	200
Al	3,612.467	80*h*	—
Fe	3,612.07	50	80
Zr	3,611.89	40	15
Co	3,611.70	—	25
Ce	3,611.65	2*w*	20*w*
Eu	3,611.58	5	40
Cs	3,611.52	—	200
Tb	3,611.33	8	50
Yb	3,611.31	50	12
Ta	3,611.13	1	25
Yt	3,611.05	60	40
Cu	3,610.81	6	25
Gd	3,610.77	6	25
Cd	3,610.51	500	1000
Se	3,610.50	(35)	—
Re	3,610.49	—	40
Ni	3,610.46	—	1000*r*
Mn	3,610.30	40	60
Fe	3,610.162	90	100
Ti	3,610.156	70	100
Cr	3,610.05	8	20
Ir	3,609.77	25	30
Ce	3,609.69	10	40
Pd	3,609.55	700*R*	1000*R*
Nd	3,609.49	30	25
Sm	3,609.484	100	60
Cr	3,609.479	12	20
Ni	3,609.314	15	200
Cu	3,609.307	5	25
V	3,609.29	30*W*	30*W*
Ne	3,609.18	(15)	—
Os	3,609.15	12	20
Fe	3,608.86	400	500
Pr	3,616.68	8	20
V	3,616.72	30	30
Hf	3,616.89	10	25
S	3,616.92	(60)	—
Re	3,617.08	—	50
P	3,617.09	(100*w*)	—
Ir	3,617.21	15	50
Fe	3,617.32	15	25
Cs	3,617.41	—	60
W	3,617.52	20	35
Fe	3,617.79	80	125
Dy	3,618.52	10	80
Hg	3,618.53	(50)	—
Se	3,618.73	(35)	—
Fe	3,618.77	400	400
V	3,618.93	100	—
Nd	3,618.97	15	20
Ir	3,619.16	3	30
Mn	3,619.28	50	75
Ni	3,619.39	150*h*	2000*R*
Os	3,619.43	25	60
Cr	3,619.46	8	30
Nb	3,619.51	200	5
Nb	3,619.73	300	3
Yb	3,619.81	100	30
Dy	3,619.96	5	25
Tm	3,619.97	20	20
Dy	3,620.176	20	80
Er	3,620.177	3	50
Cu	3,620.35	5	30
Rh	3,620.456	10	20
Gd	3,620.461	25	25
V	3,620.47	50	10
Er	3,620.94	15	25
Rn	3,621.0	(250)	—
Pr	3,621.09	2*h*	20*h*
Co	3,621.18	50*h*	15
V	3,621.21	80	15
Sm	3,621.22	10	60
Cu	3,621.24	5	20
Fe	3,621.463	100	125
Re	3,621.465	—	30
Fe	3,622.00	100	125
Tb	3,622.11	3	30
Pr	3,622.38	5	25
Eu	3,622.56	50	20*h*
V	3,622.63	1	35
Gd	3,622.81	20	20
Fe	3,623.19	80	100

	λ	I	
		J	O
Tm	3,608.77	20	100
Gd	3,608.76	125	100
Mn	3,608.49	40	60
Kr	3,607.88	(100*wh*)	—
Mn	3,607.54	40	75
Ta	3,607.41	35	70
Ti	3,607.13	5	25
Ni	3,606.85	—	100*r*
Eu	3,606.71	—	50
V	3,606.69	70	80
Fe	3,606.68	150	200
Ar	3,606.52	(1000)	—
Yb	3,606.47	60	15
Dy	3,606.13	100	200
Rh	3,605.86	30	25
Hg	3,605.80	(200)	—
Gd	3,605.62	15	20
V	3,605.59	20*h*	30
Fe	3,605.46	150	300
Co	3,605.36	—	60
Cr	3,605.33	400*R*	500*R*
Pr	3,605.05	10	20
Gd	3,604.88	12	50
Os	3,604.47	100	15
V	3,604.38	50*h*	—
Ti	3,604.284	6	20
Mg	3,604.09	(50)	—
Fe	3,603.82	12	20
Cr	3,603.74	50	15
Fe	3,603.21	80	150
Eu	3,603.20	50	100*w*
Mo	3,602.94	25	20
F	3,602.85	(60*d*)	—
Nb	3,602.56	30	30
Fe	3,602.53	30	50
Ni	*3,602.28*	*15*	*150*
Co	*3,602.08*	*35*	*200*
Cu	*3,602.03*	*25W*	*50*
Os	*3,601.83*	*20*	*60*
Cr	*3,601.67*	*30*	*50*
Zr	*3,601.19*	*15*	*400*
Yt	*3,600.73*	*300*	*100*
Ne	*3,600.17*	*(75)*	—
V	*3,600.03*	*40*	*50*
Ru	*3,599.76*	*100*	*12*
Fe	*3,599.62*	*30*	*40*
Cu	*3,599.14*	*30*	*60*
Ho	*3,598.77*	*30*	*40*
Mn	3,623.79	40	75
Ir	3,623.80	4	20
Ce	3,623.84	5	60
Zr	3,623.864	7	40
Tb	3,623.92	8	30
Lu	3,623.99	40	20
Ca	3,624.11	15	150
Cu	3,624.24	3*wh*	30*W*
Mo	3,624.46	25	25
Nd	3,624.65	6	20
Ag	3,624.71	—	25*h*
Ni	3,624.73	15	150
Ti	3,624.82	125	60
Gd	3,624.90	30	20
Co	3,624.96	—	20
Fe	*3,625.15*	*35*	*70*
Ta	*3,625.24*	*2h*	*70r*
Tb	*3,625.54*	*15*	*50*
V	*3,625.61*	*125*	*4*
Rh	*3,636.59*	*60*	*150*
Ta	*3,626.62*	*18*	*125*
Sm	*3,626.99*	*40*	*50*
Co	*3,627.81*	—	*200*
Pt	*3,628.11*	*20*	*300W*
Tb	*3,628.20*	*15*	*100*
Ir	*3,268.67*	*30*	*100*
Yt	*3,628.71*	*50*	*40*
La	*3,628.82*	*40h*	*80*
V	*3,629.31*	*2*	*50*
Dy	*3,629.44*	*50*	*100*
Gd	*3,629.52*	*60*	*40*

	λ	I J	I O		λ	I J	I O
Ti	*3,598.72*	*30*	*70*				
Os	*3,598.11*	*30*	*300*				
Cu	*3,598.01*	—	*40wh*				
Ni	*3,597.70*	*50h*	*1000r*				
Sb	*3,597.51*	*200wh*	*2*				
Rh	*3,597.15*	*100*	*200*				
Fe	*3,597.06*	*10*	*40*				
Sb	*3,596.96*	*(100)*	—				
Tb	*3,596.38*	*15*	*50*				
Rh	*3,596.19*	*50*	*200*				
Ru	*3,596.18*	*100*	*30*				
Bi	*3,596.11*	*50*	*150wh*				
Dy	*3,596.07*	*20*	*50*				
Ti	*3,596.05*	*125*	*50*				
Ta	*3,595.64*	*5*	*70*				
Mn	*3,595.12*	*25*	*50*				
Dy	*3,595.05*	*100*	*200*				
Co	*3,594.87*	—	*200W*				
Fe	*3,594.64*	*100*	*125*				
Nb	*3,593.97*	*50*	*80*				
Ne	*3,593.64*	*(250)*	—				
Ne	*3,593.53*	*(500)*	—				
Cr	*3,593.49*	*400R*	*500R*				
V	*3,593.33*	*300R*	*30*				
Ru	*3,593.02*	*150*	*60*				
Yt	*3,592.91*	*25*	*80*				
Gd	*3,592.70*	*70*	*50*				
Sm	*3,592.59*	*50*	*40*				
V	*3,592.52*	—	*40*				
Dy	*3,592.12*	*30*	*80*				
V	*3,592.02*	*300R*	*50*				
Dy	*3,591.82*	*20*	*80*				
Rb	*3,591.59*	*20*	*80*				
Dy	*3,591.42*	*100*	*200*				

Sc 3,640.74 *J* 70 *O* 50

	λ	J	O		λ	J	O
U	**3,630.73**	**20**	**8**	**Ca**	**3,630.75**	**9**	**150**
Nb	**3,630.69**	**1**	**2**	**Ce**	**3,630.79**	—	**3***w*
Sm	**3,630.67**	**5**	**5**	**W**	**3,630.82**	**8**	**9**
Hg	**3,630.65**	**(100)**	—	**Sm**	**3,630.85**	**3**	**3**
Ba	**3,630.64**	**5**	**15**	**Hf**	**3,630.87**	**6**	**15**
Nb	**3,630.62**	**15**	**5**	**Tb**	**3,630.88**	—	**15**
Eu	**3,630.50**	—	**3***d*	**Ca**	**3,630.947**	—	**10**
Dy	**3,630.46**	—	**10**	**W**	**3,630.955**	**7**	—
Ce	**3,630.42**	**1**	**6**	**Pr**	**3,630.97**	**20**	**50**
Fe	**3,630.35**	**15**	**40**	**Nd**	**3,631.02**	**8**	**10**

	λ	I J	I O		λ	I J	I O
W	3,630.32	10	10	Fe	3,631.10	10	25
Tb	3,630.28	8	30	Sm	3,631.14	15	40
Er	3,630.24	6	25	Ce	3,631.19	3	50
Ho	3,630.20	4	—	Na	3,631.27	(100)	12
Dy	3,630.18	15	15	Ti	3,631.31	—	2
Ce	3,630.15	—	5	Sc	3,631.38	(25)	—
W	3,630.09	7	1	Co	3,631.39	25	50*W*
Er	3,630.06	1	8	P	3,631.40	(50)	—
Zr	3,630.02	12	15	Tb	3,631.46	—	30
La	3,629.99	2	2	Fe	3,631.464	300	500
Os	3,629.95	12	40	Ca	3,631.51	2	—
Nd	3,629.93	9	10	Cr	3,631.69	60	10
Sb	3,629.915	15*h*	2	Ru	3,631.711	8	2
Ni	3,629.906	3*h*	4	Ir	3,631.714	2	10
Yb	3,629.90	5	2	Ho	3,631.75	4*h*	8
Er	3,629.85	1	4	Eu	3,631.78	—	3
Ce	3,629.81	—	5	Kr	3,631.87	(200*h*)	—
Eu	3,629.80	50*h*	40	J	3,631.88	(2)	—
Cu	3,629.79	2	15				
Ir	3,629.75	4	15				
Mn	3,629.74	30	100				
Pr	3,629.71	3	20				
Rn	3,629.70	(10)	—				
Fe	3,629.61	—	4				
U	3,629.59	3	2				
Gd	3,629.52	60	40				

*P** *N**

Sc 3,642.79 *J* 50 *O* 60

	λ	J	O		λ	J	O
U	3,642.78	5	2	F	3,642.80	(30)	—
Sm	3,642.74	5	25	Au	3,642.8	5*h*	10
Mo	3,642.72	2	4	Fe	3,642.81	—	20
Tb	3,642.68	8	15	W	3,642.82	5	7
Ti	3,642.67	125	300	Ce	3,642.83	1	8
Ce	3,642.62	—	6	U	3,642.85	—	6
Os	3,642.50	10	20	Th	3,642.89	2	3
Rn	3,642.5	(5)	—	Pr	3,642.92	2	8
Nd	3,642.46	6	20	Re	3,642.99	—	100
U	3,642.44	8	10	Ce	3,643.01	—	2
Mo	3,642.41	2	3	Ar	3,643.09	(100)	—
Ni	3,642.39	—	2	Fe	3,643.11	5	30
Lu	3,642.35	—	2	Pt	3,643.17	8	60
Au	3,642.34	1*wh*	5	Co	3,643.18	15	80
Ce	3,642.250	—	4	Cr	3,643.20	30	5
Th	3,642.249	4	10	Tb	3,643.26	3	15
Nd	3,642.205	—	2	W	3,643.31	5	7

	λ	I J	I O
Mo	**3,642.204**	**10**	**5**
Ce	**3,642.17**	—	**2**
Ta	**3,642.06**	**18**	**125**
U	**3,642.03**	*2h*	*6h*
F	**3,641.98**	**(60)**	—
Er	**3,641.88**	**1**	**12**
Pr	**3,641.87**	**2**	**8**
W	**3,641.85**	**5**	**7**
Th	**3,641.84**	**4**	**6**
Cr	**3,641.83**	**15**	**40**
Mo	**3,641.82**	**3**	**3**
Co	**3,641.79**	**8**	**60**
Ce	**3,641.73**	**1**	**8**
Cu	**3,641.69**	**5**	**60**
Tb	**3,641.66**	**30**	**70**
La	**3,641.656**	**4**	—
Sm	**3,641.65**	**2**	**7**
Ni	**3,461.64**	—	**6**
Pr	3,641.62	5	50
Cr	3,641.47	25	30
W	3,641.41	40	12
Mn	3,641.39	50*h*	50
Ti	3,641.33	150	60
Os	3,641.23	10	30
V	3,641.10	30*wh*	100*h*
F	3,640.89	(100)	—
Ir	3,640.87	4	30
Fe	3,640.390	200	300
Cr	3,640.388	5	30
Os	3,640.33	40	200
Ar	3,639.85	(25)	—
Tb	3,639.82	8	30
Cr	3,639.80	25	60
Pb	3,639.58	50*h*	300
Rh	3,639.51	70	125
Co	3,639.44	20	200
V	3,639.02	60	70
Pt	3,638.79	10	250
Sm	3,638.75	8	40
Tb	3,638.46	50	80
Hg	3,638.34	(100)	—
Fe	3,638.30	80	100
Ti	3,637.97	8	30
Re	3,637.84	—	50
Sb	3,637.829	60	2*h*
Nb	3,637.827	30	20
V	3,637.76	40	40
Yb	3,637.75	30	15

	λ	I J	I O
Pr	**3,643.32**	**8**	**25**
Nb	**3,643.34**	**10**	**5**
Ce	**3,643.45**	—	**5**
Mo	**3,643.47**	**20**	**1*h***
Ir	**3,643.48**	—	**2**
Dy	**3,643.50**	—	**3**
W	**3,643.514**	**6**	—
Th	**3,643.515**	**1**	**5**
Nb	**3,643.52**	**2**	**2**
Fe	**3,643.626**	**8**	**20**
Nd	**3,643.630**	**1**	**5**
Cu	**3,643.631**	—	**5**
Tm	**3,643.652**	**40**	**60**
Fe	**3,643.713**	**4**	**6**
Sm	**3,643.715**	**3**	**5**
Nb	**3,643.72**	**15**	**15**
Tb	**3,643.76**	**3**	**30**
Fe	**3,643.81**	**1**	**7**
V	**3,643.86**	**30**	**40**
Ne	**3,643.89**	**(18)**	—
Dy	**3,643.89**	**6**	**15**
Ni	**3,643.94**	—	**2**
Te	3,644.27	(25)	—
Hg	3,644.32	(40)	—
Hf	3,644.35	50	25
Ca	3,644.41	15	200
Ti	3,644.70	5	35
V	3,644.71	50	80
Ca	3,644.76	—	30
Sc	3,645.31	50	50
Tb	3,645.38	15	50
La	3,645.41	60	100
Dy	3,645.42	100	300
Pr	3,645.66	20	30
Fe	3,645.82	60	80
Gd	3,646.196	150	200*w*
Ti	3,646.20	25	70
Pr	3,646.30	15	50
W	3,646.52	35	10
Tb	3,647.06	8	30
Co	3,647.09	4	30
Co	3,647.66	8	100
Tm	3,647.73	15	30
Tb	3,647.75	—	50
Lu	3,647.77	5	100
Fe	3,647.84	400	500
Pr	3,648.30	10	30

	λ	I J	I O		λ	I J	I O
Co	3,637.32	5	30	Kr	3,648.61	(40*h*)	—
La	3,637.15	40	50	Os	3,648.806	10	100
Nb	3,636.96	30*W*	20*W*	Dy	3,648.807	30	50
Co	3,636.72	6	40	V	3,648.97	50	80
Cr	3,636.59	30	60	Cr	3,649.00	20	40
Zr	3,636.45	30	200	Fe	3,649.30	25	60
Ir	3,636.20	25	50	Co	3,649.35	4	200
Fe	3,636.19	10	40	Sm	3,649.506	30	100
Cu	3,635.92	7	50	Fe	3,649.508	100	100
V	3,635.87	25*h*	50	La	3,649.509	8	40
Ir	3,635.49	4*h*	35	Ra	3,649.55	(1000)	—
Ti	3,635.463	100	200	Ar	3,649.83	(800)	—
V	3,635.463	5	40	Fe	3,650.03	30	70
Tb	3,635.42	8	30	Mo	3,650.05	25	3
Mo	3,635.14	10	100*h*	Hg	3,650.15	500	200
Ni	3,634.94	10	50	La	3,650.17	60	100
Ru	3,634.93	100	50	Pr	3,650.18	6	30
Nd	3,634.87	15	30	N	3,650.19	(70)	—
Rn	3,634.8	(250)	—	Fe	3,650.28	50	70
Co	3,634.71	10	70	Cr	3,650.34	40	—
Pd	3,634.69	1000*R*	2000*R*	Tb	3,650.40	100	50
Ar	3,634.46	(300)	—	Gd	3,650.97	30	25
Nd	3,634.28	30	25	Kr	3,651.02	(25*h*)	—
Sm	3,634.27	25	100	Al	3,651.06	(50)	—
S	3,634.25	(35*h*)	—	Mo	3,651.11	50	2
V	3,633.91	8	35	Nb	3,651.19	400	10
Ta	3,633.79	10*h*	35	Fe	3,651.47	200	300
Ne	3,633.66	(75)	—	Sc	3,651.80	45	50
Ti	3,633.46	5	35	Tb	3,651.86	8	30
Nb	3,633.31	30	3	Re	3,651.97	—	40
Tb	3,633.29	30	30	Tb	3,652.26	8	30
Yt	3,633.12	100	50	Mo	3,652.33	25	—
Co	3,632.842	—	60	V	3,652.42	30*h*	40
Cr	3,632.839	35	80	Mo	3,652.47	25	—
Ar	3,632.68	(300)	—	Co	3,652.54	—	200*r*
Fe	3,632.56	25	30	Gd	3,652.56	25	20
Eu	3,632.17	10	30	Tl	3,652.95	50	150
V	3,632.12	70	—	Se	3,653.05	(25)	—
Fe	3,632.04	50	50	Ir	3,653.19	50	15
				Sr	3,653.27	8	30
				P	3,653.38	(100*w*)	—
				Ti	3,653.50	200	500
				Tm	3,653.61	20	30
				Os	3,653.72	10	30
				Cr	3,653.91	25	100
				Kr	3,653.97	(250*h*)	—
				Os	*3,654.49*	*15*	*100*

*P**

	λ	I J	I O
Ti	3,654.59	40	100
Gd	3,654.64	200	200 W
Hg	3,654.83	(200)	—
Tb	3,654.88	30	70
Al	3,655.00	(100)	—
Gd	3,656.16	200	200 W
Cr	3,656.26	25	80
V	3,656.71	20h	80
Os	3,656.90	30	150
Co	3,656.97	6	60
Ru	3,657.55	50	—
Rh	3,657.99	200 W	500 W
Ti	3,658.10	60	150
Tb	3,658.88	100	100
P	3,659.26	(50)	—
Ar	3,659.50	(100)	—
Fe	3,659.52	80	125
Nb	3,659.61	500	15
Ti	3,659.76	150	50
Mn	3,660.40	75	75
Ti	3,660.63	18	90
Sm	3,661.350	50	100
Ru	3,661.353	100	60
V	3,661.38	150	10
La	3,662.07	40	60
Co	3,662.16	25	100
Ti	3,662.24	100	40
Sm	3,662.25	50	50
Gd	3,662.27	200	200w
Hg	3,662.88	400	50
Hg	3,663.28	400	500
Ru	3,663.37	60	5
V	3,663.59	1wh	150
Zr	3,663.65	10	100
Ni	3,664.09	30	300
Ne	3,664.11	(250)	—
P	3,664.19	(100w)	—
Yt	3,664.61	100	100
Ir	3,664.618	15	60
Gd	3,664.621	200	200w
V	3,665.14	50h	100
Kr	3,665.33	(80)	—

Sc 3,907.48 *J* 25* *O* 125

	λ	J	O		λ	J	O
Fe	3,907.47	6	15	U	**3,907.56**	8	6
Ce	3,907.44	2	6	Ag	**3,907.59**	2*h*	3
Th	3,907.34	3	8	Ce	**3,907.64**	1	2

	λ	I J	I O
Pr	**3,907.30**	**3**	**8**
Ce	**3,907.29**	**6**	**35**
W	**3,907.20**	**5**	**6**
V	**3,907.17**	**2**	**2*h***
Gd	**3,907.125**	**100**	**100*W***
Sm	**3,907.124**	**15**	**10**
Eu	**3,907.11**	**500*R***	**1000*RW***
U	**3,907.02**	**1**	**5**
Mo	**3,906.98**	**5**	**5**
Ce	**3,906.924**	**3**	**8**
Mo	**3,906.916**	**5**	**5**
Nb	**3,906.91**	**5**	**5**
Hf	**3,906.84**	**3*d***	**3**
Sm	**3,906.805**	**5**	**6**
Th	**3,906.796**	**—**	**8**
Fe	**3,906.751**	**10**	**10**
V	**3,906.748**	**20**	**50**
Tb	**3,906.53**	**—**	**4**
Fe	**3,906.482**	**200**	**300**
Mo	**3,906.480**	**10**	**5**
Ce	**3,906.45**	**2**	**8**
Hg	**3,906.41**	**15**	**25**
Er	**3,906.32**	**12**	**25**
Co	**3,906.294**	**—**	**150**
Pt	**3,906.291**	**1*h***	**2**
Ho	**3,906.26**	**3**	**3**
Zr	**3,906.15**	**—**	**3**
Ce	**3,906.104**	**—**	**2**
Nd	**3,906.096**	**10**	**15**
Pr	**3,906.09**	**8**	**6**
Fe	**3,906.04**	**2*w***	**2*w***
Ba	**3,906.01**	**2**	**4**
Ru	**3,905.99**	**3*h***	**3**
Ti	3,904.78	35	70
Fe	3,903.90	80	100
Mo	3,902.96	500*R*	1000*R*
Fe	3,902.95	400	500
Cr	3,902.91	100	100
Gd	3,902.40	80	100
Os	3,901.71	20	150
Tm	3,900.79	50	80
Zr	3,900.52	—	100
Fe	3,899.71	300	500
Tb	3,899.19	100	200
Dy	3,898.54	—	100
Co	3,898.49	6	80*r*
Ce	3,898.27	6	80

	λ	I J	I O
Au	**3,907.650**	**3**	**5**
Tb	**3,907.65**	**—**	**3**
Fe	**3,907.67**	**1**	**2**
Ir	**3,907.75**	**—**	**4**
Cr	**3,907.78**	**10**	**30**
Tb	**3,907.79**	**3**	**3**
Nd	**3,907.84**	**12**	**20**
Ir	**3,907.90**	**2**	**10**
Tb	**3,907.91**	**—**	**5**
Fe	**3,907.94**	**60**	**100**
Ce	**3,907.96**	**—**	**2**
Pr	**3,908.03**	**50**	**100**
Tb	**3,908.08**	**—**	**20**
Ce	**3,908.09**	**1**	**8**
Gd	**3,908.15**	**3*h***	**5**
Mn	**3,908.16**	**3**	**3**
Re	**3,908.20**	**—**	**25**
Mo	**3,908.25**	**3**	**5**
Sm	**3,908.26**	**—**	**6**
V	**3,908.32**	**2*h***	**50**
U	**3,908.33**	**1**	**10**
Ce	**3,908.41**	**6**	**30**
Er	**3,908.42**	**—**	**10**
Pr	**3,908.43**	**60**	**100**
U	**3,908.47**	**10**	**8**
Th	**3,908.49**	**2**	**3**
Ce	**3,908.54**	**3**	**20**
Nb	**3,908.59**	**5**	**3**
Tb	**3,908.66**	**3**	**4**
U	**3,908.68**	**3**	**2**
Cr	**3,908.75**	**150**	**200**
Ru	**3,908.765**	**12**	**12**
Ce	**3,908.765**	**2**	**12**
Ir	**3,908.83**	**—**	**5**
Ni	**3,908.93**	**—**	**2**
Nb	**3,908.97**	**8**	**10**
Co	3,909.93	—	200*W*

	λ	I J	I O		λ	I J	I O
Fe	3,898.01	50	80				
Fe	3,897.89	60	100				
Fe	3,895.66	300	400				
Ti	3,895.25	10	70				
Co	3,894.98	3	300*R*				
Gd	3,894.71	80	150*W*				
Pd	3,894.20	200*W*	200*W*				
Co	3,894.08	100	1000*R*				
Fe	3,893.39	8	100				
Ho	*3,891.02*	*40*	*200*				
Zr	*3,890.32*	*6*	*150*				
Pr	*3,889.33*	*70*	*150*				
Cs	*3,888.65*	*10*	*150*				
Cr	*3,886.79*	*125*	*125*				
La	*3,886.37*	*200*	*400*				
Fe	*3,886.28*	*400*	*600*				
Tm	*3,883.43*	*30*	*150*				
Co	*3,881.87*	*30*	*300R*				
Os	*3,881.86*	*20*	*125*				
Fe	*3,878.57*	*300*	*300R*				
Fe	*3,878.02*	*300*	*400*				

Sc 3,911.81 *J* 30* *O* 150

	λ	I J	I O		λ	I J	I O
Ho	**3,911.80**	—	**3**	**Os**	**3,911.81**	**5**	**30**
Pr	**3,911.798**	**2**	**8**	**Cr**	**3,911.82**	—	**10**
Re	**3,911.775**	—	**3**	**Nd**	**3,911.909**	**6**	**12**
Lu	**3,911.77**	—	**3*h***	**Er**	**3,911.911**	—	**5**
Ce	**3,911.73**	—	**3**	**Th**	**3,911.914**	**3**	**8**
Fe	**3,911.70**	—	**1**	**Mo**	**3,911.94**	**5**	**5**
Dy	**3,911.68**	**1**	**5**	**Eu**	**3,911.97**	—	**4*w***
U	**3,911.67**	**18**	**18**	**Pr**	**3,911.99**	**6**	**20*d***
Gd	**3,911.66**	—	**5**	**Cr**	**3,912.00**	—	**40*wh***
Eu	**3,911.61**	—	**5**	**Fe**	**3,912.05**	**5*wh***	**5*wh***
Er	**3,911.56**	—	**8**	**Ru**	**3,912.11**	**8**	**10**
Mn	**3,911.42**	**15**	**15**	**Ta**	**3,912.13**	**3*s***	**10**
Ti	**3,911.36**	—	**7**	**Ce**	**3,912.19**	**3**	**25**
Th	**3,911.308**	**3**	**5**	**V**	**3,912.21**	**20**	**50**
Pr	**3,911.307**	**3**	**5**	**Nd**	**3,912.228**	**20**	**20**
Ce	**3,911.30**	**1**	**15**	**U**	**3,912.23**	**1**	**2**
W	**3,911.29**	**5**	**6**	**Tb**	**3,912.25**	—	**5**
Yb	**3,911.28**	—	**8**	**Pr**	**3,912.27**	**3**	**10**
U	**3,911.23**	**3**	**3**	**Th**	**3,912.28**	**15**	**15**
Ti	**3,911.19**	**5**	**40**	**Ni**	**3,912.31**	—	**2**
Nd	**3,911.17**	**25**	**25**	**U**	**3,912.408**	**8**	**6**

	λ	I	
		J	O
Ru	3,911.14	2	6
Mn	3,911.12	—	20
Mo	3,911.09	20	20
Th	3,911.003	1	5
Fe	3,911.001	2	6
U	3,910.98	3	10
Sm	3,910.92	2	1
Pt	3,910.91	2	5
Tb	3,910.85	—	5
Fe	3,910.84	10	30
La	3,910.81	5*h*	10
V	3,910.79	20	35
Ce	3,910.70	2*h*	12
Pr	3,910.572	—	10
Tb	3,910.57	—	2
Er	3,910.51	—	3
U	3,910.50	3*h*	1
Tb	3,910.40	—	5
Sm	3,910.20	1	4

	λ	I	
		J	O
Ir	3,912.410	—	2
Er	3,912.429	—	12
Eu	3,912.43	1	5
Ta	3,912.435	10*h*	15
Ce	3,912.436	5	50
Ho	3,912.44	2	4
Fe	3,912.45	1*h*	2*h*
Dy	3,912.54	—	5
Ti	3,912.59	—	15
Pr	3,912.61	5	10
Gd	3,912.745	—	5
U	3,912.745	4	2
Mn	3,912.75	5	5
Tb	3,912.78	8	5
W	3,912.82	6	7
Rh	3,912.83	1	2
Dy	3,912.85	1	5
V	3,912.886	15	40
Bi	3,912.895	2*h*	2*h*
Pr	3,912.90	80	150
Sm	3,912.97	5	20
Ni	3,912.98	—	5
Th	3,913.012	15	15
Nb	3,913.013	10	5
Ru	3,913.013	—	3
Pd	3,913.09	—	2
Nb	3,913.14	5	5
Ce	3,913.14	—	3
Fe	3,913.21	1*h*	2*h*
Eu	3,913.24	1	4
U	3,913.25	3	8
Pr	3,913.56	30	80
Fe	3,913.63	25	100
Zr	3,914.34	8	70
Li	3,915.0	—	200*wh*
Ir	3,915.38	50	150
Dy	3,915.60	—	80
Cr	3,915.84	80	125
La	3,916.04	400	400
Cr	3,916.24	60	100
Tm	3,916.47	8	80
Gd	3,916.59	100	150*w*
Fe	3,916.73	80	100
Co	3,917.11	10	80
Fe	3,917.18	70	150
Re	3,917.27	—	100*w*
Pr	3,918.86	30	100

	λ	I J	I O		λ	I J	I O
				Cr	3,919.16	125	300*r*
				Fe	3,920.26	300	500
				Cr	3,921.02	40	150
				La	3,921.53	200	400
				Zr	3,921.79	4	100
				V	3,922.43	40	80
				Co	3,922.75	—	100
				Ta	3,922.78	15	100
				Fe	3,922.914	400	600
				Ta	3,922.915	10*h*	100
				Pt	3,922.960	20*r*	100
				Pr	3,924.14	15	100
				Tb	3,925.45	200	150
				Pr	3,925.456	100	125
				Fe	3,925.65	50	80
				Fe	*3,927.92*	*300*	*500*
				Cr	*3,928.64*	*40*	*150*
				La	*3,929.22*	*300*	*400*
				Fe	*3,930.30*	*400*	*600*
				Eu	*3,930.50*	*400R*	*1000R*
				Dy	*3,931.54*	—	*200*
				Sm	*3,933.59*	*200h*	*200*
				Fe	*3,933.60*	*200*	*200*
				Ca	*3,933.67*	*600R*	*600R*
				Ir	*3,934.84*	*50*	*200*
				Pr	*3,935.82*	*50*	*125*
				Co	*3,935.97*	*15*	*400R*
				Tb	*3,939.60*	*200*	*200*
				Fe	*3,940.88*	*80*	*150*
				Cr	*3,941.49*	*60*	*200r*
				Co	*3,941.73*	—	*200wh*

Sc 4,020.40 *J* 20 *O* 50

	λ	J	O		λ	J	O
Pr	4,020.28	2	10	Mo	4,020.45	10	10
Hf	4,020.25	2*h*	6	Tb	4,020.47	3	20
Nb	4,020.24	10*h*	2	Fe	4,020.49	—	2
Pd	4,020.22	—	15*wh*	Er	4,020.52	1	20
U	4,020.17	3	3	Ce	4,020.54	—	5
*bh*Sr	4,020.1	—	2	Mo	4,020.67	—	5
Th	4,020.096	3	3	Pd	4,020.68	—	5*h*
Mn	4,020.09	5	10	U	4,020.69	2	2
Nd	4,020.06	3*d*	6*d*	Tb	4,020.74	—	2
Ir	4,020.03	100	80	Sm	4,020.77	2	3
Sm	4,019.98	15	30	Nd	4,020.87	15	15
Ce	4,019.90	3	8	Dy	4,020.897	4	10

	λ	I J	I O
Pr	4,019.84	8	15
Sm	4,019.83	8	8
Nd	4,019.81	8	10
Mo	4,019.79	—	10
Gd	4,019.73	10	15
Eu	4,019.71	—	3
Tb	4,019.66	—	2
Pb	4,019.64	6	6
Ru	4,019.55	8	12
Dy	4,019.48	2	5
Ce	4,019.480	1	6
Pr	4,019.44	3	10
Co	4,019.30	—	80
Ce	4,019.28	—	4
W	4,019.23	15	18
U	4,019.20	8	6
Ce	4,019.19	—	2
Co	4,019.140	—	5
Th	4,019.137	8	8
Re	4,019.13	—	15
Tb	4,019.12	5	40
Ni	4,019.046	—	5
Mo	4,019.046	4	2
Ce	4,019.04	4	15
U	4,018.99	15	25
Ce	4,018.92	—	2
Nd	4,018.83	10	15
Re	4,018.41	—	25
Fe	4,018.27	7	50
Os	4,018.26	4	60
Cr	4,018.20	8	35
Zr	4,018.12	—	25
Mn	4,018.10	60	80
Ti	4,017.77	8	70*h*
U	4,017.72	25	25
Eu	4,017.58	25	25*W*
Fe	4,017.15	50	80
Pr	4,016.75	20	25
Ti	4,016.28	5	30
La	4,015.393	2*h*	100
Pr	4,015.389	30	50
Ti	4,015.38	10*h*	70*h*
W	4,015.22	30	25
Ce	4,014.90	12	50
Cr	4,014.67	8	40
Fe	4,014.53	100	200
Co	4,013.94	—	300

	λ	I J	I O
Co	4,020.905	—	500*w*
Pr	4,020.991	30	40
Ru	4,020.995	12	15
Eu	4,021.01	—	5*w*
Mo	4,021.015	25	15
Zr	4,021.02	—	3*h*
Tb	4,021.12	1	10
Ce	4,021.24	—	5
U	4,021.25	6	6
Nd	4,021.33	12	12
Tm	4,021.39	10	4
U	4,021.40	3*h*	3*h*
Sm	4,021.41	1	4
Er	4,021.56	—	5
Fe	4,021.62	—	2
Tb	4,021.74	—	2
Nd	4,021.79	10	12
Ti	4,021.83	20	100
Fe	4,021.87	100	200
V	4,021.92	2	8
U	4,021.93	3	2
Er	4,021.96	—	5

	λ	I J	I O
Fe	4,013.824	—	200
Gd	4,013.817	3	25
Fe	4,013.79	40	80
Ti	4,013.58	7*h*	70*h*
Cr	4,012.47	60	70
Ti	4,012.391	50	35
Ce	4,012.388	20	60
Re	4,012.26	—	25
Nd	4,012.25	40	80
Mo	4,011.97	25	25
Eu	4,011.68	—	25
Re	4,011.51	—	35
Fe	4,009.72	100	120
Ti	4,009.66	25	60
Gd	4,009.21	2	50
Ti	4,008.93	35	80
W	4,008.75	45	45
Pr	4,008.71	50	150
Ti	4,008.06	7	50
Er	4,007.97	7	35
Zr	4,007.60	1	25
Sm	4,007.49	25	50
Fe	4,007.27	50	80
Ta	4,006.83	20	30
Ru	4,006.60	15	25
Fe	4,006.31	35	60
Ti	4,005.97	3	35
U	4,005.70	3	25
Ru	4,005.64	30	25
Tb	4,005.55	125	100*d*
Fe	4,005.25	200	250
Os	4,005.15	20	35
If	4,005.02	—	25
Re	4,004.93	—	30
Os	*4,004.02*	*6*	*50*
Ti	*4,003.81*	*70*	*50*
Os	*4,003.48*	*6*	*50*
Tb	*4,002.58*	*5*	*50*
Fe	*4,001.67*	*50*	*80*
Cr	*4,001.44*	*80*	*200*
Gd	*4,001.24*	*3*	*80*
Dy	*4,000.45*	*300*	*400*
Pr	*4,000.19*	*25*	*50*
Ce	*3,399.24*	*20*	*80*
Pr	*3,999.19*	*40d*	*50d*
Os	*3,998.93*	*12*	*80*
V	*3,998.73*	*25*	*100*
Ti	*3,998.64*	*100*	*150*

	λ	I J	I O	λ	I J	I O
Fe	3,998.05	100	150			
Co	3,997.91	20	200			
Fe	3,997.40	150	300			
Pr	3,997.05	40	100			
Os	3,996.80	10	50			
Dy	3,996.70	80	200			
Pt	3,996.57	—	50			
Tm	3,996.52	40	200			
Gd	3,996.32	100	100			
Ta	3,996.17	30h	100			
Fe	3,995.99	20	60			
La	3,995.75	300	600			
Tm	3,995.58	—	100			
Co	3,995.31	20	1000R			
Pr	3,994.83	25	300			
Nd	3,994.68	40	80			
Co	3,994.54	—	60			
Cr	3,993.97	20	60			
Ce	3,993.82	6	50			
Ba	3,993.40	50r	100R			
Cr	3,992.84	70	150			
V	3,992.80	20	60			
Ce	3,992.39	8	50			
Ir	3,992.12	60	150			
Nd	3,991.74	40	60			
Co	3,991.69	6	60			
Cr	3,991.67	50	100			
Zr	3,991.13	60	100			
Cr	3,991.12	60	200			
Yb	3,990.89	20	60			
V	3,990.57	40	125			
Fe	3,990.38	25	70			
Co	3,990.30	10	80			
Cr	3,989.99	40	80			
Ti	3,989.76	100	150			
Pr	3,989.72	125	200			
V	3,988.83	35	70			
La	3,988.52	800	1000			
Os	3,988.18	12	50			
Th	3,988.01	30	50			
Yb	3,987.99	500R	1000R			
Er	3,987.95	20	100r			

Sc 4,023.69 *J* 25* *O* 100

	λ	*I*			λ	*I*	
		J	*O*			*J*	*O*
Ce	4,023.64	—	4	Tb	4,023.716	—	6
U	4,023.60	6*h*	3*h*	Dy	4,023.722	2	10
La	4,023.59	15	50	Pr	4,023.737	—	3
Th	4,023.53	1	3	Cr	4,023.739	15	40
Cr	4,023.43	—	15	Gd	4,023.74	—	3*h*
Co	4,023.40	—	200	Nd	4,023.82	3	12
V	4,023.39	30	10	Ru	4,023.83	60	25
Ce	4,023.37	1	8	Eu	4,023.84	—	2
Re	4,023.353	—	40*W*	Ho	4,023.93	1*h*	4*v*
Gd	4,023.350	10	20	Zr	4,023.98	2	30
Zr	4,023.30	—	5	Pr	4,024.07	1	5
Sc	4,023.223	—	60	Tb	4,024.07	1	40*W*
Sm	4,023.223	25	30	Mo	4,024.09	25	30
V	4,023.174	2	7	Fe	4,024.10	2	8
U	4,023.170	6	6	Tm	4,024.24	—	7
Gd	4,023.15	10	20	Ru	4,024.30	4	7
Rh	4,023.144	5	10	Eu	4,024.34	—	3*W*
Nb	4,023.141	3	4	Ce	4,024.35	—	5
Zr	4,023.03	—	6	Pr	4,024.41	2	15
Nd	4,023.00	15	15	Dy	4,024.437	—	20
Re	4,022.96	—	25*d*	Zr	4,024.438	4	5
Ce	4,022.93	—	3	Tb	4,024.44	—	4
Eu	4,022.91	—	6*W*	Th	4,024.47	5	5
Tb	4,022.87	—	15*d*	Ce	4,024.49	5	15
Hf	4,022.84	—	2	Cr	4,024.567	1	20
W	4,022.83	5	4	Ti	4,024.573	35	80
Ce	4,022.75	—	4	Ca	4,024.68	3	3*wh*
Fe	4,022.744	—	3	Ru	4,024.69	5	12
Pr	4,022.738	6	12	Tb	4,024.70	—	4
Sm	4,022.71	5	6	Fe	4,024.74	30	120
Ru	4,022.69	3	7	Tb	4,024.781	—	7
Cu	4,022.66	25	400	Nd	4,024.785	10	20
Ce	4,022.45	—	5	Dy	4,024.91	—	10
Nb	4,022.39	10	2	Zr	4,024.918	3	25
Gd	4,022.34	—	15	Cr	4,025.012	25	100
Tb	4,022.33	—	3	U	4,025.013	3	8
Ce	4,022.27	4	15	Ni	4,025.11	—	2
Cr	4,022.26	40	80	Ti	4,025.14	25	15
Pr	4,022.20	4	8	Tb	4,025.146	—	4
Ru	4,022.16	100	40	Ce	4,025.150	2	12
W	4,022.12	10	12	Pr	4,025.19	8	15
Th	4,022.09	15	20	W	4,025.19	6	8
Ni	4,022.05	—	3	Ce	4,025.29	—	2
Nd	4,022.03	2	4				
				Pr	4,025.55	25	40
				La	4,025.88	50	50
				Mo	4,025.99	30*h*	30*h*

*P**

	λ	I J	I O
U	4,026.03	25	25
Cr	4,026.17	35	100
Mn	4,026.43	40	50
Ti	4,026.54	10	70
Ta	4,026.94	30	40
Co	4,027.04	4	200
Cr	4,027.10	30	80
Zr	4,027.20	4	100
Ti	4,027.48	3	30
Cr	4,028.02	—	35
Gd	4,028.16	8	25
Ce	4,028.41	8	35
Re	4,028.56	—	25
Zr	4,028.95	1	40
Fe	4,029.636	25	80
Re	4,029.639	—	80
Zr	4,029.68	15	40
Ta	4,029.94	5	50
Zr	4,030.04	2	35
Sr	4,030.38	—	40
Fe	4,030.49	60	120
Ti	4,030.51	18	80
Cr	4,030.68	30	40
Mn	4,030.75	20	500*r*
Cr	4,031.13	6	30
Ce	4,031.34	8	40
Tb	4,031.64	3	50
La	4,031.69	300	400
Ti	4,031.754	1	35
Pr	4,031.755	30	50
Fe	4,031.96	50	80
Tb	4,032.28	—	30
Nb	4,032.52	50	30
Fe	4,032.630	15	80
Ti	4,032.632	1	35
Ga	4,032.98	500*R*	1000*R*
Tb	4,033.04	5	125
Ta	4,033.069	10	100
Mn	4,033.073	20	400*r*
Cr	4,033.26	8	30
Re	4,033.31	—	40
Sb	4,033.54	60	70
Ir	4,033.76	25	100
Pr	4,033.86	35	50
Ti	4,033.91	3	40
Mn	4,034.49	20	250*r*
Ti	4,034.91	2	25
Sm	4,035.10	3	50

	λ	I	
		J	O
Co	4,035.55	3	150
V	4,035.63	80	40
Mn	4,035.73	60	50
Ti	4,035.83	5	50
Zr	4,035.89	2	40
Eu	4,036.11	—	50 *W*
La	4,037.21	3*h*	50
Cr	4,037.29	12	80
Gd	4,037.34	30	100
*bh*B	4,037.4	—	25
Re	4,037.51	—	30
Ce	4,037.66	3	25
Os	4,037.84	4	80
Gd	4,037.91	30	100
Cr	4,039.10	40	100
Ru	4,039.21	50	25
Pr	4,039.36	20	50
Nb	4,039.53	50	30
Ce	*4,040.76*	*5*	*70*
Ho	*4,040.84*	*30*	*150*
Ta	*4,040.87*	*5h*	*50*
Au	*4,040.94*	*40*	*50*
Mn	*4,041.36*	*50*	*100*
Os	*4,041.92*	*6*	*100*
Ce	*4,042.58*	*3*	*50*
La	*4,042.91*	*300*	*400*
K	*4,044.14*	*400*	*800*
Fe	*4,044.61*	*35*	*70*
Pr	*4,044.82*	*35*	*50*
Co	*4,045.39*	—	*400*
Ho	*4,045.43*	*80*	*200*
Fe	*4,045.81*	*300*	*400*
Dy	*4,045.98*	*12*	*150*
Hg	*4,046.56*	*300*	*200*
K	*4,047.20*	*200*	*400*
Yt	*4,047.63*	*10*	*50*
Gd	*4,047.85*	*50*	*150*
Mn	*4,048.75*	*60*	*60*
Cr	*4,048.78*	*50*	*80*
Gd	*4,049.44*	*20*	*80*
Gd	*4,049.90*	*60*	*100*
La	*4,050.08*	*60*	*60*
Pr	*4,051.15*	*30*	*50*
Ru	*4,051.40*	*200*	*125*
Gd	*4,053.20*	*80*	*100*
Gd	*4,053.65*	*40*	*100*
Ho	*4,053.92*	*200*	*400*

	λ	I J	I O		λ	I J	I O
				Gd	*4,054.73*	*20*	*80*
				Pr	*4,054.84*	*40*	*50*
				Ti	*4,055.02*	*30*	*80*
				Zr	*4,055.03*	*5*	*100*
				Ag	*4,055.26*	*500R*	*800R*
				Mn	*4,055.54*	*80*	*80*

Sc 4,670.40 *J* 300 *O* 100

	λ	J	O		λ	J	O
Cs	**4,670.28**	**(20)**	**—**	**Ce**	**4,670.44**	**—**	**2*h***
Mo	**4,670.24**	**5**	**5**	**V**	**4,670.49**	**40*r***	**60*R***
Tb	**4,670.23**	**—**	**3**	**Pr**	**4,670.54**	**—**	**10**
Te	**4,670.11**	**(30)**	**—**	**Nd**	**4,670.56**	**—**	**20**
Nb	**4,670.104**	**4*h***	**1**	**Yb**	**4,670.57**	**15**	**12**
Ce	**4,670.097**	**—**	**2**	**Os**	**4,670.69**	**—**	**6**
Er	**4,670.09**	**—**	**2**	**Ce**	**4,670.74**	**—**	**4**
Th	**4,670.03**	**1**	**3**	**Sm**	**4,670.77**	**—**	**30**
Mo	**4,670.00**	**—**	**2**	**Tb**	**4,670.82**	**—**	**5**
Ru	**4,669.98**	**—**	**40**	**Yt**	**4,670.83**	**2**	**3**
Nb	**4,669.87**	**2**	**3**	**Sm**	**4,670.833**	**—**	**30**
N	**4,669.77**	**(10)**	**—**	**Gd**	**4,670.85**	**2*h***	**3**
Sm	**4,669.65**	**40**	**50**	**Ne**	**4,670.88**	**(70)**	**—**
Ce	**4,669.638**	**—**	**3**	**Mo**	**4,670.909**	**—**	**2**
Mo	**4,669.637**	**3**	**2**	**Hf**	**4,670.91**	**2**	**5**
Er	**4,669.52**	**—**	**3**	**Ce**	**4,670.913**	**—**	**4**
Ce	**4,669.50**	**—**	**4**	**Er**	**4,671.092**	**1**	**2**
Dy	**4,669.40**	**2**	**3**	**Nd**	**4,671.094**	**—**	**20**
Tb	**4,669.39**	**—**	**3**	**Dy**	**4,671.10**	**4**	**4**
Sm	**4,669.390**	**35**	**40**	**Eu**	**4,671.18**	**1**	**30**
Se	**4,669.38**	**(10)**	**—**	**Xe**	**4,671.23**	**(200)**	**—**
Cr	**4,669.34**	**20**	**50**	**W**	**4,671.30**	**1**	**3**
U	**4,669.309**	**15**	**8**	**Ru**	**4,671.37**	**—**	**5**
V	**4,669.308**	**8**	**10**	**Eu**	**4,671.38**	**—**	**10**
Hf	**4,669.24**	**6**	**6**	**U**	**4,671.41**	**30**	**20**
Fe	**4,669.18**	**2**	**15**	**Er**	**4,671.58**	**1**	**3**
Ta	**4,669.143**	**15**	**300**	**Kr**	**4,671.61**	**(10)**	**—**
S	**4,669.14**	**(35)**	**—**	**W**	**4,671.65**	**1**	**12**
Ru	**4,669.138**	**—**	**15**	**Mn**	**4,671.688**	**5**	**100**
Nd	**4,669.13**	**—**	**5**	**Cu**	**4,671.693**	**10**	**—**
Ne	**4,669.02**	**(50)**	**—**	**Ce**	**4,671.71**	**—**	**2**
Ir	**4,668.99**	**—**	**20**	**Li**	**4,671.8**	**(4)**	**—**
U	**4,668.94**	**3**	**3**	**La**	**4,671.83**	**150**	**100**
				Sm	**4,671.84**	**—**	**2**
La	4,668.91	300*r*	200*r*				
Na	4,668.60	100	200	Pr	4,672.08	25*w*	100
Ag	4,668.48	70	200	Nb	4,672.09	100	150
Fe	4,668.14	10	125	Sm	4,674.592	40	80
Ni	4,667.77	—	100	Nd	4,674.595	10	50

	λ	I J	I O		λ	I J	I O
Ti	4,667.59	8	150	Cu	4,674.76	30*W*	200
Fe	4,667.46	20	150	Yt	4,674.848	100	80
Ni	4,666.99	—	50	Er	4,674.849	15	50
J	4,666.52	(250)	—	Rh	4,675.03	50	100
Cr	4,666.51	25	50	Ti	4,675.12	5	50
Se	4,664.98	(150)	—	Nb	4,675.37	30*w*	50*w*
Na	4,664.86	—	80	Sm	4,676.91	50	100
Cr	4,664.80	20	70	Tm	4,677.85	2	50
Pr	4,664.65	15	100	Cd	4,678.16	200*W*	200*W*
Te	4,664.34	(800)	—	Ne	4,678.22	(300)	—
Se	4,664.20	(150)	—	Br	4,678.69	(200)	—
Hf	4,664.12	100	50	Fe	4,678.85	100	150
Cr	4,663.83	15	50	Ne	4,679.13	(150)	—
Os	4,663.82	5	100	Gd	4,680.06	25	50
La	4,663.76	200	100	Zn	4,680.14	200*h*	300*w*
Sm	4,663.55	—	60	Kr	4,680.41	(500)	—
Co	4,663.41	—	700*W*	W	4,680.52	40	150
La	4,662.51	200	150	Cr	4,680.54	25	50
Eu	4,661.88	—	100	Nd	4,680.73	1	50
Eu	4,661.87	20	80*R*	Rn	4,680.83	(500)	—
Ta	4,661.12	5*h*	300	Cr	4,680.87	8	60
Ne	4,661.10	(150)	—	In	4,681.11	(200)	—
Eu	4,660.37	1	50	Sm	4,681.56	—	50
Hg	4,660.28	(200)	—	Ru	4,681.79	—	100
W	4,659.87	70	200	Ta	4,681.87	50	200
Kr	4,658.87	(2000)	—	Ti	4,681.916	100	200
Lu	4,658.02	15	100	Tm	4,681.92	2	50
Ar	4,657.94	(150)	—	In	4,682.00	250*W*	—
W	4,657.44	12	50	Ra	4,682.28	(800)	—
Co	4,657.39	35	100	Yt	4,682.32	100	60
Eu	4,656.73	1	50	Co	4,682.38	—	500
Ti	4,656.47	70	150	Gd	4,683.34	50	100
Ne	4,656.39	(300)	—	Nd	4,683.44	—	50
Cr	4,656.19	4	50	Ru	4,684.02	—	100
Ir	4,656.18	—	60	Ta	4,684.87	2	100
				Pr	4,684.94	10*w*	125
La	*4,655.50*	*300*	*150*	*He*	*4,685.75*	*(300)*	—
Te	*4,654.38*	*(800)*	—	*Ni*	*4,686.22*	*1*	*200*
Ru	*4,654.31*	—	*125*	*Te*	*4,686.95*	*(300)*	—
Cr	*4,652.16*	*150*	*200R*	*Sm*	*4,687.18*	—	*100*
Pr	*4,651.52*	*40w*	*125*	*Zr*	*4,687.80*	—	*125*
Cr	*4,651.28*	*100*	*100*	*Eu*	*4,688.23*	*2*	*100*
Cu	*4,651.13*	*40*	*250*	*Ti*	*4,691.34*	*25*	*125*
O	*4,649.15*	*(300)*	—	*Ba*	*4,691.62*	*40*	*100*
Ni	*4,648.66*	*3*	*400w*	*Ta*	*4,691.90*	*5h*	*400*
Se	*4,648.44*	*800*	—	*La*	*4,692.50*	*300*	*200*
Ru	*4,647.61*	—	*125*	*Co*	*4,693.21*	*25*	*500*
Fe	*4,647.44*	*40*	*125*				
Cr	*4,646.17*	*150*	*100*				

	λ	I J	I O		λ	I J	I O
Ne	*4,645.42*	*(300)*	—	*Ta*	*4,693.35*	*3*	*150*
Ti	*4,645.19*	*10*	*100*	*Xe*	*4,697.02*	*(300)*	—
				Gd	*4,697.49*	*4*	*100*
Ru	*4,645.09*	*30*	*100*	*Eu*	*4,698.14*	*2*	*300*
Rn	*4,644.18*	*(300)*	—				
Sm	*4,642.23*	*40*	*100*	*Co*	*4,698.38*	*8*	*300*
				Ti	*4,698.76*	*20*	*100*
				La	*4,699.63*	*200r*	*200r*

Sc 5,700.23 *O* 400*R*

Yb	**5,699.95**	—	**2**	**Cu**	**5,700.24**	—	**350**
U	**5,699.87**	—	**3**	**Sm**	**5,700.26**	—	**2**
Sm	**5,699.61**	—	**5**	**Th**	**5,700.45**	**1**	**8**
Ru	**5,699.58**	—	**10**	**Pt**	**5,700.47**	—	**2***h*
Sm	**5,699.48**	—	**2**	**Cr**	**5,700.51**	—	**8**
La	**5,699.39**	—	**5**	**Mn**	**5,700.58**	—	**3**
Mo	**5,699.28**	—	**20**	*bh***Ti**	**5,700.6**	—	**2**
Ta	**5,699.24**	—	**80***w*	**Th**	**5,700.70**	**1**	**8**
Ce	**5,699.23**	—	**40**	*Th*	**5,700.91**	**3**	**12**
Ru	**5,699.05**	—	**125**	**Sm**	**5,701.12**	—	**3**
Pt	**5,698.97**	—	**7**	**Si**	**5,701.14**	—	**15**
Er	**5,698.95**	—	**8**	**Gd**	**5,701.35**	—	**20**
Nd	**5,698.93**	**1**	**10**	**Fe**	**5,701.56**	**25**	**50**
Sm	**5,698.92**	—	**3**	**Nd**	**5,701.57**	—	**5**
Dy	**5,698.72**	—	**3**	**Mo**	**5,702.11**	—	**15**
V	**5,698.52**	**300**	**300**	**Pr**	**5,702.22**	—	**3**
*bh***F**	**5,698.4**	—	**2**	**Nd**	**5,702.24**	—	**25**
Cr	**5,698.330**	**2**	**30**	**Sm**	**5,702.25**	—	**2**
U	**5,698.327**	**1**	**2**	**Cr**	**5,702.31**	—	**20**
Mo	**5,698.27**	—	**10**	**Ru**	**5,702.36**	—	**15**
Nb	**5,698.03**	**3**	**4**	**Ce**	**5,702.39**	—	**3**
Gd	**5,697.99**	—	**6**	**Rh**	**5,702.47**	—	**8**
Pr	**5,697.98**	—	**5**	**La**	**5,702.53**	—	**5**
Nb	**5,697.90**	**3**	**5**	**Tb**	**5,702.54**	—	**15**
W	**5,697.82**	—	**35**				
*bh***Yt**	**5,697.8**	—	**5**	V	5,703.56	60	200
				V	5,706.98	—	200
				Sc	5,708.60	—	15
Ho	5,691.47	—	200	Sc	5,711.75	—	100
Na	5,688.22	—	300	Sc	5,717.25	—	20
Sc	5,686.83	—	200				
				Yb	5,720.01	8	300
				Rb	5,724.45	—	600
Sc	*5,671.80*	—	*300w*				
Sc	*5,657.87*	—	*30*				

Se 34 78.96

t_0 217°C

t_1 680°C

I.	II.	III.	IV.	V.
9.750	21.691	34.078	42.900	73.11

λ	*I*		eV
	J	*O*	
I 2,039.85	(1000)	—	6.3
I 2,062.79	(800)	—	6.3
2,591.41	(150)	—	—
I 4,730.78	(1000)	—	>2.6
I 4,739.03	(800)	—	>2.6
I 4,742.25	(500)	—	>2.6
I 5,142.14	(500)	—	—
I 5,175.98	(600)	—	—

Se 2,039.85 J (1000)

	λ	I J	I O		λ	I J	I O
W	2,039.80	10	—	Re	2,039.91	9	3
Ir	2,039.79	40	8	Al	2,039.93	(15)	—
Pt	2,039.70	20	12	Zn	2,040.00	(250)	—
Rh	2,039.67	25	—				

Se 2,062.79 J (800)

	λ	I J	I O		λ	I J	I O
Pt	2,062.78	20	20	Ir	2,063.03	80	—
Pd	2,062.56	10	—				
J	2,062.38	(900)	—				

Se 2,591.41 J (150)

	λ	I J	I O		λ	I J	I O
Te	2,591.35	(30)	—	Mn	2,591.42	12	2
Hf	2,591.33	12	10	W	2,591.49	7	12
U	2,591.25	12	18	Fe	2,591.54	100	50
Ru	2,591.24	50	—	Ru	2,581.64	1	30
Mn	2,591.23	12	—	Co	2,591.69	3	10r
Cs	2,591.17	(20)	—	W	2,591.74	6	—
Ne	2,591.15	(3)	—	Mo	2,591.77	25	2
Yb	2,591.04	5	1	Mn	2,591.82	5	—
Ir	2,591.040	4	2	Tb	2,591.83	20	—
J	2,591.02	(20)	—	Cr	2,591.85	12	100r
Pt	2,591.01	10	—				
Ru	2,590.97	100	12	Rh	2,592.16	80	—
				Fe	2,592.78	100	20
				Pd	2,593.27	100	3
Nb	2,590.94	800	15	J	2,593.47	(150)	—
Cu	2,590.53	250	1h	Ta	2,593.66	100	80d
Mn	2,588.96	80	—	Mn	2,593.73	1000R	200R
Ni	2,588.31	80	—	Nb	2,593.76	100	2h
Tm	2,588.27	80	40	Sn	2,594.42	80	60
Rh	2,587.29	100	2	Nb	2,594.74	80	3
Co	2,587.22	100h	10w	Ru	2,595.80	100	—
Se	*2,586.44*	*(15)*	—	*Pd*	*2,595.97*	*150*	*3*
Se	*2,585.25*	*(15)*	—	*Ta*	*2,596.45*	*150*	*80*
Ta	*2,584.03*	*200*	*80w*	*Rh*	*2,597.07*	*150*	*3*
Ni	*2,583.995*	*200*	—	*Fe*	*2,598.37*	*1000h*	*700*
Nb	*2,583.986*	*800wh*	*10*	*Cu*	*2,598.81*	*200*	—
Pd	*2,583.85*	*200*	—	*Nb*	*2,598.88*	*150*	—
J	*2,582.81*	*(400)*	—	*Fe*	*2,599.40*	*1000h*	*1000*

Se 4,730.78 J (1000)

	λ	I J	I O		λ	I J	I O
La	**4,730.77**	2*h*	**2**	**As**	**4,730.92**	**125**	—
Cr	**4,730.71**	**50**	**100**	**J**	**4,730.98**	**(10)**	—
Pr	**4,730.69**	**1**	**60**	**U**	**4,731.00**	**4**	**6**
Ar	**4,730.66**	**(5)**	—	**S**	**4,731.15**	**(15)**	—
Rb	**4,730.48**	**10**	—	**Ti**	**4,731.17**	**6**	**50**
V	**4,730.385**	**8**	**9**	**Xe**	**4,731.19**	**(50*h*)**	—
J	**4,730.38**	**(25)**	—	**V**	**4,731.25**	**3**	**3**
Mn	**4,730.36**	**5**	**15**	**Te**	**4,731.27**	**(70)**	—
Nb	**4,730.31**	**5*h***	**5**	**Rb**	**4,731.3**	**(8)**	—
Bi	**4,730.3**	**(25)**	—	**Hf**	**4,731.37**	**20**	**15**
Ta	**4,730.12**	**5**	**100**	**Mo**	**4,731.44**	**100**	**100**
Th	**4,729.88**	**5**	**6**	**Fe**	**4,731.49**	**1**	**5**
Te	**4,729.83**	**(50)**	—	**Er**	**4,731.59**	**1**	**8**
Bi	**4,729.82**	**2*h***	—	**U**	**4,731.60**	**50**	**40**
Cr	**4,729.72**	**6**	**30**	**Ni**	**4,731.81**	**2**	**100**
Fe	**4,729.70**	**25**	**25**	**Dy**	**4,731.851**	**10**	**30**
W	**4,729.65**	**15**	**30**	**Ir**	**4,731.855**	**5**	**8**
V	**4,729.53**	**12**	**15**	**Yb**	**4,732.02**	**15**	**1**
S	**4,729.45**	**(8)**	—	**Ar**	**4,732.08**	**(5)**	—
				Sc	**4,732.29**	**6**	**10**
La	4,728.42	300	400*r*				
Zn	4,722.16	300*h*	400*w*				
La	4,719.95	300	200*r*				
S	4,716.23	(600)	—				
Ne	*4,715.34*	*(1500)*	—				
Ne	*4,712.06*	*(1000)*	—				
Ne	*4,710.06*	*(1000)*	—				
Ne	*4,708.85*	*(1200)*	—				
Ne	*4,708.39*	*(1500)*	—				
Ar	*4,702.32*	*(1200)*	—				

*N**

Se 4,739.03 J (800)

	λ	I J	I O		λ	I J	I O
Kr	**4,739.00**	**(3000)**	—	**Mn**	**4,739.11**	**15**	**150**
Te	**4,738.67**	**(50)**	—	**Rh**	**4,739.22**	**3**	**15**
Hf	**4,738.57**	**3**	**20**	**Cl**	**4,739.42**	**(10)**	—
Dy	**4,738.50**	**2**	**2**	**P**	**4,739.49**	**(30)**	—
Cl	**4,738.41**	**(10)**	—	**Cs**	**4,739.66**	**(20)**	—
Ta	**4,738.35**	**2**	**5**	**La**	**4,739.79**	**8**	**4**
Mn	**4,738.29**	**5**	**12**	**Eu**	**4,739.96**	**2**	**2**
Gd	**4,738.13**	**50*h***	**50**	**Ra**	**4,740.07**	**(4)**	—
Co	**4,737.77**	**1**	**150**	**Ta**	**4,740.16**	**100**	**100*R***
Sc	**4,737.64**	**60*h***	**100**	**La**	**4,740.277**	**300**	**150**

	λ	I J	I O		λ	I J	I O
Pt	**4,737.56**	**2**	**4**	**U**	**4,740.285**	**10**	**3**
				Mo	**4,740.36**	**5**	**5**
				Cl	**4,740.40**	**(150)**	**—**
Ar	4,735.93	(400)	—				
Xe	4,734.15	(600)	—				
Gd	4,732.61	300	300				

*P** *N**

Se 4,742.25 *J* (500)

Ti	**4,742.11**	**1**	**15**	**Th**	**4,742.27**	**2**	**4**
Ho	**4,742.04**	**3**	**10**	**U**	**4,742.33**	**3**	**10**
Ge	**4,741.94**	**50**	**—**	**Mo**	**4,742.59**	**10**	**—**
Cd	**4,741.78**	**3**	**—**	**V**	**4,742.63**	**15**	**20**
O	**4,741.71**	**(20)**	**—**	**Br**	**4,742.70**	**(200)**	**—**
Dy	**4,741.54**	**2**	**3**	**Ti**	**4,742.79**	**40**	**100**
Fe	**4,741.53**	**1**	**12**	**N**	**4,742.90**	**(4)**	**—**
W	**4,741.52**	**2**	**12**	**Dy**	**4,743.03**	**2**	**3**
Yt	**4,741.40**	**3**	**2**	**La**	**4,743.085**	**300**	**300*r***
U	**4,741.28**	**2**	**1**	**Mo**	**4,743.085**	**10**	**10**
Sc	**4,741.02**	**60*h***	**100**	**Mo**	**4,743.61**	**3**	**3**
Se	**4,740.97**	**(600)**	**—**	**Gd**	**4,743.65**	**300**	**300**
Dy	**4,740.93**	**2**	**3**	**Th**	**4,743.70**	**2**	**3**
				Sc	**4,743.81**	**60*h***	**100**
				Ne	4,749.57	(300)	—
				Ne	4,752.73	(1000)	—
				Se	*4,763.65*	*(800)*	—
				Se	*4,765.52*	*(40)*	—
				Kr	*4,765.74*	*(1000)*	—

*P**

Se 5,142.14 *J* (500)

S	**5,141.89**	**(8)**	**—**	**Mo**	**5,142.25**	**2**	**10**
Ar	**5,141.81**	**(20)**	**—**	**Yb**	**5,142.31**	**5*h***	**—**
Fe	**5,141.75**	**100*h***	**100**	**Kr**	**5,142.7**	**(4)**	**—**
P	**5,141.49**	**(50)**	**—**	**Kr**	**5,143.05**	**(600*h*)**	**—**
Mo	**5,141.26**	**5**	**12**	**Pb**	**5,143.14**	**4*h***	**—**
Sb	**5,141.2**	**40**	**—**	**Ne**	**5,143.26**	**(5)**	**—**
Nb	**5,140.69**	**2**	**3**	**Br**	**5,143.46**	**(15)**	**—**
Nb	**5,140.57**	**5**	**10**	**C**	**5,143.49**	**15**	**—**
U	**5,140.42**	**1**	**3**	**Tl**	**5,143.58**	**5**	**—**
Cl	**5,140.38**	**(2)**	**—**				

	λ	I	
		J	O
Tm	**5,140.26**	**5**	**5**
Se	5,134.31	(35)	—
Kr	5,125.73	(400*wh*)	—
J	*5,119.28*	*(500)*	—
Se	*5,117.77*	*(25)*	—

	λ	I	
		J	O
Bi	5,144.48	300*h*	2
Al	5,144.87	(2)	—
Ne	5,144.94	(500)	—
Ne	5,145.01	(500)	—
Ar	*5,162.28*	*(500)*	—

*N**

Se 5,175.98 *J* (600)

Rh	**5,175.97**	**1**	**200**
N	**5,175.89**	**(30)**	—
Cu	**5,175.89**	**2**	—
O	**5,175.86**	**(15)**	—
Cl	**5,175.85**	**(20)**	—
Pr	**5,175.83**	**1**	**10*w***
Ba	**5,175.62**	**7**	**10**
In	**5,175.56**	**(150)**	—
In	**5,175.422**	**(300)**	—
Sm	**5,175.418**	**1**	**60**
J	**5,175.40**	**(8)**	—
In	**5,175.29**	**(400)**	—
Pr	**5,175.19**	**1**	**30*w***
J	**5,174.74**	**(2)**	—
U	**5,174.62**	**2**	**2*h***
V	**5,174.53**	**10*h***	**10*h***
N	**5,174.46**	**(5)**	—
Nb	**5,174.20**	**2**	**3**
Mo	**5,174.18**	**25*h***	**70*h***

Ar	**5,176.28**	**(10)**	—
P	**5,176.38**	**(70)**	—
V	**5,176.48**	**8**	**8**
Sb	**5,176.55**	**(50)**	—
Ni	**5,176.56**	**2**	**70**
V	**5,176.77**	**50**	**60**
Mo	**5,177.09**	**1**	**5**
La	**5,177.31**	**30**	**150**
Ba	**5,177.45**	**2**	**4**
Ar	**5,177.53**	**(40)**	—
Kr	**5,177.71**	**(6*whs*)**	—
Se	5,183.01	(15)	—
La	5,183.42	400	300
Mg	5,183.62	300	500*wh*
In	5,184.44	(300)	—
Se	5,187.68	(18)	—
La	5,188.23	500	50
Kr	*5,208.32*	*(500)*	—
Ag	*5,209.07*	*1000R*	*1500R*
Bi	*5,209.29*	*600h*	—

*P**

Si 14 28.086

t_0 1,420°C t_1 2,287°C

I.	II.	III.	IV.	V.
8.149	16.339	33.489	45.131	166.5

λ	I		eV
	J	O	
I 2,506.90	200	300	4.95
I 2,516.12	500	500	4.95
I 2,528.52	500	400	4.93
I 2,881.58	400	500	5.08
I 2,987.65	100	100	4.93
I 5,754.26	—	40	7.11
I 7,423.54	—	500	7.29

Si 2,506.90 *J* 200 *O* 300

	λ	I *J*	I *O*		λ	I *J*	I *O*
Ag	**2,506.88**	**(2*h*)**	**2**	**V**	**2,506.905**	**35**	**50*r***
Co	**2,506.877**	**3**	**10*w***	**U**	**2,506.906**	**2**	**6**
Xe	**2,506.86**	**(4)**	—	**Ru**	**2,507.014**	**80**	**60**
Ir	**2,506.601**	**2**	**10**	**Fe**	**2,507.014**	**10**	**1**
Ag	**2,506.60**	**50*h***	**15**	**U**	**2,507.07**	**2**	**6**
Fe	**2,506.57**	—	**10**	**Nb**	**2,507.137**	**2**	**3**
Kr	**2,506.56**	**(5*wh*)**	—	**Tm**	**2,507.14**	**15**	**10**
O	**2,506.56**	**(8)**	—	**Ag**	**2,507.22**	**5*h***	—
				W	**2,507.23**	**12**	—
Co	2,506.46	200*h*	50*w*				
Cu	2,506.27	500*r*	—	Pt	2,508.50	20	300
V	2,506.22	150	10	C	2,509.11	200	—
Pt	2,505.93	10	150	Rh	2,510.65	200*wh*	5
Rh	2,505.10	200	2				
Ta	2,504.45	1	200	*Fe*	*2,510.83*	*50*	*300R*
Cr	2,504.31	3	150*r*	*Ni*	*2,510.87*	*250h*	*50h*
Rh	2,503.84	150	2	*C*	*2,512.03*	*400*	—
V	2,503.02	100	7	*Ta*	*2,512.04*	*800*	*5*
				Si	*2,514.33*	*200*	*300*
Zn	*2,502.00*	*400w*	*20*	*Pd*	*2,514.48*	*200*	—
Te	*2,499.75*	*(300)*	—				
Nb	*2,499.750*	*200*	—				

Si 2,516.123 *J* 500 *O* 500

	λ	I *J*	I *O*		λ	I *J*	I *O*
				Fe	**2,516.25**	—	**2**
Xe	**2,516.12**	**(6)**	—	**Cd**	**2,516.34**	**25**	—
V	**2,516.118**	**100**	**25**	**Yb**	**2,516.36**	**5**	**2**
Re	**2,516.116**	—	**125**	**In**	**2,516.40**	**3**	—
Ta	**2,516.111**	**2**	**6**	**Cl**	**2,516.42**	**(2)**	—
Dy	**2,516.11**	—	**5**				
				Te	**2,516.49**	**(15)**	—
Mo	**2,516.109**	**25**	**25**				
vzduch	**2,516.1**	**7**	—				
Br	**2,516.05**	**(3)**	—	Si	2,519.21	200	300
Ru	**2,516.01**	**5*wh***	**20**				
Zn	**2,515.81**	**20**	**150*w***				
				Rh	*2,520.53*	*1000wh*	*10*
W	**2,515.80**	**10**	**1**				
La	**2,515.79**	**4**	—				
Ru	**2,515.76**	**8**	**6**				
Rh	**2,515.75**	**10**	**60**				
U	**2,515.73**	**2**	**6**				
Pt	2,515.58	20	500				

*P** *N**

Si 2,528.52 *J* 500 *O* 400

	λ	*I*			λ	*I*	
		J	*O*			*J*	*O*
Ba	**2,528.51**	**50***r*	—	**Sb**	**2,528.53**	**200**	**300***R*
Dy	**2,528.49**	—	**4**	**Ta**	**2,528.59**	**1**	**4**
Xe	**2,528.49**	**(3)**	—	**Lu**	**2,528.59**	**5***h*	—
V	**2,528.47**	**150***R*	**50**	**Co**	**2,528.61**	**200**	**4**
Ir	**2,528.39**	**10***h*	—	**Mn**	**2,528.70**	**2**	**12**
Th	**2,528.36**	**5**	—	**Ar**	**2,528.71**	**(5)**	—
Ar	**2,528.33**	**(10)**	—	**Ru**	**2,528.715**	—	**30**
Yb	**2,528.33**	**3**	—	**U**	**2,528.77**	**2**	**5**
Ce	**2,528.29**	**1**	**15**	**Cs**	**2,528.8**	**(8)**	—
Cr	**2,528.24**	—	**30**	**V**	**2,528.84**	**150***R*	**25**
Tb	**2,528.23**	**20**	—	**Mo**	**2,528.87**	**25**	**1**
Co	**2,528.18**	**40**	—	**Ru**	**2,528.878**	**2**	**60**
Fe	**2,528.174**	—	**6**	**Fe**	**2,528.879**	**2**	**2**
In	**2,528.17**	**5**	—	**Cl**	**2,528.88**	**(8)**	—
W	**2,528.16**	**6**	—	**W**	**2,528.91**	**15**	**3**
V	2,527.90	300*R*	35	Cu	2,529.30	600	—
Fe	2,527.43	50	200*R*	Co	2,530.13	300	40*w*
Ni	2,525.39	300*Wh*	—				
Co	2,524.96	700	50*w*				
				Fe	*2,535.60*	—	*1000*
				Hg	*2,536.52*	*2000R*	*2000R*
Si	*2,524.12*	*400*	*400*				

Si 2,881.578 *J* 400 *O* 500

	λ	*J*	*O*		λ	*J*	*O*
Ce	**2,881.578**	**2**	**40**	**Co**	**2,881.580**	**1**	**4**
Gd	**2,881.578**	—	**40**	**Dy**	**2,881.59**	—	**2***h*
Al	**2,881.46**	**(30)**	—	**In**	**2,881.61**	**2**	—
Ce	**2,881.42**	—	**4***s*	**Ir**	**2,881.64**	—	**7**
Mo	**2,881.37**	**10**	—	**O**	**2,881.70**	**(10)**	—
Ir	**2,881.36**	—	**4**	**Ir**	**2,881.74**	—	**3**
Gd	**2,881.33**	**3**	**2**	**Ce**	**2,881.77**	—	**2**
Tb	**2,881.31**	—	**10**	**Ca**	**2,881.80**	**4**	—
Ru	**2,881.28**	**3**	**30**	**U**	**2,881.91**	**2**	**4**
Rh	**2,881.254**	—	**20**	**Yb**	**2,881.92**	**3**	—
Ni	**2,881.247**	**8**	—	**Cr**	**2,881.93**	**30**	**1**
Ta	**2,881.232**	**5**	**30**	**Mo**	**2,881.94**	**3**	—
Cd	**2,881.23**	**(30)**	**50***R*	**Ti**	**2,881.95**	—	**2**
Cs	**2,881.16**	**(20)**	—	**J**	**2,882.01**	**(20)**	—
Ir	**2,881.158**	**3**	**15**	**Th**	**2,882.014**	**12**	**10**
Th	**2,881.15**	**10**	**10**	**Mo**	**2,882.04**	**3**	—
Cr	**2,881.141**	**2**	**25**	**Zr**	**2,882.09**	**3**	**2**
Na	**2,881.140**	**(60)**	**8**	**Ru**	**2,882.12**	**200**	**30**
Ce	**2,881.13**	—	**12**	**Zn**	**2,882.15**	**(25)**	—
W	**2,881.07**	**1**	**5**	**Yb**	**2,882.15**	**2**	—

	λ	I J	I O
Dy	**2,881.06**	**1***h*	**2**
U	**2,881.02**	**2**	**2**
Ho	**2,880.99**	**10**	**20**
J	2,878.64	(400)	—
Sb	2,877.91	150	250*W*
Pt	2,877.52	200*h*	40
Nb	2,876.95	500*W*	40*W*

	λ	I J	I O
W	**2,882.19**	—	**3**
Ag	**2,882.20**	**10***h*	—
V	2,882.50	200*r*	35
Nb	2,883.18	800*R*	100
Fe	2,883.70	300	—
V	2,884.78	200*r*	40
Tb	*2,891.41*	*500*	*3*
Ta	*2,891.84*	*100*	*500 W*
Pt	*2,893.86*	*25*	*500*

Si 2,987.648 *J* 100 *O* 100

	λ	I J	I O
Ir	**2,987.646**	**1**	**5**
Co	**2,987.58**	—	**2**
Nb	**2,987.55**	**10***w*	**1***w*
Ce	**2,987.54**	—	**2**
U	**2,987.52**	—	**4**
Rh	**2,987.45**	**20**	**40**
Rn	**2,987.40**	**(20)**	—
Ce	**2,987.351**	—	**8**
Mo	**2,987.35**	**25**	—
Fe	**2,987.292**	**200**	**300**
Nb	**2,987.289**	**4**	**5**
W	**2,987.287**	**15**	**8**
Cd	**2,987.2**	**(25)**	—
Co	**2,987.16**	**50**	**75***R*
U	**2,987.06**	**1**	**2**
Tb	**2,987.04**	**5**	**10**
Rh	**2,986.99**	**20**	**80**
Ta	2,986.81	100	20*d*
Tm	2,986.52	150	50
Cr	2,986.47	125	125*r*
Fe	2,986.46	60	100
Rh	2,986.20	60	150
Mo	2,986.16	50	4
Fe	2,985.55	300	80
Zr	2,985.39	3	50
Cr	2,985.32	60	10
V	2,985.17	60	1
Nb	2,985.05	50	2*h*
Fe	2,984.83	400	200*r*
Na	2,984.43	(80)	20
Ni	2,984.13	10	50*R*
Yb	2,983.98	70	10

	λ	I J	I O
Sm	**2,987.651**	—	**5***h*
Ru	**2,987.70**	**4**	**30**
Zr	**2,987.799**	**10**	**3**
U	**2,987.802**	**2**	**5**
Dy	**2,987.89**	—	**2**
Mo	**2,987.920**	**30**	**25**
J	**2,987.92**	**(12)**	—
Tl	**2,987.95**	**(10)**	—
U	**2,987.95**	**2**	**2***d*
W	**2,987.96**	**4**	**8**
Si	**2,987.99**	**(5)**	—
Ga	**2,987.99**	**6**	—
V	**2,988.02**	**80**	**10**
Ni	**2,988.05**	**4**	—
Ru	**2,988.09**	—	**8**
Ir	**2,988.19**	—	**3**
Mo	**2,988.228**	**1**	**25**
Th	**2,988.234**	**20**	**15**
Os	**2,988.26**	**1**	**8**
Fe	2,988.47	30	60
Cr	2,988.65	150	200*r*
Ru	2,988.95	100	250
Bi	2,989.03	100*wh*	250*wh*
Cr	2,989.19	90	10
Lu	2,989.27	4	50
Ta	2,989.50	15	200
Co	2,989.59	30	75*R*
Ti	2,990.16	80*Wh*	—
Nb	2,990.26	200	5
Au	2,990.28	50	—
Eu	2,990.39	100	150
Tm	2,990.54	30	80

	λ	I	
		J	O
Fe	2,983.57	400r	1000R
V	2,983.55	60	10
V	2,982.75	50	4
Ne	2,982.66	(250)	—
Nb	2,982.11	80	10w
Fe	2,982.06	90	—
Th	2,982.05	5	150
Ru	2,981.93	3	60
Cd	2,981.89	(10)	50
Fe	2,981.85	50	100
Ni	2,981.65	20	80r
Tm	2,981.49	100	60
Fe	2,981.45	200	300
Cd	2,981.34	(40)	200R
V	2,981.20	60	4
Sb	2,980.962	(125hd)	—
Ru	2,980.958	50	10
Ne	2,980.92	(50)	—
Pd	*2,980.65*	*200R*	—
Cd	*2,980.63*	*500*	*1000R*
Fe	*2,980.54*	*70*	*100*
Fe	*2,979.35*	*100*	*20*
Xe	*2,979.32*	*(200)*	—
Ti	*2,979.20*	*100wh*	—
Ta	*2,978.75*	*30*	*200r*
Ru	*2,978.641*	*150*	*50*
Th	*2,978.64*	*100h*	*1*
Ta	*2,978.18*	*150*	—
Nb	*2,977.68*	*300*	*1*
Rh	*2,977.68*	*30*	*125*
Ru	*2,976.59*	*200*	*60*
Ta	*2,976.26*	*150*	*2wh*
Fe	*2,976.13*	*60*	*100*
Te	*2,975.91*	*(100)*	—
Hf	*2,975.88*	*100*	*80*
Ta	*2,975.56*	*50*	*200*
Cr	*2,975.48*	*50*	*100r*
Ne	*2,974.71*	*(250)*	—

	λ	I	
		J	O
V	2,990.95	5	50
Ta	2,991.25	10	50
Eu	2,991.34	40	80
Ru	2,991.62	100	50
Fe	2,991.64	80	100
Cr	2,991.89	60	125r
Nb	2,991.95	100	1
Re	2,992.37	—	100
Ne	2,992.44	(150)	—
Ni	2,992.59	10	80R
Ru	2,993.27	9	60
Bi	2,993.34	100wh	200wh
J	2,993.86	(70)	—
Cr	2,994.07	50	150
Fe	*2,994.43*	*600r*	*1000R*
Ni	*2,994.46*	*10*	*125R*
Nb	*2,994.73*	*300*	*100*
Au	*2,994.99*	*100*	—
Cr	*2,995.10*	*75*	*200r*
Cr	*2,996.58*	*125*	*300r*
Cu	*2,997.36*	*30*	*300*
Pt	*2,997.97*	*200r*	*1000R*
Cr	*2,998.79*	*70*	*200r*
Ru	*2,998.89*	*100*	*50*
Fe	*2,999.51*	*300*	*500*
Pd	*2,999.55*	*100h*	—
Re	*2,999.59*	—	*125*
Fe	*3,000.45*	*80*	*100*
Cr	*3,000.89*	*125*	*150r*
Fe	*3,000.95*	*300r*	*800r*

Si 5,754.26 *O* 40

Gd	**5,754.15**	—	**25***r*
*bh***Zr**	**5,753.8**	—	**20**
Cr	**5,753.69**	—	**15**
Sm	**5,753.61**	**15**	**4**
Nd	**5,753.53**	—	**15**
Sm	**5,754.40**	—	**7***w*
Nb	**5,754.44**	**1**	**3**
W	**5,754.57**	—	**8**
Ni	**5,754.67**	—	**150***w*
V	**5,755.08**	**2**	**2***h*

	λ	I	
		J	O
W	5,753.38	—	7
Fe	5,753.14	20	40
Sm	5,753.11	—	2
Pr	5,753.022	—	5
Th	5,753.021	—	8
Re	5,752.95	—	200*w*
Co	5,752.88	—	2 *h*
Ti	5,752.84	—	5
V	5,752.74	10	10
Sm	5,752.59	—	3
Er	5,752.55	—	12
Hf	5,752.53	5	3
Ce	5,752.52	—	8
Fe	5,752.06	2*h*	8*h*
Er	5,752.023	—	4
Ru	5,752.023	—	10
Tb	5,751.89	—	10
Gd	5,751.88	—	8
U	5,751.76	—	3
Mo	5,751.40	100	125
Ho	5,751.12	—	30
V	5,750.65	—	50*W*
Nd	5,749.19	—	20
Ni	5,748.34	—	40
*bh*Zr	5,748.1	—	100
Sm	5,748.07	—	20
Tb	5,747.58	—	60
*bh*Yt	5,746.9	—	20
Ta	5,746.71	—	60
Nd	5,744.77	1	25
La	5,744.41	—	80
Ce	5,743.53	—	25
V	5,743.45	20	60
Sm	5,743.33	—	20
Nd	5,743.20	1	20*wh*
Nd	5,742.92	—	40
Bi	5,742.55	10	30
Sm	5,741.19	—	60
Sm	5,740.88	—	80
Nd	5,740.86	—	35
La	5,740.66	1	100
Re	5,740.30	—	50*w*
Ti	5,740.02	—	25
Ti	5,739.51	80	70
Er	5,739.19	—	30
Eu	5,739.00	—	300
V	5,737.06	100	100
Sm	5,736.85	—	40
Nd	5,755.20	—	2
Rh	5,755.69	—	2
Ta	5,755.81	—	40
Yb	5,755.90	—	15
Sm	5,755.93	—	4
W	5,756.09	—	12
Pr	5,756.17	1	25
Sm	5,756.40	—	15
Sm	5,756.76	—	3
Ru	5,756.83	—	7
Ti	5,756.86	12	10
Er	5,757.62	—	30
Sm	5,757.96	—	20
Sm	5,759.50	—	60
Tm	5,760.21	5	20
Nb	5,760.34	30	30
Ni	5,760.85	—	50
V	5,761.41	—	25
La	5,761.84	—	60
*bh*Ti	5,762.0	—	20
Ti	5,762.27	50	70*h*
Er	5,762.79	—	30
Fe	5,763.01	35	80
Pt	5,763.57	—	30
Sm	5,763.88	—	20
*bh*Yt	5,764.2	—	20
Tm	5,764.29	—	50
Eu	5,765.20	—	2000
Ti	5,766.35	50	70*W*
Ta	5,766.56	—	80*w*
Ta	5,767.91	—	100*w*
La	5,769.07	60	30
La	5,769.35	—	70
Hg	5,769.59	200	600
Er	5,769.93	—	20
La	5,770.01	—	25
Nd	5,770.50	—	20
Yb	5,771.67	50	30
*bh*F	5,771.9	—	100
Si	5,772.26	—	30
V	5,772.42	25	50
Ce	5,773.12	—	30
Sm	5,773.77	—	100
Ti	5,774.05	50	70*W*
Mo	5,774.55	—	20
*bh*F	5,774.8	—	100
Lu	5,775.40	5	50

	λ	I	
		J	O
Lu	5,736,55	15	150
Zr	5,735.70	—	25
W	5,735.09	25	50W
Mo	5,734.06	—	20
V	5,734.01	—	35
Gd	5,733.86	—	20
Sm	5,732.95	—	100
V	5,731.25	100	250
Eu	5,730.89	—	300
bhC	5,730.0	—	150
Sm	5,729.92	—	40
Nd	5,729.29	—	30
Nb	5,729.19	10	20
In	5,728.27	—	50
V	5,727.66	—	75
V	5,727.03	150	150
V	5,725.64	30	40
Sm	5,725.59	—	40d
Rb	5,724.95	—	50
Rb	5,724.45	—	600
bhZr	5,724.1	—	70
Mo	5,722.73	60	80
Os	5,721.93	—	80
Yb	5,720.01	8	300
Hf	5,719.18	10	40
Sm	5,719.12	20	60
Ce	5,719.03	—	40
bhZr	5,718.1	—	150
Ti	5,716.48	—	40
V	5,716.21	30	60
Ti	5,715.13	60	70
Ni	5,715.09	—	50
Ni	5,711.90	—	50
Ti	5,711.88	40	50
Sc	5,711.75	—	100
In	5,709.75	—	50wh
Ni	5,709.56	(1)	100w
Fe	5,709.38	—	100h
Si	5,708.44	2h	40
Nd	5,708.28	—	60
Pr	5,707.61	—	100w
V	5,706.98	—	200
Ho	5,706.88	—	50
Ta	5,706.28	—	50
Nd	5,706.21	—	40
Mo	5,705.72	40	40
Ta	5,704.31	—	40

	λ	I	
		J	O
Nb	5,776.07	3	30
V	5,776.68	25	50
Ta	5,776.77	—	80
Re	5,776.84	—	300w
bhF	5,777.6	—	100
Ba	5,777.66	100R	500R
Sm	5,778.33	—	50
bhZr	5,778.5	—	60
Sm	5,779.25	—	50
Pr	5,779.29	—	50
Mo	5,779.36	—	20
bhF	5,779.4	—	200
Ta	5,780.02	—	60
bhF	5,780.5	—	50
U	5,780.61	—	40
Ta	5,780.71	—	80
Os	5,780.81	—	50
Sm	5,781.89	—	100
bhF	5,782.1	—	150
Cu	5,782.13	—	1000
K	5,782.60	—	60
Eu	5,783.71	—	150s
V	5,784.38	30	50
bhF	5,784.8	—	100
Tb	5,785.18	—	40
Ti	5,785.98	60	100W
V	5,786.16	—	75
Sm	5,786.99	—	200
Nb	5,787.54	15	80
bhF	5,787.6	—	100
Cr	5,787.99	—	50wh
La	5,789.25	—	125
bhF	5,790.3	—	100
Cr	5,791.00	—	40wh
La	5,791.34	—	200
Mo	5,791.85	60	100
Rh	5,792.77	—	40
bhF	5,793.1	—	100
bhF	5,795.9	—	100
La	5,797.59	150	80
Zr	5,797.74	—	50
bhF	5,798.9	—	80
V	5,799.90	2	40
Eu	5,800.27	—	200
Ba	5,800.28	20	100
Sm	5,800.50	—	80
Os	5,800.60	—	50
bhF	5,801.8	—	80

	λ	I J	I O		λ	I J	I O
V	*5,703.56*	*60*	*200*	*K*	*5,801.96*	*20*	*50h*
				Sm	*5,802.82*	—	*80*
				Tb	*5,803.15*	—	*40*
				Nd	*5,804.02*	—	*100*
				Ti	*5,804.26*	*50*	*100h*
				bhF	*5,804.7*	—	*80*
				Ni	*5,805.23*	—	*50*

Si 7,423.54 *O* 500

	λ	J	O		λ	J	O
Ti	**7,423.17**	—	**35**	**Rh**	**7,423.68**	—	**2*h***
Zr	**7,422.75**	—	**3*h***	**Hf**	**7,423.69**	**15**	**5**
Ni	**7,422.30**	—	**600**	**Sm**	**7,423.81**	—	**3**
				Zr	**7,423.83**	—	**3*h***
				Sm	**7,424.43**	—	**3**
Co	7,417.38	—	300*W*	**Si**	**7,424.63**	—	**20**
Si	7,416.00	—	200				
Si	7,415.37	—	15				
Ni	7,409.39	—	400	Eu	7,426.57	—	500
Si	7,409.11	—	100				
Rb	7,408.17	—	500				
Si	*7,405.85*	—	*300*				
Ni	*7,493.63*	—	*600*				

Sn $^{50}_{118.59}$

t_0 231.9°C

t_1 2,270°C

I.	II.	III.	IV.	V.
7.332	14.629	30.654	40.740	81.13

λ	*I*		eV
	J	*O*	
I 2,839.99	300*R*	300*R*	4.78
I 2,863.33	300*R*	300*R*	4.32
I 3,009.15	200	300	4.33
I 3,034.12	150	200	4.30
I 3,175.02	400*R*	500	4.33
I 3,262.33	300	400	4.87
I 4,524.74	50	500	4.87
I 5,631.69	200	50	4.33

Sn 2,839.99 *J* 300*R* *O* 300*R*

	λ	*I* *J*	*I* *O*
In	**2,839.98**	**2**	—
W	**2,839.92**	**10**	—
Pd	**2,839.892**	**100**	—
U	**2,839.890**	**20**	**18**
Tm	**2,839.85**	**10**	**5**
Fe	**2,839.82**	**10*h***	—
W	**2,839.81**	**10**	**1**
Ti	**2,839.80**	**100*wh***	—
Nb	**2,839.80**	**5**	**2**
Ta	**2,839.78**	**2**	**8**
Nd	**2,839.69**	**5**	—
U	**2,839.65**	**4**	**5**
Nb	**2,839.60**	**10*h***	—
Mo	**2,839.58**	**1**	**25**
Ce	**2,839.56**	—	**5**
Na	**2,839.55**	**(20)**	**2**
Fe	**2,839.53**	**25**	**4**
Ce	**2,829.52**	—	**2**
V	**2,839.44**	**7**	**12**
Ta	2,838.24	150	2
Fe	2,838.12	150	150
Cu	2,837.55	250	—
Cd	2,836.91	80	200
C	2,836.71	200	—
Cr	2,835.63	400*r*	100
Ta	*2,833.64*	*40w*	*300w*
Pd	*2,833.07*	*80R*	*500R*
Fe	*2,832.44*	*200*	*300*
Fe	*2,831.56*	*500*	*1*
Pt	*2,830.29*	*600r*	*1000R*
Mn	**2,840.00**	—	**20**
Cr	**2,840.02**	**125**	**25**
Al	**2,840.05**	**(15)**	—
W	**2,840.10**	**2**	**9**
V	**2,840.11**	**25**	**2**
Th	**2,840.156**	**6**	**8**
Ce	**2,840.156**	—	**3**
Ir	**2,840.219**	**10**	**15**
W	**2,840.220**	**4**	**9**
Gd	**2,840.24**	**60**	**50**
Fe	**2,840.34**	**8**	—
Re	**2,840.35**	—	**40**
Ta	**2,840.39**	**50**	**2**
Fe	**2,840.42**	**20**	**125**
Cr	**2,840.44**	**6**	—
U	**2,840.47**	**2**	**3**
La	**2,840.50**	**25*h***	**3**
Ru	**2,840.54**	**8**	**60**
Ru	2,841.68	200	50
Rn	2,842.1	(150)	—
Ni	2,842.42	150	—
Ta	2,842.81	50	200
Cr	2,843.25	400*r*	125
Fe	2,843.98	300	300
Ta	2,844.25	50	400*r*
Ta	2,844.46	200	200
Ru	2,844.71	150	—
Ta	2,844.76	30	150
Ta	2,845.35	10	150
Hg	*2,847.67*	*(300)*	—
Ta	*2,848.52*	*50*	*300*
Sn	*2,850.62*	*100wh*	*80*
Ta	*2,850.98*	*150*	*400*

Sn 2,863.33 *J* 300*R* *O* 300*R*

	λ	*I* *J*	*I* *O*
Ru	**2,863.32**	**80**	**30**
Mn	**2,863.26**	—	**3**
Ce	**2,863.22**	—	**2**
Cl	**2,863.20**	**(3)**	—
U	**2,863.19**	**4*h***	**3*h***
Mo	**2,863.12**	**20**	—
Fe	**2,863.08**	**2**	**3**
Fe	**2,863.33**	—	**5**
Ce	**2,863.34**	—	**12**
Tm	**2,863.35**	—	**10**
Fe	**2,863.43**	**80**	**100**
U	**2,863.44**	**4**	**6**
Co	**2,863.54**	—	**3**
Cl	**2,863.55**	**(7)**	—

	λ	I J	I O		λ	I J	I O
V	2,863.05	7wh	20	O	2,863.57	(10h)	—
Sb	2,863.02	4wh	—	Sc	2,863.68	—	4
W	2,863.01	8s	9s	Ni	2,863.70	250	—
Ru	2,863.00	—	6	Bi	2,863.75	18	80w
La	2,862.97	15h	2	Tm	2,863.76	40	15
Rh	2,862.93	60	150	V	2,863.79	12	4
Ru	2,862.88	60	6	Mo	2,863.81	100h	30
Mo	2,862.84	—	10	Ir	2,863.84	—	15
U	2,862.80	6	6	Fe	2,863.86	100	125
Ce	2,862.78	—	15	W	2,863.88	12	10
W	2,862.77	2	6	Nd	2,863.95	2	5
Co	2,862.766	—	9h	Ni	2,864.15	300wh	—
Cr	2,862.57	300R	80	Cr	2,865.11	200R	60
Pt	2,860.68	150h	30	Ta	2,867.41	150	5h
Ta	2,858.43	300	100	Nb	2,868.52	300	15
Fe	2,858.34	200	3	Ta	2,868.65	40	150
Pd	*2,854.58*	*500h*	*4*	V	2,869.13	150r	25
Mg	*2,852.13*	*100R*	*300R*	Tm	2,869.22	300	100
				Fe	2,869.31	70	300
				Cr	*2,870.44*	*300W*	*25*
				Fe	*2,873.40*	*300*	*—*
				Fe	*2,874.17*	*200*	*300*
				Nb	*2,875.39*	*300*	*50r*

Sn 3,009.15 *J* 200 *O* 300

	λ	J	O		λ	J	O
Na	3,009.14	(20)	3	Ca	3,009.20	5	20
Fe	3,009.09	60	80	Tb	3,009.30	8	8
W	3,009.08	5	10	Yb	3,009.39	20	2
Rh	3,008.98	3	2	U	3,009.42	2d	15d
W	3,008.98	12	—	Ho	3,009.48	4h	—
Nb	3,008.96	10	—	Eu	3,009.53	—	10W
U	3,008.922	1	5	Rh	3,009.57	2	2
Tm	3,008.92	20	20	Fe	3,009.570	400	500
O	3,008.83	(10h)	—	Ir	3,009.572	—	2
Dy	3,008.821	—	5	Ru	3,009.69	—	20
Sm	3,008.82	2	3	Tb	3,009.72	10	15
Ru	3,008.80	5	50	Ce	3,009.77	—	5
Ce	3,008.79	3	40	Pd	3,009.78	10	50r
W	3,008.76	7	8	Gd	3,010.14	100	100
Ir	3,008.62	—	4	Rn	3,010.8	(100)	—
V	3,008.61	70	3	Cu	3,010.84	30	250
V	3,008.50	50	—	Ni	3,012.00	125W	800R
Th	3,008.49	12	15	Ta	3,012.54	100	125

	λ	I J	I O		λ	I J	I O
In	3,008.31	500W	—	Hf	3,012.90	100	80
Fe	3,008.14	400r	600r	Os	3,013.07	20	150
Rn	3,006.8	(300)	—	Cr	3,013.71	150	200r
Cr	3,005.06	125	300r	Cr	3,014.76	100	300r
Cr	3,003.92	150	1	V	3,014.82	100	10
Ni	3,003.63	80	500R	Cr	3,014.91	100	300r
Fe	3,003.03	100	200	Cr	3,015.19	80	200r
Yb	3,002.61	150	15	Tm	3,015.29	100	125
Ni	3,002.49	100	1000R	Cr	3,015.51	150	1
V	*3,001.20*	*200r*	*20*	*Ti*	*3,017.19*	*200*	*15*
Fe	*3,000.95*	*300r*	*800R*	*Te*	*3,017.51*	*(350)*	—
Fe	*2,999.51*	*300*	*500*	*Cr*	*3,017.57*	*200*	*300r*
Pt	*2,997.97*	*200r*	*1000R*	*Os*	*3,018.04*	*50*	*300R*
Cu	*2,997.36*	*30*	*300*	*Fe*	*3,020.49*	*300r*	*300r*
Cr	*2,996.58*	*125*	*300r*				

*N**

Sn 3,034.121 *J* 150 *O* 200

	λ	J	O		λ	J	O
Fe	**3,034.12**	—	**2**	**In**	**3,034.122**	**1**	**8**
Ag	**3,034.11**	**1*h***	**4**	**Kr**	**3,034.16**	**(2*h*)**	—
Th	**3,034.07**	**15**	**15**	**Sm**	**3,034.17**	—	**5**
Ru	**3,034.060**	**5**	**60**	**Cr**	**3,034.190**	**60**	**200*r***
Gd	**3,034.059**	**60**	**100**	**W**	**3,034.195**	**9**	**10**
U	**3,034.05**	**5**	**6**	**Pr**	**3,034.20**	**3**	—
Ir	**3,034.04**	—	**4**	**U**	**3,034.37**	**1**	**3**
Ru	**3,033.89**	**6**	—	**Co**	**3,034.43**	**2**	**80**
Yb	**3,033.86**	**3**	**1**	**Ne**	**3,034.48**	**(18)**	—
V	**3,033.82**	**90*r***	**20**	**Sm**	**3,034.49**	**2**	**5**
U	**3,033.77**	**4*d***	**4*d***	**Ca**	**3,034.53**	—	**2**
Xe	**3,033.71**	**(8)**	—	**Re**	**3,034.536**	—	**30**
Ir	**3,033.62**	**1**	**25**	**Fe**	**3,034.538**	**40**	**70**
W	**3,033.57**	**15**	**12**	**Cr**	**3,034.542**	**30**	—
Mn	**3,033.56**	**2**	**1**	**Ir**	**3,034.56**	—	**20**
Ar	**3,033.52**	**(8)**	—	**Te**	**3,034.62**	**(10)**	—
Bi	**3,033.5**	**(15)**	—	**Yb**	**3,034.63**	**6**	**2**
Zr	**3,033.46**	**2**	—	**U**	**3,034.66**	**2**	**3**
Ru	**3,033.451**	**10**	**70**	Nb	3,034.95	100*w*	1
V	**3,033.448**	**40**	**20**	Fe	3,035.74	60	100
Fe	**3,033.445**	**2**	**4**	Zn	3,035.78	100	200
Ra	3,033.44	(150)	—	Cu	3,036.10	50	200
Cr	3,032.93	100	10	Pt	3,036.45	10	200
Gd	3,032.85	100	100	Ru	3,036.47	150	50
As	3,032.84	70	125	Cr	3,037.04	100	200*r*

	λ	I J	I O
Sn	3,032.775	20	50
Nb	3,032.768	300	3
Pd	3,032.20	100h	2
Ni	3,031.87	—	200
Fe	3,031.64	200	200
Fe	3,031.21	150	150
Hf	3,031.16	90	70
Yb	3,031.11	30	100h
Os	3,030.69	40	500
Re	3,030.45	—	100
Cr	3,030.24	150	200r
Fe	3,030.15	300	300
Sb	3,029.81	200wh	100
Ti	3,029.73	150	12
Nb	3,028.44	200	50
Cr	3,028.12	125	2
Pd	3,027.91	200h	150
Gd	3,027.61	60	100
Ta	3,027.51	35w	125
Fe	*3,026.46*	*200*	*200*
Fe	*3,025.84*	*300r*	*400r*
Nb	*3,024.74*	*200*	*10w*
Bi	*3,024.63*	*50*	*250wh*
Cr	*3,024.35*	*125*	*300r*
Fe	*3,024.03*	*200*	*300*
Ti	*3,022.82*	*150 Wh*	—
Cr	*3,021.56*	*200r*	*300r*
Fe	*3,021.07*	*300r*	*700R*
Cr	*3,020.67*	*100*	*200r*
Fe	3,037.39	400r	700R
Ta	3,037.50	100	8h
Ni	3,037.93	100	800R
Ge	3,039.06	1000	1000
In	3,039.36	500R	1000R
Nb	3,039.81	300	5
Fe	3,040.43	400	400
Sb	3,040.67	(400wh)	—
Cr	3,040.85	200	500R
Os	*3,040.90*	*100*	*200*
Pt	*3,042.64*	*250R*	*200R*
Fe	*3,042.66*	*200*	*300*
Co	*3,044.00*	—	*400R*
Ni	*3,045.01*	*10*	*200*
Te	*3,047.00*	*(350)*	—
Fe	*3,047.60*	*500r*	*800r*

Sn 3,175.02 J 400R O 500

	λ	J	O
Fe	**3,174.96**	**4**	**5**
Co	**3,174.90**	—	**80**
Dy	**3,174.883**	**4**	**12**
La	**3,174.88**	**10h**	**3**
Ho	**3,174.86**	**6**	**6**
U	**3,174.84**	—	**6**
Pt	**3,174.82**	—	**2**
Ti	**3,174.80**	**100**	—
Re	**3,174.78**	—	**20w**
Yb	**3,174.76**	**4**	**2**
Mo	**3,174.67**	**1**	**4**
Tb	**3,174.66**	**15**	**15**
Mn	**3,174.65**	—	**15h**
Re	**3,174.62**	—	**30**
V	**3,174.54**	**80**	**1**
In	**3,175.03**	**3**	—
Fe	**3,175.035**	—	**1**
J	**3,175.047**	**(10)**	—
Mo	**3,175.049**	**60**	**2**
Ce	**3,175.06**	—	**10**
Fe	**3,175.08**	**2**	—
Na	**3,175.09**	**(15)**	—
Te	**3,175.11**	**(15)**	**30**
P	**3,175.14**	**(70)**	—
Ru	**3,175.15**	**100**	**20**
Xe	**3,175.25**	**(3)**	—
Ru	**3,175.298**	**40**	**30**
Cl	**3,175.30**	**(6)**	—
U	**3,175.36**	—	**8**
Fe	**3,175.447**	**200**	**200**

	λ	I J	I O
Pt	3,174.53	2	—
Sm	3,174.51	1	3
Cd	3,174.49	2	—
Th	3,174.46	5	8
Nb	3,174.44	10*h*	—
Ta	3,174.38	1	3
Th	3,174.17	10	10
Tm	3,172.82	200	200
Cr	3,172.08	200	2
La	3,171.67	1000*wh*	2*h*
Mo	3,170.35	25*r*	1000*R*
Ta	3,170.29	35	250*w*
Ti	3,168.52	300*r*	70
Tb	3,175.45	3	8
Er	3,175.52	2	10
Mo	3,175.53	1	5
Ce	3,175.58	—	2
Mo	3,175.59	3	5
Rn	3,175.6	(40)	—
Xe	3,175.64	(40)	—
Os	3,175.65	8	10
Ti	3,175.66	20*wh*	—
Kr	3,175.67	(40*wh*)	—
Nb	3,175.68	20	1
Mn	3,175.71	—	8
Ce	3,175.72	—	8
Th	3,175.73	15	12
U	3,175.74	1	4
Yb	3,175.75	5	1
Nb	3,175.78	20	1
Sm	3,175.83	2	1
Mg	3,175.84	—	5
Nb	3,175.85	50	5
Ru	3,177.05	200	60
Fe	3,177.53	300	5
Fe	3,178.01	150	300
Ca	3,179.33	400*w*	100
Fe	3,180.23	300	300
Nb	3,180.29	200	5
V	*3,183.98*	*400R*	*500R*
V	*3,185.40*	*400R*	*500R*

Sn 3,262.33 *J* 300 *O* 400

	λ	I J	I O
Cs	3,262.29	(6)	—
Os	3,262.290	50	500*R*
W	3,262.28	18	—
Fe	3,262.280	25	50
Ba	3,262.275	3	3
Dy	3,262.27	—	2
Sm	3,262.26	4	10
Mo	3,262.19	20	1
Rb	3,262.15	(30)	—
Ce	3,262.14	—	5
V	3,262.06	3	10
Dy	3,262.02	1	2
Xe	3,262.02	(3*h*)	—
Fe	3,262.013	15	30
Ir	3,262.010	2	20
Pb	3,262.353	5*h*	20*h*
Mo	3,262.353	1	8
Nd	3,262.36	2*h*	10*d*
Pr	3,262.39	—	2
Tb	3,262.40	—	2
Co	3,262.43	—	2
Hf	3,262.47	1	10
Eu	3,262.49	2*h*	5
Nb	3,262.56	5*h*	—
Sm	3,262.58	1	3
Eu	3,262.61	—	15
Mo	3,262.63	4	15
Th	3,262.67	15	12
Tb	3,262.68	3	8
Ir	3,262.72	—	3

	λ	I: J	I: O
Ba	**3,261.96**	**—**	**40**
Nb	**3,261.88**	**3**	**10**
Mo	**3,261.84**	**—**	**2**
Tb	**3,261.74**	**3**	**15**
Pd	**3,261.73**	**2*h***	**—**
U	**3,261.72**	**10**	**15**
Yb	**3,261.70**	**5**	**—**
Nb	**3,261.695**	**50**	**1**
Pt	**3,261.692**	**2**	**8**
Tm	**3,261.66**	**100**	**30**
Ce	**3,261.63**	**2*h***	**2**
Tl	**3,260.605**	**300*r***	**70**
S	**3,260.60**	**(8)**	**—**
Eu	**3,260.59**	**—**	**2*h***
Kr	**3,260.58**	**(8*h*)**	**—**
Re	**3,260.55**	**—**	**50**
Th	**3,260.54**	**10**	**8**
Ce	**3,260.52**	**—**	**2**
Ta	**3,261.511**	**35*h***	**—**
Yb	**3,261.509**	**18**	**5**
Cd	3,261.06	300	300
Nb	3,260.56	300	15
Fe	3,259.05	200	1
Pd	3,258.78	200*h*	300
Fe	3,258.77	150	—
In	3,258.56	300*R*	500*R*
In	3,256.09	600*R*	1500*R*
Fe	3,254.36	150	200
Lu	3,254.31	150	50
Co	3,254.21	—	300*R*
Nb	3,254.07	300	20
Cd	*3,252.52*	*300*	*300*
Pd	*3,251.64*	*500*	*200*
Cu	*3,247.54*	*2000R*	*5000R*
La	*3,245.12*	*300*	*400*
Os	**3,262.75**	**20**	**100**
Re	**3,262.76**	**—**	**25**
Er	**3,262.80**	**4**	**15**
Ir	**3,262.93**	**—**	**4**
Tb	**3,262.97**	**3**	**30**
Ta	**3,263.00**	**1**	**20**
Eu	**3,263.02**	**—**	**2*W***
Th	**3,263.03**	**8**	**8**
Cs	**3,263.06**	**(4)**	**—**
Fe	**3,263.07**	**—**	**1**
Ce	**3,263.071**	**—**	**10**
W	**3,263.10**	**7**	**8**
U	**3,263.11**	**8**	**25**
Rh	**3,263.14**	**40**	**200**
Co	**3,263.21**	**—**	**30**
V	**3,263.24**	**—**	**40**
Nb	3,263.37	500	3
Hg	3,264.06	(200)	—
Fe	3,265.05	150	200
Fe	3,265.62	300	300
La	3,265.67	200	300
Pd	3,267.35	200*h*	—
Sb	3,267.50	150*Wh*	150
Os	3,267.94	30	400*R*
Tm	3,269.00	150	40
Os	3,269.21	20	200
Ge	3,269.49	300	300
Fe	3,271.00	300	300
Cu	*3,273.96*	*1500R*	*3000R*
Eu	*3,280.682*	*—*	*1000R*
Ag	*3,280.683*	*1000R*	*2000R*

Sn 4,524.741 *J* 50* *O* 500

	λ	J	O
Hf	**4,524.74**	**15**	**10**
Ti	**4,524.732**	**10**	**10**
Mo	**4,524.726**	**5**	**4**
Pr	**4,524.69**	**1**	**5**
Ce	**4,524.59**	**—**	**2**
Eu	**4,524.49**	**—**	**12**
Pr	**4,524.36**	**2*w***	**10*w***
Mo	**4,524.34**	**30**	**30**
Cr	**4,524.84**	**3**	**20**
Os	**4,524.87**	**2**	**80**
Co	**4,524.93**	**—**	**3**
Ba	**4,524.95**	**30**	**80**
Tb	**4,525.01**	**—**	**3**
Fe	**4,525.15**	**50**	**100**
V	**4,525.16**	**12**	**15**
Rh	**4,525.21**	**2**	**5**

	λ	I J	I O
Gd	4,524.23	1	3
Tb	4,524.326	—	3
V	4,524.22	30	40
Tb	4,524.13	—	2
Nb	4,524.12	5*h*	5*h*
Gd	4,524.10	2	8
Ce	4,524.03	—	4
Pr	4,523.99	—	4
Sm	4,523.912	50	100
Zr	4,523.909	—	3
Re	4,523.88	—	40
Gd	4,523.84	—	8
Nd	4,523.57	2	12
Dy	4,523.50	—	2
Eu	4,522.58	—	500
*bh*Yt	4,525.3	—	5
La	4,525.303	100	100
Mo	4,525.32	8	6
Tb	4,525.34	—	2
Lu	4,525.48	—	2*h*
U	4,525.65	3	1
*bh*Yt	4,525.7	—	2
Co	4,525.79	—	5
Mo	4,525.86	3	2
Ti	4,525.934	—	2
Ru	4,525.935	—	7
Tb	4,525.94	—	3
U	4,525.96	5	8*h*
Re	4,525.99	—	30
Fe	4,528.62	200	600
Rh	4,528.72	60	500*r*
Ta	4,530.85	50	300
Co	4,530.96	8	1000*w*
Co	4,533.99	8	500
Co	*4,549.66*	—	*600*

Sn 5,631.69 *J* 200 *O* 50*

	λ	I J	I O
Tm	5,631.41	10	80
Hf	5,631.36	4	1
Te	5,630.66	(25)	—
Ar	5,630.44	(10)	—
As	5,630.42	6	—
Fe	5,630.35	2	5
Nb	5,629.83	1	3
J	5,625.70	(150)	—
Fe	5,624.55	125	150
Se	5,623.13	(300)	—
Xe	5,616.67	(150)	—
Ra	5,616.66	(250)	—
Fe	5,616.65	300	400
Ar	*5,606.73*	*(500)*	—
S	*5,606.10*	*(700)*	—
Ar	*5,597.46*	*(500)*	—
Al	*5,593.23*	*(200)*	—
Se	*5,591.16*	*(500)*	—
Sb	5,631.97	(40)	15
Mo	5,632.47	50	100
Kr	5,633.02	(100*h*)	—
Cu	5,633.14	3	—
Xe	5,633.24	(5*Wh*)	—
Fe	5,633.96	10	20
Rb	5,635.99	100	—
In	5,636.66	(150)	—
In	5,636.75	(300)	—
Ar	5,639.11	(100)	—
Sb	5,639.74	100*wh*	—
S	5,639.98	(500)	—
S	5,640.37	(500)	—
Ti	5,644.14	200	150
V	5,646.11	150	150
Ar	5,648.66	(200)	—
Te	5,649.30	(250)	—
Kr	5,649.56	(100)	—
Ar	5,650.70	(1500)	—
As	5,651.53	200	—

	λ	I	
		J	O
Br	5,589.93	(250)	—
Sn	5,588.92	(50)	2h
Ar	5,588.69	(500)	—

	λ	I	
		J	O
Ne	5,656.66	(500)	—
Ar	5,659.13	(500)	—
S	5,659.93	(600)	—
Ra	5,660.81	(1000)	—
Yt	5,662.92	400	20
N	5,666.64	(300)	—
Xe	5,667.56	(300)	—
Hg	5,677.17	(300)	—

Sr $^{38}_{87.62}$

t_0 800°C t_1 1,384°C

I.	II.	III.	IV.	V.
5.693	11.026	—	—	—

λ	*I*		eV
	J	*O*	
I 3,351.25	15	300	5.54
II 3,380.71	200	150	6.61
II 3,464.46	200	200	6.62
II 4,077.71	500	400*R*	3.04
II 4,215.52	400	300*R*	2.94
I 4,607.33	50*R*	1000*R*	2.69
I 4,832.08	8	200	4.34

Sr 3,351.246 *J* 15* *O* 300

	λ	I J	I O		λ	I J	I O
Th	**3,351.230**	**15**	**10**	**U**	**3,351.251**	**3**	**8**
Zr	**3,351.228**	—	**4**	**Pr**	**3,351.25**	**1**	**5**
Eu	**3,351.19**	—	**15*W***	**Sm**	**3,351.28**	**2**	**5**
W	**3,351.18**	**3**	**5**	**Er**	**3,351.30**	**2**	**10*w***
Co	**3,351.15**	—	**4**	**Hg**	**3,351.30**	**20*h***	**10**
Mo	**3,351.12**	**5**	**3**	**Mn**	**3,351.43**	—	**4**
Ce	**3,351.08**	—	**3**	**Ta**	**3,351.510**	**15*r***	**18*r***
Eu	**3,351.04**	**2**	**2**	**Mo**	**3,351.512**	**8**	**5**
Ta	**3,350.96**	**18*W***	**25*W***	**Ce**	**3,351.520**	—	**12**
Dy	**3,350.95**	**2**	**2**	**Fe**	**3,351.521**	**50**	**70**
Ce	**3,350.93**	—	**3**	**Co**	**3,351.54**	**2**	**35**
U	**3,350.90**	**2**	**3**	**Eu**	**3,351.56**	**2**	**12**
Rb	**3,350.89**	—	**150**	**Cr**	**3,351.596**	**8**	**35**
Sm	**3,350.88**	**4**	**9**	**Th**	**3,351.599**	**4**	**4**
Ce	**3,350.75**	—	**10**	**U**	**3,351.66**	**2**	**4**
Nb	**3,350.69**	**7*h***	**10**	**Mn**	**3,351.662**	—	**8**
Ce	**3,350.68**	—	**20**	**Ti**	**3,351.67**	**5**	**1**
Dy	**3,350.66**	**2**	**3**	**Sm**	**3,351.69**	—	**3**
Sm	**3,350.65**	—	**8**	**Os**	**3,351.73**	**12**	**50**
W	**3,350.61**	**10**	**4**	**Fe**	**3,351.74**	**60**	**80**
Ag	**3,350.56**	—	**2*h***	**Ce**	**3,351.75**	—	**3**
Ru	**3,350.549**	—	**12**	**Nd**	**3,351.80**	**1*h***	**8*d***
Ti	**3,350.548**	**1**	**10**	**U**	**3,351.83**	**2**	**5**
U	**3,350.52**	**2*h***	**8**	**Ta**	**3,351.87**	**1*h***	**2*wh***
Gd	**3,350.48**	**180**	**150**	**La**	**3,351.89**	**2**	**2**
Ho	**3,350.46**	**8**	**6**	**Ru**	**3,351.93**	**4**	**50**
Eu	**3,350.43**	—	**30**	**Cr**	**3,351.966**	**50**	**50**
Mn	**3,350.40**	—	**4**	**W**	**3,351.967**	**4**	**7**
Co	**3,350.38**	—	**4*h***	**Sc**	**3,352.048**	**3**	**12**
Ca	**3,350.36**	**3**	**15**	**Tb**	**3,352.05**	**3**	**8**
Th	**3,350.33**	**2*d***	**2**	**Hf**	**3,352.055**	**50**	**30**
Mo	**3,350.30**	**6**	**5**	**Ti**	**3,352.07**	**15**	**8**
Fe	**3,350.285**	**4**	**8**	**Ho**	**3,352.08**	**6**	**6**
U	**3,350.28**	**1*hd***	**4*d***	**Tm**	**3,352.11**	**6**	**6**
Pr	**3,350.27**	**3**	**10**	**Nd**	**3,352.22**	**2**	**8**
Er	**3,350.255**	**2**	**10**				
Ru	**3,350.251**	**2**	**10**				
				Co	3,354.38	—	200*R*
				Co	3,356.47	2	150*W*
				Lu	3,359.56	15	150
Fe	3,347.93	100	150				
Cr	3,346.74	80*r*	150*R*				
Zn	3,345.93	50	150				
Zn	3,345.57	100	500				
Zn	3,345.02	300	800				
La	3,344.56	200*wh*	300				
Re	3,344.35	—	150				
Co	3,342.73	—	150*W*				
Re	3,342.26	—	200				
La	*3,337.49*	*300wh*	*800*				

*N**

Sr 3,380.71 *J* 200 *O* 150

	λ	*I* *J*	*I* *O*		λ	*I* *J*	*I* *O*
Sm	**3,380.703**	**1**	**2**	**Cu**	**3,380.72**	**5**	**—**
U	**3,380.700**	**10**	**5**	**Nb**	**3,380.86**	**—**	**5**
W	**3,380.69**	**3**	**—**	**Th**	**3,380.883**	**2**	**5**
Pd	**3,380.67**	**2*h***	**150*w***	**Ni**	**3,380.885**	**12**	**200**
Rh	**3,380.64**	**—**	**4**	**Tb**	**3,380.89**	**—**	**8**
K	**3,380.62**	**(30)**	**—**	**U**	**3,380.90**	**3*h***	**3**
Tb	**3,380.60**	**3**	**15**	**La**	**3,380.910**	**100*h***	**200**
Ni	**3,380.57**	**100**	**600*R***	**Ru**	**3,380.915**	**2**	**10**
Gd	**3,380.52**	**2**	**2**	**Ce**	**3,380.916**	**—**	**10**
Nb	**3,380.49**	**2*h***	**3**	**Nb**	**3,380.94**	**200**	**—**
Nb	**3,380.41**	**3**	**20**	**J**	**3,380.98**	**(3)**	**—**
Ti	**3,380.28**	**150*r***	**25**	**Ir**	**3,380.99**	**1**	**15**
Eu	**3,380.25**	**10**	**15**	**Fe**	**3,380.998**	**2**	**—**
Mo	**3,380.21**	**25**	**1**	**Tl**	**3,381.00**	**(20)**	**—**
Ru	**3,380.17**	**15**	**60**	**Zn**	**3,381.04**	**(20)**	**—**
Tb	**3,380.15**	**—**	**3**	**Nd**	**3,381.05**	**4**	**4*d***
Yt	**3,380.114**	**12**	**7**	**Er**	**3,381.08**	**2**	**15**
Fe	**3,380.111**	**25**	**200**	**Kr**	**3,381.11**	**(20*wh*)**	**—**
Pr	**3,380.11**	**1**	**3**	**Mo**	**3,381.119**	**2**	**3**
Nb	**3,380.05**	**5**	**5**	**Ce**	**3,381.121**	**—**	**10**
W	**3,379.968**	**—**	**4**	**Cu**	**3,381.124**	**5*h***	**15**
Mo	**3,379.966**	**1**	**25**	**Er**	**3,381.32**	**4**	**20**
Cu	**3,379.961**	**3*h***	**2*h***	**Ti**	**3,381.33**	**—**	**5**
Ti	**3,379.93**	**5**	**2**	**Fe**	**3,381.34**	**2**	**5**
Zr	**3,379.92**	**2**	**10**	**Ir**	**3,381.370**	**—**	**3**
Ir	**3,379.83**	**2*h***	**4**	**Th**	**3,381.375**	**12**	**5**
Cr	**3,379.82**	**100**	**15**	**Pr**	**3,381.376**	**—**	**3**
Nd	**3,379.80**	**4**	**12**	**Cu**	**3,381.421**	**7**	**20**
Yb	**3,379.78**	**20**	**12**	**La**	**3,381.438**	**2**	**5**
Gd	**3,379.763**	**2**	**2**	**Ti**	**3,381.442**	**—**	**2**
Mo	**3,379.762**	**25**	**1**	**Rh**	**3,381.445**	**—**	**4**
Pr	**3,379.759**	**2**	**10**	**Ar**	**3,381.49**	**(20)**	**—**
Re	**3,379.72**	**—**	**80**	**Ce**	**3,381.491**	**—**	**20**
				Co	**3,381.50**	**—**	**100*W***
				Tb	**3,381.60**	**3**	**8**
Cr	3,379.37	100	6	**Os**	**3,381.67**	**1**	**15**
Nb	3,379.30	100	1	**U**	**3,381.69**	**—**	**6**
Fe	3,379.02	50	80				
Fe	3,378.68	80	150				
Cr	3,378.34	150	25	Sm	3,382.41	40	100
Co	3,377.06	—	100	V	3,382.53	125	—
Lu	3,376.50	10	100	Cr	3,382.68	200	35
La	3,376.33	50	100	Nd	3,382.81	10	200
V	3,376.06	60	80	Ag	3,382.89	700*R*	1000*R*
Yb	3,375.48	100	30				
				Fe	3,383.70	70	100
Ru	3,374.65	18	80	Ti	3,383.76	300*R*	70
Ni	3,374.22	6	400	Fe	3,383.98	100	200
Ar	3,373.48	(300)	—	Os	3,384.00	5	80

	λ	I J	I O
Pd	3,373.00	500wh	800r
Ti	3,372.80	400R	80
Nb	3,372.56	200	10h
Rh	3,372.25	200	300
Sc	3,372.15	150	7
Ni	3,371.99	10	400
Ti	3,371.45	15	100
Fe	3,370.79	200	300
Os	*3,370.59*	*30*	*300R*
Ne	*3,369.91*	*(700)*	*—*
Ne	*3,369.81*	*(500)*	*—*
Ni	*3,369.57*	*100*	*500R*
Fe	*3,369.55*	*200*	*300*
Rh	*3,368.37*	*50*	*300*
Co	*3,367.11*	*30*	*300R*
Sr	*3,366.33*	*10*	*100*
Ni	*3,366.17*	*12*	*400W*
Ni	*3,365.77*	*12*	*400W*
Tm	*3,362.61*	*200*	*250*
Gd	*3,362.24*	*180*	*150*
Ni	*3,361.56*	*20*	*500W*
V	*3,361.51*	*200*	*—*
Ti	*3,361.21*	*600R*	*100*
Co	3,385.22	15	250R
Hg	3,385.25	(200)	—
Ti	3,385.95	25	80
Nb	3,386.24	100	5
Os	3,387.84	15	100
Zr	3,387.87	100	100
Co	3,388.17	12	250R
Ru	3,388.71	20	80
bhSr	3,389.8	—	12
Ni	*3,391.05*	*40*	*400*
Zr	*3,391.97*	*400*	*300*
Fe	*3,392.66*	*200*	*300*
Ni	*3,392.99*	*—*	*600R*
Ar	*3,393.75*	*(250)*	*—*
Cr	*3,394.29*	*150*	*15*
Ti	*3,394.57*	*200*	*70*
Fe	*3,394.59*	*80*	*150*
Co	*3,395.37*	*50*	*400R*
Rh	*3,396.85*	*500*	*1000w*
Re	*3,399.30*	*—*	*200w*
Fe	*3,399.34*	*200*	*200*
Rh	*3,399.70*	*60*	*500*

Sr 3,464.457 *J* 200 *O* 200

	λ	J	O
W	**3,464.45**	**12**	**—**
Sm	**3,464.43**	**1**	**3**
Th	**3,464.42**	**2d**	**2d**
Yb	**3,464.37**	**50r**	**200R**
Cd	**3,464.37**	**10**	**—**
Eu	**3,464.36**	**—**	**15d**
Ne	**3,464.34**	**(75)**	**—**
Ba	**3,464.23**	**—**	**5**
Ce	**3,464.21**	**1w**	**8**
Yb	**3,464.19**	**—**	**12**
Xe	**3,464.17**	**(2h)**	**—**
V	**3,464.170**	**40**	**—**
Ce	**3,464.16**	**—**	**10**
Si	**3,464.14**	**3**	**—**
Ar	**3,464.14**	**(15)**	**—**
Gd	**3,464.139**	**3**	**4**
Sm	**3,464.08**	**10**	**30**
Mn	**3,464.04**	**(15)**	**—**
Cr	**3,464.00**	**2**	**1**
Gd	**3,463.98**	**125**	**100**
Dy	**3,464.46**	**1**	**2**
U	**3,464.47**	**—**	**4**
Fe	**3,464.49**	**1h**	**1h**
Er	**3,464.54**	**3**	**10**
Tb	**3,464.63**	**3**	**8**
S	**3,464.66**	**(8)**	**—**
U	**3,464.67**	**2**	**—**
Re	**3,464.722**	**—**	**100**
Ce	**3,464.722**	**—**	**2**
Cr	**3,464.84**	**3**	**30**
Ce	**3,464.86**	**—**	**12**
U	**3,464.87**	**—**	**2**
Os	**3,464.88**	**5**	**15**
Fe	**3,464.92**	**—**	**2**
Nd	**3,464.93**	**2h**	**10**
U	**3,464.95**	**4**	**4**
Ce	**3,464.99**	**—**	**12**
Th	**3,465.02**	**5**	**5**
Mn	**3,465.04**	**(18)**	**—**
W	**3,465.09**	**10**	**—**

	λ	I	
		J	O
Pt	3,463.94	1h	2
Ta	3,463.915	10	3
U	3,463.910	—	4
Dy	3,463.88	2	12
V	3,463.83	25	—
Nb	3,463.81	50	30
Ta	3,463.77	—	50
Ce	3,463.76	1	15s
Th	3,463.72	12	10
Nb	3,463.68	2	5
Al	3,463.63	(1)	3
Sm	3,463.611	—	2
Cr	3,463.610	1	15
Pb	3,463.6	(50)	—
Mo	3,463.57	20w	10w
W	3,463.51	25	8
Cu	3,463.499	—	6h
Co	3,463.499	—	8
Cs	3,463.42	(6)	—
V	3,463.39	2	10
Co	3,462.80	80	1000R
Tm	3,462.20	200	250
Rh	3,462.04	150	1000
Ni	3,461.65	50h	800R
Rb	3,461.57	(200)	—
Ti	3,461.50	125	80
Co	3,461.18	3	100wh
Ar	3,461.08	(300)	—
Dy	3,460.97	3	100
Pd	3,460.77	600h	300r
Re	3,460.47	—	1000W
Mn	3,460.33	500	60
Os	3,459.02	10	100
Ni	3,458.47	50h	800R
Os	3,458.38	12	200
Yb	3,458.28	100	12
Rh	3,457.93	10	125
Cr	3,457.63	125	4
V	3,457.15	150	2
Rh	3,457.07	4	100
Ti	3,456.39	125	25
Ni	3,455.27	100h	—
Co	3,455.23	10	2000R
Rh	3,455.22	12	300
Cr	3,454.99	100	—
Dy	3,454.33	10	100
Yb	3,454.07	250	40
Nb	3,453.97	100	—

	λ	I	
		J	O
Er	3,465.12	—	8d
Mn	3,465.19	—	2
Cs	3,465.20	(4)	—
Ir	3,465.22	1	12
V	3,465.249	25	2
Cr	3,465.250	30	35
Ru	3,465.29	2h	5
Dy	3,465.30	—	2
Ce	3,465.302	—	8
Cu	3,465.40	2h	10
W	3,465.408	4	4
Kr	3,465.41	(6wh)	—
Ce	3,465.42	—	8
Os	3,465.43	12	60
Nd	3,465.44	—	6wh
Sm	3,465.46	—	5
Ce	3,465.503	—	5
Ta	3,465.505	15h	—
Co	3,465.80	25	2000R
Fe	3,465.86	400	500
Cd	3,466.20	500	1000
Gd	3,467.28	100	100
Ni	3,467.50	15	300
Cd	3,467.66	400	800
Re	3,467.95	—	100w
Gd	3,468.99	150	150
Ni	3,469,49	20	300
V	3,469.52	100	2
Rh	3,469.62	10	100
Nb	3,470.25	100	3h
Rh	3,470.66	125	500
O	3,470.77	(100)	—
Rh	3,472.25	8	100
Lu	3,472.48	150	50
Ni	3,472.54	40	800R
Ne	3,472.57	(500)	—
Sb	3,473.91	300wh	3
Co	3,474.02	100	3000R
Mn	3,474.13	400	12
Nb	3,474.67	4	100
Rh	3,474.78	125	700
Sr	3,474.89	50	80
Fe	3,475.45	300	400
Fe	*3,476.70*	*200*	*300*
Yb	*3,478.84*	*300*	*40*
Rh	*3,478.91*	*100*	*500*
Nb	*3,479.56*	*200*	*5*

	λ	I J	I O		λ	I J	I O
Co	*3,453.50*	*200*	*3000R*	*Ta*	*3,480.52*	*200ws*	*70*
Ni	*3,452.89*	*50*	*600R*				
Nb	*3,452.35*	*200*	*5*	*Pd*	*3,481.15*	*2h*	*500r*
Hg	*3,451.69*	*(200)*	*—*	*Mn*	*3,482.91*	*250*	*50*
Pd	*3,451.35*	*400h*	*—*	*Co*	*3,483.41*	*10*	*300R*
				Ni	*3,483.77*	*30*	*500R*
Co	*3,449.44*	*125*	*500R*	*Pt*	*3,485.27*	*200R*	*150*
Co	*3,449.17*	*125*	*500R*				
Ni	*3,446.26*	*50h*	*1000R*				
Fe	*3,445.15*	*150*	*300*				
Fe	*3,443.88*	*200*	*400*				
Co	*3,443.64*	*100*	*500R*				

Sr 4,077.71 *J* 500 *O* 400*R*

	λ	J	O		λ	J	O
Pr	**4,077.69**	**2**	**4**	**Cu**	**4,077.716**	**—**	**5**
Mo	**4,077.682**	**10**	**8**	**Sn**	**4,077.72**	**3**	**2**
Cr	**4,077.677**	**10**	**30**	**Ta**	**4,077.721**	**2*h***	**4**
Nd	**4,077.62**	**3**	**8**	**U**	**4,077.79**	**6**	**15**
Pb	**4,077.61**	**2**	**—**	**Hg**	**4,077.81**	**150**	**150**
Rh	**4,077.57**	**4**	**5**	**Er**	**4,077.970**	**18*s***	**20*s***
Ce	**4,077.47**	**4**	**18**	**Tb**	**4,077.97**	**2**	**25**
Co	**4,077.41**	**2*h***	**100*wh***	**Dy**	**4,077.974**	**100**	**150*r***
Yt	**4,077.37**	**40**	**50**	**V**	**4,077.977**	**—**	**2**
Dy	**4,077.35**	**—**	**4**	**Pr**	**4,077.98**	**2**	**10**
La	**4,077.34**	**400**	**600**	**Ho**	**4,078.00**	**3**	**3**
Yb	**4,077.27**	**100**	**30**	**Mo**	**4,078.07**	**4**	**4**
Ti	**4,077.153**	**2**	**18**	**Re**	**4,078.214**	**—**	**10**
Nd	**4,077.150**	**4**	**10**	**W**	**4,078.124**	**6**	**7**
Cr	**4,077.089**	**10**	**35**	**Zn**	**4,078.14**	**(5)**	**—**
Nb	**4,077.088**	**5**	**3**	**Pr**	**4,078.16**	**1**	**5**
As	**4,077.08**	**10**	**—**	**Eu**	**4,078.23**	**—**	**3**
W	**4,077.06**	**5**	**6**	**Tb**	**4,078.26**	**—**	**3**
Zr	**4,077.05**	**1**	**3**	**Zr**	**4,078.31**	**—**	**10**
Ar	**4,076.96**	**(10)**	**—**	**Ce**	**4,078.32**	**4**	**15**
Eu	**4,076.95**	**—**	**10**	**Nb**	**4,078.35**	**3**	**4**
Sm	**4,076.85**	**5**	**10**	**Fe**	**4,078.36**	**40**	**80**
Ir	**4,076.81**	**—**	**2*h***	**Mo**	**4,078.381**	**3**	**5**
Fe	**4,076.80**	**1**	**8**	**Sb**	**4,078.385**	**4**	**—**
Ru	**4,076.73**	**25**	**60**	**Gd**	**4,078.46**	**10**	**15**
U	**4,076.72**	**10**	**8**	**Tb**	**4,078.47**	**1**	**5**
La	**4,076.71**	**5**	**15**	**Ti**	**4,078.474**	**50**	**125**
Ar	**4,076.64**	**(20)**	**—**	**Ce**	**4,078.51**	**1**	**5**
Fe	**4,076.637**	**50**	**80**	**Sc**	**4,078.58**	**10**	**10**
Sm	**4,076.63**	**15**	**25**	**Ce**	**4,078.61**	**—**	**5**
Co	**4,076.57**	**1**	**3*h***	**Gd**	**4,078.709**	**10**	**20**
Ir	**4,076.55**	**—**	**25**	**V**	**4,078.708**	**—**	**2**
Zr	**4,076.53**	**—**	**10**	**Tb**	**4,078.78**	**—**	**5**

	λ	I J	I O
Mo	4,076.51	5	5
Fe	4,076.50	1	2
Ti	4,076.375	1	15
Ir	4,076.370	2	2
Yt	4,076.35	8	30
Au	4,076.33	25w	4
Ce	4,076.24	1	12
U	4,076.225	3	2
Fe	4,076.222	1	2
Pr	4,076.21	1	10
Mo	4,076.19	25	25
Co	4,076.13	—	70
Nb	4,076.09	3	4
Cr	4,076.06	15	30
Er	4,076.01	—	2
Dy	4,076.01	—	2
O	4,075.87	(800)	—
Te	4,073.57	(300)	—
O	4,072.16	(300)	—
Se	4,070.16	(500)	—
Kr	4,065.11	(300)	—
Fe	4,063.60	300	400
Cu	4,062.70	20	500w
Nb	*4,058.94*	*400w*	*1000w*
Pb	*4,057.82*	*300R*	*2000R*
In	*4,056.94*	*(500)*	—
Ag	*4,055.26*	*500R*	*800R*
Ho	*4,053.92*	*200*	*400*
K	*4,047.20*	*200*	*400*
Fe	*4,045.81*	*300*	*400*
Co	*4,045.39*	—	*400*
Ar	*4,044.42*	*(1200)*	—
K	*4,044.14*	*400*	*800*
La	*4,042.91*	*300*	*400*
Xe	4,078.82	(100)	—
U	4,078.83	8	3
O	4,078.85	(70)	—
Ga	4,078.90	5	—
Ce	4,079.02	1h	15
Nb	4,079.13	3w	1h
Tb	4,079.15	—	4
Ir	4,079.17	—	3
La	4,079.18	3	25
Ta	4,079.19	4	10
Bi	4,079.21	(40w)	2h
Mn	4,079.24	40	50
W	4,079.26	3	6
Dy	4,079.27	—	6
Ce	4,079.277	—	6
Ru	4,079.277	5	12
Mo	4,079.34	4	4
Ne	4,079.359	(2)	—
Re	4,079.363	—	20
Mn	4,079.42	40	50
Nb	4,079.73	200w	500w
Ru	4,080.60	300	125
Co	4,086.31	15	400
La	4,086.71	500	500
Pd	4,087.34	100	500
Kr	4,088.33	(500)	—
Co	4,092.39	15	600W
Tm	4,094.18	30	300
In	*4,101.77*	*1000R*	*2000R*
Ho	*4,103.84*	*400*	*400*
Se	*4,108.83*	*(800)*	—
N	*4,109.98*	*(1000)*	—
Co	*4,110.53*	—	*600*

Sr 4,215.52 *J* 400 *O* 300*R*

	λ	I J	I O
U	4,215.51	2	3
Re	4,215.506	—	20
Fe	4,215.42	15	60
Na	4,215.4	—	3
W	4,215.38	5	12
Zr	4,215.31	1	4
Dy	4,215.17	8	50
Tm	4,215.53	—	10
Rb	4,215.56	300	1000R
Xe	4,215.60	(100)	—
Cl	4,215.64	(5)	—
Zr	4,215.75	3	—
In	4,215.90	5	—
N	4,215.92	(5)	—

	λ	I	
		J	O
Os	4,215.15	1	8
Pr	4,215.14	3	8
Tb	4,215.13	1*h*	30*d*
Gd	4,215.024	150	200
Se	4,215.02	(150)	—
U	4,215.01	4	4
W	4,214.96	1	5
Co	4,214.87	1	2
Nb	4,214.82	10	5
N	4,214.733	—	40
Nb	4,214.73	(25)	—
Ce	4,214.70	—	4
Xe	4,214.69	(3)	—
Nb	4,214.67	100	—
Nb	4,214.60	10	10
Ru	4,214.56	—	7
Th	4,214.55	3	—
Ir	4,214.52	—	5
Pr	4,214.48	2	3
Xe	4,213.72	(200*h*)	—
Pd	4,212.95	300*W*	500*W*
Ag	4,212.68	20*h*	150*h*
Se	4,212.58	(200)	—
Gd	4,212.02	50	150
Os	4,211.85	50	150
Se	4,211.83	(200)	—
Dy	4,211.72	15	200
Rh	4,211.14	200	15
Ag	4,210.94	30*h*	200*h*
Fe	4,210.35	200	300
Xe	4,208.48	(200*h*)	—
Eu	4,205.05	50	200*R*
Fe	*4,202.03*	*300*	*400*
Rb	*4,201.85*	*500*	*2000R*
Ar	*4,200.67*	*(1200)*	*—*
Fe	*4,199.10*	*200*	*300*
Ar	*4,198.32*	*(1200)*	*—*
Fe	*4,198.31*	*150*	*250*

	λ	I	
		J	O
Tb	4,215.95	—	3
Er	4,215.96	—	9
Fe	4,215.97	1	2
Pr	4,215.98	3	4
U	4,215.99	3*h*	5
*bh*C	4,216.0	—	—
Ba	4,216.04	(25)	—
W	4,216.06	5	1
Sc	4,216.10	—	3
Fe	4,216.19	100	200
Nb	4,216.23	10	1
Cr	4,216.36	25	60
La	4,216.55	—	10
P	4,216.56	(15)	—
Ir	4,216.58	—	8
La	4,217.554	100	200
Fe	4,217.555	100	200
Cr	4,217.63	70	150
Fe	4,219.36	200	250
P	4,222.15	(150*w*)	300
Fe	4,222.22	200	200
Xe	4,223.00	(200*h*)	—
Fe	4,224.18	80	200
Ca	*4,226.73*	*50R*	*500R*
Fe	*4,227.43*	*250*	*300*
Eu	*4,235.60*	*—*	*400r*
Fe	*4,235.94*	*200*	*300*

Sr 4,607.331 *J* 50*R** *O* 1000*R*

	λ	J	O
Co	4,607.33	—	2*h*
*bh*La	4,607.3	—	3
Ce	4,607.29	—	3

	λ	J	O
Au	4,607.34	15	30
Nd	4,607.38	—	25
Mn	4,607.62	—	50

	λ	I J	I O		λ	I J	I O
V	4,607.23	3	4	Fe	4,607.65	5	50
Ce	4,607.09	—	2	Tb	4,607.81	—	4
Mo	4,607.07	10	8	Tm	4,607.887	—	2
Pr	4,606.92	—	3	Er	4,607.89	—	2
Ru	4,606.83	—	4	Yt	4,607.97	—	2*h*
*bh*Sc	4,606.8	—	10	Ho	4,608.00	—	2
W	4,606.768	—	2	Pr	4,608.005	—	5
Nb	4,606.766	50	50	Gd	4,608.02	3	8
Ti	4,606.72	—	4	Hf	4,608.09	4	25
Gd	4,606.65	3	3	Mo	4,608.116	5	5
Er	4,606.61	1	20	Rh	4,608.119	5	15
Sm	4,606.52	—	40	Eu	4,608.15	—	2*w*
Mo	4,606.511	3	4	Os	4,608.28	—	12
Th	4,606.506	2	4	U	4,608.34	—	2
Pr	4,606.45	3*w*	30*w*	Tb	4,608.44	—	8*R*
Ce	4,606.40	15	12	Ce	4,608.49	—	6
Cr	4,606.37	3	15	Nb	4,608.58	3*h*	2
Ni	4,606.23	—	100	Gd	4,608.59	—	6
*bh*Sr	4,606.2	—	2	Hg	4,608.67	1	3
V	4,606.15	25	30	Ru	4,608.675	—	6
Gd	4,606.06	2	5	Mo	4,608.71	10	10
Dy	4,606.05	—	4				
Li	4,602.86	—	800				
Eu	4,594.02	200	500*R*				
Cs	*4,593.18*	*50R*	*1000R*				
Co	*4,581.60*	*10*	*1000w*				

Sr 4,832.075 *J* 8* *O* 200

	λ	J	O		λ	J	O
Pr	4,832.07	—	100*w*	Ti	4,832.076	1	20
Tb	4,832.03	—	3	Ta	4,832.18	—	100
Pt	4,831.96	1*h*	2	Nd	4,832.28	—	20
V	4,831.64	25	30	Ho	4,832.31	2	3
Cr	4,831.63	—	25	Sm	4,832.33	—	2
Th	4,831.60	—	4	Dy	4,832.39	2*h*	7
Pr	4,831.43	—	6	V	4,832.43	20	25
W	4,831.36	—	5	Ni	4,832.70	—	70
Pt	4,831.22	—	3	Fe	4,832.730	—	5
Tm	4,831.21	80	50	Pr	4,832.734	—	3
Ni	4,831.18	2	200	Tb	4,832.74	—	2
Er	4,831.16	—	25	Th	4,832.80	3	12
Th	4,831.13	2	10	Mo	4,832.81	10	15
Nd	4,831.08	—	5	Mo	4,832.92	10	15
Dy	4,830.88	—	2	Ru	4,833.00	1	25

	λ	I J	I O		λ	I J	I O
Tb	**4,830.84**	—	**2**	**V**	**4,833.02**	**8**	**9**
Er	**4,830.72**	—	**5**	**Pr**	**4,833.14**	—	**15***w*
Yb	**4,830.71**	—	**2**	**Sm**	**4,833.33**	—	**80**
Sm	**4,830.68**	—	**8**	**Nb**	**4,833.37**	**10**	**10**
				Rh	**4,833.47**	**2**	**2**
Mo	4,830.51	100	125	**Nd**	**4,833.51**	—	**2**
Sm	4,829.58	—	200	**Sc**	**4,833.67**	**5**	**6**
Cr	4,829.38	40	200				
Ni	4,829.03	2*h*	300*w*				
Nd	4,825.48	8	100	Gd	4,834.24	25	125
				Sm	4,834.63	—	100
Ta	4,825.43	—	150	Gd	4,835.27	10	100
La	4,824.07	150	150	Sm	4,837.65	—	100
Mn	4,823.52	80	400	Ni	4,838.65	4	150
Pr	4,822.98	10*w*	125				
Gd	4,821.71	80	150	Co	4,840.27	150	700*w*
				Ti	4,840.87	25	125
Fe	4,821.05	200*h*	200*h*	Sm	4,841.70	—	100
Ti	4,820.41	30	125	Co	4,843.46	—	300
Ta	4,819.53	—	100	Rh	4,843.99	60	100
				Sm	4,844.21	—	150
Co	*4,813.48*	*6*	*1000 W*	Ta	4,846.45	2	100
Sr	*4,811.88*	—	*40*	Sm	4,847.77	—	150
Zn	*4,810.53*	*300h*	*400w*				
Gd	*4,805.82*	*80*	*200*				
Gd	*4,801.08*	*200*	*200*	*Ni*	*4,855.41*	*1*	*400w*

Ta $^{73}_{180.948}$

t_0 2,996°C t_1 4,100°C

I.	II.	III.	IV.	V.
6.0	–	–	–	–

λ	I		eV
	J	O	
I 2,653.27	15	200	4.7
2,675.90	200	150	4.6
I 2,714.67	8	200	5.05
3.311.16	70	300	3.7
I 4,551.95	8	400	3.21
I 4,780.94	200	50	3.91

Ta 2,653.27 *J* 15* *O* 200

	λ	I			λ	I	
		J	*O*			*J*	*O*
Ru	**2,653.15**	**5**	**12**	**W**	**2,653.32**	—	**6**
Ir	**2,653.03**	**5**	**8**	**Mo**	**2,653.35**	**150**	**25**
W	**2,653.01**	**9**	**2**	**Nb**	**3,653.38**	**3**	**5**
Os	**2,652.98**	**2**	**10**	**Nb**	**2,653.47**	**2**	**3**
V	**2,652.92**	—	**18**	**La**	**2,653.49**	—	**2**
Re	**2,652.91**	—	**30**	**W**	**2,653.57**	**15**	**10**
Hf	**2,652.86**	**3**	**4**	**Cr**	**2,653.59**	**35**	**12**
U	**2,652.83**	**10**	**12**	**Eu**	**2,653.60**	**15*h***	**15**
				Hg	**2,653.68**	**40**	**80**
				Ru	**2,653.697**	**1*wh***	**20**
Rh	2,652.66	25	100				
Al	2,652.49	60	150*R*	**Co**	**2,653.703**	**40**	**5**
Hg	2,652.04	60	100	**Re**	**2,653.738**	—	**15**
Re	2,651.90	—	100	**Er**	**2,653.743**	**10**	**15**
Ru	2,651.84	9	100	**Yb**	**2,653.75**	**200**	**50**
				Ir	**2,653.76**	**5*h***	**15**
Pt	2,650.86	100	700				
Be	2,650.47	15	100				
Pb	2,650.4	80	100	Ta	2,654.01	—	15
Re	2,649.05	—	100	Ta	2,655.67	—	15
				Ta	2,656.61	2	200*R*
				Pt	2,658.17	10	100
Ta	*2,647.47*	*10*	*200*				
Pt	*2,646.89*	*100*	*1000h*				
Ta	*2,646.37*	*2*	*125*	*Ta*	*2,658.86*	*50*	*25*
Ta	*2,645.10*	*30h*	*80*	*Pt*	*2,659.45*	*500R*	*2000R*
Ta	*2,644.60*	*50W*	*20W*	*Ta*	*2,659.65*	—	*15*
				Ta	*2,661.34*	*10*	*200*
Ta	*2,643.89*	—	*50*	*Ta*	*2,661.89*	—	*60*

Ta 2,675.90 *J* 200 *O* 150

	λ	*J*	*O*		λ	*J*	*O*
U	**2,675.88**	**10**	**15**	**Nb**	**2,675.94**	**100**	**10**
W	**2,675.87**	**6**	**12**	**Au**	**2,675.95**	**100**	**250*R***
V	**2,675.761**	**2**	**12**	**V**	**2,675.97**	—	**6**
Tl	**2,675.76**	**(30)**	—	**Co**	**2,675.98**	**5**	**10*W***
Ce	**2,675.733**	—	**2**	**Co**	**2,676.01**	**10*h***	—
W	**2,675.73**	**12**	—	**Tl**	**2,676.03**	**(30)**	—
Cr	**2,675.68**	**15**	**1**	**V**	**2,676.04**	**15*h***	—
Th	**2,675.67**	**4**	**8**	**Eu**	**2,676.05**	**5**	—
La	**2,675.65**	**5**	**2**	**Tl**	**2,676.08**	—	**6**
Ne	**2,675.64**	**(150)**	—	**Rh**	**2,676.110**	**5**	**10**
Ta	**2,676.54**	—	**2**	**Fe**	**2,676.11**	**7*w***	**15*w***
Ru	**2,675.52**	**50**	—	**Ce**	**2,676.121**	—	**2**
Mn	**2,675.51**	**25**	—	**Nb**	**2,676.125**	**10*h***	**2**
				Yb	**2,676.13**	**2**	—
				Ru	**2,676.19**	**100**	**8**
Ne	2,675.24	(150)	—	**Rh**	**2,676.25**	**100**	**1**
Pt	2,674.57	10	200				

	λ	I J	I O
Rh	2,674.44	200w	1
Re	2,674.34	—	100
Nb	2,673.57	500	10
Eu	2,673.41	50	125
Mo	2,673.27	100	1
Mo	2,672.84	100	15
Mn	2,672.59	125h	15
Ta	2,672.50	30h	20
Ta	2,672.34	—	20
V	2,672.00	300R	50
Nb	2,671.93	200	20
Mo	2,671.83	100	1
Ta	2,671.63	—	30
Zn	*2,670.53*	*4*	*200*
Ta	*2,668.62*	*100*	*1*
Eu	*2,668.33*	*400*	*300w*
Ta	*2,668.07*	—	*80*
Ta	*2,667.17*	—	*20*

	λ	I J	I O
P	**2,676.28**	**(10)**	—
W	**2,676.31**	**6**	—
Mn	**2,676.33**	—	**15**
V	**2,676.35**	**18h**	—
Ru	**2,676.353**	**3**	**50**
Ce	**2,676.355**	—	**2**
Cl	2,676.95	(150)	—
Pt	2,677.15	200w	800w
Cr	2,677.16	300r	35
Cd	2,677.64	25	100
V	2,677.80	300R	70
Eu	2,678.28	100	150
V	2,678.57	150R	30
Zr	2,678.63	100	80
Ru	2,678.76	300	100
Ta	2,679.80	20	3
Fe	2,679.06	200	200
Ni	2,679.24	500wh	—
V	2,679.32	300R	70
Pd	2,679.58	100	—
Co	2,679.76	—	75W
Ti	2,679.93	12	100
Ta	2,680.06	50	30
Ta	*2,681.87*	—	*50*
V	*2,682.87*	*200R*	*50*
Rh	*2,683.56*	*200*	*1*
Zn	*2,684.16*	*6*	*300*
Ta	*2,684.28*	*2*	*150*
Ni	*2,684.41*	*500wh*	—
Fe	*2,684.75*	*400*	*3*
Eu	*2,685.65*	*100*	*150*

Ta 2,714.67 *J* 8* *O* 200

	λ	J	O
Os	**2,714.64**	**10**	**50r**
Th	**2,714.62**	**4**	**2h**
U	**2,714.58**	**8**	**10**
Ir	**2,714.55**	—	**4**
La	**2,714.54**	—	**6**
Co	**2,714.42**	**200W**	**12**
Fe	**2,714.412**	**400**	**200**
Rh	**2,714.410**	**5**	**150**
U	**2,714.29**	**2h**	**2**
Zr	**2,714.26**	**7**	**7**

	λ	J	O
Ce	**2,714.72**	—	**5**
W	**2,714.81**	**3**	**4**
Fe	**2,714.87**	**15**	**40**
Ce	**2,714.97**	—	**5**
U	**2,715.00**	**4**	**6**
V	**2,715.03**	**2**	**10**
Rh	**2,715.04**	**2**	**50**
Th	**2,715.09**	**5**	**6**
Fe	**2,715.12**	**2**	**5**

	λ	I J	I O		λ	I J	I O
V	2,714.199	100	60	Re	2,715.47	—	100
Nb	2,714.198	5	3	Eu	2,716.97	300	300
				Ta	2,717.18	—	100
				Rh	2,717.51	5	100
In	2,713.93	125*wh*	200*R*	Ta	2,718.38	—	80
*bh*B	2,713.8	—	200				
Mn	2,713.33	—	300*Wh*	Rh	2,718.54	20	150
Pt	2,713.13	10	200	Fe	2,719.02	300*r*	500*r*
Zn	2,712.49	8	300	Pt	2,719.04	100*W*	1000*w*
				Ru	2,719.51	30	100
Fe	2,711.65	50	100				
In	2,710.26	200*Rh*	800*R*				
Ta	2,710.13	3	200	Ta	2,720.76	1	150
				Fe	2,720.90	—	700r
				Ta	2,721.83	—	50
Ta	2,709.27	150	40	Fe	2,723.58	200	300
Tl	2,709.23	200R	400R				
Ta	2,706.69	1	50				
Sn	2,706.51	150R	200R				
Pt	2,705.89	200wh	1000wh				

Ta 3,311.16 *J* 70 *O* 300

	λ	J	O		λ	J	O
W	3,311.11	4	5	Ar	3,311.26	(5)	—
Er	3,311.09	—	3	Cr	3,311.30	—	8
Pd	3,311.023	2*h*	4	Ne	3,311.30	(7)	—
Ir	3,311.023	—	2	Nb	3,311.338	10	5
Cu	3,311.00	—	3	Zr	3,311.339	3	8
Dy	3,310.96	3	5	U	3,311.35	1*h*	4*d*
Ru	3,310.957	5	30	W	3,311.38	12	15
O	3,310.91	30	200	Ce	3,311.39	—	3
Nd	3,310.90	2	25	Fe	3,311.45	1	1
Ce	3,310.88	—	10	Ca	3,311.46	6	—
U	3,310.87	2	2*d*	Ce	3,311.50	1	15
Hf	3,310.86	8	15	Dy	3,311.51	—	5
Eu	3,310.80	1	3*wh*	Cs	3,311.52	(4)	—
Tb	3,310.80	3	8	Sc	3,311.71	6	3
Mo	3,310.77	20	20	U	3,311.72	12	10*d*
Nb	3,310.66	15*wh*	—	Ce	3,311.73	—	2*h*
Sm	3,310.655	10	25	Xe	3,311.80	(2)	—
Cr	3,310.647	200	2	Sr	3,311.85	4	1
Ce	3,310.63	—	5	Mn	3,311.90	—	75
U	3,310.624	—	10	Cr	3,311.93	125	6
La	3,310.62	3	2	Os	3,312.03	5	8
Tm	3,310.59	20	60	Th	3,312.08	4	3
Zr	3,310.55	—	3	Cr	3,312.081	—	10
Ir	3,310.525	12	30	Lu	3,312.11	10	100
Th	3,310.52	4*d*	3*d*				
Ag	3,310.51	—	2	Cr	3,312.18	125	5
U	3,310.50	8*d*	6*d*	Nb	3,312.60	50	40

	λ	I	
		J	O
Fe	**3,310.49**	**40**	**50**
Ar	**3,310.47**	**(3)**	**—**
Nb	**3,310.468**	**10**	**10**
Mo	**3,310.40**	**3**	**3**
Xe	**3,310.38**	**(2)**	**—**
Nd	**3,310.36**	**2**	**8**
Fe	**3,310.340**	**80**	**100**
Eu	**3,310.34**	**2**	**2*w***
Sm	**3,310.33**	**—**	**4**
Hf	**3,310.27**	**5**	**20**
Th	**3,310.25**	**15**	**8**
Tm	3,309.80	60	80
Ta	3,309.78	5	70
Ta	3,308.89	—	18*w*
P	3,308.85	(100*w*)	—
Ti	3,308.81	100	35
Ta	3,308.54	1*h*	20
V	3,308.48	80	—
Cl	3,307.90	(50)	—
Sr	3,307.53	10*s*	200
Mo	3,307.43	50	—
Fe	3,307.23	60	80
Ta	3,307.08	1*h*	35
Te	3,306.99	(150)	—
Rn	3,306.70	(100)	—
Cl	3,306.45	(40)	—
Sm	3,306.37	40	100
Fe	3,306.35	150	200
Zr	3,306.28	80	80
Fe	3,305.97	300	400
Yb	3,305.73	125	30
Nb	3,305.61	100	1
Ta	3,305.34	1*h*	15
Yb	3,304.76	40	12
Yb	3,304.56	40	15
V	3,304.47	125	—
Ta	3,304.37	15	70
Sb	3,304.11	40*wh*	—
Ta	3,304.40	3	35
La	3,303.11	150	400
Na	3,302.99	150*R*	300*R*
Zn	3,302.94	300*R*	700*R*
Ta	3,302.76	1*h*	50
Zn	3,302.59	300	800
Bi	3,302.55	—	150
Tm	3,302.45	80	125
Na	3,302.32	300*R*	600*R*

	λ	I	
		J	O
Rn	3,312.8	(100)	—
Mo	3,312.94	50	1
Eu	3,313.32	40	35
Mo	3,313.62	50	5
Th	3,313.65	50*h*	5
Cr	3,314.56	100	10
Fe	3,314.74	200	200
Pt	3,315.05	10	200
V	3,315.18	35	2
Ti	3,315.32	100	12
Cl	3,315.44	(100)	—
Ni	3,315.66	20	400*R*
Cl	3,316.86	(50)	—
V	3,316.88	60	—
Ru	3,316.90	50	5
Fe	3,317.12	80	100
V	3,317.91	80	—
Ta	3,317.93	25*W*	200
Ti	3,318.02	125	60
Ta	3,318.53	3	70
Ta	3,318.84	35	125
V	3,318.91	60	—
Fe	3,319.25	50	70
Ar	3,319.34	(300)	—
Nb	3,319.58	50	4
Dy	3,319.89	9	150
Tl	3,319.91	(35)	—
Ni	3,320.26	15	400*w*
Nb	*3,320.81*	*100*	*3*
Mo	*3,320.90*	*80*	*3*
Be	*3,321.34*	*30*	*1000r*
V	*3,321.54*	*150*	*3*
Ti	*3,321.70*	*125*	*15*
Ni	*3,322.31*	*10*	*400*
Fe	*3,322.48*	*100*	*150*
Ti	*3,322.94*	*300R*	*80*
Fe	*3,323.07*	*100*	—
Rh	*3,323.09*	*200*	*1000*
Fe	*3,323.74*	*150*	*150*
Fe	*3,324.54*	*80*	*100*
Fe	*3,325.46*	*80*	*100*
Ar	*3,325.50*	*(100)*	—
Ta	*3,325.74*	*3*	*50*
Ti	*3,326.76*	*125*	*15*
Zr	*3,326.80*	*100*	*100*
Fe	*3,328.87*	*100*	*150*
Cl	*3,329.12*	*(150)*	—
Mo	*3,329.21*	*100*	*2*

	λ	I J	I O
Pd	3,302.13	200h	1000wh
Ta	3,301.89	3	25
Pt	3,301.86	250W	300
V	3,301.65	80	—
Os	3,301.56	50	500R
Nb	3,301.49	100	1
Ta	3,299.77	10	70
V	3,298.74	80	12
Fe	3,298.13	150	200
Ta	3,297.19	1wh	18h
Cr	3,295.43	200	10
Ta	3,295.33	20w	125w
Ta	3,294.71	1	18
Nb	3,294.36	100	2
Ru	3,294.11	200	60
Ta	3,293.93	10	70
Tb	3,293.07	100	50
Te	3,329.25	(100)	—
Ti	3,329.46	200r	80
Ta	3,329.53	1	18

Ta 4,551.950 *J* 8* *O* 400

	λ	J	O
Pt	4,551.950	1h	2
W	4,551.85	10	35
V	4,551.84	8	9
Eu	4,551.7	—	2w
U	4,551.6	4	1
Rh	4,551.64	10	25
Tb	4,551.61	—	3
Pr	4,551.54	—	4
Nb	4,551.52	2	3h
Re	4,551.48	—	2h
Gd	4,551.47	2	2
Dy	4,551.37	—	3
Os	4,551.298	8	150
Ce	4,551.297	—	20
Ni	4,551.24	—	5
Pr	4,551.18	—	3
Eu	4,551.08	—	5w
Nb	4,551.03	3h	1
U	4,550.98	2	5
Gd	4,550.96	10	10
Ce	4,550.922	—	2
Tb	4,550.92	—	2
Dy	4,550.89	2	5
Pr	4,550.88	2	10
Fe	4,550.79	—	50
U	4,551.98	1	15
Ce	4,552.07	—	5
Ru	4,552.11	—	7
Er	4,552.14	1	8
Eu	4,552.19	5	5W
Pr	4,552.26	—	60
Ce	4,552.31	—	2
Pt	4,552.42	10	60
Co	4,552.44	—	25
U	4,552.45	2	2
Ti	4,552.46	50	150
La	4,552.49	—	5
W	4,552.53	3	12
Fe	4,552.55	1	10
Sm	4,552.66	40	80
Mo	4,552.80	5	5
Pr	4,552.83	—	5
Pd	4,552.89	—	2
Zr	4,553.01	—	10
Th	4,553.04	3h	3
V	4,553.05	15	20
Ce	4,553.06	—	8
Er	4,553.13	—	5
Ni	4,553.17	—	15r
Mo	4,553.22	6	12

	λ	I J	I O		λ	I J	I O
La	**4,550.78**	**3**	**4**	**Hf**	**4,553.24**	—	**5h**
Ir	**4,550.775**	—	**80**	**Pr**	**4,553.26**	—	**5h**
Pr	**4,550.769**	—	**10**	**Eu**	**4,553.3**	—	**3w**
Zr	**4,550.71**	—	**3**	**Mn**	**4,553.31**	—	**12h**
Eu	**4,550.67**	—	**20w**	**Mo**	**4,553.32**	**4**	**12**
Pr	**4,550.670**	—	**10d**				
Nb	**4,550.64**	**3**	**1**	Ta	4,553.69	2	200
				Ba	4,554.04	200	1000R
				Ru	4,554.51	200	1000R
Co	4,549.66	—	600	Cs	4,555.35	100	2000R
Fe	4,547.85	100	200	Ta	4,556.35	5	200
Ta	4,547.15	2	150				
Cr	4,545.96	125	200	Co	4,565.59	12	800W
Ir	4,545.68	4	200				
Co	4,543.81	—	500W	*Ta*	*4,565.85*	*15*	*200*
				Ta	*4,566.86*	*2h*	*100*
				Ta	*4,573.29*	*2h*	*200*
Co	*4,533.99*	*8*	*500*	*Ta*	*4,574.31*	*20*	*300*
Co	*4,530.96*	*8*	*1000w*				
Ta	*4,530.85*	*50*	*300*				
Rh	*4,528.72*	*60*	*500r*				
Fe	*4,528.62*	*200*	*600*				
Ta	*4,527.49*	*5*	*150*				
Sn	*4,524.74*	*50*	*500wh*				

Ta 4,780.94 *J* 200 *O* 50*

Ne	**4,780.88**	**(30)**	—	**Er**	**4,781.02**	**2**	**35**
W	**4,780.52**	**1**	**10**	**Yt**	**4,781.03**	**5**	**10**
Ne	**4,780.34**	**(50)**	—	**Dy**	**4,781.04**	**2**	**3**
Br	**4,780.31**	**(125)**	—	**N**	**4,781.17**	**(5)**	—
Yt	**4,780.18**	**1**	**2**	**Ne**	**4,781.24**	**(2)**	—
Co	**4,780.01**	**500**	**500w**	**Tm**	**4,781.30**	**10**	**5**
Ti	**4,779.95**	**100h**	**10**	**Cl**	**4,781.32**	**(75)**	—
N	**4,779.71**	**(15)**	—	**Co**	**4,781.43**	**2h**	**400**
U	**4,779.63**	**3**	**8**	**Ti**	**4,781.72**	**2**	**30**
Th	**4,779.60**	**2**	**2**	**Cl**	**4,781.82**	**(50)**	—
Sb	**4,779.4**	**8**	—	**Yb**	**4,781.88**	**5**	**50**
				Gd	**4,781.93**	**50**	**200**
				Ne	**4,781.95**	**(5)**	—
Br	4,776.42	(200)	—	**Hg**	**4,782.1**	**(5)**	—
Rb	4,776.41	(100)	—				
Xe	4,769.05	(100)	—				
Cl	4,768.68	(150)	—	Lu	4,785.42	200	100
Ar	4,768.67	(150)	—	Br	4,785.50	(400)	—
				Yb	4,786.60	(200)	50
Rn	4,768.59	(100)	—	Ne	4,788.93	(300)	—
Br	4,767.10	(200)	—	Cr	4,789.38	100	300
Te	4,766.03	(150)	—				
Kr	4,765.74	(1000)	—	Ne	4,789.60	(100)	—

	λ	I			λ	I	
		J	O			J	O
Se	*4,763.65*	*(800)*	—	Fe	4,791.25	200*R*	200
Kr	*4,762.43*	*(300)*	—	Xe	4,792.62	(150)	—
Ta	*4,756.51*	*10*	*150*	Cl	4,794.54	(250)	—
Ne	*4,752.73*	*(1000)*	—				
				Hg	*4,797.01*	*(300)*	—
				Cd	*4,799.92*	*300w*	*300w*
				Gd	*4,801.08*	*200*	*200*
				Ar	*4,806.07*	*(500)*	—
				Xe	*4,807.02*	*(500)*	—
				Cl	*4,810.06*	*(200)*	—
				Zn	*4,810.53*	*300h*	*400w*
				Kr	*4,811.76*	*(300)*	—

Te $^{52}_{127.60}$

t_0 452°C

t_1 1,390°C

I.	II.	III.	IV.	V.
9.007	21.543	30.611	37.817	60.27

λ	I		eV
	J	O	
I 2,142.75	—	600	—
I 2,383.25	300	500	5.8
I 2,385.76	(300)	600	5.8
4,866.22	(800)	—	—
5,755.87	250	—	—

Te 2,142.75 *O* 600

	λ	*I*			λ	*I*	
		J	*O*			*J*	*O*
Os	**2,142.73**	**3**	**4**	**Re**	**2,142.78**	**15**	**10**
Pd	**2,142.57**	—	**3**	**La**	**2,142.81**	**18*h***	**2*h***
Ta	**2,142.53**	**20*s***	**8*h***	**Ir**	**2,142.82**	**1**	**3**
				Nb	**2,142.91**	**6**	**1*h***
Zn	*2,138.56*	*500*	*800R*				

Te 2,383.25 *J* 300 *O* 500

	λ	*J*	*O*		λ	*J*	*O*
Fe	**2,383.241**	**12**	**8**	**Cr**	**2,383.33**	—	**20**
Pd	**2,383.24**	**2**	—	**Mo**	**2,383.36**	**10**	—
Os	**2,383.21**	—	**15**	**Rh**	**2,383.40**	**10**	**50**
Ag	**2,383.21**	**25**	—	**Pd**	**2,383.40**	**50*wh***	—
W	**2,383.20**	**3**	—	**V**	**2,383.436**	**4*h***	**2**
Ir	**2,383.17**	**5**	**10**	**U**	**2,383.44**	**3**	—
Fe	**2,383.055**	**2**	**6**	**Ru**	**2,383.442**	**12**	—
Mo	**2,383.05**	**4**	—	**Ir**	**2,383.45**	—	**2**
V	**2,383.00**	**80**	**8**	**Co**	**2,383.46**	**30**	**15**
W	**2,382.99**	**3**	**15**	**Re**	**2,383.48**	**5**	**25**
U	**2,382.93**	**2**	—	**Ar**	**2,383.50**	**(40)**	—
				Mo	**2,383.52**	—	**12**
				W	**2,383.54**	**4**	**2*h***
Ru	2,381.99	150	50	**Rh**	**2,383.59**	**25**	—
Tl	*2,379.69*	*200R*	*100R*				

*N**

Te 2,385.76 *J* (300) *O* 600

	λ	*J*	*O*		λ	*J*	*O*
Cr	**2,385.74**	—	**25**	**Hf**	**2,385.78**	**3**	—
Ta	**2,385.73**	—	**8**	**W**	**2,385.79**	**5**	—
V	**2,385.62**	**10*h***	—	**Co**	**2,385.816**	—	**9**
Rb	**2,385.6**	**(2)**	—	**V**	**2,385.816**	**100**	—
Rh	**2,385.60**	**10**	—	**Ir**	**2,385.86**	**3**	**20**
Fe	**2,385.58**	—	**2**	**Fe**	**2,385.95**	—	**4**
U	**2,385.55**	**2**	**3**	**Os**	**2,386.04**	**20**	**2**
Re	**2,385.50**	—	**12**	**Mo**	**2,386.06**	**18**	—
W	**2,385.49**	**9**	**3**				
Os	**2,385.49**	**2**	**15**				
				Te	2,389.79	(25)	—
Rh	**2,385.44**	**25**	**3**				
He	**2,385.42**	**(30)**	—				

*P**

Te 4,866.22 J (800)

	λ	I J	I O		λ	I J	I O
Ar	4,865.91	(10)	—	Ni	4,866.27	1	300*w*
Mo	4,865.82	8	8	Ne	4,866.48	(80)	—
Ti	4,865.62	3	—	Br	4,866.52	(4)	—
Os	4,865.60	1	80	Br	4,866.70	(20)	—
Ne	4,865.50	(100)	—	Nb	4,866.85	1	15
Hf	4,865.43	10	5	Ne	4,877.01	(70)	—
Lu	4,865.36	20	4	J	4,867.08	(3)	—
Pr	4,865.24	2	40	Ar	4,867.59	(5)	—
Te	4,865.13	(50)	—	Eu	4,867.60	3	30
Gd	4,865.04	10	400	Br	4,867.75	(10)	—
O	4,864.95	(30)	—	Ar	4,867.84	(10)	—
Ga	4,864.92	4	—				
Kr	4,864.91	(2*h*)	—				
Mo	4,864.743	3	3	Te	4,875.53	(15)	—
V	4,864.737	25*wh*	30*wh*				
Ta	4,864.66	2	20 *h*	*Ne*	*4,884.91*	*(1000)*	—
				Te	*4,885.22*	*(100)*	—
Te	4,864.10	(800)	—				
Xe	4,862.54	(400*h*)	—				
J	4,862.31	(700)	—				
H	4,861.33	(500)	—				
Kr	*4,856.60*	*(700)*	—				
Se	*4,844.96*	*(800)*	—				
Xe	*4,844.33*	*(1000)*	—				
Se	*4,840.63*	*(800)*	—				

Te 5,755.87 J (250)

	λ	I J	I O		λ	I J	I O
Kr	5,755.60	(2*wh*)	—	Pr	5,756.17	1	25
Ra	5,755.45	(25)	—	Ti	5,756.86	12	10
V	5,755.08	2	2*h*	U	5,757.34	1	2
Kr	5,755.04	(2)	—	Tm	5,758.01	5	15
Xe	5,754.60	(2*h*)	—				
Nb	5,754.44	1	3	Xe	5,758.65	(150)	—
Xe	5,754.18	(3)	—	Te	5,762.60	(15)	—
Sn	5,753.61	15	4	Te	5,763.92	(25)	—
Se	5,753.32	(25)	—	Ne	5,764.42	(700)	—
				Te	5,765.25	(70)	—
Xe	5,751.03	(200)	—	Hg	5,769.59	200	600
Ne	5,748.30	(500)	—	Te	5,770.92	(35)	—
Ar	5,739.52	(500)	—	Te	5,777.25	(15)	—
				Hg	*5,789.66*	*(500)*	—
Ne	*5,719.22*	*(500)*	—	*Hg*	*5,790.65*	*(1000)*	—
Te	*5,708.07*	*(250)*	—	*Te*	*5,803.07*	*(50)*	—
				Ne	*5,804.45*	*(500)*	—

Th $^{90}_{232.038}$

t_0 1,845°C

t_1 >3,000°C

I.	II.	III.	IV.	V.
—	—	29.5	—	—

λ	I		eV
	J	O	
3,290.59	40	—	>7.3
3,538.75	50	—	>3.5
3,601.04	10	8	>3.4
3,741.19	80	80	>3.3
II 4,019.14	8	8	3.08
4,381.86	30	30	>2.8
II 4,391.11	40	50	3.37
II 4,919.81	20	50	3.28

Th 3,290.59 *J* 30

	λ	I J	I O		λ	I J	I O
Ce	3,290.58	1	10	Yt	3,290.60	2	3*d*
Cu	3,290.54	25	25	Sm	3,290.639	5	10
W	3,290.51	5	6	Nd	3,290.643	6	10
U	3,290.47	5	6	K	3,290.65	(10*h*)	—
Ca	3,290.42	3	1*h*	Fe	3,290.72	7	15
Sm	3,290.38	8	10	Mo	3,290.82	100	40
Ce	3,290.34	1	20	Ho	3,290.96	6	8
Sm	3,290.28	4	10	Fe	3,290.99	80	125
Os	3,290.26	20	200	Tm	3,291.00	80	125
V	3,290.24	70	2	Tl	3,291.01	(40)	—
Pt	3,290.22	10	150	Hf	3,291.05	6*h*	20
Th	3,290.13	5	6	Nb	3,291.06	100	10
O	3,290.08	(18)	—	Br	3,291.1	(3)	—
Fe	3,290.04	2	3	Os	3,291.13	5	10
Nb	3,290.01	10	10	Er	3,291.27	1	5
Ar	3,289.95	(3)	—	U	3,291.33	10	12
Yt	3,289.85	10	15	Th	3,291.36	2*w*	2*w*
Mo	3,289.844	10	10	Ta	3,291.41	5	1
Ta	3,289.838	1	25	Sm	3,291.42	6*h*	6
U	3,289.75	1	2	Ar	3,291.47	(5)	—
Hf	3,289.74	3	5				
Cl	3,289.72	(12)	—	Th	3,291.74	12	10
				Cr	3,291.76	200	10
				Nb	3,292.020	100	3
V	3,289.39	70	10	Fe	3,292.023	125	150
Ho	3,289.38	20	10	Ti	3,292.07	40	70
Yb	3,289.37	1000*R*	500*R*				
Fe	3,289.35	40	—	Mo	3,292.31	300	10
Rh	3,289.14	50	150	Nb	3,292.37	20	—
				Th	3,292.52	12	10
Mo	3,289.01	30	40	Fe	3,292.59	150	300
Ti	3,288.57	20	12	Tb	3,293.07	100	50
J	3,288.35	(35)	—				
V	3,288.31	30	1	V	3,293.15	50	—
U	3,288.21	20	25	Ar	3,293.66	(25)	—
				Ru	3,294.11	200	60
Ti	3,287.65	200	40	Rh	3,294.28	25	60
O	3,287.57	(70)	—	Yb	3,294.34	20	3
Pd	3,287.25	25	300*w*				
Mo	3,287.20	25	1	Nb	3,294.36	100	2
Yb	3,286.97	30	7	Th	3,295.00	8	8
				Th	3,295.32	10	10
Fe	3,286.75	400	500	Ta	3,295.33	20*w*	125*w*
Rb	3,286.47	(60)	—	Cr	3,295.43	200	10
Nb	3,286.33	20	1				
Na	3,285.75	(100)	40	Nb	3,295.50	20	1
Tm	3,285.61	40	40	Th	3,295.53	10	10
				Fe	3,295.82	30	4
Fe	3,285.41	40	60	Nb	3,296.01	40	20
V	3,285.02	40	4	V	3,296.05	30	—
Zr	3,284.71	30	25				
Mo	3,284.62	40	—	Mo	3,296.58	25	—

	λ	I	
		J	O
Fe	3,284.59	125	200
Th	3,284.10	8h	—
Sn	3,283.51	100h	—
Nb	3,283.46	100	2
Tm	3,283.40	40	40
Sn	3,283.21	(50)	—
Cr	3,283.06	35	—
Th	3,282.97	12	8
Mo	3,282.91	30	1
Fe	3,282.89	80	80
V	3,282.53	80	12
Zn	3,282.33	300	500R
Ti	3,282.329	150	30
V	3,281.75	30	—
Mo	3,281.62	20	—
Rb	3,281.49	(20)	—
Th	3,281.41	10	8
Fe	*3,281.30*	*100*	*15*
V	*3,281.11*	*50*	*3*
Ag	*3,280.68*	*1000R*	*2000R*
Fe	*3,280.26*	*150*	*150*
Ti	*3,279.99*	*40*	*10*
Zr	*3,279.26*	*50*	*50*
Ti	*3,278.92*	*150*	*40*
Mo	*3,278.88*	*50*	*1*
Fe	*3,278.73*	*60*	*100*
Ti	*3,278.29*	*100*	*25*
Hg	*3,277.87*	*(50)*	—
Fe	*3,277.35*	*200*	*40*
Cl	*3,276.81*	*(40)*	—
Ti	*3,276.77*	*70*	*12*
Fe	*3,276.47*	*50*	*100*
Mo	*3,276.34*	*40*	*2*
Pb	*3,276.19*	*60*	—
V	*3,276.12*	*200R*	*50*
Yb	*3,275.81*	*100*	*12*
Ti	*3,275.29*	*50*	*8*
Th	*3,275.01*	*12*	*10*
Be	*3,274.64*	*(50)*	—
Fe	*3,274.45*	*60*	*80*
Na	*3,274.22*	*(40)*	*15*
Cu	*3,273.96*	*1500R*	*3000R*
Nb	*3,273.89*	*100W*	*20r*
Th	*3,273.88*	*15*	*10*
Zr	*3,273.05*	*80*	*50*
Pd	*3,272.56*	*60h*	—
Ti	*3,272.08*	*100*	*25*

	λ	I	
		J	O
Th	3,296.61	10	8
Mn	3,296.88	30	60
Nb	3,297.05	50	3
Th	3,297.37	10	6
V	3,297.52	60	—
Pb	3,297.64	30	—
Nb	3,297.66	30	1
Mo	3,297.68	60	1
Ne	3,297.74	(40)	—
Th	3,297.83	12	10
Yb	3,297.85	20	4
Sm	3,298.10	60	60
Fe	3,298.13	150	200
Mn	3,298.22	25h	50
Ru	3,298.41	25R	50
V	3,298.74	80	12
Ti	3,299.41	35	50
Th	3,299.67	8	5
Th	*3,300.49*	*30h*	—
Nb	*3,301.49*	*100*	*1*
Os	*3,301.56*	*50*	*500R*
Ru	*3,301.59*	*40*	*70*
V	*3,301.65*	*80*	—
Pt	*3,301.86*	*250W*	*300*
Pd	*3,302.13*	*200h*	*1000wh*
Na	*3,302.32*	*300R*	*600R*
Tm	*3,302.45*	*80*	*125*
Zn	*3,302.59*	*300*	*800*
Zn	*3,302.94*	*300R*	*700R*
Na	*3,302.99*	*150R*	*300R*
La	*3,302.11*	*150*	*400*
Th	*3,303.49*	*10*	*5*
Sb	*3,304.11*	*40wh*	—
V	*3,304.47*	*125*	—
Yb	*3,304.56*	*40*	*15*
Yb	*3,304.76*	*40*	*12*
Nb	*3,305.61*	*100*	*1*
Yb	*3,305.73*	*125*	*30*
Fe	*3,305.97*	*300*	*400*
Zr	*3,306.28*	*80*	*80*
Fe	*3,306.35*	*150*	*200*
Sm	*3,306.37*	*40*	*100*
Cl	*3,306.45*	*(40)*	—
Rn	*3,306.7*	*(100)*	—
Te	*3,306.99*	*(150)*	—
Fe	*3,307.23*	*60*	*80*
Mo	*3,307.43*	*50*	—
Cl	*3,307.90*	*(50)*	—

	λ	I J	I O		λ	I J	I O
				V	*3,308.48*	*80*	–
				Ti	*3,308.81*	*100*	*25*
				P	*3,308.85*	*(100w)*	–

Th 3,538.75 *J* 50

	λ	J	O		λ	J	O
U	**3,538.68**	**6*h***	**—**	**Ce**	**3,538.76**	**2**	**5**
W	**3,538.63**	**9**	**8**	**Ce**	**3,538.79**	**2**	**3**
Yt	**3,538.53**	**3**	**10**	**Pr**	**3,538.85**	**2**	**9**
Dy	**3,538.523**	**40**	**150**	**Nd**	**3,538.858**	**6**	**20**
Ho	**3,538.52**	**6**	**—**	**Sm**	**3,538.864**	**1**	**10**
Er	**3,538.519**	**9**	**18**	**Tb**	**3,538.90**	**3**	**15**
Dy	**3,538.50**	**2**	**5**	**Mo**	**3,538.92**	**4**	**3**
Tb	**3,538.50**	**3**	**3**	**U**	**3,538.98**	**3**	**1**
F	**3,538.47**	**(6)**	**—**	**Zr**	**3,539.01**	**3**	**4**
Ce	**3,538.46**	**1**	**2**	**Ce**	**3,539.09**	**10**	**100**
U	**3,538.42**	**8**	**2**	**Nb**	**3,539.12**	**15**	**1**
Ce	**3,538.41**	**1**	**2**	**Al**	**3,539.15**	**8**	**—**
Pr	**3,538.31**	**2**	**3**	**Nd**	**3,539.18**	**4**	**10**
Fe	**3,538.29**	**1*h***	**2*h***	**Fe**	**3,539.20**	**1*h***	**1*h***
Ag	**3,538.27**	**2**	**10**	**U**	**3,539.21**	**4**	**4**
Rh	**3,538.26**	**4**	**50**	**Sm**	**3,539.25**	**1**	**8**
V	**3,538.24**	**100**	**10**	**Ru**	**3,539.26**	**5**	**30**
Th	**3,538.223**	**3**	**8**	**Th**	**3,539.32**	**5**	**5**
U	**3,538.221**	**6**	**8**	**W**	**3,539.33**	**3**	**6**
Ir	**3,538.15**	**1*h***	**18**	**Ru**	**3,539.369**	**15**	**60**
Rh	**3,538.14**	**10**	**100**	**Nd**	**3,539.372**	**4**	**4**
Eu	**3,538.09**	**10*h***	**20*w***	**Dy**	**3,539.38**	**2*h***	**18**
Xe	**3,538.08**	**(2)**	**—**	**W**	**3,539.458**	**7**	**3**
Ne	**3,537.99**	**(7)**	**—**	**Mo**	**3,539.465**	**3**	**3**
Os	**3,537.988**	**4**	**1**	**Kr**	**3,539.54**	**(15)**	**—**
Ru	**3,537.95**	**25**	**70**	**Th**	**3,539.59**	**8**	**8**
Tb	**3,537.94**	**15**	**15**	**Dy**	**3,539.62**	**4**	**6**
Tm	**3,537.91**	**2**	**15**	**Pr**	**3,539.63**	**2**	**5**
Fe	**3,537.90**	**25**	**50**	**Nb**	**3,539.648**	**15**	**15**
U	**3,537.83**	**2**	**—**	**U**	**3,539.654**	**2**	**12**
Eu	**3,537.75**	**10*h***	**10**	**Tl**	**3,539.78**	**(6)**	**—**
Ca	**3,537.733**	**4**	**3**	**Tb**	**3,539.81**	**3**	**15**
Fe	**3,537.731**	**15**	**25**				
				Fe	3,540.12	60	100
Nb	3,537.62	30	2	Tb	3,540.24	50	50
Nb	3,537.47	30	30	Nb	3,540.96	500	15
Th	3,537.16	10	10	Fe	3,541.09	200	200
Mo	3,537.09	50	—	Rb	3,541.22	(100)	—
F	3,536.84	(30)	—				
				V	3,541.34	35	—
Fe	3,536.56	200	300	F	3,541.76	(100)	—
P	3,536.29	(30)	—	F	3,541.94	(60)	—

	λ	I	
		J	O
Tm	3,536.21	60	40
Sc	3,535.73	30	15
Hf	3,535.54	50	15
Tm	3,535.52	25	80
Ti	3,535.41	125	15
Kr	3,535.35	(50h)	—
Nb	3,535.30	500	300
Tm	3,534.85	30	20
Mo	3,534.69	25	10
Ti	3,533.87	35	6
V	3,533.76	40h	20
Nb	3,533,665	30	20
P	3,533.66	(30)	—
Fe	3,533.20	50	50
Mo	3,533.07	25d	—
Na	3,533.010	(200)	50
Fe	3,533.010	75	50
Hg	3,532.63	(200)	—
V	3,532.28	25h	1
Mn	3,532.12	30	50h
Mn	3,531.85	30r	40R
Er	3,531.714	25	40
Dy	3,531.712	100	100
Tb	3,531.70	50	15
Rb	3,531.60	(100)	—
V	3,530.77	100	40
K	3,530.71	(40)	—
V	3,530.45	30	—
Fe	3,530.39	25	50
P	3,530.24	(70)	—
Fe	3,529.82	80	125
Co	3,529.81	30	1000R
Tl	3,529.43	800	1000
Nb	3,528.90	40	2
Os	3,528.60	50	400R
Nb	3,528.48	50	3
Rh	3,528.02	150	1000w
Fe	*3,527.80*	*80*	*100*
Fe	*3,526.68*	*50*	*80*
Fe	*3,526.04*	*50*	*80*
V	*3,524.71*	*60*	*10*
Mo	*3,524.65*	*50h*	*5*
Ni	*3,524.54*	*100wh*	*1000R*
Fe	*3,524.24*	*50*	*60*
Hg	*3,524.19*	*(100)*	—
Tb	*3,523.66*	*50*	*30*
Nb	*3,522.36*	*50W*	—

	λ	I	
		J	O
Fe	3,542.08	100	150
Nb	3,542.56	30d	5
Zr	3,542.62	30	12
Gd	3,542.77	25	25
Ne	3,542.90	(40)	—
Hg	3,543.08	40	—
V	3,543.50	50	50
Fe	3,543.67	30	60
Rh	3,543.95	40	150
Kr	3,544.14	(30wh)	—
Kr	3,544.54	(30wh)	—
Co	3,545.04	30	2
V	3,545.20	300R	40
Ar	3,545.58	(300)	—
Fe	3,545.64	70	90
Gd	3,545.79	125	125
Ar	3,545.84	(125)	—
Mo	3,546.00	25	—
Ni	3,548.18	25	400
Mn	3,548.20	40	40h
Ar	3,548.51	(25)	—
Yt	3,549.01	50	12
W	3,549.05	25	5
Gd	3,549.36	125	125
Hg	3,549.42	(200)	—
Rh	*3,549.54*	*50*	*150*
Dy	*3,550.23*	*100*	*200*
Cr	*3,550.63*	*60*	*70*
Fe	*3,552.83*	*50*	*80*
Fe	*3,553.74*	*100*	*100*
Ar	*3,554.31*	*(300)*	—
Lu	*3,554.43*	*150*	*50*
Fe	*3,554.93*	*300*	*400*
Ar	*3,555.97*	*(100)*	—
P	*3,556.48*	*(100)*	—
Zr	*3,556.60*	*50*	*15*
Fe	*3,556.88*	*150*	*300*
Fe	*3,558.518*	*300*	*400*
Gd	*3,558.519*	*50*	*100r*
Ir	*3,558.99*	*50*	*50*
Sb	*3,559.24*	*50wh*	*2h*
Th	*3,559.46*	*10*	*10*
Nb	*3,559.60*	*100*	*2*
Os	*3,559.79*	*50*	*150*

	λ	I J	I O
Ir	*3,522.03*	*50*	—
Th	*3,521.92*	*10*	*10*
V	*3,521.84*	*80*	*20*
Rb	*3,521.44*	*(200)*	—
Fe	*3,521.26*	*200*	*300*
Sb	*3,520.474*	*(125)*	—
Ne	*3,520.472*	*(1000)*	—
Yb	*3,520.29*	*70*	*10*
V	*3,520.02*	*50*	*5*
Tl	*3,519.24*	*1000R*	*2000R*
P	*3,518.60*	*(50h)*	—
Co	*3,518.35*	*100*	*200W*
Nb	*3,517.67*	*200*	*2*

Th 3,601.04 *J* 10 *O* 8

	λ	J	O
Pr	**3,601.02**	**2**	**10**
Gd	**3,600.97**	**30**	**30**
W	**3,600.955**	**9**	—
Ho	**3,600.95**	**10**	**6**
Nd	**3,600.91**	**10**	**20**
Th	**3,600.83**	**1**	**3**
Co	**3,600.81**	—	**4**
Tb	**3,600.80**	—	**8**
J	**3,600.77**	**(5)**	—
Yb	**3,600.76**	**20**	**10**
Rh	**3,600.754**	**2**	**8**
Pr	**3,600.750**	**5**	**25**
Er	**3,600.742**	**20*w***	**30*w***
Mo	**3,600.737**	**4**	**2**
Yt	**3,600.734**	**300**	**100**
Cs	**3,600.73**	**(10)**	—
Ho	**3,600.73**	—	**6**
Ta	**3,600.70**	**1*h***	**2**
Rb	**3,600.68**	**(20)**	—
Ce	**3,600.58**	**2**	**15**
U	**3,600.49**	**2**	**1**
Tb	**3,600.44**	**50**	**8**
Th	**3,600.43**	**1**	**4**
Yb	**3,600.39**	**2**	**2**
Dy	**3,600.34**	**30**	**20**
U	**3,600.293**	—	**10**
W	**3,600.287**	**3**	**4**
Mo	**3,600.28**	**5**	**5**
Ar	**3,600.22**	**(3)**	—
J	**3,600.21**	**10**	—
Mo	**3,600.20**	**3**	**3**
La	**3,601.05**	**15**	**5**
Zr	**3,601.193**	**15**	**400**
Ti	**3,601.193**	**3**	**7**
Sm	**3,601.25**	**1**	**4**
Mn	**3,601.27**	**15*h***	**15**
Nd	**3,601.32**	**2**	**10**
Ti	**3,601.38**	—	**15**
Dy	**3,601.39**	**1*h***	**3**
F	**3,601.403**	**(30)**	—
Ir	**3,601.403**	**8**	**30**
U	**3,601.43**	**3**	—
Ru	**3,601.48**	**5**	**2**
Tb	**3,601.50**	—	**8**
Ar	**3,601.51**	**(5)**	—
W	**3,601.58**	**6**	—
Rn	**3,601.6**	**(5)**	—
Cr	**3,601.67**	**30**	**50**
Sm	**3,601.69**	**5**	**5**
Eu	**3,601.73**	—	**2*w***
Al	**3,601.74**	**15**	—
Tb	**3,601.75**	**3**	**15**
W	**3,601.798**	**4**	**5**
Pb	**3,601.8**	**(20)**	—
Os	**3,601.83**	**20**	**60**
Mo	**3,601.84**	**3**	**1**
Ho	**3,601.87**	—	**6**
Pr	**3,601.916**	—	**3**
Yt	**3,601.921**	**60**	**18**
Sm	**3,601.984**	**1**	**2**
Th	**3,601.984**	**2**	**2**
Cu	**3,602.03**	**25*W***	**50**

	λ	I	
		J	O
Ne	**3,600.17**	**(75)**	—
Nd	**3,600.12**	**10**	**15**
S	**3,600.08**	**(3)**	—
Hf	**3,600.05**	**8*h***	**2**
Tb	**3,600.04**	**3**	**15**
V	**3,600.03**	**40**	**50**
Ce	**3,599.974**	**1**	**10**
Fe	**3,599.974**	—	**5**
Bi	**3,599.94**	—	**2**
Zr	**3,599.901**	**5**	**5**
Kr	**3,599.90**	**(40*h*)**	—
S	3,599.88	(5)	—
Hf	3,599.87	8	10
U	3,599.84	18	6
Mo	3,599.829	5	3
Er	3,599.829	20*s*	30*w*
Ru	3,599.76	100	12
Ar	3,599.67	(20)	—
Nb	3,599.63	10	10
Fe	3,599.62	30	40
Pr	3,599.51	2*hd*	4*d*
Er	3,599.508	5	18
Ba	3,599.42	3	10
Cr	3,599.39	20	30
U	3,599.35	10	4
He	3,599.30	(5)	—
Nb	3,599.28	15	15
Kr	3,599.21	(25*h*)	—
Tm	3,599.16	8	4
Fe	3,599.15	5	10
Cu	3,599.140	30	60
Hf	3,599.139	10	4
Nd	3,599.07	—	10
Tb	3,598.96	3	8
U	3,598.95	3	8
Pr	3,598.91	2	8
Mo	3,598.881	15	10
W	3,598.878	7	10
Er	3,598.85	1	8*hs*
Ho	3,598.77	30	40
Re	3,598.769	—	20
Ir	3,598.76	5	25
Fe	3,598.721	2	6
V	3,598.718	10	—
Ti	3,598.716	30	70
F	3,598.70	(30)	—
W	3,598.43	12	—

	λ	I	
		J	O
Ir	**3,602.06**	—	**3**
Ca	**3,602.08**	**3**	—
Co	**3,602.084**	**35**	**200**
Fe	**3,602.085**	**5**	**20**
Cl	**3,602.10**	**(12)**	—
Kr	**3,602.12**	**(2*h*)**	—
Ni	3,602.28	15	150
W	3,602.462	12	—
Th	3,602.464	1	4
Fe	3,602.47	5	10
Os	3,602.48	10	15
Eu	3,602.50	1	6*w*
U	3,602.51	1*d*	12*d*
Tb	3,602.51	3	15
Fe	3,602.53	30	50
Nb	3,602.56	30	30
Cr	3,602.57	10	15
Zr	3,602.78	2*h*	5*h*
Dy	3,602.82	4	10
F	3,602.85	(60*d*)	—
Cs	3,602.852	(8)	—
Tb	3,602.93	—	15
Mo	3,602.941	25	20
V	3,602.943	12	10
W	3,602.97	5	1
U	3,603.13	—	4*hd*
Eu	3,603.20	50	100*w*
Fe	3,603.208	80	150
Th	3,603.208	8	8
Ce	3,603.358	—	6
Th	3,603.363	5	5
U	3,603.363	1*h*	10
Nb	3,603.43	5	4
Ce	3,603.48	—	5
Nb	3,603.53	25*wh*	2
Th	3,603.62	2	4
U	3,603.67	—	5*d*
F	3,603.72	(20)	—
Mo	3,603.724	5	4
Ce	3,603.73	—	5
Cr	3,603.74	50	15
Fe	3,603.82	12	20
Ti	3,603.84	2	15
W	3,603.91	5	5
Nb	3,603.96	5	2
*bh*Ca	3,604.00	—	4
Nd	3,604.04	6	—

	λ	I J	I O		λ	I J	I O
Nb	3,598.35	2*h*	5*d*	Th	3,604.05	4	4
Dy	3,598.26	5*h*	10	Mo	3,604.07	15	6
Ce	3,598.20	1	20	Hg	3,604.09	(50)	—
Th	3,598.12	4	6	Tb	3,604.15	—	8
Os	3,598.11	30	300	Ce	3,604.20	1	8
Au	3,598.08	10	35	U	3,604.22	6	1
Tb	3,598.06	8	15	Fe	3,604.27	—	5
Nd	3,598.03	6	10	Sm	3,604.276	20	6
Cu	3,598.01	—	40*wh*	Ti	3,604.284	6	20
W	3,597.96	4	5	Ir	3,604.31	—	5
Rh	3,597.89	3	6	V	3,604.377	50*h*	—
Tb	3,597.80	—	8	Fe	3,604.380	1	10
Er	3,597.77	—	6	Re	3,604.396	—	15*w*
U	3,597.76	6	12	U	3,604.396	6	1
Cs	3,597.73	(6)	—	Os	3,604.47	100	15
W	3,597.71	6	7	*bh*La	3,604.5	—	5*h*
Ni	3,597.70	50*h*	1000*r*	Mo	3,604.56	5	5
Os	3,597.52	10	15	Nb	3,604.64	10*w*	5*w*
Sb	3,597.51	200*wh*	2	Th	3,604.680	1	5
Al	3,597.50	(5)	—	Eu	3,604.684	1*h*	7
Cs	3,597.43	(10)	—	Mn	3,604.69	—	12
Hf	3,597.42	15	5	Er	3,604.71	2	9
V	3,597.40	5	—	Dy	3,604.85	2	4*w*
W	3,597.26	9	10	Gd	3,604.88	12	50
Nb	3,597.25	10	8	Tb	3,604.90	15	3
Ce	3,597.23	—	5	Er	3,604.901	4	12
Rh	3,597.15	100	200	Ce	3,604.93	1	6
Fe	3,597.06	10	40	Ta	3,604.98	1*h*	7
Sb	3,596.96	(100)	—	Co	3,605.01	—	15
U	3,596.88	8	6	Mo	3,605.02	5	2
Ta	3,596.86	1	7	Pr	3,605.05	10	20
Tb	3,596.80	—	8	Dy	3,605.09	2	4
U	3,596.76	8	2	Fe	3,605.21	2	12
Ce	3,596.72	1	10	Gd	3,605.25	15	15
Pd	3,596.66	1	5	U	3,605.28	—	8
Sm	3,596.64	2	4	Cr	3,605.33	400*R*	500*R*
W	3,596.61	4	5	Co	3,605.36	—	60
Co	3,596.51	—	5	Fe	3,605.458	150	300
W	3,596.43	4	5	Yt	3,605.46	6*h*	2*h*
Re	3,596.39	—	25	V	3,605.59	20*h*	30
Tb	3,596.38	15	50	Rn	3,605.6	(10)	—
Mo	3,596.35	30	—	Eu	3,605.61	—	4*w*
Lu	3,596.34	—	6	Gd	3,605.62	15	20
Re	3,596.22	—	20*h*	Ru	3,605.64	9	2
Fe	3,596.20	5	15	Th	3,605.66	6	8
Rh	3,596.194	50	200	Mn	3,605.69	—	10
Pr	3,596.193	8	25	Ce	3,605.78	—	5
W	3,596.18	15	—	Hg	3,605.80	(200)	—
Ru	3,596.179	100	30	U	3,605.82	8	8

	λ	I J	I O		λ	I J	I O
Ce	3,596.115	2	12	Rh	3,605.86	30	25
Bi	3,596.110	50	150*wh*	Ar	3,605.89	(15)	—
Dy	3,596.07	20	50	Pr	3,605.96	3	10
Ti	3,596.05	125	50	Ti	3,606.06	1	12
Rh	3,596.02	5	25	W	3,606.07	10	12
S	3,595.99	(50)	—	Er	3,606.08	4	15*d*
F	3,595.92	(15)	—	U	3,606.116	6	1
Fe	3,595.86	1	4	Tb	3,606.12	15*d*	30
Er	3,595.84	—	7	Dy	3,606.126	100	200
Os	3,595.79	5	1	Ce	3,606.13	—	5
U	3,595.75	—	5	Th	3,606.20	5*h*	—
Ta	3,595.64	5	70	Nb	3,606.27	5*wh*	2
Mo	3,595.55	5	8	Nd	3,606.28	4	12
Er	3,595.47	1	7*d*	Mo	3,606.29	10	1
Mo	3,595.40	10	—	U	3,606.32	12	8
W	3,595.39	6	8	W	3,606.34	8	10
Fe	3,595.31	7	20	Eu	3,606.38	—	5
U	3,595.24	—	5	Nd	3,606.40	2	12
Re	3,595.17	—	20*h*	Yb	3,506.47	60	15
Mn	3,595.12	25	50	Nb	3,606.49	5	3*h*
W	3,595.07	5	—	Ar	3,606.52	(1000)	—
Dy	3,595.05	100	200	Sc	3,606.62	—	4
Er	3,595.04	8	18	Br	3,606.66	(15)	—
Sm	3,594.99	—	6*d*	Mo	3,606.67	20	8
Tb	3,594.98	—	15	Fe	3,606.68	150	200
U	3,594.95	15	8	V	3,606.69	70	80
Co	3,594.87	—	200*W*	Eu	3,606.71	—	50
Tb	3,594.65	—	15	Ti	3,606.79	4	12
Fe	3,594.64	100	125	F	3,606.80	(10)	—
Dy	3,594.57	2	4	Ni	3,606.85	—	100*r*
W	3,594.53	7	5	Mo	3,606.91	30	—
S	3,594.46	(35)	—	Hg	3,606.92	(30)	—
Ir	3,594.39	30	15	Nb	3,607.01	10	1
Tb	3,594.25	8	15	Gd	3,607.066	20	15
Ne	3,594.18	(12)	—	W	3,607.066	12	10
Au	3,594.15	6	25	*bh*Sr	3,607.1	—	4
Ir	3,594.14	4	10	Ti	3,607.13	5	25
Er	3,594.128	—	9	Pr	3,607.22	2	10
Th	3,594.116	2	5	F	3,607.32	(6)	—
Ce	3,594.10	2	5	Nb	3,607.33	5	2
Ce	3,594.03	—	5	Tm	3,607.36	10	8
Cu	3,594.02	2	15	Zr	3,607.38	9	8
W	3,593.97	8	9	Th	3,607.39	4	3
Nb	3,593.97	50	80	Ta	3,607.406	35	70
Er	3,593.95	—	8	Xe	3,607.41	(5)	—
Th	3,593.88	5	5	Mo	3,607.412	4	4
Tb	3,593.75	8	30	Er	3,607.43	—	8
Sm	3,593.73	2	4	J	3,607.51	(18)	—
U	3,593.685	1	8	Mn	3,607.537	40	75

	λ	I J	I O		λ	I J	I O
Pr	3,593.685	2	4	Au	3,607.54	20	—
Ne	3,593.64	(250)	—	Tb	3,607.54	—	8
Tl	3,593.61	(10)	—	Hg	3,607.60	(18)	—
N	3,593.60	(10)	—	Ce	3,607.62	8	15
W	3,593.56	3	4	Nd	3,607.72	6	10
Nb	3,593.55	3	15	*bh*La	3,607.8	—	5*h*
Rh	3,593.530	2	10	Pr	3,607.82	3	5
Ne	3,593.526	(500)	—	Tb	3,607.86	—	8
Cr	3,593.49	400*R*	500*R*	Kr	3,607.88	100*wh*	—
Hg	3,593.48	(10)	—	Pt	3,607.89	5	—
Gd	3,593.43	15	15	Nb	3,608.01	5	5
Re	3,593.40	—	15*W*	Dy	3,608.06	2	4
Te	3,593.34	(5)	—	Rh	3,608.09	3	10
V	3,593.334	300*R*	30	Fe	3,608.15	25	15
Fe	3,593.329	2	7	La	3,608.16	6	2
K	3,593.22	(5)	—	Er	3,608.22	—	10
U	3,593.20	1	10	Tb	3,608.25	—	15
Dy	3,593.15	4	6	Cs	3,608.28	(10)	—
Ce	3,593.134	—	8	Mo	3,608.369	15	15
Ho	3,593.13	6	6	W	3,608.37	10	—
Zr	3,593.129	1	7	Cr	3,608.40	8	12
Pb	3,593.12	30	—	Pr	3,608.47	3	10
Tb	3,593.10	8	15	Mn	3,608.49	40	60
Ti	3,593.09	30	5	Eu	3,608.72	1*h*	10
Pr	3,593.04	2*h*	5	Ru	3,608.73	8	2
Ru	3,593.02	150	60	Gd	3,608.76	125	100
W	3,592.98	3	5	Tm	3,608.77	20	100
Ho	3,592.95	—	6	Ta	3,608.78	1*d*	15*r*
Eu	3,592.93	1*h*	7	Fe	3,608.86	400	500
Yt	3,592.91	25	80	K	3,608.87	(10)	—
Sm	3,592.90	—	4	F	3,608.89	(6)	—
W	3,592.85	3*d*	6*d*	U	3,608.96	10	18
U	3,592.80	6	1	Tb	3,609.06	—	8
Th	3,592.78	6	2*d*	N	3,608.09	(5)	—
Gd	3,592.70	70	50	Os	3,609.15	12	20
Fe	3,592.69	2	12	Ta	3,609.177	1	8
Mo	3,592.65	20	—	Ne	3,609.179	(50)	—
Sm	3,592.595	50	40	Th	3,609.23	4	5
Nd	3,592.59	30	20	V	3,609.29	30*W*	30*W*
V	3,592.53	—	40	Cu	3,609.307	5	25
W	3,592.42	35	9	Ni	3,609.314	15	200
Bi	3,592.4	(5)	—	Ta	3,609.357	1	8
Os	3,592.32	12	20	Nb	3,609.360	5	1
U	3,592.30	4	4	Er	3,609.44	1*h*	6*d*
Mo	3,592.24	15	—	Th	3,609.45	10	12
Ho	3,592.22	10	6	Cr	3,609.479	12	20
Fe	3,592.21	—	6	Sm	3,609.484	100	60
Dy	3,592.12	30	80	Mo	3,609.493	10	5

	λ	I	
		J	O
Nd	3,592.07	8	5
U	3,592.06	3	4
V	3,592.02	300*R*	50
Mo	3,592.01	3	5
Au	3,591.99	5	15
W	3,591.97	4	6
Gd	3,591.91	4	8
U	3,591.88	10	6
Dy	3,591.82	20	80
Mn	3,591.81	—	15
Nb	3,591.79	5	5
W	3,591.77	3	6
U	3,591.75	2	10
Ce	3,591.74	1	4
Zr	3,591.72	—	8
Mo	3,591.69	25	—
Tb	3,591.66	3	15
Os	3,591.60	5	2
Rb	3,591.59	20	80
U	3,591.56	15	4
Fe	3,591.49	—	6
Th	3,591.45	1	4
Gd	3,591.43	5	10
Dy	3,591.42	100	200
Mo	3,591.40	2	4
Tb	3,591.39	3	8
Fe	3,591.345	3	12
Eu	3,591.34	1	10
Ho	3,591.33	6	—
Ir	3,591.28	2	8
Pr	3,591.21	1	4
Nb	3,591.20	50	2
Er	3,591.16	—	7*h*
Se	3,591.03	(8)	—
Fe	3,591.00	1	4
Re	3,590.88	—	15
C	3,590.87	6	—
W	3,590.83	10	10
Si	3,590.77	5	—
Mo	3,590.74	10	10
U	3,590.73	—	4
Er	3,590.72	1	6*d*
Dy	3,590.67	20	30
Ce	3,590.60	1	50
Rh	3,590.52	3	4
U	3,590.50	10	8
Sc	3,590.475	12	18
Gd	3,590.472	15	15

	λ	I	
		J	O
Nd	3,609.495	30	25
Tm	3,609.54	25	15
Pd	3,609.548	700*R*	1000*R*
Tb	3,609.55	—	15
Ti	3,609.59	2	12
U	3,609.68	12	15
Ce	3,609.69	10	40
Co	3,609.76	3	5
Ir	3,609.77	25	30
Nd	3,609.79	10	15
Tb	3,609.88	—	8
Ta	3,609.93	18	1*h*
Nb	3,610.00	5	3
Cr	3,610.05	8	20
Ti	3,610.16	70	100
Fe	3,610.16	90	100
La	3,610.24	5	7
Ce	3,610.26	—	5
Mn	3,610.30	40	60
Xe	3,610.32	(15)	—
Th	3,610.40	4	8
Ni	3,610.46	—	1000*r*
U	3,610.487	8	4
Re	3,610.493	—	40
Sc	3,610.50	(35)	—
In	3,610.508	18	—
Cd	3,610.510	500	1000
Ta	3,610.60	15	—
Mo	3,610.62	5	3
U	3,610.69	12*d*	3*d*
Fe	3,610.70	3	10
Nb	3,610.764	5	3
Gd	3,610.766	6	25
Dy	3,610.77	2	4
Th	3,610.79	4	8
Cu	3,610.81	6	25
Pt	3,610.909	1	4
Ce	3,610.914	—	10
Ba	3,611.00	3	10
Eu	3,611.02	1	10*w*
Yt	3,611.047	60	40
Gd	3,611.049	5*h*	5
Sm	3,611.06	2	4
*bh*La	3,611.1	—	6*h*
Ta	3,611.13	1	25
Pr	3,611.16	2	5
U	3,611.24	10	5
Nb	3,611.28	5*h*	3

	λ	I J	I O
Tb	3,590.46	—	15
Eu	3,590.45	—	10wh
W	3,590.43	5	—
Nd	3,590.352	300W	400W
Ce	3,590.351	5wh	5wh
In	3,590.35	12	—
Er	3,590.35	—	25
U	3,590.32	15	5
Mo	3,590.20	15d	—
Os	3,590.11	15	30
Fe	3,590.08	3	6
Pb	3,589.92	40	—
bhZr	3,589.8	—	5
U	3,589.79	—	8
Te	3,589,78	(5)	—
V	3,589.76	600R	80
C	3,589.67	20	—
U	3,589.66	1	6
Kr	3,589.65	(70wh)	—
Sc	3,589.63	12	5
Mo	3,589.52	30d	—
Fe	3,589.46	30	50
Nb	3,589.36	100	100
F	3,589.34	(20)	—
Ru	3,589.21	100	60
Fe	3,589.107	30	70
Nb	3,589.107	20	50
Mo	3,588.949	10	10
Er	3,588.948	3	9
Fe	3,588.91	4	10
Fe	3,588.62	10	35
Bi	3,588.5	(60)	—
Ar	3,588.44	(300)	—
Er	3,588.34	—	20
Zr	3,588.32	9	9
V	3,588.14	20	1
Mo	3,588.11	15	—
Zr	3,587.984	10	10
F	3,587.980	(15)	—
Ni	3,587.93	12	200
Os	3,587.87	10	2
Pr	3,587.86	6	8
Tb	3,587.76	—	30
Fe	3,587.758	25	50
Yt	3,587.753	2	15
Er	3,587.748	3	10
Ce	3,587.64	2	10
Nd	3,587.50	20	8

	λ	I J	I O
Yb	3,611.31	50	12
Tb	3,611.33	8	50
Eu	3,611.332	—	10h
Ce	3,611.34	—	10
U	3,611.39	1	12
bhLa	3,611.4	—	4d
Tb	3,611.41	8	8
Cs	3,611.52	—	200
Eu	3,611.580	5	40
V	3,611.58	15wh	—
Ce	3,611.65	2w	20w
Co	3,611.70	—	25
Ce	3,611.73	—	5
Te	3,611.77	(5)	—
Ar	3,611.84	(5)	—
W	3,611.85	20	—
Zr	3,611.89	40	15
Dy	3,611.90	1	2
Pr	3,611.94	7	15
Mo	3,612.00	5	8
Fe	3,612.07	50	80
Eu	3,612.19	2	6w
Mo	3,612.20	20	—
Ti	3,612.25	1	15
Tb	3,612.31	—	8
Ce	3,612.32	1	8s
La	3,612.33	15h	8
Ne	3,612.35	(7)	—
Xe	3,612.37	(10)	—
Er	3,612.40	—	8
Th	3,612.43	4	8
Nd	3,612.44	20	6
Mo	3,612.45	6	6
Ce	3,612.466	—	6
Al	3,612.467	80h	—
Rh	3,612.470	50	200
Cr	3,612.609	25	35
Rn	3,612.61	(20)	—
Pr	3,612.71	2	10
Ni	3,612.74	50h	400
Dy	3,612.79	10	25
In	3,612.86	15	—
Cl	3,612.86	(10)	—
Er	3,612.87	2h	10w
Cd	3,612.875	500	800
Fe	3,612.94	4	20

	λ	I	
		J	O
Sm	3,587.47	3	15
Al	3,587.441	(80)	—
Tb	3,587.44	15	15
Fe	3,587.42	5	10
P	3,587.35	(30)	—
Pr	3,587.34	5	10
Os	3,587.31	15	60
He	3,587.25	(10)	—
Ru	3,587.20	70	5
Gd	3,587.191	15	25
Co	3,587.190	50h	200r
Pr	3,587.14	8	8
Ti	3,587.13	25	12
Rb	3,587.08	40	200
Al	3,587.06	(100)	—
Fe	3,586.99	150	200
Al	3,586.91	(500h)	—
bhSr	3,586.9	—	8
Al	3,586.80	(200wh)	—
Nb	3,586.76	20	2
Ce	3,586.75	4	10
Au	3,586.74	15	20
Al	3,586.692	(200)	—
Nb	3,586.691	20	1
Er	3,586.64	—	8d
Al	3,586.55	(200)	—
Mn	3,586.54	40	50h
Pr	3,586.53	4	10
Os	3,586.51	15	30
Pb	3,586.44	20	—
Sm	3,586.36	—	40r
Zr	3,586.294	3	20
Ta	3,586.291	3	18
Kr	3,586.25	(12h)	—
Fe	3,586.11	80	80
Mo	3,585.90	15	—
In	3,585.86	12	—
Ru	3,585.82	10h	—
Er	3,585.79	2h	10W
Tb	3,585.78	3	8
Dy	3,585.776	100	150
Fe	3,585.71	80	125
Mo	3,585.70	20	—
Cr	3,585.51	35	6
Yb	3,585.47	20	10
Te	3,585.34	(350)	—
Re	3,585.321	—	30
Fe	3,585.320.	100	150

	λ	I	
		J	O
S	3,613.03	(12)	—
Tb	3,613.06	3	15
La	3,613.08	3	12
Zr	3,613.10	40	40
Nd	3,613.24	4	10
Os	3,613.33	12	30
Gd	3,613.40	6	8
Ti	3,613.43	—	8
Fe	3,613.45	2	10
Hg	3,613.61	(40)	—
He	3,613.64	(30)	—
Cr	3,613.67	8	10
Tb	3,613.68	3	15
Ce	3,613.70	5	18
Ti	3,613.756	—	12
Cu	3,613.761	7	60
W	3,613.79	30	10
J	3,613.81	(10h)	—
Bi	3,613.82	30	—
Sc	3,613.84	70	40
Au	3,614.00	20	—
Dy	3,614.08	10	30
Fe	3,614.12	3	10
Ti	3,614.21	4	35
Cu	3,614.22	6	50
Mo	3,614.25	30	50d
Cd	3,614.45	100	60
Fe	3,614.56	6	15
Tb	3,614.63	8	30
Er	3,614.64	1	12
Mo	3,614.69	20	1
Dy	3,614.71	5	15
Zr	3,614.774	80	40
Rh	3,614.775	10	15
Rn	3,615.0	(30)	—
Th	3,615.13	10	10
Pr	3,615.16	4	40
Fe	3,615.20	1	10
Sm	3,615.24	2	40
Ti	3,615.33	10	—
Fe	3,615.35	—	9
Co	3,615.39	—	8
Kr	3,615.47	(20)	—
Nb	3,615.50	30	30
U	3,615.56	—	10
Ce	3,615.62	—	10
Cr	3,615.64	10	30
Tb	3,615.66	15	50
Fe	3,615,665	2	10

	λ	I J	I O		λ	I J	I O
Co	3,585.16	—	60	Nd	3,615.82	10	20
Dy	3,585.07	100	300				
Tb	3,585.03	50	15	Er	3,616.07	—	10
Re	3,585.02	—	15	Eu	3,616.15	10	40
				Nd	3,616.32	1h	15
Nb	3,584.97	50	30	U	3,616.33	2h	12
Gd	3,584.964	100	100	Eu	3,616.51	—	10d
Fe	3,584.960	25	30				
U	3,584.88	12	30	Fe	3,616.572	7	30
Co	3,584.80	—	25	Er	3,616.573	20	30
				Os	3,616.574	20	150
Ir	3,584.663	1h	10	Tb	3,616.58	8	8
Fe	3,584.663	60	100	Pr	3,616.68	8	20
Er	3,584.52	15h	25				
Yt	3,584.514	15	20	V	3,616.72	30	30
Ta	3,584.512	3	35	U	3,616.76	10	2
				Mo	3,616.84	15	15
Dy	3,584.43	20	50	U	3,616.888	—	8
Os	3,584.40	20	15	Hf	3,616.892	10	25
Cr	3,584.33	10wh	10wh				
Pr	3,584.257	8	15	S	3,616.92	(60)	—
Mo	3,584.256	10	5	Re	3,617.08	—	50
				P	3,617.09	(100w)	—
Sm	3,584.25	3	25d	Ir	3,617.21	15	50
Ta	3,584.21	7h	50	Ti	3,617.21	1	8
S	3,584.18	(10)	—				
W	3,584.11	9	9	Re	3,617.25	—	15
In	3,583.85	12	—	Fe	3,617.32	15	25
				Cs	3,617.41	—	60
V	3,583.70	30	60	W	3,617.52	20	35
bhCa	3,583.7	—	8	Te	3,617.55	(25)	—
P	3,583.60	(50)	—				
Rh	3,583.53	5	10	Mo	3,617.55	20	—
W	3,583.46	10	—	Nb	3,617.71	15	3
				Fe	3,617.79	80	125
Os	3,583.40	10	15	Er	3,617.82	5	12
Sm	3,583.37	4	15	Tb	3,617.88	8	8
Fe	3,583.34	15	50				
J	3,583.32	(18)	—	Re	3,617.88	—	12w
Rh	3,583.10	125	200	Pr	3,618.08	3	15
				Tb	3,618.18	3	8
Os	3,583.09	10	20	Eu	3,618.19	—	8h
Re	3,583.02	—	100w	Mo	3,618.35	20	—
V	3,582.81	25	40				
Ar	3,582.70	(30)	—	Fe	3,618.39	4	8
Sm	3,582.683	—	10	K	3,618.43	(20)	—
				Nb	3,618.44	5	10
Mo	3,582.676	1	20	W	3,618.45	12	—
Tb	3,582.63	—	8	U	3,618.49	10	2
Cr	3,582.62	12	35				
Rn	3,582.6	(18)	—	Dy	3,618.52	10	80
Mn	3,582.43	—	10	Hg	3,618.53	(50)	—
				Ce	3,618.58	3	10
Mo	3,582.42	15	—	Se	3,618.73	(35)	—
Nb	3,582.36	10	5	Fe	3,618.77	400	400
Ar	3,582.35	(50)	—				
Fe	3,582.20	30	30	Er	3,618.92	6	10w
Dy	3,582.03	10	25	V	3,618.93	100	—

	λ	I J	I O		λ	I J	I O
U	3,582.02	15	6	Nd	3,618.97	15	20
Gd	3,581.92	15	15	Pr	3,619.10	3	9
Mo	3,581.89	15	10	Ir	3,619.16	3	30
Er	3,581.842	2	10h				
Pr	3,581.840	7	10	W	3,619.27	10	10
				Mn	3,619.28	50	75
U	3,581.838	15	6	Fe	3,619.390	1	12
Mo	3,581.80	5	10	Ce	3,619.391	—	8
La	3,581.68	20h	—	Ni	3,619.392	150h	2000R
Ar	3,581.62	(15)	—				
W	3,581.24	8	8	Os	3,619.43	25	60
				Cr	3,619.46	8	30
Fe	3,581.19	600s	1000R	Nb	3,619.51	200	5
Re	3,580.97	—	40w	Pr	3,619.59	2	8
Sc	3,580.93	40	12	Nb	3,619.727	300	3
Sm	3,580.91	6	40				
Ir	3,580.86	3	15	Tb	3,619.73	8	15
				Yb	3,619.81	100	30
V	3,580.82	50	50	Ce	3,619.92	1	8
Ce	3,580.78	2	10	Dy	3,619.96	5	25
Tb	3,580.63	—	8	Tm	3,619.97	20	20
Mo	3,580.54	10	5				
Er	3,580.48	5	12d	Er	3,619.98	1	9
				Ti	3,620.02	—	10
P	3,580.35	(30)	—	Hf	3,620.04	—	8
Ti	3,580.291	5	15	U	3,620.08	2h	15
Mo	3,580.29	10	—	Dy	3,620.176	20	80
Nb	3,580.27	300	100				
Rh	3,580.26	2	10	Er	3,620.177	3	50
				Au	3,620.23	10	—
Re	3,580.13	—	80	Os	3,620.24	12	15
Au	3,580.08	15	20	Tb	3,620.29	—	15
Dy	3,580.04	30	80	Cu	3,620.35	5	30
Ta	3,580.02	—	10				
Tb	3,579.98	3	8	Rh	3,620.456	10	20
				Gd	3,620.461	25	25
Hf	3,579.90	10	15	Fe	3,620.471	2	15
Ba	3,579.67	8	10	V	3,620.472	50	10
Ta	3,579.45	35	—	P	3,620.65	(15)	—
Er	3,579.44	—	15d				
U	3,579.36	10	6	Nd	3,620.74	4	10
				Yt	2,620.941	12	2h
Tb	3,579.20	50	50	Er	3,620.945	15	25
Re	3,579.13	—	50	Rn	3,621.00	(250)	—
Ta	3,579.08	1	15	Nb	3,621.03	15	10
Mo	3,579.07	25	—				
Se	3,578.93	(12)	—	Pr	3,621.09	2h	20h
				Th	3,621.12	10	10
Co	3,578.90	2	8	Ce	3,621.15	—	8
Hg	3,578.75	(40)	—	Co	3,621.179	50h	15
Tb	3,578.70	—	8	Pr	3,621.181	1	9
Ti	3,578.687	5	25				
Cr	3,578.687	400r	500R	V	3,621.21	80	15
				Sm	3,621.22	10	60
V	3,578.64	80d	35d	Cu	3,621.24	5	20
Nb	3,578.584	1	15	Tb	3,621.39	3	8
Gd	3,578.58	—	8	Fe	3,621.463	100	125
Pr	3,578.42	3	9				

	λ	I J	I O
Tb	3,578.40	—	8
Fe	3,578.38	5	40
U	3,578.33	1	8
Ti	3,578.27	1	8
Pr	3,578.11	5	10
Co	3,578.07	—	18
Re	3,621.465	—	30
Eu	3,621.90	10	8
Fe	3,622.00	100	125
Tb	3,622.11	3	30
Ce	3,622.14	5	15
Ar	3,622.15	(15)	—
Ti	3,622.20	10	—
V	3,622.29	30	—
Th	3,622.34	4	8
Pr	3,622.38	5	25
Ce	3,622.44	1	8
Sm	3,622.51	4	10
Eu	3,622.56	50	20h
V	3,622.63	1	35
U	3,622.70	1	15
Gd	3,622.81	20	20
Mo	3,622.85	20	—
U	3,623.05	15	12
Ti	3,623.096	1	12
P	3,623.10	(15)	—
Fe	3,623.19	80	100
Mo	3,623.23	15	15
Sm	3,623.31	7	15
Fe	3,623.45	5	15
W	3,623.51	7	8
Kr	3,623.61	(30h)	—
Ru	3,623.63	10	2
Mo	3,623.70	50	—
Ce	3,623.76	1	8
Fe	3,623.77	7	35
Mn	3,623.79	40	75
Ir	3,623.80	4	20
Ce	3,623.84	5	60
Zr	3,623.86	7	40
Tb	3,623.92	8	30
Lu	3,623.91	40	20
Hf	3,624.00	20	15

Th 3,741.19 *J* 80 *O* 80

	λ	J	O
Dy	3,741.18	2	6
S	3,741.15	(3)	—
Ti	3,741.14	3	2
Mo	3,741.12	—	10h
Er	3,741.10	2	15
Os	3,741.08	15	40
Ti	3,741.06	40	150
Tb	3,741.24	8	15
Cu	3,741.24	2	50
Ta	3,741.276	1h	7h
U	3,741.280	3	—
Sm	3,741.283	25	25
Nb	3,741.30	15	2
Eu	3,741.32	5	10

	λ	I	
		J	O
Mn	**3,741.03**	**15**	**15**
Pr	**3,741.007**	**15**	**40**
Ce	**3,741.006**	—	**8**
Nd	**3,740.96**	**8**	**6**
Th	**3,740.85**	**20**	**40**
Nb	**3,740.84**	**10**	**10**
Mo	**3,740.76**	**4**	**10**
Sm	**3,740.75**	**1**	**4**
Kr	**3,740.73**	**(6)**	—
Nb	**3,740.72**	**50**	**3**
Ce	**3,740.574**	—	**2**
Mo	**3,740.571**	**3**	**3**
Nb	**3,740.54**	**2*h***	**3*h***
Br	**3,740.51**	**(4)**	—
Re	**3,740.429**	—	**20**
Ce	**3,740.427**	—	**2**
Tb	**3,740.32**	**8**	**15**
Er	**3,740.28**	**3**	**10**
Eu	**3,740.27**	**2**	**8**
Fe	**3,740.25**	**35**	**70**
V	**3,740.24**	**10**	**100**
Co	**3,740.19**	—	**60**
Ce	**3,740.13**	—	**6**
Re	**3,740.10**	—	**50*W***
Dy	**3,740.07**	**2**	**6**
Fe	**3,740.06**	**4**	**8**
Ir	**3,740.049**	—	**2**
Gd	**3,740.048**	**50**	**50**
Ce	**3,740.04**	—	**3**
Sm	**3,740.00**	—	**5**
Zn	**3,739.99**	—	**20**
In	**3,739.95**	**3**	—
Sb	**3,739.95**	**4*h***	—
Pb	**3,739.947**	**60*h***	**150**
Ce	**3,739.947**	—	**8**
Nb	3,739.79	200	100
Gd	3,739.76	50	100
Fe	3,739.53	35	80
Ni	3,739.23	10	100
Pr	3.739.19	30	80
Ti	3,738.90	10	40
Th	3,738.85	20	20
V	3,738.76	7	100
If	3,738.53	10	60
Cr	3,738.38	40	6
Fe	3,738.31	100	100
V	3,737.99	5	50

	λ	I	
		J	O
Ce	**3,741.40**	—	**8**
Nd	**3,741.43**	**15**	**20**
Fe	**3,741.48**	—	**3**
V	**3,741.50**	**8**	**80**
Os	**3,741.53**	**5**	**20**
U	**3,741.577**	**3**	**4**
Tb	**3,741.58**	**8**	**15**
Eu	**3,741.62**	—	**3**
Ti	**3,741.64**	**200**	**30**
Kr	**3,741.69**	**(200)**	—
J	**3,741.710**	**(40)**	—
W	**3,741.713**	**20**	**12**
Ce	**3,741.727**	**1**	**10**
Ir	**3,741.731**	**2**	**25**
Nb	**3,741.78**	**15**	**20**
Fe	**3,741.79**	—	**3*h***
Mo	**3,741.81**	**20**	—
Pr	**3,741.82**	**2**	**5**
Er	**2,741.85**	—	**18*d***
Tb	**3,741.89**	**8**	**8**
Hf	**3,741.94**	**2*h***	**2**
Lu	**3,742.08**	—	**2**
Ta	**3,742.09**	—	**3**
J	**3,742.13**	**(40)**	—
Nd	**3,742.16**	**4**	**6**
Zr	**3,742.21**	**3**	**3**
Ce	**3,742.22**	**3**	**6**
Eu	**3,742.23**	—	**25*W***
Th	**3,742.25**	**10**	**20**
Pr	**3,742.26**	**2**	**4**
Re	**3,742.269**	—	**15**
Ir	**3,742.27**	**2**	**25**
Ru	**3,742.280**	**100**	**70**
Mo	**3,742.285**	—	**20**
Ce	**3,742.31**	**3**	**2**
Eu	**3,742.34**	—	**3*W***
U	**3,742.35**	**4**	**6**
Nb	**3,742.39**	**50**	**30**
Tb	**3,742.43**	**3**	**8**
Fe	3,742.62	25	50
Ru	3,742.78	50	50
Th	3,742.92	8	20
Fe	3,743.363	150	200
Cu	3,743.365	40	40
Cr	3,743.58	40	40
V	3,743.61	40	10
Ar	3,743.76	(100)	—

	λ	I	
		J	O
Rh	3,737.27	10	50
Fe	3,737.13	600	1000*r*
Rh	3,737.12	1	50
Ca	3,736.901	50	12
Ni	3,736.81	15	300
Sm	3,735.97	8	50
Co	3,735.93	—	200*R*
Kr	3,735.78	(40)	—
Nd	3,735.60	50	10
Re	3,735.33	—	40
Rh	3,735.28	2	70
Ne	3,734.94	(40)	—
Fe	3,734.867	600	1000*r*
Co	3,734.867	—	60
Ir	3,734.77	30	100
Th	3,734.60	10	8
Pr	3,734.41	30	40
Co	3,734.14	—	70
Tm	3,734.13	50	150
Co	3,733.49	—	150
Fe	3,733.32	300	400
P	3,733.26	(50)	—
Pr	3,733.03	20	40*d*
Os	3,732.85	5*h*	200*R*
V	3,732.76	500*R*	70*R*
Fe	3,732.399	150	200
Co	3,732.399	—	200*r*
Cr	3,732.03	15	50
Mn	3,731.93	100	75
Fe	3,731.38	20	40
Ir	3,731.36	50	50
Sm	3,731.26	10	50
Ta	3,731.02	3	50
Fe	3,730.95	30	50
Gd	3,730.86	100	100*W*
Cr	3,730.81	12	60
Os	3,730.73	12	40
Co	3,730.48	—	200*r*
Ru	3,730.43	70	12
Fe	3,730.39	40	70
Th	3,730.38	3	15
Ti	3,729.81	150	500
Ar	3,729.29	(200)	—
Pr	3,729.11	7	40
V	3,729.03	15	80
Mn	3,728.89	100	75
Sm	*3,728.47*	*100*	*100*
Sm	3,743.86	25	50
Cr	3,743.88	40	40
Tm	3,744.07	10	100
Fe	3,744.10	20	40
P	3,744.21	(70)	—
Mo	3,744.37	80	20
Th	3,744.74	10	15
Kr	3,744.80	(150*hs*)	—
Re	3,745.44	—	40*W*
Sm	3,745.46	10	40
Co	3,745.50	—	300*R*
Fe	3,745.56	500	500
Sm	3,745.62	30	40
V	3,745.80	600	35
Fe	3,745.90	100	150
Th	3,745.98	20	15
Mo	3,746.41	40*w*	—
Os	3,746.47	20	100
Nb	3,746.91	80	20
Fe	3,746.93	25	40
Ir	3,747.20	60	100
Th	3,747.55	30	30
Dy	3,747.83	20	60
V	3,747.98	4	50
Mo	3,748.13	50	1
Ho	3,748.17	40	60
Rh	3,748.22	100	200
Fe	3,748.26	200	500
Th	3,748.30	8	10
Cr	3,748.61	30	40
Cr	3,749.00	125*R*	125*R*
Ni	3,749.04	5	50
O	3,749.47	(125)	—
Fe	3,749.49	700	1000*r*
As	3,749.77	100	—
Co	3,749.93	5	60
Th	3,750.15	5	20
Th	3,750.69	5	10
Mn	3,750.76	30	60
Pr	3,751.00	30	40
Th	3,751.11	—	10
V	3,751.23	100	4
Zr	3,751.59	40	25
Co	3,751.63	60	100
Th	3,751.74	3	10
V	3,751.78	2	50
Tm	3,751.82	5	50
Pr	3,752.29	30*d*	40*d*

	λ	I J	I O
V	3,728.34	150	20
Ru	3,728.03	150	100
Fe	3,727.62	150	200
V	3,727.34	200	40
Ne	3,727.08	(125)	—
Ru	3,726.93	150	100
Fe	3,726.92	70	100
Sm	3,726.80	3	100r
Th	3,726.73	20	30
Nb	3,726.24	100	30
Ti	3,725.16	60	150
Eu	3,724.99	50	250
Th	3,724.74	20	30
Ti	3,724.57	50	100
Fe	3,724.38	150	200
Th	3,723.70	8	10
Th	3,723.29	8	10
Ti	3,722.57	60	100
Fe	3,722.56	400	500
Ni	3,722.48	20	200
Th	3,722.19	25	35
Sm	3,721.84	50	100
Th	3,721.83	30	40
Ti	3,721.64	125	60
Kr	3,721.35	(150h)	—
Nb	3,720.46	100h	5
Th	3,720.31	10	15
Os	3,720.13	40	80
Fe	3,719.93	700	1000R
Th	3,719.44	10	30
Mn	3,718.93	100	75
Pd	3,718.91	200	300
Sm	3,718.88	5	100
Kr	3,718.63	(200h)	—
Fe	3,718.41	50	80
Th	3,718.17	10	15
Kr	3,718.02	(300h)	—
Tm	3,717.92	10	100
Th	3,717.83	10	20
Gd	3,717.49	50	100w
Ti	3,717.40	50	80
Re	3,717,29	—	150W
Nb	3,717.07	1000	8
Fe	3,716.45	100	150
Gd	3,716.37	125	150w
Os	3,752.52	100	400R
Th	3,752.57	50	40
Ti	3,752.86	80	200
Th	3,753.24	1	10
Dy	3,753.51	20	50
Ru	3,753.54	60	30
Fe	3,753.61	100	150
Ti	3,753.63	35	80
Rn	3,753.65	(50)	—
Dy	3,753.76	10	80
Th	3,754.04	5w	20w
Tb	3,755.24	100	50
Th	3,755.41	5	10
Co	3,755.45	—	100
Th	3,756.32	20r	50r
Fe	3,756.94	60	80
Dy	3,757.37	50	200
Ti	3,757.69	100	30
Th	3,757.70	5	10
Fe	3,758.23	700	700
La	3,759.08	150	400
Ti	3,759.29	400R	100
Th	3,759.31	10	20
Fe	3,760.05	100	150
Th	3,760.28	5	20
Fe	3,760.53	70	100
Th	3,761.10	8	15
Ti	3,761.32	300r	100
Tm	3,761.33	150	250
Th	3,761.47	1	10
Pr	3,761.87	100	150
Tm	3,761.92	120	200
Th	3,762.36	3	10
Th	3,762.88	20	20
V	3,763.14	6h	80
Th	3,763.33	8	15
Fe	3,763.79	400	500
Th	3,764.28	5	10
Pr	3,764.81	50	100
Rh	3,765.08	70	100
Tb	3,765.14	100	70
Fe	3,765.54	150	200

Th 4,019.137 *J* 8 *O* 8

	λ	I *J*	I *O*
Re	4,019.13	—	15
Tb	4,019.12	5	40
V	4,019.046	6	—
Ni	4,019.046	—	5
Mo	4,019.046	4	2
Ce	4,019.044	4	15
U	4,018.99	15	25
Ce	4,018.92	—	2
Ca	4,018.88	4	—
Nd	4,018.83	10	15
Bi	4,018.60	2	—
Sm	4,018.54	5	4
Se	4,018.52	(70)	—
Ce	4,018.51	—	2
Tb	4,018.44	—	2
Re	4,018.41	—	25
Eu	4,018.400	4	10
Ce	4,018.400	—	2
Zr	4,018.38	5	8
Br	4,018.33	(4)	—
W	4,018.31	3	2
U	4,018.28	3	8
Fe	4,018.27	7	50
Os	4,018.26	4	60
Ce	4,018.23	—	2
Cr	4,018.20	8	35
Zr	4,018.12	—	25
Mn	4,018.102	60	80
Ho	4,018.10	1*h*	3
Ce	4,018.07	—	2
Tb	4,017.83	1	4
Ti	4,017.77	8	70*h*
Dy	4,017.76	2	6
Rn	4,017.75	(150)	—
Eu	4,017.73	—	3
U	4,017.723	25	25
Gd	4,017.717	8	10
Er	4,017.71	—	3*d*
Ce	4,017.60	—	10*s*
Eu	4,017.58	25	25*W*
Nb	4,017.46	10	3
K	4,017.51	(15)	—
Th	4,017.49	8	8
Ni	4,017.462	—	15*Wh*
U	4,017.458	6	8
Mo	4,017.38	6	6
Co	4,019.140	—	5
Ce	4,019.19	—	2
U	4,019.20	8	6
W	4,019.23	15	18
Ce	4,019.28	—	4
Co	4,019.30	—	80
Yb	4,019.35	7	—
Pr	4,019.44	3	10
P	4,019.45	(50)	—
Ce	4,019.480	1	6
Dy	4,019.48	2	5
Se	4,019.50	(10)	—
Ru	4,019.55	8	12
Pb	4,019.64	6	6
Tb	4,019.66	—	2
Eu	4,019.71	—	3
Se	4,019.72	(20)	—
Gd	4,019.73	10	15
Mo	4,019.79	—	10
W	4,019.80	5	—
Nd	4,019.81	8	10
Sm	4,019.83	8	8
Pr	4,019.84	8	15
Ce	4,019.90	3	8
Sm	4,019.98	15	30
Ne	4,020.01	(2)	—
Ir	4,020.03	100	80
Cl	4,020.06	(15)	—
Nd	4,020.062	3*d*	6*d*
Mn	4,020.09	5	10
Th	4,020.096	3	3
*bh*Sr	4,020.1	—	2
U	4,020.17	3	3
Pd	4,020.22	—	15*wh*
Nb	4,020.24	10*h*	2
Hf	4,020.25	2*h*	6
Pr	4,020.28	2	10
W	4,020.32	2	—
Sc	4,020.40	20	50
Mo	4,020.45	10	10
Tb	4,020.47	3	20
Fe	4,020.49	—	2
Er	4,020.52	1	20
Ce	4,020.54	—	5
Mo	4,020.67	—	5
Pd	4,020.68	—	5*h*
U	4,020.69	2	2

	λ	I J	I O
W	4,017.35	5	4
Nd	4,017.31	3	6
V	4,017.29	15*h*	—
C	4,017.27	5	—
Gd	4,017.25	3	5
J	4,017.21	(25)	—
Fe	4,017.15	50	80
Fe	4,017.09	—	5
Nd	4,017.08	2	4
Zr	4,016.98	—	15
Ti	4,016.97	2	12
Co	4,016.88	5*h*	10
Tb	4,016.87	—	5
U	4,016.85	8	6
Co	4,016.83	—	5
V	4,016.82	15*wh*	—
Ru	4,016.753	—	7
Pr	4,016.748	20	25
Eu	4,016.703	—	8
Mo	4,016.702	5	5
Mn	4,016.66	5	8
W	4,016.53	12	10
Fe	4,016.43	4	15
Tb	4,016.36	1	10
U	4,016.34	8	12
Th	4,016.30	5	5
Ti	4,016.28	5	30
In	4,016.24	(50)	—
W	4,016.11	4	6
Nb	4,016.08	30	5
Mo	4,016.06	10	—
Au	4,016.050	15	10
U	4,016.046	6	5
Sm	4,016.00	—	15
J	4,015.94	(8)	—
Tb	4,015.93	1	8
Ce	4,015.88	4	20
Cu	4,015.81	—	7*wh*
Ir	4,015.79	2	4
Sm	4,015.76	4	8
Tb	4,015.62	—	4
Gd	4,015.60	—	4
Er	4,015.58	1	18
Nd	4,015.56	8	10
Tb	4,015.50	1	9*w*
La	4,015.393	2*h*	100
Pr	4,015.389	30	50
Ti	4,015.38	10*h*	70*h*

	λ	I J	I O
Tb	**4,020.74**	—	**2**
Nd	4,020.87	15	15
Dy	4,020.897	4	10
Co	4,020.90	—	500*w*
Pr	4,020.991	30	40
Ru	4,020.995	12	15
U	4,021.005	4	—
Eu	4,021.01	—	5*w*
Mo	4,021.015	25	15
Tb	4,021.12	1	10
Ce	4,021.24	—	5
U	4,021.25	6	6
Se	4,021.26	(20)	—
Nd	4,021.33	12	12
Tm	4,021.39	10	4
Sm	4,021.41	1	4
Er	4,021.56	—	5
In	4,021.66	(50)	—
Nd	4,021.79	10	12
Ti	4,021.83	20	100
Fe	4,021.87	100	200
Br	4,021.89	(4)	—
V	4,021.92	2	8
Er	4,021.96	—	5
In	4,021.99	(10)	—
Nd	4,022.03	2	4
Th	4,022.09	15	20
W	4,022.12	10	12
Ru	4,022.16	100	40
Pr	4,022.20	4	8
Cr	4,022.26	40	80
Ce	4,022.27	4	15
As	4,022.32	5	—
Gd	4,022.34	—	15
Nb	4,022.39	10	2
Ce	4,022.45	—	5
Cu	4,022.66	25	400
Ru	4,022.69	3	7
Sm	4,022.71	5	6
Pr	4,022.74	6	12
Ce	4,022.75	—	4
W	4,022.83	5	4
Tb	4,022.87	—	15*d*
Eu	4,022.91	—	6*W*
Re	4,022.96	—	25*d*
Nd	4,023.00	15	15
Zr	4,023.03	—	6

	λ	I J	I O		λ	I J	I O
Ta	4,015.25	2	8	Nb	4,023.141	3	4
U	4,015.24	—	10	Rh	4,023.144	5	10
Gd	4,015.22	—	5	Gd	4,023.15	10	20
				U	4,023.170	6	6
W	4,015.216	30	25				
Dy	4,015.18	2	10	V	4,023.174	2	7
Cl	4,015.06	(10)	—	Sm	4,023.223	25	20
Os	4,015.04	5	20	Sc	4,023.223	—	60
Cs	4,014.99	(10)	—	Se	4,023.23	(20)	—
				Zr	4,023.30	—	5
W	4,014.94	6	8				
Nb	4,014.93	8	5	Gd	4,023.350	10	20
Ce	4,014.90	12	60	Re	4,023.353	—	40*w*
U	4,014.82	2	12	Ce	4,023.37	1	8
Se	4,014.77	(70)	—	V	4,023.93	30	10
				Co	4,023.40	—	200
Dy	4,014.71	4	18				
Cr	4,014.67	8	40	Cr	4,023.43	—	15
Pr	4,014.66	2	10	Mo	4,023.56	25	—
Ir	4,014.658	2	8	Cs	4,023.58	(10)	—
Ta	4,014.63	1	6*h*	La	4,023.59	15	50
				U	4,023.60	6*h*	3*h*
Fe	4,014.53	100	200				
Sc	4,014.49	8	20	Ce	4,023.64	—	4
Pr	4,014.35	3	9	Sc	4,023.69	25	100
Br	4,014.25	(25)	—	Tb	4,023.176	—	6
Ho	4,014.18	4	2	Dy	4,023.722	2	10
				Cr	4,023.74	15	40
U	4,014.16	8	8				
Ru	4,014.15	5	12	In	4,023.76	(15)	—
Tb	4,014.09	—	4	Nd	4,023.82	3	12
Ir	4,014.00	—	4	Ru	4,023.83	60	25
Co	4,013.94	—	300	Ho	4,023.93	1*h*	4
				Zr	4,023.98	2	30
Nd	4,013.935	3	5				
In	4,013.93	(80)	—	Br	4,024.04	(20)	—
Gd	4,013.92	—	10	Pr	4,024.07	1	5
Ar	4,013.87	(200)	—	Tb	4,024.07	1	40*W*
Dy	4,013.83	2	12	Mo	4,024.09	25	30
				Fe	4,024.10	2	8
Fe	4,013.824	—	200				
Gd	4,013.817	3	25	Sb	4,024.14	4	—
P	4,013.80	(30)	—	Tm	4,024.24	—	7
Fe	4,013.795	40	80	Ru	4,024.30	4	7
J	4,013.794	(15)	—	Ce	4,024.35	—	5
				Pr	4,024.41	2	15
Ru	4,013.74	5	15				
Fe	4,013.65	1	8	Dy	4,024.437	—	20
Ti	4,013.58	7*h*	70*h*	Zr	4,024.438	4	5
Ta	4,013.54	—	5	Tb	4,024.44	—	4
Ru	4,013.50	12	15	Th	4,024.47	5	5
				Ce	4,024.49	5	15
In	4,013.49	(30)	—				
Pr	4,013.43	8	15	Cr	4,024.567	1	20
Gd	4,013.430	—	5	Ti	4,024.573	35	80
Tb	4,013.28	—	20*d*	Ru	4,024.69	5	12
Nb	4,013.27	5	3	Tb	4,024.70	—	4
				F	4,024.73	(500)	—

	λ	I	
		J	O
Th	4,013.26	5	5
Pr	4,013.23	4	10
Nd	4,013.22	5	10
Mo	4,013.21	40	1
Ta	4,013.19	1	5
W	4,013.18	5	6
U	4,013.03	—	8
In	4,012.96	(10)	—
Se	4,012.96	(150)	—
Pr	4,012.91	1	8
Tb	4,012.87	1	6
Dy	4,012.82	2	5
Eu	4,012.816	25	20
Ti	4,012.81	1	12
Nd	4,012.70	10	15
Er	4,012.58	2	4
Dy	4,012.52	—	4
Th	4,012.497	15	15
Cr	4,012.469	60	70
Tb	4,012.45	—	5
Ti	4,012.391	50	35
Ce	4,012.388	20	60
Re	4,012.26	—	25
Er	4,012.253	—	12
Zr	4,012.252	—	20
Nd	4,012.250	40	80
Nb	4,012.17	100	—
U	4,012.16	4	6
Ce	4,012.14	—	4
Ta	4,012.11	2*h*	5*h*
K	4,012.10	(20)	—
W	4,012.096	5	—
Mo	4,011.97	25	25
Mn	4,011.905	10	12
Rb	4,011.90	(15)	—
Se	4,011.88	(200)	—
W	4,011.81	3	4
U	4,011.78	5	10
Th	4,011.75	15	15
Ru	4,011.73	—	7
Sm	4,011.72	4	8
Eu	4,011.683	—	25
Te	4,011.68	(30)	—
Ce	4,011.56	3	15
Ti	4,011.534	—	4
Pr	4,011.532	1	5
Mn	4,011.531	15	15
Re	4,011.51	—	35
U	4,011.45	10	8

	λ	I	
		J	O
Fe	4,024.74	30	120
Tb	4,024.781	—	7
Nd	4,024.785	10	20
In	4,024.83	100*wh*	—
Dy	4,024.91	—	10
Zr	4,024.918	3	25
K	4,024.920	(15)	—
F	4,025.010	(150)	—
Cr	4,025.012	25	100
U	4,025.013	3	8
J	4,025.08	(30)	—
Ti	4,025.14	25	15
Tb	4,025.146	—	4
Ce	4,025.150	2	12
Pr	4,025.19	8	15
Xe	4,025.19	(15)	—
W	4,025.19	6	8
Mo	4,025.487	5	6
F	4,025.495	(300)	—
Pr	4,025.55	25	40
W	4,025.600	10	—
Dy	4,025.605	2	10
Th	4,025.612	20	20
Re	4,025.614	—	15
Cs	4,025.67	(10)	—
Cl	4,025.68	(7)	—
Tb	4,025.73	2	15
La	4,025.88	50	50
Eu	4,025.95	—	5
Hg	4,025.954	(20)	—
Mo	4,025.99	30*h*	30*h*
U	4,026.02	25	25
Nd	4,026.08	2	10
N	4,026.09	(10*h*)	—
Th	4,026.16	10	10
Cr	4,026.17	35	100
He	4,026.19	(70)	—
Nb	4,026.32	10	—
He	4,026.36	(5)	—
Tb	4,026.41	—	4
U	4,026.434	—	10
Mn	4,026.435	40	50
Al	4,026.50	(30)	—
Ru	4,026.50	1	7
Ti	4,026.54	10	70
Eu	4,026.62	—	5*W*
Nd	4,026.66	3	10
Cl	4,026.67	(4)	—
Pr	4,026.84	5	20

	λ	I J	I O		λ	I J	I O
Fe	4,011.41	1	5	Rb	4,026.90	(25)	—
V	4,011.31	2	9	Mo	4,026.92	5	5
Ce	4,011,30	—	4	Ta	4,026.94	30	40
Dy	4,011.29	8	12	U	4,027.00	8	6
Ar	4,011.23	(5)	—	Th	4,027.01	1	8
Nd	4,011.07	10	15	Co	4,027.04	4	200
W	4,011.02	5	4	Ce	4,027.05	—	5
N	4,010.99	(5)	—	Cr	4,027.10	30	80
Tb	4,010.87	—	7	Ho	4,027.20	1	5
Cu	4,010.85	—	6	Zr	4,027.205	4	100
U	4,010.82	1	10	Nb	4,027.31	5	5
Nd	4,010.76	—	4	Th	4,027.33	3	5
Tb	4,010.74	—	5	U	4,027.40	2	4
As	4,010.66	10	—	Ti	4,027.48	3	30
Pr	4,010.64	5*r*	8*r*	Gd	4,027.61	8	10
Er	4,010.54	2	6*d*	Pr	4,027.64	6	15
Cs	4,010.54	(10)	—	Th	4,027.644	3	5
Nd	4,010.45	6	20	Ce	4,027.69	3	20*s*
Eu	4,010.42	—	20*W*	Dy	4,027.787	4	15
W	4,010.37	8	9	In	4,027.79	(50*h*)	—
Mo	4,010.30	10	—	U	4,027.8	8	12
Eu	4,010.18	4	8	Ce	4,027.88	—	4
Ce	4,010.136	3	15	Nb	4,027.98	10	5
Mo	4,010.136	5	5	Ce	4,027.99	—	4
Dy	4,010.08	4	10	Cr	4,028.02	—	35
Tb	4,010.06	—	8	Gd	4,028.16	8	25
Pr	4,009.97	3	10	Yb	4,028.27	8*h*	—
C	4,009.90	10	—	Tb	4,028.31	1	8
Ir	4,009.85	2*h*	8	Dy	4,028.32	—	10
W	4,009.81	9	—	Ti	4,028.34	80	20
Hf	4,009.78	—	5	U	4,028.40	2*h*	5
Fe	4,009.72	100	120	Ce	4,028.41	8	35
Nb	4,009.71	10	5	Dy	4,028.42	8	5
Ti	4,009.66	25	60	Ru	4,028.43	2	10
Al	4,009.58	(4)	—	Au	4,028.48	10	—
Tb	4,009.54	1	10	Eu	4,028.52	—	10*W*
U	4,009.41	8	4	Re	4,028.56	—	25
Nd	4,009.367	5	8	Eu	4,028.63	—	15*w*
Mo	4,009.366	25	20	Mo	4,028.65	6	6
He	4,009.27	(10)	—	Th	4,028.66	8	8
Pr	4,009.24	2	5	Pr	4,028.69	1	5
Gd	4,009.21	2	50	Nd	4,028.75	5	10
Tb	4,009.19	—	5	W	4,028.790	10	12
U	4,009.17	15	8	S	4,028.791	(200)	—
Er	4,009.16	1	15	Ho	4,028.85	1	4
Th	4,009.07	8	10	Zr	4,028.95	1	40
Ce	4,009.06	—	12	W	4,029.02	7	5
Ir	4,008.97	—	5	Pr	4,029.04	5	12

	λ	I J	I O
Ti	4,008.93	35	80
Gd	4,008.922	3	20
U	4,008.918	—	8
Fe	4,008.873	1	5
Eu	4,008.872	2	8*W*
Br	4,008.76	(20)	—
Nd	4,008.754	10	12
W	4,008.753	45	45
Pr	4,008.71	50	150
Mo	4,008.67	20	—
Ce	4,008.66	—	8
Dy	4,008.49	—	5
Kr	4,008.48	(10*wh*)	—
Hf	4,008.46	8	5
Ce	4,008.45	—	6
Nd	4,008.42	3	5
Gd	4,008.331	10	15
Sm	4,008.330	10*h*	8
Nb	4,008.28	10	5
Ru	4,008.27	20	20
Th	4,008.22	8	10
Er	4,008.18	1	8
V	4,008.17	10	2
Sm	4,008.09	3	10
Kr	4,008.08	(25)	—
Ti	4,008.06	7	50
Mo	4,008.054	5	4
Ir	4,008.052	2	12
Mn	4,008.02	5	15
Eu	4,007.98	—	6
Er	4,007.97	7	35
Ho	4,007.96	3	4
U	4,007.93	3	8
Se	4,007.90	(150)	—
S	4,007.78	(5)	—
Pr	4,007.78	3	8
Dy	4,007.77	2	12
U	4,007.689	4	2
Eu	4,007.687	—	5*d*
In	4,007.61	(10)	—
Zr	4,007.60	1	25
Ce	4,007.59	4	15
In	4,007.54	(15)	—
Ru	4,007.53	10	20
Sm	4,007.49	25	50
Pd	4,007.47	—	5*h*
Ce	4,007.451	—	4*h*
Mo	4,007.45	5	4
Th	4,029.05	3	5
Cd	4,029.08	5	—
Ir	4,029.14	—	6
Hf	4,029.17	12	10
Cl	4,029.20	(8)	—
Ce	4,029.26	1*h*	5
Th	4,029.31	5	8
Dy	4,029.41	2	5
Rb	4,029.56	15	—
Eu	4,029.58	—	7
W	4,029.61	7	6
Fe	4,029.636	25	80
Re	4,029.639	—	80
Zr	4,029.68	15	40
Te	4,029.73	(15)	—
Pr	4,029.73	12	15
Ce	4,029.75	—	4
Yt	4,029.84	—	5
Nd	4,029.91	3	10
U	4,029.92	4	6
Ta	4,029.94	5	50
Mo	4,029.941	30	1
W	4,029.95	7	6
Eu	4,030.00	—	5
Zr	4,030.04	2	35
Se	4,030.07	(150)	—
Ru	4,030.14	—	7
Ce	4,030.15	—	5
Fe	4,030.19	4	20
Eu	4,030.20	—	10*w*
Br	4,030.29	(12)	—
Th	4,030.293	5	8
Ca	4,030.30	2*h*	10
Ce	4,030.34	4	18
Sr	4,030.38	—	40
Sm	4,030.42	3	10
Nd	4,030.47	15	20
Fe	4,030.49	60	120
Ti	4,030.51	18	80
Sc	4,030.657	2	10
Eu	4,030.66	—	5*w*
Ta	4,030.67	1*d*	10
Cr	4,030.68	30	40
Mn	4,030.755	20	500*r*
U	4,030.758	6	5
Zr	4,030.759	—	20
Th	4,030.85	8	10
Lu	4,030.86	5*h*	—

	λ	I	
		J	O
Nd	4,007.43	20	20
Yb	4,007.36	—	5
Hf	4,007.35	4*h*	5
Br	4,007.33	(10)	—
Fe	4,007.27	50	80
Ta	4,007.23	2	5
Ti	4,007.19	1	15
Dy	4,007.14	—	6
Mn	4,007.04	5	10
Th	4,007.03	20	20
W	4,007.00	6	—
Nb	4,006.99	5	—
Mo	4,006.95	5	—
Ta	4,006.835	20	30
Sm	4,006.834	5	8
Cs	4,006.77	(10)	—
Nd	4,006.761	2*d*	5*d*
Fe	4,006.761	3	7
Pr	4,006.704	5	10
Mo	4,006.70	5	5
Cd	4,006.68	5	—
Fe	4,006.63	15	20
Ru	4,006.60	15	25
Sm	4,006.58	8	8
Eu	4,006.56	—	6
Cs	4,006.54	(30)	—
Te	4,006.50	(100)	—
U	4,006.395	5	3
Nd	4,006.386	3	5
Th	4,006.386	10	10
As	4,006.34	50	—
Fe	4,006.31	35	60
Hg	4,006.27	(30)	—
Ni	4,006.14	—	10
J	4,006.10	(8)	—
Dy	4,006.07	2	10
Mo	4,006.05	20	20
Ti	4,005.97	3	35
Tb	4,005.96	1	9
Nb	4,005.93	5*h*	2
W	4,005.89	6	7
Dy	4,005.84	2	12
V	4,005.71	30	10
U	4,005.70	3	25
Ru	4,005.640	30	25
Ce	4,005.639	6	20
Br	4,005.58	(4)	—
Ir	4,005.580	3	20
Kr	4,005.57	(30*h*)	—

	λ	I	
		J	O
Al	4,030.87	(8)	—
Gd	4,030.88	—	8
Mo	4,030.91	5	3
Ru	4,031.00	12	15
Dy	4,031.08	—	7
Pr	4,031.09	8	12
Th	4,031.099	5	5
Cs	4,031.10	(10)	—
Cr	4,031.130	6	30
V	4,031.22	3	10
U	4,031.306	8	8
Nb	4,031.31	5	—
Th	4,031.327	5	5
Pb	4,031.33	5	—
Ce	4,031.34	8	40
Eu	4,031.38	3	7
Sc	4,031.40	2	10
Nd	4,031.54	3	10
Tb	4,031.64	3	50
Ce	4,031.669	—	10*d*
W	4,031.675	7	8
Er	4,031.690	—	6*w*
La	4,031.692	300	400
Ho	4,031.75	1*h*	4
Ti	4,031.754	1	35
Pr	4,031.755	30	50
U	4,031.78	2	8
Mn	4,031.79	10	8
Nd	4,031.81	15	15
V	4,031.83	3	10
Ta	4,031.96	—	5
J	4,031.96	(10)	—
Fe	4,031.965	50	80
J	4,031.09	(10)	—
Re	4,032.15	—	10
Cl	4,032.19	(4)	—
Ru	4,032.20	20	20
Ir	4,032.21	—	10
Se	4,032.22	(8)	—
Hf	4,032.27	2	5
Tb	4,032.28	—	30
Sr	4,032.379	—	20
W	4,032.385	7	6
Fe	4,032.469	1	4
Zr	4,032.471	—	5
Er	4,032.477	—	9
Dy	4,032.480	12	20
Pr	4,032.49	12	20
Mo	4,032.50	8	8

	λ	I J	I O		λ	I J	I O
Tb	4,005.55	125	100*d*	Ru	4,032.521	5	10
Th	4,005.549	30*w*	20*w*	Nb	4,032.524	50	30
W	4,005.40	10	8	Th	4,032.54	8	10
Ag	4,005.32	2	10	Tb	4,032.626	—	4
Fe	4,005.25	200	250	Fe	4,032.630	15	80
Ce	4,005.24	—	18	Ti	4,032.632	1	35
Os	4,005.15	20	35	S	4,032.81	(125)	—
Mo	4,005.12	20	—	Dy	4,032.847	—	8
Ir	4,005.02	—	25	Br	4,032.85	(20)	—
Re	4,004.932	—	30	Se	4,032.89	(10)	—
Gd	4,004.93	3	8	Ar	4,032.97	(20)	—
Zr	4,004.87	—	20	Pr	4,032.974	10	15
Fe	4,004.84	7	10	Sm	4,032.977	8	20
In	4,004.834	(10)	—	Ga	4,032.982	500*R*	1000*R*
Nd	4,004.830	3	10	Tb	4,033.04	5	125
W	4,004.75	6	—	In	4,033.066	—	4
Pr	4,004.714	25	20	Ta	4,033.069	10	100
In	4,004.709	(15)	—	Cr	4,033.072	2	15
U	4,004.617	5	3	Mn	4,033.073	20	400*r*
Eu	4,004.59	—	5	Sr	4,033.19	—	6
Ce	4,004.582	3	12	Nb	4,033.20	5	5
Er	4,004.579	—	4*d*	Cr	4,033.26	8	30
In	4,004.53	(30)	—	Re	4,033.31	—	40
Tb	4,004.50	1	5	U	4,033.43	10	12
Dy	4,004.48	—	5	Gd	4,033.49	5	10
U	4,004.41	—	6	Nd	4,033.50	5	10
Zr	4,004.40	—	10	Sb	4,033.54	60	70
Dy	4,004.33	2	8	Mn	4,033.640	5	5
Nd	4,004.26	10	15	Mo	4,033.631	6	6
Sm	4,004.24	3	8	Dy	4,033.67	4	15
U	4,004.06	20	15	P	4,033.68	(15)	—
Ce	4,004.047	—	6*h*	Eu	4,033.69	—	8*w*
Er	4,004.046	—	5	U	4,033.73	12	12
Os	4,004.02	6	50	Ir	4,033.76	25	100
Nd	4,004.01	15	20	Ce	4,033.79	—	6
Cr	4,003.92	12	30	Ar	4,033.83	(30)	—
Tb	4,003.91	1	5	Pr	4,033.86	35	50
S	4,003.89	(8)	—	Hf	4,033.88	8	5
Gd	4,003.85	8	8	Nd	4,033.90	4	10
W	4,003.84	7	—	Ti	4,033.906	3	40
Ti	4,003.81	70	50	W	4,033.913	10	—
Ce	4,003.771	18	40	U	4,034.00	4	4
Fe	4,003.767	80	30	S	4,034.01	(8)	—
Tb	4,003.76	1	8	Nd	4,034.012	2	4
Eu	4,003.71	25	18	Cr	4,034.05	—	20
Ta	4,003.70	5	15	Zr	4,034.09	2	5
Sm	4,003.69	3	6	Nd	4,034.15	2	10*d*
Co	4,003.60	—	15	Pt	4,034.17	5	—

	λ	I	
		J	O
V	4,003.54	2	9
Os	4,003.48	6	50
Sm	4,003.45	20	30
U	4,003.40	10	6
Ho	4,003.36	1	4
Cr	4,003.33	20	—
Th	4,003.32	15	15
Mn	4,003.25	5*h*	20
U	4,003.20	5	8
Nd	4,003.171	3	10
Ce	4,003.170	1	10
Th	4,003.11	10	10
Hg	4,003.10	(20)	—
Zr	4,003.096	1	20
Se	4,003.08	(60)	—
Cu	4,003.04	1*h*	40
Ce	4,002.974	1	8
Mo	4,002.97	20	—
V	4,002.94	80	6
Eu	4,002.91	—	15
Ce	*4,002.81*	*4*	*20*
Tb	*4,002.58*	*5*	*50*
Eu	*4,002.56*	*2*	*30*
Zr	*4,002.55*	—	*18*
Ti	*4,002.49*	*5*	*40*
Xe	*4,002.35*	*(40wh)*	—
U	*4,002.34*	*18*	*10*
Tb	*4,002.18*	*2*	*40w*
Mn	*4,002.16*	*5*	*15*
Se	*4,002.07*	*(60)*	—
Gd	*4,001.97*	*1*	*8*
Mn	*4,001.91*	*10*	*15*
W	*4,001.89*	*15*	—
Th	*4,001.74*	*8*	*10*
Cs	*4,001.68*	*(20)*	—
Fe	*4,001.67*	*50*	*80*
Gd	*4,001.63*	—	*15*
Ce	*4,001.55*	*5s*	*20s*
Pr	*4,001.47*	*5*	*15*
Cr	*4,001.44*	*80*	*200*
W	*4,001.37*	*9*	*9*
U	*4,001.25*	*3*	*10*
Gd	*4,001.24*	*3*	*80*
Zr	*4,001.23*	—	*9*
K	*4,001.20*	*(40)*	—
Mn	*4,001.18*	*5*	*12*
Nb	*4,001.13*	*15*	*10*
Sc	4,034.23	2*h*	8
Th	4,034.26	10	10
Pr	4,034.30	5	20
Gd	4,034.38	—	5
Mn	4,034.49	20	250*r*
Nb	4,034.52	5	10
Tm	4,034.74	10	10
Th	4,034.89	5	8
Ti	4,034.91	2	25
Cr	4,035.00	—	8
N	4,035.09	(15*h*)	—
Nb	4,035.098	3	4
Sm	4,035.101	3	50
Cr	4,035.24	—	8
Ir	4,035.33	—	6
W	4,035.35	9	10
Nd	4,035.399	2	8*d*
Gd	*4,035.403*	*5*	*8*
Pr	*4,035.42*	*1*	*10*
Ar	*4,035.47*	*(30)*	—
Co	*4,035.55*	*3*	*150*
V	*4,035.63*	*80*	*40*
Mo	*4,035.66*	*25*	*3*
Mn	*4,035.73*	*60*	*50*
Ti	*4,035.830*	*5*	*50*
Cs	*4,035.83*	*(15)*	—
Zr	*4,035.893*	*2*	*40*
Ta	*4,035.893*	*5h*	*10*
Nd	*4,036.00*	*3*	*10*
Pr	*4,036.05*	*1*	*15*
Th	*4,036.06*	*3*	*10*
J	*4,036.08*	*(50)*	—
Eu	*4,036.11*	—	*50W*
Tb	*4,036.219*	—	*10*
P	*4,036.22*	*(15h)*	—
Ba	*4,036.26*	*(10)*	—
Dy	*4,036.33*	*4*	*15*
Br	*4,036.43*	*(10)*	—
Tb	*4,036.45*	—	*10*
Ru	*4,036.49*	*3*	*10*
Cl	*4,036.53*	*(10)*	—
U	*4,036.554*	*8*	*3*
Nd	*4,036.555*	*1*	*10*
Th	*4,036.57*	*15*	*15*
La	*4,036.60*	*15*	*5*
V	*4,036.78*	*40*	*8*
Gd	*4,036.84*	*8*	*10*

	λ	I J	I O
Zr	4,001.09	—	10
Th	4,001.06	5	10
Ce	4,001.05	2	20
Pr	4,000.91	8	20
W	4,000.69	10	12
Nb	4,000.60	50	2
Nb	4,000.56	3d	8d
Mo	4,000.50	8	8
Nd	4,000.49	5d	10d
Pr	4,000.48	4	8
Tb	4,000.46	2	15
Dy	4,000.454	300	400
Fe	4,000.452	10	35
Er	4,000.452	6	35
Th	4,000.29	3	8
Fe	4,000.27	1	8
Pr	4,000.19	25	50
Gd	4,000.16	—	10h
W	4,000.09	8	—
V	4,000.08	2	8
N	4,000.98	(15)	—
U	4,000.95	8	6
Cr	3,999.68	10	40
Br	3,999.62	(8)	—
Tb	3,999.41	15	12
Nd	3,999.38	6	20
Ti	3,999.36	5	30
Ta	3,999.28	20d	30
Ce	3,999.24	20	80
V	3,999.19	40h	—
Pr	3,999.188	40d	50d
U	3,999.183	3h	10
Nb	3,999.182	10	5
Zr	3,998.97	30	30
Os	3,998.93	12	80
Cr	3,998.86	2	25
S	3,998.79	(60)	—
W	3,998.75	8	8
V	3,998.73	25	100
Nd	3,998.69	15	40
Ti	3,998.64	100	150
Mo	3,998.63	25	—
Pr	3,998.44	3	9
Sm	3,998.35	2	10
Mo	3,998.29	5	8
Ho	3,998.28	6	40
U	3,998.24	18	5
Nd	3,998.15	12	20
W	4,036.86	12	12
U	4,036.87	—	10
Sc	4,036.88	—	8
Sm	4,037.09	4	10
La	4,037.21	3h	50
Pr	4,037.22	2	8
Th	4,037.26	5	8
Xe	4,037.29	(50)	—
Cr	4,037.294	12	80
Mo	4,037.30	8	8
Gd	4,037.34	30	100
bhB	4,037.4	—	25
Re	4,037.51	—	30
Xe	4,037.59	(100)	—
Ne	4,037.61	(15)	—
Nb	4,037.659	20	—
Eu	4,037.665	—	8W
Ce	4,037.665	3	25
Cr	4,037.72	—	15
Ru	4,037.74	5	12
Mo	4,037.78	15	10
Kr	4,037.83	(30)	—
Os	4,037.84	4	80
Gd	4,037.91	30	100
Ce	4,037.96	2	18
U	4,037.987	2	8
Cr	4,037.993	—	15h
Mo	4,038.08	15	20
Nd	4,038.12	10	15
Pr	4,038.15	3	12
Se	4,038.31	(40)	—
Ti	4,038.33	1	12
Ce	4,038.34	5	10
Pr	4,038.47	10	15
Dy	4,038.53	4	15
Mo	4,038.63	10h	—
Os	4,038.64	3	9
Mn	4,038.73	15	15
Ar	4,038.82	(40)	—
Tb	4,038.86	—	10
Nb	4,039.09	10	5
Cr	4,039.10	40	100
Eu	4,039.19	—	15
Ru	4,039.21	50	25
Cr	4,039.22	3	15
Pr	4,039.36	20	50
Gd	4,039.50	—	10
Sm	4,039.50	2	10
Nb	4,039.53	50	30

	λ	I	
		J	O
Fe	3,998.05	100	150
S	3,997.97	(8)	—
Pr	3,997.96	9	20
Kr	3,997.95	(100wh)	—
Nd	3,997.93	10	20
Co	3,997.91	20	200
Th	3,997.86	10	10
Gd	3,997.77	30	20
Ce	3,997.72	2	18
Th	3,997.45	8	8
Tb	3,997.41	—	10
Fe	3,997.40	150	300
Mn	3,997.21	25	12
P	3,997.16	(70)	—
Br	3,997.13	(12)	—
V	3,997.12	40	25
U	3,997.09	2	12
Pr	3,997.05	40	100
Fe	3,996.97	20	40
Os	3,996.80	10	50
Dy	3,996.699	80	200
Tb	3,996.696	3	9
Er	3,996.695	—	25
Pr	3,996.686	4	20
Ti	3,996.65	—	12
Sc	3,996.61	10	40
Pt	3,996.57	—	50
Tm	3,996.52	40	200
Ru	3,996.51	4	10
Ce	3,996.49	—	10
Ir	3,996.45	2	8
Al	3,996.38	(10)	—
Gd	3,996.32	100	100
Ta	3,996.17	30h	100
Al	3,996.16	(18)	—
Rh	3,996.15	10	25
Th	3,996.07	10	15
Fe	3,995.99	20	60
Ru	3,995.98	30	30
U	3,995.97	8	10
Al	3,995.86	(30)	—
Tb	3,995.79	—	8
Eu	3,995.755	—	12
La	3,995.750	300	600
Re	3,995.68	—	10h
Ba	3,995.66	5	18
Rh	3,995.61	10	15
Sm	3,995.59	3	8
Tm	3,995.58	—	100

	λ	I	
		J	O
V	4,039.58	8	3
Gd	4,039.67	5	10
K	4,039.69	(15)	—
U	4,039.78	2	15
Yt	4,039.83	8	12
Cs	4,039.84	(50)	—
W	4,039.85	9	12
Ce	4,039.89	—	12
Th	4,039.90	1	8
Nd	4,039.94	—	8
Ir	4,040.08	5	40
Yb	4,040.09	8	1
Tb	4,040.10	—	8
Re	4,040.18	—	15
Ti	4,040.32	1	40
Hg	4,040.40	(20)	—
Tb	4,040.40	1	10
Ru	4,040.48	3	12
W	4,040.59	8	9
Cl	4,040.64	(9)	—
Fe	4,040.644	7	20
Ce	4,040.76	5	70
Er	4,040.78	—	20
Nd	4,040.796	40	40
Co	4,040.802	1	15
Ho	4,040.84	30	150
Ta	4,040.87	5h	50
Au	4,040.94	40	50
La	4,040.97	—	8
Ta	4,041.057	4h	40
Nd	4,041.065	5	15
Th	4,041.21	10	20
N	4,041.32	(20h)	—
Mn	4,041.36	50	100
V	4,041.60	2W	10
Br	4,041.61	(8)	—
Sm	4,041.67	10	25
Cr	4,041.79	—	20
Se	4,041.82	(8)	—
Os	4,041.92	6	100
J	4,041.94	(15)	—
Dy	4,041.99	4	12
Ru	4,042.00	1	12
Eu	4,042.06	2	10w
Ce	4,042.13	—	8
Zr	4,042.22	—	25
Cr	4,042.25	1	30
Ne	4,042.327	(10)	—

	λ	I J	I O
Co	3,995.31	20	1000R
Nd	3,995.24	—	20
Fe	3,995.20	—	10
In	3,995.16	18Wh	—
K	3,995.10	30	—
N	3,994.99	(300)	—
U	3,994.980	20wh	8
J	3,994.979	(35)	—
Os	3,994.93	5	30
Sb	3,994.90	10	—
V	3,994.89	—	35
Pr	3,994.834	25	300
Kr	3,994.83	(100)	—
Ar	3,994.81	(10)	—
Rn	3,994.76	(10)	—
Ti	3,994.70	—	25
Nd	3,994.68	40	80
Th	3,994.55	10	30
Co	3,994.54	—	60
Gd	3,994.18	6	8
Fe	3,994.12	10	25
Nd	3,994.10	2	20
Pr	3,994.01	3	20
Cr	3,993.97	20	60
Ni	3,993.95	—	30h
Eu	3,993.93	2	8
Ce	3,993.82	6	50
U	3,993.81	4	12
Th	3,993.72	5	8
Ho	3,993.71	4	10
Dy	3,993.57	—	8
Tb	3,993.55	8	30d
Ru	3,993.531	5	10
S	3,993.526	(50)	—
Ba	3,993.40	50r	100R
Sm	3,993.30	25	25
Gd	3,993.23	10	15
Pr	3,992.92	3	15
Ce	3,992.91	3	15
Cr	3,992.84	70	150
V	3,992.80	20	60
Re	3,992.72	—	20
Gd	3,992.70	15	15
Nd	3,992.57	20	30
U	3,992.54	8	10
Mn	3,992.49	75	40
Br	3,992.39	(20)	—
Ce	3,992.386	8	50
Tb	4,042.33	—	9
W	4,042.39	7	8
U	4,042.46	8	5
Ce	4,042.58	3	50
K	4,042.59	(30)	—
V	4,042.63	2	15
Ne	4,042.64	(50)	—
Sm	4,042.71	9	10
U	4,042.75	10	40
Mo	4,042.87	15	15
Sm	4,042.90	10	10
Ar	4,042.91	(80)	—
La	4,042.911	300	400
Er	4,043.01	—	10
Nd	4,043.05	1	15
Yb	4,043.06	8h	—
Th	4,043.15	3	10
Nb	4,043.16	10	3
Th	4,043.40	1	8
Cs	4,043.42	(20)	—
Cu	4,043.50	25	—
N	4,043.54	(10h)	—
Zr	4,043.58	—	24
Nd	4,043.60	5	15
Tb	4,043.66	1W	8w
Cr	4,043.70	2	30
Mo	4,043.74	8	8
Cu	4,043.75	10	—
Ti	4,043.77	—	20
Sc	4,043.80	4	12
J	4,043.88	(20)	—
Fe	4,043.90	7	25
Eu	4,043.97	—	20
Hg	4,044.10	10	5
Nb	4,044.105	10	5
K	4,044.14	400	800
W	4,044.29	12	15
Hf	4,044.39	4	10
U	4,044.416	25	18
Ar	4,044.418	(1200)	—
Tm	4,044.47	—	15
P	4,044.49	(150w)	—
Zr	4,044.56	2	25
Cl	4,044.58	(10)	—
Fe	4,044.61	35	70
Kr	4,044.67	(80)	—
Pr	4,044.818	35	50
U	4,044.824	2	10

	λ	I			λ	I	
		J	O			J	O
Th	3,992.28	10	10	Sm	4,044.95	10	10
Pr	3,992.18	10	25	Gd	4,045.01	5	20
Nd	3,992.16	8	20	Sm	4,045.05	6	10
				Mn	4,045.13	—	15
Ir	3,992.12	60	150	Mn	4,045.206	15	15
Cr	3,992.11	1	20				
Ar	3,992.06	(25)	—	Ce	4,045.209	3	8
Kr	3,991.94	(15)	—	Rn	4,045.30	(35)	—
Pr	3,991.89	10	25	Co	4,045.39	—	400
				Er	4,045.43	1wd	8wd
Co	3,991.83	—	15	Ho	4,045.43	80	200
K	3,991.77	(15)	—				
Nd	3,991.743	40	60	W	4,045.60	15	12
Th	3,991.737	3	8	Zr	4,045.61	10	10
Co	3,991.69	6	60	Ru	4,045.76	—	25
				Fe	4,045.81	300	400
Nb	3,991.68	20	15	Ag	4,045.82	2	10
Cr	3,991.67	50	100				
Mn	3,991.60	25	20	Gd	4,045.86	—	(30)
Tb	3,991.59	—	10d	Ho	4,045.95	2	10
Re	3,991.58	—	20	Ar	4,045.966	(150)	—
				Tb	4,045.97	1	25
Co	3,991.54	—	30	Dy	4,045.98	12	150
Os	3,991.49	10	40				
Mo	3,991.39	8	6	Sm	4,046.15	10	12
Dy	3,991.33	—	40	V	4,046.26	15	1
Kr	3,991.26	(10)	—	Ce	4,046.34	10	30
				Pt	4,046.45	20	—
W	3,991.22	8	9	Sc	4,046.49	—	10
Er	3,991.16	—	15				
Zr	3,991.13	60	100	Ir	4,046.54	—	8
Cr	3,991.12	60	200	Hg	4,046.56	300	200
Kr	3,991.08	(20)	—	Pr	4,046.63	3	8
				W	4,046,701	12	10
Re	3,991.04	—	25	Nd	4,046.702	2	8
Sm	3,991.02	2	10				
S	3,990.94	(40)	—	Cr	4,046.76	3	30
Yb	3,990.89	20	60	Gd	4,046.84	10	10
Mo	3,990.84	30	—	Mo	4,046.89	20	3
				Er	4,046.96	—	8
Ca	3,990.69	8	3	U	4,047.05	8	6
Re	3,990.663	—	15				
Kr	3,990.66	(15h)	—	Pr	4,047.10	12	20
Tb	3,990.63	3	12	Nd	4,047.158	10	12
V	3,990.57	40	125	Tb	4,047.16	—	9
				Te	4,047.18	(15)	—
Th	3,990.56	5	8	Cs	4,047.184	(20)	—
U	3,990.42	20	18				
W	3,990.381	10	—	K	4,047.20	200	400
Fe	3,990.379	25	70	Ce	4,047.27	2	18
Xe	3,990.33	(30wh)	—	Sm	4,047.36	6	8
				U	4,047.61	3	18
Co	3,990.30	10	80	Yt	4,047.63	10	50
Cl	3,990.19	(20)	—				
Ti	3,990.18	1w	10w	Se	4,047.77	(8)	—
U	3,990.15	10	5	Sc	4,047.79	10	25
Ce	3,990.11	2	20	Gd	4,047.85	50	150
				W	4,047.93	8	9

	λ	I J	I O		λ	I J	I O
Nd	3,990.10	20	40	Hf	4,047.96	25	8
Th	3,990.08	8	5	Os	4,048.05	2	20
Nd	3,990.02	15	20	Th	4,048.057	8	8
Sm	3,990.00	25	40	U	4,048.065	2	8
Cr	3,989.99	40	80	Pr	4,048.14	4	8
Fe	3,989.86	5	30	O	4,048.22	(10h)	—
V	3,989.80	10	—	Er	4,048.346	1	15
Ti	3,989.76	100	150	Dy	4,048.35	2	15
Pr	3,989.72	125	800	Ce	4,048.37	1	8
Mo	3,989.51	25h	—	Bi	4,048 40.	10	3h
Zr	3,989.50	1h	12	Gd	4,048.59	8	8
Ce	3,989.44	6	20	Sm	4,048.62	10	10
Ir	3,989.43	2	25	Zr	4,048.67	30	30
U	3,989.292	1h	8	Mn	4,048.75	60	60
Zr	3,989.287	1h	12	Cr	4,048.78	50	80
Gd	3,989.25	20	10	Nd	4,048.81	10	15
Zn	3,989.23	(100)	—	Gd	4,048.82	—	10W
Pr	3,989.14	1	10	Th	4,048.86	10wh	—
Fe	3,989.01	1wh	15wh	Te	4,048.89	(70)	—
Rn	3,988.98	(8)	—	Re	4,048.99	—	30
U	3,988.88	15	12	Ce	4,049.03	1	18
Th	3,988.85	10	10	U	4,049.17	3	10
V	3,988.83	35	70	Fe	4,049.33	2	10
Nd	3,988.81	6	20	Ti	4,049.40	1	35
Ta	3,988.70	10	15	Ru	4,049.41	12	15
Zr	3,988.68	—	15	Gd	4,049.44	20	80
Mn	3,988.67	15	12	Hf	4,049.45	10	10
U	3,988.64	8	8	Er	4,049.48	2	10d
Th	3,988.60	8	10	Pr	4,049.61	2	8
La	3,988.518	800	1000	U	4,049.74	2h	8
Ce	3,988.518	—	8h	Cr	4,049.78	6	30
Eu	3,988.25	2	10W	Ce	4,049.79	1h	8
Os	3,988.18	12	50	Pr	4,049.83	2	8
Nb	3,988.16	10	5	Sm	4,049.832	20	40
U	3,988.03	12	8	Nd	4,049.86	2d	10d
Pr	3,988.02	7	25	Fe	4,049.87	3	30
Th	3,988.01	30	50	J	4,049.88	(30)	—
Yb	3,987.99	500R	1000R	Gd	4,049.90	60	100
Ho	3,987.98	—	8	Sc	4,049.95	—	8
Er	3,987.95	20	100r	Cr	4,050.03	1	30
Gd	3,987.84	25	50	U	4,050.04	35	25
Eu	3,987.83	—	20w	La	4,050.08	60	60
Ir	3,987.829	—	12	J	4,050.09	(35)	—
Nd	3,987.81	6	25	Mo	4,050.09	8	6
Ru	3,987.79	50	3	S	4,050.11	(10)	—
Kr	3,987.78	(25)	—	Zr	4,050.33	10	20
Th	3,987.71	8	10	Gd	4,050.37	—	8
Ti	3,987.61	12	8	Kr	4,050.42	(50wh)	—
Mn	3,987.46	15	15				

	λ	I J	I O
Nd	*3,987.43*	*20*	*20*
Sm	*3,987.42*	*8*	*15*
Pr	*3,987.37*	*15*	*25*
Th	*3,987.23*	*8*	*10*
Gd	*3,987.22*	*100*	*100*
Pr	*3,987.17*	*3*	*15*
W	*3,987.15*	*10*	—
Co	*3,987.11*	—	*80*
Mn	*3,987.10*	*60*	*30*
Sm	*3,986.90*	*8*	*8*
U	*3,986.85*	*8*	—
Mn	*3,986.83*	*75*	*40*
Zr	*3,986.80*	—	*10*
Mg	*3,986.73*	*3*	*15w*
Th	*3,986.64*	*8*	*8*

	λ	I J	I O
Cl	*4,050.46*	*(8)*	—
Zr	*4,050.48*	—	*25*
Er	*4,050.57*	*1*	*25*
Dy	*4,050.58*	*15*	*30*
Cu	*4,050.66*	—	*30*
Ce	*4,050.81*	*1*	*12*
Hf	*4,050.88*	*1h*	*20*
Th	*4,050.89*	*5*	*15h*
Re	*4,050.93*	—	*10*
V	*4,050.96*	*4*	*15*
Rn	*4,051.00*	*(10)*	—
Th	*4,051.10*	*5*	*8*
Nd	*4,051.14*	*15*	*15*
Pr	*4,051.15*	*30*	*50*
Mo	*4,051.18*	*10*	*5*
W	*4,051.30*	*9*	*2*
Cr	*4,051.33*	*8*	*25*
V	*4,051.35*	*4*	*15*
Ru	*4,051.40*	*200*	*125*
Ce	*4,051.427*	*3*	*20*
Os	*4,051.429*	*3*	*12*
Tb	*4,051.47*	*1*	*8*

Th 4,381.86 *J* 30 *O* 30

	λ	I J	I O
Sc	**4,381.85**	—	**2**
Ce	**4,381.77**	**2**	**6**
Mn	**4,381.70**	**20**	**80**
Mo	**4,381.64**	**150**	**150**
Pr	**4,381.62**	—	**15**
Kr	**4,381.52**	**(100*h*)**	—
Th	**4,381.39**	**5**	**10**
U	**4,381.38**	**3**	**4**
Ir	**4,381.36**	—	**3**
Tb	**4,381.30**	—	**8**
Nb	**4,381.29**	**3**	**10**
Ru	**4,381.27**	—	**15**
Mn	**3,381.26**	—	**5**
Sm	**4,381.25**	—	**2**
Sc	**4,381.23**	—	**9**
Ne	**4,381.22**	**(30)**	—
Ca	**4,381.19**	**3**	—
Nb	**4,381.13**	**20**	**3**
Cr	**4,381.112**	**25**	**30**
Tm	**4,381.11**	**5**	**5**
Ce	**4,381.09**	—	**4**
Pr	**4,381.08**	—	**20*w***

	λ	I J	I O
Nd	**4,381.87**	—	**4**
Nb	**4,381.94**	—	**4**
Eu	**4,382.05**	**1**	**2**
Gd	**4,382.06**	—	**40**
U	**4,382.07**	**1**	**6**
Hg	**4,382.16**	**(10)**	—
Ce	**4,382.167**	**12**	**40**
Er	**4,382.168**	**1**	**9**
U	**4,382.34**	**5**	**18**
Mo	**4,382.41**	**20**	**10**
Pr	**4,382.42**	**20**	**30**
Tb	**4,382.45**	—	**25**
Nb	**4,382.49**	**5*h***	**3*h***
*bh***C**	**4,382.5**	—	—
Mn	**4,382.63**	—	**80*h***
Zr	**4,382.73**	—	**3**
Nd	**4,382.74**	**10**	**15**
Fe	**4,382.77**	**10**	**10**
Pr	**4,382.82**	**8*w***	**25*w***
U	**4,382.83**	—	**2**
Nb	**4,382.84**	**5**	**3**
Cr	**4,382.85**	**2**	**12**

	λ	I J	I O
Eu	**4,380.98**	**1**	**2**
Ir	**4,380.77**	—	**10**
Ce	**4,380.705**	—	**3**
Rb	**4,380.70**	**(20)**	—
Gd	4,380.64	125	100
Mo	4,380.59	15	15
V	4,380.55	10	15
Sm	4,380.42	6	25
Pr	4,380.32	20	50
Mo	4,380.29	25	30
Ce	4,380.06	2	20
Rh	4,379.92	25	60
Cl	4,379.91	(15)	—
Cr	4,379.78	2	15
Ar	4,379.74	(80)	—
*bh*La	4,379.7	—	20
O	4,379.55	15*h*	—
Xe	4,379.50	(100)	—
Bi	4,379.40	20	25
Pr	4,379.33	2	100*w*
V	4,379.24	200*R*	200*R*
Ta	4,378.82	2	40
Gd	4,378.57	—	40
W	4,378.49	12	25
Pr	4,378.26	3	15
Sm	4,378.23	100	100
Cu	4,378.20	30*w*	200*w*
La	4,378.10	30	40
Nb	4,377.96	30	10
Ne	4,377.95	(15)	—
Mo	4,377.76	200	5
Kr	4,377.71	(40*h*)	—
Cr	4,377.55	10	25
Nd	4,377.40	3	18
Te	4,377.10	(70)	—
Ir	4,377.01	4	100
Cr	4,376.80	20	20
Nd	4,376.44	4	15
Hg	4,376.19	(50)	—
J	4,376.16	(20)	—
Kr	4,376.12	(800)	—
Ar	4,375.96	(20)	—
Fe	4,375.93	200	500
Ce	4,375.92	5	40
*bh*La	4,375.8	—	30
*bh*La	4,375.7	—	15
Cr	4,375.33	30	25
Se	**4,382.87**	**(800)**	—
Ca	**4,382.93**	**2**	**5**
B	**4,382.95**	**4**	—
Ce	**4,382.96**	—	**4**
Gd	4,383.14	40	30
Eu	4,383.16	20	100*W*
Rn	4,383.30	(35)	—
La	4,383.45	50	10
*bh*La	4,383.5	—	15
Fe	4,383.55	800	1000
Xe	4,383.91	(100)	—
Br	4,384.00	(20)	—
Pr	4,384.14	25*w*	30*w*
Sm	4,384.29	50	50
Cs	4,384.43	(25)	—
Ni	4,384.54	1	25
Er	4,384.70	5	30
V	4,384.72	125*R*	125*R*
Pr	4,384.80	15	40
Sc	4,384.81	10	25
W	4,384.86	15	25
Xe	4,384.93	(30)	—
Cr	4,384.98	200	150
Ne	4,385.00	(15)	—
Te	4,385.08	(50)	—
Pr	4,385.24	5	15
Kr	4,385.27	(50*wh*)	—
P	4,385.33	(100)	—
Ru	4,385.39	40	125
Re	4,385.45	—	20
Mo	4,385.56	20	—
Ru	4,385.65	50	125
Tb	4,385.66	—	15
Nd	4,385.663	20	40
Xe	4,385.77	(70)	—
Ta	4,386.068	15	50
Tb	4,386.07	—	25
Sm	4,386.22	5	25
Ru	4,386.27	—	20
Ac	4,386.37	100	—
Tm	4,386.42	10	200
Kr	4,386.54	(300*h*)	—
Pb	4,386.58	(20)	—
Ce	4,386.83	6	15
Ti	4,386.85	80	8
V	4,387.21	12	15
U	4,387.31	4	15

	λ	I J	I O
V	4,375.30	12	20
Ta	4,375.14	5	15
Nd	4,375.04	5	30
Gd	4,374.99	—	25
Sm	4,374.97	200	200
Mn	4,374.95	20	150
Yt	4,374.93	150	150
Er	4,374.925	25*wh*	40*wh*
Nd	4,374.923	5	20
Ti	4,374.82	35	7
Rh	4,374.80	500	1000*W*
Th	4,374.79	10	15
Sc	4,374.45	25	100
Pr	4,374.41	15	50
C	4,374.28	40	—
Gd	4,374.25	20	20
Se	4,374.24	(40)	—
Ta	4,374.21	15	15
Cr	4,374.16	60	50
Gd	4,373.84	80	200
V	4,373.83	12	20
Ce	4,373.82	4	40
Pr	4,373.81	10	40
Xe	4,373.78	(50*wh*)	—
Cr	4,373.66	12	15
Co	4,373.63	—	15*wh*
Fe	4,373.57	3	50
Sm	4,373.46	50	50
Se	4,373.31	(40)	—
Cr	4,373.25	50	50
V	4,373.23	20	25
Ce	4,373.22	—	25
Rh	4,373.042	10	60
In	4,373.040	(15)	—
Cs	4,373.02	(30)	—
Te	4,373.00	(50)	—
Cl	4,372.91	(80)	—
In	4,372.87	(80)	—
In	4,372.80	(15)	—
Ta	4,372.77	2	30*h*
U	4,372.76	5	18
Pr	4,372,72	—	15*h*
Nb	4,372.64	20*h*	2
U	4,372.57	18	15
W	4,372.53	10	25
C	4,372.49	30	—
Ce	4,372.40	1	35
Ti	4,372.38	6	20
Re	4,387.41	—	25*w*
Cr	4,387.50	15	15*d*
Gd	4,387.68	—	150
Eu	4,387.88	—	200
Fe	4,387.90	35	150
He	4,387.93	(30)	—
Ti	4,388.07	5	25
Mn	4,388.08	—	60
K	4,388.13	(40)	—
Tb	4,388.25	—	20
Mo	4,388.27	20	6
Nb	4,388.36	15	10
Fe	4,388.41	50	125
J	4,388.51	(15*h*)	—
Fe	4,389.25	2	35
Kr	4,389.72	(20*h*)	—
Mn	4,389.75	—	50
Cl	4,389.76	(25)	—
W	4,389.84	6	15
Gd	4,389.88	40*h*	40

*N**

	λ	I J	I O
Xe	4,372.29	(20)	—
Ru	4,372.21	100	125
Ne	4,372.16	(30)	—
Ir	4,372.13	—	40w
Tb	4,372.04	—	20
bhLa	4,372.0	—	40
bhLa	4,371.9	—	20
Rb	4,371.80	(20)	—
U	4,371.76	1	18
Pr	4,371.61	40w	125
Rn	4,371.53	(30)	—
As	4,371.38	50	—
Ar	4,371.36	(80)	—
C	4,371.33	30	—
Cr	4,371.28	150	200
Kr	4,371.25	(20h)	—
Ru	4,371.20	5	15
Co	4,371.13	—	25wh
Hf	4,370.97	40	30
Mn	4,370.87	—	30
Yb	4,370.81	40	15
Pr	4,370.80	7	20
Ar	4,370.76	(30)	—
Os	4,370.66	3	50
Eu	4,370.46	2	60W
Ru	4,370.41	—	15
Gd	4,370.19	—	40
Sm	4,369.92	25	40
Gd	4,369.775	150	250
Fe	4,369.774	100	200
Ne	4,369.77	(70)	—
Kr	4,369.69	(200)	—
Ti	4,369.68	5	25
Th	4,369.32	5	10
O	4,369.28	(50)	—
Xe	4,369.20	(100wh)	—
Gd	4,369.17	20	50
Mo	4,369.04	25	40
Mn	4,368.88	—	50
Nd	4,368.63	15	50
Nb	4,368.43	30	15
Pr	4,368.33	90	125
O	4,368.30	(1000)	—
C	4,368.14	(30h)	—
Te	4,368.09	(30)	—
Sm	4,368.03	60	60
Nb	4,367.97	50h	2
Fe	4,367.91	70	60

	λ	I J	I O
Fe	4,367.582	50	100
Re	4,367.582	—	80
Tb	4,367.31	—	30
O	4,366.91	(100)	—
Pr	4,366.09	—	30w
Os	4,365.67	4	60
Br	4,365.60	(200)	—
La	4,364.67	50	50
Ce	4,364.66	6	30
Te	4,364.02	(400)	—
Be	4,364.00	50	—
Ar	4,363.79	(80)	—
Mo	4,363.64	200	5
Ne	4,363.52	(70)	—
Sm	4,363.45	50	60
bhB	4,363.4	—	40
Cs	4,363.27	(50)	—
Cr	4,363.13	35	25
Pr	4,362.98	20	50
Sm	4,362.91	25	60
Ne	4,362.69	(30)	—
Kr	4,362.64	(500)	—
U	4,362.05	3	30
Sm	4,362.03	150	150
Th	4,361.32	10	12
Te	4,361.27	(50)	—
Ru	4,361.21	50	40
Sm	4,361.07	20	30
Be	4,361.02	(40)	—
Gd	4,360.93	—	200
Ta	4,360.83	5	40s
Sm	4,360.71	60	100
Be	4,360.69	(35)	—
Ti	4,360.49	15	60
Tm	4,359.93	30	300
Nb	4,359.85	50	50
Pr	4,359.79	40	100
Cr	4,359.63	150	200
Ni	4,359.58	10	100
Pr	4,359.11	25	70
Ac	4,359.09	30	—
Yt	4,358.73	50	60
Re	4,358.69	—	80
Fe	4,358.50	20	70
Hg	4,358.35	500	3000w
Mo	4,358.32	40	—
N	4,358.27	(250)	—
Nd	4,358.17	20	50

Th 4,391.11 *J* 40 *O* 50

	λ	*I* *J*	*I* *O*		λ	*I* *J*	*I* *O*
Pr	**4,391.111**	—	**10**	**Eu**	**4,391.19**	—	**10***w*
Nd	**4,391.110**	**4**	**10**	**Hf**	**4,391.26**	—	**3**
Os	**4,391.08**	**3**	**15**	**Re**	**4,391.33**	—	**60***w*
Ti	**4,391.034**	**25**	**6**	**Ce**	**4,391.34**	—	**3**
Ru	**4,391.027**	—	**20**	**Eu**	**4,391.37**	—	**8***w*
Gd	**4,390.955**	**100**	**100**	**Gd**	**4,391.44**	**25**	**15**
Fe	**4,390.954**	**35**	**100**	**U**	**4,391.496**	**4**	**5**
Mo	**4,390.93**	**8***h*	**1**	*bh***La**	**4,391.5**	—	**4**
Tb	**4,390.91**	—	**30**	**Pr**	**4,391.51**	**4**	**20**
Sm	**4,390.86**	**150**	**150**	**Mo**	**4,391.53**	**15**	**15**
Ce	**4,390.81**	—	**4**	**Co**	**4,391.57**	**4**	**10***wh*
Hf	**4,390.70**	—	**3**	*bh***La**	**4,391.6**	—	**8**
Nd	**4,390.66**	**15**	**20**	**Br**	**4,391.61**	**(25)**	—
V	**4,390.61**	**3**	**10**	**Gd**	**4,391.65**	—	**3**
Mg	**4,390.58**	—	**10**	**Er**	**4,391.657**	—	**5**
U	**4,390.55**	**2**	**2**	**Ce**	**4,391.661**	**15**	**40**
Ho	**4,390.53**	—	**2**	**V**	**4,391.67**	**6**	**8**
Ce	**4,390.50**	—	**5**	**Cr**	**4,391.75**	**35**	**50**
Ru	**4,390.43**	**80**	**150***R*	**Pt**	**4,391.83**	**3**	**50**
Cl	**4,390.38**	**(8)**	—	**S**	**4,391.84**	**(30)**	—
Eu	**4,390.36**	**2**	**4***w*	**U**	**4,391.85**	**1**	**3***h*
J	**4,390.34**	**(15***h***)**	—	**Co**	**4,391.88**	**3**	**10**
Ni	**4,390.32**	—	**2**	**Ce**	**4,391.89**	—	**2**
Ce	**4,390.28**	—	**10**	**Ne**	**4,391.94**	**(150)**	—
Er	**4,390.19**	—	**6**	**Pr**	**4,391.99**	**5***w*	**30***w*
U	**4,390.16**	**3**	**1**	**J**	**4,392.00**	**(8)**	—
Ce	**4,390.15**	—	**2**	**Ce**	**4,392.03**	—	**2**
Pr	**4,390.145**	**1**	**8**	**Sm**	**4,392.07**	**2**	—
Na	**4,390.14**	—	**15**	**Gd**	**4,392.071**	**100**	**100**
Nd	**4,390.05**	**3**	**30**	**V**	**4,392.074**	**15**	**25**
Gd	**4,389.99**	**80**	**30**	**Pr**	**4,392.08**	—	**5**
V	**4,389.97**	**60***R*	**80***R*	**Nd**	**4,392.117**	**4**	**10**
Tb	**4,389.92**	—	**2**	**Mo**	**4,392.123**	**15**	**15**
Te	**4,389.92**	**(50)**	—	**Ce**	**4,392.167**	—	**2**
				Re	**4,392.17**	—	**5**
*P**				**U**	**4,392.21**	**6**	**3**
				Tb	**4,392.23**	—	**3**
				Tb	**4,392.34**	—	**3**
				Re	4,392.49	—	100
				Ir	4,392.59	4	100
				Yb	4,392.83	20	3
				Ce	4,393.19	3	35
				Xe	4,393.20	(100*wh*)	—
				Nd	4,393.35	2	15
				Sm	4,393.36	6	20
				Na	4,393.45	—	20

	λ	I J	I O
Br	4,393.56	(25)	—
U	4,393.59	6	40
Yb	4,393.75	—	25*h*
V	4,393.83	10	15
Ti	4,393.92	12	60
Re	4,394.38	—	100*r*
Tm	4,394.42	—	30
Ce	4,394.78	3	30
Os	4,394.86	6	150
Dy	4,394.98	4	25
Pr	4,395.00	15	25
Ti	4,395.03	150	50
V	4,395.23	40*R*	60*R*
Fe	4,395.29	—	80
Ne	4,395.56	(50)	—
Xe	4,395.77	(200*wh*)	—
Pr	4,395.79	2*w*	30
Ti	4,395.84	30	10
Ag	4,395.93	30	10
O	4,395.95	(80)	—
Te	4,396.00	(100)	—
Pr	4,396.12	50	80
Ag	4,396.32	—	100
Br	4,396.40	(20)	—
Mo	4,396.66	25	25
Re	4,396.79	—	25
Cr	4,397.25	5	25
Mo	4,397.29	30	30
Gd	4,397.52	5	100
Ru	4,397.80	—	150
Ne	4,397.94	(100)	—
Yt	4,398.01	100	150
Pr	4,398.26	9	25
Ta	4,398.449	10	40
Te	4,398.45	(70)	—
Hg	4,398.62	(300)	—
J	4,399.09	(20)	—
Ce	4,399.20	6	45
Pr	4,399.33	20	40
Ir	4,399.47	100	400
Cs	4,399.49	(20)	—
Ru	4,399.59	—	20
Ti	4,399.77	100	40
Kr	4,399.97	(200)	—
Pr	4,400.03	20	30
Ar	4,400.09	(300)	—
Pr	4,400.25	10	25
Sc	4,400.35	30	150

	λ	I	
		J	O
V	4,400.57	40	60
Ti	4,400.58	2	25
Nd	4,400.83	20	50
Kr	4,400.87	(100*h*)	—
P	4,400.99	(50)	—
Ar	4,401.02	(40)	—
Se	4,401.02	(100)	—
Sm	4,401.17	30	50
Fe	4,401.30	15	60
Rb	4,401.40	(30)	—
Ni	4,401.55	30	1000*W*
Gd	4,401.85	100	200
Te	4,401.89	(100)	—
Hg	4,402.06	(50)	—
Yb	4,402.30	20	6
Ta	4,402.50	20*h*	100
Ba	4,402.55	10	80
Re	4,402.60	—	30
Os	4,402.74	3	50
Mo	4,402.90	20	20
Sm	4,403.12	10	25
Gd	*4,403.14*	*100*	*100*
Sm	*4,403.36*	*50*	*50*
Pr	*4,403.60*	*40*	*100*
Ir	*4,403.78*	*10*	*300*
Ti	*4,404.27*	*30*	*50*
Kr	*4,404.33*	*(30h)*	—
Fe	*4,404.75*	*700*	*1000*
Hg	*4,404.86*	*(50)*	—
Cs	*4,405.25*	*(35)*	—
Pr	*4,405.85*	*100*	*100*
Re	*4,406.40*	—	*60*
Gd	*4,406.67*	*200*	*70*
Sr	*4,406.85*	—	*50h*
Xe	*4,406.88*	*(100wh)*	—
Fe	*4,407.72*	*50*	*100*
Mn	*4,408.08*	*5*	*60*
Gd	*4,408.26*	*150*	*100*
Fe	*4,408.42*	*60*	*125*
Pr	*4,408.84*	*100*	*125*
Kr	*4,408.89*	*(40hs)*	—
J	*4,408.96*	*(250)*	—
Ne	*4,409.30*	*(150)*	—
Sm	*4,409.34*	*100*	*100*
Ru	*4,410.03*	*80*	*150*
Kr	*4,410.37*	*(50)*	—
Mn	*4,410.49*	—	*50*

	λ	I J	I O
Te	*4,410.95*	*(50)*	—
Nd	*4,411.05*	*20*	*50*
Ti	*4,411.08*	*100*	*7*
Gd	*4,411.16*	*50*	*100*
C	*4,411.20*	*40*	—
C	*4,411.52*	*40*	—
Mn	*4,411.88*	*20*	*100*
Pr	*4,412.15*	*15*	*50*
Th	*44,12.53*	*6*	*10*
Th	*4,412.74*	*8*	*12*
Ac	*4,413.17*	*100*	—
Ne	*4,413.20*	*(50)*	—
As	*4,413.64*	*50*	—
Pr	*4,413.76*	*40*	*90*
Gd	*4,414.16*	*60*	*100*
P	*4,414.28*	*(100)*	—
P	*4,414.60*	*(70)*	—
Cd	*4,414.63*	*200*	—
Gd	*4,414.73*	*50*	*100*
Xe	*4,414.84*	*(150)*	—
Mn	*4,414.88*	*60*	*150*
O	*4,414.89*	*(300)*	—
Fe	*4,415.12*	*400*	*600*

Th 4,919.81 *J* 20 *O* 50

	λ	J	O		λ	J	O
Rh	**4,919.69**	—	**10**	**Pd**	**4,919.86**	—	**12**
Pr	**4,919.59**	—	**4**	**Ti**	**4,919.87**	**3**	**80**
Cr	**4,919.46**	—	**6**	**Sm**	**4,919.88**	—	**2**
Ca	**4,919.41**	**2**	**3**	**Ce**	**4,919.89**	—	**12*s***
Tb	**4,919.21**	—	**2**	**Ta**	**4,920.11**	**3**	**150*W***
Fe	**4,918.999**	**50**	**300**	**Co**	**4,920.26**	—	**10**
Ru	**4,918.998**		**7**	**Sm**	**4,920.37**	—	**125**
V	**4,918.985**	**2**	**3**	**Fe**	**4,920.50**	**125**	**500**
Sm	**4,918.984**	—	**125**	**Nd**	**4,920.69**	—	**60**
Ta	**4,918.88**	—	**4**	**Ce**	**4,920.78**	—	**15**
Rh	**4,918.83**	—	**5**	**Ta**	**4,920.88**	**1**	**5**
Ni	**4,918.71**	—	**40**	**Cr**	**4,920.94**	—	**50**
Gd	**4,918.66**	—	**10**	**La**	**4,920.971**	**400**	**500**
Ni	**4,918.36**	**1**	**200*W***	**Hf**	**4,920.975**	**5**	**2**
Dy	**4,918.22**	—	**4**	**Ru**	**4,921.07**	—	**40**
				Nd	**4,921.16**	—	**3**
Gd	4,916.60	—	30	**Ta**	**4,921.27**	**2**	**50**
Gd	4,915.84	—	40				
Ti	4,915.24	10	30				
Re	4,915.02	—	30	Ti	4,921.77	5	100
Ta	4,914.95	1	50	La	4,921.78	400	500
				Cr	4,922.27	40	200

	λ	I J	I O
Sm	4,914.31	—	25
Pr	4,914.03	—	60
Ni	4,913.97	—	200
Ti	4,913.62	15	125
Nd	4,913.42	—	60
Sm	4,913.26	—	150
Os	4,912.60	—	80
Ni	4,912.03	—	100
Eu	4,911.41	6	30
Ta	4,911.38	1	30
W	4,910.74	—	30
Sm	4,910.41	—	150
Gd	4,910.12	—	25
Fe	4,910.03	—	100
Fe	4,909.39	—	50
Mo	4,909.19	15	30
Fe	4,907.74	—	25
Ta	4,907.73	1	50
Mo	4,907.43	20	30
Pr	4,906.98	—	50
Cr	4,905.05	—	30
Sm	4,904.98	—	60
Lu	4,904.88	5	60
Ta	4,904.59	2	80*W*
Ni	4,904.41	1	400*W*
Co	4,904.17	—	80
Mo	4,903.81	30	80
Fe	*4,903.32*	*2*	*500*
Cr	*4,903.24*	—	*125*
Ru	*4,903.05*	—	*60*
La	*4,901.86*	—	*50*
Sm	*4,900.74*	—	*100*
La	*4,899.92*	*200*	*400*
Ti	*4,899.91*	*20*	*150*
Co	*4,899.52*	—	*400W*
Os	*4,899.21*	—	*60*
Th	*4,898.46*	—	*12*
Nd	*4,896.93*	—	*60*
Th	*4,894.95*	—	*10*
Gd	*4,894.32*	*4*	*200*
Sm	*4,894.30*	—	*60*
Sm	*4,893.34*	—	*150*
Fe	*4,891.50*	*15*	*70*
Fe	*4,890.77*	*15*	*100*
Gd	*4,889.20*	—	*60*
Re	*4,889.17*	—	*2000w*
Cr	*4,888.53*	—	*100*

	λ	I J	I O
Sm	4,922.47	—	30
Th	4,922.95	—	10
Ta	4,923.47	1	60*W*
Sm	4,923.81	—	30
Tm	4,923.83	—	30
Fe	4,923.92	50	30
Re	4,923.93	—	150
Sm	4,924.06	—	30
Nd	4,924.53	—	80
Pr	4,924.59	—	80
Fe	4,924.78	—	100
Ta	4,924.96	1	30
Fe	4,925.29	50*r*	1000*R*
Ti	4,925.41	—	25
Ni	4,925.58	—	100
V	4,925.65	20	25
Ta	4,926.00	2	60
Mo	4,926.19	12	25
Mo	4,926.43	12	25
Fe	4,927.45	6	50
Co	4,928.28	—	200*W*
Ti	4,928.34	4	100
Sm	4,929.56	—	40
Cr	4,930.18	—	35
Fe	4,930.33	—	25
Gd	4,930.71	—	80
W	4,931.56	—	30
Mo	4,933.10	15	30
Sm	4,933.30	—	25
Fe	4,933.35	30	50
Fe	4,934.02	—	40
Co	4,934.06	—	25
Ba	4,934.09	400*h*	400*h*
Mn	4,934.15	5	25
Hf	4,934.45	50	40
La	4,934.82	100	150
Er	4,935.498	—	35
Yb	4,935 50	10	200
Ni	4,935.83	1	150
Pr	4,936.00	—	50
Sm	4,936.02	—	80
Cr	*4,936.33*	*5*	*200*
Ta	*4,936.42*	—	*100s*
Th	*4,936.77*	—	*10*
Ni	*4,937.34*	—	*400w*
Sm	*4,938.10*	—	*125*
Fe	*4,938.18*	—	*100*

	λ	I	
		J	O
Ti	4,938.29	2	70
Eu	4,938.31	—	250W
Ru	4,938.43	—	60
Gd	4,938.62	2	150
Fe	4,938.82	1	300
Fe	4,939.69	2	150
Pr	4,939.73	—	100
Pr	4,940.30	—	50
Cr	4,942.49	3	125
Gd	4,942.57	—	100
Nd	4,944.83	—	50
Ni	4,945.46	—	90
Sm	4,946.31	—	60
La	4,946.47	50	100
Re	4,946.74	—	100
Th	4,947.58	2	10
Sm	4,948.63	—	125
La	4,949.77	—	200
Mo	4,950.618	30	80
Th	4,950.62	5h	10
bhF	4,950.8	—	100
Pr	4,951.36	—	150
La	4,952.07	40	50
Sm	4,952.37	—	125

Ti 22 47.90

t_0 1,800°C t_1 >3,000°C

I.	II.	III.	IV.	V.
6.835	13.6	27.5	43.237	98.84

λ	*I*		eV
	J	*O*	
3,234.52	500*R*	100	3.88
II 3,349.04	800*R*	125	4.31
II 3,361.21	600*R*	100	3.71
II 3,372.80	400*R*	80	3.68
II 3,383.76	300*R*	70	3.66
I 3,635.46	100	200	3.40
I 3,642.68	125	300	3.42
I 3,653.50	200	500	3.44
I 4,981.73	125	300	3.33
I 4,991.07	100	200	3.32
I 4,999.51	80	200	3.30
I 5,007.21	40	200	3.29

Ti 3,234.516 *J* 500*R* *O* 100

	λ	*I* *J*	*I* *O*		λ	*I* *J*	*I* *O*
V	**3,234.516**	**20**	—	**Ho**	**3,234.52**	**6**	**2**
Ar	**3,234.51**	**(100)**	—	**Fe**	**3,234.61**	**125**	**200**
Ir	**3,234.510**	—	**3**	**Rb**	**3,234.64**	**(5)**	—
Tb	**3,234.50**	**15**	**15**	**Ni**	**3,234.65**	**15**	**300**
Ce	**3,234.49**	**3**	**10**	**Ta**	**3,234.69**	**10**	**70**
Ru	**3,234.43**	**50**	**3**	**V**	**3,234.728**	—	**8**
Sm	**3,234.42**	**4**	**10**	**J**	**3,234.73**	**(3)**	—
Eu	**3,234.30**	—	**3***w*	**Os**	**3,234.731**	**10**	**100**
Ce	**3,234.27**	—	**5**	**Ru**	**3,234.798**	**1***h*	**10**
Pr	**3,234.22**	**2**	**12**	**Er**	**3,234.798**	—	**4**
Os	**3,234.20**	**12**	**150**	**Zr**	**3,234.80**	—	**2**
Mo	**3,234.18**	**5**	**8**	**Ce**	**3,234.89**	**5**	**18***s*
Ce	**3,234.161**	**8**	**40**	**Na**	**3,234.93**	**(20)**	**6**
Cs	**3,234.16**	**(6)**	—	**Ba**	**3,234.96**	**3**	—
Zr	**3,234.124**	**3**	**8**	**W**	**3,234.98**	**7**	**9**
Co	**3,234.119**	—	**2**	**Ca**	**3,234.990**	**2**	—
Cr	**3,234.06**	**150**	**10**	**Th**	**3,234.995**	**1***h*	**5**
Si	**3,234.05**	**7**	—	**Mn**	**3,235.003**	—	**30***h*
Fe	**3,233.971**	**150**	**300**	**Ce**	**3,235.01**	**1**	**10**
Mn	**3,233.968**	—	**75***wh*	**Mo**	**3,235.02**	**25**	—
Cu	**3,233.90**	**2***h*	**2***h*	**Ru**	**3,235.10**	—	**6**
Hf	**3,233.80**	**4**	**3**	**Eu**	**3,235.13**	—	**10**
Ce	**3,233.773**	—	**15**	**U**	**3,235.23**	**8**	**6**
V	**3,233.768**	**20**	**5**				
Tm	**3,233.75**	—	**10**				
				Tm	3,235.45	40	80
				Co	3,235.54	—	60
Mo	3,233.14	30	50	Re	3,235.94	—	50
Fe	3,233.05	60	100	Fe	3,236.22	200	300
Ni	3,232.96	35	300*R*	Ti	3,236.57	300*r*	70
Co	3,232.87	25	60				
Ru	3,232.75	4	50	Sm	3,236.63	40	100
				Mn	3,236.78	75	75
Li	3,232.61	500	1000*R*	Tm	3,236.80	80	100
Os	3,232.54	10	150	Co	3,237.03	—	100
Sb	3,232.50	250*wh*	150	Rh	3,237.66	20	60
Ti	3,232.28	100	30				
Os	3,232.05	20	500*R*	Ta	3,237.85	7*h*	70
				Ru	3,238.53	45	100
Os	3,231.42	12	150	Os	3,238.63	20	100
Ti	3,231.31	25	15	Ru	3,238.77	1	50
Fe	3,230.97	200	300	Ti	3,239.04	300*R*	60
Ta	3,230.85	18*w*	200				
Mn	3,230.72	75	75	Fe	3,239.44	300	400
				Ru	3,239.60	5	50
Sm	3,230.54	30	100	Sm	3,239.64	25	100
Pt	3,230.29	6	100	Ti	3,239.66	80	25
Fe	3,230.21	80	100	Ta	3,239.99	18*w*	200
Tl	3,229.75	800	2000				
Ti	3,229.42	70	15	Tm	3,240.23	80	100
				Mn	3,240.40	30	60
Ta	3,229.24	70*w*	300*w*	Mn	3,240.62	30	60

	λ	I J	I O
Os	3,229.21	5	125
Fe	3,229.12	50	80
Fe	3,228.90	40	80
Ti	3,228.60	100	30
Ru	3,228.53	150	50
Fe	3,228.25	80	100
Mn	3,228.09	100	100
Fe	3,227.75	300	200
Ta	3,227.32	10	70
Os	3,227.28	12	125
Co	3,226.985	—	80r
Ni	3,226.984	—	100
Ru	3,226.37	12	50
Ti	3,226.13	7	25
Fe	3,225.79	150	300
Nb	3,225.48	800wr	150w
Ni	3,225.02	6	300
Ti	3,224.24	150	15
Os	3,223.86	8	100
Ta	3,223.83	50W	200W
Ti	3,223.52	2	15
Ti	3,222.84	150r	20
Fe	3,222.07	100	200
Ni	3,221.65	4	300
Ti	3,221.38	6	25
Ir	3,220.78	30	100
Ti	3,220.47	25	—
Fe	3,219.81	80	100
Fe	3,219.58	125	200
Ti	3,219.21	3	15
Pd	3,218.97	8	300
Sm	3,218.60	25	100
Ti	3,218.27	150	15
Ti	3,217.94	1	12

	λ	I J	I O
Os	3,241.04	20	80
Sm	3,241.14	10	50
Ru	3,241.23	12	60
Sb	3,241.28	(350wh)	—
Ir	3,241.52	50	100
Tm	3,241.53	125	150
Ti	3,241.99	300R	60
Ta	3,242.05	15	125
Ru	3,242.16	—	80
Yt	3,242.28	100	60
Pd	3,242.70	600R	2000wh
Ta	3,242.83	10	125
Ni	3,243.06	15	400R
Mn	3,243.78	75	100
Co	3,243.84	—	100
Fe	3,244.19	200	300
La	3,245.12	300	400
Fe	3,245.98	150	200
Fe	3,246.96	70	100
Cu	3,247.54	2000R	5000R
Fe	3,248.21	150	200
Ni	3,248.46	2	150
Mn	3,248.516	100	100
Ta	3,248.522	3h	100
Ti	3,248.60	200r	25
La	3,249.35	80	300
Ti	3,249.37	20	8
Ni	3,250.74	1	125
Fe	3,251.23	150	300
Dy	3,251.26	100	100

Ti 3,349.095 *J* 800*R* *O* 125

	λ	J	O
Th	3,348.96	6	4
Mo	3,348.94	30	1
Dy	3,348.88	—	2
Sm	3,348.87	1	3
Ti	3,348.84	12	12
Nb	3,348.78	30	1
Ce	3,348.775	—	6
Th	3,348.766	4	5
Er	3,348.74	1	10
Rb	3,348.72	—	100

	λ	J	O
U	3,349.036	—	4
Nb	3,349.06	100	80
Cr	3,349.07	40	125
Hf	3,349.17	4	5
Mo	3,349.19	5	6
Sc	3,349.22	—	4
Co	3,349.222	—	3h
Cu	3,349.29	40	70
Cr	3,349.32	50	35
Th	3,349.340	10	5

	λ	I J	I O
Cs	**3,348.72**	**—**	**15**
Ru	**3,348.70**	**2**	**50**
U	**3,348.69**	**1**	**4**
Sm	**3,348.68**	**10**	**10**
Os	**3,348.66**	**15**	**30**
Nd	**3,348.59**	**—**	**2**
Ho	**3,348.58**	**4*h***	**—**
Tb	**3,348.54**	**—**	**15**
Ti	**3,348.53**	**—**	**7**
V	**3,348.37**	**20**	**—**
U	**3,348.33**	**1**	**5*d***
W	**3,348.29**	**15**	**2**
Nb	**3,348.28**	**50**	**1*h***
Ce	**3,348.19**	**—**	**8**
Nd	**3,348.16**	**2**	**10**
Er	**3,348.14**	**2**	**15**
Co	**3,348.11**	**—**	**80**
Fe	3,347.93	100	150
Ru	3,347.61	6	60
Co	3,346.93	2	100
Cr	3,346.74	80*r*	150*R*
Ti	3,346.73	60	60
Re	3,346.20	—	100
Zn	3,345.93	50	150
Zn	3,345.57	100	500
Ru	3,345.32	5	60
Zn	3,345.02	300	800
La	3,344.56	200*wh*	300
Ru	3,344.53	6	60
Ca	3,344.51	7	100
Re	3,344.35	—	150
Rh	3,344.20	20	100
Pt	3,343.90	80	100
Ti	3,343.77	70	60
Co	3,342.73	—	150*W*
Ti	3,342.71	5	10
Re	3,342.26	—	200
La	3,342.22	5	80
Ti	3,342.15	8	12
Nb	3,341.97	50	100*r*
Fe	3,341.90	80	100
Ti	3,341.87	300*R*	100
Ru	3,341.66	50	70
Hg	3,341.48	100	100
Co	3,341.34	2	60
Fe	3,340.57	100	125
Ti	3,340.34	100	80

	λ	I J	I O
W	**3,349.340**	**10**	**2**
Nb	**3,349.35**	**100**	**5**
Au	**3,349.40**	**5**	**15**
Ce	**3,349.405**	**5**	**4**
Ti	**3,349.406**	**400*R***	**100**
Tb	**3,349.42**	**50**	**30**
U	**3,349.43**	**3*d***	**4*d***
Cs	**3,349.44**	**(10)**	**—**
Cu	**3,349.46**	**2*h***	**—**
Nb	**3,349.52**	**5*h***	**30**
Co	**3,349.53**	**—**	**7**
W	**3,349.54**	**7**	**4**
Fe	**3,349.74**	**—**	**1**
J	**3,349.79**	**(10)**	**—**
La	**3,349.834**	**—**	**4**
Sm	**3,349.835**	**1**	**3**
Th	**3,349.87**	**2**	**—**
Re	**3,349.91**	**—**	**20**
Ce	**3,349.97**	**1**	**30**

*N**

	λ	I	
		J	O
Dy	3,340.01	4	80
Co	*3,339.78*	—	*150w*
Rh	*3,338.54*	*50*	*200*
Re	*3,338.17*	—	*150*
Ti	*3,337.85*	*60*	*12*
Fe	*3,337.67*	*100*	*125*
La	*3,337.49*	*300wh*	*800*
Mg	*3,336.68*	*60*	*125*
Os	*3,336.15*	*50*	*200R*
Fe	*3,335.77*	*100*	*125*
Ti	*3,335.19*	*150*	*60*
Cr	*3,334.69*	—	*150wh*
Fe	*3,334.22*	*100h*	*150h*
Co	*3,334.14*	—	*250R*
Cr	*3,333.60*	*3h*	*125*
Ti	*3,332.11*	*125*	*40*
Fe	*3,331.61*	*70*	*125*

Ti 3,361.213 *J* 600*R* *O* 100

	λ	J	O
U	**3,361.213**	—	**2**
Au	**3,361.21**	**4**	**3**
Bi	**3,361.209**	**2**	**3**
Os	**3,361.149**	**20**	**80**
Ru	**3,361.148**	—	**30**
Re	**3,361.14**	—	**25**
W	**3,361.11**	**15**	**10**
Co	**3,361.093**	—	**5**
C	**3,361.09**	**12**	—
Tm	**3,361.04**	**15**	**10**
Er	**3,361.02**	**1**	**12**
Ce	**3,360.993**	—	**3**
Ti	**3,360.990**	**1**	**10**
Fe	**3,360.93**	**1**	**7**
Nb	**3,360.902**	**50**	**3**
Pr	**3,360.900**	**2**	**8**
Ho	**3,360.86**	**4*h***	—
In	**3,360.85**	**6**	—
Rh	**3,360.80**	**1**	**2**
Ru	**3,360.79**	—	**4**
W	**3,360.74**	**2**	**4**
Gd	**3,360.72**	**25**	**25**
Ce	**3,360.71**	—	**5**
Dy	**3,360.65**	**2**	**2**
Sm	**3,360.64**	**1**	**2**
Eu	**3,361.214**	**2**	**20**
Ce	**3,361.23**	**5**	**8**
Tb	**3,361.24**	**8**	**3**
Ni	**3,361.241**	—	**10**
Ti	**3,361.26**	**50*r***	**80*r***
Co	**3,361.267**	—	**18**
Sc	**3,361.270**	**9**	**25**
Mo	**3,361.37**	**30**	**30**
Sm	**3,361.43**	**2**	**10**
Rb	**3,361.49**	**(10)**	—
V	**3,361.51**	**200**	—
Ce	**3,361.555**	—	**10**
Ni	**3,361.556**	**20**	**500*w***
Co	**3,361.558**	**2**	**80**
Pb	**3,361.58**	**15**	—
Eu	**3,361.61**	**3**	**2*h***
Ta	**3,361.64**	**50*W***	**125*W***
Er	**3,361.67**	**1**	**6**
U	**3,361.727**	**8**	**15**
Ar	**3,361.73**	**(5)**	—
Th	**3,361.739**	**4*d***	**4*d***
Kr	**3,361.74**	**(2)**	—
C	**3,361.75**	**6**	—
Ce	**3,361.76**	**1**	**25**
Cr	**3,361.770**	**100**	**10**

	λ	I	
		J	O
Ne	**3,360.63**	**(18)**	**—**
Ce	**3,360.54**	**4**	**35**
P	**3,360.49**	**(30)**	**—**
Zr	**3,360.45**	**—**	**20**
Ce	**3,360.40**	**—**	**2**
Th	**3,360.38**	**5**	**4**
Ce	**3,360.37**	**—**	**2**
U	**3,360.34**	**1**	**5**
W	**3,360.32**	**12**	**4**
Ir	**3,360.313**	**—**	**3**
Fe	**3,360.313**	**1**	**3**
Cr	**3,360.295**	**200**	**50**
Ce	**3,360.291**	**—**	**3**
Tb	**3,360.29**	**3**	**8**
Mo	**3,360.28**	**25**	**—**
Rh	3,359.89	50	100
Sc	3,359.68	25	50
Lu	3,359.56	15	150
Co	3,359.29	2	100
Re	3,359.20	—	40
Ni	3,359.11	—	60
Ru	3,359.09	20	70
Gd	3,358.63	100	100
Ta	3,358.53	25*W*	70
Cr	3,358.50	200	40
Ti	3,358.48	1	15
Nb	3,358.42	100	100
Ti	3,358.28	5	35
Mo	3,358.12	30	60*W*
Os	3,357.97	15	100
Zr	3,357.26	40	50
Ba	3,356.89	3	40*r*
Co	3,356.47	2	150*W*
V	3,356.35	60	125
Ti	3,356.20	1	10
Zr	3,356.09	40	50
Re	3,355.29	—	60*h*
Fe	3,355.23	100	100
Ti	3,354.63	20	100
Co	3,354.38	—	200*R*
Fe	3,354.06	40	40
Ti	3,354.63	20	100
Co	3,354.38	—	200*R*
Fe	3,354.06	40	40
Os	3,353.91	15	40
Sc	3,353.73	60	50
Ru	3,353.65	4	50
Re	3,353.21	—	40
Ti	3,352.94	2	25

	λ	I	
		J	O
Nd	**3,361.773**	**2**	**8**
Re	**3,361.833**	**—**	**4**
Ti	**3,361.835**	**1**	**12**
W	**3,361.852**	**2**	**6**
Ce	**3,361.853**	**—**	**8**
Ca	**3,361.92**	**10**	**125**
Sc	**3,361.93**	**9**	**25**
Fe	**3,361.95**	**1**	**6**
W	**3,361.97**	**3**	**4**
Yt	**3,362.00**	**25**	**12**
Ru	**3,362.003**	**8**	**60**
La	**3,362.04**	**3**	**10**
U	**3,362.05**	**1**	**2**
Ti	**3,362.10**	**3**	**6**
Ca	**3,362.13**	**5**	**15**
Dy	**3,362.17**	**2**	**3**
Nb	**3,362.172**	**100**	**—**
Rh	**3,362.18**	**20**	**100**
Th	**3,362.20**	**4**	**3**
Cr	**3,362.21**	**8**	**80**
*N**			

	λ	I J	I O
Ti	3,352.07	15	8
Cr	3,351.97	50	50
Ru	3,351.93	4	50
Fe	3,351.74	60	80
Os	3,351.73	12	50
Fe	3,351.52	60	70
Sr	3,351.25	15	300
Rb	*3,350.89*	—	*150*
Ti	*3,350.55*	*1*	*10*
Gd	*3,350.48*	*180*	*150*
Ca	*3,350.21*	*10*	*100*

*P**

Ti 3,372.800 *J* 400*R* *O* 80

	λ	I J	I O
Pr	**3,372.797**	**4**	**10**
Th	**3,372.794**	**4*d***	**3*d***
Pt	**3,372.791**	—	**2**
Ho	**3,372.79**	**15**	**12**
Yt	**3,372.77**	**3**	**15**
Er	**3,372.75**	**20**	**35**
Tb	**3,372.72**	—	**50**
Rn	**3,372.7**	**(18)**	—
P	**3,372.70**	**(50)**	—
Th	**3,372.696**	**2*d***	**1*d***
Ca	**3,372.68**	**3**	**1**
V	**3,372.67**	**6**	—
Ce	**3,372.65**	—	**3**
Rb	**3,372.61**	**(30)**	—
U	**3,372.60**	**1**	**4**
Nb	**3,372.56**	**200**	**10*h***
Ce	**3,372.54**	—	**8**
Rh	**3,372.53**	—	**10**
Os	**3,372.52**	**2**	**3**
Pr	**3,372.51**	**3**	**20**
Ag	**3,372.50**	**3*h***	**1**
S	**3,372.48**	**(3)**	—
Ce	**3,372.39**	—	**3**
Tb	**3,372.36**	**15**	**15**
Fe	**3,372.35**	—	**1**
Rh	**3,372.25**	**200**	**300**
Ti	**3,372.21**	**15**	**10**
W	**3,372.19**	**12**	—
Sc	**3,372.15**	**150**	**7**
Cr	**3,372.13**	**30**	—
Fe	**3,372.86**	—	**2**
Ar	**3,372.88**	**(3)**	—
Mo	**3,372.92**	**1**	**5**
Pt	**3,372.996**	—	**10**
Pd	**3,373.001**	**500*wh***	**800*r***
Os	**3,373.04**	**5**	**15**
Nd	**3,373.09**	—	**4**
Mo	**3,373.13**	**25**	**5**
S	**3,373.19**	**(5)**	—
Os	**3,373.20**	**8**	**6**
Eu	**3,373.228**	**2**	**3**
Co	**3,373.230**	—	**60**
W	**3,373.233**	**6**	**7**
Nd	**3,373.30**	—	**8**
Zr	**3,373.42**	**10**	**12**
Ce	**3,373.45**	**3*h***	**25**
Ar	**3,373.48**	**(300)**	—
F	**3,373.49**	**(15)**	—
Th	**3,373.50**	**1**	**2**
Tb	**3,373.54**	—	**8**
Ti	**3,373.58**	—	**4**
Cu	**3,373.59**	**8**	—
K	**3,373.60**	**(30)**	—
Er	**3,373.67**	**1**	**5**
U	**3,373.727**	**1**	**10**
Ce	**3,373.729**	**3**	**25**
W	**3,373.75**	**10**	**10**
Sm	**3,373.79**	**4**	**15**

*N**

	λ	I	
		J	O
Nb	**3,372.09**	**3**	**5**
Os	**3,372.083**	**8**	**40**
Fe	**3,372.080**	**7**	**40**
Mo	**3,372.06**	**2**	**5**
U	**3,372.01**	**4**	**8**
Ca	**3,372.00**	**—**	**2**
Ni	**3,371.99**	**10**	**400**
S	**3,371.90**	**(20)**	**—**
Ne	**3,371.87**	**(12)**	**—**
Ru	**3,371.86**	**18**	**70**
O	**3,371.85**	**(5*h*)**	**—**
Ce	**3,371.84**	**—**	**8**
Th	**3,371.80**	**10**	**4**
Ta	3,371.54	20	70
Ti	3,371.45	15	100
Fe	3,370.79	200	300
Os	3,370.59	30	300*R*
Ti	3,370.44	15	80
Co	3,370.33	2	80
Os	3,370.20	10	50
Ne	3,369.91	(700)	—
Ne	3,369.81	(500)	—
Ni	3,369.57	100	500*R*
Fe	3,369.55	200	300
Ti	3,369.21	25	6
Eu	3,369.05	5	40
Sc	3,368.95	20	50
Ru	3,368.45	60	100
Rh	3,368.37	50	300
Dy	3,368.12	5	60
Ni	3,367.89	—	80
Cr	3,367.53	—	50*wh*
Co	3,367.11	30	300*R*
Fe	3,366.87	15	50
Ni	3,366.81	1	60
Fe	3,366.79	25	50
Ta	3,366.66	15*w*	50
Sr	3,366.33	10	100
Ti	3,366.18	50	20
Ni	3,366.17	12	400*W*
Ni	3,365.77	12	400*w*
V	3,365.55	80	125
Cu	3,365.35	30	70
Os	3,364.12	12	100
Mo	3,363.78	30	40
Co	3,363.76	—	80*r*
Ni	3,362.81	—	100
Co	3,362.80	—	80

	λ	I J	I O		λ	I J	I O
Re	3,362.75	—	40				
Cr	3,362.71	—	40				
Tm	3,362.61	200	250				
Ru	3,362.33	*5h*	50				
Gd	3,362.24	180	150				

*P**

Ti 3,383.76 *J* 300*R* *O* 70

	λ	I J	I O		λ	I J	I O
Pr	**3,383.73**	**3**	**12**	**Tb**	**3,383.78**	**—**	**8**
Fe	**3,383.70**	**70**	**100**	**Er**	**3,383.78**	**1**	**12**
Ce	**3,383.69**	**2**	**20**	**Nb**	**3,383.80**	**5**	**15**
Mo	**3,383.55**	**1**	**5**	**Pt**	**3,383.82**	**8**	**1**
Nd	**3,383.45**	**—**	**2**	**J**	**3,383.86**	**(3)**	**—**
U	**3,383.40**	**3**	**6**	**Co**	**3,383.920**	**—**	**60**
Ce	**3,383.39**	**1**	**12**	**Ce**	**3,383.925**	**—**	**8**
Pr	**3,383.38**	**3**	**15**	**Fe**	**3,383.980**	**100**	**200**
Ce	**3,383.28**	**—**	**8**	**Mo**	**3,383.981**	**30**	**1**
Sb	**3,383.14**	**50**	**40**	**Os**	**3,384.00**	**5**	**80**
W	**3,383.12**	**18**	**—**	**Yb**	**3,384.04**	**20**	**—**
Th	**3,383.120**	**10**	**5**	**Ce**	**3,384.07**	**—**	**8**
Yt	**3,383.05**	**3**	**3**	**Er**	**3,384.09**	**1**	**6**
Zr	**3,382.90**	**—**	**3**	**Nd**	**3,384.127**	**4**	**4**
Er	**3,382.892**	**1**	**18**	**Xe**	**3,384.13**	**(20*h*)**	**—**
Ag	**3,382.891**	**700*R***	**1000**	**Hf**	**3,384.14**	**15**	**15**
Ce	**3,382.888**	**—**	**8**	**U**	**3,384.22**	**3**	**3**
Yt	**3,382.83**	**3**	**3**	**Cr**	**3,384.24**	**—**	**35**
Nd	**3,382.81**	**10**	**200**	**Er**	**3,384.25**	**1**	**6**
Tb	**3,382.80**	**8**	**15**	**W**	**3,384.34**	**6**	**7**
Cr	3,382.68	200	35	**Rn**	**3,384.4**	**(5)**	**—**
Fe	3,382.409	10	50	**U**	**3,384.45**	**8**	**10**
Sm	3,382.407	40	100	**Tl**	**3,384.47**	**(6)**	**—**
Ti	3,382.31	7	30	**Os**	**3,384.596**	**3**	**20**
Co	3,381.50	—	100*W*	**V**	**3,384.599**	**40**	**60**
Nb	3,380.94	200	—	**W**	**3,384.61**	**3**	**5**
La	3,380.91	100*h*	200	**Mo**	**3,384.62**	**25**	**30*r***
Ni	3,380.88	12	200	**Cr**	**3,384.64**	**1**	**40**
Sr	3,380.71	200	150	**Nb**	**3,384.658**	**8**	**2**
Pd	3,380.67	2*h*	150*w*	**Sm**	**3,384.663**	**8**	**10**
Ni	3,380.57	100	600*R*	**Nd**	**3,384.67**	**20**	**20**
Ti	3,380.28	150*r*	25	**Hf**	**3,384.70**	**12**	**10**
Ru	3,380.17	15	60				
Fe	3,380.11	25	200				
Cr	3,379.82	100	15	Ru	3,385.14	35	60
				Co	3,385.22	15	250*R*
Re	3,379.72	—	80	Hg	3,385.25	(200)	—
Ru	3,379.60	18	60	Ti	3,385.66	10	25
Ti	3,379.22	10	25	Ru	3,385.71	4	50

	λ	I	
		J	O
Cr	3,379.17	6	40
Re	3,379.05	—	50
Fe	3,379.02	50	80
Co	3,378.74	2	40
Fe	3,378.685	80	150
Os	3,378.679	10	50
Ru	3,378.02	12	60
V	3,377.62	30	60
Ti	3,377.48	10	15
V	3,377.39	15	35
Rh	3,377.14	10	40
Ce	3,377.13	5*s*	50
Co	3,377.06	—	100
Lu	3,376.50	10	100
La	3,376.33	50	100
V	3,376.06	60	80
Ta	3,376.05	15	35
Ti	3,375.71	—	15*h*
Ru	3,374.65	18	80
Tm	3,374.51	30	60
Ti	3,374.35	30	10
Co	3,374.30	—	60
Ni	3,374.22	6	400
Ru	3,373.98	4	60
Co	3,373.97	—	60

*P**

	λ	I	
		J	O
Os	3,385.94	10	40
Ti	3,385.95	25	80
Fe	3,387.41	8	35
Os	3,387.836	15	100
Ti	3,387.837	125	60
Zr	3,387.87	100	100
Co	3,388.17	12	250*R*
Zr	3,388.30	40	50
Ru	3,388.709	20	80
Cr	3,388.710	4	40
Ti	3,388.75	35	12
Sm	3,389.32	20	40
Ru	3,389.50	18	60
Re	3,390.24	—	40
Ti	3,390.68	10	35
V	3,390.76	15	40
Ni	3,391.05	40	400
Cr	3,391.43	150	4
Ru	3,391.89	6	50
Zr	3,391.97	400	300
Eu	3,391.99	5	40
Fe	3,392.31	80	125
Gd	3,392.53	25	40
Ru	3,392.54	40	100
Fe	3,392.66	200	300
Ti	3,392.71	8	20
Ni	3,392.99	—	600*R*
Dy	3,393.58	10	100
Ar	3,393.75	(250)	—
Ti	*3,394.57*	*200*	*70*
Fe	*3,394.59*	*80*	*150*
Co	*3,395.37*	*50*	*400R*
Eu	*3,396.58*	*10*	*100*
Rh	*3,396.85*	*500*	*1000w*
Fe	*3,396.98*	*25*	*125*
Bi	*3,397.21*	*50*	*100wh*
Tm	*3,397.50*	*50*	*100*
Ti	*3,398.63*	*6*	*20*
Re	*3,399.30*	*—*	*200w*
Fe	*3,399.34*	*200*	*200*
Zr	*3,399.35*	*40*	*100*
Rh	*3,399.70*	*60*	*500*
Ti	*3,400.16*	*1*	*10*
V	*3,400.39*	*8*	*100*
Fe	*3,401.52*	*90*	*150*
Ru	*3,401.74*	*50*	*100*
Os	*3,401.86*	*20*	*200*

	λ	I J	I O		λ	I J	I O
				Fe	*3,402.26*	*150*	*150*
				Ti	*3,402.42*	*90*	*15*
				Os	*3,402.51*	*15*	*200*
				Ti	*3,403.37*	*2*	*12*
				Cd	*3,403.65*	*500h*	*800*

Ti 3,635.463 *J* 100 *O* 200

	λ	J	O		λ	J	O
Nb	**3,635.463**	**5**	**5**	**Ir**	**3,635.49**	**4*h***	**35**
V	**3,635.463**	**5**	**40**	**Ru**	**3,635.52**	**8**	**2**
W	**3,635.434**	—	**2**	**U**	**3,635.58**	**8**	**4**
Mo	**3,635.429**	**5**	**25**	**Mo**	**3,635.61**	**3**	**4**
Tb	**3,635.42**	**8**	**30**	**Eu**	**3,635.65**	—	**7*W***
U	**3,635.40**	**5**	**6**	**Ar**	**3,635.67**	**(3)**	—
Th	**3,635.37**	**10**	**15**	**Mn**	**3,635.68**	**5**	**10**
Ho	**3,635.35**	**4**	—	**Ce**	**3,635.78**	—	**2**
Yt	**3,635.33**	**12**	**2*h***	**Nb**	**3,635.85**	**8**	**3**
Nb	**3,635.32**	**8**	**5**	**Eu**	**3,635.87**	**4**	—
Sc	**3,635.31**	**2**	**3**	**V**	**3,635.874**	**25*h***	**50**
U	**3,635.30**	—	**10**	**Ce**	**3,635.88**	—	**2**
Pr	**3,635.284**	**7**	**25**	**Cu**	**3,635.92**	**7**	**50**
Cr	**3,635.281**	**8**	**25**	**Th**	**3,635.94**	**4**	**8**
Dy	**3,635.26**	**6**	**10**	**Ca**	**3,635.95**	**2*h***	**4**
Th	**3,635.25**	**5**	**4**	**Sm**	**3,636.04**	**2**	**5**
Ti	**3,635.202**	**5**	**10**	**Ce**	**3,636.07**	**1**	**5**
Fe	**3,635.196**	**2**	**12**	**Sm**	**3,636.11**	**1**	**5**
Tb	**3,635.17**	—	**8**	**Tb**	**3,636.16**	—	**15**
Mo	**3,635.14**	**10**	**100*h***	**Th**	**3,636.167**	**3**	**3**
Au	**3,635.13**	**10**	**3**	**Sc**	**3,636.168**	—	**3**
Nd	**3,635.11**	**6**	**6**	**Fe**	**3,636.19**	**10**	**40**
U	**3,635.01**	**2**	**1**	**Ir**	**3,636.20**	**25**	**50**
Cr	**3,634.99**	**12**	**25**	**Fe**	**3,636.234**	**10**	**15**
Ni	**3,634.94**	**10**	**50**	**Sm**	**3,636.235**	**2**	**10**
Ru	**3,634.93**	**100**	**50**	**Lu**	**3,636.25**	**3**	**25**
Sm	**3,634.91**	**2**	**20**	**U**	**3,636.31**	—	**10**
Nd	**3,634.87**	**15**	**30**	**Ce**	**3,636.37**	**1**	**3**
Ar	**3,634.83**	**(5)**	—	**Zr**	**3,636.45**	**30**	**200**
Rn	**3,634.8**	**(250)**	—	**Fe**	**3,636.49**	**1**	**3**
Gd	**3,634.76**	**20**	**20**	**Pr**	**3,636.50**	**2**	**8**
Cs	**3,634.75**	**(6)**	—	**Ce**	**3,636.56**	—	**2**
Tb	**3,634.75**	—	**8**	**Th**	**3,636.57**	**4**	**4**
Co	**3,634.71**	**10**	**70**	**Cr**	**3,636.59**	**30**	**60**
Pd	**3,634.695**	**1000*R***	**2000*R***				
				Sb	3,637.83	60	2*h*
Fe	**3,634.687**	**2**	**20**	Ti	3,637.97	8	30
Ho	**3,634.68**	**4*h***	**6**	Fe	3,638.30	80	100
Er	**3,634.679**	—	**8**	Hg	3,638.34	(100)	—
Nb	**3,634.60**	—	**2**	Tb	3,638.46	50	80

	λ	I	
		J	O
Th	**3,634.58**	**4**	**8**
U	**3,634.56**	**1**	**8**
Yb	**3,634.55**	**—**	**2**
Pr	**3,634.47**	**4**	**20**
Ar	**3,634.46**	**(300)**	**—**
Nb	**3,634.44**	**15**	**5**
Ce	**3,634.43**	**—**	**2**
Kr	**3,634.42**	**(3*wh*)**	**—**
He	**3,634.37**	**(2)**	**—**
Fe	**3,634.33**	**5**	**15**
Sm	3,634.27	25	100
Ne	3,633.66	(75)	—
Ti	3,633.46	5	35
Yt	3,633.12	100	50
Ar	3,632.68	(300)	—
V	3,632.12	70	—
Fe	3,632.04	50	50
Kr	3,631.87	(200*h*)	—
Cr	3,631.69	60	10
Fe	3,631.46	300	500
P	3,631.40	(50)	—
Na	3,631.27	(100)	12
Ca	3,630.75	9	150
Sc	3,630.74	70	50
Hg	3,630.65	(100)	—
Eu	3,629.80	50*h*	40
Mn	3,629.74	30	100
Gd	3,629.52	60	40
Dy	3,629.44	50	100
Yt	3,628,71	50	40
Ir	3,628.67	30	100
Tb	3,628.20	15	100
Nb	3,628.18	50	1
Pt	3,628.11	20	300*W*
Co	3,628.81	—	200
V	3,627.712	50	4
Ti	3,627.71	12	2
Ta	3,626.62	18	125
Rh	3,626.59	60	150
Ti	3,626.08	5	25
V	3,625.61	125	4
Ti	3,624.82	125	60
Ni	3,624.73	15	150
Ca	3,624.11	15	150
Ti	*3,623.10*	*1*	*12*
Fe	*3,622.00*	*100*	*125*
Fe	*3,621.46*	*100*	*125*
Pt	3,638.79	10	250
V	3,639.02	60	70
Co	3,639.44	20	200
Rh	3,639.51	70	125
Pb	3.639.58	50*h*	300
Os	3,640.33	40	200
Fe	3,640.39	200	300
F	3,640.89	(100)	—
V	3,641.10	30*wh*	100*h*
Ti	3,641.33	150	60
Mn	3,641.39	50*h*	50
*N**			

	λ	I J	I O		λ	I J	I O
Rn	*3,621.00*	*(250)*	*—*				
Yb	*3,619.81*	*100*	*30*				
Nb	*3,619.73*	*300*	*3*				
Nb	*3,619.51*	*200*	*5*				
Ni	*3,619.39*	*150h*	*2000R*				
V	*3,618.93*	*100*	*—*				
Fe	*3,618.77*	*400*	*400*				
P	*3,617.09*	*(100w)*	*—*				
Cd	*3,614.45*	*100*	*60*				
Ti	*3,614.21*	*4*	*35*				
Ti	*3,613.76*	*—*	*12*				
Cd	*3,612.87*	*500*	*800*				
Ni	*3,612.74*	*50h*	*400*				
Rh	*3,612.47*	*50*	*200*				

Ti 3,642.675 *J* 125 *O* 300

	λ	J	O		λ	J	O
Ce	**3,642.62**	**—**	**6**	**Tb**	**3,642.68**	**8**	**15**
Os	**3,642.50**	**10**	**20**	**Mo**	**3,642.72**	**2**	**4**
Rn	**3,642.5**	**(5)**	**—**	**Sm**	**3,642.74**	**5**	**25**
Nd	**3,642.46**	**6**	**20**	**U**	**3,642.776**	**5**	**2**
U	**3,642.44**	**8**	**10**	**Sc**	**3,642.785**	**50**	**60**
Mo	**3,642.41**	**2**	**3**	**F**	**3,642.798**	**(30)**	**—**
Ni	**3,642.39**	**—**	**2**	**Au**	**3,642.8**	**5*h***	**10**
Lu	**3,642.35**	**—**	**2**	**Fe**	**3,642.81**	**—**	**20**
Au	**3,642.34**	**1*wh***	**5**	**W**	**3,642.82**	**5**	**7**
Ce	**3,642.250**	**—**	**4**	**Ce**	**3,642.83**	**1**	**8**
Th	**3,642.249**	**4**	**10**	**U**	**3,642.85**	**—**	**6**
Nd	**3,642.205**	**—**	**2**	**Th**	**3,642.89**	**2**	**3**
Mo	**3,642.204**	**10**	**5**	**Pr**	**3,642.92**	**2**	**8**
Ce	**3,642.17**	**—**	**2**	**Re**	**3,642.99**	**—**	**100**
Ta	**3,642.06**	**18**	**125**	**Ce**	**3,643.01**	**—**	**2**
U	**3,642.03**	**2*h***	**6*h***	**Ar**	**3,643.09**	**(100)**	**—**
F	**3,641.98**	**(60)**	**—**	**Fe**	**3,643.11**	**5**	**30**
Er	**3,641.88**	**1**	**12**	**Pt**	**3,643.17**	**8**	**60**
Pr	**3,641.87**	**2**	**8**	**Co**	**3,643.18**	**15**	**80**
W	**3,641.85**	**5**	**7**	**Cr**	**3,643.20**	**30**	**5**
Th	**3,641.84**	**4**	**6**	**Tb**	**3,643.26**	**3**	**15**
Cr	**3,641.83**	**15**	**40**	**W**	**3,643.31**	**5**	**7**
Mo	**4,641.82**	**3**	**3**	**Pr**	**3,643.32**	**8**	**25**
Co	**3,641.79**	**8**	**60**	**Nb**	**3,643.34**	**10**	**5**
Ce	**3,641.73**	**1**	**8**	**Ce**	**3,643.45**	**—**	**5**
Cu	**3,641.69**	**5**	**60**	**Mo**	**3,643.47**	**20**	**1*h***
Tb	**3,641.66**	**30**	**70**	**Ir**	**3,643.48**	**—**	**2**
La	**3,641.656**	**4**	**—**	**Dy**	**3,643.50**	**—**	**3**
Sm	**3,641.648**	**2**	**7**	**W**	**3,643.514**	**6**	**—**
Ni	**3,641.64**	**—**	**4**	**Th**	**3,643.515**	**1**	**5**

	λ	I			λ	I	
		J	O			J	O
Pr	**3,641.62**	**5**	**50**	**Nb**	**3,643.52**	**2**	**2**
Ce	**3,641.54**	**1**	**10**	**Fe**	**3,643.626**	**8**	**20**
				Nd	**3,643.630**	**1**	**5**
				Cu	**3,643.631**	**—**	**5**
*P**				**Tm**	**3,643.65**	**40**	**60**
				Fe	**3,643.713**	**4**	**6**
				Sm	**3,643.715**	**3**	**5**
				Nb	**3,643.72**	**15**	**15**
				Tb	**3,643.76**	**3**	**30**
				Fe	**3,643.81**	**1**	**7**
				Hf	3,644.35	50	25
				Ca	3,644.41	15	200
				Ti	3,644.70	5	35
				V	3,644.71	50	80
				Sc	3,645.31	50	50
				La	3,645.41	60	100
				Dy	3,645.42	100	300
				Fe	3,645.82	60	80
				Gd	3,646.196	150	200*w*
				Ti	3,646.200	25	70
				Co	3,647.66	8	100
				Lu	3,647.77	5	100
				Fe	3,647.84	400	500
				Os	3,648.81	10	100
				Ti	3,648.86	10	3
				Co	3,649.35	4	200
				Sm	3,649.506	30	100
				Fe	3,649.508	100	100
				Ra	3,649.55	(1000)	—
				Ar	3,649.83	(800)	—
				Hg	3,650.15	500	200
				La	3,650.17	60	100
				N	3,650.19	(70)	—
				Fe	3,650.28	50	70
				Tb	3,650.40	100	50
				Al	3,651.06	(50)	—
				Mo	3,651.11	50	2
				Nb	3,651.19	400	10
				Fe	3,651.47	200	300

*N**

Ti 3,653.50 *J* 200 *O* 500

	λ	*I*			λ	*I*	
		J	*O*			*J*	*O*
Ca	**3,653.495**	**3**	—	**W**	**3,653.52**	**1**	**5**
Au	**3,653.491**	**5**	**3**	**Mo**	**3,653.55**	**4**	**4*d***
Sm	**3,653.48**	**2**	**15**	**Th**	**3,653.59**	**2**	**3**
Ta	**3,653.39**	**1**	**3**	**Yt**	**3,653.61**	**2**	**5**
P	**3,653.38**	**(100*w*)**	—	**Tm**	**3,653.61**	**20**	**30**
W	**3,653.34**	**12**	**1**	**Nb**	**3,653.615**	**5**	**10**
Ir	**3,653.32**	—	**3**	**Re**	**3,653.615**	—	**15**
Sr	**3,653.27**	**8**	**30**	**Pr**	**3,653.65**	**1**	**4**
U	**3,653.21**	**1**	**10**	**Ce**	**3,653.67**	**8**	**18**
Os	**3,653.20**	**5*W***	**10*W***	**Os**	**3,653.72**	**10**	**30**
Ir	**3,653.19**	**50**	**15**	**Ir**	**3,653.759**	—	**2*h***
Re	**3,653.16**	—	**4*h***	**Fe**	**3,653.763**	**10**	**25**
Nd	**3,653.15**	**4**	**10**	**Ta**	**3,653.83**	**1**	**3**
Sm	**3,653.113**	**2**	**10**	**Tb**	**3,653.87**	**8**	**15**
Ce	**3,653.108**	**5**	**14**	**Mo**	**3,653.90**	**2**	**3**
Se	**3,653.05**	**(25)**	—	**Cr**	**3,653.91**	**25**	**100**
Tb	**3,652.97**	**8**	**15**	**Sr**	**3,653.928**	**3**	**15**
Tl	**3,652.95**	**50**	**150**	**Au**	**3,653.93**	**2**	**5**
Er	**3,652.88**	**4**	**20**	**Kr**	**3,653.97**	**(250*h*)**	—
W	**3,652.76**	**4**	**3**	**Fe**	**3,653.98**	**2*h***	**4**
Fe	**3,652.73**	—	**6**	**Pt**	**3,653.99**	**1*h***	**2**
Eu	**3,652.65**	**4*wh***	**5*w***	**Ce**	**3,654.09**	—	**2**
Cr	**3,652.59**	—	**3**	**U**	**3,654.13**	**1**	**5**
Er	**3,652.58**	**3**	**20**	**Nd**	**3,654.16**	**10**	**10**
Ca	**3,652.56**	**2**	**3**	**Dy**	**3,654.17**	**1**	**4**
Gd	**3,652.560**	**25**	**20**	**W**	**3,654.20**	**10**	**12**
Co	**3,652.543**	—	**200*r***	**Nb**	**3,654.23**	**5*h***	—
Th	**3,652.542**	**10**	**10**	**U**	**3,654.29**	**4**	**3**
Mo	**3,652.474**	**25**	—	**Tb**	**3,654.30**	—	**8**
Fe	**3,652.47**	—	**4**	**Cu**	**3,654.30**	**2*h***	**10*wh***
Cu	**3,652.43**	**1*h***	**10*wh***	**Pr**	**3,654.34**	**2**	**8**
V	**3,652.42**	**30*h***	**40**	**Re**	**3,654.358**	—	**8**
Ta	**3,652.41**	—	**5**	**Er**	**3,654.363**	—	**4**
Pr	**3,652.38**	**3**	**20**	**Bi**	**3,654.38**	**5*s***	**7**
Mo	**3,652.33**	**25**	—	**Ru**	**3,654.40**	**40**	**3**
Ce	**3,652.325**	—	**2**	**Nb**	**3,654.423**	**10**	**10**
Ru	**3,652.316**	**5**	**2**	**Pd**	**3,654.425**	**4**	—
Nb	**3,652.27**	**3*h***	**1**	**Co**	**3,654.446**	—	**35**
Ce	**3,652.263**	—	**3**	**Ho**	**3,654.45**	**6**	**6**
Tb	**3,652.26**	**8**	**30**	**Th**	**3,654.47**	**1**	**4**
Pt	**3,652.25**	**1*h***	**2**	**Os**	**3,654.49**	**15**	**100**
				S	**3,654.51**	**(8)**	—
*P**				**Th**	**3,654.579**	**4**	**4**
				Mo	**3,654.583**	**20**	**25**
				Ce	**3,654.592**	—	**2**
				Ti	**3,654.592**	**40**	**100**
				Pr	**3,654.61**	—	**6**

	λ	I J	I O
Gd	**3,654.64**	**200**	**200*W***
Fe	**3,654.66**	**—**	**4**
Au	**3,654.67**	**3**	**3**
W	**3,654.71**	**12**	**—**
Hg	3,654.83	(200)	—
Al	3,655.00	(100)	—
Gd	3,656.16	200	200*W*
Os	3,656.90	30	150
Rh	3,657.99	200*W*	500*W*
Ti	3,658.10	60	150
Tb	3,658.88	100	100
Ar	3,659.50	(100)	—
Nb	3,659.61	500	15
Ti	3,659.76	150	50
Mn	3,660.40	75	75
Ti	3,660.63	18	90
Ru	3,661.35	100	60
V	3,661.38	150	10
Ti	3,662.24	100	40
Gd	3,662.27	200	200*w*
Hg	3,662.88	400	50
Hg	3,663.28	400	500
Ru	3,663.37	60	5
V	3,663.59	1*wh*	150
Ni	3,664.09	30	300
Ne	3,664.11	(250)	—
P	3,664.19	(100*w*)	—
Yt	3,664.61	100	100
Gd	3,664.62	200	200*w*
Kr	3,665.33	(80)	—
Kr	*3,669.01*	*(150h)*	*—*
V	*3,669.41*	*300*	*20W*
Fe	*3,669.52*	*150*	*200*
Fe	*3,670.07*	*200*	*200*
Ar	*3,670.64*	*(300)*	*—*
Ti	*3,671.67*	*70*	*150*
Yb	*3,675.09*	*200*	*50*
Ar	*3,675.22*	*(300)*	*—*
Tb	*3,676.35*	*200*	*100*
V	*3,676.68*	*150h*	*300*

Ti 4,981.73 *J* 125 *O* 300

	λ	I J	I O		λ	I J	I O
Sm	**4,981.71**	—	**50**	**Mo**	**4,981.83**	**3**	**10**
Re	**4,981.54**	—	**15**	**Ir**	**4,981.88**	—	**3**
Tl	**4,981.35**	**(15)**	—	**Dy**	**4,981.96**	—	**2*h***
Nd	**4,981.28**	—	**10**	**Pr**	**4,982.00**	—	**2**
Th	**4,980.95**	—	**8**	**Ra**	**4,982.03**	**(10)**	—
Nd	**4,980.89**	—	**2*h***	**Ce**	**4,982.13**	—	**8**
Hg	**4,980.82**	**(6)**	—	**Yt**	**4,982.14**	**50**	**8**
Er	**4,980.71**	—	**2**	**Fe**	**4,982.51**	—	**200**
Re	**4,980.69**	—	**4**	**W**	**4,982.60**	**5**	**40**
Tm	**4,980.67**	**3**	**5**	**Tb**	**4,982.78**	—	**2**
Hg	**4,980.57**	**(70)**	—	**Kr**	**4,982.83**	**(50*h*)**	—
Tb	**4,980.54**	—	**4**	**Na**	**4,982.84**	**100**	**200*wh***
Pr	**4,980.52**	—	**3*r***	**Os**	**4,982.89**	—	**6**
Pt	**4,980.38**	—	**4**	**Nd**	**4,982.90**	—	**2**
Sc	**4,980.37**	**8**	**6**	**Fe**	**4,983.26**	—	**100*h***
Ru	**4,980.35**	—	**60**	**Sm**	**4,983.37**	—	**15**
Sm	**4,980.26**	—	**4**	**Ru**	**4,983.449**	—	**7**
Th	**4,980.19**	—	**4**	**Sc**	**4,983.451**	**4**	**6**
Dy	**4,980.17**	—	**2*h***				
Ni	**4,980.161**	**1**	**500*W***	*N**			
Tb	**4,980.16**	—	**5**				
Cu	**4,980.01**	**4**	—				
Br	4,979.76	(125)	—				
Mo	4,979.11	30	100				
Kr	4,978.89	(100*h*)	—				
Rn	4,978.84	(300)	—				
Ti	4,976.82	—	15				
Se	4,975.66	(300)	—				
Ti	4,975.36	4	80				
Ne	4,974.76	(50)	—				
Ne	4,973.54	(100)	—				
Fe	4,973.11	—	100				
Ti	4,973.05	2	35				
Xe	4,972.71	(200*h*)	—				
Li	4,971.99	—	500				
Co	4,971.96	—	150				
Xe	4,971.71	(100*wh*)	—				
Ni	4,971.35	—	100				
La	4,970.39	100	125				
Cl	4,970.12	(50)	—				
P	4,969.64	(150)	—				
Gd	4,969.16	—	100				
O	4,968.76	(100)	—				
Ti	4,968.58	1	40				
O	4,967.86	(80)	—				

	λ	I J	I O
Nb	4,967.78	50	150
O	4,967.40	(50)	—
Co	4,966.59	—	100
Fe	4,966.10	1	300
V	4,965.40	40	—
Nb	4,965.37	15	100
Ar	4,965.12	(40)	—
Gd	4,965.06	10*h*	100
Ti	4,964.75	—	25
Kr	*4,960.25*	*(100h)*	—
Fe	*4,957.61*	*150*	*300*
Ne	*4,967.12*	*(150)*	—
Ne	*4,957.05*	*(1,000)*	—
Ar	*4,956.75*	*(100)*	—
Ne	*4,955.38*	*(150)*	—
La	*4,949.77*	—	*200*
Ti	*4,948.19*	—	*12*

Ti 4,991.07 *J* 100 *O* 200

	λ	I J	I O
Ce	**4,990.67**	—	**10**
Fe	**4,990.46**	—	**5**
U	**4,990.12**	—	**3**
Os	**4,990.10**	—	**6**
Th	**4,990.03**	—	**5**
Ar	**4,989.945**	**(80)**	—
Nd	**4,989.937**	—	**35**
U	**4,989.92**	**3*h***	—
W	**4,989.87**	—	**5**
Sm	**4,989.44**	—	**60**
Ti	4,989.15	4	100
Nb	4,988.97	10	150
Fe	4,988.96	—	100*h*
Xe	4,988.77	(150*h*)	—
Co	4,988.04	—	500*R*
J	4,986.93	(36)	—
La	4,986.83	100	150
As	4,985.60	50	—
Fe	4,985.56	—	100
Fe	4,985.26	—	100
Ni	4,984.13	1	500*W*
Fe	4,983.85	—	200*h*
*P**			

	λ	I J	I O
Xe	**4,991.17**	**(50*wh*)**	—
N	**4,991.22**	**(5)**	—
Fe	**4,991.277**	—	**80**
La	**4,991.284**	**80**	**100**
Tb	**4,991.40**	—	**3**
Hg	**4,991.5**	—	**15**
U	**4,991.54**	—	**2**
Sb	**4,991.66**	**(10)**	—
Pr	**4,991.77**	—	**3**
Sc	**4,991.92**	**8**	**8**
Se	**4,992.03**	**(12)**	—
Sm	**4,992.031**	—	**80**
*bh***F**	**4,992.1**	—	**50**
U	**4,992.12**	—	**2**
Mo	**4,992.24**	**3**	**8**
Pr	**4,992.31**	—	**2**
Tb	**4,992.32**	—	**2**
Er	**4,992.35**	—	**2**
Hf	**4,992.36**	**2**	**3*h***
Ce	**4,992.40**	—	**12**
Nb	**4,992.47**	**2**	**10**
Ru	**4,992.74**	—	**25**
Se	**4,992.75**	**(300)**	—
*N**			

Ti 4,999.51 *J* 80 *O* 200

	λ	*I*			λ	*I*	
		J	*O*			*J*	*O*
La	4,999.47	300	400	Ru	4,999.55	—	6
Fe	4,999.12	—	4	Hf	4,999.68	40	30
Gd	4,999.08	—	10*h*	Ir	4,999.74	2	6
Dy	4,999.00	—	2*h*	Mo	4,999.91	25	50
Nd	4,998.55	—	10	Th	4,999.94	1	8
Kr	4,998.54	(5*wh*)	—	Sm	5,000.05	—	25
Ne	4,998.50	(10)	—	Ni	5,000.33	—	150*w*
Dy	4,998.47	—	2*h*	Er	5,000.38	—	15
Gd	4,998.39	2	25	Ne	5,000.39	(3)	—
Ni	4,998.23	—	150	Nd	5,000.44	—	10
Ta	4,998.18	—	15*W*	*bh*F	5,000.5	—	10
Ce	4,998.13	—	18	W	5,000.504	—	7
Pt	4,997.97	—	6	Hf	5,000.54	2	2
Tb	4,997.93	—	5	Pr	5,000.59	—	3
				Sb	5,000.6	20	—
Fe	4,997.80	300	20	*bh*F	5,000.6	—	80
Ti	4,997.10	4	50	Nb	5,000.70	1	3
La	4,996.82	30	25	Te	5,000.87	(25)	—
Fe	4,995.63	60	3	Nb	5,000.95	5	30
Cl	4,995.52	(60)	—	Al	5,000.97	(15)	—
J	4,995.15	(25)	—	Xe	5,001.01	(15)	—
Ne	4,994.93	(150)	—	Ti	5,001.010	2	80
N	4,994.36	(30)	—	W	5,001.09	—	12
Fe	4,994.133	—	200	N	5,001.13	(150)	—
Lu	4,994.13	400	250	Lu	5,001.14	—	100
Bi	4,993.92	(20)	—	Sm	5,001.23	—	40
S	4,993.51	(150)	—				

*P** *N**

Ti 5,007.21 *J* 40 *O* 200

	λ	*J*	*O*		λ	*J*	*O*
Ar	5,007.09	(2)	—	W	5,007.23	2	15
Yt	5,006.97	2	8	Er	5,007.24	—	15
Ar	5,006.84	(2)	—	Fe	5,007.29	—	25*h*
Cu	5,006.79	30	—	*bh*Mg	5,007.3	—	12
*bh*F	5,006.6	—	30	Co	5,007.30	—	5
W	5,006.16	7	40	N	5,007.32	(150)	—
Fe	5,006.13	5	300	Dy	5,007.32	—	5
Ir	5,005.78	—	2	Mo	5,007.62	2	10
Fe	5,005.72	—	200	Pr	5,007.63	—	4
Ce	5,005.70	—	10	Nb	5,008.04	1	5
Se	5,005.69	(8)	—	*bh*F	5,008.1	—	2
V	5,005.60	12	12	Th	5,008.19	1	8
Pb	5,005.43	4	20	U	5,008.22	25	30
				J	5,008.36	(5)	—
				Tb	5,008.63	—	10

	λ	I	
		J	O
Ne	5,005.33	(50)	—
Ne	5,005.16	(500)	—
N	5,005.14	(500)	—
Fe	5,004.79	100	3
Ar	5,004.32	(20)	—
Br	5,002.70	(40)	—
V	5,002.33	90	90
F	5,001.98	(30)	—
Fe	5,001.87	40	300
N	5,001.47	(200)	—
*P**			

	λ	I	
		J	O
U	**5,008.69**	—	**4**
Ar	5,009.35	(100)	—
Yb	5,009.52	50	20
Tm	5,009.76	50	50
Cu	5,009.83	20	—
Sb	5,010.42	(40)	—
N	5,010.62	(100)	—
Ne	5,011.00	(25)	—
Fe	5,012.07	—	300
Cu	5,012.61	20	—
Xe	5,012.83	(50*h*)	—
Ba	5,013.08	50	—
Eu	5,013.14	1	125
Kr	5,013.29	(100)	—
Ti	5,014.19	15	40
Ti	5,014.24	30	100
La	5,014.45	30*h*	2
Mo	5,014.59	20	30
V	5,014.62	125	125
Fe	5,014.96	—	500
Gd	5,015.06	—	100*W*
He	5,015.67	(100)	—
Ti	5,016.17	15	100
N	5,016.39	(70)	—
Ar	5,017.16	(60)	—
Ni	5,017.59	1	100*w*
Fe	5,018.44	50	80
O	5,018.78	(30)	—
Se	5,019.32	(25)	—
O	5,019.34	(50)	—
V	5,019.86	35	—
Ti	5,020.03	80	100
O	5,020.13	(70)	—
Br	5,020.58	(30)	—
Cu	5,021.28	20	—
Kr	5,021.88	(100)	—
Fe	5,022.25	—	150
Kr	5,022.40	(200)	—
Ti	5,022.866	18	100
Ne	5,022.870	(25)	—
Eu	5,022.90	—	125
Fe	5,023.48	300	10
Ti	*5,024.84*	*15*	*100*
Ti	*5,025.58*	*8*	*100*
N	*5,025.66*	*(100)*	—

	λ	I	
		J	O

	λ	I	
		J	O
Xe	5,028.28	(200)	—
Sm	5,028.44	—	200
Eu	5,029.48	—	500w
Fe	5,030.78	125	1
Sc	5,031.02	200h	50
Se	5,031.26	(40)	—
Ne	5,031.35	(250)	—
Ar	5,032.02	(60)	—
Kr	5,033.85	(100wh)	—
Tm	5,034.21	100	100
Ni	5,035.37	5	300w
Ti	5,035.91	30	125
Ti	5,036.47	25	125
Ne	5,037.75	(500)	—
Ti	5,038.40	20	100
Br	5,038.77	(60)	—
Nb	5,039.04	30	200
Ti	5,039.95	25	125
Ti	5,040.62	40	40
P	5,040.74	(70)	—
Hf	5,040.82	150	100

Tl 81 204.37

t_0 303.5°C t_1 1,450°C

I.	II.	III.	IV.	V.
6.106	20.423	29.8	50.8	–

λ	I		eV
	J	*O*	
I 2,767.87	300*R*	400*R*	4.44
I 2,918.32	200*R*	400*R*	5.18
I 3,229.75	800	2000	4.77
I 3,519.24	1000*R*	2000*R*	4.49
I 3,775.72	1000*R*	3000	3.28
I 5,350.46	2000*R*	5000*R*	3.28
I 6,549.77	50	300	5.14

Tl 2,767.87 *J* 300*R* *O* 400*R*

	λ	*I*			λ	*I*	
		J	*O*			*J*	*O*
U	**2,767.783**	**2*h***	**2**	**Bi**	**2,767.88**	—	**4**
Ag	**2,767.783**	**1*h***	**5**	**Ar**	**2,767.99**	**(2)**	—
Re	**2,767.75**	—	**6**	**Mo**	**2,768.09**	—	**10**
Cl	**2,767.74**	**(4)**	—	**K**	**2,768.1**	**(2)**	—
Rh	**2,767.73**	**4**	**100**	**Fe**	**2,768.11**	**8**	**35**
Ir	**2,767.65**	**5*h***	**6**	**Nb**	**2,768.128**	**100**	**10**
Cr	**2,767.54**	**8*d***	**30**	**V**	**2,768.131**	**18**	**12**
Fe	**2,767.523**	—	**300**	**Cr**	**2,768.15**	**10**	—
Ag	**2,767.523**	**200**	**30**	**U**	**2,768.17**	**2*d***	**3*d***
Fe	**2,767.50**	**400*wh***	**10**	**Rh**	**2,768.23**	**4**	**50**
Cd	**2,767.49**	**(2)**	—	**Yb**	**2,768.27**	**2**	—
U	**2,767.41**	**4**	**5**	**Re**	**2,768.28**	—	**3**
La	**2,767.40**	**8**	**4**	**Co**	**2,768.29**	—	**9**
				Ta	**2,768.307**	**2**	**10**
				V	**2,768.314**	—	**8**
Rh	2,766.54	150	5	**W**	**2,768.33**	**20**	**10**
Cr	2,766.540	300*r*	40	**Fe**	**2,768.337**	**1**	—
Cu	2,766.37	25	500	**Ce**	**2,768.337**	**3*w***	**5**
V	2,765.67	200*h*	50				
Ru	2,765.44	150	50				
Cr	2,764.35	6	200*r*	V	2,768.56	150*R*	35
Pd	2,763.09	30*r*	300*r*	Ni	2,768.78	250*wh*	—
				Ru	2,768.93	200	60
				Cu	2,768.67	400	5*h*
Ni	*2,759.02*	*500wh*	—	Cr	2,769.91	40	400*r*
				Zn	2,770.86	25	300
				Zn	2,770.98	150	300
				Pt	2,771.67	15	500
				Fe	2,772.11	300	300
				Ru	2,772.45	150	—
				Lu	2,772.58	150*h*	5
				Ho	*2,774.70*	*300*	—

Tl 2,918.32 *J* 200*R* *O* 400*R*

	λ	*J*	*O*		λ	*J*	*O*
Tm	**2,918.27**	**50**	**25**	**Fe**	**2,918.36**	**25**	**40**
W	**2,918.25**	**8**	**12**	**Au**	**2,918.37**	**25**	—
Zr	**2,918.24**	**15**	**15**	**Ca**	**2,918.42**	**2*s***	—
V	**2,918.21**	**15*h***	—	**Gd**	**2,918.44**	**3**	**2**
Fe	**2,918.16**	**2**	**3**	**Ru**	**2,918.50**	**60**	—
B	**2,918.15**	**2**	—	**Nb**	**2,918.56**	**10**	—
Ir	**2,918.035**	—	**12**	**Ir**	**2,918.57**	**2**	**18**
Fe	**2,918.027**	**100**	**125**	**Hf**	**2,918.58**	**8**	**30**
V	**2,917.93**	**2*h***	**12**	**W**	**2,918.63**	**20**	**8**
Dy	**2,917.92**	—	**2**	**Ce**	**2,918.66**	—	**30*s***

	λ	I J	I O		λ	I J	I O
Ce	**2,917.89**	—	**2**	**Dy**	**2,918.67**	**1**	**2**
Fe	**2,917.87**	—	**2**	**U**	**2,918.68**	**10**	**5**
Os	**2,917.83**	**3**	**10**	**Ti**	**2,918.78**	**10*wh***	—
Rn	**2,917.8**	**(3)**	—	**Ce**	**2,918.781**	—	**10**
Th	**2,917.79**	**6**	**4**	**Fe**	**2,918.82**	**8**	**15**
Ru	**2,917.774**	**2**	**60**	**Mo**	**2,918.83**	**25**	**15**
Ir	**2,917.770**	—	**10**	**U**	**2,918.87**	**2**	**3**
J	**2,917.73**	**(15)**	—	**Re**	**2,918.88**	—	**10**
Ce	**2,917.69**	—	**2**	**Nb**	**2,918.92**	**10*h***	**1**
Nb	2,917.05	100*r*	10*w*	Ag	2,920.04	100*wh*	—
Ta	2,914.12	30	200	V	2,920.38	125*r*	20
Sn	2,913.542	125*wh*	100*wh*	Tl	2,921.52	100*R*	200*R*
Pt	2,913.542	25	300	Pd	2,922.49	25	200
Ne	2,913.17	(150)	—	V	2,923.62	150*r*	50*r*
Pt	2,912.26	25	300	V	2,924.02	300*R*	70*r*
Fe	2,912.16	150	150				
Lu	*2,911.39*	*300*	*100*	*V*	*2,924.64*	*200r*	*60*
V	*2,911.06*	*200r*	*30*	*Fe*	*2,926.59*	*400*	*150*
Os	*2,909.06*	*400*	*500R*	*Ru*	*2,927.54*	*200*	*50*
V	*2,908.82*	*400R*	*70r*	*Nb*	*2,927.81*	*800r*	*200*
Nb	*2,908.24*	*200*	*20r*	*Pt*	*2,929.79*	*200w*	*800R*
Eu	*2,906.68*	*300*	*300W*				

Tl 3,229.75 *J* 800 *O* 2000

	λ	I J	I O		λ	I J	I O
Zr	**3,229.72**	—	**2*h***	**Fe**	**3,229.791**	**1**	**1**
Mn	**3,229.711**	—	**2**	**Mo**	**3,229.794**	**4**	**25**
Mo	**3,229.710**	**50**	—	**Cr**	**3,229.866**	**8**	—
Be	**3,229.69**	—	**15**	**Nd**	**3,229.872**	**10**	—
W	**3,229.66**	**8**	**9**	**Fe**	**3,229.873**	**10**	**10**
V	**3,229.61**	**8**	**15**	**Ta**	**3,229.88**	**50**	**35*h***
Ce	**3,229.60**	—	**10**	**Ar**	**3,229.91**	**(3)**	—
Sm	**3,229.595**	**2**	**8**	**Er**	**3,229.93**	**2**	**10**
Fe	**3,229.593**	**5**	**10**	**Dy**	**3,229.94**	**2**	**8*d***
Tb	**3,229.59**	—	**8**	**Ag**	**3,229.99**	**4*h***	**1**
Nb	**3,229.56**	**50**	**5**	**J**	**3,229.993**	**(15)**	—
Te	**3,229.52**	**(25)**	—	**Fe**	**3,229.994**	**20**	**20**
U	**3,229.502**	**25**	**18**	**Ce**	**3,230.02**	—	**3**
Rn	**3,229.5**	**(5)**	—	**Tb**	**3,230.03**	**8**	**15**
Ne	**3,229.50**	**(7)**	—	**Hf**	**3,230.06**	—	**10**
Ti	**3,229.42**	**70**	**15**	**Ce**	**3,230.08**	**1**	**10**
Mg	**3,229.371**	—	**25**	**Er**	**3,230.12**	**1**	**4**
Dy	**3,229.37**	**2**	**10**	**Tm**	**3,230.15**	**20**	**10**
Ce	**3,229.363**	**3**	**25**	**Ne**	**3,230.16**	**(18)**	—
Mo	**3,229.36**	**5**	—	**Fe**	**3,230.21**	**80**	**100**

	λ	I J	I O		λ	I J	I O
Ir	3,229.28	2	35	Nb	3,230.24	30	2
Ta	3,229.24	70*w*	300*w*	Pt	3,230.287	6	100
Os	3,229.21	5	125	Ce	3,230.288	—	8
Cr	3,229.20	—	35	Dy	3,230.32	—	2
Ti	3,229.193	60	30	Os	3,230.40	5	40
Tb	3,229.19	8	15	Si	3,230.43	3	—
Fe	3,229.123	60	80	Sm	3,230.54	30	100
Ce	3,229.122	1	25	Rh	3,230.55	(15)	—
Rb	3,229.11	—	10*h*	Ho	3,230.57	6	6
U	3,229.07	—	6	J	3,230.572	(10)	—
Xe	3,229.03	(3*h*)	—	Eu	3,230.58	—	4
F	3,229.00	(2*h*)	—	Er	3,230.585	15	25
Dy	3,228.97	1	9	W	3,230.59	10*s*	1
Th	3,228.969	15	12				
W	3,228.96	10	1	Li	3,232.61	500	1000*R*
Nb	3,228.95	3	1	Ti	3,234.52	500*r*	100
Tm	3,228.91	30	15				
Fe	3,228.90	40	80	*Pd*	*3,242.70*	*600R*	*2000wh*
				Tl	*3,243.68*	*(12)*	—
Nb	3,225.48	800*wr*	150*w*				

Tl 3,519.24 *J* 1000*R* *O* 2000*R*

	λ	J	O		λ	J	O
P	3,519.22	(15)	—	Nb	3,519.33	3	2
Bi	3,519.18	—	10	Cr	3,519.45	—	6*h*
Os	3,519.176	8	15	Rh	3,519.54	2	40
V	3,519.167	—	10	Zr	3,519.60	10	100
Pr	3,519.13	2	10	Ru	3,519.63	30	70
Er	3,519.09	2*rh*	20*rh*	Nb	3,519.649	20	5
Ce	3,519.08	4	25	Ho	3,519.65	4*h*	—
Ru	3,518.98	3	30	Th	3,519.69	6	3*d*
Tb	3,518.96	—	15	Ce	3,519.74	—	18
U	3,518.95	—	3	Tb	3,519.76	15	50
Os	3,518.94	5	20	Ni	3,519.766	30	500*h*
Ir	3,518.91	—	2*h*	Dy	3,519.770	1*h*	2
Eu	3,518.90	—	5	Nd	3,519.77	2	10*d*
Th	3,518.89	3	3*d*	Co	3,519.82	25*h*	2*h*
Fe	3,518.88	2	10	Ho	3,519.92	10*h*	10
Zr	3,518.87	1*h*	4	Ce	3,519.93	—	3
Hf	3,518.75	15	5	Ti	3,519.94	—	5
Ce	3,518.73	—	3	U	3,519.96	12	6
Os	3,518.72	30	200	Os	3,519.998	20	30
Dy	3,518.71*m*	2	4	Ar	3,520.00	(15)	—
Ca	3,518.706	2*h*	—	V	3,520.02	50	5
Fe	3,518.685	1	7	Cu	3,520.031	10	30
Th	3,518.685	—	3	Er	3,520.034	4	12
Sm	3,518.682	—	5	Nb	3,520.05	20	20

	λ	I J	I O		λ	I J	I O
Ni	**3,518.62**	**8**	**90**	**Sb**	**3,520.07**	**2*h***	**4**
P	**3,518.60**	**(50*h*)**	**—**	**Co**	**3,520.08**	**—**	**100*W***
Mo	**3,518.57**	**3**	**1**	**Ce**	**3,520.10**	**1**	**2**
W	**3,518.55**	**5**	**—**	**Ru**	**3,520.13**	**40**	**60**
Pr	**3,518.51**	**—**	**4**	**Eu**	**3,520.14**	**4**	**4**
Ce	**3,518.49**	**—**	**3**	**Ho**	**3,520.16**	**10*h***	**8**
Eu	**3,518.48**	**10**	**20**	**Mo**	**3,520.20**	**20**	**1**
W	**3,518.478**	**7**	**10**	**Ce**	**3,520.247**	**—**	**4**
W	**3,518.478**	**—**	**10**	**Ti**	**3,520.253**	**28**	**10**
U	**3,518.47**	**—**	**2**	**Yb**	**3,520.29**	**70**	**10**
Th	**3,518.404**	**2**	**6**	**Zr**	**3,520.30**	**—**	**2**
Cr	**3,518.40**	**1**	**8**				
Tb	**3,518.38**	**—**	**8**	Ne	3,520.47	(1000)	—
Hf	**3,518.38**	**3**	**—**	Ni	3,524.54	100*wh*	1000*R*
Ce	**3,518.37**	**1**	**12**	Rh	3,528.02	150	1000*w*
Co	**3,518.35**	**100**	**200*W***	Tl	3,529.43	800	1000
Sm	**3,518.30**	**2**	**1**	Co	3,529.81	30	1000*R*
Mo	**3,518.22**	**15**	**6**				
Nb	**3,518.178**	**4**	**3**				
Er	**3,518.176**	**8**	**20**				
Yb	**3,518.179**	**30**	**6**				
Pd	3,516.94	500*R*	1000*R*				
Ni	3,515.05	50*h*	1000*R*				
Ni	3,510.34	50*h*	900*R*				
Co	*3,502.28*	*20*	*2000R*				

Tl 3,775.72 *J* 1000*R* *O* 3000

	λ	J	O		λ	J	O
V	**3,775.719**	**2**	**30**	**Bi**	**3,775.74**	**—**	**2**
Rh	**3,775.716**	**3**	**8**	**Sm**	**3,775.85**	**—**	**2**
Hg	**3,775.70**	**(2)**	**—**	**Ca**	**3,775.86**	**2**	**2**
Eu	**3,775.69**	**4*W***	**4*W***	**Fe**	**3,775.87**	**—**	**2**
Er	**3,775.66**	**2**	**8**	**Th**	**3,775.94**	**10**	**20**
Mo	**3,775.65**	**20**	**15**	**Ce**	**3,775.986**	**15**	**12**
U	**3,775.61**	**5**	**12**	**U**	**3,775.991**	**12**	**10**
Ni	**3,775.57**	**40*h***	**500*h***	**Ba**	**3,776.0**	**7**	**3**
Ce	**3,775.56**	**—**	**4**	**Dy**	**3,776.01**	**—**	**2**
Nd	**3,775.50**	**20**	**10**	**Tb**	**3,776.02**	**—**	**8**
Re	**3,775.464**	**—**	**10**	**Ti**	**3,776.059**	**60**	**8**
Zr	**3,775.461**	**1*h***	**7**	**Sm**	**3,776.06**	**—**	**15**
Sm	**3,775.45 9**	**3**	**10**	**Pr**	**3,776.09**	**3**	**8**
Ar	**3,775.45**	**(10)**	**—**	**Mo**	**3,776.10**	**5**	**5**
Nb	**3,775.448**	**10**	**5**	**Ce**	**3,776.15**	**5**	**5**
W	**3,775.446**	**7*s***	**8**	**V**	**3,776.157**	**2**	**50**

	λ	I J	I O		λ	I J	I O
U	**3,775.44**	**4**	**4**	**Nb**	**3,776.161**	**5**	**2**
Eu	**3,775.41**	—	**4*W***	**Eu**	**3,776.20**	—	**10*w***
Ce	**3,775.36**	**4**	**2**	**Os**	**3,776.25**	**15**	**50**
Th	**3,775.32**	**5**	**20**	**Hg**	**3,776.26**	**(30)**	—
U	**3,775.262**	**3**	**6**	**Th**	**3,776.28**	**1**	**8**
Tb	**3,775.26**	**8**	**15**	**Cd**	**3,776.32**	**3**	—
Nd	**3,775.24**	**8*d***	**15*d***	**Nd**	**3,776.34**	**12**	**15**
Pr	**3,775.211**	**4**	**10**	**Fe**	**3,776.46**	**70**	**125**
Ir	**3,775.206**	—	**8**	**Tb**	**3,776.49**	**100**	**100**
V	**3,775.19**	**5**	**25**	**Eu**	**3,776.52**	**4*W***	**4*W***
P	**3,775.02**	**(30)**	—	**Mn**	**3,776.53**	**25**	**25**
Tb	**3,774.96**	—	**8**	**Mo**	**3,776.55**	**8**	**5**
Nd	**3,774.93**	**6**	**10**	**Yt**	**3,776.56**	**12**	**12**
U	**3,774.91**	**1**	**10**	**Nb**	**3,776.60**	**3**	**3**
Dy	**3,774.84**	**1*h***	**4**	**Ce**	**3,776.61**	**3**	**6**
Fe	**3,774.83**	**40**	**100**	**W**	**3,776.69**	**2*h***	—
Si	**3,774.74**	**4**	—	**Ce**	**3,776.71**	—	**2**
Sm	**3,774.68**	—	**15*W***	**U**	**3,776.74**	**5*h***	**3**
V	**3,774.67**	**12**	**1**	**S**	**3,776.80**	**(3)**	—
Ti	**3,774.65**	**1**	**2**	**V**	**3,776.88**	**2**	**10**
Mn	**3,774.64**	—	**20**	**Ir**	**3,776.93**	**2**	**10**
Os	**3,774.62**	**12**	**60**	**Os**	**3,776.99**	**20**	**150**
Co	**3,774.60**	—	**200*W***	**Fe**	**3,777.06**	**7**	**12**
Pr	**3,774.59**	**2**	**10**	**Co**	**3,777.078**	**2**	**2**
Ar	**3,774.54**	**(3)**	—	**Zr**	**3,777.080**	**10**	**15**
Hg	**3,774.52**	**(30)**	—	**Sm**	**3,777.088**	**10**	**30**
Nb	**3,774.44**	**5**	**3**	**Er**	**3,777.089**	**10*h***	**25**
				Ce	**3,777.091**	**15**	**6**
				Kr	3,778.09	(500*h*)	—
				Kr	3,783.13	(500*h*)	—
				Tl	3,793.95	(25)	—
				Mo	3,798.25	1000*R*	1000*R*

Tl 5,350.46 *J* 2000*R* *O* 5000*R*

	λ	J	O		λ	J	O
W	**5,350.445**	—	**18**	**Ar**	**5,350.58**	**(20)**	—
Er	**5,350.44**	—	**12**	**Sm**	**5,350.62**	—	**15**
Te	**5,350.41**	**(8)**	—	**Nb**	**5,350.74**	**50**	**150**
Gd	**5,350.406**	—	**25**	**Zr**	**5,350.90**	—	**2**
Eu	**5,350.40**	—	**60*h***	**Nb**	**5,351.04**	**2**	**3**
Re	**5,350.39**	—	**2*w***	**Ti**	**5,351.08**	**60**	**50**
V	**5,350.38**	**3*h***	—	**Th**	**5,351.13**	—	**8**
Zr	**5,350.35**	**5**	**4**	**N**	**5,351.21**	**(30)**	—
Sc	**5,350.30**	—	**3**	**Yb**	**5,351.32**	**3**	**50**
Zr	**5,350.09**	**5**	**4**	**Sm**	**5,351.56**	**3**	**15*d***

	I	I J	I O		λ	I J	I O
Xe	5,350.03	(2)	—	Eu	5,351.67	—	150
U	5,349.92	3	5	W	5,351.90	—	20
Mn	5,349.88	—	20	Zr	5,351.91	—	3
W	5,349.86	—	6	J	5,351.95	(2)	—
Mo	5,349.79	4	6	Co	5,352.05	—	500*w*
Fe	5,349.74	—	4*h*	*bh*F	5,352.1	—	10
Sc	5,349.70	—	6	Dy	5,352.12	—	4
J	5,349.65	(15)	—	Er	5,352.21	—	8
Ba	5,349.621	—	7	Os	5,352.25	—	20
Gd	5,349.62	—	8	Sm	5,352.28	—	2
Nd	5,349.58	—	4	U	5,352.32	3*h*	3*h*
Ta	5,349.57	—	30	Mo	5,352.347	4	10
Eu	5,349.474	—	2	Tb	5,352.35	—	10
Ca	5,349.474	12	12	Pr	5,352.40	2	80
Cs	5,349.31	(15)	—				
Rh	5,349.308	—	20				
Sc	5,349.29	—	30				
Nd	5,349.26	—	2				
Re	5,349.25	—	3				
Ne	5,349.21	(150)	—				
Cs	5,349.16	(25)	—				
Sm	5,349.14	—	25				
Lu	5,349.12	2	25				
Ta	5,349.093	—	80				
Co	5,349.087	—	80				
Cs	5,348.95	(25)	—				
W	5,348.949	—	30				
Mo	5,348.81	1*h*	4*h*				
Sm	5,348.75	—	20				
Gd	5,348.69	—	20				
Ne	5,341.09	(1000)	—				

Tl 6,549.77 *J* 50* *O* 300

*bh*Zr	6,549.7	—	2	Sm	6,549.77	—	100*d*
Nd	6,549.54	—	10	U	6,549.88	—	4
Gd	6,549.24	—	25	Hf	6,550.01	10	1
La	6,549.18	—	3	Yb	6,550.19	—	10
Eu	6,549.14	—	20*W*	Nd	6,550.191	—	8*wh*
Hf	6,548.72	10	1	Sr	6,550.25	10	100
Dy	6,548.28	—	3	Zr	6,550.54	—	18
				Hg	6,550.97	—	20
				Cu	6,550.98	—	3
Fe	6,546.24	50	150	Sm	6,551.04	—	5
Sm	6,542.76	—	200				
				Tm	6,551.19	—	10

	λ	I (J)	I (O)
Tl	6,552.63	(10)	200
Ti	6,556.07	—	150
Co	6,563.42	5	200*w*
Eu	*6,567.87*	—	*600*
Sm	*6,569.31*	—	*500d*

U 92 238.03

t_0 1,133°C t_1 3,500°C

I.	II.	III.	IV.	V.
4	—	—	—	—

λ	I		eV
	J	O	
3,552.17	12	8	<3.5
II 3,670.07	18	15	3.49
3,672.58	15	8	<3.4
II 3,859.58	30	20	3.24
II 4,090.14	40	25	3.24
II 4,241.67	50	40	3.49
II 5,492.97	50	60	2.26

U 3,552.17 *J* 12 *O* 8

	λ	*I* *J*	*I* *O*
Te	3,552.15	(15)	—
Fe	3,552.12	6	10
Ir	3,552.10	2	20
Pr	3,552.07	2	10
Ce	3,552.067	2	5
Al	3,552.00	(2)	—
Dy	3,551.99	—	4
Nb	3,551.961	5	2
Pr	3,551.96	—	8*hd*
Tb	3,551.96	—	15
Zr	3,551.95	40	30
Th	3,551.87	3	2*d*
As	3,551.82	10	—
Er	3,551.795	1	5*s*
Yt	3,551.790	2	6
Ce	3,551.77	1	10
U	3,551.755	3	2
Nd	3,551.748	6	15
Co	3,551.67	—	2
Ce	3,551.66	1	10
Ho	3,551.60	4	—
Re	3,551.595	—	25
Dy	3,551.59	6	10
W	3,551.540	10	—
V	3,551.537	12	25
Ni	3,551.53	12	50
Ce	3,551.43	1	10
Pr	3,551.40	2	8
Th	3,551.39	1	2*d*
Pt	3,551.37	15	—
Re	3,551.31	—	40*W*
Er	3,551.29	2	20
Eu	3,551.28	1*h*	3*d*
W	3,551.276	5	6
U	3,551.269	5	3
Dy	3,551.16	1	25
P	3,551.15	(30)	—
Fe	3,551.120	—	2
Ta	3,551.118	18	—
Nb	3,551.11	10*h*	5
Rn	3,551.10	(5)	—
Pr	3,551.05	2*d*	5*d*
U	3,551.04	6	6
Tb	3,551.03	3	30
W	3,551.028	10	7
Mo	3,550.96	6	4
J	3,552.19	(18)	—
Sm	3,552.195	—	2
Nb	3,552.22	1	2
Ce	3,552.24	—	2
Sm	3,552.28	3	15
Er	3,552.317	—	15
W	3,552.323	43	5
Yb	3,552.342	15	15
Rh	3,552.36	(3)	—
Ce	3,552.37	2	2
Mo	3,552.40	3	3
Ba	3,552.45	8	1
Eu	3,552.52	2	6
Zr	3,552.66	4	5
U	3,552.67	10	5
Pr	3,552.68	6	15
Yt	3,552.69	4	6
Hf	3,552.70	35	20
Mo	3,552.720	5	3
Co	3,552.721	—	20
Mn	3,552.721	—	12
Ce	3,552.73	10	18*s*
Th	3,552.75	1	2*d*
J	3,552.76	(5)	—
V	3,552.81	10	7
Fe	3,552.83	50	80
Cr	3,552.85	5*wh*	2*h*
Tb	3,552.92	—	8
Ce	3,552.93	—	3
Cr	3,552.95	—	3*wh*
Co	3,552.99	—	20
U	3,552.01	2	5
Pd	3,552.08	15*Wh*	100*R*
Th	3,553.11	6	5
Co	3,553.16	—	5*h*
Dy	3,553.19	2	3
Er	3,553.20	15	25
V	3,553.27	30	80
Pr	3,553.34	2	4
Th	3,553.38	4	4
Sr	3,553.40	—	4
Ta	3,553.41	2	7
U	3,553.44	—	15
Ni	3,553.48	10	50
Kr	3,553.49	(20*h*)	—
Mg	3,553.51	—	8

	λ	I	
		J	O
Os	3,550.94	10	6
W	3,550.84	8	8
U	3,550.82	20	12
La	3,550.81	4	10
Os	3,550.71	10	15
W	3,550.68	6	7
Cr	3,550.635	60	70
Ru	3,550.630	5	4
Co	3,550.59	—	200
V	3,550.50	10	—
Zr	3,550.46	4	35
Nb	3,550.45	30	40
Th	3,550.29	4	4
Ru	3,550.27	8	4
Nb	3,550.24	5	4
Dy	3,550.23	100	200
Sm	3,550.222	2	25
Er	3,550.222	8	12
Pr	3,550.22	8*rd*	30*rd*
Tm	3,550.16	15	5
Rh	3,549.99	2	10
Re	3,549.89	—	40
Fe	3,549.87	5	15
Xe	3,549.86	(10)	—
Er	3,549.86	20*s*	25
Yb	3,549.82	20	10
Dy	3,549.80	—	6
Th	3,549.741	4	4
Zr	3,549.736	3	15
S	3,549.72	(8)	—
Pr	3,549.72	1	5
Os	3,549.68	12	20
Mg	3,549.61	—	4
Er	3,549.55	1	7
Rh	3,549.54	50	150
Pr	3,549.53	3	10
Zr	3,549.51	10	10
Hg	3,549.42	(200)	—
Nd	3,549.38	20	—
Gd	3,549.36	125	125
Ir	3,549.29	2	8
Cr	3,549.28	—	5*wh*
Nb	3,549.26	8	8
Dy	3,549.25	2	4
U	3,549.20	1	12
Ce	3,549.12	1	6
W	3,549.053	25	5
Ta	3,549.046	3	35
Dy	3,553.55	—	4
Tb	3,553.46	—	8
Nd	3,553.57	30	30
Au	3,553.572	10	20
Ar	3,553.58	(15)	—
Nb	3,553.62	15	20*W*
Re	3,553.65	—	25*W*
Fe	3,553.741	100	100
Gd	3,553.744	—	15
U	3,553.81	2	1
Mo	3,553.83	1	10
Ru	3,553.848	10	4
Pr	3,553.849	2	4
Cr	3,553.97	6	6
Nd	3,553.99	4	10
Hf	3,554.00	—	5
Xe	3,554.04	(10)	—
Os	3,554.05	10	8
Zr	3,554.07	6	10
Fe	3,554.120	20	50
Nb	3,554.125	30	—
Sm	3,554.17	3	25
Mo	3,554.19	10	5
W	3,554.21	8	9
Er	3,554.29	—	9
Ar	3,554.31	(300)	—
Pr	3,554.38	3	5
He	3,554.39	(7)	—
Lu	3,554.43	150	50
Fe	3,554.51	2	4
Nb	3,554.52	20	15
Os	3,554.54	5	8
Ce	3,554.63	2	10
Nb	3,554.66	20	15
Er	3,554.67	1	6*d*
Dy	3,554.83	4	4
Fe	3,554.93	300	400
Ce	3,554.99	—	25
V	3,555.14	3	10
Ce	3,555.16	2	5
W	3,555.17	15	6
Pr	3,555.20	4*rd*	10*rd*
Dy	3,555.27	—	6
Tb	3,555.29	3	15
U	3,555.32	6	15
Eu	3,555.37	1*h*	4*d*
Kr	3,555.54	(8*wh*)	—
Ir	3,555.60	2	10

	λ	I J	I O		λ	I J	I O
Yt	3,549.01	50	12	U	3,555.68	2	4
				Tb	3,555.70	—	8
Mo	3,548.95	—	6	Os	3,555.71	8	10
Eu	3,548.93	—	4	Nd	3,555.72	15	100
U	3,548.85	3*h*	4				
Ce	3,548.83	1	8	Ta	3,555.73	2*h*	10
Tb	3,548.82	3	15	V	3,555.74	1	15
				W	3,555.752	7	8
Hf	3,548.81	2	5	Th	3,555.755	3	12
Kr	3,548.71	(6)	—	Cr	3,555.788	6	20
Ir	3,548.65	1*h*	25				
U	3,548.62	—	8	Ce	3,555.788	1	4
Ar	3,548.51	(25)	—	Tm	3,555.82	—	6
				Co	3,555.94	6	—
Eu	3,548.51	1*h*	4*w*	Dy	3,555.96	4	6
Pt	3,548.49	8	—	Ar	3,555.97	(100)	—
Tm	3,548.48	20	8				
Co	3,548.44	6	18	Os	3,555.973	12	30
U	3,548.42	—	5	Nb	3,556.017	10	5
				W	3,556.02	6	—
W	3,548.26	7	5	Ce	3,556.10	—	8
Er	3,548.23	1	8*d*	Cr	3,556.13	4	20
Mn	3,548.202	40	40*h*				
Dy	3,548.202	2	8	Ti	3,556.18	1	15
Ni	3,548.18	25	400	V	3,556.25	2	40
				Mo	3,556.33	25	—
Ce	3,548.17	5	10*h*	Ce	3,556.36	1	5
Sr	3,548.093	—	50	P	3,556.48	(100)	—
Pr	3,548.09	5*d*	4*d*				
Nb	3,548.089	5*d*	5	Zr	3,556.60	50	15
Fe	3,548.089	—	9	Tb	3,556.67	3	15
				Fe	3,556.70	1	4
Eu	3,548.03	1*h*	4	Sm	3,556.73	3	35
Mn	3,548.029	15	40*h*	Mo	3,556.76	40	15
Fe	3,548.02	7	10				
Cr	3,547.98	—	5*wh*	V	3,556.80	40*wh*	30
Dy	3,547.89	—	4	Eu	3,556.85	—	10*w*
				Fe	3,556.88	150	300
Os	3,547.88	8	1	Ce	3,556.89	—	6
Mn	3,547.80	15	40*h*	Ar	3,556.91	(10)	—
Ba	3,547.77	6	10				
Zr	3,547.68	12	200	Mo	3,556.94	25	2*d*
Dy	3,547.53	2	6	Gd	3,557.06	25	25
				Er	3,557.11	1	4
Er	3,547.51	1*h*	25	Pr	3,557.14	5	15
W	3,547.47	6	8	V	3,557.16	—	8
Fe	3,547.20	8	25				
U	3,547.19	12	10	Ir	3,557.17	30	25
Eu	3,547.11	—	15*W*	W	3,557.21	8	8
				La	3,557.25	8	10
Ti	3,547.03	12	30	Fe	3,557.30	—	15
Ce	3,547.00	3	15	Sm	3,557.36	2	30
V	3,546.96	20	—				
Dy	3,546.84	10	50	Sm	3,557.45	1	15
Co	3,547.71	—	8	Th	3,557.46	5	5
				Ce	3,557.49	1	6
U	3,546.68	8	6	Tm	3,557.79	20	40
Ce	3,546.66	—	8	Rb	3,557.80	(15)	—

	λ	I J	I O
*bh*Sr	3,546.6	—	4
Er	3,546.58	—	8*W*
U	3,546.55	3	5
Ti	3,546.54	1	4
Tb	3,546.52	8	50
W	3,546.488	7	3
Nb	3,546.487	5	5
Lu	3,546.38	—	7
U	3,546.38	15	1
Pr	3,546.28	2	6
Th	3,546.26	3	5
Ce	3,546.19	2	20
Os	3,546.133	5	8
U	3,546.132	6	12
Ir	3,546.10	—	5
Tb	3,546.05	3	15
Nb	3,546.03	8	5
Pr	3,546.02	1	8
Mo	3,546.00	25	—
Ho	3,545.97	20	20
Ce	3,545.91	1	5
Ar	3,545.84	(125)	—
Gd	3,545.79	125	125
Ce	3,545.78	1	8
Dy	3,545.74	2	6
U	3,545.67	8	3
Fe	3,545.64	70	90
Ce	3,545.60	1	10
Ar	3,545.58	(300)	—
J	3,545.48	(10)	—
Tb	3,545.40	3	8
V	3,545.34	—	35
Th	3,545.29	3	5
W	3,545.23	10	12
V	3,545.20	300*R*	40
Co	3,545.04	30	2
Ti	3,544.99	2	10
Sm	3,544.983	2	5
Yt	3,544.980	5	7
W	3,544.97	6	7
Cu	3,544.963	6	35
Pr	3,544.96	10	30
Th	3,544.95	2	5
Tb	3,544.93	—	15
Ho	3,544.91	—	8
W	3,544.80	5	7
Eu	3,544.77	1	10*W*
Ba	3,544.71	5	20
Er	3,557.81	2*W*	10*Wd*
Ne	3,557.84	(12)	—
U	3,557.842	1	12
W	3,557.92	12	2
Ta	3,557.98	2*h*	35
Pr	3,558.007	3	10
Nb	3,558.012	5	5
J	3,558.013	(7)	—
Er	3,558.02	1	10
Mo	3,558.10	15	15
Gd	3,558.19	25	25
Dy	3,558.20	4	15
Ti	3,558.518	7	15
Fe	3,558.518	300	400
Gd	3,558.519	50	100*r*
Cr	3,558.519	—	20
Sc	3,558.54	40	15
Cr	3,558.58	8*h*	—
Ce	3,558.71	1	8
Er	3,558.74	1	10
Tb	3,558.77	8	15
Co	3,558.78	—	40*W*
Os	3,558.80	2	4*h*
Pr	3,558.81	2	9
Sm	3,558.87	—	5
Re	3,558.95	—	50*W*
Zr	3,558.96	2	8
Ir	3,558.99	50	50
Yb	3,559.076	—	5
Pr	3,559.081	4	8
Fe	3,559.083	2	4
W	3,559.09	4	6
Sm	3,559.10	20	40*d*
Nb	3,559.12	8	8
Sb	3,559.24	50*wh*	2*h*
Dy	3,559.27	4	10
Ce	3,559.33	—	6
Tb	3,559.39	8	8
Th	3,559.46	10	10
Fe	3,559.509	25	50
Ar	3,559.51	(15)	—
Nb	3,559.60	100	2
Yb	3,559.61	—	10
W	3,559.71	8	10
Mo	3,559.72	25	—
Tb	3,559.76	8	15
Cr	3,559.78	10	15
Os	3,559.79	50	150
Mo	3,559,88	5	5

	λ	I J	I O
U	3,544.66	—	4
Nb	3,544.65	15	20
Fe	3,544.63	6	50
Mo	3,544.62	2	40*h*
Os	3,544.58	10	12
Kr	3,544.54	(30*wh*)	—
W	3,544.46	12	1
F	3,544.39	(6)	—
Tb	3,544.36	3	8
Er	3,544.35	1	10
Dy	3,544.211	10*h*	25
U	3,544.209	6	3*h*
Eu	3,344.15	1	10*W*
Kr	3,544.14	(30*wh*)	—
Ho	3,544.07	—	8
Nb	3,544.02	10	8
Pr	3,544.00	10*r*	40*r*
Te	3,543.96	(10)	—
Rh	3,543.95	40	150
Eu	3,543.88	4	15*W*
Tb	3,543.86	15	50
Os	3,543.72	12	15
W	3,543.71	7	8
Fe	3,543.67	30	60
Ce	3,543.52	—	10
V	3,543.50	50	50
Nd	3,543.47	—	8
Fe	3,543.39	2	10
Nd	3,543.352	2	10
U	3,543.35	1*d*	12*rd*
Ce	3,543.28	2	10
Co	3,543.26	—	35
Tb	3,543.23	8	30
Ar	3,543.16	(10)	—
U	3,543.156	8	2
Mo	3,543.115	3	4
W	3,543.107	12	4
Hg	3,543.08	40	—
Nb	3,542.98	10	10
Er	3,542.98	3	7*d*
Ne	3,542.90	(40)	—
Dy	3,542.86	2	6
Gd	3,542.77	25	25
U	3,542.72	6	2
Os	3,542.71	10	150
V	3,542.66	2	15
W	3,542.65	3	4
Pr	3,542.63	2	8
Zr	3,542.62	30	12
Nb	3,559.901	10	—
Er	3,559.902	7	15
P	3,559.92	(30)	—
Ni	3,559.93	—	5
Th	3,559.96	8	6
W	3,560.07	5	5
Dy	3,560.149	20	40
Mo	3,560.15	8	6
Sm	3,560.27	3	25
Co	3,560.306	3	18*h*
U	3,560.308	2*h*	4
Yb	3,560.33	50	20
Tl	3,560.40	(12)	—
Nb	3,560.48	10	—
Eu	3,560.57	1*h*	7*W*
V	3,560.60	50	10
Fe	3,560.70	15	50
Yb	3,560.727	100	9
Nd	3,560.729	20	30
Tl	3,560.77	(25)	—
Ce	3,560.80	2	300
Os	3,560.85	100	150*R*
Th	3,560.86	4	10
Co	3,560.89	25	200
Tm	3,560.92	—	8
Ar	3,561.04	(15)	—
Nb	3,561.14	10	5
Ir	3,561.17	2	5
J	3,561.18	(40)	—
Hg	3,561.20	6*h*	10
Ne	3,561.23	(12)	—
W	3,561.25	4	8
Pr	3,561.27	4	25
Er	3,561.273	—	12*s*
Mo	3,561.36	20	1
Ir	3,561.37	—	5
U	3,561.41	—	12
W	3,561.45	10	—
Ce	3,561.54	—	6
Ti	3,561.57	20	10
Sm	3,561.58	4	15
Nd	3,561.59	2	8
Hf	3,561.66	35	20
Pr	3,561.68	3	6
Nb	3,561.69	10	4
Tb	3,561.74	200	200
Ni	3,561.75	12	70
U	3,561.80	30	12

	λ	I J	I O
Ag	3,542.61	5	30
Ir	3,542.575	10*h*	10
U	3,542.570	—	8
Nb	3,542.56	30*d*	5
Ti	3,542.55	1	7
Sm	3,542.46	2	9
Pr	3,542.39	3	8
U	3,542.38	1	4
Er	3,542.36	4	10*d*
Dy	3,542.33	20	90
W	3,542.27	6	6
Fe	3,542.23	1	5
Ce	3,542.18	1	4
Mo	3,542.17	5	5
Eu	3,542.16	5*W*	20*W*
Fe	3,542.08	100	150
F	3,541.94	(60)	—
Rh	3,541.912	10	50
Ce	3,541.911	1	4
Os	3,541.906	5	20
Nb	3,541.90	15	10
U	3,541.89	6	2
Ta	3,541.88	15	35
Dy	3,541.86	—	3
F	3,541.76	(100)	—
Tb	3,541.75	3	8
Ce	3,541.66	2	10
W	3,541.65	10	10
Ru	3,541.641	10	60
Nd	3,541.626	4*d*	10*d*
Th	3,541.62	6	8
Sm	3,541.37	1	10
Eu	3,541.34	2	12
V	3,541.336	35	—
Dy	3,541.29	—	6
U	3,541.28	8	—
Nb	3,541.25	5	50
Rb	3,541.22	(100)	—
Rn	3,541.20	(10)	—
Ru	3,541.15	—	4
Fe	3,541.09	200	200
Ru	3,541.05	—	10
Nb	3,540.96	500	15
Ta	3,540.82	1	18
Ce	3,540.79	1	5
Ho	3,540.76	8	6
W	3,540.73	12	10
Fe	3,540.71	4	10

	λ	I J	I O
Fe	3,561.82	—	4
Nb	3,561.88	8	—
Er	3,561.89	—	8
Ti	3,561.91	12	5
U	3,562.05	6	—
Ce	3,562.09	—	6
Co	3,562.10	—	15
Mo	3,562.11	5	5
V	3,562.14	8	12
K	3,562.15	(15)	—
Eu	3,562.18	1	4
Pr	3,562.22	4	10
Cr	3,562.28	12	20
Os	3,562.34	20	50
P	3,562.47	(30)	—
Cr	3,562.475	5	10
Zr	3,562.48	—	8
W	3,562.519	12	—
Tb	3,562.52	—	8
Pr	3,562.55	4	10
Dy	3,562.67	1	5
Ir	3,562.82	2	6
Cr	3,562.88	—	10
Pb	3,562.89	20	—
Tb	3,562.90	8	30
Co	3,562.92	—	10
Ne	3,562.94	(15)	—
Pr	3,562.98	3	8
Mo	3,563.14	15	15
Dy	3,563.15	100	200
Er	3,563.16	6	15*W*
Ar	3,563.26	(100)	—
Th	3,563.38	3*h*	10
V	3,563.396	8	15
U	3,563.397	6	—
Er	3,563.402	1	12
Nd	3,563.403	10	50
W	3,563.45	7	9
Nb	3,563.510	30	30
Er	3,563.55	—	5
Nb	3,563.62	15	10
U	3,563.66	5	12
Dy	*3,563.70*	*30*	*50*
Tb	*3,563.74*	—	*8*
Mo	*3,563.75*	*10*	*10*
Pr	*3,563.77*	*8*	*30*
Tm	*3,563.88*	*3*	*15*

	λ	I J	I O
Ir	3,540.69	2h	5
Dy	3,540.67	2	10
V	3,540.53	4	25
U	3,540.46	15	6
Tb	3,540.24	50	50
Ru	3,540.22	1	12
Fe	3,540.12	60	100
Tl	3,540.05	(20)	—
Re	3,539.95	—	15
Pr	3,539.92	5	25
Zr	3,539.91	7	8
Sm	3,539.89	—	10
Tb	3,539.81	3	15
U	3,539.654	2	12
Nb	3,539.648	15	15
Er	3,539.60	—	9
Th	3,539.59	8	8
Kr	3,539.54	(15)	—
Dy	3,539.38	2h	18
Ru	3,539.37	15	60
Re	3,539.33	—	25
Ru	3,539.26	5	30
Sm	3,539.25	1	8
Nd	3,539.18	4	10
Nb	3,539.12	15	1
Ce	3,539.09	10	100
Tb	3,538.90	3	15
Sm	3,538.864	1	10
Mg	3,538.86	—	8
Nd	3,538.858	6	20
Pr	3,538.85	2	9
Th	3,538.75	50	—
W	3,538.63	9	8
Yt	3,538.53	3	10
Dy	3,538.52	40	150
Er	3,538.52	9	18
U	3,538.42	8	2
Rh	3,538.26	4	50
V	3,538.24	100	10
Th	3,538.223	3	8
U	3,538.221	6	8
Ir	3,538.15	1h	18
Rh	3,538.14	10	100
Eu	3,538.09	10h	20W
Ru	3,537.95	25	70
Tb	3,537.94	15	15
Tm	3,537.91	2	15

	λ	I J	I O
bhCa	3,564.0	—	20
Gd	3,564.05	10	10
Nb	3,564.08	20	2
Nd	3,564.09	4	30
Fe	3,564.12	2	15
Rh	3,564.13	1	25
Dy	3,564.24	10	25
Ar	3,564.27	(100)	—
Cr	3,564.29	6	15
Ti	3,564.397	1	10
Er	3,564.40	1h	8d
Fe	3,564.53	15	30
Ti	3,564.54	2	12
Cr	3,564.71	8	20
Re	3,564.73	—	20
Ta	3,564.79	2	18
Co	3,564.951	—	150W
Cr	3,564.953	1	8
Nb	3,565.06	15h	10
Fe	3,565.38	300	400
Fe	3,565.59	4	10
Ta	3,565.63	50	—
Tb	3,565.74	15	15
U	3,565.75	15h	1
Ne	3,565.84	12	—
Tm	3,565.90	50	40
Au	3,565.93	15	20
Ti	3,565.99	25	6
Ce	3,566.03	6W	10W
Mo	3,566.05	10	10
Th	3,566.08	3	10
Pr	3,566.089	6	10
Er	3,566.090	—	12
Ti	3,566.095	2h	8
Zr	3,566.10	10	20
Cr	3,566.16	12Wh	80Wh
V	3,566.18	100	25
Nd	3,566.34	—	8
Ni	3,566.37	100Wh	2000R
P	3,566.42	(70)	—
Sm	3,566.471	2	10
Tm	3,566.474	20	60
Ir	3,566.49	—	8
U	3,566.60	10	30
Pd	3,566.63	—	60
Sb	3,566.64	25Wh	—
Ta	3,566.71	5	50
Sm	3,566.83	3	20
Fe	3,567.04	15	50

	λ	I	
		J	O
Fe	3,537.90	25	50
Mn	3,537.89	—	12
Eu	3,537.75	10h	10
Fe	3,537.73	15	25
Dy	3,537.67	—	9
Nb	3,537.62	30	2
Fe	3,537.48	8	25
Nb	3,537.475	30	30
Pr	3,537.47	3	8
Re	3,537.466	—	80W
W	3,537.45	12	12
U	3,537.445	12	5
Ce	3,537.437	3	10
Mo	3,537.27	15	15
Cr	3,537.23	1	15
Th	3,537.16	10	10
Ce	3,537.13	1	10
Tb	3,537.11	3	15
Mo	3,537.09	50	—
U	3,537.06	1	8
Zr	3,536.94	5	8
F	3,536.84	(30)	—
Ce	3,536.70	2	10
Tb	3,536.62	3	30
Hf	3,536.62	2	10
Tm	3,536.57	20	60
Ru	3,536.567	—	50
Dy	3,536.56	—	10
Fe	3,536.557	200	300
Ce	3,536.48	2	8
Tb	3,536.32	8	15
Ta	3,536.30	2	20
P	3.536.29	(30)	—
W	3,536.27	15	3
Tm	3,536.21	60	40
Fe	3,536.19	10	40
Re	3,536.16	—	8
Dy	3,536.024	10	125
Er	3,536.023	12	20S
Ru	3,535.83	12	60
Ir	3,535.79	10	12
Sc	3,535.73	30	15
Cd	3,535.69	15	5
Sm	3,535.63	5	25
Ce	3,535.57	1	10
W	3,535.55	10	12
Hf	3,535.54	50	15
Tm	3,535.52	25	80
Ti	3,535.41	125	15

	λ	I	
		J	O
Nb	3,567.10	30W	5
S	3,567.17	(40)	—
Tb	3,567.35	15	15
Tm	3,567.36	—	10
Hf	3,567.364	10	20
Fe	3,567.38	2	10
Ar	3,567.66	(300)	—
Sc	3,567.70	40	15
Lu	3,567.84	7	100
Tb	3,567.86	—	8
Nb	3,567.995	50	2
Ir	3,567.995	5	15
W	3,568.04	10	10
Ce	3,568.13	1	10
Re	3,568.23	—	40
Sm	3,568.26	50	40
Fe	3,568.42	4	20
Nb	3,568.506	50	10
Tb	3,568.51	50	50
Ne	3,568.53	(25)	—
Fe	3,568.82	7	15
Zr	3,568.87	—	12
Nd	3,568.88	12	6
V	3,568.94	15	20
Fe	3,568.979	35	50
Tb	3,568.98	15	15
W	3,568.987	9	9
Dy	3,568.994	—	100
Hf	3,569.04	50	20
U	3,569.06	20	12
Nd	3,569.229	4	20
W	3,569.235	7d	8d
Er	3,569.25	—	8
Ce	3,569.32	2	8
Co	3,569.38	100	400R
Nb	3,569.47	15	20
Mn	3,569.493	8	25h
Zr	3,569.494	2h	15
Nd	3,569.50	—	12
Pr	3,569.56	5	10
Dy	3,569.67	10	40
Os	3,569.77	30	100
Mn	3,569.80	5	12
Mn	3,570.04	—	20R
Tb	3,570.07	—	8
La	3,570.09	10	10
Fe	3,570.097	300	300
Mn	3,570.10	15	20R

	λ	I	
		J	O
Ta	3,535.40	18Wh	15
Ru	3,535.37	8	30
Kr	3,535.35	(50h)	—
Ar	3,535.33	(15)	—
Nb	3,535.30	500	300
Zr	3,535.15	—	10
Ru	3,535.052	1	10
Ce	3,535.046	—	8
Mg	3,535.04	—	8
Ir	3,534.98	—	10
Dy	3,534.963	—	125
Er	3,534.962	9	20
Sm	3,534.91	—	10
Tm	3,534.85	30	20
Re	3,534.83	—	25
V	3,534.73	5	20
Mo	3,534.69	25	10
Er	3,534.55	2	15d
Nd	3,534.54	4	15d
Sm	3,534.53	—	10
Pr	3,534.52	4	25
Ce	3,534.44	—	10
U	3,534.33	2	8
Nb	3,534.22	15	1
Tb	3,534.21	3	8
Al	3,534.20	15	—
Nb	3,534.12	15h	15
Ce	3,534.05	10	35
Ti	3,533.87	35	6
Tb	3,533.86	8	30
V	3,533.76	40h	20
Pr	3,533.749	5	20
Cu	3,533.746	15	50
V	3,533.68	10h	40
Nb	3,533.665	30	20
P	3,533.66	(30)	—
Tb	3,533.60	—	15
U	3,533.57	20	10
Ce	3,533.56	3	12
Os	3,533.41	20	40
Co	3,533.36	—	200W
Zr	3,533.22	3	30
Fe	3,533.20	50	50
Ce	3,533.11	—	8
Mo	3,533.07	25d	—
P	3,533.06	(15)	—
Na	3,533.010	(200)	50
Fe	3,533.010	75	50

	λ	I	
		J	O
Rh	3,570.18	150	400R
Re	3,570.256	—	30
Fe	3,570.258	15	50
P	3,576.33	(30d)	—
Tb	3,570.34	—	8
Yb	3,570.56	15	2
Pr	3,570.56	4	8
Ru	3,570.59	60	12
Mo	3,570.647	20	15
W	3,570.655	15	15
Er	3,570.76	1	10
Tb	3,571.03	—	15
V	3,571.04	20	35
Pd	3,571.15	40h	40h
V	3.571.21	5	10
Fe	3,571.23	6	40
Ne	3,571.26	(12)	—
Tb	3,571.35	—	8
Yt	3,571.43	2	15
Th	3,571.57	8	8
V	3,571.65	35	40
Er	3,571.73	1	15d
Pr	3,571.79	3	10
Ta	3,571.85	3	18
Ni	3,571.87	40h	1000R
Gd	3,571.94	20	25
Fe	3,571.996	80	100
Pt	3,571.999	15	—
Tb	3,572.07	8	30
Ar	3,572.29	(300)	—
Th	3,572.399	8	8
Nd	3,572.405	6	25
Ce	3,572.43	5	12
Zr	3,572.47	80	60
W	3,572.48	35	10
Sc	3,572.52	50	30
Mo	3,572.59	20	—
Kr	3,572.68	(15Wh)	—
Pb	3,572.73	20	200
Cr	3,572.748	5	25
Nd	3,572.751	10	10
Er	3,573.077	—	10
Zr	3,573.084	9	10
Nd	3,573.18	4	10
Ho	3,573.22	10	8
Fe	3,573.40	20	50
W	3,573.41	7	8
Ta	3,573.44	70W	15

	λ	I J	I O
Cr	3,532.89	2	8
Ce	3,532.88	1	10
Ru	3,532.81	12	60
Os	3,532.80	20	100
Tb	3,532.70	8	30
N	3,532.65	(15)	—
Hg	3,532.63	(200)	—
Sm	3,532.54	4	20
Zr	3,532.46	—	8
V	3,532.28	25h	1
Ir	3,532.27	3	12
Ta	3,532.21	2	15
Mn	3,532.12	30	50h
Mn	3,532.00	8	50h
Mn	3,531.85	30r	40R
Eu	3,531.83	2	12
Ru	3,531.79	1	8
Ho	3,531.74	20	10
Er	3,531.714	25	40
Dy	3,531.712	100	100
Tb	3,531.70	50	15
U	3,531.64	1	15
Rb	3,531.60	(100)	—
Ce	3,531.59	2	18
Ta	3,531.58	3	35
Ru	3,531.39	9	60
Er	3,531.26	1	12
Eu	3,531.14	10	10
Os	3,531.13	12	15
U	3,531.11	20	8
Cr	3,531.08	—	15
Ce	3,530.95	1	10
V	3,530.87	20	1
Pr	3,530.84	2	9
Nb	3,536.82	15	3
V	3,530.77	100	40
W	3,530.76	7	8
Ir	3,530.74	2	10
K	3,530.71	(40)	—
La	3,530.66	6	15
Tb	3,530.64	—	8
Ce	3,530.63	—	10
Sm	3,530.60	4	8
Ni	3,530.59	—	30
Ti	3,530.58	1	15
V	3,530.45	30	—
Fe	3,530.389	25	50
Cu	3,530.386	20	50
V	3,573.52	20h	20
Tb	3,573.57	—	15
Cr	3,573.64	15	60
J	3,573.67	18	—
Ce	3,573.70	3	10
Ir	3,573.72	100	8
Ti	3,573.74	40	20
Fe	3,573.836	15	20
Dy	3,573.838	80	80
Er	3,573.843	4	10
Mo	3,573.88	25	20
Sm	3,573.888	—	10h
Fe	3,573.888	30	40
Cr	3,574.04	15	50
Tm	3,574.06	30	20
Os	3,574.08	5	30
U	3,574.11	8	12
Tb	3,574.13	—	8
Dy	3,574.16	100	200
Ti	3,574.24	3	15
Nd	3,574.34	8	20
V	3,574.35	40	—
La	3,574.43	8	30
Ru	3,574.58	15	4
Ne	3,574.64	(18)	—
Gd	3,574.738	25	25
Er	3,574.743	1	10
U	3,574.76	1	12
V	3,574.77	12	25
Ho	3,574.78	20	10
Cr	3,574.80	15	8
Cr	3,574.93	6	8
Pr	3,574.958	4	10
Co	3,574.96	25	200
Nd	3,574.98	8	30
Fe	3,575.12	5	15
V	3,575.129	15	20
Nb	3,575.129	10	10

	λ	I J	I O		λ	I J	I O
Tb	*3,530.38*	*—*	*15*				
Eu	*3,530.36*	*1*	*12W*				
P	*3,530.24*	*(70)*	*—*				
Zr	*3,530.22*	*1*	*10*				
Os	*3,530.06*	*20*	*100*				
Ce	*3,530.02*	*3*	*18*				
Sm	*3,530.00*	*—*	*15*				
Fe	*3,529.820*	*80*	*125*				
Re	*3,529.816*	*—*	*10*				
Co	*3,529.81*	*30*	*1000R*				
Tb	*3,529.76*	*3*	*15*				
V	*3,529.73*	*2*	*20*				
W	*3,529.56*	*20*	*10*				
Tl	*3,529.43*	*800*	*1000*				
Ru	*3,529.28*	*3*	*30*				
Ce	*3,529.27*	*2*	*8*				
Re	*3,529.21*	*—*	*25*				
Nd	*3,529.19*	*2*	*8*				

U 3,670.07 *J* 18 *O* 15

3,672.58 *J* 15 *O* 8

	λ	I J	I O		λ	I J	I O
Nd	3,671.45	10	—	—	—	—	—
Tb	3,671.42	3	3	*bh*Sr	3,671.5	—	4
Pb	3,671.39	70	—	Pb	3,671.503	7	50
Pr	3,671.370	2	8	Ce	3,671.51	—	2
Nb	3,671.368	5	3	Fe	3,671.52	—	2
Ce	3,671.32	1	5	U	3,671.539	1	4
Er	3,671.31	1	12	Th	3,671.543	4	6
Zr	3,671.27	30	40	Pr	3,671.56	2	8
Ru	3,671.219	4	10	Ho	3,671.65	4	6
Gd	3,661.216	100	150*W*	Mo	3,671.65	3	5
V	3,671.20	70	100	Nd	3,671.66	6	10
Zr	3,671.04	—	3	Ti	3,671.67	70	150
Ar	3,671.01	(3)	—	Dy	3,671.697	1*h*	6
Nd	3,670.91	—	4	Fe	3,671.700	1	3
Os	3,670.89	20	200	Nb	3,671.73	5	2
Eu	3,670.83	1*h*	4	Ag	3,671.82	7*h*	—
Sm	3,670.82	50	100	Pr	3,671.91	5	20
Fe	3,670.81	6	20	As	3,671.92	15	—
Pr	3,670.794	2	8	Ce	3,671.94	1	10
W	3,670.786	5	—	Cu	3,671.95	3	20
W	3,670.77	6	7	Pt	3,671.99	10	80
Yb	3,670.68	10	2	La	3,672.02	3	25
Ce	3,670.672	—	6	Ru	3,672.07	1	4
Mo	3,670.668	25	—	S	3,672.14	(20)	—
Sm	3,670.66	5	10	Ce	3,672.17	1	10

	λ	I J	I O		λ	I J	I O
Ar	3,670.64	(300)	—	W	3,672.18	4	5
Th	3,670.63	1	5	U	3,672.19	—	5
Tb	3,670.55	3	15	Eu	3,672.21	—	7*W*
U	3,670.53	—	6	Sm	3,672.213	5	6
Re	3,670.522	—	20	Ce	3,672.26	—	2
Mn	3,670.517	15	25	Hf	3,672.27	3	25
Nd	3,670.51	4	10	Ho	3,672.30	2	6
Sm	3,670.50	3	10	Tb	3,672.30	3	8
Si	3,670.49	4	—	Er	3,672.303	2	18
Ce	3,670.490	1	10	Th	3,672.304	1	4
Ni	3,670.43	20	150	Dy	3,672.31	100	100
Li	3,670.40	—	5	Nd	3,672.36	12	30
Re	3,670.36	—	15	Ru	3,672.38	4	6
Cl	3,670.29	(8)	—	Re	3,672.400	—	20
Ho	3,670.29	6	6	V	3,672.403	40*h*	100
U	3,670.265	2	3	Nb	3,672.44	5	5
Pr	3,670.263	3	8	Tb	3,672.53	1*d*	2*d*
La	3,670.23	3	2	Xe	3,672.57	(10)	—
Ca	3,670.21	4	2	Nb	3,662.576	3	3
U	*3,670.072*	*18*	*15*	*U*	*3,672.579*	*15*	*8*
Ce	3,670.072	—	2*W*	W	3,672.59	18	1
Fe	3,670.071	200	200	Pr	3,672.62	1	7
Th	3,670.062	4*d*	3*d*	Zr	3,672.66	—	5
Co	3,670.058	—	20	Dy	3,672.67	4	10
Nb	3,670.05	15	1	Fe	3,672.71	2	4
Fe	3,670.03	—	100	Ce	3,672.79	5	15
Re	3,670.025	—	15	Mo	3,672.82	20	20
Nd	3,670.022	2	4	W	3,672.95	2*h*	4*d*
Ce	3,669.962	—	3	Hg	3,673.02	(5)	—
Th	3,669.758	1*d*	5*d*	U	3,673.06	—	6
W	3,669.92	7	—	Fe	3,673.09	4	10
Xe	3,669.91	(10)	—	Tm	3,673.14	15	10
Sm	3,669.89	2	8	Er	3,673.142	1	12
Mn	3,669.84	30	30	Ir	3,673.143	—	10
Pr	3,669.82	2	9	Dy	3,673.15	5	25
Eu	3,669.81	1*Wh*	8*Wh*	Eu	3,673.20	2	20
Re	3,669.77	—	25	Mo	3,673.22	3	4
Fe	3,669.76	—	2	Nb	3,673.23	10	5
Nb	3,669.74	8	5	Ar	3,673.16	(5)	—
Ce	3,669.72	—	2	Th	3,673.262	10	10
Yb	3,669.71	80	50	Ce	3,673.265	—	3
Rn	3,669.70	(5)	—	V	3,673.40	80*h*	150
Ar	3,669.62	(10)	—	Rn	3,673.41	(15)	—
Tb	3,669.62	—	30	Ca	3,673.45	3	5
Fe	3,669.523	150	200	Pr	3,673.49	2	7
Pr	3,669.523	3	9	Nd	3,673.54	20	20
Ho	3,669.52	4	6	Rh	3,673.56	—	2
Ru	3,669.49	70	50	Ce	3,673.64	1	10*S*

	λ	I J	I O
Re	**3,669.43**	**—**	**15**
Th	**3,669.413**	**4**	**4**
V	**3,669.408**	**300**	**20*W***
Mn	**3,669.40**	**3**	**3**
Dy	**3,669.39**	**—**	**2**
U	**3,669.36**	**3**	**—**
Nb	**3,669.35**	**3*h***	**2**
Mo	**3,669.34**	**8**	**8**
W	**3,669.33**	**6**	**4**
Ce	**3,669.31**	**—**	**3**
La	**3,669.27**	**3*h***	**—**
Ni	**3,669.24**	**10**	**150**
Tb	**3,669.21**	**3**	**8**
Fe	**3,669.16**	**30**	**50**
Eu	**3,669.13**	**1*h***	**7*W***
Ce	**3,669.08**	**1**	**2**
Ho	**3,669.05**	**6**	**6**
S	**3,669.049**	**(60)**	**—**
Er	**3,669.02**	**—**	**4**
Kr	**3,669.01**	**(150*h*)**	**—**
Nb	**3,669.007**	**10**	**10**
Ti	**3,668.97**	**40**	**100**
Dy	**3,668.91**	**3**	**6**
Fe	**3,668.893**	**1**	**5**
Sm	**3,668.887**	**—**	**2**
Pr	**3,668.83**	**40**	**100**
Nd	**3,668.79**	**4**	**25**
Kr	**3,668.74**	**(10)**	**—**
Er	**3,668.738**	**4*Wh***	**1**
Ru	**3,668.729**	**2**	**4**
Ce	**3,668.72**	**2**	**12*S***
W	3,668.66	12	10
K	3,668.63	(10)	—
Nb	3,668.62	5	5
P	3,668.59	(50)	—
Tb	3,668.50	3	15
Er	3,668.491	3	25
Yt	3,668.489	20	7
Mo	3,668.488	5	5
Zr	3,668.45	9	10
U	3,668.43	—	4
Gd	3,668.31	10	5
Fe	3,668.214	4	15
Hf	3,668.207	1	10
Ir	3,668.18	—	8
Th	3,668.15	2	8
Tm	3,668.08	20	80
Tb	**3,673.73**	**3**	**8**
Ce	**3,673.75**	**—**	**5**
Th	**3,673.80**	**10**	**10**
Cl	**3,673.83**	**(18)**	**—**
Mo	**3,673.87**	**4**	**3**
Fe	**3,673.90**	**2**	**6**
Fe	3,674.043	7	7
Pt	3,674.045	4	80
Ce	3,674.047	—	5
Tb	3,674.05	3	15
Sm	3,674.053	10	50
Nd	3,674.055	10	30
Gd	3,674.06	30	100*W*
Er	3,674.087	2	20
Dy	3,674.093	50	100
U	3,674.13	—	8
Pr	3,674.14	2	9
Ce	3,674.150	1	6
Ni	3,674.15	50*r*	200
W	3,674.16	5*d*	4*d*
Eu	3,674.18	—	15*W*
Ho	3,674.36	2	6
Fe	3,674.41	5	12
Dy	3,674.45	1	7
Ce	3,674.47	—	4
W	3,674.58	12	10
Nd	3,674.64	6	15
Eu	3,674.67	—	25
Nb	3,674.678	4	20
V	3,674.685	20	2
Zr	3,674.72	40	100
Rh	3,674.76	4	10
Fe	3,674.768	25	40
Ho	3,674.77	15	8
Nb	3,674.78	5	20
Ta	3,674.83	3*h*	20
Pr	3,674.88	2	5
Th	3,674.89	1	5
Au	3,674.90	5	4
W	3,674.97	8	—
Ir	3,674.98	50	100
Tb	3,675.03	3	8
U	3,675.08	6	4
Yb	3,675.09	200	50
Th	3,675.14	1	5
Ar	3,675.22	(300)	—
U	3,675.25	—	8

	λ	I J	I O
Tb	3,668.07	3	15
Er	3,668.04	1	6*d*
Cl	3,668.03	(20)	—
Cr	3,668.029	4	15
Mo	3,668.002	3	4
Fe	3,667.999	10	60
Ce	3,667.98	15	80*s*
U	3,667.975	8	15
Mo	3,667.97	6	10
Sm	3,667.91	10	25
V	3,667.74	25*h*	80
W	3,667.72	10	10
Ta	3,667.68	—	10
Pr	3,667.67	4	10
Nb	3,667.66	10	3
Sm	3,667.54	—	10
Cr	3,667.45	25	—
Tb	3,667.35	—	15
Cd	3,667.32	10	—
Ce	3,667.28	—	4*v*
Fe	3,667.26	25	80
W	3,667.18	7	8
Pr	3,667.14	2	9
U	3,667.13	5	8
Ho	3,667.05	10	6
Nb	3,667.00	15	8
Th	3,666.98	2	5
Sm	3,666.95	2*h*	10
Mo	3,666.93	6	6
Rh	3,666.91	4	10
Ta	3,666.89	—	15
Dy	3,666.85	3	6
Eu	3,666.84	4*W*	5*W*
W	3,666.81	4	5
Hf	3,666.77	4	25
Mo	3,666.71	10	8
Ho	3,666.65	4	6
Cr	3,666.64	15	20
Sm	3,666.56	2	8
Nd	3,666.56	4	10
Hg	3,666.55	—	9
Sc	3,666.54	8	15
Nb	3,666.53	10	8
Th	3,666.38	2	4
Ce	3,666.35	—	5
Dy	3,666.31	1	5
Os	3,666.309	15	40
Sm	3,666.26	3	6
Fe	3,666.25	7	20
Ru	3,675.262	3	6
Sc	3,675.262	2	5
Ca	3,675.31	2	10*h*
Mo	3,675.35	25	25*r*
Ce	3,675.36	1	8
Os	3,675.45	15	40
Ta	3,675.49	—	18
V	3,675.50	10	40
Pr	3,675.518	—	4
Ce	3,675.524	—	5
W	3,675.55	10	12
Th	3,675.57	10	10
Mn	3,675.67	—	20
V	3,675.70	70	100
Hf	3,675.74	6	10
Tb	3,675.78	8	30
Pr	3,675.91	2	6
Mo	3,675.98	4	10
Re	3,676.00	—	20
Dy	3,676.01	2	6
K	3,676.05	(10)	—
Ce	3,676.16	1	12
Mo	3,676.23	10	8
P	3,676.26	(100*W*)	—
W	3,676.311	5	6
Nb	3,676.312	20	10
Fe	3,676.314	100	200
Sm	3,676.315	3	8
Cr	3,676.32	15	40
Tb	3,676.35	200	100
Ru	3,676.41	2*h*	5
*bh*Ca	3,676.5	—	8
Er	3,676.51	2	20
Co	3,676.55	35	100
U	3,676.560	15	3
Dy	3,676.56	30	6
Re	3,676.57	—	8
Ho	3,676.59	4	6
Eu	3,676.62	2	20*W*
Ir	3,676.66	2	15
Ru	3,676.67	4	8
V	3,676.68	150*h*	300
Th	3,676.69	4	6
W	3,676.80	10	10
Sm	3,676.84	5	10
Eu	3,676.87	1	15
Cu	3,676.878	1*h*	25*Wh*
Fe	3,676.879	2	10
Ta	3,676.89	1*h*	5

	λ	I J	I O
Rh	3,666.215	30	70
U	3,666.210	8	6
Cr	3,666.17	2	12
Ta	3,666.10	—	20
Ce	3,666.02	—	15
Cr	3,665.98	15	20
W	3,665.88	8*s*	9
Tm	3,665.81	20	40
Nd	3,665.751	4	10
Mo	3,665.747	15	15
Cu	3,665.735	5	20
Th	3,665.731	5	6
Tb	3,665.60	3	30
Ce	3,665.50	—	4
Dy	3,665.40	—	4
Sm	3,665.38	—	6
Ta	3,665.35	—	10
Hf	3,665.346	25	20
Kr	3,665.33	(80)	—
La	3,665.22	4	4
Er	3,665.21	—	8
U	3,665.207	1	5
Dy	3,665.20	5	7
Th	3,665.185	4	5
Nd	3,665.180	12	25
Nb	3,665.16	5*h*	4*h*
Eu	3,665.15	—	4*W*
V	3,665.14	50*h*	100
Ir	3,665.07	2	5
Ce	3,665.05	—	8
Ce	3,664.95	—	8
Cr	3,664.94	40	1
Mo	3,664.813	40	20
Rn	3,664.81	(25)	—
Ce	3,664.73	1	8
Nb	3,664.695	30	30
Fe	3,664.694	3	12
Nd	3,664.65	10	20
Tb	3,664.64	8	50
Pr	3,664.640	3	10
Dy	3,664.63	6	10
Gd	3,664.621	200	200*W*
Ir	3,664.618	15	60
Yt	3,664.61	100	100
Fe	3,664.54	8	35
Er	3,664.44	20*h*	40
Mo	3,664.30	5	5
Tb	3,664.28	—	8
Sc	3,664.25	2	4
Ru	3,676.95	3	8
Mn	3,676.96	100	60
Nb	3,677.080	8	5
V	3,677.085	70	25
Ce	3,677.17	—	8
Sm	3,677.25	—	15
Fe	3,677.31	30	40
U	3,677.39	—	10
W	3,677.41	10	—
Ho	3,677.62	6	6
Fe	3,677.63	60	80
Cr	3,677.68	35	6
Mo	3,677.70	6	6
Pr	3,677.72	2	10
Th	3,677.74	4	5
Ti	3,677.770	1	8
Sm	3,677.775	10	25
Nb	3,677.78	5*h*	5
*bh*B	3,677.8	—	50
Cr	3,677.888	70	3
Th	3,677.89	8	70
Tm	3,677.980	20	30
Ru	3,677.982	1	5
Os	3,678.02	10	10
Th	3,678.03	10	10
Ru	3,678.06	2	4
Nb	3,678.078	10	1
Sm	3,678.081	2	15
Eu	3,678.12	2	8*W*
S	3,678.13	(10)	—
Sm	3,678.17	—	15*d*
Nd	3,678.19	12	10
Ce	3,678.22	—	8
Ca	3,678.23	2*h*	15
Ru	3,678.25	—	5
Ar	3,678.27	(10)	—
Eu	3,678.27	—	50
Ir	3,678.29	—	10
Pr	3,678.30	—	6
Ru	3,678.31	15	6
Sc	3,678.34	15	5
Re	3,678.39	—	6
Th	3,678.47	1	4
Dy	3,678.52	1	5
Sc	3,678.54	15	3
Mo	3,678.61	4	6
J	3,678.611	(7)	—
Kr	3,678.66	(7*h*)	—
Nb	3,678.72	3*h*	5
U	3,678.75	15	6

	λ	I J	I O
P	3,664.19	(100*W*)	—
Ne	3,664.11	(250)	—
Ni	3,664.09	30	300
Sm	3,664.01	—	15
Fe	3,663.95	2	4
U	3,663.93	—	10
Rb	3,663.86	(15)	—
Ta	3,663.84	—	15
W	3,663.82	6	7
Nb	3,663.74	8*h*	—
Th	3,663.71	10	10
Ce	3,663.70	1	10*s*
Mo	3,663.66	4	4
Zr	3,663.65	10	100
V	3,663.59	1*wh*	150
Fe	3,663.46	7	25
Eu	3,663.44	4*W*	4*W*
Kr	3,663.44	(20)	—
Ru	3,663.37	60	5
W	3,663.36	9	8
Mo	3,663.30	8	8
Mg	3,663.28	400	500
Fe	3,663.27	3	8
Cr	3,663.21	20	35
Th	3,663.20	5	4
Nb	3,663.18	3*h*	5
W	3,663.15	6	7
Tb	3,663.12	15	50
Ta	3,663.10	—	18*d*
Pt	3,663.095	2	50
Hg	3,663.09	3	5*d*
Nd	3,663.033	2	6
Re	3,663.029	—	5
Mo	3,662.990	10	8
Ce	3,662.990	1	8*s*
Ho	3,662.98	4	6
Eu	3,662.94	5*W*	15*W*
Sm	3,662.90	10	25
Mg	3,662.88	400	50
Er	3,662.87	—	10
Fe	3,662.85	5	30
Cr	3,662.84	8	25
Rb	3,662.78	15	—
Sm	3,662.69	10	25
U	3,662.66	2	15*r*
Tb	3,662.64	—	8
Ba	3,662.53	5	10
Ce	3,662.49	—	8
Cr	3,662.38	4*h*	6

	λ	I J	I O
Tb	3,678.78	3	30
Ce	3,678.85	—	8
Tm	3,678.864	40	50
Fe	3,678.864	50	100
Zr	3,678.90	5	6
Er	3,678.96	—	12
Fe	3,678.98	1	5
Rn	3,679.00	(30)	—
Cr	3,679.07	5	15
*bh*B	3,679.10	—	200
Ce	3,679.16	1	6
Ho	3,679.19	4	6
Sm	3,679.25	1	25
Nd	3,679.40	2*h*	4*h*
Ce	3,679.42	4	12
Eu	3,679.49	1	15
Tb	3,679.56	—	15
Kr	3,679.561	(50)	—
W	3,679.607	9	12
Nb	3,679.608	10*h*	—
Kr	3,679.611	(50)	—
F	3,679.67	(15)	—
Ti	3,679.673	12	5
Ho	3,679.70	4*h*	6
Th	3,679.71	12	12
Cr	3,679.82	8	40
Fe	3,679.91	300	500
Pr	3,679.98	2	5
Sm	3,680.00	4	20
Hg	3,680.01	(40)	—
Ar	3,680.06	(10)	—
Ru	3,680.063	—	4
Ce	3,680.08	2	10
Er	3,680.10	—	6
V	3,680.11	50*h*	125
Mn	3,680.15	5	5
Re	3,680.22	—	15
Ti	3,680.27	6*h*	—
Kr	3,680.37	(100*Wh*)	—
Zr	3,680.374	2	9
Ce	3,680.42	—	4
Mo	3,680.60	20	20
Mo	3,680.68	20	20
Eu	3,680.76	3	4
Fe	3,680.80	7	12
Ce	3,680.85	—	6
W	3,680.86	5*d*	4*d*
U	3,680.88	1	20
Sm	3.680.96	5	25
Rh	3,681.04	25	20

	λ	I J	I O
Ta	3,662.34	15	15
U	3,662.331	10	10
Eu	3,662.33	—	8
Er	3,662.275	1	15
Ho	3,662.27	10	20
Gd	3,662.268	200	200*W*
Nd	3,662.26	30	30
Sm	3,662.254	50	50
Tb	3,662.25	—	8
Dy	3,662.24	—	6
Ti	3,662.237	100	40
Th	3,662.19	4	4
Co	3,662.16	25	100
Mo	3,662.15	8	6
Zr	3,662.14	5	5
Re	3,662.12	—	10
La	3,662.07	40	60
Nb	3,662.05	3	5
Er	3,662.04	1	9
Ni	3,661.951	6	50
S	3,661.95	(8)	—
Ce	3,661.91	—	6
U	3,661.90	—	12*r*
Rh	3,661.86	5	10
Mo	3,661.78	10	10
J	3,661.77	(40*d*)	—
Dy	3,661.75	4	6
Ce	3,661.73	2	10
Ir	3,661.71	30	50
Xe	3,661.70	(10*Wh*)	—
Nd	3,661.691	4	10
Ta	3,661.688	3	20*r*
Sm	3,661.687	4	6
Nb	3,661.68	5*h*	5
Gd	3,661.67	20	10
Pr	3,661.624	5	25
Th	3,661.624	5*h*	10
Ru	3,661.58	5	1
U	3,661.48	1*h*	5
V	3,661.38	150	10
Fe	3,661.37	1	10
Ru	3,661.352	100	60
Sm	3,661.350	50	150
Nd	3,661.34	25	40
Zr	3,661.33	1	5
Os	3,661.25	10	10
W	3,661.24	7*d*	3
Zr	3,661.20	—	18

	λ	I J	I O
N	3,681.10	(10)	—
Th	3,681.19	5	6
Fe	3,681.23	2	6
Ta	3,681.24	1	15
Ti	3,681.27	—	15
Re	3,681.31	—	15
Ce	3,681.38	3	10
K	3,681.52	(30)	—
Mo	3,681.55	3	4
Os	3,681.57	8	30
Zr	3,681.646	—	5
Fe	3,681.651	2	6
Nb	3,681.68	10	1
Cr	3,681.69	8	18
Tb	3,681.71	—	8
Sm	3,681.719	6	30
Mo	3,681.725	25	8
Pr	3,681.86	5	10
Nd	3,681.87	1	6
Th	3,681.89	8	8
Mo	3,681.95	1	4
U	3,682.04	15	6
Sm	3,682.075	3	5
Ce	3,682.077	2	10*s*
Mn	3,682.087	40	25
W	3,682.092	20	25
Fe	3,682.209	300	400
Sm	3,682.212	2	10
Hf	3,682.236	30	25
Ne	3,682.243	(75)	—
Tb	3,682.26	30	50
Th	3,682.36	4	4
*bh*Zr	3,682.4	—	30
Eu	3,682.44	1	12
U	3,682.46	—	8
Ag	3,682.47	4	50
Th	3,682.49	1	4
Dy	3,682.52	2*Wh*	5*h*
Sm	3,682.54	—	8*d*
Ce	3,682.647	1	8
Ho	3,682.65	4	6
Er	3,682.71	4	18
Gd	3,682.74	20	10
Au	3,682.86	10*h*	20
Nb	3,682.95	10*h*	—
Pt	3,682.98	2	8
Co	3,683.05	—	200*R*
Ta	3,683.058	1	18
Fe	3,683.058	100	200
V	3,683.13	60	100

	λ	I	
		J	O
Mo	3,661.08	5	5
Mf	3,661.05	25	10
Kr	3,661.00	(15)	—
Nd	3,660.97	2	10
Mo	3,660.92	6	6
Zr	3,660.91	1	4
Ru	3,660.81	8	2
Er	3,660.78	1	10
Tb	3,660.75	—	30
U	3,660.74	8	1
Co	3,660.69	2	5
Ce	3,660.64	10	40
Ti	3,660.63	18	90
W	3,660.61	9	8
Eu	3,660.60	—	30*d*
Re	3,660.53	—	25
Ar	3,660.44	(15)	—
Tb	3,660.44	3	15
Mn	3,660.40	75	75
Pr	3,660.375	20	40
W	3,660.370	4	5
Nb	3,660.366	30	20
Fe	3,660.33	1	5
W	3,660.17	3	6
Ce	3,660.16	4	12
Th	3,660.12	2	4
Pr	3,660.07	3	10
Ce	3,659.97	5	20
Nd	3,659.95	4	15
Ne	3,659.93	(7)	—
Ti	3,659.76	150	50
Th	3,659.63	1	4
Sm	3,659.62	6	10
Nb	3,659.61	500	15
U	3,659.59	2	10
Er	3,659.55	1	12*W*
W	3,659.54	3	6
Fe	3,659.52	80	125
Th	3,659.51	12	10
Ar	3,659.50	(100)	—
Br	3,659.50	(25)	—
Tb	3,659.45	3	15
Pt	3,659.42	2	5
Mo	3,659.36	30	30
Cu	3,659.35	4	8
W	3,659.32	6	10
P	3,659.26	(50)	—
Ce	3,659.23	6	20
U	3,659.16	1	15

	λ	I	
		J	O
Tm	3,683.20	20	10
Tb	3,683.26	—	15
Eu	3,683.27	1*h*	18*W*
W	3,683.31	7	8
Ag	3,683.32	10	—
Th	3,683.23	4	5
W	3,683.39	7	8
Ag	3,683.45	2	4
Pb	3,683.471	50	300
Zn	3,683.471	(15)	20
Er	3,683.472	—	25
Mn	3,683.474	—	12
In	3,683.520	15	—
Ir	3,683.523	—	
As	3,683.58	15	—
U	3,683.59	8*d*	2*d*
W	3,683.62	6	—
Pr	3,683.846	2	4
Eu	3,683.85	1*h*	8*W*
W	3,683.94	9	10
Er	3,684.01	—	6
Fe	3,684.11	200	300
Sm	3,684.120	5	20
Gd	3,684.124	150	200*W*
Mo	3,684.22	25*d*	1
Ce	3,684.24	—	6
Er	3,684.28	3	15
Nd	3,684.29	6	10
Lu	3,684.32	1	15
Mo	3,684.327	1	5
Ir	3,684.327	5	6
V	3,684.33	10	40
Co	3,684.48	—	200*W*
Mn	3,684.52	5	5
Sm	3,684.54	6	6
Os	3,684.55	8	15
U	3,684.62	5	5
Nd	3,684.64	1	8
W	3,684.66	9	10
Cu	3,684.67	4	12
Tb	3,684.81	8	15
Er	3,684.86	1	10*W*
Mn	3,684.87	15	15
Yt	3,684.90	8	5
Nd	3,684.909	10	10
Hg	3,684.91	(18)	—
Pr	3,684.92	1	8
Cu	3,684.93	3	7
Co	3,684.96	—	8

	λ	I J	I O
Hf	3,659.045	8	3
Pr	3,659.038	8	10
U	3,659.01	8	2
Mo	3,658.92	20	—
Tb	3,658.88	100	100
Er	3,658.84	1*W*	10*W*
Th	3,658.81	1	5
Ta	3,658.78	3	35
Ce	3,658.77	—	6
Re	3,658.74	—	4
U	3,658.68	—	6
Nb	3,658.60	5*h*	5*h*
Fe	3,658.55	1	5
Mn	3,658.52	5	5
W	3,658.37	3	6
Pr	3,658.35	1	7
Mo	3,658.33	25	4
V	3,658.264	10	1
Ce	3,658.258	—	10
Tb	3,658.22	3	15
Pr	3,658.21	2	15
Th	3,658.173	6	8
Cr	3,658.168	20	—
Mo	3,658.15	5	5
Ti	3,658.10	60	150
Ce	3,658.09	—	5
Th	3,658.07	10	10
Nd	3,657.991	2*h*	4*h*
Rh	3,657.987	200*W*	500*W*
Co	3,657.91	5	18
Fe	3,657.904	4	20
Mn	3,657.902	10	12
Nb	3,657.897	5	4
W	3,657.88	15	5
U	3,657.82	—	10
Ce	3,657.68	—	5
Pr	3,657.625	1	10
Ir	3,657.616	2	5
W	3,657.59	25	10
Ru	3,657.55	50	—
Ta	3,657.495	2	15
V	3,657.492	5	20
Pr	3,657.42	4	25
Mo	3,657.35	30	30
U	3,657.322	15	6
Sm	3,657.316	1	15
Ta	3,657.27	1*h*	18
Rn	3,657.20	(10)	—
Re	3,657.14	7	20
W	3,685.02	7	9
Ru	3,685.07	3	8
Nb	3,685.13	2*W*	10*W*
Ho	3,685.16	15	6
Er	3,685.17	—	6*W*
Ti	3,685.19	700*R*	150
Mn	3,685.21	—	15
Pr	3,685.26	8	60
Rn	3,685.30	(8)	—
Ce	3,685.52	—	6
Cr	3,685.55	3	4
Mn	3,685.56	12	12
W	3,685.60	4	7
Zr	3,685.63	—	5
Pr	3,685.69	2	7
Ne	3,685.736	(75)	—
Dy	3,685.74	6	6
U	3,685.77	1	15
Nd	3,685.80	20	30
Th	3,685.87	3	4
Xe	3,685.90	(40)	—
Ru	3,685.951	5	10
Ti	3,685.955	—	40
Fe	3,686.00	125	150
Ce	3,686.04	—	8
Mo	3,686.11	5	5
Kr	3,686.15	(80*Wh*)	—
Ra	3,686.16	(10)	—
Tm	3,686.16	5	5
Ta	3,686.18	2	35
Se	3,686.21	(35)	—
Fe	3,686.258	4	10
V	3,686.262	100	100
Ce	3,686.263	—	5
Gd	*3,686.34*	*200*	*150W*
Re	*3,686.46*	—	*8*
Pr	*3,686.478*	*6*	*25*
Co	*3,686.484*	—	*8*
J	*3,686.547*	*(70)*	—
Cu	*3,686.555*	*25*	—
Cr	*3,686.80*	*5h*	*20h*
Ta	*3,686.82*	*1h*	*20W*
Pr	*3,687.04*	*20d*	*60d*
Ir	*3,687.08*	*4*	*40*
Er	*3,687.100*	—	*10*
Fe	*3,687.101*	*7*	*15*
S	*3,687.13*	*(15)*	—

	λ	I J	I O
Nb	3,657.11	5	5
J	3,657.06	(15)	—
Co	3,656.97	6	60
Cl	3,656.95	(15)	—
U	3,656.93	8	8
Os	3,656.90	30	150
Ta	3,656.89	2h	15
Zr	3,656.88	—	10
Dy	3,656.85	2	4
Cu	3,656.79	3	4
Tb	3,656.77	3	8
Ce	3,656.75	—	10s
V	3,656.71	20h	80
Th	3,656.70	1	6
W	3,656.684	5	7
Ir	3,656.677	—	5
Ag	3,656.64	8h	—
U	3,656.63	—	8
bhCa	3,656.6	—	12
Tb	3,656.48	3	15
Ir	3,656.42	—	6
Pr	3,656.27	1	9
Cr	3,656.26	25	80
Fe	3,656.23	5	15
Th	3,656.20	10	10
W	3,656.17	3	8
Gd	3,656.16	200	200W
Nd	3,655.96	4	10
Cu	3,655.86	7	20
Ce	3,655.85	12	25
Mo	3,655.79	25	—
Sn	3,655.78	25h	30
Yb	3,655.73	—	8
Fe	3,655.67	2h	15
Dy	3,655.60	2	10
Fe	3,655.47	25	25
Ar	3,655.29	(15)	—
Pr	3,655.12	2	9
U	3,655.11	4	8
Mo	3,655.08	20	8
Al	3,655.00	(100)	—
Al	3,654.98	(18)	—
Ce	3,654.97	3	15
Re	3,654.93	—	15
U	3,654.89	—	25
Tb	3,654.88	20	70
Rh	3,654.87	10	40
Tb	3,687.15	3	15
Pr	3,687.20	15d	50d
Nd	3,687.29	10	20
Ti	3,687.35	—	10
Pt	3,687.43	3	35
Fe	3,687.46	300	400
V	3,687.47	15	20
Cr	3,687.54	5	12h
Fe	3,687.66	15	15
Er	3,687.758	3	10
Gd	3,687.759	200	200W
Eu	3,687.79	50W	40W
			8
Ce	3,687.80	2	10
Sm	3,687.98	—	50
Nb	3,687.97	300W	20W
Nd	3,688.05	6	15
W	3,688.069	7	12
V	3,688.069	200R	200
Tb	3,688.15	15	30
Ir	3,688.17	—	10
Nb	3,688.18	20	5
J	3,688.21	(125)	—
Rn	3,688.30	(40)	—
Mo	3,688.31	150	4
Nd	3,688.33	8	10
Ni	3,688.41	15	150
Sm	3,688.42	8	20
Eu	3,688.44	500W	1000W
Cl	3,688.44	(15)	—
Cr	3,688.46	6	18
Ba	3,688.47	—	12
Fe	3,688.48	8	40
Re	3,688.65	—	10
Ce	3,688.66	1	10
Th	3,688.76	12	12
Re	3,688.89	—	10
Mo	3,688.97	5	20
Os	3,689.06	30	200
Tb	3,689.12	8	15
Ce	3,689.16	—	8
U	3,689.20	1	10
Pb	3,689.31	40	—
Pr	3,689.40	10	40
Eu	3,689.42	—	10W
Fe	3,689.46	150	200
Re	3,689.52	—	100W
Cr	3,689.63	5	8

	λ	I J	I O		λ	I J	I O
Sm	3,654.85	1	9	Ce	3,689.68	—	8
Mg	3,654.83	(200)	—	Nb	3,689.69	15	25
Gd	3,654.64	200	200W	Pr	3,689.71	10	20
Ti	3,654.59	40	100	Tb	3,689.72	3	8
				Ta	3,689.73	2	35
Mo	3,654.58	20	25				
Os	3,654.49	15	100	W	3,689.88	8	10
Co	3,654.45	—	35	Ti	3,689.91	40	100
Nb	3,654.42	10	10	Pr	3,689.98	3	9
Ru	3,654.40	40	3	Ru	3,690.03	100	5
				Sm	3,690.08	—	10
Re	3,654.36	—	8				
Pr	3,654.34	2	8	Nd	3,690.09	—	8
Cu	3,654.30	2h	10Wh	Th	3,690.119	10	10
Tb	3,654.30	—	8	Ce	3,690.122	—	8
W	3,654.20	10	12	W	3,690.26	8	9
				V	3,690.28	125	200
Nd	3,654.16	10	10				
Kr	3,653.97	(250h)	—	Pd	3,690.34	1000W	300h
Sr	3,653.93	3	15	Mo	3,690.36	25	—
Cr	3,653.91	25	100	Re	3,690.38	—	8
Tb	3,653.87	8	15	Tb	3,690.44	—	8
				Pr	3,690.45	2	15
Fe	3,653.76	10	25				
Os	3,653.72	10	30	Fe	3,690.46	6	15
Ce	3,653.67	8	18	Th	3,690.49	10	10
Re	3,653.615	—	15	Nd	3,690.54	4	8
Nb	3,653.615	5	10	Yb	3,690.58	50	10
				Mo	3,690.59	5	10
Tm	3,653.61	20	30				
Ti	3,653.50	200	500	Kr	3,690.65	(30)	—
Sm	2,653.48	2	15	Rh	3,690.70	50	125
P	3,653.38	(100W)	—	Co	3,690.723	10	60
Sr	3,653.27	8	30	Os	3,690.724	20	15
				Fe	3,690.73	60	80
U	3,653.21	1	10				
Os	3,653.20	5W	10W	Ar	3,690.90	(300)	—
Ir	3,653.19	50	15	Sm	3,690.93	2	15
Nd	3,653.15	4	10	Tb	3,691.15	50	50
Sm	3,653.113	2	10	Nb	3,691.18	15	5
				Re	3,691.38	—	12
Ce	3,653.108	5	15				
Se	3,653.05	(25)	—	Pr	3,691.48	3	9
Tb	3,652.97	8	15	Re	3,691.50	—	100W
Tl	3,652.95	50	150	Eu	3,691.55	—	20d
Er	3,652.88	4	20	Sm	3,691.57	3	25W
				Yb	3,681.69	—	10
Er	3,652.58	3	20				
Gd	3,652.56	25	20	Ta	3,691.879	15	—
Co	3,652.543	—	200r	W	3,691.88	20d	4d
Th	3,652.542	10	10	Th	3,691.881	2	8
Mo	3,652.47	25	—	Tb	3,692.02	8	15
				Th	3,692.08	4	8
Cu	3,652.43	1h	10Wh				
V	3,652.42	30h	40	Re	3,692.126	—	15
Pr	3,652.33	25	—	Ti	3,692.134	1	12
Mo	3,642.33	25	—	Sm	3,692.221	40	90
Tb	3,652.26	8	30	V	3,692.225	150R	200R

	λ	I	
		J	O
Th	3,652.17	10	10
U	3,652.07	8	12
Re	3,651.97	—	40
Tb	3,651.86	8	30
Sc	3,651.80	45	50
Hf	3,651.77	1	15
Re	3,651.661	—	25W
Cr	3,651.660	18	1
Nd	3,651.59	6h	10
Fe	3,651.47	200	300
Mo	3,651.35	2	10
Co	3,651.26	—	20
Nb	3,651.19	400	10
Gd	3,651.12	15	10
Mo	3,651.11	50	2
Fe	3,651.10	3	10
Al	3,651.09	(18)	—
Al	3,651.06	(50)	—
Pr	3,651.04	2	9
Kr	3,651.02	(25h)	—
W	3,651.00	12	10
Sm	3,650.98	10	25
Gd	3,650.97	30	25
Tb	3,650.93	8	8
Ce	3,650.88	3	12
Nb	3,650.81	15	15
Th	3,650.77	10	10
Nd	3,650.69	6	15
Nd	3,650.415	6	12
Er	3,650.414	2	15
Tb	3,650.40	100	50
Os	3,650.38	4	15
Cr	3,650.34	40	—
Ru	3,650.32	12	3
Fe	3,650.28	50	70
N	3,650.19	(70)	—
Pr	3,650.18	6	30
La	3,650.174	60	100
Sm	3,650.168	5	25
Hg	3,650.14	500	200
Ce	3,650.12	1	10
Mo	3,650.05	25	3
Fe	3,650.03	30	70
Cr	3,649.86	4	15
Nb	3,649.85	20	20
Ar	3,649.83	(800)	—
Th	3,649.74	3	8
Ce	3,649.73	1	10
Cd	3,649.60	15	20

	λ	I	
		J	O
Rh	3,692.36	150Wd	50hd
O	3,692.44	(50)	—
Th	3,692.57	2	10
Mo	3,692.64	150	3
Ho	3,692.65	15	10
Er	3,692.652	12	20
Ir	3,692.69	4	15
U	3,692.75	—	10
Sm	3,692.76	6	20
Nd	3,692.77	8	12
Mn	3,692.81	50	50
Sm	3,692.90	2	10
Tb	3,692.95	8	30
Fe	3,693.03	7	15
Ta	3,693.05	3h	25r
Cr	3,693.09	5	10
Co	3,693.11	15	80
Nd	3,693.13	8	20
Pr	3,693.363	8	30
Co	3,693.364	—	18
Nb	3,693.37	10	8
Ce	3,693.42	1	10
Co	3,693.48	10	35
Xe	3,693.49	(40)	—
Tb	3,693.56	—	30
Ru	3,693.59	5	20
Mn	3,693.67	60	50
U	3,693.70	18	4
Ce	3,693.71	—	8
In	3,693.79	(25)	—
Eu	3,693.82	1	25W
In	3,693.90	(35)	—
Ni	3,693.93	—	50
Sm	3,694.00	150	100
Fe	3,694.010	300	400
Gd	3,694.011	25	25
In	3,694.03	(40)	—
Tb	3,694.12	—	8
Er	3,694.19	15	25d
Ne	3,694.197	(250)	—
Yb	3,694.203	1000R	500R
Ho	3,694.24	20	10
Sm	3,694.31	8	15
Mo	3,694.42	20	—
Ti	3,694.45	20	80
W	3,694.51	20	10
Ta	3,694.52	18	7h
V	3,694.62	5	60

	λ	I J	I O
Ra	3,649.55	(1000)	—
La	3,649.509	8	40
Fe	3,649.508	100	100
Sm	3,649.506	30	100
Tb	3,649.41	3	8
Pr	3,649.40	3	10
Co	3,649.35	4	200
Fe	3,649.30	25	60
Th	3,649.25	10	10
Hf	3,649.10	5	20
Pb	3,649.00	(20)	—
Cr	3,648.997	20	40
V	3,648.97	50	80
Dy	3,648.807	30	50
Os	3,648.806	10	100
Kr	3,648.61	(40h)	—
Mo	3,648.607	10	10
Cr	3,648.53	8	10
Gd	3,648.47	—	20
Th	3,648.42	4	8
Cu	3,648.38	7	10
Os	3,648.304	5	10
Pr	3,648.299	10	30
Nd	3,648.19	8	12
Co	3,648.14	2	20
Ce	3,647.95	5	10
Fe	3,647.84	400	500
Lu	3,647.77	5	100
Ce	3,647.751	3	10
Tb	3,647.75	—	50
Tm	3,647.73	15	30
Ir	3,647.70	—	20
Co	3,647.06	8	100
Th	3,647.65	8	8
W	3,647.52	9	8
Hf	3,647.49	2	20
Fe	3,647.43	15	20
Cr	3,647.394	20	2
Co	3,647.388	—	8h
V	3,647.34	—	20
Nb	3,647.30	20	10
Sm	2,647.27	3	8
Tm	3,647.24	20	10
Co	3,647.09	4	30
Tb	3,647.06	8	30
Ce	3,646.96	5	15
V	3,646.85	20h	—
Eu	3,646.75	5h	20

	λ	I J	I O
Nb	3,694.67	10	10
Pr	3,694.69	4h	40
Tb	3,694.75	8	50
Dy	3,694.75	30	20
Tm	3,694.76	20	20
Nd	3,694.79	20	30
Er	3,694.81	6	15
Sm	3,694.82	3	15
Ce	3,694.91	2	15S
Mo	3,694.94	30	40
Fe	3,695.05	150	200
U	3,695.21	—	8
V	3,695.33	70h	125
Ta	3,695.38	7S	15
Rh	3,695.52	10	15
Bi	3,695.55	50	—
Tb	3,695.69	—	8
Cr	3,695.858	3	12
V	3,695.865	100r	150
Nb	3,695.90	200	4
Th	3,695.98	8	8
Ir	3,696.16	2	20
Er	3,696.25	2	18
Tb	3,696.30	8	15
Eu	3,696.44	1	12d
Ar	3,696.51	(20)	—
Hf	3,696.51	2	8
Mn	3,696.57	50	100
Ru	3,696.59	15	50
Yt	3,696.62	20h	6
Pr	3,696.66	2	9
Nb	3,696.68	50	2
Gd	3,696.76	15	10
Tb	3,696.85	8	30
Dy	3,696.87	2	10h
Ti	3,696.88	3	20
Er	3,696.92	1	12
Mo	3,697.03	25	1
U	3,697.13	3h	10
Nd	3,697.16	10	20
Dy	3,697.25	4	10
Nb	3,697.39	20	10
Fe	3,697.43	60	100
Zr	3,697.458	20	20
W	3,697.46	9	10
Nd	3,697.54	10	20
Tm	3,697.58	3	8
Ce	3,697.66	1	10

	λ	I J	I O
Eu	3,646.66	—	10W
Ce	3,646.65	3	10
Re	3,646.63	—	10
W	3,646.52	35	10
Tb	3,646.46	3	8
Pr	3,646.30	15	50
U	3,646.22	—	10
Ti	3,646.200	25	70
Gd	3,646.196	150	200W
Cr	3,646.16	8	18
Pd	3,645.97	—	15
Er	3,645.94	2	15
V	3,645.90	20	—
Fe	3,645.82	60	80
Nd	3,645.78	1	8
Pr	3,645.66	20	30
Gd	3,645.634	20	20
Nd	3,645.626	6	15
Tb	3,645.60	—	8
W	3,645.599	20	4
V	3,645.596	1	15
Pr	3,645.54	8	20
Fe	3,645.49	7	15
Ce	3,645.45	2	10W
Dy	3,645.42	100	300
La	3,645.414	60	100
Ho	3,645.41	8	8
Er	3,645.40	12	25S
Sm	3,645.39	6	8
Tb	3,645.38	15	50
Sc	3,645.31	50	50
Ir	3,645.30	2h	25
Sm	3,645.29	7	9
Cu	3,645.23	5	20
Fe	3,645.22	—	9
Co	3,645.19	3	60
Eu	3,645.17	5Wh	10W
Fe	3,645.08	8	20
U	3,645.03	15	8
Tb	3,644.93	—	15
Fe	3,644.80	6	20
Ca	3,644.76	—	30
V	3,644.71	50	80
Ti	3,644.70	5	35
Cr	3,644.69	18	5
Nd	3,644.66	4	10
Pr	3,644.54	3	8
Nd	3,644.43	15	—

	λ	I J	I O
Re	3,697.70	—	150W
Tb	3,697.72	3	8
Gd	3,697.74	200W	200W
Ru	3,697.76	6	8
Nb	3,697.85	50	50
Cr	3,698.00	40	—
Pr	3,698.07	12	25
Ir	3,698.10	8	30
Th	3,698.11	2	8
Ce	3,698.12	1	12
Mo	3,698.166	5	15h
Zr	3,698.167	80	50
Tb	3,698.17	3	8
Dy	3,698.17	30	10
Ti	3,698.18	8	10
Rh	3,698.26	10	15
Th	3,698.30	10	12
Ce	3,698.36	2	10
Hf	3,698.39	25	10
Tb	3,698.45	—	8
Yb	3,698.59	15	3
Rh	3,698.595	15	20
Fe	3,698.605	20	40
Ce	3,698.66	1	20
W	3,698.72	9	9
Os	3,698.83	12	20
Er	3,698.97	—	10
Co	3,699.02	—	8h

	λ	I J	I O
Ca	*3,644.41*	*15*	*200*
Hf	*3,644.35*	*50*	*25*
Hg	*3,644.32*	*(40)*	—
Ce	*3,644.30*	*1*	*8*
Te	*3,644.27*	*(25)*	—
U	*3,644.24*	*2*	*18*
Tb	*3,644.13*	*3*	*15*
Sm	*3,643.99*	—	*8*
Dy	*3,643.89*	*6*	*15*
Ne	*3,643.89*	*(18)*	—
V	*3,643.86*	*30*	*40*
Tb	*3,643.76*	*3*	*30*
Nb	*3,643.72*	*15*	*15*
Tm	*3,643.65*	*40*	*60*
Fe	*3,643.63*	*8*	*20*
Mo	*3,643.47*	*20*	*1h*
Pr	*3,643.32*	*8*	*25*
Tb	*3,643.26*	*3*	*15*
Cr	*3,643.20*	*30*	*5*
Co	*3,643.18*	*15*	*80*
Pt	*3,643.17*	*8*	*60*
Fe	*3,643.11*	*5*	*30*
Ar	*3,643.09*	*(100)*	—
Re	*3,642.99*	—	*100*
Pr	*3,642.92*	*2*	*8*
Ce	*3,642.83*	*1*	*8*
Fe	*3,642.81*	—	*20*
Au	*3,642.80*	*5h*	*10*
F	*3,642.798*	*(30)*	—
Sc	*3,642.78*	*50*	*60*
Sm	*3,642.74*	*5*	*25*
Tb	*3,642.68*	*8*	*15*
Ti	*3,642.67*	*125*	*300*

U 3,859.58 *J* 30 *O* 20

	λ	J	O
Eu	**3,859.52**	—	**4*d***
Nd	**3,859.422**	**12**	**10**
Pr	**3,859.416**	**3**	**3**
Au	**3,859.37**	**3**	—
V	**3,859.34**	**8**	**20**
Al	**3,859.33**	**(10)**	—
W	**3,859.29**	**12**	**10**
Eu	**3,859.22**	—	**2*wh***
Fe	**3,859.216**	**100**	**100**
Mg	**3,859.21**	—	**2**
Pr	**3,859.14**	**4**	**4**
Nd	**3,859.670**	**25**	**25**
Ca	**3,859.675**	**5*h***	**6**
Ru	**3,859.71**	**15**	**6**
Ce	**3,859.79**	—	**3**
Ta	**3,859.80**	**2**	**10**
Th	**3,859.83**	**10**	**5**
Fe	**3,859.91**	**600**	**1000*r***
Ce	**3,859.94**	**5**	**3**
W	**3,859.98**	**30**	**15**
Sm	**3,860.14**	**1**	**5**
S	**3,860.15**	**(8)**	—

	λ	I	
		J	O
U	**3,859.01**	**4**	**8**
Nb	**3,858.95**	**50**	**20**
Cr	**3,858.89**	**20*Wh***	**35*Wh***
W	**3,858.85**	**6**	**5**
Mo	**3,858.84**	**15**	—
Ca	**3,858.81**	**2*h***	—
Kr	**3,858.78**	**(5*Wh*)**	—
Sm	**3,858.74**	—	**20**
Ru	**3,858.69**	**4**	**4**
V	**3,858.685**	**15**	**50**
U	**3,858.682**	**6**	**3**
Ta	**3,858.61**	**2*h***	**10*h***
Yb	**3,858.555**	**3**	—
Nd	**3,858.55**	**3**	**3**
Xe	**3,858.53**	**(10)**	—
Sm	**3,858.514**	**10*h***	**10*d***
U	**3,858.508**	**2*h***	**2**
Dy	**3,858.40**	—	**5**
Er	**3,858.39**	**2**	**4**
Mo	**3,858.32**	**5**	**5**
Hf	**3,858.31**	**5*Wh***	**5*Wh***
Ni	**3,858.301**	**70*h***	**800*r***
Co	**3,858.299**	**4**	**18**
Pr	**3,858.26**	**8**	**20**
U	**3,858.18**	**2**	—
Ti	**3,858.14**	**7**	**40**
Dy	**3,858.09**	—	**6**
Nd	3,857.85	20	20
Tm	3,857.84	5	15
Cr	3,857.63	25	50
Ru	3,857.55	25	50
Kr	3,857.32	(20*Wh*)	—
Mo	3,857.20	60	—
Nd	3,857.14	20*h*	30
Os	3,857.09	15	150
Ce	3,857.02	2	10
Nd	3,856.95	15	20
Co	3,856.80	2	80
Nb	3,856.68	5	20
V	3,856.67	—	20
Mn	3,856.53	30	15
Rh	3,856.51	20	50
Ru	3,856.46	8	50
Fe	3,856.37	300	500
Cr	3,856.28	15	20
O	3,856.16	(18)	—
Ir	3,856.07	4	25
Ce	**3,860.18**	**4**	**2**
Sm	**3,860.28**	**5**	**4**
La	**3,860.21**	**2**	—
Ce	**3,860.40**	**3**	**6**
Cu	**3,860.46**	**7**	**30**
Sm	**3,860.62**	**10**	**25**
Ce	**3,860.626**	**15**	**8**
Zr	**3,860.627**	—	**5**
V	**3,860.658**	—	**20**
U	**3,860.630**	**15*h***	**1**
S	**3,860.64**	**(15)**	—
Eu	**3,860.72**	**6**	**5*d***
Ru	**3,860.723**	**4**	**20**
In	**3,860.73**	**(5)**	—
Rb	**3,860.80**	**(5)**	—
Cl	**3,860.83**	**(150)**	—
Re	**3,860.857**	—	**3**
Nb	**3,860.857**	**10**	**5**
Hf	**3,860.910**	**6**	**6**
Fe	**3,860.915**	**2**	**1**
Nd	**3,860.94**	**4**	**4**
Cl	**3,860.99**	**(100)**	—
Ce	**3,860.993**	**2**	**2**
Sm	**3,861.059**	**4**	**15*d***
W	**3,861.062**	**10**	**10**
Ta	**3,861.077**	**1**	**3*h***
Hg	**3,861.08**	**(10*h*)**	—
Ti	**3,861.084**	**3**	**10*h***
U	3,861.12	—	5
Co	3,861.16	15	300*R*
Eu	3.861.18	30*W*	30*W*
Sm	3,861.183	7	15
Pr	3,861.31	5	15
Cl	3,861.34	(50)	—
Fe	3,861.342	50	80
Sm	3,861.59	8	15
V	3,861.61	3*h*	10*h*
Ho	3,861.68	20	40
U	3,861.73	6	8
Ti	3,861.74	3	10
Cu	3,861.75	2	50
Sm	3,861.79	6	10
Cl	3,861.88	(20)	—
Ir	3,861.94	10	15
U	3,862.05	1	8
Re	3,862.12	—	10
V	3,862.22	20	[illegible]
Er	3,862.41	3	15

	λ	I J	I O
Re	3,855.93	—	20
Sm	3,855.89	15	5
V	3,855.84	200	200
Sc	3,855.62	15	15
Gd	3,855.581	6	15
Tb	3,855.58	—	10
Ru	3,855.578	15	4
Cr	3,855.57	30	30
W	3,855.54	9	10
Nb	3,855.50	50*h*	—
Nb	3,855.45	—	10
Tb	3,855.38	—	10
V	3,855.37	50*r*	50*r*
Cr	3,855.29	35	35
La	3,854.91	40	—
Pr	3,854.90	30	80*W*
Cr	3,854.79	15*d*	20*d*
Cl	3,854.75	(30)	—
Ru	3,854.73	5*Wh*	10
Os	3,854.70	12	30
U	3,854.65	30	20
Th	3,854.55	20	20
U	3,854.23	—	20
Cr	3,854.22	15	40
Sm	3,854.20	25	200*Wh*
Pb	3,854.053	100	—
Ho	3,854.05	20	10
Ti	3,853.73	3	12
Pr	3,853.49	8	10
Nd	3,853.48	20	40
Mn	3,853.47	20	25
Os	3,853.44	15	100
Nb	3,853.38	20	10
Lu	3,853.29	—	10
Cr	3,853.18	10	20
Ce	3,853.16	3	25
Ti	3,853.05	4	18
Dy	3,853.04	—	100
Nd	3,852.90	8	10
Pr	3,852.80	50	100
Fe	3,852.57	100	150
Gd	3,852.50	8	100
Sc	3,852.392	15	15
Ce	3,852.387	25	8
Nd	3,852.38	50	60
Cr	3,852.22	12	60
Ru	3,852.14	10	12
V	3,852.10	10	20
Mo	3,851.99	15	10

	λ	I J	I O
Ce	3,862.46	10	15
Sm	3,862.48	3	10
Nd	3,862.487	12	20
V	3,862.490	—	10
Cr	3,862.55	20	25
Ru	3,862.65	60	2
Cu	3,862.76	—	10
Ti	3,862.83	4	30*h*
Nb	3,862.93	5	10
Nb	3,863.05	20	3
U	3,863.08	1	10
Re	3,863.15	—	10
Nd	3,863.33	4	10
Nb	3,863.38	10	15
Th	3,863.39	20	20
U	3,863.40	20	5
Sm	3,863.407	8	15*d*
Nd	3,863.408	20	20
Co	3,863.61	—	30
Fe	3,863.74	30	60
Re	3,863.75	—	15
V	3,863.866	15	25
Zr	3,863.874	6	200
Bi	3,863.90	(100)	—
Sm	3,864.05	2	10
Eu	3,864.10	10*W*	7*W*
Mo	3,864.11	500*R*	1,000*R*
Bi	3,864.20	150*R*	—
V	3,864.300	—	30
U	3,864.305	10	8
Zr	3,864.335	20	50
W	3,864.336	10	12
O	3,864.42	(18)	—
U	3,864.48	10	8
La	3,864.49	150	100
Ti	3,864.50	6	15
Cl	3,864.60	(15)	—
Hf	3,864.75	20	2
V	3,864.86	50*r*	100*r*
Nb	3,865.02	200*h*	—
Nb	3,865.039	—	10*d*
Os	3,865.045	5	10
Mo	3,865.15	20	—
Sm	3,865.24	—	10
Ru	3,865.40	4	10
Be	3,865.43	—	30
Dy	3,865.45	2	10
Pr	3,865.458	125*r*	200*r*

	λ	I	
		J	O
Re	3,851.98	—	20
Ra	3,851.90	(25)	—
Sm	3,851.88	8	10
Nd	3,851.75	15	8
U	3,851.73	2	15
Cl	3,851.69	(30)	—
F	3,851.67	(200)	—
Nd	3,851.66	—	20
Pr	3,851.62	150*W*	200
Cl	3,851.42	(75)	—
Cl	3,851.20	(100)	—
Rb	3,851.20	(20)	5
V	3,851.17	15	60
Gd	3,850.98	6	10
Co	3,850.95	—	100
Pr	3,850.825	15	50
J	3,850.82	(15)	—
Fe	3,850.820	75	200
Mo	3,850.819	25	—
Ar	3,850.57	(400)	—
Ru	3,850.43	10	50
Sb	3,850.23	20	—
Nd	3,850.227	2	10
V	3,850.16	2	15*W*
Cr	3,850.04	40*r*	40*r*
F	3,849.99	(600)	—
Fe	3,849.97	400	500
Os	3,849.94	20	125
Xe	3,849.87	(25*Wh*)	—
Ni	3,849.58	15*h*	—
Cr	3,849.53	20	20
Dy	3,849.40	—	20
Cr	3,849.36	30*h*	40*h*
V	3,849.32	25	60
Zr	3,849.25	4	10
Hf	3,849.18	15	15
La	3,849.013	150	200
Sm	3,849.012	4	10
Er	3,849.00	2	12
Cr	3,848.98	50*d*	80*d*
Sm	3,848.81	10	150*d*
Tb	3,848.75	200	100
*bh*B	3,848.7	—	200
U	3,848.62	8	10
Ce	3,848.60	2	20
Nd	3,848.52	20	10
Ti	3,848.310	2	10
Nd	3,848.309	—	40

	λ	I	
		J	O
Sr	3,865.46	—	50
Os	3,865.47	200	125
Be	3,865.52	—	10
Fe	3,865.53	400	600
Eu	3,865.56	—	30*W*
Ir	3,865.64	20	25
Mn	3,865.66	12	12
U	3,865.92	25	20
Nd	3,865.98	30	35
Ti	3,866.028	5	12
Be	3,866.03	—	15
Ti	3,866.44	10	40*h*
Os	3,866.48	5	10
Nd	3,866.52	2	10
Dy	3,866.59	—	25
V	3,866.74	50	5
Nd	3,866.80	2	15
Ce	3,866.82	6	12
Gd	3,866.98	—	10
Fe	3,867.22	100	150
V	3,867.35	3	15
He	3,867.48	(15)	—
U	3,867.51	—	10
Pr	3,867.55	2	20
S	3,867.56	(150)	—
Se	3,867.60	(50)	—
V	3,867.602	35	70
Gd	3,867.619	—	10
Sm	3,867.623	8	10
Ru	3,867.84	35	60
Nb	3,867.917	20	30
Fe	3,867.923	8	30
W	3,867.97	35	30
Ti	3,868.40	8	50
Dy	3,868.46	—	50
Ar	3,868.53	50	—
Cl	3,868.62	(40)	—
Os	3,868.69	10	30
Dy	3,868.81	—	60
Nd	3,869.04	4	20
Mo	3,869.08	30	25
N	3,869.10	(15)	—
Tl	3,869.19	(18)	—
Ti	3,869.29	8	15
Th	3,869.36	8	10
Dy	3,869.43	—	25
Ir	3,869.47	2	15
Fe	3,869.56	80	100

	λ	I	
		J	O
Mo	3,848.30	20	25
Mg	3,848.24	10	10
Nd	3,848.23	10*d*	10*d*
Ir	3,848.16	2	10
Ta	3,848.05	5	30
Tm	3,848.02	250	400
Er	3,847.89	1	18*d*
Yb	3,847.86	—	40*W*
Nd	3,847.85	50	60
Eu	3,847.84	—	12*W*
W	3,847.49	15	18
N	3,847.38	(10)	—
V	3,847.33	70*h*	100
Nd	3,847.25	10	20
Mo	3,847.248	25	25
F	3,847.09	(800)	—
Zr	3,847.01	4	10
*bh*B	3,847.0	—	100
Dy	3,846.99	2	10
Nd	3,846.97	20	20
Fe	3,846.80	100	125
Nd	3,846.71	30*d*	30*d*
Ir	3,846.69	—	12
Ho	3,846.68	10	10
Ru	3,846.676	10	12
Pr	3,846.60	30	70*d*
U	3,846.56	—	10
Ti	3,846.45	3	15
Fe	3,846.415	75	50
Os	3,846.411	12	15
Dy	3,846.36	—	25
U	3,846.243	1	10
Th	3,846.242	8	10
W	3,846.21	20	20
Bi	3,846.03	100	—
Fe	3,846.001	1	10
Sm	3,845.997	4	10
La	3,845.997	50	40
Nd	3,845.99	25	30
Kr	3,845.98	(15)	—
Mo	3,845.95	20	20
Nb	3,845.90	30	10
Cl	3,845.82	(30)	—
Bi	3,845.80	(100)	—
Fe	3,845.70	6	10
Cl	3,845.68	(75)	—
Tb	3,845.61	15	10
Co	3,845.47	100	500*R*

	λ	I	
		J	O
Ti	3,869.61	4	10
Eu	3,869.73	—	20*Wh*
Tb	3,869.75	15	15
Dy	3,869.87	—	100
Re	3,869.92	—	40
Rh	3,870.01	3	15
Ti	3,870.13	5	12
Cr	3,870.27	3*W*	80*W*
Er	3,870.35	1	10
Ca	3,870.51	—	15
Co	3,870.53	8	70
U	3,870.56	8	8
V	3,870.58	10	35
U	3,871.04	1	30
V	3,871.08	35	60
Nb	3,871.19	20	15
Ru	3,871.21	1*h*	10
Au	3,871.35	20	25
U	3,871.38	20*h*	3*h*
Zr	3,871.390	20	10
B	3,871.39	20*h*	—
Ce	3,871.40	25	8
Mo	3,871.45	50	—
La	3,871.631	15	200
Dy	3,871.635	—	20
Fe	3,871.75	60	100
Sm	3,871.78	25	6
U	3,871.878	12	8
Mo	3,871.882	25	—
Mn	3,872.06	—	10
Dy	3,872.117	150	300
Mn	3,872.125	10	10
Rh	3,872.39	3	50
Fe	3,872.50	300	300
Hf	3,872.55	20	6
Ca	3,872.56	—	30
Th	3,872.728	20	30
Ce	3,872.729	15	8
V	3,872.75	2*h*	10*h*
W	3,872.83	10	12
Yb	3,872.86	5	12
Tb	3,873.00	—	10
U	3,873.09	2	12
Co	3,873.11	80	500*R*
Ir	3,873.15	10	25
Mn	3,873.196	12	12
Sm	3,873.200	3	20
Ti	3,873.21	7	40
Sm	3,873.47	3	10

	λ	I	
		J	O
V	3,845.454	—	20
Sc	3,845.454	10	10
Zr	3,845.44	20	20
Cl	3,845.42	(50)	—
Fe	3,845.17	60	100
Hg	3,845.15	(30)	—
Th	3,845.02	10W	20W
Ir	3,845.01	2h	15
V	3,844.89	7	25
Gd	3,844.58	15	15
Pr	3,844.56	10	50
Kr	3,844.45	(50Wh)	—
V	3,844.44	50h	100
Mn	3,843.98	100	75
Co	3,843.69	—	60
Os	3,843.66	12	80
V	3,843.51	15	50
Re	3,843.45	—	50
Cl	3,843.26	(100)	—
Fe	3,843.259	100	125
Zr	3,843.02	40	40
Sc	3,843.00	20	25
V	3,842.99	—	20
Nd	3,842.988	40	30
U	3,842.986	12	12
As	3,842.82	50	—
Nd	3,842.69	12	30
Tb	3,842.49	50	40
Pr	3,842.36	40	80
In	3,842.27	(35)	—
Co	3,842.05	20	400R
Dy	3,842.02	—	20
Th	3,841.96	20	20
Nd	3,841.95	30	30
V	3,841.89	12	35
Ti	3,841.75	12	20
Pb	3,841.62	60	—
Bi	3,841.60	(25)	—
Co	3,841.46	—	60
Dy	3,841.32	—	100
Os	3,841.29	10	30
Cr	3,841.28	80	150
Lu	3,841.18	8	100
Mn	3,841.08	50	50
Fe	3,841.05	400	500
Pr	3,841.01	10	30
Ag	3,840.80	3h	20
La	3,840.71	70	50

	λ	I	
		J	O
Ru	3,873.52	45	30
V	3,873.62	12	35
Sm	3,873.718	3	12
Os	3,873.724	20	20
K	3,873.75	20	—
Fe	3,873.76	80	125
Th	3,873.82	5	10
Co	3,873.95	80	400R
Dy	3,873.99	—	100
U	3,874.04	15	15
Er	3,874.12	1W	18d
Ti	3,874.14	4	12
Tb	3,874.18	200	200
V	3,874.34	2	10
U	3,874.36	—	10
W	3,874.41	12	12
Pr	3,874.45	4	20
U	3,874.46	2	10
Cr	3,874.53	12	70
Hg	3,874.98	(30h)	—
V	3,875.07	50	70r
Sm	3,875.17	10	50
Tb	3,875.11	8	20
Re	3,875.25	—	40
Ti	3,875.26	8	35h
U	3,875.34	6	12
Kr	3,875.44	(150Wh)	—
Nb	3,875.76	50	10
Ca	3,875.81	—	50
Nd	3,875.87	4	30
V	3,875.90	3	40
Fe	3,876.04	15	40
C	3,876.05	40	—
Ru	3,876.08	3	20
V	3,876.09	30	50
U	3,876.13	2	15
Pr	3,876.18	30	80
C	3,876.19	125	—
Cs	3,876.39	—	300
C	3,876.41	60	—
Lu	3,876.63	100	50
C	3,876.67	40	—
V	3,876.74	2	25
Os	3,876.77	50	300
Co	3,876.83	40	300W
Re	3,876.89	—	60
Hf	3,877.10	30	4
Sm	3,877.18	5	30

	λ	I J	I O
Ce	3,840.45	35	30
V	3,840.440	20	40
Fe	3,840.439	300	400
Zn	3,840.34	(50)	3
Os	3,840.30	20	150
Tb	3,840.26	—	50
Nd	3,839.81	20	30
Mn	3,839.78	125	100
Ru	3,839.69	30	50
Gd	3,839.64	—	25
U	3,839.63	2	30
V	3,839.38	5	30
Hg	3,839.26	(50)	—
Fe	3,839.259	75	100
Sm	3,839.19	3	25
Ta	3,839.03	5	30
V	3,839.00	10	60
Nd	3,838.98	30	20
Nd	3,838.72	25	50
Ce	3,838.54	3	35
Nd	3,838.33	25	40
Mg	3,838.26	200	300
Tm	3,838.20	60	80
U	3,838.15	10	8
U	3,837.83	8h	12
Kr	3,837.82	(30)	—
Kr	3,837.70	(30)	—
Mo	3,837.29	100W	—
Fe	3,837.142	6	25
P	3,837.14	(30)	—
Ru	3,836.70	40	8
Ta	3,836.30	3	30
Nd	3,836.541	100	80
Kr	3,836.54	(30Wh)	—
U	3,836.520	15	6
Sc	3,836.519	25	25
Sm	3,836.515	25	50
Gd	3,836.514	60	100Wh
Dy	3,836.51	40	100
Th	3,836.505	50W	50W
Er	3,836.505	10	40
Au	3,836.48	30	30
Fe	3,836.33	60	100
Re	3,836.32	—	40
Sm	3,836.12	4	20
Nd	3,836.11	15	40
Ti	3,836.08	30	15
Cr	3,836.07	8	25

	λ	I J	I O
Pr	3,877.22	80W	125W
Se	3,877.28	(50)	—
Os	3,877.31	10	20
Rh	3,877.34	5	20
Sm	3,877.47	8	50
Nb	3,877.56	20	50
Fe	3,878.02	300	400
Os	3,878.572	12	50
Fe	3,878.575	300	300R
Nd	3,878.58	2	30
Fe	3,878.68	—	30
V	3,878.71	100	35
Co	3,878.74	—	70r
Dy	3,879.05	2	25
Pr	3,879.21	80	100
Cr	3,879.22	15	60
V	3,879.23	3	35
Nb	3,879.35	300	5
Nd	3,879.54	5	30
U	3,879.55	—	18
Er	3,879.66	2W	20d
V	3,879.661	5	50
U	3,879.71	2	20
Er	3,880.04	2W	20Wd
Nd	3,880.38	4	25
Pr	3,880.47	60	80
Er	3,880.66	6Wd	25Wd
Sm	3,880.75	30	40
Hf	3,880.82	30	20
Cr	3,881.21	18	60
W	3,881.39	20	20
U	3,881.46	20	30
Sm	3,881.77	—	20
Cr	3,881.856	6	50
Os	3,881.858	20	125
Co	3,881.87	30	300R
Re	3,881.89	—	25
Dy	3,882.00	—	25
Ti	3,882.15	8	25
O	3,882.19	(35)	—
Ti	3,882.33	7	20
U	3,882.36	18	18
Sm	3,882.50	10	25
Ti	3,882.89	10	35
Tm	3,883.13	10	100
Nb	3,883.14	30	30
Fe	3,883.289	40	70
Cr	3,883.292	80	60

	λ	I	
		J	O
Os	3,836.06	20	150
Zr	3,835.96	5	25
V	3,835.56	12	50
H	3,835.40	(40)	—
Er	3,835.26	2d	25d
V	3,835.18	—	30
Nb	3,835.17	20	20
Ru	3,835.05	6	50
Pr	3,834.92	15	30
Cr	3,834.73	12	25
Ar	3,834.68	(800)	—
In	3,834.65	(35)	—
In	3,834.61	(40)	—
Sm	3,834.60	15	25
In	3,834.56	(40)	—
Mn	3,834.36	75	75r
Re	3,834.23	—	50
Fe	3,834.225	400	400
V	3,834.224	12	20
Rh	3,833.89	50	25
Ir	3,833.88	—	50
Mn	3,833.86	75	75
Mo	3,833.75	25	80
Ta	3,833.74	200	40
Re	3,833.70	—	30
Nd	3,833.60	12	20
Fe	3,833.31	60	100
Pr	3,833.042	20	20
Th	3,833.037	10W	20W
Nd	3,833.03	30	60
U	3,833.02	15	20r
Yt	3,832.89	80	30
Ni	3,832.87	—	25
V	3,832.835	5	25
Pb	3,832.83	50	—
Re	3,832.41	—	20
Mg	3,832.31	200	250
Tl	3,832.30	(30)	—
Pd	3,832.29	150	150
Os	3,832.18	10	30
U	3,831.86	3h	15
Nb	3,831.84	300	5
Ru	3,831.79	50	60
Ni	3,831.69	20	300
U	3,831.46	25	25
Hf	3,831.13	25	25
Cr	3,831.032	25	40
Nd	3,831.030	30	60d
Pr	3,830.72	60	100

	λ	I	
		J	O
U	3,883.33	18	10
Zn	3,883.34	2h	50
Tm	3,883.43	30	150
U	3,883.64	—	11
Cr	3,883.66	20	30
V	3,883.89	5	40
Ru	3,884.02	6	20
Fe	3,884.36	35	80
V	3,884.46	1	30
Co	3,884.61	—	100
Ru	3,884.676	6	20
U	3,884.681	20	8h
Eu	3,884.76	5Wh	50Wh
Th	3,884.83	15	20
V	3,884.84	70	4
P	3,885.17	(150)	—
Pr	3,885.19	40W	100W
Cr	3,885.22	50	40
Sm	3,885.28	50	50
Co	3,885.29	4	70
Zr	3,885.42	6	25
Nb	3,885.44	100	50
Fe	3,885.51	60	100
Nb	3,885.68	30	15
Os	3,885.75	8	20
V	3,885.77	1	20
Fe	3,886.28	400	600
La	3,886.37	200	400
V	3,886.59	15	25
Os	3,886.75	8	30
Cr	3,886.79	125	125
Mo	3,886.82	30	30
Tb	3,886.83	8	20
Yb	3,887.31	40	6
Tm	3,887.35	8	80
U	3,887.45	—	20
Re	3,887.48	—	20
Tb	3,887.67	3	25
U	3,887.70	10	20
Nd	3,887.87	20	25
Re	3,887.95	—	20
U	3,888.206	6	12
Tb	3,888.21	—	30
Bi	3,888.23	2	40
He	3,888.646	(1000)	—
Cs	3,888.65	10	150
Fe	3,888.82	15	40
Ho	3,888.95	20	40

	λ	I J	I O		λ	I J	I O
N	*3,830.39*	*150*	—	*Dy*	*3,888.99*	—	*20*
Sm	*3,830.30*	*10*	*50*	*H*	*3,889.05*	*(60)*	—
Tb	*3,830.29*	*8*	*30*	*Pr*	*3,889.33*	*70*	*150*
V	*3,830.27*	*4*	*40*	*Mn*	*3,889.45*	*50*	*25*
bhB	*3,830.2*	—	*50*	*Ta*	*3,889.46*	*40W*	—
Cr	*3,830.03*	*50*	*150W*				
Nd	*3,830.00*	*8*	*20*				
Re	*3,829.81*	—	*25*				
U	*3,829.80*	—	*15*				
Ne	*3,829.77*	*(40)*	—				
Mn	*3,829.68*	*60*	*60*				
Nd	*3,829.63*	*30*	*40*				

U 4,090.135 *J* 40 *O* 25

	λ	J	O		λ	J	O
Tb	4,090.13	—	2	Er	4,090.135	—	6*d*
Fe	4,090.08	—	2	Nb	4,090.16	5	5
Pd	4,090.07	—	4	Eu	4,090.20	—	6*d*
Mn	4,089.94	20	80	Tm	4,090.29	3	10
Re	4,089.92	—	25	Cr	4,090.30	15	30
Pr	4,089.87	2	5	W	4,090.31	2	4
As	4,089.87	3	—	Co	4,090.35	—	20*h*
Ce	4,089.85	—	6	U	4,090.37	1	6
Eu	4,089.745	1*h*	9	Dy	4,090.40	—	4
Zr	4,089.743	10*h*	12	La	4,090.41	—	3
Ce	4,089.742	—	6	Gd	4,090.42	20	100
Mo	4,089.731	6	6	Ce	4,090.47	1	15
Br	4,089.73	(8)	—	Zr	4,090.51	10	15
Yb	4,089.685	7	50	V	4,090.58	25	60
Nd	4,089.678	4	10	Mn	4,090.61	53	12
Ir	4,089.64	2	15	Br	4,090.62	(5)	—
La	4,089.61	5	40	W	4,090.64	5	6
J	4,089.55	(3)	—	Pr	4,090.74	8	15
Dy	4,089.511	—	7	Gd	4,090.75	—	5
Tb	4,089.51	1	10	Os	4,090.769	3	3
Pr	4,089.47	1	3	Eu	4,090.770	—	3
Re	4,089.43	—	2*h*	Er	4,090.771	—	9
Nb	4,089.40	5	—	Ce	4,090.775	2	—
W	4,089.38	5	6	Zr	4,090.79	2	25
U	4,089.37	—	4	Sb	4,090.86	2	—
Tb	4,089.33	1	20	Mo	4,090.88	20	10
O	4,089.27	(60*h*)	—	Ce	4,090.947	2	6
P	4,089.25	30	—	Mo	4,090.953	20	—
Fe	4,089.22	2	10	Ir	4,090.97	—	3
Br	4,089.20	5	—	Nd	4,090.99	5	5*d*
Ce	4,089.16	—	4	W	4,091.04	3	5
				Ce	4,091.058	—	3

	λ	I	
		J	O
Ce	4,088.85	2	15
Os	4,088.44	3	100*h*
Ac	4,088.37	100	—
Kr	4,088.33	(500)	—
Co	4,088.30	—	50
U	4,088.25	18	25
Gd	4,087.71	100	80
Er	4,087.64	1	20
U	4,087.39	12	6
Pd	4,087.34	100	500
Dy	4,087.213	4	15
Pr	3,087.206	3	50
O	4,087.14	(40*h*)	—
Fe	4,087.10	5	50
La	4,086.71	500	500
Tb	4,086.60	—	15
Co	4,086.31	15	400
Pr	4,086.24	2	25
Mo	4,086.02	15	15
Ir	4,085.93	15*h*	15
Ag	4,085.87	25	—
Ta	4,085.80	80*W*	5
Zr	4,085.66	2	25
Gd	4,085.65	80	100
Ru	4,085.43	50	40
Pr	4,085.34	8	15
Fe	4,085.32	70	100
Ce	4,085.23	5	20
O	4,085.12	(70)	—
Th	4,085.04	15	15
Cr	4,085.02	1	15
Fe	4,085.01	30	80
O	4,084.66	(30)	—
Fe	4,084.50	80	120
Sm	4,084.40	—	80
Mo	4,084.38	40	40
Zr	4,084.30	1	30
Au	4,084.12	12	25
U	4,084.09	2	10
F	4,083.92	(40)	—
O	4,083.91	(40*h*)	—
Fe	4,083.78	2	15
Yt	4,083.71	10	50
Mn	4,083.63	60	80
Re	4,083.59	—	25
Re	4,083.37	—	25
Pr	4,083.34	20	20
Ce	4,083.23	6	35*d*
Mn	4,082.94	60	80

	λ	I	
		J	O
Ru	**4,091.062**	**20**	**20**
Pr	**4,091.11**	—	**3**
U	4,091.52	12	6
P	4,091.53	(30)	—
U	4,091.64	—	12
Os	4,091.82	12	100
Se	4.091.95	(70)	—
Cr	4,092.17	2	25
Sm	4,092.26	25	10
Mn	4,092.388	20	30
Co	4,092.391	15	500*W*
V	4,092.41	6	15
Ir	3,092.61	20	60
Pr	4,092.63	5	15
Ca	4,092.633	2	15
V	4,092.69	12	20
Gd	4,092.712	40	100
Ce	4,092.715	2	18
Co	4,092.85	10	25
O	4,092.94	(80)	—
Hf	4,093.16	20	25
Ce	4,093.28	—	15
K	4,093.70	(20)	—
Gd	4,093.72	15	15
Ce	4,093.95	3	20
U	4,093.99	3*h*	12
Tm	4,094.18	30	300
Zr	4,094.27	1	30
P	4,094.43	(30)	—
Gd	4,094.49	15	25
Th	4,094.75	15	15
Mo	4,094.91	30	—
Ca	4,094.93	7	15
Pr	4,094.97	12	50
Mn	4,095.25	15	15
V	4,095.49	15	40
U	4,095.75	**25**	**18**
Fe	4,095.98	40	80
Dy	4,096.11	2	15
Pr	4,096.34	8	20
U	4,096.35	5	20
O	4,096.53	(30)	—
Mo	4,096.810	15	20
Eu	4,096.815	2	15
Pr	4,096.82	15	30
Ru	4,097.03	15	15
O	4,097.24	(70)	—

	λ	I	
		J	O
V	4,082.93	6	15
Ru	4,082.79	10	15
Rh	4,082.78	50	100
Co	4,082.60	—	50
Sm	4,082.59	25	10
Ti	4,082.46	25	60
Nd	4,082.43	15	15
Ar	4,082.40	(30)	—
Sc	4,082.396	10	25
Ir	4,082.38	3	25
Pr	4,081.90	30	75
Cr	4,081.74	2	25
Mo	4,081.44	50	50
Re	4,081.43	—	30
Ir	4,081.40	2	15
W	4,081.30	25	2
U	4,081.26	4	10
Tb	4,081.23	1	30
Ce	4,081.22	8	40
Zr	4,081.21	7	150
Mo	4,081.08	30	—
Pr	4,081.02	25	50
U	4,080.61	20	12
Ru	4,080.60	300	125
Cu	4,080.553	—	30W
Sm	4,080.547	10	15
Hf	4,080.44	15	15
Fe	*4,080.22*	*10*	*60*
Ne	*4,080.15*	*(50)*	—
P	*4,080.04*	*(150)*	—
Ir	*4,079.90*	—	*25*
Fe	*4,079.84*	*40*	*80*
Pr	*4,079.79*	*35*	*50*
Nb	*4,079.73*	*200W*	*500W*
Ti	*4,079.72*	*7*	*40*
Mn	*4,079.42*	*40*	*50*
Mn	*4,079.24*	*40*	*50*
Bi	*4,079.21*	*(40W)*	*2h*
La	*4,079.18*	*2*	*25*
O	*4,078.85*	*(70)*	—
Xe	*4,078.82*	*(100)*	—
Ti	*4,078.47*	*50*	*125*
Fe	*4,078.36*	*40*	*80*
Dy	*4,077.974*	*100*	*150r*
Tb	*4,077.97*	*2*	*25*
Hg	*4,077.81*	*150*	*150*
U	*4,077.79*	*6*	*15*

	λ	I	
		J	O
Tb	4,097.45	5	20
Rh	4,097.52	10	25
Cr	4,097.65	—	20h
U	4,097.74	1	10
Ru	4,097.79	125	25
Se	4,097.91	(60)	—
Cr	4,097.96	—	20
U	4,098.03	12	12
Nd	4,098.17	5	15
Cr	4,098.18	—	20h
Mo	4,098.183	15	20
Fe	4,098.19	40	100
Pr	4,098.41	12	20
Ca	4,098.53	3	15
Gd	4,098.61	6	25h
Kr	4,098.72	(250)	—
Mo	4,098.74	20	20
Ne	4,098.77	(50)	—
Xe	4,098.89	(50h)	—
Gd	4,098.91	100	100
Ce	4,098.98	1	15
Cr	4,099.02	8	30
Ti	4,099.17	8	25
U	4,099.27	1	20r
Tb	4,099.46	1	25W
La	4,099.54	100	100
V	4,099.80	12	23
N	4,099.94	(150)	—
Ir	*4,100.15*	*3h*	*100*
Os	*4,100.30*	*3*	*60*
Fe	*4,100.74*	*30*	*80*
Pr	*4,100.75*	*50*	*200*
Tb	*4,100.90*	*2*	*50d*
Nb	*4,100.92*	*200W*	*300W*
Te	*4,101.07*	*(50)*	—
Ho	*4,101.09*	*40*	*40*
Cr	*4,101.16*	*2*	*30*
Fe	*4,101.27*	*10*	*40*
H	*4,101.73*	*(100)*	—
Ru	*4,101.74*	*60*	*20*
Ce	*4,101.772*	*6*	*35*
In	*4,101.773*	*1000R*	*1000R*
U	*4,101.90*	*1*	*18*
Mo	*4,102.15*	*25*	*30*
V	*4,102.16*	*15*	*30*
Yt	*4,102.38*	*30*	*150*
Tb	*4,102.52*	*2*	*25W*
W	*4,102.70*	*30*	*35*

	λ	I J	I O		λ	I J	I O
Sr	4,077.71	500W	400r	Mn	4,102.96	20	100
Cr	4,077.68	10	30	O	4,103.01	(50)	—
Co	4,077.41	2h	100Wh	F	4,103.08	(150)	—
Yt	4,077.37	40	50	U	4,103.12	1h	18
La	4,077.34	400	600	F	4,103.22	(30)	—
Yb	4,077.27	100	30	Dy	4,103.31	50	50
Cr	4,077.09	10	35	Tb	4,103.37	—	30
Ru	4,076.73	25	60	F	4,103.52	(300)	—
U	4,076.72	10	8	F	4,103.72	(50)	—
Fe	4,076.64	50	80	Ho	4,103.84	400	400
Sm	4,076.63	15	25	Hg	4,103.87	(50)	—
Ir	4,076.55	—	25	F	4,103.871	(50)	—
Yt	4,076.35	8	30	Dy	4,103.88	4	30
Mo	4,076.19	25	25	Tb	4,103.90	2	30
Co	4,076.13	—	70	Ar	4,103.91	(200)	—
Cr	4,076.06	15	30	Fe	4,104.13	25	100
C	4,076.00	80	—	Cu	4,104.23	1h	30
O	4,075.87	(800)	—	Re	4,104.422	—	30
Sm	4,075.83	40	40	Ce	4,104.425	1	30
Cu	4,075.59	—	40	Co	4,104.43	2	30
Mn	4,075.254	5	25	O	4,104.73	(50)	—
Mo	4,075.246	25	25	Co	4,104.75	—	50
C	4,074.89	40	—	Cr	4,104.87	8	35
Cr	4,074.86	10	25	La	4,104.88	2	40
Fe	4,074.79	40	80	Ce	4,104.996	3	40
Os	4,074.68	6	80	O	4,105.001	(125)	—
C	4,074.53	50	—	Ir	4,105.20	2	25
U	4,074.49	10	10	U	4,105.31	3	10
W	4,074.36	45	50	Mh	4,105.36	20	50
Fe	4,073.775	20	80	Tm	4,105.84	30	300
Gd	4,073.771	50	100	Ce	4,106.13	2	30
Ce	4,073.73	3	30	Nd	4,106.58	15	25
Te	4,073.57	(300)	—	Sm	4,106.60	5	100
Ce	4,073.48	8	50	U	4,106.93	10	25
Gd	4,073.21	80	80	Ce	4,107.42	8	30
Dy	4,073.11	15	80	Mo	4,107.47	40	30
Zr	4,072.70	3	100	Rh	4,107.49	8	25
In	4,072.40	200Wh	—	Fe	4,107.492	100	120
Ar	4,072.40	(40)	—	Ru	4,107.84	20	25
O	4,072.16	(300)	—	Os	4,107.97	—	25
Ar	4,072.01	(150)	—	U	4,108.35	—	10
Ce	4,071.81	5	30	Cr	4,108.40	4	30
Fe	4,071.74	200	300	Ho	4,108.63	40	100
Os	4,071.56	3	30	Se	4,108.83	(800)	—
U	4,071.11	25	15	F	4,109.17	(100)	—
Fe	4,070.78	20	50	P	4,109.19	(70)	—
J	4,070.75	(150)	—	Kr	4,109.23	(100hs)	—
Ir	4,070.68	10	30	Nd	4,109.45	30	30
Gd	4,070.40	10	40	Cr	4,109.58	10	40
Gd	4,070.29	—	80				

	λ	I			λ	I	
		J	O			J	O
				Xe	*4,109.71*	*(60)*	—
				V	*4,109.79*	*20r*	*40r*
				Fe	*4,109.81*	*100*	*120*

U 4,241.67 *J* 50 *O* 40

	λ	J	O		λ	J	O
Ce	**4,241.64**	—	**2**	**Zr**	**4,241.69**	**2**	**100**
Na	**4,241.6**	—	**3**	**Pd**	**4,241.70**	—	**2*h***
Os	**4,241.52**	—	**9**	**Ce**	**4,241.74**	—	**3**
Co	**4,241.51**	—	**3*h***	**Au**	**4,241.77**	**30**	**40**
Nb	**4,241.449**	**5**	**3**	**N**	**4,241.80**	**(100*h*)**	—
W	**4,241.448**	**10**	**30**	**Re**	**4,241.82**	—	**2**
Ce	**4,241.40**	—	**5**	**Dy**	**4,241.83**	—	**2**
Re	**4,241.39**	—	**30**	**Co**	**4,241.89**	—	**3*h***
Cl	**4,241.38**	**(60)**	—	**Hf**	**4,241.93**	**10**	**2**
V	**4,241.32**	**8**	**15**	**Cs**	**4,241.97**	**(10)**	—
Pr	**4,241.30**	**2**	**10**	**Ce**	**4,242.009**	—	**10**
Gd	**4,241.28**	—	**10**	**Zr**	**4,242.013**	—	**4**
Ce	**4,241.24**	—	**3**	**Tm**	**4,242.15**	**100**	**500**
Nd	**4,241.21**	**4**	**12**	**Pb**	**4,242.20**	**2**	—
Zr	**4,241.202**	**2**	**100**	**Ne**	**4,242.20**	**(5)**	—
La	**4,241.20**	**15*h***	—	**Tb**	**4,242.25**	—	**4**
Cr	**4,241.19**	—	**10**	**Ce**	**4,242.259**	—	**5**
Re	**4,241.16**	—	**20**	**Cu**	**4,242.26**	—	**20**
U	**4,241.112**	**2**	**10**	**U**	**4,242.29**	**3**	**6**
Fe	**4,241.112**	**4**	**1**	**Cr**	**4,242.38**	**50**	**4**
Tb	**4,241.08**	—	**4**	**Mg**	**4,242.47**	—	**4**
Ru	**4,241.05**	**20**	**100**	**Pb**	**4,242.47**	**10**	—
Pr	**4,241.02**	**12**	**50**	**Eu**	**4,242.47**	**1*h***	**6**
Eu	**4,240.90**	**1**	**2**	**Tb**	**4,242.55**	**2*h***	**12**
Mo	**4,240.83**	**25**	**30**	**U**	**4,242.57**	**3**	**3**
Dy	**4,250.81**	—	**2**	**Fe**	**4,242.59**	—	**3**
*bh***Ca**	**4,240.8**	—	**3**	**Rb**	**4,242.60**	**(150)**	—
Dy	**4,240.75**	—	**3**	**Zr**	**4,242.617**	—	**3**
Al	**4,240.75**	**(15)**	—	**Ba**	**4,242.617**	**5**	**10**
Ca	**4,240.74**	**8*h***	—	**Nb**	**4,242.63**	**20**	**10**
Cr	**4,240.70**	**30**	**200**	**Ce**	**4,242.723**	**3**	**15**
Gd	**4,240.68**	—	**4**	**Th**	**4,242.725**	**2**	**5**
Ce	**4,240.67**	—	**2**				
U	**4,240.59**	**10**	**10**				
				Ru	4,243.06	40	100
				Mo	4,243.14	25	—
Fe	4,240.37	5	30	As	4,243.26	100	—
Zr	4,240.34	1	100	Pr	4,243.53	8	20
Mo	4,246.28	20	25	Gd	4,243.84	100	60
Mo	4,240.08	20	25				
Ce	4,239.91	4	35	Re	4,244.14	—	30
				U	4,244.372	25	25

	λ	I	
		J	O
Dy	4,239.87	4	50
Fe	4,239.85	15	40
Nd	4,239.83	10	20
U	4,239.74	10	10
Fe	4,239.73	10	30
Mn	4,239.72	50	100
Zr	4,239.31	5	100
Cr	4,238.96	15	100
Fe	4,238.82	100	200
Gd	4,238.78	200	200
Re	4,238.58	—	20
La	4,238.38	300	500
Xe	4,238.25	(200*h*)	—
Fe	4,238.04	15	80
Ti	4,237.89	8	20
Cr	4,237.71	—	70
Sm	4,237.66	50	60
Ar	4,237.23	(40)	—
Mo	4,237.16	10	20
N	4,236.98	(30*h*)	—
Br	4,236.88	(25)	—
Sm	4,236.74	50	60
Ru	4,236.67	—	20
Kr	4,236.64	(100*h*)	—
U	4,236.45	6	10
Re	4,236.24	—	25
Pr	4,236.21	3*W*	20*W*
U	4,236.04	10	12
Ce	4,236.02	1	30
Yt	4,235.94	30	60
Fe	4,235.94	200	300
Eu	4,235.60	—	400
Cl	4,235.49	(25)	—
J	4,235.47	(25)	—
Tb	4,235.34	1*h*	20
Mn	4,235.29	100	80
Mn	4,235.14	—	80
U	4,234.69	8	10
Zr	4,234.63	1	25
Sm	4,234.57	40	60
U	4,234.53	15	18
Cr	4,234.51	2	60
W	4,234.35	7	25
Ce	4,234.21	2	30
Nd	4,234.20	10	20
Cl	4,234.09	(50)	—
Co	4,234.00	—	100*W*
Ne	4,233.86	(30)	—
Fe	4,233.61	150	250
W	4,244.373	20	40
Rb	4,244.44	25	—
P	4,244.55	(30)	—
Sm	4,244.70	80	100
Ra	4,244.72	(40)	—
Mo	4,244.80	80	4
Ru	4,244.83	—	25
Fe	4,245.26	40	80
Gd	4,245.34	—	25
Ta	4,245.35	15	30
Xe	4,245.38	(200*h*)	—
Ti	4,245.51	3	20
Dy	4,245.92	4	25
Mo	4,246.02	30	30
Fe	4,246.09	30	80
F	4,246.16	300*h*	—
U	4,246.26	2	30
Tm	4,246.38	4	20
Gd	4,246.55	3	150
Mo	4,246.62	25	—
Ce	4,246.71	4	30
Ru	4,246.73	—	20
Re	4,246.81	—	25
Sc	4,246.83	500	80
P	4,246.88	(150*W*)	70
U	4,247.14	8	10
Dy	4,247.36	4	30
Nd	4,247.37	20	50
Fe	4,247.43	100	200
Pr	4,247.66	35	60
Se	4,248.00	(100)	—
Fe	4,248.23	40	150
Cr	4,248.34	2	30
Ce	4,248.68	8	60
Cr	4,248.71	2	35
Cu	4,248.96	15	80
Ti	4,249.12	3*h*	60*h*
Eu	4,249.43	—	20
Pr	4,249.48	2	20
Sm	4,249.54	40	50
P	4,249.57	(100)	—
La	4,249.99	50	100
Fe	4,250.13	150	250
Kr	4,250.58	(150)	—
Ne	4,250.68	(50)	—
Mo	4,250.69	125	5
Fe	4,250.79	250	400
Te	4,251.15	(70)	—

	λ	I J	I O
O	4,233.32	(100)	—
Re	4,233.28	—	20
Fe	4,233.17	100	100
U	4,233.13	1	12
Re	4,232.96	—	20
Cr	4,232.87	—	60
Mo	4,232.59	100	125
Ce	4,232.57	2	20
Hf	4,232.44	6	20
U	4,232.39	5*h*	12*h*
Nd	4,232.38	15	40
Ru	4,232.32	—	40
Cr	4,232.22	5	70
Tb	4,232.19	3	20
Cs	4,232.188	(25)	—
Ce	4,232.05	2	20
U	4,232.04	15	15
Dy	4,232.03	2	20
Yb	4,231.99	2	20
Ce	4,231.745	5	30
Te	4,231.74	(30)	—
U	4,231.68	—	25
Ne	4,231.60	50	—
La	*4,230.95*	*50*	*150*
Cr	*4,230.48*	*8*	*70*
Ru	*4,230.31*	—	*60*
Sm	*4,229.70*	*30*	*40*
Ru	*4,229.31*	—	*40*
Nb	*4,229.15*	*100*	*50*
Te	*4,228.46*	*(50)*	—
U	*4,228.43*	*1*	*10*
Hg	*4,227.87*	*70)*	—
Zr	*4,227.76*	*8*	*150*
Ce	*4,227.75*	*5*	*40*
Re	*4,227.46*	—	*200W*
Fe	*4,227.43*	*250*	*300*
Hg	*4,227.29*	*(100)*	—
In	*4,227.16*	*(50h)*	—
Gd	*4,227.14*	*20*	*50*
Cs	*4,227.10*	*(50)*	—
Cr	*4,226.76*	*30*	*125*
Ce	*4,226.734*	*30*	*50*
Ca	*4,226.728*	*50R*	*500R*
Gd	*4,226.57*	*50*	*200*
Tb	*4,226.44*	—	*50*
Fe	*4,226.43*	*25*	*80*
Rn	*4,226.06*	*(50)*	—
Ar	4,251.18	(800)	—
Yt	4,251.20	8	25
U	4,251.33	6	10
Pr	4,251.49	15*W*	40*W*
Xe	4,251.57	(50*Wh*)	—
Ti	4,251.61	3	20
Gd	4,251.74	10	300
Sm	4,251.79	200	200
Mo	4,251.87	60	60
Co	*4,252.31*	—	*150*
U	*4,252.43*	*20*	*15*
Kr	*4,252.67*	*(50hs)*	—
Nb	*4,252.97*	*50*	*30*
Ce	*4,253.36*	*3*	*40S*
Gd	*4,253.37*	*4*	*50*
Gd	*4,253.62*	*50*	*50*
Nb	*4,253.69*	*40*	*25*
O	*4,253.74*	*(50h)*	—
O	*4,253.98*	*(100h)*	—
Cr	*4,254.35*	*1000R*	*5000R*
Ho	*4,254.43*	*20*	*100*
Kr	*4,254.85*	*(100h)*	—
Nb	*4,255.44*	*50*	*30*
Ce	*4,255.78*	*6*	*40*
Ti	*4,256.036*	*15*	*80*
Zr	*4,256.037*	—	*60*
Te	*4,256.10*	*(50)*	—
Sm	*4,256.40*	*150*	*150*
Nd	*4,256.48*	*20*	*40*
La	*4,256.91*	—	*50*
Re	*4,257.59*	—	*125W*
Mn	*4,257.66*	*40*	*100*
Ti	*4,258.53*	*7*	*70*
Sm	*4,258.56*	*40*	*50*
Ir	*4,259.11*	*10*	*200*
Ar	*4,259.36*	*(1200)*	—
Kr	*4,259.44*	*(80hs)*	—
U	*4,259.45*	*3*	*12*
Bi	*4,259.62*	*60Wh*	—
Fe	*4,260.48*	*300*	*400*
Os	*4,260.85*	*200*	*200*
Te	*4,261.08*	*(300)*	—
Cr	*4,261.35*	*50*	*125*
Ti	*4,261.60*	*8*	*70*
Hg	*4,261.88*	*(70)*	—
Gd	*4,262.09*	*10*	*150*
Cr	*4,262.13*	*8*	*40*

	λ	I	
		J	O
Fe	4,225.96	30	80
Gd	4,225.85	50	150
Te	4,225.70	(50)	—
Fe	4,225.46	20	80
Pr	4,225.33	40	50
Sm	4,225.32	30	40
Dy	4,225.15	8	40
Ti	4,224.79	8	40
Fe	4,224.52	15	60
Cr	4,224.51	12	60
U	4,224.43	10	8
Fe	4,224.18	80	200
Br	4,223.88	(80)	—
Sm	4,223.71	20	50
Xe	4,223.00	(200h)	—
Pr	4,222.98	40	125
O	4,222.78	(50)	—
Cr	4,222.73	15	100
Ce	4,222.60	18	80
U	4,222.37	8	18
Fe	4,222.22	200	200
P	4,222.15	(150W)	200
U	4,221.80	8	12
Cr	4,221.57	35	80
Dy	4,221.10	8	60
Re	4,221.08	—	100
J	4,220.96	(80)	—
Ru	4,220.67	—	60

U 5,492.97 *J* 50 *O* 60

	λ	J	O
Th	5,492.64	—	8
W	5,492.59	—	4h
Sm	5,492.50	2	6
Pr	5,492.37	1	4
W	5,492.32	50	50
Nd	4,492.30	—	8
Mo	5,492.17	8	15
Ar	5,492.06	(40)	—
Rh	5,991.87	—	5
As	5,491.81	12	—
Er	5,491.72	—	12
bhZr	5,491.7	—	20
Pr	5,491.671	—	3h
Ti	5,491.67	—	3
Gd	5,491.65	—	6
Y	5,493.18	—	3
Hf	5,493.22	6	1
Nd	5,493.34	—	4
Gd	5,493.38	—	8
La	5,493.45	20	15
Ar	5,493.49	(20)	—
J	5,493.50	(20)	—
Fe	5,493.51	—	4
Sm	5,493.72	—	80
Mo	5,493.80	12h	20h
Nd	5,494.01	1	15
J	5,494.07	(15)	—
SB	5,494.31	—	3
Ne	5,494.41	(50)	—
Fe	5,494.46	—	2

	λ	I J	I O		λ	I J	I O
J	**5,491.57**	**(100)**	—	**U**	**5,494.67**	**1**	**8*h***
Yt	**5,491.54**	—	2*h*	**Ta**	**5,494.78**	—	**50**
Kr	**5,491.43**	(4*h*)	—	**Ni**	**5,494.89**	—	**5**
Gd	**5,491.37**	—	**6**	**Br**	**5,495.06**	**(150)**	—
Kr	**5,491.33**	(2*h*)	—	**Xe**	**5,495.07**	**(10*Wh*)**	—
Sm	**5,491.31**	—	**2**	**Eu**	**5,495.17**	—	**125**
Os	**4,491.*a*8**	—	**8**				
U	**5,491.24**	**6**	**8**				
Ce	**5,491.15**	—	**10**	N	5,495.70	(70)	—
La	**5,491.07**	—	**10**	Ar	5,495.87	(1,000)	—
				U	5,496.44	1	12
Nb	**5,491.06**	**1**	**3**	J	5,496.60	(30)	—
Kr	**5,490.94**	**(50)**	—	In	5,496.69	(30*h*)	—
Ti	**5,490.84**	—	**3**				
				In	5,496.90	(30*h*)	—
				J	5,496.92	(900)	—
Ti	5,490.15	2	70	In	5,497.38	(30)	—
Ar	5,490.13	(60)	—	Yt	5,497.40	40	20
Ta	5,490.11	—	60	Fe	5,497.52	5	150
Co	5,489.65	—	150*W*				
Br	5,488.79	(70)	—	In	5,497.55	(70)	—
				In	5,497.64	(50)	—
Eu	5,488.65	—	500	As	5,497.98	200	—
Ra	5,488.32	(25)	—	In	5,498.05	(30)	—
Ti	5,488.20	30	30	Sm	5,498.21	—	80
Te	5,488.07	(50)	—				
Fe	5,487.14	5	50	Ta	5,499.44	—	60*S*
				Kr	5,499.54	(50)	—
U	5,487.02	8	12	P	5,499.71	(150)	—
Sr	5,486.12	8	40	U	5,500.70	—	10
Er	5,485.93	—	30	Kr	5,500.71	(50)	—
Sm	5,485.42	—	40				
Sc	5,484.62	—	60	La	5,501.34	50	200
				Fe	5,501.47	—	150
Ru	5,484.32	—	60	Ra	5,501.98	(250)	—
Co	5,483.96	—	150*W*	P	5,502.83	(30*h*)	—
P	5,483.55	(70)	—	In	5,503.15	(30)	—
Co	5,483.34	—	500*W*				
U	5,482.55	18	12	W	5,503.45	1	45
				La	5,503.81	—	100
La	5,482.27	50	25	Ti	5,503.90	3	60
Ra	5,482.13	(100)	—	U	5,504.15	10	12
Sc	5,481.99	—	60	Sr	5,504.17	25	60
Yb	5,481.94	2	50				
Ti	5,481.43	1	35	Nb	5,504.58	3*h*	30*W*
				J	5,504.72	(60)	—
Mn	5,481.40	—	50	Ra	5,505.50	(25)	—
U	5,481.22	25	30	Yb	5,505.501	2	40
Sr	5,480.84	30	100*h*	Mn	5,505.87	—	40
La	5,480.74	40	5				
U	5,480.27	25	15	La	5,506.00	—	50
				Ar	5,506.11	(500)	—
N	5,480.10	(30)	—	Mo	5,506.49	100	200*r*
Br	5,479.99	(30)	—	In	5,506.71	(30)	—
Ru	5,479.40	—	40	Br	5,506.78	(300)	—
Te	5,479.13	(50)					
Yb	5,478.52	50	10	Fe	5,506.782	10	150

	λ	I	
		J	O
Pt	5,478.49	2	50
W	5,477.80	20	25
P	5,477.74	(30)	—
Ti	5,477.71	2	70
Os	5,477.27	—	30
Co	5,477.08	—	40
Ni	5,476.91	8	400W
Lu	5,476.69	1000	500
Fe	5,476.58	—	80
Pt	5,475.77	2	60
Ta	5,475.54	—	40
Fe	5,474.92	—	100
Ti	5,474.23	50	30
Fe	5,473.92	—	100
S	5,473.63	(750)	—
Ar	5,473.44	(500)	—
Mo	5,473.37	25	50
*bh*C	5,473.3	—	70
Xe	5,472.61	(500)	—
Pb	5,472.40	(25)	—
Eu	5,472.33	—	1000*s*
Ag	5,471.55	100	500*h*
Ti	5,471.21	25	25
Co	*5,469.30*	—	*125*
Kr	*5,468.17*	*(200hs)*	—
Sm	*5,466.73*	—	*80*
Yt	*5,466.47*	*20*	*150*
Br	*5,466.23*	*(150)*	—
U	*5,465.69*	*8*	*12*
Ag	*5,465.49*	*500R*	*1000R*
J	*5,464.61*	*(900)*	—
Sb	*5,464.08*	*(100)*	—
Fe	*5,463.28*	—	*100*
Ta	*5,461.29*	—	*80*
P	*5,460.85*	*(100)*	—
Hg	*5,460.74*	*(2000)*	—
Xe	*5,460.39*	*(200)*	—
La	*5,458.69*	*50*	*5*
Ar	*5,457.37*	*(200)*	—
Cl	*5,457.02*	*(75)*	—
Cl	*5,456.27*	*(50)*	—
Fe	*5,455.61*	*30*	*300*
La	*5,455.15*	*1*	*200*
Ru	*5,454.82*	—	*100*
Co	*5,454.55*	—	*300W*
S	*5,453.88*	*(750)*	—
Sm	*5,453.02*	—	*100*
Eu	*5,452.96*	—	*1000S*

	λ	I	
		J	O
S	5,507.01	(25)	—
In	5,507.11	(30)	—
P	5,507.13	(70)	—
In	5,507.33	(70)	—
Ne	5,507.34	(25)	—
V	5,507.75	60	60
Pr	5,509.15	2	50
Sm	5,509.66	—	40
S	5,509.67	(25)	—
Yt	5,509.30	40	30
La	5,510.34	—	200
U	5,510.43	—	10
Eu	5,510.55	—	300*s*
Ru	5,510.71	—	100
In	5,510.80	(30)	—
In	5,510.88	(70)	—
V	5,511.18	25	25
U	5,511.50	2	30
Ce	5,512.08	—	50*s*
Sm	5,512.10	—	80
Ti	5,512.53	12	125
Rb	5,512.57	30	—
O	5,512.70	(70*h*)	—
In	5,512.82	(150)	—
In	5,512.92	(100)	—
In	5,513.00	(70)	—
In	5,513.06	(70)	—
In	5,513.10	(50)	—
In	5,513.16	(30)	—
Sc	5,514.21	—	60
Ti	5,514.35	10	70
Tb	5,514.54	—	50
Ti	5,514.542	15	80
Fe	5,514.63	10	50
W	5,514.70	8	50W
P	5,514.77	(30)	—
Sm	*5,516.14*	—	*200*
Ta	*5,518.91*	—	*100W*
Ba	*5,519.11*	*60*	*200wh*
In	*5,519.36*	*(500)*	—
Sc	*5,520.50*	—	*80*
Se	*5,522.42*	*(750)*	—
Rb	*5,522.79*	*100*	—
Kr	*5,522.94*	*(60)*	—
Ce	*5,522.99*	—	*100*
In	*5,523.00*	*(50)*	—
In	*5,523.287*	*(50)*	—

	λ	I J	I O
Ar	5,451.65	(500)	—
Eu	5,451.53	—	1,000S
P	5,450.65	(100)	—
Xe	5,450.45	(100)	—
Te	5,449.82	(75)	—
Yb	5,449.29	100	20

	λ	I J	I O
Co	5,523.292	—	300W
Os	5,523.53	—	100
In	5,523.61	(50)	—
In	5,523.86	(70)	—
In	5,523.91	(100)	—
Hf	5,524.35	50	40
Ar	5,524.93	(300)	—
Eu	5,526.62	—	60
Sc	5,526.81	300wh	100
Yt	5,527.54	15	100
U	5,527.85	40	25
Mg	5,528.46	30	60
N	5,530.27	(50)	—
Co	5,530.77	—	500
Xe	5,531.07	(300)	—
Re	5,532.66	—	100
Mo	5,533.05	100	200
Ne	5,533.68	(75)	—
Ar	5,534.45	(60)	—
Rh	5,535.04	1	80
N	5,535.39	(70)	—
Ba	5,535.55	200R	1000R
La	5,535.67	100	50
In	5,535.94	(70)	—
Br	5,536.30	(50)	—
In	5,536.55	(70)	—

V 23 50.942

t_0 1,710°C t_1 3,000°C

I.	II.	III.	IV.	V.
6.738	14.2	29.5	48.5	64

λ	*I*		eV
	J	*O*	
II 3,093.11	400*R*	100*R*	4.40
II 3,102.30	300*R*	70	4.36
II 3,110.71	300*R*	70	4.33
II 3,118.38	200*R*	70	4.31
II 3,125.28	200*R*	80	4.29
I 3,183.41	100*R*	200*R*	3.91
I 3,183.98	400*R*	500*R*	3.90
I 3,185.40	400*R*	500*R*	3.96
I 4,379.24	200*R*	200*R*	3.13
I 4,389.97	60*R*	80*R*	3.10
I 4,408.51	20*R*	30	3.08

V 3,093.108 *J* 400*R* *O* 100*R*

	λ	*I*	
		J	*O*
Dy	3,093.108	5	15
Th	3,093.05	6	12
U	3,093.01	20	20
Rn	3,093.0	(18)	—
Cl	3,093.00	(4)	—
Ta	3,092.994	1	18
Mg	3,092.991	20	125
Mo	3,092.92	10*d*	—
Nd	3,092.915	6	8
Ne	3,092.91	(4)	—
Nb	3,092.89	5	1*h*
Al	3,092.84	18	50*R*
Ce	3,092.82	1	4
Fe	3,092.78	30	50
Na	3,092.73	(200)	50
Ce	3,092.724	—	4
V	3,092.720	50*r*	100*r*
U	3,092.720	3	5
Al	3,092.713	1000	1000
Yt	3,092.712	—	8
Mo	3,092.70	2	20
Yb	3,092.56	30	—
Sc	3,092.52	4	3
T*a*	3,092.44	15	50
Sc	3,092.42	—	3
Xe	3,092.41	(10)	—
Ir	3,092.401	—	5
Fe	3,092.399	5	4
Cd	3,092.39	15	10
Ru	3,091.87	5	50
Fe	3,091.58	200	300
V	3,091.44	—	10
Mg	3,091.08	10	80
Hg	3,090.60	(200)	—
Os	3,090.49	15	80
Os	3,090.30	12	100
Co	3,090.25	1	80
Ru	3,090.23	6	50
Os	3,090.08	15	100
Ru	3,089.80	5	60
V	3,089.63	10	—
Co	3,089.59	—	100*r*
Ru	3,089.14	12	60
V	3,089.13	2	30
Re	3,088.77	—	60
*bh*B	3,088.6	—	100
Tm	3,093.12	60	30
Ce	3,093.24	—	6
Si	3,093.28	6	—
Au	3,093.3	5	—
Zr	3,093.32	—	3
Ce	3,093.34	—	12
Fe	3,093.36	40	70
U	3,093.37	3*d*	4*d*
Ar	3,093.41	(50)	—
Yb	3,093.44	2	—
Ru	3,093.45	—	4
Ca	3,093.46	3	2
Rh	3,093.48	3	2
Cr	3,093.49	100	1
W	3,093.51	10	12
Bi	3,093.58	8	10*w*
Os	3,093.59	15	125
Ce	3,093.61	—	18
Re	3,093.65	—	60
Mo	3,093.68	—	10
Cd	3,093.74	2	3
Er	3,093.75	1	6
Yt	3,093.77	15*d*	9*d*
V	3,093.79	—	30
Fe	3,093.806	40	50
Ti	3,093.813	—	5
Dy	3,093.82	1	10

	λ	I J	I O
Os	3,088.27	12	60
V	3,088.11	—	30
Ir	3,088.04	2	50
Ti	3,088.02	500*R*	70
Co	3,087.81	—	60
Os	3,087.75	10	50
Mo	3,087.62	200	30
V	3,087.06	2	25
Co	3,086.78	—	200*R*
V	3,086.50	40	2
Co	3,086.40	2	80
Os	3,086.27	10	50
V	3,086.21	30	1
Ru	3,086.07	6	60
Mo	3,085.61	25	125
V	*3,084.38*	—	*40*
Rh	*3,083.96*	*2*	*150*
Fe	*3,083.74*	*500*	*500*
V	*3,083.54*	—	*60*
V	*3,083.21*	*50*	*2*
Co	*3,082.62*	*50*	*150R*
V	*3,082.52*	*30*	*3*
Re	*3,082.43*	—	*100w*
Al	*3,082.15*	*800*	*800*
V	*3,082.11*	*2h*	*80r*
V	*3,082.01*	*2h*	*15*
Gd	*3,082.00*	*60*	*100*
V	*3,081.25*	*50*	*5*
V	*3,081.00*	*30*	*2*
Cd	*3,080.93*	*100*	*150*
Ni	*3,080.75*	*60*	*200*
V	*3,080.33*	*1h*	*15*
Mn	*3,079.63*	*50*	*125*
V	*3,078.95*	*15*	—
Ti	*3,078.64*	*500R*	*60*
Os	*3,078.38*	*15*	*125*
Os	*3,078.11*	*15*	*125*

V 3,102.299 *J* 300*R* *O* 70

	λ	J	O
K	**3,102.25**	—	**20*R***
Sn	**3,102.23**	**(2)**	—
W	**3,102.22**	**12**	**2**
Dy	**3,102.19**	**1*h***	**4**
Sc	**3,102.149**	**4**	**3**
Sm	**3,102.299**	**4**	**8**
Ir	**3,102.35**	—	**4**
Hf	**3,102.358**	**3*h***	**15**
Ce	**3,102.358**	—	**12**
Ca	**3,102.359**	**4**	**2*h***

	λ	I J	I O
Hf	3,102.147	1	6
Fe	3,102.144	2	2
Ir	3,102.144	—	4
B	3,102.09	5	—
Yb	3,102.07	8	1
K	3,102.03	—	50*R*
La	3,101.99	4*h*	—
Sm	3,101.93	8	10
Gd	3,101.925	3	5
Nb	3,101.917	10	2
Mo	3,101.915	—	5
Dy	3,101.91	2	4
Ni	3,101.88	150	400*R*
U	3,101.86	4	6
Ra	3,101.80	(75)	—
Ce	3,101.79	—	20
Ta	3,101.72	2	7
Re	3,101.703	—	3
U	3,101.699	8	8
Th	3,101.69	8	10
Mn	3,101.56	50	50
Ni	3,101.55	150	1000*R*
Os	3,101.53	20	125
Hf	3,101.40	90	60
Mo	3,101.34	10	80
Pr	3,101.27	2	40
V	3,100.93	100	20
Mo	3,100.87	2	40
Ru	3,100.84	50	70
Fe	3,100.666	100	100
Re	3,100.666	—	100
Gd	3,100.51	80	100
Fe	3,100.304	100	100
Mn	3,100.302	60	60
Fe	3,099.97	40	40
Cu	3,099.93	10	60
Fe	3,099.90	60	60
Co	3,099.67	—	50
Ru	3,099.28	60	70
Os	3,099.26	8	40
Ni	3,099.11	50	200
Tm	3,098.59	60	80
Co	3,098.20	5	100*r*
Fe	3,098.19	60	70
Eu	3,097.46	5	200*w*
Ti	3,097.19	150	20
Ni	3,097.12	50	200
Fe	3,102.361	2	2
U	3,102.39	10*d*	12*d*
Ru	3,102.40	—	20
Co	3,102.41	4	60
Ce	3,102.43	—	10
Ti	3,102.52	—	5
Rh	3,102.53	—	10
Gd	3,102.557	25	25
Ce	3,102.563	—	15*h*
U	3,102.61	5	6
Ar	3,102.63	(3)	—
Fe	3,102.64	5	6
Ag	3,102.66	—	3
Th	3,102.67	12	15
Ho	3,102.68	4	—
Er	3,102.69	2	15
Os	3,102.716	8	20
J	3,102.72	(2)	—
Xe	3,102.73	(2)	—
Fe	3,102.872	20	30
Pb	3,102.874	10	—
Tm	3,102.88	15	20
U	3,102.90	3	6
Tb	3,102.97	15	30
Ti	3,102.975	2	2
Ce	3,103.01	—	2
Ta	3,103.25	15	70
Co	3,103.74	2	80
Ti	3,103.80	200	20
Co	3,103.98	—	60*r*
V	3,103.99	—	15
La	3,104.59	50	200
Os	3,104.98	15*d*	200*d*
Ru	3,105.28	40	50
Ru	3,105.41	1	50
Ni	3,105.47	35	200
Os	3,105.99	20	150
Ti	3,106.23	150	25
Ru	3,106.84	3	50
Co	3,107.04	3	70
V	3,107.14	—	10
Os	3,107.38	10	40
Ru	3,107.71	5	60
V	3,108.70	50	3
Re	3,108.81	—	125
Os	3,108.98	15	125
Hf	3,109.12	100	50

	λ	I	
		J	O
Mn	3,097.06	40	75*w*
Mg	3,096.90	25	150
Co	3,096.70	3	60
Ru	3,096.57	60	70
Cr	3,096.53	—	35
Co	3,096.40	3	60
V	3,095.90	—	15
Cr	3,095.86	3	125
Co	3,095.72	2	60
Ta	3,095.39	18*w*	70*w*
Re	3,095.06	—	40
V	*3,094.69*	—	*40*
Mo	*3,094.66*	*25*	*150*
Nb	*3,094.18*	*1000*	*100*
Cu	*3,093.99*	*50*	*150*

*P**

	λ	I	
		J	O
V	3,109.37	70	1
Os	3,109.38	20	125
Co	3,109.51	1*h*	60
Dy	3,109.77	20	40

*N**

V 3,110.706 *J* 300*R* *O* 70

	λ	J	O
Mn	**3,110.68**	**35**	**35**
Ti	**3,110.67**	**100**	**10**
Ar	**3,110.66**	**(3)**	—
Mo	**3,110.64**	—	**3**
Ti	**3,110.620**	**18**	**8**
Os	**3,110.618**	**10**	**30**
Pr	**3,110.58**	**1**	**10**
Ru	**3,110.55**	**6**	**60**
U	**3,110.52**	**8**	**3**
Fe	**3,110.281**	**30**	**40**
Ce	**3,110.278**	**1**	**30**
Eu	**3,110.24**	—	**20**
Sc	**3,110.238**	**5**	**3**
Ir	**3,110.227**	**1**	**4**
U	**3,110.226**	**3**	**3**
Sm	**3,110.193**	**4**	**15**
Fe	**3,110.188**	**6**	**6**
Ta	**3,110.10**	**1**	**3**
Ti	**3,110.09**	**35**	**3**
U	**3,110.04**	**3**	**6**
Th	**3,110.022**	**15**	**15**
Co	**3,110.021**	**2**	**60**
Sm	**3,109.96**	**10*h***	**1**

*P**

	λ	J	O
Fe	**3,110.71**	—	**2**
Dy	**3,110.76**	**5*h***	**10**
U	**3,110.78**	**1**	**3**
Nb	**3,110.80**	**5**	**1*h***
Ta	**3,110.81**	**70*w***	—
Co	**3,110.82**	**2**	**60**
U	**3,110.83**	**8**	**6**
Fe	**3,110.841**	**10**	**20**
Be	**3,110.842**	—	**20**
Mo	**3,110.844**	**30**	—
Cr	**3,110.860**	**8**	**25**
Re	**3,110.862**	—	**100**
Er	**3,110.87**	**3**	**12**
Hf	**3,110.874**	**40**	**30**
Zr	**3,110.878**	**8**	**10** *u*
Bi	**3,110.88**	**(5)**	—
Be	**3,110.96**	—	**15**
Ce	**3,110.998**	—	**2**
Cr	**3,111.00**	—	**8**
Bi	**3,111.02**	**(7)**	—
Os	**3,111.09**	**20**	**100**
W	**3,111.12**	**15**	**9**
Zr	**3,111.160**	**1**	**2**
Ce	**3,111.165**	—	**20**
Bi	**3,111.20**	**(10)**	—
Mo	**3,111.22**	**15**	**1**
Ce	**3,111.23**	—	**10**

	λ	I J	I O
Ti	3,111.28	8	15
Pr	3,111.337	2	15
Co	3,111.339	3	20
Cr	3,111.34	—	6
Pb	3,111.38	2	—
Bi	3,111.41	(15)	—
Rb	3,111.43	(30)	—
Eu	3,111.432	—	200
Nb	3,111.450	5	4
Kr	3,111.45	(2*h*)	—
U	3,111.46	3	3

*N**

V 3,118.38 *J* 200*R* *O* 70

	λ	I J	I O
Cu	3,118.355	1*h*	5
W	3,118.354	7	10
Cs	3,118.35	(4)	—
Os	3,118.33	20	150
Co	3,118.25	2	60
Re	3,118.193	—	200
Pb	3,118.19	2	—
Ti	3,118.130	1	10
Mn	3,118.128	—	4*h*
Os	3,118.12	15	80
Ru	3,118.07	50	50
Ne	3,118.02	(12)	—
Pt	3,118.01	15*h*	1
Te	3,117.99	(25)	—
U	3,117.98	8*d*	6*d*
Rh	3,117.96	10	4
Ba	3,117.94	—	2
Sm	3,117.899	3	5
Ti	3,117.899	—	4
Dy	3,117.898	1*h*	1
Os	3,117.891	10	30
Sc	3,117.890	—	2
Fe	3,117.888	3	3
Ag	3,117.88	3*h*	2
Ir	3,117.86	—	5
Ar	3,117.85	(3)	—
Yb	3,117.80	30	7
Ce	3,117.77	—	3
Fe	3,117.762	2	2
Ir	3,117.760	—	5
Sm	3,117.72	3	8
Pb	3,117.7	(100)	—
Lu	3,118.43	5	40
Er	3,118.44	—	6
Ho	3,118.51	10	8
Gd	3,118.60	1	2
Cr	3,118.65	200	35
Ru	3,118.68	3	50
U	3,118.74	2	2
W	3,118.79	7	8
Mo	3,118.81	10	—
Ti	3,118.82	15	2
		9	
Er	3,118.83	3	12
Ir	3,118.836	2	2
Ce	3,118.839	—	2
Sm	3,118.86	—	2
Cd	3,118.91	2	3
Pb	3,118.92	—	5*h*
Rb	3,119.00	(20)	—
Gd	3,119.02	1	2
Pr	3,119.03	2	15
U	3,119.049	6	6
Er	3,119.05	2	8
Fe	3,119.49	80	100
As	3,119.60	50	100
Ti	3,119.80	150	4
Mn	3,120.34	—	50
Cr	4,120.37	150	40
Fe	3,120.43	80	100
V	3,120.73	80	12
Ir	3,120.76	2	50
Fe	3,120.87	50	80
V	3,121.14	200*r*	60

	λ	I J	I O		λ	I J	I O
Ti	**3,117.67**	**200**	**15**	Re	3,121.37	—	100
Ca	**3,117.65**	**2**	**10**	Co	3,121.41	6	150*r*
Fe	**3,117.640**	**10**	**20**	Co	3,131.566	3	60*r*
				Pr	3,121.571	4	50
Th	**3,117.636**	**12**	**12**	V	3,121.749	—	12
Eu	**3,117.63**	—	**2*h***				
				Rh	3,121.755	—	150
				Ir	3,121.78	1	35
Ta	3,117.44	10	70	Xe	3,121.87	(150)	—
V	3,116.78	25	—	Mo	3,122.00	150	5
Fe	3,116.633	—	150	Fe	3,122.30	20	70
As	3,116.63	150	—				
Fe	3,116.59	150	—	Au	3,122.78	5	500*h*
				V	3,122.89	300*r*	12
Os	3,116.47	15	50	Ti	3,123.07	15	35
Cu	3,116.35	12	50	Rh	3,123.70	2	150
Hg	3,116.24	(100)	—	Ru	3,124.17	8	60
Ta	3,115.86	18*w*	50				
Mn	3,115.46	25	50				
Bi	3,115.42	500	—				
Rh	3,114.91	2	100				
Os	3,114.81	12	50				
Ni	3,114.12	50	300				
Pd	3,114.04	500*w*	400*w*				
Ta	3,113.90	35*w*	50				
V	3,113.57	100	7				
Co	3,113.48	4	100				
Ru	3,113.40	—	50				
V	3,112.92	—	12				
Ru	3,112.68	3	50				
V	3,112.124	1	20				
Mo	3,112.124	10	40				
Ru	3,111.91	5	50				

*P** *N**

V 3,125.28 *J* 200*R* *O* 80

	λ	I J	I O		λ	I J	I O
Fe	**3,125.27**	**2**	**5**	**Cs**	**3,125.3**	**(4)**	—
Ir	**3,125.25**	—	**2*h***	**Ce**	**3,125.358**	—	**2**
Zr	**3,125.19**	**2**	**3**	**W**	**3,125.358**	**9**	**10**
Er	**3,125.17**	**2**	**15**	**Yb**	**3,125.44**	**2**	**1**
Th	**3,125.16**	**8**	**10**	**Cl**	**3,125.44**	**(6)**	—
Eu	**3,125.10**	**1*h***	**5**	**Th**	**3,125.46**	**10**	**10**
V	**3,125.00**	**50**	**4**	**Cr**	**3,125.47**	**4**	**8**
Cr	**3,124.98**	**125**	**20**	**Re**	**3,125.52**	—	**30**
Ta	**3,124.97**	**20**	**50**	**Ti**	**3,125.55**	—	**2**
Os	**3,124.94**	**10*wh***	**150*h***	**Pb**	**3,125.6**	**(50)**	—
Sc	**3,124.935**	**1*h***	**2**	**Er**	**3,125.65**	**2**	**10**
Sm	**3,124.927**	**3**	**9**	**Fe**	**3,125.654**	**300**	**400**
Ir	**3,124.927**	—	**4**	**Ti**	**3,125.656**	—	**5**

	λ	I	
		J	O
Tm	3,124.90	30	15
U	3,124.900	10	12
Fe	3,124.89	7	15
Eu	3,124.87	5	4*w*
Ce	3,124.86	—	3
Ge	3,124.82	80	200
W	3,124.73	3	8
Ru	3,124.61	2	50
Nd	3,124.57	10	12
Tb	3,124.54	8	8
W	3,124.50	5	6

	λ	I	
		J	O
Hg	3,125.663	150	200
Th	3,125.71	3	8
La	3,125.72	3	—
Ce	3,125.76	—	20
Nb	3,125.89	15	1
Te	3,125.91	(10)	—
Zr	3,125.918	5	8
J	3,125.92	(10)	—
Cl	3,125.96	(5)	—
Ru	3,125.963	12	70
Sm	3,126.00	—	5
Tm	3,126.01	—	25
Kr	3,126.02	(6*h*)	—
Mo	3,126.03	10	—
Cu	3,126.11	20	80
Fe	3,126.17	70	150
Ne	3,126.19	(150)	—
V	3,126.21	100*R*	60
Co	3,126.72	—	70
Co	3,127.25	—	100
Ce	3,127.53	—	40
Ta	3,127.76	100	18*w*
Ru	3,127.91	100	10
V	3,128.28	60	2
Dy	3,128.41	10	40
V	3,128.69	70	3
Cr	3,128.699	150	30
Cu	3,128.701	15	70
Re	3,128.95	—	100*W*
Os	3,129.23	15	60
Ni	3,129.31	—	125
Fe	3,129.33	60	100
Na	3,129.37	(60)	35
Co	3,129.48	2	40
Ta	3,129.55	7	50
Ru	3,129.60	1	50
Ru	3,129.84	4	60
Ta	3,129.95	8	50
V	3,130.27	200*r*	50
Be	3,130.42	200	200
Ta	3,130.58	35	100*W*
Eu	3,130.74	100	100*W*
Nb	3,130.786	100	100
Rh	3,130.790	2	60
Ti	3,130.80	100	25
Be	3,131.07	150	200
Os	3,131.11	30	125

	λ	I J	I O		λ	I J	I O
				Tm	3,131.26	500	400
				Hg	3,131.55	300	400
				Hf	3,131.81	10	40
				Hg	3,131.83	100	200
				Co	3,132.22	—	40
				Fe	3,132.51	40	70
				V	3,132.594	20	80*r*
				Mo	3,132.594	300*R*	1000*R*
				Ta	3,132.64	25	250*W*
				Cd	*3,133.17*	*300*	*200*
				V	*3,133.33*	*200r*	*50*
				Tm	*3,133.89*	*200*	*200*
				Ni	*3,134.108*	*150*	*1000R*
				Fe	*3,134.111*	*125*	*200*
				Hf	*3,134.72*	*125*	*80*
				V	*3,134.93*	*150r*	*30*
				Dy	*3,135.37*	*50*	*100*
				V	*3,136.51*	*200*	*20*
				Co	*3,137.33*	—	*150r*
				Rh	*3,137.71*	—	*100*
				V	*3,138.06*	*70*	—
				Pt	*3,139.39*	*80*	*300*
				V	*3,139.74*	*150*	*15*
				Fe	*3,139.91*	*40*	*70*
				Co	*3,139.94*	*10*	*150r*

V 3,183.406 *J* 100*R* *O* 200*R*
3,183.982 *J* 400*R* *O* 500*R*

Th	**3,183.79**	**8**	**8**	**Eu**	**3,183.81**	**8**	**8*h***
Lu	**3,183.73**	**6**	**3**	**Ce**	**3,183.84**	—	**2**
Tb	**3,183.64**	**3**	**15**	**Hg**	**3,183.85**	**6**	**6**
J	**3,183.58**	**(10)**	—	**Tb**	**3,183.88**	**3**	**15**
Fe	**3,183.57**	**3**	**5**	**Sm**	**3,183.92**	**40**	**60**
Th	**3,183.55**	**6**	**8**	**Ba**	**3,183.95**	—	**2**
Ce	**3,183.52**	—	**40**	**Ti**	**3,183.968**	**2**	**12**
W	**3,183.51**	**6**	**9**	**Th**	**3,183.971**	**5**	**12**
Er	**3,183.41**	**4**	**15**	**W**	**3,183.98**	**6**	**1**
V	***3,183.406***	***100R***	***200R***	***V***	***3,183.982***	***400R***	***500R***
Cr	**3,183.325**	**150**	**6**	**Ce**	**3,183.99**	—	**4**
Mo	**3,183.322**	—	**6*d***	**Ti**	**3,184.02**	**10**	**5**
Sm	**3,183.30**	**2**	**4**	**W**	**3,184.04**	**7**	**10**
Tb	**3,183.29**	**3**	**15**	**Ti**	**3,184.09**	**10**	**3**

	λ	I J	I O		λ	I J	I O
Ni	**3,183.25**	**—**	**25**	**Fe**	**3,184.11**	**2**	**5**
Dy	**3,183.20**	**3**	**7**	**Ho**	**3,184.13**	**4*h***	**—**
Ba	**3,183.16**	**—**	**5**	**Ag**	**3,184.15**	**20**	**—**
Fe	**3,183.11**	**50**	**7**	**Sm**	**3,184.19**	**—**	**3**
Ce	**3,183.09**	**—**	**5**	**Dy**	**3,184.20**	**1**	**2**
Ni	**3,183.04**	**—**	**5**	**Ce**	**3,184.21**	**—**	**20**
Mo	**3,183.03**	**—**	**10**	**Eu**	**3,184.22**	**—**	**10*W***
Fe	**3,182.97**	**70**	**125**	**Nb**	**3,184.223**	**150**	**5**
Cd	**3,182.91**	**(8)**	**—**	**Ce**	**3,184.33**	**—**	**2**
Re	**3,182.87**	**—**	**100**	**Os**	**3,184.337**	**5**	**10**
Zr	**3,182.86**	**18**	**12**	**Cr**	**3,184.341**	**30**	**—**
W	**3,182.85**	**8**	**9**	**Ni**	**3,184.37**	**3**	**150**
U	**3,182.83**	**8**	**6**	**Mo**	**3,184.40**	**1**	**3**
Zn	**3,182.82**	**(3)**	**—**	**U**	**3,184.406**	**2**	**6**
Os	**3,182.803**	**8**	**40**	**Ba**	**3,184.41**	**—**	**3**
Ir	**3,182.798**	**—**	**2**	**W**	**3,184.415**	**9**	**10**
V	**3,182.675**	**35**	**—**	**Ta**	**3,184.55**	**18**	**70**
Re	**3,182.667**	**—**	**15**	**Mo**	**3,184.57**	**4**	**6**
Ce	**3,182.66**	**—**	**20**	**Pd**	**3,184.59**	**2**	**—**
Th	**3,182.65**	**10**	**10**	**Ce**	**3,184.622**	**—**	**12**
V	**3,182.59**	**35**	**—**	**Fe**	**3,184.622**	**40**	**60**
Ta	**3,182.571**	**18**	**70*r***				
Ti	**3,182.57**	**40*wh***	**—**				
Os	**3,182.567**	**15**	**100**				
U	**3,182.55**	**5**	**6**				
Pr	**3,182.45**	**3**	**15**				
Fe	3,182.06	80	80				
Os	3,181.88	12	100				
Ti	3,181.84	50*wh*	—				
Fe	3,181.52	70	80				
Ta	3,180.95	35	100				
Fe	3,180.75	100	100				
Cr	3,180.70	150	30				
Nb	3,180.29	200	5				
Fe	3,180.23	300	300				
V	3,179.41	25	1				
B	3,179.35	100	5				
Ca	3,179.33	400*w*	100				
Cr	3,179.28	10*h*	100				
Ru	3,179.26	50*r*	50				
Mn	3,178.49	50	150				
Os	3,178.06	20	150				
Fe	3,178.01	150	300				
V	3,177.68	20	—				
Fe	3,177.53	300	5				
Co	3,177.27	—	100				
Ru	3,177.05	200	60				

	λ	I	
		J	O
Pb	3,176.54	100	—
Mo	3,176.33	50	—
Nb	3,175.85	50	5
Fe	3,175.45	200	200
Ru	3,175.15	100	20
P	3,175.14	(70)	—
Mo	3,175.05	60	2
Sn	3,175.02	400*hr*	500*h*
Ti	*3,174.80*	*100*	—
V	*3,174.54*	*80*	*1*
V	*3,174.08*	*35*	*5*
Tm	*3,173.58*	*100*	*50*
Nb	*3,173.20*	*100*	*2*
Ar	*3,172.96*	*(150)*	—
Tm	*3,172.82*	*200*	*200*
V	*3,172.23*	*25*	—
Cr	*3,172.08*	*200*	*2*
Fe	*3,172.07*	*100*	*100*
V	*3,171.14*	*30*	—
La	*3,171.67*	*1000wh*	*2h*
Fe	*3,171.35*	*80*	*100*
Mo	*3,170.35*	*25r*	*1000R*
Ta	*3,170.29*	*35*	*250w*
Ti	*3,168.52*	*300r*	*70*
V	*3,168.14*	*40*	*12*
Fe	*3,167.86*	*100*	*2*
Ru	*3,167.47*	*100*	*5*
V	*3,167.44*	*150R*	*5*
Os	*3,166.51*	*20*	*200*

V 3,185.396 *J* 400*R* *O* 500*R*

	λ	J	O
Os	**3,185.33**	**12**	**150**
Fe	**3,185.32**	**25**	—
Si	**3,185.28**	**3**	—
Er	**3,185.25**	**2**	**15**
W	**3,185.20**	**5**	**6**
U	**3,185.14**	**3**	**4**
Mo	**3,185.10**	**8**	**20**
Ag	**3,185.09**	**4*h***	—
Zr	**3,185.07**	—	**2**
W	**3,185.05**	**10**	**1**
Th	**3,184.898**	**10**	**10**
Ru	**3,184.897**	—	**4**
Fe	**3,184.896**	**150**	**200**
Cu	**3,184.84**	**8**	—
U	**3,185.397**	—	**4**
Ce	**3,185.41**	—	**4**
Ru	**3,185.44**	—	**12**
Tm	**3,185.48**	**40**	**20**
Rn	**3,185.5**	**(10)**	—
Ce	**3,185.506**	—	**2**
Tl	**3,185.51**	**(6)**	—
Cd	**3,185.55**	**(15)**	—
Eu	**3,185.56**	**10*wh***	**10*wh***
Re	**3,185.563**	—	**200**
Rh	**3,185.59**	**20**	**100**
U	**3,185.71**	**10**	**12**
Mo	**3,185.711**	**3**	**8**
Sm	**3,185.84**	**1**	**3**

	λ	I J	I O
P	**3,184.82**	**(50)**	**—**
Dy	**3,184.78**	**5**	**18**
U	**3,184.76**	**1**	**3**
Re	**3,184.75**	**—**	**150**
Ce	**3,184.72**	**—**	**10**
Ir	**3,185.89**	**—**	**2*h***
Co	**3,185.95**	**—**	**40**
Yt	**3,185.96**	**2*h***	**7**
Sm	**3,186.015**	**5**	**40**
Cu	**3,186.017**	**3**	**—**
Ru	**3,186.04**	**25**	**80**
Ir	**3,186.11**	**—**	**3**
Ce	**3,186.13**	**—**	**40**
Ar	**3,186.19**	**(3)**	**—**
Tb	**3,186.23**	**3**	**8**
P	**3,186.24**	**(50)**	**—**
Ti	3,186.45	80	150
Fe	3,186.74	300	20
V	3,186.86	10	1
Os	3,186.98	15	100
Fe	3,187.29	60	—
Mo	3,187.59	50	1
V	3,187.71	100*R*	35
Tl	3,187.74	(50)	—
He	3,187.74	(200)	—
Cr	3,188.01	60*h*	150*h*
Ru	3,188.34	50	60
Co	3,188.37	2*h*	100
V	3,188.51	100*R*	35
Fe	3,188.57	100	150
Fe	3,188.82	100	150
Rh	3,189.05	20	100
Nb	3,189.28	300*r*	10*w*
Os	3,189.46	15	125
Na	3,189.78	(60)	35
Ru	3,189.98	50	50
V	3,190.68	150*R*	50
Ti	3,190.87	200*r*	40
Hg	3,191.03	(100)	—
Nb	3,191.10	300*w*	100*w*
Rh	3,191.19	50	300
Nb	3,191.43	200	3
Fe	3,191.66	150	200
Lu	3,191.80	60*h*	3
Ti	3,191.99	20	100
Mo	3,192.10	50	3
V	3,192.71	20	1
Fe	3,192.80	8	150
Yb	3,192.88	80	12
La	3,193.02	60	15
V	3,193.19	20	1
Fe	3,193.23	70	100
Fe	3,193.80	50	10

	λ	I J	I O
V	3,193.92	20	100
J	3,193.95	(100)	—
Mo	3,193.97	50r	1000r
Nb	3,194.27	150w	2w
Nb	3,194.98	300	30
Ru	3,195.15	100	—
V	3,195.48	25h	—
Fe	3,196.08	150	10
V	3,196.56	20	5
Fe	3,196.93	300	500
Ni	3,197.11	—	300
V	3,198.01	30r	100r
Fe	3,199.52	200	300
V	3,199.82	10	25
Ti	3,199.91	150	200
Ar	3,200.39	(100)	—
Fe	3,200.47	150	150
Ru	3,200.26	100	2
V	3,202.38	20r	100r

V 4,379.24 *J* 200*R* *O* 200*R*

	λ	I J	I O
Hf	4,379.17	4h	10h
Ho	4,379.15	—	3
Nd	4,379.11	1	2
Ce	4,379.08	—	3
Br	4,378.97	(7)	—
Pr	4,378.84	—	3
Ta	4,378.822	2	40
Ce	4,378.818	—	4
Tb	4,378.70	—	9w
Pr	4,378.63	—	3
Ce	4,378.58	—	5
Gd	4,378.57	—	40
U	4,378.53	2	6
W	4,378.49	12	25
Cu	4,378.43	2	—
O	4,378.41	(10h)	—
Ce	4,378.407	—	3
Er	4,378.34	—	9
Cr	4,378.32	1	5
Pr	4,378.26	3	15
Sm	4,378.23	100	100
Cu	4,378.20	30w	200w
Th	4,378.13	1	4
La	4,378.10	30	40
Ag	4,379.25	1	5
Pb	4,379.26	—	4
Yt	4,379.32	—	4
Pr	4,379.33	2	100w
Bi	4,379.4	20	25
Dy	4,379.41	—	2
Xe	4,379.44	(5wh)	—
Ne	4,379.50	(100)	—
Gd	4,379.52	—	2
Nb	4,379.525	3	2
O	4,379.55	(15h)	—
Pd	4,379.56	2h	6
Tb	4,379.60	—	3w
bhLa	4,379.6	—	10
Ho	4,379.64	2	—
U	4,379.644	2	2
bhLa	4,379.7	—	20
Ar	4,379.74	(80)	—
Zr	4,379.776	8	10
Br	4,379.78	(6)	—
Cr	4,379.782	2	15
Eu	4,379.81	2	5w
Ce	4,379.84	—	2
Nd	4,379.88	—	4
Cl	4,379.91	(15)	—

	λ	I J	I O
Nb	4,377.96	30	10
Mo	4,377.76	200	5
Te	4,377.10	(70)	—
Ir	4,377.01	4	100
Kr	4,376.12	(800)	—
Fe	4,375.93	200	500
V	4,375.30	12	20
Sm	4,374.97	200	200
Mn	4,374.95	20	150
Yt	4,374.93	150	150
Rh	4,374.80	500	1000 *W*
Sc	4,374.45	25	100
Cr	4,374.16	60	50
Gd	4,373.84	80	200
V	4,373.23	20	25
Cl	4,372.91	(80)	—
In	4,372.87	(80)	—
Ru	4,372.21	100	125
Pr	4,371.61	40*w*	125
Ar	4,371.36	(80)	—
Cr	4,371.28	150	200
Gd	4,369.775	150	250
Fe	4,369.774	100	200
Ne	4,369.77	(70)	—
Kr	4,369.69	(200)	—
Xe	4,369.20	(100*wh*)	—
V	4,368.60	9	15
Pr	4,368.33	90	125
O	4,368.30	(1000)	—
Sm	4,368.03	60	60
Fe	4,367.91	70	60
Fe	4,367.582	50	100
Re	4,367.582	—	80
O	*4,366.91*	*(100)*	—
Br	*4,365.60*	*(200)*	—
V	*4,364.22*	*10*	*15*
Te	*4,364.02*	*(400)*	—
Mo	*4,363.64*	*200*	*5*
V	*4,363.52*	*12*	*20*
Kr	*4,362.64*	*(500)*	—
V	*4,361.40*	*9*	*12*
Gd	*4,360.93*	—	*200*
V	*4,360.58*	*12*	*15*
Tm	*4,359.93*	*30*	*300*
Cr	*4,359.63*	*150*	*200*
Hg	*4,358.35*	*500*	*3000w*
N	*4,358.27*	*(250)*	—
V	*4,355.94*	*20*	*25*
Kr	*4,355.48*	*(3000)*	—
Rh	**4,379.92**	**25**	**60**
Ce	**4,380.06**	**2**	**30**
Co	**4,380.07**	**3**	**5*wh***
Kr	**4,380.11**	**(2)**	—
W	**4,380.115**	**2**	**8**
Eu	**4,380.17**	—	**3*w***
Dy	**4,380.23**	**2**	**5**
U	**4,380.27**	**6**	**12**
Mo	**4,380.29**	**25**	**30**
Pr	**4,380.32**	**20**	**50**
Ce	**4,380.33**	—	**4**
Sm	**4,380.42**	**6**	**25**

V 4,389.97 *J* 60*R* *O* 80*R*

	λ	I	
		J	O
Tb	**4,389.92**	—	**2**
Te	**4,389.92**	**(50)**	—
Gd	**4,389.88**	**40*h***	**40**
Ni	**4,389.870**	**5**	**5**
La	**4,389.868**	—	**5**
W	**4,389.84**	**6**	**15**
Tb	**4,389.81**	—	**3**
Ce	**4,389.807**	—	**4**
Dy	**4,389.79**	**2**	**3**
Cl	**4,389.76**	**(25)**	—
Yb	**4,389.76**	**10**	**1**
Mn	**4,389.75**	—	**50**
Kr	**4,389.72**	**(20*h*)**	—
Sc	**4,389.60**	—	**10**
Mo	**4,389.57**	**6**	**6**
Pr	**4,389.51**	—	**8**
In	**4,389.48**	**5*h***	—
Tb	**4,389.40**	—	**3**
Cl	**4,389.32**	**(6)**	—
Fe	**4,389.25**	**2**	**35**
Eu	**4,389.22**	—	**2*w***
Ce	**4,389.11**	—	**5**
Eu	**4,389.07**	**2**	**5**
Tb	**4,389.04**	—	**6**
Gd	**4,388.991**	**4**	**4**
Ru	**4,388.990**	—	**12**
Sm	**4,388.987**	**8**	**10**
Kr	**4,388.90**	**(3*h*)**	—
Cs	**4,388.76**	**(10)**	—
Fe	4,388.41	50	125
Mo	4,388.27	20	6
K	4,388.13	(40)	—
Mn	4,388.08	—	60
He	4,387.93	(30)	—
Fe	4,387.90	35	150
Eu	4,387.884	—	200
Gd	4,387.678	—	150
V	4,387.21	12	15
Ti	4,386.85	80	8
Pb	4,386.58	(20)	—
Kr	4,386.54	(300*h*)	—
Tm	4,386.42	10	200
Ac	4,386.37	100	—
Ta	4,386.07	15	50
Xe	4,385.77	(70)	—
Nd	4,385.66	20	40

	λ	I	
		J	O
Gd	**4,389.99**	**80**	**30**
Nd	**4,390.05**	**3**	**30**
Na	**4,390,14**	—	**15**
Pr	**4,390.145**	**1**	**8**
Ce	**4,390.15**	—	**2**
U	**4,390.16**	**3**	**1**
Er	**4,390.19**	—	**6**
Ce	**4,390.28**	—	**10**
Ni	**4,390.32**	—	**2**
J	**4,390.34**	**(15*h*)**	—
Eu	**4,390.36**	**2**	**4*w***
Cl	**4,390.38**	**(8)**	—
Ru	**4,390.43**	**80**	**150*R***
Ce	**4,390.50**	—	**5**
Ho	**4,390.53**	—	**2**
U	**4,390.55**	**2**	**2**
Mg	**4,390.58**	—	**10**
V	**4,390.61**	**3**	**10**
Nd	**4,390.66**	**15**	**20**
Hf	**4,390.70**	—	**3**
Ce	**4,390.81**	—	**4**
Sm	**4,390.86**	**150**	**150**
Tb	**4,390.91**	—	**30**
Mo	**4,390.93**	**8*h***	**1**
Fe	**4,390.954**	**35**	**100**
Gd	**4,390.955**	**100**	**100**
Ru	**4,391.027**	—	**20**
Ti	**4,391.034**	**25**	**6**
Os	**4,391.08**	**3**	**15**
Nd	**4,391.110**	**4**	**10**
Pr	**4,391.111**	—	**10**
Th	**4,391.114**	**40**	**50**

	λ	I J	I O		λ	I J	I O
Ru	4,385.65	50	125				
Mo	4,385.56	20	—				
Ru	4,385.39	40	125				
P	4,385.33	(100)	—				
Kr	4,385.27	(50*wh*)	—				
Te	4,385.08	(50)	—				
Cr	4,384.98	200	150				
Xe	4,384.93	(30)	—				
Pr	4,384.80	15	40				
V	4,384.72	125*R*	125*R*				
Er	4,384.70	5	30				
Sm	4,384.29	50	50				
Xe	4,383.91	(100)	—				
Fe	4,383.55	800	1000				
La	4,383.45	50	10				
Rn	4,383.30	(35)	—				
Eu	4,383.16	30	100*W*				
Gd	4,383.14	40	30				
Se	4,382.87	(800)	—				
Mn	4,382.63	—	80*h*				
Mo	4,381.64	150	150				
Kr	4,381.52	(100*h*)	—				
Ne	4,381.22	(30)	—				
Gd	4,380.64	125	100				
V	4,380.55	10	15				

*P**

V 4,408.51 *J* 20*R* *O* 30

	λ	J	O		λ	J	O
Fe	**4,408.42**	**60**	**125**	**W**	**4,408.71**	**4**	**8**
Ce	**4,408.34**	**—**	**3**	**Ce**	**4,408.77**	**—**	**4**
W	**4,408.28**	**12**	**25**	**Tb**	**4,408.79**	**—**	**3**
Gd	**4,408.26**	**150**	**100**	**Hf**	**4,408.81**	**—**	**8**
V	**4,408.20**	**—**	**30**	**Nd**	**4,408.82**	**5**	**15**
Pr	**4,408.16**	**2**	**10**	**Pr**	**4,408.84**	**100**	**125**
Mn	**4,408.08**	**5**	**60**	**Ir**	**4,408.86**	**—**	**4**
Dy	**4,408.05**	**4**	**3**	**Ce**	**4,408.869**	**1**	**12**
U	**4,407.96**	**12**	**12**	**U**	**4,408.87**	**2**	**—**
Be	**4,407.91**	**(35)**	**20**	**Kr**	**4,408.89**	**(40*hs*)**	**—**
Ce	**4,407.82**	**—**	**3**	**Th**	**4,408.891**	**2**	**6**
Fe	**4,407.72**	**50**	**100**	**J**	**4,408.96**	**(250)**	**—**
Ti	**4,407.68**	**10**	**2**	**Ce**	**4,409.05**	**—**	**4**
V	**4,407.64**	**9*R***	**15*h***	**Ti**	**4,409.22**	**8**	**2**
Br	**4,407.62**	**(12)**	**—**	**Gd**	**4,409.26**	**10**	**10**
Dy	**4,407.54**	**—**	**2**	**Ne**	**4,409.30**	**(150)**	**—**
Sm	**4,407.52**	**7**	**10**	**Ce**	**4,409.301**	**—**	**2**
Mo	**4,407.44**	**30**	**—**	**J**	**4,409.31**	**(15)**	**—**
				Sm	**4,409.337**	**100**	**100**

	λ	I	
		J	O
Nd	4,407.07	8	15
Xe	4,406.88	(100*wh*)	—
Mo	4,406.87	10	8
Sr	4,406.85	—	50*h*
Ba	4,406.84	(3)	20
Eu	4,406.78	—	15*w*
Ir	4,406.76	2	25
Gd	4,406.67	200	70
Pr	4,406.66	8	25
V	4,406.64	30	40
Se	4,406.58	(70)	—
Pd	4,406.55	2	30
Re	4,406.40	—	60
V	4,406.15	15*W*	20*W*
Pr	4,405.85	100	100
Ti	4,405.68	6	20
Te	4,405.49	(15)	—
Ce	4,405.47	2	18
Tb	4,405.41	—	15
Eu	4,405.27	—	15*W*
Cs	4,405.25	(35)	—
Ba	4,405.23	(20)	—
Pr	4,405.14	20	25
V	4,405.01	6	12
Ti	4,404.90	10	15
Hg	4,404.86	(50)	—
Fe	4,404.75	700	1000
Pr	4,404.71	4	25*w*
Mo	4,404.55	8	15
As	4,404.53	15	—
Kr	4,404.33	(30*h*)	—
Ti	4,404.27	30	50
Os	4,404.21	1	18
Mo	4,404.18	15*h*	1
W	4,403.95	9	20
Cs	4,403.85	(20)	—
Ir	4,403.78	10	300
V	4,403.67	15	20
Pr	4,403.60	40	100
J	4,403.55	(20)	—
Cr	4,403.50	25	15*d*
Cr	4,403.37	6	15*d*
Sm	4,403.36	50	50
Mo	4,403.34	20	—
Pr	4,403.28	10	25
Tb	4,403.19	—	20
Gd	4,403.14	100	100
Sm	4,403.12	10	25
Sm	4,403.05	10	15
Yb	**4,409.34**	**10**	**6**
Er	**4,409.359**	**2**	**35**
Ho	**4,409.36**	**1**	**3**
Dy	**4,409.38**	**8**	**30**
Mo	**4,409.44**	**12**	**10**
Tb	**4,409.51**	—	**40*w***
Ti	**4,409.519**	**10**	**3**
Nd	**4,409.520**	**2**	**8**
Cr	**4,409.55**	—	**5**
*bh***La**	**4,409.6**	—	**4**
Ne	**4,409.62**	**(20)**	—
Zr	**4,409.63**	—	**3**
Eu	**4,409.64**	—	**6*W***
U	**4,409.67**	**4**	**4**
Pr	**4,409.68**	**2*w***	**25*w***
Mo	4,409.95	15	15
Ru	4,410.03	80	150
C	4,410.06	30	—
Cs	4,410.208	(20)	—
Nb	4,410.213	30	15
Cr	4,410.30	8	25
Kr	4,410.37	(50)	—
Mn	4,410.49	—	50
Ni	4,410.52	4	25
Fe	4,410.71	—	20
Te	4,410.95	(50)	—
Nd	4,411.05	20	50
Yb	4,411.07	—	20
Ti	4,411.08	100	7
Cr	4,411.09	12	15*d*
Gd	4,411.16	50	100
Ir	4,411.18	2	40
C	4,411.20	40	—
La	4,411.204	10*h*	4
Pr	4,411.33	—	25
Sb	4,411.50	10*h*	—
C	4,411.52	40	—
Sm	4,411.58	5	25
Mo	4,411.70	25	25
Te	4,411.78	(50)	—
Sm	4,411.83	10	15
Mn	4,411.88	20	100
Ti	4,411.936	12	8
P	4,411.94	(15*h*)	—
Ce	4,412.02	2	20
V	4,412.14	20	25
Pr	4,412.15	15	50

	λ	I	
		J	O
Cl	4,403.03	(12)	—
Mo	4,402.90	20	20
Os	4,402.74	3	50
Re	4,402.60	—	30
Ba	4,402.55	10	80
Ta	4,402.50	20*h*	100
Mo	4,402.49	15	15
Yb	4,402.30	20	6
Hg	4,402.06	(50)	—
Te	4,401.89	(100)	—
Gd	4,401.85	100	200
Ni	4,401.55	30	1000*W*
Tb	4,401.54	—	20
Rb	4,401.4	(30)	—
Fe	4,401.30	15	60
Nb	4,401.172	10	—
Sm	4,401.166	30	50
Se	4,401.02	(100)	—
Ar	4,401.02	(40)	—
P	4,400.99	(50)	—
Pb	4,400.89	10	—
Kr	4,400.87	(100*h*)	—
Ni	4,400.870	3	15
Nd	4,400.83	20	50
Gd	4,400.76	8	15
Ti	4,400.582	2	25
Os	4,400.579	1	18
V	4,400.57	40	60
Sc	4,400.355	30	150
Nb	4,400.354	10	5
Fe	4,400.351	1	20
Pr	4,400.25	10	25
Ar	4,400.09	(30)	—
Pr	4,400.03	20	30
Kr	4,399.97	(200)	—
Se	4,399.94	(12)	—
Sm	4,399.86	15	20
Cr	4,399.82	3	20
Ti	4,399.77	100	40
Br	4,399.72	(10)	—
Ca	4,399.64	10	—
U	4,399.63	6	15
Ru	4,399.59	—	20
Cs	4,399.49	(20)	—
Ir	4,399.47	100	400
Kr	4,399.39	(15*hs*)	—
Pr	4,399.33	20	40
Ce	4,399.20	6	35
W	4,412.20	10	20
Mo	4,412.22	25	—
Cr	4,412.250	10	35
As	4,412.25	10	—
Nd	4,412.26	15	40
Ne	4,412.28	(20)	—
Cd	4,412.31	10	—
J	4,412.37	(25)	—
Ti	4,412.43	1	15
Ne	4,412.54	(15)	—
Mo	4,412.77	20	30
U	4,413.14	4	15
Se	4,413.16	(20)	—
Ac	4,413.17	100	—
Ce	4,413.19	2	35
Ne	4,413.20	(50)	—
Ne	4,413.56	(15)	—
As	4,413.64	50	—
V	4,413.68	10	15
Pr	4,413.76	40	90
Ce	4,413.80	1	20
Cr	4,413.87	15	25
Gd	4,414.16	60	100
P	4,414.28	(100)	—
Cr	4,414.35	—	15*h*
Pr	4,414.40	4	20
V	4,414.54	5	10
P	4,414.60	(70)	—
Cd	4,414.63	200	—
Gd	4,414.73	50	100
Xe	4,414.84	(150)	—
Mn	4,414.88	60	150
O	4,414.89	(300)	—
W	4,415.07	6	15
Fe	4,415.12	400	600
U	4,415.24	12	12
Sc	4,415.56	25	100
Cu	4,415.60	—	40*w*
In	4,415.69	15*h*	—
Cd	4,415.70	20	1
Ta	4,415.74	10	40
Re	4,415.82	—	40
Se	4,415.85	(10)	—
Xe	4,416.07	(80*wh*)	—
Th	4,416.24	8	15
Tb	4,416.26	—	30
V	4,416.47	7	15*w*
Ti	4,416.54	10	70

	λ	I	
		J	O
Se	4,399.17	(15)	—
Cl	4,399.14	(15)	—
J	4,399.09	(20)	—
Ce	4,398.79	3	20
Hg	4,398.62	(300)	—
Mo	4,398.49	12	6
Te	4,398.45	(70)	—
Ta	4,398.449	10	40
Ti	4,398.31	10	3
Pr	4,398.26	9	25
Nd	4,398.03	5	15
Yt	4,398.01	100	150
Cs	4,397.99	(10)	—
Ne	4,397.94	(100)	—
Ru	4,397.80	—	150
Gd	4,397.52	5	100
Sm	4,397.35	5	20
Nd	4,397.31	2	15
Mo	4,397.29	30	30
Os	4,397.26	3	18
Cr	4,397.25	5	25
Cs	4,396.91	(15)	—
Re	4,396.79	—	25
Mo	4,396.66	25	25
Ce	4,396.58	1h	20
Br	4,396.40	(20)	—
Ag	4,396.32	—	100
Pr	4,396.12	50	80
Te	4,396.00	(100)	—
O	4,395.95	(80)	—
Ag	4,395.93	30	10
Ti	4,395.84	30	10
Pr	4,395.79	2w	30
Xe	4,395.77	(200wh)	—
Ne	4,395.56	(50)	—
Fe	4,395.29	—	80
V	4,395.23	40R	60R
Ti	4,395.03	150	50
Os	4,394.86	6	150
Ce	4,394.78	3	30
Tm	4,394.42	—	30
Re	4,394.38	—	100r
Ti	4,393.92	12	60
V	4,393.83	10	15
U	4,393.59	6	40
Xe	4,393.20	(200wh)	—
Ce	4,393.19	3	35
Nd	4,416.56	4	15
Ne	4,416.82	(50)	—
J	4,416.84	(15)	—
Kr	4,416.884	(20)	—
Nd	4,416.884	8	20
Ce	4,416.90	4h	25s
O	4,416.97	(150)	—
Kr	4,417.24	(40)	—
Eu	4,417.25	—	60w
Ti	4,417.28	20	80
Re	4,417.29	—	20
P	4,417.30	(30)	—
Hf	4,417.35	50	25
Sm	4,417.58	80	80
Ti	4,417.72	80	40
Hf	4,417.91	1	25
bhLa	4,418.1	—	25
bhLa	4,418.2	—	50
Ti	4,418.34	20	10
U	4,418.47	—	15
Kr	4,418.76	(50)	—
Ce	4,418.78	10	40
Gd	4,419.04	200	200
Pr	4,419.06	30	80
Ce	4,419.30	1	18
Sm	4,419.34	20	40
Nb	4,419.44	20	10
Er	4,419.61	4	25
Pr	4,419.67	50	100
Mn	4,419.78	20	100
V	4,419.93	20	30
Zr	4,420.46	1	20
W	4,420.468	10	30
Os	4,420.468	100	400R
Sm	4,420.529	200	200
Er	4,420.534	—	15
P	4,420.64	(70)	—
Ru	4,420.84	—	40
Ar	4,420.90	(40)	—
Sm	4,421.136	150	150
Te	4,421.14	(70)	—
Pr	4,421.23	35w	100
Gd	4,421.24	8	100
Ne	4,421.38	(30)	—
Ru	4,421.46	—	60
Ne	4,421.56	(50)	—
V	4,421.57	20h	30h

	λ	I J	I O
V	4,393.09	9	12
Yb	4,392.83	20	3
Ir	4,392.59	4	100
Re	4,392.49	—	100
V	4,392.074	15	25
Gd	4,392.071	100	100
Pr	4,391.99	5w	30w
Ne	4,391.94	(150)	—
S	4,391.84	(30)	—
Pt	4,391.83	3	50
Cr	4,391.75	35	50
Ce	4,391.66	15	40
Re	4,391.33	—	60w
Se	4,421.68	(20)	—
Ti	4,421.76	15	60
Ti	4,421.95	35	6
Gd	4,422.41	40	100
Er	4,422.47	20	30
Ne	4,422.52	(300)	—
Fe	4,422.57	125	300
Yt	4,422.59	60	60
Kr	4,422.70	(100hs)	—
Hf	4,422.74	25	15
Ti	4,422.82	25	80
Tb	4,423.11	—	35
V	4,423.21	25w	40
P	4,423.55	(30)	—
Mo	4,423.62	40	40
J	4,423.76	(80)	—
P	4,423.9	(30)	—
Ar	4,423.99	(80)	—
Cr	4,424.28	35	25
Sm	4,424.342	300	300
Nd	4,424.343	50	50
V	4,424.56	15	20
Pr	4,424.59	35	90
Ne	4,424.80	(300)	—
Kr	4,425.19	(100)	—
Hg	4,425.22	(30)	—
Ne	4,425.40	(150)	—
Ca	4,425.44	20	100
Cs	4,425.66	(20)	—
V	4,426.00	15h	25h
Ar	4,426.01	(300)	—
Ti	4,426.05	25	80
Rb	4,426.1	(30)	—
Se	4,426.12	(20)	—
Gd	4,426.15	2	50
Ir	4,426.27	10	400w
Mo	4,426.67	30	30
Ti	4,427.10	60	126
Fe	4,427.312	200	500
V	4,427.312	15	20
As	4,427.38	200	—
La	4,427.57	50	30
Sm	4,427.58	25	25
Ne	4,427.75	(30)	—
bhLa	4,428.1	—	30
P	4,428.15	(70)	—
bhLa	4,428.2	—	40
J	4,428.22	(35)	—
Ru	4,428.46	—	125

P*

	λ	I	
		J	O
V	4,428.51	20	25
Ne	4,428.54	(100)	—
Pr	4,429.24	125	200
Ce	4,429.27	5	35
V	4,429.80	25	30
La	4,429.90	300	200
Ar	4,430.18	(100)	—
Ti	4,430.37	15	35
Lu	4,430.48	2	30
V	4,430.50	5	10
Fe	4,430.62	8	200
Gd	4,430.63	40	150
Rb	4,430,7	(20)	—
Ne	4,430.90	(50)	—
Ar	4,431,02	(80)	—
S	4,431.02	(20)	—
Se	4,431.37	3	50
Kr	4,431.67	(500)	—
J	4,431.730	(20)	—
As	4,431.73	200	—
Gd	4,431.77	1	60
Pr	4,431.89	5w	40
Ba	4,431.90	30	60
Cr	4,432.17	15	30
Se	4,432.33	(60)	—
Pr	4,432.34	10w	80
S	4,432.41	(50)	—
O	4,432.412	1	30

W 74 183.85

t_0 1,927°C

t_1 3,370°C

I.	II.	III.	IV.	V.
7.98	–	–	–	–

λ	I		eV
	J	O	
II 2,397.09	30	18	5.56
II 2,589.17	25	15	5.55
I 2,944.40	20	30	4.57
I 2,946.98	18	20	4.57
II 3,613.79	30	10	5.24
I 4,008.75	45	45	3.45
I 4,294.61	50	50	3.25
I 4,302.11	60	60	3.24

W 2,397.09 *J* 30 *O* 18

	λ	*I* *J*	*I* *O*
Co	**2,397.03**	—	**6**
V	**2,396.98**	**3***wh*	—
Ru	**2,396.94**	**8**	**3**
Rh	**2,396.93**	**5**	—
Re	**2,396.81**	**7**	**30**
Co	**2,396.774**	—	**90**
Os	**2,396.77**	**12**	**100**
Nb	**2,396.770**	**3***h*	—
Ru	2,396.71	80	60
Pt	2,396.69	30	2
Ni	2,396.63	2	12
Rh	2,396.55	200	2
V	2,396.48	5	10
Er	2,396.38	30	—
Ni	2,396.37	3	12
Cr	2,396.37	3	30
Ta	2,396.30	—	80
Co	2,396.23	—	10
W	2,396.22	8	6
Pt	2,396.17	18	25
Nd	2,396.12	15	—
Ir	2,396.09	5	20
Re	2,396.04	3	10
W	2,395.89	—	6
Ir	2,395.886	8	15
Cr	2,395.79	2*h*	25
W	2,395.71	8	5
Fe	2,395.62	100*Wh*	50
Ni	2,395.61	4	10
W	2,395.47	—	8
Os	2,395.39	10	10
Br	2,395.34	(25)	—
Nb	2,395.32	2	15
W	2,395.30	—	6
Mo	2,395.25	—	12
Sb	2,395.20	15	50
W	2,395.10	12	5
V	2,395.09	5*h*	10
B	2,395.07	15	—
Ta	2,395.04	25	—
Mo	2,394.74	9	18
*bh*C	2,394.6	—	20
Ir	2,394.56	—	10
Ni	2,394.52	20	18
W	2,394.45	5	7
Ir	2,394.33	—	30

	λ	*I* *J*	*I* *O*
Zn	**2,397.17**	**(2)**	—
Nb	**2,397.18**	**2**	—
J	**2,397.21**	**(12)**	—
Hf	**2,397.21**	**2***h*	—
Ir	**2,397.22**	**15**	**3**
Zr	**2,397.23**	—	**12**
La	**2,397.24**	**7***h*	—
Er	**2,397.25**	**3**	**7**
Co	**2,397.27**	—	**4**
Br	**2,397.30**	**(3)**	—
U	**2,397.32**	**20**	—
Re	**2,397.36**	**8**	**30**
Co	**2,397.39**	**25**	**4**
Ta	**2,397.42**	**2**	—
W	**2,397.43**	—	**5**
Os	2,397.61	3	20
W	2,397.72	3	12
Cr	2,397.77	35	5
V	2,397.78	—	30
W	2,397.98	8	8
Yb	2,398.01	5	10
V	2,398.13	—	10
Os	2,398.18	3	25
V	2,398.27	—	15
Co	2,398.37	18	2
In	2,398.38	(25)	—
Nb	3,398.48	30	10
*bh*B	2,398.5	—	200
Ca	2,398.56	20	100*R*
V	2,398.68	—	10
Ir	2,398.75	150	10
Pd	2,398.83	15*h*	—
Rh	2,398.95	—	10
W	2,399.04	2*h*	9
Cr	2,399.06	2	50
Lu	2,399.14	50	10
In	2,399.18	—	10
Fe	2,399.24	30	20
W	2,399.33	8	—
Hg	2,399.38	10	20*wh*
Ir	2,399.57	2*h*	15
Cr	2,399.58	—	40
Pb	2,399.583	12	35
Os	2,399.65	—	20
La	2,399.66	20*h*	—
Re	2,399.67	5	10

	λ	I J	I O		λ	I J	I O
Os	2,394.29	10	45	V	2,399.68	150	—
				Hg	2,399.74	15	10
V	2,394.27	30	6	Ru	2,399.75	—	12
Nd	2,394.20	20	—	V	2,399.96	2	30
W	2,394.17	8	5				
Cr	2,394.01	50	—	Cu	2,400.11	100	5
Ru	2,393.97	—	50	W	2,400.336	6	—
				Fe	2,400.338	25	4
Ir	2,393.93	25	—	W	2,400.50	10	1
Re	2,393.91	20*w*	18				
Co	2,393.90	15	8				
Os	2,393.86	5	30	*Co*	*2,400.56*	—	*30*
Hf	2,393.83	100	80	*Rh*	*2,400.71*	*30*	—
				Co	*2,400.84*	*2*	*30*
Zr	2,393.82	2*h*	9	*Bi*	*2,400.88*	*100*	*200R*
Zn	2,393.81	—	15*h*	*V*	*2,400.90*	*50h*	*1*
Pb	2,393.79	1000	2500				
Re	2,393.67	—	20	*Pt*	*2,401.00*	*9*	*25*
				Co	*2,401.11*	*2*	*30*
				Os	*2,401.13*	*8*	*25*
V	*2,393.57*	*500*	—	*W*	*2,401.29*	*3*	*12*
W	*2,393.42*	*2h*	*10*	*Co*	*2,401.60*	*2*	*30*
Hf	*2,393.36*	*80*	*50*				
Ru	*2,393.25*	*1*	*80*	*Ni*	*2,401.84*	*10*	*40r*
Hf	*2,393.18*	*40*	*20*	*W*	*2,401.86*	*15*	*2h*
				Pt	*2,401.87*	*30*	*300*
W	*2,392.93*	*15*	*8*	*Mo*	*2,401.93*	*40*	*10*
V	*2,392.90*	—	*25*	*Pb*	*2,401.95*	*40*	*50*
Cr	*2,392.89*	*2h*	*40*				
Ru	*2,392.424*	*6*	*80*	*Co*	*2,402.17*	—	*30r*
Rh	*2,392.42*	*40*	—	*W*	*2,402.44*	*20*	*15*
				Ag	*2,402.57*	*30wh*	—
Re	*2,392.38*	*7*	*20*	*Ru*	*2,402.72*	*150r*	*100*
Cr	*2,392.37*	*2h*	*25*	*Te*	*2,403.00*	*(100)*	—
Lu	*2,392.19*	*100*	*30*				
Al	*2,392.15*	*(30)*	—	*W*	*2,403.07*	*10*	*2h*
Re	*2,391.30*	*7*	*20*	*Pt*	*2,403.09*	*50*	*400*
				W	*2,403.22*	*12*	*3*
V	*2,391.26*	*3w*	*25*	*V*	*2,403.25*	*35*	*1*
Ir	*2,391.25*	*25*	—	*Cu*	*2,403.33*	*300*	*100*
Ir	*2,391.18*	—	*50*				
Os	*2,390.95*	*3*	*30*	*U*	*2,403.42*	*30*	—
W	*2,390.89*	*10*	*4*	*W*	*2,403.47*	*10*	—
				Mo	*2,403.61*	*25*	*25*
V	*2,390.78*	*3*	*25*	*Te*	*2,403.66*	*(30)*	—
Rh	*2,390.62*	*50w*	*10*	*Os*	*2,403.85*	*3*	*25*
Ir	*2,390.67*	*6*	*40*				
Ag	*2,390.54*	*80h*	—				
Re	*2,390.44*	*7*	*18*				
W	*2,390.37*	*20*	*10*				

W 2,589.17 *J* 25 *O* 15

	λ	I *J*	I *O*		λ	I *J*	I *O*
Ir	**2,589.12**	—	**2**	**Ge**	**2,589.19**	**6**	**6**
Br	**2,589.10**	**(6)**	—	**Nb**	**2,589.27**	**2**	**3**
Kr	**2,589.08**	**(30)**	—	**Cs**	**2,589.3**	**(2)**	—
Zr	**2,589.07**	**10**	**5**	**Ir**	**2,589.38**	—	**5**
Th	**2,589.063**	**10**	**20**	**Os**	**2,589.39**	**2**	**8**
Ce	**2,589.057**	—	**5**	**Ru**	**2,589.41**	**60**	—
Cr	**2,589.05**	**2**	—	**In**	**2,589.5**	**2**	—
Ru	**2,589.04**	—	**20**	**Os**	**2,589.51**	**1**	**5**
Nb	**2,588.965**	**20**	**1**	**Ce**	**2,589.56**	—	**2**
Ir	**2,588.964**	—	**3**	**Ru**	**2,589.57**	**3**	**60**
Mn	**2,588.957**	**80**	—	**U**	**2,589.59**	**4**	**10**
Zr	**2,588.95**	—	**15**				
Ga	**2,588.92**	**5**	—				
U	**2,588.91**	**2**	**5**	Zr	2,589.653	—	12
Ce	**2,588.88**	—	**3**	W	2,589.654	12	2*d*
				Mn	2,589.71	50*h*	10
Ru	**2,588.86**	**3**	—	Ru	2,589.78	5	8
Fe	**2,588.79**	**20**	**1**	Ta	2,589.81	25	8
Mo	**2,588.78**	**30**	**10**				
				Au	2,590.04	50	30
				Ti	2,590.26	1	30
J	2,588.68	(40)	—	Cu	2,590.53	250	1*h*
W	2,588.55	8	—	Fe	2,590.54	35	2
Ni	2,588.31	80	—	Co	2,590.59	—	75*W*
Mg	2,588.28	8	8				
Tm	2,588.27	80	40	Os	2,590.75	8	75
				N	2,590.91	(25)	—
Cr	2,588.20	1	8	Nb	2,590.94	800	15
Ru	2,588.19	—	10	Ru	2,590.97	100	12
W	2,588.09	12	—	J	2,591.02	(20)	—
*bh*B	2,588.0	—	50				
Fe	2,587.999	—	40	Ru	2,591.12	—	50
				Re	2,591.14	—	15
Nb	2,587.96	20	3	Cs	2,591.17	(20)	—
Fe	2,587.95	50	—	Ru	2,591.24	50	—
Pt	2,587.79	—	10	U	2,591.25	12	18
Pr	2,587.76	15	—				
W	2,587.758	2	12	Fe	2,591.26	—	20
				Te	2,591.35	(30)	—
Os	2,587.49	2	8	Se	2,591.41	(150)	—
Nb	2,587.40	30	—	Re	2,591.43	—	15
W	2,587.32	8	9	W	2,591.49	7	12
Mo	2,587.31	30*wh*	1*h*				
Pd	2,587.289	15*h*	—	Fe	2,591.54	100	50
				Re	2,591.59	—	30
Rh	2,587.287	100	2	Ru	2,591.64	1	30
Ni	2,587.25	50	—	Co	2,591.69	3	10*r*
Co	2,587.22	100*h*	10*w*	W	2,591.74	6	—
Re	2,587.14	—	40*w*				
U	2,587.07	15*h*	6	Mo	2,591.77	25	2
				Tb	2,591.83	20	—
Er	2,587.04	1	10	Cr	2,591.85	12	100*r*
Re	2,587.00	—	20	W	2,591.970	6	—
Al	2,586.95	(50)	—	Mo	2,591.975	1	40

	λ	I	
		J	O
W	2,586.94	30	75
Re	2,586.79	—	100
J	2,586.734	(40)	—
W	2,586.733	6	—
W	2,586.64	6	10
Ra	2,586.61	(50)	—
Se	2,586.44	(15)	—
W	2,586.34	12	8
Nb	2,586.10	40*wh*	—
Ru	2,586.08	—	12
Te	2,586.04	(15)	—
W	2,585.96	20	8
Mo	2,585.95	30	15
Fe	2,585.88	100	70
Ru	2,585.73	—	50
Ru	2,585.65	—	10
Ta	2,585.61	—	30
Tl	2,585.59	2	30*R*
W	2,585.43	2	8
Ru	2,585.341	—	20
Co	2,585.336	—	50*W*
Se	2,585.25	(15)	—
W	2,585.22	3	9
V	2,584.96	100	2
U	2,584.90	4	10
Re	2,584.77	—	25
Ta	*2,584.69*	*—*	*40*
Fe	*2,584.54*	*30*	*100*
W	*2,584.38*	*10*	*15*
Mn	*2,584.31*	*15*	*150w*
W	*2,584.23*	*6*	*5*
Ru	*2,584.14*	*3*	*50*
Pd	*2,584.13*	*75*	*—*
Mn	*2,584.11*	*5w*	*15*
Cr	*2,584.10*	*25*	*1*
Ta	*2,584.03*	*200*	*80w*
Ni	*2,583.995*	*200*	*—*
Nb	*2,583.986*	*800wh*	*10*
Pd	*2,583.85*	*200*	*—*
Fe	*2,583.75*	*—*	*250*
W	*2,583.66*	*1*	*6*
Zr	*2,583.65*	*—*	*15*
W	*2,583.51*	*9*	*—*
Mn	*2,583.28*	*—*	*15*
Ti	*2,583.22*	*1*	*15*
W	*2,583.21*	*9*	*12*
Co	*2,583.18*	*40w*	*—*

	λ	I	
		J	O
Ru	2,592.02	6	60
Ir	2,592.06	20	100
Cd	2,592.14	—	30
Rh	2,592.16	80	—
Nb	2,592.20	3	20
Fe	2,592.29	—	12
Mn	2,592.30	2	12
Ce	2,592.34	—	10
Ta	2,592.44	—	12*h*
Kr	2,592.48	(60)	—
Ge	2,592.54	15	20
U	2,592.57	6	12
Er	2,592.58	1	10
Cu	2,592.63	50	1000
Fe	2,592.78	100	20
Mo	2,592.79	15	3
Re	2,592.87	—	50*w*
Mn	2,592.94	3	150
V	2,593.05	50*h*	—
Ta	2,593.08	1	150
Pd	2,593.27	100	3
Mo	2,593.378	20	3
W	2,593.382	4	12
J	2,593.47	(150)	—
Fe	2,593.52	—	25
U	2,593.57	6	18
Ti	*2,593.64*	*2*	*20*
Ta	*2,593.66*	*100*	*80d*
Ru	*2,593.700*	*—*	*20*
Mo	*2,593.705*	*40*	*20*
Fe	*2,593.726*	*70*	*15*
Mn	*2,593.729*	*1000R*	*200R*
Nb	*2,593.76*	*100*	*2h*
Na	*2,593.83*	*—*	*20R*
Na	*2,593.93*	*—*	*15R*
Mo	*2,594.00*	*—*	*15*
Fe	*2,594.04*	*2*	*20*
Fe	*2,594.15*	*2*	*20*
Pd	*2,594.27*	*25wh*	*—*
Ru	*2,594.39*	*35wh*	*—*
Sn	*2,594.42*	*80*	*60*
W	*2,594.54*	*15*	*4*
Nb	*2,594.74*	*80*	*3*
W	*2,594.80*	*10*	*—*
Ru	*2,594.852*	*4*	*60*
Re	*2,594.853*	*—*	*40*
V	*2,595.10*	*70h*	*—*

	λ	I J	I O		λ	I J	I O
Ru	*2,583.04*	*40*	*30*	*Ne*	*2,595.21*	*(50)*	—
V	*2,583.01*	*50*	—	*Re*	*2,595.24*	—	*60*
Mn	*2,582.96*	*50*	—	*Ta*	*2,595.26*	*2*	*50*
W	*2,582.82*	*8*	—	*Mo*	*2,595.40*	*20*	*25*
J	*2,582.81*	*(400)*	—	*W*	*2,595.57*	*20*	*2*
Re	*2,582.77*	—	*15*	*Ta*	*2,595.59*	*50h*	*40h*
Fe	*2,582.58*	*80*	*25*	*Ag*	*2,595.63*	*40wh*	—
Hf	*2,582.54*	*35*	*25*	*Ru*	*2,595.64*	—	*20*
W	*2,582.52*	*12*	—	*Mn*	*2,595.65*	*25*	—
Zn	*2,582.49*	*40*	*300*	*W*	*2,595.761*	*10*	*1*
Zn	*2,582.44*	—	*100*	*Mn*	*2,595.761*	*25*	*200*
Fe	*2,582.30*	—	*50*	*Ru*	*2,595.80*	*100*	—
Co	*2,582.24*	*500wh*	*50w*	*Ir*	*2,595.83*	*25h*	*3h*
Mo	*2,582.16*	*3*	*25*	*Pd*	*2,595.97*	*150*	*3*
Bi	*2,582.14*	*5*	*35*	*Pt*	*2,596.00*	*20*	*200*
Os	*2,581.96*	*5*	*80*	*Ta*	*2,596.12*	*10wh*	*30*
Ru	*2,581.91*	*2*	*30*	*W*	*2,596.34*	*10*	—
Ti	*2,581.72*	*30*	*4*	*Re*	*2,596.40*	—	*25*
Rh	*2,581.69*	*150*	*1*	*Ta*	*2,596.45*	*150*	*80*
W	*2,581.50*	—	*8*	*Ti*	*2,596.58*	*4*	*40*
Re	*2,581.42*	—	*25*	*Ta*	*2,596.61*	—	*20*
W	*2,581.20*	*20*	*7*	*W*	*2,596.67*	*3s*	*12*
Ru	*2,581.14*	*2*	*60*	*Ba*	*2,596.68*	—	*40*
Yb	*2,581.12*	*100*	*20*	*Mo*	*2,596.77*	—	*20*
Fe	*2,581.11*	*25*	*1*	*Re*	*2,596.78*	—	*25*
W	*2,581.06*	—	*12*	*Mn*	*2,596.83*	*50r*	—
Os	*2,581.05*	*4*	*25*	*W*	*2,596.86*	*15s*	*1*
Co	*2,580.84*	*4*	*50W*	*Re*	*2,596.94*	—	*25*
Ti	*2,580.82*	—	*30*	*Th*	*2,597.05*	*5*	*20*
Ru	*2,580.80*	*3*	*50*	*Rh*	*2,597.07*	*150*	*3*
Ag	*2,580.74*	*150wh*	*1*	*Ru*	*2,597.15*	*50h*	—
Eu	*2,580.60*	—	*25w*	*Al*	*2,597.18*	*(50)*	—
W	*2,580.49*	*7*	*12*	*Mo*	*2,597.22*	—	*20*
				Ru	*2,597.33*	—	*30*
				Mo	*2,597.38*	*30*	*2*
				W	*2,597.47*	*7*	—
				U	*2,597.69*	*15*	*25*
				W	*2,597.73*	*4*	*10*
				W	*2,597.95*	*12*	—
				Re	*2,597.96*	—	*20*

W 2,944.395 *J* 20 *O* 30

	λ	I J	I O		λ	I J	I O
Ce	**2,944.35**	—	**18**	**Fe**	**2,944.40**	**600**	**70**
Re	**2,944.32**	—	**10**	**Yb**	**2,944.46**	**3**	—
Bi	**2,944.29**	**4**	**5**	**Ho**	**2,944.50**	**20**	**10**
Mo	**2,944.21**	**2**	**25**	**Fe**	**2,944.51**	—	**2**

	λ	I J	I O
Zr	**2,944.20**	**3**	**2*h***
Cl	**2,944.20**	**(4)**	**—**
U	**2,944.19**	**12**	**8**
Ru	**2,944.18**	**—**	**12**
Ga	**2,944.17**	**15*r***	**10**
Cs	**2,944.1**	**(2)**	**—**
Er	**2,944.07**	**—**	**12**
Mo	**2,943.989**	**3**	**—**
Ce	**2,943.987**	**—**	**6**
W	**2,943.96**	**4**	**5**
Ru	**2,943.92**	**5**	**50**
Ni	**2,943.914**	**20**	**50*r***
Mn	**2,943.908**	**2**	**2**
U	**2,943.895**	**25**	**10**
Cd	**2,943.89**	**(5)**	**—**
Ir	**2,943.87**	**5*d***	**4*d***
V	**2,943.83**	**—**	**7*h***
Sm	**2,943.79**	**2**	**8**
Ta	**2,943.77**	**5**	**10**
Ga	2,943.64	20*r*	10
Fe	2,943.57	6	12
Co	2,943.484	—	30
Ru	2,943.481	—	30
Mo	2,943.380	25	1
Re	2,943.380	—	15
Tm	2,943.36	15	6
W	2,943.33	6	7
V	2,943.20	25*r*	30
Ir	2,943.151	20	30
Co	2,943.15	100*wh*	—
Re	2,943.14	—	60
Ti	2,943.13	60*wh*	—
Ar	2,942.90	(100)	—
Th	2,942.86	10	10*s*
Mo	2,942.850	1	10
Os	2,942.848	8	30
Pt	2,942.75	3	20
Mn	2,942.74	1	10
Fe	2,942.63	5	10
W	2,942.61	10	2
W	2,942.44	3	8
V	2,942.35	20*h*	80*r*
Pr	2,942.30	10	—
J	2,942.27	(20)	—
W	2,942.26	10	2
Ru	2,942.25	100	30
Te	2,942.16	(100*h*)	—

	λ	I J	I O
Dy	**2,944.56**	**1*h***	**2**
V	**2,944.57**	**300*r***	**50**
Xe	**2,944.61**	**(2*h*)**	**—**
U	**2,944.637**	**8**	**2**
Nb	**2,944.644**	**2*h***	**4**
Hf	**2,944.71**	**1**	**20**
Eu	**2,944.72**	**—**	**3**
Pt	**2,944.754**	**2**	**15**
V	**2,944.755**	**—**	**2*h***
Ce	**2,944.77**	**—**	**4**
Mo	**2,944.82**	**50*h***	**2**
Tb	**2,944.87**	**—**	**5**
Te	**2,944.96**	**(5)**	**—**
Tl	**2,945.04**	**25**	**50**

	λ	I	
		J	O
W	2,942.139	10	6
Ta	2,942.137	40	150
U	2,942.12	12	8
Mg	2,942.11	2*h*	20
Xe	2,942.10	(10*h*)	—
Ho	2,942.05	10	—
Ti	2,941.99	150	100
U	2,941.92	30	15
Cr	2,941.88	25	12
Tb	2,941.70	10	3
Re	2,941.56	—	15
J	2,941.55	(12)	—
Nb	2,941.54	300	50
V	2,941.49	150*r*	12
Se	2,941.48	(15)	—
V	2,941.37	300*r*	40
U	2,941.343	12	8
Fe	2,941.343	300	600
W	2,941.24	4	8
Mo	2,941.22	40	2
Ir	2,941.08	5	20
In	2,941.05	(80)	—
Mn	2,941.04	1	25
Mo	2,940.978	1	10
Cr	2,940.978	10	—
W	2,940.95	3	5
Eu	2,940.82	—	10
Ce	2,940.78	—	15
Hf	2,940.77	12	60
Fe	2,940.59	80	200
Ir	2,940.54	10	15
Yb	2,940.51	25	3
Mn	2,940.48	—	40 *Wh*
Eu	2,940.46	—	15
Mn	2,940.39	—	40 *Wh*
U	2,940.37	12	6
Ru	2,940.36	3	50
Cr	2,940.22	30	—
Ta	3,940.21	50	150
W	2,940.20	18	4
Mo	2,940.10	40	2
Ta	2,940.06	40*w*	100
Tb	2,940.03	3*h*	10
Ru	2,939.94	3	30
Mn	2,939.90	—	12
W	2,939.74	12	2
Ru	2,939.69	5	12
Ce	2,939.54	—	12

	λ	I	
		J	O
Fe	2,939.51	30	3
Cr	2,939.45	20	6
In	2,939.36	10	—
Mn	2,939.30	—	50
Ho	2,939.29	10	—
Ta	2,939.28	40*h*	200
Ir	2,939.27	15	20
W	2,939.18	2	8
Ru	2,939.13	—	12
Fe	2,939.08	20	80
W	2,939.04	4	9
Hg	2,939.03	(10)	—
W	2,938.85	9	1
Pt	2,938.81	2	15
Ir	2,938.76	—	10
In	2,938.71	10	—
Ti	2,938.70	100*wh*	—
V	2,938.67	2	12
Mo	2,938.59	10	1
Ta	2,938.56	—	10
Ag	2,938.55	200*wh*	200
Mg	2,938.54	—	25
Cs	2,938.5	(20)	—
W	2,938.499	6	8
Ir	2,938.47	12	18*h*
Ta	2,938.43	3	50
Mo	2,938.300	30	1
Bi	2,938.298	300*w*	300*w*
V	2,938.25	60	2
Mn	*2,937.92*	—	*25 Wh*
Fe	*2,937.81*	*150*	*300*
Hf	*2,937.79*	*100*	*50*
Na	*2,937.72*	*(40)*	*6*
Nb	*2,937.71*	*20wh*	—
V	*2,937.69*	*10*	*20w*
Mo	*2,937.665*	*1*	*20*
W	*2,937.661*	*10*	*6d*
Ru	*2,937.55*	—	*20*
Ru	*2,937.34*	—	*20*
Ti	*2,937.32*	*5*	*35*
W	*2,937.14*	*12*	*8*
V	*2,937.04*	*25*	*2*
J	*2,937.00*	*(20)*	—
Cr	*2,936.93*	*18*	*4*
Fe	*2,936.90*	*500r*	*700r*
Mo	*2,936.78*	*25*	*2*
Ho	*2,936.77*	*1000R*	—
W	*2,936.67*	*20*	*10*

	λ	I	
		J	O

	λ	I J	I O
Nb	2,936.66	30	—
Ir	2,936.62	—	40
Mg	2,936.54	—	20
Re	2,936.50	—	25
U	2,936.45	20	12
Ti	2,936.17	100wh	—
Ru	2,936.02	—	20
W	2,936.002	3	10
Tm	2,935.997	300	80
V	2,935.87	15h	30
Xe	2,935.86	(30h)	—
W	2,935.75	10d	—
Mn	2,935.66	—	20
W	2,935.62	2	8
Ru	2,935.52	80	10
W	2,935.35	10	5
Ta	2,935.16	2	20
Cr	2,935.14	40	8
W	2,934.99	12	15
bhB	2,934.9	—	100
Ta	2,934.85	4	40
Os	2,934.64	5	30
Cr	2,934.49	—	25h
V	2,934.40	50h	10
Mo	2,934.30	50h	30
Ag	2,934.23	200h	10
Ru	2,934.18	—	30
Mn	2,934.02	—	25
Cr	2,933.97	40	2h
V	2,933.83	35	2
Ta	2,933.55	150	400
Ru	2,933.24	150	20
Ir	2,933.14	5	20
Mn	2,933.06	15	80
W	2,932.86	10	1
Ne	2,932.72	(75)	—
Cr	2,932.70	25	2
Ta	2,932.69	80w	400
Nb	2,932.66	80	1h
In	2,932.62	300	500
U	2,932.61	25	10
Ar	2,932.60	(20)	—
Tm	2,932.59	25	5
Th	2,932.52	25wh	—
V	2,932.32	80	12
Au	2,932.19	40	8
Nb	2,932.13	50	—
Rh	2,931.94	20	80

W 2,946.98 *J* 18 *O* 20

	λ	I			λ	I	
		J	O			J	O
Ir	**2,946.97**	**10**	**20*h***	**Ru**	**2,946.99**	**12**	**60**
Ta	**2,946.91**	**10**	**150**	**Dy**	**2,947.06**	—	**2**
Nb	**2,946.90**	**30**	**3**	**Hg**	**2,947.08**	**(25)**	—
Ce	**2,946.86**	—	**4**	**Yb**	**2,947.11**	**2**	—
Cr	**2,946.842**	**30**	**5**	**Hf**	**2,947.130**	**15**	**15**
Tm	**2,946.84**	**30**	**10*d***	**Fe**	**2,947.13**	**2**	**4**
Dy	**2,946.78**	—	**2**	**Se**	**2,947.13**	**(4)**	—
Yb	**2,946.75**	**8**	**1**	**Ce**	**2,947.15**	—	**4**
Nd	**2,946.72**	**10**	—	**Sm**	**2,947.19**	**3**	**8**
Mo	**2,946.69**	**25**	**3**	**Dy**	**2,947.22**	**1*h***	**2**
Er	**2,946.61**	**5**	**8**	**Ca**	**2,947.25**	**2*h***	—
Re	**2,946.58**	—	**10**	**F**	**2,947.27**	**(2*h*)**	—
V	**2,946.53**	**5**	**10**	**Mo**	**2,947.28**	**25**	**2**
W	**2,946.51**	—	**5**	**Ne**	**2,947.297**	**(150)**	—
Pd	**2,946.44**	**2**	—	**Eu**	**2,947.300**	—	**30**
W	**2,946.43**	**12**	—	**Fe**	**2,947.36**	**20**	**30**
Mo	**2,946.422**	**6**	**10**	**Ir**	**2,947.37**	**2**	**12**
Sm	**2,946.420**	—	**6**	**Ta**	**2,947.377**	**2*h***	**1**
U	**2,946.41**	**2**	**3**	**W**	**2,947.38**	**10**	**12**
Ce	**2,946.38**	—	**8**	**Na**	**2,947.44**	**(40)**	**6**
Dy	**2,946.32**	—	**2**	**Ni**	**2,947.45**	**10**	—
				Cr	**2,947.50**	**25**	—
Yb	2,946.29	15	3				
Ta	2,946.26	2	10	Fe	2,947.66	100	10
Nb	2,946.12	20	4	W	2,947.72	8	—
Mo	2,946.01	40	20	Ta	2,947.80	2	10
Yt	2,945.95	100	2	Fe	2,947.88	200	600*r*
				Tm	2,948.01	20	15
Yb	2,945.90	60	10	V	2,948.07	70	2
U	2,945.89	12	8				
Nb	2,945.88	100	2	U	2,948.09	12	10
Ho	2,945.83	70*h*	3	Tm	2,948.15	10	15
Tb	2,945.70	3	10	Cd	2,948.16	35	—
				Os	2,948.23	5	12
Fe	2,945.70	5	10	Ti	2,948.25	30	100
Na	2,945.69	(20)	2				
W	2,945.67	1	4	Ce	2,948.38	—	10
Ru	2,945.668	300	60	Yt	2,948.40	5*h*	20
Mo	2,945.66	2	20	Fe	2,948.43	70	80
				Pb	2,948.72	125	—
Ti	2,945.47	100*wh*	—	Fe	2,948.726	7	10
Zr	2,945.46	10*h*	4				
Er	2,945.28	—	15	Tl	2,948.73	(30)	—
He	2,945.104	(100)	—	Zr	2,948.94	20	12
Ru	2,945.101	50	6	Fe	2,948.95	4	10
				Ne	2,949.04	(10)	—
Fe	2,945.05	30	100	Th	2,949.07	8	10
				Re	2,949.098	—	20*h*
				J	2,949.105	(30)	—
				W	2,949.12	4	5

*P**

	λ	I J	I O
V	2,949.17	80	6
Ho	2,949.19	20	—
Mn	2,949.20	30	100
Re	2,949.26	—	15
Ne	2,949.32	(15)	—
Cr	2,949.44	18	—
Ru	2,949.50	12	80
Te	2,949.52	(100)	—
Os	2,949.53	10	30
Kr	2,949.54	(15*h*)	—
V	2,949.63	15*r*	30
Rb	2,949.68	(10)	—
Fe	2,949.70	5	10
Lu	2,949.73	1	20*h*
Ir	2,949.76	10	25
Re	2,949.88	—	20
Ru	2,949.96	20	4
Kr	2,950.21	(30*h*)	—
Te	2,950.21	(10)	—
Fe	2,950.24	2300	700
Ce	2,950.30	—	12
Yb	2,950.32	10	2
V	2,950.35	100*r*	25
W	2,950.44	20	—
Ir	2,950.48	—	12
La	2,950.49	50	3
Ru	2,950.54	2	30
Hf	2,950.68	12	15
Ir	2,950.77	2	12
Re	2,950.83	—	25
Nb	2,950.88	200	150
Ir	2,951.22	3	10*h*
Na	2,951.23	(100)	40
Tm	2,951.26	150	30
Ce	2,951.29	—	10
W	2,951.41	3	5
Zr	2,951.479	15	15
Te	2,951.48	(15)	—
Fe	2,951.56	4	10
V	2,951.562	9	—
Lu	2,951.69	80	20
W	2,951.79	3	6
Cd	2,951.82	25	—
Ta	2,951.92	200	400*w*
V	2,952.07	150*R*	35
Ti	2,952.08	25*wh*	—
Ru	2,952.25	25	4
Se	2,952.28	(12)	—

	λ	I J	I O
W	2,952.29	30	12s
Os	2,952.34	8	15
Na	2,952.39	(12)	—
Cr	2,952.46	20	—
Ru	2,952.50	2	60
Eu	2,952.68	150	200w
Ru	2,952.69	10	—
Ir	2,952.73	—	10
Ta	2,952.99	100h	30h
Mn	2,953.01	—	10
Ho	2,953.11	10	20
Sm	2,953.19	4	12
Yt	2,953.28	12h	3
Cr	*2,953.36*	*50*	*4*
Fe	*2,953.49*	*50*	*100*
Tm	*2,953.59*	*30*	*5*
Cr	*2,953.71*	*25*	*6*
Fe	*2,953.78*	*80*	*5*
Fe	*2,953.94*	*150*	*400r*
Mo	*2,953.95*	*20*	*1*
Ru	*2,954.10*	*20*	*6*
V	*2,954.33*	*20*	*30*
Re	*2,954.34*	—	*25*
Au	*2,954.39*	*50*	—
Pd	*2,954.394*	*60h*	—
Ru	*2,954.49*	*20*	*100*
Fe	*2,954.65*	*70*	*100*
Co	*2,954.74*	*100*	*2*
Ti	*2,954.76*	*150wh*	—
W	*2,954.90*	*8*	*10*
W	*2,954.98*	*8*	*2*
Hg	*2,955.13*	*(100)*	—
Mo	*2,955.16*	*25*	*1*
Ru	*2,955.36*	*1*	*50*
Co	*2,955.386*	—	*30*
Ar	*2,955.39*	*(40)*	—
Rh	*2,955.41*	—	*20*
V	*2,955.58*	*60*	*1*
Gd	*2,955.60*	*25*	*2*
Ne	*2,955.73*	*(40)*	—
Lu	*2,955.78*	*60h*	*2*
Zr	*2,955.783*	*30*	*10*
V	*2,955.80*	*2*	*20*
Mo	*2,955.84*	*30*	*2*
Ce	*2,955.94*	—	*25*
Mo	*2,956.05*	*25*	*10*
U	*2,956.06*	*60*	*10*
Mn	*2,956.10*	*1*	*20*

	λ	I	
		J	O
Rb	2,956.12	(70)	—
Ti	2,956.13	25	125
Tb	2,956.21	40	10
Cr	2,956.60	18	—
W	2,956.67	10	8
Fe	2,956.70	8h	25h
Ti	2,956.80	8	35
Ta	2,956.84	100	1h
Mo	2,956.90	30	1
In	2,957.01	25	50
Tm	2,957.05	20	3
W	2,957.26	10	—
Ir	2,957.364	—	20
Fe	2,957.365	300	300
J	2,957.49	(30)	—
V	2,957.52	125r	20
Ta	2,957.60	30	100
Co	2,957.67	1	50
W	2,957.92	7	7
Ru	2,958.00	5	60
Hf	2,958.02	5	30
Kr	2,958.35	(20wh)	—
J	2,958.36	(25)	—
V	2,958.60	40	1
W	2,958.72	2	6
Ti	2,958.99	150wh	—
Pt	2,959.09	3	20
Fe	2,959.34	25	60
Te	2,959.43	(25)	—
Eu	2,959.48	—	30

W 3,613.79 *J* 30 *O* 10

	λ	J	O		λ	J	O
Th	3,613.78	3	5	J	3,613.81	(10h)	—
Cu	3,613.761	7	60	Bi	3,613.82	30	—
Eu	3,613.76	1wh	2h	Sc	3,613.84	70	40
Ce	3,613.701	5	18	U	3,613.997	2	—
Pr	3,613.70	2wh	6dh	Au	3,614.00	20	—
Tb	3,613.68	3	15	Th	3,614.01	8	6
Cr	3,613.67	8	10	Ce	3,614.03	1	5
Mo	3,613.642	4	3	Dy	3,614.083	10	30
He	3,613.641	(30)	—	Eu	3,614.084	2d	2d
Hg	3,613.61	(40)	—	Fe	3,614.12	3	10
Sm	3,613.59	2	3	Ti	3,614.21	4	35
W	3,613.55	6	—	Cu	3,614.22	6	50
Sm	3,613.52	2	2	Mo	3,614.253	30	50d
Nb	3,613.452	3	5	W	3,614.255	2	5
Fe	3,613.452	2	10	Cd	3,614.450	100	60

	λ	I J	I O
Zr	**3,613.451**	**1**	**7**
U	**3,613.403**	**3*h***	**4**
Gd	**3,613.400**	**6**	**8**
Mo	**3,613.37**	**5**	**5**
Ho	**3,613.33**	**8**	**6**
Os	**3,613.329**	**12**	**30**
Nd	**3,613.244**	**4**	**10**
Nb	**3,613.239**	**5**	**2**
Cr	**3,613.18**	**8**	**5**
U	**3,613.152**	**2**	**3**
Fe	**3,613.148**	**2**	**6**
Zr	**3,613.10**	**40**	**40**
La	**3,613.08**	**3**	**12**
Dy	**3,613.06**	**4**	**6**
Xe	**3,613.06**	**(8)**	—
Tb	**3,613.06**	**3**	**15**
S	**3,613.03**	**(12)**	—
Fe	**3,612.94**	**4**	**20**
Te	**3,612.90**	**(5)**	—
Cd	**3,612.875**	**500**	**800**
Er	**3,612.87**	**2*h***	**10*w***
Th	**3,612.869**	**1**	**5**
Cl	**3,612.86**	**(10)**	—
In	**3,612.86**	**15**	—
Dy	**3,612.79**	**10**	**25**
Ca	**3,612.77**	**3**	—
Ni	**3,612.74**	**50*h***	**400**
Pr	**3,612.71**	**2**	**10**
U	**3,612.67**	**8**	**6**
Nb	**3,612.65**	**4**	**3**
Rn	3,612.61	(20)	—
Cr	3,612.609	25	35
Rh	3,612.470	50	200
Al	3,612.467	80*h*	—
Nd	3,612.44	20	6
La	3,612.33	15*h*	8
Mo	3,612.20	20	—
Fe	3,612.07	50	80
Zr	3,611.89	40	15
W	3,611.85	20	—
V	3,611.58	15*wh*	—
Yb	3,611.31	50	12
Yt	3,611.05	60	40
Ta	3,610.60	15	—
Cd	3,610.510	500	1000
In	3,610.508	18	—
Se	3,610.50	(35)	—
Xe	3,610.32	(15)	—

	λ	I J	I O
Ir	**3,614.454**	**2*h***	**2**
Rh	**3,614.51**	**2**	**4**
Fe	**3,614.56**	**6**	**15**
Tb	**3,614.63**	**8**	**30**
Er	**3,614.64**	**1**	**12**
Mo	**3,614.69**	**20**	**1**
Dy	**3,614.707**	**5**	**15**
Fe	**3,614.715**	**1*h***	**6**
Sm	**3,614.74**	**1**	**4**
Zr	**3,614.774**	**80**	**40**
Rh	**3,614.775**	**10**	**15**
W	**3,614.80**	**2**	**6**
Eu	**3,614.86**	**2**	—
Fe	**3,614.87**	**2**	—
Rn	3,615.0	(30)	—
Kr	3,615.47	(20)	—
Nb	3,615.50	30	30
W	3,615.54	5	6
Tb	3,615.66	15	50
Er	3,616.573	20	30
Os	3,616.574	20	150
V	3,616.72	30	30
Mo	3,616.84	15	15
S	3,616.92	(60)	—
P	3,617.09	(100*w*)	—
Ir	3,617.21	15	50
Fe	3,617.32	15	25
W	3,617.52	20	35
Te	3,617.55	(25)	—
Mo	3,617.554	20	—
Nb	3,617.71	15	3
Fe	3,617.79	80	125
Mo	3,618.35	20	—
K	3,618.43	(20)	—
W	3,618.45	12	—
Hg	3,618.53	(50)	—
Se	3,618.73	(35)	—
Fe	3,618.77	400	400
V	3,618.93	100	—
Nd	3,618.97	15	20
W	3,619.27	10	10
Mn	3,619.28	50	75
Ni	3,619.39	150*h*	2000*R*
Os	3,619.43	25	60
Nb	3,619.51	200	5
Nb	3,619.73	300	3
Yb	3,619.81	100	30
Tm	3,619.97	20	20

	λ	I	
		J	O
Mn	3,610.30	40	60
Fe	3,610.162	90	100
Ti	3,610.156	70	100
Ta	3,609.93	18	1*h*
Ir	3,609.77	25	30
Pd	3,609.55	700*R*	1000*R*
Tm	3,609.54	25	15
Nd	3,609.49	30	25
Sm	3,609.48	100	60
Ni	3,609.31	15	200
V	3,609.29	30*W*	30*W*
Ne	3,609.18	(50)	—
Fe	3,608.86	400	500
Tm	3,608.77	20	100
Gd	3,608.76	125	100
Mn	3,608.49	40	60
W	3,608.37	10	—
Mo	3,608.369	15	15
Fe	3,608.15	25	15
Kr	3,607.88	(100*wh*)	—
Hg	3,607.60	(18)	—
Au	3,607.54	20	—
Mn	3,607.537	40	75
J	3,607.51	(18)	—
Ta	3,607.41	35	70
W	3,607.066	12	10
Gd	3,607.066	20	15
Hg	3,606.92	(30)	—
Mo	3,606.91	30	—
V	3,606.69	70	80
Fe	3,606.68	150	200
Mo	3,606.67	20	8
Br	3,606.66	(15)	—
Ar	3,606.52	(1000)	—
Yb	3,606.47	60	15
W	3,606.34	8	10
Dy	3,606.13	100	200
Tb	3,606.12	15*d*	30
W	3,606.07	10	12
Ar	3,605.89	(15)	—
Rh	3,605.86	30	25
Hg	3,605.80	(200)	—
Gd	3,605.62	15	20
V	3,605.59	20*h*	30
Fe	3,605.46	150	300
Cr	3,605.33	400*R*	500*R*
Gd	3,605.25	15	15
Tb	3,604.90	15	3

	λ	I	
		J	O
Dy	3,620.18	20	80
Gd	3,620.46	25	25
V	3,620.47	50	10
P	3,620.65	(15)	—
W	3,620.83	8	—
Er	3,620.94	15	25
Rn	3,621.0	(250)	—
Nb	3,621.03	15	10
Co	3,621.18	50*h*	15
V	3,621.21	80	15
Fe	3,621.46	100	125
Fe	3,622.00	100	125
Ar	3,622.15	(15)	—
V	3,622.29	30	—
Eu	3,622.56	50	20*h*
Gd	3,622.81	20	20
Mo	3,622.85	20	—
U	3,623.05	15	12
P	3,623.10	(15)	—
W	3,623.13	5	—
Fe	3,623.19	80	100
Mo	3,623.23	15	15
W	3,623.51	7	8
Kr	3,623.61	(30*h*)	—
Mo	3,623.70	50	—
Mn	3,623.79	40	75
Lu	3,623.99	40	20
Hf	3,624.00	20	15
Ca	3,624.11	15	150
W	3,624.27	10	—
Mo	3,624.46	25	25
Ni	3,624.73	15	150
Ti	3,624.82	125	60
Gd	3,624.90	30	20
Fe	3,625.15	35	70
Nb	3,625.17	15	8
Ru	3,625.20	30	4
Gd	3,625.27	15	15
W	*3,625.40*	*10*	*10*
V	*3,625.61*	*125*	*4*
Rh	*3,626.59*	*60*	*150*
Ru	*3,626.74*	*40*	*3*
Sm	*3,626.99*	*40*	*50*
W	*3,627.24*	*10*	*12*
Mo	*3,627.35*	*30*	*—*
V	*3,627.71*	*50*	*4*
Nb	*3,628.18*	*50*	*1*

	λ	I J	I O
Os	3,604.47	100	15
V	3,604.38	50*h*	—
Sm	3,604.28	20	6
Hg	3,604.09	(50)	—
Mo	3,604.07	15	6
W	3,603.91	5	5
Cr	3,603.74	50	15
F	3,603.72	(20)	—
Sb	3,603.53	25*wh*	2
Fe	3,603.21	80	150
Eu	3,603.20	50	100*w*
W	3,602.97	5	1
Mo	3,602.94	25	20
F	3,602.85	(60*d*)	—
Nb	3,602.56	30	30
Fe	3,602.53	30	50
W	3,602.46	12	—
Co	*3,602.08*	*35*	*200*
Yt	*3,601.92*	*60*	*18*
Cr	*3,601.67*	*30*	*50*
W	*3,601.58*	*6*	*—*
F	*3,601.40*	*(30)*	*—*
Gd	*3,600.97*	*30*	*30*
W	*3,600.95*	*9*	*—*
Yt	*3,600.73*	*300*	*100*
Tb	*3,600.44*	*50*	*8*
Dy	*3,690.34*	*30*	*20*
Ne	*3,600.17*	*(75)*	*—*
V	*3,600.03*	*40*	*50*
Kr	*3,599.90*	*(40h)*	*—*
Ru	*3,599.76*	*100*	*12*
Fe	*3,599.62*	*30*	*40*
Cu	*3,599.14*	*30*	*60*
W	*3,598.88*	*7*	*10*
Ho	*3,598.77*	*30*	*40*
Ti	*3,598.72*	*30*	*70*
F	*3,598.70*	*(30)*	*—*
W	*3,598.43*	*12*	*—*
Os	*3,598.11*	*30*	*300*
W	*3,597.71*	*6*	*7*
Ni	*3,597.70*	*50h*	*1000r*
Sb	*3,597.51*	*200wh*	*2*
W	*3,597.26*	*9*	*10*
Rh	*3,597.15*	*100*	*200*
Sb	*3,596.96*	*(100)*	*—*
Mo	*3,596.35*	*30*	*—*
Rh	*3,596.19*	*50*	*200*

	λ	I J	I O
W	*3,628.38*	*20d*	*3*
Ir	*3,628.67*	*30*	*100*
Yt	*3,628.71*	*50*	*40*
La	*3,628.82*	*40h*	*80*
W	*3,628.93*	*15*	*—*
Dy	*3,629.44*	*50*	*100*
Nb	*3,629.47*	*30*	*—*
Gd	*3,629.52*	*60*	*40*
Mn	*3,629.74*	*30*	*100*
Eu	*3,629.80*	*50h*	*40*
W	*3,630.09*	*7*	*1*
W	*3,630.32*	*10*	*10*
Hg	*3,630.65*	*(100)*	*—*
Sc	*3,630.74*	*70*	*50*
W	*3,630.82*	*8*	*9*
W	*3,630.95*	*7*	*—*
Na	*3,631.27*	*(100)*	*12*
P	*3,631.40*	*(50)*	*—*
Fe	*3,631.46*	*300*	*500*
Cr	*3,631.69*	*60*	*10*
Kr	*3,631.87*	*(200h)*	*—*
W	*3,631.95*	*10*	*15*
Fe	*3,632.04*	*50*	*50*
V	*3,632.12*	*70*	*—*
Fe	*3,632.56*	*25*	*30*
Ar	*3,632.68*	*(300)*	*—*
W	*3,632.71*	*8*	*9*
Cr	*3,632.84*	*35*	*80*
Yt	*3,633.12*	*100*	*50*
Tb	*3,633.29*	*30*	*30*
Nb	*3,633.31*	*30*	*3*
Ne	*3,633.66*	*(75)*	*—*
S	*3,634.25*	*(35h)*	*—*
Nd	*3,634.28*	*30*	*25*
Ar	*3,634.46*	*(300)*	*—*
Pd	*3,634.69*	*1000R*	*2000R*
Rn	*3,634.8*	*(250)*	*—*
Ru	*3,634.93*	*100*	*50*
Ti	*3,635.46*	*100*	*200*
Zr	*3,636.45*	*30*	*200*
Cr	*3,636.59*	*30*	*60*

	λ	I J	I O		λ	I J	I O
W	*3,596.18*	*15*	*—*				
Ru	*3,596.179*	*100*	*30*				
Bi	*3,596.11*	*50*	*150wh*				
Ti	*3,596.05*	*125*	*50*				
S	*3,595.99*	*(50)*	*—*				
W	*3,595.39*	*6*	*8*				
W	*3,595.07*	*5*	*—*				
Dy	*3,595.05*	*100*	*200*				
Fe	*3,594.64*	*100*	*125*				
W	*3,594.53*	*7*	*5*				
S	*3,594.46*	*(35)*	*—*				
Ir	*3,594.39*	*30*	*15*				
W	*3,593.971*	*8*	*9*				
Nb	*3,593.966*	*50*	*80*				
Ne	*3,593.64*	*(250)*	*—*				
Ne	*3,593.53*	*(500)*	*—*				
Cr	*3,593.49*	*400R*	*500R*				
V	*3,593.33*	*300R*	*30*				
Pb	*3,593.12*	*30*	*—*				
Ti	*3,593.09*	*30*	*5*				
Ru	*3,593.02*	*150*	*60*				
Gd	*3,592.70*	*70*	*50*				
Sm	*3,592.595*	*50*	*40*				
Nd	*3,592.595*	*30*	*20*				
W	*3,592.42*	*35*	*9*				
Dy	*3,592.12*	*30*	*80*				
V	*3,592.02*	*300R*	*50*				
Dy	*3,591.42*	*100*	*200*				
Nb	*3,591.20*	*50*	*2*				
W	*3,590.83*	*10*	*10*				

W 4,008.753 *J* 45 *O* 45

	λ	J	O		λ	J	O
Pr	**4,008.71**	**50**	**150**	**Nd**	**4,008.754**	**10**	**12**
U	**4,008.70**	**1**	**2**	**Br**	**4,008.76**	**(20)**	**—**
Mo	**4,008.67**	**20**	**—**	**Eu**	**4,008.872**	**2**	**8** *W*
Ce	**4,008.66**	**—**	**8**	**Fe**	**4,008.873**	**1**	**5**
Dy	**4,008.49**	**—**	**5**	**U**	**4,008.918**	**—**	**8**
Kr	**4,008.48**	**(10*wh*)**	**—**	**Gd**	**4,008.922**	**3**	**20**
Hf	**4,008.46**	**8**	**5**	**Ti**	**4,008.93**	**35**	**80**
Ce	**4,008.45**	**—**	**6**	**Ir**	**4,008.97**	**—**	**5**
Nd	**4,008.42**	**3**	**5**	**Ce**	**4,009.06**	**—**	**12**
Gd	**4,008.331**	**10**	**15**	**Th**	**4,009.07**	**8**	**10**
Sm	**4,008.330**	**10*h***	**8**	**Er**	**4,009.16**	**1**	**15**
Nb	**4,008.28**	**10**	**5**	**U**	**4,009.17**	**15**	**8**
Ru	**4,008.27**	**20**	**20**	**Tb**	**4,009.19**	**—**	**5**
Th	**4,008.22**	**8**	**10**	**Gd**	**4,009.21**	**2**	**50**
Er	**4,008.18**	**1**	**8**	**Pr**	**4,009.24**	**2**	**5**

	λ	I	
		J	O
V	4,008.17	10	2
Sm	4,008.09	3	10
Kr	4,008.08	(25)	—
Ti	4,008.06	7	50
Mo	4,008.054	5	4
Ir	4,008.052	2	12
Mn	4,008.02	5	15
Eu	4,007.98	—	6
Er	4,007.97	7	35
Ho	4,007.96	3	4
Co	4,007.94	—	3
U	4,007.93	3	8
Se	4,007.90	(150)	—
S	4,007.78	(5)	—
Pr	4,007.78	3	8
Dy	4,007.77	2	12
Tb	4,007.75	—	3
U	4,007.689	4	2
Eu	4,007.687	—	5*d*
La	4,007.66	2*h*	3
In	4,007.61	(10)	—
Zr	4,007.60	1	25
Ce	4,007.59	4	15
In	4,007.54	(15)	—
Ru	4,007.53	10	20
Sm	4,007.49	25	50
Pd	4,007.470	—	5*h*
U	4,007.469	2	2
Ce	4,007.451	—	4*h*
Mo	4,007.45	5	4
Nd	4,007.43	20	20
Yb	4,007.36	—	5
Hf	4,007.35	4*h*	5
Br	4,007.33	(10)	—
Fe	4,007.27	50	80
Ta	4,007.23	2	4
Ti	4,007.19	1	15
Dy	4,007.14	—	6
W	4,007.00	6	—
Ta	4,006.83	20	30
Ru	4,006.60	15	25
Cs	4,006.54	(30)	—
Te	4,006.50	(100)	—
As	4,006.34	50	—
Fe	4,006.31	35	60
Hg	4,006.27	(30)	—
Ti	4,005.97	3	35
W	4,005.89	6	7
He	4,009.270	(10)	—
Tb	4,009.27	—	2
Mo	4,009.366	25	20
Nd	4,009.367	5	8
Zr	4,009.387	—	3
S	4,009.39	(3)	—
U	4,009.41	8	4
Tb	4,009.54	1	10
Th	4,009.546	1	3
Ce	4,009.55	1*h*	2
Al	4,009.58	(4)	—
Ti	4,009.66	25	60
Nb	4,009.71	10	5
Fe	4,009.72	100	120
Sr	4,009.75	2	—
Hf	4,009.78	—	5
Er	4,009.785	—	3
W	4,009.81	9	—
Ir	4,009.85	2*h*	8
C	4,009.90	10	—
Pr	4,009.97	3	10
Ni	4,009.98	—	3
Tb	4,010.06	—	8
Dy	4,010.08	4	10
Mo	4,010.136	5	5
Ce	4,010.136	3	15
Eu	4,010.18	4	8
Ca	4,010.26	2*h*	2
Mo	4,010.30	10	—
U	4,010.35	2*h*	2
W	4,010.37	8	9
W	4,011.02	5	4
Re	4,011.51	—	35
Te	4,011.68	(30)	—
Eu	4,011.683	—	25
Se	4,011.88	(200)	—
Mo	4,011.97	25	25
W	4,012.10	5	—
Nb	4,012.17	100	—
Nd	4,012.25	40	80
Re	4,012.26	—	25
Ce	4,012.388	20	60
Ti	4,012.391	50	35
Cr	4,012.47	60	70
Eu	4,012.82	25	20
Se	4,012.96	(150)	—
W	4,013.18	5	6

	λ	I	
		J	O
V	4,005.71	30	10
Ru	4,005.64	30	25
Kr	4,005.57	(30*h*)	—
Tb	4,005.55	125	100*d*
Th	4,005.549	30*w*	20*w*
W	4,005.40	10	8
Fe	4,005.25	200	250
Os	4,005.15	20	35
Ir	4,005.02	—	25
Re	4,004.93	—	30
W	4,004.75	6	—
Pr	4,004.71	25	20
In	4,004.53	(30)	—
Os	4,004.02	6	50
Cr	4,003.92	12	30
W	4,003.84	7	—
Ti	4,003.81	70	50
Ce	4,003.771	18	40
Fe	4,003.767	80	30
Eu	4,003.71	25	18
Os	4,003.48	6	50
Sm	4,003.45	20	30
Se	4,003.08	(60)	—
Cu	4,003.04	1*h*	40
V	4,002.94	80	6
Tb	4,002.58	5	50
Eu	4,002.56	2	30
Ti	4,002.49	5	40
Xe	4,002.35	(40*wh*)	—
Tb	4,002.18	2	40*w*
Se	4,002.07	(60)	—
W	4,001.89	15	—
Fe	4,001,67	50	80
Cr	4,001.44	80	200
W	4,001.37	9	9
Gd	4,001.24	3	80
K	4,001.20	(40)	—
W	4,000.69	10	12
Nb	4,000.60	50	2
Dy	4,000.454	300	400
Fe	4,000.452	10	35
Er	4,000.452	6	35
Pr	4,000.19	25	50
W	4,000.09	8	—
Cr	3,999.68	10	40
Ti	3,999.36	5	30
Ta	3,999.28	20*d*	30
Ce	3,999.24	20	80
V	3,999.191	40*h*	—
Mo	4,013.21	40	1
In	4,013.49	(30)	—
Ti	4,013.58	7*h*	70*h*
Fe	4,013.79	40	80
P	4,013.80	(30)	—
Gd	4,013.817	3	25
Fe	4,013.824	—	200
Ar	4,013.87	(200)	—
In	4,013.93	(80)	—
Co	4,013.94	—	300
Br	4,014.32	(25)	—
Fe	4,014.53	100	200
Cr	4,014.67	8	40
Se	4,014.77	(70)	—
Ce	4,014.90	12	60
W	4,014.94	6	8
W	4,015.22	30	25
Ti	4,015.38	10*h*	70*h*
Pr	4,015.389	30	50
La	4,015.393	2*h*	100
W	4,016.11	4	6
In	4,016.24	(50)	—
Ti	4,016.28	5	30
Pr	4,016.75	20	25
Fe	4,017.15	50	80
J	4,017.21	(25)	—
Eu	4,017.58	25	25*W*
U	4,017.72	25	25
Rn	4,017.75	(150)	—
Ti	4,017.77	8	70*h*
Mn	4,018.10	60	80
Zr	4,018.12	—	25
Cr	4,018.20	8	35
Os	4,018.26	4	60
Fe	4,018.27	7	50
Re	4,018.41	—	25
Se	4,018.52	(70)	—
U	4,018.99	15	25
Tb	4,019.12	5	40
W	4,019.23	15	18
Co	4,019.30	—	80
P	4,019.45	(50)	—
Sm	4,019.98	15	30
Ir	4,020.03	100	80
Sc	4,020.40	20	50
Co	4,020.90	—	500*w*
Pr	4,020.99	30	40
Mo	4,021.01	25	15
In	4,021.66	(50)	—

	λ	I	
		J	O
Pr	3,999.188	40*d*	50*d*
Zr	3,998.97	30	30
Os	3,998.93	12	80
Cr	3,998.86	2	25
S	3,998.79	(60)	—
W	3,998.75	8	8
V	3,998.73	25	100
Nd	3,998.69	15	40
Ti	3,998.64	100	150
Mo	3,998.63	25	—
Ho	3,998.28	6	40
W	3,998.16	6	7
Fe	3,998.05	100	150
Kr	3,997.95	(100*wh*)	—
Co	3,997.91	20	200
Gd	3,997.77	30	20
W	3,997.76	6	5
Fe	3,997.40	150	300
Mn	3,997.21	25	12
P	3,997.16	(70)	—
W	3,997.13	7	6
V	3,997.12	40	25
Pr	3,997.05	40	100
Fe	3,996.97	20	40
Os	3,996.80	10	50
Dy	3,996.70	80	200
Er	3,996.69	—	25
Sc	3,996.61	10	40
Pt	3,996.57	—	50
Tm	3,996.52	40	200
Gd	3,996.32	100	100
Ta	3,996.17	30*h*	100
Rh	3,996.15	10	25
Fe	3,996.99	20	60
Ru	3,995.98	30	30
Al	3,995.86	(30)	—
La	3,995.75	300	600
Tm	3,995.58	—	100
Co	3,995.31	20	1000*R*
K	3,995.10	30	—
N	3,994.99	(300)	—
J	3,994.98	(35)	—
Os	3,994.93	5	30
V	3,994.89	—	35
Pr	3,994.834	25	300
Kr	3,994.83	(100)	—
Ti	3,994.70	—	25
Nd	3,994.68	40	80

	λ	I	
		J	O
Ti	4,021.83	20	100
Fe	4,021.87	100	200
W	4,022.12	10	12
Ru	4,022.16	100	40
Cr	4,022.26	40	80
Cu	4,022.66	25	400
Re	4,022.96	—	25*d*
Sm	4,023.223	25	30
Sc	4,023.223	—	60
Re	4,023.35	—	40*w*
V	4,023.39	30	10
Co	4,023.40	—	200
Mo	4,023.56	25	—
La	4,023.59	15	50
Sc	4,023.69	25	100
Cr	4,023.74	15	40
Ru	4,023.83	60	25
Zr	4,023.98	2	30
Tb	4,024.07	1	40*W*
Mo	4,024.09	25	30
Ti	4,024.57	35	80
F	4,024.73	(500)	—
Fe	4,024.74	30	120
In	4,024.83	100*wh*	—
Zr	4,024.92	3	25
F	*4,025.010*	*(150)*	*—*
Cr	*4,025.012*	*25*	*100*
J	*4,025.08*	*(30)*	*—*
W	*4,025.19*	*6*	*8*
F	*4,025.49*	*(300)*	*—*
W	*4,025.60*	*10*	*—*
La	*4,025.88*	*50*	*50*
Cr	*4,026.17*	*35*	*100*
He	*4,026.19*	*(70)*	*—*
Mn	*4,026.43*	*40*	*50*
Ti	*4,026.54*	*10*	*70*
Co	*4,027.04*	*4*	*200*
Cr	*4,027.50*	*30*	*80*
Zr	*4,027.20*	*(50h)*	*—*
In	*4,027.79*		
Ti	*4,028.34*	*80*	*20*
W	*4,028.790*	*10*	*12*
S	*4,028.791*	*(200)*	*—*
W	*4,029.02*	*7*	*5*
W	*4,029.61*	*7*	*6*
Fe	*4,029.636*	*25*	*80*
Re	*4,029.639*	*—*	*80*
Ta	*4,029.94*	*5*	*50*

	λ	I	
		J	O
Th	3,994.55	10	30
Co	3,994.54	—	60
Fe	3,994.12	10	25
Cr	3,993.97	20	60
Ni	3,993.95	—	30*h*
W	3,993.90	6	7
Ce	3,993.82	6	50
Tb	3,993.55	8	30*d*
S	3,993.53	(50)	—
Ba	3,963.40	50*r*	100*R*
Sm	3,993.30	25	25
Cr	3,992.84	70	150
V	3,992.80	20	60
W	3,992.75	6	7
Nd	3,992.57	20	30
W	3,992.50	7	—
Mn	*3,992.49*	*75*	*40*
Ce	*3,992.39*	*8*	*50*
Ir	*3,992.12*	*60*	*150*
Nd	*3,991.74*	*40*	*60*
Co	*3,991.69*	*6*	*60*
Cr	*3,991.67*	*50*	*100*
W	*3,991.22*	*8*	*9*
Zr	*3,991.13*	*60*	*100*
Cr	*3,991.12*	*60*	*200*
Yb	*3,990.89*	*20*	*60*
V	*3,990.57*	*40*	*125*
W	*3,990.381*	*10*	—
Fe	*3,990.379*	*25*	*70*
Co	*3,990.30*	*10*	*80*
Cr	*3,989.99*	*40*	*80*
Ti	*3,989.76*	*100*	*150*
Pr	*3,989.72*	*125*	*200*
Zn	*3,989.23*	*(100)*	—
V	*3,988.83*	*35*	*70*
La	*3,988.52*	*800*	*1000*
Os	*3,988.18*	*12*	*50*
Th	*3,988.015*	*30*	*50*
W	*3,988.007*	*6*	*7*
Yb	*3,987.99*	*500R*	*1000R*
Er	*3,987.95*	*20*	*100r*
Gd	*3,987.84*	*25*	*50*
Ru	*3,987.79*	*50*	*3*
Gd	*3,987.22*	*100*	*100*
W	*3,987.15*	*10*	—
Co	*3,987.11*	—	*80*
Mn	*3,987.10*	*60*	*30*

	λ	I	
		J	O
Se	*4,030.07*	*(150)*	—
Fe	*4,030.49*	*60*	*120*
Ti	*4,030.51*	*18*	*80*
Mn	*4,030.75*	*20*	*500r*
Tb	*4,031.64*	*3*	*50*
W	*4,031.67*	*7*	*8*
La	*4,031.69*	*300*	*400*
Pr	*4,031.75*	*30*	*50*
Fe	*4,031.96*	*50*	*80*
W	*4,032.38*	*7*	*6*
Fe	*4,032.63*	*15*	*80*
S	*4,032.81*	*(125)*	—
Ga	*4,032,98*	*500R*	*1,000R*
Tb	*4,033.04*	*5*	*125*
Ta	*4,033.069*	*10*	*100*
Mn	*4,033.073*	*20*	*400r*
Sb	*4,033.54*	*60*	*70*
Ir	*4,033.76*	*25*	*100*
Pr	*4,033.86*	*35*	*50*
W	*4,033.91*	*10*	—
Mn	*4,034.49*	*20*	*250r*
Sm	*4,035.10*	*3*	*50*
W	*4,035.35*	*9*	*10*
Co	*4,035.55*	*3*	*150*
V	*4,035.63*	*80*	*40*
Mn	*4,035.73*	*60*	*50*
Ti	*4,035.83*	*5*	*50*
J	*4,036.08*	*(50)*	—
Eu	*4.036.11*	—	*50W*
W	*4,036.86*	*12*	*12*
La	*4,037.21*	*3h*	*50*
Xe	*4,037.29*	*(50)*	—
Cr	*4,037.294*	*12*	*80*
Gd	*4,037.34*	*30*	*100*
Xe	*4,037.59*	*(100)*	—
W	*4,037.75*	*6*	—
Os	*4,037.84*	*4*	*80*
Gd	*4,037.91*	*30*	*100*
Cr	*4,039.10*	*40*	*100*
Ru	*4,039.21*	*50*	*25*
Pr	*4,039.36*	*20*	*50*
W	*4,039.42*	*6*	*7*
Nb	*4,039.53*	*50*	*30*
Cs	*4,039.84*	*(50)*	—
W	*4,039.85*	*9*	*12*
W	*4,040.59*	*8*	*9*
Ce	*4,040.76*	*5*	*70*
Ho	*4,040.84*	*30*	*150*

	λ	I	
		J	O
Mn	3,986.83	75	40
Fe	3,986.17	8	125
Li	3,985.79	—	100
Fe	3,985.39	40	125
Mn	3,985.24	100	75
Ru	3,984.86	70	60
Cr	3,984.34	60	80
Dy	3,984.23	—	80
W	3,984.18	8	—
Hg	3,983.98	(400)	—
Fe	3,983.96	125	200
Cr	3,983.91	60	200
Dy	3,983.66	8	150
W	3,983.29	25	12
Sm	3,983.14	60	100
W	3,982.96	6	7
Yt	3,982.59	100	60
Ti	3,982.48	30	80
Pr	3,982.06	100	125
Dy	3,981.94	100	150
Tb	3,981.88	200	80
Fe	3,981.77	100	150
Ti	3,981.76	70	100
Rn	3,981.68	(150)	—
W	3,981.26	7	6d
Cr	3,981.23	50	100
W	3,980.64	15	12
Cr	3,979.80	20	80
Co	3,979.52	12	150w
Ru	3,979.42	60	60
Gd	3,979.34	30	100
W	3,979.29	12	12
Ta	3,979.28	3h	50h
Sm	3,979.19	50	50
V	3,979.14	8	50
Ce	3,978.89	50	50
Cr	3,978.68	40	80
Co	3,978.65	—	100
Dy	3,978.57	15	200
Ru	3,978.44	70	60
Fe	3,977.74	150	300
Os	3,977.23	40	300
Mn	3,977.08	100	50
Tb	3,976.82	200	150
Cr	3,976.66	300	300
Nb	3,976.51	80h	—
Ir	3,976.31	70	10
Pr	3,976.29	10	50
Ta	4,040.87	5h	50
Au	4,040.94	40	50
Mn	4,041.36	50	100

W 4,294.61 *J* 50 *O* 50

	λ	I *J*	I *O*		λ	I *J*	I *O*
Mo	**4,294.60**	**6**	**6**	**Sn**	**4,294.65**	**(3)**	**—**
Rb	**4,294.57**	**2**	**—**	**U**	**4,294.651**	**1**	**3*h***
S	**4,294.43**	**(80)**	**—**	**Eu**	**4,294.67**	**2**	**4**
Rb	**4,294.362**	**3**	**—**	**Pr**	**4,294.70**	**10**	**25**
Tb	**4,294.36**	**1*h***	**10**	**Nd**	**4,294.73**	**2**	**15**
Te	**4,294.25**	**(70)**	**—**	**O**	**4,294.74**	**(30*h*)**	**—**
Fe	**4,294.13**	**400**	**700**	**Ce**	**4,294.76**	**—**	**10**
Ti	**4,294.12**	**80**	**60**	**Sc**	**4,294.77**	**20**	**20**
P	**4,294.11**	**(15)**	**—**	**Hf**	**4,294.787**	**2**	**25**
W	**4,294.10**	**10**	**20**	**Ru**	**4,294.791**	**—**	**20**
Tb	**4,294.05**	**—**	**5**	**Zr**	**4,294.792**	**1**	**40**
Rb	**4,293.99**	**40**	**—**	**Dy**	**4,294.94**	**15**	**25**
Os	**4,293.95**	**6**	**60**	**Ar**	**4,294.97**	**(20)**	**—**
Mo	**4,293.88**	**20**	**20**	**W**	**4,295.005**	**4**	**5**
Eu	**4,293.87**	**—**	**6*w***	**Sb**	**4,295.008**	**2*h***	**—**
Pb	**4,293.84**	**(7)**	**—**	**Ho**	**4,295.01**	**—**	**3**
Mo	**4,293.75**	**4**	**10**	**Dy**	**4,295.038**	**—**	**20**
Sm	**4,293.73**	**1**	**3**	**Er**	**4,295.039**	**—**	**15**
Sn	**4,293.61**	**(2)**	**—**	**Th**	**4,295.09**	**1**	**4**
Pr	**4,293.58**	**8**	**20**	**U**	**4,295.10**	**1**	**15**
Cr	**4,293.56**	**8**	**50**	**Pr**	**4,295.11**	**5**	**12**
Rb	**4,293.48**	**2**	**—**	**Zr**	**4,295.15**	**—**	**4**
				Kr	**4,295.21**	**(8 *h*)**	**—**
				Nd	**4,295.23**	**2**	**10**
Te	4,293.35	(70)	—	**U**	**4,295.346**	**10**	**10**
Mo	4,293.21	100	125				
Pr	4,293.13	8	25	**Tb**	**4,295.35**	**—**	**10**
Kr	4,292.92	(600)	—	**Mo**	**4,295.37**	**—**	**5**
Zn	4,292.88	25	25	**Eu**	**4,295.44**	**5**	**20**
				Sm	**4,295.45**	**3**	**2**
Nb	4,292.48	50	30	**Dy**	**4,295.58**	**2**	**8**
Sm	4,292.18	60	100				
Mo	4,292.13	80	100	**Nb**	**4,295.62**	**5**	**3**
Cr	4,291.96	20	35	**Ca**	**4,295.63**	**2**	**2*h***
Dy	4,291.93	—	25	**W**	**4,295.69**	**3**	**—**
				Sm	**4,295.72**	**5**	**15**
V	4,291.82	30*r*	40				
Cl	4,291.76	(50)	—				
Re	4,291.65	—	25	Ti	4,295.756	40	100
Fe	4,291.47	20	125	Cr	4,295.757	40	125
Br	4,291.40	(150)	—	Ni	4,295.89	2*h*	100
				Lu	4,295.95	3	30
Re	4,291.18	—	100*w*	La	4,296.05	200	200
Ti	4,290.94	30	70				
Ne	4,290.40	(100)	—	Gd	4,296.08	4	200
Fe	4,290.38	5	35	V	4,296.11	25	30
Ti	4,290.23	60	35	Nb	4,296.16	30	10
				Gd	4,296.29	40	40
Zr	4,290.21	20	40	Xe	4,296.40	(200*h*)	—
Mo	4,290.18	25*h*	30*h*				
W	4,290.14	3	8	Ce	4,296.68	25	40
Ce	4,289.94	25	50	Sm	4,296.75	50	100

	λ	I	
		J	O
Gd	4,289.90	100	40
Tb	4,289.73	—	30
Cr	4,289.72	800*r*	3000*R*
Ce	4,289.45	4	25
Ca	4,289.36	20	35
Ti	4,289.07	50	125
Rh	4,288.71	100	400
Ce	4,288.67	—	30
Mo	4,288.64	100	80
P	4,288.52	(50)	—
Te	4,288.51	(50)	—
Cs	4,288.35	(35)	—
Fe	4,288.15	6	50
Pt	4,288.06	1*h*	75
Ni	4,288.00	—	150
Ti	4,287.88	30	10
Ti	4,287.40	50	100
Nb	4,286.99	30	15
La	4,286.97	300	400
Sm	4,286.64	60	100
Ir	4,286.62	3	200
Kr	4,286.49	(40)	—
Ta	4,286.38	20*h*	80
W	4,286.21	2	6
Gd	4,286.12	20	60
W	4,286.01	8	15
Ti	4,286.009	40	100
Os	4,285.89	3	30
Te	4,285.84	(70)	—
Gd	4,285.83	20	60
Co	4,285.79	—	125
Sm	4,285.48	200	200
Fe	4,285.44	50	125
Ce	4,285.37	8	25
Tb	4,285.13	3	25
Ti	4,284.99	20	40
Cr	4,284.72	8	40
Ni	4,284.68	10	25
Mo	4,284.60	80*h*	125*h*
Nd	4,284.52	20	25
Ru	4,284.33	20	25
Cr	4,284.22	30	—
Mn	4,284.08	20	80
W	4,283.81	5	10
Pr	4,283.78	1	25*w*
Sm	4,283.50	6	30
Kr	*4,282.97*	(*100*)	—
Rh	4,296.77	20	40
Cr	4,297.05	15	100
Gd	4,297.18	4	100
Se	4,297.31	(40)	—
Ru	4,297.71	50	60
Cr	4,297.74	30	125
Pr	4,297.76	40	50
V	4,298.03	20	25
Fe	4,298.04	400	100
Zn	4,298.33	25	25
W	4,298.41	3	6
Gd	4,298.42	1	25
Ti	4,298.66	50	125
Pr	4,298.92	30	30
Ca	4,298.99	18	30
F	4,299.18	(150)	—
Ti	4,299.23	20	70
Fe	4,299.24	400	500
Nb	4,299.60	30	20
Ti	4,299.64	10	60
Cr	4,299.72	50	100
Ti	4,300.05	100	40
Ar	4,300.10	(1200)	—
Mn	4,300.20	5	60
Ce	4,300.33	15	40
La	4,300.43	20	25
Kr	4,300.49	(200)	—
Cr	4,300.51	20	100
Ti	4,300.56	20	125
Cs	4,300.64	(30)	—
Ar	4,300.66	(30)	—
Zn	4,300.81	(25)	—

	λ	I	
		J	O
Ti	4,282.71	25	70
Pr	4,282.44	40	75
Fe	4,282.41	300	600
Se	4,282.10	(100)	—
Ti	4,281.38	20	80
Mn	4,281.10	50	100
Sm	4,281.00	30	50
Sm	4,280.78	200	200
Gd	4,280.50	100	200
Cr	4,280.40	50	80
Se	4,280.36	(150)	—
La	4,280.26	30	50
Pr	4,280.10	30	60
Sm	4,279.95	40	50
Sm	4,279.67	100	100
Mo	4,279.02	100	2
Tb	4,278.51	100	200
W	4,278.41	3	10
Ti	4,278.23	15	50
Ar	4,277.55	(80)	—
Cs	4,277.10	(50)	—
Tb	4,276.75	2h	50w
W	4,276.746	10	15
Te	4,276.68	(50)	—
Ti	4,276.43	20	50
W	4,276.03	4	10
La	4,275.64	500	40
Ne	4,275.56	(70)	—
W	4,275.49	10	15
O	4,275.47	(50h)	—
F	4,275.21	(100h)	—
Cu	4,275.13	30	80
Tl	4,274.98	(100)	—
W	4,274.94	5	10
Cr	4,274.80	800r	4000R
Ne	4,274.66	(50)	—
Ti	4,274.58	40	100
W	4,274.55	12	20
Gd	4,274.17	—	100
Kr	4,273.97	(1000)	—
W	4,273.69	5	10
Te	4,273.40	(70)	—
Li	4,273.28	100h	200r
W	4,272.31	3	8
Pr	4,272.27	35	50
Ar	4,272.17	(1200)	—

W 4,302.11 *J* 60 *O* 60

	λ	*I*	
		J	*O*
Pr	4,302.10	5*w*	60
U	4,302.09	4	5
Ti	4,301.934	50	25
Mo	4,301.932	10	10
U	4,301.72	2	6
Ho	4,301.62	1	3
Er	4,301.604	2	25
Ir	4,301.603	10	200
Zr	4,301.59	—	4
Pr	4,301.58	30	20
Eu	4,301.56	1	5
Ca	4,301.54	5	—
Tb	4,301.54	—	2*h*
Ce	4,301.533	—	10
Kr	4,301.53	(40)	—
U	4,301.47	15	15
Sm	4,301.28	1	7
Mo	4,301.264	20	20
Er	4,301.260	—	4
Nd	4,301.25	8	15
Nb	4,301.21	2	1
Tb	4,301.18	—	5
Cr	4,301.178	25	100
V	4,301.17	10	—
Ru	4,301.15	—	5
Ho	4,301.09	2*h*	2
Ti	4,301.088	50	150
Re	4,301.05	—	3
Co	4,301.03	—	3
Nb	4,300.99	30	30
Bi	4,302.14	50*wh*	2*h*
V	4,302.15	7	8
Fe	4,302.19	10	50
As	4,302.26	5	—
Yt	4,302.29	8	30
Pt	4,302.43	2	—
Kr	4,302.45	(10)	—
Ca	4,302.527	25	50
U	4,302.530	—	2
Dy	4,302.57	—	2
Ce	4,302.65	—	10
*bh*Sr	4,302.7	—	4
Dy	4,302.72	—	10
Cr	4,302.77	2	40
Zr	4,302.89	1	100
Nb	4,302.91	10*h*	2
Tb	4,302.94	—	10*d*
Ta	4,302.978	40*W*	125*W*
Ti	4,302.979	—	10
Dy	4,303.03	—	8
Pr	4,303.14	5	20
Fe	4,303.17	15	12
Co	4,303.24	2	15
Ne	4,303.25	(30)	—
W	4,303.33	15	—
Nd	4,303.57	40	100
Pr	4,303.59	60	100
O	4,303.78	(60*h*)	—
Cl	4,304.07	(40)	—
Gd	4,304.90	100	100
K	4,304.937	(40)	—
Sm	4,304.944	100	100
Re	4,305.32	—	30
Sr	4,305.447	—	40
Cr	4,305.453	20	150
Fe	4,305.455	50	100
N	4,305.46	(30)	—
V	4,305.48	20	25
W	4,305.63	4	10
Mn	4,305.66	—	50
Pr	4,305.76	90	150
Ti	4,305.92	150	300
Ir	4,305.95	2	25
La	4,306.00	—	30
Pr	4,306.08	15	25
V	4,306.21	20	30

*P**

	λ	I J	I O
Ne	4,306.24	(70)	—
Gd	4,306.35	80	200
Ce	4,306.72	15	30
U	4,306.78	4	40*r*
Tl	4,306.80	(40)	—
W	4,306.87	15	20
V	4,307.18	20	30
Pr	4,307.24	2*w*	30
Cl	4,307.42	(75)	—
Cr	4,307.49	1	35
Ru	4,307.59	50	20
W	4,307.64	12	12
Pr	4,307.67	10	30
Ca	4,307.74	20	45
Rn	4,307.76	(400)	—
Ti	4,307.905	100	100
Fe	4,307.906	800*R*	1000*R*
Bi	4,308.18	12	50
W	4,308.50	4	8
Dy	4,308.62	12	100
Er	4,308.63	3	30
Pr	4,308.89	5	25
W	4,308.95	4	10
Sm	4,309.00	150	200
Fe	4,309.04	20	40
Cl	4,309.06	(50)	—
K	4,309.08	(40)	—
Se	4,309.09	(25)	—
Ba	4,309.32	(80)	—
Fe	4,309.38	70	125
Lu	4,309.57	2	25
Yt	4,309.63	50	50
Ce	4,309.74	4	25
V	4,309.79	20	30
W	4,309.88	2	8
W	4,310.26	3	12
Xe	4,310.51	(200*h*)	—
Ir	4,310.59	8	150
Ce	4,310.70	3	30*s*
Gd	4,310.99	—	100
Mo	4,311.02	30	1
Ag	4,311.07	25	5
W	4,311.10	3	8
Pr	4,311.102	10*w*	50
Nb	4,311.26	100	30
Os	4,311.40	9	150
Ir	4,311.50	10	300
Ce	4,311.59	3	25

	λ	I	
		J	O
Ti	4,311.653	7	25
Mo	4,311.653	40	2
W	4,312.35	3	10
Cr	4,312.47	1	30
Mn	4,312.55	20	100
Ti	4,312.87	100	35
Gd	*4,313.85*	*80*	*200*
Sc	*4,314.08*	*150*	*50*
Gd	*4,314.40*	*20*	*100*
Ti	*4,314.80*	*20*	*100*
Fe	*4,315.09*	*300*	*500*
As	*4,315.86*	*50*	—
La	*4,315.90*	*3h*	*50*
Pr	*4,316.057*	*8*	*50*
Gd	*4,316.061*	*60*	*150*
W	*4,316.81*	*7*	*15*
O	*4,317.16*	*(150)*	—
Kr	*4,317.81*	*(500wh)*	—
Kr	*4,318.55*	*(400)*	—
W	*4,318.57*	*3*	*9*
Ti	*4,318.64*	*50*	*100*
Ca	*4,318.65*	*20*	*60*
Tb	*4,318.85*	*30*	*150*
Sm	*4,318.93*	*300*	*300*
Sr	*4,319.12*	*20*	*50*
Te	*4,319.26*	*(50)*	—
Sm	*4,319.53*	*15*	*50*
Kr	*4,319.58*	*(1000)*	—
Cr	*4,319.64*	*20*	*100*
O	*4,319.65*	*(150)*	—
Se	*4,320.39*	*(100)*	—
Gd	*4,320.53*	*20*	*60*
Cr	*4,320.59*	*6*	*125*
Ce	*4,320.72*	*8*	*50*
Sc	*4,320.74*	*40*	*50*
Gd	*4,321.11*	*20*	*50*
Mn	*4,321.16*	*5*	*60*
Cr	*4,321.24*	*2*	*70*
Cr	*4,321.62*	*3*	*70*
Ti	*4,321.66*	*25*	*70*
Se	*4,322.19*	*(60)*	—
Gd	*4,322.195*	*25*	*50*
La	*4,322.50*	*150*	*150*
Eu	4,322.57	—	60
W	*4,322.75*	*3*	*12*
Kr	*4,322.98*	*(150wh)*	—
Sm	*4,323.28*	*100*	*125*
Mn	*4,323.40*	—	*50*
Cr	*4,323.52*	*15*	*100*

	λ	I J	I O
Pr	4,323.55	35	100
Gd	4,324.07	—	100
As	4,324.10	50	—
W	4,324.59	7	—

Xe $^{54}_{131.30}$

t_0 −111.6°C t_1 −108.0°C

I.	II.	III.	IV.	V.
12.127	21.204	32.115	∼46	∼76

λ	*I*	eV
II 2,979.32	(200)	17.35
III 3,624.05	(600)	—
II 3,922.53	(500)	—
I 4,500.98	(500)	11.07
I 4,624.28	(1000)	11.00
I 4,671.23	(1000)	10.97

Xe 2,979.32 *J* (200)

	λ	*I*			λ	*I*	
		J	*O*			*J*	*O*
Hf	2,979.28	1	18	Fe	2,979.35	100	20
Ti	2,979.20	100*wh*	—	Os	2,979.43	6	12
U	2,979.19	2	4	Tm	2,979.44	10	3
Zr	2,979.184	10	10	In	2,979.58	10	—
Pt	2,979.183	2	1	Ho	2,979.63	40	20
W	2,979.17	5	—	Na	2,979.66	(40)	35
V	2,979.11	35	—	Yb	2,979.67	2	—
Fe	2,979.09	30	—	Sc	2,979.68	10	2*h*
J	2,979.06	(12)	—	Ru	2,979.72	40	30
Ar	2,979.05	(40)	—	W	2,979.721	6	6
Na	2,979.050	(5)	—	Cr	2,979.74	60	10
Nb	2,978.94	50	1	Ne	2,979.816	(50)	—
Yb	2,978.90	3	—	Pt	2,979.806	8	1
Kr	2,978.87	(25)	—	Kr	2,979.81	(20)	—
Fe	2,978.85	7	—	Yb	2,979.85	4	1
Ta	2,978.75	30	200*r*	W	2,979.86	10	10
Ru	2,978.641	150	50	Fe	2,979.87	2	4
Th	2,978.64	100*h*	1	Mo	2,979.873	20	3
W	2,978.640	1	4	Nb	2,979.875	30	2
				Ru	2,979.86	80	60
Ta	2,978.18	150	—				
Nb	2,977.68	300	1	Cd	2,980.63	500	1000*R*
Ru	2,976.59	200	60	Pd	2,980.65	200*R*	—
Ta	2,976.26	150	2*wh*	Sb	2,980.96	(125*hd*)	—
Te	2,975.91	(100)	—	Fe	2,981.45	200	300
				Tm	2,981.49	100	60
Hf	2,975.88	100	80				
Ne	2,974.71	(250)	—	Ne	2,982.66	(250)	—
Nb	2,974.10	200	5	Fe	2,983.57	400*r*	1000*R*
Fe	2,973.24	400*R*	500*R*	Fe	2,984.83	400	200*r*
Fe	2,973.13	400*R*	500*R*	Fe	2,985.55	300	80
Nb	2,972.57	100	40				
				Fe	*2,987.29*	*200*	*300*
				Nb	*2,990.26*	*200*	*5*
Fe	*2,970.10*	*200*	*400*				
Te	*2,967.21*	*(300)*	—				
Fe	*2,966.90*	*600r*	*1000R*				

Xe 3,624.05 *J* (600)

	λ	*J*	*O*		λ	*J*	*O*
Hf	3,624.00	20	15	Zn	3,624.07	3	10
Lu	3,623.99	40	20	Ca	3,624.11	15	150
Th	3,623.98	6	4	Ta	3,624.17	3*s*	3*h*
Zr	3,623.96	1	2	Ce	3,624.18	1	10
Tb	3,623.92	8	30	Cu	3,624.24	3*wh*	30*W*
Zr	3,623.86	7	40	Dy	3,624.25	10	10
Ce	3,623.84	5	60	W	3,624.27	10	—

	λ	I J	I O		λ	I J	I O
Ir	**3,623.80**	**4**	**20**	**Er**	**3,624.28**	**1**	**10**
Mn	**3,623.79**	**40**	**75**	**Fe**	**3,624.307**	**2**	**10**
Fe	**3,623.77**	**7**	**35**	**Nd**	**3,624.32**	**2**	**6**
Ce	**3,623.76**	**1**	**8**	**Co**	**3,624.33**	**4**	**8**
Eu	**3,623.72**	**4*d***	**4*d***	**Nb**	**3,624.36**	**3**	**3**
Mo	**3,623.70**	**50**	—	**Mo**	**3,624.46**	**25**	**25**
Ru	**3,623.63**	**10**	**2**	**W**	**3,624.468**	**1**	**3**
Kr	**3,623.61**	**(30*h*)**	—	**Th**	**3,624.474**	**1**	**4**
Rn	**3,623.53**	**(3)**	—	**Cs**	**3,624.56**	**(4)**	—
W	**3,623.51**	**7**	**8**	**U**	**3,624.57**	**1**	**4**
Fe	**3,623.45**	**5**	**15**	**Er**	**3,624.60**	**1**	**10**
Au	**3,623.44**	**3**	—	**Sc**	**3,624.621**	**2*h***	**2**
Sm	**3,623.31**	**7**	**15**	**Mo**	**3,624.625**	**5**	**5**
Mo	**3,623.23**	**15**	**15**	**Nd**	**3,624.65**	**6**	**20**
Fe	**3,623.19**	**80**	**100**	**P**	**3,624.68**	**(5)**	—
Eu	**3,623.13**	**2**	**3*d***	**Ni**	**3,624.73**	**15**	**150**
W	**3,623.13**	**5**	—	**Tb**	**3,624.80**	**3**	**15**
P	**3,623.10**	**(15)**	—	**Fe**	**3,624.81**	**2**	**12**
Ti	**3,623.096**	**1**	**12**	**Ti**	**3,624.82**	**125**	**60**
U	**3,623.05**	**15**	**12**	**Fe**	**3,624.89**	**2**	—
Th	**3,623.03**	**1**	**3**	**Th**	**3,624.898**	**8**	**8**
				Gd	**3,624.901**	**30**	**20**
				Ca	**3,624.96**	**3**	—
Nb	3,619.73	300	3				
Fe	3,618.77	400	400	U	3,625.09	3	—
Cd	3,612.87	500	800	Fe	3,625.15	35	70
				Nb	3,625.17	15	8
				Ru	3,625.20	30	4
Xe	*3,612.37*	*(10)*	—				
Pd	*3,609.55*	*700R*	*1000R*				
Ar	*3,606.52*	*(1000)*	—	Ar	3,634.46	(300)	—
				Pd	3,634.69	1000*R*	2000*R*

Xe 3,922.53 *J* (500)

	λ	I J	I O		λ	I J	I O
Zr	**3,922.44**	—	**2**	**Mo**	**3,922.66**	**3**	**3**
V	**3,922.431**	**40**	**80**	**W**	**3,922.68**	**2**	—
U	**3,922.427**	**8**	**6**	**Sm**	**3,922.69**	**4**	**8**
Ta	**3,922.42**	**50**	**5**	**P**	**3,922.71**	**(50)**	—
Sm	**3,922.38**	**60**	**60**	**Tb**	**3,922.74**	**8**	**50**
Nb	**3,922.35**	**10**	**5**	**Ta**	**3,922.78**	**15**	**100**
W	**3,922.33**	**3**	**6**	**Mn**	**3,922.908**	**5*w***	**5**
Mo	**3,922.32**	**10**	**10**	**Fe**	**3,922.914**	**400**	**600**
Rb	**3,922.26**	**(10)**	—	**Ta**	**3,922.915**	**10*h***	**100**
Pr	**3,922.24**	**3**	**4**	**Pt**	**3,922.960**	**20*r***	**100**
Th	**3,922.22**	**3**	**8**	**Pr**	**3,922.961**	**3*h***	**4**
Rh	**3,922.19**	**8**	**15**	**K**	**3,923.053**	**(20)**	—
In	**3,922.16**	**(10)**	—	**U**	**3,923.054**	**6**	**15**

	λ	I	
		J	O
Tb	3,922.09	8	20
In	3,922.08	(10)	—
Sm	3,922.039	6	40
U	3,922.037	2	—
Os	3,922.03	10	30
V	3,921.90	20	35
Zr	3,921.79	4	100
Mo	3,921.77	1	3
Mn	3,921.76	20*wh*	20*h*
Ce	3,921.73	1	25
Cs	3,921.69	(4)	—
Kr	3,921.68	(6*h*)	—
Nb	3,921.67	2	3
U	3,921.55	8	8
Mo	3,921.54	20	—
La	3,921.53	200	400
Ti	3,921.42	6	40
Pr	3,921.41	2	10
Nb	3,921.35	10*h*	—
Cu	3,921.27	1*h*	40
U	3,921.24	6	8
Cr	3,921.02	40	150
Fe	3,920.26	300	500
La	3,916.04	400	400
Xe	3,907.91	(50*h*)	—
Eu	*3,907.11*	*500R*	*1000RW*
Mo	*3,902.96*	*500R*	*1000R*
Ce	3,923.109	4	15
Nd	3,923.110	1	3
Mn	3,923.327	20	10
Tb	3,923.33	3	6
Gd	3,923.34	4	6
Br	3,923.35	(15)	—
Ho	3,923.37	2*h*	5
Se	3,923.38	(8)	—
U	3,923.38	2*h*	—
Cu	3,923.44	1*h*	2
Zr	3,923.46	1	3
Ru	3,923.47	100	60
S	3,923.48	(200)	—
Sc	3,923.50	5	3
Pr	3,923.56	8	6
Sm	3,923.68	2	10
Mo	3,923.748	20	10
Ca	3,923.75	2	2
Pr	3,923.82	3	9
Hf	3,923.90	12	12
U	3,923.92	10	8
In	3,923.94	(2)	—
J	3,924.01	(5)	—
Se	3,924.02	(2)	—
Fe	3,927.92	300	500
La	3,929.22	300	400
Fe	3,930.30	400	600
Eu	3,930.50	400*R*	1000*R*
J	3,931.01	(400)	—
Rn	3,931.82	(250)	—
Ca	3,933.67	600*R*	600*R*
Xe	*3,938.92*	*(10)*	—
J	*3,940.24*	*(500)*	—
Xe	*3,943.57*	*(10)*	—
Al	*3,944.03*	*1000*	*2000*
Ar	*3,947.50*	*(1000)*	—
Xe	*3,948.16*	*(60)*	—
Xe	*3,948.72*	*(10)*	—
Ar	*3,948.98*	*(2000)*	—
La	*3,949.11*	*800*	*1000*

Xe 4,500.98 *J* (500)

	λ	*I*	
		J	*O*
In	**4,500.95**	**(50)**	—
Gd	**4,500.88**	**3**	**3**
In	**4,500.77**	**(30)**	—
Er	**4,500.75**	**20**	**20**
Gd	**4,500.68**	**2**	**5**
In	**4,500.63**	**(15)**	—
Se	**4,500.57**	**(10)**	—
Mo	**4,500.523**	**5*h***	—
Pr	**4,500.518**	**2*w***	**20**
Cr	**4,500.29**	**30**	**50**
La	**4,500.22**	**4**	**15**
Ne	**4,500.18**	**(50)**	—
Tl	**4,500.03**	**(8)**	—
Th	**4,599.98**	**6**	**12**
Ca	**4,499.90**	**10*h***	—
Ne	**4,499.84**	**(5)**	—
Nb	**4,499.80**	**10**	**5**
Kr	4,489.88	(400*h*)	—
Ne	4,488.09	(300)	—
Xe	*4,485.95*	*(10)*	—
Xe	*4,480.86*	*(200wh)*	—
Te	*4,478.73*	*(800)*	—
Cr	**4,501.11**	**30**	**40**
P	**4,501.2**	**(15)**	—
Ti	**4,501.27**	**100**	**60**
Mo	**4,501.29**	**25**	**25**
Sm	**4,501.37**	**9**	**5**
U	**4,501.50**	**1**	**4**
Cs	**4,501.52**	**(35)**	—
Cr	**4,501.79**	**6**	**15**
Nd	**4,501.81**	**8**	**50**
Pr	**4,501.83**	**1**	**20**
U	**4,501.94**	**4**	**2**
V	**4,501.95**	**15**	**20**
La	**4,502.16**	**10*h***	—
Mn	**4,502.22**	**40**	**125**
U	**4,502.25**	**3**	**2**
N	**4,502.27**	**(5)**	—
Kr	4,502.35	(600)	—
Rn	4,508.48	(250)	—
Ar	4,510.73	(1000)	—
In	4,511.32	4000*R*	5000*R*
Xe	*4,521.86*	*(50h)*	—
Ar	*4,522.32*	*(800)*	—

Xe 4,624.28 *J* (1000)

	λ	*I*	
		J	*O*
Mo	**4,624.24**	**25**	**25**
Th	**4,624.14**	**2**	**3**
S	**4,624.11**	**(20)**	—
Dy	**4,624.10**	**2**	**3**
Cl	**4,623.96**	**(6)**	—
Th	**4,623.89**	**3**	**3**
Se	**4,623.77**	**(150)**	—
W	**4,623.687**	**3**	**12**
Mo	**4,623.68**	**2**	—
Mo	**4,623.46**	**15**	**15**
U	**4,623.43**	**8**	**5**
Ti	**4,623.094**	**40**	**125**
Cs	**4,623.091**	**(20)**	—
Sb	**4,623.07**	**20**	—
Ta	**4,622.96**	**1**	**50**
Kr	4,619.15	(1000)	—
Xe	4,617.50	(50)	—
V	**4,624.41**	**15**	**20**
Dy	**4,624.42**	**2**	**3**
Gd	**4,624.43**	**8**	**15**
Cr	**4,624.57**	**6**	**15**
U	**4,624.71**	**8**	**6**
Er	**4,624.78**	**1**	**3**
Ce	**4,624.90**	**10**	**8**
Th	**4,625.02**	**3**	**4**
Fe	**4,625.05**	**12**	**100**
W	**4,625.17**	**1**	**5**
Pr	**4,625.30**	**1**	**4**
Se	**4,625.37**	**(8)**	—
Ar	**4,625.46**	**(10)**	—
Rn	**4,625.48**	**(500)**	—
Er	**4,625.56**	**2**	**3**
Ar	4,628.44	(1000)	—
Xe	4,633.30	(25)	—
Kr	4,633.88	(800)	—

	λ	I			λ	I	
		J	O			J	O
Xe	4,615.50	(100)	—				
Kr	4,615.28	(500)	—				
Xe	4,615.06	(50*h*)	—				
Xe	4,611.89	(700)	—				
Xe	*4,603.03*	*(300h)*	—				

Xe 4,671.23 *J* (2000)

	λ	J	O		λ	J	O
Eu	**4,671.18**	**1**	**30**	**W**	**4,671.30**	**1**	**3**
Dy	**4,671.10**	**4**	**4**	**U**	**4,671.41**	**30**	**20**
Er	**4,671.09**	**1**	**2**	**Er**	**4,671.58**	**1**	**3**
Hf	**4,670.91**	**2**	**5**	**Kr**	**4,671.61**	**(10)**	—
Ne	**4,670.88**	**(70)**	—	**W**	**4,671.65**	**1**	**12**
Gd	**4,670.85**	**2*h***	**3**	**Mn**	**4,671.688**	**5**	**100**
Yt	**4,670.83**	**2**	**3**	**Cu**	**4,671.693**	**10**	—
Yb	**4,670.57**	**15**	**12**	**Li**	**4,671.8**	**(4)**	—
U	**4,670.49**	**40*r***	**60*R***	**La**	**4,671.83**	**150**	**100**
Sc	**4,670.40**	**300*wh***	**100**	**Mo**	**4,671.90**	**30**	**30**
Cs	**4,670.28**	**(20)**	—	**Tm**	**4,671.98**	**20**	**15**
Mo	**4,670.24**	**5**	**5**	**Pr**	**4,672.08**	**25*w***	**100**
Te	**4,670.11**	**(30)**	—	**Kr**	**4,672.09**	**(2*wh*)**	—
Nb	**4,670.10**	**4*h***	**1**	**Nb**	**4,672.091**	**100**	**150**
Th	**4,670.03**	**1**	**3**	**Br**	**4,672.11**	**(4)**	—
Nb	**4,669.87**	**2**	**3**	**Br**	**4,672.2**	**100**	—
N	**4,669.77**	**(10)**	—	**Xe**	**4,672.20**	**(50*h*)**	—
				Br	**4,672.56**	**(12)**	—
				Nd	**4,672.69**	**2**	**5**
Xe	4,668.49	(50)	—				
Xe	4,666.28	(25*h*)	—				
Kr	4,658.87	(2000)	—	Xe	4,674.56	(25)	—
				Xe	4,676.46	(100*wh*)	—
Xe	*4,653.00*	*(25)*	—				
Xe	*4,651.94*	*(100)*	—	*Xe*	*4,690.97*	*(100)*	—
				Xe	*4,693.34*	*(10h)*	—
				Xe	*4,697.02*	*(300)*	—
				Xe	*4,698.01*	*(150h)*	—

*P**

Y $^{39}_{88.905}$

t_0 1,490°C t_1 2,500°C

I.	II.	III.	IV.	V.
~ 6.6	12.4	20.5	—	77

λ	I		eV
	J	O	
II 3,242.28	100	60	4.01
II 3,600.73	300	100	3.62
II 3,633.12	100	50	3.41
II 3,710.29	150	80	3.52
II 3,774.33	100	12	3.41
I 4,643.70	100	50	2.67
I 4,674.85	100	80	2.72

Y 3,242.280 *J* 100 *O* 60

	λ	*I*	
		J	*O*
Fe	**3,242.27**	**1**	**3**
Th	**3,242.26**	**5**	**8**
Se	**3,242.19**	**(25)**	—
Ru	**3,242.165**	—	**80**
Zr	**3,242.163**	**2*h***	**1**
Ce	**3,242.13**	—	**12**
Ta	**3,242.05**	**15**	**125**
Sm	**3,242.03**	**4**	**40**
W	**3,242.02**	**10**	**10**
U	**3,241.991**	**8**	**10**
Ti	**3,241.986**	**300*R***	**60**
Os	**3,241.98**	**12**	**15**
Tb	**3,241.94**	**3**	**8**
Be	**3,241.83**	**(50)**	**5**
Nb	**3,241.82**	**10**	**2**
Si	**3,241.80**	**6**	—
Os	**3,241.796**	**10**	**15**
Pd	**3,241.79**	**2**	—
Ca	**3,241.76**	**2*h***	**1**
Fe	**3,241.68**	**2**	—
Be	**3,241.65**	**(15)**	—
U	**3,241.59**	**1**	**3**
Sm	**3,241.57**	**6**	**20**
Co	**3,251.55**	—	**5**
Th	**3,241.534**	**3*h***	**3**
Tm	**3,241.53**	**125**	**150**
Ir	**3,241.52**	**50**	**100**
Rn	**3,241.50**	**(40)**	—
Re	**3,241.46**	—	**25**
Os	**3,241.43**	**10**	**15**
Sb	3,241.28	(350*Wh*)	—
Ru	3.241.23	12	60
Sm	3,241.14	10	50
Os	3,241.04	20	80
Cr	3,240.95	2	35
Mo	3,240.71	60	3
Mn	3,240.62	30	60
Mn	3,240.40	30	60
Tm	3,240.23	80	100
Pt	3,240.20	6	40
Pb	3,240.19	—	30*h*
Ta	3,239.99	18*w*	200
Ti	3,239.66	80	25
Sm	3,239.64	25	100
Ru	3,239.60	5	50
Fe	3,239.44	300	400

	λ	*I*	
		J	*O*
Cs	**3,242.28**	**(10)**	—
Dy	**3,242.285**	—	**4**
Ir	**3,242.32**	—	**8**
Nb	**3,242.41**	**5**	**1**
Nd	**3,242.45**	**2**	**4**
Sm	**3,242.48**	**2**	**4**
Dy	**3,242.51**	—	**4**
Nb	**3,242.53**	**10**	**3**
Ce	**3,242.54**	—	**2**
Ta	**3,242.57**	**1**	**7**
Ir	**3,242.61**	—	**3**
Pd	**3,242.70**	**600*R***	**2000*wh***
Zr	**3,242.76**	—	**2*h***
Ce	**3,242.788**	—	**3**
Se	**3,242.79**	**(10)**	—
Ta	**3,242.83**	**10**	**125**
Ru	**3,242.85**	—	**20**
Pb	**3,242.86**	**40**	—
Lu	**3,242.93**	**4*h***	—
Hf	**3,243.00**	**8**	**6**
Th	**3,243.03**	**8**	**8**
Ni	**3,243.06**	**15**	**400*R***
Fe	**3,243.11**	**20**	**50**
Pd	**3,243.13**	**60*h***	—
Fe	3,243.40	20	70
Ru	3,243.50	12	70
Fe	3,243.72	60	—
Mn	3,243.78	75	100
Co	3,243.84	—	100
Cr	3.244.11	4	30
Fe	3,244.19	200	300
La	3,245.12	300	400
Ta	3,245.28	2	70
Pr	3,245.46	4	40
Fe	3,245.98	150	200
Fe	3,246.48	25	40
Ce	3,246.67	3	35
Ta	3,246.90	—	35*h*
Fe	3,246.96	70	100
Co	3,247.00	—	35
Co	3,247.18	—	80
Nb	3,247.57	100*w*	50*w*
Eu	3,247.53	5	50*W*
Cu	3,247.540	2000*R*	5000*R*
Mn	3,247.542	—	125
Mo	3,247.62	20	30

	λ	I	
		J	O
Ti	3,239.04	300R	60
Ru	3,238.77	1	50
Cr	3,238.76	200	6
Os	3,238.63	20	100
Ru	3,238.53	45	100
Cr	3,238.09	20	30
Nb	3,238.02	200	20
V	3,237.87	100h	30
Ta	3,237.85	7h	70
Fe	3,237.82	100	1
Cr	3,237.73	30	40
Nb	3,237.68	50	2
Rh	3,237.66	20	60
Re	3,237.51	—	30
Mn	3,237.41	—	30h
Mn	3,237.07	25	40
Co	3,237.03	—	100
Tm	3,236.80	80	100
Mn	3,236.78	75	75
Ce	3,236.73	8	35
Sm	3,236.63	40	100
Ti	3,236.57	300r	70
Nb	3,236.40	200	10
Fe	3,236.22	200	300
Re	3,235.94	—	50
Dy	3,235.90	8	35s
Co	3,235.54	—	60
Tm	3,235.45	40	80
Mn	3,235.00	—	30h
Os	3,234.73	10	100
Ta	3,234.69	10	70
Ni	3,234.65	15	300
Fe	3,234.61	125	200
Ti	3,234.52	500r	100
Ar	3,234.51	(100)	—
Ru	3,234.43	50	3
Os	3,234.20	12	150
Ce	3,234.16	8	40
Cr	3,234.06	150	10
Fe	3,233.971	150	300
Mn	3,233.968	—	75wh
Fe	*3,233.05*	*60*	*100*
Ni	*3,232.96*	*35*	*300R*
Co	*3,232.87*	*25*	*60*
Li	*3,232.61*	*500*	*1000R*
Os	*3,232.54*	*10*	*150*
Sb	*3,232.50*	*250wh*	*150*

	λ	I	
		J	O
Fe	3,248.21	150	200
Ni	3,248.46	2	150
Mn	3,248.516	100	100
Ta	3,248.522	3h	100
Ti	3,248.60	200r	25
Nb	3,248.93	50	5
Fe	3,249.19	35	70
La	3,249.35	80	300
Ni	3,249.44	—	30
V	3,249.57	30	40
Tm	3,249.84	40	10
Ru	3,249.93	3	30
Co	3,250.00	—	60
Ru	3,250.01	3	30
Cd	3,250.17	100	—
Nb	3,250.27	100	5h
Pt	3,250.35	8	40
Ta	3,250.358	3	70
Sm	3,250.361	10	50
Fe	3,250.63	40	60
Ni	3,250.74	1	125
Mo	3,250.75	100	2
V	3,250.78	50	10
Fe	*3,251.23*	*150*	*300*
Dy	*3,251.26*	*100*	*100*
Te	*3,251.37*	*(150)*	—
Pd	*3,251.64*	*500*	*200*
Ti	*3,251.91*	*150*	*50*
Pt	*3,251.98*	*1*	*100*
Fe	*3,252.44*	*40*	*90*
Cd	*3,252.52*	*300*	*300*
Ti	*3,252.91*	*200r*	*60*
Fe	*3,252.93*	*50*	*80*
Mn	*3,252.95*	*50*	*75*
Fe	*3,253.60*	*80*	*100*
Nb	*3,254.07*	*300*	*20*
Co	*3,254.21*	—	*300R*
Ti	*3,254.25*	*125*	*35*
Lu	*3,254.31*	*150*	*50*
Fe	*3,254.36*	*150*	*200*
Sm	*3,254.38*	*15*	*100*
Os	*3,254.91*	*12*	*60*
Ge	*3,255.34*	*100Wh*	—
Fe	*3,255.89*	*100*	*20*
In	*3,256.09*	*600R*	*1500R*
Mn	*3,256.14*	*50*	*75*
Ta	*3,256.77*	*1*	*100*
Os	*3,256.92*	*12*	*80*

	λ	I J	I O		λ	I J	I O
Ti	*3,232.28*	*100*	*30*	*bhB*	*3,257.0*	—	*100*
Os	*3,232.05*	*20*	*500R*	*Ar*	*3,257.58*	*(100)*	—
V	*3,231.95*	*100*	*8*	*Fe*	*3,257.59*	*100*	*100*
Os	*3,231.42*	*12*	*150*	*Co*	*3,258.02*	—	*60*
				Tm	*3,258.04*	*60*	*125*
Fe	*3,230.97*	*200*	*300*				
Ta	*3,230.85*	*18w*	*200*	*Mn*	*3,258.41*	*40*	*75*
Mn	*3,230.72*	*75*	*75*	*In*	*3,258.56*	*300R*	*500R*
Sm	*3,230.54*	*30*	*100*	*Fe*	*3,258.77*	*150*	—
Pt	*3,230.29*	*6*	*100*	*Pd*	*3,258.78*	*200h*	*300*
				Re	*3,258.85*	—	*100*
Fe	*3,230.21*	*80*	*100*				
Tl	*3,229.75*	*800*	*2000*	*Fe*	*3,259.05*	*200*	*1*
Ta	*3,229.24*	*70w*	*300w*				
Os	*3,229.21*	*5*	*125*				
Fe	*3,229.12*	*50*	*80*				
Fe	*3,228.90*	*40*	*80*				
Ti	*3,228.60*	*100*	*30*				
Ru	*3,228.53*	*150*	*50*				
Fe	*3,228.25*	*80*	*100*				
Mn	*3,228.09*	*100*	*100*				
Fe	*3,227.75*	*300*	*200*				
Ta	*3,227.32*	*10*	*70*				
Os	*3,277.28*	*12*	*125*				
Co	*3,226.985*	—	*80r*				
Ni	*3,226.984*	—	*100*				
Ca	*3,225.90*	*10*	*80*				
Fe	*3,225.79*	*150*	*300*				
Nb	*3,225.48*	*800wr*	*150w*				

Y 3,600.734 *J* 300 *O* 100

Cs	**3,600.73**	**(10)**	—	**Mo**	**3,600.737**	**4**	**2**
Ho	**3,600.73**	—	**6**	**Er**	**3,600.742**	**20*w***	**30*w***
Ta	**3,600.70**	**1*h***	**2**	**Pr**	**3,600.750**	**5**	**25**
Rb	**3,600.68**	**(20)**	—	**Rh**	**3,600.754**	**2**	**8**
Ce	**3,600.58**	**2**	**15**	**Yb**	**3,600.76**	**20**	**10**
U	**3,600.49**	**2**	**1**	**J**	**3,600.77**	**(5)**	—
Tb	**3,600.44**	**50**	**8**	**Tb**	**3,600.80**	—	**8**
Th	**3,600.43**	**1**	**4**	**Co**	**3,600.81**	—	**4**
Yb	**3,600.39**	**2**	**2**	**Th**	**3,600.83**	**1**	**3**
Dy	**3,600.34**	**30**	**20**	**Nd**	**3,600.91**	**10**	**20**
U	**3,600.293**	—	**10**	**Ho**	**3,600.95**	**10**	**6**
W	**3,600.23**	**3**	**4**	**W**	**3,600.955**	**9**	—
Mo	**3,600.28**	**5**	**5**	**Gd**	**3,600.97**	**30**	**30**
Ar	**3,600.22**	**3**	—	**Pr**	**3,601.02**	**2**	**10**
J	**3,600.21**	**(10)**	—	**Th**	**3,601.04**	**10**	**8**
Mo	**3,600.20**	**3**	**3**	**La**	**3,601.05**	**15**	**5**
Ne	**3,600.17**	**(75)**	—	**Zr**	**3,601.193**	**15**	**400**

	λ	I	
		J	O
Nd	3,600.12	10	15
S	3,600.08	(3)	—
Hf	3,600.05	8*h*	2
Tb	3,600.04	3	15
V	3,600.03	40	50
Ce	3,599.974	1	10
Fe	3,599.974	—	5
Bi	3,599.94	—	2
Zr	3,599.901	5	5
Kr	3,599.90	(40*h*)	—
S	3,599.88	(5)	—
Hf	3,599.87	8	10
U	3,599.84	18	6
Mo	3,599.829	5	3
Er	3,599.829	20*s*	30*W*
Ho	3,599.81	4	—
Ru	3,599.76	100	12
Th	3,599.73	1	3
Ar	3,599.67	(20)	—
Nb	3,599.63	10	10
Fe	3,599.623	30	40
Sb	3,599.62	4	—
Cu	3,599.14	30	60
Ti	3,598,72	30	70
Os	3,598.11	30	300
Ni	3,597.70	50*h*	1000*r*
Sb	3,597.51	200*wh*	2
Rh	3,597.15	100	200
Sb	3,596.96	(100)	—
Tb	3,596.39	15	50
Rh	3,596.19	50	200
Bi	3,596.11	50	150*wh*
Dy	3,596.07	20	50
Ti	3,596.05	125	50
Ta	3,595.64	5	70
Mn	3,595.12	25	50
Dy	3,595.05	100	200
Co	3,594.87	—	200*W*
Fe	3,594.64	100	125
Nb	3,593.97	50	80
Ne	3,593.64	(250)	—
Ne	3,593.53	(500)	—
Cr	3,593.49	400*R*	500*R*
V	3,593.33	300*R*	30
Ru	3,593.02	150	60
Yt	3,592.91	25	80
Gd	3,592.70	70	50

	λ	I	
		J	O
Ti	3,601.193	3	7
Sm	3,601.25	1	4
Mn	3,601.27	15*h*	15
Nd	3,601.32	2	10
Ti	3,601.38	—	15
Dy	3,601.39	1*h*	3
F	3,601.403	(30)	—
Ir	3,601.403	8	30
U	3,601.42	3	—
Ru	3,601.48	5	2
Tb	3,601.50	—	8
Ar	3,601.51	(5)	—
W	3,601.58	6	—
Rn	3,601.60	(5)	—
Cr	3,601.67	30	50
Sm	3,601.69	5	5
Eu	3,601.73	—	2*w*
Al	3,601.74	15	—
Tb	3,601.75	3	15
W	3,601.798	4	5
Pb	3,601.80	(20)	—
Os	3,601.83	20	60
Mo	3,601.84	3	1
Ho	3,601.87	—	6
Cu	3,602.03	25*W*	50
Co	3,602.08	35	200
Ni	3,602.28	15	150
Fe	3,602.53	30	50
Eu	3,603.20	50	100*w*
Fe	3,603.21	80	150
Gd	3,604.88	12	50
Cr	3,605.33	400*R*	500*R*
Co	3,605.36	—	60
Fe	3,605.46	150	300
Hg	3,605.80	(200)	—
Dy	3,606.13	100	200
Ar	3,606.52	(1000)	—
Fe	3,606.68	150	200
V	3,606.69	70	80
Eu	3,606.71	—	50
Ni	3,606.85	—	100*r*
Ta	3,607.41	35	70
Mn	3,607.54	40	75
Mn	3,608.49	40	60
Gd	3,608.76	125	100
Tm	3,608.77	20	100
Fe	3,608.86	400	500

	λ	I J	I O		λ	I J	I O
Dy	3,592.12	30	80	Ni	3,609.31	15	200
V	3,592.02	300*R*	50	Sm	3,609.48	100	60
Dy	3,591.82	20	80				
Rb	3,591.59	20	80	Pd	3,609.55	700*R*	1000*R*
Dy	3,591.42	100	200	Ti	3,610.156	70	100
				Fe	3,610.162	90	100
Ce	3,590.60	1	50				
Nd	3,590.35	300*W*	400*W*				
V	3,589.76	600*R*	80				
Fe	3,589.46	30	50				
Nb	3,589.36	100	100				
Ar	*3,588.44*	*(300)*	—				
Ni	*3,587.93*	*12*	*200*				
Co	*3,587.19*	*50h*	*200r*				
Rb	*3,587.08*	*40*	*200*				
Fe	*3,586.99*	*150*	*200*				
Al	*3,586.91*	*(500h)*	—				
Al	*3,586.80*	*(200wh)*	—				
Al	*3,586.69*	*(200)*	—				
Al	*3,586.55*	*(200)*	—				
Dy	*3,585.78*	*100*	*150*				
Fe	*3,585.71*	*80*	*125*				
Te	*3,585.34*	*(350)*	—				
Fe	*3,585.32*	*100*	*150*				
Dy	*3,585.07*	*100*	*300*				
Gd	*3,584.96*	*100*	*100*				
Fe	*3,584.66*	*60*	*100*				
Rh	*3,583.10*	*125*	*200*				
Re	*3,583.02*	—	*100w*				
Fe	*3,581.19*	*600r*	*1000R*				
Nb	*3,580.27*	*300*	*100*				
Cr	*3,578.69*	*400r*	*500R*				
Dy	*3,577.99*	*50*	*150*				

Y 3,633.12 *J* 100 *O* 50

	λ	J	O		λ	J	O
Fe	**3,633.08**	**3**	**10**	**Hf**	**3,633.18**	**8**	**6**
Ce	**3,633.07**	**—**	**2**	**Au**	**3,633.24**	**15**	**1**
Xe	**3,633.06**	**(6)**	**—**	**Dy**	**3,633.26**	**—**	**25**
Dy	**3,633.00**	**3**	**10**	**Mo**	**3,633.29**	**20**	**—**
Nb	**3,632.999**	**3**	**5**	**U**	**3,633.29**	**15*d***	**8*d***
Fe	**3,632.98**	**8**	**12**	**Tb**	**3,633.29**	**30**	**30**
Co	**3,632.842**	**—**	**60**	**Nb**	**3,633.31**	**30**	**3**
Cr	**3,632.839**	**35**	**80**	**Co**	**3,633.33**	**—**	**8**
Er	**3,632.785**	**1**	**5**	**Pr**	**3,633.35**	**2**	**7**
Ce	**3,632.782**	**—**	**3**	**Th**	**3,633.36**	**3**	**3**
Ne	**3,632.75**	**(4)**	**—**	**Ce**	**3,633.40**	**—**	**8**
Dy	**3,632.73**	**4**	**10**	**Ti**	**3,633.46**	**5**	**35**

	λ	I J	I O
U	**3,632.713**	**2**	**3**
W	**3,632.708**	**8**	**9**
Hf	**3,632.69**	**2**	**5**
Ar	**3,632.68**	**(300)**	—
Th	**3,632.63**	**6**	**6**
Au	**3,632.56**	**2*wh***	—
Cu	**3,632.558**	**3**	**25**
Ir	**3,632.557**	—	**2**
Fe	**3,632.556**	**25**	**30**
Kr	**3,632.49**	**(4)**	—
J	**3,632.48**	**(2)**	—
Hg	**3,632.38**	**(10)**	—
Ce	**3,632.30**	—	**6**
Eu	**3,632.175**	**10**	**30**
U	**3,632.17**	**3*hd***	—
V	**3,632.12**	**70**	—
Ce	**3,632.11**	**2**	**10**
Fe	**3,632.04**	**50**	**50**
Ti	**3,632.00**	—	**6**
Kr	3,631.87	(200*h*)	—
Cr	3,631.69	60	10
Fe	3,631.464	300	500
Tb	3,631.46	—	30
P	3,631.40	(50)	—
Co	3,631.39	25	50*W*
Se	3,631.38	(25)	—
Na	3,631.27	(100)	12
Ce	3,631.19	3	50
Sm	3,631.14	15	40
Fe	3,631.10	10	25
Pr	3,630.97	20	50
Ca	3,630.75	9	150
Sc	3,630.74	70	50
Hg	3,630.65	(100)	—
Fe	3,630.35	15	40
Tb	3,630.28	8	30
Er	3,630.24	6	25
Os	3,629.95	12	40
Eu	3,629.80	50*h*	40
Mn	3,629.74	30	100
Gd	3,629.52	60	40
Dy	3,629.44	50	100
V	3,629.31	2	50
Sr	3,629.14	—	30
La	3,628.82	40*h*	80
Yt	3,628.71	50	40
Er	3,628.70	12*h*	25
Nd	**3,633.47**	**4**	**12**
Zr	**3,633.49**	**8**	**8**
Kr	**3,633.54**	**(3*h*)**	—
Er	**3,633.541**	**2**	**15**
W	**3,633.57**	**3**	**4**
Br	**3,633.64**	**(6)**	—
Tb	**3,633.66**	—	**8**
Ne	**3,633.665**	**(75)**	—
Th	**3,633.70**	**2**	**1*d***
Nb	**3,633.71**	**3**	**8**
Dy	**3,633.77**	**1**	**4**
Ta	**3,633.79**	**10*h***	**35**
Fe	**3,633.83**	**3**	**7**
Ir	**3,633.862**	—	**3**
Ce	**3,633.865**	—	**2**
V	**3,633.91**	**8**	**35**
Ru	**3,633.92**	**5**	**2**
Cr	**3,634.01**	**6**	—
Zr	**3,634.15**	**2**	**25**
Th	**3,634.21**	**6**	**5**
He	**3,634.23**	**(15)**	—
S	**3,634.25**	**(35*h*)**	—
Dy	**3,634.27**	**3*h***	**3**
Sm	**3,634.271**	**25**	**100**
Nd	3,634.28	30	25
Ar	3,634.46	(300)	—
Pd	3,634.69	1000*R*	2000*R*
Co	3,634.71	10	70
Rn	3,634.80	(250)	—
Nd	3,634.87	15	30
Ru	3,634.93	100	50
Ni	3,634.94	10	50
Cr	3,634.99	12	25
Mo	3,635.15	10	100*h*
Cr	3,635.281	8	25
Pr	3,635.284	7	25
Tb	3,635.42	8	20
Mo	3,635.43	5	25
V	3,635.463	5	40
Ti	3,635.463	100	200
Ir	3,635.49	4*h*	35
V	3,635.87	25*h*	50
Cu	3,635.92	7	50
Fe	3,636.19	10	40
Ir	3,636.20	25	50
Lu	3,636.25	3	25
Zr	3,636.45	30	200

	λ	I	
		J	O
Ir	3,628.67	30	100
Tb	3,628.20	15	100
Nb	3,628.18	50	1
Pt	3,628.11	20	300*W*
Co	3,627.81	—	200
V	3,627.71	50	4
Sm	3,626.99	40	50
Ta	3,626.62	18	125
Rh	3,626.59	60	150
Tb	3,626.50	15	30
Ir	3,626.29	10	30
Ti	3,626.08	5	25
Ir	3,625.71	7	30
V	3,625.61	125	4
Tb	3,625.54	15	50
Ta	3,625.24	2*h*	70*r*
Fe	3,625.15	35	70
Ti	3,624.82	125	60
Ni	3,624.73	15	150
Ag	3,624.71	—	25*h*
Mo	3,624.46	25	25
Cu	3,624.24	3*wh*	30*W*
Ca	3,624.11	15	150
Tb	3,623.92	8	30
Zr	3,623.86	7	40
Ce	3,623.84	5	60
Mn	3,623.79	40	75
Fe	3,623.77	7	35
Mo	3,623.70	50	—
Fe	3,623.19	80	100
V	3,622.63	1	35
Eu	3,622.56	50	20*h*
Pr	3,622.38	5	25
Tb	3,622.11	3	30
Fe	3,622.00	100	125
Fe	*3,621.46*	*100*	*125*
Sm	*3,621.22*	*10*	*60*
Rn	*3,621.00*	*(250)*	—
Er	*3,620.177*	*3*	*50*
Dy	*3,620.176*	*20*	*80*
Yb	*3,619.81*	*100*	*30*
Nb	*3,619.73*	*300*	*3*
Nb	*3,619.51*	*200*	*5*
Os	*3,619.43*	*25*	*60*
Ni	*3,619.39*	*150h*	*2000R*
Mn	*3,619.28*	*50*	*75*
V	*3,618.93*	*100*	—

	λ	I	
		J	O
Cr	3,636.59	30	60
Co	3,636.72	6	40
Re	3,637.06	—	25
La	3,637.15	40	50
Co	3,637.32	5	30
V	3,637.76	40	40
Sb	3,637.83	60	2*h*
Re	3,637.84	—	50
Ti	3,637.97	8	30
Fe	3,638.30	80	100
Hg	3,638.34	(100)	—
Tb	3,638.46	50	80
Sm	3,638.75	8	40
Pt	3,638.79	10	250
V	3,639.02	60	70
Co	3,639.44	20	200
U	3,639.49	—	25
Rh	3,639.51	70	125
Pb	3,639.58	50*h*	300
Cr	3,639.80	25	60
Tb	3,639.82	8	30
Os	3,640.33	40	200
Cr	3,640.388	5	30
Fe	3,640.390	200	300
Ir	3,640.87	4	30
F	3,640.89	(100)	—
V	3,641.10	30*wh*	100*h*
Os	3,641.23	10	30
Ti	3,641.33	150	60
Mn	3,641.39	50*h*	50
Cr	3,641.47	25	30
La	3,641.52	4	25
Pr	3,641.62	5	50
Tb	3,641.66	30	70
Cu	3,641.69	5	60
Co	3,641.79	8	60
Cr	3,641.83	15	40
F	3,641.98	(60)	—
Ta	3,642.06	18	125
Ti	3,642.67	125	300
Sm	3,642.74	5	25
Sc	3,642.78	50	60
Re	3,642.99	—	100
Ar	3,643.09	(100)	—
Fe	3,643.11	5	30
Pt	3,643.17	8	60
Co	3,643.18	15	80
Pr	3,643.32	8	25

	λ	I	
		J	O
Fe	3,618.77	400	400
Dy	3,618.52	10	80
Fe	3,617.79	80	125
Cs	3,617.41	—	60
Ir	3,617.21	15	50
P	3,617.09	(100w)	—
Re	3,617.08	—	50
Os	3,616.57	20	150
Tb	3,615.66	15	50
Cd	3,614.45	100	60
Mo	3,614.25	30	50d
Cu	3,614.22	6	50
Cu	3,613.76	7	60
Cd	3,612.87	500	800
Ni	3,612.74	50h	400
Rh	3,612.47	50	200
Fe	3,612.07	50	80
Cs	3,611.52	—	200
Tb	3,611.33	8	50
Cd	3,610.51	500	1000
Ni	3,610.46	—	1000r
Mn	3,610.30	40	60
Fe	3,610.162	90	100
Ti	3,610.156	70	100

	λ	I	
		J	O
Tm	3,643.65	40	60
Tb	3,643.76	3	30
V	3,643.86	30	40
Hf	3,644.35	50	25
Ca	3,644.41	15	200
V	3,644.71	50	80
Co	3,645.19	3	60
Sc	3,645.31	50	50
Tb	3,645.38	15	50
La	3,645.41	60	100
Dy	3,645.42	100	300
Fe	3,645.82	60	80
Gd	3,646.196	150	200w
Ti	3,646.20	25	70
Pr	3,646.30	15	50
Co	3,647.66	8	100
Tb	3,647.75	—	50
Lu	3,647.77	5	100
Fe	3,647.94	400	500
Os	3,648.806	10	100
Dy	3,648.807	30	50
V	3,648.97	50	80
Fe	3,649.30	25	60
Co	3,649.35	4	200
Sm	3,649.506	30	100
Fe	3,649.508	100	100
Ra	3,649.55	(1000)	—
Ar	3,649.83	(800)	—
Fe	3,650.03	30	70
Hg	3,650.15	500	200
La	3,650.17	60	100
Fe	3,650.28	50	70
Tb	3,650.40	100	50
Nb	3,651.19	400	10
Fe	3,651.47	200	300
Sc	3,651.80	45	50
Co	3,652.54	—	200r
Tl	3,652.95	50	150
P	3,653.38	(100w)	—
Ti	3,653.50	200	500
Cr	3,653.91	25	100
Kr	3,653.97	(250h)	—
Os	3,654.49	15	100
Ti	3,654.59	40	100
Gd	3,654.64	200	200W
Hg	3,654.83	(200)	—
Tb	3,654.88	30	70
Al	3,655.00	(100)	—
Gd	3,656.16	200	200W

Y 3,710.290 *J* 150 *O* 80

	λ	*I*			λ	*I*	
		J	*O*			*J*	*O*
Eu	**3,710.29**	**1***wh*	**20***w*	**Er**	**3,710.290**	**8***wh*	**15***wh*
W	**3,710.289**	**5**	**4**	**Sm**	**3,710.30**	**15**	**25**
Ce	**3,710.25**	**1**	**2**	**U**	**3,710.308**	**1**	**8**
Ti	**3,710.19**	—	**6**	**Yb**	**3,710.31**	**6**	**4**
U	**3,710.17**	**5**	**1**	**Ru**	**3,710.32**	—	**4**
Mo	**3,710.14**	**15**	**20**	**F**	**3,710.36**	**(10)**	—
Cr	**3,710.09**	**2**	**4**	**Nb**	**3,710.448**	**20**	**15**
Dy	**3,710.08**	**5**	**20**	**P**	**3,710.45**	**(30***h***)**	—
Pr	**3,710.012**	**2**	**6**	**Sb**	**3,710.52**	**5**	**2**
Eu	**3,710.01**	—	**5**	**U**	**3,710.53**	**4**	**3**
Ti	**3,709.96**	**25**	**80**	**La**	**3,710.59**	**2**	**3**
Re	**3,709.94**	—	**40**	**Cr**	**3,710.60**	**2**	**2**
Ce	**3,709.93**	**10**	**25**	**Dy**	**3,710.73**	**2**	**4**
Ar	**3,709.90**	**(5)**	—	**Ho**	**3,710.75**	**8***h*	**8**
U	**3,709.88**	**2**	**2**	**Ca**	**3,710.753**	**3**	—
Nb	**3,709.74**	**2**	**2**	**Cs**	**3,710.77**	**(4)**	—
Th	**3,709.67**	**3**	**5**	**U**	**3,710.78**	**2**	**4**
Pr	**3,709.665**	—	**3**	**Ta**	**7,710.79**	—	**7***wh*
Fe	**3,709.665**	**4**	**4**	**Sm**	**3,710.87**	**7**	**100***r*
Ne	**3,709.64**	**(40)**	—	**Eu**	**3,710.88**	**30***h*	**35**
Ce	**3,709.59**	—	**5**	**U**	**3,710.91**	**8**	**3**
Fe	**3,709.533**	**6**	**6**	**Lu**	**3,710.95**	—	**3**
Sm	**3,709.526**	**6**	**10**	**Ce**	**3,711.00**	—	**2**
J	**3,709.51**	**(3)**	—	**Na**	**3,711.07**	**(60)**	**8**
U	**3,709.46**	**3***d*	—	**Pr**	**3,711.10**	**3**	**8**
Nb	**3,709.42**	**10**	**5**	**V**	**3,711.12**	**80**	—
V	**3,709.33**	**25**	—	**Ta**	**3,711.15**	—	**5**
Tb	**3,709.30**	—	**15**	**Ir**	**3,711.153**	—	**7**
Ce	**3,709.29**	—	**25**	**Pr**	**3,711.218**	**2**	**3**
Zr	**3,709.26**	**30**	**50**	**Fe**	**3,711.225**	**50**	**80**
Fe	**3,709.249**	**400**	**600**	**Cr**	**3,711.28**	**12**	—
Ag	**3,709.248**	**3**	**10**	**Ho**	**3,711.30**	**4**	**6**
Nb	**3,709.247**	**30**	**5**	**Th**	**3,711.31**	**20**	**30**
Os	**3,709.14**	**1**	**4**	**Nb**	**3,711.338**	**20**	**20**
Ru	**3,709.10**	—	**5**	**Mn**	**3,711.340**	—	**10**
				Fe	**3,711.41**	**25**	**50**
Co	3,708.82	—	100	**J**	**3,711.44**	**(18)**	—
V	3,708.72	60	100	**W**	**3,711.48**	**3**	**6**
Sm	3,708.66	25	50	**Mo**	**3,711.51**	**5**	**5**
Ti	3,708.65	3	50	**Re**	**3,711.53**	—	**15**
Fe	3,707.925	60	80				
				Sm	**3,711.545**	**10**	**10**
Nb	3,707.918	100*W*	3*W*	**Nd**	**3,711.545**	**1**	**2**
Fe	3,707.82	50	80				
Ti	3,707.53	10	100				
Fe	3,707.05	100	150	Tb	3,711.74	30	200
Os	3,706.56	15	50	Co	3,712.18	8	40
				Gd	3,712.71	250	200*W*
Ti	3,706.23	125	30	Sm	3,712.76	100	100

	λ	I	
		J	O
Mn	3,706.08	—	75
P	3,706.05	(150*w*)	—
V	3,706.03	50	50
La	3,705.82	80*h*	50
Fe	3,705.57	500	700
V	3,705.03	70	100
V	3,704.70	150*R*	200*R*
Fe	3,704.46	100	125
Ti	3,704.29	25	70
Co	3,704.06	35	300*r*
Tb	3,703.92	100	70
V	3,703.58	100*R*	200*R*
Y	3,703.32	10*w*	4
Os	3,703.25	30	100
Re	3,703.24	—	40
Tb	3,702.85	200	50
Mo	3,702.55	150	10
Ti	3,702.29	20	60
Co	3,702.24	—	200
Fe	3,702.03	30	50
Mn	3,701.73	30	60
Tm	3,701.36	80	150
Fe	3,701.09	200	300
Ru	3,700.99	20	50
Sm	3,700.93	4	50
Rh	3,700.91	150*d*	150*d*
Sm	3,700.59	30	50
V	3,700.34	100	10
Tm	3,700.26	80	150
Pt	3,699.91	5	80
Gd	3,699.75	250	200*W*
Pr	3,699.51	10	40
Fe	3,698.60	20	40
Zr	3,698.17	80	50
Nb	3,697.85	50	50
Gd	*3,697.74*	*200w*	*200w*
Re	*3,697.70*	—	*150w*
Fe	*3,697.43*	*60*	*100*
Mn	*3,696.57*	*50*	*100*
Nb	*3,695.90*	*200*	*4*
V	*3,695.86*	*100r*	*150*
V	*3,695.33*	*70h*	*125*
Fe	*3,695.05*	*150*	*200*
Ti	*3,694.45*	*20*	*80*
Yb	*3,694.203*	*1000R*	*500R*
Ne	*3,694.197*	*(250)*	—
Fe	*3,694.01*	*300*	*400*
Os	3,712.84	12	50
Cr	3,712.95	125	12*s*
Nb	3,713.01	80*h*	100
Rh	3,713.02	100*r*	100
Ne	3,713.08	(250)	—
Eu	3,713.46	2	50
La	3,713.54	60	200
Gd	3,713.58	80	100*W*
Os	3,713.73	20	100
V	3,713.96	10	60
Pr	3,714.06	20	50
La	3,714.86	40	60
Eu	3,714.91	2	40
Ti	3,715.40	2*h*	40
V	3,715.47	400*R*	70
La	3,715.52	50	100
Fe	3,715.91	50	80
Gd	3,716.37	125	150
Fe	3,716.45	100	150
Nb	3,717.07	1000	8
Re	3,717.29	—	150*W*
Ti	3,717.40	50	80
Gd	3,717.49	50	100*w*
Tm	3,717.92	10	100
Kr	3,718.02	(300*h*)	—
Y	3,718.11	4	12
Fe	3,718.41	50	80
Kr	3,718.63	(200*h*)	—
Sm	3,718.88	5	100
Pd	3,718.91	200	300
Mn	3,718.93	100	75
Sm	3,719.45	10	50
Gd	3,719.46	40	40
Os	3,719.52	12	40
Fe	3,719.93	700	1000*R*
Os	3,720.13	40	80
Ti	3,720.38	10	40
Nb	3,720.46	100*h*	5
Kr	3,721.35	(150*h*)	—
Ti	3,721.64	125	60
Th	3,721.83	30	40
Sm	3,721.84	50	100
V	3,722.00	20	70
Gd	3,722.03	30	50
V	3,722.19	40	40
Ni	3,722.48	20	200
Fe	3,722.56	400	500
Ti	3,722.57	60	100

	λ	I J	I O		λ	I J	I O
Sm	3,694.00	150	100	Eu	3,722.61	10	40W
Mo	3,692.64	150	3	Sb	3,722.79	50	40 Ws
Rh	3,692.36	150wd	500hd	Fe	3,724.38	150	200
V	3,692.225	150R	200R	Ti	3,724.57	50	100
Sm	3,692.221	40	90	Eu	3,724.99	50	250
Re	3,691.50	—	100w	Ti	3,725.16	60	150
Ar	3,690.90	(300)	—	Sm	3,726.80	3	100r
Fe	3,690.73	60	80	Fe	3,726.925	70	100
Rh	3,690.70	50	125	Ru	3,726.926	150	100
Pd	3,690.34	1000w	300h	V	3,727.34	200	40
V	3,690.28	125	200	Fe	3,727.62	150	200
Ti	3,689.91	40	100	Ru	3,728.03	150	100
Re	3,689.52	—	100W	Sm	3,728.47	100	100
Fe	3,689.46	150	200	Mn	3,728.89	100	75
Os	3,689.06	30	200	V	3,729.03	15	80
Eu	3,688.44	500W	1000W	Ar	3,729.29	(200)	—
Ni	3,688.41	15	150	Ti	3,729.81	150	500
Mo	3,688.31	150	4	Fe	3,730.39	40	70
J	3,688.21	(125)	—	Co	3,730.48	—	200r
V	3,688.07	200R	200	Gd	3,730.86	100	100W
Nb	3,687.97	300w	20w	Co	3,732.399	—	200r
Gd	3,687.76	200	200w	Fe	3,732.399	150	200
Fe	3,687.46	300	400	V	3,732.76	500R	70R
Gd	3,686.34	200	150W	Os	3,732.85	5h	200R
V	3,686.26	100	100	Fe	3,733.32	300	400
Fe	3,686.00	125	150	Co	3,733.49	—	150
				Tm	3,734.13	50	150
				Ir	3,734.77	30	100
				Fe	3,734.87	600	1000r

Y 3,774.33 *J* 100 *O* 12*

Nd	3,774.32	30	20	Nb	3,774.38	5	2
Cl	3,774.25	(25)	—	Os	3,774.40	15	60
Sr	3,774.22	2h	1	Nb	3,774.44	5	3
Th	3,774.21	8	8	Hg	3,774.52	(30)	—
W	3,774.14	18	—	Ar	3,774.54	(3)	—
Sm	3,774.13	5h	20	Pr	3,774.59	2	10
V	3,774.11	3	20	Os	3,774.62	12	60
Nd	3,774.08	10d	20d	Ti	3,774.65	1	2
Eu	3,774.08	10W	50W	V	3,774.67	12	1
Pr	3,774.064	50	100	Si	3,774.75	4	—
U	3,774.06	12w	—	Fe	3,774.83	40	100
Br	3,773.83	(3)	—	Dy	3,774.84	1h	4
V	3,773.81	2	—	U	3,774.91	1	10
Os	3,773.78	5	12	Nd	3,774.93	6	10
Th	3,773.760	20	20	P	3,775.02	(30)	—

	λ	I	
		J	O
J	3,773.758	(10)	—
W	3,773.705	18	20
Fe	3,773.699	10	40
Cl	3,773.68	(20)	—
Nb	3,773.62	3	3
U	3,773.55	5	2
Hg	3,773.54	(2)	—
Pr	3,773.48	3	8
Yb	3,773.46	6	2
Ce	3,773.44	3	5
U	3,773.425	10	20
Kr	3,773.424	(50)	—
Sm	3,773.420	4	15
Sm	3,773.34	1	15
Mo	3,773.325	3	2
Dy	3,773.319	3	8
Ce	3,773.21	2	8
Pr	3,773.20	5	9
Nd	3,773.175	10	10*d*
Ru	3,773.170	4	12
Au	3,773.16	10	—
Nb	3,773.15	8	5
La	3,773.12	20	2
Th	3,773.07	2	5
Dy	3,773.06	3	8
V	3,772.97	40	2
V	3,770.97	60	30
Sm	3,770.73	50	25
Gd	3,770.70	60	50
Ar	3,770.37	(400)	—
Ni	3,769.45	50*h*	2
P	3,768.70	(50)	—
Cr	3,768.24	60	60
Ru	3,767.35	50	50
Fe	3,767.19	400	500
Ne	3,766.29	(75)	—
Fe	3,765.54	150	200
Tb	3,765.14	100	70
Rh	3,765.08	70	100
Pr	3,764.81	50	100
Fe	3,763.79	400	500
Ta	3,763.44	60	12
Tm	3,761.92	120	200
Pr	3,761.87	100	150
Tm	3,761.33	150	250
Ti	3,761.32	300*r*	100

	λ	I	
		J	O
V	3,775.19	5	25
Pr	3,775.21	4	10
Nd	3,775.24	8*d*	15*d*
Tb	3,775.26	8	15
U	3,775.262	3	6
Th	3,775.32	5	20
Ce	3,775.36	4	2
U	3,775.44	4	4
W	3,775.446	7*s*	8
Nb	3,775.448	10	5
Ar	3,775.45	(10)	—
Sm	3,775.459	3	10
Zr	3,775.461	1*h*	7
Nd	3,775.50	20	10
Ni	3,775.57	40*h*	500*h*
U	3,775.61	5	12
Mo	3,775.647	20	15
Er	3,775.665	2	8
Eu	3,775.69	4*W*	4*W*
Hg	3,775.70	(2)	—
Tl	3,775.72	1000*R*	3000
Ti	3,776.06	60	8
Fe	3,776.46	70	125
Tb	3,776.49	100	100
Ne	3,777.16	(75)	—
Ru	3,777.59	50	60
Kr	3,778.09	(500*h*)	—
Sm	3,778.13	100	40
Fe	3,779.45	70	100
La	3,780.67	50	50
Ar	3,780.84	(50)	—
Ar	3,781.36	(300)	—
Nb	3,781.38	200	5
Pr	3,781.64	50*d*	100*d*
Mo	3,782.07	100	—
Os	3,782.19	200	400*R*
Gd	3,782.28	50	25
Kr	3,783.13	(500*h*)	—
Ta	3,784.25	50*w*	150
Fe	3,785.95	80	125
Ru	3,786.05	100	70
Fe	3,786.18	60	100
Mo	3,786.36	125	2
Fe	3,786.68	50	125
Nd	3,787.16	50	60*d*
Fe	3,787.88	300	500

	λ	I J	I O		λ	I J	I O
Fe	*3,760.05*	*100*	*150*	*Y*	*3,788.70*	*30*	*30*
Ti	*3,759.29*	*400R*	*100*	*Fe*	*3,790.09*	*100*	*200*
La	*3,759.08*	*150*	*400*	*Mn*	*3,790.21*	*125*	*100*
Fe	*3,758.23*	*700*	*700*	*Ru*	*3,790.51*	*150*	*70*
Ti	*3,757.69*	*100*	*30*	*La*	*3,790.82*	*300*	*400*
Tb	*3,755.24*	*100*	*50*	*B*	*3,792.50*	*(500)*	—
Fe	*3,753.61*	*100*	*150*	*Bi*	*3,792.80*	*500h*	—
Os	*3,752.52*	*100*	*400R*	*Os*	*3,793.91*	*300*	*125*
V	*3,751.23*	*100*	*4*	*La*	*3,794.77*	*200*	*400*
As	*3,749.77*	*100*	—	*Fe*	*3,795.00*	*400*	*500*
Fe	*3,749.59*	*700*	*1000r*	*Tm*	*3,795.76*	*150*	*250*
O	*3,749.47*	*(125)*	—	*Gd*	*3,796.39*	*150*	*150w*
Cr	*3,749.00*	*125R*	*125R*	*Fe*	*3,797.52*	*200*	*300*
Fe	*3,748.26*	*200*	*500*	*Mo*	*3,798.25*	*1000R*	*1000R*
Rh	*3,748.22*	*100*	*200*	*Fe*	*3,798.51*	*300*	*400*
Y	*3,747.55*	*15*	*12*	*Ru*	*3,798.90*	*100*	*70*
				Pd	*3,799.19*	*150*	*200w*
				Rh	*3,799.31*	*100*	*25*
				Ru	*3,799.35*	*100*	*70r*
				Fe	*3,799.55*	*300*	*400*
				Ir	*3,800.12*	*100*	*150*
				Sn	*3,801.00*	*150h*	*200h*

Y 4,643.70 *J* 100 *O* 50

	λ	J	O		λ	J	O
Er	**4,643.69**	**15**	**50**	**Co**	**4,643.72**	—	**15**
Nb	**4,643.68**	**3*h***	**3**	**Br**	**4,644.01**	**(4)**	—
Ca	**4,643.66**	**4*h***	—	**Ba**	**4,644.10**	**(10)**	—
U	**4,643.62**	—	**12**	**U**	**4,644.14**	**3**	**1**
Br	**4,643.52**	**(25)**	—	**Tb**	**4,644.15**	—	**2**
Pr	**4,643.50**	**5*w***	**60*w***	**Rn**	**4,644.18**	**(300)**	—
Fe	**4,643.47**	**2**	**35**	**Ce**	**4,644.20**	**6**	**6**
Dy	**4,643.46**	—	**3**	**Eu**	**4,644.24**	—	**50**
Nb	**4,643.31**	**3**	**3**	**Co**	**4,644.32**	—	**70**
Lu	**4,643.29**	—	**8*h***	**Ir**	**4,644.36**	—	**3**
Tb	**4,643.27**	—	**2*h***	**Er**	**4,644.37**	—	**3**
Sb	**4,643.19**	**15**	—	**Th**	**4,644.371**	—	**3*w***
Er	**4,643.19**	—	**2**	**V**	**4,644.44**	**5**	**6**
Rh	**4,643.184**	**10**	**15**	**In**	**4,644.436**	**(125)**	—
Ne	**4,643.182**	**(5)**	—	**Yb**	**4,644.543**	**3**	**30**
Ce	**4,643.17**	—	**5**	**Tm**	**4,644.57**	—	**20**
W	**4,643.15**	**8*w***	**12**	**In**	**4,644.65**	**(60)**	—
La	**4,643.12**	**5*h***	**8**	*bh***Zr**	**4,644.7**	—	**40**
Tm	**4,643.11**	—	**35**	**Zr**	**4,644.829**	—	**5**
N	**4,643.106**	**(100)**	—	**Ne**	**4,644.833**	**(40)**	—
Ce	**4,643.08**	—	**2**	**Nd**	**4,644.84**	**2*w***	**4**
Tm	**4,642.94**	—	**25**	**U**	**4,644.91**	**3**	**2**

	λ	I	
		J	O
Ca	**4,642.88**	4*h*	—
Mn	**4,642.81**	—	50
Tb	**4,642.79**	—	3
Dy	**4,642.78**	3*h*	2
Eu	**4,642.77**	—	15*w*
Mo	**4,642.70**	6	8
Br	**4,642.63**	(4)	—
Ru	**4,642.60**	—	7
Tm	**4,642.58**	—	10
W	**4,642.56**	8	30
U	**4,642.48**	2*h*	1
Ru	**4,642.38**	—	7
Sm	4,642.23	40	100
Ar	4,642.15	(80)	—
Tb	4,641.98	—	40
Os	4,641.831	3*h*	30
O	4,641.827	(150)	—
P	4,641.72	(50)	—
Te	4,641.19	(70)	—
Nd	4,641.10	—	80
J	4,640.88	(50)	—
V	4,640.73	20	25
*bh*Zr	4,640.6	—	150
Ne	4,640.44	(70)	—
Ir	4,640.08	2*wh*	40
V	4,640.07	15	25
Ti	4,639.95	15	60
Ti	4,639.67	15	40
Pr	4,639.55	2	50
Ti	4,639.37	18	80
Gd	4,639.01	3	25
O	4,638.865	(70)	—
In	4,638.861	70	3
Os	4,638.62	—	25
In	4,638.24	(125)	—
In	4,638.10	(200)	—
Fe	4,638.02	10	80
*bh*Zr	4,637.8	—	100
Fe	4,647.52	10	100
P	4,637.16	(50)	—
Ne	4,636.97	(50)	—
Gd	4,636.655	12	25
Se	4,636.65	(150)	—
Ne	4,636.63	(70)	—
Ne	4,636.12	(70)	—
Pr	4,635.692	—	40
Ru	4,635.690	—	125
W	**4,644.93**	—	8
Re	**4,644.95**	—	15
Nd	**4,644.98**	—	5
Mn	**4,645.03**	—	40
Ru	4,645.09	30	100
Ti	4,645.19	10	100
Tb	4,645.26	—	60*W*
La	4,645.28	40	25
Sm	4,645.40	—	40
Ne	4,645.42	(300)	—
Lu	4,645.47	2	25*h*
Pr	4,646.06	8	50
Cr	4,646.17	150	100
V	4,646.395	30	40
Nd	4,646.399	—	50
U	4,646.60	40	25
Sm	4,646.68	—	50
Nd	4,646.69	4	60
Cr	4,646.81	3	35
Pr	4,646.99	—	25
Sb	4,647.32	(80)	—
Fe	4,647.54	40	125
La	4,647.509	50	10
Te	4,647.51	(50)	—
Ru	4,647.61	—	125
Mo	4,647.81	25	25
Cr	4,648.13	6	40
Lu	4,648.21	2	25*h*
Se	4,648.44	(800)	—
Ni	4,648.66	3	400*w*
Lu	4,648.85	—	25
Cr	4,648.87	3	50
Nb	4,648.95	20	50
P	4,649.05	(50)	—
O	4,649.15	(300)	—
Cu	4,649.27	60	—
Cr	4,649.46	3	60
Sm	4,649.49	—	25
Nd	4,649.67	—	40
Ne	4,649.90	(70)	—
Ti	4,650.02	4	60
Nd	4,650.23	—	25
O	4,650.85	(70)	—
Cu	4,651.13	40	250
Cr	4,651.28	100	100
Pr	4,651.52	40*w*	125
Th	4,651.56	15	30

	λ	I	
		J	O
V	4,635.18	25	30
Os	4,634.77	—	30
Tm	4,634.24	10	80
Nd	4,634.23	—	50
Cr	4,634.09	80h	5
Zr	4,633.98	—	35
Kr	4,633.88	(800)	—
Mo	4,633.10	25h	25h
Ta	4,633.06	3h	150
Fe	4,632.92	4	70
J	4,632.32	(50)	—
Pr	4,632.28	1	40
Cr	4,632.18	8	25
Tb	4,632.06	—	30
Os	4,631.83	5	100
U	4,631.62	3	30
Re	4,630.84	—	50
Te	4,630.57	(50)	—
N	4,630.55	(300)	—
Sm	4,630.21	—	40
As	4,630.14	200	—
Nb	4,630.11	20	30
Co	*4,629.38*	*5*	*600W*
Ti	*4,629.34*	*7*	*70*
Co	*4,628.94*	—	*125*
Pr	*4,628.75*	*50w*	*200*
Ar	*4,628.44*	*(1000)*	—
Ne	*4,628.31*	*(150)*	—
As	*4,627.80*	*200*	—
Mn	*4,627.74*	—	*50*
Mo	*4,627.47*	*80*	*80*
In	*4,627.38*	*(150)*	—
Eu	*4,627.22*	*15*	*50*
Eu	*4,627.12*	—	*(300R)*
Tm	*4,626.55*	*3*	*50*
Mn	*4,626.54*	*15*	*80*
Mo	*4,626.47*	*80*	*100*
Tm	*4,626.31*	*20*	*50*
Cr	*4,626.19*	*125*	*100*
Co	*4,625.78*	—	*200*
Rn	*4,625.48*	*(500)*	—
Eu	*4,625.30*	—	*50w*
Fe	*4,625.05*	*12*	*100*
Xe	*4,624.28*	*(1000)*	—
Se	*4,623.77*	*(150)*	—
Ti	*4,623.09*	*40*	*125*
Co	*4,623.04*	—	*150*
Xe	4,651.94	(100)	—
Cr	4,652.16	150	200R
Re	4,652.32	—	30
Tm	4,653.01	15	25
Eu	4,653.48	—	25w
Ne	4,653.70	(50)	—
Ru	4,654.31	—	125
Te	4,654.38	(800)	—
Nd	4,654.73	—	30d
Cr	4,654.75	8	70
Co	4,654.85	—	25
Tm	4,655.08	—	35
Hf	4,655.19	4	50
O	4,655.36	(50)	—
La	4,655.50	300	150
In	4,655.657	(50)	—
Ni	4,655.661	—	40
In	4,655.79	(100)	—
Ir	4,656.18	—	60
Cr	4,656.19	4	50
Ne	4,656.39	(300)	—
Ti	4,656.47	70	150
Eu	4,656.73	1	50
S	4,656.74	(80)	—
Co	4,657.39	35	100
W	4,657.44	12	50
Ar	*4,657.94*	*(150)*	—
Lu	*4,658.02*	*15*	*100*
P	*4,658.11*	*(100)*	—
Kr	*4,658.87*	*(2000)*	—
W	*4,659.87*	*70*	*200*
Hg	*4,660.28*	*(200)*	—
Eu	*4,660.37*	*1*	*50*
Ne	*4,661.10*	*(150)*	—
Ta	*4,661.12*	*5h*	*300*
O	*4,661.65*	*(125)*	—
Eu	*4,661.87*	*20*	*80R*
Eu	*4,661.88*	—	*100*
La	*4,662.51*	*200*	*150*
Co	*4,663.41*	—	*700W*
Sm	*4,663.55*	—	*60*
La	*4,663.76*	*200*	*100*
Os	*4,663.82*	*5*	*100*
Cr	*4,663.83*	*15*	*50*
Hf	*4,664.12*	*100*	*50*
Se	*4,664.20*	*(150)*	—
Te	*4,664.34*	*(800)*	—

	λ	I J	I O		λ	I J	I O
Ta	*4,622.96*	*1*	*50*	*Pr*	*4,664.65*	*15*	*100*
Br	*4,622.75*	*(200)*	—	*Cr*	*4,664.80*	*20*	*70*
Cr	*4,621.96*	*40*	*50*	*Na*	*4,664.86*	—	*80*
Hf	*4,620.86*	*4*	*50*	*Se*	*4,664.98*	*(150)*	—
In	*4,620.24*	*(200)*	—				
La	*4,619.88*	*200*	*150*	*Cr*	*4,666.51*	*25*	*50*
bhZr	*4,619.8*	—	*80*	*J*	*4,666.52*	*(250)*	—
Cr	*4,619.55*	*30*	*50*	*Ni*	*4,666.99*	—	*50*
Ta	*4,619.51*	*10*	*300*	*Ne*	*4,667.36*	*(100)*	—
Fe	*4,619.30*	*8*	*100*	*Fe*	*4,667.46*	*20*	*150*
Kr	*4,619.15*	*(1000)*	—	*Ti*	*4,667.59*	*8*	*150*
Se	*4,618.77*	*(100)*	—	*Ni*	*4,667.77*	—	*100*
Ti	*4,617.27*	*100*	*200*	*Fe*	*4,668.14*	*10*	*125*
In	*4,617.16*	*(200)*	—	*Ag*	*4,668.48*	*70*	*200*
Os	*4,616.78*	*6*	*150*	*Na*	*4,668.60*	*100*	*200*
Ir	*4,616.39*	*5*	*200*	*La*	*4,668.91*	*300r*	*200r*
Cr	*4,616.14*	*200*	*300r*	*Ta*	*4,669.14*	*15*	*300*
				Cr	*4,669.34*	*20*	*50*
				Sm	*4,669.65*	*40*	*50*
				Sc	*4,670.40*	*300wh*	*100*
				V	*4,670.49*	*40r*	*60R*
				Xe	*4,671.23*	*(2000)*	—

Y 4,674.848 *J* 100 *O* 80

	λ	J	O		λ	J	O
Pr	**4,674.80**	—	**25**	**Er**	**4,674.849**	**15**	**50**
Cu	**4,674.76**	**30*W***	**200**	**Ce**	**4,674.88**	—	**2*h***
Ru	**4,674.65**	—	**20**	**Cs**	**4,674.89**	**(10)**	—
Ho	**4,674.62**	**3**	**4**	**N**	**4,674.98**	**(5)**	—
Dy	**4,674.61**	—	**3**	**Rh**	**4,675.03**	**50**	**100**
Nd	**4,674.595**	**10**	**50**	**Pr**	**4,675.04**	—	**5**
Sm	**4,674.592**	**40**	**80**	**Tm**	**4,675.08**	—	**25**
Xe	**4,674.56**	**(25)**	—	**W**	**4,675.09**	**1**	**8**
Ru	**4,674.52**	—	**7**	**Ti**	**4,675.12**	**5**	**50**
Ce	**4,674.49**	—	**3**	**Tb**	**4,675.18**	—	**2**
Re	**4,674.30**	—	**15**	**Lu**	**4,675.29**	**1**	**4**
Pr	**4,674.24**	—	**3**	**Tm**	**4,675.31**	—	**35**
U	**4,674.23**	**8**	**8**	**Ce**	**4,675.312**	—	**2**
Tm	**4,674.21**	**8**	**10**	**Nb**	**4,675.37**	**30*w***	**50*w***
Nd	**4,674.18**	—	**5**	**Re**	**4,675.39**	—	**2*h***
Tb	**4,674.04**	—	**2**	**Hf**	**4,675.45**	**10**	**10**
Os	**4,674.01**	—	**3**	**Pr**	**4,675.47**	—	**3**
W	**4,673.99**	**1**	**2**	**Eu**	**4,675.48**	—	**20*W***
Nd	**4,673.97**	—	**5**	**Nd**	**4,675.52**	—	**5**
Kr	**4,673.80**	**(3)**	—	**J**	**4,675.53**	**(50)**	—
In	**4,673.77**	**(5)**	—	**Ir**	**4,675.54**	**2**	**15**
U	**4,673.74**	**3*h***	—	**Se**	**4,675.55**	**(10)**	—
O	**4,673.71**	**(30)**	—	**Er**	**4,675.62**	**4**	**15**

	λ	I	
		J	O
Th	**4,673.66**	**2**	**6**
Ba	**4,673.62**	**5**	**40**
Dy	**4,673.615**	**8**	**10**
Tb	**4,673.61**	—	**4**
Nb	**4,673.59**	**5**	**2**
Cu	**4,673.55**	**6**	—
Be	**4,673.46**	**(100)**	—
As	4,672.70	50	—
Xe	4,672.20	(50*h*)	—
Be	4,673.20	100	—
Nb	4,672.09	100	150
Pr	4,672.08	25*w*	100
La	4,671.83	150	100
Mn	4,671.69	5	100
Xe	4,671.23	(2000)	—
Ho	**4,675.63**	**1**	**3**
Ni	**4,675.639**	—	**8**
Br	**4,675.64**	**(4)**	—
Mo	**4,675.70**	**4**	**2**
Sb	**4,675.74**	**(15)**	—
P	**4,675.78**	**(70)**	—
Dy	**4,675.81**	**2**	**7**
J	**4,676.05**	**(5)**	—
Th	**4,676.054**	—	**2**
Pr	**4,676.18**	**8*h***	**10**
O	**4,676.25**	**(125)**	—
Nd	**4,676.26**	**2*h***	**2**
U	**4,676.30**	**3*h***	**1**
*bh***Y**	**4,676.3**	—	**5**
Xe	4,676.46	(100*wh*)	—
Sm	4,676.91	50	100
J	4,676.94	(80)	—
Tm	4,677.85	2	50
Ta	4,678.02	2	40
Cd	4,678.16	200*W*	200*W*
Ne	4,678.22	(300)	—
Ne	4,678.50	(50)	—
Br	4,678.69	(200)	—
Fe	4,678.85	100	150
P	4,678.94	(100)	—
Ne	4,679.13	(150)	—
Gd	4,680.06	25	50
Zn	4,680.14	200*h*	300*w*
Ne	4,680.36	(100)	—
Kr	4,680.41	(500)	—
W	4,680.52	40	150
Cr	4,680.54	25	50
Nd	4,680.73	1	50
Rn	4,680.83	(500)	—
Cr	4,680.87	8	60
In	4,681.11	(200)	—
Ne	4,681.20	(50)	—
Sm	4,681.56	—	50
Ru	4,681.79	—	100
Ta	4,681.87	50	200
Ti	4,681.916	100	200
Tm	4,681.92	2	50
In	4,682.00	250*W*	—
Ra	4,682.28	(800)	—
Yt	4,682.32	100	60
Co	4,682.38	—	500
Gd	4,683.34	50	100

*P**

	λ	I	
		J	O
Nd	4,683.44	—	50
Ru	4,684.02	—	100
Nd	4,684.04	—	40
Ta	4,684.87	2	100
Pr	4,684.94	10w	125
In	4,685.22	(100)	—
Eu	4,685.25	—	60W
Ta	4,685.27	2	80
He	4,685.75	(300)	—
Ni	4,686.22	1	200
Te	4,686.95	(300)	—
Sm	4,687.18	—	100
Ne	4,687.67	(100)	—
Zr	4,687.80	—	125
Pr	4,687.81	1	50
Eu	4,688.23	2	100
Zr	4,688.45	—	50
Sm	4,688.73	—	50
Ta	4,688.84	—	40
Cr	4,689.37	35	80
Xe	4,690.97	(100)	—
Kr	4,691.28	(100)	—
Ti	4,691.34	25	125
Fe	4,691.41	10	80
Ba	4,691.62	40	100
Ta	4,691.90	5h	400
Os	4,692.06	3	80
La	4,692.50	300	200
Co	4,693.21	25	500
Ta	4,693.35	3	150
Kr	4,694.44	(200h)	—
Xe	4,697.02	(300)	—
Gd	4,697.49	4	100
Xe	4,698.01	(150h)	—
Eu	4,698.14	2	300
Co	4,698.38	8	300
Ti	4,698.76	20	100
O	4,699.21	(100)	—
La	4,699.63	200r	200r
Mn	4,701.16	5	100
Ta	4,701.32	2h	150
Ni	4,701.34	—	100
Ni	4,701.54	—	150
Ar	4,702.316	(1200)	—
Gd	4,702.323	100	50
Tb	4,702.42	—	80
Ne	4,702.53	(150)	—
La	4,703.28	300r	200r
Ni	4,703.81	—	200
Ne	4,704.39	(1500)	—

Zn $^{30}_{65.30}$

t_0 419.4°C

t_1 907°C

I.	II.	III.	IV.	V.
9.392	17.960	39.7	—	—

λ	*I*		eV
	J	*O*	
I 2,138.56	500	800*R*	5.80
II 2,502.00	400	20	10.95
II 2,557.96	300	10	10.95
I 3,282.33	300	500*R*	7.78
I 3,302.59	300	800	7.78
I 3,302.94	300*R*	700*R*	7.78
I 3,345.02	300	800	7.78
I 4,680.14	200	300	6.66
I 4,722.16	300	400	6.66
I 4,810.53	300	400	6.66
I 6,362.35	500	1000	7.74

Zn 2,138.56 *J* 500 *O* 800*R*

	λ	I			λ	I	
		J	O			J	O
Nb	**2,138.55**	**—**	**2**	**Ir**	**2,138.57**	**—**	**15**
As	**2,138.53**	**—**	**2**	**Ni**	**3,138.58**	**15**	**10**
Cu	**2,138.51**	**—**	**25***wh*	**Fe**	**2,138.59**	**—**	**8**
Os	**2,138.40**	**4**	**3**	**Os**	**2,138.61**	**—**	**8**
Yb	**2,138.32**	**10**	**—**	**Rh**	**2,138.66**	**25**	**—**
				Os	**2,138.75**	**1**	**3**
Zn	2,136.46	(10)	—				
				Te	*2,142.75*	—	*600*
Cu	*3,135.98*	*500w*	*25*				

Zn 2,502.001 *J* 400 *O* 20*

	λ	J	O		λ	J	O
Rn	**2,502.00**	**(7)**	**—**	**W**	**2,502.08**	**6**	**2**
Yb	**2,501.99**	**20**	**5**	**W**	**2,502.18**	**6**	**2**
Ta	**2,501.98**	**12**	**20**	**Cs**	**2,502.20**	**(2)**	**—**
U	**2,501.90**	**2**	**2**	**Mo**	**2,502.222**	**20**	**—**
W	**2,501.900**	**10**	**10**	**Os**	**2,502.29**	**1**	**10***s*
Ru	**2,501.89**	**30**	**30**	**Tl**	**2,502.30**	**20**	**—**
O	**2,501.80**	**(35)**	**—**	**Ru**	**2,502.38**	**4***h*	**20**
Fe	**2,501.69**	**4**	**20**	**Fe**	**2,502.39**	**60**	**3**
V	**2,501.613**	**30**	**35**				
Mo	**2,501.609**	**3**	**2**				
				Rh	2,505.10	200	2
Nb	2,501.40	150	—				
Te	2,499.75	(300)	—	*Cu*	*2,506.27*	*500r*	—
Nb	2,499.750	200	—				
B	*2,497.73*	*400*	*500*				

Zn 2,557.958 *J* 300 *O* 10*

	λ	J	O		λ	J	O
Nb	**2,557.94**	**100***h*	**—**	**U**	**2,557.960**	**2**	**4**
Rh	**2,557.92**	**50**	**1**	**Ru**	**2,558.016**	**10**	**4**
Ni	**2,557.87**	**80**	**—**	**W**	**2,558.023**	**6**	**—**
Br	**2,557.82**	**(4)**	**—**	**Sn**	**2,558.05**	**6**	**30**
Os	**2,557.77**	**1**	**10**	**Os**	**2,558.09**	**2**	**12**
Hg	**2,557.75**	**(5)**	**—**	**Po**	**2,558.10**	**(20)**	**—**
Al	**2,557.71**	**(40)**	**—**	**Mn**	**2,558.29**	**12**	**—**
Ta	**2,557.709**	**100**	**50**	**Ta**	**2,558,34**	**5***wh*	**—**
Yb	**2,557.70**	**4**	**—**				
Ru	**2,557.697**	**5***wh*	**30**				
				Ni	2,560.30	500*h*	—

	λ	I J	I O
W	2,557.56	6	12
Mn	2,557.54	50	1
B	2,557.52	6	—
Nb	2,556.94	200	5
Co	2,556.76	150	50*w*
Ni	2,555.11	1000*h*	—
Ta	2,554.62	100	50
Pd	2,561.02	200	—
J	2,561.49	(150)	—
Zn	*2,564.45*	*(25)*	—
Ni	*2,566.08*	*600h*	—
J	*2,566.26*	*(300)*	—

Zn 3,282.333 *J* 300 *O* 500*R*

	λ	I J	I O
Ti	3,282.329	150	30
Ir	3,282.327	2	2
Ce	3,282.32	—	2
Pr	3,282.30	1	6
Gd	3,282.26	5	5*r*
Co	3,282.23	—	3
Cs	3,282.10	(3)	—
Kr	3,282.08	(15*h*)	—
Co	3,282.04	—	4
B	3,282.01	12	4
Ce	3,282.00	—	3
Ho	3,281.98	15	12
Pt	3,281.97	3	10
Ce	3,281.95	—	3
W	3,281.94	10	12
Tb	3,281.92	—	8
Ir	3,281.90	—	3
Ni	3,281.88	—	20
Zr	3,281.87	—	2*h*
Sm	3,281.77	—	3
V	3,281.75	30	—
Lu	3,281.74	5	60
Ba	3,281.73	—	8
Ar	3,281.72	(15)	—
Rh	3,281.701	1	5
Cu	3,281.696	5	—
U	3,281.65	—	2
Si	3,281.64	3	—
Mo	3,281.62	20	—
Gd	3,281.51	5	5
Co	3,281.588	—	7
Ce	3,281.587	—	3
U	3,281.55	1	5
Ba	3,281.50	—	25
Rb	3,281.49	(20)	—
Nd	3,281.487	6	4
U	3,282.36	—	2
Fe	3,282.44	1	2
Mo	3,282.48	8	2
Yt	3,282.519	5	3
Eu	3,282.522	2	3*W*
Ag	3,282.525	1*h*	3
V	3,282.53	80	12
U	3,282.54	6	10
Th	3,282.61	12	7
Te	3,282.67	(10)	—
Ni	3,282.70	—	100
Cu	3,282.72	15*W*	25
Zr	3,282.739	—	10
Nd	3,282.78	2	8
Dy	3,282.79	1	2
Ni	3,282.827	1	25
Zr	3,282.834	10	10
Be	3,282.890	—	8
Fe	3,282.892	80	80
Mo	3,282.91	30	1
Th	3,282.97	12	8
Cr	3,283.06	35	—
Ho	3,283.07	4*h*	—
Tb	3,283.10	8	30
U	3,283.104	4	6
Pr	3,283.12	2	15
Ce	3,283.17	—	3
P	3,283.20	(15)	—
Pt	3,283.209	—	8
Yt	3,283.21	4	3
Sn	3,283.21	(50)	—
Fe	3,286.75	400	500
Pd	3,287.25	25	300*w*
Ti	3,287.65	200	40
Yb	3,289.37	1000*R*	500*R*

	λ	I J	I O
Ca	**3,281.48**	**4**	**1*h***
Th	**3,281.41**	**10**	**8**
Th	**3,281.40**	**15**	**50**
Ag	3,280.683	1000*R*	2000*R*
Eu	3,280.682	—	1000*R*
Fe	3,280.26	150	150
Ti	3,278.92	150	40
Fe	3,277.35	200	40
V	3,276.12	200*R*	50
Cu	3,273.96	1500*R*	3000*R*
Fe	*3,271.00*	*300*	*300*
Ge	*3,269.49*	*300*	*300*
Fe	*3,265.62*	*300*	*300*

	λ	I J	I O
Yb	3,289.85	—	1000
Os	3,290.26	20	200
Mo	*3,292.31*	*300*	*10*
Zn	*3,299.39*	*(15)*	—

Zn 3,302.59 *J* 300 *O* 800

3,302.941 *J* 300*R* *O* 700*R*

	λ	I J	I O
Ta	3,302.76	1*h*	50
Mo	3,302.72	25	—
Zr	3,302.67	6	10
Pr	3,302.66	2	15
Tb	3,302.64	—	8
Nb	3,302.62	10	1
Zn	***3,302.59***	***300***	***800***
J	3,302.56	(10)	—
Bi	3,302.55	—	150
Kr	3,302.54	(10)	—
B	3,302.51	10	—
U	3,302.492	1	3
J	3,302.489	(5)	—
Dy	3,302.47	2	4
Tm	3,302.45	80	125
Yb	3,302.44	5	7
J	3,302.43	(2)	—
Na	3,302.32	300*R*	600*R*
Ta	3,302.33	1*h*	3
Kr	3,302.28	(4*h*)	—
U	3,302.26	2*h*	1*h*
Re	3,302.23	—	30
Cr	3,302.19	1*h*	50*h*
Nb	3,302.18	10	5
Y	3,302.17	2	5

	λ	I J	I O
V	3,302.82	4	6
Fe	3,302.86	5	1
Cr	3,302.88	8	30
Ce	3,302.91	—	10*W*
Ca	*3,302.94*	*6*	*4*
Zn	***3,302.941***	***300R***	***700R***
Na	3,302.99	150*R*	300*R*
J	3,303.08	(25)	—
Ir	3,303.09	—	3
La	3,303.11	150	400
Mo	3,303.113	2	4
Re	3,303.21	—	30
Ce	3,303.22	—	10
Mn	3,303.28	—	40
Nb	3,303.32	30	1
W	3,303.33	3	7
Mo	3,303.34	5	25
U	3,303.37	3*h*	4
Fe	3,303.47	5	5
Th	3,303.49	10	5
Cu	3,303.52	2	—
Mo	3,303.54	20	—
Fe	3,303.57	10	70
U	3,303.597	12	10

	λ	I J	I O
Pd	3,302.13	200*h*	1000*wh*
Pt	3,302.12	—	2*h*
Ti	3,302.10	20	8
Sm	3,302.09	4	8
Dy	3,302.02	—	2
Eu	3,301.95	2	25
Er	3,301.93	2	15
Ru	3,301.91	8	30
Ce	3,301.905	—	10
Pr	3,301.898	—	10
Ta	3,301.895	3	25
Ar	3,301.87	(10)	—
Pt	3,301.861	250*W*	300
Al	3,301.86	2	—
W	3,301.85	10	1
Ir	3,301.76	—	2
U	3,301.754	3	10
Kr	3,301.75	(5*h*)	—
Sr	3,301.73	10	100
Dy	3,301.72	—	2
Mo	3,301.71	4	4
Er	3,301.68	—	3
Sm	3,301.67	8	15
Ce	3,301.652	—	3
Th	3,301.651	1*h*	6
V	3,301.651	80	—
U	3,301.650	1	6
K	3,301.60	(10)	—
Re	3,301.600	—	50
Ru	3,301.59	40	70
Eu	3,301.57	1	15*w*
O	3,301.56	(10)	—
Os	3,301.559	50	500*R*
Dy	3,303.60	—	2
Sm	3,303.60	—	5
Ir	3,303.63	—	4
W	3,303.66	—	6
Ce	3,303.67	—	5
Cs	3,303.72	(4)	—
Re	3,303.75	—	40
Ce	3,303.77	—	8
U	3,303.84	—	6
Co	3,303.88	—	60r
Zn	3,305.96	(20)	—
Fe	3,305.97	300	400
Fe	3,306.35	150	200
Te	3,306.99	(150)	—
Cr	3,310.65	200	2
Ar	*3,319.34*	*(300)*	—
Be	*3,321.34*	*40*	*1,000R*

*P**

Zn 3,345.020 *J* 300 *O* 800

	λ	J	O
Xe	3,344.97	(2*h*)	—
Ti	3,344.93	—	8
W	3,344.90	12	3
Th	3,344.88	8	8
U	3,344.87	6	6
Ru	3,344.80	—	5
Ar	3,344.79	(3)	—
Zr	3,344.786	15	15
Ce	3,344.761	8	50
P	3,344.758	—	2
Sm	3,345.02	3*h*	1
Nd	3,345.088	2	12
W	3,345.089	8	9
Ta	3,345.11	3*h*	3
Cr	3,345.15	2	15
Ce	3,345.23	—	3
Eu	3,345.31	2	3
Ru	3,345.317	5	60
K	3,345.32	(30)	—
Mn	3,345.35	—	15

	λ	I J	I O		λ	I J	I O
Er	**3,344.75**	**1**	**8**	**Cr**	**3,345.37**	**1**	**18**
Mo	**3,344.746**	**40**	**50**	**Dy**	**3,345.37**	**2**	**3**
Rb	**3,344.73**	**(15)**	—	**Ce**	**3,345.44**	—	**20**
Ag	**3,344.66**	**2*h***	—	**Be**	**3,345.45**	—	**2**
Ti	**3,344.63**	—	**2**	**Er**	**3,345.46**	**1**	**6**
Pr	**3,344.564**	**1**	**8**	**Ne**	**3,345.49**	**(7)**	—
U	**3,344.561**	—	**4**	**Rn**	**3,345.50**	**(2)**	—
La	**3,344.560**	**200*wh***	**300**	**Nd**	**3,345.52**	**2**	**10**
Ce	**3,344.552**	—	**4*w***	**U**	**3,345.54**	**4**	**12**
Yt	**3,344.55**	**2*h***	**3**	**Zn**	**3,345.572**	**100**	**500**
Ru	**3,344.53**	**6**	**60**	**Mo**	**3,345.572**	**2**	**5**
Ca	**3,344.513**	**7**	**100**	**Fe**	**3,345.68**	—	**1**
Cr	**3,344.51**	**2**	**20**	**Nd**	**3,345.71**	**4**	**20**
Tb	**3,344.50**	**3**	**8**	**Kr**	**3,345.73**	**(4)**	—
Dy	**3,344.49**	**2**	**3**	**Dy**	**3,345.79**	**2**	**3**
Ho	**3,344.46**	**6**	**4**	**Tm**	**3,345.85**	**15**	**6**
W	**3,344.44**	**7**	**8**	**W**	**3,345.86**	**15**	**10**
Ne	**3,344.43**	**(18)**	—	**U**	**3,345.89**	**1*h***	**12**
Er	**3,344.36**	**1**	**12**	**V**	**3,345.90**	**125**	—
Sm	**3,344.349**	**10**	**40**	**Te**	**3,345.93**	**(35)**	—
Re	**3,344.347**	—	**150**	**Zn**	**3,345.934**	**50**	**150**
Th	**3,344.34**	**2**	**2**	**In**	**3,345.94**	**(5)**	—
Ce	**3,344.33**	—	**8**	**Gd**	**3,345.99**	**15**	**20**
U	**3,344.32**	**1**	**4**				
Nb	**3,344.245**	**10**	—				
				Ti	3,349.03	800*R*	125
Co	**3,344.245**	—	**4**	Ti	3,349.41	400*R*	100
Ir	**3,344.22**	**1**	**10**	Gd	3,350.48	180	150
Rh	**3,344.20**	**20**	**100**				
Sm	**3,344.172**	—	**4**				
Er	**3,344.17**	—	**4**	*Ti*	*3,361.21*	*600R*	*100*
Rn	**3,344.10**	**(1)**	—				
Ti	3,341.87	300*R*	100				
Cr	3,339.80	150	25				
V	3,337.85	150	2				
La	3,337.49	300*wh*	800				

Zn 4,680.14 *J* 200 *O* 300

	λ	J	O		λ	J	O
Ce	**4,680.13**	—	**6**	**Fe**	**4,680.30**	—	**9**
Gd	**4,680.06**	**25**	**50**	**Pr**	**4,680.32**	—	**2**
Sm	**4,680.04**	—	**10**	**Ne**	**4,680.36**	**(100)**	—
Pr	**4,679.83**	—	**3**	**Kr**	**4,680.41**	**(500)**	—
V	**4,679.77**	**3**	**4**	**Ce**	**4,680.458**	—	**2**
Tb	**4,679.62**	—	**2**	**Ir**	**4,680.465**	—	**2**
Pr	**4,679.50**	—	**2**	**Pr**	**4,680.47**	—	**4**
Eu	**4,679.48**	—	**4*w***	**Sc**	**4,680.48**	**4*h***	**5**

	λ	I J	I O
Re	4,679.47	—	20
Xe	4,679.45	(3h)	—
Gd	4,679.18	25	25
Ne	4,679.13	(150)	—
Dy	4,679.12	2	3
Pr	4,679.11	3	8
Er	4,679.07	2	6
W	4,679.041	2	15
Pr	4,679.039	—	5
J	4,678.98	(8)	—
P	4,678.94	(100)	—
Nd	4,678.91	1	2h
Fe	4,678.85	100	150
Tb	4,678.81	—	2
Br	4,678.69	(200)	—
Ne	4,678.22	(300)	—
Cd	4,678.16	200W	200W
Xe	4,676.46	(100wh)	—
O	4,676.25	(125)	—
Yt	4,674.85	100	80
Cu	4,674.76	30W	200
Be	4,673.46	(100)	—
Be	4,672.20	100	—
Nb	4,672.09	100	150
La	4,671.83	150	100
Xe	4,671.23	(2000)	—
Sc	4,670.40	300wh	100
Ta	4,669.14	15	300
La	4,668.91	200r	200r
Na	4,668.60	100	200
Ag	4,668.48	70	200
Ti	4,667.59	8	150
Fe	4,667.46	20	150
Ne	4,667.36	(100)	—
J	4,666.52	(250)	—
Te	4,664.34	(800)	—
La	4,663.76	200	100
Co	4,663.41	—	700W
La	4,662.51	200	150
Ta	4,661.12	5h	300
Hg	4,660.28	(200)	—
Kr	4,658.87	(2000)	—
Ne	4,656.39	(300)	—
La	4,655.50	300	150
Te	4,654.38	(800)	—

	λ	I J	I O
W	4,680.52	40	150
Cr	4,680.54	25	50
Gd	4,680.65	—	3
Th	4,680.652	3	4
Nd	4,680.73	1	50
Rn	4,680.83	(500)	—
Cr	4,680.87	8	60
V	4,680.89	3	4
Ce	4,680.99	—	3
Dy	4,681.00	—	4
Mo	4,681.05	8	6
Te	4,681.06	(15h)	—
Eu	4,681.06	1	15
In	4,681.11	(200)	—
U	4,681.18	4	2
W	4,681.19	1	10
Ne	4,681.20	(50)	—
Th	4,681.21	2	2
Nd	4,681.25	—	5
S	4,681.32	(5)	—
Ru	4,681.39	—	10
Os	4,681.43	—	2
Eu	4,681.51	1	10
Er	4,681.558	—	3
Sm	4,681.560	—	50
Ta	4,681.87	50	200
Ti	4,681.92	100	200
In	4,682.00	250W	—
Ra	4,682.28	(800)	—
Yt	4,682.32	100	60
Co	4,682.38	—	500
In	4,685.22	(100)	—
He	4,685.75	(300)	—
Ni	4,686.22	1	200
Te	4,686.95	(300)	—
Ne	4,687.67	(100)	—
Xe	4,690.97	(100)	—
Ta	4,691.90	5h	400
La	4,692.50	300	200
Co	4,693.21	25	500
Kr	4,694.44	(200h)	—
Xe	4,697.02	(300)	—
Co	4,698.38	8	300
La	4,699.63	200r	200r
Ar	4,702.32	(1200)	—
La	4,703.28	300r	200r

	λ	I J	I O		λ	I J	I O
				Ne	*4,704.39*	*(1500)*	—
				Hg	*4,704.63*	*(200)*	—
				Br	*4,704.86*	*(250)*	—
				O	*4,705.32*	*(300)*	—
				As	*4,707.82*	*200*	—
				Ne	*4,708.85*	*(1200)*	—

Zn 4,722.159 *J* 300 *O* 400

	λ	J	O		λ	J	O
Ne	**4,722.15**	**(5)**	—	**Kr**	**4,722.16**	**(3)**	—
La	**4,722.14**	**2*h***	—	**Bi**	**4,722.19**	**5**	**10**
Th	**4,722.11**	—	**2**	**Sr**	**4,722.28**	—	**30**
Er	**4,722.02**	—	**4**	**Bi**	**4,722.55**	**100**	**1000**
Pr	**4,721.91**	—	**3**	**Ti**	**4,722.62**	**8**	**80**
Ir	**4,721.88**	—	**2**	**Sm**	**4,722.63**	—	**3*h***
Rn	**4,721.76**	**(150)**	—	**Pr**	**4,722.67**	—	**5**
Hf	**4,721.71**	**4**	**10**	**Er**	**4,722.70**	—	**12**
Pr	**4,721.67**	—	**3**	**Ne**	**4,722.71**	**(15)**	—
Ar	**4,721.62**	**(10)**	—	**U**	**4,722.73**	**50**	**40**
Ne	**4,721.54**	**(70)**	—	**Hg**	**4,722.80**	**(5)**	—
V	**4,721.51**	**12**	**15**	**Bi**	**4,722.83**	**5**	**10**
Pr	**4,721.471**	—	**2**	**V**	**4,722.86**	**15**	**20**
Gd	**4,721.466**	**20*h***	**20**	**Ta**	**4,722.88**	—	**200**
Cl	**4,721.43**	**(25)**	—	**Sm**	**4,722.95**	—	**3*h***
Co	**4,721.41**	**100*h***	**8**	**Mo**	**4,723.06**	**20**	**10**
Sm	**4,721.40**	—	**20**	**W**	**4,723.09**	—	**4**
U	**4,721.32**	**2*h***	—	**Cr**	**4,723.10**	**8**	**125**
Os	**4,721.284**	—	**12**	**Nb**	**4,723.15**	**3*h***	**1**
Cl	**4,721.28**	**(6)**	—	**Dy**	**4,723.167**	**2**	**2**
V	**4,721.25**	**2**	**3**	**Ti**	**4,723.168**	**7**	**40**
Dy	**4,721.23**	**5**	**12**	**Ru**	**4,723.22**	—	**6**
Er	**4,721.06**	—	**5**	**Er**	**4,723.24**	—	**3**
Xe	**4,721.00**	**(2*wh*)**	—	**Mo**	**4,723.31**	**4**	**4**
Rh	**4,721.000**	**5**	**8**	**Mo**	**4,723.449**	**4**	**4**
Ir	**4,720.93**	—	**2*wh***	**Th**	**4,723.452**	**4**	**5**
Ru	**4,720.92**	—	**15**				
Nd	**4,720.91**	—	**4**	Gd	4,723.73	200	100
Sc	**4,720.83**	**5**	**4**	Yb	4,726.07	200	45
Yb	**4,720.78**	**1**	**7**	Ar	4,726.91	(200)	—
				Co	4,727.94	—	300
La	4,719.95	300	200*r*	La	4,728.42	300	400*r*
Cr	4,718.43	150	200				
La	4,716.44	200	100	Se	4,730.78	(1000)	—
S	4,716.23	(600)	—	Gd	4,732.61	300	300
Ni	4,715.78	2	200	Xe	4,734.15	(600)	—
				Gd	4,735.76	150	150
Ne	4,715.34	(1500)	—	Ar	4,735.93	(400)	—
Ni	4,714.42	8	1000				

	λ	I J	I O		λ	I J	I O
Eu	4,713.61	2	400	*Kr*	*4,739.00*	(3000)	—
La	4,712.93	150	100*r*	*Se*	*4,739.03*	*(800)*	—
Ne	4,712.06	(1000)	—	*La*	*4,740.28*	*300*	*150*
				Eu	*4,740.52*	*2*	*500*
Ne	4,710.06	(1000)	—	*Se*	*4,740.97*	*(600)*	—
				Se	*4,742.25*	*(500)*	—
				La	*4,743.08*	*300*	*300r*
				Gd	*4,743.65*	*300*	*300*
				Ne	*4,749.57*	*(300)*	—
				Co	*4,749.68*	*100h*	*500*

*P**

Zn 4,810.53 *J* 300 *O* 400

Kr	**4,810.51**	**(3)**	—	**Nb**	**4,810.60**	**10**	**100**
Nd	**4,810.509**	**2*wh***	**2**	**Ne**	**4,810.63**	**(100)**	—
Rh	**4,810.49**	**25**	**15**	**Cr**	**4,810.73**	—	**30**
Ce	**4,810.39**	—	**6**	**U**	**4,810.89**	—	**6**
N	**4,810.29**	**(5)**	—	**Mo**	**4,811.062**	**50**	**50**
Dy	**4,810.24**	—	**2**	**Pr**	**4,811.064**	—	**5**
Tb	**4,810.20**	—	**3*r***	**Ti**	**4,811.08**	**1**	**12**
Re	**4,810.10**	—	**4**	**Hf**	**4,811.14**	**2**	**6**
Ne	**4,810.063**	**(150)**	—	**V**	**4,811.14**	**4**	—
Cl	**4,810.06**	**(200)**	—	**Re**	**4,811.21**	—	**4**
Ru	**4,809.70**	—	**4**	**Nb**	**4,811.30**	**2**	**15**
W	**4,809.65**	**1**	**2**	**Nd**	**4,811.34**	**60**	**60**
Se	**4,809.61**	**(4)**	—	**Er**	**4,811.47**	—	**2**
Ne	**4,809.50**	**(10)**	—	**Cl**	**4,811.57**	**(12)**	—
Zr	**4,809.468**	—	8	**Au**	**4,811.62**	**15**	**50**
Ir	**4,809.467**	—	**3**	**U**	**4,811.70**	—	**4*h***
Nb	**4,809.37**	**8**	**5**	**Pr**	**4,811.73**	—	**7**
Cr	**4,809.289**	—	**12**	**Kr**	**4,811.76**	**(300)**	—
Eu	**4,809.289**	**2**	**10**	**Sr**	**4,811.88**	—	**40**
Hf	**4,809.18**	**10**	**8**	**Ni**	**4,812.00**	—	**10**
Cl	**4,809.05**	**(9)**	—	**As**	**4,812.01**	**10**	—
La	**4,809.01**	**150**	**150**				
Sm	**4,809.00**	—	**2**				
				Co	4,813.48	6	1000*W*
				Ge	4,814.80	200	—
Xe	4,807.02	(500)	—	S	4,815.51	(800)	—
Ar	4,806.07	(500)	—	Br	4,816.71	(300)	—
Gd	4,805.82	80	200	Ne	4,817.64	(300)	—
La	4,804.04	150	150				
Gd	4,801.08	200	200	Ne	4,818.79	(150)	—
				Cl	4,819.46	(200)	—
Cr	4,801.03	70	200	Fe	4,821.05	200*h*	200*h*
Cd	4,899.92	300*w*	300*w*	Ne	4,821.92	(300)	—
Hg	4,897.01	(300)	—	Xe	4,823.41	(150*h*)	—
				Mn	4,823.52	80	400

	λ	I J	I O		λ	I J	I O
Co	*4,792.86*	*5*	*600W*	La	4,824.07	150	150
Ne	*4,788.93*	*(300)*	—	Kr	4,825.18	(300)	—
Br	*4,785.50*	*(400)*	—	Ra	4,825.91	(800)	—
Mn	*4,783.42*	*60*	*400*				
Co	*4,781.43*	*2h*	*400*				
				Ne	*4,827.34*	*(1000)*	—
Co	*4,780.01*	*500*	*500w*	*Ne*	*4,827.59*	*(300)*	—
				Xe	*4,829.71*	*(400)*	—
				Te	*4,831.29*	*(800)*	—
				Kr	*4,832.07*	*(800)*	—
				Ne	*4,837.31*	*(500)*	—
				Co	*4,840.27*	*150*	*700w*
				Se	*4,840.63*	*(800)*	—

Zn 6,362.35 *J* 500 *O* 1000

	λ	J	O		λ	J	O
Sc	**6,362.29**	**5**	**2**	**In**	**6,362.37**	**(12)**	—
In	**6,362.13**	**(40)**	—	*bh***Zr**	**6,362.5**	—	**5**
Hg	**6,362.10**	**(15)**	—	**Re**	**6,362.69**	—	**2**
Nd	**6,362.09**	—	**10**	**Xe**	**6,362.80**	**(3*wh*)**	—
Sc	**6,361.801**	**8**	**10*R***	**Cr**	**6,362.87**	**8**	**150**
Er	**6,361.80**	—	**4**	**In**	**6,362.90**	**(40)**	—
In	**6,361.74**	**(20)**	—	**Re**	**6,362.94**	—	**2**
*bh***Sc**	**6,361.7**	—	**2**	**In**	**6,362.96**	**(40)**	—
In	**6,361.49**	**(20)**	—	**Si**	**6,363.10**	**1**	**3**
Nd	**6,361.43**	—	**10**	**Sm**	**6,363.16**	—	**3**
V	**6,361.27**	**2**	**5**	**Gd**	**6,363.22**	—	**3**
W	**6,361.08**	**1**	**3**	**J**	**6,363.26**	**(15)**	—
				Ru	**6,363.41**	—	**4**
				Rh	**6,363.77**	—	**2**
Xe	6,356.35	(300)	—				
Eu	6,350.04	—	600*W*	*Ne*	*6,382.99*	*(1000)*	—
Ra	*6,336.90*	*(500)*	—				
Ne	*6,334.43*	*(1000)*	—				

Zr $^{40}_{91.22}$

t_0 1,900°C t_1 >2,990°C

I.	II.	III.	IV.	V.
6.951	14.03	24.10	33.972	—

λ	I		eV
	J	O	
II 3,391.98	400	300	3.82
II 3,438.23	200	250	3.69
I 3,547.68	12	200	3.56
I 3,601.19	15	400	3.59
I 4,687.80	—	125	3.40
I 4,739.48	—	100	3.26
I 4,772.31	—	100	3.22

Zr 3,391.98 *J* 400 *O* 300

	λ	*I*	
		J	*O*
Nd	3,391.95	—	6
Ru	3,391.89	6	50
Ce	3,391.88	—	3
Mo	3,391.85	30	1
Sm	3,391.84	3	2
Rh	3,391.73	1	3*d*
Tb	3,391.72	3	15
Th	3,391.717	3	1
Ce	3,391.593	—	10
Nb	3,391.591	5	1
Lu	3,391.55	2	10
Mo	3,391.54	—	4
W	3,391.53	10	10
Cr	3,391.43	150	4
U	3,341.40	—	6
Cr	3,391.37	—	12
Nb	3,391.33	3	5
Os	3,391.29	5	2
Rn	3,391.27	(8)	—
Re	3,391.24	—	2*h*
Dy	3,391.13	3*d*	3*d*
Sm	3,391.12	2	8
Yb	3,391.10	40	10
W	3,391.100	10	10
Ni	3,391.05	40	400
Hg	3,391.04	2*h*	—
U	3,391.04	2	2
Co	3,388.17	12	250*R*
Zr	3,387.87	100	100
Hg	3,385.25	(200)	—
Co	3,385.22	15	250*R*
Fe	3,383.98	100	200
Ti	3,383.76	300*R*	70
Ag	3,382.89	700*R*	1000*R*
Nd	3,382.81	10	200
Cr	3,382.68	200	35
Ni	*3,380.57*	*100*	*600R*
Zr	*3,374.73*	*30*	*20*
Ni	*3,374.22*	*6*	*400*
Zr	*3,373.42*	*10*	*12*
Pd	*3,373.00*	*500wh*	*800r*
Ti	*3,372.80*	*400R*	*80*
Rh	*3,372.25*	*200*	*300*
Ni	*3,371.99*	*10*	*400*

	λ	*I*	
		J	*O*
Er	3,391.989	12	30
Dy	3,391.99	—	4
Eu	3,391.992	5	40
Tb	3,392.01	8	15
Fe	3,392.014	6	20
Cu	3,392.02	—	7
Th	3,392.04	15	10
Ho	3,392.05	8	6
Mo	3,392.17	2	15
Ti	3,392.197	1	2
Rn	3,392.20	(2)	—
Ce	3,392.26	—	3
Ar	3,392.31	(3)	—
Fe	3,392.313	80	125
Nb	3,392.34	30	20*r*
Re	3,392.38	—	3*h*
Ir	3,392.47	4*h*	2
Gd	3,392.53	25	40
Ru	3,392.54	40	100
Ir	3,392.59	—	3
K	3,392.63	(10)	—
Fe	3,392.66	200	300
Ti	3,392.71	8	20
V	3,392.73	30	1
Ne	3,392.78	(20)	—
Ce	3,392.784	—	10
Ca	3,392.81	2	1
Ar	3,392.81	(100)	—
Hf	3,392.811	3	20
Cl	3,392.87	(15)	—
La	3,392.94	3	4
Th	3,392.97	3	3
Ni	3,393.99	—	600*R*
Zr	3,393.12	25	30
Ar	3,393.75	(250)	—
Ti	3,394.57	200	70
Fe	3,394.59	80	150
Co	3,395.37	50	400*R*
Zr	3,396.33	10	12
Rh	3,396.85	500	1000*w*
Re	3,399.30	—	200*w*
Fe	3,399.34	200	200
Zr	3,399.35	40	100
Rh	3,399.70	60	500
Fe	3,401.52	90	150
Os	3,401.86	20	200

	λ	I J	I O		λ	I J	I O
				Fe	3,402.26	150	150
				Os	3,402.51	15	200
				Zr	3,402.87	10	10
				Cd	3,403.65	500*h*	800
				Zr	3,403.68	15	15
				Pd	3,404.58	1000*R*	2000*R*
				Zr	3,404.83	35	40
				Co	3,405.12	150	2000*R*
				Fe	3,407.46	400	400
				Zr	3,408.08	9	10
				Co	3,409.18	125	1000*R*
				Ni	3,409.58	—	300
				Zr	3,410.25	50	50
				Rh	3,412.27	60	300
				Co	3,412.34	200	1000*R*
				Co	3,412.63	40	1000*R*

Zr 3,438.230 *J* 200 *O* 250

	λ	J	O		λ	J	O
W	3,438.21	3	7	Ce	3,438.235	—	3
Ir	3,438.09	—	6	Hf	3,438.235	25	25
Ce	3,438.07	—	15	Fe	3,438.306	3*h*	10
Sm	3,438.05	3	9	Pr	3,438.31	1*h*	25*w*
Y	3,437.953	—	3	Eu	3,438.31	—	5
Fe	3,437.952	7	15	Er	3,438.32	1	12
U	3,437.93	10	6	In	3,438.34	(50)	—
V	3,437.87	1*h*	2	Ru	3,438.37	35	70
Ce	3,437.81	—	10	U	3,438.41	3	3
V	3,437.77	2	4	Nb	3,438.42	50*wh*	1
Xe	3,437.73	(2*wh*)	—	Hf	3,438.43	3	12
Re	3,437.72	—	100	Er	3,438.473	1*h*	9
Co	3,437.69	—	150*Wh*	Nd	3,438.474	1	6
U	3,437.67	—	6	In	3,438.52	(10)	—
Er	3,437.64	1	10	Tb	3,438.57	8	15
Tm	3,437.64	15	10	Os	3,438.61	10*h*	4
Sm	3,437.61	2*h*	3	U	3,438.70	3	2
Pr	3,437.60	—	6	Co	3,438.71	—	80*W*
Ir	3,437.50	3	30	Yb	3,438.72	80	4
Sb	3,437.49	8	2*h*	Te	3,438.73	(10)	—
Ta	3,437.37	300*wh*	7	W	3,438.81	6	9
Pb	3,437.36	10	—	Yb	3,438.84	100	20
Ce	3,437.32	—	12	Mo	3,438.87	20	20
Th	3,437.31	2	6	Kr	3,438.88	(3*h*)	—
Ni	3,437.28	40	600*R*	Ce	3,438.881	—	3
W	3,437.220	10	3	Co	3,438.91	—	30
Mo	3,437.216	25	25	U	3,438.92	—	3

	λ	I	
		J	O
Ru	3,436.74	150	300R
Rh	3,434.89	200r	1000r
Ni	3,433.56	50wh	800R
Pd	3,433.45	500h	1000h
Cr	3,433.31	150	30
Co	3,433.04	150	1000R
Nb	3,432.70	100	10
Co	3,431.575	40	500R
Zr	3,431.569	8	10
Bi	3,431.23	(150)	—
Bi	3,430.83	(200)	—
Zr	3,430.53	50	50
Zr	3,430.29	—	10
Tm	3,429.97	100	100
Ru	3,428.31	100	100
Nb	3,426.57	200	5
Nb	3,425.42	300	30r
Tm	3,425.08	300	200
Zr	3,424.82	9	15
Re	3,424.60	—	300W
Ni	3,423.71	25	600R
Pd	3,421.24	1000R	2000R
Cr	3,421.21	200	50
Dy	3,438.952	9	25
Th	3,438.953	12	10
W	3,438.966	4	5
Nd	3,438.97	—	6
Ne	3,438.97	(4)	—
Mn	3,438.974	20	20
Sm	3,438.98	—	8
Ta	3,439.00	70W	—
Tb	3,439.05	—	15
Nb	3,439.210	12	30
Gd	3,439.212	35	60
Mo	3,439.23	2	2
Rb	3,439.34	(200)	—
Zr	3,440.45	—	10
Rh	3,440.53	100	2
Fe	3,440.61	300	500
Zr	3,440.97	2	10
Fe	3,440.99	200	300
Pd	3,441.40	2h	800h
Tm	3,441.51	80	150
Co	3,442.93	15	400R
Co	3,443.64	100	500R
Fe	3,443.88	200	400
Ti	3,444.31	150	60
Fe	3,445.15	150	300
Sm	3,445.62	10R	150R
Ni	3,446.26	50h	1000R
Zr	3,446.61	1	15
K	3,446.72	100R	150R
Ta	3,446.91	150W	2W
Zr	3,447.36	3	150w
K	3,447.701	75R	100R
Ne	3,447.703	(150)	—
Co	3,449.17	125	500R
Co	3,449.44	125	500R
Zr	3,449.91	—	10
Pd	3,451.35	400h	—
Hg	3,451.69	(200)	—
Nb	3,452.35	200	5
Ni	3,452.89	50	600R
Co	3,453.50	200	3000R
Yb	3,454.07	250	40
Rh	3,455.22	12	300
Co	3,455.23	10	2000R
Zr	3,455.91	1	12
Zr	3,457.18	—	10

Zr 3,547.68 *J* 12* *O* 200

	λ	*I*	
		J	*O*
U	**3,547.533**	**3**	**1**
Dy	**3,547.53**	**2**	**6**
Er	**3,547.51**	**1*h***	**25**
W	**3,547.47**	**6**	**8**
Th	**3,547.41**	**2*d***	**1*d***
Mo	**3,547.40**	**4**	**3**
Ca	**3,547.38**	**3**	**2**
Eu	**3,547.37**	—	**3*w***
Th	**3,547.34**	**1*d***	**2*d***
Fe	**3,547.20**	**8**	**25**
U	**3,547.19**	**12**	**10**
Eu	**3,547.11**	—	**15*w***
Ti	**3,547.03**	**12**	**30**
Ce	**3,547.00**	**3**	**15**
Ru	**3,546.98**	**5**	**2**
V	**3,546.96**	**20**	—
Dy	**3,546.84**	**10**	**50**
Sm	**3,546.82**	—	**3**
Co	**3,546.71**	—	**8**
U	**3,546.68**	**8**	**6**
Ce	**3,546.66**	—	**8**
*bh***Sr**	**3,546.6**	—	**4**
Gd	3,545.79	125	125
Rh	3,543.95	40	150
Os	3,542.71	10	150
Fe	3,542.08	100	150
Fe	3,541.09	200	200
Fe	3,540.12	60	100
Ce	3,539.09	10	100
Dy	3,538.52	40	150
Rh	3,538.14	10	100
Fe	*3,536.56*	*200*	*300*
Nb	*3,535.30*	*500*	*300*
Co	*3,533.36*	—	*200w*
Zr	*3,533.22*	*3*	*30*
Co	*3,529.81*	*30*	*1000R*
Tl	*3,529.43*	*800*	*1000*
Co	*3,529.03*	—	*200R*
Os	*3,528.60*	*50*	*400R*
Rh	*3,528.02*	*150*	*1000w*
Ni	*3,527.98*	*15*	*200*
Co	*3,526.85*	*25*	*300R*

	λ	*I*	
		J	*O*
Ba	**3,547.77**	**6**	**10**
Mn	**3,547.80**	**15**	**40*h***
Ce	**3,547.81**	—	**3**
Os	**3,547.88**	**8**	**1**
Dy	**3,547.89**	—	**4**
Cr	**3,547.98**	—	**5*wh***
Fe	**3,548.02**	**7**	**10**
Mn	**3,548.029**	**15**	**40*h***
Eu	**3,548.03**	**1*h***	**4**
Fe	**3,548.089**	—	**9**
Nb	**3,548.089**	**5*h***	**5**
Pr	**3,548.09**	**5*d***	**4*d***
Sr	**3,548.093**	—	**50**
Re	**3,548.11**	—	**3**
Nb	**3,548.131**	**5**	**2*h***
Th	**3,548.133**	**1*d***	**2*d***
Ce	**3,548.17**	**5**	**10*h***
Ni	**3,548.18**	**25**	**400**
Dy	**3,548.202**	**2**	**8**
Mn	**3,548.202**	**40**	**40*h***
Er	**3,548.23**	**1**	**8*d***
W	**3,548.259**	**7**	**5**
U	**3,548.260**	—	**3**
Zr	**3,548.40**	—	**2*h***
U	**3,548.42**	—	**5**
Mo	**3,548.439**	**2*h***	**3*h***
Co	**3,548.444**	**6**	**18**
Tm	**3,548.48**	**20**	**8**
Eu	**3,548.51**	**1*h***	**4*w***
U	**3,548.62**	—	**8**
Zr	**3,548.64**	**2**	**3**
Ir	**3,548.65**	**1*h***	**25**
Sr	**3,548.66**	—	**2**
Dy	**3,548.726**	**2**	**3**
Mo	**3,548.730**	**3**	**2**
Cr	**3,548.731**	**4*wh***	**2*wh***
Gd	3,549.36	125	125
Zr	3,549.51	10	10
Rh	3,549.54	50	150
Zr	3,549.74	3	15
Dy	3,550.23	100	200
Zr	3,550.46	4	35
Co	3,550.59	—	200
Zr	3,551.95	40	30
Pd	3,553.08	15*wh*	100*R*
Fe	3,553.74	100	100

	λ	I J	I O
Fe	3,554.93	300	400
Nd	3,555.72	15	100
Fe	3,556.88	150	300
Fe	*3,558.52*	*300*	*400*
Ce	*3,560.80*	*2*	*300*
Co	*3,560.89*	*25*	*200*
Tb	*3,561.74*	*200*	*200*
Dy	*3,563.15*	*100*	*200*
Fe	*3,565.38*	*300*	*400*
Zr	*3,566.10*	*10*	*20*
Ni	*3,566.37*	*100wh*	*2000R*
Zr	*3,568.87*	—	*12*

Zr 3,601.193 *J* 15* *O* 400

	λ	I J	I O
La	**3,601.05**	**15**	**5**
Th	**3,601.04**	**10**	**8**
Pr	**3,601.02**	**2**	**10**
Gd	**3,600.97**	**30**	**30**
Ho	**3,600.95**	**10**	**6**
Nd	**3,600.91**	**10**	**20**
Th	**3,600.83**	**1**	**3**
Co	**3,600.81**	—	**4**
Tb	**3,600.80**	—	**8**
Yb	**3,600.76**	**320**	**10**
Rh	**3,600.754**	**2**	**8**
Pr	**3,600.760**	**5**	**25**
Er	**3,600.742**	**20*w***	**30*w***
Mo	**3,600.737**	**4**	**2**
Y	**3,600.734**	**300**	**100**
Cs	**3,600.73**	**(10)**	—
Ho	**3,600.73**	—	**6**
Ta	**3,600.70**	**1*h***	**2**
Ce	**3,600.58**	**2**	**15**
U	**3,600.49**	**2**	**1**
Tb	**3,600.44**	**50**	**8**
Th	**3,600.43**	**1**	**4**
Yb	**3,600.39**	**2**	**2**
Dy	**3,600.34**	**30**	**20**
U	**3,600.293**	—	**10**
W	**3,600.287**	**3**	**4**
Mo	**3,600.28**	**5**	**5**
Mo	**3,600.20**	**3**	**3**
Nd	**3,600.12**	**10**	**15**
Hf	**3,600.05**	**8*h***	**2**
Tb	**3,600.04**	**3**	**15**
Ti	**3,601.93**	**3**	**7**
Sm	**3,601.25**	**1**	**4**
Mn	**3,601.27**	**15*h***	**15**
Nd	**3,601.32**	**2**	**10**
Ti	**3,601.38**	—	**15**
Dy	**3,601.39**	**1*h***	**3**
Ir	**3,601.40**	**8**	**30**
Ru	**3,601.48**	**5**	**2**
Tb	**3,601.50**	—	**8**
Cr	**3,601.67**	**30**	**50**
Sm	**3,601.69**	**5**	**5**
Eu	**3,601.73**	—	**2*w***
Tb	**3,601.75**	**3**	**15**
W	**3,601.798**	**4**	**5**
Os	**3,601.83**	**20**	**60**
Mo	**3,601.84**	**3**	**1**
Ho	**3,601.87**	—	**6**
Pr	**3,601.916**	—	**3**
Y	**3,601.921**	**60**	**18**
Sm	**3,601.984**	**1**	**2**
Th	**3,601.984**	**2**	**2**
Cu	**3,602.03**	**25*W***	**50**
Ir	**3,602.06**	—	**3**
Co	**3,602.084**	**35**	**200**
Fe	**3,602.085**	**5**	**20**
Ni	**3,602.28**	**15**	**150**
Cr	3,605.33	400*R*	500*R*
Fe	3,605.46	150	300
Dy	3,606.13	100	200
Fe	3,606.68	150	200

	λ	I	
		J	O
Os	3,598.11	30	300
Ni	3,597.70	50*h*	1000*r*
Rh	3,597.15	100	200
Rh	3,596.19	50	200
Dy	3,595.05	100	200
Co	3,594.87	—	200*w*
Cr	3,593.49	400*R*	500*R*
Dy	3,591.42	100	200
Nd	3,590.35	300*W*	400*W*
Zr	*3,586.29*	*3*	*20*
Fe	*3,581.19*	*600r*	*1000R*
Cr	*3,578.69*	*400r*	*500R*
Fe	3,608.86	400	500
Ni	3,609.31	15	200
Pd	3,609.55	700*R*	1000*R*
Ni	3,610.46	—	1000*R*
Cd	3,610.51	500	1000
Cs	3,611.52	—	200
Zr	3,611.89	40	15
Rh	3,612.47	50	200
Ni	*3,612.74*	*50h*	*400*
Cd	*3,612.87*	*500*	*800*
Zr	*3,613.10*	*40*	*40*
Zr	*3,614.77*	*80*	*40*
Fe	*3,618.77*	*400*	*400*
Ni	*3,691.39*	*150h*	*2000R*

Zr 4,687.80 *O* 125

	λ	J	O
Nb	4,687.78	5	5
W	4,687.65	1	12
Ce	4,687.64	—	2
Pr	4,687.23	—	3
Sm	4,687.18	—	100
U	4,686.922	—	8
Ti	4,686.921	2	8
V	4,686.920	12	15
Eu	4,686.85	—	3*W*
Ce	4,686.81	—	2
Eu	4,686.66	—	2*W*
Th	4,686.59	1	2
Os	4,686.49	—	2
Tb	4,686.42	—	3
Gd	4,686.40	5*h*	10
W	4,686.383	—	8
Hf	4,686.38	—	6
Ni	4,686.22	1	200
Ta	4,685.27	2	80
Eu	4,685.25	—	50*W*
Pr	4,684.94	10*w*	125
Ta	4,684.87	2	100
Ru	4,684.02	—	100
Gd	4,683.34	50	100
Co	4,682.38	—	500
Y	4,682.32	100	60
Ti	4,681.92	100	200
Pr	4,687.81	1	50
Er	4,687.83	—	3
Re	4,687.86	—	15
Sm	4,688.13	—	4
Gd	4,688.14	10	20
Ho	4,688.20	2	2
Mo	4,688.22	30	25
Ce	4,688.229	—	2
Eu	4,688.23	2	100
Hf	4,688.39	4	30
Ti	4,688.394	3	10
Zr	4,688.45	—	50
Co	4,688.48	—	20
Eu	4,688.50	1	3*w*
Yb	4,688.51	5*h*	1
Nd	4,688.557	1	5
Pr	4,688.559	—	4
Tb	4,688.63	—	15
Er	4,688.63	2	5
La	4,688.66	20	6
Sm	4,688.73	—	50
Ta	4,688.84	—	40
Ce	4,688.89	—	6
Gd	4,688.91	—	5
Pr	4,689.04	—	3
U	4,689.07	40	30
Tb	4,689.10	—	3
Nb	4,689.16	2	3

	λ	*I*	
		J	*O*
Ta	4,681.87	50	200
Ru	4,681.79	—	100
Cr	4,680.87	8	60
W	4,680.52	40	150
Zn	4,680.14	200*h*	300*w*
Fe	4,678.85	100	150
Cd	4,678.16	200*W*	200*W*
Sm	4,676.91	50	100
Rh	4,675.03	50	100
Y	4,674.85	100	80
Cu	4,675.76	30*W*	200
Sm	4,674.59	40	80
Nb	*4,672.09*	*100*	*150*
Pr	*4,672.08*	*25w*	*100*
Ta	*4,669.14*	*15*	*300*
La	*4,668.91*	*300r*	*200r*
Na	*4,668.60*	*100*	*200*
Ag	*4,668.48*	*70*	*200*
Fe	*4,668.14*	*10*	*125*
Ti	*4,667.59*	*8*	*150*
Fe	*4,667.46*	*20*	*150*
Co	*4,663.41*	—	*700W*
La	*4,662.51*	*200*	*150*
W	*4,659.87*	*70*	*200*

	λ	*I*	
		J	*O*
Th	**4,689.17**	**6**	**10**
Cr	4,689.37	35	80
Ti	4,691.34	25	125
Fe	4,691.41	10	80
Ba	4,691.62	40	100
Ta	4,691.90	5*h*	400
Os	4,692.06	3	80
La	4,692.50	300	200
Eu	4,692.63	2	60*w*
Co	4,693.21	25	500
Ta	4,693.35	3	150
Pr	4,695.77	2	60
Gd	4,697.488	4	100
Cu	4,697.490	5*w*	50*w*
Eu	4,698.14	2	300
Co	4,698.38	8	300
Cr	4,698.46	12	60
Ti	4,698.76	20	100
La	4,699.63	200*r*	200*r*
Mn	4,701.16	5	100
Ta	4,701.32	2*h*	150
Ni	4,701.34	—	100
Ni	4,701.54	—	150
Tb	4,702.42	—	80
La	*4,703.28*	*300r*	*200r*
Ni	*4,703.81*	—	*200*
Sm	*4,704.41*	—	*200*
Cu	*4,704.60*	*50*	*200*
Ta	*4,706.09*	*2*	*200*
Mo	*4,707.22*	*125*	*125*
Cr	*4,708.04*	*150*	*200*
Ru	*4,709.48*	*80*	*150*
Mn	*4,709.71*	*15*	*150*
Zr	*4,710.07*	—	*60*
*N**			

Zr 4,739.48 *O* 100

Ta	**4,739.44**	—	**2**
Pr	**4,739.32**	—	**3**
Rh	**4,739.22**	**3**	**15**
U	**4,739.193**	—	**8**
Tb	**4,739.19**	—	**2**
Eu	**4,739.17**	—	**80**
Mn	**4,739.11**	**15**	**150**
Ce	**4,739.53**	—	**10**
Sm	**4,739.56**	—	**10**
Mg	**4,739.59**	—	**5**
Pt	**4,739.76**	—	**5**
La	**4,739.79**	**8**	**4**
Hf	**4,739.81**	—	**12**
Pr	**4,739.886**	—	**4**

	λ	I	
		J	O
Pr	**4,738.62**	—	**8***d*
Hf	**4,738.57**	**3**	**20**
Dy	**4,738.50**	**2**	**2**
U	**4,738.44**	—	**3**
Ru	**4,738.40**	—	**10**
Os	**4,738.351**	—	**20**
Ta	**4,738.347**	**2**	**5**
Mn	**4,738.29**	**5**	**12**
W	**4,738.16**	—	**8**
Gd	**4,738.131**	**50***h*	**50**
Hf	**4,738.13**	—	**4**
Os	**4,738.04**	—	**12**
Co	4,737.77	1	150
Sc	4,737.64	60*h*	100
Cr	4,737.35	80	200
*bh*Zr	4,736.9	—	60
Fe	4,736.78	50	125
Pr	4,736.69	1	125
Mo	4,736.64	10	10
Eu	4,736.61	—	60
Gd	4,735.76	150	150
Co	4,734.83	—	150
Gd	4,734.44	40	100
Pr	4,734.18	8*w*	100*w*
Sc	4,734.09	60*h*	100
Tm	4,733.32	5	80
Gd	4,732.61	300	300
Ni	4,732.46	—	100
Zr	4,732.33	—	25
Ni	4,731.81	2	100
Mo	4,731.44	100	100
Ru	4,731.33	—	60
Cr	4,730.71	50	100
Pr	4,730.69	1	60
Ta	4,730.12	5	100
Sc	4,729.25	50*h*	100
Ir	4,728.86	3	150
Sc	4,728.77	25	50
Gd	4,728.57	100	150
Sm	4,728.44	—	150
La	4,728.42	300	400*r*
Co	4,727.94	—	300
Mn	4,727.48	20	150
Cr	4,727.15	20	80
Ba	4,726.45	30	80
Sm	4,726.03	—	100

	λ	I	
		J	O
Nd	**4,739.888**	—	**3**
Gd	**4,739.91**	—	**10***h*
Tb	**4,739.93**	—	**8**
Eu	**4,739.96**	**2**	**2**
Ta	**4,740.163**	**100**	**100***R*
Ni	**4,740.165**	—	**15**
Ir	**4,740.26**	—	**2***wh*
La	**4,740.277**	**300**	**150**
U	**4,740.285**	**10**	**3**
Ru	**4,740.33**	—	**7**
Mo	**4,740.36**	**5**	**5**
*bh***Zr**	**4,740.5**	—	**8**
Th	**4,740.517**	**15**	**20**
Eu	**4,740.524**	**2**	**500**
Nb	**4,740.61**	**3**	**3**
Dy	**4,740.93**	**2**	**3**
Sc	4,741.02	60*h*	100
Sm	4,741.73	—	80
Ti	4,742.79	40	100
La	4,743.08	300	300*r*
Gd	4,743.65	300	300
Sc	4,743.91	60*h*	100
Os	4,743.89	—	60
Pr	4,744.16	—	80
Pr	4,744.92	10*w*	100
Cr	4,745.31	2	80
Sm	4,745.67	—	250
Co	4,746.11	—	100
Pr	4,746.93	25*w*	100
Re	4,748.38	—	50*w*
La	4,748.73	200	100
Co	4,749.68	100*h*	500
Nb	4,749.70	50	100
Sm	4,750.72	—	60
Cr	4,752.08	40	100
Ni	4,752.43	—	150
Tb	4,752.52	80	100
Sc	4,753.15	40	80
Mn	4,754.04	60	400
Co	*4,754.36*	*2*	*200*
Ni	*4,754.77*	—	*100*
Sm	*4,755.37*	—	*100*
Cr	*4,756.11*	*100*	*300*
Ta	*4,756.51*	*10*	*150*
Ni	*4,756.52*	*3*	*250*
Co	*4,756.72*	—	*100*

	λ	I			λ	I	
		J	O			J	O
Cr	*4,724.42*	*10*	*125*	*Ru*	*4,757.84*	—	*125*
Gd	*4,723.73*	*200*	*100*	*Pr*	*4,757.94*	*5w*	*100w*
Cr	*4,723.10*	*8*	*125*	*Ti*	*4,758.12*	*60*	*125*
Ta	*4,722.88*	—	*200*				
Bi	*4,722.55*	*100*	*1000*	*Ti*	*4,759.28*	*8*	*100*
				Mo	*4,760.19*	*125*	*125*
Zn	*4,722.16*	*300h*	*400w*	*Sm*	*4,760.26*	—	*150*
La	*4,719.95*	*300*	*200r*	*Mn*	*4,762.38*	*40*	*100*
Sm	*4,719.84*	—	*125*	*Ni*	*4,762.63*	*1h*	*150*
Cr	*4,718.43*	*150*	*200*				
Sm	*4,718.35*	—	*100*	*Zr*	*4,762.78*	—	*10*
				Ni	*4,763.95*	*1*	*150*
Sm	*4,717.74*	—	*100*	*Cr*	*4,764.29*	*35*	*200*
La	*4,716.44*	*200*	*100*	*La*	*4,766.89*	*10*	*100*
Sm	*4,716.11*	—	*150*	*Co*	*4,767.14*	—	*100*
Ni	*4,715.78*	*2*	*200*				
Sm	*4,715.27*	—	*100*	*Gd*	*4,767.25*	*25*	*100*
				Cr	*4,767.86*	*8*	*100*
Ni	*4,714.42*	*8*	*1000*	*Co*	*4,768.08*	*10*	*300*
Eu	*4,713.61*	*2*	*400*				
Sm	*4,713.07*	—	*100*				
La	*4,712.93*	*150*	*100r*				
Zr	*4,711.92*	—	*15*				
Ti	*4,710.19*	*25*	*100*				

Zr 4,772.31 *O* 100

Pr	**4,772.30**	—	**5**	**Pt**	**4,772.32**	—	**5**
Nd	**4,772.26**	—	**8**	**Sm**	**4,772.47**	—	**2**
Tb	**4,772.03**	—	**3**	**Pr**	**4,772.53**	—	**4**
Dy	**4,771.93**	**2**	**8**	**W**	**4,772.54**	—	**12**
Pr	**4,771.87**	—	**10**	**V**	**4,772.58**	**3**	**3**
Nb	**4,771.85**	**5*h***	**3*h***	**Eu**	**4,772.66**	—	**5*w***
Re	**4,771.80**	—	**5**	**U**	**4,772.70**	**18**	**6**
Nb	**4,771.73**	—	**4**	**Nb**	**4,772.81**	**1**	**3**
Eu	**4,771.72**	**1**	**3*w***	**Fe**	**4,772.82**	**4**	**10**
Tb	**4,771.68**	—	**2**	**Nd**	**4,772.88**	—	**10**
Mn	**4,771.672**	—	**8**	**Tb**	**4,772.97**	—	**3**
Sm	**4,771.667**	—	**8**	**Eu**	**4,773.01**	—	**2*W***
Cr	**4,771.61**	—	**18**	**Gd**	**4,773.09**	—	**20**
Rh	**4,771.56**	—	**2**	**Ru**	**4,773.15**	—	**15**
Sc	**4,771.44**	**3**	**2**	**Nb**	**4,773.25**	**10**	**10**
Tb	**4,771.30**	—	**5**	**Mo**	**4,773.29**	**10**	**10**
*bh***Zr**	**4,771.3**	—	**4**	**Ni**	**4,773.41**	—	**15**
Co	**4,771.11**	—	**500*w***	**Er**	**4,773.43**	—	**2**
Ti	**4,771.10**	**1**	**10**	**Mo**	**4,773.437**	**20**	**20**
Nd	**4,771.06**	—	**2**	**U**	**4,773.440**	—	**4**
Mo	**4,770.87**	**8**	**8**	**Pr**	**4,773.52**	—	**10**
Dy	**4,770.81**	**2**	**3**	**Dy**	**4,773.69**	**2**	**2**
Rh	**4,770.79**	**15**	**2**	**Hf**	**4,773.71**	**4**	**25**

	λ	I	
		J	O
Eu	**4,770.78**	—	**15**
W	**4,770.77**	—	**6**
Ta	4,768.98	5	150
*P**			

	λ	I	
		J	O
Mn	4,774.10	—	50
Sm	4,774.14	—	100
Co	4,776.32	—	300
Sm	4,777.84	—	100
Ir	4,778.16	3	50
Co	4,778.25	—	100
Sc	4,779.35	40	80
Co	4,780.01	500	500*w*
Ta	4,780.93	200	50
Co	4,781.43	2*h*	400
Sm	4,781.84	—	60
Yb	4,781.88	5	50
Gd	4,781.93	50	200
Sm	4,783.10	—	150
Pr	4,783.35	10*w*	125
Mn	4,783.42	60	400
Gd	4,784.64	50	100
Zr	4,784.92	—	40
Co	4,785.07	—	50
Lu	4,785.42	200	100
Sm	4,785.87	—	100
Ni	4,786.54	2	300*W*
Yb	4,786.60	200	50
Fe	4,786.81	—	150
Pd	*4,788.17*	*4h*	*200h*
Zr	*4,788.67*	—	*10*
Cr	*4,789.38*	*100*	*300*
Fe	*4,789.65*	—	*100*
Cr	*4,790.34*	*1*	*100*
Fe	*4,791.25*	*200R*	*200*
Re	*4,791.42*	—	*200w*
Sm	*4,791.58*	—	*150*
Gd	*4,791.60*	—	*150*
Cr	*4,792.51*	*40*	*200*
Au	*4,793.60*	*60*	*200W*
Co	*4,792.86*	*5*	*600W*
Os	*4,793.99*	*6*	*300*
Co	*4,795.85*	—	*100*
Cr	*4,796.37*	*1h*	*125*
Co	*4,796.37*	—	*100*
Cd	*4,799.92*	*300w*t	*300w*
Cr	*4,801.03*	*70*	*200*
Gd	*4,801.08*	*200*	*200*
Gd	*4,802.58*	*40h*	*100*

TABLE XVII
COINCIDENCE OF ANALYTICAL LINES OF THE ELEMENTS OF RARE EARTHS

Dy 66 162.50

t_0 — t_1 —

I.	II.	III.	IV.	V.
6.82	—	—	—	—

λ	I		eV
	J	O	
II 3,944.692	150	300	
I 4,211.719	15	200	

Dy 3,944.692 *J* 150 *O* 300

	λ	*I*			λ	*I*	
		J	*O*			*J*	*O*
Yt	3,944.687	—	3	Re	3,944.73	—	30
U	3,944.621	2	8	Sm	3,944.74	5	5
Pr	3,944.617	3	8	Fe	3,944.75	1	4
Eu	3,944.59	—	5*W*	Ru	3,944.78	—	4
Sb	3,944.58	8	—	W	3,944.80	6	7
Er	3,944.423	—	12	Ce	3,944.83	—	5
Nd	3,944.421	—	20	Fe	3,944.896	8	15
Ir	3,944.37	4	20	Pr	3,944.899	12	30
Re	3,944.35	—	15	Ce	3,944.92	—	6
F	3,944.33	(20)	—	Co	3,944.95	2	50*h*
Ga	3,944.29	2	—	O	3,945.04	(20)	—
Ar	3,944.27	(50)	—	Hg	3,945.09	(100)	—
Th	3,944.25	1	3	Fe	3,945.13	10	30
Tb	3,944.20	—	6	Eu	3,945.16	—	5
Ru	3,944.19	4	10	V	3,945.17	—	2
La	3,944.15	2	—	W	3,945.21	6	5
Pr	3,944.14	5	9	Mo	3,945.25	6	10
U	3,944.130	15	8	Co	3,945.33	15	200
Ni	3,944.126	—	5*wh*	Hf	3,945.36	6*h*	8
Ce	3,944.09	—	8	Ce	3,945.364	—	2
Se	3,944.07	(20)	—	Tb	3,945.39	—	8
Ir	3,944.06	—	2	Pr	3,945.42	4	12
Al	3,944.03	1000	2000	Kr	3,945.48	(5*h*)	—
Rh	3,944.019	4	5	Cr	3,945.49	15	50
Re	3,944.016	—	15	Ce	3,945.507	1	3
Er	3,944.014	—	2	Th	3,945.515	15	20
Eu	3,944.01	—	8*h*	J	3,945.52	(2)	—
Ce	3,943.89	15	40	Gd	3,945.54	150	200*W*
U	3,943.82	5	35	Cu	3,945.570	2	—
Pb	3,943.80	5	—	Ru	3,945.572	100	50
Pr	3,943.75	8	10	Ba	3,945.61	—	5*h*
Th	3,943.69	10	0	F	3,945.65	(10)	—
Nb	3,943.67	50	20	Pr	3,945.66	10	30
V	3,943.664	18	50	Eu	3,945.67	1	8*w*
Tb	3,943.66	—	8	U	3,945.73	3	2
Gd	3,943.63	20	20	Gd	3,945.74	—	3
Sm	3,943.62	3	8	Ce	3,945.82	—	2
Cr	3,943.61	4	18	Th	3,945.825	8	10
Eu	3,943.60	—	3	Re	3,945.91	—	40
Fe	3,943.59	—	2*h*	Ce	3,945.92	—	2
Xe	3,943.579	(10)	—	As	3,945.93	10	—
Mo	3,943.51	4	5	Dy	3,945.94	—	3
U	3,943.498	10	6	Cr	3,945.97	7	50
Ce	3,943.497	1	6	W	3,945.98	5	1
As	3,943.47	3	—	Br	3,946.01	(3)	—
Pr	3,943.37	3	5	Hf	3,946.012	4*h*	—
Fe	3,943.35	8	40	Ce	3,946.048	—	2

	λ	I J	I O		λ	I J	I O
Gd	**3,943.25**	**40**	**40**	**Ca**	**3,946.05**	—	**2**
Sm	**3,943.23**	**25**	**40**	**Ar**	**3,946.10**	**(25)**	—
Ir	**3,943.20**	—	**6**	**Th**	**3,946.15**	**20**	**20**
Er	**3,943.19**	—	**10**	**Ce**	**3,946.17**	—	**6**
				Eu	**3,946.18**	—	**4*w***

Dy 4,211.719 *J* 15 *O* 200

	λ	J	O		λ	J	O
Er	**4,211.718**	**5**	**30**	**Tb**	**4,211.72**	**2**	**25**
U	**4,211.68**	**10**	**10**	**Ti**	**4,211.728**	**6**	**30**
Ca	**4,211.64**	**2**	—	**Ho**	**4,211.73**	**3**	**5**
U	**4,211.62**	—	**18**	**Mn**	**4,211.75**	**20**	**30**
Ce	**4,211.58**	—	**3**	**Se**	**4,211.83**	**(200)**	—
Th	**4,211.52**	**8**	**8**	**Os**	**4,211.855**	**50**	**150**
Cr	**4,211.35**	**30**	**100**	**Pr**	**4,211.858**	**25*d***	**50*d***
Zr	**4,211.33**	—	**12**	**Cu**	**4,211.86**	**6**	**1*h***
Te	**4,211.32**	**(15)**	—	**Zr**	**4,211.87**	**15**	**18**
U	**4,211.31**	**6**	**10**	**Ce**	**4,211.91**	**2**	**1**
Nd	**4,211.29**	**15**	**30**	**Gd**	**4,212.02**	**50**	**150**
Dy	**4,211.25**	**2**	**4**	**Nb**	**4,212.04**	**3**	**4**
Pr	**4,211.24**	**2**	**5**	**Ru**	**4,212.06**	**80**	**125**
Ho	**4,211.24**	**1*h***	**3**	**Zr**	**4,212.16**	—	**3**
*bh***Sr**	**4,211.2**	—	**3**	**Hg**	**4,212.22**	**(30)**	—
Tb	**4,211.16**	**1*h***	**2**	**Tb**	**4,212.23**	—	**3*w***
Rh	**4,211.14**	**200**	**15**	**U**	**4,212.26**	**10**	**12**
Mo	**4,211.02**	**12**	**15**	**Sc**	**4,212.34**	—	**3**
Nd	**4,210.98**	**5*h***	**25**	**C**	**4,212.36**	**(5)**	—
Ag	**4,210.94**	**30*h***	**200*h***	**Ce**	**4,212.39**	**2**	**1**
Th	**4,210.92**	**3**	**8**	**Na**	**4,212.40**	—	**3**
U	**4,210.87**	**12**	**12**	**U**	**4,212.48**	**3**	**5**
Th	**4,210.77**	**1**	**3**	**Sc**	**4,212.49**	—	**4**
Cr	**4,210.75**	—	**12**	**Hg**	**4,212.53**	**(50)**	—
Eu	**4,210.67**	**1**	**2**	**Nb**	**4,212.534**	**3**	**2**
				Tb	**4,212.54**	—	**5**
Kr	**4,210.67**	**(25*wh*)**	—	**Se**	**4,212.58**	**(200)**	—
				Zr	**4,212.62**	—	**15**
				Cr	**4,212.660**	**8**	**80**
				Si	**4,212.66**	**2**	—
				Ag	**4,212.68**	**20*h***	**150*h***
				Pr	**4,212.70**	**2**	**5**
				U	**4,212.73**	**44**	**5**
				Nb	**4,212.75**	**1**	**2**

Er $^{68}_{167.27}$

t_0 —　　　　　　　　　　　　t_1 —

I.	II.	III.	IV.	V.
—	—	—	—	—

λ	*I*		eV
	J	*O*	
I 4,007.97	7	35	
4,151.11	3	20	

Er 4,007.97 *J* 7 *O* 35

	λ	I	
		J	*O*
Ho	4,007.96	3	4
Co	4,007.94	—	3
U	4,007.93	3	8
Se	4,007.90	(150)	—
S	4,007.78	(5)	—
Pr	4,007.78	3	8
Dy	4,007.77	2	12
Tb	4,007.75	—	3
U	4,007.689	4	2
Eu	4,007.687	—	5*d*
La	4,007.66	2*h*	3
In	4,007.61	(10)	—
Zr	4,007.60	1	25
Ce	4,007.59	4	15
In	4,007.54	(15)	—
Ru	4,007.53	10	20
Sm	4,007.49	25	50
Pd	4,007.470	—	5*h*
U	4,007.469	2	2
Ce	4,007.451	—	4*h*
Mo	4,007.45	5	4
Nd	4,007.43	20	20
Yb	4,007.36	—	5
Hf	4,007.35	4*h*	5
Br	4,007.33	(10)	—
Fe	4,007.27	50	80
Ta	4,007.23	2	4
Ti	4,007.19	1	15
Dy	4,007.14	—	6
Mn	4,007.04	5	10
Th	4,007.03	20	20
Ce	4,007.02	—	2
W	4,007.00	6	—
Nb	4,006.99	5	—
Sm	4,006.98	—	3
Gd	4,006.97	—	3
Mo	4,006.95	5	—
Ta	4,006.835	20	30
Sm	4,006.834	5	8
Cs	4,006.77	(10)	—
Nd	4,006.761	2*d*	5*d*
Fe	4,006.761	2	7
Pr	4,006.704	5	10
Mo	4,006.70	5	5
Cd	4,006.68	5	—
Fe	4,006.63	15	20
Ru	4,006.60	15	25
Eu	4,007.98	—	6
Mn	4,008.02	5	15
Ir	4,008.052	2	12
Mo	4,008.054	5	4
Ti	4,008.062	7	50
Kr	4,008.08	(25)	—
Sm	4,008.09	3	10
V	4,008.17	10	2
Er	4,008.18	1	8
Th	4,008.22	8	10
Ru	4,008.27	20	20
Nb	4,008.28	10	5
Sm	4,008.330	10*h*	8
Gd	4,008.331	10	15
Nd	4,008.42	3	5
Ce	4,008.45	—	6
Hf	4,008.46	8	5
Kr	4,008.48	(10*wh*)	—
Dy	4,008.49	—	5
Ce	4,008.66	—	8
Mo	4,008.67	20	—
U	4,008.70	1	2
Pr	4,008.71	50	150
W	4,008.755	45	45
Nd	4,008.754	10	12
Br	4,008.76	(20)	—
Eu	4,008.872	2	8*W*
Fe	4,008.873	1	5
U	4,008.918	—	8
Gd	4,008.922	3	20
Ti	4,008.93	35	80
Ir	4,008.97	—	5
Ce	4,009.06	—	12
Th	4,009.07	8	10
Er	4,009.165	1	15
U	4,009.170	15	8
Tb	4,009.19	—	5
Gd	4,009.21	2	50
Pr	4,009.24	2	5
He	4,009.270	(10)	—
Tb	4,009.27	—	2
Mo	4,009.37	25	20
Nd	4,009.38	5	8
Zr	4,009.387	—	3
S	4,009.39	(3)	—
U	4,009.41	8	4
Tb	4,009.54	1	10

	λ	I	
		J	O
Sm	4,006.58	8	8
Eu	4,006.56	—	6
Cs	4,006.54	(30)	—
Te	4,006.50	(100)	—
U	4,006.395	5	3
Nd	4,006.386	3	5
Th	4,006.386	10	10
Th	4,009.546	1	3
Ce	4,009.55	1*h*	2
Al	4,009.58	(4)	—

Er 4,151.11 *J* 3 *O* 20

	λ	J	O
Re	4,151.05	—	2*h*
Pr	4,151.01	3*d*	6*d*
Zr	4,150.97	10	25
Ti	4,150.96	15	35
Ce	4,150.91	3	18
Tb	4,150.88	—	4
Mo	4,150.85	20*h*	—
U	4,150.82	3	3
Ho	4,150.8	—	2
Re	4,150.793	—	2
Nd	4,150.788	3	3
Er	4,150.78	2	7*w*
B	4,150.71	2	—
Os	4,150.70	1*h*	8
V	4,150.672	8	10
Ne	4,150.67	(30)	—
Gd	4,150.61	—	5
Ti	4,150.55	1	10
Re	4,150.53	—	10
Tb	4,150.53	1*h*	12
Co	4,150.43	2	3
U	4,150.41	8	8
Ce	4,150.40	1	2
Ni	4,150.37	—	3
Ru	4,150.30	9	20
Fe	4,150.26	2	50
La	4,150.24	—	2
Nb	4,150.12	20	15
Tm	4,150.08	—	3
Pr	4,150.04	2	5
U	4,150.03	2	5
Th	4,149.99	12	10
Re	4,149.97	—	40*w*
Tb	4,151.13	1	6
Ho	4,151.14	4	4
P	4,151.19	(15)	—
Sm	4,151.21	—	5*d*
U	4,151.24	1	6
Ag	4,151.26	2*h*	—
Cs	4,151.27	(20)	—
Ir	4,151.35	2	30
Pd	4,151.36	—	2
Mo	4,151.37	2	4
N	4,151.46	(1000)	—
Eu	4,151.55	1*h*	4
U	4,151.57	3*h*	2
Gd	4,151.59	—	3
Eu	4,151.66	—	4
Nd	4,151.68	10	5
Ce	4,151.72	—	3
Mo	4,151.88	15	20
Ir	4,151.951	—	2
La	4,151.955	300	200
Fe	4,151.955	1	4
Ce	4,151.97	8	30
Nb	4,152.03	8	10
Sm	4,152.07	3	3
Pd	4,152.12	—	2
Fe	4,152.17	5	70

Eu $^{63}_{151.96}$

t_0 1,200°C t_1 —

I.	II.	III.	IV.	V.
5.67	11.24	—	—	—

λ	*I*		eV
	J	*O*	
II 3,907.11	500*R*	1000*RW*	
II 4,129.74	50*R*	150*R*	
II 4,205.05	50	200*R*	

Eu 3,907.11 *J* 500*R* *O* 1000*RW*

	λ	I			λ	I	
		J	O			J	O
U	3,907.02	1	5	Sm	3,907.124	15	10
Mo	3,906.98	5	5	Gd	3,907.125	100	100*W*
Cs	3,906.93	(20)	—	Sn	3,907.15	4	—
Ce	3,906.924	3	8	V	3,907.17	2	2*h*
Mo	3,906.916	5	5	J	3,907.19	(18)	—
Nb	3,906.91	5	5	W	3,907.20	5	6
Hf	3,906.89	3*d*	3	Ce	3,907.29	6	35
Sm	3,906.805	5	6	Pr	3,907.30	3	8
Th	3,906.796	—	8	Th	3,907.34	3	8
Fe	3,906.751	10	10	O	3,907.44	(18)	—
V	3,906.748	20	50	Ce	3,907.445	2	6
Tb	3,906.53	—	4	Fe	3,907.47	6	15
Fe	3,906.482	200	300	Sc	3,907.48	25	125
Mo	3,906.480	10	5	V	3,907.52	2	—
Ce	3,906.45	2	8	U	3,907.56	8	6
Hg	3,906.41	15	25	Ag	3,907.59	2*h*	3
Er	3,906.32	12	25	Ce	3,907.64	1	2
Co	3,906.294	—	150	Au	3,907.650	3	5
Pt	3,906.291	1*h*	2	Tb	3,907.65	—	3
Ho	3,906.26	3	3	Fe	3,907.67	1	2
Kr	3,906.25	(150)	—	Sb	3,907.74	(8)	—
Zr	3,906.15	—	3	Ir	3,907.75	—	4
Ce	3,906.104	—	2	Cr	3,907.78	10	30
Nd	3,906.096	10	15	Tb	3,907.79	3	3
Pr	3,906.093	8	6	Nb	3,907.84	12	20
Fe	3,906.04	2*w*	2*w*	Ir	3,907.90	2	10
Ba	3,906.01	2	4	Xe	3,907.91	(50*h*)	—
Ru	3,905.99	3*h*	6	Tb	3,907.912	—	5
W	3,905.97	7	8	Fe	3,907.94	60	100
Dy	3,905.95	1	6	Ce	3,907.96	—	2
Ce	3,905.92	—	3	Pr	3,908.03	50	100
U	3,905.90	—	8	Tb	3,908.08	—	20
Nd	3,905.89	30	40	Ce	3,908.09	1	8
Yb	3,905.88	10	2	Gd	3,908.15	3*h*	5
Xe	3,905.85	(5)	—	Mn	3,908.16	3	3
Ce	3,905.85	—	3	Re	3,908.20	—	25
Tb	3,905.83	—	2	Mo	3,908.25	3	5
Ho	3,905.78	6	30	Sm	3,908.26	—	6
J	3,905.72	(10)	—	V	3,908.32	2*h*	50
Cr	3,905.659	4	2	U	3,908.33	1	10
Eu	3,905.657	—	3	Ce	3,908.41	6	30
Gd	3,905.65	50	50	Er	3,908.42	—	10
Tb	3,905.61	—	10	Pr	3,908.43	60	100
				U	3,908.47	10	8
				Th	3,908.49	2	3
				Ce	3,908.54	3	20
				Nb	3,908.59	5	3

Eu 4,129.74 *J* 50*R* *O* 150*R*

	λ	*I*			λ	*I*	
		J	*O*			*J*	*O*
Ar	4,129.70	(10)	—	Nd	4,129.87	—	2
Mo	4,129.69	4	8	Ce	4,129.92	—	3
U	4,129.66	4	4	Nb	4,129.93	30	15
Eu	4,129.62	—	15	W	4,129.98	6	5
Tm	4,129.54	2	1	Sm	4,129.99	—	10
U	4,129.46	5	6	Fe	4,130.04	3	20
Ho	4,129.44	1	3	Ir	4,130.100	—	15
Er	4,129.433	2	10	Mo	4,130.105	8	10
Nb	4,129.426	20	15	Ce	4,130.12	2	—
Dy	4,129.425	8	20	Tb	4,130.14	—	10
Tb	4,129.42	1*w*	10	V	4,130.145	2	3
Ta	4,129.38	40	200	Cl	4,130.21	(4)	—
Cr	4,129.37	—	10	Ce	4,130.23	4	—
O	4,129.34	(15)	—	Mo	4,130.26	8	—
Sm	4,129.225	12	5	Tb	4,130.35	1	3
Fe	4,129.22	1	5	Gd	4,130.38	10	200
Ce	4,129.18	1	5	Dy	4,130.42	2	12
Ti	4,129.17	7	15	Ca	4,130.43	2	—
Se	4,129.15	(200)	—	Zr	4,130.45	—	4
Pr	4,129.148	15	20	Re	4,130.46	—	10
J	4,129.13	(2)	—	Ne	4,130.51	(20)	—
Dy	4,129.13	2	10	Ba	4,130.66	60*wh*	50*r*
Ca	4,129.12	2	2*h*	Ce	4,130.71	8	25
Ce	4,129.10	—	3	Nd	4,130.72	8	15
Os	4,128.96	3	60	Ir	4,130.73	—	2
Ce	4,128.90	1*h*	3	U	4,130.74	3*h*	—
U	4,128.89	—	4	Pr	4,130.770	20	25
Rh	4,128.87	150	300	P	4,130.77	(30)	—
V	4,128.86	5	9	Mo	4,130.85	15	—
Mo	4,128.83	40	40	Cl	4,130.86	(20)	—
U	4,128.76	2	2	S	4,130.95	(15)	—
Fe	4,128.73	2*h*	2	Si	4,130.96	25*w*	—
Tb	4,128.72	—	4	Xe	4,131.01	(10)	—
Nd	4,128.70	10	10	Dy	4,131.04	4	12
J	4,128.69	(35)	—	Th	4,131.049	5*d*	8*d*
Ar	4,128.65	(20)	—	Os	4,131.049	1	12
Ce	4,128.48	—	3	Ne	4,131.054	(70)	—
Tb	4,128.46	—	3	Ce	4,131.10	8	30
Ir	4,128.43	2	10	Tb	4,131.11	1	8
Cr	4,128.40	6	10	Mn	4,131.12	40	50
				Ru	4,131.22	—	10
Pd	4,128.37	—	5*h*	Ti	4,131.25	2	15
Ce	4,128.36	4	10*s*	Nd	4,131.31	4	8
U	4,128.34	20	18	Ce	4,131.347	—	2
Ho	4,128.32	3	—	U	4,131.353	6	6
Yt	4,128.30	30	150				
				Cr	4,131.36	20	30
Mo	4,128.28	25	50	Mn	4,131.430	—	10
Dy	4,128.24	4	20	Th	4,131.433	8	8

	λ	I J	I O
Pb	4,128.21	5	—
Mn	4,128.14	15*s*	15
Nd	4,128.13	4	8
Sm	4,128.12	2	5
Si	4,128.11	20*w*	—
Eu	4,128.10	—	2
Re	4,128.09	—	6
Ca	4,128.08	4	2
Ne	4,128.072	(30)	—
V	4,128.071	20*r*	30*r*
Ce	4,128.068	3	10

Eu 4,205.05 *J* 50 *O* 200*R*

	λ	I J	I O
Dy	4,205.03	2	7
Eu	4,204.91	—	5
Gd	4,204.84	—	25
Sm	4,204.813	10	8
Mo	4,204.809	20	25
Ce	4,204.74	—	15
Yt	4,204.70	15	15
Ca	4,204.62	2	—
Mo	4,204.61	10	15
Pr	4,204.58	1	5
Os	4,204.56	—	12
Cl	4,204.54	(18)	—
Sc	4,204.538	2*h*	4
Re	4,204.53	—	25*w*
Cr	4,204.47	30	80
W	4,204.41	10	20
Tl	4,204.40	(2)	—
U	4,204.37	10	15
Nd	4,204.35	3*d*	10*d*
Nb	4,204.32	3	4
Kr	4,204.31	(3*wh*)	—
Ce	4,204.30	3	1
V	4,204.200	6	2
Cr	4,204.198	6	50
La	4,204.04	25	200
Cl	4,205.07	(10)	—
In	4,205.08	(50)	—
V	4,205.09	20	5
*bh*Ca	4,205.1	—	6
In	4,205.15	(30)	—
Ce	4,205.16	2*h*	6
Sc	4,205.19	1	10
In	4,205.217	(15)	—
Os	4,205.222	1	9
Tb	4,205.23	—	2
Nd	4,205.25	4	5
Nb	4,205.31	15	15
Sm	4,205.36	4	8
Tl	4,205.4	(2)	—
Xe	4,205.404	(10)	—
Fe	4,205.55	6	50
W	4,205.56	12	7
Nd	4,205.59	15	20
U	4,205.63	—	3
Dy	4,205.64	2	5
Tb	4,205.64	—	3
N	4,205.65	(5*h*)	—
Pr	4,205.72	1	5
Sm	4,205.77	2	4
Ce	4,205.79	—	4
Mo	4,205.81	1	2
Ta	4,205.88	30	100
Ce	4,205.89	1*h*	4
Zr	4,205.919	3	10
Er	4,205.92	—	6
Tm	4,206.00	5	20
Ru	4,206.02	40	100
Br	4,206.07	(2)	—
Hg	4,206.10	(30)	—

Gd $^{64}_{157.25}$

t_0 —

t_1 —

I.	II.	III.	IV.	V.
6.16	—	—	—	—

λ	*I*		eV
	J	*O*	
II 4,098.61	6	25*h*	
II 4,184.264	150	150	

Gd 4,098.61 *J* 6 *O* 25*h*

	λ	*I*	
		J	*O*
Tb	4,098.60	2	2
Ca	4,098.53	3	15
Nd	4,098.51	5	10
W	4,098.49	3	4
Ce	4,098.47	—	4
Pr	4,098.41	12	20
Cl	4,098.40	(12)	—
Sc	4,098.34	—	6
O	4,098.24	(5*h*)	—
Nb	4,098.22	5	5
Fe	4,098.19	40	100
Mo	4,098.183	15	20
Cr	4,098.18	—	20*h*
Nd	4,098.17	5	15
Ce	4,098.15	—	8
Er	4,098.103	—	8
Os	4,098.10	3	9
Gd	4,098.04	—	10
U	4,098.03	12	12
Cr	4,097.96	—	20
Se	4,097.91	(60)	—
Br	4,097.88	(2)	—
Sb	4,097.87	4	—
Yb	4,097.86	8	2
Ru	4,097.79	125	25
Th	4,097.75	—	5
U	4,097.74	1	10
Mo	4,097.71	2	3
Th	4,097.69	3	5
W	4,097.67	5	6
Cr	4,097.65	—	20*h*
Nb	4,097.64	3*h*	4
Ce	4,097.62	—	2
Rh	4,097.52	10	25
Tb	4,097.45	5	20
U	4,097.40	2	1
Th	4,097.339	3	5
Ce	4,097.338	—	4
Ir	4,097.25	—	3
O	4,097.24	(70)	—
Dy	4,097.24	—	2
Hf	4,097.24	10	4
Co	4,097.203	—	2
Bi	4,097.2	3	4
Ta	4,097.19	3*h*	5
Ar	4,097.15	(5)	—
Fe	4,097.11	1	4

	λ	*I*	
		J	*O*
Pr	4,098.65	3	8
Kr	4,098.72	(250)	—
La	4,098.73	4	—
Cl	4,098.74	(8)	—
Mo	4,098.74	20	20
Ne	4,098.74	(50)	—
Tb	4,098.83	—	3
Xe	4,098.89	(50*h*)	—
Pd	4,098.90	—	2
Gd	4,098.91	100	100
Nd	4,098.936	8	8
Th	4,098.939	10	10
Sm	4,098.95	—	8
Ce	4,098.98	1	15
Cr	4,099.02	8	30
W	4,099.026	6	7
Ca	4,099.029	1	2*h*
Eu	4,099.05	—	8*d*
Nb	4,099.07	5	5
Zr	4,099.08	1	3
Tb	4,099.14	1	9
Ti	4,099.17	8	25
Dy	4,099.21	—	4
U	4,099.269	1	20*r*
Pd	4,099.27	4*h*	—
Zr	4,099.31	—	10
Ce	4,099.39	—	6
S	4,099.44	(8)	—
Td	4,099.46	1	25*W*
Ar	4,099.47	(5)	—
Ir	4,099.48	—	4
La	4,099.54	100	100
Kr	4,099.71	(3)	—
Eu	4,099.72	5	12*d*
Ce	4,099.75	2	12
V	4,099.80	12	25
Pr	4,099.88	1	4
Dy	4,099.89	2	6
U	4,099.93	5	5
N	4,099.94	(150)	—
Sm	4,099.95	—	4
He	4,100.00	(2)	—
Ir	4,100.15	3*h*	100
Fe	4,100.17	1	10
Ho	4,100.2	—	4
Pr	4,100.22	10	15
Nd	4,100.24	8	10

	λ	I J	I O		λ	I J	I O
Ru	4,097.03	15	15	Eu	4,100.25	—	2
Cl	4,096.98	(6)	—	Gd	4,100.26	5	15
Fe	4,096.97	—	2	Os	4,100.300	3	60
V	4,096.94	2	7	Ne	4,100.30	(5*h*)	—
Ce	4,096.926	2	—				
Er	4,096.926	—	8				

Gd 4,184.264 *J* 150 *O* 150

	λ	I J	I O		λ	I J	I O
Se	4,184.26	(25)	—	Ca	4,184.28	8	2
Lu	4,184.25	200	100	Tb	4,184.30	2	10
Pr	4,184.24	1	4	Ti	4,184.33	20	8
Mo	4,184.17	4	4	Ce	4,184.38	1	3
U	4,184.15	2	2	Mo	4,184.39	4	4
Os	4,184.13	2	25	Nb	4,184.44	50	20
Se	4,184.08	(10)	—	Kr	4,184.473	(20)	—
Ce	4,184.07	4	3	Ni	4,184.475	—	6
Er	4,184.06	—	4	U	4,184.50	3*h*	1
Te	4,183.99	(30)	—	Ce	4,184.63	—	2
W	4,183.83	5	12	Tb	4,184.66	—	2
Ce	4,183.80	—	3	Fe	4,184.895	80	100
Sm	4,183.76	15	10	Cr	4,184.895	10	35
Dy	4,183.73	8	15	Tb	4,184.95	—	2
W	4,183.67	—	10	Nd	4,184.980	10	15
Gd	4,183.62	—	35	Eu	4,184.985	—	7
Dy	4,183.61	—	8	Er	4,184.99	—	7
U	4,183.60	5	2	Kr	4,185.12	(50)	—
Th	4,183.57	8	8	Pr	4,185.15	2	5
V	4,183.43	20	3				
Nb	4,183.38	10*h*	—				
Sm	4,183.33	2	2				
Zr	4,183.32	1	40				
Ti	4,183.29	7	20				
U	4,183.27	1	10				
Ir	4,183.21	4	40				

Hf $^{72}_{178.49}$

t_0 2,230°C t_1 3,200°C

I.	II.	III.	IV.	V.
5.5	—	—	—	—

λ	I		eV
	J	O	
II 3,134.72	125	80	—
II 4.093.161	20	25	—

Hf 3,134.72 *J* 125 *O* 80

	λ	I *J*	I *O*		λ	I *J*	I *O*
Eu	3,134.70	1	3*W*	O	3,134.79	(100)	—
U	3,134.69	2	5	Cs	3,134.8	(4)	—
Ti	3,134.65	—	2	Ru	3,134.802	100	10
Ce	3,134.60	—	4	Nd	3,134.897	30	40
Ca	3,134.56	2	1	Rb	3,134.90	(10)	—
Th	3,134.43	12	10	Cr	3,134.92	—	10
Se	3,134.42	(70)	—	V	3,134.93	150*r*	30
Fe	3,134.405	1*h*	3*h*	Mg	3,135.0	2*h*	4
Ho	3,134.40	4*h*	6	Ag	3,135.02	8*h*	—
Nb	3,134.34	15	2	Gd	3,135.04	2	5
O	3,134.32	(10)	—	Re	3,135.068	—	15
Cr	3,134.31	50	3	Ti	3,135.069	—	2
Tb	3,134.26	8	8	Kr	3,135.10	(8)	—
Sm	3,134.19	1	6	Yt	3,135.17	18	10
Fe	3,134.111	125	200	Ce	3,135.18	—	15
Ni	3,134.108	150	1000*R*	V	3,135.19	—	8
Re	3,134.02	—	30	Ir	3,135.23	—	2*h*
				Cr	3,135.34	25	1
				Tb	3,135.35	8	15
				Pr	3,135.351	2	10
				Fe	3,135.36	100	1
				Dy	3,135.37	50	100
				Nb	3,135.40	10	2
				Fe	3,135.45	3	10
				Rh	3,135.47	—	5

Hf 4,093.161 *J* 20 *O* 25

	λ	*J*	*O*		λ	*J*	*O*
Mo	4,093.161	5	5	Zr	4,093.163	—	12
W	4,093.14	4	5	U	4,093.285	2	1
Sc	4,093.13	8	6	Ce	4,093.285	—	15
Nd	4,093.10	2*h*	2	Tb	4,093.35	—	2
Co	4,093.05	—	10	Th	4,093.39	8	10
Sm	4,093.04	3	3	V	4,093.50	5	9
Eu	4,092.95	—	5	J	4,093.51	(10)	—
O	4,092.94	(80)	—	U	4,093.61	1	3
Er	4,092.90	—	6	Ir	4,093.63	—	2
Pr	4,092.85	2	5	Dy	4,093.65	—	5
Co	4,092.848	10	25	K	4,093.70	(20)	—
Ce	4,092.715	2	18	Gd	4,093.72	15	15
Gd	4,092.712	40	100	Ru	4,093.78	1	5
V	4,092.69	12	20	Pr	4,093.79	1	8
Nd	4,092.65	—	4	Ce	4,093.95	3	20
Ca	4,092.633	2	15	U	4,093.99	3*h*	12
Pr	4,092.63	5	15	Tb	4,094.04	—	7
Ir	4,092.61	20	60	Sm	4,094.042	10	10

	λ	I	
		J	O
*bh*Sr	4,092.6	—	4
Cl	4,092.53	(8)	—
Dy	4,092.43	—	4
V	4,092.41	6	15
W	4,092.40	4	5
Co	4,092.391	15	600*W*
Mn	4,092.388	20	30
Fe	4,092.287	1	6
U	4,092.286	1*h*	2
Pt	4,092.261	1	4
Sm	4,092.257	25	10
Tb	4,092.19	—	12
Cr	4,092.17	2	25
Ce	4,092.08	1	10
Se	4,091.95	(70)	—
V	4,091.94	5	8
Xe	4,091.88	(2*h*)	—
Ce	4,091.85	—	6
U	4,091.84	4	3
Os	4,091.82	12	100
Dy	4,091.77	4	8
Er	4,091.770	—	6
Gd	4,091.754	—	5*h*
Ec	4,091.75	—	2
Ta	4,091.642	—	2
U	4,091.636	—	12
Ce	4,091.59	—	3
Fe	4,091.56	3	8
Dy	4,091.533	4	9
P	4,091.53	(30)	—
U	4,091.52	12	6
Yb	4,091.50	5	2
Tm	4,094.18	30	300
Ce	4,094.20	—	3
Zr	4,094.27	1	30
V	4,094.28	5	10
Rn	4,094.3	(10)	—
Eu	4,094.309	1*h*	4
Ce	4,094.312	2	—
Mo	4,094.35	2	3
P	4,094.43	(30)	—
Tb	4,094.45	5	10
Mo	4,094.47	4	5
Gd	4,094.49	15	25
Ce	4,094.59	—	2
Nd	4,094.61	2	8
U	4,094.62	5	6
Er	4,094.66	—	6
Pb	4,094.68	5	—
Th	4,094.75	15	15
Sc	4,094.84	6	8

Ho $^{67}_{164.930}$

t_0 —　　　　t_1 —

I.	II.	III.	IV.	V.
—	—	—	—	—

λ	*I*		eV
	J	*O*	
2,936.77	1000*R*	—	—
I 4,103.84	400	400	—

Ho 2,936.77 *J* 1000*R*

	λ	*I*			λ	*I*	
		J	*O*			*J*	*O*
Pd	2,936.77	1	2	U	2,936.776	4	6
Te	2,936.77	(5)	—	Mo	2,936.781	25	2
Ca	2,936.72	4*h*	—	Fe	2,936.90	500*r*	700*r*
Bi	2,936.7	(3)	—	Cr	2,936.93	18	4
Ir	2,936.68	5	15*h*	J	2,937.00	(20)	—
W	2,936.67	20	10	V	2,937.04	25	2
Nb	2,936.66	30	—	W	2,937.14	12	8
Th	2,936.47	12	12	U	2,937.15	2	3
Nb	2,936.46	2*h*	1	Yb	2,937.18	10	2
U	2,936.453	20	12	Ti	2,937.32	5	35
Fe	2,936.449	5	5	Nb	2,937.33	15	2
Zr	2,936.31	15	10	U	2,937.35	4	5
Th	2,936.19	10	10				
Ti	2,936.17	100*wh*	—				
Br	2,936.16	(5)	—				

Ho 4,103.84 *J* 400 *O* 400

	λ	*J*	*O*		λ	*J*	*O*
Er	4,103.81	—	12	Hg	4,103.87	(50)	—
Yt	4,103.80	2	7	F	4,103.871	(50)	—
Re	4,103.78	—	2*h*	Dy	4,103.878	4	30
Mo	4,103.75	5	5	Eu	4,103.88	—	5
Ce	4,103.73	—	4	Tb	4,103.90	2	30
F	4,103.724	(50)	—	Ar	4,103.91	(200)	—
Tm	4,103.72	10	5	*bh*Ca	4,104.1	—	4
Ir	4,103.67	—	10	Sm	4,104.12	10	5
Sm	4,103.65	2	3	Fe	4,104.13	25	100
Os	4,103.62	3	20	Nb	4,104.17	30	3
Mn	4,103.55	10	12	Cu	4,104.226	1*h*	30
F	4,103.52	(300)	—	Nd	4,104.227	4	10
Mn	4,103.463	—	12	Hf	4,104.233	4	8
Tb	4,103.46	—	8	Ir	4,104.235	2	20
Ce	4,103.45	—	4	Ta	4,104.24	5	6
Eu	4,103.41	3	2*W*	Ho	4,104.30	—	2
Gd	4,103.410	—	5*h*	Rb	4,104.31	20	—
vzduch	4,103.4	20	—	Er	4,104.34	—	3
Tb	4,103.37	—	30	Th	4,104.39	10	10
Dy	4,103.312	50	50	V	4,104.40	7	15
Er	4,103.311	1	18	Re	4,104.422	—	30
Ho	4,103.30	3	3	Ce	4,104.425	1	30
F	4,103.22	(30)	—	Co	4,104.43	2	30
Tb	4,103.21	—	4	U	4,104.44	2	2
Pr	4,103.20	1	2	Pr	4,104.48	2	5
U	4,103.12	1*h*	18	Nd	4,104.54	5	8
Xe	4,103.10	(5*h*)	—	Rb	4,104.66	2	—
F	4,103.08	(150)	—	O	4,104.73	(50)	—

	λ	I	
		J	O
O	4,103.01	(50)	—
Mn	4,102.96	20	100
Si	4,102.95	10	12
J	4,102.88	(10)	—
U	4,102.86	6	5
Ce	4,102.724	1	2
Eu	4,102.72	3	10
Ti	4,102.71	1	3
W	4,102.70	30	35
Nd	4,102.56	5	10
Br	4,102.53	(10)	—
Tb	4,102.52	2	25*w*
Ho	4,102.40	3	—
Yt	4,102.38	30	150
Ce	4,102.36	2	18
*bh*Sr	4,102.3	—	2
Ru	4,102.285	10	15
Zr	4,102.281	—	10
U	4,102.21	4	1
N	4,102.18	(5)	—
V	4,102.159	15	30
Re	4,102.159	—	4*h*
Mo	4,102.15	25	30

	λ	I	
		J	O
Co	4,104.75	—	50
Sb	4,104.77	3	—
V	4,104.778	5	12
U	4,104.779	2*h*	2
Br	4,104.80	(2)	—
Rb	4,104.82	3	—
Pr	4,104.86	1	4
Cr	4,104.87	8	35
La	4,104.88	2	40
U	4,104.94	—	6
Xe	4,104.95	(20)	—
Fe	4,104.95	4*h*	4*h*
Gd	4,104.99	—	5
Ce	4,104.996	3	40
O	4,105.001	(125)	—
Ta	4,105.02	8	12
Dy	4,105.05	4	10
Mo	4,105.08	20	15
V	4,105.17	12*r*	20*r*
Ir	4,105.20	2	25
U	4,105.31	3	10
Ce	4,105.343	—	2
Th	4,105.343	12	12
Mn	4,105.36	20	50
Tb	4,105.38	1	15
Os	4,105.42	—	6
Br	4,105.47	(2)	—
Ce	4,105.49	—	2
Mo	4,105.53	10	10

Lu $^{71}_{174.97}$

t_0 — t_1 —

I.	II.	III.	IV.	V.
~5.0	—	—	—	—

λ	*I*		eV
	J	*O*	
II 4,184.25	200	100	5.11
I 4,518.57	40	300	>2.7

Lu 4,184.25 *J* 200 *O* 100

	λ	*I*			λ	*I*	
		J	*O*			*J*	*O*
Pr	4,184.24	1	4	Se	4,184.26	(25)	—
Mo	4,184.17	4	4	Gd	4,184.264	150	150
U	4,184.15	2	2	Ca	4,184.28	8	2
Os	4,184.13	2	25	Tb	4,184.30	2	10
Se	4,184.08	(10)	—	Ti	4,184.33	20	8
Ce	4,184.07	4	3	Ce	4,184.38	1	3
Er	4,184.06	—	4	Mo	4,184.39	4	4
Te	4,183.99	(30)	—	Nb	4,184.44	50	20
W	4,183.83	5	12	Kr	4,184.473	(20)	—
Ce	4,183.80	—	3	Ni	4,184.475	—	6
Sm	4,183.76	15	10	U	4,184.50	3*h*	1
Dy	4,183.73	8	15	Ce	4,184.63	—	2
W	4,183.67	—	10	Tb	4,184.66	—	2
Gd	4,183.62	—	35	Fe	4,184.895	80	100
Dy	4,183.61	—	8	Cr	4,184.895	10	35
U	4,183.60	5	2	Tb	4,184.95	—	2
Th	4,183.57	8	8	Nd	4,184.980	10	15
V	4,183.43	20	3	Eu	4,184.985	—	7
Nb	4,183.38	10*h*	—	Er	4,184.99	—	7
Sm	4,183.33	2	2	Kr	4,185.12	(50)	—
Zr	4,183.32	1	40	Pr	4,185.15	2	5
Ti	4,183.29	7	20				
U	4,183.27	1	10				
Ir	4,183.21	4	40				

Lu 4,518.57 *J* 40 *O* 300

	λ	*J*	*O*		λ	*J*	*O*
Dy	4,518.54	4	6	U	4,518.59	3	1
Mo	4,518.44	5	5	Cr	4,518.63	—	6
Hf	4,518.29	4	10	Er	4,518.64	1*h*	40*w*
Ce	4,518.28	—	4	Th	4,518.65	2	2
Tb	4,518.21	—	2	Mo	4,518.67	4*h*	4*h*
U	4,518.08	4	1	Eu	4,518.68	5	8*w*
Ti	4,518.03	60	100	Pr	4,518.69	—	2
Ce	4,518.02	—	12	Ti	4,518.70	10	30
Er	4,517.94	—	2*w*	Os	4,518.89	—	15
Tm	4,517.87	—	15	Dy	4,518.97	1	2
Ru	4,517.82	—	60	Hf	4,519.03	8	1
Nd	4,517.78	—	2*h*	Rb	4,519.07	10	—
Eu	4,517.76	—	20*w*	*bh*Ca	4,519.1	—	3
Ne	4,517.74	(100)	—	Pr	4,519.11	—	3
Pr	4,517.59	15*w*	40*w*	W	4,519.17	3	12
V	4,517.57	4	5	Cl	4,519.19	(18)	—
Fe	4,517.53	3	30	U	4,519.22	3	2
In	4,517.42	10	—	Th	4,519.27	—	6*w*
Mo	4,517.41	8	12	Co	4,519.29	—	40

	λ	I	
		J	O
Ir	4,517.372	—	4
W	4,517.370	2	9
Eu	4,517.36	—	6
Ne	4,517.29	(15)	—

	λ	I	
		J	O
*bh*Zr	4,519.3	—	20
Er	4,519.44	—	5
Sm	4,519.53	—	2
Eu	4,519.56	—	3
Tm	4,519.58	1	50
Mo	4,519.586	40	—
Ce	4,519.952	—	8
Pr	4,519.629	—	9
Sm	4,519.63	80	150
Gd	4,519.66	100	150
Xe	4,519.69	(2)	—
Re	4,519.74	—	40*d*
Th	4,519.75	1	4
U	4,519.77	1	2
Cr	4,519.828	—	10
Dy	4,519.83	4	5
Os	4,519.86	—	10

Nd $^{60}_{144.24}$

t_0 840°C t_1 –

I.	II.	III.	IV.	V.
~6.31	–	–	–	–

λ	I		eV
	J	O	
II 3,951.15	30	40	3.92
II 4,303.57	40	100	–

Nd 3,951.15 *J* 30 *O* 40

	λ	*I*			λ	*I*	
		J	*O*			*J*	*O*
Th	3,951.101	2	2	W	3,951.168	12	5
Cr	3,951.097	8	50	Fe	3,951.168	125	150
Mo	3,950.99	15	15	Ru	3,951.21	6	10
Xe	3,950.92	(125)	—	Eu	3,951.31	—	3
Eu	3,950.85	—	7	Zr	3,951.33	—	5
Sr	3,950.81	—	8*h*	Mo	3,951.35	4	4
Ce	3,950.80	1	6	Ce	3,951.42	—	4
Tb	3,950.78	—	5	La	3,951.43	2*h*	4
Hf	3,950.76	—	10*h*	Er	3,951.49	1	2
Nd	3,950.75	6	10	P	3,951.50	(70)	—
U	3,950.70	1	3	Th	3,951.52	20	20
Pr	3,950.66	4	10	Mo	3,951.53	4	4
Br	3,950.61	(30)	—	U	3,951.55	8	1
Re	3,950.610	—	30*d*	Yt	3,951.60	8	6
Mo	3,950.50	—	8	Xe	3,951.61	(3*wh*)	—
U	3,950.48	—	8	Ce	3,951.62	1	8
Ce	3,950.424	3	10	Tb	3,951.70	—	4
Nd	3,950.412	10	20	Co	3,951.72	2	4*h*
Tb	3,950.416	—	20*d*	Cr	3,951.76	5	40
Ru	3,950.41	10	10	Hf	3,951.83	4	15
Dy	3,950.40	—	50	Se	3,951.84	(25)	—
Th	3,950.39	30	30	Pr	3,951.843	5	12
Yt	3,950.36	100	60	U	3,951.85	4	1
Er	3,950.35	104	30	Tb	3,951.87	—	6
Zr	3,950.29	—	2*h*	W	3,951.876	8	7
Mo	3,950.26	4	4	Sm	3,951.885	3	10
V	3,950.23	10	20	Nd	3,951.89	—	8
Ru	3,950.21	15	12	Pb	3,951.94	50*h*	—
U	3,950.132	6	3	Ir	3,951.95	8	20
Tb	3,950.126	—	4	Mn	3,951.96	50	40
W	3,950.12	12	9	V	3,951.97	50	35
Ru	3,950.00	8	10	O	3,951.99	(10)	—
Cl	3,949.96	(10)	—	Gd	3,952.01	60	100
Fe	3,949.957	100	150	Pr	3,952.02	4	5
Nb	3,949.93	3	4	Tb	3,952.05	—	2
Ca	3,949.92	2*r*	—	C	3,952.08	6	—
J	3,949.90	(20)	—	Ce	3,952.11	1	10
Tb	3,949.87	—	6	Ta	3,952.16	3	8
Sm	3,949.845	2	2	Nd	3,952.19	20	30
Eu	3,949.837	—	5	N	3,952.21	(10)	—
Ce	3,949.82	—	10	Eu	3,952.25	—	3*W*
Os	3,949.78	10	50	F	3,952.26	(3)	—
Cr	3,949.62	—	6	W	3,952.262	4	5
Eu	3,949.59	—	50*W*	U	3,952.263	6	3
Cr	3,949.58	1	5	Mo	3,952.27	2	4
Ca	3,949.57	2	—	Se	3,952.28	—	3
				Ru	3,952.29	1	4

	λ	I J	I O
Co	3,952.33	4	40
Pr	3,952.357	3	8
Rn	3,952.36	(25)	—
Nb	3,952.37	50	3
Cr	3,952.40	18	60
U	3,952.447	2	4
Th	3,952.451	1	3
W	3,952.52	7	8
Ce	3,952.54	30	60
Se	3,952.60	(5)	—
Fe	3,952.51	50	80
Ir	3,952.616	3	20
Tb	3,952.620	—	2*w*
Ru	3,952.68	30	20
Fe	3,952.70	1	8
Ar	3,952.74	(15)	—
Os	3,952.77	15	40

Nd 4,303.57 *J* 40 *O* 100

	λ	I J	I O
Ta	4,303.54	15	1
W	4,303.533	—	4
V	4,303.527	7	15
Gd	4,303.46	6	6
W	4,303.33	15	—
U	4,303.32	6	6
Ne	4,303.25	(30)	—
Co	4,303.24	2	15
Fe	4,303.17	15	12
Pr	4,303.14	5	20
Dy	4,303.03	—	8
Ti	4,302.979	—	10
Ta	4,302.978	40*W*	125*W*
Tb	4,302.94	—	10*d*
Nb	4,302.91	10*h*	2
Zr	4,302.89	1	100
Cr	4,302.77	2	40
Dy	4,302.72	—	10
*bh*Sr	4,302.7	—	4
Ce	4,302.65	—	10
Dy	4,302.57	—	2
U	4,302.530	—	2
Ca	4,302.527	25	50
Kr	4,302.45	(10)	—
Pr	4,303.59	60	100
Hf	4,303.596	6	10
Dy	4,303.60	—	3
O	4,303.78	(60*h*)	—
Er	4,303.81	—	12
Nb	4,303.88	10	3
Mo	4,303.89	10	—
Ne	4,303.95	(5)	—
Ti	4,303.96	—	8
C	4,304.0	(10)	—
Mo	4,304.020	10	12
Tb	4,304.02	—	12
Cl	4,304.07	(40)	—
La	4,304.12	2	3
Se	4,304.13	(10)	—
U	4,304.14	1	15*r*
Li	4,304.15	2	—
Ce	4,304.278	2*h*	15
Tb	4,304.28	—	4
Hf	4,304.407	2	6
Re	4,304.411	—	5*w*
Nd	4,304.44	12	20
U	4,304.47	6	8
Mo	4,304.55	5	4
Dy	4,304.58	—	3
Pr	4,304.65	1	4
Zr	4,304.68	—	15
Nb	4,304.69	5	3

Pr 59 140,907

t_0 940°C t_1 —

I.	II.	III.	IV.	V.
5.76	—	—	—	—

λ	I		eV
	J	O	
II 4,179.422	40	200	—
II 4,222.98	40	125	2.99

Pr 4,179.422 *J* 40 *O* 200

	λ	I			λ	I	
		J	*O*			*J*	*O*
V	4,179.419	10	20	U	4,179.43	1	10
Er	4,179.401	—	5	Cr	4,179.459	8	—
Eu	4,179.40	1*h*	2	Dy	4,179.46	—	3
Ba	4,179.37	—	6	Cu	4,179.51	6	1*h*
Ar	4,179.31	(20)	—	Hf	4,179.53	8	5
Ce	4,179.29	1	10	Kr	4,179.58	(20*wh*)	—
Cr	4,179.26	40	100	Nd	4,179.585	10	10
Te	4,179.24	(70)	—	U	4,179.63	1	2
Co	4,179.23	2	15	Br	4,179.64	(40)	—
Ce	4,179.08	—	8	N	4,179.68	(2*h*)	—
V	4,179.06	2	—	Th	4,179.72	8	8
Ge	4,179.04	25*wh*	—	Ca	4,179.74	2	—
U	4,179.00	12	15	Re	4,179.75	—	2*h*
Tb	4,178.97	2*h*	50*d*	Nb	4,179.755	20	10
Fe	4,178.87	10	10	Tb	4,179.80	1	6
Pr	4,178.641	10	18	Ce	4,179.806	2	1
Nd	4,178.639	6	10	Zr	4,179.809	8	15
Er	4,178.602	—	8*s*	Ti	4,179.88	—	4
Re	4,178.597	—	3	Ac	4,179.93	60	—
Dy	4,178.59	2	8	Mo	4,179.94	5	8
Tb	4,178.56	—	2	Cr	4,179.959	1	25
Mo	4,178.535	2	5	Th	4,179.965	8	8
Nd	4,178.532	—	5	Tm	4,189.02	2	5
U	4,178.52	8	8	Xe	4,180.10	(500)	—
Nd	4,178.44	5	—	Na	4,180.2	—	3
Nb	4,178.41	10	1	W	4,180.24	5	10
Ar	4,178.39	(20)	—	Er	4,180.28	2	6*d*
Ce	4,178.390	2	1	U	4,180.31	1*h*	12
P	4,178.36	(300*w*)	—	Dy	4,180.33	2	2
				Sm	4,180.38	2	10
				Tb	4,180.40	1*h*	20
				Pr	4,180.40	2	5

Pr 4,222.98 *J* 40 *O* 125

	λ	*J*	*O*		λ	*J*	*O*
K	4,222.97	(40)	—	Gd	4,222.98	—	10
Mo	4,222.96	10	15	Xe	4,223.00	(200*h*)	—
U	4,222.94	2	1	N	4,223.04	(25)	—
Tb	4,222.91	—	6*W*	Tl	4,223.05	(25)	—
Ce	4,222.88	—	3	Sm	4,223.06	2	3
Na	4,222.8	—	3	Lu	4,223.09	4	1
O	4,222.78	(50)	—	Ce	4,223.15	—	3
U	4,222.74	—	6	Ir	4,223.16	2*h*	15
Cr	4,222.73	15	100	Nd	4,223.21	3	15
Tb	4,222.71	—	9	U	4,223.30	3	1*h*

	λ	I			λ	I	
		J	O			J	O
Ar	4,222.67	(20)	—	Tb	4,223.32	1	10
Tm	4,222.67	4	10	J	4,223.33	(15)	—
Ce	4,222.60	18	80	Sb	4,223.34	4	—
Mo	4,222.41	15	20	Dy	4,223.38	—	2
Br	4,222.40	(4)	—	Cr	4,223.470	—	15
U	4,222.37	8	18	Ho	4,223.47	—	4
Eu	4,222.31	—	5	Pr	4,223.51	3	12
Ho	4,222.26	1*h*	3	In	4,223.57	5	—
Dy	4,222.221	—	10	Ho	4,223.64	2	—
Fe	4,222.221	200	200	U	4,223.65	1	6
Kr	4,222.20	(20*h*)	—	Pt	4,223.69	3	—
P	4,222.15	(150*w*)	300	Sm	4,223.71	20	50
U	4,222.15	3*h*	1	Er	4,223.72	2	9
Nd	4,222.06	2	5	Tb	4,223.82	—	3
W	4,222.05	8	15	Br	4,223.88	(80)	—
Tb	4,222.02	—	3	Ce	4,223.884	1	20
Tm	4,221.95	2	2	Mo	4,223.92	4	5
Ba	4,223.96	3	5	Lu	4,223.99	—	—

Sm $^{62}_{150.35}$

t_0 1,350°C t_1 –

I.	II.	III.	IV.	V.
5.6	~11.4	–	–	–

λ	I		eV
	J	O	
II 4,424.342	300	300	3.28
II 4,434.32	200	200	3.17

Sm 4,424.342 *J* 300 *O* 300

	λ	I	
		J	*O*
Ce	4,424.31	—	6
Cr	4,424.28	35	25
Mo	4,424.20	3	5
Ce	4,424.14	—	3
Gd	4,424.10	10	25
Cr	4,424.07	2	10
Rh	4,424.047	2	5
Cs	4,424.046	(10)	—
Ar	4,423.99	(80)	—
Th	4,423.94	4*w*	8*w*
Pr	4,423.93	1	5
V	4,423.91	6	7
P	4,423.9	(30)	—
La	4,423.898	5	5
Nb	4,423.87	5	1
Tb	4,423.86	—	2
Re	4,423.82	—	2
W	4,423.78	5	12
J	4,423.76	(80)	—
U	4,423.74	8	6
K	4,423.72	(10)	—
Pr	4,423.685	—	2
Ce	4,423.678	3	25
Mo	4,423.62	40	40
P	4,423.55	(30)	—
Ce	4,423.44	—	12
Sm	4,423.38	2	9
Eu	4,423.35	—	4*W*
Pr	4,423.32	—	10*w*
Cr	4,423.318	6	15
Na	4,423.31	—	3
U	4,423.29	6	6
V	4,423.21	25*w*	40
Tb	4,423.11	—	35
Nd	4,424.343	50	50
Ti	4,424.39	2	15
Tb	4,424.46	—	3
Ce	4,424.54	—	3
V	4,424.56	15	20
Er	4,424.57	—	10
Pr	4,424.59	35	90
Nb	4,424.66	5*h*	—
Ir	4,424.75	—	3
Ru	4,424.78	—	25
Ne	4,424.80	(300)	—
W	4,424.91	2	8
Ta	4,424.96	3*h*	10
Sm	4,424.99	2	3
Gd	4,425.01	4	8
Ce	4,425.12	—	2
Cr	4,425.13	1	15
Br	4,425.14	(12)	—
Kr	4,425.19	(100)	—
U	4,425.20	5	2
Pr	4,425.220	—	2
Hg	4,425.22	(30)	—
Tb	4,425.25	—	2*h*
Ce	4,425.33	—	3*s*
Ne	4,425.40	(150)	—
U	4,425.41	—	8
Ca	4,425.44	20	100
Sb	4,425.48	2*h*	—
Pr	4,425.50	—	3

Sm 4,434.32 *J* 220 *O* 200

	λ	I	
		J	*O*
J	4,434.23	(20)	—
Mo	4,434.23	6*h*	—
Tb	4,434.06	—	2
W	4,434.05	—	3
Ti	4,434.00	50	100
Mg	4,433.99	—	8
Cr	4,433.97	1	10
Ir	4,433.91	—	10
U	4,433.89	12	15
Sm	4,433.88	200	200
Ce	4,434.37	—	3
Ir	4,434.47	—	2
Tb	4,434.48	—	20
Hf	4,434.52	—	6*h*
V	4,434.60	12	20
Eu	4,434.80	2	20*w*
Pr	4,434.85	3*w*	25
Se	4,434.92	(40)	—
Ir	4,434.93	—	3
U	4,434.94	—	8

	λ	I			λ	I	
		J	O			J	O
Ar	4,433.83	(20)	—	Ce	4,434.952	—	8
Fe	4,433.79	2	30	Th	4,434.953	—	4
Se	4,433.74	(25)	—	Mo	4,434.953	80	80
Ce	4,433.725	1	12	Ir	4,434.958	—	3
Ne	4,433.721	(70)	—	Ca	4,434.960	25	150
Mn	4,433.68	—	12	Te	4,434.96	(70)	—
Tb	4,433.65	—	4	Tb	4,435.00	—	4
W	4,433.642	—	3	Dy	4,435.02	—	2
Gd	4,433.638	12	20	Rn	4,435.05	(200)	—
Ti	4,433.58	2	15	Ne	4,435.094	(5)	—
Mo	4,433.50	125	4	Nd	4,435.095	4	15
N	4,433.48	(5*h*)	—	Fe	4,435.15	3	70
Ne	4,433.40	(10)	—	Pr	4,435.25	—	5
Sm	4,433.35	—	4	W	4,435.44	—	6
Rh	4,433.32	5	15	Ce	4,435.47	—	2
Ce	4,433.23	—	3*h*	Eu	4,435.53	—	2,000
Fe	4,433.221	20	150				
Eu	4,433.22	1	6*w*				

Tb $^{65}_{158.924}$

t_0 327°C t_1 —

I.	II.	III.	IV.	V.
~6.74	—	—	—	—

λ	*I*		eV
	J	*O*	
3,899.19	100	200	—
4,278.51	100	200	—

Tb 3,899.19 *J* 100 *O* 200

	λ	I			λ	I	
		J	*O*			*J*	*O*
Dy	3,899.15	—	20	K	3,899.24	(10)	—
V	3,899.13	5*h*	12	Nb	3,899.25	15	10
Cu	3,899.11	—	2*wh*	Ce	3,899.27	—	2
U	3,899.10	3	5	Mn	3,899.33	25	12
Zr	3,899.09	1	5	Ce	3,899.38	2	6
Fe	3,899.038	8	20	Eu	3,899.45	3	5*w*
Er	3,899.036	1	2	Lu	3,899.54	3*h*	—
Fe	3,899.01	—	6	Tb	3,899.54	8	15
Ce	3,899.94	3	15	Pr	3,899.55	8	10
Pr	3,898.840	9	15	Ho	3,899.64	6	—
U	3,898.837	6	6	Eu	3,899.708	—	4
F	3,898.83	(20)	—	Fe	3,899.709	300	500
Th	3,898.79	1	2	Ti	3,899.712	2	15
W	3,898.782	6*d*	—	W	3,899.78	5	—
Ta	3,898.781	5*h*	10	Ce	3,899.859	2*h*	2
Eu	3,898.78	—	20*W*	Ar	3,899.86	(100)	—
Pt	3,898.75	2*h*	20	Hf	3,899.94	6	15
Tb	3,898.73	—	4	Cs	3,900.09	(4)	—
F	3,898.72	(3)	—	Pr	3,900.108	1	3
U	3,898.69	—	6	K	3,900.11	(10)	—
La	3,898.604	5	30	Eu	3,900.169	—	12*w*
Ce	3,898.600	—	4	V	3,900.175	2*h*	50
Ba	3,898.58	—	5	Ce	3,900.20	2*h*	10
Nb	3,898.557	5	5	Nd	3,900.226	30	30
Zr	3,898.557	—	2	Mo	3,900.227	3	1
No	3,898.55	8	6	U	3,900.33	12	6
Dy	3,898.54	—	100	Dy	3,900.39	—	3
Er	3,898.53	10	30	Os	3,900.394	12	50
U	3,898.514	3	6	Tb	3,900.40	—	2
Sm	3,898.508	4	—	Eu	3,900.42	2	8*w*
Eu	3,898.50	—	2*W*	Zr	3,900.517	—	100
Ti	3,898.494	5	20	Eu	3,900.519	—	12*w*
Co	3,898.494	6	80*r*	Fe	3,900.519	3*d*	60
Gd	3,898.48	10	10	Nb	3,900.53	10	30
Th	3,898.477	8	10	Ti	3,900.54	50*h*	30
J	3,898.44	(25)	—	Th	3,900.58	3	8
Ce	3,898.43	—	2	Ar	3,900.63	(10)	—
Mn	3,898.362	—	30	Hf	3,900.65	15	2
Ru	3,898.361	6	12	Mo	3,900.66	—	2
Nb	3,898.285	200	3	Al	3,900.68	(200)	—
V	3,898.278	—	10				
Ce	3,898.27	6	80				
Eu	3,898.26	8*Wh*	5*W*				
V	3,898.14	—	40				
Mg	3,898.12	8	5				
V	3,898.02	2	10				
Ce	3,898.014	—	5				

	λ	I J	I O
Fe	3,898.013	50	80
W	3,897.91	15	12
Fe	3,897.895	60	100
Au	3,897.89	25	30
K	3,897.87	(60)	—
Tb	3,897.85	8	15
Br	3,897.73	(3)	—
Eu	3,897.73	—	20*W*
Pr	3,897.72	3	8
U	3,897.71	8	6

Tb 4,278.51 *J* 100 *O* 100

	λ	J	O
W	4,278.41	3	10
Th	4,278.32	2	8
Ca	4,278.28	4	—
Ce	4,278.25	1	10
Ti	4,278.23	15	50
Gd	4,278.20	6	6
U	4,278.17	4	5
Pr	4,278.04	15	35
In	4,277.99	2	—
Si	4,277.95	3	1
W	4,277.91	1	6
Hg	4,277.80	(10)	—
Cr	4,277.79	—	25*wh*
Tb	4,277.76	—	8
Dy	4,277.76	—	2
Yb	4,277.73	2	25
Eu	4,277.64	—	4
Ar	4,277.55	(80)	—
U	4,277.54	4	5
F	4,277.51	(40*h*)	—
Lu	4,277.50	3	30
Nb	4,277.496	1	5
Mo	4,277.41	15	15
Zr	4,277.37	1	4
S	4,278.54	(30)	—
Mo	4,278.58	2	5
Rh	4,278.60	10	25
Dy	4,278.61	—	3
Pr	4,278.62	2	10
Mn	4,278.68	5	15
Ru	4,278.69	—	7
Ti	4,278.81	3	25
As	4,278.82	10	—
Ne	4,278.85	(5)	—
Ce	4,278.87	5	20
V	4,278.88	8*h*	2
F	4,278.89	(20*h*)	—
V	4,278.92	15	—
W	4,278.93	4	4
Pr	4,278.99	3	15
Mo	4,279.02	100	8
Ta	4,279.06	10	30
Ce	4,279.17	—	2
Re	4,279.20	—	3
Eu	4,279.25	—	8
Ne	4,279.28	(15)	—
Yt	4,279.3	15	3
Ce	2,279.32	—	4
U	4,279.33	—	10
Mo	4,279.39	—	3
Fe	4,279.480	1	5
Te	4,279.48	(15)	—
Nb	4,279.499	5	5
Sm	4,279.501	2	4

Tm $^{69}_{168.934}$

t_0 —

t_1 —

I.	II.	III.	IV.	V.
—	—	—	—	—

λ	*I*		eV
	J	*O*	
II 3,761.33	150	250	3.29
4,187.62	30	300	—

Tm 3,761.33 *J* 150 *O* 250

	λ	*I*	
		J	*O*
Ti	3,761.32	300*r*	100
Ce	3,761.18	—	3
Sm	3,761.14	2*h*	6
Eu	3,761.134	10	10
Nb	3,761.126	20	15
Tb	3,761.12	—	15
Th	3,761.10	8	15
U	3,761.04	8	4
Re	3,761.02	—	2*h*
Yb	3,761.00	6	3
Pr	3,760.96	3	5
Nd	3,760.94	4*d*	10*d*
Gd	3,760.93	10	10
Mo	3,760.885	10	6
U	3,760.884	15	10
Ir	3,760.84	—	2*h*
Ce	3,760.82	—	2
Rn	3,760.80	(10)	—
V	3,760.79	5	30
Eu	3,760.76	—	10*W*
Nb	3,760.760	8*h*	2
Gd	3,760.71	25	25
Sm	3,760.70	40	25
Ce	3,760.69	—	6
Nb	3,760.644	4	5
W	3,760.639	6	7
Fe	3,760.53	70	100
Rh	3,760.405	2	6
Ce	3,760.399	3	5
Co	3,760.39	—	30
W	3,760.38	9	—
Er	3,760.36	1*d*	15*d*
Nd	3,760.360	6*d*	10*d*
U	3,760.345	2	1
Eu	3,760.34	10*W*	10W
Pr	3,760.30	1	2
Th	3,760.28	5	20
Os	3,760.27	10	20
V	3,760.23	40	6
Ta	3,760.21	2*d*	8
Ce	3,760.18	—	2
Sm	3,760.17	—	5
Tb	3,760.13	3	15
W	3,760.126	10	15
Pr	3,760.08	8	20
Fe	3,760.052	100	150
Sm	3,760.049	10	25
Ta	3,761.35	—	8
Pr	3,761.38	2	8
Dy	3,761.40	—	2*h*
Fe	3,761.41	8	20
V	3,761.44	7	40
Ce	3,761.45	—	2
Th	3,761.47	1	10
Ru	3,761.51	45	12
Ir	3,761.55	—	5
Nd	3,761.57	10	20
U	3,761.60	6	2
Pr	3,761.61	6	10
W	3,761.62	8	7
Cr	3,761.70	8	4
Ca	3,761.72	4	2
Mo	3,761.75	6	6
P	3,761.81	(30)	—
Pr	3,761.867	100	150
Cr	3,761.868	10*h*	5*h*
Ti	3,761.89	15	12
Tm	3,761.92	120	200
Ir	3,761.94	4	2
Ce	3,761.95	—	2
U	3,761.96	8	2
Sr	3,762.00	3	1
Xe	3,762.05	(3*h*)	—
Mo	3,762.09	30	10
U	3,762.11	6	10
Ir	3,762.208	—	2
Fe	3,762.209	3	7
Gd	3,762.21	3	10
Ce	3,762.22	—	3
Xe	3,762.26	(5)	—
Ce	3,762.280	5	4
Dy	3,762.28	1*h*	3
Ti	3,762.30	1	10
Eu	3,762.33	—	9*w*
Tb	3,762.34	—	3
Nd	3,762.361	4	10
Th	3,762.363	3	10
Pr	3,762.37	3	5
Si	3,762.42	3	—
Nb	3,762.45	20*h*	5
Hg	3,762.51	25	2
O	3,762.51	(12)	—
Zr	3,762.513	—	2*h*
Yb	3,762.55	10	4

	λ	I J	I O		λ	I J	I O
Ru	3,760.03	50	20	Pr	3,762.56	2	8
				Nd	3,762.59	20	15
				Ni	3,762.62	—	2
				Ca	3,762.64	3	2
				U	3,762.66	2	2
				Ce	3,762.67	3	2

Tm 4,187.62 *J* 30 *O* 300

	λ	J	O		λ	J	O
Mo	4,187.61	3	3	Hf	4,187.66	10	8
Fe	4,187.59	1	3	Pr	4,187.79	2	2
Zr	4,187.56	3	9	Fe	4,187.80	150	200
Zr	4,187.47	—	4	Rn	4,187.81	(35)	—
Ce	4,187.323	15	35	U	4,187.87	5	—
La	4,187.316	40	50	Re	4,187.92	—	6
Co	4,187.25	3	50	Ir	4,187.97	—	8
Tb	4,187.16	—	15	Er	4,187.98	—	7*wd*
Fe	4,187.04	200	250	U	4,188.066	8	10
U	4,186.98	4	10	P	4,188.07	(30*h*)	—
Yb	4,186.90	10*h*	—	Gd	4,188.099	—	20
Ce	4,186.86	—	2	Tb	4,188.10	1*h*	15
Ho	4,186.84	3	3	Sm	4,188.12	25	10
Dy	4,186.81	12	100*w*	Hf	4,188.24	—	8
U	4,186.79	5	6	Mo	4,188.32	80	100
Zr	4,186.78	—	3	Ce	4,188.38	1*h*	5
Er	4,186.71	2	8	Tb	4,188.51	2*h*	20
Zr	4,186.69	3	3	Th	4,188.59	3	1
Tb	4,186.60	—	2	Ce	4,188.66	—	3
Ce	4,186.599	25	80				

Yb $^{70}_{173,04}$

t_0 (1,800°C) t_1 –

I.	II.	III.	IV.	V.
6.25	12.11	–	–	–

λ	I		eV
	J	*O*	
II 3,694.203	1000*R*	500*R*	3.35
I 3,987.994	500*R*	1000*R*	3.10

Yb 3,694.203 *J* 1000*R* *O* 500*R*

	λ	*I*			λ	*I*	
		J	*O*			*J*	*O*
Yt	3,694.20	—	4	Ho	3,694.24	20	10
Ne	3,694.197	(250)	—	La	3,694.27	10	4
Er	3,694.19	15	25*d*	Sm	3,694.31	8	15
Ce	3,694.17	—	4*w*	Pr	3,694.321	1	3
Tb	3,694.12	—	8	U	3,694.325	1	6
Mn	3,694.11	5	5	Th	3,694.35	1	5
In	3,694.03	(40)	—	Dy	3,694.36	4	6
Gd	3,694.011	25	25	Mo	3,694.42	20	—
Fe	3,694.010	300	400	Ti	3,694.447	20	80
Sm	3,694.00	150	100	Ir	3,694.454	—	2
Th	3,693.94	3	2	W	3,694.51	20	10
Ni	3,693.93	—	50	Ta	3,694.52	18	7*h*
In	3,693.90	(35)	—	Ag	3,694.62	1*h*	2*h*
Dy	3,693.84	1	4	V	3,694.622	5	60
Eu	3,693.82	1	25*w*	Nb	3,694.67	10	10
In	3,693.79	(25)	—	Pr	3,694.69	4*h*	40
Nb	3,693.76	2	1	Tb	3,694.75	8	50
Ce	3,693.71	—	8	Dy	3,694.75	30	20
U	3,693.705	18	4	Tm	3,694.76	20	20
Th	3,693.698	4	6	Nd	3,694.792	20	30
Mn	3,693.67	60	50	Nb	3,694.795	5	—
Ru	3,693.59	5	20	Ho	3,694.81	4	—
Tb	3,693.56	—	30	Er	3,694.811	6	15
Pr	3,693.493	1	20	U	3,694.816	3*wh*	—
Xe	3,693.49	(40)	—	Sm	3,694.817	3	15
Co	3,693.48	10	35	Th	3,694.89	1	5
Ce	3,693.42	1	10	Ce	3,694.91	2	15*s*
Mo	3,693.375	3	5	Mo	3,694.945	30	40
Nb	3,693.368	10	8	W	3,694.946	—	2
Co	3,693.364	—	18	Rh	3,694.95	2	3
Pr	3,693.363	8	30	Cr	3,694.97	6*h*	—
U	3,693.32	6	4	Fe	3,695.05	150	200
Mo	3,693.23	—	3	Er	3,695.09	—	2
Nd	3,693.13	8	20	V	3,695.16	2	—
Co	3,693.11	15	80	La	3,695.2	2	2*h*
Cr	3,693.09	5	10	U	3,695.21	—	8
Ta	3,693.05	3*h*	35*r*	Ce	3,695.237	—	2
Fe	3,693.03	7	15	W	3,695.244	4	5
Tb	3,692.95	8	30	Th	3,695.29	1	5
				V	3,695.33	70*h*	125
				Cu	3,695.36	—	6
				Ta	3,695.38	7*s*	15
				Ce	3,695.42	—	2
				Pr	3,695.46	—	4

Yb 3,987.994 *J* 500*R* *O* 1000*R*

	λ	*I*			λ	*I*	
		J	*O*			*J*	*O*
Ce	3,987.990	—	5	Ho	3,988.00	6	—
Ho	3,987.98	—	8	W	3,988.007	6	7
Er	3,987.95	20	100*r*	Th	3,988.015	30	50
Re	3,987.93	—	3	Pr	3,988.02	7	25
Cr	3,987.89	—	3	U	3,988.03	12	8
Gd	3,987.84	25	50	Nb	3,988.16	10	5
Eu	3,987.83	—	20*w*	Os	3,988.179	12	50
Ir	3,987.829	—	12	Ar	3,988.18	(5)	—
Nd	3,987.81	6	25	Dy	3,988.21	—	4
Tb	3,987.80	—	4	Eu	3,988.25	2	10*W*
Ru	3,987.79	50	3	U	3,988.29	1	2
Kr	3,987.78	25	—	Te	3,988.33	(5)	—
U	3,987.72	2	2	Ce	3,988.518	—	8*h*
Th	3,987.71	8	10	La	3,988.518	800	1000
Ce	3,987.70	—	2	Eu	3,988.58	—	5
Tb	3,987.67	—	6	Th	3,988.60	8	10
Er	3,987.663	—	4	U	3,988.64	8	8
Ir	3,987.657	—	4	Cr	3,988.66	—	5
Ti	3,987.61	12	8	Mn	3,988.67	12	12
Ho	3,987.55	—	3	Zr	3,988.68	—	15
W	3,987.52	4	5	Ta	3,988.70	10	15
Ce	3,987.51	—	2*s*	Nd	3,988.81	6	20
Pb	3,987.5	5	—	V	3,988.83	35	70
Yt	3,987.497	3	2	Th	3,988.85	10	10
Mn	3,987.46	15	15	Co	3,988.878	—	2
Nd	3,987.43	20	20	U	3,988.885	15	12
Sm	3,987.42	8	15	Dy	3,988.89	—	5
Ir	3,987.38	—	5	Ir	3,988.978	—	4
Mo	3,987.373	5	4	Rn	3,988.98	(8)	—
Pr	3,987.371	15	25	Fe	3,989.01	1*wh*	15*wh*
Tb	3,987.29	3	3	Sc	3,989.06	2	3
Eu	3,987.26	—	3*W*	U	3,989.11	1	2
Th	3,987.23	9	10	Pr	3,989.14	1	10
Gd	3,987.22	100	100	Br	3,989.23	(3)	—
Pr	3,987.17	3	15	Zn	3,989.23	(100)	—
W	3,987.15	10	—	Gd	3,989.25	20	10
Co	3,987.11	—	80	Zr	3,989.287	1*h*	12
Mn	3,987.098	60	30	U	3,989.292	1*h*	8
Eu	3,987.098	—	2	Ir	3,989.43	2	25
Kr	3,987.09	(5*whs*)	—	Ce	3,989.44	6	20
Re	3,987.08	—	6	Zr	3,989.50	1*h*	12
Sm	3,987.07	2	2	Tb	3,989.506	—	5
Dy	3,987.06	—	7	Mo	3,989.506	25*h*	—
Ce	3,987.058	—	2	Ti	3,989.57	—	2*h*
U	3,987.053	4	2				
Cu	3,987.01	3	—				
Mo	3,986.98	1	4				

	λ	I	
		J	O
Tb	3,986.94	—	3
Eu	3,986.92	—	6
Sm	3,986.90	8	8
U	3,986.85	8	—
Mn	3,986.83	75	40
Zr	3,986.80	—	10
Br	3,986.75	(3)	—
Mg	3,986.73	3	15*w*
Er	3,986.69	—	7*d*
Th	3,986.64	8	8
Eu	3,986.62	—	10
Br	3,986.54	(3)	—
Ho	3,986.50	—	5
U	3,986.44	1	2
Ce	3,986.39	1	15
Ir	3,986.38	2	2

7408357
3 1378 00740 8357